Methods in Enzymology

Volume 168

HORMONE ACTION

Part K

Neuroendocrine Peptides

METHODS IN ENZYMOLOGY

Methods in Enzymology

Volume 168

Hormone Action

Part K

Neuroendocrine Peptides

EDITED BY

P. Michael Conn

DEPARTMENT OF PHARMACOLOGY
THE UNIVERSITY OF IOWA
COLLEGE OF MEDICINE
IOWA CITY, IOWA

ACADEMIC PRESS, INC.
Harcourt Brace Jovanovich, Publishers
San Diego New York Berkeley Boston
London Sydney Tokyo Toronto

ACADEMIC PRESS, INC.
San Diego, California 92101

United Kingdom Edition published by
ACADEMIC PRESS LIMITED
24-28 Oval Road, London NW1 7DX

LIBRARY OF CONGRESS CATALOG CARD NUMBER: 54-9110

ISBN 0-12-182069-6 (alk. paper)

PRINTED IN THE UNITED STATES OF AMERICA
89 90 91 92 9 8 7 6 5 4 3 2 1

Table of Contents

Section I. Preparation of Chemical Probes

Section II. Equipment and Technology

Section III. Preparation and Maintenance of Biological Materials

Section IV. Quantitation of Neuroendocrine Substances

Section V. Use of Chemical Probes

Section VI. Localization of Neuroendocrine Substances

Contributors to Volume 168

Article numbers are in parentheses following the names of contributors. Affiliations listed are current.

S. AGRAWAL (54), *Medical Research Council (MRC), Laboratory of Molecular Biology, Cambridge CB2 2QH, England*

M. C. AGUILA (51), *Department of Physiology, Neuropeptide Division, The University of Texas Southwestern Medical Center at Dallas, Dallas, Texas 75235*

WILFRIED ALLAERTS (5), *Laboratory of Cell Pharmacology, University of Leuven, School of Medicine, B-3000 Leuven, Belgium*

ROBERTA ALLEN (32), *Autonomic Physiology Laboratory, Department of Medicine, University of California, San Diego, Medical Center, San Diego, California 92103*

P. C. ANDREWS (6), *Department of Biochemistry, Purdue University, West Lafayette, Indiana 47907*

H. ARAI (54), *Department of Psychopharmacology, Psychiatric Research Institute of Tokyo, Setagaya-ku, Tokyo 156, Japan*

WILLIAM McD. ARMSTRONG (9), *Department of Physiology and Biophysics, Indiana University, School of Medicine, Indianapolis, Indiana 46223*

FRANK BALDINO, JR. (55), *Cephalon, Inc., West Chester, Pennsylvania 19380*

WILLIAM A. BANKS (45), *Veterans Administration, Medical Center, and Tulane University School of Medicine, New Orleans, Louisiana 70146*

C. WAYNE BARDIN (42), *The Population Council, Center for Biomedical Research, New York, New York 10021*

AYALLA BARNEA (50), *Departments of Obstetrics and Gynecology and of Physiology, The University of Texas Southwestern Medical Center at Dallas, Dallas, Texas 75235*

ROBERT C. BAXTER (20), *Department of Endocrinology, Royal Prince Alfred Hospital, Camperdown, New South Wales, Australia 2050*

MARGERY C. BEINFELD (1), *Department of Pharmacology, St. Louis University Medical School, St. Louis, Missouri 63104*

NIRA BEN-JONATHAN (9, 15), *Department of Physiology and Biophysics, Indiana University, School of Medicine, Indianapolis, Indiana 46223*

M. BERG (35), *Protein Chemistry, Rorer Biotechnology Inc., Rockville, Maryland 20850*

THOMAS A. BICSAK (31), *Department of Reproductive Medicine, University of California, San Diego, La Jolla, California 92093*

A. E. BISHOP (57), *Department of Histochemistry, Royal Postgraduate Medical School, University of London, Hammersmith Hospital, London W12 0NN, England*

J. EDWIN BLALOCK (3), *Department of Physiology and Biophysics, The University of Alabama at Birmingham, Birmingham, Alabama 35294*

M. BLUM (43), *Fishberg Research Center for Neurobiology, The Mount Sinai Medical Center, New York, New York 10029*

KENNETH L. BOST (3), *Department of Physiology and Biophysics, The University of Alabama at Birmingham, Birmingham, Alabama 35294*

THOMAS BROCK (24), *Departments of Medicine and Pharmacology, The Neurobiology Curriculum and the Biological Sciences Research Center, The University of North Carolina at Chapel Hill, School of*

Medicine, Chapel Hill, North Carolina 27599

ELAINE R. BROWN (44), *Department of Anatomy and Neurobiology, Washington University School of Medicine, St. Louis, Missouri 63110*

MARVIN R. BROWN (32), *Autonomic Physiology Laboratory, Departments of Medicine and Surgery, University of California, San Diego, Medical Center, San Diego, California 92103*

J. PETER H. BURBACH (28, 29), *Rudolf Magnus Institute for Pharmacology, Medical Faculty, University of Utrecht, 3521 GD Utrecht, The Netherlands*

JOHN C. CAMBIER (22), *Department of Pediatrics, National Jewish Center for Immunology and Respiratory Medicine, Denver, Colorado 80206*

CAROLYN A. CAMPEN (42), *The Clayton Foundation Laboratories for Peptide Biology, The Salk Institute, La Jolla, California 92037*

PETER CARMELIET (5), *Laboratory of Cell Pharmacology, University of Leuven, School of Medicine, B-3000 Leuven, Belgium*

MARK S. CARTER (44), *Department of Anatomy and Neurobiology, Washington University School of Medicine, St. Louis, Missouri 63110*

MARIE-FRANCOISE CHESSELET (55), *Department of Pharmacology, The Medical College of Pennsylvania, Philadelphia, Pennsylvania 19129*

MILLIE M. CHIEN (22), *Department of Pediatrics, National Jewish Center for Immunology and Respiratory Medicine, Denver, Colorado 80206*

C. CHRISTODOULOU (54), *Medical Research Council (MRC), Laboratory of Molecular Biology, Cambridge CB2 2QH, England*

BIBIE M. CHRONWALL (56), *Division of Structural and Systems Biology, Kansas City School of Basic Life Sciences, University of Missouri, Kansas City, Missouri 64801, and Experimental Therapeutics Branch, National Institutes of Health, Bethesda, Maryland 20892*

P. MICHAEL CONN (18), *Department of Pharmacology, The University of Iowa, College of Medicine, Iowa City, Iowa 52242*

ANNE Z. CORRIGAN (42), *The Clayton Foundation Laboratories for Peptide Biology, The Salk Institute, La Jolla, California 92037*

DAVID W. CRABB (48), *Departments of Medicine and Biochemistry, Indiana University, School of Medicine, Indianapolis, Indiana 46223*

JEAN D. CREMINS (44), *Department of Anatomy and Neurobiology, Washington University School of Medicine, St. Louis, Missouri 63110*

THOMAS L. CROXTON (9), *Department of Physiology and Biophysics, Indiana University, School of Medicine, Indianapolis, Indiana 46223*

MICHAEL D. CULLER (37), *Reproductive Neuroendocrinology Section, Laboratory of Molecular and Integrative Neuroscience, National Institute of Environmental Health Sciences, National Institutes of Health, Research Triangle Park, North Carolina 27709*

KRISTINE D. DAHL (30), *Veterans Administration, Medical Center, Seattle, Washington 98108*

CARL DENEF (5), *Laboratory of Cell Pharmacology, University of Leuven, School of Medicine, B-3000 Leuven, Belgium*

ROBERT A. DESHARNAIS (59), *Laboratory of Populations, The Rockefeller University, New York, New York 10021*

JACK E. DIXON (6, 48), *Department of Biochemistry, Purdue University, West Lafayette, Indiana 47907*

CHARLES H. EMERSON (26, 40), *Department of Medicine, Division of Endocrinology, University of Massachusetts, School of Medicine, Worcester, Massachusetts 01655*

P. C. EMSON (54), *Medical Research Council (MRC) Group, Agricultural and Food*

Research Council (A.F.R.C.), Institute of Animal Physiology and Genetics Research (I.A.P.G.R.), Babraham, Cambridge CB2 4AT, England

JOHN ENG (19), *Solomon A. Berson Research Laboratory, Veterans Administration, Medical Center, Bronx, New York 10468, and Mount Sinai School of Medicine, City University of New York (CUNY), New York, New York 10029*

LAUREL A. FISHER (32), *Department of Pharmacology, The University of Arizona, College of Medicine, Tucson, Arizona 85724*

GARY FISKUM (35), *Department of Biochemistry, The George Washington University School of Medicine and Health Sciences, Washington, D.C. 20037*

EDWARD D. FRENCH (7), *Department of Pharmacology, The University of Arizona, College of Medicine, Tucson, Arizona 85724*

M. J. GAIT (54), *Medical Research Council (MRC), Laboratory of Molecular Biology, Cambridge CB2 2QH, England*

MARVIN C. GERSHENGORN (13), *Department of Medicine, Division of Endocrinology and Metabolism, Cornell University Medical College, New York, New York 10021*

MICHEL GOEDERT (34), *Medical Research Council (MRC), Laboratory of Molecular Biology, Cambridge CB2 2QH, England*

HECTOR A. GONZALEZ (27), *Servicio de Dermatologia, Hospital Clinico de la Universidad de Chile, Santos Dumont 999, Santiago, Chile*

ELI HAZUM (36, 52), *Department of Biochemistry, Glaxo Research Laboratories, Research Triangle Park, North Carolina 27709*

CARLA M. HEKMAN (31), *Department of Basic and Clinical Research, Scripps Clinic and Research Foundation, La Jolla, California 92037*

KLAUS HERMANN (39), *Dermatology Clinic, University of Munich, Munich, Federal Republic of Germany*

AARON J. W. HSUEH (30, 31), *Department of Reproductive Medicine, University of California, San Diego, La Jolla, California 92093*

JANE HUMM (24, 25), *Departments of Medicine and Pharmacology, The Neurobiology Curriculum and the Biological Sciences Research Center, The University of North Carolina at Chapel Hill, School of Medicine, Chapel Hill, North Carolina 27599*

W. C. HYMER (21), *Department of Molecular and Cell Biology, Paul M. Althouse Laboratory, The Pennsylvania State University, University Park, Pennsylvania 16802*

XIAO-CHI JIA (30), *Department of Reproductive Medicine, University of California, San Diego, La Jolla, California 92093*

DIANE JOLLEY (42), *The Clayton Foundation Laboratories for Peptide Biology, The Salk Institute, La Jolla, California 92037*

ABBA J. KASTIN (45), *Veterans Administration, Medical Center, and Tulane University School of Medicine, New Orleans, Louisiana 70146*

WOJCIECH KEDZIERSKI (27), *Department of Obstetrics and Gynecology, The University of Texas Southwestern Medical Center at Dallas, Dallas, Texas 75235*

MARCIA E. KENDALL (21), *Department of Molecular and Cell Biology, Paul M. Althouse Laboratory, The Pennsylvania State University, University Park, Pennsylvania 16802*

KEITH M. KENDRICK (11), *Department of Behavioural Physiology, Agricultural Food and Research Council (A.F.R.C.), Institute of Animal Physiology and Genetics Research (I.A.P.G.R.), Babraham, Cambridge CB2 4AT, England*

USHIO KIKKAWA (23), *Department of Biochemistry, Kobe University School of Medicine, Kobe 650, Japan*

JOHN S. KIZER (24, 25), *Departments of Medicine and Pharmacology, The Neurobiology Curriculum and the Biological*

Sciences Research Center, The University of North Carolina at Chapel Hill, School of Medicine, Chapel Hill, North Carolina 27599

JEANNE B. KOGER (25), *Departments of Medicine and Pharmacology, The Neurobiology Curriculum and the Biological Sciences Research Center, The University of North Carolina at Chapel Hill, School of Medicine, Chapel Hill, North Carolina 27599*

JAMES E. KRAUSE (44), *Department of Anatomy and Neurobiology, Washington University School of Medicine, St. Louis, Missouri 63110*

RUDOLPH G. KRAUSE II (58), *Medical Products Department, E. I. du Pont de Nemours & Company, Incorporated, Experimental Station Laboratory, Wilmington, Delaware 19898*

RICHARD M. KRIS (35), *Protein Chemistry, Rorer Biotechnology Inc., Rockville, Maryland 20850*

EDMUND F. LAGAMMA (47), *Department of Pediatrics and Department of Neurobiology and Behavior, State University of New York (SUNY) at Stony Brook, Stony Brook, New York 11794*

LAWRENCE H. LAZARUS (33), *Laboratory of Molecular and Integrative Neuroscience, Peptide Neurochemistry Group, National Institute of Environmental Health Sciences, National Institutes of Health, Research Triangle Park, North Carolina 27709*

DENIS A. LEONG (17), *Departments of Internal Medicine and Neuroscience, University of Virginia Medical Center, Charlottesville, Virginia 22908*

JON E. LEVINE (10), *Department of Neurobiology and Physiology, Northwestern University, Evanston, Illinois 60208*

MICHAEL E. LEWIS (55, 56, 58), *Cephalon, Inc., West Chester, Pennsylvania 19380*

CAROL D. LINDEN (35), *Department of Biochemistry, The George Washington University School of Medicine and Health Sciences, Washington, D.C. 20037*

BIN LIU (28), *Rudolf Magnus Institute for Pharmacology, Medical Faculty, University of Utrecht, 3521 GD Utrecht, The Netherlands*

WALTER C. LOW (15), *Department of Physiology and Biophysics, Indiana University, School of Medicine, Indianapolis, Indiana 46223*

CRAIG A. MCARDLE (18), *Institute for Hormone and Fertility Research (IHF), D-2000 Hamburg 54, Federal Republic of Germany*

JOSEPH T. MCCABE (59), *Laboratory of Neurobiology and Behavior, The Rockefeller University, New York, New York 10021*

S. M. MCCANN (51), *Department of Physiology, Neuropeptide Division, The University of Texas Southwestern Medical Center at Dallas, Dallas, Texas 75235*

RICHARD MCCLINTOCK (42), *The Clayton Foundation Laboratories for Peptide Biology, The Salk Institute, La Jolla, California 92037*

MARGARET R. MACDONALD (44), *Department of Anatomy and Neurobiology, Washington University School of Medicine, St. Louis, Missouri 63110*

PHILIPPE MAERTENS (5), *Laboratory of Cell Pharmacology, University of Leuven, School of Medicine, B-3000 Leuven, Belgium*

THOMAS F. J. MARTIN (14), *Department of Zoology, The University of Wisconsin, Madison, Wisconsin 53706*

GONZALO MARTÍNEZ DE LA ESCALERA (16), *Instituto de Investigaciones Biomédicas, Universidad Nacional Autónoma de México, 04510, Mexico DF., Mexico*

ANNICK MIGNON (5), *Laboratory of Cell Pharmacology, University of Leuven, School of Medicine, B-3000 Leuven, Belgium*

CAROLYN D. MINTH (48), *Department of Biochemistry, Purdue University, West Lafayette, Indiana 47907*

TERRY W. MOODY (35), *Department of Biochemistry, The George Washington Uni-*

versity School of Medicine and Health Sciences, Washington, D.C. 20037

ICHIRO MURAI (15), *Department of Physiology and Biophysics, Indiana University, School of Medicine, Indianapolis, Indiana 46223*

ANDRÉS NEGRO-VILAR (37, 38), *Reproductive Neuroendocrinology Section, Laboratory of Molecular and Integrative Neuroscience, National Institute of Environmental Health Sciences, National Institutes of Health, Research Triangle Park, North Carolina 27709*

SIMON NEUBORT (41), *Neurology Research Center, State of New York, Department of Health, Helen Hayes Hospital, West Haverstraw, New York 10993*

YASUTOMI NISHIZUKA (23), *Department of Biochemistry, Kobe University School of Medicine, Kobe 650, Japan*

THOMAS L. O'DONOHUE[1] (1), *Central Nervous System Disease Research, Monsanto-G. D. Searle and Company, St. Louis, Missouri 63198*

KOUJI OGITA (23), *Department of Biochemistry, Kobe University School of Medicine, Kobe 650, Japan*

REINHARD A. PALOVCIK (8), *Department of Neuroscience, University of Florida, College of Medicine, Gainesville, Florida 32610*

DONALD W. PFAFF (59), *Laboratory of Neurobiology and Behavior, The Rockefeller University, New York, New York 10021*

M. IAN PHILLIPS (8, 39), *Department of Physiology, University of Florida, College of Medicine, Gainesville, Florida 32610*

J. M. POLAK (57), *Department of Histochemistry, Royal Postgraduate Medical School, University of London, Hammersmith Hospital, London W12 0NN, England*

JOHN C. PORTER (27), *Departments of Obstetrics and Gynecology and of Physiology, The University of Texas Southwestern Medical Center at Dallas, Dallas, Texas 75235*

KEITH D. POWELL (10), *Department of Neurobiology and Physiology, Northwestern University, Evanston, Illinois 60208*

MOHAN K. RAIZADA (39), *Department of Physiology, University of Florida, College of Medicine, Gainesville, Florida 32610*

DENNIS D. RASMUSSEN (12), *Department of Reproductive Medicine, University of California, San Diego, La Jolla, California 92093*

JOSEPH R. REEVE, JR. (46), *Department of Medicine, UCLA School of Medicine, University of California, Los Angeles, Los Angeles, California 90024*

CATHERINE RIVIER (42), *The Clayton Foundation Laboratories for Peptide Biology, The Salk Institute, La Jolla, California 92037*

JEAN RIVIER (42), *The Clayton Foundation Laboratories for Peptide Biology, The Salk Institute, La Jolla, California 92037*

WIM ROBBERECHT (5), *Laboratory of Cell Pharmacology, University of Leuven, School of Medicine, B-3000 Leuven, Belgium*

WADE T. ROGERS (58), *Engineering Physics Laboratory, E. I. du Pont de Nemours & Company, Incorporated, Experimental Station Laboratory, Wilmington, Delaware 19898*

JOSEPH SCHLESSINGER (35), *Rorer Biotechnology Inc., Rockville, Maryland 20850*

JAMES S. SCHWABER (56, 58), *Medical Products Department, E. I. du Pont de Nemours & Company, Incorporated, Experimental Station Laboratory, Wilmington, Delaware 19898*

JEFFREY SCHWARTZ (4), *Medical Research Centre, Prince Henry's Hospital, Melbourne, Australia 3004*

CAROLYN D. SCOTT (20), *Department of Endocrinology, Royal Prince Alfred Hospital, Camperdown, New South Wales, Australia 2050*

[1] Deceased.

MARK S. SHEARMAN (23), *Department of Biochemistry, Kobe University School of Medicine, Kobe 650, Japan*

KAREN C. SWEARINGEN (16), *Department of Biology, Mills College, Oakland, California 94613*

LUC SWENNEN (5), *Laboratory of Cell Pharmacology, University of Leuven, School of Medicine, B-3000 Leuven, Belgium*

JAMES P. TAM (2), *The Rockefeller University, New York, New York 10021*

GEORGE R. UHL (53), *Gene Neuroscience Unit and Departments of Neurology and Neuroscience, Addiction Research Center, National Institute on Drug Abuse and The Johns Hopkins School of Medicine, Baltimore, Maryland 21224*

WYLIE VALE (4, 42), *The Clayton Foundation Laboratories for Peptide Biology, The Salk Institute, La Jolla, California 92037*

GUY VALIQUETTE (41), *Department of Medicine, State of New York, Department of Health, Helen Hayes Hospital, West Haverstraw, New York 10993, and Department of Neurology, College of Physicians and Surgeons of Columbia University, New York, New York 10032*

HUBERT H. M. VAN TOL (29), *Rudolf Magnus Institute for Pharmacology, Medical Faculty, University of Utrecht, 3521 GD Utrecht, The Netherlands*

JOAN M. VAUGHAN (42), *The Clayton Foundation Laboratories for Peptide Biology, The Salk Institute, La Jolla, California 92037*

JOSEF K. VOGLMAYR (42), *Division of Reproductive Biology, Florida Institute of Technology, Melbourne, Florida 32901*

JOHN H. WALSH (46), *Department of Medicine, UCLA School of Medicine, University of California, Los Angeles, Los Angeles, California 90024*

PAULUS S. WANG (27), *Department of Physiology, National Yang-Ming Medical College, Shih-Pai, Taipei, Taiwan 11121, Republic of China*

RICHARD I. WEINER (16), *Department of Obstetrics, Gynecology and Reproductive Sciences, School of Medicine, University of California, San Francisco, San Francisco, California 94143*

W. WETSEL (38), *Reproductive Neuroendocrinology Section, Laboratory of Molecular and Integrative Neuroscience, National Institute of Environmental Health Sciences, National Institutes of Health, Research Triangle Park, North Carolina 27709*

JEFFREY D. WHITE (47), *Division of Endocrinology, Department of Medicine and Department of Neurobiology and Behavior, State University of New York (SUNY) at Stony Brook, Stony Brook, New York 11794*

JOHN T. WILLIAMS (7), *Institute for Advanced Biomedical Research, The Oregon Health Sciences University, Portland, Oregon 97201*

WILLIAM E. WILSON (33), *Laboratory of Molecular and Integrative Neuroscience, Peptide Neurochemistry Group, National Institute of Environmental Health Sciences, National Institutes of Health, Research Triangle Park, North Carolina 27709*

IRENE WINICOV (13), *Department of Medicine, Division of Endocrinology and Metabolism, Cornell University Medical College, New York, New York 10021*

ROSALYN S. YALOW (19), *Solomon A. Berson Research Laboratory, Veterans Administration, Medical Center, Bronx, New York 10468, and Mount Sinai School of Medicine, City University of New York (CUNY), New York, New York 10029*

W. SCOTT YOUNG III (49), *Laboratory of Cell Biology, National Institute of Mental Health, National Institutes of Health, Bethesda, Maryland 20892*

Preface

This volume supplements Volumes 103 and 124 of *Methods in Enzymology,* which also deal with neuroendocrine peptides, and is considerably less archival than its predecessors. In many cases the methods presented include shortcuts and conveniences not included in the sources from which they were taken. The techniques are described in a context that allows comparisons to other related methodologies. The authors have been encouraged to do this in the belief that such comparisons are valuable to readers who must adapt extant procedures to new systems. Also, so far as possible, methodologies have been presented in a manner that stresses their general applicability and potential limitations. Special attention has been paid to statistics and means of data analysis and assessment.

Although some topics for various reasons are not covered, the three volumes dealing with neuroendocrine peptides provide a substantial and current overview of the extant methodology in the field and a view of its rapid development.

This volume does not include a cross-index since the recent, preceding volumes included comprehensive indexes. Since publication of Volume 124, most ancillary information is contained in *Methods in Enzymology* Volumes 139 [*Cellular Regulators* (*Part A: Calcium- and Calmodulin-Binding Proteins*)] and 141 [*Cellular Regulators* (*Part B: Calcium and Lipids*)].

Particular thanks go to the authors for their attention to meeting deadlines and for maintaining high standards of quality, to the series editors for their encouragement, and to the staff of Academic Press for their help and timely publication of the volume.

P. MICHAEL CONN

METHODS IN ENZYMOLOGY

VOLUME I. Preparation and Assay of Enzymes
Edited by SIDNEY P. COLOWICK AND NATHAN O. KAPLAN

VOLUME II. Preparation and Assay of Enzymes
Edited by SIDNEY P. COLOWICK AND NATHAN O. KAPLAN

VOLUME III. Preparation and Assay of Substrates
Edited by SIDNEY P. COLOWICK AND NATHAN O. KAPLAN

VOLUME IV. Special Techniques for the Enzymologist
Edited by SIDNEY P. COLOWICK AND NATHAN O. KAPLAN

VOLUME V. Preparation and Assay of Enzymes
Edited by SIDNEY P. COLOWICK AND NATHAN O. KAPLAN

VOLUME VI. Preparation and Assay of Enzymes (*Continued*)
Preparation and Assay of Substrates
Special Techniques
Edited by SIDNEY P. COLOWICK AND NATHAN O. KAPLAN

VOLUME VII. Cumulative Subject Index
Edited by SIDNEY P. COLOWICK AND NATHAN O. KAPLAN

VOLUME VIII. Complex Carbohydrates
Edited by ELIZABETH F. NEUFELD AND VICTOR GINSBURG

VOLUME IX. Carbohydrate Metabolism
Edited by WILLIS A. WOOD

VOLUME X. Oxidation and Phosphorylation
Edited by RONALD W. ESTABROOK AND MAYNARD E. PULLMAN

VOLUME XI. Enzyme Structure
Edited by C. H. W. HIRS

VOLUME XII. Nucleic Acids (Parts A and B)
Edited by LAWRENCE GROSSMAN AND KIVIE MOLDAVE

VOLUME XXVII. Enzyme Structure (Part D)
Edited by C. H. W. HIRS AND SERGE N. TIMASHEFF

VOLUME XXVIII. Complex Carbohydrates (Part B)
Edited by VICTOR GINSBURG

VOLUME XXIX. Nucleic Acids and Protein Synthesis (Part E)
Edited by LAWRENCE GROSSMAN AND KIVIE MOLDAVE

VOLUME XXX. Nucleic Acids and Protein Synthesis (Part F)
Edited by KIVIE MOLDAVE AND LAWRENCE GROSSMAN

VOLUME XXXI. Biomembranes (Part A)
Edited by SIDNEY FLEISCHER AND LESTER PACKER

VOLUME XXXII. Biomembranes (Part B)
Edited by SIDNEY FLEISCHER AND LESTER PACKER

VOLUME XXXIII. Cumulative Subject Index Volumes I–XXX
Edited by MARTHA G. DENNIS AND EDWARD A. DENNIS

VOLUME XXXIV. Affinity Techniques (Enzyme Purification: Part B)
Edited by WILLIAM B. JAKOBY AND MEIR WILCHEK

VOLUME XXXV. Lipids (Part B)
Edited by JOHN M. LOWENSTEIN

VOLUME XXXVI. Hormone Action (Part A: Steroid Hormones)
Edited by BERT W. O'MALLEY AND JOEL G. HARDMAN

VOLUME XXXVII. Hormone Action (Part B: Peptide Hormones)
Edited by BERT W. O'MALLEY AND JOEL G. HARDMAN

VOLUME XXXVIII. Hormone Action (Part C: Cyclic Nucleotides)
Edited by JOEL G. HARDMAN AND BERT W. O'MALLEY

VOLUME XXXIX. Hormone Action (Part D: Isolated Cells, Tissues, and Organ Systems)
Edited by JOEL G. HARDMAN AND BERT W. O'MALLEY

VOLUME XL. Hormone Action (Part E: Nuclear Structure and Function)
Edited by BERT W. O'MALLEY AND JOEL G. HARDMAN

VOLUME XLI. Carbohydrate Metabolism (Part B)
Edited by W. A. WOOD

VOLUME XLII. Carbohydrate Metabolism (Part C)
Edited by W. A. WOOD

VOLUME XLIII. Antibiotics
Edited by JOHN H. HASH

VOLUME XLIV. Immobilized Enzymes
Edited by KLAUS MOSBACH

VOLUME XLV. Proteolytic Enzymes (Part B)
Edited by LASZLO LORAND

VOLUME XLVI. Affinity Labeling
Edited by WILLIAM B. JAKOBY AND MEIR WILCHEK

VOLUME XLVII. Enzyme Structure (Part E)
Edited by C. H. W. HIRS AND SERGE N. TIMASHEFF

VOLUME XLVIII. Enzyme Structure (Part F)
Edited by C. H. W. HIRS AND SERGE N. TIMASHEFF

VOLUME XLIX. Enzyme Structure (Part G)
Edited by C. H. W. HIRS AND SERGE N. TIMASHEFF

VOLUME L. Complex Carbohydrates (Part C)
Edited by VICTOR GINSBURG

VOLUME LI. Purine and Pyrimidine Nucleotide Metabolism
Edited by PATRICIA A. HOFFEE AND MARY ELLEN JONES

VOLUME LII. Biomembranes (Part C: Biological Oxidations)
Edited by SIDNEY FLEISCHER AND LESTER PACKER

VOLUME LIII. Biomembranes (Part D: Biological Oxidations)
Edited by SIDNEY FLEISCHER AND LESTER PACKER

VOLUME LIV. Biomembranes (Part E: Biological Oxidations)
Edited by SIDNEY FLEISCHER AND LESTER PACKER

VOLUME LV. Biomembranes (Part F: Bioenergetics)
Edited by SIDNEY FLEISCHER AND LESTER PACKER

VOLUME LVI. Biomembranes (Part G: Bioenergetics)
Edited by SIDNEY FLEISCHER AND LESTER PACKER

VOLUME LVII. Bioluminescence and Chemiluminescence
Edited by MARLENE A. DELUCA

VOLUME LVIII. Cell Culture
Edited by WILLIAM B. JAKOBY AND IRA PASTAN

VOLUME LIX. Nucleic Acids and Protein Synthesis (Part G)
Edited by KIVIE MOLDAVE AND LAWRENCE GROSSMAN

VOLUME LX. Nucleic Acids and Protein Synthesis (Part H)
Edited by KIVIE MOLDAVE AND LAWRENCE GROSSMAN

VOLUME 61. Enzyme Structure (Part H)
Edited by C. H. W. HIRS AND SERGE N. TIMASHEFF

VOLUME 62. Vitamins and Coenzymes (Part D)
Edited by DONALD B. MCCORMICK AND LEMUEL D. WRIGHT

VOLUME 63. Enzyme Kinetics and Mechanism (Part A: Initial Rate and Inhibitor Methods)
Edited by DANIEL L. PURICH

VOLUME 64. Enzyme Kinetics and Mechanism (Part B: Isotopic Probes and Complex Enzyme Systems)
Edited by DANIEL L. PURICH

VOLUME 65. Nucleic Acids (Part I)
Edited by LAWRENCE GROSSMAN AND KIVIE MOLDAVE

VOLUME 66. Vitamins and Coenzymes (Part E)
Edited by DONALD B. MCCORMICK AND LEMUEL D. WRIGHT

VOLUME 67. Vitamins and Coenzymes (Part F)
Edited by DONALD B. MCCORMICK AND LEMUEL D. WRIGHT

VOLUME 68. Recombinant DNA
Edited by RAY WU

VOLUME 69. Photosynthesis and Nitrogen Fixation (Part C)
Edited by ANTHONY SAN PIETRO

VOLUME 70. Immunochemical Techniques (Part A)
Edited by HELEN VAN VUNAKIS AND JOHN J. LANGONE

VOLUME 71. Lipids (Part C)
Edited by JOHN M. LOWENSTEIN

VOLUME 72. Lipids (Part D)
Edited by JOHN M. LOWENSTEIN

VOLUME 73. Immunochemical Techniques (Part B)
Edited by JOHN J. LANGONE AND HELEN VAN VUNAKIS

VOLUME 74. Immunochemical Techniques (Part C)
Edited by JOHN J. LANGONE AND HELEN VAN VUNAKIS

VOLUME 75. Cumulative Subject Index Volumes XXXI, XXXII, XXXIV–LX
Edited by EDWARD A. DENNIS AND MARTHA G. DENNIS

VOLUME 76. Hemoglobins
Edited by ERALDO ANTONINI, LUIGI ROSSI-BERNARDI, AND EMILIA CHIANCONE

VOLUME 77. Detoxication and Drug Metabolism
Edited by WILLIAM B. JAKOBY

VOLUME 78. Interferons (Part A)
Edited by SIDNEY PESTKA

VOLUME 79. Interferons (Part B)
Edited by SIDNEY PESTKA

VOLUME 80. Proteolytic Enzymes (Part C)
Edited by LASZLO LORAND

VOLUME 81. Biomembranes (Part H: Visual Pigments and Purple Membranes, I)
Edited by LESTER PACKER

VOLUME 82. Structural and Contractile Proteins (Part A: Extracellular Matrix)
Edited by LEON W. CUNNINGHAM AND DIXIE W. FREDERIKSEN

VOLUME 83. Complex Carbohydrates (Part D)
Edited by VICTOR GINSBURG

VOLUME 84. Immunochemical Techniques (Part D: Selected Immunoassays)
Edited by JOHN J. LANGONE AND HELEN VAN VUNAKIS

VOLUME 85. Structural and Contractile Proteins (Part B: The Contractile Apparatus and the Cytoskeleton)
Edited by DIXIE W. FREDERIKSEN AND LEON W. CUNNINGHAM

VOLUME 86. Prostaglandins and Arachidonate Metabolites
Edited by WILLIAM E. M. LANDS AND WILLIAM L. SMITH

VOLUME 87. Enzyme Kinetics and Mechanism (Part C: Intermediates, Stereochemistry, and Rate Studies)
Edited by DANIEL L. PURICH

VOLUME 88. Biomembranes (Part I: Visual Pigments and Purple Membranes, II)
Edited by LESTER PACKER

VOLUME 89. Carbohydrate Metabolism (Part D)
Edited by WILLIS A. WOOD

VOLUME 90. Carbohydrate Metabolism (Part E)
Edited by WILLIS A. WOOD

VOLUME 91. Enzyme Structure (Part I)
Edited by C. H. W. HIRS AND SERGE N. TIMASHEFF

VOLUME 92. Immunochemical Techniques (Part E: Monoclonal Antibodies and General Immunoassay Methods)
Edited by JOHN J. LANGONE AND HELEN VAN VUNAKIS

VOLUME 93. Immunochemical Techniques (Part F: Conventional Antibodies, Fc Receptors, and Cytotoxicity)
Edited by JOHN J. LANGONE AND HELEN VAN VUNAKIS

VOLUME 94. Polyamines
Edited by HERBERT TABOR AND CELIA WHITE TABOR

VOLUME 95. Cumulative Subject Index Volumes 61–74, 76–80
Edited by EDWARD A. DENNIS AND MARTHA G. DENNIS

VOLUME 96. Biomembranes [Part J: Membrane Biogenesis: Assembly and Targeting (General Methods; Eukaryotes)]
Edited by SIDNEY FLEISCHER AND BECCA FLEISCHER

VOLUME 97. Biomembranes [Part K: Membrane Biogenesis: Assembly and Targeting (Prokaryotes, Mitochondria, and Chloroplasts)]
Edited by SIDNEY FLEISCHER AND BECCA FLEISCHER

VOLUME 98. Biomembranes (Part L: Membrane Biogenesis: Processing and Recycling)
Edited by SIDNEY FLEISCHER AND BECCA FLEISCHER

VOLUME 99. Hormone Action (Part F: Protein Kinases)
Edited by JACKIE D. CORBIN AND JOEL G. HARDMAN

VOLUME 100. Recombinant DNA (Part B)
Edited by RAY WU, LAWRENCE GROSSMAN, AND KIVIE MOLDAVE

VOLUME 101. Recombinant DNA (Part C)
Edited by RAY WU, LAWRENCE GROSSMAN, AND KIVIE MOLDAVE

VOLUME 102. Hormone Action (Part G: Calmodulin and Calcium-Binding Proteins)
Edited by ANTHONY R. MEANS AND BERT W. O'MALLEY

VOLUME 103. Hormone Action (Part H: Neuroendocrine Peptides)
Edited by P. MICHAEL CONN

VOLUME 104. Enzyme Purification and Related Techniques (Part C)
Edited by WILLIAM B. JAKOBY

VOLUME 105. Oxygen Radicals in Biological Systems
Edited by LESTER PACKER

VOLUME 106. Posttranslational Modifications (Part A)
Edited by FINN WOLD AND KIVIE MOLDAVE

VOLUME 107. Posttranslational Modifications (Part B)
Edited by FINN WOLD AND KIVIE MOLDAVE

VOLUME 108. Immunochemical Techniques (Part G: Separation and Characterization of Lymphoid Cells)
Edited by GIOVANNI DI SABATO, JOHN J. LANGONE, AND HELEN VAN VUNAKIS

VOLUME 109. Hormone Action (Part I: Peptide Hormones)
Edited by LUTZ BIRNBAUMER AND BERT W. O'MALLEY

VOLUME 110. Steroids and Isoprenoids (Part A)
Edited by JOHN H. LAW AND HANS C. RILLING

VOLUME 111. Steroids and Isoprenoids (Part B)
Edited by JOHN H. LAW AND HANS C. RILLING

VOLUME 112. Drug and Enzyme Targeting (Part A)
Edited by KENNETH J. WIDDER AND RALPH GREEN

VOLUME 113. Glutamate, Glutamine, Glutathione, and Related Compounds
Edited by ALTON MEISTER

VOLUME 114. Diffraction Methods for Biological Macromolecules (Part A)
Edited by HAROLD W. WYCKOFF, C. H. W. HIRS, AND SERGE N. TIMASHEFF

VOLUME 115. Diffraction Methods for Biological Macromolecules (Part B)
Edited by HAROLD W. WYCKOFF, C. H. W. HIRS, AND SERGE N. TIMASHEFF

VOLUME 116. Immunochemical Techniques (Part H: Effectors and Mediators of Lymphoid Cell Functions)
Edited by GIOVANNI DI SABATO, JOHN J. LANGONE, AND HELEN VAN VUNAKIS

VOLUME 117. Enzyme Structure (Part J)
Edited by C. H. W. HIRS AND SERGE N. TIMASHEFF

VOLUME 118. Plant Molecular Biology
Edited by ARTHUR WEISSBACH AND HERBERT WEISSBACH

VOLUME 119. Interferons (Part C)
Edited by SIDNEY PESTKA

VOLUME 120. Cumulative Subject Index Volumes 81–94, 96–101

VOLUME 121. Immunochemical Techniques (Part I: Hybridoma Technology and Monoclonal Antibodies)
Edited by JOHN J. LANGONE AND HELEN VAN VUNAKIS

VOLUME 122. Vitamins and Coenzymes (Part G)
Edited by FRANK CHYTIL AND DONALD B. MCCORMICK

VOLUME 123. Vitamins and Coenzymes (Part H)
Edited by FRANK CHYTIL AND DONALD B. MCCORMICK

VOLUME 124. Hormone Action (Part J: Neuroendocrine Peptides)
Edited by P. MICHAEL CONN

VOLUME 125. Biomembranes (Part M: Transport in Bacteria, Mitochondria, and Chloroplasts: General Approaches and Transport Systems)
Edited by SIDNEY FLEISCHER AND BECCA FLEISCHER

VOLUME 126. Biomembranes (Part N: Transport in Bacteria, Mitochondria, and Chloroplasts: Protonmotive Force)
Edited by SIDNEY FLEISCHER AND BECCA FLEISCHER

VOLUME 127. Biomembranes (Part O: Protons and Water: Structure and Translocation)
Edited by LESTER PACKER

VOLUME 128. Plasma Lipoproteins (Part A: Preparation, Structure, and Molecular Biology)
Edited by JERE P. SEGREST AND JOHN J. ALBERS

VOLUME 129. Plasma Lipoproteins (Part B: Characterization, Cell Biology, and Metabolism)
Edited by JOHN J. ALBERS AND JERE P. SEGREST

VOLUME 130. Enzyme Structure (Part K)
Edited by C. H. W. HIRS AND SERGE N. TIMASHEFF

VOLUME 131. Enzyme Structure (Part L)
Edited by C. H. W. HIRS AND SERGE N. TIMASHEFF

VOLUME 132. Immunochemical Techniques (Part J: Phagocytosis and Cell-Mediated Cytotoxicity)
Edited by GIOVANNI DI SABATO AND JOHANNES EVERSE

VOLUME 133. Bioluminescence and Chemiluminescence (Part B)
Edited by MARLENE DELUCA AND WILLIAM D. MCELROY

VOLUME 134. Structural and Contractile Proteins (Part C: The Contractile Apparatus and the Cytoskeleton)
Edited by RICHARD B. VALLEE

VOLUME 135. Immobilized Enzymes and Cells (Part B)
Edited by KLAUS MOSBACH

VOLUME 136. Immobilized Enzymes and Cells (Part C)
Edited by KLAUS MOSBACH

VOLUME 137. Immobilized Enzymes and Cells (Part D)
Edited by KLAUS MOSBACH

VOLUME 138. Complex Carbohydrates (Part E)
Edited by VICTOR GINSBURG

VOLUME 139. Cellular Regulators (Part A: Calcium- and Calmodulin-Binding Proteins)
Edited by ANTHONY R. MEANS AND P. MICHAEL CONN

VOLUME 140. Cumulative Subject Index Volumes 102–119, 121–134

VOLUME 141. Cellular Regulators (Part B: Calcium and Lipids)
Edited by P. MICHAEL CONN AND ANTHONY R. MEANS

VOLUME 142. Metabolism of Aromatic Amino Acids and Amines
Edited by SEYMOUR KAUFMAN

VOLUME 143. Sulfur and Sulfur Amino Acids
Edited by WILLIAM B. JAKOBY AND OWEN GRIFFITH

VOLUME 144. Structural and Contractile Proteins (Part D: Extracellular Matrix)
Edited by LEON W. CUNNINGHAM

VOLUME 145. Structural and Contractile Proteins (Part E: Extracellular Matrix)
Edited by LEON W. CUNNINGHAM

VOLUME 146. Peptide Growth Factors (Part A)
Edited by DAVID BARNES AND DAVID A. SIRBASKU

VOLUME 147. Peptide Growth Factors (Part B)
Edited by DAVID BARNES AND DAVID A. SIRBASKU

VOLUME 148. Plant Cell Membranes
Edited by LESTER PACKER AND ROLAND DOUCE

VOLUME 149. Drug and Enzyme Targeting (Part B)
Edited by RALPH GREEN AND KENNETH J. WIDDER

VOLUME 150. Immunochemical Techniques (Part K: *In Vitro* Models of B and T Cell Functions and Lymphoid Cell Receptors)
Edited by GIOVANNI DI SABATO

VOLUME 151. Molecular Genetics of Mammalian Cells
Edited by MICHAEL M. GOTTESMAN

VOLUME 152. Guide to Molecular Cloning Techniques
Edited by SHELBY L. BERGER AND ALAN R. KIMMEL

VOLUME 153. Recombinant DNA (Part D)
Edited by RAY WU AND LAWRENCE GROSSMAN

VOLUME 154. Recombinant DNA (Part E)
Edited by RAY WU AND LAWRENCE GROSSMAN

VOLUME 155. Recombinant DNA (Part F)
Edited by RAY WU

VOLUME 156. Biomembranes (Part P: ATP-Driven Pumps and Related Transport: The Na,K-Pump)
Edited by SIDNEY FLEISCHER AND BECCA FLEISCHER

VOLUME 157. Biomembranes (Part Q: ATP-Driven Pumps and Related Transport: Calcium, Proton, and Potassium Pumps)
Edited by SIDNEY FLEISCHER AND BECCA FLEISCHER

VOLUME 158. Metalloproteins (Part A)
Edited by JAMES F. RIORDAN AND BERT L. VALLEE

VOLUME 159. Initiation and Termination of Cyclic Nucleotide Action
Edited by JACKIE D. CORBIN AND ROGER A. JOHNSON

VOLUME 160. Biomass (Part A: Cellulose and Hemicellulose)
Edited by WILLIS A. WOOD AND SCOTT T. KELLOGG

VOLUME 161. Biomass (Part B: Lignin, Pectin, and Chitin)
Edited by WILLIS A. WOOD AND SCOTT T. KELLOGG

VOLUME 162. Immunochemical Techniques (Part L: Chemotaxis and Inflammation)
Edited by GIOVANNI DI SABATO

VOLUME 163. Immunochemical Techniques (Part M: Chemotaxis and Inflammation)
Edited by GIOVANNI DI SABATO

VOLUME 164. Ribosomes
Edited by HARRY F. NOLLER, JR., AND KIVIE MOLDAVE

VOLUME 165. Microbial Toxins: Tools for Enzymology
Edited by SIDNEY HARSHMAN

VOLUME 166. Branched-Chain Amino Acids
Edited by ROBERT HARRIS AND JOHN R. SOKATCH

VOLUME 167. Cyanobacteria
Edited by LESTER PACKER AND ALEXANDER N. GLAZER

VOLUME 168. Hormone Action (Part K: Neuroendocrine Peptides)
Edited by P. MICHAEL CONN

VOLUME 169. Platelets: Receptors, Adhesion, Secretion (Part A) (in preparation)
Edited by JACEK HAWIGER

VOLUME 170. Nucleosomes (in preparation)
Edited by PAUL M. WASSARMAN AND ROGER D. KORNBERG

VOLUME 171. Biomembranes (Part R: Transport Theory: Cells and Model Membranes) (in preparation)
Edited by SIDNEY FLEISCHER AND BECCA FLEISCHER

VOLUME 172. Biomembranes (Part S: Transport: Membrane Isolation and Characterization) (in preparation)
Edited by SIDNEY FLEISCHER AND BECCA FLEISCHER

VOLUME 173. Biomembranes [Part T: Cellular and Subcellular Transport: Eukaryotic (Nonepithelial) Cells] (in preparation)
Edited by SIDNEY FLEISCHER AND BECCA FLEISCHER

VOLUME 174. Biomembranes [Part U: Cellular and Subcellular Transport: Eukaryotic (Nonepithelial) Cells] (in preparation)
Edited by SIDNEY FLEISCHER AND BECCA FLEISCHER

VOLUME 175. Cumulative Subject Index Volumes 135–139, 141–167 (in preparation)

VOLUME 176. Nuclear Magnetic Resonance (Part A: Spectral Techniques and Dynamics) (in preparation)
Edited by NORMAN J. OPPENHEIMER AND THOMAS L. JAMES

VOLUME 177. Nuclear Magnetic Resonance (Part B: Structure and Mechanism) (in preparation)
Edited by NORMAN J. OPPENHEIMER AND THOMAS L. JAMES

VOLUME 178. Antibodies, Antigens, and Molecular Mimicry (in preparation)
Edited by JOHN J. LANGONE

Section I

Preparation of Chemical Probes

[1] Strategy and Methodology for Development of Antisera against Procholecystokinin

By MARGERY C. BEINFELD and THOMAS L. O'DONOHUE[1]

Introduction

Recent recombinant DNA studies have indicated that cholecystokinin (CCK),[2] like many other biologically active peptides, originates as a large precursor in which CCK is flanked by both amino- and carboxyl-terminal segments. Like many other peptides, it is uncommon for antisera against the processed peptide to detect their respective high molecular weight precursors. The reason for this is not completely clear, although at least two possibilities exist: (1) the processed peptide may contain posttranslationally modified amino acids (like sulfated tyrosine, amidated carboxyl-terminal amino acid, or acetylated serine residue) which are not present in the precursor but which are modified after cleavage of the precursor. The processed peptide antisera may recognize these modified amino acids and fail to recognize the precursor. (2) The tertiary structure of the precursor in solution may be such that the processed peptide is not on the surface of the precursor where it is accessible to interaction with the antisera.

General Strategy and Rationale

From the proposed sequence of rat prepro-CCK determined by recombinant DNA techniques,[2] four peptides were obtained, and their relationship to the sequence of prepro-CCK is shown in Fig. 1. The single-letter amino acid code was utilized, and the nomenclature of the company that synthesized them (Penninsula Labs) has been adopted. In this nomenclature, D-10-Y signifies a 10-amino-acid peptide with an amino-terminal aspartic acid and a carboxyl-terminal tyrosine. These peptides were conjugated to bovine serum albumin (BSA) and were used to raise antisera in rabbits.

The following considerations influenced the selection of these sequences as antigens: (1) the sequences do not straddle the presumed cleavage sites, indicated by the arrows in Fig. 1 (an exception was made

[1] Deceased.

[2] R. J. Deschenes, L. J. Lorenz, R. S. Haun, B. A. Roos, K. J. Collier, and J. E. Dixon, *Proc. Natl. Acad. Sci. U.S.A.* **81,** 726 (1984).

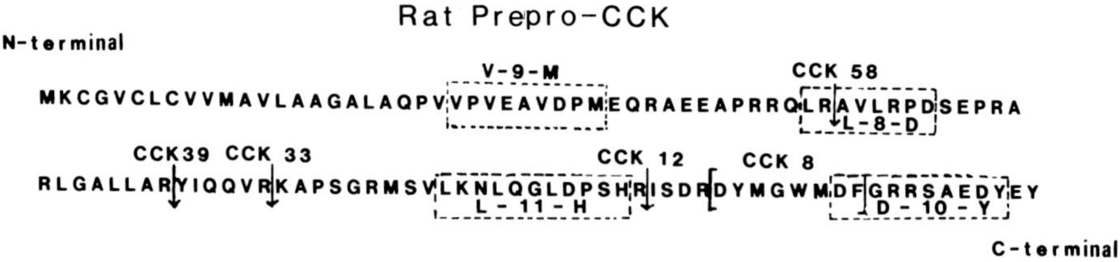

Rat Prepro-CCK

N-terminal

V-9-M CCK 58

MKCGVCLCVVMAVLAAGALAQPVVPVEAVDPMEQRAEEAPRRQLRAVLRPDSEPRA

L-8-D

CCK39 CCK 33 CCK 12 CCK 8

RLGALLARYIQQVRKAPSGRMSVLKNLQGLDPSHRISDRDYMGWMDFGRRSAEDYEY

L - 11 - H D - 10 - Y

C-terminal

FIG. 1. Sequence of rat prepro-CCK.

for L-8-D), (2) the signal sequence (assumed to include about 20 amino acids from the amino terminus) is not included, (3) as many proline residues as possible are included, (4) serine residues are excluded, and (5) in the case of D-10-Y, a little of CCK8 is included (two amino acids) so that it will cross-react with carboxyl-terminal-extended CCK peptides.

Proline residues were included because proline is considered to be a helix-breaking amino acid. Therefore, the proline-rich regions of the precursor might be more likely to be found on the surface where they would be accessible to antibody binding. Serine residues were excluded because in the precursor they may be acetylated or glycosylated posttranslationally and thus might prevent the antisera from detecting them.

Methods

Synthetic Peptides

V-9-M, L-8-D, and D-10-Y are synthesized and purified by Penninsula Labs. The peptide I-11-H, which is the porcine equivalent of L-11-H, is available from Bachem (Torrance, CA). Neither V-9-M or L-8-D can be labeled with the chloramine-T method,[3] since they lack a tyrosine or histidine. Additional tyrosine-extended peptides are made for the purpose of labeling during the synthesis of V-9-M and L-8-D. When the synthesis of V-9-M and L-8-D is complete, but before they are cleaved from the resin, their resins are divided in half and a tyrosine is added to one-half of the resin, creating Y-10-M and Y-9-D along with V-9-M and L-8-D. Fortunately, the antisera raised against V-9-M and L-8-D cross-react completely with these tyrosine-extended peptides so that it is possible to use them (when labeled) in the V-9-M and L-8-D radioimmunoassays (RIAs).

Iodination

All of the synthetic peptides used as tracers, Y-10-M, Y-9-D, I-11-H, and D-10-Y, are iodinated with the chloramine-T method.[3] The three

[3] W. M. Hunter and F. C. Greenwood, *Biochem. J.* **89,** 114 (1963).

tyrosine-containing peptides can be labeled to very high specific activity, with incorporation of greater than 90% of the iodine into the peptide. The I-11-H peptide incorporates much less iodine under the same reaction conditions, but it still works as a tracer in the I-11-H RIA. The reaction mixture consists of 2 μg of peptide in 75 μl of 0.25 *M* phosphate buffer, pH 7.5, 1 mCi ^{125}I in a 1.5-ml Eppendorf tube. Ten microliters of chloramine-T (2 mg/ml) is added, mixed, and, after 15 sec, 20 μl sodium metabisulfite, 4 mg/ml, is added and the reaction mixed again. The iodinated peptide is separated from the free iodine by Sephadex G-10 chromatography in a 10-ml disposable plastic pipet eluted with standard assay buffer: 50 m*M* Tris–HCl, 0.2% sodium azide, 1% BSA, pH 7.5. The trace is stored at −20° for up to 2 months.

Peptide Conjugation to BSA

The peptides are conjugated to BSA with a peptide-to-BSA molar ratio of 10 : 1 with a final concentration of about 0.1 mg/ml peptide, 6.8 mg/ml BSA, 0.2 *M* borate buffer, pH 8.5. A small amount of iodinated peptide (~500,000 cpm) is added so that the degree of conjugation may be tested. Glutaraldehyde (22 m*M*, freshly diluted) is added dropwise to a final concentration of 5 m*M*, and the solution (final volume ~10 ml) is placed on an aliquot mixing rocker overnight at room temperature. A small aliquot (about 100 μl) is diluted to 1.0 ml with water in an Eppendorf tube, and the degree of incorporation of labeled peptide into BSA is tested by addition of 40% trichloroacetic acid (TCA) dropwise until a precipitate is obtained. The tube is centrifuged and the pellet and supernatant counted. Typically, the degree of incorporation of peptide into BSA ranges from about 56 to 86%.

Production of Antisera

The initial immunization is 100 μg equivalent of peptide conjugated to BSA emulsified in an equal volume of Freund's complete adjuvant and injected in multiple intradermal sites in the backs of at least two male New Zealand White rabbits. The booster injections at about monthly intervals consist of 50 μg peptide equivalents emulsified in Freund's incomplete adjuvant. Rabbits are bled prior to immunization and after three peptide injections. The rabbits are then immunized, bled, and tested monthly. The best antisera are usually obtained after more than three peptide injections.

Methods of Radioimmunoassay

The titer of the antisera is determined by its binding to the related radiolabeled peptide in standard assay buffer. Antibody bound and free peptide are separated by precipitation with 25% polyethylene glycol (PEG, Sigma, MW 8000) with the addition of 0.2 ml outdated human plasma (obtained from the Red Cross), followed by centrifugation and decanting of the supernatant. This method of separation routinely gives less than a 10% blank.

Results

Over the course of several years, we succeeded in obtaining usable antisera against all of the synthetic peptides depicted in Fig. 1. In general, they did not have particularly high titers, ranging from 1 : 600 to 1 : 60,000 final dilution in the RIA. They were relatively sensitive, and all of them could detect 20 pg or less of their respective antigen. The antiserum against V-9-M detects a high molecular weight peptide which we think may be pro-CCK as well as smaller peptides which appear to be products of the processing of pro-CCK.[4] We have utilized the antisera to characterize the different forms of CCK precursor and its products in rat brain[4–6] and duodenum, and to determine that some of the non-CCK8 products of pro-CCK are enriched in synaptic vesicles along with CCK8.[7,8] These studies have allowed us to formulate a model of how pro-CCK is processed. Current biosynthesis studies in progress in our laboratory are utilizing these reagents to evaluate this model of pro-CCK processing.

Another more rational strategy for selecting the peptide sequences, which are the most hydrophilic and thus most likely to be on the surface of pro-CCK, is to apply the analysis of Hopp and Woods.[9] A computer program called ANTIGEN is available from BCTIC (Biomedical Computing Technology Information Center, Vanderbilt Medical Center), which runs on various personal computers that will calculate and plot the results of this analysis. When the sequence of prepro-CCK was entered into this program, the computer verified that the regions selected were indeed quite hydrophilic, and the CCK8 sequence was found to be quite hydrophobic, which may explain why conventional CCK8 antisera do not detect pro-CCK effectively.

Disadvantages

In a sense these antisera represent "first generation" reagents. Because of the way they were developed, they have certain disadvantages:

[4] M. C. Beinfeld, *Brain Res.* **344,** 351 (1985).
[5] M. C. Beinfeld, *Biochem. Biophys. Res. Commun.* **127,** 720 (1985).
[6] M. C. Beinfeld, *Peptides* **6,** 857 (1985).
[7] L. R. Allard and M. C. Beinfeld, *Neuropeptides* **6,** 239 (1985).
[8] L. R. Allard and M. C. Beinfeld, *Reg. Peptides* **12,** 59 (1985).
[9] T. P. Hopp and K. R. Woods, *Proc. Natl. Acad. Sci. U.S.A.* **78,** 3824 (1981).

(1) because they were raised against short peptide fragments modeling the much longer actual fragments found in tissue, the RIAs do not display exact parallel displacement with tissue extract volume, so that exact quantitation of these peptides in tissue extracts is not possible; (2) because they are based on the rat sequence alone, they may be fairly species specific; (3) because they were selected for their ability to bind trace in solution and not their ability to stain tissue, they do not work well in immunohistochemical staining assays. Now that we know the sequence of pro-CCK in at least three species and have a clear idea where pro-CCK is cleaved during processing, we plan to synthesize some of the actual intermediates we think are made during the processing of pro-CCK and use them as antigens to produce even better "second generation" antisera.

[2] High-Density Multiple Antigen–Peptide System for Preparation of Antipeptide Antibodies

By JAMES P. TAM

Introduction

Synthetic peptides conjugated to protein carriers have been shown to induce antibodies reactive with their cognate sequences in the native proteins.[1-3] Antipeptide antibodies of predetermined specificities are useful laboratory reagents for confirming new proteins from recombinant DNA, exploring biosynthetic pathways and precursors, and probing structural functions of proteins.[2] Because of the advantage of being conveniently available through chemical synthesis, synthetic peptide antigens can also be used for producing vaccines and for passive immunoprophylaxis.[1-10] A convenient and versatile approach to the synthesis of a

[1] M. Sela and R. Arnon, *in* "New Developments with Human and Veterinary Vaccines" (A. Mizrahi, I. Hertman, M. A. Klingberg, and A. Kohn, eds.), p. 315. Liss, New York, 1980.

[2] R. A. Lerner, *Nature (London)* **299,** 592 (1982).

[3] H. Langbeheim, R. Arnon, and M. Sela, *Proc. Natl. Acad. Sci. U.S.A.* **73,** 4636 (1976).

[4] J. L. Bittle, R. A. Houghten, H. Alexander, T. M. Shinnick, J. G. Sutcliffe, R. A. Lerner, D. J. Rowlands, and F. Brown, *Nature (London)* **298,** 30 (1982).

[5] C. Carelli, F. Audibert, J. Gaillard, and L. Chédid, *Proc. Natl. Acad. Sci. U.S.A.* **79,** 5392 (1982).

[6] A. M. Prince, H. Ikram, and T. P. Hopp, *Proc. Natl. Acad. Sci. U.S.A.* **79,** 579 (1982).

[7] F. Zavala, J. P. Tam, M. R. Hollingdale, A. H. Cochrane, I. Quakyi, R. S. Nussenzweig, and V. Nussenzweig, *Science* **228,** 1436 (1985).

peptide–antigen carrier suitable for generating antipeptide antibodies known as the "multiple antigen peptide" (MAP) system is described. The MAP utilizes a simple "scaffolding" of a low number of sequential levels (n) of a trifunctional amino acid as the core carrier and 2^n peptide antigens to form a macromolecule with a high density of peptide antigens. A suitable MAP model is an octabranching MAP consisting of a core matrix made up of three levels of lysine and eight amino ends for anchoring peptide antigens. The MAP containing both the carrier and antigenic peptides is synthesized in a single manipulation by the solid-phase method.[11] After a simple purification scheme, the MAP, which usually has a molecular weight greater than 10,000, was directly used as an immunizing agent. Thus, such a design completely eliminated the conventional step of conjugation of the peptide to a carrier.

Concept and Design

The basic idea makes use of a limited sequential propagation of a trifunctional amino acid (or similar homologs) to form a core that serves as a low-molecular-weight carrier. The trifunctional amino acid, Boc-Lys(Boc), was found to be suitable since both N^{α}- and N^{ε}-amino groups are available as reactive ends. Sequential propagation of Boc-Lys(Boc) will generate 2^n reactive ends. The first level coupling of Boc-Lys(Boc) will produce 2 reactive amino ends as a bivalent MAP (Fig. 1A). The sequential generation of a second, third, and fourth level with Boc-Lys-(Boc) will produce a MAP containing 4 (tetravalent), 8 (octavalent), and 16 (hexadecavalent) reactive amino ends to which peptide antigens are attached (Fig. 1B). The synthesis of the antigenic peptide on the lysinyl core matrix (Fig. 2) proceeded in two ways. For short peptide antigens that contained 10 residues or less, it was necessary to extend the antigenic peptide from the lysinyl core by a linker of simple tri- or tetrapeptide of glycine, alanine, or β-alanine. For those antigenic peptides longer than 10 residues, however, a linker was not used.

Essentially, the MAP consists of three structural features (Fig. 1B): (1) a simple amino acid such as glycine or β-alanine in a benzyl ester (or

[8] R. Arnon, M. Sela, M. Parent, and L. Chédid, *Proc. Natl. Acad. Sci. U.S.A.* **77,** 6769 (1980).

[9] B. Morein, B. Sundquist, S. Hoglund, K. Dalsgarad, and A. Osterhaus, *Nature* (*London*) **398,** 457 (1984).

[10] R. DiMarchi, G. Brooke, C. Gale, V. Cracknell, T. Doel, and N. Mowat, *Science* **232,** 639 (1986).

[11] R. B. Merrifield, *J. Am. Chem. Soc.* **85,** 2149 (1963).

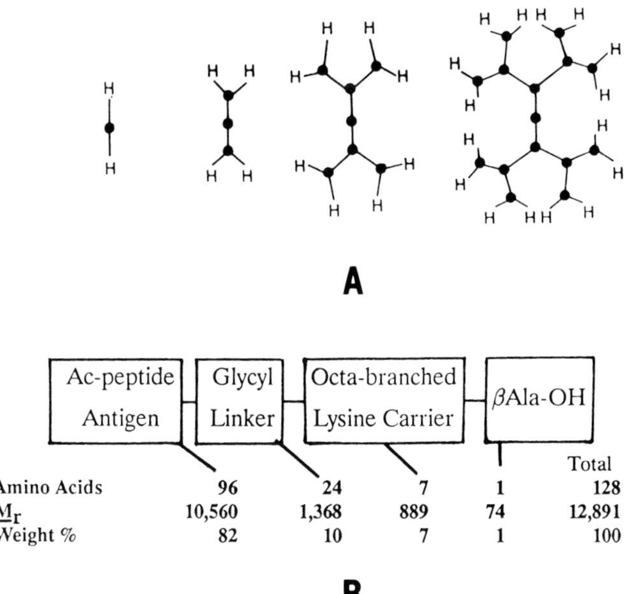

FIG. 1. Schematic representation of the MAP. (A) Four different generations of the lysinyl core with the first generation (lysine itself) as a divalent core, a second generation with tetravalency, a third generation with octavalency, and a fourth generation with hexadecavalency. (B) Structural features and composition of an octameric MAP containing an average peptide of 12 residues.

benzyhydrylamine) linkage to the solid-phase matrix to initiate the synthesis and as internal standard (1 amino acid), (2) an inner core of two to four levels of trifunctional amino acids (7 to 11 amino acids), and (3) an outer surface core of acetylated synthetic peptide attached to the inner core matrix with or without a peptide extender (36 to 288 amino acids). Thus, one major characteristic of the MAP is that the core matrix is small, and the bulk is formed by a high density of peptide antigens layered around the core matrix. This design is in strong contrast to the conventional peptide–carrier conjugate which comprises a large protein carrier and a low density of peptide antigens. In an octabranched MAP, the peptide antigen accounts for more than 80% of the total weight of the MAP (Fig. 1B). It is perhaps important to emphasize the MAP has the appearance and molecular weight of a small protein. Furthermore, the MAP is oligomeric and contains noncationic peptidyl lysine amide linkages on both the N^{α} and N^{ε} termini of lysine. Such a design differs markedly from the conventional polylysinyl conjugate, which is cationic and polymeric in lysinyl residues. Moreover, the dendritic and short peptide chains on the

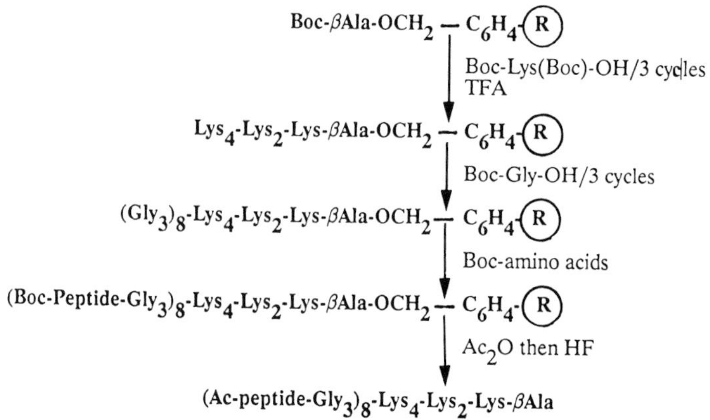

FIG. 2. Synthesis of antigenic peptides on lysinyl core matrix.

MAP are likely to be mobile, which may contribute to the enhanced immunogenicity of the MAP.[12]

Methods

General Procedures for the Synthesis of Maps

The synthesis of an octabranched matrix core with peptide antigen is carried out manually by a stepwise solid-phase procedure[11] on Boc-βAla-OCH_2-Pam resin[13] with a typical scale of 0.5 g of resin (0.05 mmol and a resin substitution level of 0.1 mmol/g for the present synthesis which should be lower if a higher branching lysine core matrix is used). After removal of the Boc group by 50% trifluoroacetic acid (TFA) in CH_2Cl_2 and neutralization of the resulting salt by diisopropylethylamine (DIEA), the synthesis of the first level of the carrier core is achieved using a 4 equivalent excess of preformed symmetrical anhydride of Boc-Lys(Boc) (0.2 mmol) in dimethylformamide (DMF, 12 ml/g) followed by a second coupling via dicyclohexylcarbodiimide (DCC) alone in CH_2Cl_2. The second and third levels are synthesized by the same protocol with 0.4 and 0.8 mmol, respectively, of preactivated Boc-Lys(Boc) to give, after deprotection, the octabranched core matrix containing eight functional amino groups. All subsequent couplings of the peptide–antigen sequence require 1.6 mmol of preactivated amino acids.

[12] E. Westhof, D. Altschuh, D. Moras, A. C. Bloomer, A. Mondragon, A. Klug, and M. H. V. Van Regenmortel, *Nature (London)* **311,** 123 (1984).

[13] A. R. Mitchell, S. B. H. Kent, M. Engelhard, and R. B. Merrifield, *J. Org. Chem.* **43,** 2845 (1978).

The protecting groups for the synthesis of the peptide antigens are as follows: *tert*-butyloxycarboxyl (Boc) group for the α-amino terminus and benzyl alcohol derivatives for most side chains of trifunctional amino acids, i.e., Asp(OBzl), Glu(OBzl), Lys(2ClZ), Ser(Bzl), Thr(Bzl), and Tyr(BrZ), plus Arg(Tos) and His(Dnp). Because of the geometric increase in weight gain and volume, a new volume ratio of 30 ml solvent/g resin is used. Deprotection by TFA (20 min) is preceded by two TFA prewashes for 2 min each. Neutralization by DIEA was in CH_2Cl_2 (5% DIEA), and there is an additional neutralization in DMF (2% DIEA). For all residues except Arg, Asn, Gln, and Gly, the first coupling is done with the preformed symmetric anhydride in CH_2Cl_2, and the second coupling is performed in DMF; each coupling is allowed to proceed for 1 hr. The coupling of Boc-Asn and Boc-Gln is mediated by the preformed 1-hydroxybenzotriazole ester in DMF. Boc-Gly and Boc-Arg are coupled with water-soluble DCC alone to avoid the risk of formation of dipeptide and lactam, respectively. All couplings are monitored by a quantitative ninhydrin test[14] after each cycle, and, if needed, a third coupling of symmetrical anhydride in *N*-methylpyrrolidinone at 50° for 1 hr is used.[15] After deprotection, the peptide chains are capped on their α-amino group by acetylation in acetic anhydride/DMF (3 mmol) containing 0.3 mmol of *N*,*N*-dimethylpyridine.

After completion of the MAP, protected peptide–resin (0.3 g) is treated with 1 *M* thiophenol in DMF for 8 hr (3 times and at 50° if necessary to complete the reaction) to remove the N^{im}-dinitrophenyl protecting group of His (when present). The branched peptide–oligolysine matrix is removed from the cross-linked polystyrene resin support by treating with 50% TFA/CH_2Cl_2 (10 ml) for 5 min to remove the N^{α}-Boc group and then by using the low–high HF method[16] or the low–high TFMSA method[17] of cleavage to give the crude MAP. The crude peptide and resin are then washed with cold ether–mercaptoethanol (99 : 1, v/v, 30 ml) to remove *p*-thiocresol and *p*-cresol, and the peptide is extracted into 100 ml of 8 *M* urea, 0.2 *M* dithiothreitol in 0.1 *M* Tris–HCl buffer, pH 8.0. To remove all remaining aromatic by-products generated in the cleavage step, the peptide is dialyzed in Spectra Por 6 tubing (MW cutoff 1000) by equilibration in a deaerated and N_2-purged solution containing 8 *M* urea, 0.1 *M* NH_4HCO_3–$(NH_4)_2CO_3$, pH 8.0, with 0.1 *M* mercaptoethanol at 0° for 24 hr. The dialysis is continued in 8 *M* and then in 2 *M* urea, all in 0.1 *M*

[14] V. K. Sarin, S. B. H. Kent, J. P. Tam, and R. B. Merrifield, *Anal. Biochem.* **117,** 147 (1981).

[15] J. P. Tam, *Proc. Am. Pept. Symp., 9th,* p. 305 (1985).

[16] J. P. Tam, W. F. Heath, and R. B. Merrifield, *J. Am. Chem. Soc.* **105,** 6442 (1983).

[17] J. P. Tam, W. F. Heath, and R. B. Merrifield, *J. Am. Chem. Soc.* **108,** 5242 (1986).

NH_4HCO_3–$(NH_4)_2CO_3$ buffer, pH 8.0, for 12 hr and then sequentially in distilled water and 1 *M* acetic acid to remove all the urea. The lyophilized MAP is then purified batchwise by high-performance gel permeation or ion-exchange chromatography. All of the purified materials prepared in our laboratory gave satisfactory amino acid analyses.

Immunization Procedure

Rabbits (New Zealand White, two for each antigen) are immunized by subcutaneous injection of the MAP (1 mg in 1 ml phosphate-buffered saline) in Freund's complete adjuvant (1 : 1) on day 0 and in Freund's incomplete adjuvant (1 : 1) on days 21 and 42 and bled on day 49. Inbred 6- to 8-week-old mice are immunized in the footpad with 80 μg of MAP in Freund's complete adjuvant (1 : 1) on day 0 and in Freund's incomplete adjuvant 4 times every 3 weeks and bled 1 week after the last boosting. The antisera are used without any purification.

Immunological Assay

An enzyme-linked immunosorbent assay (ELISA) is used to test all antisera for their ability to react with the MAP used for immunization. Peptide antigen (0.5 μg/well) in carbonate–bicarbonate buffer (pH 9.0) is incubated at 4° overnight in a 96-well microtiter plate. Rabbit antisera (serially diluted in 10 m*M* phosphate-buffered saline) is then incubated with the antigen for 2 hr at 20°. Goat anti-rabbit IgG horseradish peroxidase conjugate is then added before incubation for an additional hour. The bound conjugate is reacted with chromogen (*o*-dianisidine dihydrochloride at 1 mg/ml in 10 m*M* phosphate buffer, pH 5.95) for 0.5 hr, and the absorbance of each well is determined with a micro-ELISA reader.

Purification and Characterization

Since the MAP was found to have the unusual ability to aggregate, the crude MAP after cleavage from the resin support is purified by extensive dialysis under basic and strongly denaturing conditions with 8 *M* urea and mercaptoethanol to remove the undesirable aromatic additives of the cleavage reactions, such as *p*-cresol and thiocresol, which tend to adhere strongly to the peptides. In addition, the base treatment under such conditions converts any strong acid-catalyzed *O*-acyl rearrangement product of serinyl peptides to the *N*-acyl peptides and converts any residual level of Met(O) to Met and His(Dnp) to His. The crude MAPs may be further purified by high-performance gel permeation or ion-exchange chromatog-

TABLE I
IMMUNOLOGICAL RESPONSE TO MAPs

Peptide	Sequence	Test animal	Half-maximal response ($\log_{10}$)[a] Preimmune	Immune	Reactive to native protein[b]
IG-11	IEDNEYTARQG	Rabbit	<0.5	4.6	+ (A)
FA-14	FEPSEAEISHTQKA	Mouse	<0.5	3.6	+ (A)
YP-13	YIQHKLQEIRHSP	Rabbit	<0.5	5.5	− (A)
NP-16	$(NANP)_4$	Mouse	<0.5	2.2	+ (B)
DV-9	DGISAAKDV	Rabbit	<0.5	4.0	+ (B)

[a] Half-maximal response of the antiserum in the dilution versus absorbance curve in ELISA.

[b] Detection either by immunoprecipitation of the labeled protein and SDS gel electrophoresis of the precipitate (A) or by immunoblotting experiment (B).

raphy. Purification by C_4 or C_8 reversed-phase high-performance chromatography also gives satisfactory results. In most cases, however, the MAPs may be used directly without further purification.

Immunological Responses to MAP

Five MAPs with peptides containing 9 to 16 residues were tested in animals (Table I). Two MAPs, IG-11 and YP-13, are internal sequences of 11 and 13 residues related, respectively, to tyrosine protein kinases, $p60^{src}$[18] and *ros*.[19] Similarly, peptides FA-14, NP-16, and DV-9, correspond, respectively, to T cell receptor,[20] circumsporozoite protein of *Plasmodium falciparum*,[7] and G_o protein.[21] Antipeptide antibodies based on the conventional approach to IG-11, NP-16, and DV-9 are known.[22–24]

[18] A. Czernilofsky, A. Levison, H. Varmus, J. M. Bishop, E. Tisher, and H. Goodman, *Nature (London)* **287,** 198 (1980).

[19] W. S. Neckameyer and L.-H. Wang, *J. Virol.* **53,** 879 (1985).

[20] Y. Yanagi, Y. Yoshikai, K. Leggett, S. P. Clark, I. Aleksander, and T. W. Mak, *Nature (London)* **308,** 145 (1984).

[21] J. B. Dame, J. L. Williams, T. F. McCutchan, J. L. Weber, R. A. Wirtx, W. T. Hockmeyer, W. L. Maloy, J. D. Haynes, I. Schneider, D. Roberts, G. S. Sanders, E. P. Reddy, C. L. Diggs, and L. H. Miller, *Science* **225,** 593 (1984).

[22] T. Tanabe, T. Nukada, Y. Nishikawa, K. Sugimoto, H. Suzuki, H. Takahashi, M. Noda, T. Haga, A. Ichiyama, K. Kangawa, N. Minamino, H. Matsuo, and S. Numa, *Nature (London)* **315,** 242 (1985).

[23] T. Y. Wong and A. R. Goldberg, *Proc. Natl. Acad. Sci. U.S.A.* **80,** 2529 (1983).

[24] S. M. Mumby, R. A. Kahn, D. R. Manning, and A. G. Gilman, *Proc. Natl. Acad. Sci. U.S.A.* **83,** 265 (1986).

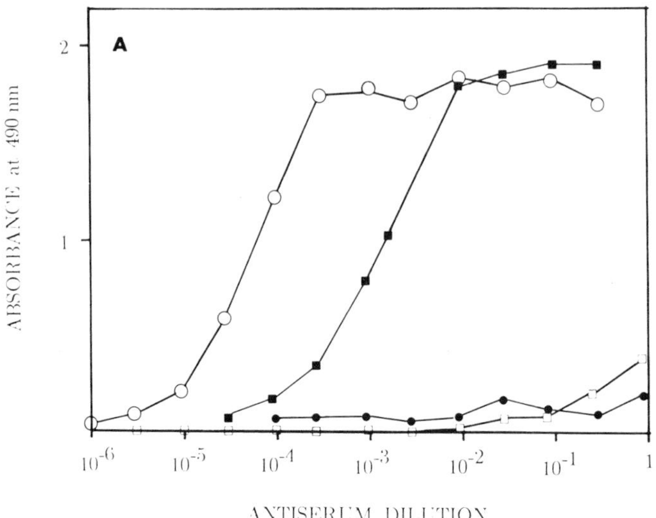

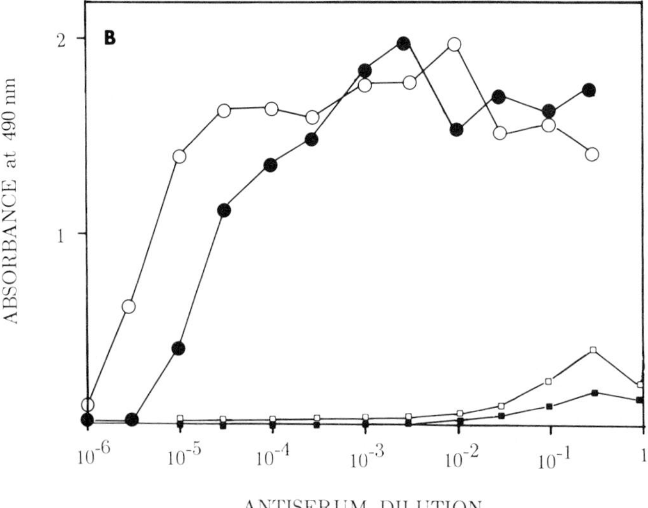

FIG. 3. (A) Immunoreactivities by ELISA of antisera derived from the MAP IG-11, with MAP IG-11 as antigenic substrate for ELISA: primary immunization (■), after two boostings (○). Controls with MAP IG-11 as antigenic substrate: antisera to MAP YP-13 (●) and MAP IG-11 antiserum preincubated with excess of IG-11 monomeric peptide (□). There is no reactivity with the control preimmune antiserum. (B) Antisera raised to the MAP YP-13 (Table I) studied by ELISA versus MAP YP-13. Primary immunization (●), after a single boosting (○); controls, antiserum to MAP IG-11 (□) and preimmune antiserum (■).

As shown in Table I, all five MAPs were found to be strong immunogens and produced high specific antibody titers after boostings. Both YP-13 and IG-11 MAPs were found to produce high titer antibody responses in rabbits with antisera dilution to 10^4-fold (Fig. 3A and B). Two MAPs, FA-14 and NP-16, were found to produce only a moderate response in mice (BALB/c and C57BL/6J, respectively) after the first inoculation, and the antibody titer increased approximately 10- to 50-fold after boostings.

Comment

The synthesis of MAPs is initiated on a Boc-βAla-OCH_2-Pam resin[13] with a loading of 0.1 mmol/g resin (Fig. 2). A low loading of the resin substitution between 0.03 and 0.1 mmol is highly desirable and necessary since it increased geometrically with each addition of Boc-Lys(Boc) and the synthesis would be difficult if the conventional loading of 0.3–0.8 mmol/g were used. For the synthesis of an octabranched MAP, Boc-Lys(Boc) is coupled sequentially to the resin ($n = 3$) to give an octavalent core matrix of 0.8 mmol/g of loading. In some cases, a triglycyl linker is added as an extender before initiation of synthesis of the peptide antigens.

Several points regarding the unusual properties of the MAP during its synthesis should be emphasized. (1) The synthesis generally requires a long coupling time (1–2 hr/coupling), and DMF is a more suitable coupling solvent than CH_2Cl_2 in preventing peptide aggregation. The peptide resin should not be dried at any stage of the synthesis since resolvation of a dried resin containing MAPs was found to be extremely difficult. (2) The efficiency of coupling is closely monitored.[14] Synthetic errors such as deletion peptides would be amplified and would be difficult to correct by conventional purification methods. (3) Cleavage of the MAP from the resin is achieved with an improved acid-deprotection method with either HF or TFMSA[16,17] in dimethyl sulfide to avoid many attendant strong-acid-catalyzed side reactions. This point needs to be stressed since it is important to have clean chemistry in order to arrive at a high-quality product.[4] The synthesis of MAPs can be carried out using other solid supports such as polyamide resins and other chemistry such as the Fmoc-*tert*-butyl strategy.[25]

Acknowledgments

This work was supported in part by Grant CA 36544 and AID.

[25] E. Atherton, C. J. Logan, and R. C. Shepard, *J. Chem. Soc., Perkin Trans. 1*, p. 538 (1981).

[3] Preparation and Use of Complementary Peptides

By KENNETH L. BOST and J. EDWIN BLALOCK

Introduction

Previous experiments have shown that peptides encoded by complementary strands of nucleic acids (designated "complementary" peptides) using the same reading frame have the ability to bind one another.[1-3] Not only could peptides complementary to corticotropin (ACTH), γ-endorphin, or luteinizing hormone-releasing hormone (LHRH) bind the appropriate hormone, the immune system recognized the complementary peptides as being antigenically similar to the respective hormone receptor binding site.[1,4-7] Stated differently, antibodies directed against a peptide complementary to a particular hormone would also bind to that hormone's receptor binding site. While the ability to immunoaffinity purify a receptor for a specific hormone is a valuable technique, the concept of complementary peptide sequences binding one another is not limited to this and has also been applied to antibody–antibody interactions[8] and to peptides which bind ribonuclease S peptide.[3] These studies suggest the general applicability of this technique, and one purpose of this chapter is to consider some potential applications of the methodology. The primary intent, however, is one of a technical nature. While it may appear simple to construct complementary peptides, there are many considerations that on first examination are not obvious. In this chapter we present a composite of the technical considerations we have generated to date relative to constructing complementary peptides which will bind a given sequence.

Antiparallel and Parallel Complementary Peptides

Owing to the strides being made in peptide chemistry, it is not difficult to obtain synthetic peptides of specified sequences. Furthermore, increased interest in molecular biology has expanded the number of nucleo-

[1] K. L. Bost, E. M. Smith, and J. E. Blalock, *Proc. Natl. Acad. Sci. U.S.A.* **82,** 1372 (1985).
[2] J. E. Blalock and K. L. Bost, *Biochem. J.* **234,** 679 (1986).
[3] Y. Shai, M. Flashner, and I. M. Chaiken, *Biochemistry* **26,** 669 (1987).
[4] K. L. Bost and J. E. Blalock, *Mol. Cell. Endocrinol.* **44,** 1 (1986).
[5] D. J. J. Carr, K. L. Bost, and J. E. Blalock, *J. Neuroimmunol.* **12,** 329 (1986).
[6] J. J. Mulchahey, J. D. Neill, L. D. Dion, K. L. Bost, and J. E. Blalock, *Proc. Natl. Acad. Sci. U.S.A.* **83,** 9714 (1986).
[7] T. J. Gorcs, P. E. Gottschall, D. H. Coy, and A. Arimura, *Peptides (N.Y.)* **7,** 1137 (1986).
[8] L. R. Smith, K. L. Bost, and J. E. Blalock, *J. Immunol.* **138,** 7 (1987).

tide and amino acid sequences available for study. Thus, both the sequence information and the technology are available for constructing complementary peptides; however, there are several considerations when deciding which peptide is appropriate to synthesize. Table I is an inclusive list of the possible amino acids encoded by complementary codons in both the 5′ to 3′ and 3′ to 5′ directions, wherein lies the first question. In

TABLE I
COMPLEMENTARY AMINO ACIDS

Amino acid	Codon (5′ to 3′)	Complementary codon (3′ to 5′)	Amino acid encoded by the complementary codon	
			5′ to 3′ direction	3′ to 5′ direction
Isoleucine	AUA	UAU	Tyr	Tyr
	AUC	UAG	Asp	Stop
	AUU	UAA	Asn	Stop
Methionine	AUG	UAC	His	Tyr
Leucine	CUA	GAU	Stop	Asp
	CUC	GAG	Glu	Glu
	CUG	GAC	Gln	Asp
	CUU	GAA	Lys	Glu
	UUA	AAU	Stop	Asn
	UUG	AAC	Gln	Asn
Valine	GUA	CAU	Tyr	His
	GUC	CAG	Asp	Gln
	GUG	CAC	His	His
	GUU	CAA	Asn	Gln
Phenylalanine	UUC	AAG	Glu	Lys
	UUU	AAA	Lys	Lys
Lysine	AAA	UUU	Phe	Phe
	AAG	UUC	Leu	Phe
Asparagine	AAC	UUG	Val	Leu
	AAU	UUA	Ile	Leu
Glutamine	CAA	GUU	Leu	Val
	CAG	GUC	Leu	Val
Histidine	CAC	GUG	Val	Val
	CAU	GUA	Met	Val
Glutamic acid	GAA	CUU	Phe	Leu
	GAG	CUC	Leu	Leu

(*continued*)

TABLE I (*continued*)

Amino acid	Codon (5′ to 3′)	Complementary codon (3′ to 5′)	Amino acid encoded by the complementary codon	
			5′ to 3′ direction	3′ to 5′ direction
Aspartic acid	GAC	CUG	Val	Leu
	GAU	CUA	Ile	Leu
Tyrosine	UAC	AUG	Val	Met
	UAU	AUA	Ile	Ile
Threonine	ACA	UGU	Cys	Cys
	ACC	UGG	Gly	Trp
	ACG	UGC	Arg	Cys
	ACU	UGA	Ser	Stop
Proline	CCA	GGU	Trp	Gly
	CCC	GGG	Gly	Gly
	CCG	GGC	Arg	Gly
	CCU	GGA	Arg	Gly
Alanine	GCA	CGU	Cys	Arg
	GCC	CGG	Gly	Arg
	GCG	CGC	Arg	Arg
	GCU	CGA	Ser	Arg
Serine	UCA	AGU	Stop	Ser
	UCC	AGG	Gly	Arg
	UCG	AGC	Arg	Ser
	UCU	AGA	Arg	Arg
	AGC	UCG	Ala	Ser
	AGU	UCA	Thr	Ser
Arginine	AGA	UCU	Ser	Ser
	AGG	UCC	Pro	Ser
	CGA	GCU	Ser	Ala
	CGC	GCG	Ala	Ala
	CGG	GCC	Pro	Ala
	CGU	GCA	Thr	Ala
Glycine	GGA	CCU	Ser	Pro
	GGC	CCG	Ala	Pro
	GGG	CCC	Pro	Pro
	GGU	CCA	Thr	Pro
Cysteine	UGC	ACG	Ala	Thr
	UGU	ACA	Thr	Thr
Tryptophan	UGG	ACC	Pro	Thr

which direction should a complementary peptide be encoded? Initially, the usual 5′ to 3′ translational direction would seem to be the only logical possibility. However, since the middle base of the codon specifies the hydropathic nature of the amino acid,[2] peptides encoded in either direction using the same RNA sequence and the same reading frame will be similar with respect to their hydropathicity. Stated simply, the middle base is always the second base regardless of the direction of reading. Since the middle base determines the hydropathic nature of an amino acid, one would predict that peptides encoded in the 5′ to 3′ or 3′ to 5′ direction might have similar binding characteristics if a similar hydropathic nature was important. In one case studied,[2] we found that peptides encoded by the RNA complementary to the mRNA for ACTH in either the 5′ to 3′ or 3′ to 5′ direction had almost identical abilities to bind ^{125}I-ACTH. In this particular instance, ^{125}I-ACTH binding to the antiparallel or parallel encoded peptides was similar; however, this may not always be the case.

A second reason to consider 3′ to 5′ encoded complementary peptides is the finding that regions of complementarity between known receptor–ligand pairs were found to occur in the 3′ to 5′ direction.[9] One region of complementarity between interleukin 2 and its receptor was subsequently shown to represent a binding-site sequence for this receptor–ligand pair.[10] While the biological mechanisms by which these 3′ to 5′ complementarities occur in nature is unclear at present, additional sequence information on pairs of interacting proteins will be necessary to understand this occurrence.

In conclusion, when synthesizing complementary peptides based on nucleotide sequences, consideration should be given to peptides encoded in both the 5′ to 3′ and 3′ to 5′ directions. Presently, adequate experimental evidence is not available to determine which peptide will be the most appropriate for each application of the methodology.

Consensus Sequences

It is possible to generate a complementary peptide from a primary amino acid sequence without prior knowledge of the nucleotide sequence. This process is analogous to constructing multiple oligonucleotide probes from primary sequence information, but it is more restricted in the sense that a single amino acid (e.g., glutamine) encoded by two different codons

[9] K. L. Bost, E. M. Smith, and J. E. Blalock, *Biochem. Biophys. Res. Commun.* **128,** 1373 (1985).

[10] D. A. Weigent, P. D. Hoeprich, K. L. Bost, T. K. Brunck, W. E. Reiher, and J. E. Blalock, *Biochem. Biophys. Res. Commun.* **139,** 367 (1986).

TABLE II
AMINO ACID SEQUENCES FOR SUBSTANCE P AND ITS 3′ TO 5′ COMPLEMENTARY PEPTIDE

Substance P amino acid sequence[11]										
1	2	3	4	5	6	7	8	9	10	11
Arg	Pro	Lys	Pro	Gln	Gln	Phe	Phe	Gly	Leu	Met
3′ to 5′ possible amino acid complements (see Table I)										
1	2	3	4	5	6	7	8	9	10	11
Ser	Gly	Phe	Gly	Val	Val	Lys	Lys	Pro	Asp	Tyr
Ser	Gly	Phe	Gly	Val	Val	Lys	Lys	Pro	Glu	
Ala	Gly		Gly					Pro	Asp	
Ala	Gly		Gly					Pro	Glu	
Ala									Asn	
Ala									Asn	
Consensus complementary sequence based on 3′ to 5′ possibilities										
1	2	3	4	5	6	7	8	9	10	11
Ala	Gly	Phe	Gly	Val	Val	Lys	Lys	Pro	Asn	Tyr
Actual complementary sequence based on nucleotide sequence read 3′ to 5′										
1	2	3	4	5	6	7	8	9	10	11
Ala	Gly	Phe	Gly	Val	Val	Lys	Lys	Pro	Asn	Tyr

(e.g., CAA and CAG) is complemented by only one amino acid (e.g., leucine in the 5′ to 3′ direction and valine in the 3′ to 5′ direction) (see Table I).

An example of constructing a complementary peptide directly from primary sequence information is given in Table II. From the published amino acid sequence for substance P,[11] all the possible 3′ to 5′ complementary amino acids were listed from Table I. Obviously, only the amino acids at positions 1 and 10 are questionable as to the complementary amino acid which should be used, and, in fact, only six peptides would have to be synthesized to cover all the possible combinations. However, several educated guesses can be made to further restrict the amino acids that should be used. First, depending on the species one is working with, preferred codon usage tables[12] can be utilized to determine the probability of which particular codon would be used for each amino acid for, in this example, substance P. The corresponding complementary codons could then be weighted appropriately. Second, the frequency of occurrence of a particular complementary amino acid at a single position can also be taken into account. For example, at position 1 it is likely that alanine

[11] M. M. Chang and S. E. Leeman, *J. Biol. Chem.* **245,** 4784 (1970).
[12] R. Grantham, C. Gautier, and M. Gouy, *Nucleic Acids Res.* **8,** 1893 (1980).

would be the appropriate choice as a complement since it occurs four out of six times. Third, it can also be appropriate to use the set of possible complementary amino acids encoded in both the 5′ to 3′ and 3′ to 5′ directions to select the most likely amino acid. Once again, this can be done without dramatically changing the hydropathicity at a single position since the middle base specifies the hydropathic nature of an amino acid.

As an example, it is not clear in Table II which complementary amino acid should occupy position 10. If one considers the possible complementary amino acids in the 5′ to 3′ direction for leucine given in Table I (i.e., Glu, Gln, Lys, Gln) with the possible 3′ to 5′ complementary amino acid (i.e., Asp, Glu, Asp, Glu, Asn, Asn), an amine would be the most likely choice based simply on the number of amines versus the number of acids. Therefore, the consensus complementary peptide for substance P would have alanine in position 1 and asparagine in position 10, using the rationale described above. When compared with the complementary peptide sequence generated by reading the nucleotide sequence complementary to the nucleotide sequence for substance P in the 3′ to 5′ direction, the consensus sequence is identical (see Table II).

It should be pointed out that the same rationale can be used to generate consensus complementary peptides using 5′ to 3′ encoded complementary amino acids. However, the number of possibilities increases in this direction when compared to the possible complementary amino acids in the 3′ to 5′ direction. For example, alanine is complemented by cysteine, glycine, arginine, or serine in the 5′ to 3′ direction but only by arginine in the 3′ to 5′ direction. The restricted nature of the set of 3′ to 5′ complementary amino acids results from the first base of the sense codon complementing the first base of the antisense codon. In the 5′ to 3′ direction, the third base of the sense codon complements the first base of the antisense codon. Owing to the degeneracy of the genetic code, 5′ to 3′ complementary amino acids are less restricted.

Experimentally, two peptides were synthesized based on the consensus complementary sequence for substance P. One peptide had asparagine at position 10, which also corresponds to the nucleotide-derived complement (see Table II), and the second peptide had glutamic acid at position 10. While either peptide coated onto microtiter wells was capable of binding tritiated substance P, the glutamic acid-containing peptide routinely bound more radiolabel than the asparagine-containing peptide. Figure 1 shows the ability of the glutamic acid-containing complementary peptide to bind radiolabel and the ability of this binding to be blocked by unlabeled substance P or the analog [D-Pro^2-D-Phe^7-D-Trp^9]substance P. Thus, here is another example of a 3′ to 5′ encoded complementary peptide binding its radiolabeled ligand.

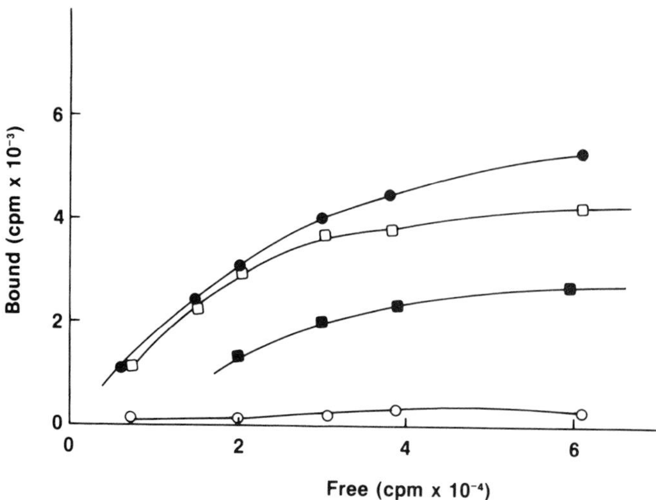

FIG. 1. Binding of tritiated substance P to its 3′ to 5′ complementary peptide. The peptide NH_2-Tyr-Glu-Pro-Lys-Lys-Val-Val-Gly-Phe-Gly-Ala-COOH was placed in carbonate buffer (pH 8.6) at 0.5 mg/ml and coated onto microtiter wells (0.1 ml/well) overnight at 4°. After blocking within 1% bovine serum albumin (BSA) in phosphate-buffered saline (PBS) for 1 hr, varying concentrations of ^{3}H-labeled substance P were added in PBS containing 0.5% BSA and 0.02% Tween 20 for 2 hr. The amount of ^{3}H-labeled substance P was determined in the presence of 0.5% BSA (●), a 500-fold excess of soluble complementary peptide (□), a 500-fold excess of soluble complementary peptide which had been cross-linked to BSA via glutaraldehyde (■), or a 500-fold excess unlabeled substance P (○). The ability of a 500-fold excess of the analog [D-Pro^2-D-Phe^7-D-Trp^9]substance P to block binding was indistinguishable from that of unlabeled substance P.

Problematic Sequences: Stop Codons

Certain codon sequences can present problems in attempts to construct complementary peptides. For example, the 5′ to 3′ complementary codon to the UUA codon for leucine is UAA which is a stop codon. If the complementary peptide to be synthesized is of sufficiently small size that terminating the peptide and starting a second one are not appropriate, an amino acid can be substituted for the stop codon. Routinely, we select this substitution using rules similar to those for generating consensus complementary peptides. For example, in the 5′ to 3′ direction, leucine is complemented by the amino acids Glu, Gln, Lys, and Gln. In the 3′ to 5′ direction, the codon UUA is complemented by AAU encoding Asn. Thus, on a frequency basis, this should be an amine, and two possibilities for substituting for this UAA stop codon would be Gln or Asn.

Solid-Phase Binding Assays Using Complementary Peptides

Choice of Complementary Peptide

In addition to the considerations discussed above relative to selecting an appropriate complementary peptide to synthesize, there are several additional factors that should be noted when conducting solid-phase binding assays. First, the length of the complementary peptide generally directly influences the affinity of binding to a particular ligand. For example, the complementary peptide to ACTH 1–24 bound this hormone with a K_D of 0.3 nM,[2] whereas smaller pairs of complementary peptides had dissociation constants in the micromolar range (see Table III). Often it is difficult to demonstrate specific binding using very small peptides (e.g., hexamers), probably due to limited secondary structure in solution and to the restrictions imparted to a small peptide bound to a solid support. Second, peptides with opposite amino to carboxy orientations have been shown to have similar[3] or differing (K. L. Bost and J. E. Blalock, unpublished observations) abilities to bind a ligand. The significance of this observation is presently under investigation. Third, especially with short peptides, the amount of nonspecific binding can sometimes be reduced by neutralizing the dipole moment of the complementary peptide. This can be accomplished by amidating carboxy termini and acetylating amino termini to reduce charge–charge interactions.

TABLE III

DISSOCIATION CONSTANTS FOR PAIRS OF COMPLEMENTARY PEPTIDES

Complementary peptide pair	Number of amino acids	Dissociation constant (M)	Ref.
ACTH 1–24/5′ to 3′ complementary peptide	24	0.3×10^{-9}	2
ACTH 1–24/3′ to 5′ complementary peptide	24	0.3×10^{-9}	2
γ-Endorphin/5′ to 3′ complementary peptide	17	2×10^{-5}	5
LHRH/5′ to 3′ complementary peptide	10	$\sim 1 \times 10^{-4}$	6[a]
Ribonuclease S peptide/5′ to 3′ complementary peptide	20	1.3×10^{-6}	3
Ribonuclease S peptide/5′ to 3′ inverted complementary peptide	20	1.2×10^{-6}	3
Substance P/3′ to 5′ complementary peptide	9	6×10^{-6}	[b]

[a] The dissociation constant was estimated from the concentration of peptide necessary to cause a 50% inhibition of a biological assay.

[b] K. L. Bost and J. E. Blalock, unpublished observations.

Solid-Phase Binding Assays

The reader is referred to other publications[13,14] which describe general aspects of solid-phase binding assays. We typically coat microtiter wells with 10–50 μg of the peptide in an appropriate buffer overnight at 4°. At this point, wells are washed and blocked with an appropriate irrelevant protein [e.g., 1% bovine serum albumin (BSA) for 2 hr]. Reaction times with radiolabeled ligands are typically 2 hr in buffer such as phosphate-buffered saline (PBS) with 0.5% BSA and 0.05% Tween 20 after which unbound radiolabel is washed out with a similar buffer.

There are several variables in solid-phase binding assays using microtiter plates which should be optimized for each system. The first variable is the microtiter plates which are used as a solid support. Polyvinyl plates have been used extensively because of their high protein-binding capacity. Recently, however, high-binding polystyrene plates like Nunc-immuno plates (Interlab, Newbury Park, CA) or Immulon plates (Dynatech, Chantilly, VA) have become available, and these are routinely used in our laboratory. These polystyrene plates have uniform binding characteristics which minimize well-to-well variations. A second variable is the method used to coat peptides to wells. Surprisingly, this is a very important variable which becomes more critical as the size of the peptide to be coated is decreased. There is no reason to assume that all the orientations which a peptide assumes on adsorbing to a solid support will be conducive for binding a radiolabeled ligand. Therefore, we routinely determine optimal methods for coating a particular peptide prior to performing binding assays. Typically, proteins are coated in carbonate buffer, around pH 9.0[1] or in phosphate-buffered saline, pH about 7.0[2]; however, this is not always optimal for peptides. Activating plates by pretreatment with 20 mM glutaraldehyde for 1 hr, followed by two 0.15 M NaCl washes and subsequent addition of the peptide for coating, can be an effective alternative.[15] Conversely, microtiter plates are commercially available (Micro Membranes, Inc., Newark, NJ) which allow covalent coupling of proteins or peptides to the surface. As a final consideration, the binding ability of some peptides can be markedly enhanced by coupling to a carrier protein (see the following section).

Nonspecific binding of the radiolabeled ligand can usually be controlled by the number of washing steps and by the composition of the washing buffer. An optimal washing buffer should permit specific binding

[13] D. N. Orth, this series, Vol. 37, p. 22.

[14] G. E. Trivers, C. C. Harris, C. Rougeot, and F. Dray, this series, Vol. 103, p. 409.

[15] L. M. Kuo and R. J. Robb, *J. Immunol.* **137,** 1538 (1986).

(i.e., binding which is blockable by unlabeled ligand and an appropriate antagonist) to occur but should minimize nonspecific binding. We typically use phosphate-buffered saline with 0.5% BSA and 0.05% Tween 20; however, depending on the assay, ovalbumin, bacitracin, and 1% fetal calf serum have been used alone or in combination to give optimal binding results.

An alternative method to using microtiter plates is that described by Shai *et al.*[3] Here peptides are immobilized on silica beads, and binding interactions are analyzed using high-performance affinity chromatography. The advantages of using this method over solid-phase binding assays using microtiter plates has been previously reviewed.[16–18]

Solid-Phase versus Solution Binding

One final observation relative to solid-phase binding assays should be made. Earlier, reference was made to the ability of ^{3}H-labeled substance P to bind microtiter wells previously coated with its complementary peptide. This binding could be blocked by unlabeled substance P or the antagonist [D-Pro^2-D-Phe^7-D-Trp^9]substance P but could not be effectively blocked with a 500-fold excess of soluble complementary peptide (see Fig. 1). On coupling the glutamic acid-containing complementary peptide for substance P to BSA, this conjugate was able to effectively block binding (~50%), whereas BSA was without effect. Thus, peptides or proteins interacting with a third surface may not have similar binding properties when placed in solution. Therefore, if binding of complementary peptide pairs in solution is weak, one should consider coupling a peptide to a solid support or a carrier protein. Furthermore, the binding ability of some peptides may be enhanced by coupling to a carrier protein prior to immobilization on a solid support.

Use of Complementary Peptides

Antireceptor Antibodies

Polyclonal antibodies against the ACTH, endorphin, and LHRH receptor binding sites have been produced in rabbits by immunization with the appropriate complementary peptides.[1,4–7] The ability to generate antireceptor antibodies for all three hormone receptors supports the fidelity of

[16] H. E. Swaisgood and I. M. Chaiken, *J. Chromatogr.* **327,** 193 (1985).

[17] H. E. Swaisgood and I. M. Chaiken, *Biochemistry* **25,** 4148 (1986).

[18] I. M. Chaiken, *J. Chromatogr.* **376,** 11 (1986).

this methodology. Before discussing the technical aspects, there are some general considerations which need to be addressed. First, the responses of individual rabbits to a particular antigen is quite heterogeneous; thus, some rabbits produce high titers whereas others do not. It would be advisable to immunize multiple animals and screen them, using immunoassays,[13,14] for the best producer. Second, producing antireceptor antibodies results in physiological responses in the immunized animal[7] (K. L. Bost and J. E. Blalock, unpublished observations). In other words, animals are making anti-self antibodies. It is important to realize that these anti-self antibody responses may not only be relatively low-titer responses but also may be transient. Thus, it is necessary to screen each bleed to determine the highest antireceptor titer.

Previous publications[19,20] have dealt with antibody production from a technical point of view. Since most peptides are not very immunogenic, coupling to a carrier protein is necessary prior to immunization. While there are a large variety of commercially available cross-linking agents (Pierce Chemical, Rockford, IL), the choice of cross-linker can affect the antireceptor antibody titer. Table IV shows that mice immunized with keyhole limpet hemocyanin (KLH) conjugated to the 5′ to 3′ complementary peptide for ACTH 1–24 (HTCA) via glutaraldehyde[1] routinely gave higher titers than carbodiimide coupling,[10] but not necessarily higher antireceptor antibody titers. In addition to pointing out that the method for cross-linking a peptide to a carrier protein can affect the antibody response, these results also show that antipeptide and antireceptor titers do not always correlate. In other words, only a portion of the anticomplementary peptide antibodies produced should be expected to cross-react with the receptor binding site. For this reason, it is important to screen sera not only against the immunogen but also against the receptor. Furthermore, biological assays to determine the agonistic or antagonistic properties of the antibodies are extremely helpful. For example, antibodies against HTCA were assayed for their ability to bind Y-1 adrenal cells[1] and for their ability to induce rounding and steroidogenesis of these ACTH receptor-positive cells.[1]

Purification of antireceptor antibodies is most easily accomplished by removing antibodies against the carrier protein[1] or by affinity purification using peptide-conjugated affinity columns. Peptides are routinely conjugated to Affi-Gel 10 or Affi-Gel 15 (see Technical Bulletin 1099, Bio-Rad, Richmond, CA), and this peptide-conjugated affinity column is then used

[19] B. A. L. Hurn and S. M. Chantler, this series, Vol. 70, p. 104.

[20] G. Galfre and C. Milstein, this series, Vol. 73, p. 1.

TABLE IV
ANTIBODY PRODUCTION TO THE 5′ TO 3′ COMPLEMENTARY PEPTIDE FOR ACTH 1–24 (HTCA) IN BALB/c MICE

Mouse[a]	Coupling method[b]	Adjuvant[c]	Antibody titer^{-1}: Anti-HTCA[d]	Antibody titer^{-1}: Anti-ACTH receptor[e]
1	Glutaraldehyde	Freund's	4800	1000
2	Glutaraldehyde	Ribi	6400	2000
3	Carbodiimide	Freund's	2400	800
4	Carbodiimide	Ribi	2400	1600

[a] Three mice were injected per group, and representative mice from each group are shown here. Mice were injected twice with 100 μg of peptide with 10 days between injections. Blood was taken for antibody screening 10 days after the final immunization.

[b] HTCA was coupled to keyhole limpet hemocyanin (KLH) with glutaraldehyde[1] or carbodiimide[10] as previously described. Coupling efficiency for glutaraldehyde was approximately 60%, whereas for carbodiimide it was approximately 15%. The amount of peptide each mouse received was adjusted accordingly.

[c] Freund's incomplete adjuvant (Sigma Chemical Co., St. Louis, MO) or Ribi MPL + TDM adjuvant (Ribi Immunochem Research, Inc., Hamilton, MT) was used.

[d] Antibody titers were determined against HTCA coated onto microtiter plates.

[e] Antibody titers were determined against Y-1 adrenal cells fixed to microtiter plates as previously described.[1]

to specifically purify the antibodies. The advantages of using Affi-Gel 10 or 15 include rapid, highly efficient coupling plus the presence of a spacer arm to prevent steric hindrance. Purified antireceptor antibodies can then be used for immunoaffinity purification of receptors from the surface of receptor-positive cells.[21,22] Solubilization of receptors from cells, which would be necessary prior to immunoaffinity procedures, has been dealt with extensively.[23,24]

Antigen–Antibody Interactions

Previously, we have shown that antibodies made against pairs of complementary peptides have an idiotype–antiidiotype relationship.[8] As judged by radioimmunoassay, antibodies against ACTH could bind antibodies against its 5′ to 3′ encoded complementary peptide, HTCA, and this binding could be blocked with either ACTH or HTCA. Furthermore,

[21] M. Wilchek, T. Miron, and J. Kohn, this series, Vol. 104, p. 3.

[22] G. J. Calton, this series, Vol. 104, p. 381.

[23] L. M. Hjelmeland and A. Chrambach, this series, Vol. 104, p. 305.

[24] J. V. Renswoude and C. Kempf, this series, Vol. 104, p. 329.

antibodies directed against β-endorphin could bind antibodies against the 5′ to 3′ encoded complementary peptide for γ-endorphin. In each case, the antibodies were binding one another at or near their antigen binding sites as evidenced by the ability of the appropriate ligand to block binding. By definition, these antibodies had an idiotype–antiidiotype relationship. Thus, this method allows one to generate antiidiotypic antibodies in a predetermined rather than a random manner. Whether HTCA sequences or their analogs can be found within the antigen combining sites of anti-ACTH antibodies and vice versa is presently under investigation. In the one case studied to date, however, we have found that complementary sequences located within the hypervariable region (i.e., antigen binding region) of a monoclonal antibody and its protein antigen specify the contact points for these interacting proteins (K. L. Bost and J. E. Blalock, unpublished observations). Thus, it may be that all the possible antigens or their peptide equivalents to which an individual can respond are encoded in the nucleic acids complementary to those encoding antibody binding sites.

Future Directions

There are many potential situations in which complementary peptides can be applied, and future studies will determine the uses. Initial studies suggest that the complementary peptide to ACTH, when injected *in vivo*, can partially inhibit stress-induced steroid production by binding ACTH (K. L. Bost and J. E. Blalock, unpublished observations). It may also be possible to use such binding peptides as substitutes for antibodies in radioimmunoassays or enzyme-linked immunosorbent assays. Along a more theoretical vein, it may be possible to identify peptide equivalents of nonpeptide ligands. If binding sites for these nonpeptide ligands could be identified from receptors or specific antibodies, then sequences complementary to these binding sites may represent peptide analogs similar to the morphinelike peptide Tyr-Gly-Gly-Phe.

[4] Fluorescent and Cytotoxic Analogs of Corticotropin-Releasing Factor: Probes for Studying Target Cells in Heterogeneous Populations

By JEFFREY SCHWARTZ and WYLIE VALE

Introduction

Since the discovery of corticotropin-releasing factor (CRF),[1] cells in numerous tissues, including pituitary,[2-5] central nervous system,[6-7] adrenal, prostate, and spleen,[8] have been identified as CRF-target cells. Conventional methods, including autoradiography, radiolabeled ligand binding, and studies of biological activity, have been employed to identify CRF-target cells. We recently adapted methods, previously used with other peptide hormones,[9,10] to fluorescently label CRF and thereby identify CRF-target cells.[11]

In the pituitary CRF-target cells constitute fewer than 10% of the total. In order to more precisely study the action of these cells we developed methods, using fluorescence labeling, to sort CRF-target cells from mixed populations for further studies. In order to study interactions among pituitary cells and the roles of CRF-target cells, we also adapted methods, previously used with other peptide hormones (Ref. 12, for example), to

[1] W. Vale, J. Spiess, C. Rivier, and J. Rivier, *Science* **213,** 1394 (1981).

[2] P. C. Wynn, G. Aguilera, J. Morell, and K. J. Catt, *Biochem. Biophys. Res. Commun.* **110,** 602 (1983).

[3] P. Leroux and G. Pelletier, *Endocrinology (Baltimore)* **114,** 14 (1984).

[4] E. B. De Souza, M. H. Perrin, J. Rivier, W. Vale, and M. J. Kuhar, *Brain Res.* **296,** 202 (1984).

[5] M. H. Perrin, Y. Haas, J. E. Rivier, and W. W. Vale, *Endocrinology (Baltimore)* **118,** 1171 (1986).

[6] M. R. Brown and L. A. Fisher, *Brain Res.* **280,** 75 (1983).

[7] E. B. De Souza, T. R. Insel, M. H. Perrin, J. Rivier, W. W. Vale, and M. J. Kuhar, *J. Neurosci.* **5,** 3189 (1985).

[8] J. R. Dave, L. E. Eiden, and R. L. Eskay, *Endocrinology (Baltimore)* **112,** 813 (1985).

[9] E. Hazum, P. Cuatrecasas, J. Marian, and P. M. Conn, *Proc. Natl. Acad. Sci. U.S.A.* **77,** 6692 (1980).

[10] Z. Naor, D. Atlas, R. N. Clayton, D. S. Forman, A. Amsterdam, and K. J. Catt, *J. Biol. Chem.* **256,** 3049 (1981).

[11] J. Schwartz, N. Billestrup, M. Perrin, J. Rivier, and W. Vale, *Endocrinology (Baltimore)* **119,** 2376 (1986).

[12] T.-M. Chang and D. M. Neville, *J. Biol. Chem.* **252,** 1505 (1977).

synthesize a cytotoxic analog of CRF. This analog can be used to selectively eliminate CRF-target cells and, thus, to study physiological responses in the absence of these cells.

This chapter describes the syntheses of the fluorescent and cytotoxic analogs of CRF, the use of the fluorescent analog to study CRF-target cells by fluorescence microscopy and to sort these cells by fluorescence-activated cell sorting, and the use of the cytotoxic analog of CRF in studies where elimination of CRF-target cells is indicated.

Reagents

Methanol, HPLC grade (Burdick and Jackson)
Dichloromethane, HPLC grade (Fisher)
Triethylamine (Aldrich)
Acetic acid, 50% in distilled H_2O
Trifluoroacetic acid (Pierce)
Acetonitrile, HPLC grade (Burdick and Jackson)
HEPES dissociation buffer (HDB): 137 m*M* NaCl, 5 m*M* KCl, 0.7 m*M* Na_2HPO_4, 25 m*M* HEPES, 10 m*M* glucose, pH adjusted to 7.3
Collagenase–DNase: 0.4% collagenase (Worthington, Type II), 80 μg/ml of DNase II (Sigma), 0.4% bovine serum albumin (BSA), 0.2% glucose in HDB
Viokase, 0.25% (Gibco) in HDB
β-PJ medium: prepared from powdered SFRE-199-2/Earle's salts (Kansas City Biologicals) and other components as specified in Ref. 13
β-PJ incubation medium (IM): β-PJ medium plus 0.1% crystalline BSA
β-PJ culture medium: β-PJ medium plus the following hormones and growth factors added immediately prior to use: insulin (5 mg/liter), transferrin (5 mg/liter), parathyroid hormone (0.5 μg/liter), T_3 (30 p*M*), fibroblast growth factor (1 μg/liter); the culture medium usually includes fetal bovine serum
Poly(D-lysine) (Sigma), 20 μg/ml of distilled H_2O, filter sterilized
Propidium iodide (Sigma), 1 mg/ml of distilled H_2O, filter sterilized
Glutaraldehyde, EM grade (Electron Microscopy Services)
Sodium chloride–sodium phosphate buffer: 0.14 *M* NaCl, 5 m*M* sodium phosphate, pH 7.4
Gelonin (Pierce)
NP-40 (Calbiochem), 0.5% solution in water

[13] W. Vale, J. Vaughan, G. Yamamoto, T. Bruhn, C. Douglas, D. Dalton, C. Rivier, and J. Rivier, this series, Vol. 103, p. 565.

A Fluorescent Analog of CRF

Synthesis of Fluorescent CRF

A fluorescent analog of CRF is synthesized by conjugating the fluorochrome fluorescein to [Nle^{21,38},Arg^{36}]rCRF. This analog of CRF is used as a starting material because, unlike native species of CRF, it contains no methionine residues and has only one free amine. It is completely bioactive but will not contain oxidized methionine after chemical modification, and chemistry involving free amine groups will be limited to the amino terminus. Advantage can be taken of the latter property, and when [Nle^{21,38},Arg^{36}]rCRF is mixed with excess fluorescein isothiocyanate (FITC, Research Organics, Cleveland, OH) in a solution of methanol, dichloromethane, and triethylamine (ratios by volume, 400 : 200 : 1) in the dark, the sole significant product is [fluorescein-Ser^{1},Nle^{21,38},Arg^{36}]rCRF.

In this procedure, reaction progress is followed by reverse-phase HPLC. At room temperature the reaction is typically complete after 3–4 hr. Sample equipment and procedures for the chromatography are as follows: 5 μm C_{18} 0.46 × 25 cm column Vydac (Hesperia, CA); two Beckman (Fullerton, CA) Model 100A dual-piston pumps; Axiom (Calabasas, CA) 710 system controller; Kratos (Ramsey, NY) Spectroflow 773 absorbance detector; Rheodyne (Cotati, CA) 7125 injector with a 2-ml loop; Shimadzu (Kyoto, Japan) C-R3A Chromatopac integrator–recorder; loading 1% of the total reaction volume and eluting under the following conditions: buffer A, 0.1% trifluoroacetic acid; buffer B, 0.1% trifluoroacetic acid–60% acetonitrile; flow rate, 1.8 ml/min; back-pressure, 2200 psi; gradient, 10% B to 95% B in 30 min. After equilibrium is reached, the reaction is stopped by drying under N_2. The product is dissolved in 50% acetic acid and is separated from unreacted FITC and other contaminants by gel chromatography [0.9 × 30 cm, BioGel P-2 (Bio-Rad Laboratories, Richmond, CA), equilibrated and eluted with 50% acetic acid]. Further purification is accomplished by HPLC (buffer system: 0.1% trifloroacetic acid, 43.2% acetonitrile), collecting fractions corresponding to the major absorbance peak at 210 n*M* during elution under isocratic conditions. Once the desired degree of purity is achieved, the product is immediately lyophilized, because it will degrade over time in solution. [Fluorescein-Ser^{1},Nle^{21,38},Arg^{36}]rCRF is kept in the dark at −20° until use.

Labeling of Cells

Dissociation of Rat Cells. CRF-target cells among dissociated anterior pituitary cells can be easily labeled with the fluorescent CRF analog.

Anterior pituitary cells from male Sprague–Dawley rats, weighing 200–250 g, can be dissociated as described in an earlier volume of this series.[13] Briefly, the freshly removed anterior pituitaries are placed in a solution of 0.4% collagenase (Worthington, Type II) in HEPES dissociation buffer (HDB) in a water-jacketed spinner flask (Celstir, Wheaton Instruments, Millville, NJ), with the impeller set to spin at 100–200 rpm. The tissue fragments are gently drawn in and out of a siliconized Pasteur pipet every 30–40 min throughout the entire procedure. At a point where most of the fragments have dissociated and the remainder are small and thready, cells in the collagenase–DNase solution are transferred to a sterile plastic tube and centrifuged at 475 *g* for 8 min. The pelleted cells are then resuspended in 30 ml of 0.25% Viokase solution, returned to the suspension flask, and incubated and stirred until fragments and threads disappear, but for no longer than 10 min. The Viokase solution and cells are then transferred to a sterile plastic tube, centrifuged as before, and resuspended in incubation medium with 2% fetal bovine serum. Cells are washed by repeated centrifugation and resuspension in 40 ml of culture medium (3–5 times).

Dissociation of Bovine Cells. Bovine anterior pituitary cells can be prepared by a similar method. In this instance, anterior pituitary tissue is minced into cubes (~1 mm^3) with sterile scalpel blades prior to dissociation in the collagenase. Addition of Viokase can be omitted as this does not significantly increase the yield of cells.

Dissociation of bovine anterior pituitary cells by this method, it should be noted, rarely results in complete dissociation of the cells, and considerable amounts of undesired material are also generated. Nevertheless, the following procedure is quite effective in providing large numbers of viable cells devoid of much debris: Following dissociation, the cells are washed in β-PJ containing 4% BSA. After resuspension in β-PJ, the cells are filtered through nylon mesh (44-μm pore size, Small Parts, Miami, FL). A useful apparatus for this procedure can be easily made by placing the mesh over the top of a silanized 100-ml beaker, held in place by an elastic made of rubber tubing. It is helpful to form a depression in the middle of the mesh prior to sterilizing the apparatus by autoclave. In order to separate the pituitary cells from erythrocytes and debris, the cells are layered on a density gradient consisting of one layer Ficoll (Histopaque 1119, Sigma, St. Louis, MO) and one layer Percoll (Pharmacia, Piscataway, NJ; 40.5% Percoll in HDB solution isotonic to cells). After centrifugation for 15 min at 1700 *g*, the pituitary cells are found at the Percoll–Ficoll interface. The cells should then be washed 3 times by centrifugation and resuspended in the appropriate medium for the next step.

Labeling. The labeling of dissociated bovine cells can be done as soon as they are dissociated. Rat cells, as prepared above, are best labeled after

an overnight recovery period, possibly because of an effect of proteases on CRF-binding sites. Dissociated cells can be maintained overnight in the same type of spinner suspension flask as described in the dissociation procedure. The cells are suspended in B-PJ culture medium, containing nystatin but no serum. Taking into account the total volume of the vessel and the volume occupied by the medium, carbon dioxide is added to the flask to approximately 10% of the volume of air above the medium, and the flask is tightly closed. The cells are stirred at 100–200 rpm while warm water (37°) is circulated through the jacket of the flask.

In order to label cells for fluorescence-activated cell sorting, the cells are incubated in an HDB solution, containing 0.4% BSA, 0.2% glucose. To this are added 70 n*M* (bovine cells) or 150 n*M* (rat cells) [fluorescein-Ser^1,$Nle^{21,38}$,Arg^{36}]rCRF (fluorescent CRF) and 1 μg/ml propidium iodide to label dead cells. In order to facilitate solubilizing the fluorescent CRF, it should be first dissolved in a minute volume of 50% acetic acid, which is then added to the HDB solution. After a 60-min incubation at room temperature, the binding of fluorescent CRF is close enough to equilibrium to permit further studies. The medium for labeling dissociated cells for fluorescence microscopy can be either the HDB solution described above or β-PJ, without serum or added glucocorticoids. The same concentrations of fluorescent CRF and propidium iodide should be used in either case.

It is also possible to label cells cultured on microscope slides (using, for example, LabTek tissue culture slides, Miles Scientific, Naperville, IL). This may be desirable, for example, to test the effect of some type of treatment in culture on CRF binding. The slides should be coated with an agent that promotes adhesion, such as poly(D-lysine), prior to seeding with cells. At the appropriate time, the fluorescent CRF can be added to the medium bathing the cells, provided that the medium does not contain any component that competes with or inhibits the binding of CRF. Whenever fluorescent CRF is added to media, the media should contain 0.1% BSA to minimize loss of the peptide due to adhesion to noncellular surfaces.

Fluorescence Microscopy

The incubation procedures prior to fluorescence microscopy should reflect the needs of the study to be performed. For example, if viable cells are to be viewed, and the receptors visualized solely on the surface of the cells, then the incubation with the fluorescent CRF should be performed at 4° with an incubation period of at least 60 min. If, for example, the cells are to be labeled and viability is not essential, then conditions that maxi-

mize binding of CRF to its receptors should be employed and propidium iodide need not be added. In any event, prior to viewing the cells should be washed free of unbound fluorescent CRF and propridium iodide. Suspended cells can be washed once by centrifugation (5 min at 475 *g* is sufficient) followed by resuspension in HDB and recentrifugation. After removing the HDB supernatant, the pellet of cells can be resuspeneded in the residual volume of buffer and deposited by pipet onto a microscope slide. Cells cultured on slides can be washed by simply aspirating the medium and adding a wash buffer. Simple buffers, such as HDB, are desirable for the wash because no component interferes with light absorption or transmission. Care must be exercised when using HDB, though, because of the possibility of loosening the cells from the glass.

With fluorescent CRF, the best viewing is that which can be accomplished within minutes of washing. The reasons for this include that the viability of cells decreases following repeated manipulations and that the fluorochrome is subject to photobleaching. A mounting solution may be used provided it does not decrease the transmission of the fluorescent light and is not itself fluorescent. We usually do not use any special mounting solution and find that when a coverslip is placed directly on the cells in a minimum of buffer the results are easily seen and quite striking. The drawback of not using any mounting solution is that the viewing must be complete before the buffer dries.

Slides thus prepared may be viewed with any suitably equipped fluorescence microscope. In our studies we have used a Leitz (Wetzlar) Dialux 22 Microscope (Leitz, Rockleigh, NJ) equipped with filters passing excitation light between 450 and 490 nm, a dichroic mirror reflecting below 510 nm, and a barrier filter of 515 nm. Cells labeled with fluorescent CRF appear green, while dead cells contain red nuclei or cytoplasm. Specific fluorescence due to binding of fluorescent CRF to cells can often be distinguished from autofluorescence of certain cells by comparing the appearance of fluorescent cells at different wavelengths of excitation and fluorescent light. Labeled cells only appear bright when excited at the proper wavelength (~490 nm) and viewed at the proper fluorescence wavelengths (~517 nm) for fluorescein.

Flow Microfluorimetry

The availability of flow-cytometric equipment in recent years has enabled further study and characterization of fluorescently labeled cells. Dissociated cells, labeled as described above, can be analyzed by flow cytometry. With flow cytometry, as with microscopy, a medium should be selected for the wash that contains nothing to interfere with excitation or fluorescence detection. Multiple parameters such as forward and 90°

light scatter can be monitored as well as fluorescence, to provide information on the size and likely presence of secretory granules in the fluorescent and nonfluorescent cells. In systems capable of detecting fluorescence of propidium iodide, dead cells can also be counted and the fluorescein fluorescence associated with these cells accounted for.

Figure 1 illustrates typical flow-cytometric data obtained with rat cells incubated with fluorescent CRF. These data were obtained using the Salk Institute Flow Microfluorimeter. The laser was tuned to 488 n*M* at 600 mW output. Fluorescein fluorescence was collected through a 520-nm bandpass filter, propidium iodide (PI) fluorescence through a 600-nm barrier filter. Fluorescence signals from the fluorescein channel were compressed with a three-decade logarithmic amplifier. Forward angle light scatter signals were routed through a time–amplitude converter (Los Alamos National Laboratory Drawing, 4Y-223131) for pulse width measurement as an estimate of cell diameter. Analog signals from all parameters were digitized and stored as list mode data files in a DEC micro-11/73 computer (Digital Equipment Corporation) for subsequent analysis. Figure 1 is a bivariate isocontour display relating forward light scatter, time–amplitude conversion (essentially cell size), to fluorescein fluorescence intensity (64 × 64 channel array). Dead cells (i.e., any cells with signals on scale in the PI channel) are disregarded. The panels represent flow cytometric data obtained from dissociated rat anterior pituitary cells suspended in HDB, 0.4% BSA, 0.2% glucose, 1 μg/ml PI labeled with PI only (top), 175 n*M* fluorescent CRF (middle), and 175 nm fluorescent CRF in the presence of 175 μM unlabeled rCRF (bottom). Each panel represents approximately 25,000 cells. Each channel in which more than three events appears is mapped as a dot. Contour lines were drawn at 25, 50, and 75% of the population.

The box (Fig. 1) represents a population of rat anterior pituitary cells that is labeled by fluorescent CRF (i.e., CRF-target cells, middle panel). This labeling is competitively inhibited to a large extent by the presence of excess unlabeled CRF, and this is why the overall fluorescence of this population is markedly decreased in the bottom panel. Any residual elevated fluorescence observed in the bottom panel is likely due to low affinity (nonspecific) binding. The boxes as illustrated can be used to quantitate the fraction of cells fluorescently labeled over background fluorescence. The boxes also represent typical "windows" for cell sorting (i.e., the sorter is instructed to deflect all "events" occurring in the window). Size determinations can also be made; the cells that are labeled in the present example appear to be smaller than the average in that sample.

Labeling of cells with [fluorescein-Ser^{1},Nle^{21,38},Arg^{36}]rCRF is possible because, despite the presence of a fluorochrome, this analog of CRF is bioactive, and thus binds to specific CRF receptors on cells. Once bound

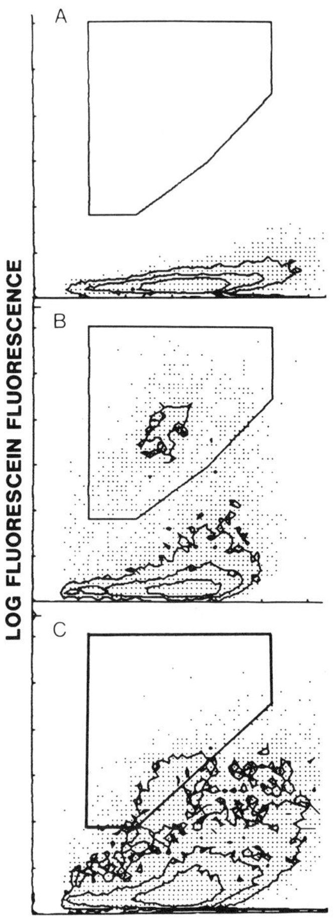

FIG. 1. Flow microfluorometric analysis of dissociated rat anterior pituitary cells. Cells were dissociated and maintained overnight as described in the text. On the following day they were divided into three portions in centrifuge tubes, washed three times in β-PJ incubation medium, preincubated at 37° for 1 hr, then washed once in HDB, 0.1% BSA. The cells were resuspended in cold (4°) HDB, 0.1% BSA containing (A) 1 μg/ml propidium iodide, (B) 1 μg/ml propidium iodide plus 175 n*M* fluorescent CRF or (C) 1 μg/ml propidium iodide plus 175 n*M* fluorescent CRF plus 175 μ*M* unlabeled rCRF. The cells were incubated 90 min at 4°, washed once with HDB, 0.1% BSA (4°), and analyzed by flow microfluorimetry as described in the text.

to cells, the analog acts as does CRF and stimulates the secretion of adrenocorticotropin (ACTH).[11] Flow-cytometric and fluorescence microscopic studies indicate that the binding can be blocked by the presence of excess unlabeled CRF but not inactive fragments or other peptides,[11] and the fluorescent label dissociates from cells after prolonged periods in a medium without fluorescent CRF.

The above described methods have been used with some success to label other CRF-target cells. Cultured dissociated rat cerebral cortex (Fig. 2) and hypothalamic cells (see Ref. 14 for methods of cell culture) have been labeled with fluorescent CRF. Rat brain and pituitary slices have also been labeled. In pituitary slices, the pattern of areas with heavy fluorescence corresponds to that in published studies with autoradiographic methods.[4] The possibilities of using fluorescent CRF to label cells in tissue slices are still being explored.

Cell Sorting

CRF-target cells have been successfully sorted from mixed populations of dissociated rat and bovine anterior pituitary cells. The procedures can be divided into three steps: labeling, sorting, and recovery. For labeling, cells can be prepared as described above for fluorescence microscopic studies of dissociated cells. Extreme care should be exercised in order to minimize the amount of debris and erythrocytes present, because the sorting apparatus cannot distinguish between these and pituitary cells, and it is the total number of "events" that determines the efficiency of the procedure. Propidium iodide need not be present during the incubation with fluorescent CRF. Rather, it should be added later in the suspension buffer, because cells will be dying throughout the process and all those that expire prior to sorting ought to be excluded. After the incubation with fluorescent CRF, the cells should be centrifuged and resuspended in *cold* (4°) buffer or medium to slow the dissociation of fluorescent CRF from the cells. Prior to sorting, the cells should be carefully passed through sterile nylon mesh (44- or 52-μm pore size) in order to remove aggregations of cells which might clog the flow apparatus and to minimize the number of "passenger" cells that might accompany the desired sorted cells.

The procedures employed to sort cells obviously depend on the available equipment. In our studies the Salk Institute cell sorter was used, and the procedures we have followed are described as an example. The apparatus is similar to that described for the flow microfluorimeter with additional hardware to enable the sorting. Sorting windows are established as illustrated in Fig. 1 and described below. Sample flow rate in the sorter is

[14] W. J. Shoemaker, R. A. Peterfreund, and W. Vale, this series, Vol. 103, p. 347.

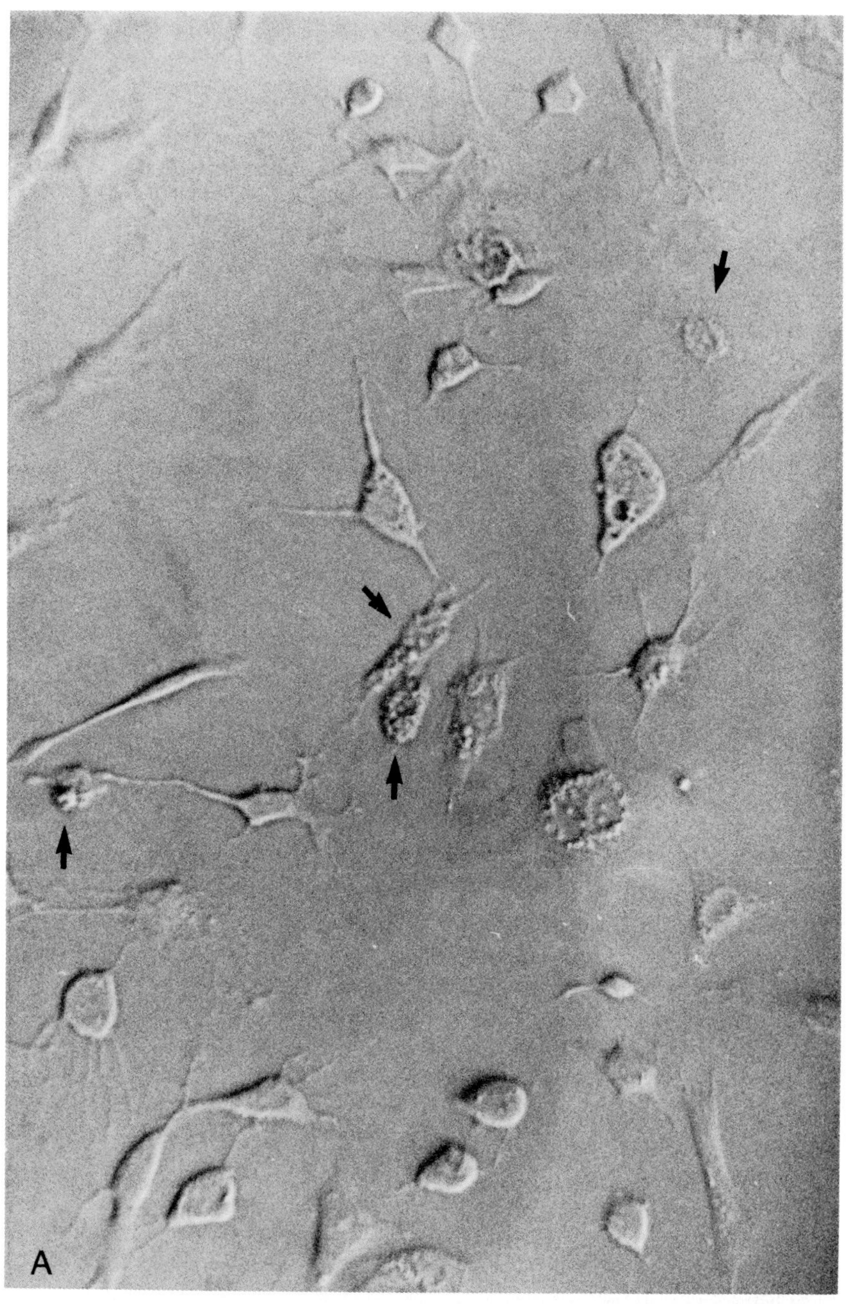

FIG. 2. Cultured dissociated rat cerebral cortex cells labeled with fluorescent CRF. Cells were cultured for 1 week on LabTek tissue culture slides under conditions described in Ref. 14. On the day of observation they were washed three times in β-PJ incubation medium, preincubated 1 hr at 37°, then incubated for 1 hr at 4° in HDB, 0.1% BSA, containing 70 nM fluorescent CRF (no propidium iodide was added to permit black and white photomicrog-

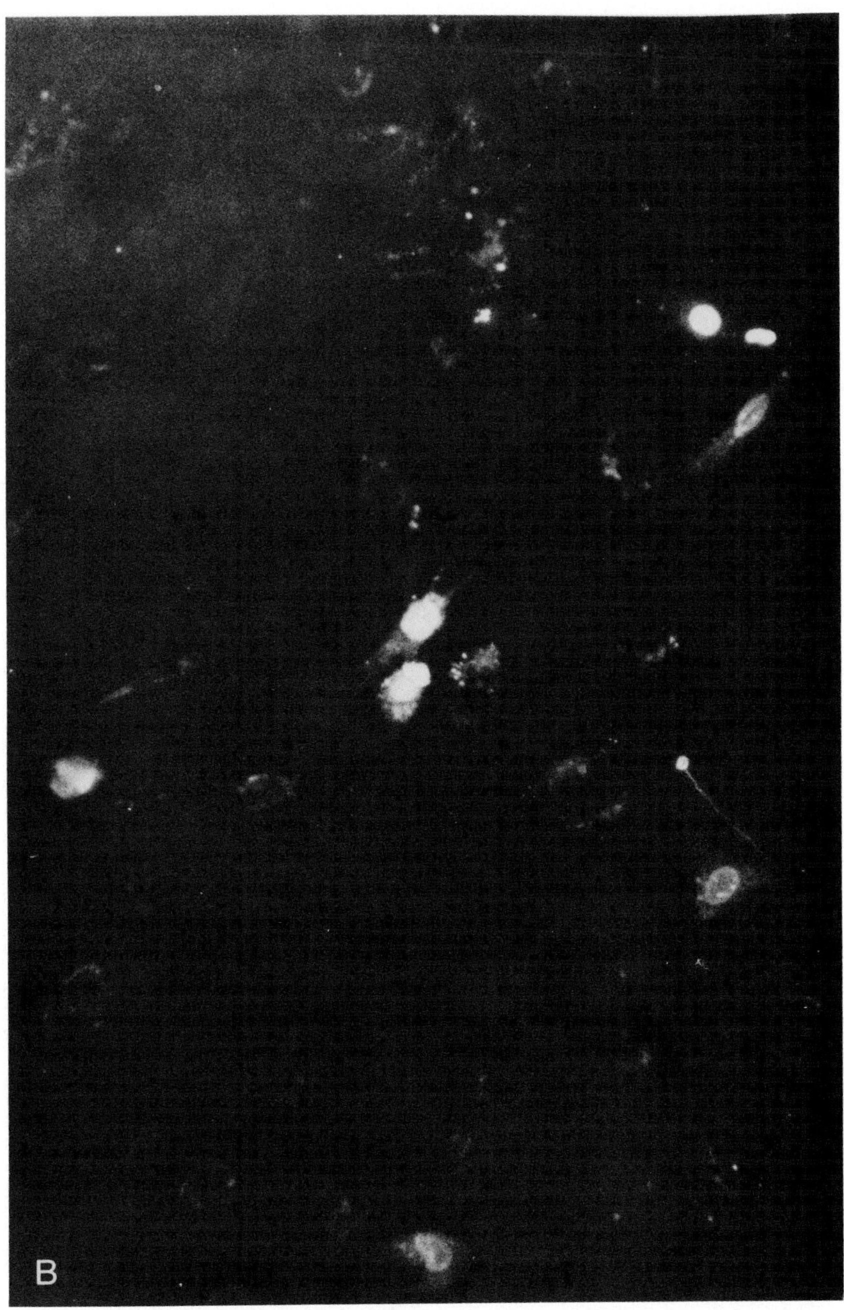

raphy). Cells were washed once in HDB, 0.1% BSA and examined under conditions described in the text. (A) Phase-contrast photomicrograph showing all cells in the field. (B) Same field, fluorescence photomicrograph showing only cells labeled with fluorescent CRF. Labeled cells are indicated in photomicrograph A by arrows. (A and B) ×264.

typically 2000 events per second, an event being either a cell or particle that scatters light. Events within the sorting window boundaries are deflected by charging the flow stream while the event moves into a droplet prior to passage through a static field where deflection occurs. There are generally three droplets per deflection (i.e., the droplets preceding and following the droplet containing the selected event are also deflected). With droplets generated at 40 KHz, the additional droplets rarely contain contaminating cells, the frequency of cells being approximately 1 in every 20 droplets.

Sorting parameters for bovine cells are typically the following. (1) Cells emitting fluorescent light at the wavelength of propidium iodide, that is, dead cells, are electronically gated out. (2) Particles too small to be cells or too large to be single cells, as determined by forward light scatter, are electronically gated out. (3) A sorting window is circumscribed around the remaining "events" on the basis of fluorescence, encompassing the most highly fluorescent (severalfold increase in fluorescence) over background (see Fig. 1). This typically includes 10–15% of the total. This percentage is higher than the observed percentage of labeled cells as assessed by fluorescence microscopy. This is because, unlike fluorescence microscopy, the cell sorter cannot distinguish between autofluorescence and fluorescein fluorescence, nor can it distinguish between cells and other material that has bound fluorescent CRF nonspecifically and that happens to scatter light as a cell. Also, the window is made to include more cells because as the procedure continues, all the cells that bind fluorescent CRF will become less brightly fluorescent as the analog dissociates from the cell. Thus while 15% of the events may be within the window at the beginning of a typical "sort," only 9% may be there after 45 min.

During recovery the cells should be collected into sterile, Ca^{2+}-free medium to minimize aggregation and adhesion of cells to surfaces. Depending on the design of the sorting apparatus, a number of receptacles can be used for the recovery of cells. The decision of which type of receptacle to use must take into account a number of factors, including the small number of cells that will be recovered, the need to wash the cells extensively to remove any traces of fluorescent CRF, and the need to avoid aggregation and adhesion of cells. Cells can be collected into a sterile centrifuge tube, but extreme care must be exercised to avoid loss of cells caused by adhesion to the inner surface of the tube. If possible, the cells should be sorted into a sterile petri dish (100 mm). By the time the sort has been completed, most cells will have settled to the bottom. The cells can be washed by carefully aspirating and replacing the medium. The cells can be resuspended by gentle agitation and then aliquoted.

Two other receptacles have been adapted for use with sorted cells, tissue culture slides (e.g., LabTek) and Transwell tissue culture inserts (Costar, Cambridge, MA). With these, the cells are aliquoted as they are sorted, and thus care must be exercised that the same number of the same type of cell (the fraction of cells in a sorting window will change during the course of a sorting procedure due to dissociation of the fluorescent CRF from the cells) is collected into each receptacle. Cells can be collected into tissue culture slides as described above for petri dishes. An interesting procedure has been developed for use of the Transwell inserts. The insert is removed from the tissue culture dish, and enough collection medium is placed in it to cover the membrane on the bottom. Surface tension prevents the medium from going through the membrane. The insert is placed inside the top of a sterile 50-ml centrifuge tube (Corning), which can be positioned in the sorting apparatus. When placed in the path of the cell sorter effluent the insert/centrifuge tube is placed such that the main outflow of the sorter, containing the undesired components, flows through one of the three large vent holes of the insert, while the desired droplets are deflected toward the membrane. After the sort, the cells in the inserts can be washed by placing them in wells containing medium, pulling them out and letting most of the medium drip through the membrane, and placing them in wells with fresh medium again.

After the sorted cells have been thoroughly washed, they can be cultured for 3–4 days of recovery and then used in experiments. These cells respond to CRF by increasing ACTH secretion by the same factor as their controls. As assessed by immunocytochemistry, the percentage of cells containing ACTH increases from 6.1% to over 80% of the total after sorting.[11]

A Cytotoxic Analog of CRF

A cytotoxic analog of CRF can be synthesized by conjugating [$Nle^{21,38}$,Arg^{36}]rCRF to gelonin. Gelonin is a potent ribosome-inactivating protein isolated from the seeds of *Gelonium multiforum* that is nontoxic to intact cells, but cytotoxic in gelonin–concanavalin A and gelonin–immunoglobulin conjugates.[15,16] In a simple procedure, glutaraldehyde can be coupled to [$Nle^{21,38}$,Arg^{36}]rCRF, and then, after extensive rinsing to wash out any free glutaraldehyde, the glutaraldehyde-activated CRF analog can be bound to gelonin.

The bioactive CRF analog [$Nle^{21,38}$,Arg^{36}]rCRF is reacted with an ex-

[15] F. Stirpe, S. Olsnes, and A. Pihl, *J. Biol. Chem.* **255,** 6947 (1980).

[16] W. A. Blattler, B. S. Kuenzi, J. M. Lambert, and P. D. Senter, *Biochemistry* **24,** 1517 (1985).

cess of glutaraldehyde. [$Nle^{21,38}$,Arg^{36}]rCRF (4 mg, 720 nmol) is placed in 440 μl sodium chloride–sodium phosphate buffer, pH 7.4. Acetonitrile (150 μl) is added to dissolve the peptide. Glutaraldehyde (10 μl of 25% solution, 25 μmol) is then added. The solution is allowed to react at room temperature for 5 min. Cold (4°) sodium chloride–sodium phosphate buffer is added to precipitate the peptide, and the mixture is quickly washed very carefully twice by ultrafiltration (1000 MW cutoff filter, Amicon ultrafiltration cell, Danvers, MA), with sodium chloride–sodium phosphate buffer, to remove all glutaraldehyde.

The mixture, containing the reacted peptide, is concentrated to a volume of approximately 300 μl by ultrafiltration. Acetonitrile is added until the suspension dissolves. Gelonin (2 mg, 67 nmol) is added, and the clear solution is stirred at room temperature overnight. The reaction product is washed thoroughly by ultrafiltration with a 30,000 MW cutoff filter in place in sodium chloride–sodium phosphate buffer.

In a recently synthesized batch of gelonin–CRF conjugate, the molecular ratios of methionine, tyrosine, and glycine to norleucine obtained in the analysis indicated a ratio of 20 mol CRF analog per mole gelonin. Analysis of the molecular ratios of all the amino acids in the reaction product conformed almost exactly to the ratios calculated for 20 mol [$Nle^{21,38}$,Arg^{36}]rCRF per mole gelonin. Radioimmunoassay of the gelonin–[$Nle^{21,38}$,Arg^{36}]rCRF conjugate indicated 30% potency compared both to [$Nle^{21,38}$,Arg^{36}]rCRF and to rat CRF (as assessed by the amount of peptide required to displace 50% of radiolabeled CRF). In experiments designed to assess the biological activity of the cytotoxic conjugate by measuring ACTH secretion in cells in response to acute exposure, the EC_{50} of the conjugate was shifted to the left of that for sheep CRF (oCRF). However, when calculated on the basis of 20 mol of [$Nle^{21,38}$,Arg^{36}]rCRF per mole of the cytotoxic conjugate, there appeared to be no shift in the EC_{50}. Yet, the amount of ACTH present in the media of cells exposed to the gelonin–[$Nle^{21,38}$,Arg^{36}]rCRF conjugate was greater than the maximum amount secreted in response to oCRF, suggesting that something more than secretion was occurring.

Usually cytotoxicity is measured by the inhibition of incorporation of radiolabeled nucleotides or amino acids into DNA or protein, or by similar methods. Since corticotropes constitute a very small fraction of anterior pituitary cells, however, small differences in overall incorporation of radioactivity into protein and DNA cannot be unquestionably measured. Instead, efficacy of the gelonin–CRF conjugate as a cytotoxin was assessed by its effects on ACTH content and CRF-stimulated ACTH secretion. The specificity was assessed by its effects on LH content and secretion.

In order to test the efficacy and specificity of the cytotoxic conjugate the following procedure was employed, using dissociated rat anterior pituitary cells. This procedure can also be used to eliminate the ACTH secretory response to CRF in cultured cells for purposes of further studies. At least 6 hr after dissociation, 2.7×10^6 cells in 10 ml β-PJ in each of seven petri dishes (100 × 20 mm, Falcon, Oxnard, CA) are treated with β-PJ vehicle, oCRF (2, 10, 50 n*M* final concentrations), the CRF–gelonin conjugate analog (2, 10, 50 n*M* final concentrations), or unconjugated [$Nle^{21,38}$,Arg^{36}]rCRF and gelonin in appropriate concentrations to reflect the content of the 2 or 10 n*M* cytotoxic conjugate. Twelve hours later the cells are washed three times in β-PJ, incubated 1 hr at 37°, washed twice more, and plated at 3×10^5 cells per well in Linbro 24-well tissue culture plates (Flow Laboratories, McLean, VA) in β-PJ. The cells are cultured at 37° for 3 days by which time the cells firmly attach to the plates. Acute (4 hr) ACTH and luteinizing hormone (LH) secretory responses can be assessed as previously described.[13,17,18] ACTH and LH contents of cells are measured following solubilization with NP-40 (0.05% final concentration). Studies of the content of cells typically use those cells from all five pretreatment groups that were not exposed to either CRF or gonadotropin-releasing hormone (GnRH) in the acute secretion experiments.

Results from studies performed according to the above procedures indicated a specific irreversible effect of the cytotoxic conjugate on ACTH-containing cells.[19] The conjugate at all concentrations studied decreased ACTH content markedly and eliminated the ACTH secretory response to CRF. There were no significant effects on LH content or secretory response, except for a small nonspecific dimunition of the LH secretory response to GnRH in cells pretreated with the cytotoxic conjugate at 50 n*M*. Control responses indicated that the effects of the conjugate on ACTH content and secretion were specific to the conjugate itself and not its components.

Subsequent studies have demonstrated that the cytotoxic conjugate does not act by preventing adhesion of cells to cell culture surfaces. Thus, an alternative method for treating cells with the cytotoxic conjugate involves first culturing them in dishes, whose surfaces may be precoated with poly(D-lysine) or another agent to speed adhesion of the cells. Once the cells are firmly attached, the cytotoxic conjugate can be added. Fol-

[17] W. Vale, G. Grant, M. Amoss, R. Blackwell, and R. Guillemin, *Endocrinology (Baltimore)* **91,** 562 (1972).

[18] W. Vale, J. Vaughan, M. Smith, G. Yamamoto, J. Rivier, and C. Rivier, *Endocrinology (Baltimore)* **113,** 1121 (1983).

[19] J. Schwartz, B. Penke, J. Rivier, and W. Vale, *Endocrinology (Baltimore)* **121,** 1454 (1987).

lowing the treatment, the cytotoxic conjugate can be removed by aspiration and repeated washing with fresh medium. The cells can be returned to culture for subsequent experiments.

Recent studies with CRF-responsive neoplastic cell lines, using the cytotoxic and fluorescent CRF, indicate a toxic effect of the conjugate on these cells as well. Studies of the efficacy of the cytotoxic conjugate *in vivo* have thus far been inconclusive. It should be possible to develop procedures to enable the reliable elimination of CRF-target cells *in vivo*.

Summary

The procedures briefly outlined above described the syntheses of two conjugates of CRF. In each case a molecule with a specific biochemical or physical property was conjugated to an analog of CRF to enable association of the particular property with only CRF-target cells. Thus, by conjugating gelonin or fluorescein to CRF it becomes possible to selectively impair the cellular function of, or fluorescently label, CRF-target cells among mixed populations. These methodologies are still in the early stages of development, and it can be expected that they will become more refined and improved as workers with specialized experience (in synthetic methods, microscopy, cell culture, etc.) continue this work.

Acknowledgments

This research was supported by National Institutes of Health Grants AM26741 and HL06808. Research conducted in part by The Clayton Foundation for Research, California Division. W. Vale is a Clayton Foundation Investigator. The authors wish to thank Dr. J. Rivier for supplying [$Nle^{21,38}$,Arg^{36}]rCRF and for incalculable assistance in synthesizing the fluorescent and cytotoxic conjugates. We also gratefully acknowledge the contributions of Joe Trotter, Botond Penke, Robert Galyean, Ron Kaiser, John Dykert, Gayle Yamamoto, Anne Corrigan, Mary Tam, Carolyn Campen, Nils Billestrup, Gotfryd Kupryszewski, Marilyn Perrin, Yaira Haas, Eric Widmaier, and Erica Nishimura and the manuscript preparation by Bethany Connor and Susan McCall.

Section II

Equipment and Technology

[5] Cell-to-Cell Communication in Peptide Target Cells of Anterior Pituitary

By CARL DENEF, PHILIPPE MAERTENS, WILFRIED ALLAERTS, ANNICK MIGNON, WIM ROBBERECHT, LUC SWENNEN, and PETER CARMELIET

Introduction

There is growing evidence in support of the hypothesis that in various endocrine tissues specific cells are responsive to peptides which are locally produced and secreted and that these peptides have a local regulatory role.[1] Various morphological and cell biological characteristics of the anterior pituitary also suggest that cell-to-cell communication may play an important role in the regulation of hormone release and cellular differentiation in this tissue.[1] Experimental evidence in support of the existence of cell-to-cell communication has recently become available although mainly by *in vitro* approaches. For example, application of antiserum against vasoactive intestinal peptide (VIP) in pituitary cell monolayer cultures appears to lower basal prolactin (PRL) release,[2] and since VIP is synthesized in pituitary cells,[3] possibly in lactotropes, VIP may function as a paracrine or autocrine factor on lactotropes. The finding that opiate receptor blockers and anti-β-endorphin antisera influence release of luteinizing hormone (LH) in pituitary cell monolayer cultures has suggested the existence of a communication between corticotropes and gonadotropes through β-endorphin.[4]

We have studied intercellular communication in the anterior pituitary by preparing highly enriched populations of the various pituitary cell types and then measuring the influence of one of these cell types on the activity of the other(s) in 3-dimensional cell cultures (also known as reaggregates, aggregates, or aggregate cell cultures). The existence of regulatory signals from one cell type to another was estimated from the change in secretory response induced by coaggregation of both cell types. Secre-

[1] C. Denef, *Clin. Endocrinol. Metab.* **15,** 1 (1986).

[2] T. C. Hagen, M. A. Arnaout, W. J. Scherzer, D. R. Martinson, and T. L. Garthwaite, *Neuroendocrinology* **43,** 641 (1986).

[3] M. A. Arnaout, T. L. Garthwaite, D. R. Martinson, and T. C. Hagen, *Endocrinology (Baltimore)* **119,** 2052 (1986).

[4] M. S. Blank, A. Fabbri, K. J. Catt, and M. L. Dufau, *Endocrinology (Baltimore)* **118,** 2097 (1986).

tory responses were followed as a function of time in a perifusion system in which aggregates remained functional for periods up to several weeks.

Three-Dimensional Pituitary Cell Cultures

Cell Dispersion

Rats are sacrificed by decapitation (Harvard animal decapitator No. 135, Harvard Apparatus, Millis, MA) between 8 and 9 a.m. After washing the heads in water and 70% ethanol, the cranium is immediately opened within the confines of a laminar air flow hood, and the brains are removed. The neurointermediate lobe is, except for the 14-day-old female rats, dissected free from the anterior pituitary *in situ*. The anterior pituitary is then removed and collected in a 55-mm petri dish in 5 ml Dulbecco's modified Eagle's medium (DMEM) (prepared from powder medium without $NaHCO_3$ and containing in 1 liter medium 25 mM HEPES and 1 g glucose; Gibco, Grand Island, NY). One liter of the medium is supplemented with 0.3% bovine serum albumin (BSA) (Fraction V, Serva, Heidelberg, FRG), 3.7 g $NaHCO_3$, 35 mg penicillin G (Sigma Chemical Co., St. Louis, MO), and 50 mg streptomycin sulfate (Sigma), pH 7.4. This medium is further denoted as DMEM–0.3% BSA. The pituitary lobes are washed with 5 ml of DMEM–0.3% BSA in order to remove surface blood and connective tissue, cut into 0.5- to 1.0-mm blocks with a sterile razor blade, and transferred to a small 10-ml Erlenmeyer flask.

After the blocks of tissue settle, the medium is replaced by 2 ml 0.5% bovine pancreatic trypsin (Type III, Sigma) in DMEM–0.3% BSA. The trypsin solution is freshly prepared or not older than 5 days. All subsequent incubations are done at 37° in a shaking water bath at 60 rpm. After a 15-min incubation in the trypsin solution, bovine pancreatic deoxyribonuclease (DNase Type DN-EP, Sigma) (4 μg in 2 ml DMEM–0.3% BSA) is added for 1 min to avoid coating of the tissue blocks by nucleohistone material released from damaged cells. DNase and trypsin are removed and replaced for 10 min by 2 mg/2 ml soybean trypsin inhibitor (Sigma, Type I-S) in DMEM–0.3% BSA. Intercellular Ca^{2+} is removed by incubating subsequently for 5 and 15 min with, respectively, 2 ml 2 mM and 4 ml 1 mM EDTA in Ca^{2+}- and Mg^{2+}-free Earle's balanced salt solution supplemented with 0.3% BSA and 20 mM HEPES with corrections of NaCl to 5.98 g/liter and $NaHCO_3$ to 1 g/liter, pH 7.4 (EBSS–0.3% BSA).

After 3 washings in 5 ml EBSS–0.3% BSA, cells are transferred to a conical tube in 0.75–1 ml EBSS–0.3% BSA in which they are mechanically dispersed: the initial cell suspension is obtained by very gently aspirating and expelling the tissue blocks 3–4 times with a flame-polished

Pasteur pipet (diameter minimum 0.7 mm). Undissociated tissue is allowed to sediment to the bottom of the tube. The supernatant, containing dispersed cells, is removed and transferred to a conical tube containing 5 μg DNase in 4.5 ml DMEM–0.3% BSA. A volume of 750 μl EBSS–0.3% BSA is added to the undissociated tissue blocks, which are mechanically dispersed again following the same procedure. This procedure (adding the supernatant to the DNase solution; dispersing the resting pellet) is repeated maximally 5–7 times, unless all blocks are dispersed earlier. Resting undissociated tissue is discarded.

The cell suspension is filtered through a 50-μm nylon mesh (Nybolt, Swiss Silk Bolting Cloth Mfg. Co. Ltd., Zurich, Switzerland), and the cells are centrifuged (10 min, 120 *g*) through a layer of 3% BSA in DMEM, pH 7.4. The pellet is resuspended in culture medium, and a fraction is taken for counting the cells with a Coulter Counter (Model Z_B, Coulter Electronics Ltd., Harpenden, Herts, England). The viability, estimated by the trypan blue exclusion test, was always better than 95% in our laboratory. Cell yield per pituitary was $0.8–1.2 \times 10^6$, $1.5–2.1 \times 10^6$, and $3–4.5 \times 10^6$ cells for 14-day-old female, adult male, and adult female (random cycle) rats, respectively. When examined immediately after dispersion, cell preparations should be essentially free of clumps or reaggregated cells.

The incubation volumes mentioned are sufficient for dispersion of 25 anterior lobes of adult male rats, 10 anterior lobes of adult females, and 40 pituitaries of 14-day-old females. All glassware used during the cell preparation must be previously siliconized. Parameters which deteriorate yield and viability are a rise in pH above 7.6 (which in the current method is avoided by the addition of HEPES to the media), a low concentration or too old solution of trypsin (appearance of nondispersed small clumps), and too heavy shaking or handling of the tissue blocks throughout the procedure (partial dispersion into single cells and subsequent loss during medium changes).

Culture Medium

In order to better standardize culture conditions and to study the influence of the hormonal environment on the secretory activity of reaggregates, serum-free defined culture medium is used. The serum-free culture medium, essentially the same as described previously,[5] is prepared from a special mixture composed for our laboratory by Gibco (Paisley, Scotland). It consists of a DMEM–Ham's F12 mixture (1 : 1), supplemented with 15 mM HEPES buffer, 15 mM TES buffer, 20 μM ethanol-

[5] C. Denef, M. Baes, and C. Schramme, *Endocrinology* (*Baltimore*) **114,** 1371 (1984).

amine, 25 nM sodium selenite, and 5 mg/liter insulin. To 1 liter of this medium are added the following: 1 g $NaHCO_3$, 35 mg penicillin (Sigma), 50 mg streptomycin (Sigma), 5 g BSA (Fraction V, analytical grade, Serva), 40 mg human transferrin (Serva or Gibco, Grand Island, NY), 1 mg catalase (65,000 U/mg) (Boehringer, Mannheim, FRG), and 10 mmol ethanol (J. T. Baker, Deventer, Holland). The latter two substances are added to scavenge superoxide anions and peroxides which can form in the absence of serum. High-quality water (Milli-RO 15, Milli-Q UF system, Millipore, Molsheim, France) is used. The pH of the medium is adjusted to 7.4, and osmolarity is 320 mOs. Triiodothyronine (T_3) (Serva) and the glucocorticoid dexamethasone (Serva) are added according to the experimental design. Dexamethasone (Dex) is stored as a 1 mM stock solution in absolute ethanol at 4°. T_3 is dissolved at 0.5 μM in saline and stored as a 0.5 ml stock solution at $-25°$. Further dilutions of Dex and T_3 are made in culture medium.

Cell Aggregation

The suspended cells are seeded at a density of 1 or 2 $\times$ 10^6 cells/2 ml culture medium in a 35-mm untreated petri dish with vents (Becton Dickinson, Cockeysville, MD). The dishes are subjected to continuous gyratory shaking at 65 rpm, amplitude 25 mm diameter on an Infors HT shaker (Bottmingen, Switzerland) in a humidified CO_2–air incubator (1.2% CO_2) at 37° to allow reaggregation. After 2 days in culture, aggregates from 4–6 dishes are collected and transferred in 6 ml new culture medium to 55-mm untreated petri dishes with vents (Gosselin Plastic, Hazebrouck, France). Thereafter medium is renewed every 3–4 days. Care should be taken to control the pH of the medium in the dishes when culture medium is renewed. When the number of cells/dish exceeds 6 $\times$ 10^6 in 6 ml, the pH drops to 7.2 within 3 days. When aggregates are composed of enriched populations of lactotropes, or even of cells of the total population of adult female rat anterior pituitary, the number of cells/dish/6 ml should not exceed 3–4 $\times$ 10^6.

The aggregation process using cells from immature rats and serum-supplemented culture medium is described in our original paper.[6] In the defined culture medium without serum and using cells from adult rats, the aggregation process is essentially similar. However, some additional aspects, not published before are worthwhile to mention here.

The aggregation process starts immediately after placing the cell suspension in the 35-mm petri dish on the gyratory shaker. Within 1 day, as

[6] B. Vanderschueren, C. Denef, and J.-J. Cassiman, *Endocrinology (Baltimore)* **110,** 513 (1982).

examined under an inverted microscope, the suspended cells reassociate into aggregates resembling bunches of grapes. Only a negligible number of cells remain dispersed. At this stage there is no smoothing of the aggregate surface, and the overall shape is rather irregular. At the aggregate surface, stringlike appendices of a few cells, laterally attached to each other, also often occur at this stage. These appendices become less frequent at day 2, at which time the cellular surface also becomes a little smoother. The smoothing process continues with the aging of the aggregates as the result of morphological alterations of the cells, adjusting their contours to each other. This feature is especially observed in the smaller pituitary cell types, obtained by velocity sedimentation at unit gravity (below), that are immunocytochemically identified as lactotropes, corticotropes, and folliculostellate cells.[7]

From the second day on, another process results in the enlargement of the aggregates. This enlargement is not the result of steady growth but is achieved by agglomeration of several aggregates 100–150 μm in size into larger structures with an average size of 300 μm. Some aggregates are linked to each other, resulting in a short chain of 3 to 4 units or a closed ring of 5 to 6 units. After the fifth day these agglomerates become more compact, so that the original aggregates are merged into the overall structure. The morphology of these structures remains relatively stable from day 5 up to several weeks. The overall shape is oblong if the agglomeration occurred in a chainlike way or rounded off in the ringlike agglomeration. If the proportion of folliculostellate cells is considerably larger, e.g., in an enriched population obtained by velocity sedimentation at unit gravity (see below),[7] the aggregation process is somewhat different. The initial size of such aggregates on day 1 is considerably larger and reaches 300–500 μm in diameter if 1×10^6 cells/2 ml medium are used. Further alterations of the morphology of folliculostellate cell-enriched aggregates consist of smoothing and compacting of the primary aggregates. No ringlike agglomeration has been observed, probably because of the bigger size of the primary aggregates. Eventually, two or three aggregates are linked to each other. The final shape of the aggregates is a very smooth, oblong or regularly rounded off structure. The overall size is 2 or 3 times larger than in the "normal" aggregates. The size of the primary aggregates diminishes to 50–100 μm when enriched populations of lactotropes or somatotropes are used. Also, the smoothing and agglomeration of these primary aggregates are less pronounced. The size of the aggregates can be decreased by increasing the frequency of gyratory shaking and/or decreasing the volume of culture medium in the dishes.

[7] M. Baes, W. Allaerts, and C. Denef, *Endocrinology (Baltimore)* **120,** 685 (1987).

Cellular Organization and Topographical Distribution of Cell Types in the Aggregates

According to our previous electron microscopic study of aggregates from immature rat pituitary cultured in serum-supplemented medium,[6] cells are organized in a 3-dimensional structure, areas of close apposition alternating with more dilated intercellular spaces. Cells had a round to oval shape, and there was no evidence for sorting out of different cell types or specific polarization. A similar picture was seen with aggregates from adult rats cultured in serum-free defined medium.[8] We here report the topographical distribution of lactotropes, corticotropes, gonadotropes, and folliculostellate cells in the aggregates as studied by light microscopy of sections immunostained for PRL, adrenocorticotropic hormone (ACTH), LH plus follicle-stimulating hormone (FSH), and S-100 protein (staining folliculostellate cells), respectively.

Methods

The aggregates are fixed in Zamboni fluid[9] [4% paraformaldehyde and 15% (v/v) picric acid], prepared exactly as described previously.[10] Fixation is carried out at room temperature for 4–5 hr. After fixation, aggregates are rinsed with phosphate-buffered saline (PBS) before being processed by a dehydration procedure. For preliminary staining experiments, samples are dehydrated in 100% ethanol (2 changes of 10 min), followed by 5 min butanol (Merck, Darmstadt, FRG), prior to embedding in Paraplast (Sherwood Medical Industries, St. Louis, MO) for 45 min at 60°. For the proper immunohistochemical staining of thin sections, embedding in the polyhydroxy aromatic acrylic resin LR White (London Resin Company, P.O. Box 29, Woking, Surrey, England)[11] is used. Therefore a partial dehydration procedure is performed, by successive steps of 25% ethanol (5 min), 50% ethanol (5 min), and 70% ethanol (2 changes of 15 min) before an intermediate step of infiltration with resin diluted 1 : 1 with 70% ethanol (2 changes of 15 min).[12] Then this mixture is removed as completely as possible and replaced by fresh undiluted resin. After 1 hr, an additional replacement of resin is performed before overnight incubation at room temperature. Samples are embedded in gelatin capsules,

[8] M. Baes and C. Denef, *Endocrinology (Baltimore)* **120,** 280 (1987).

[9] L. Zamboni and C. De Martino, *J. Cell. Biol.* **35,** 148A (1967).

[10] C. Denef, E. Hautekeete, A. De Wolf, and B. Vanderschueren, *Endocrinology (Baltimore)* **103,** 724 (1978).

[11] B. Causton, *in* "Immunolabeling for Electron Microscopy" (J. M. Polak and I. M. Varndell, eds.), p. 29. Elsevier, Amsterdam, 1984.

[12] B. G. Timms, *Am. J. Anat.* **175,** 267 (1986).

sealed, and polymerized at 50° for 24 hr. Thin sections of 1 μm are cut with a glass knife on an ultramicrotome (LKB 2128 Ultratome, LKB Instruments, Gent, Belgium). For preliminary staining experiments 5-μm-thick Paraplast sections are used, obtained with a rotary microtome with steel knife (Lipshaw, Detroit, MI).

Prior to immunolabeling according to the PAP immunoperoxidase technique, sections are pretreated with 0.1% bovine pancreatic trypsin (Type III, Sigma) for 10 min at 37° in Tris–HCl buffer (pH 7.8).[13] After rinsing the sections in 2 changes of PBS for 5 min each, a standard PAP technique is followed. Primary antisera are diluted 1 : 500 to 1 : 1000, and incubation proceeds overnight at 4°. Goat anti-rabbit immunoglobulins are used for 1 hr at room temperature (diluted 1 : 50), whereas the PAP complex is applied for 20 min (room temperature) at a dilution of 1 : 100. The final immunoreactive product is visualized for 10 min with 3,3′-diaminobenzidine–4HCl (Serva) in Tris–HCl buffer (pH 7.6). Staining intensity is further enhanced by incubation with 1% osmium tetroxide for 10 min.[14] ACTH and rat PRL, LH, and FSH antisera are obtained from Dr. A. F. Parlow through the National Pituitary Hormone Program (NIADDK, Bethesda, MD). The S-100 antiserum used to stain folliculostellate cells is obtained from Dakopatts (Glostrup, Denmark).

Results

Lactotropes. Lactotropes were found randomly distributed throughout the aggregates, at the center as well as at the periphery. In aggregates of adult female rats they have oval or polygonal shapes, with a slightly eccentric nucleus and clearly granulated cytoplasm. This corresponds to the fact that lactotropes bear the largest secretory granules of all the anterior lobe cell types, the mature granules being 600–900 nm.[15] With lower proportions of lactotropes in the primary aggregates, clustered groups of 3 to 4 lactotropes were sometimes found.

Corticotropes. Corticotropes were also found at the center and the periphery of the aggregates. As described by others,[15-17] corticotropes have a stellate shape and form cytoplasmic extensions between neighboring cells. In our aggregates, however, their shape was polygonal rather

[13] C. S. Morris and E. Hitchcock, *J. Clin. Pathol.* **38,** 481 (1985).

[14] E. J. Gosselin, C. C. Cate, O. S. Pettengill, and G. D. Sorenson, *Am. J. Anat.* **175,** 135 (1986).

[15] M. G. Farquhar, E. H. Skutelsky, and C. R. Hopkins, *in* "The Anterior Pituitary" (A. Tixier-Vidal and M. G. Farquhar, eds.), p. 83. Academic Press, New York, 1975.

[16] P. K. Nakane, *J. Histochem. Cytochem.* **18,** 9 (1970).

[17] P. K. Nakane, *in* "The Anterior Pituitary" (A. Tixier-Vidal and M. G. Farquhar, eds.), p. 45. Academic Press, New York, 1975.

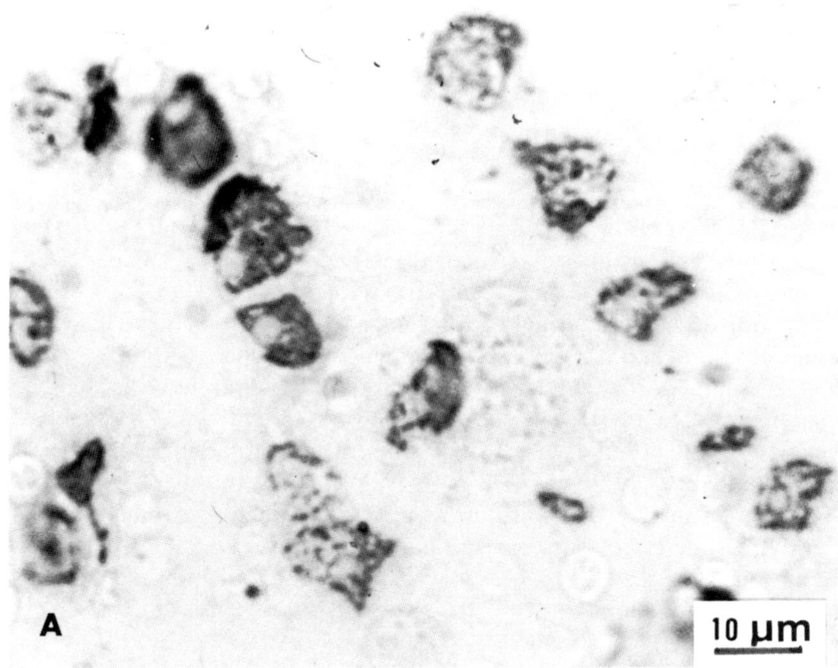

FIG. 1. Semithin LR White section (A, B, D) of aggregates immunostained with human ACTH antiserum diluted 1 : 1000 (A); with rat FSH antiserum and LH antiserum diluted 1 : 500 and 1 : 1000, respectively (B); or with S-100 protein antiserum diluted 1 : 700 (D). Arrows indicate cytoplasmic extensions of folliculostellate cells (S-100). Paraplast section (C) of aggregates immunostained with S-100 protein antiserum diluted 1 : 1000. The aggregates were prepared from rat pituitary cells separated by unit gravity sedimentation: A, gradient fraction 3 from adult females; B, gradient fraction 6 from adult females; C, gradient fraction 2 from 14-day-old females; and D, gradient fraction 3 from adult females.

than stellate (Fig. 1A). In aggregates prepared from small cells separated by velocity sedimentation of pituitary cells from 14-day-old female rats, after 1 week patches of ACTH-positive cells were found at the periphery of the aggregates. After 5–6 weeks in culture, these patches developed into buttonlike protrusions consisting entirely of ACTH-positive cells. Whether these structures are the result of proliferation of ACTH cells or whether they are a manifestation of the migratory capacities of these cells could not be precluded. Ohtsuka *et al.*[18] found proliferation by mitosis of

[18] Y. Ohtsuka, H. Ishikawa, T. Omoto, Y. Takasaki, and F. Yoshimura, *Endocrinology (Baltimore)* **18,** 133 (1971).

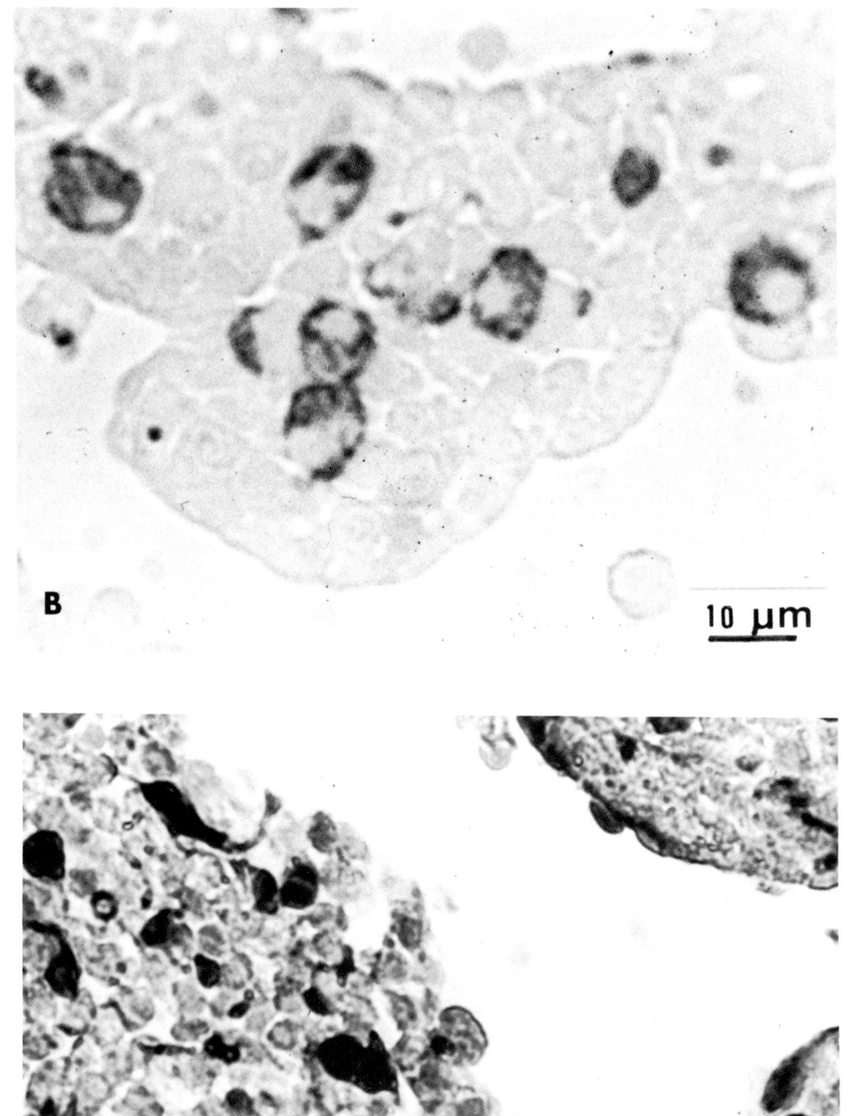

FIG. 1B and C.

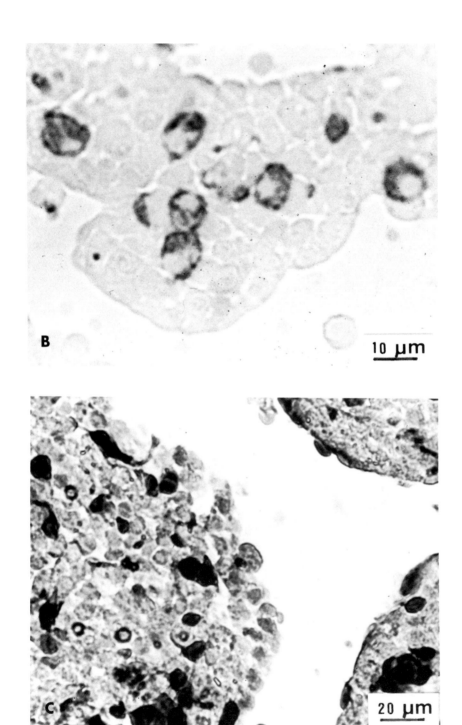

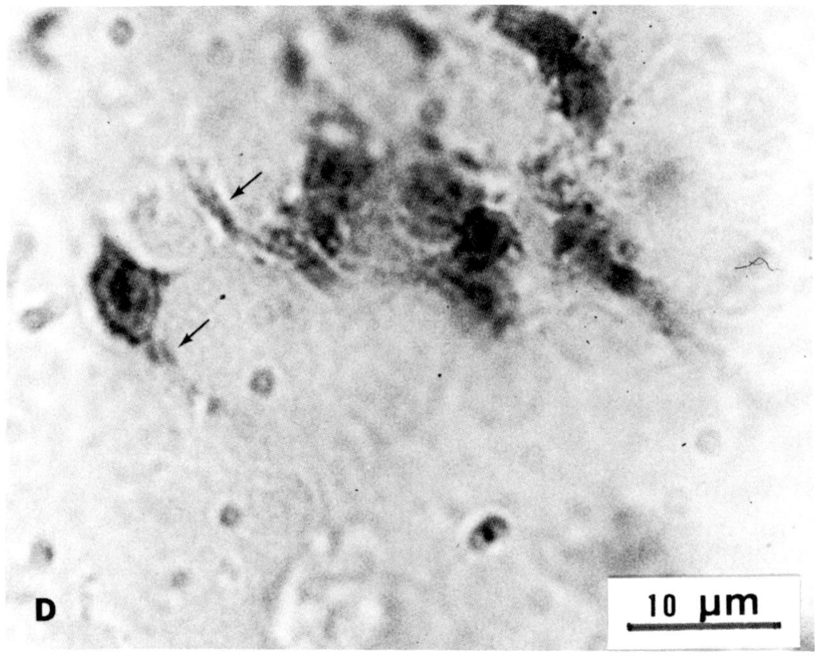

FIG. 1D. See legend on p. 54.

cultured chromophobes and differentiation into acidophils producing large amounts of ACTH but neither growth hormone (GH) nor PRL.

Gonadotropes. Gonadotropes were found mostly at the center of aggregates (Fig. 1B), sometimes with two or three cells linked to each other. Aggregates of adult females stained simultaneously for LH and FSH showed oval as well as polygonal cells, corresponding to the type A and B cells described by Nakane.[16,17]

Folliculostellate Cells. Folliculostellate cells (FS cells) were distributed all over the aggregates and displayed morphological characteristics different from FS cells of the intact pituitary.[19,20] On 5-μm-thick sections a random distribution of the small polygonal or pyramidal cell bodies was observed (Fig. 1C). At higher magnification and using LR White sections of 1 μm, we observed long and slender cytoplasmic extensions protruding among the granulated cells (Fig. 1D). Some FB cells were also found at the periphery of the aggregates, with extensions bordering the aggregate,

[19] E. Vila-Porcile, *Z. Zellforsch. Mikrosk. Anat.* **129,** 328 (1972).

[20] T. Nakajima, H. Yamaguchi, and K. Takahashi, *Brain Res.* **191,** 523 (1980).

and occasionally a FS cell was observed linking the borders of two agglomerated aggregates. At the ultrastructural level, intercellular spaces resembling pituitary follicles with microvillous projections into the lumen have been reported previously.[6]

Ultrastructure of the Cells

We have shown by electron microscopy[6,8] that the various pituitary cell types looked very healthy and retained their specific characteristics. Most cells were well granulated. Plasma membranes running in close parallel formed specialized junctions (tight junctions, adherens junctions, and septa-containing junctions) very much like those in the intact pituitary. Intercellular spaces were usually empty except for some microfibrils and basal lamina-like material. There was no proliferation of fibroblasts.

Perifusion of Aggregates

Secretory responses of the aggregates can easily be measured in a perifusion apparatus, which allows the dynamics of these responses to be followed as a function of time.

Secretory Response Studies during Short-Term Perifusion

This perifusion system is intended to follow secretion over a period up to 6 hr and has been described previously.[6,21] Our up-to-date system is shown schematically in Fig. 2. Eight acrylic chambers (9 mm diameter; 20 mm high) are mounted in an acrylic water jacket (37°). The top and bottom of each chamber are fitted airtight but removable. A nylon mesh (25-μm pore size) inserted in the bottom of the chamber retains the aggregates in the chambers. The peristaltic pump (Minipuls 2, Gilson Medical Electronics, Villiers-le-Bel, France; eight channels) pumps the perfusion medium to the chambers, the inlet being constructed so that the medium is forced to flow along the side wall of the chambers in order to gain the desired temperature, to let air bubbles in the tubing escape, and to avoid mechanical disturbance of the aggregates. A vent at the top of each chamber allows adjustment of the amount of medium inside the chamber. Before use the perifusion chambers, connecting polyethylene tubings, Tygon pump tubings, and plastic containers used to hold the perifusion medium are rinsed for 20 min with 1% BSA in water in order to minimize adsorption of test substances or secreted hormones.

[21] C. Denef and M. Andries, *Endocrinology (Baltimore)* **112,** 813 (1983).

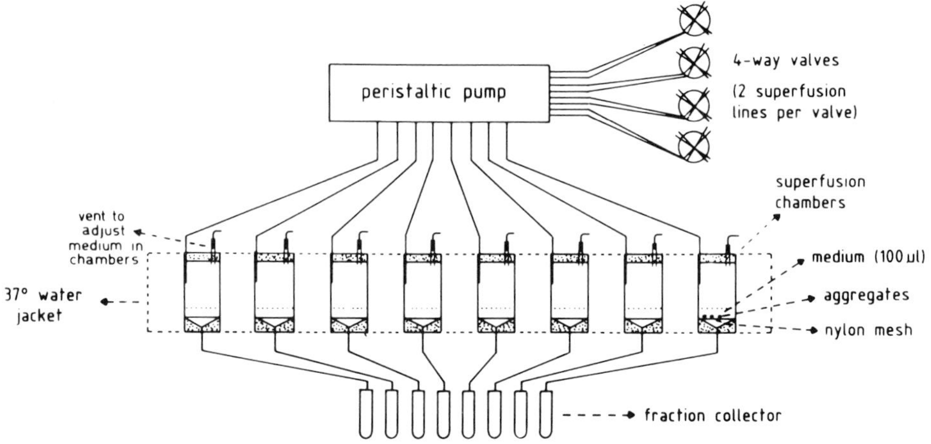

FIG. 2. Short-term perifusion system.

The perifusion medium consists of DMEM (Gibco) supplemented with 1 g/liter $NaHCO_3$ and 15 m*M* each of HEPES and TES buffers (Sigma, tissue culture tested) adjusted to pH 7.5. Aggregates, suspended in the medium, are gently introduced in the chamber with a plastic pipet. It is mandatory that they remain covered with medium throughout the loading process. During perifusion the aggregates remain covered with approximately 100 μl perifusion medium. The air above the medium buffers possible pulsations of the peristaltic pump. Flow rates used range between 0.125 and 1 ml/min. The average dead volume in the system is 1 ml. Dead time is estimated from the retention time of a known amount of one of the pituitary hormones injected into the system.

After a 2.5-hr equilibration period (to stabilize basal hormone release), the aggregates are exposed to rectangular pulses of the test substances. Control medium always contains the same vehicle in which the test substances are diluted. Each of the eight-port valves with zero dead volume fittings (Valco Instruments Co., Houston, TX) allows switching over from control medium to medium with test substances for two perifusion chambers simultaneously. The eluates are collected by a fraction collector, capable of collecting 8 fractions simultaneously (Gilson, Model 202), in 1- to 4-min fractions in polystyrene tubes containing 100 μl of 2% BSA in saline and are stored at $-20°$ until radioimmunoassay (RIA) of the hormones under study. Under these conditions and with a minimum of 10 ml perifusion medium in the holding plastic containers (50 ml tubes, Nunc, Roskilde, Denmark), the pH in the system remains stable. It should be noted that mechanical disturbance (e.g., shaking the table) or exposure of the aggregates for even seconds to air flow elicits an artificial burst of PRL and GH release, drastically lowering the magnitude of the PRL and GH response to stimulatory agents administered up to 2 hr later. Another important note is that the system should be thoroughly rinsed with water and, with the exception of the acrylic materials, sterilized with 70% ethanol after use. Acrylic material can be sterilized with 7% peracetic acid.

Secretory Response Studies during Long-Term Perifusion

Recently, we succeeded in maintaining aggregates for several weeks in perifusion. This perifusion system is shown schematically in Fig. 3. The acrylic perifusion chamber has a conical shape, maximal diameter and height being 10 mm, and is mounted in a 37° water jacket. To avoid infection the entire system is built in such a way that it can easily be sterilized and can function closed, contacts with air being through sterile filters. The system is sterilized with diluted peracetic acid (see above) and then thoroughly rinsed with water, after which it is coated with 1% BSA in water (see above). An electrically driven solenoid three-way valve allows switching over from control medium to perifusion medium with test substance. Perifusion medium is the serum-free chemically defined culture medium described above. It is kept in holding flasks with screw caps containing fittings for connecting polyethylene tubing and a sterile air filter to keep an atmospheric pressure inside the flasks. Holding flasks remain in the dark in a small refrigerator at 4°. The flow rate of the

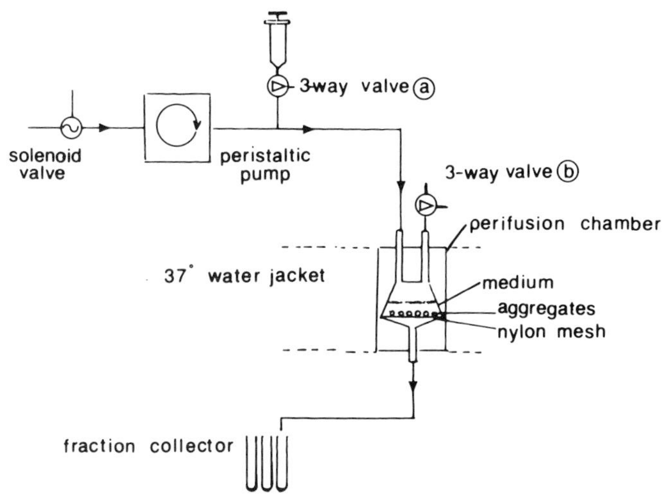

FIG. 3. Continuous perifusion system.

peristaltic pump (Pharmacia Fine Chemicals, Uppsala, Sweden) is 1 ml/hr during basal conditions and 3 ml/hr during the experimental sessions. The three-way valve (a) is intended to remove air from the tubing in case a new holding flask has to be connected. The three-way valve (b) allows introduction of the aggregates as well as adjustment of the amount of medium covering the aggregates, retained in the bottom of the chamber by a nylon mesh (25-μm pore size). Timings for switching over from control to test substance and vice versa by the solenoid valve are programmed by a microprocessor so that the system can run automatically. This allows trains of pulses with preset pulse periods and durations to be given over long periods of time.

Aggregates maintained under continuous perifusion conditions retained excellent secretory responsiveness. After 4 weeks basal PRL release was at the same level as during the first days of perifusion. Basal LH release declined gradually to about 10% of initial values after 4 weeks but, with 1×10^7 aggregated pituitary cells (from 14-day-old female rats) loaded in the chamber, remained detectable. As shown in Fig. 4, the gonadotropes remained highly sensitive to gonadotropin-releasing hormone (GnRH), a 15-min pulse of 0.1 nM eliciting a more than 10-fold rise in LH release after 25 days of perifusion. Addition of 500 nM dopamine caused a rapid inhibition of PRL release, whereas removal of the catecholamine, after long-term exposure to it, provoked an immediate rise of PRL release. As far as tested on the eighth day of perifusion, LH release remained responsive without change in magnitude to successive 15-min pulses of 0.03 nM GnRH, imposed every hour over 24 hr (data not shown).

The continuous perifusion system can also be used for long-term collection of bioactive material from aggregates, including not only pituitary hormones such as PRL and GH, which are secreted at a relatively high basal rate, but also specific peptides that are produced in the pituitary and have potential paracrine action. We previously showed that GnRH is capable of stimulating PRL release from aggregates containing gonadotropes and lactotropes and that aggregates composed mainly of gonadotropes release a substance(s) with PRL-releasing activity when incubated in a static incubation system.[21] We here show that similar material is secreted continuously in the continuous perifusion system for several weeks (Fig. 5). Perifusion was at a flow rate of 2.5 ml/hr, and a 12-min pulse of GnRH at 0.3 nM was given every 12 hr. The perifusion chamber contained 10^7 aggregated cells of a population consisting of 70% gonadotropes obtained by velocity sedimentation at unit gravity.[10] The gonadotrope-conditioned medium was collected up to a volume of about 5 ml, the pH was measured, and the sample was bioassayed for PRL-releasing

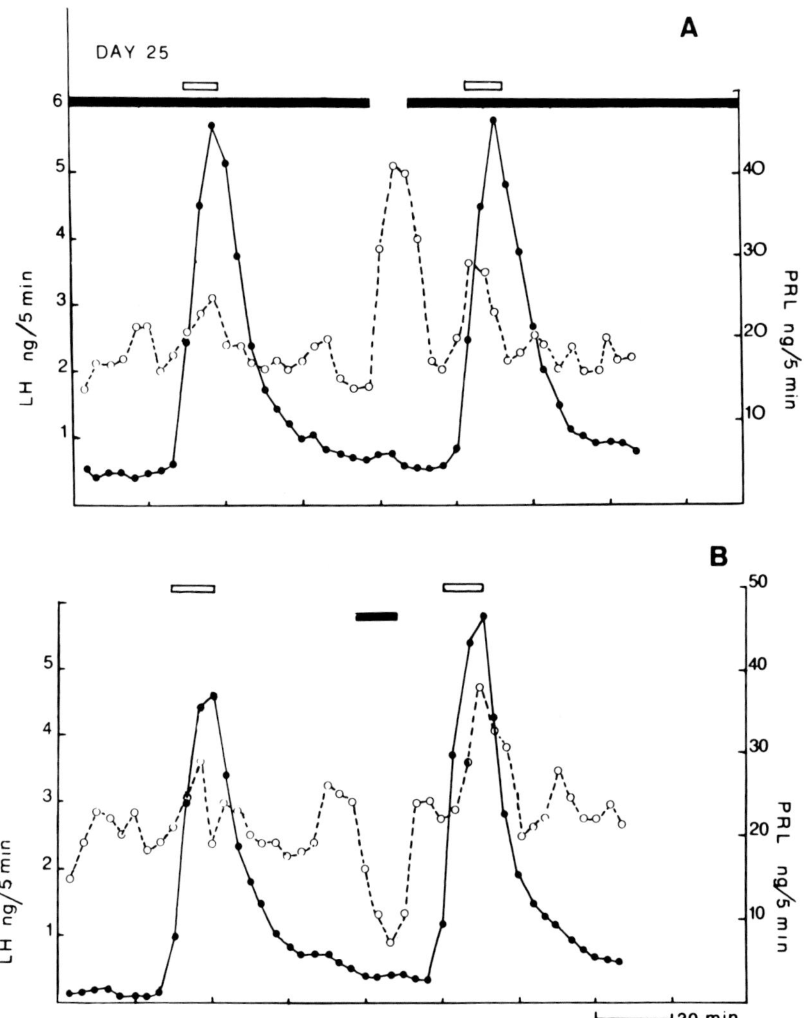

FIG. 4. Effect of 0.1 n*M* GnRH (A, B) (□), 500 n*M* dopamine (B) (■), and dopamine withdrawal (A) on LH (●) and PRL (○) secretion in pituitary cell aggregates from 14-day-old female rats after 25 days in continuous perifusion. In A dopamine was added to the system from day 0 of perifusion as well as during the cell dispersion and cell aggregation procedure.

activity in the short-term type perifusion system containing aggregates consisting of an enriched population of lactotropes. The latter were from adult female rats obtained by velocity sedimentation at unit gravity[22] and

[22] C. Denef, L. Swennen, and M. Andries, *Int. Rev. Cytol.* **76,** 225 (1982).

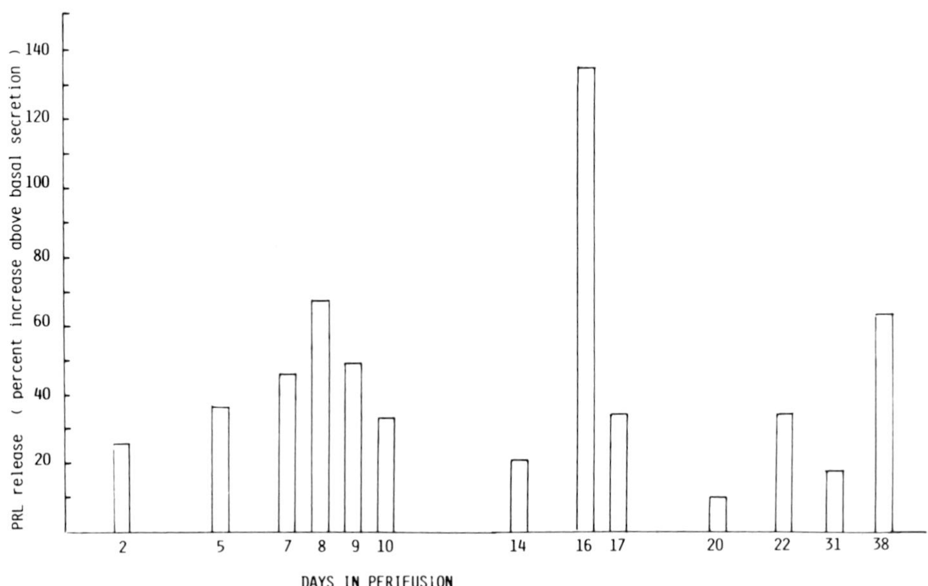

FIG. 5. Release of a substance(s) with PRL-releasing activity by gonadotrope-rich aggregates from 14-day-old female rat pituitary in the continuous perifusion system.

cultured in defined medium supplemented with 0.05 n*M* T_3 and 4 n*M* dexamethasone. The perifusion medium was the same defined culture medium used to produce the gonadotrope-conditioned medium, and the pH of the perifusion medium was adjusted to that of the gonadotrope-conditioned medium. Control medium processed in a similar way as the gonadotrope-conditioned medium did not contain any PRL-releasing activity. The low levels of PRL present in the gonadotrope-conditioned medium were negligible compared to PRL released from the lactotrope-rich aggregates in the bioassay system.

Separation of Anterior Pituitary Cell Types

As different anterior pituitary cell types show differences in their cell size and their buoyant density, the main tool for fractionating cells so far has been sedimentation at unit gravity or centrifugation. There are several reports of enrichment according to cell size of gonadotropes, somatotropes, and thyrotropes by means of sedimentation at unit gravity in an albumin gradient. Also, centrifugal elutriation has been successfully applied to separate cells according to size. A few studies reported enrichment of pituitary cell types after isopycnic centrifugation through a density gradient of dextran, BSA, metrizamide, and Percoll. The topic of methods for separation of pituitary as well as other cell types has been extensively reviewed.[22–25] The best results are obtained if an animal model is used in which the proportional number of the desired cell type is already high in the initial cell suspension or if the size of the cell type differs markedly from the others, which is the case for somatotropes in adult male rats,[26] gonadotropes in 14-day-old female rats,[10] and lactotropes in adult female rats.[26] Castration will lead to increased numbers of gonadotropes, chemical thyroidectomy to increased numbers and enlarging of thyrotropes, etc. None of the mentioned techniques gives an absolute purification. It is to be expected, however, that further improvement of the enrichment can be achieved when the different methods are combined. A few examples already exist in the literature.[27,28]

We illustrate in this chapter how lactotropes and somatotropes can be purified from both adult and immature rats by combining sedimentation at unit gravity with isopycnic centrifugation through a Percoll density gradient and by sequential application of two unit gravity sedimentation procedures. Both techniques offer the advantage that cells can be separated under sterile conditions with a high degree of reliability and using equipment which is inexpensive.

Velocity Sedimentation at Unit Gravity

Enzymatically dispersed pituitary cells are allowed to sediment by gravitational force through a linear gradient of BSA in medium that is optimal for the cells being separated. We use an acrylic chamber of the Hymer type[23] of 14 or 23 cm diameter, permitting separation of 40 million and 100 million cells. The system is shown schematically in Fig. 6. Details of the separation in a chamber of 14 cm diameter have been published.[10] The characteristics of separation are listed in Table I. In our experience a good cell dispersion procedure is extremely important for obtaining a satisfactory cell recovery (range 53–73%). However, cell aggregation is also improved by the quality of the cells, and, since clustering is a func-

[23] W. C. Hymer and J. M. Hatfield, this series, Vol. 103, p. 257.

[24] W. C. Hymer and J. M. Hatfield, *in* "Cell Separation: Methods and Selected Applications" (T. G. Pretlow and T. P. Pretlow, eds.), Vol. 3, p. 163. Academic Press, New York, 1984.

[25] T. B. Pretlow, E. E. Weir, and J. G. Zettergren, *Int. Rev. Exp. Pathol.* **14,** 19 (1975).

[26] L. Swennen, M. Baes, C. Schramme, and C. Denef, *Neuroendocrinology* **40,** 78 (1985).

[27] G. Snyder and W. C. Hymer, *Endocrinology* (*Baltimore*) **96,** 792 (1975).

[28] B. Scheikl-Lenz, J. Sandow, A. W. Herling, L. Träger, and H. Kuhl, *Acta Endocrinol.* (*Copenhagen*) **113,** 211 (1986).

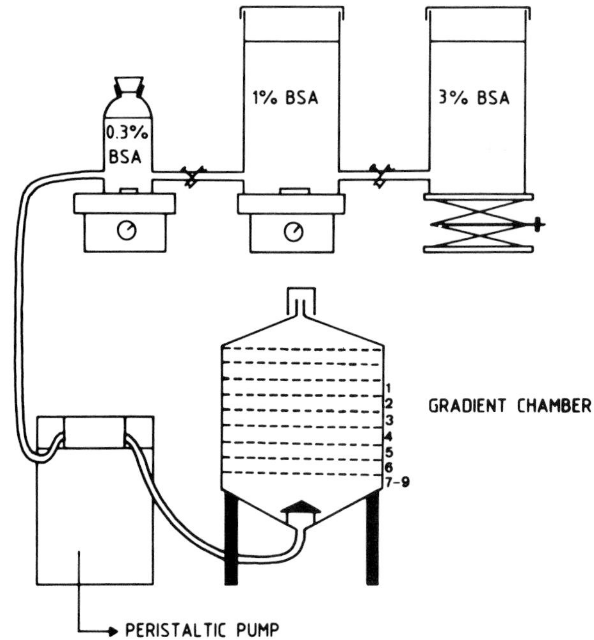

FIG. 6. Velocity sedimentation gradient at unit gravity.

tion of the time during which cells remain at high concentrations, it is very important to work as fast as possible and with a cell suspension diluted to a concentration not exceeding 10^6 cells/ml.

Isopycnic Sedimentation through a Percoll Density Gradient

Isopycnic centrifugation is used to separate cells with different densities. Cells are sedimented in a gradient with sufficient force and for a sufficient period of time for them to arrive at their buoyant density levels in the gradient. At these locations in the gradient, no additional sedimentation will occur, and the cells will be separated according to their respective densities. The buoyant density of a cell is affected by the environment, especially the osmolality of the medium. As Percoll (Pharmacia) can be made isosmotic, cells can band in Percoll at their physiological densities.

It is necessary to use Percoll solutions of physiological pH and ionic strength. In our laboratory, Percoll is diluted in DMEM with 25 m*M* HEPES (supplemented with 3.7 g $NaHCO_3$, 35 mg penicillin, and 50 mg streptomycin for 1 liter). In order to make Percoll isotonic for cells, it is common to dilute 9 parts of 100% Percoll with 1 part of a 10× concen-

TABLE I
CHARACTERISTICS OF THE SEPARATION OF ANTERIOR PITUITARY CELLS IN TWO DIFFERENT TYPES OF GRADIENT CHAMBERS

Parameter	Type 1	Type 2
Chamber diameter	14 cm	23 cm
Chamber constant *a*	15.38 cm^3/mm	42.5 cm^3/mm
Number of cells separable	40×10^6	100×10^6
Vessel volume		
A	80 ml	200 ml
B	600 ml	1600 ml
C	600 ml	1600 ml
Sedimentation height	6.5 cm	6.5 cm
Number of fractions	11	11
Content/fraction	100 ml	270 ml
Flow rate during gradient introduction		
First 5 min	7.5 ml/min	18 ml/min
Next 5 min	15 ml/min	35 ml/min
Resting minutes	46 ml/min	112 ml/min
Flow rate during sucrose introduction	22 ml/min	53 ml/min

trated DMEM solution (or other physiological buffers) and use this as a stock solution. It has been reported, however,[29] that this medium is hypertonic because of the significant volume occupation by the Percoll particles. Since the osmolality of this working solution is approximately 25% higher than the value obtained with the dilution of concentrated DMEM in water, the usual dilution factor must be multiplied by 1.25 to give a correct dilution to be used to obtain a truly isotonic Percoll working solution.

First a 70% working solution is prepared by mixing 7 parts of a 100% Percoll solution with 2.125 parts of a 4× concentrated DMEM solution and 0.875 parts reagent grade water. This working solution is adjusted by dropwise addition of 1 *N* HCl to a final pH of 7.35 ± 0.05. This isosmotic Percoll working solution is diluted with DMEM to produce solutions of known density within the range limits required.

A hyperbolic gradient is produced spanning the range between the limit density solutions (e.g., 25–70% Percoll) as follows. Twenty-five milliliters of 70% Percoll is added at a rate of 0.5 ml/min to 15 ml 25% Percoll solution, thus generating the hyperbolic gradient. A two-channel peristaltic pump (Minipuls 2, Gilson) is used to generate the gradient. The Alto, CA), the heaviest solution being carefully pumped below the lighter solution, thus generating the hyperbolic gradient. A two-channel peristaltic pump (Minipuls 2, Gilson) is used to generate the gradient. The shape of the gradient may be changed by altering the volumes and concen-

[29] R. Vincent and D. Nadeau, *Anal. Biochem.* **141,** 322 (1984).

trations of the gradient-forming chambers. For this purpose a Fortran computer program is used. The theoretical hyperbolic density distribution of this gradient is controlled by means of density marker beads (Pharmacia) and appeared to be exactly as expected. The density is linearly related to the percentage of Percoll and can, for our working solutions, be calculated from the expression density = 0.0012813(% Percoll) + 1.00778. The density gradient is prepared freshly or 24 hr before the experiment and left at room temperature.

A three-layer density gradient is used for all cell separations. The top layer is always 1 ml 25% Percoll. The middle layer is 25 ml of the actual gradient from 25 to 70% Percoll (densities 1.0404–1.0970). The bottom layer always contains 70% Percoll.

Cells to be separated ($5–15 \times 10^6$ in 1–3 ml) are carefully layered on the top of the 25% Percoll. This is done using a siliconized Pasteur pipet, mounted on a Pipetus pipet-aid, keeping the tip on the wall of the tube just above the surface of the liquid. The gradient is then centrifuged at 954 *g* for 20 min at room temperature in special Ertalon sleeves in a Beckman J-6B centrifuge. The rotor accelerated in 2 min to reach 500 rpm and decelerated in 9 min from 500 rpm to rest. After centrifugation, the tube contents are carefully fractionated. This is done by means of a fraction recovery system (MSE Scientific Instruments Ltd., Crawley, England) which holds the centrifuge tube securely. The sample is removed through a puncture hole in the bottom of the tube, made by screwing a needle into the tube, and the removed fractions are diluted 5× with DMEM and centrifuged through a layer of 3% BSA in DMEM at 678 *g* for 10 min at room temperature to free the cells from Percoll. Cells in the resuspended pellets are counted with a Coulter counter. Samples of each fraction are taken for immunocytochemical cell staining, using the PAP procedures as described in detail in our previous papers,[7,8,10,26] and the remainder of the cells are cultured as aggregates.

Improved Purification of Somatotropes and Lactotropes from Adult Male Rats by Sequential Velocity Sedimentation at Unit Gravity and Isopycnic Sedimentation

Velocity sedimentation of anterior pituitary cells obtained from adult male rats yields a population of 70% somatotropes in the bottom of the gradient[26] whereas only about 30% lactotropes are found in gradient fraction 3.[26] To date, lactotropes from male rats have not been isolated to a high degree of enrichment. Since they are functionally and morphologically different from those of females, it is important that highly enriched lactotrope populations from male rats become available in order to study their potential role in cell-to-cell communication.[8]

TABLE II
ISOPYCNIC SEDIMENTATION ON PERCOLL GRADIENTS OF A SOMATOTROPE-ENRICHED POPULATION FROM ADULT MALE RATS OBTAINED BY VELOCITY SEDIMENTATION AT UNIT GRAVITY[a]

	Fraction			
	1	2	3	4
Cell type	d = 1.043–1.056	d = 1.056–1.069	d = 1.069–1.075	d = 1.075–1.085
ACTH	7.4 ± 1.4 (5)	8.1 ± 2.0 (6)	2.2 ± 0.6 (6)	1.1 ± 0.6 (6)
PRL	8.2 ± 1.8 (5)	9.5 ± 2.0 (5)	2.7 ± 0.7 (7)	1.0 ± 0.3 (7)
FSH–LH	2.1 ± 0.6 (5)	13.7 ± 2.6 (6)	3.7 ± 2.0 (6)	0.4 ± 0.1 (4)
GH	24.8 ± 2.0 (6)	49.9 ± 4.7 (7)	79.6 ± 3.9 (7)	91.0 ± 1.8 (7)
TSH	5.4 ± 0.9 (5)	11.2 ± 2.1 (6)	3.0 ± 0.5 (6)	2.8 ± 0.4 (7)
S-100	55.1 ± 3.3 (3)	11.4 ± 2.6 (3)	0.7 ± 0.2 (3)	0.3 ± 0.2 (3)

[a] Values represent mean % ± SEM of each cell type counted in each fraction, with the number of independent experiments in parentheses.

Dispersed anterior pituitary cells from adult male rats (3 months of age) are first separated on a velocity sedimentation gradient at unit gravity. Cells from the combined fractions 6 to 9 (5–15 × 10^6 cells) are transferred to the Percoll gradient and sedimented as described above. The fractions collected and the distribution of cell types are shown in Table II. Somatotropes made up over 90% of the population in Percoll fraction 4 (density 1.075–1.085). As shown in Table III, when cells recovered from fraction 3 of the unit gravity sedimentation gradient (5–15 × 10^6 cells)

TABLE III
ISOPYCNIC SEDIMENTATION ON PERCOLL GRADIENTS OF A PITUITARY CELL POPULATION FROM ADULT MALE RATS ISOLATED IN FRACTION 3 OF A VELOCITY SEDIMENTATION GRADIENT AT UNIT GRAVITY[a]

	Fraction			
	1	2	3	4
Cell type	d = 1.043–1.058	d = 1.058–1.063	d = 1.063–1.075	d = 1.075–1.085
ACTH	7.3 ± 2.0 (3)	7.2 ± 2.8 (3)	2.6 ± 1.0 (8)	0.7 ± 0.4 (3)
PRL	31.1 ± 12.2 (5)	45.8 ± 15.1 (4)	77.3 ± 4.0 (8)	87.6 ± 4.3 (3)
FSH–LH	3.3 ± 2.2 (3)	3.1 ± 1.9 (3)	1.3 ± 0.3 (8)	0.6 ± 0.3 (3)
GH	19.2 ± 2.0 (2)	11.7 ± 1.1 (2)	8.2 ± 1.6 (7)	6.3 ± 3.9 (3)
TSH	6.1 ± 1.6 (3)	3.7 ± 1.1 (3)	3.6 ± 0.6 (8)	0.0 ± 0.0 (3)
S-100	50.0 ± 9.3 (3)	33.3 ± 18.6 (3)	7.1 ± 2.3 (7)	0.4 ± 0.4 (4)

[a] Values represent mean % ± SEM of each cell type counted in each fraction, with the number of independent experiments in parentheses.

were sedimented on the Percoll gradient, a population consisting of 77 and 87% lactotropes was obtained in Percoll fractions 3 (density 1.064–1.075) and 4 (density 1.075–1.085), respectively. Recovery of all the cells from the gradient was 76.3 ± 2.0%. Cells from all Percoll fractions readily formed aggregates.

Improved Purification of Somatotropes from 14-Day-Old Female Rats by Sequential Sedimentation at Unit Gravity and Isopycnic Sedimentation

Gonadotropes are about the only anterior pituitary cell type which can be enriched by unit gravity sedimentation using pituitary cells from 14-day-old female rats: this cell type makes up about 70% of cells found in fractions 7–9 of this type of gradient.[10] However, it is clear that other enriched cell populations from developing animals are of interest to evaluate cell-to-cell communication in the immature rat.[30]

Fractions 3 and 4 from the unit gravity sedimentation gradient (type 2 chamber) which contained only 20 and 27% somatotropes, respectively, were pooled for further separation on the Percoll gradient. These pooled fractions contained on the average $14.7 \pm 1.3 \times 10^6$ cells. Cells were collected in 7 fractions with density boundaries of 1.049, 1.055, 1.068, 1.074, 1.082, 1.086, and 1.097. Cells from fractions 1 and 2 were pooled as were cells from fractions 4 and 5. Recovery from fractions 6 and 7 was negligible. Total cell recovery from the Percoll gradient was 51.9 ± 5.0%. The composition of the different fractions is given in Table IV. Somatotropes were mainly found in fraction 4–5, where they represent 83.2 ± 1.2% of cells. There is also significant enrichment of corticotropes and lactotropes in Percoll fraction 1–2. Cells from all Percoll fractions readily formed aggregates.

Other investigators have used mature animals for separating pituitary cells on density gradients. Snyder and Hymer,[27] using two discontinuous gradients of BSA, reported a somatotrope enrichment of 85%; they used 250-g animals. Hall *et al.*[31] separated anterior pituitary cells from 150- to 200-g rats using a Percoll density gradient and obtained one fraction containing 90% somatotropes and another containing 70% somatotropes. The latter fraction was mainly contaminated by lactotropes. An interesting technique was reported by Scheikl-Lenz *et al.*,[28] who obtained a 90% somatotrope fraction by combining centrifugal elutriation with density gradient sedimentation. Again, these authors used only adult rats. The sequential combination of unit gravity sedimentation and Percoll centrifu-

[30] W. Robberecht and C. Denef, *Endocrinology (Baltimore)* **122,** 1496 (1988).

[31] M. Hall, S. L. Howell, D. Schulster, and M. Wallis, *J. Endocrinol.* **94,** 257 (1982).

TABLE IV
ISOPYCNIC SEDIMENTATION ON PERCOLL GRADIENTS OF A PITUITARY CELL POPULATION FROM 14-DAY-OLD FEMALE RATS ISOLATED IN COMBINED FRACTIONS 3 AND 4 OF A VELOCITY SEDIMENTATION GRADIENT AT UNIT GRAVITY[a]

	Fraction		
Cell type	1 + 2 d = 1.042–1.055	3 d = 1.055–1.068	4 + 5 d = 1.069–1.082
ACTH	56.3 ± 3.8 (4)	23.8 ± 2.0 (3)	2.3 ± 0.8 (4)
PRL	60.1 ± 1.0 (3)	24.9 ± 3.3 (2)	4.1 ± 1.4 (3)
FSH–LH	0.9 ± 0.1 (4)	15.6 ± 3.9 (4)	3.0 ± 0.5 (5)
GH	1.0 ± 0.1 (3)	14.1 ± 5.5 (3)	83.1 ± 1.2 (4)
TSH	5.4 ± 0.3 (2)	15.7 ± 1.4 (2)	1.8 ± 0.3 (3)

[a] Values represent mean % ± SEM of each cell type counted in each fraction, with the number of independent experiments in parentheses.

gation using immature rats results in an enrichment of somatotropes which is comparable to the enrichment reported in the literature for adult animals; the point of interest, however, is the fact that this technique allows somatotropes to be obtained from developing animals.

As unit gravity sedimentation gradient fraction 4 contained considerably more gonadotropes than fraction 3,[10] we tried to use the latter fraction alone in order to obtain a somatotrope-enriched population depleted of gonadotropes. In these experiments $10.5 \pm 1.0 \times 10^6$ cells of fraction 3 of the velocity sedimentation gradient were loaded on the Percoll gradient. Cells were collected in the same fractions (Table IV); 82.0 ± 9.0% of the cells was recovered. In the somatotrope-rich population (Percoll fraction 4–5), there were only 0.9 ± 0.5% gonadotropes, considerably fewer than when BSA fractions 3 and 4 were combined. The enrichment of somatotropes was comparable.

Improved Purification of Lactotropes from Adult Female Rats by a Two-Step Velocity Sedimentation Gradient Procedure

Separation of anterior pituitary cells obtained from adult female rats yields only a 70% lactotrope population with a considerable amount of other pituitary cell types still present.[26] Preliminary results by sequential sedimentation of a velocity sedimentation gradient, followed by an isopycnic sedimentation gradient on Percoll, resulted in only slightly better enrichment of lactotropes (80%) and did not eliminate other pituitary cells from the most enriched lactotrope fraction.

TABLE V
DISTRIBUTION OF PITUITARY CELL TYPES FROM ADULT FEMALE RAT FRACTIONS AFTER A TWO-STEP UNIT GRAVITY SEDIMENTATION PROCEDURE[a]

Fraction	Cell number fraction ($\times 10^3$)	Cell type				
		ACTH	PRL	FSH–LH	GH	TSH
4	589	1.0	92.1	0.3	7.3	0.3
5	1000	1.6	93.4	0.5	6.2	0.2
6	742	1.6	86.0	1.0	1.2	0.2
7–9	683	6.1	79.1	4.0	0.8	5.0

[a] Values represent % of each cell type in the gradient fractions.

In another approach, freshly dispersed anterior pituitary cells (7×10^7 cells) obtained from adult female rats (3 months old, 200 g) are separated by a velocity sedimentation gradient at unit gravity. Fractions 3, 4, and 5 of this gradient (2.6×10^7 cells), which contained lactotropes ranging from 30% in fraction 3 to 70% in fraction 5, are allowed to form aggregates and then kept in culture for 3 weeks. The serum-free defined culture medium is used, without any supplementations of dexamethasone or T_3. After 3 weeks, the aggregates are redispersed using the same method as described for intact glands. Redispersed cells (5.6×10^6) are then placed again on a velocity sedimentation gradient under the same conditions as the first gradient procedure. Identification by immunocytochemistry of the cells obtained from fractions 4 to 9 of this second gradient is shown in Table V. As can be seen, fractions 4 and 5 consisted of over 90% lactotropes. Corticotropes, gonadotropes, and thyrotropes had increased in size and/or density during the 3-week culture period as they sedimented deeper in the gradient, while the somatotropes behaved in the opposite way. Although yield in the various fractions was only $1–2 \times 10^6$ cells, the number of cells was sufficient to allow culture as reaggregates.

Value of the Present Technologies to the Study of Intercellular Communication

We have shown that with a combination of the methods described in this chapter it is possible to distinguish whether a secretagogue affects secretion by its target cell *in vitro* through a direct effect or whether its primary action also involves another cell type, which in turn transmits a signal to the target cell.[7,8,21] This interacting cell type could be called an "intercell" in comparison to interneurons in the nervous system. Evidence for the participation of intercells can be obtained by comparing the secretory responses of aggregates that consist of a highly enriched population of the target cell type with the secretory responses of aggregates that are composed of a mixture of the enriched target cell population and a highly enriched population of the suspected intercell. The higher the proportion of the intercell in the coaggregates, the larger will be the change in secretory response induced. Absence of the secretory response in aggregates consisting of greater than 90% of the target cell indicates only that the target cell response is fully mediated by the intercell. The exact mode of action of the intercell, however, cannot be determined; it may function as an informative cell, simply transducing a (stimulatory or inhibitory) signal, or may exert a trophic action on the target cell so that the latter becomes responsive to the secretagogue.

The 3-dimensional configuration and tissuelike organization of the cells in the aggregate most likely favor the expression of these communication systems. As has been shown for brain cells, aggregates express various functional and morphogenetic capabilities which are very similar to the *in vivo* capabilities of the tissue.[32–34] These data support the relevance of our findings using pituitary cell aggregates. Furthermore, the perfusion system mimics *in vivo* conditions *in vitro* far better than do static incubation systems. The fact that pituitary cell aggregates remain functional during weeks in a continuous perifusion system is a further indication of the validity and reliability of continuous perifusion as an *in vitro* system. Moreover, this system is of considerable economical value as various experimental designs can be accommodated in sequence on successive days instead of starting a new primary culture each time.

A major question in cell-to-cell communication is whether it is based on the secretion of paracrine (or autocrine) factors. If so the observed effect should be mimicked by medium previously conditioned by the presence of a highly enriched population of the intercell. Again, a continuous perifusion system which allows collecting of the conditioned medium, in sufficiently large quantities so that paracrine factors can eventually be extracted and characterized from it, can be expected to be of great value. On the other hand, release and possible paracrine role of peptides such as opioids and angiotensin II, already identified as intrinsic substances in the gonadotropes,[1] can be studied using the perifused aggregates. The use of peptide receptor blockers or peptide antiserum will provoke alterations in the secretory response of a particular target cell if the particular peptide is involved in cell-to-cell communication.

[32] P. Honegger and D. Lenoir, *Brain Res.* **199,** 425 (1980).

[33] B. B. Garber and A. A. Moscona, *Dev. Biol.* **27,** 217 (1972).

[34] P. Linser and A. A. Moscona, *Proc. Natl. Acad. Sci. U.S.A.* **76,** 6476 (1979).

[6] Application of Fast Atom Bombardment Mass Spectrometry to Posttranslational Modifications of Neuropeptides

By P. C. ANDREWS and JACK E. DIXON

Introduction

Fast atom bombardment mass spectrometry (FABMS) is a method that has undergone an extraordinary increase in use for analysis of proteins and peptides. This has occurred despite the limited availability of the instruments due to the expense of both the instrument itself and the personnel and facilities necessary to operate them adequately. Fast atom bombardment is a method used to vaporize nonvolatile compounds directly from a solution without prior derivatization.[1] The peptides are dissolved in a liquid matrix (commonly glycerol, thioglycerol, or a dithiothreitol/dithioerythritol mixture) and desorbed from the surface of the matrix by a beam of atoms (usually xenon) with an energy of 4–10 keV.[2] Desorption may be effected equally well with a beam of high-energy cesium atoms.

Many of the examples cited in this chapter are from our laboratory or represent collaborative efforts with other researchers.[3] An attempt has been made to cite some of the many examples from the literature; however, this review is not meant to be exhaustive, and we recognize that many fine studies will not be cited. A number of excellent reviews have

[1] M. Barber, R. S. Bordoli, R. D. Sedgwick, and A. N. Tyler, *J. Chem. Soc., Chem. Commun.*, p. 325 (1981).

[2] H. R. Morris, M. Panico, M. Barber, R. S. Bordoli, R. D. Sedgwick, and A. Tyler, *Biochem. Biophys. Res. Commun.* **101,** 623 (1981).

[3] It is impossible to thank all of our collaborators, but the authors wish to especially acknowledge the following individuals for their generous donations of time and interest: Dr. R. Cotter, Dr. C. Fenselau, D. Heller, Dr. M. Hermodson, Dr. T. D. Lee, and Dr. D. Smith. We also wish to thank L. Lofland for typing the manuscript and M. Poling for preparation of the figures. The spectra in Figs. 1 and 3 were obtained at the Middle Atlantic Mass Spectrometry Facility, Johns Hopkins University, Baltimore, MD. A Kratos MS-50 mass spectrometer equipped with a 23-kilogauss magnet and a postacceleration detector was used. The spectrum in Fig. 2 was obtained at the Beckman Research Institute, City of Hope, Duarte, CA, using a JEOL HX100HF double focusing mass spectrometer having a mass range of 4500 at 5 kV. Postacceleration ion detection was achieved using an off-axis detector having a conversion dynode at −20 kV. All other spectra were obtained at Purdue University using a Kratos MS-50 mass spectrometer with a 23 kilogauss magnet. P. C. A. was supported by NSF grant ECE 8613167 and a grant from the American Diabetes Association.

recently appeared describing various aspects of FABMS of peptides.[4–13] We encourage those interested to make use of these reviews as further sources for citations. This chapter describes the application of FABMS to the analysis of posttranslational modifications associated with neuropeptides. Emphasis will be placed on the use of chemical and enzymatic methods in conjunction with FABMS as a way to obtain solutions to specific peptide structure problems.

Neuropeptides may undergo a vast array of posttranslational modifications. Precursors to neuropeptides are exposed to many of the cotranslational and posttranslational modifications to which proteins are subjected in the endoplasmic reticulum and the Golgi apparatus. These modifications may include disulfide bond formation, glycosylation, hydroxylation, phosphorylation, sulfation, acylation, alkylation, and proteolysis among many others.[14] Bioactive peptides may also undergo a unique series of processing events associated with secretory granules. These events include carboxy-terminal amidation and specific proteolytic events. These processing events are usually crucial for biological activity. Occurrence of a posttranslational processing event at a particular site may affect subsequent processing events. Knowledge of the precursor sequence is insufficient to predict proteolytic processing events as well as most other posttranslational modifications. While the amino acid sequences for many precursors to small neuropeptides and peptide hormones have been deduced from their cDNA sequences, the posttranslational processing steps leading to formation of the final, physiologically active products are rarely known. This chapter briefly describes the problems associated with classic methods of protein structure analysis with respect to posttranslational modifications and then discusses specific examples in which FABMS has been effectively used.

[4] S. A. Carr and K. Beimann, this series, Vol. 106, p. 30.

[5] R. M. Caprioli, *Mass Spectrom. Rev.* **6,** 237 (1987).

[6] H. R. Morris, G. W. Taylor, M. Panico, A. Dell, A. T. Etienne, R. A. McDowell, and M. B. Judkins, *in* "Methods in Protein Sequence Analysis" (M. Elzinga, ed.), p. 243. Humana Press, Clifton, New Jersey, 1982.

[7] D. R. Marshak and B. A. Fraser, *in* "Brain Peptides Update" (J. Martin, M. Brownstein, and D. Krieger, eds.), Vol. 1, p. 9. Wiley, New York, 1987.

[8] D. R. Marshak and B. A. Fraser, *in* "HPLC in Biotechnology" (W. Hancock, ed.). Wiley, New York, 1988.

[9] A. L. Burlingame, T. A. Baille, and P. J. Derrick, *Anal. Chem.* **58,** 165R (1986).

[10] K. Biemann and F. A. Martin, *Mass Spectrom. Rev.* **6,** 1 (1987).

[11] T. D. Lee, *in* "Microcharacterization of Peptides: A Practical Manual" (J. E. Shively, ed.), p. 403. Humana Press, Clifton, New Jersey, 1986.

[12] C. Fenselau and R. J. Cotter, *Chem. Rev.* **87,** 501 (1987).

[13] R. M. Caprioli, *Biochemistry* **27,** 513 (1988).

[14] F. Wold, *Annu. Rev. Biochem.* **50,** 783 (1981).

FABMS as a Complement to Classic Methods of Analysis

Neuropeptides have historically been defined by their specific biological activities. More recently, cDNA sequences have been used to deduce the primary structure of precursors to putative biologically active peptides. If the ultimate structure (after posttranslational modifications) is to be determined, it is, of course, necessary to purify the peptide in question and to determine its structure directly.

Recent methods for amino acid analysis[15,16] and Edman degradation[17,18] have increased the sensitivity of these methods dramatically. However, identification of modified amino acids by either of these methods is still a difficult problem. In the case of amino acid analysis, a number of posttranslational modifications are labile to the hydrolysis conditions and so may not be observed. Moreover, many commonly used derivatization methods require a free amino group for resolution and/or detection. Once the modification is released from the amino acid by hydrolysis, a convenient "handle" is not always available for analysis, which may thus require specialized analytical methods. Likewise, Edman degradation involves alternating cycles of base and strong acid which may result in destruction of the more labile modified amino acids. Many derivatives of modified amino acids also exhibit limited solubility in the organic solvents employed in the extraction steps of automated Edman degradation following cleavage of the phenylthiocarbamyl (PTH) amino acid. Inefficient extraction of these residues results in the absence of any detectable amino acid in those cycles. Branch points found within a peptide (disulfide bonds, etc.) also cause the same problem. Edman degradation also requires a free amino terminus and suffers from the intrinsic problem that sequence information at the carboxy-terminal end of the peptide is frequently difficult to obtain.

Peptides having a monoalkyl amino acid at the amino terminus (e.g., *N*-methyl amino acids) will usually lose that residue in the coupling step,[19–21] producing a new amino terminus which will in turn react with phenyl isothiocyanate to form the normal phenylthiocarbamyl derivative which is stable under the coupling conditions. Thus, two amino acid

[15] G. Tarr, *in* "Microcharacterization of Peptides: A Practical Manual" (J. E. Shively, ed.), p. 155. Humana Press, Clifton, New Jersey, 1986.

[16] R. L. Heinrikson and S. C. Meredith, *Anal. Biochem.* **136,** 65 (1984).

[17] R. M. Hewick, M. W. Hunkapillar, L. E. Hood, and W. J. Dryer, *J. Biol. Chem.* **256,** 7990 (1981).

[18] D. H. Hawke, D. C. Harris, and J. E. Shively, *Anal. Biochem.* **147,** 315 (1985).

[19] L. S. Frost, M. Carpenter, and W. Paranchych, *Nature (London)* **271,** 87 (1978).

[20] J. Y. Chang, *FEBS Lett.* **91,** 63 (1978).

[21] J. Hemple, K. Nilsson, K. Larsson, and J. Jornvall, *FEBS Lett.* **194,** 333 (1986).

derivatives will be released in the first cycle, one of which is an unusual amino acid. Since the cleavage of the *N*-alkyl amino acid during coupling is rarely complete, the degradation is often out of phase from the start. If an automated sequencer has an initial double coupling step followed by an ethyl acetate wash, it is possible to miss the first residue altogether. *N*-Alkyl amino acids have been observed only at amino termini. They can be detected by amino acid analysis, depending on the detection method, but are easily missed. The mass of a *N*-alkylated peptide will, however, provide a clear indication of modification. For example, the mass of a peptide containing a *N*-methyl residue will be 14 mass units higher than predicted from its sequence deduced from cDNA. This example illustrates one advantage of having independent methods of analysis.

FABMS does not require prior degradation of the peptide nor does it involve exposure to harsh chemical conditions prior to analysis. Because FABMS is a relatively mild desorption process, little fragmentation occurs, resulting in spectra which are relatively easy to interpret and which reflect the total mass of the peptide, including the modified residue(s). This also means that mixtures of peptides may be analyzed directly without going through time-consuming and potentially low-yield purification steps.

Identification of posttranslational modifications in neuropeptides has been compounded by the low levels of material normally available. FABMS is comparable in sensitivity to modern methods of sequence and amino acid analyses. Generally, 10 pmol to 1.0 nmol of peptide is required for analysis, depending on the size and structure of the peptide and the configuration of the instrument. The sensitivity of FABMS indicates that it should be considered as a standard microanalytical tool along with amino acid analysis and Edman degradation. Other useful characteristics of FABMS include the ability to determine the mass of a peptide with great accuracy if necessary and the absence of a requirement for a free amino terminus in order to provide useful information.

In some cases, fragmentation ions derived from a peptide are observed in sufficient intensity to allow partial or complete assignment of structure.[2,6,22,23] Most frequently, however, only the molecular ion is observed. Tandem mass spectrometric methods, of course, will provide useful fragmentation information which can be related to the sequence more easily.[24–26]

[22] A. M. Buko, L. R. Phillips, and B. A. Fraser, *Biomed. Mass Spectrom.* **10,** 408 (1983).

[23] D. H. Williams, C. V. Bradley, S. Santikarn, and G. Bojesen, *Biochem. J.* **201,** 105 (1982).

[24] D. M. Desiderio and I. Katakuse, *Anal. Biochem.* **129,** 425 (1983).

[25] K. B. Tomer, F. W. Crow, M. L. Gross, and K. D. Kopple, *Anal. Chem.* **56,** 880 (1984).

[26] D. F. Hunt, J. R. Yates, J. Shabanowitz, S. Winston, and C. R. Hauer, *Proc. Natl. Acad. Sci. U.S.A.* **83,** 6233 (1986).

Although FABMS is a powerful tool for peptide structure analysis, it does suffer some minor drawbacks. One of the more important is an inability to distinguish between Leu and Ile without tandem mass spectrometry. It is also very difficult to distinguish Lys (128.095 Da) from Gln (128.059 Da). Both Edman degradation and amino acid analysis make these distinctions without difficulty. High salt concentrations in the sample will interfere with analysis either by completely suppressing ionization or by shifting the molecular ions from the protonated form (MH^+) completely to the salt cationated forms (e.g., MNa^+), making interpretation difficult. A number of other problems, discussed later in this chapter using specific examples, include selective suppression of ionization in peptide mixtures, interpretation of mass spectra when more than one posttranslational modification is present or more than one peptide in a proteolytic map displays an anomalous mass, and the paucity of fragmentation patterns suitable for complete structure analysis.

Design of Experiments and Interpretation of Results

Determination of the mass of a peptide does not provide much direct structural information by itself. However, in conjunction with an amino acid analysis or with sequence information it becomes a very rich source of information. For example, the number of amidated residues or the identity of amino-terminal blocking groups may often be deduced, in the absence of other modifications, from the observed mass and composition data. A hypothetical case is illustrated by considering the structure of luteinizing hormone-releasing hormone (LHRH), which has the structure <Glu-His-Trp-Ser-Tyr-Gly-Leu-Arg-Pro-Gly-NH_2 (<Glu is a pyroglutamyl residue). This structure was, of course, elucidated before FABMS was available. The mass range predicted for LHRH from its amino acid composition is 1198.6–1200.6 Da. The 2-Da range indicates the ambiguity regarding the amidation of the two carboxyl groups (γ-carboxyl group of Glu and the carboxy terminus) which would be deamidated during the acid hydrolysis prior to analysis. The mass determined by FABMS is 1181.6, or 19 Da lower than predicted for the nonamidated peptide (1200.6 Da). Loss of water (18 Da) from formation of an internal amide bond and amidation of one of the carboxyl groups would account for the mass discrepancy. The most likely model would be cyclization of an amino-terminal glutamine to a pyroglutamyl residue (loss of water) and amidation of the carboxy terminus (loss of 1 Da). Identification of a blocked amino terminus and mapping appropriate proteolytic digests using FABMS would provide further support for this structure.

An excellent example of this approach can be found in a recent study describing the structure of lamprey LHRH.[27] FABMS provided key information regarding the number of amides and verification of the nature of the amino-terminal blocking group. Proteolytic mapping experiments similar to those described later in this chapter were used to localize the modification sites. In addition, some sequence information obtained from fragmentation ions was consistent with other analytical methods utilized.

In a similar manner to the examples cited above, the number of amidated residues may be determined for other small peptides, and the probable identity of single side chain modifications may be determined. When more than one modification is present, the ability to identify the probable nature of the modifications from the mass of the intact peptide becomes more difficult.

The sequences of many neuropeptides are known only from the translated cDNA sequence of the precursor. The putative neuropeptide sequence within the prohormone may have been identified by sequence similarity with another known peptide or simply because it is flanked by potential prohormone conversion sites.[28-30] Although some recognition sites for specific posttranslational processing events have been identified, whether they will be utilized in a particular protein must be determined experimentally.

A number of selected mass discrepancies between calculated and experimental masses associated with various posttranslational modifications are listed in Table I along with some of their possible interpretations. This list is, of course, incomplete both in mass discrepancies and in possible assignments. However, it serves to illustrate a number of points. The first is simply that posttranslational modifications leading to a decrease in mass are less common than those which result in a mass increase. The second point is that modifications which cause no net mass change can occur (e.g., isopeptide bond formation). The most important point, however, is that a particular mass discrepancy, in the absence of other analytical data, is usually not sufficient to identify the type of modification. For example, although a mass discrepancy of +14 suggests that a methylation has occurred, the list of possible methylation sites in peptides is long and

[27] N. M. Sherwood, S. A. Sower, D. R. Marshak, B. A. Fraser, and M. J. Brownstein, *J. Biol. Chem.* **261,** 4812 (1986).

[28] P. K. Lund, R. H. Goodman, P. L. Dee, and J. F. Habener, *Proc. Natl. Acad. Sci. U.S.A.* **79,** 345 (1982).

[29] H. Nawa, T. Hirose, H. Takashima, S. Inayama, and S. Nakanishi, *Nature (London)* **306,** 32 (1983).

[30] P. C. Andrews, K. A. Brayton, and J. E. Dixon, *in* "Regulatory Peptides" (J. Polak, ed.). Birkhauser/Verlag, Basel, 1988.

TABLE I
SELECTED MASS DIFFERENCES BETWEEN EXPERIMENTAL AND CALCULATED MASSES FOR MH^+

Monoisotopic mass difference (Da)	Possible modification(s)
−31.97207	Loss of sulfur: lanthionine
−18.0156	Loss of water: pyroglutamic acid, dehydroalanine, internal amide linkage
−0.9840	Amidation of carboxyl group
0	Isopeptide bond
+0.9840	Hydrolysis of amide bond (Gln, Asn, etc.)
+2.0156	Two-proton reduction: disulfide bond reduction
+14.0156	Addition of methylene: methylation
+15.9949	Addition of oxygen: oxidation, hydroxylation
+18.0156	Addition of water: cleavage of internal peptide bond (e.g., insulin consists of two peptides linked by a disulfide bond)
+27.9949	Formylation
+28.0313	Dimethylation: dimethyllysine, two separate methylation sites
+31.9898	Addition of two oxygen atoms: dioxidation, dihydroxylation
+42.0106	Acetylation
+42.0469	Trimethylation: trimethyllysine, multiple methylation sites
+43.9898	Carboxylation: γ-carboxyglutamic acid
+46.99	Selenocysteine
+70.0055	Pyruvic acid
+79.9663	Phosphorylation
+79.9568	Sulfation
+126.9047	Iodination
+210.1984	Myristic acid
+238.2297	Palmitic acid
+266.2688	Stearic acid

may include Glu or Asp methyl esters, methyllysine, methylhistidine, methylarginine, or an N-terminal methylation. This presupposes that the residues composing the peptide are accurately known so that a Thr for Ser or other substitution may be excluded.

A mass discrepancy of +28 is subject to even greater ambiguity because it may be interpreted as dimethylation at a single site (e.g., dimethyllysine) or as monomethylation at two separate sites. Further uncertainty arises from this mass discrepancy since it can also be due to a formylated peptide. The difference in mass between a dimethylated and a formylated peptide (0.03 Da) is too small to be convincingly determined solely from the molecular ion for peptides with mass greater than 1000 Da. Often, several methods of analysis are required to unambiguously identify the nature and site of a posttranslational modification. Phosphorylation and sulfation also produce similar changes in mass and require additional chemical methods of analysis or more detailed mass spectral analysis. Even highly phosphorylated peptides may be observed by FABMS in both positive and negative modes. A recent study described the location of 8 phosphoseryl residues in a 23-residue peptide from the fragmentation ions.[31] It would have been extremely difficult to obtain this information using classic methods of peptide structure analysis.

Carboxy-Terminal Amidation

For a neuropeptide sequence deduced from the cDNA, the glutamine and asparagine residues will be identified conclusively from the specific codons for those amino acids. In addition to these amide-containing residues, numerous neuropeptides possess carboxy-terminal amides. Whether a peptide is amidated at the carboxy terminus is frequently in doubt. The amide nitrogen donor for a carboxy-terminally amidated peptide is a glycyl residue.[32,33] The presence of Gly at the carboxy terminus of a putative neuropeptide does not necessarily indicate that the isolated peptide will be amidated. Edman degradation cannot identify a carboxy-terminal amide, unambiguous identification of carboxy-terminal amino acids in peptides is often a problem, and Gly is a common problem residue for amino acid analysis due to contamination of the peptide with Gly from exogenous sources (fingerprints, buffers, etc.). For these reasons, the glycine-extended form of a neuropeptide Y homolog from anglerfish (aPY) was confirmed by FABMS (Fig. 1).[34] The average mass was found to be 4221.3 Da in good agreement with the calculated value of 4221.6 Da for the glycine-extended peptide, indicating that the isolated form of aPY was not amidated at the carboxy terminus.

A similar question arose regarding the nature of the carboxy-terminal residues of three glucagonlike peptides from the anglerfish proglucagon gene II product (GLP-II) (Fig. 2B).[35] The experimental and theoretical masses of the molecular ions for the various peptides are listed in Table II. Partial sequence analysis by automated Edman degradation identified the amino termini and verified that the peptides were derived from gene II. Although only part of the protein sequences was determined, the

[31] C. Fenselau, D. N. Heller, M. S. Milo, and H. B. White, *Anal. Biochem.* **150,** 309 (1985).

[32] A. F. Bradbury, M. D. A. Finnie, and D. G. Smyth, *Nature* (*London*) **298,** 686 (1982).

[33] B. A. Eipper, R. E. Mains, and C. C. Glembotski, *Proc. Natl. Acad. Sci. U.S.A.* **80,** 5144 (1983).

[34] P. C. Andrews, D. Hawke, J. E. Shively, and J. E. Dixon, *Endocrinology* (*Baltimore*) **116,** 2677 (1985).

[35] P. C. Andrews, D. H. Hawke, T. D. Lee, K. Legesse, B. D. Noe, and J. E. Shively, *J. Biol. Chem.* **261,** 8128 (1986).

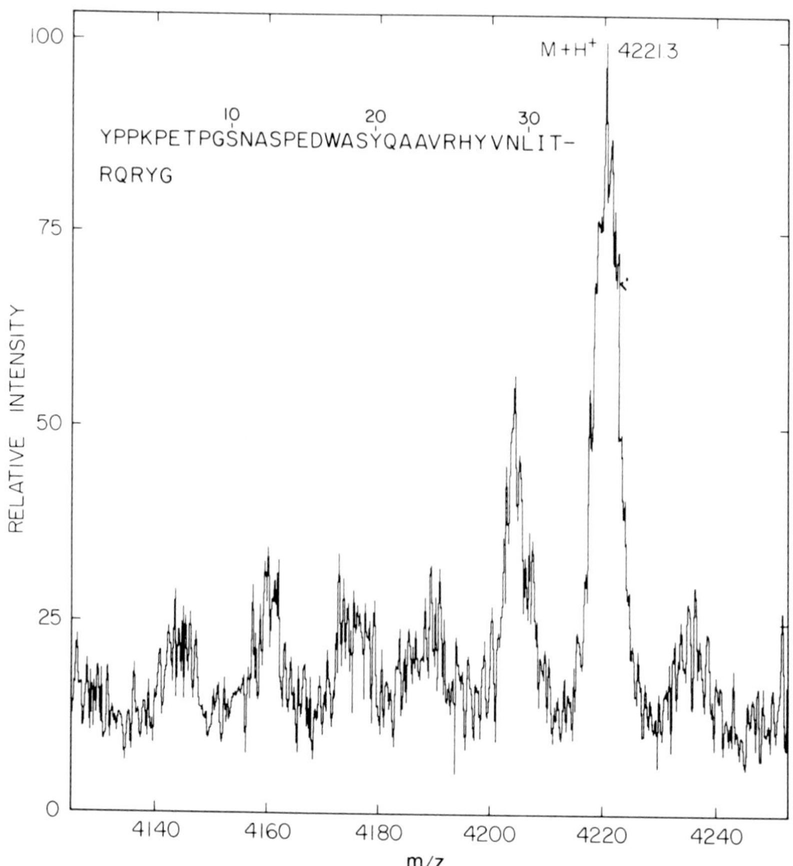

FIG. 1. Fast atom bombardment mass spectrum of aPY (5.4 μg). The average molecular ion is indicated. The spectrum shown represents a single scan. The inset indicates the structure of aPY. See footnote 3 for instrumentation details.

TABLE II
MASSES OF ANGLERFISH GLUCAGONLIKE PEPTIDES

Anglerfish preproglucagon II residue numbers[a]	Mass of monoisotopic MH^+ (Da)	
	Observed	Calculated
89–122	3815.1	3814.9
89–119	3373.5	3373.6
89–118	3315.6	3315.6

[a] Numbering from initiator Met.

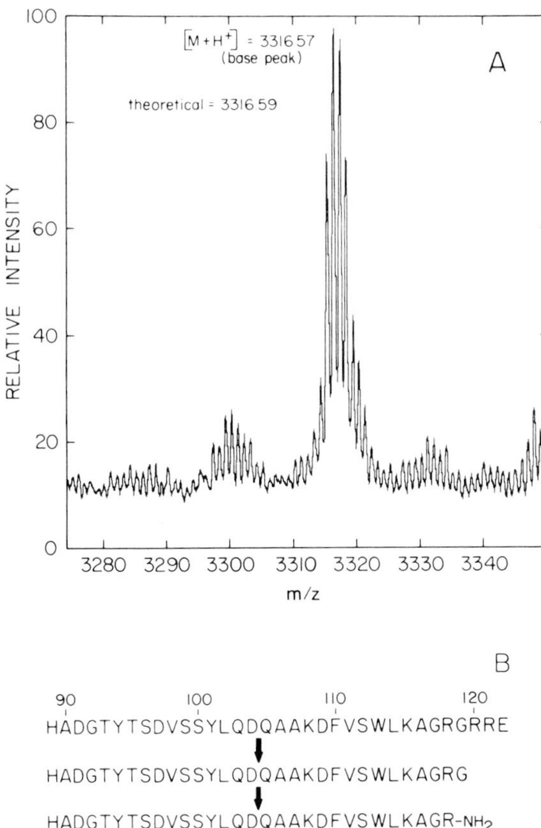

FIG. 2. (A) Mass spectrum of the molecular ion region of amidated GLP-II. The theoretical and experimental masses of the base peak are indicated. The resolution was approximately 3500 Da. See footnote 3 for details regarding instrumentation. (B) Structures of the three forms of GLP-II isolated from anglerfish pancreas. Arrows indicate the probable order of processing.

masses obtained were found to be consistent with those predicted from the cDNA-deduced sequences. In particular, one peptide (residues 89–118) was found to have a mass consistent with a carboxy-terminal arginineamide (Fig. 2A). It should be noted that the monoisotopic mass indicated in Table II is not the base peak in Fig. 2A due to the natural isotopic distribution (see Ref. 36 for discussion). The first major peak in the mass envelope corresponds to the monoisotopic mass. Generally, for peptides less than 4000 Da the monoisotopic peak is readily distinguished.

[36] J. A. Yergey, R. J. Cotter, D. Heller, and C. Fenselau, *Anal. Chem.* **56,** 2262 (1984).

Similarly, the masses of the other two forms of GLP-II indicated that a glycine-extended form (89–119) was also present, as was a form containing an unprocessed prohormone conversion site (89–122). This approach of identifying a carboxy-terminally amidated peptide by its 1-Da mass discrepancy has been used recently for a number of peptides.[27,37–39] Because the carboxy-terminal amide group is determined by difference, all other carboxyl and carboxamido groups must be individually accounted for either experimentally or by deduction from the cDNA sequence.

Clearly, when the cDNA for the precursor to a neuropeptide is known, the mass of the peptide can provide a very precise confirmation of structure. Any significant discrepancy between the calculated and the experimental masses indicates a structural difference.

Application of FABMS to Specific Posttranslational Modifications

To illustrate how FABMS may aid structural analysis of peptides we review several specific structural problems associated with biologically active peptides which were addressed in our laboratory and in collaboration with others. We hope these examples illustrate the many applications of FABMS to peptide structure determination.

O-Glycosylated Peptides

Characterization of posttranslationally modified side chains by FABMS is a particularly important problem because difficulties are frequently encountered in the analysis of their PTH derivatives and amino acid compositions. No single chemical method exists which will identify more than a few amino acid modifications. O-glycosylation sites, in particular, are difficult to predict since no unique recognition sequence occurs as has been identified for sites of N-glycosylation. Two O-glycosylated forms of a 22-residue somatostatin shown to be O-glycosylated at Thr-5 (Fig. 3B)[40] were examined by FABMS. Carbohydrate analysis indicated that both forms contained 1 mol of *N*-acetylgalactosamine and 1 mol of galactose per mole peptide. One of the forms was subsequently demonstrated to contain sialic acid. A rapid confirmation of this structure

[37] H. R. Morris, M. Panico, T. Etienne, J. Tippins, S. I. Girgis, I. MacIntyre, *Nature (London)* **308,** 746 (1984).

[38] J. L. Witten, M. H. Schaffer, M. O'Shea, J. C. Cook, M. E. Hemling, and K. L. Rinehart, *Biochem. Biophys. Res. Commun.* **124,** 350 (1984).

[39] B. S. Rothman, D. H. Hawke, R. O. Brown, T. D. Lee, A. A. Deghan, J. E. Shively, and E. Meyer, *J. Biol. Chem.* **261,** 1616 (1986).

[40] P. C. Andrews, M. H. Pubols, M. A. Hermodson, B. T. Sheares, and J. E. Dixon, *J. Biol. Chem.* **259,** 13267 (1984).

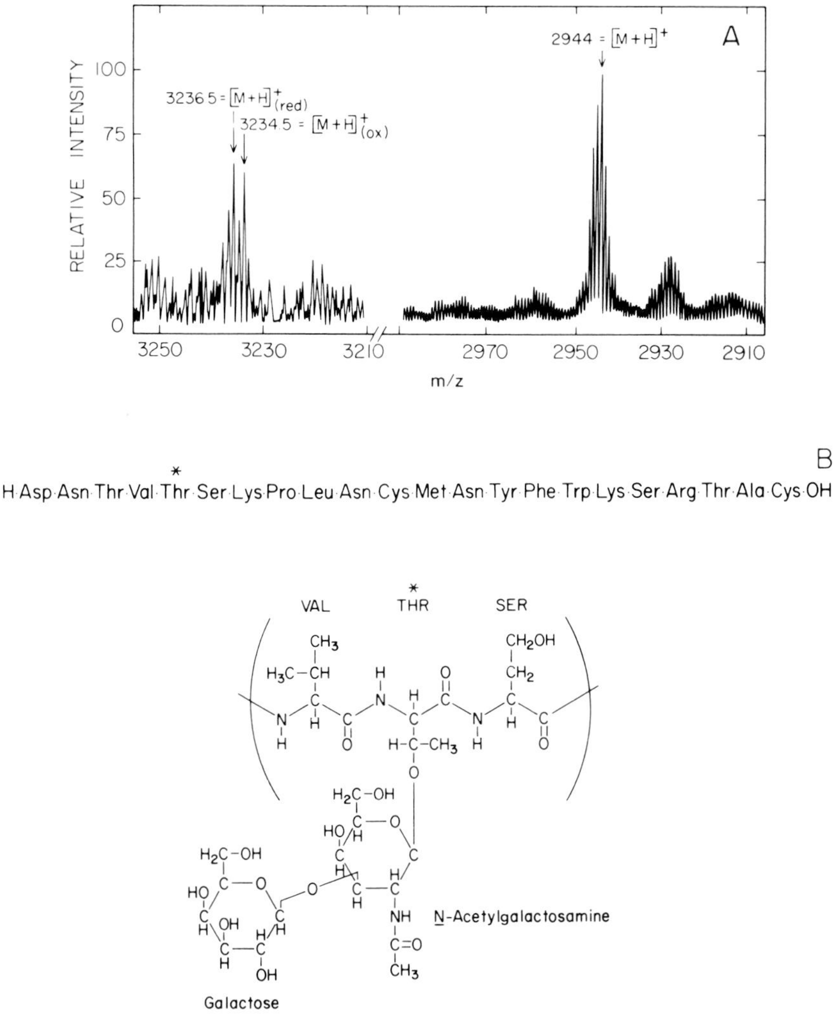

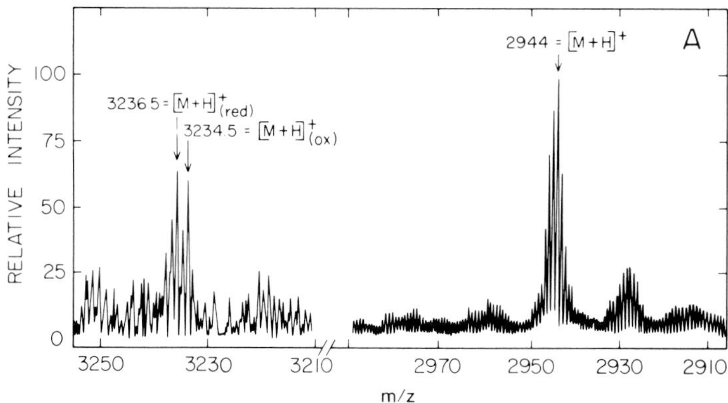

B

H·Asp·Asn·Thr·Val·Thr*·Ser·Lys·Pro·Leu·Asn·Cys·Met·Asn·Tyr·Phe·Trp·Lys·Ser·Arg·Thr·Ala·Cys·OH

VAL THR* SER

N-Acetylgalactosamine

Galactose

FIG. 3. (A) Molecular ion regions of the fast atom bombardment mass spectrum of O-glycosylated catfish somatostatin-22. The theoretical average masses are indicated next to the mass peaks. Masses for both the peptide with an intact disulfide bond, $[M + H]^{+}_{(ox)}$, and the reduced peptide, $[M + H]^{+}_{(red)}$, are shown. The spectrum was obtained as in Fig. 1. (B) Structure of the O-glycosylated 22-residue somatostatin from catfish pancreas. The site of O-glycosylation is indicated by an asterisk (Thr-5). One possible structure for the carbohydrate moiety is also indicated and is based on the structure of piscine antifreeze glycoprotein.

was obtained by FABMS of a mixture of the two forms (Fig. 3A). One form had a mass of 2944 for the molecular ion, which corresponded to the calculated mass of the peptide plus 1 mol *N*-acetylgalactosamine and 1 mol of galactose. The larger form (3234.5 Da) had a mass 290.5 Da higher. This mass difference corresponds to the mass of *N*-acetylneuraminic acid (less water of condensation). Thus, a single analysis confirmed the glycosylation, indicated the number of sialic acid residues, and identified the sialic acid as being *N*-acetylneuraminic acid. The number and nature of the sialic acid moieties were recently examined for a series of peptides containing more complex, biantennary N-linked carbohydrates.[41] In this case, N-glycosylated peptides derived from human fibrinogen were found to contain one or two residues of *N*-acetylneuraminic acid apiece.

It is worthwhile to note that somatostatin contains a single disulfide bond. Slow reduction of this disulfide bond occurs during analysis by the action of the atom source on the sample. The reduction is time dependent and results in an increase in mass of 2 Da. The mass envelope for the sialylated peptide displays an obvious split owing to partial reduction of the disulfide. The mass envelope at m/z 2944 is also skewed toward higher mass but not to the same degree. The use of this phenomenon as a diagnostic test for disulfide-containing peptides has been described.[42,43]

Analysis of Hydroxylated Peptides

Identification and location of a hydroxylysyl residue in a 28-residue somatostatin (SS-28) required a significant amount of work in two different laboratories[44,45] including an effort to resolve the hydroxylated peptide from the nonhydroxylated form which was also present. FABMS of the intact SS-28 verified the presence of a hydroxylated residue[45] but could not indicate which of the three lysyl residues in SS-28 was hydroxylated. It was possible, however, to utilize FABMS analysis of a tryptic digest of partially hydroxylated SS-28 to assign the hydroxylation site (Fig. 4). Although this site had previously been identified using classic biochemical approaches,[44,45] this study serves to emphasize the ease with which a proteolytic map may be analyzed using FABMS. The key feature of this mass spectrum is the pair of mass peaks at m/z 756.9 and 772.9, 16 Da

[41] R. R. Townsend, D. N. Heller, C. C. Fenselau, and Y. C. Lee, *Biochemistry* **23,** 6389 (1984).

[42] R. Yazdanparast, P. C. Andrews, D. L. Smith, and J. E. Dixon, *Anal. Biochem.* **153,** 348 (1986).

[43] H. R. Morris and P. Pucci, *Biochem. Biophys. Res. Commun.* **126,** 1122 (1985).

[44] J. Spiess and B. D. Noe, *Proc. Natl. Acad. Sci. U.S.A.* **82,** 277 (1985).

[45] P. C. Andrews, D. Hawke, J. E. Shively, and J. E. Dixon, *J. Biol. Chem.* **259,** 15021 (1984).

apart. These represent the fragment FFWK (residues 19–23) from SS-28 in both the hydroxylated and nonhydroxylated forms. Weak sodium and potassium adducts were visible for both peptides, indicating that the ions observed were molecular ions and not fragmentation ions.

The spectrum of the tryptic digest is rich in other structural information. Note that, while the mass spectrum shown was not run as soon as the sample was exposed to the xenon atom source, quite a strong signal was observed for residues 14–18 and 24–28 linked by a disulfide bond. Some reduction had occurred, however, resulting in the mass at m/z 505.7 corresponding to residues 14–18. This, of course, provided verification that the two peptides are joined via a disulfide bond. While the major site of tryptic cleavage seemed to be after Arg-13, some cleavage occurred after Lys-14, resulting in formation of residues 15–18 and 24–28 linked by a disulfide bond. The mass at m/z 349.5 corresponding to residues 15–18 was not observed owing to overlap with matrix ions in the low mass region. Two of the higher mass matrix ions are indicated in Fig. 4. In this case, ions overlapping the entire sequence of SS-28 could be identified from a single proteolytic digest.

Peptide Mapping by FABMS

FABMS is perhaps the single most powerful method for analyzing proteolytic digests. Interpretation of the data is usually straightforward for small peptides (<50 residues). The use of very specific proteases as in the example in the preceding section results in spectra which are not complex. Use of less specific proteases such as α-chymotrypsin or subtilisin should be avoided with larger peptides because the large number of overlapping fragments can be very difficult to assign. For the same reason, care should be taken to use only well-characterized batches of trypsin or other specific proteases. Proteases or chemical cleavages of low specificity can, however, be very useful for the analysis of smaller peptides. It is important when dealing with mixtures of peptides to distinguish between molecular ions and fragment ions. Molecular ions will usually form cationated species when low concentrations of salt are present. Fragmentation ions are not usually observed as cationated forms.

The suppression of sample desorption by high salt concentration may best be avoided by using volatile buffers such as *N*-ethylmorpholinium acetate, a buffer which is appropriate for many proteases. An alternative buffer, *N*-ethylmorpholinium trifluoroacetate, forms a syrup on drying the digest but does not seem to interfere with desorption. Nonvolatile salts may be removed using one of the many commercial reversed-phase cartridges designed for desalting peptides. Care should be taken, however, that the recommended prewash with organic solvents is carried out

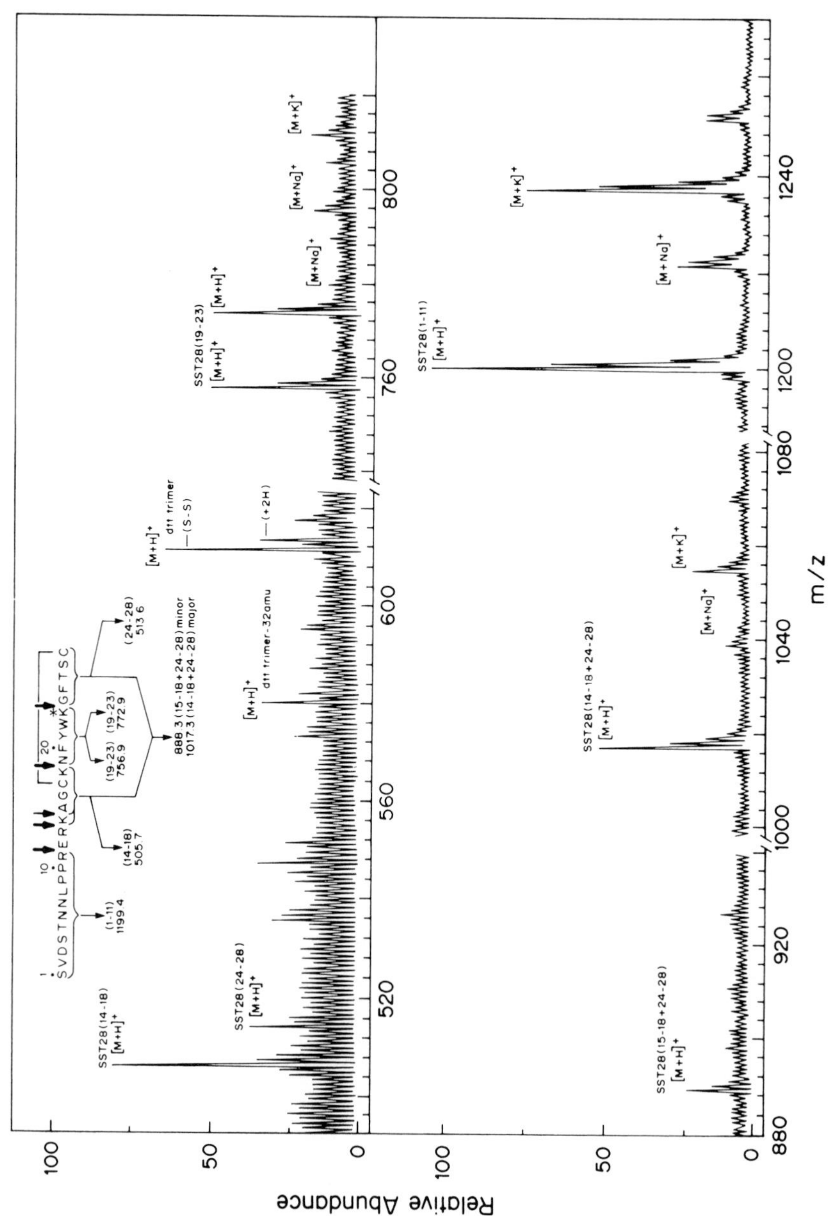
Relative Abundance
m/z
SVDSTNNLPPRERKAGCKNFFWKTFTSC
(1-11) 1199.4
(14-18) 505.7
(19-23) 756.9
(19-23) 772.9
(24-28) 513.6
888.3 (15-18+24-28) minor
1017.3 (14-18+24-28) major
SST28(14-18) [M+H]+
SST28(24-28) [M+H]+
dtt trimer-32amu [M+H]+
dtt trimer (S-S) [M+H]+
(+2H)
SST28(19-23) [M+H]+
[M+H]+
[M+Na]+
[M+K]+
SST28(15-18+24-28) [M+H]+
SST28(14-18+24-28) [M+H]+
SST28(1-11) [M+H]+
520
560
600
760
800
880
920
1000
1040
1080
1200
1240

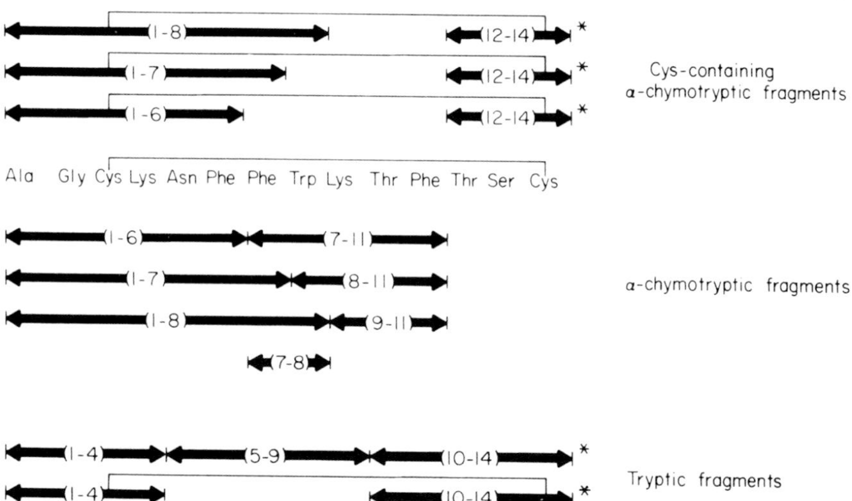

FIG. 5. Proteolytic fragments of both normal and modified somatostatin-14, observed by FABMS, are indicated by double-ended arrows. Disulfide bonds are indicated by thin lines. Fragments that are methylated in the modified somatostatin are indicated by asterisks.

(extensively) to avoid contamination of the peptides by the organic phase coating of the cartridge material. This type of contamination is not detectable by most other methods of analysis but can be disastrous for samples analyzed by FABMS. Similar contamination may also be observed for many silica-based reversed-phase HPLC columns, particularly with new columns. The use of large-pore polystyrene-based reversed-phase columns avoids this type of contamination.

Figure 5 compares schematically the data for tryptic and chymotryptic digests of the 14-residue somatostatin (SS-14) found in most species. While both enzymes cleaved SS-14 at the sites expected from their primary specificities, the chymotryptic digest produced a series of overlap-

FIG. 4. Fast atom bombardment mass spectrum of trypsinized anglerfish somatostatin-28. The identities of the tryptic fragments are indicated just above the appropriate molecular ion, $[M + H]^+$. The corresponding sodium $[M + Na]^+$ and potassium $[M + K]^+$ adducts are also indicated. The position of the hydroxylysyl residue is indicated by an asterisk (inset). Major tryptic cleavage sites are indicated by bold arrows and the minor cleavage site by a small arrow. Theoretical masses of the noncharged tryptic fragments are indicated below the sequence. The intact disulfide bond occurring between Cys-17 and Cys-28 is indicated by the line connecting these residues above the peptide sequence. Spectra were obtained on a Kratos MS-50 mass spectrometer using xenon as the atom source. Approximately 200 pmol of the digests was placed on the probe tip in a dithioerythritol/dithiothreitol matrix after dissolution in dilute acid. CsI cluster ions were used for calibration.

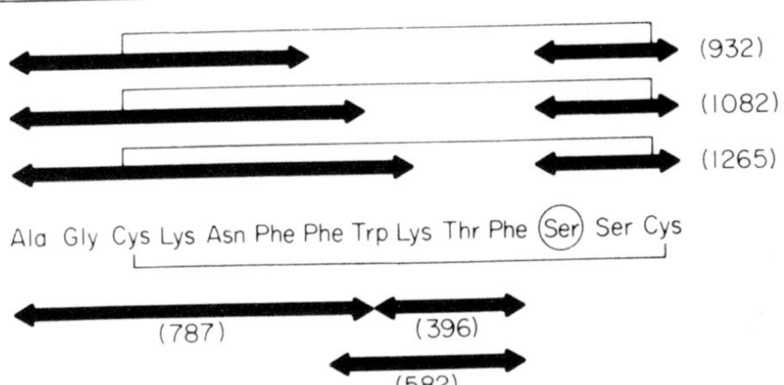

FIG. 6. Schematic diagram of chymotryptic mapping data for lamprey somatostatin-14. Experimental masses of the molecular ions are indicated in parentheses. Disulfide bonds are indicated by thin lines. The site of the Ser for Thr substitution is also indicated.

ping fragments. The use of chymotrypsin provides more structural information in this case than does trypsin.

This approach was used to map a natural homolog of mammalian SS-14 isolated from lamprey pancreas which contained a Ser for Thr substitution (Fig. 6).[46] The chymotryptic digest unambiguously indicated the site of the Ser for Thr substitution as being in position 12. A tryptic digest would not have been able to distinguish between substitutions at Thr-10 or Thr-12. Several of the following examples also involve mapping peptide structures from proteolytic digests of varying complexity.

Analysis of Artifacts Introduced by Peptide Isolation Procedures

A similar approach to that described above was used to identify the sites of artifactual methyl ester formation in two natural peptides purified by reversed-phase HPLC using 0.1% trifluoroacetic acid (TFA) in methanol as the elutrope. Partial methylation of carboxyl groups occurred as a result of storage in the methanol at $-20°$. The first peptide was SS-14, which was shown to be methylated at the carboxy terminus by a chymotryptic map (Fig. 5) similar to the ones described above. Because the SS-14 was partially methylated, fragments containing the methyl group appeared as pairs of ions 14 mass units apart (Table III). Again, by difference, only the carboxy-terminal tripeptide could be methylated.

A more complex situation arose with the second peptide (Fig. 7), which corresponds to the carboxy-terminal portion of pro-aPY.[47] Again, only partial methylation has occurred. Of the side chain carboxyl groups, only those of glutamyl residues were esterified to a significant extent.

[46] P. C. Andrews, H. G. Pollock, W. M. Elliott, J. H. Youson, and E. M. Plisetskaya, submitted for publication.

[47] P. C. Andrews and J. E. Dixon, *J. Biol. Chem.* **261,** 8674 (1986).

TABLE III
MOLECULAR IONS OBSERVED FOR α-CHYMOTRYPSIN AND TRYPSIN DIGESTS OF PARTIALLY METHYLATED SOMATOSTATIN-14

Residue numbers	Theoretical monoisotopic mass (Da)	Experimental mass (Da)
1–8, 12–14[a]	1293.55	1294
1–8, 12–14	1279.52	1280
1–7, 12–14[a]	1107.5	1107.5
1–7, 12–14	1093.44	1094
1–6, 12–14[a]	960.4	960
1–6, 12–14	946.37	946.5
1–8	972.43	972.5
1–7	785.35	785.5
1–6	639.28	638.5
1–4, 10–14[a]	946.4	946.5
1–4, 10–14	931.38	932
5–9	741.36	741
10–14[a]	571.23	571.5
10–14	557.21	557
1–4	377.17	377

[a] Ion-containing methyl ester at carboxy terminus.

These observations were confirmed by automated Edman degradation. No evidence for PTH-Asp methyl ester was found, while a significant amount of PTH-Glu methyl ester was observed in addition to the expected PTH-Glu. A methylated form of the carboxy-terminal tryptic fragment suggested that the carboxy terminus was also methyl esterified. Interestingly, the carboxy-terminal tryptophan was oxidized (mass +32 Da) but Trp-9 did not appear to be. This example represents two types of structural artifacts (methyl ester formation and Trp oxidation) which could not easily be identified by classic methods but which contribute to peptide heterogeneity and may affect activity. Methyl esterification is also a natural posttranslational modification. It should be noted that peptides containing oxidized forms of methionine are also, of course, apparent by FABMS.

FIG. 7. Summary of tryptic digest of the carboxy-terminal peptide from pro-aPY analyzed by FABMS. The theoretical monoisotopic masses are indicated.

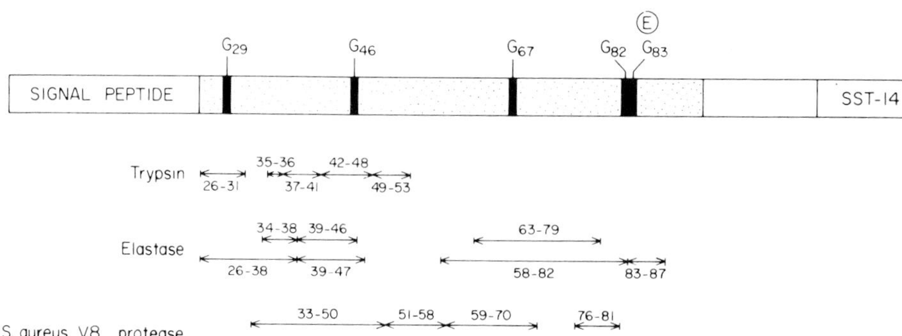

FIG. 8. Summary of proteolytic maps of anglerfish preprosomatostatin I (residues 26–92, stippled region). Arrows indicate fragments for which molecular ions were observed for digestions with trypsin, elastase, and *Staphylococcus aureus* V8 protease. The positions of all five glycyl residues deduced from the cDNA sequence are indicated by solid bars. The site of the Glu for Gly substitution is indicated (circled residue).

FABMS Used in the Analysis of Proteins and Precursors to Neuropeptides

When proteolytic maps of peptides larger than about 40 residues are examined, a number of minor problems associated with FABMS begin to become apparent and become more significant as the sizes of the peptides examined become larger. The first problem is that desorption of certain peptides in a mixture seems to be suppressed by the presence of the other peptides. This behavior is related, in part, to differing hydrophobicities.[48] It is not clear what other physical properties are responsible for the suppression; however, the problems may be alleviated in a number of ways. First, several matrices are available,[1,2,14,38,49] and qualitative and quantitative differences in desorption may be observed using them. Second, the pattern of ions observed will often shift after a period of time in the source. Thus, an immediate scan followed by a second scan a few minutes later will sometimes reveal more ions than a single scan. A mass scan in the negative mode may also identify ions not apparent in the positive mode.

A more systematic approach is to perform two or three separate digests of the protein using different proteases with the expectation that proteolytic fragments not seen in one digest will overlap fragments in a different digest which are desorbed. This approach is illustrated diagrammatically in Fig. 8 for a portion of anglerfish preprosomatostatin I corre-

[48] S. Naylor, A. J. Findeis, B. W. Gibson, and D. H. Williams, *J. Am. Chem. Soc.* **108,** 6359 (1986).

[49] J. L. Gower, *Biomed. Mass Spectrom.* **12,** 191 (1985).

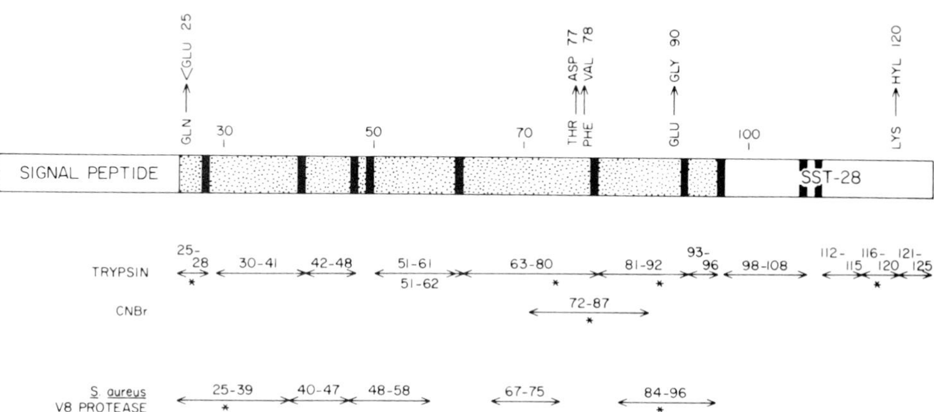

FIG. 9. Schematic diagram of anglerfish preprosomatostatin II. Solid bars indicate Arg residues, and the nonhormone portion of prosomatostatin II is indicated by stippling. The fragments identified by the mapping experiments are summarized below the preprohormone. Asterisks indicate fragments whose masses deviate from those predicted by the structures deduced from the cDNA. The site and nature of the discrepancies with the sequence deduced from the cDNA are indicated. The cyclization of Gln-26 to a pyroglutamyl residue is indicated by Gln → <Glu.

sponding to residues 26–92 (stippled region).[50] Although ions corresponding to less than one-half the peptide were observed after trypsin digestion, ions corresponding to overlapping proteolytic fragments for most of the peptide were obtained from elastase and *Staphylococcus aureus* V8 protease digests. The point behind this mapping experiment was to resolve a Glu for Gly discrepancy between the experimental amino acid composition and that deduced from the cDNA. The question was which of the five candidate Gly residues had been replaced by Glu. The Glu for Gly substitution was found to be located near the carboxy terminus at position 83.

A much more difficult situation arose with the analogous peptide from anglerfish prosomatostatin II,[51] which is the precursor to the hydroxylated 28-residue somatostatin described above. The peptide, residues 25–96 of preprosomatostatin II (schematically indicated as the stippled region in Fig. 9), was found to be resistant to Edman degradation, indicating a blocked amino terminus. Moreover, the experimental amino acid composition deviated significantly from that deduced from the cDNA. It was not possible to determine the site of signal cleavage in preprosomatostatin II from the composition due to uncertainty in the data.

Table IV summarizes the results of a tryptic analysis. The masses listed in Table IV are for the peptides, not their molecular ions. Note that

[50] P. C. Andrews and J. E. Dixon, *Biochemistry* **26,** 4853 (1987).

[51] P. C. Andrews, R. Nichols, and J. E. Dixon, *J. Biol. Chem.* **262,** 12692 (1987).

TABLE IV
ANALYSIS OF PROSOMATOSTATIN FRAGMENTS BY FABMS

Residue numbers	Calculated mass[a] (Da)	Observed mass (Da)	Post-Edman mass (Da)	N-Terminal residue
Tryptic digest				
25–28	530	NO[b]	512.5	—
29–41	1573.7	NO[b]	1445	Glu
42–48	980.5	980	852	Gln
49–50	245.2	NO[b]	NO[b]	—
51–61	1204.6	1205	1253	Ser
51–62	1360.7	1360.5	NO[b]	—
63–80	2032.0	1998.5	1926.5	Ala
81–92	1265.5	1193[c]	1065	Glu
93–96	505.2	505[c]	NO[b]	—
Staphylococcus aureus V8 protease digest				
25–39	1816.8	1798	1798	Blocked
40–47	1093.6	1093.5	980	Leu
48–58	1186.6	1187	1031	Arg
67–75	1000.5	1000.5	887	Leu
84–96	1423.6	1341.5	1227	Asp

[a] Calculated nominal mass.
[b] NO, Not observed.
[c] Observed in negative mode only.

two tryptic fragments corresponding to residues 63–80 and 81–92 deviated from their theoretical masses. Molecular ions corresponding to three predicted fragments were not observed. It was already known that the mass of the amino-terminal fragment would not correspond to the calculated mass because of the presence of a blocking group. In addition, the ions corresponding to residues 63–80 and 81–92 could not be assigned with certainty. For this reason, one cycle of manual Edman degradation was performed on the digest as described by Tarr[52] and the digest analyzed again by FABMS. This type of analysis is a form of subtractive Edman degradation in which the peptides rather than the PTH-amino acids are analyzed. Edman analysis immediately verified the previously assigned masses for residues 42–48 and 51–61. Note that although a Ser residue (87.1 Da) was removed from the amino terminus of the 51–61 fragment, the mass actually increased by 48 Da owing to the addition of one molecule of phenylisothiocyanate (135 Da) to the ε-amino group of Lys-61.

[52] G. E. Tarr, this series, Vol. 47, p. 335.

Edman cleavage also allowed the assignment of the signal at *m/z* 1998.5 to the tryptic fragment 63–80 and the fragment at *m/z* 1193 to residues 81–92. The difference in calculated and experimental masses for the 81–92 fragment (−72.5 Da) is equivalent to a Gly for Glu substitution. This observation was compatible with the discrepancy in the amino acid composition. Unfortunately, the 81–92 fragment contains three Glu residues. The amino-terminal Glu could be excluded by the results of the subtractive Edman degradation. Distinguishing between the other two sites required further digests. The mass discrepancy for the fragment corresponding to residues 63–80 could not be explained by substitution of a single amino acid. Again, further analysis was required.

One anomalous ion was observed which could not be assigned to a reasonable tryptic fragment. It was tentatively assigned to residues 25–28 on the assumption that the amino-terminal Gln had cyclized to pyroglutamic acid, resulting in the blocked amino terminus. Subsequent mass information (described below) confirmed this assignment. It should be noted that ions corresponding to two tryptic fragments (25–28 and 29–41) which could not be observed in the initial digest were observed after one cycle of Edman degradation. In general, we have found that after one cycle of Edman degradation a significant increase in sensitivity occurs. This increase might be due to removal during the solvent extraction steps of impurities which suppress ionization. Changes in the physical properties of the peptides after removal of the amino-terminal residue might also be responsible. Smaller peptides may occasionally be lost during the solvent extraction steps following cleavage of the phenylthiocarbamyl residues from the peptide.

In order to obtain more complete information on the structure of the prosomatostatin II fragment, further mapping by FABMS was performed using cleavage by cyanogen bromide and *S. aureus* V8 protease (Table IV). The data are summarized with the results of the tryptic mapping experiments in Fig. 9. Mass information for the entire 72 residues was obtained when data from all the mapping experiments were considered. The presence of a pyroglutamyl residue at the amino terminus was confirmed by the *S. aureus* V8 protease-produced fragment having a mass of 1798 Da (residues 25–39) which did not change in mass after one cycle of manual Edman degradation. The mass of the cyanogen bromide fragment, residues 72–87 (1698.0 Da), deviated from the cDNA-deduced mass (1731.7 Da) by 33.7 Da, indicating the substitutions responsible for the −72.5 Da mass discrepancy of the tryptic fragment (residues 81–92) lay outside residues 72–87. By the process of elimination, the substitution must lie between residues 88 and 92. The only candidate site in this region is Glu-90, thus suggesting a Gly for Glu substitution at this site. The *S.*

aureus V8 protease fragment corresponding to residues 84–96 deviated from the cDNA sequence by −72.5 Da, also consistent with a Gly for Glu substitution at position 90. This assignment was later confirmed by automated Edman degradation of a tryptic fragment corresponding to residues 81–92.

The mass data obtained from the proteolytic and cyanogen bromide digests indicate that the second mass discrepancy (33.5 Da) lies within the overlap between the tryptic fragment (residues 63–80) and the CNBr fragment (i.e., between residues 72 and 80). The mass of the *S. aureus* V8 protease fragment, corresponding to residues 67–75, was not anomalous, further limiting the site of substitution to five residues, 76–80. All the mapping data localizing the mass discrepancy of 33.5 Da to residues 76–80 are compatible with the substitution of Asp and Val residues for Thr-77 and Phe-78 (or vice versa). These substitutions were compatible with the amino acid composition data. In order to determine the order of the substitutions, Edman sequence analysis of a peptide corresponding to residues 63–96 was performed.

The inability to determine the order of residues 78 and 79 from mass spectral data alone arose from the lack of a cleavage which would distinguish the two residues. Presumably, the use of an Asp-specific protease[53] would have solved this problem. The sequence of a cDNA clone for anglerfish preprosomatostatin II was determined in order to ascertain the reason for the discrepancy. The redetermined sequence indicated an Asp-77 and Val-78, consistent with the protein data and suggesting a cDNA sequence error. Interestingly, residue 90 was unambiguously a glutamyl residue according to the cDNA, suggesting that the discrepancy at residue 90 might be due to microheterogeneity.

This last example indicates both the advantages and the disadvantages of having a cDNA sequence available. The assumption cannot be made that the cDNA sequence is infallible, nor that the cDNA clone sequenced corresponds exactly to the isolated protein. As an aside, four out of the eight prohormones whose products we have extensively examined have at least one sequence discrepancy relative to the cDNA sequences. This represents an admittedly small selection; however, it does urge caution in interpretation of peptide structure solely on the basis of cDNA sequence. The advantages of being able to interpret the ion data based on a protein sequence deduced from the cDNA far outweigh any possible problems concerning sequence errors or microheterogeneity.

This example also illustrates the usefulness of manual Edman degradation in conjunction with FABMS. This is particularly important when

[53] G. R. Drapeau, *J. Biol. Chem.* **255,** 839 (1980).

multiple mass discrepancies exist, thus making mass assignments difficult. More than one cycle of Edman degradation may, of course, be run if needed to verify the mass assignments. A peptide having a blocked amino terminus will be resistant to Edman degradation and may be identified on analysis of the mass spectrum. The use of manual Edman degradation as an adjunct to FABMS has been applied to a series of small peptides.[54] It has also been used effectively in the analysis of proteolytic maps by FABMS of Gly-tRNA synthetase (subunit molecular mass 35 and 65 kDa).[55]

An alternative approach to verification of mass assignments is to subject aliquots of the digest to specific chemical modification reagents in an approach similar to the use of Edman degradation or chemical reduction of disulfide-containing peptides described above. A large number of reagents of varying degrees of specificity are available for this purpose. Sun and Smith have recently described the use of performic acid to identify Cys-containing peptides.[56] The mass of a cysteine-containing peptide increases by 48 Da on performic acid treatment, while two cysteine residues result in a mass increase of 96 Da. Peptides containing Met and Trp residues also exhibit shifts in mass different from those observed for Cys. Using this approach, Sun and Smith were able to unambiguously assign all six Cys-containing peptides in a trypsin digest of CNBr-cleaved ribonuclease A.[56]

When not all the expected ions are observed in a mixture, a more time-intensive, but nevertheless effective, approach is to partially or completely resolve the fragments by HPLC prior to FABMS. This is particularly useful for larger proteins. Data from an example of this approach for a tryptic digest of a small (29-residue) peptide are summarized in Fig. 10. Glucagon I, isolated from anglerfish pancreas, was digested with trypsin and the fragments resolved by HPLC. FABMS of the resolved peptides provided masses consistent with the cDNA sequence except that two peaks corresponding to residues 18–24 were observed with one having the correct theoretical mass (878.0 Da), the second being 1 mass unit higher (878.9 Da). The fragment having the higher mass represents a form containing a deamidated Gln residue which apparently formed during isolation of the peptide. The use of FABMS for analysis of digests resolved by HPLC is likely to gain increased use with the introduction of liquid chromatography mass spectrometry (LCMS) systems that allow effluent

[54] C. V. Bradley, D. H. Williams, and M. R. Hanley, *Biochem. Biophys. Res. Commun.* **104,** 1223 (1982).

[55] B. W. Gibson and K. Biemann, *Proc. Natl. Acad. Sci. U.S.A.* **81,** 1956 (1984).

[56] Y. Sun and D. L. Smith, *Anal. Biochem.* **172,** in press (1988).

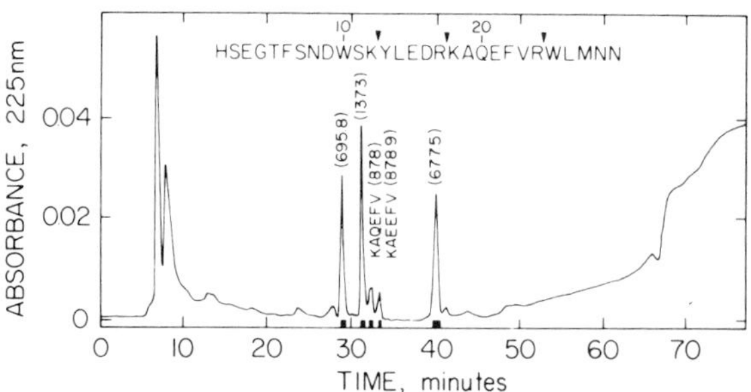

FIG. 10. Reversed-phase HPLC on a Zorbax ODS column of a tryptic digest of anglerfish glucagon I. The structures of the fragments discussed in the text are indicated. The major sites of tryptic cleavage are indicated by arrowheads. Experimental masses (Da) are indicated in parentheses. The gradient was from 0 to 50% acetonitrile in 60 min, then to 70% at 70 min. The flow rate was 0.7 ml/min, and both solvents contained 0.1% TFA. The structure of anglerfish glucagon I is indicated at the top of the figure.

from HPLC to be introduced directly into the source of a mass spectrometer.[57–60]

Assignment of Disulfide Bonds in Proteins

One of the more elegant uses of FABMS is in the assignment of disulfide bonds in proteins.[61–64] Knowledge of the disulfide bond connectivity in a protein puts constraints on the tertiary structure, thus providing useful information on the spatial relationships between portions of the peptide backbone. Before the assignment of disulfide bonds may be made, however, the primary sequence must be known. Two major experimental hurdles must be overcome in order to assign all disulfide bonds unambiguously. The first is the cleavage of the peptide backbone between each Cys residue. This frequently requires the use of a number of reagents. Native proteins (disulfide bonds intact) are normally much more resistant to proteolysis than the reduced and alkylated forms. For this reason, cyanogen

[57] M. L. Vestal, *Science* **226,** 275 (1984).

[58] R. M. Caprioli, T. Fan, and J. S. Cottrell, *Anal. Chem.* **58,** 2949 (1986).

[59] R. M. Caprioli, W. T. Moore, B. DaGue, and M. Martin, *J. Chromatogr.*, in press (1988).

[60] R. M. Caprioli, *Biomed. Environ. Mass Spectrom.*, in press (1988).

[61] A. M. Buko and B. A. Fraser, *Biomed. Mass Spectrom.* **12,** 577 (1985).

[62] H. R. Morris and P. Pucci, *Biochem. Biophys. Res. Commun.* **126,** 1122 (1985).

[63] R. Yazdanparast, P. C. Andrews, D. L. Smith, and J. E. Dixon, *J. Biol. Chem.* **262,** 2507 (1987).

[64] T. Takao, M. Yoshida, Y. Hong, S. Aimota, and Y. Shimonishi, *Biomed. Mass Spectrom.* **11,** 549 (1984).

bromide cleavage of the protein is frequently the first step in a digestion protocol. Initial cleavage at Met residues with cyanogen bromide introduces a few "nicks" into the Gordian knot of a native protein, making it more susceptible to unraveling by subsequent proteolytic digestion. Because digests of peptides containing multiple disulfide bonds are complex, it is advisable to avoid use of the less specific proteases in the primary digests for reasons described earlier. Even when specific cleavage reagents are utilized, unique assignment of ions may not be possible in all cases and may require special methods for verification.

Extreme care should be exercised to avoid reagents and conditions which may promote disulfide bond reduction or interchange. Particular care should be taken to be certain that the buffers and proteases used for the proteolytic digests are devoid of thiols. To assure that disulfide bonds remain intact, long-term (greater than 2 hr) proteolytic digests should be avoided. It is also useful to run a blank proteolytic digest with protease but no protein substrate in order to identify background ions from fragments derived from the protease. These ions are rare but may be observed on occasion. If it is not possible to introduce cuts between all Cys residues, it may be necessary to isolate or at least partially purify those fragments by HPLC and then to introduce further peptide backbone cleavages using less specific proteases, partial acid hydrolysis, or other nonspecific cleavage methods.

The second major problem is the recognition of disulfide bond-containing peptides in the spectrum. These peptides may be identified by following the time-dependent reduction in intensity of disulfide-containing peptides in the source with concomitant increase in intensity of the reduced product peptide(s). Alternatively, the digest may be split and one-half chemically reduced before FABMS. Ions exhibiting a reduction-dependent difference will correspond to disulfide-containing peptides. Running one cycle of Edman degradation will result in loss of at least two amino acid equivalents from peptides containing interchain disulfide bonds.

Assignment of the disulfide bonds in a "model" protein, lysozyme, will serve to illustrate the method. Table V indicates the masses observed for peptides in a crude tryptic digest of cyanogen bromide-cleaved native lysozyme. An initial search of the lysozyme sequence, using PROSEAR,[65] for the observed fragment masses indicated that in only two cases could unique mass assignments be made if the specificities of the

[65] PROSEAR and PROFRAG are modules of PROCOMP, a program designed to process protein structure data. Developed for the IBM XT, AT, and compatibles by P. C. Andrews (in preparation).

TABLE V
PEPTIDES IDENTIFIED IN THE TRYPTIC DIGEST OF CNBr-CLEAVED HEN EGG LYSOZYME

Residue numbers	Calculated monoisotopic mass of MH$^+$ (Da)	Experimental mass of MH$^+$ (Da)	Calculated fragments within 1 Da	Fragments compatible with cleavage sites	Post-Edman mass of MH$^+$ (Da)	N-Terminal residue(s)	Fragments compatible with N-terminal amino acid
1–5	606.4	607[a]	1	1	478[a]	Lys	1
2–5	478.3	478[a]	2	1	379	Val	1
6–12	660.3	660[a,b,c,d]	2	1	557[a,d]	Cys	1
15–21	874.4	874[b]	5	1	737[a]	His	1
22–33	1268.6	1269[b,d]	2	1	1347[d]	Gly	1
34–45	1428.6	1429[b]	2	1	1282	Phe	1
46–61	1753.8	1754[b]	4	1	1640[a,b]	Asn	1
62–68	936.4	936[b,d]	4	1	—	—	—
69–73	517.3	517[b]	3	1	416	Thr	1
74–96	2337.1	2337[a,b,d]	1	1	2358[d]	Asn	1
98–105	744.3	744[a,b,c]	4	1	631	Leu/Ile	1
106–112	902.5	902[b]	2	1	789[a]	Asn	1
113–114	289.1	289	8	1	—	—	—
117–125	1045.5	1046[a,b]	3	1	988	Gly	1
126–128	335.1	335[b,d]	6	1	—	—	—
62–68, 74–96	3268.5	3270[e]			3107[e]		
22–33, 115–116	1515.7	1516[a,b,e]			—[e]		
6–12, 126–128	992.4	992[e]			—		

[a] The sodium adduct was also observed.
[b] The potassium adduct was also observed.
[c] Both homoserine and homoserine lactone forms were observed.
[d] Peptides produced by reduction of disulfide bonds.
[e] One or more of the component peptides resulting from chemical reduction of the disulfide bond(s) were also observed.

cleavage reagents were ignored. Mass spectra obtained both before and after chemical reduction served to identify the disulfide-containing peptides. The computer search for matches corresponding to the remainder of the peptides was performed for reduced lysozyme. When the branch points occurring at disulfide bonds are included, the number of possible assignments increases dramatically. For example, the tryptic fragment with an experimental mass for the molecular ion of 1754 ± 1 Da could be assigned to residues 7–22 (1753.9 Da), 45–60 (1753.8 Da), 46–61 (1753.8 Da), or 51–64 (1754.8 Da). Only one of these four potential fragments (residues 46–61) is consistent with tryptic cleavage of a cyanogen bromide digest. Use of the additional criterion of the known specificity of the cleavage reagents was sufficient to provide unique mass assignments for all the ions indicated in Table V.

Inappropriate cleavage can occur even with the most specific reagents and enzymes, leading to difficulties in interpretation. For example, even low levels of α-chymotrypsin contamination of trypsin may affect digests, especially at high protease to substrate ratios. Trypsin itself has intrinsic chymotryptic activity. In addition, cyanogen bromide will provide only partial cleavages when followed by Ser, Thr, or Cys.[66,67] For these reasons it is advisable to perform one or more cycles of manual Edman degradation on aliquots of the digest prior to FABMS when the mass assignments are in doubt. This approach identified the amino-terminal residues of most of the fragments observed (Table V) and allowed independent confirmation of the mass assignments.

Disulfide-containing peptides were identified by chemical reduction of an aliquot of the digest prior to FABMS. Ions which disappeared on reduction with concomitant appearance of product peptides were considered to correspond to cystine-containing peptides. Identification of the mass corresponding to residues 22–33/115–116 (1516 Da) confirms the connection between Cys-30 and Cys-115. Likewise, the mass at 992 Da corresponding to residues 6–12/126–128 confirms the Cys-6 to Cys-127 disulfide bond. It was not possible to introduce specific cleavages between Cys-76 and Cys-80 or between Cys-80 and Cys-94 (Fig. 11), so the connectivity for these four Cys residues could not be ascertained. The peptide containing these cystines (62–68/74–96) was purified from the digest using reversed-phase HPLC and subjected to partial acid hydrolysis. FABMS of the hydrolysate identified three disulfide-containing peptides (Fig. 12), allowing unambiguous assignment of the Cys-64 to Cys-80 and

[66] N. Doyen and C. Lapresle, *Biochem. J.* **177,** 251 (1979).

[67] W. A. Schroeder, J. B. Shelton, and J. R. Shelton, *Arch. Biochem. Biophys.* **130,** 551 (1969).

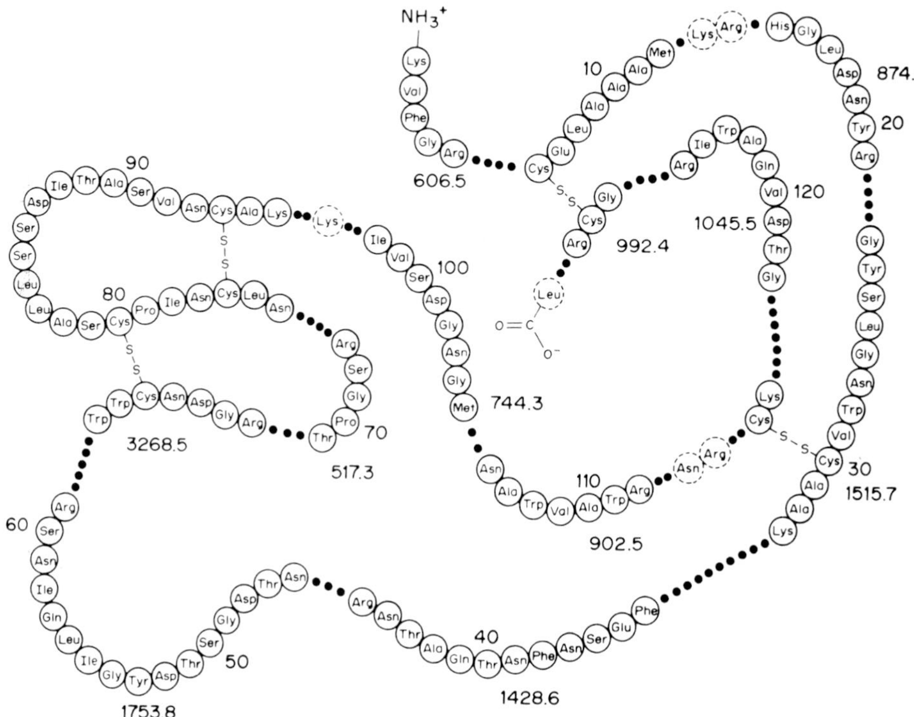

FIG. 11. Peptide products produced by digestion of hen egg white lysozyme with CNBr followed by trypsin. Numbers represent the molecular ions (MH^+) of each tryptic fragment. Methionyl residues are indicated as Met, not as homoserine. Molecular ions were not assigned to the peptides encircled with dashed lines. Dots indicate cleavage sites introduced by cyanogen bromide or trypsin.

the Cys-76 to Cys-94 disulfide bonds. Figure 11 summarizes the disulfide assignments for lysozyme and indicates the trypsin and cyanogen bromide cleavage sites.

The most difficult situation in assignment of disulfide bonds, of course, is one in which two adjacent Cys residues occur in a peptide. One approach to this problem might be to isolate a fragment containing this sequence, subject it to partial acid hydrolysis in hopes of introducing a cleavage of the Cys—Cys peptide bond, and then analyzing the fragments by FABMS. An alternative is to introduce a cut close to the amino-terminal side of the pair and then to run a series of manual Edman degradation cycles until the Cys—Cys peptide bond has been cleaved.[68] The

[68] F. M. Greer, M-Scan, Berkshire, England, personal communication, and P. C. Andrews, unpublished results.

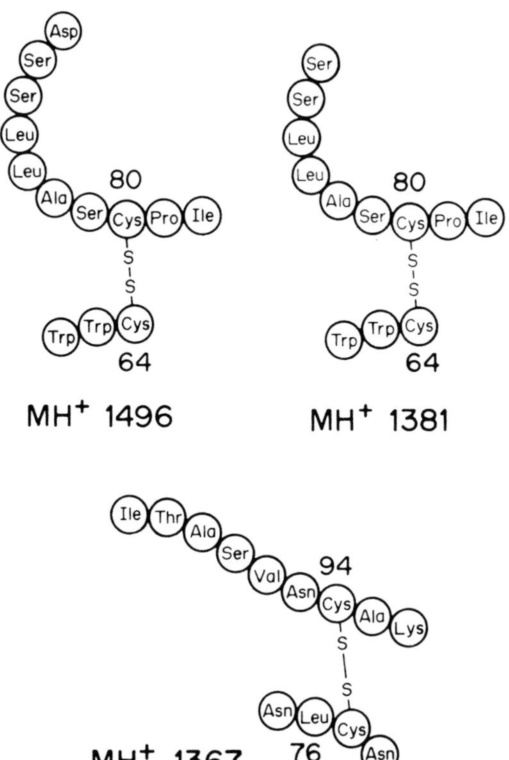

FIG. 12. Three of the disulfide-containing peptides from partial acid hydrolysis of the 62–68/74–96 tryptic peptide of hen egg white lysozyme. The experimental masses of the molecular ions are indicated.

number of Edman cycles should be kept to a minimum due to low recovery of disulfide bonds. When the amino-terminal residue of a peptide is cystine, one cycle of manual Edman degradation results in destruction of most of that disulfide bond.[63,69]

Assignments of ions to fragments is greatly facilitated by computer programs which calculate masses for peptide digests and which perform automatic mass assignments. The program used for many of the calculations described in this chapter is PROCOMP and related modules.[65]

Sequence Information from Fragmentation Ions

FABMS will sometimes produce sequence information via fragmentation of the peptide. For small peptides having blocked amino termini,

[69] P. C. Andrews, unpublished results.

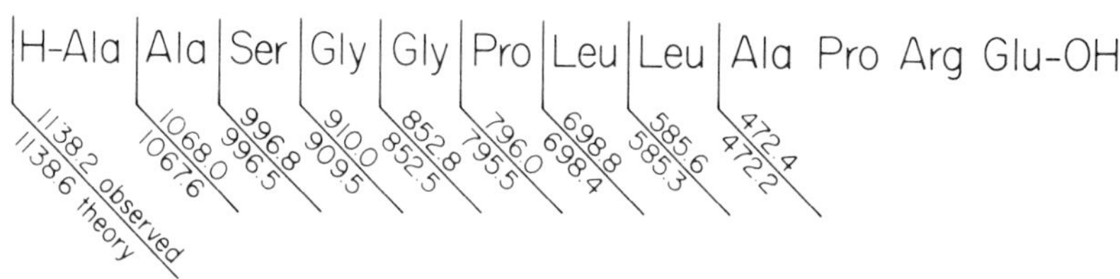

FIG. 13. Fragmentation ions observed for the dodecapeptide preceding somatostatin-14 in anglerfish prosomatostatin I.

FABMS is the method of choice for structure determination. However, extensive fragmentation of peptides sufficient to assign all or even most of the residues does not occur reliably. Peptides less than 20 residues in length are more likely to provide sequence information from fragmentation than are larger peptides. Substitution of krypton or argon for xenon and adjusting the accelerating voltage may increase the number of fragment ions observed.[22] Tandem mass spectrometers are better suited for obtaining complete structural information and have been the subject of several articles.[25,26,70] Figure 13 illustrates part of the fragmentation observed for a dodecameric peptide produced during the proteolytic processing of prosomatostatin. The ion series shown represents cleavage of the peptide bond. Only those ions resulting from cleavage of a single bond are indicated in Fig. 13. An ion series (not shown) arising from cleavage of two peptide bonds allowed the remainder of the residues to be ordered. Ion series generated by fragmentation of N—C or C—C bonds in the peptide backbone may also be observed.

Alternatively, sequence information may be obtained by use of appropriate chemical methods in conjunction with FABMS. For example, carboxy-terminal sequencing using carboxypeptidases and analysis of the truncated peptides by FABMS rather than analysis of the released amino acids is a sensitive and effective method for obtaining carboxy-terminal sequence information.[54,71,72] Aminopeptidases may be used in an equivalent manner but have few advantages over automated Edman degradation.[71,72] Subtractive Edman degradation, discussed above, is particularly appropriate when working with mixtures of peptides. Finally, partial internal sequence information may be obtained from overlapping peptide fragments produced by partial digests using proteases having broad substrate specificity (see the somatostatin-14 example above).

[70] K. Biemann and H. A. Scoble, *Science* **237,** 992 (1987).

[71] R. Self and A. Parente, *Biomed. Mass Spectrom.* **10,** 78 (1983).

[72] R. M. Caprioli and T. Fan, *Anal. Biochem.* **154,** 596 (1986).

Summary

FABMS is a powerful and sensitive analytical technique capable of providing structural information unattainable by standard methods of peptide analysis. Many posttranslational modifications are undetectable by other routine analytical methods. In addition, FABMS is capable of providing information regarding posttranslational modifications at levels of peptide comparable to those required for other methods of analysis (10–1000 pmol). FABMS has had the effect on protein structure analysis that structure determination of any neuropeptide might now be considered incomplete without some form of mass spectrometric analysis.

Much of the recent explosive increase in the use of mass spectrometry for solving problems in peptide structure analysis can be traced to improvements in methods capable of producing molecular ions from nonvolatile species.[1,2] With the development of these methods, it can be expected that refinements of existing methods and new ionization methods will continue to increase the mass range and sensitivity available for peptide structure determination. For a brief review of other mass spectrometric methods applicable to peptides, see Delgass and Cooks.[73]

[73] W. N. Delgass and R. G. Cooks, *Science* **235,** 545 (1987).

[7] Electrophysiological Analysis of Opioid Peptides: Extracellular and Intracellular Approaches

By EDWARD D. FRENCH and JOHN T. WILLIAMS

Introduction

The isolation of opiate-like substances from mammalian brain[1,2] launched a flurry of research activity aimed at elucidating their physiological role in both peripheral and central tissues. However, to assess and compare the effects of both exogenous and endogenous opiates and to understand their function in the physiology of pain perception, reward, learning and memory, hormonal regulation, etc., it became necessary to utilize techniques which allowed the direct, controlled application of these substances onto well-defined neuronal populations while monitoring

[1] J. Hughes, *Brain Res.* **88,** 295 (1975).

[2] J. Hughes, T. W. Smith, H. W. Kosterlitz, L. A. Fothergill, B. A. Morgan, and H. R. Morris, *Nature (London)* **258,** 577 (1975).

changes in cellular excitability. Two methods of drug delivery, microiontophoresis/micropressure (both for *in vivo* and *in vitro* studies) and superfusion (mainly for *in vitro*), are usually combined with either extracellular or intracellular recording of the electrical responses of neurons to the applied drugs. The aim of this chapter is to survey the techniques currently available for the study of opioid responses in neurons, and to provide the interested investigator with a basis for selecting the most appropriate set of methods for a particular intended line of inquiry.

Possibly the most important recent advance in methods for the study of opioids has been the development of agonists and antagonists specific to opioid receptor subtypes (Table I). The techniques in this chapter are predicated on the use of these selective agents, which can eliminate ambiguities in interpretation of drug effects. The discussion is divided into two main sections, dealing first with *in vivo,* then *in vitro* preparations. It should be realized that techniques which are used in one type of preparation can to some extent also be used in the other. For example, microiontophoresis is primarily carried out *in vivo* although it can be used effectively, and often for a different purpose, *in vitro*. Both preparations have clear advantages and disadvantages (Table II), yet they can be complementary in understanding the actions of opioids.

In Vivo Techniques

Preparation of the Animal

Although a variety of anesthetic agents can be used in *in vivo* preparations, we have found halothane and chloral hydrate to be the most acceptable. Halothane is administered as a gaseous mixture and as such requires an air or oxygen supply, flow meter for setting the delivery volume, an anesthetic vaporizer, and a means to remove and vent it safely out of the laboratory. The last item is extremely important given the hepatic toxicity of this agent. The advantage of halothane is that the depth of anesthesia can be quickly and accurately controlled and maintained at a fixed level throughout a recording period of many hours.[3]

Generally, animals prepared for halothane are fitted with a tracheal breathing tube made of polyethylene tubing. The external portion of this tube is then attached to stainless steel tubing which enters a T adapter. One arm of the T is used for introduction of the anesthetic, and the other opens into a larger cylinder (a 10-ml syringe barrel) attached to a vacuum line (Fig. 1A). This permits the volatilized halothane (0.75–1% at 1 liter/min) to pass through the tube where it is inspired by the animal's sponta-

[3] E. D. French and G. R. Siggins, *Regul. Pept.* **1,** 127 (1980).

TABLE I
SPECIFIC AGONISTS AND ANTAGONISTS TO OPIOID SUBTYPE RECEPTORS

Compound	pK_i μ	pK_i δ	pK_i κ	Reference
Agonist				
[D-Ala[2],MePhe[4],Glyol[5]] enkephalin	9.6	6.7	6.0	*a, b*
[D-Pen[2],D-Pen[5]]enkephalin	6.6	9.0	4.9	*a, c*
U-50488	6.7	5.1	9.1	*a, d*
Dynorphin A (1–13) amide	9.2	8.8	10.7	*d*
Tifluadom	8.1	6.9	10.1	*d*
	pA_2 μ	pA_2 δ	pA_2 κ	
Antagonist				
Naloxone (not selective)	8.6			*e*
		8.1		*f*
	8.9	8.4	8.1	*b*
ICI 174864	5.2			*e*
		7.6		*f*
	>5.3	7.5	>5.3	*g*

[a] From H. I. Mosberg, J. R. Omnaas, and A. Goldstein, *Mol. Pharmacol.* **31,** 599 (1987).
[b] From J. Magnan, S. J. Paterson, A. Tavani, and H. W. Kosterlitz, *Naunyn-Schmiedeberg's Arch. Pharmacol.* **319,** 197 (1982).
[c] From Pelton *et al.*[35]
[d] From I. F. James and A. Goldstein, *Mol. Pharmacol.* **25,** 337 (1984).
[e] From Williams and North.[25]
[f] From Mihara and North.[21]
[g] From Cotton *et al.*[34]

neous breathing and then evacuated without requiring that the in- and outflow rates be perfectly balanced. Using this arrangement, we can adjust to a light depth of anesthesia as evidenced by the presence of vibrassal movements. Although halothane is normally vaporized with oxygen we have not found any differences in the status of the animals using a mixture of halothane and air, which is considerably less expensive and potentially less dangerous.

We have also found that chloral hydrate (350 mg/kg, i.p.) is a very acceptable alternative to halothane and has the advantage of being injectable and not requiring the costly equipment or safety precautions men-

TABLE II
ADVANTAGES AND DISADVANTAGES OF *in Vivo* AND *in Vitro* PREPARATIONS

In vivo preparation
  Advantages
    1. Neurons little affected by preparation
    2. Major physiological systems remain intact
    3. Neurons can be identified physiologically
    4. Afferent and efferent pathways are left in tact such that neuronal circuitry can be investigated
    5. Chronic drug exposure may more closely resemble clinical situations
  Disadvantages
    1. Drug application is limited to local (e.g., iontophoresis) or systemic routes with the concentration at the recording site usually indeterminable
    2. Specificity of agonist–antagonist interactions are difficult to assess since locally or systemically applied drugs often have indirect effects
    3. Localizing the site of drug actions (pre- or postsynaptic) is problematic
    4. Possible artifacts produced by anesthetic agents
    5. Difficult to alter the extracellular milieu
    6. Lack of stability for intracellular recordings

In vitro preparation
  Advantages
    1. Drugs can be applied in known concentrations
    2. The extracellular ionic composition, temperature, and pH can be accurately controlled or selectively altered
    3. Absence of anesthetics
    4. Stability for long-term intracellular recording
    5. Neurons and processes can often be visualized, thus facilitating accurate placement of stimulating and recording electrodes
    6. Neurons can be isolated from presynaptic influences
    7. Synaptic potentials and the cellular properties appear similar to the *in vivo* situation
    8. Possible to use techniques (e.g., voltage clamp) to isolate and define selective conductances
    9. All of the advantages above in tissue taken from chronic drug-treated animals
  Disadvantages
    1. Normal physiological processes may be altered by slicing
    2. Neuronal circuitry incomplete and activation of fiber inputs greatly limited
    3. Spontaneous activity possibly different than *in vivo*
    4. Neuronal responses to drugs may differ from those *in vivo*
    5. Contribution or importance of other tissues in chronic treatments is absent

tioned above. However, chloral hydrate can stimulate secretion of mucus in the airways which can progress to highly labored (intercostal) breathing and even death. Thus, a polyethylene (PE) tracheal breathing tube (for spontaneous breathing of room air) is highly recommended for routine use with injectable anesthetics. If mucus-related difficulties do arise, a smaller PE tubing attached to a 5-ml syringe can be inserted into the tracheal tube

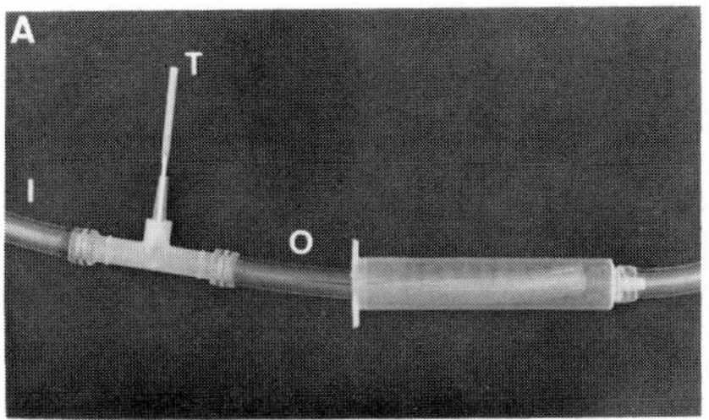

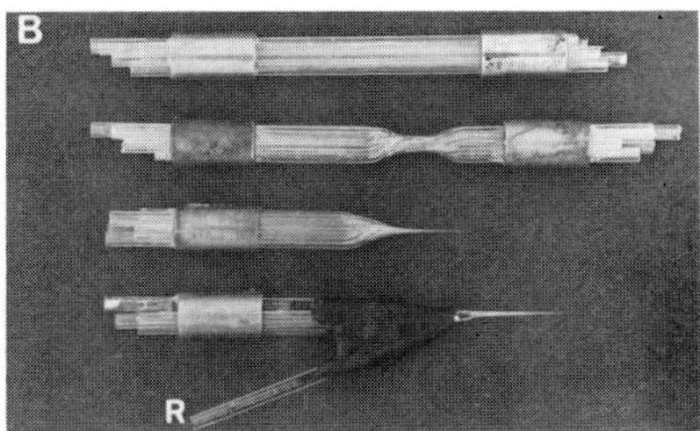

FIG. 1. (A) Breathing apparatus for administering the volatile anesthetic halothane. I and O indicate the in- and outflow routes. The tracheal tube (T) allows the animal to spontaneously breathe the passing anesthetic mixture. (B) Multibarrel micropipet for recording and iontophoresis. The five capillary tubes (top) are secured by brass rings glued to each end, then heated and twisted. The twisted blank is then repositioned, heated, and pulled to the desired shape and shank length. The bottom part of figure shows a five-barrel micropipet to which a glass microelectrode has been affixed. In this case the tip of the recording electrode (R) extends 25 μm beyond the micropipet. The entire assembly is strengthened with a waxlike substance.

to clear the upper air passages. Note, however, that mechanical irritation of the airways will also result in an increased mucous response and should be kept to a minimum. Another shortfall with chloral hydrate is that the depth of anesthesia is cyclical. To minimize this we routinely administer 50 mg/kg chloral hydrate at 40-min intervals. Generally, it has been our experience that the response of various neuronal populations to the opioids did not differ as a function of the anesthetic agent used.

The surgical preparation of the brain for recording is rather straightforward yet can easily induce tissue trauma. A dissecting microscope with a long working distance objective should be used throughout the surgery and the lowering of the electrode into the brain. Care must be exercised when drilling away the skull and incising the underlying dura mater. A 25-gauge hypodermic needle bent at a slight angle about 5 mm from the tip is ideally suited for lifting and cutting the dura. If the cortex is damaged, however, it will quickly become swollen. This can alter the calculation of depth measurements which are crucial when performing tests in small nuclei. Also, brain pulsations can occur and markedly interfere with any long-term recording of single-unit activity. In such cases, a small slab of agar held in place by a pressor foot can dampen the pulsations.

Extracellular Recording

Although a variety of commercial iontophoretic/micropressure units are available, we have had extensive experience with the Medical Systems Neurophore BH-2 which provides low noise–high compliance constant current with automatic current neutralization, clock devices for timing drug applications, digital readouts of eject/retain currents and resistances of the drug barrels, analog outputs of applied currents, connectors for interfacing with other lab triggering or gating devices, and direct plug-in compatibility for pressure modules.

The instrumentation for recording from multibarrel electrodes is identical to that using a single extracellular electrode. Ideally, the amplified signal is filtered of extraneous noise (1–10 kHz bandpass) and passed through a discriminating device to isolate out single-unit action potentials. These can then be integrated over time, recorded on a paper or magnetic recorder, or fed into a computer. We use an Apple II+ with software for acquisition, storage, and analysis of interspike interval and poststimulus (drug) time histograms (Klaus Liebold, Research Institute of Scripps Clinic, La Jolla, CA). In addition to the filtered signal we routinely monitor an unfiltered dc trace on the ratemeter record which represents the electrode tip potential.[3]

Iontophoresis/Micropressure

A number of published articles have provided extensive discussions of both the theoretical framework and practical aspects for iontophoresis.[4–9] While this part of the chapter cannot and is not intended to be an exposition of all these works, it does detail methods that draw on the refinement of past techniques which have evolved from the practicalities dictated by hands-on "spritzing."

Electrode Construction. Multibarrel electrodes must serve equally well the two distinct processes of recording single neuronal action potentials while simultaneously permitting the controlled application of minute amounts of drugs. Oftentimes when the size of the electrode tip is such that the ohmic resistances of the drug barrels are within a range that permits easy current ejection of compounds, the recording properties are such that single units cannot be adequately isolated from multiunit activity. In the final analysis only trial and error will find the best mix for the

[4] D. R. Curtis, *in* "Physical Techniques in Biological Research" (W. H. Nastuk, ed.), p. 144. Academic Press, New York, 1964.

[5] G. C. Salmoiraghi and F. E. Bloom, *Science* **144,** 493 (1964).

[6] G. C. Salmoiraghi and C. N. Stefanis, *Int. Rev. Neurobiol.* **10,** 1 (1967).

[7] K. Krnjevic, *in* "Methods in Neurochemistry" (R. Fried, ed.), p. 129. Dekker, New York, 1971.

[8] F. E. Bloom, *Life Sci.* **14,** 1819 (1974).

[9] R. D. Purves, "Microelectrode Methods for Intracellular Recording and Ionophoresis." Academic Press, New York, 1981.

shape, shank length, and geometry of the micropipet tip to meet the experimenter's needs.

We have used two different types of multibarrel pipette. One version begins as a commercially available prefabricated five-barrel glass blank (R & D Scientific Glass, Spencerville, Maryland).[5] The longer center capillary is fixed in the lower portion of a vertical electrode puller (Narishige) and the fused capillary end inserted into a melted glass rod held by the puller's upper chuck. The blank is heated and pulled to the desired shaft length, which is dictated by the depth of the area to be recorded. Generally, the longer the electrode and thinner the shaft the more prone it is to develop current passing problems. The tip of the pulled electrode is broken back to a diameter of 4–8 μm by touching it to a glass syringe plunger under microscopic control. "Bumping" the electrode before filling it with solutions allows more precise fracturing of the glass tip, and peptide solutions (often in very small quantity) are put only into electrodes with acceptable tip diameters. The micropipet is then placed in a holder, submerged in a beaker of distilled water which is then brought to a boil, and allowed to cool to room temperature. This process facilitates the diffusion of fluid to the tip of all the barrels. The water is removed by a syringe with a 30-gauge needle and replaced with a drug solution. The electrode is placed tip down inside a Teflon (or similar material) holder with a machined out center well having a diameter that permits the electrode to be suspended by the upper flanged barrels.[4] This unit is put into a centrifuge cup and spun at 1000 rpm for 10–20 min at 10°C, after which the electrodes are placed back in a holder with the tips submerged in water until needed.

More recently, we have used a somewhat different five-barrel electrode which we fabricate in the lab from fiber-filled glass capillary tubes (1.5 mm o.d., 1.1 mm i.d., WP-Instruments, New Haven, CT). These are held together with a brass or stainless steel ring cemented (cyanoacyrlate; Krazy Glue) at each end (Fig. 1B). The tubes are staggered to accommodate the attachment of a micropressure tubing and to lessen the likelihood of salt bridges forming between the different drug and recording/current balance barrels. A small amount of petroleum jelly applied to the tops of the barrels will help prevent cross-talk. The glass blanks are held by the metal collars in the chucks of the vertical electrode puller. The blank is heated, and when sufficiently pliable the lower chuck is twisted 360° and let drop approximately 1–1.5 cm before turning the heating coil off. The twisted portion is then recentered in the coil, heated, and pulled (primary pull by gravity, secondary by magnet) to the desired length and tip shape. These electrodes can then be bumped, filled with drug solutions, and used immediately. Also, only 5 μl is needed to fill a barrel. The ease of construction, low cost, rapid filling and small volumes are obvious advantages with this micropipet.

Another type of microelectrode which can be fabricated in the lab is a combined single recording microelectrode–micropipet ensemble. This combination electrode can, in some cases, make for better isolation of single-unit action potentials and for smaller ejection currents and pressure. The pulled recording electrode must be bent at an angle to permit it to fit alongside and slightly beyond the tip of the micropipet; this can enable recording from the cell soma while applying drugs to the dendritic field. Combining the two elements requires a means by which each can be fixed in space and aligned in the same planes for gluing. This takes two microscopes set at 90° angles, micropositioners, and a stable surface to minimize vibration during the time it takes for the epoxy cement to dry. Application of the glue must take place under microscopic control to ensure that the expoxy does not come into contact with the micropipet orifices. The entire structure can be further reinforced with melted wax or similar substance (Pyseal, Fisher Scientific Co.). Our experience has found that this type of electrode can be highly effective when confined to the more superficial structures (<4 mm) of the brain. With the deeper lying areas, it is not uncommon that pressure exerted by the tissue separates the recording tip from the micropipet, often giving negative findings during iontophoretic drug tests. Also, we have found that if the micropipet tips are "bumped" too large they tend to get clogged with tissue debris and are blocked to current or pressure ejection, or there is greater passive diffusion of drugs out of the barrels thus requiring higher backing currents. Although this electrode has certain advantages over the standard five-barrel, the facilities and time required for its manufacture and its limited usefulness in actual experimental situations may make its cost-to-benefit ratio too high for many investigators.

Drug Solutions. The ejection of test substances from micropipets can be accomplished by the application of current or pressure through the drug barrel.[10,11] For most substances, the concentration (100 mM to 1 M) and charge-to-mass ratio are sufficiently large to ensure current passage and the ejection of the charged moiety. With peptides, however, the small charge-to-mass ratio and need to use lower concentrations (endorphins, 3 mM; enkephalins, 30 mM) dictate that these compounds be dissolved in NaCl (165 mM) and expelled electroosmotically in the sodium or chloride hydration shell.[4] It is imperative that low retaining currents (~5 nA) be

[10] M. Sakai, B. E. Swartz, and C. D. Woody, *Neuropharmacology* **18,** 209 (1979).

[11] P. Bevan, C. M. Bradshaw, R. Y. K. Pun, N. T. Slater, and E. Szabadi, *Experientia* **37,** 296 (1981).

used in these instances to prevent the ejection of peptides via the counter-ion current. Iontophoretic application also requires that the pH of the drug solution be selected to ensure a charged drug species (pK_a values will determine this). Usually, this requires a pH near 4 or 8. Although extremes in the pH may produce nonspecific effects on neuronal firing patterns,[12,13] we have not found this to be the case at the pH values mentioned above. Nevertheless, tests with solutions of comparable pH are recommended for values less than 4 or greater than 8.

Micropressure is another means by which substances can be ejected at the single neuronal level. There are certain distinct advantages to this technique including lack of current artifacts, ability to use small concentrations of drug or solutions of organic noncharged molecules (e.g., ethanol, benzodiazepines), and neutral pH values.[10,14] However, pressure artifacts such as the actual movement of the fine electrode tip with the on- and off-set of the pressure head can occur as evidenced by a change in action potential amplitude. Also, different micropipets may have markedly different ejection properties given the fact that flow is proportional to the third power of the tip diameter.[9] If retaining currents are not employed, drugs also can diffuse out of the pipet. While this can be circumvented by an adaptor which allows pressure and retaining current in the same barrel, this necessitates using solutions with current carrying properties, and therefore reintroduces some disadvantages. Also, the interaction of opposing pressure fronts may be of some practical concern when assessing or interpreting agonist/antagonist relationships.

Iontophoretic Drug Tests in Vivo. At the outset it is important to establish the criteria by which any given drug test is included in the sample. We routinely specify that neuronal firing rates be changed by $>20\%$ at least twice, no current unbalance, and no dc change >10 mV.[3,15] With antagonists the agonist response must also be reduced by at least 50%. Interest in determining the relative potency of various compounds or differences in neuronal sensitivity to agonists with iontophoretic techniques can be quite problematic. Since the relationship between passage of current and ejection of a charged molecule is best described by Hittorf's law, the amount of drug delivered is determined by the amount and time of current application, valence of the ion, Faraday's constant, and the transport number of the compound. This last entry varies according to a molecule's solubility, polarity, dissociation constant, and medium into

[12] D. R. Curtis, J. W. Phillis, and J. C. Watkins, *J. Physiol. (London)* **158,** 296 (1961).

[13] R. C. A. Frederickson, L. M. Jordan, and J. W. Phillis, *Brain Res.* **35,** 556 (1971).

[14] M. R. Palmer, S. M. Wuerthele, and B. J. Hoffer, *Neuropharmacology* **19,** 931 (1980).

[15] W. Zieglgansberger, E. D. French, G. R. Siggins, and F. E. Bloom, *Science* **205,** 415 (1979).

which it is ejected.[4,8,9] Thus, any meaningful intersubstance potency comparisons must be made on the basis of transport numbers. Since this value can change dramatically from one micropipet to another, efforts to compile useful potency ratios can be extremely tedious.[11,16–18]

In the absence of such measurements we were able to assess the effects of the opioid peptides, Met-enkephalin and β-endorphin, on neuronal activity in a number of structures comprising the limbic system.[3,15] Although we did find that the opioids were generally depressant throughout the central nervous system, there were instances of excitatory responses. In most cases both effects were sensitive to blockade by the opiate antagonist naloxone. Also, in the majority of cells, the peptides required much larger currents to elicit responses comparable to those of morphine. However, with the higher ejection currents, we frequently found that the electrode tip potential was markedly changed even in the presence of automatic current neutralization (Fig. 2). The direct application of current necessary to produce comparable tip potentials would in some cases directly alter neuronal firing. With current controls, however, the geometric relationship of the current carrying electrode to the cell may not be the same as that of the drug barrel,[8] and the passing of either positive or negative current may not produce the same conductance changes as those associated with drug applications.[19] Therefore, current neutralization and rejection of tests with a tip potential change of greater than 10 mV is the preferred approach.

In regions with little or no spontaneous activity, it is necessary to concurrently apply an excitant, such as glutamic acid or acetylcholine, to drive firing rates sufficient for drug interaction measurements. In all such tests it is imperative that the compounds be applied at regular and timed intervals to avoid possible false-positives resulting from the phenomenon of barrel "warm-up."[4,8] Also, when using the excitant glutamic acid, care must be taken to limit the duration of application since this excitatory amino acid can readily lead to a depolarization blockade of activity. This is evident by the decline in action potential amplitude with increased firing. Moreover, we have found that higher currents (15–20 nA) are needed for retaining glutamate than for other substances to prevent even a small leakage of glutamate out of the pipet which can eventually result in depolarization inactivation. The inclusion of an excitant also serves as an additional control for the specificity of action of antagonists and helps

[16] B. J. Hoffer, N. Neff, and G. R. Siggins, *J. Neuropharmacol.* **10,** 175 (1971).

[17] C. M. Bradshaw, R. Y. K. Pun, N. T. Slater, and E. Szabadi, *J. Pharmacol. Methods* **5,** 67 (1981).

[18] D. A. Hosford, H. J. Haigler, and R. S. Turner, *J. Neurosci. Methods* **4,** 135 (1981).

[19] W. Zieglgansberger and J. Champagnat, *Brain Res.* **160,** 95 (1979).

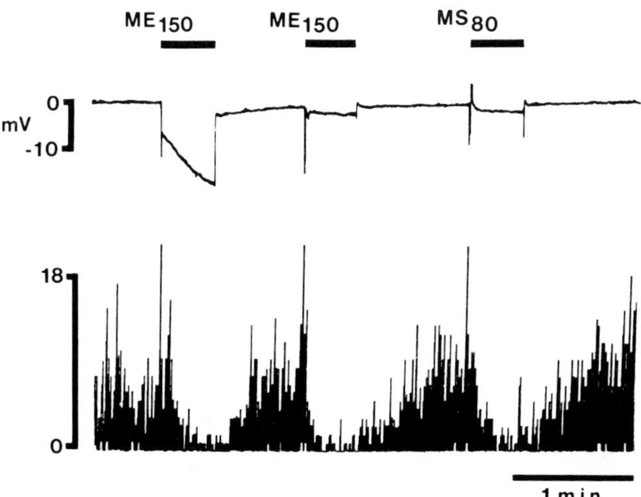

FIG. 2. Opioid-induced inhibition of a spontaneously active central amygdala nucleus neuron. With the first application a large change in the electrode tip potential occurs even in the presence of complete current neutralization. Changes of more than 10 mV with the subsequent drug ejections would have been grounds for excluding this test from the sample. Since the next Met-enkephalin (ME) and morphine (MS) applications showed no further dc imbalance, the observed inhibitions were considered drug- rather than current-mediated effects. Numbers over bars refer to nanoamperes (nA) of ejection current and length of bars duration of application. Ordinate on bottom trace is in spikes per second. Note also the artifactual activity coincident with the onset of current.

control for nonspecific local anesthetic effects, like that ascribed to naloxone.[20] An example of this is shown in Fig. 3, where the inclusion of acetylcholine helped to ensure specificity of the antagonism of Met-enkephalin by naloxone.

In order to determine if an observed iontophoretic effect is mediated through an indirect (presynaptic) or direct (postsynaptic) site of action, it becomes necessary to test the drug under conditions of altered transmitter function. This is most easily accomplished by specific antagonists or selective lesions of major afferent inputs. For example, the potential contribution of acetylcholine to the opioid-induced excitation of hippocampal neurons[3,5,20] was assessed in animals with electrolytic lesions of the septal nuclei or during concurrent application of the muscarinic cholinergic blocker scopolamine (Fig. 3). The continued excitation of pyramidal neurons under these circumstances strongly suggested that the opioid-

[20] R. A. Nicoll, G. R. Siggins, N. Ling, F. E. Bloom, and R. Guillemin, *Proc. Natl. Acad. Sci. U.S.A.* **74,** 2584 (1977).

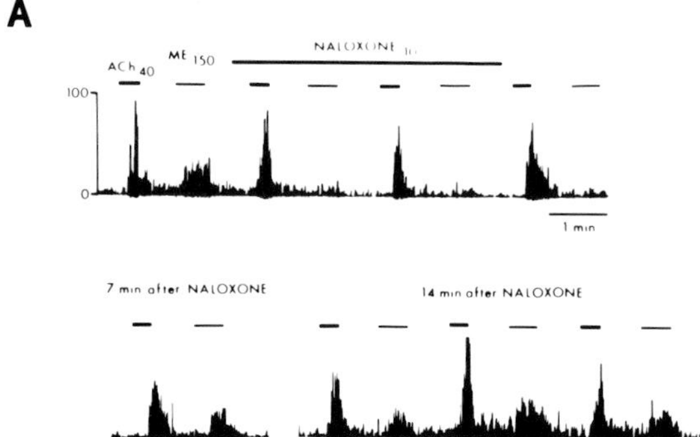
A
NALOXONE 10
ACh 40
ME 150
100
0
1 min
7 min after NALOXONE
14 min after NALOXONE

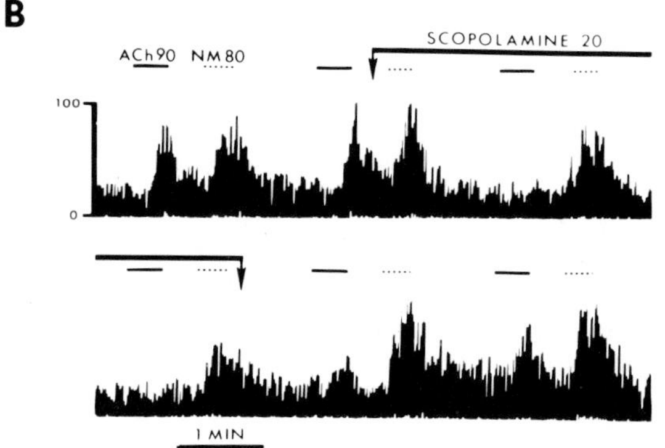
B
SCOPOLAMINE 20
ACh90 NM80
100
0
1 MIN

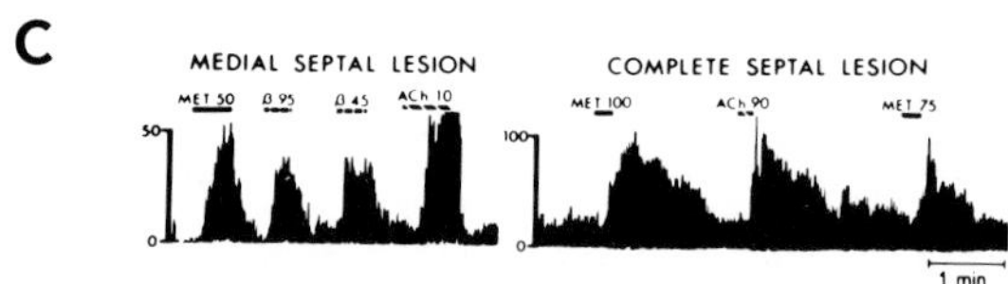
C
MEDIAL SEPTAL LESION
COMPLETE SEPTAL LESION
50
0
100
0
1 min

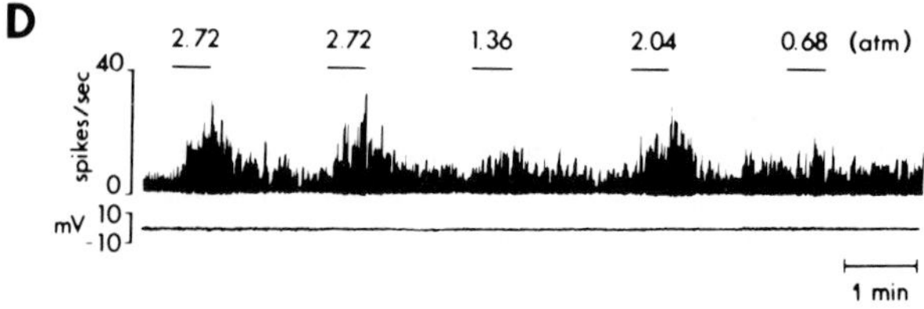
D
2.72
2.72
1.36
2.04
0.68 (atm)
40
spikes/sec
0
mV 10
-10
1 min

induced effects were not cholinergic dependent. In fact, further iontophoretic experiments in this region under conditions of altered γ-aminobutyric acid (GABA) transmission revealed that the hippocampal excitations were mediated indirectly through opioid-induced inhibition of inhibitory (presumably GABA) interneurons.[15]

In Vitro Techniques

The *in vitro* approach is best suited to study the actions of opiates at the cellular level. Such studies include quantitative identification of the receptor subtype and the ionic conductance and second messenger system(s) which mediate the change(s) in membrane properties produced by opioids on single neurons. The minimum requirements necessary to carry out these experiments are 2-fold: the first is the absolute necessity to be able to make long-term stable intracellular recordings; the second is the choice of a preparation that has a large proportion of cells that show a stable, nondesensitizing, and reproducible response to opioids over an extended period of time. A third consideration is the choice of ligands for such a study. Highly selective ligands for μ and δ subtype opioid receptors have been developed {Tyr-D-Ala^2-Gly-Phe-Glyol (DAGO) for the μ, and [D-Pen^2, D-Pen^5]enkephalin (DPDPE) for the δ}. The endogenous peptide dynorphin is selective for the κ-subtype opioid receptor. With the proper use of these compounds and the demonstration that naloxone in concentrations of 10–100 nM shifts the dose–response curves in a parallel manner, any opioid action can be ascribed to one or another of the opioid receptor subtypes (see Table I).

Two *in vitro* preparations in which the actions of opioids have been extensively studied are the locus coeruleus (LC) of rat brain and the

FIG. 3. Opioid-induced excitations of hippocampal pyramidal neurons under different test situations. (A) Excitatory action of Met-enkephalin (ME) and acetylcholine (ACh) in the hippocampus and the blockade of the action of ME by iontophoretic naloxone. Regularly spaced applications throughout eliminate potential barrel "warm-up" artifacts, and inclusion of a nonopiate excitant helps control against possible nonspecific depressant effects of naloxone. (B and C) Examination of the possibility that opioid effects in the hippocampus were mediated through ACh by using a cholinergic antagonist (B) or elimination of hippocampal cholinergic afferents by medial or complete septal nucleus lesions (C). NM and β refer to normorphine and β-endorphin, respectively. (D) Excitatory action of ME applied by micropressure. The concentration of ME in the pipet was 30 mM in normal saline at pH 7. Between ejections, diffusion of peptide from the pipet was minimized by a backing current of -5 nA. The line beneath the record shows that no imbalance of tip potential is observed during pressure applications. From French and Siggins.[3]

submucous plexus of guinea pig ileum.[21–27] The isolation of the tissue, mechanical stabilization, identification of the neuron and opioid-induced response for each are described below.

Preparation of Tissues

Brain Slices. Adult rats (150–250 g) are anesthetized with halothane and killed. Halothane is the anesthetic of choice since its effects are rapid in onset and short in duration. No differences in the properties of LC neurons were found between animals treated with halothane and those which were decapitated directly.

The scalp is split, the skull peeled away, and the brain rapidly removed rostral to caudal. The brain is placed on a glass plate and two transverse sections made at either end of the pons. The block of brainstem is mounted rostral surface up on a glass slide, fixed with cyanoacrylate adhesive, and immersed in oxygenated physiological saline at 4°. Slices (300 μm) are cut transversely using a vibratome (Lancer) and placed on a nylon mesh in a small chamber (500 μl) into the bottom of which enters physiological saline (1.5 ml/min) preheated to 37° and saturated with 95% O_2–5% CO_2. The perfusion solution is drained from the top of the chamber either by a gravity-driven siphon or actively drawn out. The slice is totally immersed in perfusion solution and immobilized by placing an electron microscope grid on top of its surface, held in place by small platinum weights. The time which elapses between killing the rat and setting the slice in the recording chamber is not as important as the care in handling the tissue during the brain removal and slicing procedure. Neurons in such slices retain unchanged membrane properties, synaptic potentials, and drug sensitivities for up to 15 hr following removal from the animal.[28] Other methods for the use of brain slices have been presented elsewhere.[29]

The LC is identifiable in the living slice as a slightly translucent area about 1 mm from the midline close to the ventral surface of the fourth

[21] S. Mihara and R. A. North, *Br. J. Pharmacol.* **88,** 315 (1986).

[22] R. A. North, J. T. Williams, A. Surprenant, and M. J. Christie, *Proc. Natl. Acad. Sci. U.S.A.* **84,** 5487 (1987).

[23] C. M. Pepper and G. Henderson, *Science* **209,** 394 (1980).

[24] J. T. Williams, T. M. Egan, and R. A. North, *Nature (London)* **299,** 74 (1982).

[25] J. T. Williams and R. A. North, *Mol. Pharmacol.* **26,** 489 (1984).

[26] R. A. North and J. T. Williams, *J. Physiol. (London)* **364,** 265 (1985).

[27] J. T. Williams, M. J. Christie, and R. A. North, *J. Pharmacol. Exp. Ther.* **243,** 397 (1987).

[28] J. T. Williams, R. A. North, S. A. Shefner, S. Nishi, and T. M. Egan, *Neuroscience* **13,** 137 (1984).

[29] R. Dingledine (ed.), "Brain Slices." Plenum, New York, 1984.

ventricle; in most studies the recording site is selected in a slice close to the caudal extent of the nucleus. Usually only a single slice from one animal contains enough LC for recording.

Submucous Plexus. The submucous plexus is obtained form the small intestine of guinea pigs (150–250 g). The ileum is opened and pinned in a petri dish with the mucosal surface up. Under a dissection microscope (×10–60) the mucosa is stripped away from segments of the small intestine. Beneath the mucosa is the submucous plexus sheath which is gently separated from the underlying circular smooth muscle. The neurons of the plexus, mesenteric arterioles, and a fine connective tissue network make up the sheath which is pinned flat in the bath. The ganglia of the plexus contain 4–30 neurons which can be visualized at 320× using Normarski optics. The ganglia are interconnected with nerve tracts on which stimulating microelectrodes can be placed. The tissue is superfused at 2 ml/min with heated (35–37°) Krebs solution (Table III).

Intracellular Recordings

Intracellular recordings are made after an incubation period of 30–60 min. Electrodes containing potassium chloride (2 *M*) having a resistance of 30–60 MΩ have been used successfully in the LC. Higher resistance electrodes (60–90 MΩ) are necessary for recording from submucous plexus neurons. These cells are equivalent in diameter to locus coeruleus neurons but are very flat and seem to be more sensitive to the type of electrode which is used. Most experiments are carried out with KCl-filled electrodes because their electrical properties are generally better than those filled with other salt solutions. In some experiments however, potassium methyl sulfate (2 *M*) is used in order to investigate chloride conductances which occur at resting as well as with synaptic potentials.

Voltage clamp experiments in the LC are performed with a single-electrode voltage clamp amplifier (Axoclamp II) and electrodes with resistances of 40 MΩ or less.[28] Fast currents involved in postsynaptic events and other voltage-dependent currents (e.g., I_{Ca} and $I_{K,A}$) can also be resolved. In experiments where fast currents are expected, recording electrodes are coated to within 100 μm of the tip with sylgard encapsulating resin under a dissection microscope. The coated electrode is then cured by passing the tip through a coil heated to about 100–150° for a few seconds. This insulating resin reduces the capacitance of the electrodes, allowing a higher switching frequency. With these electrodes, and a low level of superfusion solution, a switching frequency of 3–10 kHz can be used successfully. Sylgard-coated electrodes should be used routinely for voltage clamp experiments.

TABLE III
CONTENT OF SUPERFUSION SOLUTION FOR *in Vitro* PREPARATIONS

	Concentration (m*M*)[a]									
Brain slice[b]	NaCl	KCl	NaH_2PO_4	$MgCl_2$	$CaCl_2$	$NaHCO_3$	Glucose	Choline chloride	Sodium isothionate	EGTA
Normal	126	2.5	1.2	1.3	2.4	25	10	0	0	0
Low Na^+	0	2.5	1.2	1.3	2.4	25	10	126	0	0
Low Cl^-	0	2.5	1.2	1.3	2.4	25	10	0	126	0
Ca^{2+}-free	126	2.5	1.2	1.3	0	25	10	0	0	0.5

[a] Other ions such as tetraethylammonium, barium, cesium, rubidium, and potassium are simply added to the normal solution from a 1 *M* stock solution of their chloride salt.

[b] Submucous plexus has 5 m*M* KCl and is otherwise the same.

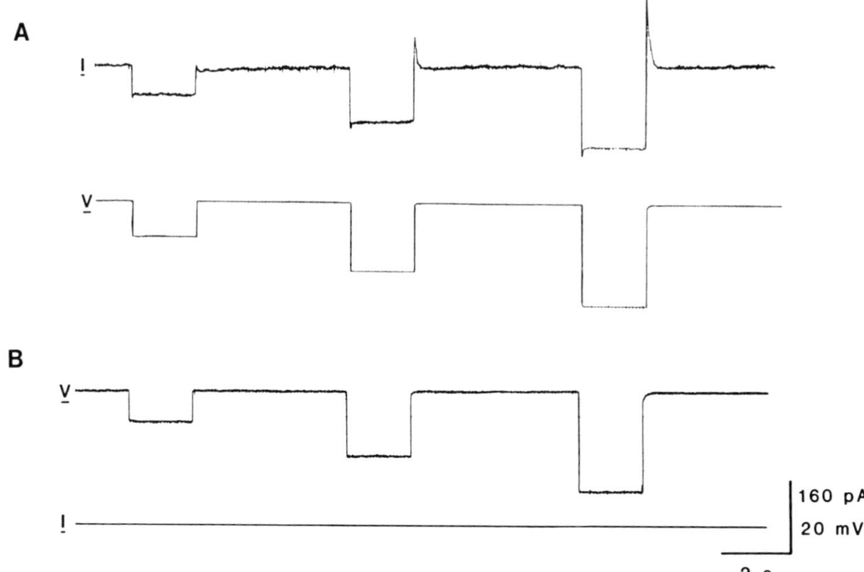

FIG. 4. Voltage clamp experiment with two electrodes in a single LC neuron. Electrode 1 (A) is connected to the single-electrode voltage clamp amplifier (SEVC); membrane current (*I*) and voltage (*V*) are recorded. Electrode 2 (B) is connected to another amplifier to measure membrane potential (*V*). Potential steps applied by the SEVC are identical to those measured by electrode 2.

The potential at the head stage of the amplifier is monitored at all times with a separate oscilloscope to verify that the switch clamp is set correctly. That is, the head stage potential is horizontal for at least 20% of the cycling period before the membrane potential is sampled and held for the next cycle. This simple procedure allows the investigator to determine at all times the reliability of the voltage clamp, eliminating the possible artifacts of clamping the electrode rather than the cell.[30] The only way to unambiguously demonstrate that the single electrode is correctly clamping the neuron under study is to use a second microelectrode to record the membrane voltage while under single-electrode voltage clamp (Fig. 4).

Voltage clamp experiments are used to measure currents in two conditions: at steady-state (time-independent currents) and during the approach to steady-state following voltage commands (time-dependent currents). Steady-state current–voltage plots can be constructed directly on

[30] T. G. Smith, H. Lecar, S. J. Redman, and P. W. Gage, *in* "Voltage and Patch Clamping with Microelectrodes." Williams & Wilkins, Baltimore, Maryland, 1985.

an X/Y plotter with the use of slow ramp potentials. The speed of the ramp is determined by constructing current–voltage plots at various speeds and comparing the currents flowing with those at the end of very long (several seconds) step commands to the same potential. A ramp speed which produced a current–voltage plot identical to that which follows prolonged step commands to many different potentials is used. Time-dependent currents are studied with step voltage commands to various membrane potentials. The currents which flow during the voltage command and those which continue to flow at the end of the voltage command (tail currents) can be plotted for analysis. The direction, amplitude, time course, and voltage range of time-dependent currents can then be analyzed. These data, along with the use of agents which block specific ion channels and changing the driving force of different ions by manipulating the extracellular ion content, can be used to determine which ions carry both steady-state and transient currents (see below). This type of analysis is used to describe the neuron under control conditions and in the presence of opioid agonists and antagonists.

An important consideration when using voltage clamp analysis is knowing to what extent the far-reaching extensions of the neuron are under voltage control, that is, the space clamp of the neuron under study. This is particularly true in the central nervous system where dendritic arborization can be extensive. In the locus coeruleus the minimum value for the reversal potential of the opioid current is about -105 mV, but in some cells reversals occur only at much more negative potentials. Equivalent circuit models in which opioids increase a potassium conductance having certain properties (inward rectification) have indicated that the failure to reverse the enkephalin current arises from inadequate voltage control of a part of the cell on which the enkephalin is acting (either dendrites or electrotonically coupled cells). The methods used to determine the extent of the dendritic load on neurons uses the shape of the membrane time constant.[31] A single exponential would be predicted by a sphere. Deviations from exponential can be used to determine the dendritic load on the cell. Since this method relies on the shape of the electrotonic potential to estimate the dendrite to soma conductance ratio (rho), it is subject to error if there are voltage-dependent conductance changes in the voltage range through which the membrane potential passes during the electrotonic potential. Therefore, knowledge of the passive and active membrane properties of the neurons is required.

[31] J. J. B. Jack, D. Noble, and R. W. Tsien, "Electric Current Flow in Excitable Cells." Oxford Univ. Press (Clarendon), London and New York, 1975.

Application of Drugs

Determination of concentrations of antagonists can be based on binding data from the literature. Generally, antagonist binding dissociation constants (K_D values) closely approximate the effective concentration for pharmacological antagonism. Concentrations of antagonists should be chosen that are known to be specific. Determination of the concentration range in which agonists act is more difficult since the binding literature generally comes from preparations where conditions are vastly different from those in the *in vitro* tissue slice. Agonist binding is much more sensitive to the ionic composition, substrates, and the state of second messenger systems than antagonist binding. The choice of agonist concentration is therefore based on effect, on the ability of selective antagonists to block the agonists effects, and on a comparison of the concentrations employed with those that are effective in other pharmacological systems. One advantage of the use of several different preparations is the possibility of testing the specificity of a given compound. For example, the action of DAGO and DPDPE were tested in the locus coeruleus and the submucous plexus of the guinea pig ileum.[22] In the locus coeruleus, DAGO hyperpolarized the neurons by about 20 mV with an EC_{50} of about 100 nM whereas DPDPE at 2 μM had no action. In the submucous plexus, however, DPDPE produced a 20-mV hyperpolarization with an EC_{50} of about 20 nM, whereas DAGO was ineffective (Fig. 5).

Actions of Opioids

LC neurons were identified electrophysiologically by the presence of spontaneous activity (0.5–2 Hz) which continued throughout the period of recording, an action potential which was about 80 mV in amplitude with a duration of 1.3 msec, and a distinct shoulder on the falling phase.[28] The properties of these cells were easily separable from neurons in the surrounding nuclei. Superfusion of opioid agonists decreased the firing rate and caused a membrane hyperpolarization in all cells which had the properties of typical locus coeruleus neurons. The hyperpolarization produced by opioids could be repeated for periods of several hours on a single cell without any significant rundown (tolerance) of the response. Additionally, in the continued presence of opioids the cells remained hyperpolarized until the opioid was washed out. This homogeneous population of neurons which respond repetitively to opioids is ideal for the study of the ionic mechanism and the classification of the receptor subtype which mediates the hyperpolarization.

In the submucous plexus, two groups of neurons are distinguished by

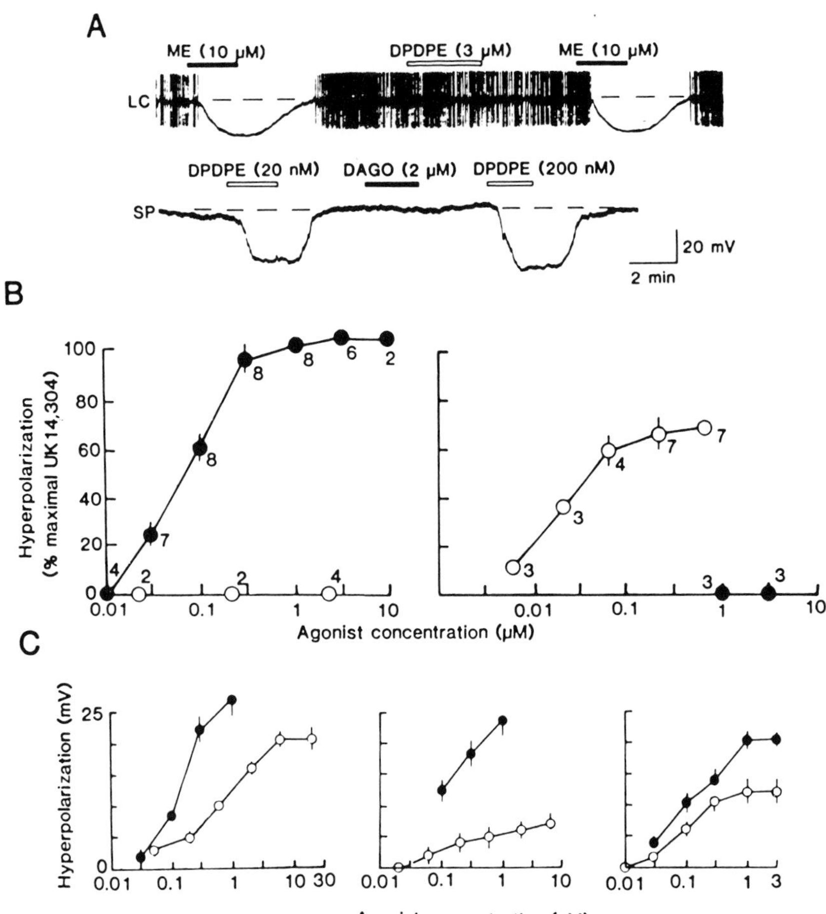

FIG. 5. Selectivity of opioid agonists on neurons of the locus coeruleus and submucous plexus. (A) Intracellular recordings of membrane potential in a locus coeruleus (LC) neuron (top) and a submucous plexus (SP) neuron (bottom). Note the spontaneous action potentials in the locus coeruleus. Full action potential amplitude is not shown. Superfusion of [Met⁵]enkephalin caused a membrane hyperpolarization in the locus coeruleus neurons while DPDPE had no effect. In the submucous plexus, however, DPDPE hyperpolarized the membrane and DAGO was ineffective. (B) Dose–response curves to DAGO (left) and DPDPE (right) summarized from many cells in locus coeruleus (LC) and submucous plexus (SMP). In order to compare the two preparations the dose–response curves are normalized to the percentage of the maximal response to a full α_2-adrenoceptor agonist [UK 14,304; 5-bromo-6-(2-imidazolin-2-ylamino)quinoxaline]. The amplitude of the hyperpolarization produced by a maximal concentration of UK 14,304 was identical in both the locus coeruleus and submucous plexus (20–25 mV). ●, LC; ○, SMP. (C) Dose–response curves in the locus coeruleus (LC) and submucous plexus (SMP) to other less selective opioid peptides, [Met⁵]enkephalin (left), metorphamide (middle), and β-endorphin (right). ●, LC; ○, SMP. From North *et al.*[22]

the presence or absence of excitatory postsynaptic potentials which are mediated by nicotinic cholinergic receptors. Neurons which have the nicotinic excitatory postsynaptic potential also have an inhibitory postsynaptic potential which is mediated by noradrenaline released from sympathetic nerve fibers and acting on α_2-adrenoceptors.[32,33] These neurons are also hyperpolarized by opioid agonists.[21] The hyperpolarizing response to opioids can be robust and reproducible but is often smaller than that caused by noradrenaline and tends to decrease in amplitude with repeated application of the same concentration of agonist.

Receptor Subtypes. It is now possible to distinguish the subtypes of opioid receptors using selective agonists and antagonists. The use of agonists alone (in the correct concentration range) can be used to separate the major receptor subtypes. Table I indicates the most selective agents at each of the receptor subtypes. The search for selective antagonists at the various opioid receptors continues. Naloxone shows virtually no discrimination between μ and δ subtypes, and it has about a 10-fold lower affinity for the κ-opioid receptors than μ receptors. Separation between μ and δ receptors is possible with ICI 174/864, which has about a 30-fold greater affinity for the δ subtype.[34] A somatostatin analog, CTP, antagonizes μ receptors but also retains some somatostatin activity, rendering it useless in preparations which are also sensitive to somatostatin.[35]

Antagonist Affinities. Classic pharmacological methods (e.g., Schild plot) can be used to measure antagonist affinity at the single-cell level (reviewed in detail in Ref. 36). The first requirement for such studies is an easily measurable response which is reproducible over an extended period of time. The opioid-induced hyperpolarization in the LC and submucous plexus meets this requirement. The choice of agonist used should not affect the analysis of the antagonist's affinity except for the possibility of interference from metabolism, diffusion, or uptake of the compound. We have found, for example, that in the locus coeruleus the dose response for [Met^5]enkephalin was shifted to the left 3- to 30-fold when peptidase inhibitors were included in the superfusion solution.[27] It is possible over the wide agonist concentration range which is used in a Schild analysis that the metabolism of enkephalin could account for deviation from a strictly competitive antagonism.

The most important dose response for a Schild analysis is that which is obtained in the absence of any antagonist since this serves as the refer-

[32] R. A. North and A. Surprenant, *J. Physiol. (London)* **358,** 17 (1985).

[33] S. Mihara, Y. Katayama, and S. Nishi, *Neuroscience* **16,** 1057 (1985).

[34] R. Cotton, M. G. Giles, J. S. Shaw, and D. Timms, *Eur. J. Pharmacol.* **97,** 331 (1984).

[35] J. T. Pelton, K. Gulya, V. Hruby, S. P. Duckles, and H. I. Yamamura, *Proc. Natl. Acad. Sci. U.S.A.* **82,** 236 (1985).

[36] T. P. Kenakin, *Pharmacol. Rev.* **36,** 165 (1984).

ence point for comparing subsequent dose–response curves in the presence of the blocker.[21] Ideally, at the end of the experiment the antagonist is washed out and the original dose response is reproduced. In the slice preparation, however, it is often difficult to wash out antagonists. It seems that the antagonist binds to both nonspecific as well as receptor sites during its application, and with washout it comes off nonspecific sites and binds to the receptor. Thus, the agonist response can remain blocked long after the washout. In the face of these diffusion barriers other control procedures can be used to monitor the condition of the cell. One is to check the response to an agonist which acts on another receptor before and after exposure to the antagonist. For example, check the action of noradrenaline before and after the testing with multiple concentrations of naloxone. If there is no change in sensitivity or magnitude of the noradrenaline effect, it may be safe to assume that the properties of the cell have not changed drastically during the course of the naloxone tests. We have used such an approach in the locus coeruleus where the responses mediated by α_2-adrenoceptors and μ-opioid receptors utilize the same potassium conductance and each agonist gives the same maximum response amplitude.[37]

Characterization of Ion Conductances. Ion conductances are characterized by the following: (1) the voltage range in which the channels are opened or closed, (2) the time course for channels opening and closing, (3) the type of channel blockers which prevent ion flux or ion selectivity of the channel, and (4) the agonist type which activates a receptor that is linked to the channel either directly or indirectly through a second messenger system. Ion conductances induced by opioids can affect a channel which is already operable in the membrane by changing its voltage dependence, activation/inactivation, or probabilities of channel opening or closing. One approach to determine which ion conductances are affected by opioids is to determine the conductances in the cell at rest and then to compare the properties of these channels in the absence and presence of opioids.

Calcium currents. Calcium currents are most commonly studied after treatment of the slice with tetrodotoxin (TTX,1 μM) and reducing potassium conductances with addition of tetraethylammonium (10 mM) or $BaCl_2$ (2 mM) to the superfusion solution or with intracellular CsCl (1–2 M CsCl in recording electrode). Voltage steps from different holding potentials activate subtypes of voltage-dependent calcium currents. These include low threshold (−70 mV) transient conductances, a higher threshold (−45 mV) inactivating conductance, and a high threshold nonin-

[37] M. J. Christie, J. T. Williams, and R. A. North, *Mol. Pharmacol.* **32,** 633 (1987).

activating conductance. One set of calcium channels is easily distinguishable by their sensitivity to dihydropyridines (nifedipine and Bay K 8644). In the LC, when potassium currents are partially blocked, a calcium current occurs with steps from −55 mV to less negative potentials, but it inactivates or is overcome within 10–25 msec by an outward potassium current. Thus, calcium currents can only be studied in isolation following complete suppression of potassium currents by substitution of barium for potassium or by substitution of tetraethylammonium chloride (TEA) for sodium chloride.

Barium currents. Barium currents are studied by substituting $BaCl_2$ (2.5 m*M*) for $CaCl_2$ in the superfusion solution. Voltage steps identical to those described above are used to activate barium currents.

Potassium currents. $I_{K(rest)}$ is commonly studied by measuring the slope of the steady-state current–voltage plot in a potential range that is linear near the resting potential (in the LC between −90 and −70 mV). In order to determine to what extent other ion conductances contribute to this resting conductance, current–voltage plots are made in controls and following superfusion with solutions which are calcium free and contain TTX (1 μM), cobalt (1 m*M*), TEA (10 m*M*), and CsCl (2 m*M*). This cocktail is aimed at blocking all active currents and leaves only the resting conductance.

$I_{K,A}$ is studied by stepping the membrane potential from −90 mV to less negative potentials. An outward current activates rapidly at the onset of the depolarization at potentials less negative then −65 mV and inactivates within 200 msec following the voltage step.[38,39] This potassium current is blocked more or less selectively by 4-aminopyridine (4-AP) and selectively by dendrotoxin.[40]

$I_{K,Ca}$ can also be studied by measuring the calcium-sensitive outward current that follows the repolarization of a voltage step from a holding potential near −60 mV to a potential of about −30 mV.[41,42] This outward current reverses at the potassium equilibrium potential and shifts to less negative values with the addition of potassium to the extracellular solution. The calcium-sensitive component can be determined by subtracting the outward current recorded after the addition of $CoCl_2$ (1–2 m*M*,

[38] M. A. Rogawski, *Trends NeuroSci.* (*Pers. Ed.*) **8,** 214 (1985).

[39] J. A. Conner and C. F. Stevens, *J. Physiol.* (*London*) **213,** 1 (1971).

[40] J. V. Halliwell, I. B. Othman, A. Pelchen-Matthews, and J. O. Dolly, *Proc. Natl. Acad. Sci. U.S.A.* **83,** 493 (1986).

[41] B. Hille, "Ionic Channels of Excitable Membranes." Sinauer, Sunderland, Massachusetts, 1984.

[42] K. Morita, R. A. North, and T. Tokimasa, *J. Physiol.* (*London*) **329,** 341 (1982).

apamin, scorpion toxin) to the superfusion solution from that recorded in control.

$I_{K,IR}$ is studied using steady-state current–voltage plots looking specifically in the potential range around the potassium equilibrium potential. An increase in conductance at potentials more negative than the potassium equilibrium potential suggests the presence of the inward rectifier whose activation can be blocked by addition of CsCl (1 mM), RbCl (1 mM), or $BaCl_2$ (1 mM) to the superfusion solution. $I_{K,IR}$ also increases with elevation of extracellular potassium levels. The time dependence of activation of the inward rectifier is often as rapid as the settling time of the step hyperpolarizing command potentials. The time course of activation of the inward rectifier varies in tissues from very fast (<10 msec) to about 100–200 msec.[43–46]

Sodium currents. Sodium currents are first studied by decreasing calcium currents [removal of calcium from the superfusion solution and addition of EGTA (500 μM)] or by addition of $MgCl_2$ (10 mM), $CoCl_2$ (2 mM), or $CdCl_2$ (200 μM) to the superfusion solution. Inward currents which remain are tested for their sensitivity to TTX and the changes that occur after substitution of choline chloride for NaCl in the superfusion solution which reduces the extracellular content of sodium ions from 152 to 26 mM. TTX-sensitive persistent sodium currents in LC[28] and cortical neurons[47] and a TTX-sensitive inactivating sodium current which is responsible for action potential generation have been described.

Opioid currents. The activation of μ- and δ-opioid receptors increased potassium conductance in both the locus coeruleus (μ) and the submucous plexus (δ). The opioid current was measured by subtraction of the steady-state current–voltage plot in the presence of opioids from that obtained in controls over the potential range from -130 to -50 mV. The opioid conductance increased with membrane hyperpolarization. In high potassium solutions the opioid conductance increased at a given voltage, and the slope of the G_{enk}–potential plot also increased. In this respect the opioid conductance was unlike other potassium conductances in the membrane. The opioid conductance was similar to the inward rectifier in that solutions containing Ba (10–100 μM), Cs (1–2 mM), and Rb (2 mM)

[43] D. Noble, *Trends NeuroSci. (Pers. Ed.)* **8,** 499 (1985).

[44] A. Constanti and M. L. Galvin, *J. Physiol. (London)* **335,** 153 (1983).

[45] S. Hagiwara, "Membrane Potential Dependent Ion Channels in Cell Membranes." Raven, New York, 1983.

[46] C. E. Stafstrom, P. C. Schwindt, M. C. Chubb, and W. E. Crill, *J. Neurophysiol.* **53,** 153 (1985).

[47] G. K. Aghajanian and Y. Y. Wang, *Brain Res.* **371,** 390 (1986).

all blocked the increase in conductance found with membrane hyperpolarization.

Second messengers. At the present time there are only indirect methods to investigate the second messenger involved in opioid action. At least two systems have been suggested; cyclic adenosine monophosphate[48] and a pertussis toxin-sensitive G protein.[47] Forskolin and the addition of high concentrations of cAMP (8-bromo-cAMP and dibutyryl-cAMP) produce a small depolarization in the locus coeruleus. The hyperpolarization or outward current produced by opioids was not affected by forskolin or the cAMP analogs. In the submucous plexus, forskolin in concentrations of 30–300 n*M* depolarizes the cells through a decrease in potassium conductance, but the opioid conductance is not affected[49] (A. Surprenant, personal communication).

The involvement of a pertussis toxin-sensitive G protein has been suggested in studies where animals were injected intracerebroventricularly with the toxin and 2–4 days later brain slices were made and the action of opioids tested. In pertussis toxin-treated animals, there was no effect by morphine or clonidine.[47] Such experiments are suggestive, but there are some problems with this approach. Acute administration of pertussis toxin alone for periods of 1–8 hr *in vitro* had no effect. Pertussin toxin-sensitive proteins are found on many different cells in the brain. Some cells may be indirectly affected by either interruption or excess synaptic transmission or through other neurohumoral actions. Also, the diffusion of toxin from intraventricular injections may be limited to structures close to the ventricles.

Another approach to determine whether a G protein is involved is to inject stable analogs of GTP (GTPγS, 10 m*M*) into the neurons. Using this method in the submucous plexus, opioids caused a membrane hyperpolarization which was only partially reversible, and with repeated application the membrane potential reached a level near the potassium equilibrium potential.[22] In the locus coeruleus cells recorded with GTPγS-filled electrodes, the component of the opioid-induced hyperpolarization was irreversible, but opioids caused a further reversible hyperpolarization. In voltage clamp experiments, the opioid current was found to be voltage independent. That is, the outward current observed at −60 mV was of the same amplitude as that measured at −120 mV. The interpretation of this observation was that diffusion of GTPγS from the electrode was probably affecting the area near the soma but there was also a distal component not affected.

[48] R. Andrade and G. K. Aghajanian, *J. Neurosci.* **5,** 2359 (1985).

[49] S. Mihara, R. A. North, and A. Surprenant, *J. Physiol. (London)* **390,** 335 (1987).

The use of compounds contained within intracellular recording electrodes is not satisfactory in several ways. Very often anything which is included within the electrode changes its electrical properties significantly. Since very often the single-electrode voltage clamp is also used in conjunction with these experiments difficulties frequently arise. With these electrodes it is almost impossible to get any kind of control response to a given drug. It often takes a period of 5–15 min following an impalement of a neuron before the membrane properties have stabilized to the point where a meaningful drug test can be performed. By that time there can be significant diffusion of the contents of the electrode solution into the cell. It may happen that the contents of the electrode change the properties of the cell so drastically that it is not possible to assess the quality of the impalement. This is especially true of a compound such as GTPγS which affects all G proteins and therefore many other cellular processes. When the concentration of the drug contained within the electrode is decreased, controls can be obtained, but the experiment is often equivocal because too little drug can be injected into the cell.

In spite of the problems which go along with both the pertussis toxin experiments and the use of compounds within the recording electrode the results strongly suggest that opioids act through a G protein to increase a potassium conductance. The opioid-activated potassium conductance is not the only one affected by these agents; the α_2-adrenoceptor and somatostatin-activated potassium conductances are similarly affected in both the locus coeruleus and the submucous plexus.[22] This observation suggests that these compounds act through a similar second messenger system which opens a common potassium channel. That a common potassium conductance is affected by α_2-adrenoceptors and opioid receptors has been demonstrated in both the locus coeruleus and the submucous plexus.[21,26] When opioids were applied in a concentration which caused a maximum increase in conductance, superfusion of α_2-adrenoceptor agonists produced no further increase in conductance.[21,26] Such occlusion studies suggest that the potassium conductance increased by opioids is identical to that increased by α_2-adrenoceptor agonists. The voltage dependence of the opioid and α_2-adrenoceptor-mediated currents are also identical. In fact, the opioid and α_2-adrenoceptor currents in the locus coeruleus and submucous plexus are also identical in their voltage dependence.[22]

In summary, the similarities between both the μ- and δ-opioid receptor-mediated increase in potassium conductance is remarkable. What is even more interesting is that this potassium conductance is identical to that which mediates other receptor-mediated responses in the locus coeruleus and the submucous plexus as well as a number of other tissues.

Such tissues include the action of acetylcholine on M-muscarinic receptors in the heart.[50,51] This is a site in which the combination of whole cell recording and inside-out and outside-out patch recordings have shown without question the G protein link.[52] Experiments similar to those done in the heart on the muscarinic receptor are the next step toward the understanding at the cellular level of the acute and chronic actions of opioids.

Acknowledgments

We would like to thank Dr. William Colmers for critical evaluation of the manuscript. EDF and JTW are currently supported by U.S. Public Health Administration Grants DA03876 and DA04523.

[50] M. Soejima and A. Noma, *Pfluegers Arch.* **400,** 424 (1984).
[51] P. J. Pfaffinger, J. M. Martin, D. D. Hunter, N. M. Nathanson, and B. Hille, *Nature (London)* **317,** 536 (1985).
[52] A. Yatani, J. Codina, A. M. Brown, and L. Birnbaumer, *Science* **235,** 207 (1987).

[8] Dose–Response Testing of Peptides by Hippocampal Brain Slice Recording

By M. IAN PHILLIPS and REINHARD A. PALOVCIK

Introduction

There has been tremendous progress in peptide research with the discovery of numerous peptides in the brain which were once thought to belong exclusively to peripheral organs. Immunocytochemistry, HPLC, chromatography, radioimmunoassay, and, more recently, molecular probes of cDNA and mRNA have been used to establish the existence and localization of some 50 neuropeptides. Many studies have demonstrated that receptors for these peptides also exist in the brain. Membrane homogenization procedures with radioisotope-labeled synthetic peptide agonists and antagonists have defined the K_D and V_{max} of peptide receptors in the brain. The recent autoradiography methods applied to brain slices and amenable to Scatchard analysis have been used to show the distribution of binding sites in the brain. These methodologies, however, do not show that peptides in their binding sites produce any biological effect.

To demonstrate and study the biological action of peptides in brain tissue under controlled conditions, several techniques are available. Biochemical techniques are described elsewhere in this volume. The electrophysiological techniques available to measure membrane potential changes include microiontophoresis, pressure ejection, and perfusion.

Microiontophoresis is very useful in ejecting peptides and producing responses which can be measured by extracellular recordings. There are problems, however, because the amount of peptide ejected is difficult to quantify and data are expressed as nanoamperes of current used to eject the peptide. Although the transport number can be defined by ejecting radioisotope-labeled peptide and quantifying the amount of radioactivity released, the cumbersome procedure should be applied each time recordings are made. This is because the resistance at the pipet tips vary with each penetration into tissue.

A puff of compressed air to pressure eject peptides from the pipettes has been used as an alternative to microiontophoresis. It has the advantage that a known volume and concentration of peptide can be released. Also, no electrophysiological controls are necessary as with microiontophoresis such as switching from anodal to cathodal current ejection. However, pressure ejection can cause mechanical effects on tissue. Mechanical movement or vibration caused by the sudden volume release can be amplified through the recording electrode and mistaken for a response. Such an artifact cannot always be controlled by the same amount of pressure in a pipet filled with saline or cerebrospinal fluid (CSF) because the tips are not in the absolute identical location, and, as a result, a different mechanical artifact may occur or be absent. Either effect could lead to misinterpretation of the results.

In this chapter, we present an alternative approach which no doubt has limitations but offers two important advantages over those methods. One is a constant perfusion of known peptide concentration and the second is the advantage of intracellular recording during perfusion. The hippocampal brain slice is used for this procedure because it is stable, spontaneously active, and sensitive to a variety of peptides.

Hippocampal Slices

Hippocampal slices have been used for recording for several years. To test different doses of peptides, however, there is a need for great stability while perfusing slices, and this means addition of known concentrations during intracellular recording. Maintaining slices under constant perfusion for long periods while allowing for the unobtrusive addition and removal of several drugs during maintained intracellular recording is the problem to be solved. Numerous chambers for successfully maintaining central nervous system slices *in vitro* have been described in the literature.[1–13] Some of these designs are available commercially. To carry out dose–response curves, however, the design described here has several advantages over the previously described apparatus or commercial models. (1) It allows us to keep a number of brain slices alive while adding and removing any of a large number of substances at fixed concentrations over a period of up to 30 hr, (2) the design provides excellent stability which makes possible the long duration and stable intracellular recordings, (3) the design is simple, (4) it gives consistency and accuracy of results, and (5) the modular construction enables the quick replacement of components should any malfunction occur. The system described here has been designed to maximize viability of slices and stability of intracellular recordings for recording the longer duration actions of neuropeptides as well as short latency, short duration effects of fast-acting peptides and neurotransmitter substances.

We have evaluated this design with regard to reliability and validity of the results obtained. Stable recordings and repeatable dose–response curves have been attained for angiotensin II, insulin, and GnRH (gonadotropin-releasing hormone). Desensitization appears to be minimized because drugs "wash by" the slices rather than being in continuous contact with them, and the design also permits the administration of very low doses with reproducible results. The dose–response curves for angiotensin II, insulin, GnRH, and carbachol based on the recording correlate with receptor-binding studies of these substances in the hippocampus and

[1] R. Dingledine (ed.), "Brain Slices," p. 381. Plenum, New York, 1984.

[2] C. F. Dore and C. D. Richards, *J. Physiol. (London)* **239,** 83 (1974).

[3] R. Dingledine, J. Dodd, and J. S. Kelly, *J. Physiol. (London)* **269,** 13 (1977).

[4] C. D. Richards and W. J. B. Tegg, *Br. J. Pharmacol.* **59,** 526P (1977).

[5] W. F. White, J. V. Nadler, and C. W. Cotman, *Brain Res.* **152,** 591 (1978).

[6] H. L. Haas, B. Schaerer, and M. Vosmansky, *J. Neurosci. Methods* **1,** 323 (1979).

[7] T. J. Teyler, *Brain Res. Bull.* **5,** 391 (1980).

[8] D. A. Brown and J. V. Halliwell, *in* "Electrophysiology of Isolated Mammalian CNS Preparations" (G. A. Kerkut and H. V. Wheal, eds.), p. 285. Academic Press, New York, 1981.

[9] I. A. Langmoen and P. Andersen, *in* "Electrophysiology of Isolated Mammalian CNS Preparations" (G. A. Kerkut and H. V. Wheal, eds.), p. 51. Academic Press, New York, 1981.

[10] K. Lee, M. Oliver, F. Schottler, and G. Lynch, *in* "Electrophysiology of Isolated Mammalian CNS Preparations" (G. A. Kerkut and H. V. Wheal, eds.), p. 189. Academic Press, New York, 1981.

[11] R. A. Nicoll and B. E. Alger, *J. Neurosci. Methods* **4,** 153 (1981).

[12] C. D. Richards, *in* "Electrophysiology of Isolated Mammalian CNS Preparations" (G. A. Kerkut and H. V. Wheal, eds.), p. 107. Academic Press, New York, 1981.

[13] J. F. Koerner and C. W. Cotman, *J. Neurosci. Methods* **7,** 243 (1983).

in some cases, produce responses of lower doses than the binding kinetics would predict.

Preparation of Tissue

We have used male Sprague–Dawley rats (100–250 g) which are kept in a constant temperature, 12 hr light–12 hr dark rat room. The hippocampal slices are prepared after rapid dissection of the brain at room temperature (22–24°). The cortex is peeled away from the hippocampus, which can be clearly seen, and the dorsal hippocampus removed and placed on a tissue chopper. Slices are cut at 400 μm thickness with the center 3 mm of the hippocampus, perpendicular to its length (Fig. 1A). Slices are transferred to a nylon mesh grid. This can be done either with a camel hair brush or by drawing the slices through a wide-mouth pipet. Slices are continuously perfused at 30 ml/hr from beneath the grid. The temperature is maintained at 34°, and humidified 95% O_2, 5% CO_2 is passed over the slices. The slices remain viable for 30–40 hr. The time from the dissection to the chamber is typically 3.5–4.0 min. The slices are allowed to recover for at least 2 hr under continuous perfusion before recordings are made. The slices are laid out on the nylon mesh so that the CA1 region is visible under the binocular microscope (Fig. 1B).

Intracellular electrodes are filled with 2 *M* potassium citrate and advanced into the slice until a change in membrane potential is seen; then the electrode is maintained in position for stable membrane potential and intracellular recordings. Preparations with long-term (1 hr or more) stable baselines are used for collecting data.

Perfusion Medium

Slices are continuously perfused with the perfusion medium and collected into the perfusion medium during transfer from the tissue chopper to the nylon grids. The perfusion medium used is artificial CSF (ACSF) containing 124 m*M* NaCl, 5 m*M* KCl, 1.25 m*M* KH_2PO_4, 1.3 m*M* $MgSO_4$, 26 m*M* $NaHCO_3$, 10 m*M* glucose, and 2.5 m*M* $CaCl_2$. An alternative to ACSF is cell culture medium. Dulbecco's modified Eagle's medium (DME) contains a number of amino acids and vitamins in addition to inorganic salts and glucose. It is obtainable commercially from Sigma Chemical Co. However, ACSF has been effective in our hands. The chamber allows for continuous perfusion with the addition of substances at fixed concentration during recording. Slices are 1 mm above the flowing perfusion medium. During testing the level of medium rises to touch the nylon grid. Surface tension makes a meniscus between the slice and grid and the perfusion medium. This method minimizes the effects of slight

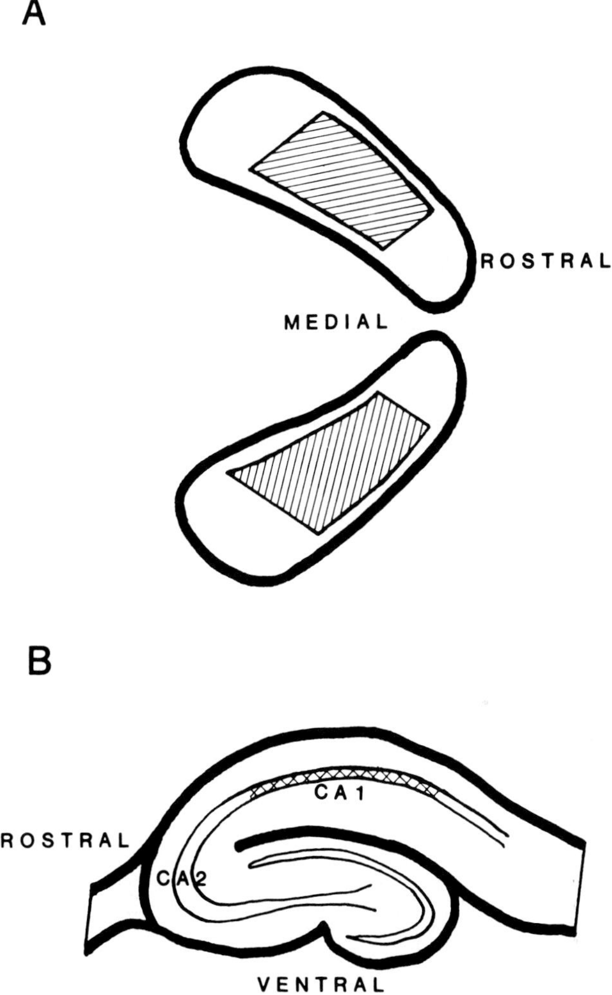

FIG. 1. The hippocampus. (A) Diagram of appearance of hippocampi after they have been dissected out of the brain (dorsal aspect). The cross-hatched area indicates the CA1 area from which the sections are cut. (B) A hippocampal slice (cross section). Slices are cut 400 μm thick.

changes in the level of medium and prevents mechanical artifacts when substances are injected. An adjustable overflow drainage system ensures a constant level of perfusion. Substances are injected in the perfusion medium as vehicle from one of 10–18 hypodermic syringes (1 ml) connected by PE-10 tubing as injection lines which enter the chamber ports.

To deliver drugs, the stems of the syringes are moved by turns on a drive bolt which controls the rate and steadiness of the injection.

The Chambers

The design of the external chamber is similar to that used by other workers. There are separate lower and upper chambers (see Fig. 2). The lower chamber contains distilled water and a heater to humidify the bubbled 95% O_2, 5% CO_2 and preheat the perfusion media. The upper chamber contains the humidified O_2–CO_2 atmosphere, ground wires, injectors, line ports, and a specially designed inner chamber that holds the slices.

Lower Chamber. The lower chamber is 17 cm in diameter and 7 cm high. The purpose of the lower chamber is to humidify an O_2–CO_2 mixture and to preheat the perfusion media to 37°. The lower chamber consists of a large Plexiglas tube section with top and bottom covers. Ports are drilled into the sides and cover to accommodate 1/8 in. × 10–32 tubing fittings, through which perfusion media passes to the inner chamber. All inflow and drainage of perfusion media, and O_2–CO_2, are channeled through

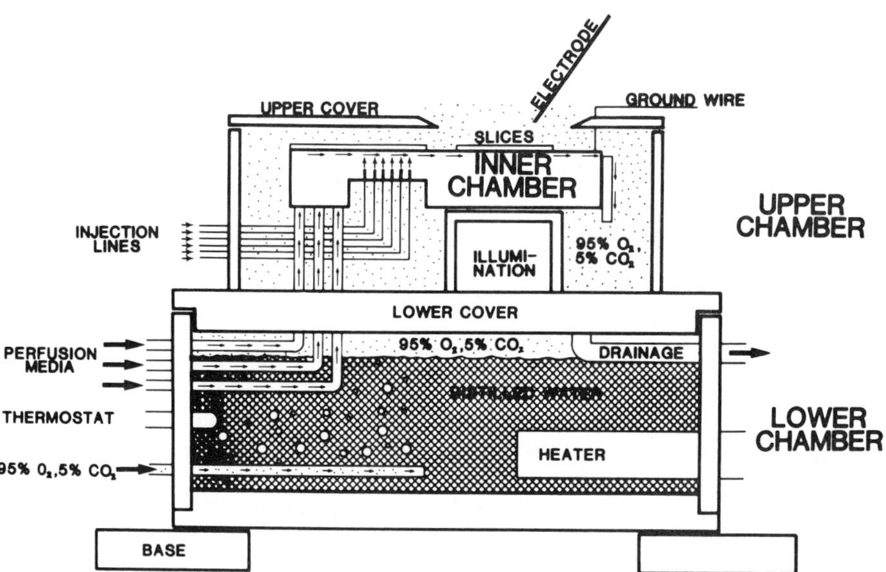

FIG. 2. Schematic of the brain chamber. The lower chamber supplies heat and humidified O_2/CO_2 to the upper chamber. Perfusion media are heated by circulating in the lower chamber. Slices sit on a nylon grid atop the inner chamber, which is interchangeable for different experiments. Peptides to be injected are circulated through the upper chamber and then flow into the perfusion medium just before it reaches the slices on the inner chamber. The base of the lower chamber is firmly fastened to the recording platform.

such fittings. The O_2–CO_2 is bubbled through the lower chamber distilled water to humidify it. The gases then pass to the upper chamber and over the slices.

Upper Chamber. The upper chamber functions to house the humidified, heated O_2–CO_2 atmosphere which flows over the slices. The upper chamber also holds the inner chamber (Fig. 3A), through which the perfusion medium passes and on which the slices rest. The upper chamber is made from a circular Plexiglas tube (14 cm diameter), which is glued to the lower chamber cover. A support for the inner chamber is also glued to the lower chamber cover. Different configurations of inner chambers can then be attached to this support to accommodate different experiments. The inner chamber support may also contain a source of illumination (fiber optics), or the preparation may be illuminated from above or both. Injection tubes are channeled into the upper chamber through the wall or cover and are then connected to the inner chamber (see Fig. 2).

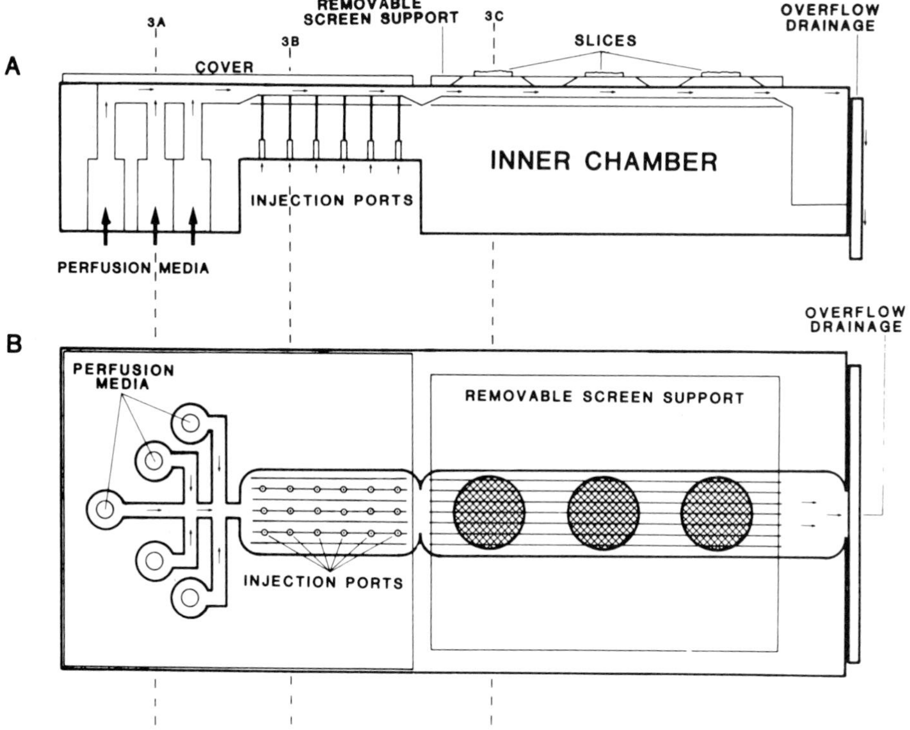

FIG. 3. Side view (A) and top view (B) of the inner chamber showing narrow channels milled into the top of the inner chamber. An adjustable overflow drainage plate regulates the height of the meniscus in the inner chamber.

Inner Chamber. The inner chamber design (see Fig. 3) is critical for maintaining the slices in a stable condition for addition and removal of drugs and changing perfusion media. It consists of a small block of Plexiglas (5 cm × 10 cm × 12 mm) onto whose upper surface channels and ports are milled in the appropriate configuration. The actual layout of these milled channels may vary depending on experimental requirements. An important feature is that the perfusion medium meniscus (surface) at the outflow be at least 1 mm below the meniscus at the slices. With the outflow meniscus 1 mm below the slices, volume can vary, within limits, without changing the vertical position of the slices or disrupting the tip of the electrode. Another reason for having the slices sit 1 mm above the perfusion medium meniscus is that the flow may be increased up to 50 ml/hr without affecting slice viability. This allows for a quicker "wash by" of the injected drugs. A screen-covered Plexiglas insert is fastened to the top of this inner chamber. The thickness of this cover is 1 mm, or, if 1 mm Plexiglas is not readily available, the shape of the remaining cover thickness may be milled into the top of the inner chamber to the appropriate depth. Additional channels are then milled onto the surface of the inner chamber. This milling should be performed carefully to cut smoothly and

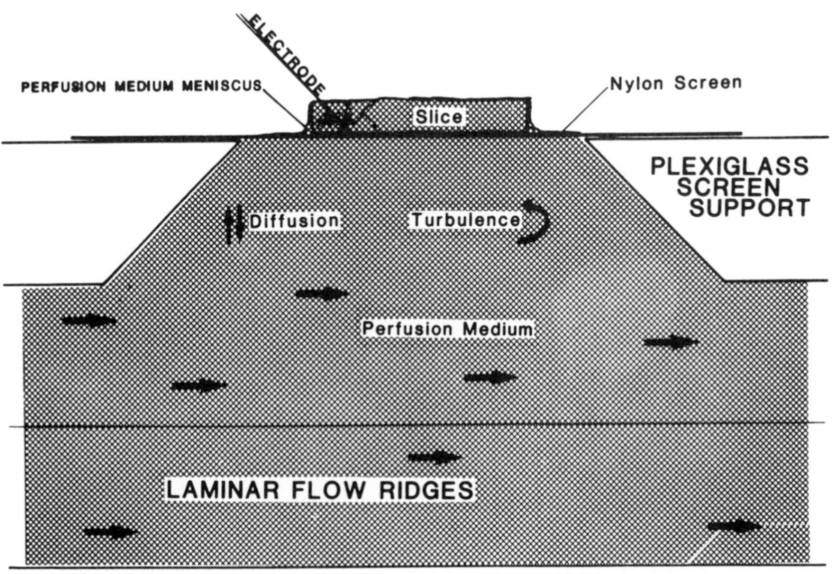

FIG. 4. The slice support and perfusion system. The nylon screen is stretched tightly and glued to the Plexiglas screen support. Flowing perfusion medium is indicated by arrows. Peptides reach the slices primarily by diffusion but also by some flow turbulence.

not to heat Plexiglas excessively. Heat can produce toxic compounds that may subsequently poison the slices.

The channels are aligned in a three-part configuration. This consists of a part where the perfusion medium enters the inner chamber, an area where drugs are injected into the perfusion medium, and a section where the perfusion medium contacts the slices (Fig. 3). Since the bulk of the perfusion medium flows past the slices (1 mm beneath them), the slices are subjected to a minimum of disruption due to flow turbulence, and drugs reach the slices primarily by diffusion (Fig. 4). The concentration of peptide arriving at the slice depends on rate of perfusion and duration of peptide injection. We have determined the concentration of peptides by sampling in the vicinity of the slice and measuring by radioimmunoassay. The concentration at the slice is the concentration predicted from that which is injected with a peak at 1 min postinjection which represents the time taken to arrive at the slice from the injector (Fig. 5). In our experiments, volume and duration of injection and rate of perfusion are held constant and concentration varied by altering concentration of the injectate. We typically inject 100 μl of drug solution into a chamber whose total volume is 1.0 ml. This produces a 10-fold dilution of the injectate when it reaches the slices.

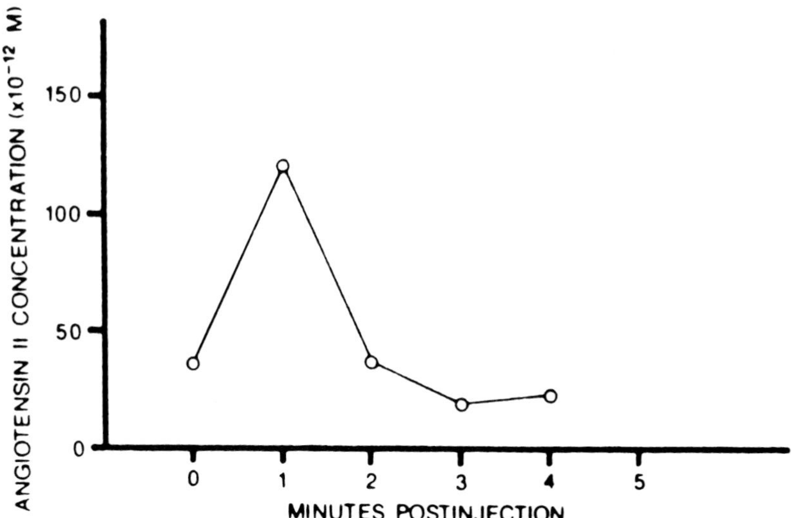

FIG. 5. Determination of angiotensin II concentration in the vicinity of the slice using radioimmunoassay. Angiotensin II in ACSF was injected into the flowing perfusion medium to achieve a bath concentration of 10^{-10} *M*. At 1-min intervals, 10 μl was sampled in the vicinity of the slice. Samples were diluted to 100 μl, and the angiotensin II concentration was determined by radioimmunoassay. The results show peak concentration and clearance time.

Several other inner chamber designs can be used based on the above principle. For example, a spare can be used to maintain numerous slices without exposing all slices to drugs tested. Another chamber can be used for rapid removal of slices for freezing in liquid nitrogen during any part of the electrophysiological response. Later, biochemical analyses can be performed on the tissue. One feature of the chamber design is that since peptides reach the slices about 1.5 min after the injection, any hydraulic or vibrational artifact will be clearly seen on the record at the time of injection. The response to the drug can, therefore, be evaluated independent of artifacts, if present, since they are sufficiently separated in time.

Reliability and Validity. The present slice chamber has been evaluated for reliability and validity of the results obtained. The apparatus is considered reliable when we can obtain a reproducible response to a fixed dose of peptide. We have demonstrated repeatable patterns of intracellular neuronal firing rates to a fixed dosage of carbachol, angiotensin, insulin, and GnRH. Insulin (10^{-9} *M*) is capable of inhibiting most hippocampal pyramidal neurons. Angiotensin is very potent in exciting pyramidal neurons. Doses as low as 10^{-12} *M* have been found to repeatedly excite some neurons, whereas ACSF alone and doses of carbachol below 10^{-8} *M* produce no effects.

There are other factors which interfere with the repeatability of results which are not related to the apparatus. Some of these are as follows: (1) habituation, desensitization, or tachyphylaxis; (2) potentiation; (3) long-term alterations; and (4) dropout of cells due to the progressive mortality of the preparation. These sorts of effects are generally more prevalent with repeated high doses. It is generally possible, however, to find a dose low enough that still produces a repeatable effect. Theoretically, one may postulate that the tissue is never precisely the same once a drug has been applied. It is, therefore, important to do complete dose–response studies to see where the lower threshold lies and at what dose one starts to get nonrepeatable results. The present design facilitates the acquisition of nearly complete dose–response curves.

Peptide Action

Angiotensins

To establish the validity of the method for peptide injection, two tests were carried out. The first aspect of chamber functioning to be tested was whether concentration delivered to the slices was actually that projected from the design. Phenol red was injected through the injection ports and its progress visually observed through the chamber. The dye flowed

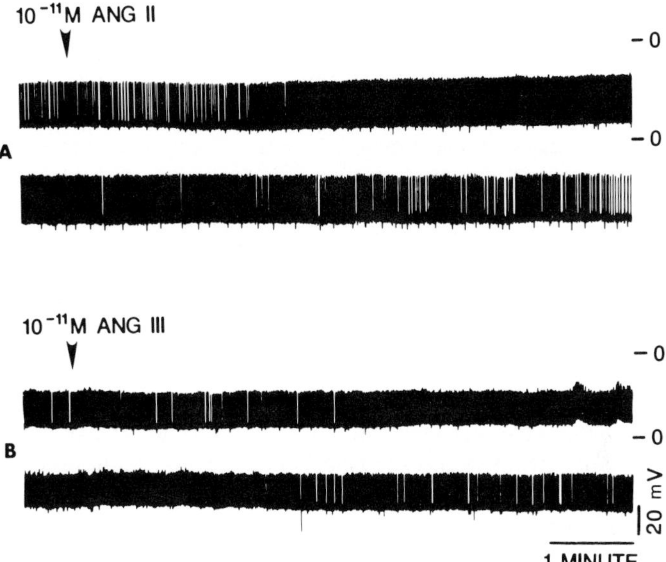

FIG. 6. Electrophysiological responses to (A) Ang II (10^{-11} M) and (B) Ang III (10^{-11} M) of a cell recorded in the CA1 pyramidal layer of hippocampal slice *in situ*.

smoothly through the apparatus to the slices once it entered the perfusion medium.

In order to measure the concentration of peptide delivered to the slice, we injected angiotensin II (Ang II) through the injector into the perfusion medium and collected 10-μl samples in the vicinity of the slice at fixed times after injection. These samples were diluted to 100 μl and the amount of Ang II measured by radioimmunoassay.[14] The results are depicted in Fig. 5. The concentrations reported elsewhere in this chapter reflect the peak of the curve, or maximum concentration, as in Fig. 5, to which the bottom surfaces of the slices are exposed. Responses to Ang II could be obtained with lower doses (10^{-12} M) than previously reported (Fig. 6). Responses were quantified by sampling the firing rate before and 10 sec after the drugs reached the slices. The firing rate for these samples was averaged over a 1-min duration of the record. In most cases, spacing of injections depended on cessation of previously elicited activity. The transit time, for drugs to reach the slices postinjection, was also determined by injecting a 2×10^{-6} M carbachol solution. Subsequent latencies were determined by subtracting injection to carbachol effect time from injection to peptide effect time, effectively removing transit time.

[14] M. I. Phillips and B. K. Stenstrom, *Circ. Res.* **56,** 212 (1985).

Intracellular recordings were obtained from 70 pyramidal neurons in the CA1 region of the hippocampus. Of these, 58 cells were tested with Ang II, of which 58.3% showed increases in firing, 6.7% showed decreases, and 35% had no response. Ang III was tested on 31 cells, of which 61.5% increased firing rate, 7.7% decreased, and 30.8% exhibited no change. No significant differences were found between proportions of cells responding to Ang II versus Ang III [$\chi^2(df = 2) = 0.422, p > 0.25$].

There were 16 cells in which Ang II and Ang III were tested on the same cell. Whether a cell responded to Ang II correlated highly with response of the same cell to Ang III for a change in firing rate ($r = 0.98$, $n = 16$, $t = 18.43$, $p < 0.05$) and polarization ($r = 0.59$, $n = 8$, $t = 1.79$, $p < 0.10$). The correlation between the effects of Ang II and Ang III on polarization was not significant ($r = 0.125$, $n = 16$, $t = 0.47$, $p > 0.10$), showing that Ang II and Ang III are able to bind to the same receptors but are not equally effective in eliciting a physiological response. Percent increase in firing rate was correlated with magnitude of membrane potential change for Ang III ($r = 0.56$, $n = 11$, $t = 2.03$, $p < 0.05$) but not for Ang II ($r = 0.08$, $n = 15$, $t = 0.29$, $p > 0.10$).

The effects of Ang II on hippocampal slice neurons obtained in this intracellular study were largely in agreement with results we obtained with extracellular recordings.[15] Ang II excited most cells over a broad dose range (10^{-12} to 10^{-6} *M*) (Fig. 7). Very few cells were also found to be inhibited by Ang II. Ang III exhibited a response pattern similar to that found for Ang II. The latencies and durations of effects to Ang III were similar to Ang II, but both could be distinguished from those of GnRH which had two distinct types of excitatory effect.[16]

Carbachol

In order to determine if the substances injected into the perfusion medium would alter neuronal activity in a predictable manner, we injected carbachol in a range of concentrations from 10^{-10} to 10^{-4} *M*. With concentrations of 10^{-7} *M* and below, there was no detectable effect of carbachol on the extracellular firing rate. At 10^{-6} *M*, there appeared to be a slight increase in firing rate, which increased progressively with increasing doses. At a 2-fold higher dose, 2×10^{-6} *M*, we found a burst of excitatory activity consisting of an increase in the rate of action potentials. At 10^{-6} *M* there was an initial large burst followed by an inhibitory pause with subsequent excitation lasting for several minutes. This pattern is repeatable with correspondingly greater intensity at higher doses.

[15] R. A. Palovcik and M. I. Phillips, *Brain Res.* **323,** 345 (1984).

[16] R. A. Palovcik and M. I. Phillips, *Neuroendocrinology* **44,** 137 (1986).

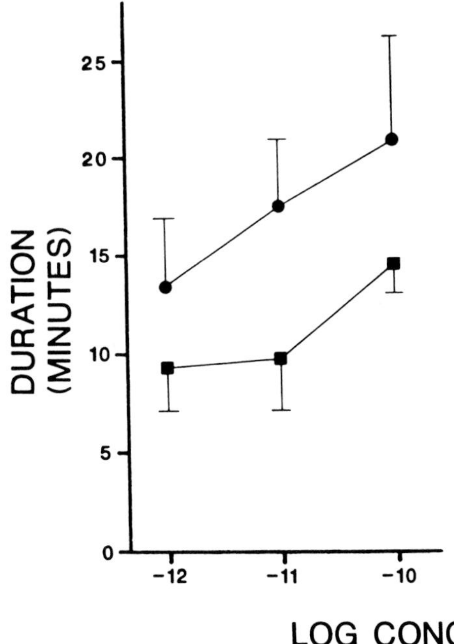

FIG. 7. Dose–response curve for duration of effect of increasing doses of Ang II in hippocampal slices. (●) Change in firing rate; (■) change in membrane potential.

A similar pattern of responding was produced by carbachol in this dose range during intracellular recording from other neurons. The amount of depolarization produced by carbachol was directly proportional to the dose. We can, therefore, test the effects of known doses of drugs on neurons using our slice chamber. The doses used above have proven effective within a physiological range for cholinergic agonists, which is 10^{-6} to 10^{-4} *M*. The unusually low dose effects of angiotensin are unlikely to be artifacts because they could not be achieved with other peptides or carbachol under very similar circumstances.

Insulin

Various insulins including the deoctapeptide or proinsulin were injected into the perfusion medium and took 1 min to reach the slices. Extracellular recordings were obtained with 10 μm tip electrodes filled with perfusion medium and lowered into the CA1 pyramidal cell layer. Single pyramidal cell spikes were selected by triggering on the largest spike (3–5 mV) that followed the typical hippocampal pyramidal bursting pattern. When a stable pattern of spontaneous firing was established, insulin was injected into the chamber. The predominant effect of insulin was to decrease the frequency of spontaneous spikes over a wide dose range. The effect of insulin was maintained for up to 30 min depending on the dose. Significant inhibition was seen at 10^{-9} *M* and at higher doses. The inhibitory effect was dose dependent, and the IC_{50} was in the nanomolar range, indicating that at half-maximum inhibitory response, insulin had a similar affinity constant to ^{125}I-labeled insulin binding to its receptors.[17] There was a much smaller inhibitory response to proinsulin (10%) and no inhibition to the deoctapeptide. Again, this fits with the binding data where controls are proportionately less effective in binding to insulin receptors.

The results are in keeping with the electrophysiological response being the biological response following insulin receptor binding. Of course, it is not possible to specify that the actual binding sites are on the same cells that gave the electrophysiological response, but the presence of binding in the hippocampus for insulin and the demonstration of electrophysiological responses to insulin in the hippocampus suggest a connection. Further proof would require detection of binding sites, possibly with autoradiography, of cells that are filled with a detectable dye after recording. With dye-filling techniques and electron microscopy, this should be possible.

GnRH

The majority of cells tested with GnRH (n = 40) responded with an increase in firing (44%). The rest were either inhibited (14%), showed no effect (40%), or were excited and inhibited (2%). What was interesting about the response to GnRH was the response all the time. With GnRH, a short latency, short duration excitation was produced by low doses (10^{-11} *M*), but at a higher dose there were longer latency, long-lasting excitations (latency = 314.6 ± 12.6, duration = 45.4 ± 4.3 min) (Fig. 8). When low Ca^{2+} (0.5 m*M*), high Mg^{2+} (10 m*M*) perfusion medium was given, the responses to GnRH were still produced. Therefore, the responses observed were not the result of Ca^{2+} as a second messenger or synaptic release of another transmitter. The results indicate that GnRH has a direct action on the cells recorded and the pattern of responding with a long duration, long latency is quite distinctly different from any pattern we observed with angiotensin or insulin. The presence of long duration effects with a single application of a peptide after the peptide has been cleared from the perfusing medium is also different from any result we have obtained with neurotransmitters. It is these profound electrophysio-

[17] R. A. Palovcik, M. I. Phillips, M. S. Kappy, and M. K. Raizada, *Brain Res.* **309,** 187 (1984).

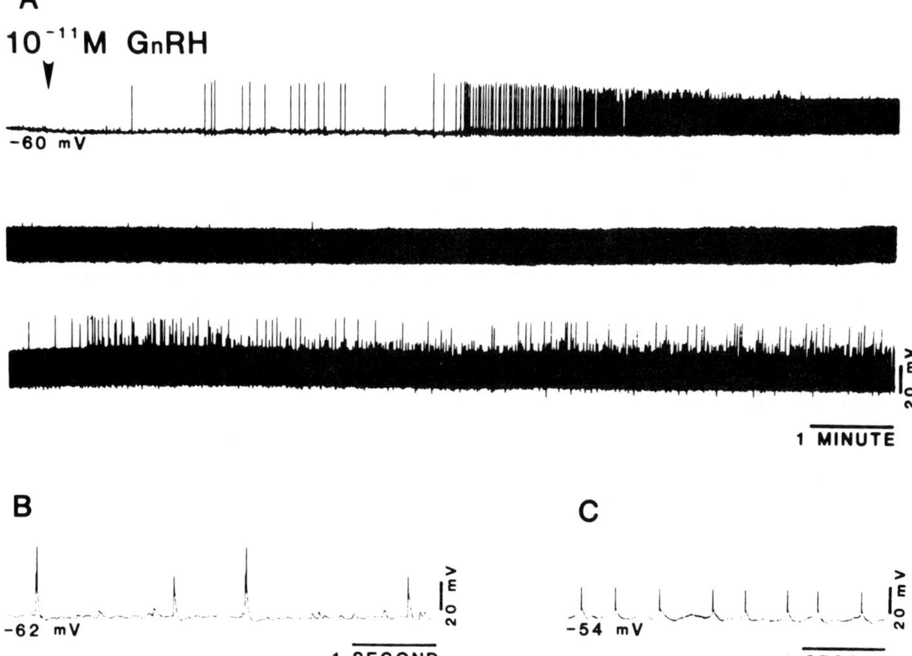

FIG. 8. Long latency, long duration response to GnRH by a pyramid cell in the CA1 region of the medial hippocampus of a rat. (A) Intracellular recording in a CA1 cell in a hippocampal brain slice suspended in a buffered medium (124 mM NaCl, 5 mM KCl, 1.25 mM KH_2PO_4, 1.3 mM $MgSO_4$, 26 mM $NaHCO_3$, 2.5 mM $CaCl_2$, and 10 mM glucose). Recording was made using a 40-MΩ electrode filled with 2 M potassium citrate. The record shows that 2 min after application of 10^{-11} M GnRH to the perfusion medium of the tissue chamber, activity increased and depolarization occurred. Control injections of perfusion medium had no effect. The GnRH was cleared by flow within 3 min. The activity lasted for several minutes at a high rate (10/sec). The total duration of the response was greater than 1 hr. The record is from the same cell with a 23-sec break between the first and second traces and a 30-min break between the second and third traces. (B) Faster speed record of the spikes during the early part of the response reveals that the tall spikes were multiple action potentials arising from large depolarizations. (C) Action potentials recorded during the fast firing period 10 min after application of GnRH. The depolarizing after potentials are absent in this case. During the latter stages of recovery, the response pattern returned to those of B. The long duration of firing was found in 25% of neurons tested (n = 50). Other responses observed were short duration responses (21%), inhibition (12%), and no response (42%).

logical responses that should be studied in order to find out more about the unique properties of peptides in controlling specific behaviors.

Summary

The brain slice chamber described offers a method of studying, with intracellular electrodes, the relationship of response to dose of peptides. By raising the level of the slices 1 mm above the level of flowing perfusion medium, we can test substances in known concentrations, free from artifacts, during long duration, stable intracellular recordings. Manipulation of Ca^{2+}/Mg^{2+} ratios in the medium can help to define synaptic and second messenger mediation of the responses. The addition of substances to the perfusion medium in this system could be combined with iontophoresis and/or micropressure techniques. Pathways in the slices may also be stimulated electrically and analyzed for the involvement of various synaptic transmitters. The results with the method so far show distinct differences among the peptides studied. Thus, there are several advantages to this method in establishing the physiological role of peptides in the brain.

Acknowledgment

This work is supported by National Institutes of Health Grant RO1-HL 27334.

[9] Patch Clamp Recording from Anterior Pituitary Cells Identified by Reverse Hemolytic Plaque Assay

By THOMAS L. CROXTON, WILLIAM McD. ARMSTRONG, and NIRA BEN-JONATHAN

Introduction

The process of exocytotic secretion is accomplished via a rise in intracellular free Ca^{2+} levels. Since cytosolic Ca^{2+} can both control and be controlled by the conductances of the cell membrane to other ions, changes in the activities of several different ion channels are often associated with secretion. Attempts to study the electrophysiology of secretion by anterior pituitary cells have been hampered, until recently, by two technical difficulties. The first difficulty results from the fact that the anterior pituitary consists (in most species) of a mixture of cell types which are morphologically indistinguishable, obscuring the identity of any particular cell studied. The second results from the fact that anterior pituitary cells are small and not electrically coupled, making impalements with microelectrodes difficult to perform and interpret.

Notable but slow progress has been made in the study of pituitary electrophysiology. The methods used in previous investigations (reviewed by Douglas and Taraskevich[1]) demonstrate both the ingenuity of earlier workers and the need for better techniques, particularly for identification of cell type. Extensive data have been collected from cell lines derived from pituitary tumors,[2,3] but the relevance of these data to secretion by normal cells is uncertain. Some workers have used unidentified cells in hemipituitaries or in culture and inferred the cell type on the basis of responsiveness to secretagogues.[4] Some have enriched the fraction of cells of a particular type by sedimentation.[5,6] Others have exploited the apparent anatomical separation of certain cell types in particular species.[7,8] More recently, immunological techniques, including the reverse hemolytic plaque assay[9,10] and a fluorescent staining procedure,[11] have been used successfully.

Electrophysiological techniques used in studies of the anterior pituitary have included the recording of action potentials with microsuction electrodes,[4] impalements with conventional microelectrodes to measure membrane potentials,[5,8] and patch clamp recording.[3,9,10,12] Radioactive lipophilic anions,[13] fluorescent dyes,[14,15] and radioactive isotopes[16,17] have also been used to measure membrane potentials, intracellular Ca^{2+}, and

[1] W. W. Douglas and P. S. Taraskevich, *in* "The Electrophysiology of the Secretory Cell" (A. M. Poisner and J. M. Trifaro, eds.), p. 63. Elsevier, Amsterdam, 1985.

[2] S. Ozawa, *in* "The Electrophysiology of the Secretory Cell" (A. M. Poisner and J. M. Trifaro, eds.), p. 221. Elsevier, Amsterdam, 1985.

[3] B. S. Wong, H. Lecar, and M. Adler, *Biophys. J.* **39,** 313 (1982).

[4] P. S. Taraskevich and W. W. Douglas, *Proc. Natl. Acad. Sci. U.S.A.* **74,** 4064 (1977).

[5] J. M. Israel, C. Denef, and J. D. Vincent, *Neuroendocrinology* **37,** 193 (1983).

[6] R. Limor, D. Ayalon, A. M. Capponi, G. V. Childs, and Z. Naor, *Endocrinology (Baltimore)* **120,** 497 (1987).

[7] P. S. Taraskevich and W. W. Douglas, *Nature (London)* **276,** 832 (1978).

[8] W. T. Mason and D. W. Waring, *Neuroendocrinology* **41,** 258 (1985).

[9] T. L. Croxton, N. Ben-Jonathan, and W. McD. Armstrong, *Biophys. J.* **49,** 217a (1986).

[10] C. J. Lingle, S. Sombati, and M. E. Freeman, *J. Neurosci.* **6,** 2995 (1986).

[11] C. Marchetti, G. V. Childs, and A. M. Brown, *Am. J. Physiol.* **252,** E340 (1987).

[12] W. T. Mason and D. W. Waring, *Neuroendocrinology* **43,** 205 (1986).

[13] M. C. Gershengorn, E. Geras, M. J. Rebecchi, and B. G. Rubin, *J. Biol. Chem.* **256,** 12445 (1981).

[14] W. Schlegel and C. B. Wollheim, *J. Cell Biol.* **99,** 83 (1984).

[15] D. L. Clapper and P. M. Conn, *Biol. Reprod.* **32,** 269 (1985).

[16] J. A. Williams, *J. Physiol. (London)* **260,** 105 (1976).

[17] S. Saith, R. J. Bicknell, and J. G. Schofield, *FEBS Lett.* **148,** 27 (1982).

ion fluxes, respectively. Measurements using radioactive isotopes or fluorescent dyes are usually performed in cell suspensions and measure a mean value; because of this, they have been used primarily with the homogeneous populations of tumor cell lines. However, recent techniques for imaging Ca^{2+} levels within single cells[18] may prove to be of great value in the study of identified normal anterior pituitary cells. Patch clamp recording is a technique that is easily applied to individual cultured cells. It does not involve impalement of the cell membrane and is thus free of many of the difficulties implicit in work with conventional microelectrodes. Patch clamp recording also differs from conventional microelectrode recording in that an electrical potential is applied and transmembrane current is measured. In this chapter we describe the identification of living pituitary cells of a particular type by a reverse hemolytic plaque assay and the study of the ionic conductances of their cell membranes by the method of patch clamp recording.

Culture and Identification of Anterior Pituitary Cells

Antiserum Production

The reverse hemolytic plaque assay requires the use of an antiserum against the hormone secreted by the cell of interest. We have prepared antisera against rat prolactin (PRL) and rat luteinizing hormone (LH) in adult female New Zealand rabbits. A successful immunization can yield 50 ml or more of serum, sufficient for use with many thousands of culture dishes. For immunization of two rabbits, approximately 500 μg of hormone is added to 1 ml of saline, small amounts of 0.1 *M* NaOH are added, if needed, to dissolve the hormone, and the solution is drawn into a 5-cm^3 glass syringe. One milliliter of Freund's complete adjuvant (Sigma, F-4258) is then drawn into a matching syringe, and the two syringes are attached to opposite ends of a 22-gauge microemulsifying needle (Popper & Sons, Inc., New Hyde Park, NY 11040). Beginning with the aqueous solution, the suspension is forced from syringe to syringe until it is very viscous and a small drop added to water does not disperse. The suspension is injected intradermally into about a dozen sites in a shaved quarter of each rabbit's back using 26-gauge needles. Injection sites are inspected daily, and necrotic wounds are drained and treated with a topical antiseptic.

Booster injections are given 4 and 10 weeks after the initial injection, using half the amount of hormone and Freund's incomplete adjuvant

[18] D. A. Williams, K. E. Fogarty, R. Y. Tsien, and F. S. Fay, *Nature* (*London*) **318,** 558 (1985).

(Sigma, F-5506). Blood is drawn from an ear vein 6 weeks after the initial injection, and the serum is tested for hormone binding (see below). Twelve weeks after the initial injection (or later if additional boosters are required to obtain an acceptable titer) rabbits are anesthetized and bled from the heart to obtain the maximum amount of blood.

The antibody titer of each antiserum is determined by standard radioimmunoassay techniques. Briefly, serial dilutions of antiserum in phosphate-buffered saline plus 1% egg white are incubated overnight at 4° with radioactive iodinated hormone. Anti-rabbit γ-globulin is added for an additional 24-hr incubation, the solution is centrifuged and decanted, and the radioactivity in the pellet is counted. Initial dilutions of 1 : 6,000 for our PRL antiserum and 1 : 48,000 for LH antiserum yielded 30% binding of counts.

Anterior Pituitary Cell Culture

The procedure for preparation of primary cultures of rat anterior pituitary cells is similar to that described by Ben-Jonathan *et al.*[19] One or two adult Wistar rats are sacrificed by decapitation, and their anterior pituitary glands are placed in Ca^{2+}-, Mg^{2+}-free buffer containing 125 m*M* NaCl, 5 m*M* KCl, 1.25 m*M* KH_2PO_4, 13.6 m*M* $NaHCO_3$, 14 m*M* glucose, 1× MEM amino acids (Gibco Laboratories, Grand Island, NY 14072), 1× MEM vitamins (Gibco), and 0.002% phenol red. The buffer is adjusted to pH 7.3–7.4 with 0.5 *M* NaOH. Under a laminar flow hood, each pituitary is cut into 6–8 pieces with a scalpel blade and transferred to a siliconized scintillation vial containing 1.5 ml of the above buffer plus 1 drop 0.5 *M* NaOH and 3 mg trypsin (Cooper Biomedical, Malvern, PA). The vial is flushed with 95% O_2/5% CO_2 and incubated in a shaking water bath at 37° for 33 min. One milligram of deoxyribonuclease (Sigma, D-5025) is then added, and the incubation is continued for an additional 2 min. The tissue fragments are transferred to a conical plastic centrifuge tube and are washed twice with 1 ml of buffer containing lima bean trypsin inhibitor (0.75 mg/ml, Cooper Biomedical) and 3 times with 1 ml of buffer.

The tissue fragments are next suspended in 2 ml of buffer and dispersed by gently drawing 50–100 times into a siliconized Pasteur pipet. The cell suspension is then filtered through nylon mesh and diluted to a total volume of 6 ml. The yield, determined by counting with a hemocytometer, is typically 3–4 $\times 10^6$ cells/rat. A volume containing 30,000 cells (about 50 μl) is pipetted into a number of 35-mm polystyrene culture dishes (Corning, 25000) not coated with poly(L-lysine). These are placed in a humidified, 5% CO_2 incubator. After 30 min of incubation to allow cell

[19] N. Ben-Jonathan, E. Peleg, and M. T. Hoefer, this series, Vol. 103, p. 249.

attachment, 2 ml of Dulbecco's culture medium is gently added to each dish. This medium is prepared by mixing 43 ml of Dulbecco's medium (Gibco), 5 ml horse serum, 1.25 ml newborn calf serum, 0.5 ml 100× MEM nonessential amino acids, 0.125 ml 10 mg/ml gentamicin, 0.125 ml 10,000 U/ml Mycostatin, and sufficient 0.5 *M* NaOH to adjust the pH to 7.4. Some of the culture medium is refrigerated in a sterile bottle for use following the reverse hemolytic plaque assay.

Reverse Hemolytic Plaque Assay

The reverse hemolytic plaque assay was adapted to the identification of anterior pituitary cells by Neill and Frawley[20] and has been used by several groups in combination with electrophysiological techniques.[9,10,21,22] Our procedure differs from that employed by others in that red blood cells are removed after the assay and the pituitary cells are returned to the incubator for 24–72 hr before electrophysiological experiments. This allows greater flexibility in the timing of experiments and permits recovery of the pituitary cells from any trauma associated with the plaque assay. The approach described below involves several modifications of the procedure previously described.[23]

The general scheme of the hemolytic plaque assay is outlined in Fig. 1. Staphylococcal protein A is first conjugated to sheep red blood cells (RBCs) by incubation with Cr^{3+}. The RBCs are then spread uniformly on the culture plate containing the pituitary cells and incubated with antiserum against the desired hormone. Antibodies attach to protein A via their Fc portion and bind the secreted hormone via their Fab portions. This binding results in activation of added guinea pig complement and lysis of RBCs in the immediate vicinity of an actively secreting cell. The hemolytic plaque formed is easily detected by light microscopy.

Preparations for the assay are as follows. Four milliliters of sheep blood (Colorado Serum Company, Denver, CO 80216) is centrifuged at 800 *g* for 10 min at room temperature, and the plasma and upper layer of cells are discarded. The pellet is then resuspended in 3 ml of saline using a large-bore Pasteur pipet and centrifuged as before. This washing is repeated an additional 2 times. The following solutions are then added to a test tube: 1500 μl saline, 200 μl packed RBCs, 200 μl 0.5 mg/ml protein A, and 500 μl 0.2 mg/ml $CrCl_3$. The tube is then covered with Parafilm and incubated at room temperature for 1 hr, vortexing at 15-min intervals.

[20] J. D. Neill and L. S. Frawley, *Endocrinology (Baltimore)* **112,** 1135 (1983).

[21] M. Hiriart and D. R. Matteson, *Biophys. J.* **51,** 250a (1987).

[22] K. A. Gregerson and G. S. Oxford, *Biophys. J.* **51,** 431a (1987).

[23] P. F. Smith, E. H. Luque, and J. D. Neill, this series, Vol. 124, p. 443.

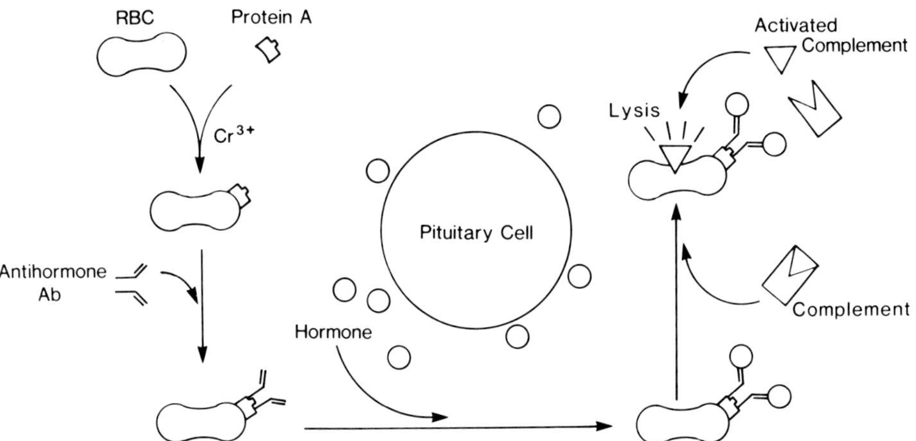

FIG. 1. Steps involved in reverse hemolytic plaque formation. See the text for explanation.

After incubation, the cells are centrifuged for 5 min and washed with saline. Slight pink coloration of the wash fluid is acceptable, but significant hemolysis at this point indicates that plaques will not form properly. This problem can be solved by aging the $CrCl_3$ solution or by reducing the concentration of Cr^{3+} in the incubation suspension. The RBCs are washed 2 additional times in 3 ml of Dulbecco's modified Eagle medium (Gibco) to which 0.1% bovine serum albumin has been added (DMEM–BSA). The final cell pellet is transferred to a plastic scintillation vial using DMEM–BSA and is diluted to a total volume of 10 ml. This RBC plus protein A suspension can be stored refrigerated for up to 1 week.

The reverse hemolytic plaque assay is performed 24 hr after the initial cell plating. A suspension containing all the reagents necessary for hemolytic plaque formation is made immediately prior to use. Two milliliters of the RBC plus protein A suspension prepared as above is centrifuged at 800 *g* for 5 min, and the supernatant is discarded. Eighty microliters of the packed cells is mixed with other reagents as listed in Table I to prepare 400 μl of the desired hemolysis suspension. Adjustments in the final concentrations of antiserum and complement may be required with different antisera. This volume is sufficient for reverse hemolytic plaque assays of six culture dishes.

Two 2 × 20 mm strips of double-sided Scotch tape (3M, #665) are attached to parallel edges of a #2 glass coverslip (not etched) for later use. A culture dish is dumped and inverted on a paper towel. A coverslip onto which a labeled grid is etched (Bellco Glass, Inc., #1916-92525) is taped to the bottom of the dish using four $\frac{1}{2} \times \frac{1}{8}$ in. pieces of autoclave tape. The

TABLE I
COMPOSITION OF HEMOLYSIS SUSPENSIONS USED FOR IDENTIFICATION OF ANTERIOR PITUITARY LACTOTROPES AND GONADOTROPES

Reagents	Volume (μl) PRL	LH
Antiserum (1 : 5)[a]	40 (1 : 50)[b]	80 (1 : 25)[b]
RBC plus protein A suspension	80	80
Complement (1 : 10)[a]	80 (1 : 50)[b]	60 (1 : 67)[b]
GnRH, 10^{-7} M[a]	—	80 (20 nM)[b]
DMEM–BSA	200	100
Total volume	400	400

[a] Initial concentration.
[b] Final concentration.

dish is righted and rinsed with 1 ml of DMEM–BSA. A corner of paper towel is used to remove the final drop of wash fluid from the dish. The unetched coverslip prepared above is placed in the bottom of the culture dish, tape side down, to form a Cunningham chamber over the cultured cells (see Fig. 2). Sixty microliters of the hemolysis suspension is drawn into a pipet and slowly expelled near one edge of the chamber, filling it

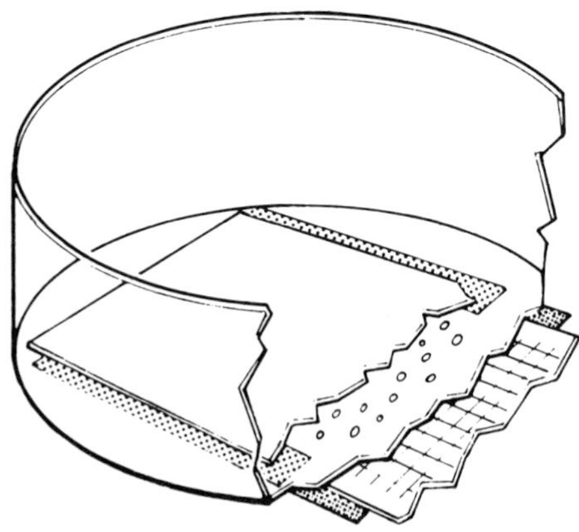

FIG. 2. Cutaway view of a Cunningham chamber used in the reverse hemolytic plaque assay. The layers represented are (from top to bottom) glass coverslip, thin space between two strips of tape filled with a suspension of all reagents necessary for plaque formation, polystyrene culture dish with attached anterior pituitary cells, and glass coverslip with etched, labeled grid.

uniformly. The dish is then covered and placed in a humidified 5% CO_2 incubator. This procedure is repeated with the next dish.

After 30 min of incubation, each dish is examined with an inverted microscope at low power. The locations of several hemolytic plaque-forming cells are recorded in a notebook (see Fig. 3). The dish is then moved to a laminar flow hood, and the chamber is gently flushed with 60 μl of cell culture medium. A corner of paper towel is used to soak up the displaced RBC-rich solution, and 1 ml of culture medium is carefully added. The unetched coverslip is slowly removed from the dish with a pair of forceps, an additional 2 ml of culture medium is added, and the dish is returned to the incubator.

Validation of the Technique

Specificity of the plaque assay for cells of the desired type may be established by both direct and indirect means. An important indirect test is to establish the specificity of the antiserum. Cross-reactivity of antisera with other anterior pituitary hormones is determined in radioimmunoassays in which varying amounts of test hormones are added to the first incubation. The relative ability of each hormone to displace the iodinated hormone from the antiserum is determined by comparison of plots of bound counts versus the logarithm of added hormone concentration. Thirty percent displacement of bound counts from our LH antiserum required 1 ng of NIADDK rLH RP-2 but about 100 ng of NIADDK rFSH RP-2 or NIAMD rTSH I-3. Our PRL antiserum showed no detectable cross-reactivities with 50-ng quantities of corticotropin (ACTH), LH (NIADDK rLH RP-2), follicle-stimulating hormone (FSH) (NIADDK rFSH RP-2), or thyroid-stimulating hormone (TSH) (NIADDK rTSH RP-2). If required, the cross-reactivity of an antiserum could be reduced by preabsorption with the cross-reacting material.

Verification of specificity can also be accomplished by direct means. One approach is exemplified by our use of secretagogues to evaluate the specificity of the reverse hemolytic plaque assay for gonadotropes. First, we established that plaques form readily when 10^{-8} *M* gonadotropin-releasing hormone (GnRH) is included in the hemolysis suspension and do not form in its absence. Second, inclusion of 10^{-6} *M* thyrotropin-releasing hormone (TRH) in the hemolysis suspension instead of GnRH resulted in no plaque formation. Finally, plaque assays performed with the same culture dishes on sequential days revealed that (1) cells which form plaques when stimulated with GnRH do not form plaques subsequently in the presence of TRH and (2) dishes which show no plaque formation when TRH is included do show an appropriate density of

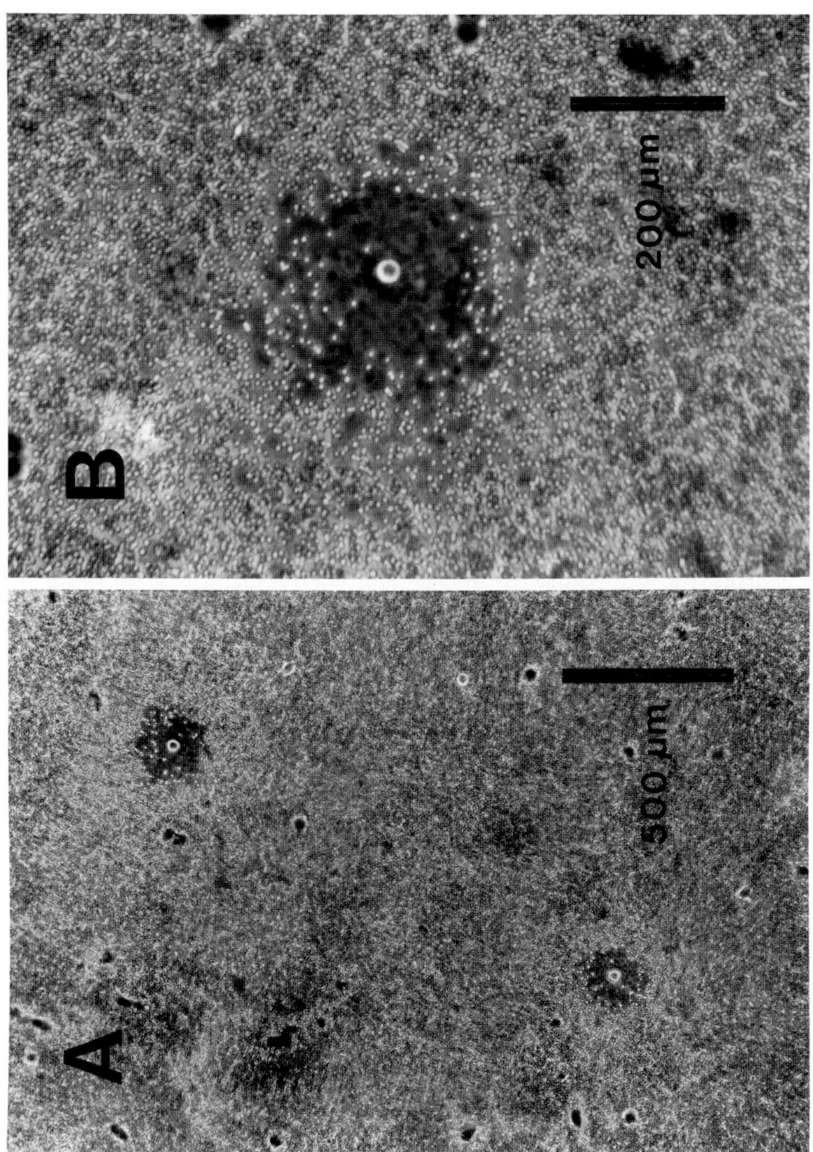

FIG. 3. (A) Reverse hemolytic plaque assay for gonadotropes. Many cells that did not form plaques are seen. The attached grid is not visible in this focal plane. (B) A gonadotrope surrounded by red blood cell ghosts in a reverse hemolytic plaque.

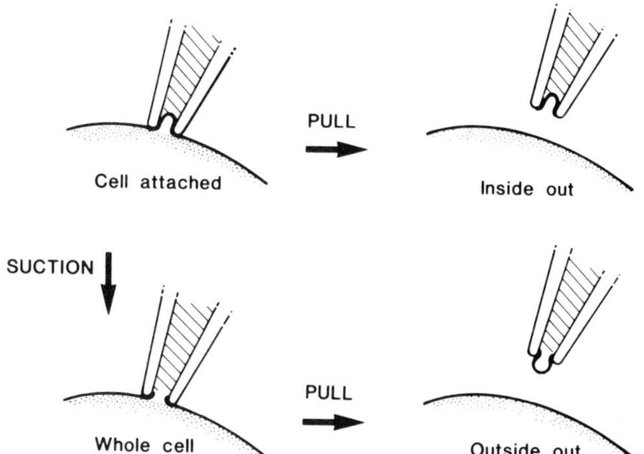

FIG. 4. Patch clamp configurations. The cell-attached configuration is obtained initially. Others are obtained as diagrammed by applying strong suction to the pipet interior and by pulling the pipet away from the cell.

plaques when later incubated with GnRH. Taken together, these data indicate that the plaque-forming cells are indeed gonadotropes.

To evaluate reproducibility, we have also performed sequential reverse hemolytic plaque assays using GnRH on both days. Of 34 cells recorded as nonplaque forming on the first day, 16 (47%) were dead or missing on the second day and none of the 18 (53%) cells still present formed plaques. Of 63 cells recorded as plaque forming on the first day, 20 (32%) were not found or were dead by the end of the second day, 40 (63%) formed plaques, and 3 (5%) were apparently healthy but did not form plaques. Loss of 10–20% of identified cells is often seen and probably occurs as the unattached RBCs are washed out of the chamber. This does not cause significant problems since we record the locations of at least six cells per dish and use only one cell per dish for patch clamp experiments.

Patch Clamp Recording

Choice of Patch Clamp Configuration

Patch clamp recording[24] can be carried out in four configurations which are summarized in Fig. 4. Each configuration provides a different

[24] O. P. Hamill, A. Marty, E. Neher, B. Sakmann, and F. J. Sigworth, *Pfluegers Arch.* **391,** 85 (1981).

perspective of cellular electrophysiology. The cell-attached configuration (also called on-cell configuration) is obtained when a gigaohm seal forms between a patch pipet and the cell membrane. Withdrawing the pipet from the cell at this stage will usually yield an intact patch of membrane across the pipet tip that has a surface area of a few square micrometers. This is called the inside-out configuration since the inside (cytosolic) surface of the membrane faces outward from the pipet. If the cell-attached patch is subjected to strong suction or high voltage, the small patch of membrane can be ruptured to produce the whole-cell configuration. Finally, if the pipet is withdrawn while in the whole-cell configuration, the membrane near the pipet will sometimes reform into an intact bilayer with the extracellular face of the membrane facing out of the pipet. This is the outside-out configuration.

The whole-cell configuration does not usually resolve individual ion channel currents but allows measurement of macroscopic currents through the greatest portion of the cell membrane. It can also be used in a current clamp mode to measure something analogous to the membrane potential of the intact cell. Although the physical structure of the cell is preserved, the cytosol is essentially replaced by the pipet solution. Diffusible modulators of channel activity may be washed out, and changes in the ionic composition of the intracellular fluid are unavoidable. Thus, normal function of the cell is not reliably observed in whole-cell recording.

The inside-out and outside-out configurations allow high resolution measurement of currents through individual channels in small pieces of cell membrane. These detached configurations are ideal for characterization of ion channel conductances, selectivities, and gating kinetics and for quantitating the effects of physical and chemical stimuli on these parameters. By recording with different applied potentials and with different bath compositions, a very complete characterization of a channel can be obtained from a single patch.

Unlike the other three, the cell-attached configuration measures membrane currents under conditions of presumably normal cellular function. However, interpretation of cell-attached data is complicated by the following: (1) measured currents must pass through two membranes in series and changes in channel activity in either membrane will produce a change in the measured current, (2) the actual voltage across either membrane cannot be controlled or readily measured, (3) neither the cytosolic nor pipet solutions can be easily changed to evaluate ion selectivity of channels in the small membrane patch, and (4) substances which might alter channel gating cannot be directly applied to either face of the small membrane patch.

In any given study, the configuration of choice will depend on the cellular mechanisms hypothesized, the stability of patches obtained with the cells of interest, and the capabilities of the data recording system. The cell-attached configuration is particularly useful in initial experiments designed to select channels that show a change in activity when the cell is stimulated for further study. However, since the patch pipet shields channels in the small patch from any substances infused into the bath, the presence of channels that open in direct response to added secretagogue may not be revealed with this configuration. Another consideration is the susceptibility of cell-attached recording to artifacts since damage to the cell can produce changes in channel activities. Results of such experiments should be analyzed with caution and reported only if fully reproducible.

The whole-cell configuration is particularly useful for the study of low conductance channels since a summation current is measured. Many Ca^{2+} channels were first discovered in this way, and even carrier-mediated processes have been studied.[25] However, experiments with the whole-cell configuration must be carefully designed to eliminate, by ion substitution and pharmacological blockade, all membrane currents except the one of interest.

Detached patches are probably most useful for quantitating the characteristics of selected channels observed previously in cell-attached or whole-cell experiments. The use of detached configurations alone in a "shotgun" approach can be frustrating because of the variety of channels found and the volume of data generated. Studies with the detached configurations often require a data recording system with a high sample frequency and a large capacity. Of the two detached configurations, the inside out is obtained with greater yield and is generally more stable. However, the outside-out configuration is required to observe directly the effects of extracellular messengers on the kinetics of channel gating.

Instrumentation and Equipment

A number of excellent patch clamp electrometers are commercially available that perform the basic functions of voltage clamping, current-to-voltage conversion, and amplification. This and an oscilloscope are all the electronics needed to observe single-channel gating. However, the serious use of patch clamp recording requires much additional instrumentation to record, display, and analyze the signal obtained. Although a number of manufacturers offer systems which perform these functions, we have often found it necessary to modify available equipment to obtain features that were either unavailable or prohibitively expensive. A block diagram

[25] P. Jauch, O. H. Petersen, and P. Lauger, *J. Membr. Biol.* **94,** 99 (1986).

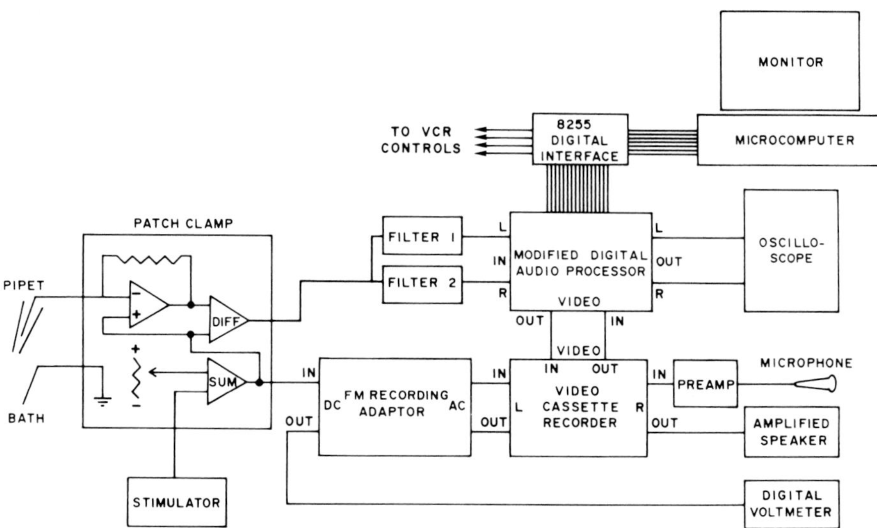

FIG. 5. Data recording system. The output of the patch clamp difference amplifier is the pipet current and that of the sum amplifier is the pipet potential. These signals are processed by a modified digital audio processor and an FM recording adaptor for recording on video tape. On playback, the digitized current signal is transferred to microcomputer memory for data analysis.

of our system is given in Fig. 5. A stimulator (Grass Instruments, Quincy, MA, Model SD9B) is connected to the stimulus input of a List EPC-7 patch clamp electrometer (Medical Systems Corp., Greenvale, NY 11548) to provide square voltage pulses used to monitor seal resistance while the gigaseal is forming. The current monitor output of the electrometer is passed through two 6-pole Bessel filters constructed from resistor tunable modules (Frequency Devices, Haverhill, MA, 01830, 736LT-4). The use of two analog filters allows us to record both with a cutoff frequency chosen to give an excellent signal-to-noise ratio (800, 1200, or 2000 Hz) and with a cutoff frequency that produces less distortion of fast-gating channel events (3200, 5000, or 8000 Hz).

Both filter outputs are delivered to a modified digital audio processor (Unitrade Inc., Philadelphia, PA 19130, DAS 501) that incorporates the additional circuitry designed by Bezanilla.[26] This unit performs analog-to-digital conversion of each input channel at a rate of 44,100 samples per second and converts the serial bit stream to a video-compatible signal that may be recorded by a standard video cassette recorder (VCR). An analog

[26] F. Bezanilla, *Biophys. J.* **47,** 437 (1985).

output, obtained by reconversion of the digital signal, is also available from this unit. We use this output to monitor the experiment on an oscilloscope. Our VCR (Sony, SL-HF 450) allows two audio channels to also be recorded. One is used to record experimental notes spoken into a microphone, while the other records the applied pipet potential via an FM recording adaptor (A. R. Vetter Co., Rebersburg, PA). To minimize electrical interference, the computer is turned off during experiments.

On playback, the pipet voltage is shown on a digital voltmeter, the spoken notes are heard from an amplified speaker, and the current signal is displayed on an oscilloscope and a strip chart recorder. Data for analysis are obtained from the parallel digital output of the digital audio processor. These are transferred to the memory of an IBM XT microcomputer where they may be accessed by programs written in Microsoft FORTRAN. A hardware interface is required for transfer of the 16-bit parallel data words to the 8-bit data bus of the computer. We have recently published details for the construction of an interface which facilitates automated analysis of the recorded data.[27] By recognizing a sample of previously recorded data and beginning the transfer of data to memory immediately after this marker is found, the assembly language driver of our interface is able to store consecutive records. This allows analysis of tens of megabytes of consecutive data points using a computer system with as little as 256 kbytes of memory. In addition, output lines are connected via optoisolators to the play, stop, and rewind buttons of the VCR to permit software control of these functions. The tape positioning necessary for transfer of consecutive records is performed automatically. Two data records are stored on the VCR tape (differing in the filter cutoff frequency) and either one can be chosen for storage in the computer by switch selection of the clock signal synchronous with that data.

The patch clamp setup used in our laboratory is diagrammed in Fig. 6. Experiments are performed in a grounded Faraday cage on a vibration isolation table (Ehrenreich Photo-optical Industries, Garden City, NY 11530, Model 78300). The culture plate is held on the platform of an inverted microscope within a grounded metal block. A lightly chlorided silver wire, attached to this block but insulated from it by several layers of autoclave tape, extends to the bottom of the culture dish and is connected to the signal ground of the patch clamp head stage. A grounded aluminum box with cutouts for entry of pipets and light is placed on the microscope stage to provide further electrical shielding. The condensor stage of the microscope has been removed to provide increased depth of field.

A 3-dimensional hydraulic micromanipulator (Narishige, MO-103) is

[27] T. L. Croxton, S. J. Stump, and W. McD. Armstrong, *Biophys. J.* **52,** 653 (1987).

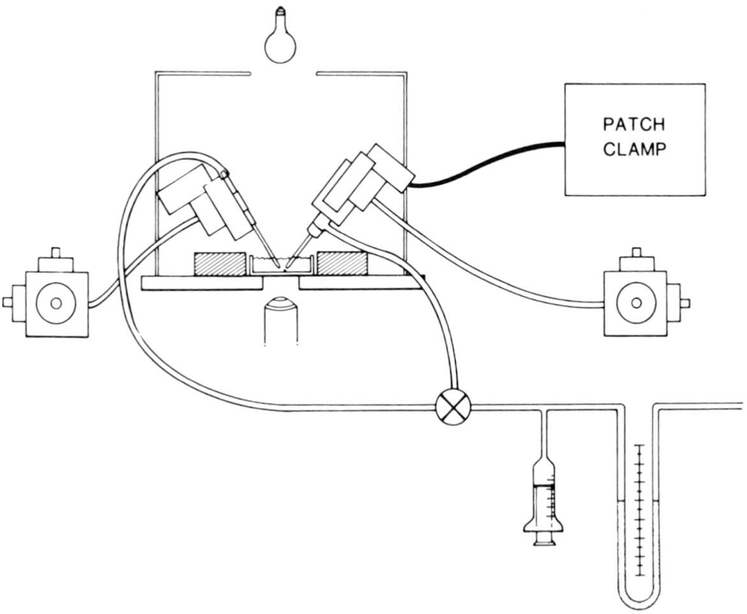

FIG. 6. Patch clamp setup. A culture dish containing pituitary cells is placed on the stage of an inverted microscope. A patch clamp pipet (right) is attached to the patch clamp head stage which is mounted on a 3-dimensional hydraulic micromanipulator. An infusion pipet (left) is mounted directly on a second micromanipulator. A syringe and water manometer pressure system is connected via a stopcock to the interior of either pipet.

used to position the electrometer head stage and the attached patch pipet. A second micromanipulator is used for positioning an infusion pipet near a cell. The controls for both micromanipulators reside on a separate table. An air syringe/water manometer pressure system is connected via stopcocks to the patch and infusion pipets.

The second micromanipulator can instead be used for changing the bath solution. Two L-shaped lengths of glass tubing are glued together in an inverted T, and the upper ends of this assembly are connected via polyethylene tubing to two 20-cm^3 syringes mounted on an infusion/withdrawal pump (Harvard Apparatus, Model 940). Before the experiment, this system is flushed and filled with the solution that will be used to replace the bath solution. At the appropriate time, the inverted T is lowered into the bath near the periphery, and the pump is turned on. Approximately 18 ml of fluid is injected at a rate of about 4 ml/min, and an equal volume is withdrawn simultaneously. The pump is then switched off, and the delivery tubing is raised above the solution surface. Thus, an effective replacement of the 1 ml volume of bath is achieved with minimal mechanical disruption of the patch and without leaving external volumes of solution in electrical continuity with the bath.

Preparation of Patch Clamp Pipets

Techniques for manufacturing patch clamp pipets have been described in detail by others.[24,28] A number of specialized devices for pipet preparation are commercially available, but satisfactory results can be obtained with minor modifications of equipment already present in most electrophysiological laboratories. We describe herein a few adaptations of equipment which might be useful to individuals not previously involved in patch clamp experiments.

Hematocrit tubing (Thomas Scientific, #2412B62) is washed with 95% ethanol, rinsed with distilled water, and soaked overnight in dilute $K_2Cr_2O_7$/HCl. They are then rinsed extensively with distilled water, dried at 105°, and stored in a clean Coplin jar. Pulling is performed immediately prior to use in two stages using a vertical puller (David Kopf Instruments, Tujunga, CA 91042, Model 700C) which has been modified to allow switch selection of two different heater currents. This yields more reproducible results than those obtained if the heater powerstat is readjusted between the first and second pulls for each pipet. The chain connecting the solenoid to the vertical slide is disconnected, and a 3-position toggle switch is mounted on the front panel. The wire between the wiper arm of the heater powerstat and the heater transformer is disconnected, and the heater transformer is reconnected to the common contact of the added switch. The switch contacts corresponding to the up and down switch positions are connected to the wiper arms of the heater and solenoid powerstats, respectively. A jumper is installed across the solenoid activating microswitch, and the solenoid itself is disconnected. In this way, the switch selects a heater current (20 A) set by the heater control (up), no current (center), or a second heater current (14–15 A) set by the solenoid control (down).

For the first pull, a semicircular 40-mm stop constructed from brass tubing (i.d. 9/16 in.) is placed around the vertical slide, the heat selector is switched up, and the start button is pushed. As the vertical slide falls, the heat selector is switched to the middle position. The heater is allowed to cool for a few seconds, the pipet is repositioned so that the narrowed portion is centered on the heating coil, and the stop is removed. A piece of plastic which serves as a windscreen is then lowered over the face of the puller, and the heat selector is flipped down to begin the second pull. The

[28] D. P. Corey and C. F. Stevens, *in* "Single-Channel Recording" (B. Sakmann and E. Neher, eds.), p. 53. Plenum, New York, 1983.

heat of the second pull is adjusted to yield pipets with a steep taper and a tip diameter of 1–2 μm. Infusion pipets are pulled similarly but with tip diameters of 3–5 μm.

The tapered portion of the patch pipet, except for the tip, is coated with Sylgard 184 (Corning) in a jig similar to that described by Corey and Stevens.[28] We apply the Sylgard in several overlapping steps, beginning near the tip, to obtain a thick but uniform coating. Pipets are then heat polished under a microscope (100×). The pipet is first mounted in a holder connected to a mechanical 3-dimensional micromanipulator, and the pipet tip is positioned in the focal plane near the center of the field. A V-shaped filament of 36-gauge platinum wire, mounted to a holder on the microscope stage and heated to dull orange color, is brought to 5–10 μm from the pipet tip for a few seconds to smooth the glass of the tip.

Experimental Procedures

The bath solution we use for patch clamp experiments contains Hanks' salts, 1 ml 0.5% phenol red/250 ml, 5 m*M* HEPES, and sufficient 0.5 *M* Tris to adjust the pH to 7.4. This solution contains (in m*M*) 137 Na^+, 138 Cl^-, 5.7 K^+, 1.3 Ca^{2+}, 0.4 Mg^{2+}, 0.8 phosphate, and 5.6 glucose. The same bath solution is used to fill the patch pipet in cell-attached studies and as diluent for any substances to be applied near the cell with the infusion pipet. Pipets are back filled through a 0.2-μm syringe filter and a $2\frac{1}{2}$ in. 32-gauge needle (Popper & Sons, Inc.). Air bubbles that occasionally remain in the tip are dislodged by sharp tapping with a wooden stick.

The patch clamp electrometer is initially set to SEARCH mode (maintaining a near zero average pipet current), and the gain is reduced. The stimulator is set to deliver 100-mV pulses at a rate of 16/sec, and stimulus scaling by the electrometer is set to 0.001. A culture dish is removed from the incubator, and the culture medium is removed with a Pasteur pipet. The dish is gently washed 5 times with 1-ml portions of the bath solution which are delivered through a 0.2-μm syringe filter in a Swinney holder. The dish is then filled with 1 ml of filtered solution and is placed on the microscope stage. The cells are brought into focus at low power, and the dish is moved to position a selected cell in midfield.

After the culture dish is in place, the patch pipet is filled and mounted in the pipet holder of the head stage. The head stage is moved with the hydraulic micromanipulator until the shadow of the pipet is seen through the microscope. The aluminum box is then placed over the metal block containing the dish, the microscope objective is switched to 40×, and the gain of the electrometer is increased to 100 mV/pA. About 5 cm of water pressure is applied to the interior of the patch pipet, and the pipet is lowered into the solution. The resistance of the patch pipet (typically 2–5 MΩ) is noted from the amplitude of current produced by the voltage stimulus. The micromanipulator for the infusion pipet is adjusted so that the shadow of that pipet is within the microscope field. Finally, the patch pipet is lowered to a position just above the cell to be studied, and the pipet potential offset is adjusted to zero.

Maintaining a few centimeters of water pressure, the patch pipet is brought into contact with the cell. This is done by simultaneous movement of two of the micromanipulator controls. Contact is seen as slight dimpling of the cell surface and a 2- to 3-fold increase in pipet tip resistance. Pressure is released, and 3–5 cm water suction is applied. The ideal response is a rise in resistance over several seconds to a plateau of about 100 MΩ followed in 5–20 sec by a sudden rise to over 5 GΩ. The increase in resistance is accompanied by a significant decrease in the amplitude of current noise. As the resistance increases, the stimulus scaling of the patch clamp is increased to maintain a pulse current of 3–30 pA. Once a gigaseal has formed, the suction is released, the stimulus scaling is switched to zero, the mode is changed to VOLTAGE CLAMP, and the VCR is started.

If the behavior described above is not seen, a gigaseal can still be obtained by using higher applied suction, albeit with the risk of damage to the cell. Damage may be indicated by a decrease in resistance or by an irreversible stimulation of channel activity. We have found that most cells can tolerate 20–30 cm water suction, but it is best to use as little suction as possible and to limit the time that suction is applied to periods of 15–30 sec. Gigaseals often form as suction is being applied or released. Seals that form under mild conditions tend to have higher resistances and last longer than those that require application of more suction.

During cell-attached recording it is useful to have some means of stimulating the cell with releasing hormones or drugs. Dropwise addition of substances to the bath is usually unsatisfactory, since diffusion to the vicinity of the cell is very slow and the associated vibration is likely to activate or damage the cell. A simple and effective alternative is the use of an infusion pipet positioned near the cell. We have found that gentle, yet effective, delivery can be accomplished by using a pipet with a tip diameter near 4 μm and applying about 10 cm water pressure to the pipet interior. Micrometer-drive devices that deliver known volumes could also be used, but these can introduce significant electrical noise if connected via electrolyte-filled tubing. In either case, the infusion pipet should be kept above the surface of the bath until just before use and should be raised into the air immediately afterward to minimize leakage. Since the volume delivered is very small in comparison to the total bath volume, the local concentrations of added substances decay after the pipet is removed, and reversible processes can be studied repetitively in a single cell.

Selected Experimental Results

We have performed patch clamp recording from identified rat lactotropes and gonadotropes. Selected results are presented here to illustrate ways in which patch clamp recording can be used to investigate the role of ion conduction in hormone secretion by anterior pituitary cells. The methods for analysis of patch clamp data are varied, and for discussion of these the reader is referred to other sources.[29] Cell-attached recordings from lactotropes show different degrees of spontaneous channel activity. Figure 7 shows a record with moderate activity. The pipet was maintained at zero potential and contained the same Hanks' solution as the bath (see above). A variety of electrical events are seen, including openings and closings of large, slow channels, bursts of activity from faster channels, and spikes (possibly resulting from action potentials).

Experiments with lactotropes have also been performed in which the pipet was filled with 150 m*M* KCl and the patch was detached to the inside-out configuration. These experiments have revealed the presence in lactotropes of a large conductance anion-selective channel and a variety of K^+-selective channels.[9] Figure 8 demonstrates a standard method of analysis in which the single-channel current is plotted as a function of applied potential, commonly called the $I-V$ curve. The slope of this plot is the single-channel conductance which, for this channel, was near 200 pS. The potential at which the single-channel current is zero can be used to probe ionic selectivity of the channel. In this experiment the pipet voltage required to eliminate current through the channel was clearly negative, indicating a selectivity for K^+ over either Na^+ or Li^+.

It is also possible to calculate from single-channel records the proportion of time that a channel is open. An example of this analysis is shown in Fig. 9 for an inside-out patch from a lactotrope. In this experiment both the pipet and the bath contained 140 m*M* KCl. The bath solution was buffered with EGTA to a calculated Ca^{2+} activity of 1 μM. The $I-V$ curve obtained (Fig. 9A) shows a conductance near 200 pS. The potential of zero current is near zero as would be expected for a K^+-selective channel

[29] F. J. Sigworth, *in* "Single-Channel Recording" (B. Sakmann and E. Neher, eds.), p. 301. Plenum, New York, 1983.

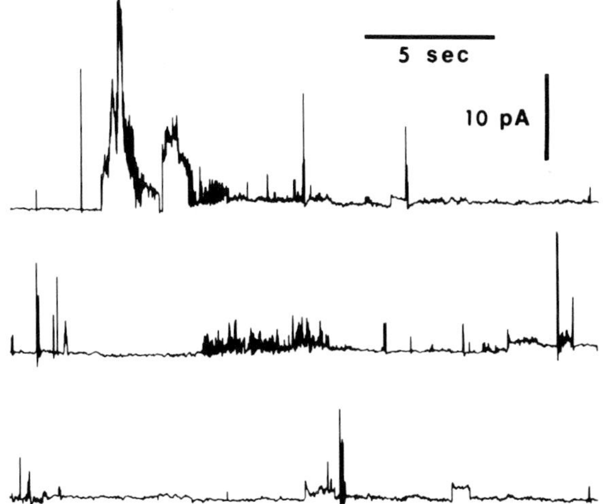

FIG. 7. Cell-attached patch clamp record from a rat pituitary lactotrope. Upward deflections correspond to inward currents through the membrane patch. The three traces shown are consecutive portions of a single record.

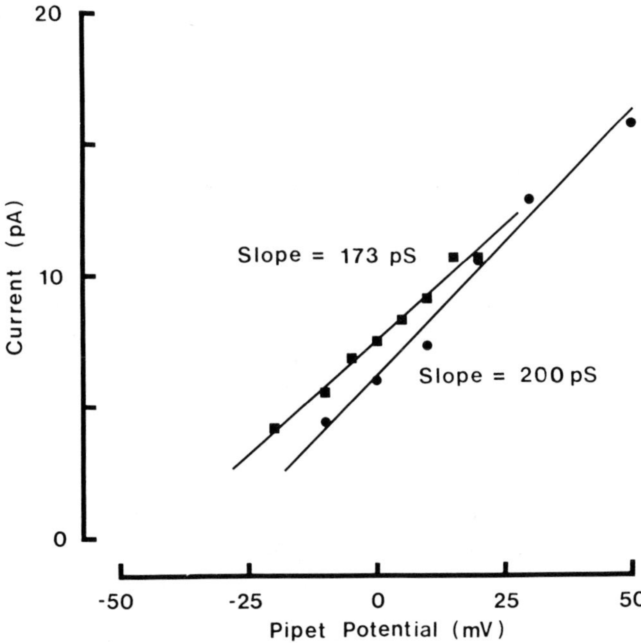

FIG. 8. Current–voltage (I–V) relation for a K^+-selective ion channel from a rat pituitary lactotrope. Data were obtained from a single patch in the inside-out configuration. The patch pipet contained 50 mM K^+, and the bath contained either 5.7 mM K^+ and 137 mM Na^+ (●) or 5.7 mM K^+ and 144 mM Li^+ (■). The intercepts of the fitted lines were −31 and −43 mV, respectively.

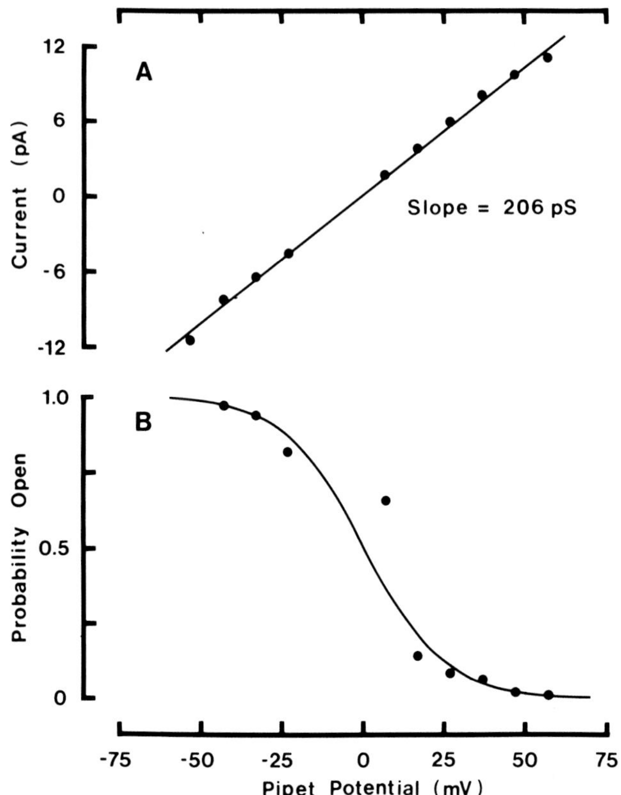

FIG. 9. (A) Current–voltage (I–V) relation for an ion channel from a rat pituitary lactotrope. Data were obtained from a patch in the inside-out configuration. The pipet and bath each contained 140 mM KCl, and the Ca^{2+} activity of the bath solution was buffered to 10^{-6} M with EGTA. The intercept of the fitted line was -0.5 mV. (B) Fraction of time (probability) open for the same channel as in A. The point at $V = 10$ mV was excluded from the fit shown by the smooth curve.

under these conditions. Thus, this channel is likely of the same type as that in Fig. 8. In Fig. 9B, the proportion of time that the channel was open (P_o) is plotted as a function of applied potential. A strong dependence on applied potential is seen which is well represented by the relationship $\ln[(1 - P_o)/P_o] = aV + b$. The smooth curve is drawn using coefficients obtained by fitting with this linearized form. This channel was open 50% of the time at transmembrane potentials near zero. Taken together, the data of Figs. 8 and 9 indicate the presence in lactotropes of a K^+-selective channel with large conductance that is opened by depolarization of the cell membrane.

Figure 10 shows a cell-attached recording from a rat gonadotrope. Initially, rounded outward deflections are seen at apparently random intervals. During the time period indicated by the bar, 2 n*M* GnRH was infused near the cell. This eliminated the rounded outward deflections and induced highly regular oscillations in membrane current. Over a period of several minutes, the outward phase of these oscillations lengthened, and the amplitude of the inward phase increased. The oscillations then progressed into a series of very large inward pulses, after which the process became irregular. The ionic basis and significance of these induced currents are currently under investigation.

Summary

The study of hormone secretion by anterior pituitary cells is complicated by the presence of multiple cell types. For unambiguous interpretation of data it is necessary to identify the cells from which measurements are made. We have described a reliable experimental approach involving the identification of cultured cells of a particular type with a reverse hemolytic plaque assay. The electrical characteristics of individual identi-

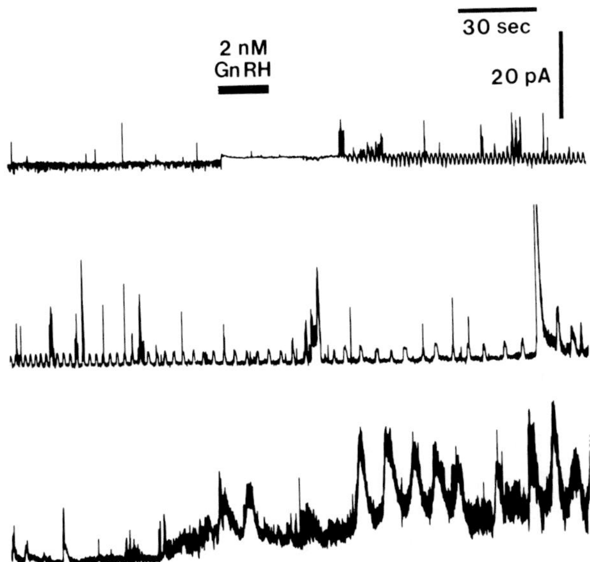

Fig. 10. Cell-attached patch clamp record from a rat pituitary gonadotrope. The bath and patch pipet each contained Hanks' salt solution (see the text). Upward deflections correspond to inward currents through the membrane patch. The three traces shown are consecutive portions of a single record.

fied cells can then be studied using patch clamp recording. This electrophysiological approach is well suited to the study of complex systems in cultured cells. Although this combined approach requires some expertise in a variety of techniques, it is workable and should yield valuable information regarding the role of ion channels in the cellular control of hormone secretion by the anterior pituitary.

Acknowledgments

The studies reported herein were supported by U.S. Public Health Service Grants DK36575 and DK07554.

[10] Microdialysis for Measurement of Neuroendocrine Peptides

By JON E. LEVINE and KEITH D. POWELL

Introduction

The hypothalamohypophyseal portal plexus is a diminutive and physically inaccessible system that poses extraordinary technical difficulties for the study of hypophysiotropic hormone secretion. As a consequence, hypothalamic releasing factor secretion rates have often been assessed by indirect methods, such as measurement of tissue content or extrapolations from anterior pituitary secretory patterns. Studies of neurohormone release in portal vessel plasma of anesthetized, immobilized animals have often yielded valuable data, yet information provided by this approach is ultimately limited since anesthetics and acute surgical trauma most likely alter the rate of neurohormone release, and because momentary releasing factor–pituitary hormone relationships cannot be assessed due to the transection of the pituitary stalk.

To circumvent these problems, we introduced in 1980 the use of a modified push–pull perfusion system[1] for the determination of hypothalamic peptide release in conscious, freely moving animals. Notable advantages of the latter technique have included the ability to sample neuropeptide release in a physiological context, and to establish and analyze endocrine correlates of hypothalamic releasing factor activity in unanesthetized animals. These advantages notwithstanding, there have remained

[1] J. E. Levine and V. D. Ramirez, *Endocrinology (Baltimore)* **107,** 1782 (1980).

technical difficulties and limitations inherent in the use of the push–pull perfusion method which can prevent high experimental success rates and preclude its uninterrupted use in chronic experiments. For these reasons, we have recently developed a microdialysis method for the measurement of neuropeptide levels in the extracellular fluid of the anterior pituitary gland. This microdialysis system incorporates many of the positive attributes of the push–pull perfusion system into a new approach that permits automated sampling of neuropeptide levels by a more efficient and reliable means over prolonged experimental periods.

The microdialysis approach[2] is rapidly gaining recognition as a powerful tool in the measurement of neurotransmitter release at specific brain loci. The method has been applied successfully in the determination of catecholamine,[3,4] amino acid,[2,5] and substance P[2] release patterns at various intracerebral sites. One group[6] has also used the approach for monitoring oxytocin release in the olfactory bulb. Our pituitary microdialysis system has recently proved useful and advantageous in estimating release profiles of luteinizing hormone-releasing hormone (LHRH), a hypothalamic releasing hormone that is known to govern the secretion of luteinizing hormone (LH) from gonadotropic cells of the anterior pituitary gland. The purpose of this chapter is to describe the design, construction, and operation of a pituitary microdialysis system, and to demonstrate the applicability of the approach for determining hypophysiotropic hormone release patterns in unanesthetized animals. Since comprehensive reviews of intracerebral dialysis are available in this volume[7] and elsewhere,[2] the following is intended to serve only as a practical guide for our specific application of the microdialysis approach in the measurement of neuroendocrine peptide levels in extracellular fluid of the anterior pituitary gland.

Principles of Microdialysis

The basic principle of peptide sampling by microdialysis is similar to that which has been described[8] for push–pull perfusion techniques: a moving pool of fluid (infusate) is brought into contact with a relatively

[2] U. Ungerstedt, *in* "Measurement of Neurotransmitter Release in Vivo" (C. A. Marsden, ed.), p. 81. Wiley, New York, 1984.

[3] T. Zetterstrom, T. Sharp, C. A. Marsden, and U. Ungerstedt, *J. Neurochem.* **41,** 1769 (1983).

[4] R. L'Heureux, T. Dennis, O. Curet, and B. Scatton, *J. Neurochem.* **46,** 1794 (1986).

[5] M. Sandberg and S. Lindstrom, *J. Neurosci. Methods* **9,** 65 (1983).

[6] K. M. Kendrick, *Brain Res.* (in press).

[7] K. M. Kendrick, this volume [11].

[8] J. E. Levine and V. D. Ramirez, this series, Vol. 124, p. 466.

stationary pool of fluid (extracellular fluid), and exchange of solutes (peptides) occurs between the two pools by diffusion. The direction and magnitude of the exchange for a particular neuroendocrine factor is largely determined by the concentration gradient which exists between the two pools. A neurohormone of higher concentration in extracellular fluid will diffuse along its gradient and can be measured in the perfusates/dialyzates to give an estimate of its original concentration in the extracellular fluid, providing the relative exchange rate is known. Conversely, substances may be added to the infusate so as to produce a concentration gradient which favors its delivery to the extracellular space. The major difference between microdialysis and push–pull perfusion is that the former approach makes use of a semipermeable membrane as a physical boundary between infusate and extracellular fluid, while the latter method allows free contact between these two pools. One important result of this dissimilarity is that no "washing" of tissue by infusate occurs with the microdialysis technique, and thus cell bodies, cell processes, and extracellular compartments are protected from physical perturbation by the procedure. A second consequence is that there is no need to actively withdraw dialyzate from the dialyzed tissue, since under the proper conditions the fluid will already follow this path of least resistance. Thus, potentially deleterious pressure imbalances within the system are virtually eliminated. The process of sample collection is also greatly simplified without the requirement for a withdrawal pump.

Special Considerations in Measurement of Hypothalamic Releasing Factors

Selection of Probe Design. Microdialysis probes designed as a horizontal tube,[4] a semicircular loop,[3] or concentric vertical tubes[5] have been used effectively by several groups in the measurement of neurotransmitter release, the choice of probe being made in most cases on the basis of the surgical accessibility and physicochemical characteristics of the target tissue. For the measurement of hypothalamic releasing factors the same consideration must be given to the overall geometry of the median eminence or anterior pituitary, and to their surrounding structures. In the rat, the vertical, concentric tube design appears to comprise the most viable option owing to the constraints imposed by the small size of these target structures and their limited accessibility via the surrounding bony structures. For the same reasons the use of the conventional dorsal stereotaxic approach is also advisable.

The Microdialysis Membrane. An important feature of any microdialysis system is the dialysis membrane used as the interface between extracellular fluid and infusate. When attempting to monitor peptide levels the choice of membrane composition becomes even more critical, since secondary structure and distribution of charge about the peptide molecule can become determining factors in the ability of a membrane material to pass a particular peptide. Indeed, the latter factors make it quite difficult in many cases to predict whether a membrane possessing a given molecular weight cutoff limit will, in fact, be permeable to a peptide with a molecular weight below that limit. We have determined that the Amicon Vitafiber membrane (Amicon Corporation, Danvers, MA) with a 10,000 or greater MW cutoff is generally suitable for use in the estimation of LHRH levels. Membranes with the 100,000 MW cutoff may sometimes give unsatisfactory exchange rates because they are more easily damaged during probe construction and operation. Commercially available probes (Carnegie-Medicin, Solna, Sweden) have proved most satisfactory for this use, as they provide reliable and reproducible exchange rates and because they are less likely to be damaged during insertion and operation. Our own probes constructed with PVDF (Spectrum, Inc., Los Angeles, CA) have also yielded adequate exchange rates, although the diameter of this tubing is too large for the present application. Probes constructed with cellulose membranes have yielded much less consistent results. Clearly, the empirical determination of optimal membrane type for each individual application is a necessary step in the development of useful probes.

Selection of Target Area. The hypothalamic releasing factors are released from neurovascular terminals in the median eminence and conveyed to the anterior pituitary gland by the hypothalamohypophyseal portal vessel system. After diffusion from the secondary portal plexus into the extracellular spaces of the anterior pituitary, peptide molecules can bind to specific membrane receptors on pituitary cells to regulate the synthesis and secretion of specific pituitary hormones. Given this cascade of neuroendocrine events, it follows logically that physiological changes in peptidergic hormone secretory rates must be reflected by fluctuations in peptide levels in (1) the neurovascular spaces of the median eminence, (2) portal vessel plasma, and (3) the extracellular fluid of the anterior pituitary gland. The distribution of peptidergic axon terminals within the median eminence is nonuniform and usually quite limited in size, and is therefore better approached with a focal perfusion technique such as push–pull perfusion. In addition, sealing of the tip of most microdialysis probes makes them less suitable for measurement of peptides in the median eminence, since exchange through this plane is prevented. In the extracellular spaces of the anterior pituitary, however, the hypothalamic releasing factors are more evenly distributed throughout a larger tissue area, and therefore an elongated exchange surface can be employed. The use of a greater exchange surface, in turn, allows for greater absolute recovery and detectability of peptide. The latter factor provides perhaps the best argument in favor of the use of microdialysis probes in the anterior pituitary, since microdialysis probes can be designed which have longer, cylindrical exchange surfaces. An additional advantage of probe placement in the anterior pituitary is that peptides recovered from pituitary dialyzates are, by definition, hypothalamic releasing factors; peptides recovered from median eminence perfusates or dialyzates may be, in part, of synaptic as well as neurovascular origin.

Design, Construction, and Operation of Pituitary Microdialysis Probes

Physical Requirements. Microdialysis probes that we currently use for pituitary dialysis procedures are designed with particular consideration given to the dimensions of the rat pituitary gland. To maximize the total dialysis exchange surface, a probe tip is constructed of a 2.0-mm-long cylinder of dialysis tubing, this distance corresponding to the maximal dorsoventral depth of anterior pituitary tissue. A second important consideration is the protection of the microdialysis probe as it is advanced through guide cannula, meninges, and pituitary tissue. To shield the delicate dialysis membrane during this process, a tungsten wire is affixed to the tip of the probe so that it sustains the impact of any contact between probe and tissue.

Probe Specifications. The microdialysis probe assembly (Fig. 1) consists of (1) an outer stainless steel cannula (29 gauge), (2) a cylinder of dialysis tubing (2.0 mm in length, 300 μm o.d.; Amicon Vitafiber membrane), (3) a fused silica infusion tube (25 μm i.d., 150 μm o.d.; Anspec Co., Ann Arbor, MI), (4) a fused silica outflow tube (100 μm i.d., 300 μm o.d.), (5) a short length of PE 20 tubing which provides a sleeve for the inflow tube and a housing for the outflow tube, (6) a plastic hub which serves as a male fitting that can be secured within the female fitting on a chronically implanted guide cannula, (7) a hook-shaped tungsten wire (120 μm) that is glued to the stainless steel cannula and the tip of the dialysis tube, and (8) a guide cannula (19 gauge) and stainless steel stylette (28 gauges).

Probe Construction. The microdialysis probe assembly is constructed with the aid of a dissecting microscope, according to the design depicted in Fig. 1. The 29-gauge stainless steel outer cannula is buffed to a fine taper at both ends and acid-etched (successively placed in methanol, a solution of 4% sulfuric acid and 4% hydrochloric acid, a solution of 2% hydrofluoric acid and 12% nitric acid, and a distilled water rinse). A hub

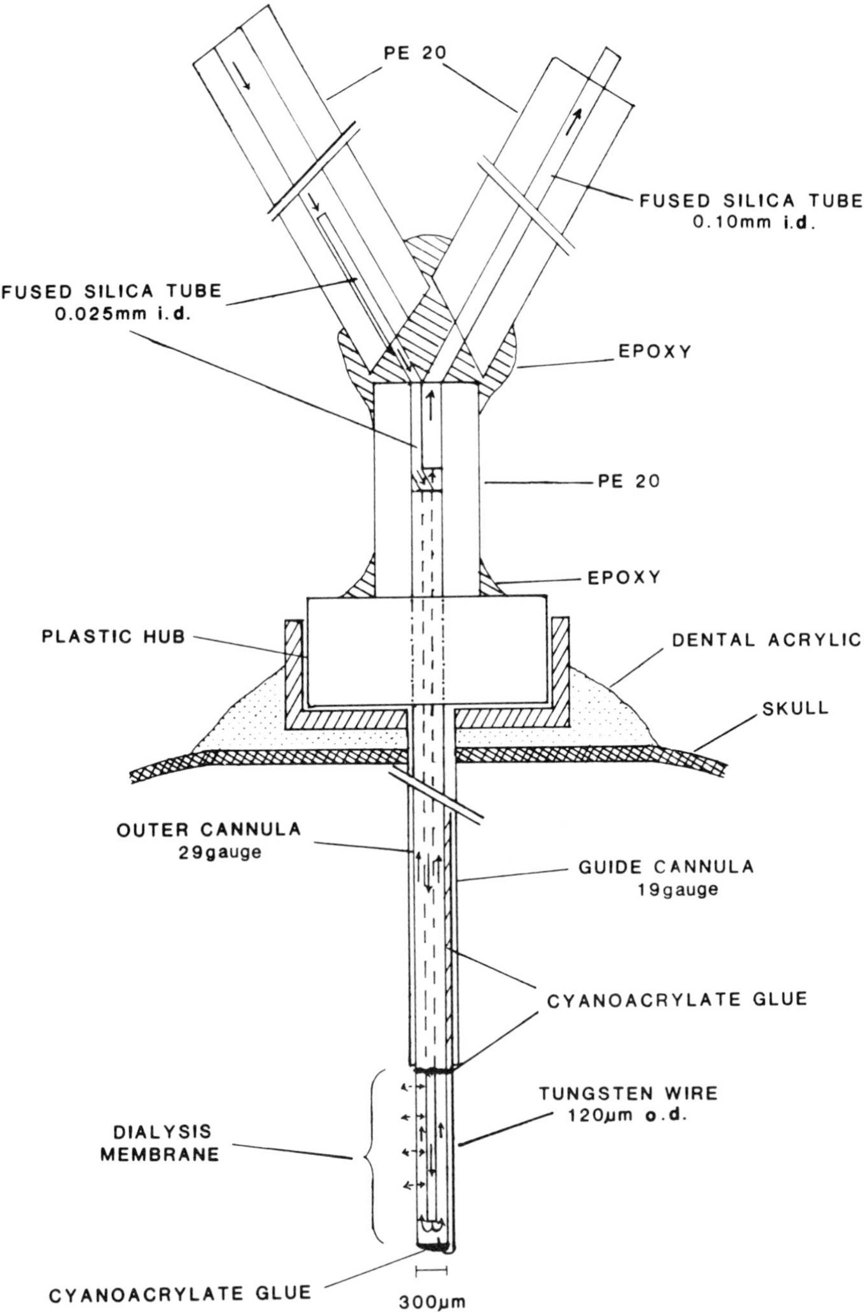

FIG. 1. Schematic diagram of the pituitary microdialysis probe. For details of construction, see the text.

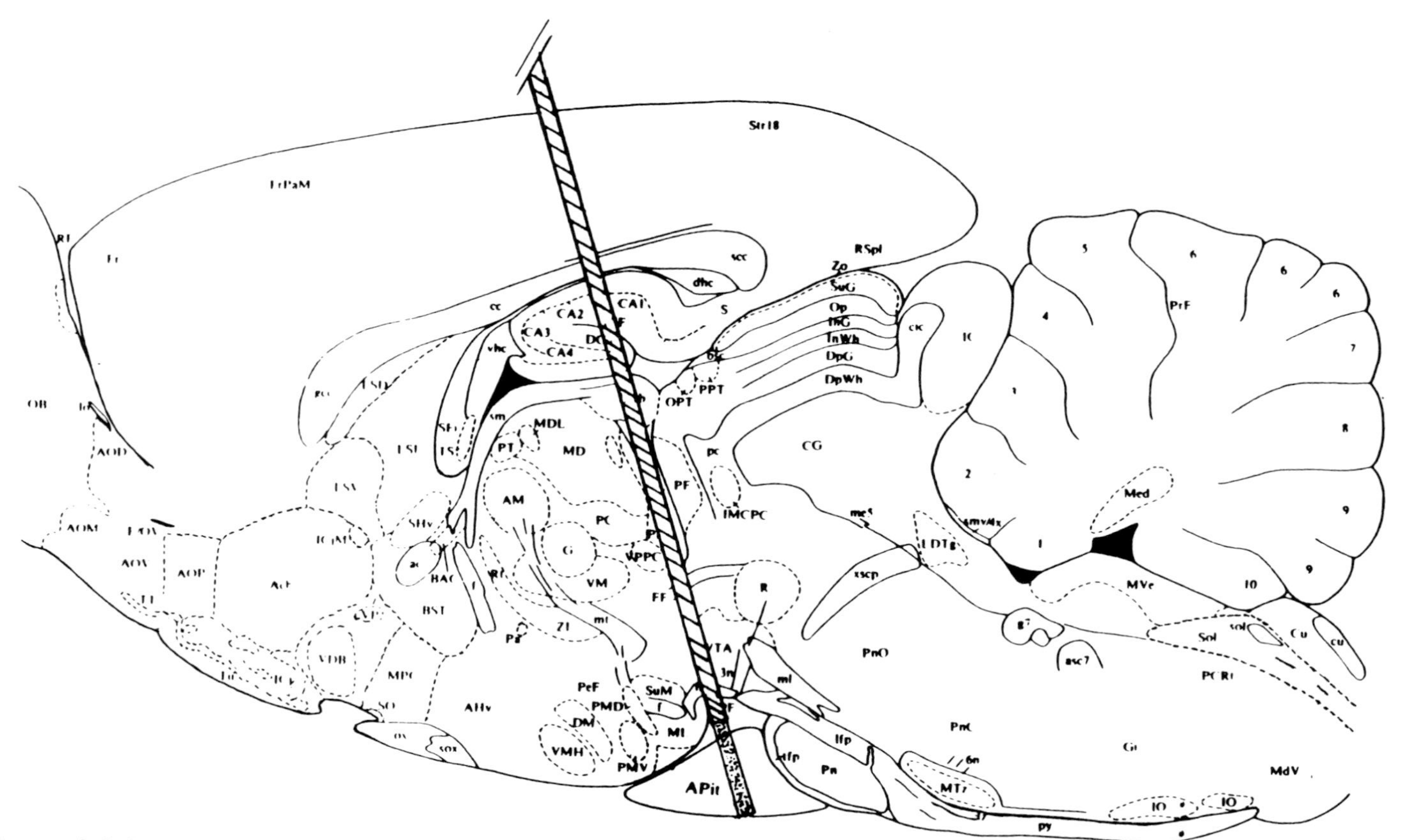

Lateral 0.9 mm

(the end of a plastic 1-ml syringe) is slid along the outer cannula until 2.5 cm of the cannula protrudes from the end of the hub; the hub is then glued in place with 5-min epoxy. An inner infusion assembly is constructed from a short length of PE 20, a fused silica infusion tube, and a fused silica outflow tube, as portrayed in Fig. 1. The infusion tube is cut so that 1.8 mm of the tube protrudes past the end of the outer cannula when the infusion assembly is inserted into it. With the infusion assembly removed from the outer cannula, a 2.0 mm section of dialysis tubing is cut from the hollow fiber bundle and secured to the end of the outer cannula with a minimum of cyanoacrylate glue, the glue being carefully applied with a fine microdissecting wire. A tungsten support wire is carefully hooked over the open tip of the dialysis membrane, and the length of the wire is secured to the outer cannula (but not to the dialysis tube) with cyanoacrylate glue. The tip of the dialysis cylinder is also sealed with an extremely fine drop of glue, without allowing significant capillary movement of the glue upward within the cylinder before drying. The infusion assembly (filled with dialysis medium) is then reinserted within the outer cannula as shown in Fig. 1 and fixed in place with epoxy. One section of PE 20 tubing (also filled with dialysis medium) is slid onto the end of the inflow tube and a second onto the entire length of the outflow tube, and each is secured with epoxy to the infusion assembly. Upon completion, each probe should be tested to verify its suitability with respect to consistency of flow and exchange rate. When not in use, probes should be stored with tips immersed in a sterile solution.

To match each microdialysis probe that is constructed, guide cannulae are fashioned from stainless steel hypodermic needles (19-gauge) and fitted with 28-gauge stylettes. A single guide cannula is implanted in each experimental animal by conventional stereotaxic means so that only the stylette penetrates into pituitary tissue (anterior 2.0 mm, ventral 11.6 mm, lateral 0.9 mm; Fig. 2). The guide tube is cemented to skull screws with dental acrylic.

Automated Collection System. A microliter syringe pump of high quality (syringe infusion pump Model 22, Harvard Apparatus, South Natick, MA, or CMA Syringe Pump, Bioanalytical Systems, West Lafayette, IN) is used to infuse 0.30–2.00 μl/min of a Krebs–Ringer phosphate medium through the dialysis system. To carry out this procedure in an unanesthetized rat, a remote sampling system (Fig. 3) is used which allows the animal freedom about the cage while eliminating the need for a swivel

FIG. 2. Schematic representation of microdialysis probe placement. Parasagittal rat brain section drawn after coordinates of G. Paxinos and C. Watson, "The Rat Brain: In Stereotaxic Coordinates." Academic Press, New York, 1982.

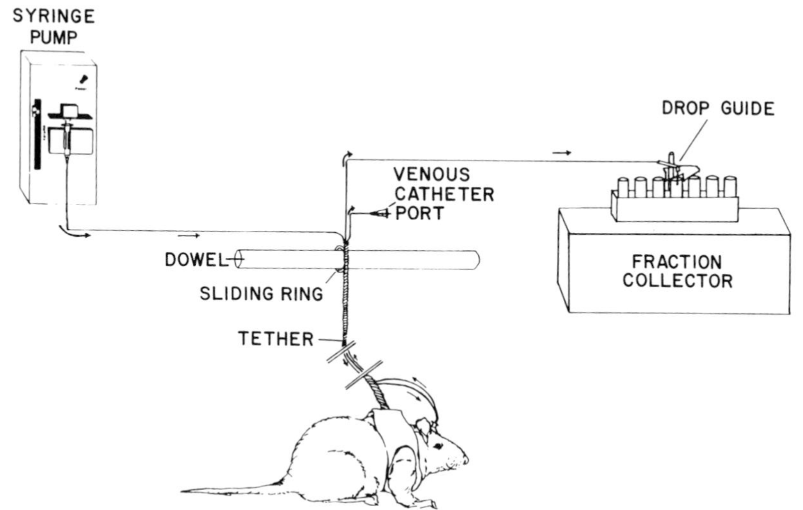

FIG. 3. Diagrammatic representation of the automated microdialysis collection system.

device. Each rat is fitted with a tether and jacket, the tether being attached at the top of the cage to a ring which can slide along the length of a fixed dowel. The dialysis inflow and outflow tubing, as well as a venous catheter if in use, are conducted through the tether to the top of the cage. With this arrangement, the rat is free to move about the cage but is prevented from circling and twisting the tubing. Outflow tubing is connected to a homemade drop guide device fitted to a fraction collector (Retriever III Fraction Collector, I.S.C.O., Lincoln, NE). By means of a small hinge and wire guide, this device ensures that (1) sample drops are collected at the center of collection tubes, (2) the end of the outflow tube briefly touches the inside of the collection tube during advancement, and (3) the outflow tube is lifted and placed in the next tube during advancement to prevent sample loss on the exterior of collection tubes.

Operation of the Microdialysis System. Throughout the day of experimentation the rat is unrestrained except for the tether, ring, and dowel system, and laboratory rat chow and tap water are available *ad libitum*. To initiate experiments, the stylette is removed from the guide cannula, and the microdialysis probe is carefully inserted. If necessary, animals may be etherized just prior to insertion of the probe, and a dissecting microscope can be used to better visualize the advancement of the probe through the guide cannula. With the start of the infusion pump, fluid is pumped to the tip of the probe, and from there it is automatically delivered via the outflow tube to the fraction collector. At a flow rate of 1 μl/

min, the delivery time from rat to collection tube is approximately 2 min, and the total volume of fluid held in the outflow tubing is 2.1 μl. Samples are stored at $-20°$ for subsequent peptide radioimmunoassay. To prevent loss of sample by adsorption to collection tubes or sample handling, the radioimmunoassay procedures are carried out in the collection tubes. When required, blood samples are collected from an atrial catheter and processed for pituitary hormone radioimmunoassays. Following experiments, probe placements and tissue damage are examined by standard histological techniques.

Precautions and Potential Problems. Four potential technical problems in the use of the microdialysis probe that merit particular attention are as follows.

1. Proper handling of the dialysis membrane. The dialysis membrane is easily damaged and must be handled with great care during excision from the hollow fiber bundle and gluing to the outer cannula. Micromanipulation of the microdialysis tubing should be carried out with little or no contact allowed between the critical regions of the membrane and dissecting instruments. We have found that the easiest method of maneuvering the membrane tube is to gently move it about a clean glass dissection surface with forceps. The dialysis membrane should also never be allowed to dry after it is initially wetted, since this can drastically affect the permeability properties of the material.

2. Surgical precautions. Since pituitary size can vary significantly as a function of age or endocrine status, it is advisable to carefully determine the proper stereotaxic coordinates for each experimental application, and to adjust the size of the dialysis cylinder accordingly. It is also recommended that stereotaxic coordinates be selected which allow for avoidance of the main blood vessels of the Circle of Willis during implantation of the guide cannula and stylette.

3. Flow rate. A flow rate should be selected which provides the best compromise between relative recovery (concentration in dialyzate/concentration in test fluid) and absolute recovery (total amount of substance in dialyzate sample). Relative recoveries decrease with increased flow rate, while absolute recoveries increase within a range of flow rates for a given probe. At sufficiently high speeds fluid is forced out of the tip of the probe into the surrounding medium due to the buildup of pressure within the system, and absolute recovery is diminished as a result. A flow rate should be chosen which (a) yields a suitable relative recovery rate, (b) produces an absolute recovery rate which allows for easy detection of the peptide in the radioimmunoassay, and (c) does not produce outward fluid flow from the dialysis chamber to extracellular fluid. A flow rate of 0.67

μl/min has proved suitable in experiments using the probe described above. A gas-tight syringe should be used with syringe pumps to deliver this flow rate with precision and consistency.

4. Pressure in the microdialysis system. As noted above, the development of pressure within the tip of the probe can greatly affect probe performance by causing outward flow of infusate through the dialysis membrane. This situation can be avoided by the use of outflow tubing of greater internal diameter than the inflow tubing. This reduces resistance to flow in the outflow side of the system and thus decreases development of "back pressure." The choice of outflow tubing internal diameter, however, must also be made with consideration to possible loss of resolution between samples owing to an increase in the dead space within the outflow line.

Validation of Probe Performance

Probes constructed according to the preceding methods were tested *in vitro* for relative and absolute recovery of LHRH at various flow rates, and for sensitivity to changes in LHRH concentration. Probe tips were immersed in solutions of radioiodinated or unlabeled LHRH at various concentrations, and dialyzates of these solutions were counted in a gamma counter or processed through the LHRH radioimmunoassay. The results of a test of one probe are depicted in Fig. 4. In this test, a probe was placed successively in solutions containing 7,904 and 16,714 cpm/15 μl at the times indicated, and serial 10-min dialyzate fractions were collected at 1.5 μl/min. As depicted in Fig. 4, the relative recovery rate was approximately 14% and did not vary throughout the experiment. Following the change of test solutions, the amount of LHRH recovered in the dialyzate immediately increased in proportion to the increase in radiolabeled LHRH concentration in the surrounding medium. In other tests the relative LHRH recovery rate at similar flow rates has varied between 4 and 20% for homemade and commercially obtained probes.

In Vivo Application of Pituitary Microdialysis

Male rats (250–500 g body weight) were anesthetized with ketamine (116 mg/kg) and thiamylal sodium (20 mg/kg) and fitted stereotaxically with guide cannulae and stylettes. Animals were allowed to recover from surgery for a minimum of 4 days. Pituitary microdialysis was carried out in each animal for at least 3 hr and, in one case, continuously for 4 days. The flow rate in these experiments was 0.67 μl/min. In some experiments,

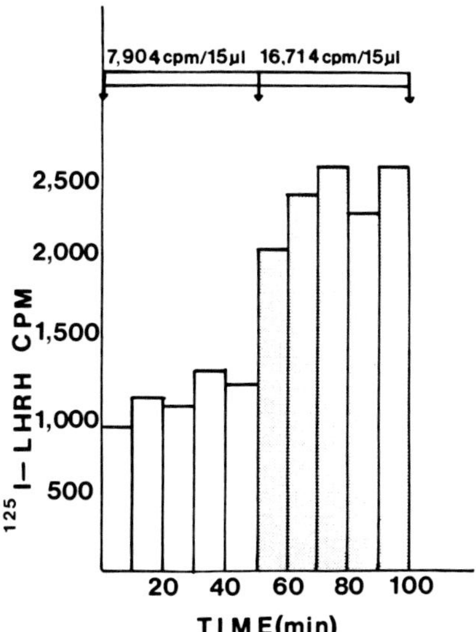

FIG. 4. Analysis of dialysis exchange rates *in vitro* at two different concentrations of synthetic radiolabeled LHRH. Exchange rates at both concentrations were 11.5–16.2%. Arrows indicate the duration of exposure of the microdialysis probe tip to the indicated concentrations.

the microdialysis procedures were interrupted while the animal was etherized and subjected to bilateral castration.

Levels of immunoreactive LHRH in pituitary microdialyzates typically fluctuated between 1 and 4 pg/10 min fraction, although peaks as high as 10 pg/10 min were noted. Patterns of LHRH levels in successive dialyzates were remarkably similar to profiles of LHRH in median eminence[9–11] or pituitary push–pull perfusates,[12] inasmuch as fluctuations (pulses) were evident that exceeded severalfold the amount of variation in the LHRH radioimmunoassay. Results from a 7-hr microdialysis session in a testes-intact male rat are depicted in Fig. 5. A cluster of LHRH pulses occurs at 120–240 min into the microdialysis session, this profile being typical for *in vivo* LHRH[10–12] and LH[13] release patterns in intact male rats. Figure 6 depicts the LHRH levels in a single male rat in two different

[9] J. E. Levine and V. D. Ramirez, *Endocrinology (Baltimore)* **111,** 1439 (1982).

[10] D. Dluzen and V. D. Ramirez, *J. Endocrinol.* **107,** 331 (1985).

[11] M. T. Duffy and J. E. Levine, *Abstr. Annu. Meet. Soc. Neurosci., 16th,* Abstr. 322.9 (1986).

[12] D. Dluzen and V. D. Ramirez, *Neuroendocrinology* **45,** 328 (1987).

[13] G. B. Ellis and C. Desjardins, *Endocrinology (Baltimore)* **110,** 1618 (1982).

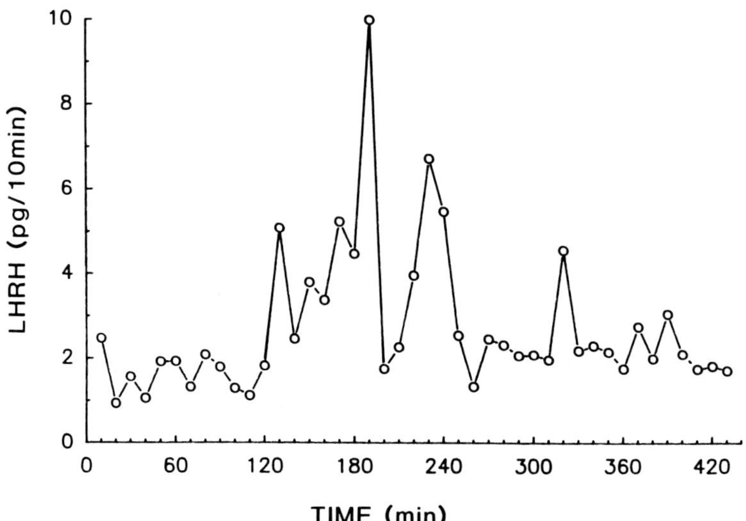

FIG. 5. Profile of LHRH levels throughout a 7-hr pituitary microdialysis experiment in a testes-intact male rat. Flow rate was 0.67 μl/min.

experimental sessions starting at 0 and 98 hr, respectively, after castration. The data demonstrate that the absolute levels of LHRH in this rat are unchanged at 4 days following castration. In current experiments we are attempting to determine if other characteristics of LHRH levels in the microdialyzates (e.g., pulse frequency) are significantly changed at various times following the removal of gonadal feedback.

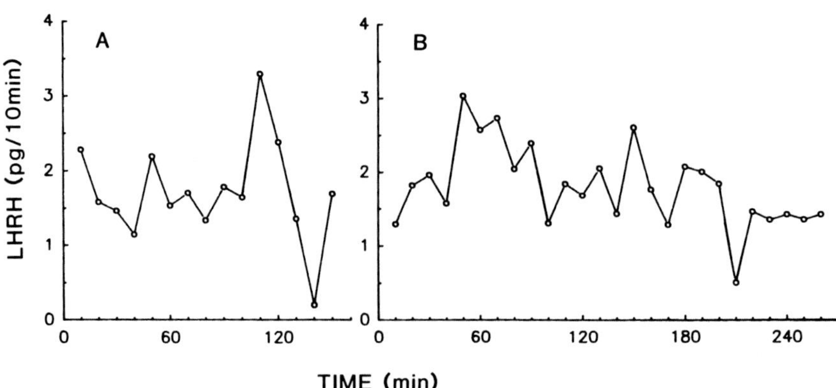

FIG. 6. Profile of LHRH in the extracellular fluid as determined by pituitary microdialysis procedures in an individual male rat (A) immediately following castration and (B) at 4 days following castration.

The average level of LHRH in microdialyzates obtained in these experiments was 2.3 pg/10 min, which is equivalent to 1.7 fmol/10 min. The latter figure is in the range of concentration values obtained in experiments where substance P was measured in striatal dialyzates.[14] Using a relative recovery rate of approximately 14% at the 0.67 μl/min flow rate, we calculate that the average LHRH concentration in the extracellular fluid of the anterior pituitary of the male rat is 1.7 fmol/6.7 μl $\times$ 7.1 = 1.8 n*M*. This value is in reasonable agreement reported EC_{50} values for LHRH stimulation of LH secretion in male rat pituitary cell cultures (reported EC_{50} values ranging between 0.8[15] and 20.0 n*M*[16]). The K_a for LHRH binding to receptors on male rat pituitary membranes (6.2×10^8 M^{-1} [17]) also appears to be in similar range. Comparison of the mean LHRH level in dialyzates versus pituitary push–pull perfusates[12] shows the former to be somewhat higher, a finding that is predicted on the basis of the difference between the two techniques in exchange rate and total exchange surface area.

Advantages and Disadvantages of Pituitary Microdialysis

In Table I we compare the characteristics of pituitary microdialysis and push–pull perfusion of the hypothalamus[10,11] or anterior pituitary gland.[12] For some applications, i.e., when it is desirous to collect peptides at the site of release, push–pull perfusion would appear to be the method of choice for reasons stated in the foregoing discussion. Push–pull perfusion would also prove more advantageous if a suitable microdialysis membrane could not be found for use in the microdialysis procedures, this situation being quite likely for larger peptides possessing hydrophobic/hydrophilic properties that are incompatible with the use of any available membrane. Indeed, the major potential problem in the use of microdialysis procedures is the identification of a membrane material that will allow sufficient exchange of a peptide of interest. A second potential drawback of the microdialysis procedure is the difficulty with which probes are constructed and maintained. Probe failure can occur due to harsh handling, partial or total drying of the dialysis membrane, or by inadvertant outflow of dialyzate through the membrane. The latter condition occurs when pressure within the dialysis system is unacceptably high, and it

[14] E. Brodin, N. Lindefors, and U. Ungerstedt, *Acta Physiol. Scand. Suppl.* **515,** 17 (1983).

[15] D. E. Suter and N. B. Schwartz, *Endocrinology (Baltimore)* **117,** 855 (1985).

[16] F. Kamel, J. A. Balz, C. L. Kubajak, and V. A. Schneider, *Endocrinology (Baltimore)* **120,** 1651 (1987).

[17] R. N. Clayton, K. Channabasavaiah, J. M. Stewart, and K. J. Catt, *Endocrinology (Baltimore)* **110,** 1108 (1982).

TABLE I
DIRECT COMPARISON OF PUSH–PULL PERFUSION OF MEDIOBASAL HYPOTHALAMUS OR PITUITARY GLAND AND MICRODIALYSIS OF PITUITARY TISSUE

Hypothalamic perfusion	Pituitary perfusion	Pituitary microdialysis
Samples at site of release	Samples at site of action	Samples at site of action
Samples neurosecretion and synaptic release	Samples presumed neurosecretion	Samples presumed neurosecretion
Samples from a population of axons	Samples net signal to anterior pituitary	Samples net signal to anterior pituitary
Subacute sampling periods (hours)	Subacute sampling periods (hours)	Chronic sampling periods (days)
Occlusions, hydrostatic pressure changes possible	Occlusions, hydrostatic pressure changes possible	Occlusions, hydrostatic pressure changes not possible
o.d. of outer cannula = 0.51 mm	o.d. of outer cannula = 0.51 mm	o.d. of outer cannula = 0.30 mm
Peptidases possible in perfusate	Peptidases possible in perfusate	Peptidases likely excluded from dialyzate
Infusion and withdrawal, tedious pump-balancing necessary	Infusion and withdrawal, tedious pump-balancing necessary	Infusion only, no pump-balancing necessary
Flow rate 5–20 μl/min	Flow rate 5–20 μl/min	Flow rate 0.5–2 μl/min
Streaming medium contacts tissue	Streaming medium contacts tissue	Moving fluid does not contact tissue
Experimental success rate 50–80%	Experimental success rate 50–80%	Experimental success 95–100% *if* the probe is operative
Easy to assemble	Easy to assemble	Very difficult to assemble

results in damage to the dialysis membrane structure. Purchase of commercial probes (Carnegie Medicin) or meticulous care in the assembly and handling of homemade devices are strongly advised.

The proper use of an acceptable microdialysis probe offers several extremely important advantages in the *in vivo* monitoring of hypothalamic releasing factor levels. Certainly one of the greatest potential advantages of a pituitary microdialysis system is its use in chronic sampling of hypothalamic releasing factors. Since many important neuroendocrine phenomena occur over periods of days or weeks, it is critically important that measurements of hypothalamic releasing factors be made within these time domains to clearly understand their role in these processes. The regulation of LH secretion by LHRH during the course of an entire estrous cycle and progressive changes in LHRH secretory profiles at various time points following the removal of gondal feedback mechanisms are two situations where the characterization of peptide levels over a prolonged period would allow powerful within-subject comparisons to be made under different endocrine conditions, and thereby yield a more complete and detailed physiological picture.

Other major advantages of the microdialysis approach relate to its reduced invasiveness compared to more conventional monitoring procedures. Like push–pull perfusion, the use of microdialysis obviates the need of anesthesia, acute surgical stress, and transection of the pituitary stalk. In contrast to local perfusion procedures, the microdialysis method produces virtually no disturbance of surrounding tissue during experiments, produces implantation lesions that are reduced in size, and once in operation it is not subject to deleterious occlusions and/or pressure imbalances. Furthermore, the need for postcollection treatment of samples, e.g., acid or methanol extraction, is eliminated as peptidases are excluded from the dialyzate by the microdialysis membrane. A final important advantage of the microdialysis system is that it can be totally automated, as described in a preceding section. A welcome result is that the labor intensity of the method is reduced and the rate of data acquisition increased.

Conclusions

Our preliminary data suggest that pituitary microdialysis may offer a valuable new approach in the study of hypothalamic releasing factor release. It is a method of reduced invasiveness and holds great promise as a reliable technique for automated, continuous sampling of hypothalamic releasing factors over prolonged experimental periods. Combined with atrial catheterization and radioimmunoassay of peripheral pituitary hormone levels, the pituitary microdialysis approach should also be of great value in the analysis of dynamic relationships between hypothalamic peptide release and the secretion of the pituitary hormones governed by these factors. It may even become possible through the choice of an appropriate dialysis membrane to measure both a hypothalamic releasing factor and its target pituitary hormone in the same microdialyzate samples, thus eliminating the need for collection and processing of blood samples. Although there remain obstacles to overcome in some applications of the microdialysis method, such as adsorption of some peptides by certain dialysis membrane materials, we anticipate that technical developments in the near future will minimize these problems and encourage more widespread use of the approach in neuroendocrine research.

[11] Use of Microdialysis in Neuroendocrinology

By KEITH M. KENDRICK

Introduction

Techniques for *in vivo* sampling of neurotransmitter release into the extracellular space in the brain have mainly evolved out of a need to measure short-term changes in such release in specific brain regions. Although the sampling of neurotransmitter "overflow" into the extracellular space is, at best, an indirect sign of changes in synaptic activity, it remains, potentially, one of the most powerful methods of demonstrating such changes in specific neurotransmitter systems and correlating them with behavioral, endocrine, or other events.

The most widely used method for sampling neurotransmitter release *in vivo* is the push–pull cannula technique first introduced by Gaddum in 1961.[1] The application of this technique to neuroendocrinology has been reviewed recently,[2,3] and it has been used successfully to measure the *in vivo* output of a number of substances including acetylcholine, amino acids, monoamines and their metabolites, and peptides. Push–pull perfusion is, however, subject to a number of drawbacks. These include the fact that the tissue surrounding the probe is damaged, to varying degrees, by the necessity to push liquid out into it. This can lead to blood contaminated samples and to blockage in the pull line by small pieces of damaged tissue. Such blockages increase tissue damage, and inconsistent volumes may be collected over sampling periods, by producing an inbalance in the flow rates in the push and pull lines. This potential hazard often compels the experimenter to monitor constantly the animal undergoing sampling, especially where a conscious preparation is used. A second problem with the technique is that it is not possible to calibrate the system so that an estimate can be made of the actual concentrations of substances in the extracellular space of the region sampled. Other *in vivo* sampling methods include cortical-cup perfusions,[4] which are restricted to the surfaces of the brain and cannot be used on subcortical structures, and more recently

[1] J. H. Gaddum, *J. Physiol.* (*London*) **155,** 1 (1961).

[2] J. E. Levine and V. D. Ramirez, this series, Vol. 124, p. 466.

[3] V. D. Ramirez, *in* "*In Vivo* Perfusion and Release of Neuroactive Substances: Methods and Strategies" (A. Bayón and R. Drucker-Colín, eds.), p. 249. Academic Press, New York, 1985.

[4] F. Moroni and G. Pepeu, *in* "Measurement of Neurotransmitter Release *in Vivo*" (C. A. Marsden, ed.), p. 63. Wiley, New York, 1984.

METHODS IN ENZYMOLOGY, VOL. 168

voltammetry[5,6] has provided a sophisticated, if somewhat nonspecific, method of rapid monitoring of monoamine and ascorbic acid release.

With the potential problems inherent in the above methods of *in vivo* sampling, it is perhaps not surprising that the majority of neurochemical studies have confined themselves to simple analysis of homogenized tissue as their main tool for investigating neurotransmitter release. Also, more recently, the development of a number of *in vitro* sampling techniques,[2,7] using tissue slices, has provided a further attractive alternative. However, an *in vivo* sampling technique has been developed which is an extension of the push–pull method but without many of its disadvantages. This technique, called microdialysis, or intracranial dialysis when applied to the brain, involves the perfusion of fluid inside a semipermeable membrane instead of directly in the tissue. The idea was first put forward by Delgado *et al.* in 1971[8] but was adapted by Ungerstedt and Pycock[9] using hollow dialysis fibers perfused by physiological fluid. Although the technique has been around for more than a decade, microdialysis has, to date, been used virtually exclusively by neuropharmacologists and particularly to investigate the release of striatal dopamine. Nevertheless, microdialysis has been used to measure release of acetylcholine,[10,11] amino acids,[12–36]

[5] R. N. Adams and C. A. Marsden, *in* "Handbook of Psychopharmacology" (L. L. Iversen, S. P. Iversen, and S. H. Synder, eds.), Vol. 15, p. 1. Plenum, New York, 1982.

[6] C. A. Marsden, M. P. Brazell, and N. T. Maidment, *in* "Measurement of Neurotransmitter Release *in Vivo*" (C. A. Marsden, ed.), p. 127. Wiley, New York, 1984.

[7] E. Gallardo and V. D. Ramirez, *Proc. Soc. Exp. Biol. Med.* **155,** 79 (1977).

[8] J. M. R. Delgado, F. V. Defeudis, R. H. Roth, D. K. Ryugo, and B. M. Mitruka, *Arch. Int. Pharmacodyn. Ther.* **198,** 9 (1971).

[9] U. Ungerstedt and C. Pycock, *Bull. Schweiz. Akad. Med. Wiss.* **30,** 44 (1974).

[10] S. Consolo, C. F. Wu, F. Fiorentini, H. Ladinsky, and A. Vezzani, *J. Neurochem.* **48,** 1459 (1987).

[11] G. Damsma, B. H. C. Westerink, J. B. de Vries, C. J. Van den Berg, and A. S. Horn, *J. Neurochem.* **48,** 1523 (1987).

[12] H. Benveniste, J. Drejer, A. Schousboe, and N. H. Diemer, *J. Neurochem.* **43,** 1369 (1984).

[13] L. Brodin, U. Tossman, U. Ungerstedt, and S. Grillner, *Neurosci. Lett. Suppl.* **26,** S419 (1982).

[14] S. P. Butcher and A. Hamberger, *J. Neurochem.* **48,** 713 (1987).

[15] S. P. Butcher, M. Sandberg, H. Hagberg, and A. Hamberger, *J. Neurochem.* **48,** 722 (1987).

[16] J. Drejer, H. Benveniste, N. H. Diemer, and A. Schousboe, *J. Neurochem.* **45,** 145 (1985).

[17] A. Hamberger, C.-H. Karlsson, A. Lehmann, and B. Nystrom, *in* "Glutamine, Glutamate and GABA in the Central Nervous System: Neurology and Neurobiology" (L. Hertz, E. Kvamme, E. McGeer, and A. Schousboe, eds.), Vol. 7, p. 473. Liss, New York, 1983.

[18] A. Hamberger, S. P. Butcher, H. Hagberg, I. Jacobson, A. Lehmann, and M. Sandberg,

monoamines and metabolites,[21,22,35,37–63] neuropeptides such as luteinizing hormone-releasing hormone,[64,65] oxytocin,[22,23,49,66] substance K,[25,67] substance P,[25,35,67–72] and purines[35,73–75] from a number of brain regions, and it is therefore a potentially useful technique in the field of neuroendocri-

in "Excitatory Amino Acids" (P. J. Roberts, J. Storm-Mathison, and H. F. Bradford, eds.). Macmillan, New York, in press.

[19] A. Hamberger and B. Nystrom, *Neurochem. Res.* **9,** 1181 (1984).

[20] I. Jacobson and A. Hamberger, *Brain Res.* **294,** 103 (1984).

[21] R. D. Johnson and J. B. Justice, *Brain Res. Bull.* **10,** 567 (1983).

[22] K. M. Kendrick, C. De la Riva, and B. A. Baldwin, *Neurosci. Lett. Suppl.* **29,** S50 (1987).

[23] K. M. Kendrick, E. B. Keverne, C. Chapman, and B. A. Baldwin, *Brain Res.* **442,** 171 (1988).

[24] J. Korf and K. Venema, *J. Neurochem.* **45,** 1341 (1985).

[25] N. Lindefors, U. Tossman, J. Segovia, E. Brodin, and U. Ungerstedt, *Brain Res.*, in press.

[26] M. Sandberg, S. P. Butcher, and H. Hagberg, *J. Neurochem.* **44,** 42 (1986).

[27] M. Sandberg and S. Lindstrom, *J. Neurosci. Methods* **9,** 65 (1983).

[28] J. Segovia, U. Tossman, M. Herrera-Marschitz, M. Garcia-Munoz, and U. Ungerstedt, *Neurosci. Lett.* **70,** 364 (1986).

[29] U. Tossman, S. Eriksson, A. Delin, L. Hagenfeldt, D. Law, and U. Ungerstedt, *J. Neurochem.* **41,** 1046 (1983).

[30] U. Tossman, G. Jonsson, and U. Ungerstedt, *Acta Physiol. Scand.* **127,** 533 (1986).

[31] U. Tossman, J. Segovia, and U. Ungerstedt, *Acta Physiol. Scand.* **127,** 547 (1986).

[32] U. Tossman and U. Ungerstedt, *Eur. J. Pharmacol.* **123,** 295 (1986).

[33] U. Tossman and U. Ungerstedt, *Acta Physiol. Scand.* **128,** 9 (1986).

[34] U. Tossman, T. Wieloch, and U. Ungerstedt, *Neurosci. Lett.* **62,** 231 (1985).

[35] U. Ungerstedt, *in* "Measurement of Neurotransmitter Release *in Vivo*" (C. A. Marsden, ed.), p. 81. Wiley, New York, 1984.

[36] A. M. Young and H. F. Bradford, *J. Neurochem.* **47,** 1399 (1986).

[37] R. D. Blakely, S. A. Wages, J. B. Justice, Jr., J. G. Herndon, and D. B. Neill, *Brain Res.* **308,** 1 (1984).

[38] G. Di Chiara and A. Imperato, *Ann. N.Y. Acad. Sci.* **473,** 367 (1986).

[39] W. H. Church and J. B. Justice, Jr., *Anal. Chem.* **59,** 712 (1987).

[40] W. H. Church, J. B. Justice, Jr., and D. B. Neill, *Brain Res.* **412,** 397 (1987).

[41] W. H. Church, J. B. Justice, Jr., and L. D. Byrd, *Eur. J. Pharmacol.* **139,** 345 (1987).

[42] I. Fairbrother and G. W. Arbuthnott, *Neurosci. Lett. Suppl.* **29,** S52 (1987).

[43] L. Hernandez, B. G. Stanley, and B. G. Hoebel, *Life Sci.* **39,** 2629 (1986).

[44] L. Hernandez, L. Paez, and C. Hamlin, *Pharmacol. Biochem. Behav.* **18,** 159 (1982).

[45] P. H. Hutson, G. S. Sarna, B. J. Sahakian, C. T. Dourish, and G. Curzon, *Ann. N.Y. Acad. Sci.* **473,** 321 (1986).

[46] P. M. Hutson and G. Curzon, *Neurosci. Lett. Suppl.* **29,** S51 (1987).

[47] A. Imperato and G. Di Chiara, *J. Neurosci.* **4,** 966 (1984).

[48] A. Imperato and G. Di Chiara, *J. Neurosci.* **5,** 297 (1985).

[49] K. M. Kendrick, E. B. Keverne, C. Chapman, and B. A. Baldwin, *Brain Res.* **439,** 1 (1988).

[50] K. M. Kendrick and G. Leng, *Brain Res.* **440,** 402 (1988).

[51] C. Routledge and C. A. Marsden, *Ann. N.Y. Acad. Sci.* **473,** 537 (1986).

[52] T. Sharpe, T. Ljungberg, T. Zetterstrom, and U. Ungerstedt, *Pharmacol. Biochem. Behav.* **24,** 1755 (1986).

nology. Microdialysis probes can also be used to introduce substances into tissue without the necessity of pumping liquid out into it[76] and to sample concentrations of substances from physiological fluids, from blood, for example.[77]

In this review I discuss the principles of microdialysis and present an overview of the different probe designs and their *in vitro* performances over a range of substances including physiological amino acids, monoamines and their metabolites, neuropeptides, and purines. I also present a brief review of a few *in vivo* neuroendocrine experiments using

[53] T. Sharpe, N. T. Maidment, M. P. Brazell, T. Zetterstrom, G. W. Bennett, and C. A. Marsden, *Neuroscience* **12,** 1213 (1984).

[54] T. Sharpe, T. Zetterstrom, T. Ljungberg, and U. Ungerstedt, *Brain Res.* **401,** 322 (1987).

[55] T. Sharpe, T. Zetterstrom, and U. Ungerstedt, *J. Neurochem.* **47,** 113 (1986).

[56] R. E. Strecker, T. Sharp, P. Brundin, T. Zetterstrom, U. Ungerstedt, and A. Bjorklund, *Neuroscience* **22,** 169 (1987).

[57] S. A. Wages, W. H. Church, and J. B. Justice, Jr., *Anal. Chem.* **58,** 1649 (1986).

[58] T. Zetterstrom, M. Herrera-Marschitz, and U. Ungerstedt, *Brain Res.* **376,** 1 (1986).

[59] T. Zetterstrom, P. Brundin, F. H. Gage, T. Sharpe, O. Isacson, S. B. Dunnett, U. Ungerstedt, and A. Bjorkland, *Brain Res.* **362,** 344 (1986).

[60] T. Zetterstrom, T. Sharpe, C. Marsden, and U. Ungerstedt, *J. Neurochem.* **41,** 1769 (1983).

[61] T. Zetterstrom, T. Sharpe, and U. Ungerstedt, *Naunyn-Schmiedeberg's Arch. Pharmacol.* **334,** 117 (1986).

[62] T. Zetterstrom, T. Sharpe, and U. Ungerstedt, *Neurosci. Lett. Suppl.* **29,** S48 (1987).

[63] T. Zetterstrom and U. Ungerstedt, *Eur. J. Pharmacol.* **97,** 29 (1984).

[64] J. E. Levine and K. D. Powell, this volume [10].

[65] K. D. Powell and J. E. Levine, *Biol. Reprod. Suppl.* **36,** 214 (1987).

[66] B. A. Baldwin, C. Chapman, K. M. Kendrick, and E. B. Keverne, *J. Physiol.* (*London*) **388,** 8P (1987).

[67] N. Lindefors, E. Brodin, and U. Ungerstedt, *J. Pharmacol. Methods,* in press.

[68] E. Brodin, B. Linderoth, B. Gazelius, and U. Ungerstedt, *Neurosci. Lett.* **76,** 357 (1987).

[69] E. Brodin, N. Lindefors, and U. Ungerstedt, *Acta Physiol. Scand. Suppl.* **515,** 17 (1983).

[70] N. Lindefors, E. Brodin, E. Theodorsson-Norheim, and U. Ungerstedt, *Regul. Pept.* **10,** 217 (1985).

[71] N. Lindefors, E. Brodin, and U. Ungerstedt, *in* "Substance P" (E. Scrabenec and D. Powell, eds.), p. 239. Boole Press, Dublin, 1984.

[72] N. Lindefors, Y. Yammamoto, T. Pantaleo, H. Lagercrantz, E. Brodin, and U. Ungerstedt, *Neurosci. Lett.* **69,** 94 (1986).

[73] H. Hagberg, P. Andersson, J. Lacarewicz, I. Jacobson, S. Butcher, and M. Sandberg, *J. Neurochem.* **49,** 227 (1987).

[74] K. M. Kendrick, B. A. Baldwin, T. R. Cooper, and D. F. Sharman, *Neurosci. Lett.* **70,** 272 (1986).

[75] T. Zetterstrom, L. Vernet, U. Ungerstedt, U. Tossman, B. Jonzon, and B. B. Fredholm, *Neurosci. Lett.* **29,** 111 (1982).

[76] M. Ruggeri, K. Fuxe, U. Ungerstedt, A. Harfstrand, and L. F. Agnati, *Neurosci. Lett. Suppl.* **26,** S26 (1986).

[77] U. Ungerstedt, *Neurosci. Suppl.* **22,** 1986P (1987).

microdialysis techniques. A further review of the application of microdialysis to neuropeptide measurement can be found in this volume.[64]

Principles of Microdialysis

Ungerstedt has likened the technique of microdialysis to implantation of an artificial blood vessel into tissue.[78] Essentially, it involves implanting into the brain, or other tissue, a probe which has a small length of semipermeable tubing attached at the end. Liquid (either 0.9% NaCl or artificial cerebrospinal fluid) is then pumped through the probe. The idea is that chemical substances will diffuse in the direction of lowest concentration from the brain, or other tissue, into the fluid flowing around the inside of the dialysis tubing. The sampling area of these probes is thought to be approximately 1 mm around the membrane.[78] Unlike push–pull, however, by increasing the length of the membrane the amount of tissue sampled can be substantially increased. Changes in the amount of substance that is recovered in the fluid flowing inside the dialysis probe will therefore be directly proportional to the changes in their concentration in the tissue immediately outside the dialysis membrane. An estimate, of the actual concentration of the substance of interest in the extracellular fluid, can be made by calculating a recovery factor for each probe using *in vitro* calibration tests. This recovery estimate (called "relative recovery"[35]) will depend on a number of factors including the molecular weight cutoff of the membrane used, its size and wall thickness, and the speed of perfusion. Other factors such as substances binding to the membrane, or degrading rapidly, may also reduce the probe's recovery, and this may rule out the technique for reliably measuring release of some large molecular weight, hydrophobic peptides.

Probe Design and Dialysis Membranes

There are a number of different probe designs which have been described in the literature; however, the most important factor in deciding on which probe to use is the performance of the membrane to be attached to it. Table I gives a list of some of the membranes available, substances measured by them *in vivo,* and suppliers. Figure 1 is a photograph of a number of different probes which have been used, and Fig. 2 shows their designs schematically. The membrane is either attached to the probe in a straight length or in a U-shape. The U-shape design (Figs. 1D–G, 2D and E) is the easiest to construct and simply consists of two lengths of stainless-steel tubing joined at one end by a length of dialysis tubing (Figs. 1F

[78] U. Ungerstedt, *Curr. Sep.* **7,** 43 (1986).

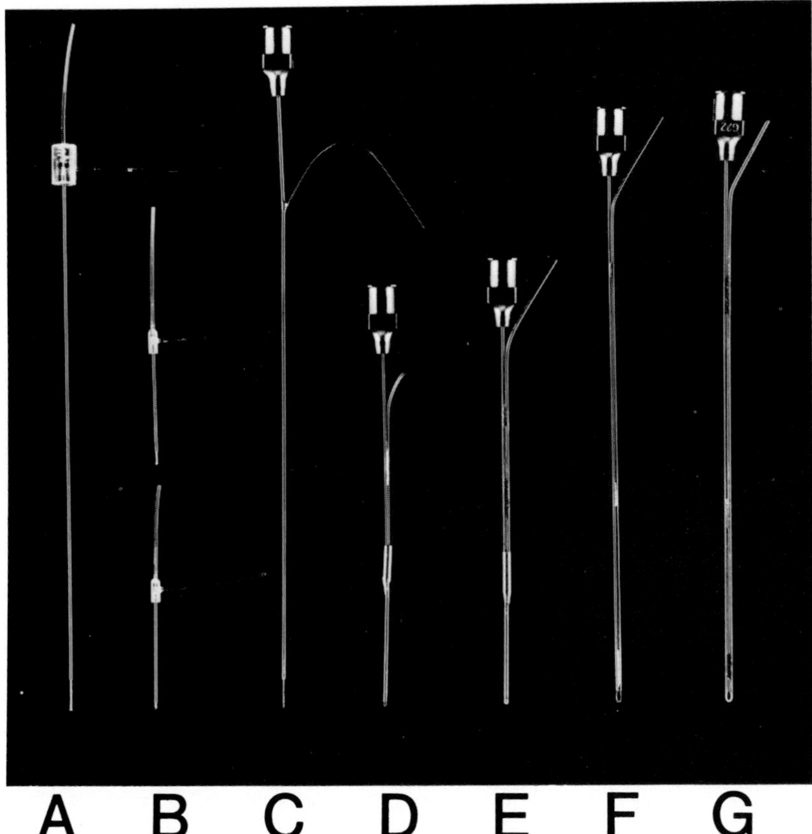

FIG. 1. Microdialysis probes. (A) Concentric design probe supplied by Carnegie Medicin for use in large animal (with a 5 mm membrane length). (B) Same as A but designed for use in rodents (2 mm membrane length above and 1 mm below). (C) Concentric probe with 5 mm cellulose membrane (0.32 mm o.d., supplied by Enka Glanzstoff) for use in large animals. (D) Loop design probe with 2 mm of cellulose membrane exposed (membrane is 0.25 mm o.d., supplied by Medicell Int.) designed for rats. (E) Same as D but with a 2 mm length of Amicon Vitafiber membrane exposed (Amicon Type 3×50, 50,000 MW cutoff). (F) Loop design probe with membrane (5 mm, Medicell Int.) attached to two lengths of stainless-steel tubing, for use in large animals. (G) Same as F but with a 5 mm length of Amicon Vitafiber membrane.

and G and 2E) or a continuous loop inside one length of stainless-steel tubing (Figs. 1D and E and 2D). Lengths of exposed membrane are usually between 2 and 5 mm although lengths of up to 10 mm have been used. To maintain the rigidity of the membrane loop, usually a length of nylon or metal is inserted into it (see Fig. 2D and E). A second type of design

TABLE I
MICRODIALYSIS MEMBRANES AND SUBSTANCES MEASURED BY THEM *in Vivo*

Membrane material	o.d. (mm)	MW cutoff	Supplier	Substances measured	Brain region sampled
Cellulose	0.20	6,000	B.R.I.[a] or B.A.S.[b]	Amino acids	Striatum[21]
				Monoamines	Striatum,[21,37,43] N. accumbens, lat. hypothalamus, N. paraventricular[43]
Cellulose	0.25	5,000	Medicell Int.[c] or Dow Corp.[d]	Acetylcholine	Striatum[10]
				Amino acids	Preoptic area[22]
				Monoamines	N. accumbens,[54,60] frontal cx.,[46] lat. hypothalamus,[45,51] preoptic area,[22] striatum,[35,53–56,58–60] N. supraoptic[50]
				Purines	Striatum,[75] zona incerta[74]
Cellulose	0.32	5,000	Enka Glanz.[e] or COBE Labs.[f]	Amino acids	Hippocampus,[26] striatum,[14,15,36] N. lat. geniculate,[27] olfactory bulb[20]
				Monoamines	Striatum[39–42,57]
				Purines	Striatum[73]
Copolymer[g]	0.31	15,000	Dasco[h]	Acetylcholine	Hippocampus[10]
Copolymer[i]	0.50	20,000	Carnegie Med.[j] or B.A.S.[b]	Amino acids	N. accumbens, hippocampus, hypothalamus, cerebellum, occipital cx., septum, olfactory tubercle, frontal cx., ventral tegmental area,[30] globus pallidus,[28,30,31] spinal cord,[13] striatum,[25,30,31] substantia nigra[25,30]
				Monoamines	N. accumbens,[76] striatum[38,47,48]

				Substance K	N. accumbens,[67] N. solitary tract,[72] striatum,[25] substantia nigra[25]
				Substance P	N. accumbens,[67] striatum,[25] substantia nigra[25]
Acrylic copolymer	0.30	50,000	Amicon[k]	Amino acids	Hippocampus,[12,16] olfactory bulb,[23] striatum[29,32–35]
				Monamines	Olfactory bulb,[49] striatum,[63] substantia nigra[49]
				Purines	Olfactory bulb,[49] substantia nigra[49]
				Substance P	Striatum[35,69,70]
				Oxytocin	Olfactory bulb,[22,23,49,66] substantia nigra[22,49,66]
Polysulfone	0.30	100,000	Amicon[k]	LHRH	Anterior pituitary[64,65]

[a] Brain Research Instruments, 207 Hartley Av., Princeton, NJ 08540 (suppliers of complete concentric probes—2 or 4 mm membrane lengths—as well as membrane).

[b] Biological Analytical Systems, 2701 Kent St., Lafayette, IN 47906.

[c] Medicell Int., 239 Liverpool Rd., London N1 1LX, England.

[d] Dow Corp., Midland, MI 48640.

[e] Enka Glanzstoff AG, Werk Wuppertal-Barmen, Product Group Dialyzing Membrane, Oder Str. 28, P.O. Box 20 09 16, D-5600 Wuppertal 2, Federal Republic of Germany.

[f] Cuprophan Type B4AH, COBE HF 130 dialyzer; COBE Labs., Gloucester, England.

[g] Polyacrylonitrile/sodium methallyl sulfonate.

[h] Dasco, Bologna, Italy.

[i] Polycarbonate/polyether.

[j] Carnegie Medicin, Roslagsvagen 101, S-104 05 Stockholm, Sweden (available only as part of concentric probe: 1, 2, 4, 5, and 10 mm membrane lengths).

[k] Amicon Corp., 17 Cherry Hill Dr., Danvers, MA 01923; Model 3×50 Vitafiber filtration cartridge for the 50,000 MW cutoff membrane and 3S100 for the 100,000.

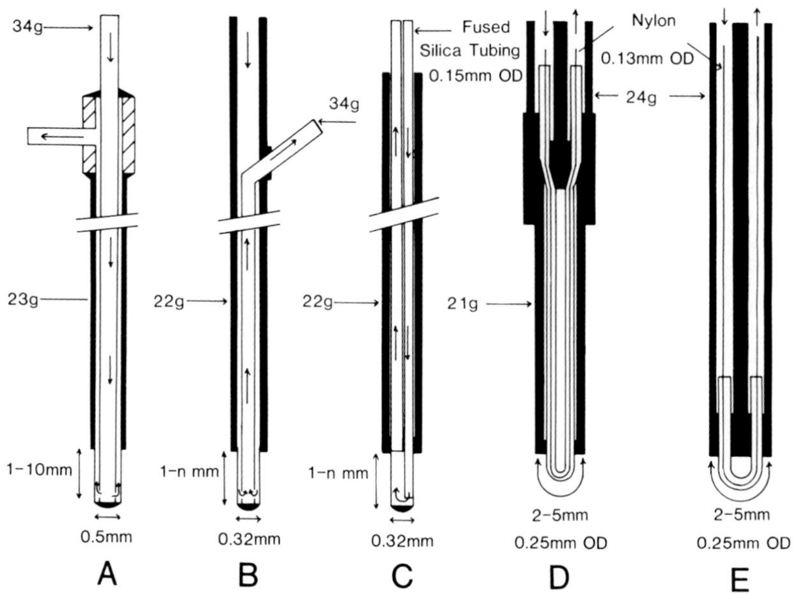

FIG. 2. Designs of some microdialysis probes. (A) Concentric Carnegie Medicin probe[78] (Fig. 1A and B). (B) Concentric design but using a 0.32 mm o.d. cellulose membrane (Enka Glanzstoff; Fig. 1C). (C) Widely used concentric design employing the same membrane as B but with the inflow and outflow tubes in glass.[26,39,57] (D) Loop design with cellulose membrane (Medicell Int.) inside a single stainless-steel tube[50] (Fig. 1D). The same design using the Amicon Vitafiber membrane (Fig. 1E) requires 19- or 20-gauge tubing at the bottom and 22-gauge tubing at the top. This general design has been widely used.[27,35] (E) Loop design with cellulose membrane (Medicell Int.) inside two lengths of stainless-steel tubing[74] (Fig. 1F). The same design, but using the Amicon Vitafiber membrane, requires 22-gauge tubing[23,49] (Fig. 1G). In all cases the direction of flow is indicated by arrows, and membranes are cemented in place using epoxy resin or similar substances.

(not illustrated) is a long, straight probe, which is inserted horizontally through an animal's head. This latter probe was designed for bilateral sampling of the striatum[10,34,38,47,48,63,75] and hippocampus[12,16,17] and offers considerable advantages in detecting release of substances since a large amount of tissue is sampled.

The final, most elegant, design is a concentric probe with a straight length of membrane (Figs. 1A–C and 2A–C). The probe is either constructed so that liquid flows down a fine central tubing and back up a large outer tubing (or vice versa), similar to a push–pull cannula (Fig. 2A and B), or two glass tubes are inserted into the membrane (one ending at the top of the membrane—the outflow—and the other ending close to the bottom of the membrane—the inflow). The two main advantages of concentric probe designs are that they can be made smaller than U-shape probes (down to 26-gauge outer tubing, for example[43]) and they do not cause so much damage if they twist around while being inserted into the brain (which can happen in chronic preparations when probes are being lowered into the brain through guide tubes). Two of these concentric designed probes are available commercially (produced by B.R.I. and Carnegie Medicin; see Table I).

Pumps

Most syringe pumps, capable of producing flow rates between 0.5 and 20 μl/min, are suitable for use with microdialysis probes. In our laboratory we have used a number of different models of Harvard and Sage pumps. We have also used a battery-driven miniature syringe pump (Graseby Medical, Watford, England, Model MS16A). Carnegie Medicin supplies a pump specially designed for use with microdialysis probes (Model CMA/100). The concentric design probes have a greater back pressure than other probe designs, and this necessitates the use of good quality pumps and pressure-resistant connections between pump and probe, preferably in Teflon.

In Vitro Recoveries of Microdialysis Probes

Before using microdialysis probes to measure release of substances in brain or other tissue, it is essential first to assess their performance in recovering those substances *in vitro*. While such *in vitro* recoveries may not precisely reflect *in vivo* conditions, they do provide a basis for estimating extracellular concentrations of substances in the brain and, in conjunction with tissue homogenate data, allow some estimate of the relationship between extracellular and intracellular concentrations. We have carried out such *in vitro* testing on a number of different probes including those manufactured by Carnegie Medicin (2 and 5 mm membrane lengths, Fig. 1A and B), concentric probes with the 0.32 mm o.d. Enka Glanzstoff cellulose membrane (2 and 5 mm lengths, Fig. 1C), the U-shape design with the 0.25 mm o.d. Medicell cellulose membrane (2 and 5 mm lengths, Fig. 1D and F) and the 0.30 mm o.d. Amicon 3×50 Vitafiber membrane (2 and 5 mm lengths, Fig. 1E and G). Recoveries of amino acids, monoamines and metabolites, neuropeptides, and purines are presented in Table II.[79,80] The recovery figures are not influenced by the

[79] P. Lindroth and K. Mopper, *Anal. Chem.* **51,** 1667 (1979).

[80] K. M. Kendrick, E. B. Keverne, B. A. Baldwin, and D. F. Sharman, *Neuroendocrinology* **44,** 149 (1986).

TABLE II

MEAN *in Vitro* RECOVERIES (%) OF MICRODIALYSIS PROBES[a]

Substances	20,000 MW, 0.5 mm o.d. 5 mm	20,000 MW, 0.5 mm o.d. 2 mm	50,000 MW, 0.3 mm o.d. 5 mm	50,000 MW, 0.3 mm o.d. 2 mm	5,000 MW, 0.32 mm o.d. 5 mm	5,000 MW, 0.32 mm o.d. 2 mm	5,000 MW, 0.25 mm o.d. 5 mm	5,000 MW, 0.25 mm o.d. 2 mm
Amino acids								
Alanine	40.5	20.5	10.9	5.5	16.9	7.8	15.5	6.5
Arginine	40.5	20.3	11.6	6.0	13.0	6.2	12.8	6.2
Aspartate	40.2	21.2	11.0	5.7	15.0	7.5	11.9	5.7
GABA	33.3	17.2	7.8	3.9	12.9	6.5	9.6	4.9
Glutamate	40.5	20.2	9.9	5.0	12.1	6.4	11.3	5.9
Glutamine	38.0	20.1	10.5	5.5	13.0	7.0	12.4	6.3
Glycine	42.6	22.1	12.6	6.8	17.6	9.0	17.2	10.5
Histidine	36.1	18.7	9.5	5.0	12.3	6.6	12.2	6.3
Isoleucine	37.0	18.0	8.3	4.5	10.8	5.4	9.0	4.8
Leucine	36.0	17.5	9.0	4.8	10.6	5.5	9.0	5.4
Lysine	34.7	17.5	8.0	4.2	10.3	5.4	8.5	4.6
Methionine	36.8	18.5	8.4	4.2	11.3	6.0	9.0	4.6
Phenylalanine	33.2	17.4	8.7	4.3	10.5	5.4	8.9	4.7
Serine	35.7	18.3	11.6	6.0	12.5	6.3	12.3	6.2
Taurine	42.4	23.9	14.0	7.2	16.8	8.9	15.1	8.8
Tyrosine	38.1	19.6	14.8	7.6	16.5	8.4	15.6	8.0
Valine	33.5	17.4	9.3	5.0	12.3	6.0	12.0	5.9
Monoamines								
Adrenaline	22.5	11.8	7.0	3.4	9.0	4.5	8.8	4.5
Dopamine	24.0	12.5	7.5	3.7	11.1	5.8	10.9	5.7
DOPAC	30.3	15.0	8.6	4.5	13.5	6.7	13.0	6.5
HVA	30.9	15.1	8.0	4.0	13.0	6.6	12.8	6.3
Noradrenaline	24.5	12.8	7.1	3.7	10.0	4.9	9.7	4.8
MHPG	30.4	16.0	8.8	4.5	13.5	6.9	13.0	6.4
VMA	30.0	15.1	8.6	4.4	12.8	6.6	12.6	6.4
5-HT	23.5	12.5	7.3	3.7	10.8	5.5	10.6	5.4
5-HIAA	30.2	15.2	8.6	4.4	12.9	6.6	12.6	6.4
Neuropeptides								
Angiotensin II	19.0	9.4	3.6	1.7	2.0	1.0	1.9	0.9
AVP	18.3	9.1	3.9	1.9	1.2	0.6	1.1	0.6
β-Endorphin	3.0	1.4	0.2	0.1	—	—	—	—
Bombesin	16.6	8.1	3.3	1.5	1.3	0.7	1.2	0.6
CCK-8	12.7	6.2	2.5	1.3	0.8	0.4	0.7	0.3
Dynorphin 1–17	6.5	3.3	1.1	0.6	0.2	0.1	0.2	0.1
LHRH	15.6	8.0	3.9	1.9	1.0	0.5	1.0	0.5
[Leu]enkephalin	20.9	10.5	6.0	3.1	6.0	3.0	5.8	3.0
[Met]enkephalin	24.8	13.0	6.5	3.3	6.6	3.3	6.4	3.2
Neurotensin	12.0	6.3	2.6	1.3	0.7	0.4	0.6	0.3
NPY	1.5	0.7	0.1	—	—	—	—	—
Oxytocin	16.4	8.6	4.0	2.0	1.1	0.6	0.9	0.5
Substance K	18.0	9.1	4.0	2.0	1.2	0.6	1.2	0.5
Substance P	15.5	7.5	3.4	1.6	0.8	0.4	0.8	0.4
TRH	19.4	11.8	4.5	2.3	3.9	2.0	3.8	1.9
Purines								
Uric acid	35.0	18.5	12.0	5.9	15.0	7.6	14.8	7.5

[a] Flow rate was 2.0 μl/min. In each case percentage relative recoveries are mean values for three individual probes. In most cases variation among probes was not greater than 20% for a substance, but the 50,000 MW cutoff Amicon Vitafiber probe varied as much

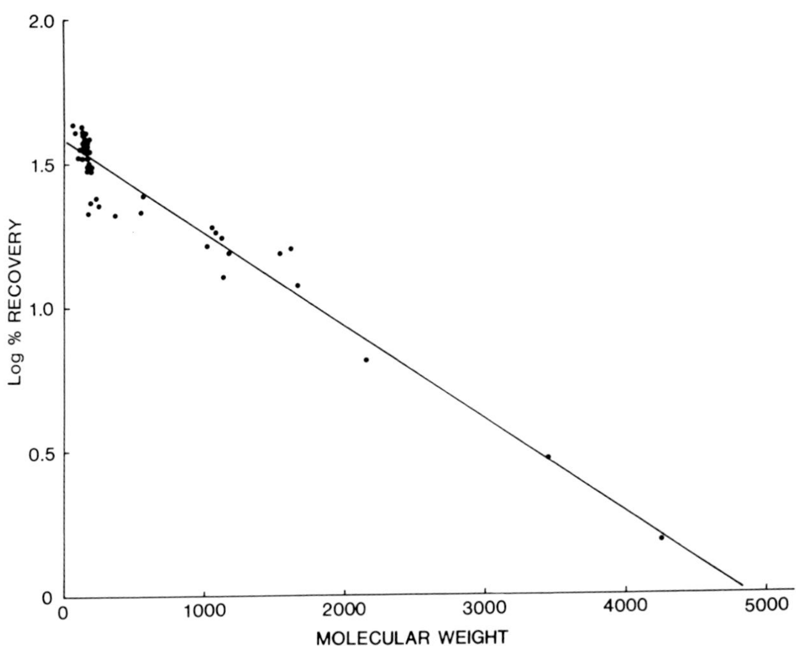

FIG. 3. Linear relationship between the log % recovery (*in vitro*) shown by a 5 mm Carnegie Medicin probe and the molecular weight of the substance sampled (indicating an exponential relationship between the two factors). Recoveries of 42 different substances are plotted. Recovery is minimal at approximately 5000 MW, even though the membrane's nominal cutoff is 20,000. A 2.0 μl/min flow rate was used.

as 40% for recovery of the neuropeptides. Within probe recoveries usually varied less than 10%. Where no figure is shown recoveries were below 0.1%. The solution passed through the probes was 0.9% NaCl, with ascorbic acid and EDTA (1 μg/ml) added for the cellulose and Amicon Vitafiber probes, for recovery of monoamines and metabolites and uric acid (to help prevent oxidation of the substances). This addition was not essential, however, for the Carnegie Medicin 20,000 MW cutoff probe. For the neuropeptides, 0.5% bovine serum albumin was added to help prevent peptides from sticking to the membrane and metal parts of the probe. This was particularly important for the Amicon Vitafiber membrane. For amino acids, probes were placed in a 25 or 50 μM standard solution, and concentrations were measured by precolumn derivatization with *o*-phthalaldehyde followed by high-performance liquid chromatography (HPLC) with fluorescence detection.[23,79] For monoamines and metabolites and uric acid, probes were placed in a 25 or 50 nM standard solution and concentrations measured by HPLC with electrochemical detection.[49,80] Recoveries of neuropeptides were calculated mainly with the probes placed in high concentrations of peptide (10 or 20 μM) and concentrations calculated using HPLC with UV detection. However, AVP and oxytocin recoveries were also calculated with the probes placed in a 2.5, 5, or 10 nM concentration of the peptide (see Fig. 5) and peptide concentrations measured using radioimmunoassay (the percentage recovery of these two peptides was not greatly altered, however, with the different concentrations).

number of substances present in the medium external to the probe; however, conducting *in vitro* tests at temperatures lower than 37° may reduce the recoveries shown by the cellulose membranes.[57]

From Table II it can be seen that the Carnegie Medicin probes have considerably better recoveries for all substances measured, although recoveries for the higher molecular weight neuropeptides (β-endorphin and neuropeptide Y) are low. Recovery figures among probes can vary as much as 20% (especially with neuropeptides), and so *in vitro* tests should be conducted on all probes used in a study, particularly where an individual animal is sampled on more than one occasion. Figure 3 shows that there is an exponential relationship between recovery and molecular weight for the 20,000 MW cutoff, Carnegie Medicin probe. Figure 3 also shows that the membrane is only potentially useful in picking up substances of up to approximately 5000 MW. The same exponential relationship between molecular weight and recovery is also shown by the Amicon Vitafiber membrane and the two cellulose membranes used. In each case the membranes reliably pick up substances up to around 25–30% of their MW cutoff.

Probes with the cellulose membrane perform reasonably well with amino acids and monoamines and metabolites but not with neuropeptides, which is not surprising given their lower molecular weight cutoff. Similar recovery figures for amino acids and monoamines and metabolites have been reported for other cellulose membranes.[21,43] The Amicon Vitafiber membrane is the least efficient recovering amino acids and monoamines and metabolites but is better with neuropeptides than the cellulose membrane. This reduced performance of the 50,000 MW cutoff membrane for low molecular weight substances illustrates the fact that there may be a trade-off of reduced recoveries for low MW substances when using high MW cutoff membranes (owing to the fewer number of pores per mm^2). Also, although there are some membranes available with MW cutoffs greater than 50,000, they often leak when liquid is pumped through them and can only be used with a push–pull arrangement.

All the probes recover substances with 0.9% NaCl or artificial cerebrospinal fluid; however, there is some advantage in recovering certain peptides through adding 0.5% bovine serum albumin[23] to try to prevent them from sticking to the membrane or to the metal of the probe. The use of bacitracin (0.03%)[25,67] or aprotinin (Sigma) in the dialyzate may also help reduce enzymatic neuropeptide degradation, even though one of the advantages of the dialysis technique is that most high molecular weight enzymes are excluded from entering the dialyzate. The cellulose and Amicon Vitafiber membranes show reduced recoveries of monoamines and metabolites (particularly noradrenaline, adrenaline, dopamine, and 5-hydroxytryptamine) over time if ascorbic acid and EDTA (~1 μg/ml) are not added to prevent their oxidation. The metal used on the Carnegie Medicin probes is specially coated to reduce oxidation of monoamines, and therefore antioxidant or similar additives do not have to be used.

Figure 4 shows that, for all membranes, recovery is exponential across different flow rates, in the range of 1–8 μl/min. At higher flow rates, however, increased pressure produced within the probe may adversely affect recovery. Similarly at very low flow rates (<0.1 μl/min) recovery is affected due to sample dispersion.[57] The flow rate chosen is clearly related to the measurement method of the substance of interest and also to the time resolution required. Thus, with high-performance liquid chromatography (HPLC), where small volumes are used for detection, low flow rates tend to be used (0.2–2.0 μl/min), whereas for neuropeptides, where larger volumes are required for radioimmunoassay, higher flow rates are used (2–10 μl/min).

Figure 5 shows that the recoveries of substances by the microdialysis probes are not altered by changes in their concentration in the medium surrounding them. Figure 6 shows that the probes also respond rapidly to sudden changes in the external concentration of substances. Thus, in principle, the concentration of substance recovered from the microdialysis probe should quantitatively reflect changes in concentrations of substances in the extracellular space of the brain.

Where the same probe is used repeatedly *in vivo*, efficiency of passage of substances from the external medium into the probe may be correspondingly reduced as a result of a coating of protein on the membrane. This may affect recoveries of substances (particularly higher MW substances like peptides) but can be largely remedied by soaking the membrane in a weak acetic acid solution (~1%) overnight. It is prudent, however, to repeatedly test probes for recoveries of substances when used on a number of occasions. Most of the membranes used in microdialysis probes will either disintegrate or show reduced recoveries if they are not kept moist. Thus, completed probes should normally be kept in distilled water when not in use.

In Vivo Experiments Using Microdialysis

Table I lists the brain regions sampled, and the substances measured, in *in vivo* microdialysis experiments. For acute preparations microdialysis probes can be directly implanted stereotaxically into the brain region of interest. For chronic preparations, however, animals are usually implanted with guide tubes through which dialysis probes can be passed.

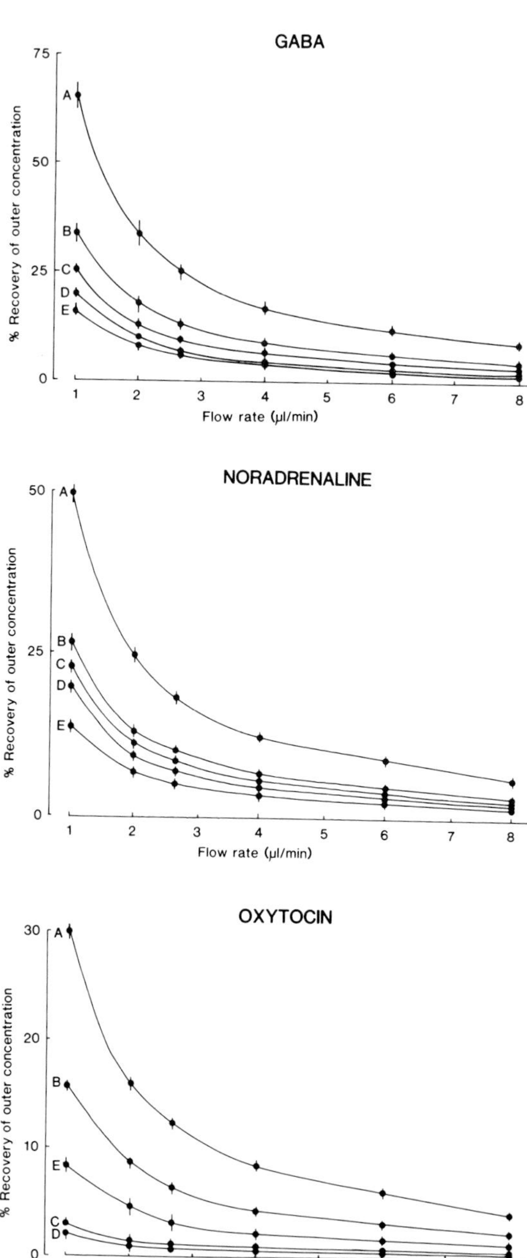

FIG. 4. The effect of flow rate on *in vitro* recoveries of γ-amino-*n*-butyric acid (GABA), noradrenaline, and oxytocin by (A) 5 mm and (B) 2 mm membrane length Carnegie Medicin probes, (C) a 5 mm cellulose membrane probe (0.32 mm o.d. type), (D) probe as C but with 0.25 mm o.d. membrane, and (E) a 5 mm Amicon Vitafiber membrane probe.

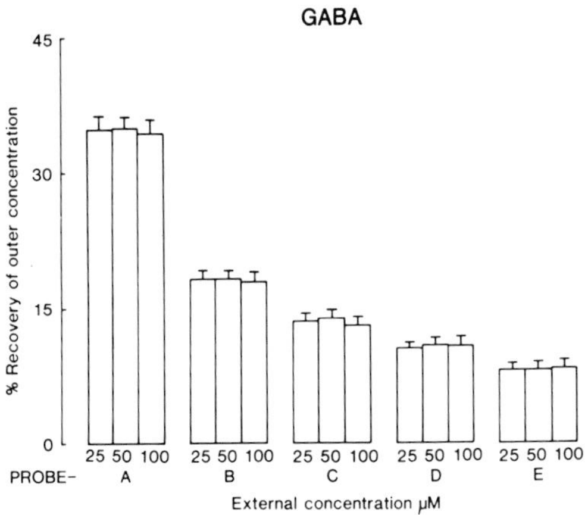

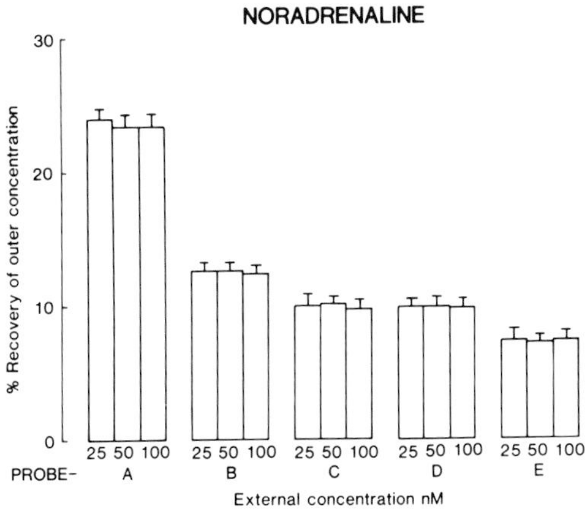

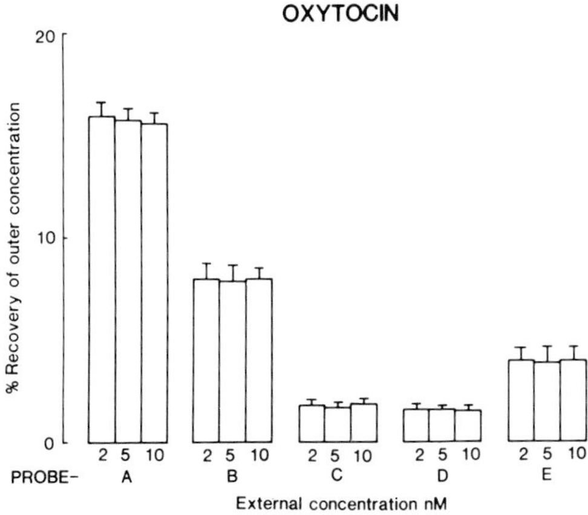

FIG. 5. Mean ± SEM recoveries of GABA, noradrenaline, and oxytocin are not influenced by changes in their external concentration. Histograms show recoveries of three probes of each type. Probes A, B, C, D, and E are as in Fig. 4. A flow rate of 2.0 μl/min was used.

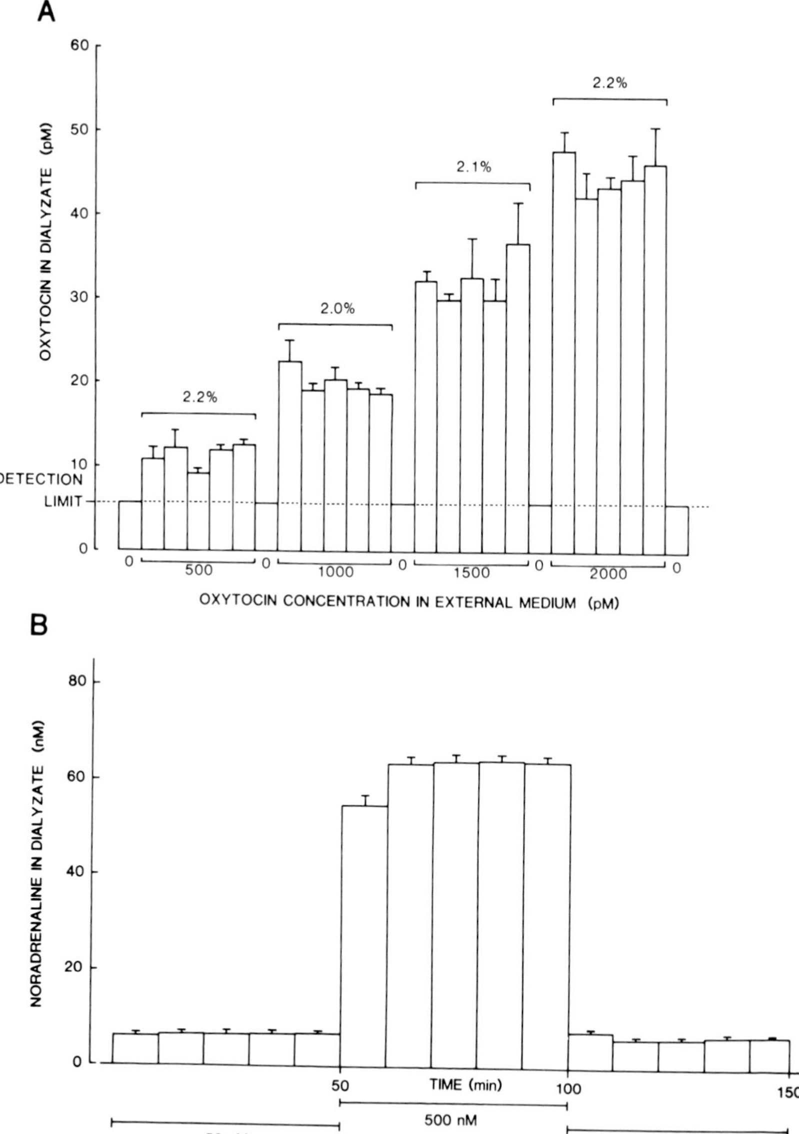

FIG. 6. Mean ± SEM recoveries of (A) oxytocin, for three 5 mm Amicon Vitafiber membrane probes placed in ascending concentrations of peptide (with 0.9% NaCl stages between each change in concentration) (sample duration was 50 min and flow rate 3 μl/min), and (B) noradrenaline, for three 5 mm cellulose (0.25 mm o.d.) membrane probes subjected to a 10-fold increase in the external concentration followed by a 10-fold decrease (sample duration was 15 min and flow rate was 2.0 μl/min). In both cases the concentration picked up by the dialysis probes reliably reflects corresponding changes in the external concentration, and percentage relative recoveries are not altered.

Guide tubes are normally used since, if the dialysis probe is permanently implanted in the brain for 48 hr or so, glial formation starts to occur in the tissue surrounding the membrane, and the amount of substances picked up is correspondingly reduced. Using guide tubes (or in some cases probes permanently implanted for short periods), chronic intracranial dialysis experiments have been carried out in rats,[10,11,21,37–40,43–48,52,54,58,60] rabbits,[17,20] cats,[27] and sheep.[22,23,49,66] Continuous sampling for up to 5 days has been reported.[43] Larger animals, such as sheep, have the advantage of being able to carry battery-driven infusion pumps on their backs, thereby avoiding the problems of using fluid swivels or similar equipment to connect a moving animal to a pump.

Typically, concentrations of substances take approximately 1–1.5 hr to return to basal levels after damage release caused by implantation of a microdialysis probe. Such damage to blood vessels caused by inserting the probe in the brain may cause the probe to sample substances from the bloodstream as well as from brain. No evidence for such pickup from the bloodstream has been found,[33,35,49] and it would therefore appear that any holes in the blood–brain barrier produced by insertion of the probes are rapidly sealed. Experiments measuring substances in the brain which also have high circulating concentrations in the blood, however, should confirm that the blood–brain barrier is intact.

Histological analysis of brain sites sampled by microdialysis show that there is no appreciable damage in the tissue immediately surrounding the membrane. This is particularly well illustrated by sections taken through the brain with the dialysis membrane *in situ*.[10,20] Equally, the finding that repeated insertion of microdialysis probes into the same brain site, over a period of months, has only minor effects on basal pickup of substances[74] further attests to the fact that such sampling does very little damage to the brain area where the membrane is situated.

Very few *in vivo* neuroendocrine-related experiments using microdialysis have been published so far; however, a small number of relevant studies in this field, both on rats and sheep, have been completed.

Oxytocin Release from the Sheep Olfactory Bulb and Substantia Nigra

In the sheep, cerebrospinal fluid concentrations of oxytocin increase during parturition, vaginocervical stimulation, and suckling,[80] and intracerebroventricular infusions of oxytocin to estrogen-primed ewes stimulate maternal behavior.[81] We have used microdialysis to investigate release of oxytocin from brain regions, known to contain oxytocin-

[81] K. M. Kendrick, E. B. Keverne, and B. A. Baldwin, *Neuroendocrinology* **46,** 56 (1987).

immunoreactive terminals, which have previously been implicated in the control of maternal behavior. Experiments using a microdialysis probe with the Amicon Vitafiber membrane (see Figs. 1H and 2E) measured simultaneous release of oxytocin and monoamines and their metabolites from the olfactory bulb and substantia nigra of conscious sheep during parturition and suckling.[49,66] Animals were surgically prepared with guide tubes, and dialysis sampling took place inside in their home pens. The animals were free to move around at all times, and a 0.9% NaCl solution containing 0.5% bovine serum albumin was passed through the probes (at 3 μl/min) using a battery-driven syringe pump (Graseby Medical MA16A) attached to the wool on their backs. Samples were collected every 50 min, and concentrations of monoamines and metabolites were measured by HPLC with electrochemical detection and oxytocin by radioimmunoassay.

Mean concentrations of oxytocin and dopamine were significantly raised in both the olfactory bulb and substantia nigra during parturition and suckling but not during separation of the ewes from their lambs. Figure 7 shows oxytocin and dopamine concentrations in dialysis samples taken over a 15-hr period (including parturition and suckling) from the olfactory bulbs and substantia nigra of an individual animal. The estimated extracellular concentration of the peptide was between 5 and 10% of the concentration measured in homogenates of the two brain regions. Intravenous infusions of oxytocin (which raised plasma levels of the peptide above 2 n*M*) did not influence the concentrations of the peptide picked up by the dialysis probe. Oxytocin was also not detectable in dialysis samples taken from the cortex.

In a further experiment, using the same microdialysis sampling technique, it was found that a 15-min period of vaginocervical stimulation produced significant increases in concentrations of oxytocin and of the amino acid transmitters aspartate, GABA, and glutamate in dialysis samples taken from the olfactory bulbs. Figure 8 shows concentrations of these substances in an individual animal in samples taken before, during, and after vaginocervical stimulation. As an *in vivo* comparison of microdialysis probes, the concentrations of oxytocin in dialysis samples taken from the olfactory bulbs of the same sheep were compared using a cellulose (Medicell, 0.25 mm o.d.), Amicon Vitafiber, or Carnegie Medicin membrane probe (all with a 5 mm length of membrane). Mean $\pm$ SEM concentrations of oxytocin in dialysis samples (1.5 μl/min, flow rate) were 165.5 $\pm$ 20.0 p*M* for the Carnegie Medicin probe compared to 36.5 $\pm$ 6.5 p*M* for the Amicon Vitafiber probe and 15.5 $\pm$ 3.0 p*M* for the cellulose membrane. The relationship among the *in vivo* pickup of the three probes closely resembles that for the *in vitro* recoveries shown in Table II.

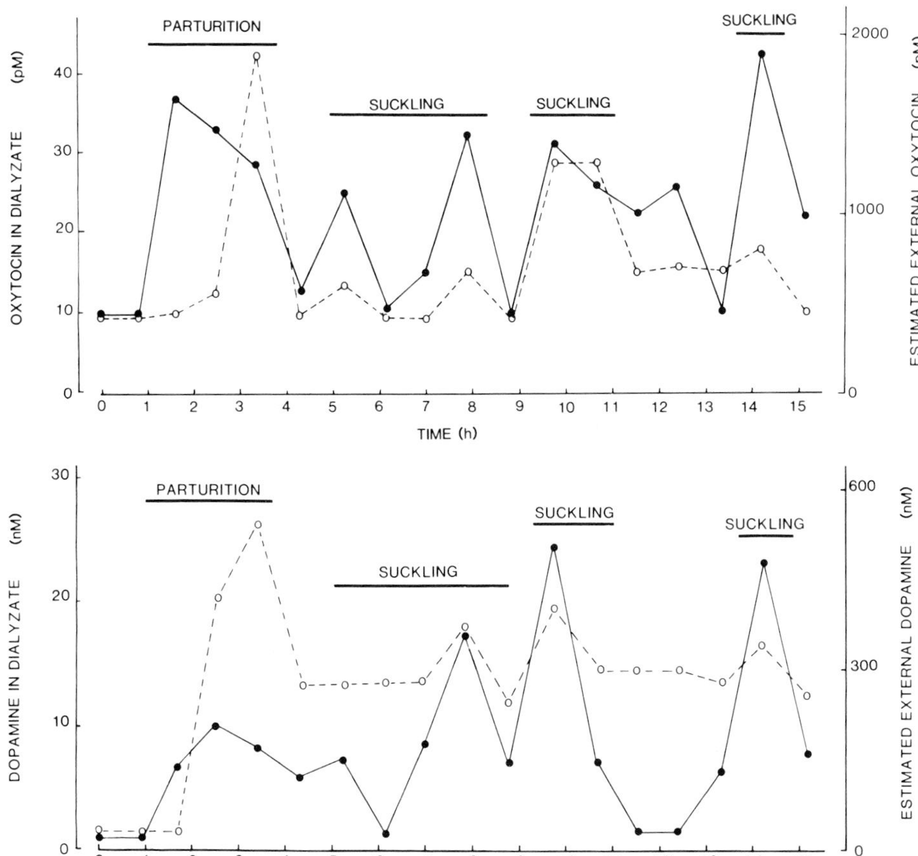

FIG. 7. Concentrations of oxytocin and dopamine in sequential 50-min samples taken simultaneously from the olfactory bulb (○---○) and substantia nigra (●——●) of a sheep over a 15-hr period. The scale on the left shows the substance concentration in the dialyzate and that on the right the estimated external concentration (based on *in vitro* recoveries). Black bars indicate the occurrence of parturition and suckling (for suckling, the black bars do not indicate that the behavior was continuous but that it occurred for at least 10 min during a particular sample).

Hemorrhage-Induced Noradrenaline and 5-HT Release from the Rat Supraoptic Nucleus

A role for the ascending noradrenergic pathways from the brain stem to the supraoptic nucleus (SON) in the control of vasopressin release has received much experimental support, although no evidence exists that

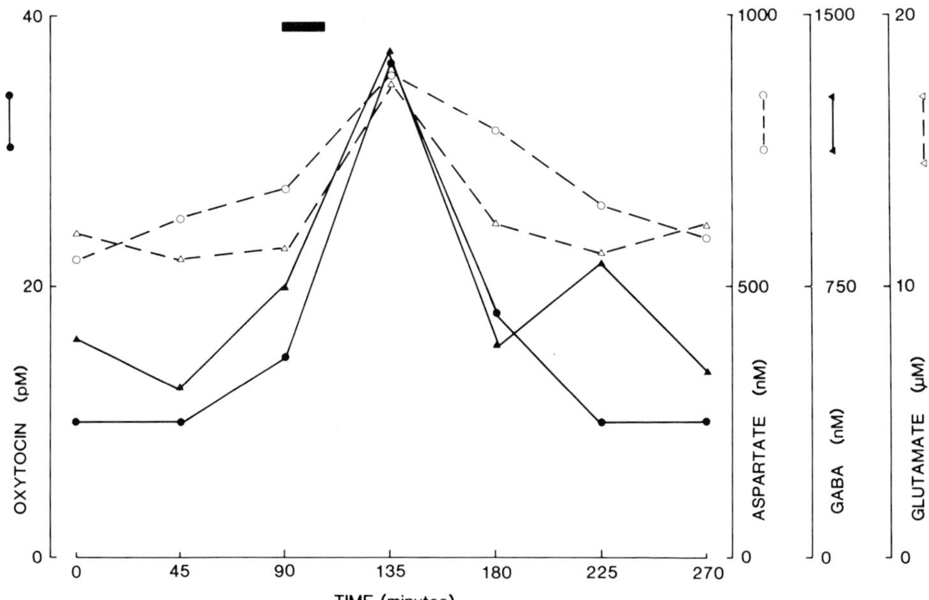

FIG. 8. Mean concentrations of oxytocin, aspartate, GABA, and glutamine in sequential dialysis samples (45-min duration) taken from the olfactory bulb of a conscious sheep. The black bar indicates a 15-min period of vaginocervical stimulation.

noradrenaline (NA) is actually released in the SON during an event (such as hemorrhage) where vasopressin is released. We have used the microdialysis technique to demonstrate release of NA and 5-hydroxytryptamine (5-HT) from the region of the SON of a urethane-anesthetized rat during a 3- to 4-ml hemorrhage.[50] A 2 mm loop of cellulose membrane (Medicell, 0.25 mm o.d.) was used with the probe (design shown in Figs. 1D and 2D). Figure 9 shows that mean concentrations of NA and 5-HT are significantly increased in dialysis samples taken from the region of the SON but not in those from a nearby control site, the olfactory tubercle. The purine metabolite, uric acid, is also increased significantly in the SON and not the olfactory tubercle during hemorrhage, and this may reflect a purine neuromodulation of noradrenergic cells.

FIG. 9. Mean ± SEM concentrations of (A) noradrenaline and (B) 5-hydroxytryptamine in dialysis samples (15-min duration) taken from urethane-anesthetized rats before, during, and after a 3- to 4-ml hemorrhage. Data are from 16 animals [9 SON (supraoptic nucleus) and 7 OT (olfactory tubercle)]. The hemorrhage typically caused a 40 mm Hg fall in blood pressure.

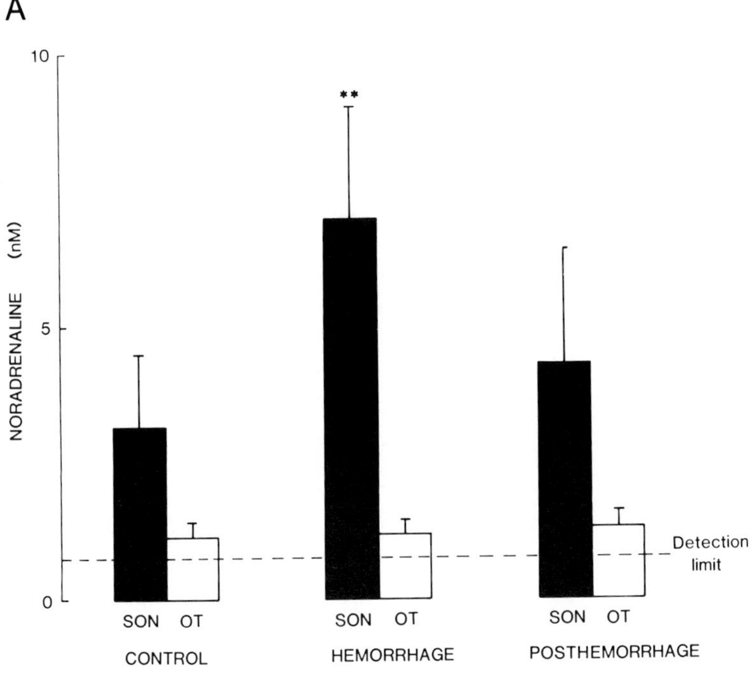
A
10
5
0
NORADRENALINE (nM)
**
Detection
limit
SON OT
CONTROL
SON OT
HEMORRHAGE
SON OT
POSTHEMORRHAGE

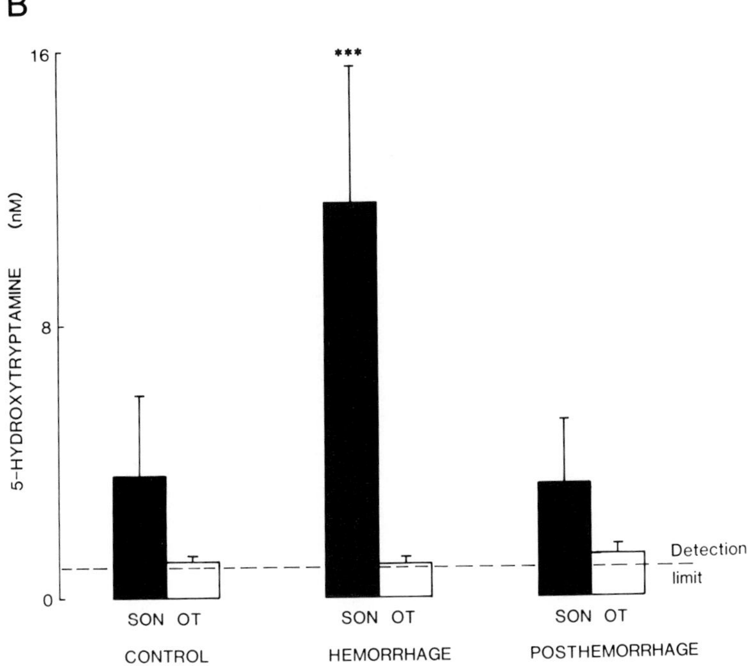
B
16
8
0
5-HYDROXYTRYPTAMINE (nM)
***
Detection
limit
SON OT
CONTROL
SON OT
HEMORRHAGE
SON OT
POSTHEMORRHAGE

Other Neuroendocrine-Related Experiments

An *in vivo* experiment using microdialysis (2 mm cellulose loop of 0.25 mm o.d. membrane) in the rat has demonstrated increased release of adrenaline from the hypothalamus following stimulation of the rostral ventrolateral medulla.[51] Such increases were associated with increases in blood pressure and therefore also provide evidence that adrenergic pathways from the brain stem to the hypothalamus may play a role in regulation of blood pressure.

Experiments using microdialysis have simultaneously introduced the peptide cholecystokinin (CCK-8) into the nucleus accumbens of the rat while monitoring dopamine release.[76] CCK-8 had a pronounced, dose-dependent releasing action on dopamine release in the anterior but not the caudal part of the nucleus accumbens.

Finally, recent experiments using an Amicon 100,000 MW cutoff Vitafiber membrane have successfully measured basal luteinizing hormone-releasing hormone (LHRH) release (2–15 pg/10 min) from the anterior pituitary.[64,65] The authors report that their microdialysis probe picked up 10 times as much peptide as a push–pull cannula in this region.[65] The recoveries of LHRH shown by their probe are less than those of the Carnegie Medicin probe but greater than that of the 50,000 MW cutoff Amicon Vitafiber membrane probe. These experiments are described in detail elsewhere in this volume.[64]

Future Application of Microdialysis Techniques to Neuroendocrinology

Microdialysis is now firmly established as an alternative *in vivo* sampling technique to the push–pull cannula. It has been used to sample the *in vivo* release of a variety of different substances from a large number of different brain regions and should prove to be of enormous potential value to neuroendocrinologists requiring an *in vivo* sampling technique with the minimum amount of damage to the tissue sampled. It also provides the opportunity, unique to *in vivo* sampling techniques, to estimate the actual concentrations of substances in the extracellular space surrounding the probe.

The limiting factor in microdialysis experiments, as with all *in vivo* sampling techniques, is the sensitivity of the assays used to measure the substances sampled. However, the increasing reliability of microbore HPLC detection of monoamines and metabolites, or amino acids, and the availability of increasingly sensitive peptide radioimmunoassays have considerably improved the situation. The improvement in the pickup of substances by new membranes, used in microdialysis probes over the last few years, has also increased the potential application of the technique.

A number of experiments have used the microdialysis technique in conjunction with voltammetry in recent years[53,82,83] in an attempt to define more precisely what is being measured by the voltammetry electrodes. This combination of technologies offers considerable scope in future research.

While microdialysis sampling has been mainly confined to the brain and spinal cord, it can also be utilized to sample release from peripheral endocrine organs such as the adrenal, ovary, and testis. It has recently been used, for example, to sample substances from muscle.[84] Equally, substances could be sampled from physiological fluids, such as blood or cerebrospinal fluid, where direct sampling, by removal of these fluids, may be impracticable or liable to compromise interpretation of results. For example, microdialysis probes have recently been used to measure serum concentrations of quinolinic acid by implanting a probe (Carnegie Medicin) in the jugular vein of a rat.[77] Such direct sampling of substances in the blood or cerebrospinal fluid also obviates the necessity to use complex and time-consuming extraction methods normally required for these fluids.

The potential use of microdialysis probes as chemical releasers remains to be fully exploited. They offer considerable advantages over other methods of chemical release into the brain since no liquid actually passes into the brain tissue. Microdialysis probes can be used to investigate the effects of one substance on the release of another, by pumping one substance through the probe (so that it diffuses from the probe out into the surrounding tissue) while simultaneously measuring the release of another substance (which diffuses from the tissue into medium pumped through the probe). A number of different substances can be pumped through the probe during the same experiment using a liquid-switching valve placed between the syringe pump and the microdialysis probe. Another application of microdialysis probes as chemical releasers is for long-term treatment of small brain regions with substances, for example, where the behavioral or endocrine effects of such treatment need to be monitored over long periods or where the experimenter wishes to release a substance into the brain only for particular periods during each hour or day.

[82] J. B. Justice, Jr. and D. B. Neill, *Ann. N.Y. Acad. Sci.* **473,** 170 (1986).

[83] C. A. Marsden, K. F. Martin, C. Routledge, M. P. Brazell, and N. T. Maidment, *Ann. N.Y. Acad. Sci.* **473,** 106 (1986).

[84] H. Askmark, S-M. Aquilonius, L. Hallander, and U. Ungerstedt, *Scand. Congr. Neurol. 26th, June 11–14, 1986* (Abstr.).

[12] *In Vitro* Perifusion of Human Hypothalamic and Pituitary Tissue

By Dennis D. Rasmussen

Introduction

In vitro perifusion of hypothalamic or pituitary tissue for the purpose of investigating mechanisms of hypothalamohypophyseal neuroendocrine regulation is a methodology which complements *in vivo* as well as alternative *in vitro* techniques, and it circumvents many of their disadvantages. For example, although methods for collection of hypothalamohypophyseal portal blood have been developed which allow direct quantification of regulatory factor (neurotransmitter, neuromodulator, releasing factor, etc.) release into the portal flow, these techniques usually require use of systemic anesthesia (which must be assumed to alter normal interneuronal dynamics within the hypothalamus) and permit only the activity of a relatively small population of neurons which secrete into the portal flow to be monitored. Even more recently developed methods which allow assay of the hypothalamohypophyseal portal blood,[1] the extracellular fluid of the median eminence and arcuate nucleus,[2-4] or the third ventricular cerebrospinal fluid,[5] all without anesthesia, share common characteristics which also make them insufficient for a critical investigation of neuroendocrine interactions.

One problem concerns localization of the true site (or sequential sites) of action mediating the response to experimental manipulations. With these and other *in vivo* preparations it is difficult or impossible to determine whether a response is mediated by a primary action exerted directly within the hypothalamus or through intermediary effects (e.g., pituitary, adrenal, ovarian, blood flow, metabolic), possibly by different but interacting neuroendocrine systems. Although such interactions may certainly be physiologically important, they confound the critical fundamental analyses of variables directly regulating secretion of the regulatory factor

[1] I. J. Clarke and J. T. Cummins, *Endocrinology (Baltimore)* **111,** 1737 (1982).
[2] J. E. Levine, F. P. Kwok-Yuen, V. D. Ramirez, and G. L. Jackson, *Endocrinology (Baltimore)* **111,** 1449 (1982).
[3] J. E. Levine and V. D. Ramirez, *Endocrinology (Baltimore)* **111,** 1439 (1982).
[4] J. E. Levine and V. D. Ramirez, *Endocrinology (Baltimore)* **107,** 1782 (1980).
[5] D. A. Van Vugt, W. D. Diefenbach, E. Alston, and M. Ferin, *Endocrinology (Baltimore)* **117,** 1550 (1985).

under evaluation. In addition, reliable isolation and analysis of a regulatory factor after it has been secreted into the cerebrospinal fluid, or especially the blood, often presents formidable problems. With *in vivo* models it is also impossible to adequately identify and/or isolate and control confounding neural, endocrine, and metabolic variables. Finally, it is not possible to use these techniques with humans.

With the pituitary, the same problems of excluding extraneous sites of action, isolating from interfering plasma factors, and eliminating confounding variables apply, with the additional problem of detecting possibly very small and acute hormonal changes after they have been diluted in the large circulating plasma and tissue distribution volume. In contrast, *in vitro* perifusion of isolated hypothalamic and pituitary tissue allows investigation of interactions between neuroendocrine factors directly and exclusively at the desired pituitary or hypothalamic site. The concentration, pattern (pulsatile, constant, gradient), and duration (acute, chronic) of administration of one or several potentially interacting modulators can be easily controlled, and a defined culture medium can be used in which the metabolic stability of the modulators is much greater than when injected *in vivo,* and is readily verifiable, with no interference of anesthetics. Furthermore, in contrast to cell culture techniques, neuronal networks and cell–cell contacts are maintained and secretory products and metabolites are removed without prolonged contact with the cells, more accurately simulating *in vivo* conditions.

The development, characterization, and utilization of the *in vitro* perifusion technique with rat hypothalamic and pituitary tissue has been thoroughly described in a previous volume of this series.[6,7] Accordingly, in this chapter we focus on the adaptation and characterization of this methodology as applied specifically to the human.

Method

Human hypothalamic or pituitary tissue is maintained in approximately 0.5- or 0.1-ml perifusion chambers, respectively, constructed from the barrels of 3-ml (0.5) or 1-ml (0.1) disposable syringes and immersed in a 37° water bath (Fig. 1). Medium 199 (Grand Island Biological Co., Grand Island, NY) which has been bubbled with 95% O_2–5% CO_2 for 1 hr and contains 0.1% crystalline bovine serum albumin (Pentex, Miles Laboratories, Elkhart, IN) with pH adjusted to 7.3 is kept on ice and pumped at approximately 100 μl/min by a low pulsation peristaltic pump (Rainin

[6] J. E. Levine and V. D. Ramirez, this series, Vol. 124, p. 466.

[7] A. Negro-Vilar and M. D. Culler, this series, Vol. 124, p. 67.

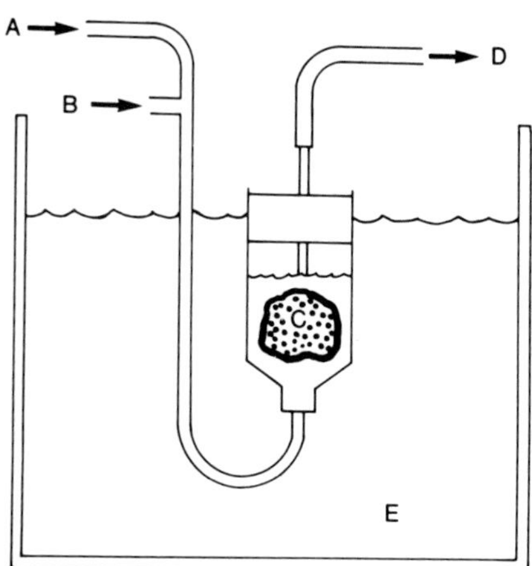

FIG. 1. Perifusion chamber. A, Inlet; B, injection port or alternate inlet; C, tissue; D, outlet; E, water bath.

Rabbit, Rainin Instrument Co., Woburn, MA) through 5 ft of gas-permeable 0.020-inch i.d. × 0.037-inch o.d. Silastic medical-grade tubing (Dow Corning Corp., Midland, MI) which is enclosed in a chamber flushed with 95% O_2–5% CO_2, as described by Takahashi *et al.*[8] The medium is then delivered (A) through 2 ft of 0.020-inch i.d. × 0.060-inch o.d. Tygon tubing (Norton Performance Plastics, Akron, OH), submerged in the 37° water bath, into the bottom of the chamber. The medium flows up and around the tissue (C), and exits the chamber through a 22-gauge needle pierced through the rubber seal (from the disposable syringe plunger) which closes the chamber. The volume of the chamber is regulated by adjusting the position of the needle. The perifusate effluent is then delivered (D) by Tygon tubing through a hole drilled in the wall of a small refrigerator to an automated fraction collector (FC-80 Microfractionator, Gilson Medical Electronics, Middleton, WI) which is maintained at 4°.

An injection port (B) allows bolus administration of small volumes (e.g., 10–25 μl) of test substances into the flow of the medium just before it enters the chamber. Repeated pulses or more prolonged administration of test substances is accomplished with two peristaltic pumps regulated by a microprocessor-based timer (ChronTrol, Lindburg Enterprises, San Diego, CA), each delivering media of different composition through a T connector. Alternatively, a commercially available automatic pulsatile delivery system (APS10, Endotronics, Inc., Coon Rapids, MN) can be

[8] J. S. Takahashi, H. Hamm, and M. Menaker, *Proc. Natl. Acad. Sci. U.S.A.* **77,** 2319 (1980).

employed to deliver a test substance in pulses of varying waveforms in the medium flow, as described by Negro-Vilar and Culler.[7]

Hypothalamus

In our early studies, hypothalami from adult men and women were obtained at autopsy within 24 hr after death. The mediobasal hypothalamus (MBH) was removed with sagittal cuts in the lateral sulci, transverse cuts through the rostral edges of the mammillary bodies and optic chiasm, and a horizontal cut at a depth of 3–4 mm. After removing the optic chiasm, the MBH was bisected at the midline and each half-MBH was sliced by three coronal cuts into four sections. The four pieces of an individual half-MBH were separated by stainless steel screens in the perifusion chamber. During studies of both gonadotropin-releasing hormone GnRH[9] and β-endorphin (βEND)[10] release, we found that hypothalamic tissue obtained within 12 hr postmortem initially released relatively large quantities of these peptides, and that this large initial release declined within 1.5 hr to basal levels (Fig. 2).

Functional βEND and GnRH secretory capacity of this tissue obtained within 12 hr postmortem was confirmed by the increased release of both peptides in response to stimulation by 56 m*M* potassium (Fig. 2). Since equimolar sodium did not alter the release of either peptide (Fig. 2), the response to potassium was apparently due to specific membrane depolarization rather than nonspecific response to monovalent ions or osmotic changes. Functional viability was also confirmed by increased release of both GnRH and βEND in response to veratridine (Fig. 2), an alkaloid which depolarizes cells by inactivating the sodium conductance mechanism of functional membrane sodium channels, thereby increasing sodium permeability.[11,12] Furthermore, the stimulation of both GnRH and βEND release by potassium was suppressed by perifusion with calcium-free medium (Fig. 2), consistent with the requirement for calcium in the normal active secretory response to depolarization in many neuronal and hormonal systems.[13]

The increased release of both GnRH and βEND in response to 56 m*M*

[9] D. D. Rasmussen, J. H. Liu, P. L. Wolf, and S. S. C. Yen, *J. Clin. Endocrinol. Metab.* **62,** 479 (1986).

[10] D. D. Rasmussen, J. H. Liu, P. L. Wolf, and S. S. C. Yen, *Neuroendocrinology* **45,** 197 (1987).

[11] P. M. Conn and D. C. Rogers, *Endocrinology (Baltimore)* **107,** 2133 (1980).

[12] M. Ohta, T. Narohashi, and R. F. Keeler, *J. Pharmacol. Exp. Ther.* **184,** 143 (1973).

[13] R. B. Kelly, J. W. Deutsch, S. S. Carlson, and J. A. Wagner, *Annu. Rev. Neurosci.* **2,** 299 (1979).

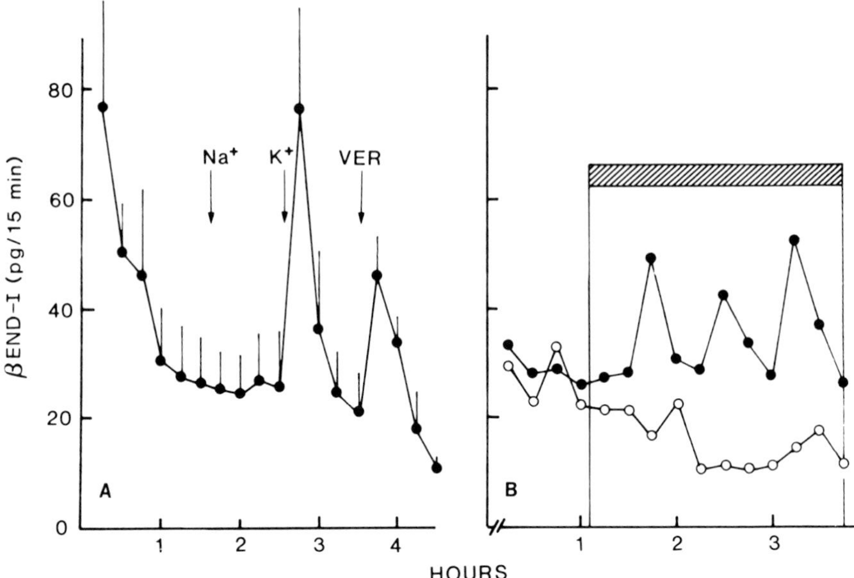

FIG. 2. Immunoreactive (I) β-endorphin (βEND) release from adult human half-MBHs. (A) The mean (+SE) βEND-I release from six half-MBHs (four male, two female) in response to 5-min pulses of 56 mM Na^+, 56 mM K^+, or 50 μM veratridine (VER). (B) βEND-I release, following a 1.5-hr stabilization period, from two matching half-MBHs (female) in response to 5-min pulses of 56 mM K^+ during control perifusion (●) or with calcium-free medium containing 1 mM of the calcium chelator EGTA (○), as indicated by the hatched box. (From Rasmussen *et al.*,[10] reproduced with permission of Karger, Basel.)

potassium but not sodium, the stimulation by the sodium channel activator veratridine, and the suppressed response to potassium in the absence of calcium demonstrated that the fundamental secretory properties of the human hypothalamic tissue obtained within 12 hr of death were satisfactorily preserved. In contrast, tissue obtained 12–24 hr postmortem did not reliably respond to these physiological challenges in a consistent fashion, but rather released undetectable levels of these peptides or responded erratically. In further studies with human hypothalamic tissue obtained within 12 hr postmortem, dopamine stimulated the release of GnRH and βEND, and these responses to dopamine were prevented by administration of the dopamine receptor antagonist haloperidol but not the adrenergic receptor antagonist phentolamine,[9,10] demonstrating that membrane receptor mechanisms in this tissue remained functional.

We have also used this methodology to investigate regulation of GnRH release from human fetal hypothalamic tissue. Although the neuroendocrine activity of the fetal hypothalamus may differ from that of the adult, compelling evidence indicates that the human hypothalamic–pituitary system forms a well-differentiated functional unit by midgestation.[14–16] Norepinephrine, dopamine, and serotonin are all present in the fetal hypothalamus by 12 weeks of gestation,[17] and all of the hypothalamic nuclei[17–20] and the tuberoinfundibular dopaminergic system[21] are well differentiated by 15 weeks. An intact hypothalamohypophyseal portal system is established as early as 11.5 weeks gestation,[22] and GnRH-containing neurons in the MBH with axon terminals in contact with portal capillaries of the median eminence are present by 16 weeks gestation.[23,24] Furthermore, hypothalamic control of fetal gonadotrope activity appears to be operative, as evidenced by parallel increases in hypothalamic GnRH content and fetal pituitary gonadotropin secretion,[14,25] the fetal pituitary response to exogenous GnRH,[26] and low levels of serum luteinizing hormone (LH) and follicle-stimulating hormone (FSH) which are found in anencephalic infants.[14]

These neuroanatomical and functional data suggest that the human fetal MBH may provide a useful model for the investigation of hypothalamic GnRH release in humans. Accordingly, we investigated the release of GnRH from human fetal hypothalamic tissues obtained at autopsy immediately after prostaglandin $F_{2\alpha}$- and urea-induced termination and delivery at 21–23 weeks gestation.[27,28] The time between induction and delivery was less than 6 hr in all cases. The MBH was removed with sagittal cuts in the hypothalamic sulci, coronal cuts through the rostral edges of the mammillary bodies and optic chiasm, and a horizontal cut at

[14] S. L. Kaplan, M. M. Grumbach, and M. L. Aubert, *Recent Prog. Horm. Res.* **32,** 161 (1976).

[15] A. Decherney and F. Naftolin, *Clin. Obstet. Gynecol.* **23,** 749 (1980).

[16] P. D. Gluckman, M. M. Grumbach, and S. L. Kaplan, *Endocrinol. Rev.* **2,** 363 (1981).

[17] M. Hyppa, *Neuroendocrinology* **9,** 257 (1972).

[18] M. S. Gilbert, *J. Comp. Neurol.* **62,** 81 (1934).

[19] J. W. Papez, *Res. Publ.–Assoc. Res. Nerv. Ment. Dis.* **20,** 31 (1940).

[20] H. Kuhlenbeck, "The Human Diencephalon—A Summary of Development, Structure, Function and Pathology." Karger, Basel, 1954.

[21] A. Nobin and A. Björklund, *Acta Physiol. Scand. Suppl.* **388,** 1 (1973).

[22] J. A. Thliveris and R. W. Currie, *Am. J. Anat.* **157,** 441 (1980).

[23] C. Bugnon, B. Bloch, and D. Fellmann, *C. R. Hebd. Seances Acad. Sci., Ser. D* **282,** 1625 (1976).

[24] C. Bugnon, B. Bloch, and D. Fellmann, *Brain Res.* **128,** 249 (1977).

[25] T. M. Siler-Khodr and G. S. Khodr, *Am. J. Obstet. Gynecol.* **130,** 795 (1978).

[26] G. V. Groom and A. R. Boyns, *Fed. Eur. Biochem. Soc.* **33,** 57 (1973).

[27] D. D. Rasmussen, J. H. Liu, P. L. Wolf, and S. S. C. Yen, *J. Clin. Endocrinol. Metab.* **57,** 881 (1983).

[28] D. D. Rasmussen, J. H. Liu, W. H. Swartz, V. S. Tueros, and S. S. C. Yen, *Clin. Endocrinol. (Oxford)* **25,** 127 (1986).

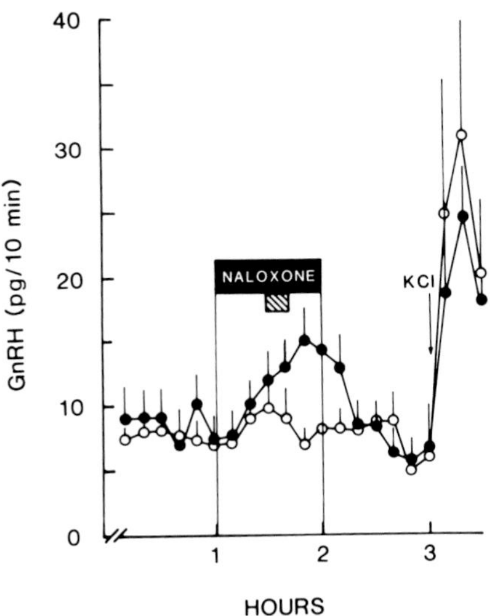

FIG. 3. GnRH release from paired human fetal half-MBHs in response to a 1-hr infusion of 40 μM naloxone with a superimposed 10-min pulse of medium alone (●) or 40 μM βEND in medium (○), as indicated by the hatched box. Each point represents the mean (+SE) of four experiments (two male and two female). (From Rasmussen *et al.*,[27] reproduced with permission of Karger, Basel.)

a depth of 3 mm, and the optic chiasm and tracts were trimmed from the tissue. Each MBH was quartered and the two tissue pieces of each half-MBH were perifused together in one chamber, separated by stainless steel screens. Perifusion was initiated within 60–90 min of delivery.

GnRH release by this midgestational fetal hypothalamic tissue followed a pattern similar to that of the adult tissue, i.e., high initial release rate declining to a lower basal rate, with functional viability confirmed by the rapid release of GnRH in response to a depolarizing dose of KCl after 5 hr of perifusion but no response to equimolar administration of NaCl.[27] As with the adult MBH tissue, dopamine administration stimulated GnRH release from the fetal human hypothalamus in a dose-dependent and apparently dopamine receptor-mediated fashion.[28] Also, the fetal MBH tissue responded to administration of the opiate receptor antagonist naloxone with increased release of GnRH, and this response to naloxone was inhibited by simultaneous administration of equimolar βEND (Fig. 3).[27] The naloxone-induced GnRH release from the human fetal MBH was consistent with and proportionately similar in magnitude to the *in vitro* response of adult rat MBHs evaluated in a similar perifusion system.[29]

[29] M. M. Wilkes and S. S. C. Yen, *Life Sci.* **28,** 2355 (1981).

Furthermore, this increased release of hypothalamic GnRH in response to opiate receptor blockade is consistent with the demonstration by Schulz *et al.*[30] that an increase in circulating LH followed *in vivo* administration of βEND antiserum into the arcuate nucleus of female rats. Accordingly, these studies provide strong evidence that perifusion of fetal hypothalamic tissue may be used for elucidation of hypothalamic mechanisms in the control of pituitary function in humans.

Although these studies indicate that perifused adult and fetal human hypothalamic tissues remain functionally viable with at least fundamental secretory capacity as well as receptor-mediated mechanisms intact under the conditions which have been described, they do not resolve whether the tissue is functioning in a relatively "normal" physiological manner. It has previously been demonstrated that all of the necessary neural elements comprising the GnRH "pulse generator" are not only resident within the MBH but capable of functioning independent of neural innervation from the remainder of the brain *in vivo*.[31-33] Thus, if the human MBH tissue which is perifused *in vitro* is indeed functioning in a "normal" physiological manner, it should be possible to demonstrate that the *in vitro* MBH releases GnRH in a pulsatile manner. We have addressed this issue by modifying the perifusion conditions to accommodate resolution of acute changes in GnRH release from the human fetal hypothalamus, i.e., the intact MBH (as opposed to several MBH pieces, but trimmed more closely to extend laterally only approximately 2 mm from the midline and only approximately 2 mm in depth) was perifused in a chamber containing only 150 μl of medium (as opposed to approximately 300 μl) in addition to the tissue, the flow rate of the medium perfusing the chamber was increased from 100 to 150 μl/min, the fraction collection interval was decreased from 15 to 10 min, the perifusate fractions were lyophilized so that when reconstituted in a smaller volume of assay buffer the entire fraction could be utilized for the radioimmunoassay determinations, and these reconstituted perifusate fractions were assayed in triplicate instead of duplicate.

When MBHs were perifused for 13 hr, each released GnRH in a distinctly pulsatile pattern (Fig. 4). Significant pulses, as determined using the PULSAR pulse detection algorithm developed by Merriam and Wachter[34] and indicated by asterisks in Fig. 4, occurred at intervals of 60.6 ±

[30] R. Schulz, A. Wilhelm, K. M. Pirke, C. Gramsch, and A. Herz, *Nature (London)* **294,** 757 (1981).

[31] C. A. Blake and C. H. Sawyer, *Endocrinology (Baltimore)* **94,** 730 (1974).

[32] M. Ferin, J. L. Antunes, E. Zimmerman, I. Dyrenfurth, A. G. Frantz, A. Robinson, and P. W. Carmel, *Endocrinology (Baltimore)* **101,** 1611 (1977).

[33] L. C. Krey, W. R. Butler, and E. Knobil, *Endocrinology (Baltimore)* **96,** 1073 (1975).

[34] G. R. Merriam and K. W. Wachter, *Am. J. Physiol.* **243,** E310 (1982).

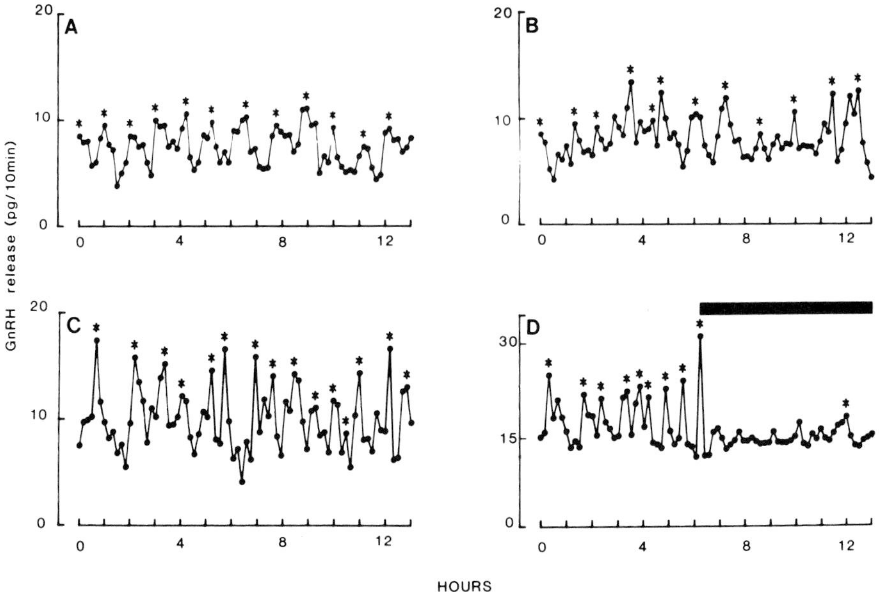

FIG. 4. GnRH release from fetal human MBHs *in vitro*. (A–D) Each represents the perifusion of a single MBH. Asterisks signify significant pulses, as described in the text. D shows GnRH release during perifusion with control medium and after addition of verapamil (50 μM) and nifedipine (5 μM), as indicated by the solid bar.

4.3 min. In another perifusion, addition of the membrane calcium channel blockers verapamil (50 μM) and nifedipine (5 μM) dramatically suppressed this pulsatile GnRH release (Fig. 4). In order to further confirm that the fluctuations in GnRH levels during these perifusions represented pulsatile GnRH release rather than fluctuations arising from experimental procedures or assay variability, we then conducted another perifusion with GnRH (15 pg/10 min fraction) added to the medium perfusing an empty 150-μl chamber. When the perifusate fractions were collected, frozen, lyophilized, reconstituted, assayed, and analyzed identically to the experimental samples, no significant GnRH pulses were detected (data not shown). In a final experiment, an MBH obtained 3.25 hr postmortem from a 65-year-old man was bisected longitudinally at the midline of the median eminence, and each intact half-MBH was perifused (200 μl/min) in chambers containing only 200 μl of medium. Both half-MBHs released GnRH in a significantly pulsatile manner (Fig. 5), with a frequency of 1 peak/79 $\pm$ 19 min. Thus, both fetal and adult human MBHs release GnRH in a pulsatile fashion under these conditions. Furthermore, the frequency of pulsatile GnRH release *in vitro* is similar to the fre-

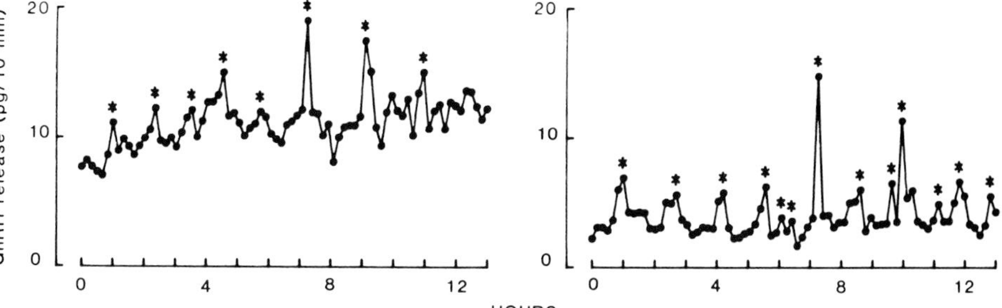

FIG. 5. GnRH release from adult human half-MBHs *in vitro*. Asterisks signify significant pulses, as described in the text.

quency of pulsatile LH, and presumably GnRH, secretion *in vivo*,[35] strongly suggesting that this *in vitro* tissue is indeed functioning relatively "normally."

Pituitary

The perifusion of human fetal and adult pituitary tissue provides a methodology for investigating the direct regulation of human pituitary hormone secretion. An advantage of this technique which had not been fully appreciated until recently is the ability to investigate very rapid and transient changes in hormone secretion, i.e., changes which would be unresolvable when diluted in the large circulating plasma and tissue distribution volume. We have recently investigated these changes with human fetal and adult anterior pituitaries or quarter-pituitaries, obtained under the conditions described for the hypothalamic tissue. When fetal pituitaries were perifused at a high flow rate (0.4 ml/min) in a small chamber (50 μl) with perifusate fractions collected at 2-min intervals, it became apparent the LH was released in a high-frequency pulsatile fashion (Fig. 6), with a pulse interval of 12.7 ± 1.7 min.[36] The corresponding lack of pulsatile fluctuations during control LH infusions (Fig. 6) and the concordance in detection of significant LH pulses in separate assays of the same perifusate samples (data not shown)[36] confirmed that these LH fluctuations were not due to either experimental or assay variability. In addition, the functional viability of the gonadotropes was confirmed by a dramatic

[35] S. S. C. Yen, C. C. Tsai, F. Naftolin, G. Vandenberg, and L. Ajabor, *J. Clin. Endocrinol. Metab.* **34,** 671 (1972).

[36] M. Gambacciani, J. H. Liu, W. H. Swartz, V. S. Tueros, S. S. C. Yen, and D. D. Rasmussen, *Neuroendocrinology* **45,** 402 (1987).

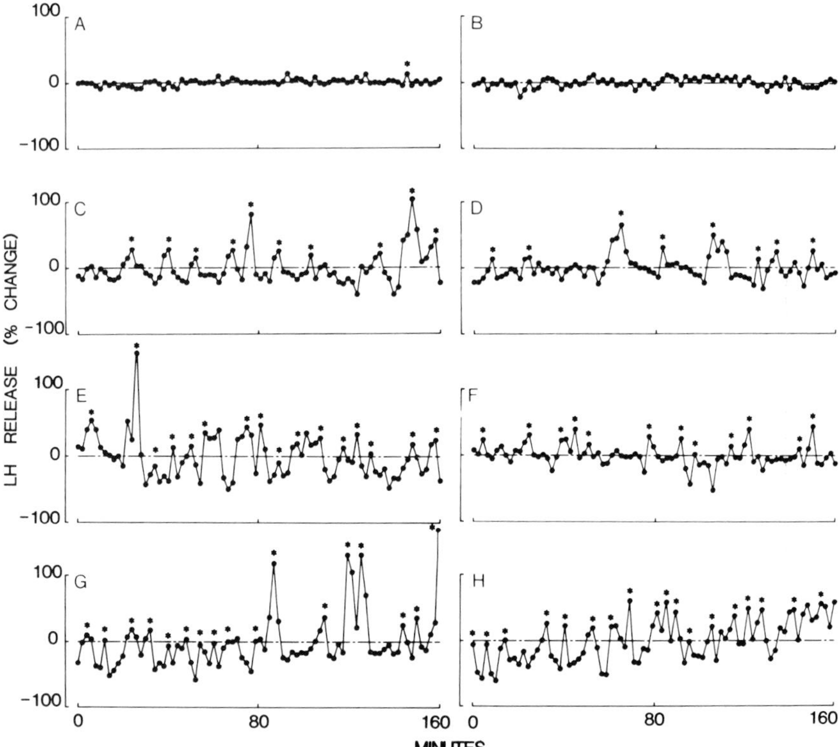

FIG. 6. LH concentrations during perifusions of empty control perifusion chambers with human LH added to the medium (A and B) and during perifusions of human fetal anterior pituitaries with medium alone (C–H), expressed as percentage change from the overall mean. (A) Control, (B) control, (C) 22-week male, (D) 22-week male, (E) 21-week female, (F) 22-week female, (G) 23-week male, (H) 23-week male. Asterisks signify significant pulses, as described in the text. (From Gambacciani *et al.*,[36] reproduced with permission of Karger, Basel.)

LH secretory response to GnRH stimulation at the end of the perifusions, and by the demonstration that the pulsatile LH release was calcium dependent (data not shown).[36]

We have also shown that adrenocorticotropin hormone (ACTH) is released in an intrinsically pulsatile fashion,[37] with pulse intervals of 11.3 ± 0.8 min. This similarity of the periodicities of LH and ACTH from the human pituitary, together with the similar previously demonstrated

[37] M. Gambacciani, J. H. Liu, W. H. Swartz, V. S. Tueros, D. D. Rasmussen, and S. S. C. Yen, *Clin. Endocrinol.* (*Oxford*) **26,** 557 (1987).

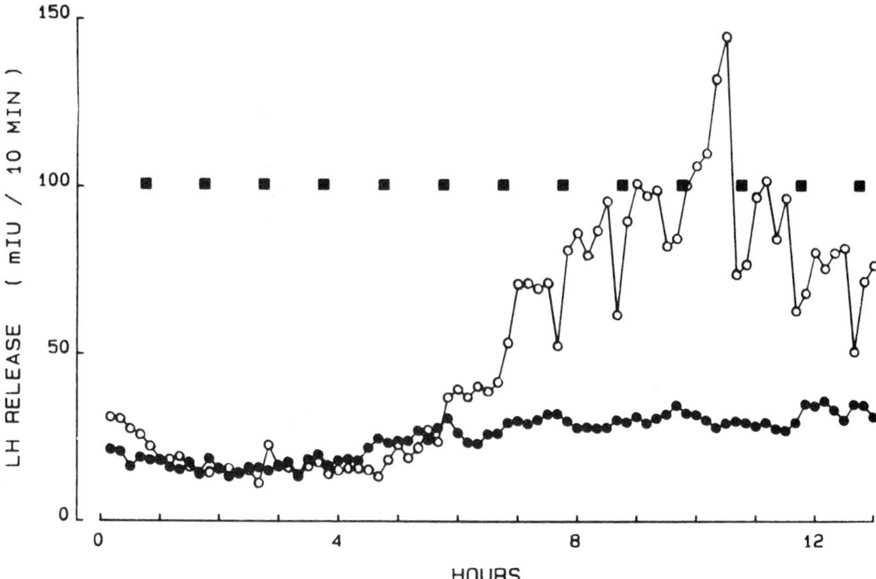

FIG. 7. LH release from two matching halves of a human fetal anterior pituitary perifused with 10-min pulses of either 10 n*M* GnRH (○) or control (●) medium administered every hour, as indicated by the squares.

pulsatile release of prolactin and growth hormone from perifused monkey hemipituitaries,[38] suggests that the basal secretion of these hormones may be intrinsically pulsatile, possibly entrained by a common mechanism.

An additional advantage of using this perifusion technique with human pituitary tissue is the ability to administer discrete brief stimuli repeatedly over a long period of time. Figure 7 shows the results of perifusions of the two halves of a human fetal pituitary, with 10-min pulses of either GnRH (10 n*M*) or control medium administered every hour for 13 hr. It is apparent that approximately 6 hr of pulsatile GnRH administration was required before the tissue was capable of responding with increased LH secretion, presumably reflecting induction of the response mechanism.

Conclusion

In vitro perifusion of human hypothalamic and pituitary tissue provides a valuable model for the investigation of hypothalamohypophyseal

[38] J. K. Stewart, D. K. Clifton, D. J. Koercker, A. D. Rogol, T. Jaffee, and C. J. Goodner, *Endocrinology (Baltimore)* **116,** 1 (1985).

regulation and interactions which cannot otherwise be directly evaluated in the human. As with other *in vitro* techniques, this methodology provides a complementary approach to *in vivo* investigations in both humans and laboratory animals.

Acknowledgments

Preparation of this chapter was supported by a grant from the National Institutes of Health, HD22608-01, to DDR.

[13] Transient Permeabilization of Endocrine Cells: Inositol Lipid Metabolism

By IRENE WINICOV and MARVIN C. GERSHENGORN

Introduction

The technique of transient permeabilization by hypoosmotic shock treatment (HOST) in the presence of ATP, previously used to incorporate aequorin into endocrine cells,[1] has been extended to the study of inositol lipid metabolism. Phosphatidylinositol (PtdIns) synthesis has been studied in GH_3 cells, cloned pituitary cells that secrete prolactin in response to thyrotropin-releasing hormone (TRH). A permeabilization approach to the study of PtdIns synthesis has the potential advantage of allowing the uptake and incorporation of *myo*-[^{3}H]inositol to be studied as dissociated events. The major advantage of permeabilization by the HOST method followed by resealing is to gain access to the intracellular milieu of large numbers of intact cells. This permits biochemical measurements to be made that are not possible in the small number of cells entered by microinjection.[2] This method, in contrast to permeabilization by membrane dielectric breakdown[3] or by detergents,[4] permits the cells to reseal and then allows for experimentation in intact cells. Cells after HOST permeabilization and resealing are viable and show normal growth in culture (see below). The demonstration by Borle and Snowdowne[1] that aequorin could be incorporated into GH_3 cells shows that substances of molecular weight

[1] A. B. Borle and K. W. Snowdowne, *Science* **217,** 252 (1982).

[2] J. E. Brown, L. J. Rubin, A. J. Ghalayini, A. P. Tarver, R. F. Irvine, M. J. Berridge, and R. E. Anderson, *Nature (London)* **311,** 160 (1984).

[3] P. F. Baker and D. E. Knight, *Nature (London)* **276,** 620 (1978).

[4] M. C. Gershengorn, E. Geras, V. Spina Purrello, and M. J. Rebecchi, *J. Biol. Chem.* **259,** 10675 (1984).

much greater than allowed by isotonic ATP treatment (800–900 exclusion limit in one well-characterized procedure[5]) may be loaded into treated cells by this method.

We found that permeabilized and resealed GH_3 cells incorporated [^{3}H]inositol into PtdIns in a concentration-dependent manner under basal conditions. The rate of incorporation was much more rapid than when [^{3}H]inositol was added to intact GH_3 cells.[6,7] Stimulation by TRH caused a rise in the specific radioactivity of PtdIns. Basal and TRH-stimulated increases in specific radioactivities of PtdIns were similar in plasma membrane and endoplasmic reticulum fractions isolated from permeabilized and resealed cells.

Methods

Materials

myo-[2-^{3}H]Inositol is obtained from New England Nuclear. Tissue culture supplies are from Gibco. ATP (potassium salt) is from Sigma, as are all other reagents unless specified. Solvents are HPLC grade from Burdick and Jackson. K6 silica gel thin-layer chromatography plates are obtained from Whatman.

Solutions

Balanced salt solution (BSS) contains 135 m*M* NaCl, 4.5 m*M* KCl, 0.5 m*M* $MgCl_2$, 1.5 m*M* $CaCl_2$, 5.6 m*M* glucose, and 10 m*M* HEPES (*N*-2-hydroxethylpiperazine-*N*′-2-ethanesulfonic acid), pH 7.4; phosphate-buffered saline (PBS) solution contains 0.1 *M* NaCl, 50 m*M* sodium phosphate, pH 7.4; CMF–PBS is 137 m*M* NaCl, 4 m*M* KCl, 1.1 m*M* glucose, 2 m*M* sodium phosphate, pH 7.4; HOST medium consists of 3 m*M* ATP (dipotassium salt) in 3 m*M* HEPES, pH 7.4. Additions, for example, 1 to 100 m*M* inositol, are made after 2 min of cell exposure to HOST medium. An ATP stock solution (100 m*M*, adjusted to pH 7.4 with KOH) is stored as aliquots at −20° and used within 1 month of preparation. Recovery solution contains 2 *M* KCl with 3 m*M* HEPES, pH 7.4.

Permeabilization and Resealing

The HOST permeabilization procedure uses a relatively high extracellular concentration of ATP (common to many permeabilization

[5] T. H. Steinberg, A. S. Newman, J. A. Swanson, and S. C. Silverstein, *J. Biol. Chem.* **262,** 9994 (1987).

[6] M. J. Rebecchi, R. N. Kolesnick, and M. C. Gershengorn, *J. Biol. Chem.* **258,** 227 (1983).

[7] M. J. Rebecchi and M. C. Gershengorn, *Biochem. J.* **216,** 287 (1983).

schemes[8,9]) combined with low quantities of divalent cations. It is distinct from most other methods in that it takes place in hypotonic medium (at reduced temperature) and is terminated by removal of ATP after a fixed interval. This is important since long-term exposure to millimolar levels of ATP has been shown to be deleterious to growth of some cells, for example, murine erythroleukemia cells.[9] Indeed, we have found that prolonged exposure of GH_3 cells to ATP in HOST or to ATP in isotonic KCl medium causes decreased viability.

The method is essentially that of Snowdowne and Borle.[10] GH_3 cells are grown in monolayer culture in Ham's F10 medium supplemented with 15% horse serum and 2.5% fetal calf serum.[11] Cells are harvested with 0.02% EDTA, centrifuged at 180 *g* for 5 min, then resuspended, and incubated (in suspension) in growth medium for 30–60 min at 37° to allow for recovery prior to the experiment. They are then washed twice with PBS (or CMF–PBS). Two milliliters of chilled HOST medium is added to the cell pellet. After the cells are resuspended, other factors are added. After 2–3 min of incubation on ice, 0.14 ml of chilled recovery solution (2 *M* KCl, 3 m*M* HEPES, pH 7.4) is added, and the cells are incubated for an additional 15 min. Cells are then centrifuged, washed, and used.

To validate the method, in particular, to show that the resealed cells are capable of reestablishing themselves in long-term culture, the growth of HOST cells is compared to control cells (cells harvested and resuspended in growth medium at 37°) and sham-treated cells (cells harvested and resuspended in 0.13 *M* KCl, 3 m*M* HEPES, pH 7.4, for 17 min, on ice). Solutions are sterilized by filtration (0.20-μm filters, Nalgene). Equal volumes of cell suspensions are seeded into culture dishes (10×35 mm) at densities of $0.005–0.05 \times 10^6$ cells/cm^2. After overnight attachment, the medium is aspirated. Adherent cells are counted, and this time point is designated day 1. Cells are counted in a hemocytometer, and their viability is determined by trypan blue exclusion. Fresh medium (2 ml per dish) is added at 3-day intervals. Figure 1 is a composite of three growth experiments, performed in duplicate. The growth rate of cells from each of the three treatments (HOST, sham, and control) were identical. The initial lower cell numbers in HOST versus control or sham-treated plates appears to reflect a lower plating efficiency for these cells or a loss of less than 10% of the cells.

[8] J. P. Bennett, S. Cockcroft, and B. D. Gomperts, *J. Physiol.* (*London*) **317,** 355 (1981).

[9] S. B. Chahwala and L. C. Cantley, *J. Biol. Chem.* **259,** 13717 (1984).

[10] K. W. Snowdowne and A. B. Borle, *Am. J. Physiol.* **246,** E198 (1984).

[11] T. F. J. Martin and A. H. Tashjian, Jr., *in* "Biochemical Actions of Hormones" (G. Litwack, ed.), Vol. 4, p. 269. Academic Press, New York, 1977.

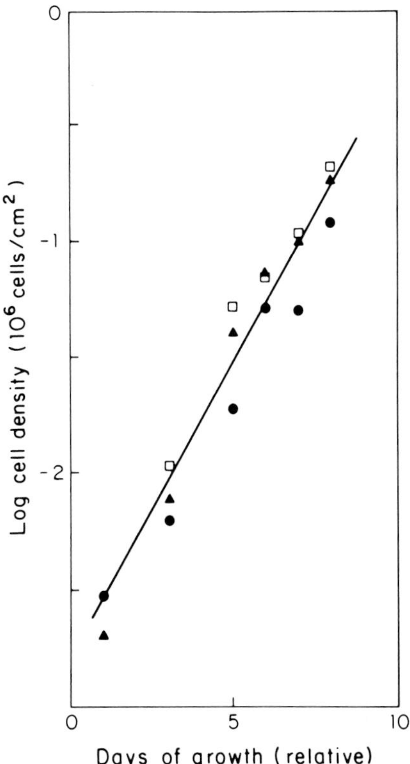

FIG. 1. Viability/growth after HOST (●) treatment. GH_3 cells are harvested and incubated as described in *Methods*. After centrifugation (180 *g*, 5 min) the pellet (60×10^6 cells) is suspended in 3.5 ml sterile PBS, and 1-ml aliquots are pipetted into sterile plastic 12×7.5 cm capped culture tubes (Falcon). The cells are centrifuged again (1 min, 1200 *g*), and the supernatant is aspirated off. "HOST" cells are then resuspended with 0.75 ml of 3 m*M* ATP, 3 m*M* HEPES, pH 7.4, and incubated on ice for 2 min. Recovery solution (0.052 ml 2 *M* KCl with 3 m*M* HEPES, pH 7.4) is added, and incubation on ice is continued for another 15 min. "Sham" (▲) cells are resuspended with 0.8 ml ice-cold 130 m*M* KCl, 3 m*M* HEPES, pH 7.4, and incubated on ice for 17 min. Control (□) cells are resuspended with 0.8 ml Ham's F10 culture medium (37°) and incubated at room temperature for 17 min. The treatment is terminated by centrifugation (1 min, 1200 *g*), and pellets are resuspended in culture medium. Aliquots are removed, counted in a hemocytometer, and tested for viability by trypan blue exclusion. Equal volumes of cells are seeded at densities which range from 5×10^3 to 50×10^3 cells/cm^2 in several experiments. Data from several experiments are combined after adjustment for differences in initial cell density. Trypan blue exclusion before seeding is 98–100% for control and sham cells and 90–95% for HOST cells.

The length of HOST and the ratio of cells to volume of HOST medium are important variables in determining the long-term viability of the treated cells. At ratios of 0.05–0.06 ml HOST medium to 10^6 cells, greater than 90% of treated cells exclude trypan blue 30 min after resealing and the growth of HOST cells is indistinguishable from growth of sham-treated or control cells (see Fig. 1). Permeabilization under more adverse conditions, for example, 3–4 min HOST with a ratio of 0.2 ml HOST medium/10^6 cells, results in a reduction of plating efficiency and initial cell viability (data not shown). However, the cells which did attach appear to grow at a rate parallel to control and sham-treated cells after an initial lag period. Thus, the requirements to support long-term growth at optimal rates appear to be more stringent than those needed for short-term cell survival.[1,10]

Time Course of Permeabilization and Uptake of [^{3}H]Inositol

The time courses of permeabilization and uptake of [^{3}H]inositol are compared so that the optimal time for complete equilibration of [^{3}H]inositol between the medium and the intracellular compartments could be determined. Table I presents data from a representative experiment. Trypan blue permeability is measured prior to addition of recovery solution and reaches 100% at 10 min. As expected because of its smaller size and lack of charge, uptake of [^{3}H]inositol is faster than that of trypan blue and reaches a plateau level within 3 min. Loss of [^{3}H]inositol from prelabeled cells (Table I) shows a similar time course during permeabilization. The maximal decrease occurs at or before 2 min of HOST. Hence, [^{3}H]inositol uptake could routinely be performed for as short a period as was needed to completely permeabilize the cells as monitored with trypan blue. A period of 2–3 min is used in all experiments.

Time Course and Concentration Dependence of Incorporation of [^{3}H]Inositol into PtdIns

Uptake and incorporation of [^{3}H]inositol into intact GH_3 cell lipids is relatively slow. To attain labeling of PtdIns to isotopic steady state requires 18–24 hr.[7] Thus, plasma membranes isolated from acutely labeled (10–45 min; 1–100 m*M* extracellular [^{3}H]inositol) intact GH_3 cells stimulated by TRH fail to show increased labeling of inositol lipids.[12] Since some of this initial delay in labeling of inositol lipids in intact cells is due to the slow equilibration of the large intracellular pool of unlabeled inositol, HOST permeabilization of cells in the presence of labeled inositol is

[12] A. Imai and M. C. Gershengorn, *Nature (London)* **325,** 726 (1987).

TABLE I
TIME COURSE OF PERMEABILIZATION AND EQUILIBRATION OF [^{3}H]INOSITOL

Time (min)	Permeable (%)	[^{3}H]Inositol uptake[a] (dpm)	[^{3}H]Inositol loss[b] (dpm)
2	55	1888 ± 68	6911
5	85	1661 ± 326	6732
10	100	1069 ± 147	6087
60	100	1469 ± 246	5775

[a] Cells are harvested and incubated as described in *Methods*, then washed twice with CMF–PBS, and finally resuspended in 0.185 ml of CMF–PBS. Fifty microliters of cell suspension is then pipetted into a series of test tubes to which have been added 0.5 ml HOST supplemented with [^{3}H]inositol. Cells are incubated in HOST medium for the indicated times. A portion (5 μl) is then withdrawn for trypan blue staining and counting, and 0.035 ml of recovery solution is added to the remainder. Cells are incubated on ice for 15 min further, then centrifuged. The pellets are resuspended in 0.2 ml BSS, then spun through silicone oil in a Beckman microfuge at 800 g for 45 sec, after which ^{3}H counts are obtained. There are approximately 0.7×10^6 cells per point.

[b] Cells are resuspended in 1 ml BSS containing [^{3}H]inositol (20 μCi/ml) and incubated at 37° for 45 min. Cells are washed twice with CMF–PBS and finally resuspended in 0.15 ml CMF–PBS. Aliquots of the suspension (5 μl) are pipetted into test tubes on ice, and 0.5 ml of HOST medium is added. After 2 min, HOST medium is supplemented with 1 mM unlabeled inositol. Samples are incubated in an ice bath for the indicated times. Recovery solution is added, and samples are incubated for 15 min further at 4°. Cells are centrifuged, and the pellet is resuspended with 0.2 ml of BSS and then spun through silicone oil. At $t = 0$ (prior to HOST) a total of 17,000–18,000 dpm [^{3}H]inositol has been taken up per point. The residual ^{3}H radioactivity could be accounted for as [^{3}H]PtdIns.

performed as a prelude to studying incorporation of [^{3}H]inositol into PtdIns. In the representative experiment presented in Table II, 0.28 nmol of [^{3}H]inositol is incorporated into PtdIns (per 10^6 cells) under "basal" conditions. Basal levels are constant over a time period ranging from 30 min to several hours. These cells respond to TRH with increased labeling of PtdIns, reaching a plateau of additionally incorporated [^{3}H]inositol of 0.78 nmol into PtdIns per 10^6 cells. For comparison, intact GH_3 cells contain approximately 1 nmol of PtdIns per 10^6 cells. Hence, in permeabilized and resealed GH_3 cells stimulated by TRH, it appears that virtually all cellular PtdIns (0.28 + 0.78 nmol) is labeled to isotopic steady state within 1 hr.

TABLE II
TIME COURSE OF INCORPORATION OF $[^3H]$INOSITOL INTO PtdIns[a]

Time (min)	$[^3H]$PtdIns (nmol)/10^6 cells Basal	TRH
30	0.26	0.57
60	0.31	0.96
120	0.22	0.76
240	0.28	0.76

[a] GH_3 cells (50×10^6) are harvested and incubated in medium as described in *Methods*, then washed twice with BSS. Cells are pelleted and resuspended in 2 ml of HOST medium. After 2 min, 100 mM $[^3H]$inositol (15 μCi/ml) is added. After 3 min on ice, 0.14 ml recovery solution is added, and the cells are further incubated for 15 min (on ice). Cells are then washed and resuspended in 2 ml of BSS with 100 mM $[^3H]$inositol (15 μCi/ml) and 10 mM LiCl. Samples (1.4×10^6 cells per ml) are pipetted into test tubes and incubated at 37° in the presence or absence of 1 μM TRH (5 μl as 1000× solution). Reactions are terminated at the specified time intervals by the addition of 1 ml of chloroform/methanol/concentrated HCl (100 : 100 : 1). PtdIns is extracted and resolved from other lipids as previously described.[7]

A similar experimental approach is used to show that PtdIns synthesis stimulated by TRH can occur at the plasma membrane as well as in the more widely accepted intracellular location, the endoplasmic reticulum.[13] [We found PtdIns synthase (CDPdiacylglycerol–inositol 3-phosphatidyltransferase) activity in both plasma membrane and endoplasmic reticulum fractions isolated from GH_3 cells.[12,14]] The HOST method is used to allow rapid equilibration of the intracellular pool of inositol so that TRH stimulation of *de novo* synthesis of PtdIns can be measured within a time scale during which any transport of PtdIns from the endoplasmic reticulum to the plasma membrane would play only a minor role. Under these conditions (15 min of incubation after the end of the recovery period from

[13] L. E. Hokin, *Annu. Rev. Biochem.* **54,** 205 (1985).
[14] A. Imai and M. C. Gershengorn, *J. Biol. Chem.* **262,** 6457 (1987).

HOST/labeling followed by 5 min in the presence TRH), TRH-stimulated increases in the incorporation of [^{3}H]inositol into PtdIns are found in both the endoplasmic reticulum and plasma membrane.[12] The similarity in the specific radioactivity of PtdIns in both the endoplasmic reticulum and the plasma membrane is consistent with the hypothesis that independent synthesis of PtdIns occurs at both intracellular sites.

Conclusion

Permeabilization by hypoosmotic shock treatment is a gentle method for rapid labeling of large quantities of clonal endocrine GH_3 cells with [^{3}H]inositol, a substrate for the synthesis of PtdIns. Further work will be required to establish the mechanism for the basal and TRH-stimulated increases in labeling of PtdIns, which could occur through enzyme-mediated base exchange as well as through PtdIns synthase activity. Both activities have been found in GH_3 cell membranes.[12,14]

The HOST method may be used for studies of other metabolites in the inositol lipid pathway. For example, preliminary studies in our laboratory have shown increased incorporation of label into CDP-diglyceride after HOST permeabilization in the presence of [^{3}H]CTP. The true strength of this method is that the treated cells grow normally in culture, making it possible to study the metabolism of a wide range of introduced substrates under a variety of growth conditions.

[14] Cell Cracking: Permeabilizing Cells to Macromolecular Probes

By THOMAS F. J. MARTIN

Introduction

There are a number of cellular processes which have resisted *in vitro* biochemical analysis since a high degree of structural integrity appears to be required. The regulated exocytotic secretion of hormones, neurotransmitters, and enzymes by endocrine, neural, and exocrine cells is a process well-defined by ultrastructural but not biochemical techniques. A variety of permeabilization techniques have been applied to secretory cells to attempt to render the exocytotic apparatus accessible to biochemical probes. These techniques have included high-voltage field-induced permeabilization,[1] reversible ATP-induced[2] or chelator-induced[3] permeabilization, membrane disruption with detergents such as digitonin[4] or saponin[5] or with Sendai virus,[2] and the use of pore-forming hydrophobic peptides such as streptolysin O[6] and staphylococcal α toxin.[7] In addition, various attempts to demonstrate regulated secretion in broken cells have been reported, but with few exceptions (e.g., sea urchin eggs[8] and *Paramecium*[9]) these experiments appear to be unreproducible.

In this chapter, we describe a minimally disruptive, mechanical shear technique (so-called cell cracking) for rendering cells permeable to macromolecular probes. For cracked GH_3 pituitary cells, we describe the preservation of ATP-dependent, Ca^{2+}-dependent prolactin (PRL) release and the finding that a cytosolic protein is required for regulated secretion.

Methods

Assembly and Use of Ball Homogenizer. The cell cracking method employs a stainless steel ball homogenizer shown in Figs. 1 and 2. Cell suspensions are gently forced through the narrow clearance established between a bored hole and an appropriately sized steel or tungsten carbide ball. Cells passed once through the chamber emerge structurally intact in appearance but rendered permeable to high molecular weight probes. Permeabilization apparently results from a large plasma membrane opening induced by deformation of the cell during passage through a restricted opening. The ball homogenizer was introduced by Balch and Rothman[10] for isotonic homogenization of cultured cells for the isolation of intact organelles. For this purpose, multiple passes of cells through the chamber are required to thoroughly homogenize cells. In contrast, we have found that a single pass through an appropriate clearance permeabilizes but does not homogenize cells.

Construction of the ball homogenizer is shown schematically in Fig. 2 and described in the figure legend. It is a modification of the apparatus described by Balch and Rothman[10] based on a design from the European

[1] D. E. Knight and M. C. Scrutton, *Biochem. J.* **234,** 497 (1986).
[2] B. D. Gomperts and J. M. Fernandez, *Trends Biochem. Sci.* **10,** 414 (1985).
[3] I. Schulz, T. Kimura, H. Wakasugi, W. Haase, and A. Kribben, *Philos. Trans. R. Soc. London, Ser. B.* **296,** 105 (1981).
[4] L. A. Dunn and R. W. Holz, *J. Biol. Chem.* **258,** 4989 (1983).
[5] J. C. Brooks and S. Treml, *J. Neurochem.* **40,** 468 (1983).
[6] T. W. Howell and B. D. Gomperts, *Biochim. Biophys. Acta* **927,** 177 (1987).
[7] G. Ahnert-Hilger, S. Bhakdi, and M. Gratzl, *J. Biol. Chem.* **260,** 12730 (1985).
[8] J. H. Crabb and R. C. Jackson, *J. Cell Biol.* **104,** 2263 (1985).
[9] M. Momayezi, C. J. Lumpert, H. Kersken, V. Gras, H. Plattner, M. H. Krinks, and C. B. Klee, *J. Cell. Biol.* **105,** 181 (1987).
[10] W. E. Balch and J. E. Rothman, *Arch. Biochem. Biophys.* **240,** 413 (1985).

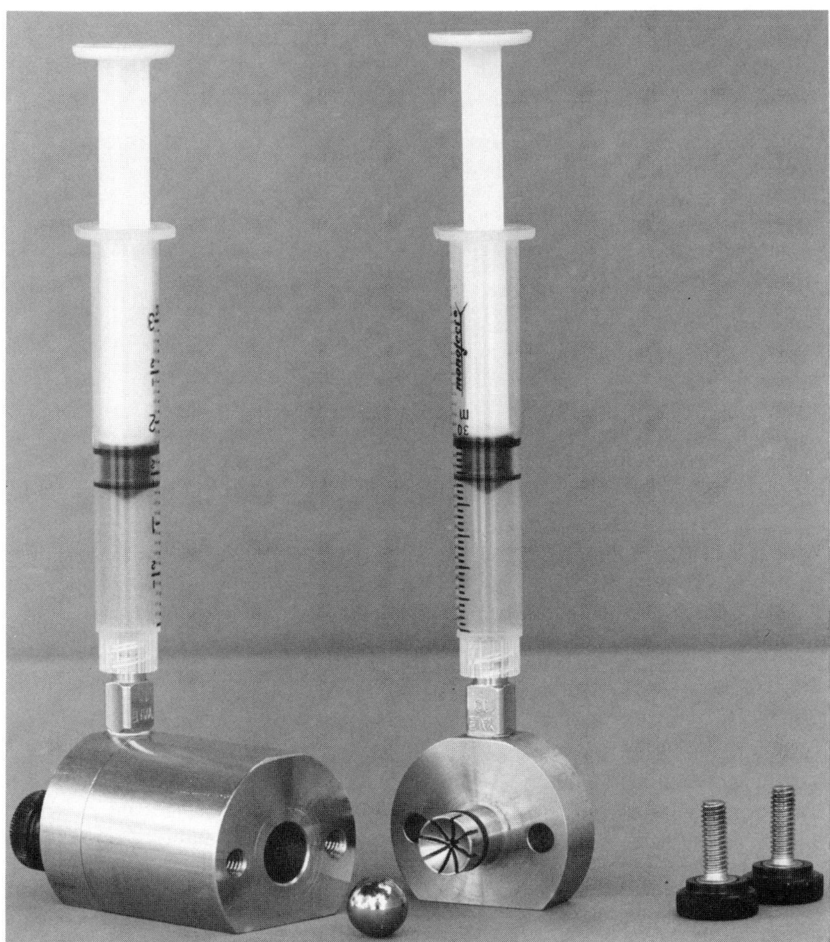

FIG. 1. Stainless steel ball homogenizer for cell cracking. Cell suspensions are gently pushed through the homogenizer using 3-ml disposable syringes at inlet and outlet Luer-Lok fittings.

Molecular Biology Laboratory, Heidelberg. The precise bore of the hole relative to the size of the ball needs to be empirically determined for each cell type. Figure 3 shows the basic empirical experiment which defines the appropriate size of ball to be used. Separate aliquots of GH_3 cell suspensions were passed gently through the homogenizer fitted with a series of balls which establish the nominal clearances shown on the abscissa. GH_3 cells have a diameter of approximately 15 μm, and a nominal clearance of 2 μm was found to be optimal. This was achieved with either an 8.01-mm bore chamber and 8.008-mm ball in a metric homogenizer, or with a

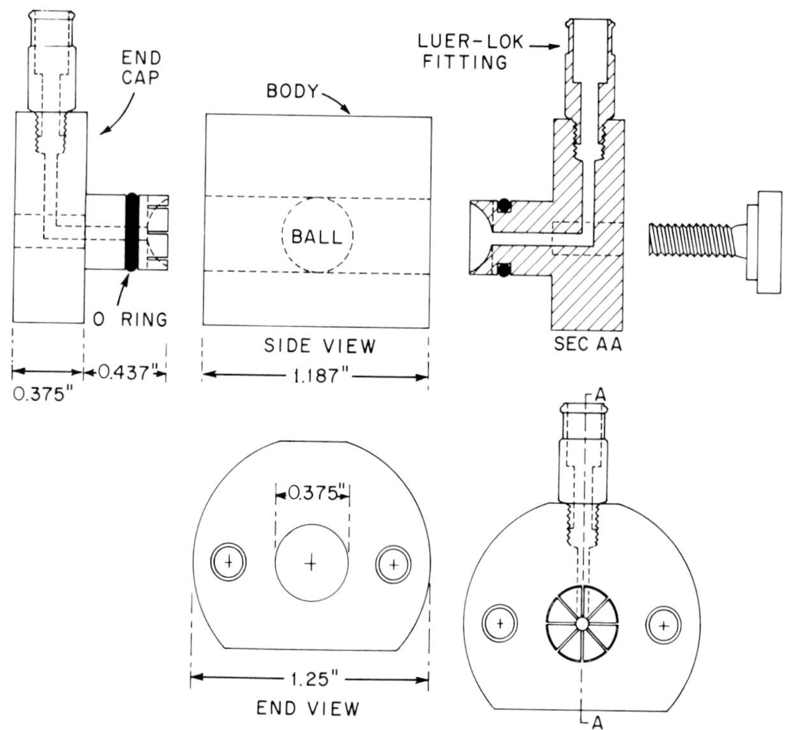

FIG. 2. Schematic for construction of ball homogenizer. From $1\frac{1}{4}$ inch diameter stainless steel round stock, cut one piece at $1\frac{1}{4}$ inch and two pieces at 1 inch. Finish to 1.187 and 0.812 inch, respectively. Drill and bore center hole in body to 0.374 inch and hone to 0.375 ± 0.0002 inch or to accommodate precision ball. Machine end caps by turning a center post to 0.374 inch diameter by 0.437 inch length. Spot with 3/8 inch ball end mill, 3/16 inch deep, and drill 1/16-inch hole 5/8 inch deep. Groove for O ring (1/16 × 1/16 × 3/16 inch). Eight slots, 0.025 × 0.125 inch, on center post prevent ball from sealing hole. Assemble parts and mill top and bottom flat. Drill 1/16-inch hole in end caps until they intersect horizontal hole. Secure syringe adapters. Drill and tap eight 32 × 1/2 inch holes, 7/8 inch apart, in body ends and clearance holes in end caps. Secure end caps to body with thrumbscrews (eight 32 × 5/8 inch). Precision grade balls are available from Industrial Tectonics Inc. (Ball Division, P.O. Box 1128, Ann Arbor, MI 48106). For the homogenizer shown here, Grade 25 tungsten carbide balls from 0.3747 to 0.3755 inch in 0.0001-inch increments were used.

0.3750-inch bore and 0.3749-inch ball in a nonmetric homogenizer (see the legend to Fig. 3). Cell permeabilization was assessed by trypan blue dye uptake. Optimal fitting of a ball in the bored housing results in 95–99% trypan blue staining. The homogenizer was used at room temperature with a cold suspension of cells using two 2.5-ml aliquots in sequence. Prechilling the homogenizer on ice resulted in suboptimal clearance and

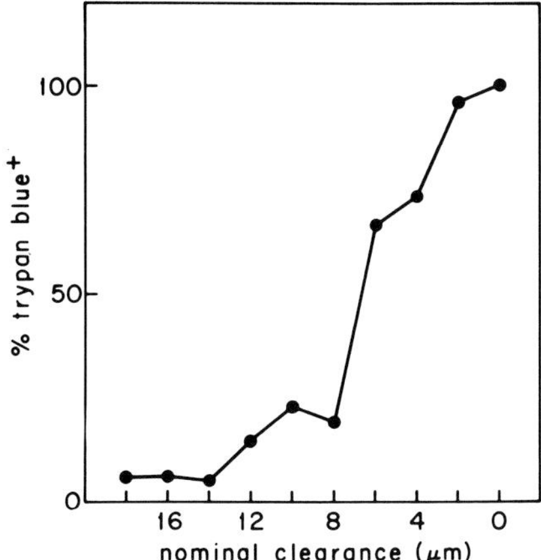

FIG. 3. Optimizing clearance for cell cracking. Suspensions of GH_3 cells were passed once through a ball homogenizer fitted with a size range of balls. Cells were then counted in a hemocytometer after mixing an equal volume of cells with 0.4% trypan blue in 0.15 *M* NaCl. In this experiment, a metric ball homogenizer constructed by the EMBL, Heidelberg, was utilized. Nominal clearances were established with the use of 8.01- or 8.02-mm bored chambers with stainless steel balls 8.002, 8.004, 8.006, 8.008, and 8.010 mm in diameter. Similar results were obtained with the nonmetric apparatus described in Figs. 1 and 2; nonmetric balls differ in diameter by 0.0001 inch or approximately 0.0025 mm.

an increase in cell debris. Empirical assessment of the chamber–ball combination must be done at well-defined temperatures to render permeabilization reproducible. Cell density was found to be a less important factor such that reproducible permeabilization of GH_3 cells was obtained using cell suspensions over a 10-fold density range (10^6–10^7 cells/ml).

Characterization of Cracked GH_3 Cells. Preliminary biochemical criteria have indicated excellent preservation of cellular organelle markers. Cells cracked by a single pass through the homogenizer sediment at low speed (800 *g*), as do intact cells. In the experiment depicted in Fig. 4, GH_3 cell suspensions were passed through the homogenizer once or repeatedly. Subsequently, cracked (single pass) and homogenized (multiple pass) cells were sedimented at indicated speeds and pellets analyzed to assess the extent of cellular disruption. Following a single pass, 60–70% of the cellular PRL was found to be associated with fast sedimenting (800 *g*) cracked cells. In contrast, successive passes increasingly converted the cells to a homogenate such that by the seventh pass most (>80%) of the PRL sedimented with smaller organelles at 5000–100,000 *g* (P-5 to P-100). Little (<7%) of the PRL was found in the high-speed supernatant fraction (S-100). These results confirm that the ball homogenizer can also be uti-

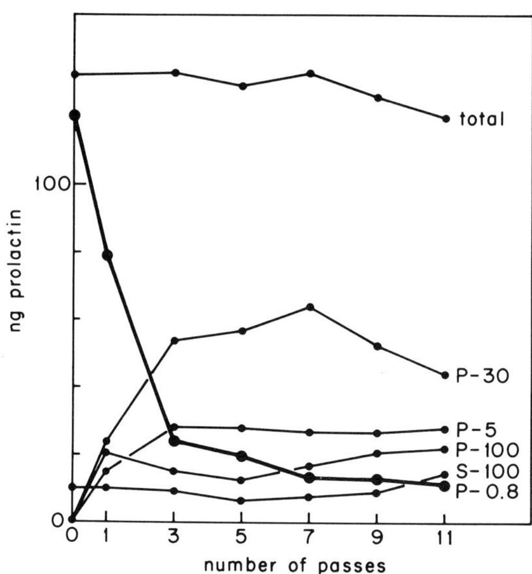

FIG. 4. Localization of PRL in GH_3 cells cracked or homogenized in ball homogenizer. Suspensions of GH_3 cells were passed through the ball homogenizer the indicated number of times. Homogenates were centrifuged successively at 800, 5000, 30,000, and 100,000 *g*. Each pellet (P-0.8, P-5, P-30, and P-100) was resuspended in detergent-containing buffer [40 m*M* potassium phosphate, pH 7.2, 0.15 *M* NaCl, 2 m*M* phenylmethylsulfonyl fluoride (PMSF), 10 μg/ml leupeptin, 0.5% NP-40, and 5 mg/ml deoxycholate]. The final supernatant (S-100) and detergent-solubilized pellets were assayed for PRL content by radioimmunoassay.

lized for isotonic homogenization and organelle isolation. In other studies it has been demonstrated that 60–70% of other organelle markers remained associated with cracked cells prepared by a single pass (800 *g*); these include lysosomal (*N*-acetylglucosaminidase) and Golgi (UDPgalactosyltransferase) markers.

Preliminary immunocytochemical studies indicated that cracked cells exhibit immunostaining properties resembling those of intact cells when assessed with antibodies directed against PRL and tubulin.[11] Transmission electron microscopy demonstrated that the cracked cells contain a full complement of organelles including mitochondria and endoplasmic reticulum; however, strikingly absent was the granular cytoplasmic background. Recent studies[11,12] using a Hitachi S-900 scanning electron microscope indicated excellent preservation of surface structure in cracked cells. In addition, large (1 μm) openings in the surface membrane were present in cracked but not in intact cells.

[11] W. V. Welshons, unpublished observations.

[12] T. F. J. Martin and J. H. Walent, manuscripts in preparation.

Several methods have been used to characterize the degree of permeabilization of cracked GH_3 cells. Staining at the light microscope level with trypan blue (M_r 960) was observed. With the fluorescence microscope, nuclear staining with ethidium bromide (M_r 394) was evident. In addition, a fluorescein-conjugated wheat germ agglutinin (M_r 33,000) was found to visualize the nuclear membrane of cracked but not intact cells.[12] These studies establish the degree of permeabilization for cracked cells. Additional evidence has indicated that large (M_r >100,000) cytoplasmic macromolecules are released from cracked cells (see below).

Functional Properties of Cracked Cells for Studies of Ca^{2+}-Activated Secretion. Permeabilized GH_3 cells have been used to study Ca^{2+} regulation of PRL secretion. Cracked cells have been utilized in either of two formats: as freshly cracked cells with diluted cytosol factors residually present or as cellular "ghosts" which have been washed free of cytosol by centrifugation.

In order to assess PRL release by cracked cells, intact GH_3 cells must be extensively washed and preincubated to reduce background values. This is done in a manner similar to that used for intact cell secretion studies.[13] Monolayer cultures are washed 3 times in F10 medium (37°) containing 0.1% bovine serum albumin, incubated in the same for 30–60 min and followed by 3 additional washes. Cells are harvested by incubation (37° for 5 min) in Hanks' Mg^{2+}-free, Ca^{2+}-free buffer containing 0.1 m*M* EDTA, and washed several times in cold buffer lacking Ca^{2+} and containing 100 μM EGTA. Washed cells are resuspended in a potassium glutamate buffer (20 m*M* sodium HEPES, pH 7.2, 5 m*M* glucose, 2 m*M* EGTA, 0.1% bovine serum albumin, 0.12 *M* potassium glutamate and 20 m*M* NaCl) previously found to support Ca^{2+}-dependent PRL release from electropermeabilized GH_3 cells.[14-16] Cells in chilled potassium glutamate buffer (10^6–10^7 cells/ml) are cracked by passage through the ball homogenizer, and the cracked cell suspension is distributed to chilled reaction tubes containing Ca^{2+} and MgATP in a total volume of 0.2 ml. Alternatively, chilled suspensions of cracked cells are washed 2–3 times (800 *g* for 4 min) to prepare cytosol-free cellular ghosts. Ghosts are resuspended in potassium glutamate buffer and distributed into reaction tubes. Following warming to 30° for 5–15 min, reactions are chilled and sedimented at 100,000 *g* for 60–90 min. Centrifugation was conducted in an airfuge for a small number of samples or in a 25 rotor (Beckman) which contains 100 positions for 1-ml polycarbonate tubes. The clear supernatant overlying a

[13] T. F. J. Martin and J. A. Kowalchyk, *Endocrinology (Baltimore)* **115,** 1527 (1984).

[14] S. A. Ronning and T. F. J. Martin, *Biochem. Biophys. Res. Commun.* **130,** 524 (1985).

[15] S. A. Ronning and T. F. J. Martin, *J. Biol. Chem.* **261,** 7834 (1986).

[16] S. A. Ronning and T. F. J. Martin, *J. Biol. Chem.* **261,** 7840 (1986).

small pellet is removed and assayed for PRL content by radioimmunoassay. The PRL content of cell pellets is determined by radioimmunoassay following solubilization of the pellets in detergent-containing buffer (see the legend to Fig. 4).

PRL release by cracked cells at 30° has been found to be Ca^{2+} and MgATP dependent. MgATP at 2–4 mM was found to be optimal. PRL release was found to be minimal at 10^{-9}–10^{-8} M free Ca^{2+} and maximal at 10^{-6} M Ca^{2+} with half-maximal stimulation at 10^{-7} to 3×10^{-7} M. Results from a representative experiment using cracked cells are shown in Table I. In contrast, ghosts prepared from cracked cells by washing failed to release PRL even in the presence of optimal Ca^{2+} and MgATP (Table I). If, however, cytosol fractions were added to the incubation, Ca^{2+}- and MgATP-dependent PRL release was observed (Table I). PRL release with Ca^{2+} and MgATP at 30° was linear for 15 min, and maximal PRL release

TABLE I

PRL RELEASE FROM CRACKED GH_3 CELLS[a]

Cell type	PRL released (ng/10^6 cells)
Cracked cells	
No incubation	3.8 ± 0.08
10^{-9} M Ca^{2+}	4.9 ± 0.06
10^{-6} M Ca^{2+}	10.2 ± 0.40
Ghosts	
No incubation	2.0 ± 0.16
10^{-6} M Ca^{2+}, no cytosol	4.6 ± 0.36
10^{-9} M Ca^{2+}, with cytosol	4.3 ± 0.40
10^{-6} M Ca^{2+}, with cytosol	13.4 ± 0.90

[a] Suspensions of GH_3 cells in potassium glutamate buffer were cracked as described in the text and dispersed into reaction tubes. A portion of the cracked cells was washed twice in chilled buffer by centrifugation (800 g, 4 min). Resultant ghosts were resuspended to original density in buffer and delivered to reaction tubes. Incubations in a final volume of 0.2 ml were conducted at 30° for 15 min with 10^{-9} or 10^{-6} M Ca^{2+}, 4 mM MgATP, and 25 μg crude rat brain cytosol where indicated. PRL content of supernatants obtained from high-speed centrifugation of chilled reactions was determined by radioimmunoassay. The mean and range of duplicate reactions are shown.

from cracked cells or ghosts corresponded to approximately 30% of the total PRL pool.

Preparation of GH_3 cell ghosts has provided an assay for characterization of the cytosolic factor required for Ca^{2+}- and MgATP-dependent PRL release. High-speed supernatants from several rat tissues (liver, brain) have been utilized as a source for this factor. Rat brain cytosol was optimal at less than 25 μg of crude cytosol protein for PRL release from 0.5×10^6 cell ghosts incubated in 0.2 ml. Preliminary characterization of rat brain factor has indicated that conventional protein chromatographic purification strategies are feasible. A tentative size estimate established for the rat brain factor by chromatography on a calibrated S-300 Sephacryl column is M_r 225,000.[12]

Summary

The ball homogenizer described here can render cells of diameter greater than approximately 10 μm permeable to macromolecular probes. This approach may be useful for studying a variety of cellular processes which require structural integrity for function, such as secretion. Preservation of cellular morphology is sufficient, such that the approach may be useful for immunocytochemical studies or for immunoneutralization studies with antibodies directed against antigens involved in a variety of cell functions.

Acknowledgments

The work described here was supported by National Science Foundation Grant DCB-8512441. The author gratefully acknowledges contributions of J. H. Walent to biochemical studies of the cytosolic factor and of W. V. Welshons to immunocytochemical and ultrastructural characterization of the cracked cells. In addition, the skillful assistance of the University of Wisconsin Zoology Department machine shop (D. W. Hoffman and R. J. Ganje) is gratefully acknowledged.

[15] Microsurgical Techniques for Studying Functional Correlates of Hypothalamohypophyseal Axis

By Ichiro Murai, Walter C. Low, and Nira Ben-Jonathan

Given the complexity of the neural and vascular connections between the hypothalamus and the pituitary gland, and the inability to maintain adult brain tissue in culture, microsurgical techniques are indispensable for elucidating dynamic interactions between hypothalamic neuroactive substances and pituitary hormone secretion. Although microsurgical techniques have been developed for a number of species, we focus on the rat, which has been used extensively for investigating neuroendocrine mechanisms. Emphasis will be placed on microsurgical methods that have been adapted for use in unanesthetized, freely moving rats. These include posterior pituitary lobectomy (LOBEX), hypophysectomy (HYPOX), pituitary stalk section, and neural transplantation.

Structural and Anatomical Features of Hypothalamohypophyseal Complex

The Hypothalamus

The hypothalamus extends from the optic chiasma to the mamillary bodies and forms the floor and lateral walls of the third ventricle. It is composed of the following: (1) clusters of large perikarya, the magnocellular nuclei, with axon terminals in the neural lobe; (2) smaller perikarya, the parvicellular nuclei, with neuroendocrine or autonomic functions; (3) afferent and efferent neuronal tracts connecting the hypothalamus to other parts of the brain; (4) diffuse networks of fine nerve fibers providing intrahypothalamic connections; and (5) several types of support cells, i.e., the ependyma, which have nutritional or transport functions. The ventral portion of the hypothalamus, the median eminence, contains the primary capillaries of the portal vessels and a rich network of nerve terminals which do not form typical synapses.[1]

The Posterior Pituitary

The posterior pituitary is composed of two juxtaposed, but distinct, structures: the neural lobe, derived from the neural ectoderm, and the

[1] A. Björklund, R. Y. Moore, A. Nobin, and V. Stenevi, *Brain Res.* **51,** 171 (1973).

intermediate lobe, which originates from the buccal ectoderm. The neural lobe contains axon terminals of the supraopticohypophyseal and tuberohypophyseal nerve tracts, which carry oxytocin/vasopressin and dopamine, respectively. It also receives direct adrenergic innervation from the superior cervical ganglia.[2] The posterior pituitary contains sinusoids and specialized neuroglial cells, the pituicytes, which do not appear to have a direct secretory function. In the rat, which, unlike the human, has a well-defined intermediate lobe, cells occur in layers and are interposed by nerve endings.

The Anterior Pituitary

The anterior pituitary is connected to the hypothalamus by the pituitary stalk. It lacks a major direct innervation except for a few nonmyelinated fibers which accompany blood vessels. The anterior pituitary in the rat accounts for 85–90% of the total weight of the gland. Recent immunocytochemical and electron microscopy techniques have identified six to seven cell types based on size, intracellular organelles, and hormone content.[3] New evidence suggests the presence of dual hormone-secreting cells[4] as well as nonsecretory follicular cells[5] which might function in ion transport and/or paracrine interactions.

The Hypophyseal Portal Vasculature

The hypophyseal portal vasculature consists of primary capillaries in the median eminence, pituitary stalk, and neural lobe. These are supplied by the superior, middle, and inferior hypophyseal arteries, respectively, which are branches of the internal carotid artery. Blood reaches the anterior lobe from the median eminence via the long portal vessels, which run along the pituitary stalk, and from the neural lobe via the short portal vessels, which bridge the avascular cleft of the intermediate lobe.[6] About 25–30% of the total blood flow to the rat anterior pituitary is supplied by the short portal vessels.[7] The frequently used term "portal veins" is inappropriate because these vessels have fenestrated (capillarylike) endo-

[2] J. M. Saavedra, *Neuroendocrinology* **40,** 281 (1985).

[3] M. O. Dada, G. T. Campbell, and C. A. Blake, *J. Endocrinol.* **101,** 87 (1984).

[4] D. A. Leong, S. K. Lau, Y. N. Sinha, D. L. Kaiser, and M. O. Thorner, *Endocrinology (Baltimore)* **116,** 1371 (1985).

[5] M. Baes, W. Allaerts, and C. Denef, *Endocrinology (Baltimore)* **120,** 685 (1987).

[6] P. M. Daniels and M. M. L. Prichard, *Acta Endocrinol. (Copenhagen), Suppl.* **201,** 1 (1975).

[7] J. C. Porter, R. S. Mical, N. Ben-Jonathan, and J. G. Ondo, *Recent Prog. Horm. Res.* **29,** 161 (1973).

thelia. Blood is drained from the secondary capillary plexus within the anterior lobe into the systemic circulation by the Y-shaped pituitary veins. It is still controversial whether portal blood flows in one direction, i.e., toward the anterior pituitary, or whether it can go in several directions, depending on the state of vasoconstriction in the various vascular beds.[8]

Surgical Manipulation of the Pituitary Gland

In most mammals, the pituitary gland is enclosed in a deep recess of the sphenoid bone. In contrast, the rat pituitary rests on a flat surface of the basisphenoid bone and is completely enclosed by a fold of the dura mater, the diaphragma sellae, through which the pituitary stalk penetrates. These features make the rat pituitary relatively accessible for surgical manipulation. The surgical methods described herein are based on an initial exposure of the pituitary gland by the parapharyngeal approach, described in detail by Porter.[9] Emphasis will be placed on the numerous modifications which we have introduced in order to adapt this technique for a chronic preparation.

Exposure of the Pituitary by the Parapharyngeal Approach

The choice of anesthesia is critical. First, the anesthetic agent should be sufficiently potent to maintain a deep level of anesthesia. Second, it should last for 30–60 min, the approximate duration of surgery. Third, side effects such as suppression of respiration and/or other sympathetic activity should be minimal. Fourth, rats should recover from anesthesia within a short time and have minimal anesthesia-induced endocrine alterations. Although no anesthetic agent is ideal, we have found that Brevital (sodium methohexital) administered intraperitoneally (i.p.) (50–70 mg/kg), is suitable for this purpose.

The anesthetized rat is placed in a supine position on an operating board. If blood samples are to be collected, implantation of a jugular vein cannula should be performed next. An L-shaped incision is made along the midline of the neck, extending 3.5 cm caudal from the lower jaw to a point above the manubrium. Jugular cannulation is done according to Harms and Ojeda,[10] exteriorizing the cannula through the back of the neck. The advantage of this method is that the jugular vein is not occluded, and stress during blood withdrawal or injection of substances is minimal. The rat's head is then secured in a metal holder with adjustable

[8] R. M. Bergland and R. B. Page, *Science* **208,** 18 (1979).

[9] J. C. Porter, this series, Vol. 39, p. 166.

[10] P. G. Harms and S. R. Ojeda, *J. Appl. Physiol.* **36,** 391 (1974).

ear bars. The head holder, operating board, dissecting microscope, and tissue retractors are similar to those described by Porter.[9] The operating field should be well illuminated, and any commercially available fiber optic device can be used.

Instead of transecting the trachea as is done in acute surgery, such as portal blood collection, the trachea is intubated. The instruments used for this procedure are shown in Fig. 1A. The operating board is rotated, pointing the rat's nose toward the operator, and a thin fiber optic illuminator is placed inside the oral cavity. The jaws are opened and the tongue is pressed against the upper jaw with a modified spatula bent at a shallow angle which functions as a laryngoscope. The tracheal cannula is made from a 4.5-cm-long polyethylene tubing (PE 205) glued to a cut yellow disposable pipet tip which can fit onto a respirator. The cannula is placed over a guiding 19-gauge needle with a smooth round tip. Upon visualizing the vocal cords (Fig. 2A), the cannula is gently inserted into the trachea and connected, if necessary, to a rodent respirator supplied with either air or oxygen. Accumulated mucus may be cleared by applying gentle suction via a small vinyl catheter inserted into the tracheal cannula.

The sternohyoid and omohyoid muscles are exposed by blunt dissection with two forceps. Two single-prong retractors are placed on the

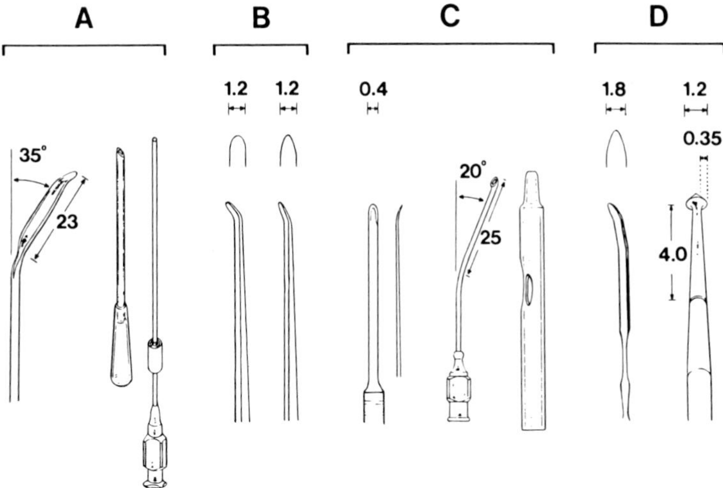

FIG. 1. Instruments which have been adapted for the different surgical procedures. Units are in millimeters, except for degrees of angles; all descriptions are from left to right. (A) For tracheal intubation: spatula used as a laryngoscope; tracheal cannula; guiding needle. (B) For scraping muscles: two slightly different spatulas. (C) For posterior pituitary lobectomy: microknife for cutting the dura (frontal and side views); aspiration needle; plastic holder with a ventilation hole. (D) For pituitary stalk section: spatula for separating the dura; cutting burr.

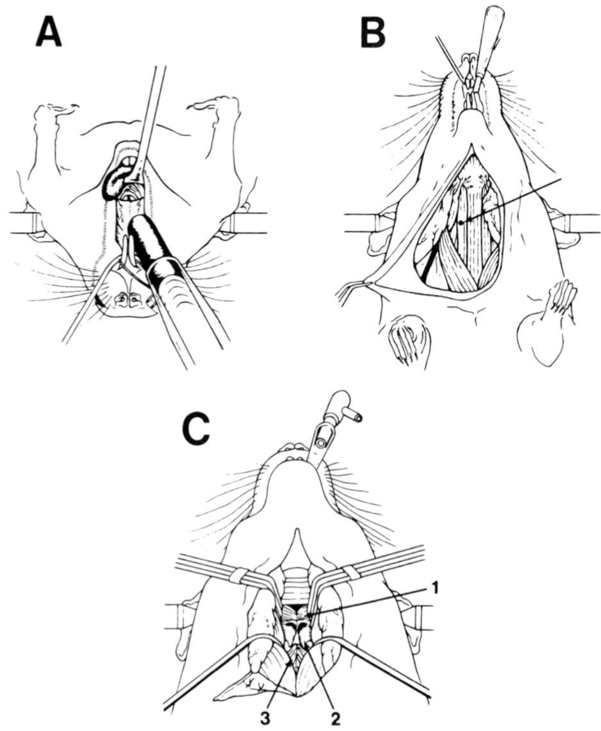

FIG. 2. Procedure for tracheal intubation and exposure of the occipital bone. (A) Inside view of oral cavity which is exposed by the use of an intubation spatula and illuminated by a thin fiber optic extension. (B) L-Shaped incision used for both the cannulation of the jugular vein and the separation of the sternohyoid and omohyoid muscles. Arrow and dot show the site for placing the single-prong retractors. (C) Exposure of the occipital bone by using two pairs of retractors. Arrow 1 designates the muscle covering the basiphenoid bone and part of the occipital bone. Arrow 2 shows the location of the occipital bone. Arrow 3 designates the "triangle" as referred to in the text.

surface of the sternohyoid muscle (Fig. 2B) at the interaural line 2–3 mm from either side of the trachea, and they are pushed down until the top of the muscle covering the first vertebra is seen. This point is referred to as a "triangle" with white lines. The tips of the retractors are further pushed down until they reach the surface of the occipital bone (Fig. 2C). The retractors are then pulled posteriorly and laterally along the "triangle" and are secured tightly by fasteners to the operating board. A second pair of three-prong retractors is used to further expose the occipital bone (Fig. 2C). Note that this exposure is much more restricted than that used for portal blood collection.[9] When performed by a skilled operator, no bleeding should occur up to this point.

Utilizing a 1.2-mm-wide spatula with a smooth tip bent at a shallow angle (Fig. 1B), a restricted portion of the occipital bone is scraped free of muscle and connective tissue. Scraping is done carefully and should not exceed rostrally 1 mm below the suture between the basisphenoid and occipital bones (see the legend to Fig. 2C). Puncturing of the pharynx, bleeding, and/or breathing difficulties can result if the area rostral to the suture is exposed. The exact location of the suture line (Fig. 3A) can be easily recognized after few preliminary trials. At this stage, the rat is surgically prepared for either LOBEX, HYPOX, or pituitary stalk transection.

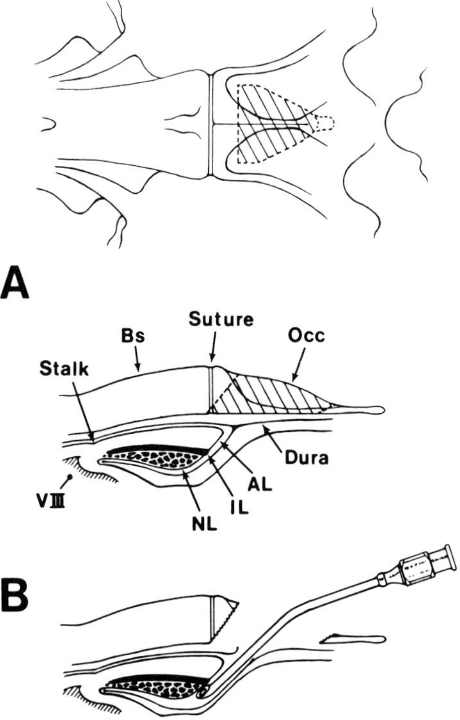

FIG. 3. Surgical approach for posterior pituitary lobectomy. (A) Ventral view (top) showing the basisphenoid and occipital bones and the location and shape of the drilling hole (hatched area). Sagittal view (bottom) of the same region as above showing the different components of the pituitary and surrounding structures. Bs, Basisphenoid bone; Occ, occipital bone; AL, anterior lobe; IL, intermediate lobe; NL, neural lobe; VIII, third ventricle. Note the location of the suture line between the basisphenoid and occipital bones and the hatched area showing the site of drilling. (B) Final approach for removal of the posterior pituitary showing the position of the aspiration needle directly beneath the posterior pituitary.

Posterior Pituitary Lobectomy

A number of important neuroendocrine parameters can be investigated by performing LOBEX. For example, LOBEX induces a prolonged diabetes insipidus, thus creating an excellent animal model for testing the efficacy of synthetic agonists/antagonists of vasopressin. The role played by oxytocin during lactation and its importance in the induction of labor can be studied in the LOBEX rat under controlled conditions. By performing LOBEX with or without adrenal demedullation, the source(s) of circulating β-endorphin and/or enkephalin and their possible peripheral physiological functions can be investigated. Moreover, removal of the posterior pituitary can resolve the question whether functional relationships exist between the two lobes of the pituitary.

Utilizing the technique of LOBEX, we have obtained interesting, and sometimes unexpected, results. For example, we have shown that dopamine from the posterior lobe participates in the inhibition of prolactin (PRL) release.[11] LOBEX in lactating rats completely abolishes the suckling-induced rise in PRL, indicating that the posterior pituitary contains a PRL-releasing factor (PRF).[12] Cycling rats subjected to LOBEX on the morning of proestrus have a normal LH surge but a blunted PRL rise. This suggests that input from the posterior lobe (possibly via PRF) contributes to the PRL surge on proestrus (unpublished observations). Moreover, LOBEX performed on estrus results in an increased frequency of luteinizing hormone (LH) pulsatility, indicating that the posterior lobe exerts an inhibitory influence over hypothalamic secretion of LH-releasing hormone (LHRH) (unpublished observations).

Since the rat neurointermediate lobe is separated from the anterior lobe by a cleft, LOBEX can be performed with relative ease. Some investigators have performed LOBEX by forcing a suction needle through the anterior lobe.[13] This might damage the anterior pituitary, however, resulting in some loss of pituitary function. To circumvent this problem, we have developed a technique which is less destructive but more difficult, since it requires a blind surgical approach during the insertion of the aspiration needle.

As shown in Fig. 3, a V-shaped and angled 3.0 × 2.5 mm hole is drilled in the occipital bone by operating a dental burr with a 1-mm head (Storz, #N-1566) at a maximal speed (14,000 rpm). The shape and location of the hole were carefully chosen to prevent unintentional scraping of the muscle during drilling and to allow for the subsequent insertion of the aspira-

[11] I. Murai and N. Ben-Jonathan, *Neuroendocrinology* **43,** 453 (1986).

[12] I. Murai and N. Ben-Jonathan, *Endocrinology* (*Baltimore*) **121,** 205 (1987).

[13] P. E. Smith, *Am. J. Physiol.* **99,** 345 (1932).

tion needle. After drilling, the hole is slightly enlarged by picking pieces of bone with a microhook (Storz, #E-858). At this stage, the caudal portion of the anterior pituitary (~0.5 mm) and rostral portion of the pons with the basilar artery can be seen. It should be noted that the posterior border of the anterior lobe in young rats is located slightly more caudally than that in adults (300 g body weight), and drilling should be adjusted accordingly.

A small, incomplete incision is made in the dura with a microknife (Storz, #E-153), about 0.2 mm below the posterior border of the pituitary. Since occasionally the dura at this site is transversed by small veins, the incision should be placed below the veins. However, these veins can break during enlargement of the incision, resulting in bleeding from lateral sites. Bleeding should be stopped by applying small pieces of Gelfoam. The incision is enlarged bilaterally with an angled microknife (V. Mueller, #AU-12362), without causing leakage of cerebrospinal fluid (CSF). If some leakage does occur, it should be stopped by applying slight pressure with an ear swab. The dura is then cut completely by pointing a microknife (Fig. 1C) in an angle down toward the posterior pituitary without touching either lobe. The small separation which has been created between the posterior lobe and the dura is enlarged with a shallow-angled thin spatula (Fig. 1B). Although the above procedure appears complicated, it permits the removal of the posterior lobe with minimal leakage of CSF and without damaging the pons, hypothalamus, or anterior pituitary by the aspiration needle.

The aspiration needle is made from a 20-gauge needle bent at a 20° angle with a flat and smooth tip (Fig. 1C). The needle is attached to a plastic holder (Fig. 1C), fashioned from a 1-ml plastic syringe with a hole made in the middle. The holder is connected via rubber tubing to a vacuum line and the pressure is controlled by an adjustable clamp. The aspiration needle is gently inserted directly beneath the posterior lobe (Fig. 3B), using a blind approach. Suction is applied by placing the index finger over the ventilation hole. The posterior lobe is dislodged in one piece and held by suction to the end of the needle. Upon withdrawal of the aspiration needle, the posterior lobe is inspected and removed. Sham operation consists of insertion and withdrawal of the needle without application of suction.

After confirming that no bleeding or leakage of CSF has occurred, a piece of Gelfoam is placed over the exposed tissue, and the site is sprinkled with antibiotic powder. The retractors are removed and the skin is sutured with No. 4 silk thread. Accumulated mucus is cleared, and the tracheal cannula is slightly withdrawn in order to test if the laryngeal reflex has been restored. The tracheal cannula is removed, and the rat is placed under oxygen. The rat is fully awake 1.5–2 hr after the initial anesthesia. Ample water should be provided since LOBEX rats begin to drink immediately after recovering from anesthesia. Since anesthesia might affect hormone secretion for several hours, an overnight recovery period is preferred if permitted by the experimental design. No special postoperative care is required, and LOBEX rats look and behave normally except for polyuria and polydipsia.

Postexperimental inspection of the surgical area is very important. Rats are divided into short-term (6–48 hr) and long-term (more than 4 days) experimental groups. In the short-term group, the brain is examined under a dissecting microscope for possible damage to the pons, hypothalamus, stalk, or anterior pituitary. The site which previously contained the posterior lobe is inspected for completeness of posterior lobectomy and for the existence of a blood clot whose size is larger than that of a normal posterior lobe. This is usually caused by insufficient suppression of lateral bleeding after cutting the dura. The dorsal surface of the anterior lobe is examined for the presence of dark-colored spots. These are usually caused by either an incorrect angle of insertion of the aspiration needle or by applying too much suction. In the long term group, the anterior pituitary is inspected for the presence of a pale color and/or atrophy. These could be due either to the presence of large blood clots or to damage of the stalk. Since abnormal findings are often associated with some loss of pituitary function, such animals should be eliminated from the study.

Hypophysectomy and Pituitary Transplantation

HYPOX and pituitary transplantation to remote sites have been extensively used in studies aimed at understanding the complex interactions between hypothalamic, pituitary, and peripheral hormones. For example, the question whether estrogen activates the hypothalamic dopaminergic system directly, or indirectly via increasing PRL secretion, can be best resolved by utilizing HYPOX rats. Several sites have been used successfully for pituitary transplantation, including the anterior chamber of the eye, different regions of the brain, the testes, and the kidney. Pituitary transplantation can be performed in HYPOX rats, for studying pituitary function in the absence of hypothalamic input and under low circulating levels of target organ hormones. On the other hand, pituitary transplantation into host rats with intact pituitaries is a useful procedure for obtaining a prolonged elevation of plasma PRL levels.

HYPOX can be performed either by an intraaural or by a parapharyngeal approach.[14] The advantage of the intraaural method is that it does not

[14] H. B. Waynforth, "Experimental and Surgical Technique in the Rat," p. 143. Academic Press, New York, 1980.

require a dissecting microscope or special microsurgical instruments, and it can be quickly performed under ether anesthesia. The major disadvantage is that it is based on a completely blind approach, and, thus, possible damage to nearby structures can be assessed only at the completion of the experiment. On the other hand, the method described below affords confirmation of successful HYPOX at the time of surgery.

Tracheal intubation and exposure of the occipital bone are performed exactly as described for LOBEX. Drilling is done at the same location and direction as for LOBEX, except that the hole is smaller and slightly oval (2.3 mm wide, 1.5 mm long). A small, complete incision of the dura is made with a microknife, and the entire pituitary gland is removed by suction. The plastic holder is the same as described before (Fig. 1C), but the aspiration needle is made from a bent 19-gauge needle with a flat and smooth tip. Care should be taken not to break the dura beneath the posterior pituitary. Removal of the pituitary results in bleeding through the hole in the dura, originating primarily from the broken bilateral veins draining the pituitary. Bleeding is suppressed with ear swabs. Postoperative care includes maintenance of the rats at a temperature of 27–28°, provision of drinking water with 5% glucose, and softening of the dry food with either water or milk. Postoperative care can be terminated within 24 hr.

For transplantation of the pituitary under the kidney capsule, rats are anesthetizied with Brevital as described before. Unilateral or bilateral incisions are made 0.5 cm caudal to the costal border of the thorax. The kidney is exposed and the capsule is nicked with small scissors. A thin spatula (Fig. 1B) is used to enlarge the space and create a small pocket between the capsule and the body of the kidney. Up to two or three pituitaries can be inserted and should be pushed away from the site of incision toward the inferior pole. The peritoneal cavity and then the skin are sutured with No. 4 silk thread. Since there is some immunological rejection of transplanted pituitaries from outbred strains,[15] littermates should be used as pituitary donors if possible. No special postoperative care is required, and the transplanted pituitaries are revascularized within 2–3 weeks.

Pituitary Stalk Transection

Transection of the pituitary stalk without damaging nearby brain and pituitary structures is a difficult and delicate operation. The reasons for this are as follows: (1) the stalk is located in a deep site and in close proximity to important nerves, blood vessels (e.g., internal carotid arteries), and vital hypothalamic sites, allowing only a limited space for surgi-

[15] R. A. Adler, *Endocrinol. Rev.* **7**, 302 (1986).

cal maneuvers; (2) the stalk is very delicate, and its length (without stretching) from the orifice in the dura near the median eminence to the anterior pituitary is no more than 1 mm; (3) portal vessels can regenerate within about 1 week, necessitating the insertion of an appropriate barrier between the cut ends of the stalk to prevent revascularization.

The stalk-sectioned rat provides an excellent animal model for differentiating between hypothalamic and anterior pituitary sites of action of hormones, neurotransmitters, or drugs. Whereas most investigators have utilized the parapharyngeal approach,[16] a subtemporal method has also been described.[17] The advantage of the latter is that the pituitary remains protected by the dura without being touched by surgical instruments, and the stalk can be slightly elongated by retraction of the temporal lobe. The disadvantages are that a large part of the skull has to be removed, severe leakage of CSF and bleeding are to be expected, and recovery time is prolonged. On the other hand, the parapharyngeal technique, as modified by us, results in minimal bleeding and fast recovery.

After exposure of the occipital bone as described for LOBEX, drilling is accomplished in two stages (Fig. 4). First, an angled hole is drilled in the occipital bone with a 1-mm-wide dental burr, as shown in Fig. 4A. The hole is wider laterally (3.2 mm) and shorter caudally (1.5 mm) than that made for LOBEX. A 0.2-mm-thick spatula with a round tip (Fig. 1D) is used to create a small space between the basisphenoid bone and the dura covering the anterior pituitary. A small and soft vinyl sheet (3.0×8.5 mm) is then inserted between the bone and the anterior pituitary to protect the dura during the next stage of drilling (Fig. 4B). Since the stalk is located approximately 4 mm rostral to the posterior border of the anterior pituitary, no more than a 4.5-mm length of the sheet should be inserted frontal to the latter point. The above measurements are typical of 300 g rats and should be slightly adjusted for younger rats.

Muscle and connective tissue attached to the basisphenoid bone (only in the midline area) are gently pushed forward and upward. The operating board is elevated 10°, and drilling continues, utilizing a smaller dental burr with a 0.7 mm head (Storz, #N-1625A), until the stalk becomes visible. Throughout the drilling, a thin spatula with 1-mm gradations is used to verify the extent of forward drilling necessary for exposing the stalk. Drilling should not cause breakage of the transverse sinus in the basisphenoid bone, which is located about 4 mm rostral to the suture line (Fig. 4). However, if bleeding from the sinus does occur, it can be stopped by packing the sinus with bone wax, followed by compression with an ear swab.

[16] S. E. Brolin, *Acta Physiol. Scand.* **6,** 336 (1943).

[17] J. H. Adams, P. M. Daniel, and M. M. L. Prichard, *Q. J. Exp. Physiol.* **48,** 217 (1963).

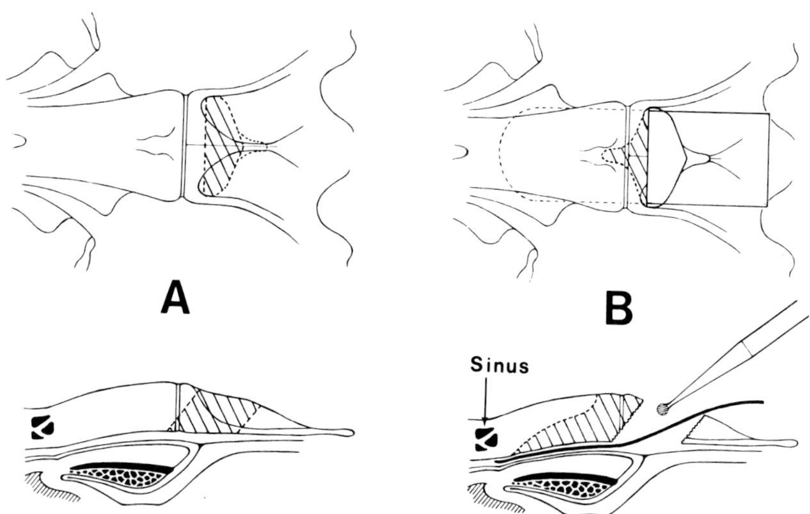

FIG. 4. Drilling to be done in preparation for pituitary stalk section. (A) Ventral (top) and sagittal (bottom) views of the first stage of drilling (hatched area). (B) Ventral (top) and sagittal (bottom) views of the second stage of drilling (hatched area). Note the locations of the transverse sinus of the basisphenoid bone and the placement of the vinyl sheet for protection during drilling.

After removing the vinyl sheet, the stalk and the dura are simultaneously cut by using a specially made cutting burr (Figs. 1D and 5A) which is operated at a medium speed. During this step, the operator cannot see the stalk, which is obstructed from view by the top of the cutting burr. A line which has been etched 4 mm from the tip of the burr serves as a marker during drilling by aligning it with the posterior border of the pituitary. Complete transection of the stalk is accomplished by slight lateral movements of the cutting burr. The advantage of using a cutting burr rather than a microknife or a microhook is that it can be manipulated in the limited space available without damaging the hypothalamus.

The transection of the stalk results in some bleeding and leakage of CSF. The surgical area is cleaned and dried by ear swabs, and the hole in the dura is slightly enlarged bilaterally by a microhook (Stortz, #E-858). A smooth and flat spatula, with a tip bent at a 90° angle (0.9 mm long and 1.7 mm wide) is placed on the cut edge of the stalk and is slightly pulled toward the operator to verify the completion of stalk transection. This manipulation also aids in creating a small space for insertion of the vinyl barrier. Using delicate forceps, a thin and flexible vinyl barrier is placed between the cut ends of the stalk, covering part of the ventral portion of

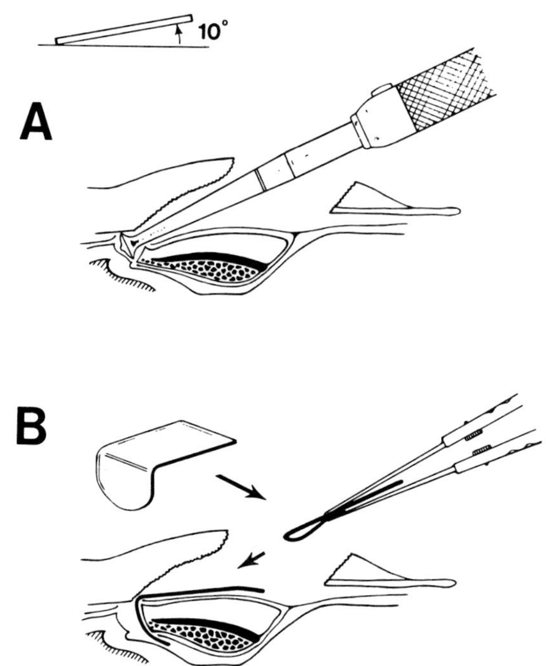

FIG. 5. Surgical approach for pituitary stalk section. (A) Transection of the pituitary stalk by using a cutting burr. (B) Placement of a vinyl barrier to separate the cut ends of the stalk in order to prevent revascularization. Note the elevation of the surgical board (top) which is necessary for both procedures.

the anterior pituitary (Fig. 5B). The size of the barrier is 6.3 × 3.0 mm, and it is bent prior to insertion at a 90° angle 2.8 mm from the edge by using a razor blade holder. Sham operation includes drilling and visualization of the stalk.

Closure of the surgical area, removal of the tracheal cannula, and placement of the rat under oxygen are as described for LOBEX. Postoperative care is the same as for HYPOX rats. On occasion, stalk-sectioned rats may look healthy for 9–12 hr, have some deteriorating conditions during the next 24 hr, but, in general, return to normal within 48 hr. Postexperimental inspection of stalk-sectioned rats should reveal significant atrophy of the anterior pituitary within 1–2 weeks and severe reduction in the size of the neural lobe, without much effect on the intermediate lobe. The hypothalamus should be inspected for any possible damage, and the median eminence area should be examined for completion of stalk transection without signs of revascularization.

Water Consumption of LOBEX and Stalk-Sectioned Rats

We have compared water consumption of LOBEX and stalk-sectioned rats for 13–14 days after surgery (Fig. 6). It is interesting that within 1 day, water consumption increases significantly in both LOBEX and stalk-sectioned rats whereas it is reduced in Sham-operated rats. During the next 4 days, water consumption in stalk-sectioned rats reduces to near normal levels, whereas in LOBEX rats it remains elevated. This indicates that prestored vasopressin from the posterior pituitary of the stalk-sectioned rat can be released for a limited period until it is exhausted. During the second week after surgery, water consumption in both LOBEX and stalk-sectioned rats is similar, and significantly higher than that in controls.

Neural Transplantation in the Hypothalamus

Neural grafting studies typically adopt one of two strategies: either the replacement of missing or defective neurons of the hypothalamus, or the reinstatement of afferent systems that innervate the hypothalamus. The first involves the grafting of *homotypic* neurons into the site of missing or defective neurons of the host brain. The second involves the use of *heterotypic* grafts from globally diverging systems of the brain stem, i.e., the serotonergic neurons of the raphe, noradrenergic neurons of the locus

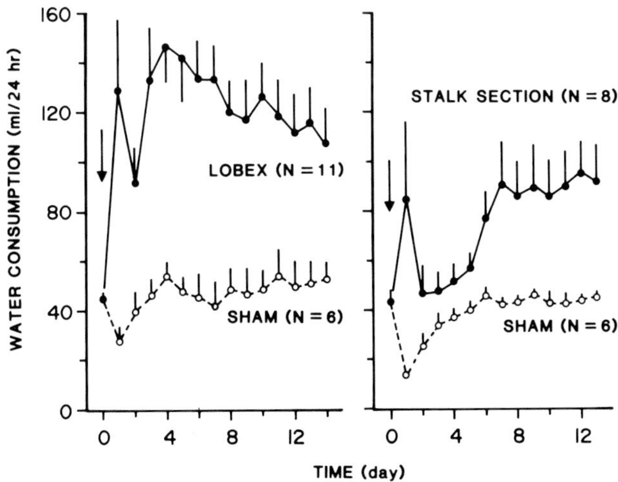

FIG. 6. Comparison of daily water consumption following posterior pituitary lobectomy or pituitary stalk section.

coeruleus, or dopaminergic neurons of the ventral mesencephalon. These are implanted directly into the target site rather than to their site of origin.

Transplant recipients belong to one of several models of hypothalamic dysfunction. Some are experimentally induced by electrolytic or chemical lesions of specific neuronal groups or afferent fiber pathways.[18,19] Others, e.g., hypogonadal mice[20] or Brattleboro rats,[21] are the result of inherited disorders that cause deficiencies of certain hormones. Three general approaches to neural transplantation have been established: (1) injection of cell suspensions into the brain parenchyma, (2) injection of solid tissue pieces into the ventricles, and (3) grafting of solid tissue pieces into aspirated cavities. The advantages and disadvantages of each will be discussed accordingly.

Preparation of the Transplant Recipient

Transplant recipients are anesthetized with ketamine–HCl (100 mg/kg body weight, i.p.). Since the effects of ketamine are short lasting, the surgical procedure should be completed within 30–60 min. The scalp is shaved, and the rat is placed in a stereotaxic apparatus (Kopf), with the incisor bar positioned at 5.0 mm above the interaural line. The scalp is swabbed with alcohol, and a midline incision is made with a scalpel blade. The scalp is retracted and immobilized with hemostats, and the connective tissue attached to the skull is removed by scraping with the edge of a scalpel blade. Openings through the skull are made with a dental burr. The size and shape of the opening depends on the method of transplantation. With cell suspension transplants or solid tissue grafts into the ventricles, small openings of approximately 1 mm in diameter are made. Transplants that are placed into aspirated cavities require larger openings ($\sim 3 \times 3$ mm).

Drilling through the bone continues until a thin translucent layer remains. Forceps are used to circumscribe the boundaries of the opening, detach the thin layer from the rest of the skull, and expose the dura and the underlying cortical tissue. A small incision is made in the dura for the insertion of a syringe needle or a guide cannula for the cell suspension or intraventricular methods of transplantation. For the cavitation approach, a cross incision is made in the dura, and the dura is folded over the skull to

[18] G. Jonsson, *Handb. Chem. Neuroanat.* **1,** 463 (1983).

[19] J. T. Coyle and R. Schwarcz, *Handb. Chem. Neuroanat.* **1,** 508 (1983).

[20] D. T. Krieger and M. J. Gibson, *in* "Neural Transplants, Development and Function" (J. R. Sladek and D. M. Gash, eds.), p. 187. Plenum, New York, 1984.

[21] D. M. Gash, P. H. Warren, L. B. Dick, J. R. Sladek, and J. R. Ison, *Ann. N.Y. Acad. Sci.* **394,** 672 (1982).

expose the underlying cortical tissue. The cavity is made by aspiration, the details of which are described in a later section.

Age and Dissection of Donor Fetal Tissue

As indicated in Table I, the appropriate age of fetal donor tissue is dependent on the area of brain required for transplantation and the method of transplantation. The optimal donor age occurs when the neurons are undergoing a period of proliferation and differentiation. The range of donor ages for the cell suspension method is more restrictive than that for solid tissue grafts and requires tissue at an early embryonic age. This may be due to the inability of neurons that have established extensive processes to survive the dissociation process.

Donor tissue is obtained by cesarean delivery from timed-pregnant females. Rats are anesthetized with equithesin (3.3 ml/kg body weight, i.p.), and the abdominal area is shaved and swabbed with alcohol. Fetuses are removed from the pregnant dam using sterilized instruments and gloves. Large forceps and scissors are used to make a midline incision in the abdominal skin, and a second pair of forceps and scissors is used to cut through the musculature and expose the uterine horns. A third set is then used to remove the uterine horn with its string of fetuses. Fetuses are placed in a sterilized plastic petri dish, where they remain viable for up to 2 hr. Some protocols require that the fetuses be removed individually, starting at one end of the uterine horn. Hemostats are used to clamp arteries to prevent excessive blood loss after the removal of each fetus.

Each fetus is placed on a sterile microscope slide, and the surrounding uterine musculature is cut to expose the amniotic sac. The sac is pierced with iridectomy scissors and separated from the fetus. The umbilical cord is then cut, separating the fetus from the placenta. The fetus is

TABLE I
APPROPRIATE AGE OF FETAL DONOR TISSUE FOR NEURAL TRANSPLANTATION

	Gestational age (days)	
Donor tissue	Cell suspension	Solid grafts
Hypothalamus	ND[a]	16–19
Locus coeruleus	13–15	15–19
Raphe nucleus	13–15	16–19
Ventral tegmentum	13–16	15–17

[a] ND, Not determined.

transferred to a second sterilized microscope slide for the brain dissection. Using another set of instruments, incisions are made in the scalp of the fetal cranium to expose the brain. The brain is gently removed from the base of the skull and inverted in a solution of CEM 2000 tissue culture medium to expose the ventral surface.

The Fetal Hypothalamus. With the fetal brain positioned ventral side up (Fig. 7A), bilateral incisions are made 0.5 mm lateral from the midline extending 1 mm rostral and 1 mm caudal to the optic recess. An anterior incision is made transverse to the midline just caudal to the bifurcation of the anterior cerebral artery. A second incision is made at the optic recess, and a third is made 1 mm posterior to the optic recess. The anterior block of tissue predominantly contains the anlage of the preoptic area, superchiasmatic, and supraoptic nuclei. The posterior block of tissue predominantly contains the anlage of the ventral hypothalamus and periventricular nuclei.

The Fetal Locus Coeruleus. The locus coeruleus is most easily dissected by a dorsal approach. A midsagittal incision is made through the tectum and cerebellum to expose the fourth ventricle. Transverse inci-

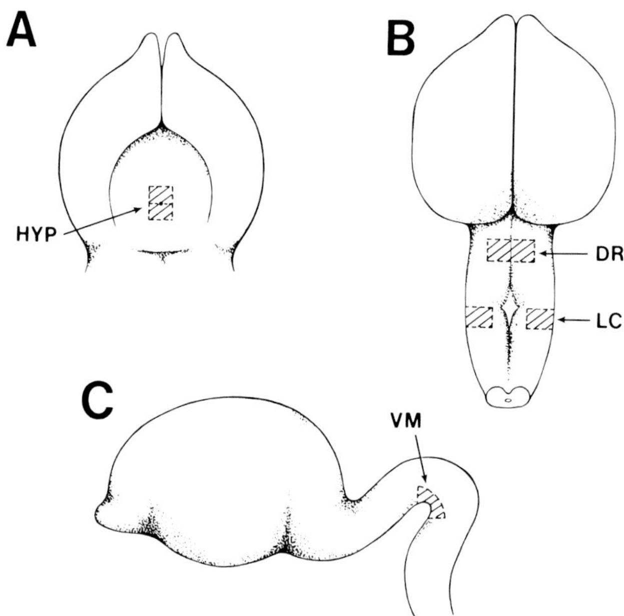

FIG. 7. Dissection of fetal brain for transplantation. (A) Ventral view of fetal brain showing the hypothalamic (HYP) region. (B) Dorsal view showing the dorsal raphe nuclei (DR) and locus coeruleus (LC). Overlying tectum and cerebellum have been removed. (C) Sagittal view showing the ventral mesencephalon (VM).

sions are made bilaterally along the roof of the ventricle and caudal to the cerebellar anlage, extending 1 mm from the lateral edges (Fig. 7B). Bilateral transverse incisions are also made just rostral to the pontine flexure approximately 1 mm caudal to the first incision. These incisions also extend 1 mm from the lateral borders. Sagittal cuts are then made to excise tissue blocks of approximately 1 mm^3.

The Fetal Raphe Nucleus. The raphe is approached by making an incision along the midline of the tectum to expose the underlying tegmentum (Fig. 7B). The dorsal raphe nuclei are situated near the midline of the tegmentum and are dissected by making bilateral transverse incisions 1 mm in length from the midline in the dorsal mesencephalic flexure. A second transverse incision is made 1 mm caudal to the first, and the tissue bounded by the midline and transverse cuts is excised as two blocks, each approximately 1 mm^3.

The Fetal Ventral Tegmentum. The ventral tegmentum is most easily dissected from the ventral surface of the brain. Transverse incisions are made 1 mm rostral and 1 mm caudal to the ventral mesencephalic flexure, extending 1 mm lateral from the midline (Fig. 7C). Lateral cuts joining the transverse incisions are made to excise a block of approximately $2 \times 2 \times 1$ mm.

Cell Suspension Transplantation

Transplantation of cell suspensions enhances the survival of neurons implanted into the parenchyma of the brain, and it is useful in situations that call for multiple grafts with minimal damage to the brain. The disadvantage is that, for homotypic grafts, it is difficult to distinguish between transplanted cells and those of the host unless the transplanted cells are labeled.[22] In addition, this approach is not amenable to studies that require electrical stimulation and recording since the transplants are not directly visible. However, this can be overcome when used in conjunction with brain slice techniques.[23]

The cell suspension method involves the dissociation of the dissected fetal brain tissue into a cell suspension and the stereotaxic injection of the suspension into the parenchyma. CEM 2000 medium is initially used for the collection of fetal tissue. The medium is gassed with 95% O_2, 5% CO_2 for approximately 10 min, and is then sterilized by vacuum filtration using a Nalgene 0.2-μm filter. Tissue collected from one to two litters is added

[22] J. Wells, B. P. Vietje, D. G. Wells, M. Boucher, and R. P. Bodony, *Brain Res.* **383,** 333 (1986).

[23] M. Segal, A. Björklund, and F. H. Gage, *in* "Neural Grafting in the Mammalian Central Nervous System" (A. Björklund and U. Stenevi, eds.), p. 389. Elsevier, Amsterdam, 1986.

to 25 ml of CEM 2000. Tissue is gently shaken for 2 min and then washed in Ham's F12, followed by a wash in a working solution of 0.05% dispase and 0.15% collagenase; DNase (0.015%) is added if the tissue becomes too sticky.

The tissue and media are poured into a Tekmar lab bag and placed into a Stomacher 80 blender for 5–10 min to gently dissociate the tissue pieces. The contents of the bag are inspected for the presence of tissue debris adhering to threads of DNA. The debris is filtered with a 57-μm-mesh Nytex screen. The cell suspension is then pipetted into a centrifuge tube and spun at 200 *g* for 5 min. The supernatant is removed, and the pellet is resuspended with 5 ml of CEM 2000 to inactivate any remaining enzyme. The suspension is again spun, the pellet resuspended in 10 ml of CEM 2000, and a sample is taken to determine cell number and viability using trypan blue.[24] The remaining cell suspension is centrifuged and the pellet is resuspended in a volume of Ca^{2+}- and Mg^{2+}-free Hanks' balanced salt solution to yield approximately 50,000–100,000 cells/μl for transplantation.

Injections are made into predetermined sites in the host hypothalamus using the stereotaxic coordinates of Pellegrino *et al.*[25] Transplant recipients are prepared as described before. A Hamilton syringe with a 23-gauge needle is mounted to a micromanipulator and attached to the stereotaxic apparatus. Two microliters of the cell suspension is drawn into the syringe, and the needle is positioned in the desired dorsoventral location in the parenchyma. The cell suspension is injected at a rate of 1 μl/min, after which the needle is kept in place for an additional 3–5 min to allow for the diffusion of the suspension into the host tissue. If the injection rate is too fast, or the injection volume too large, the suspension will flow back up the needle tract and away from the desired injection site. After removing the injection needle, Gelfoam is placed in the skull, the scalp sutured, and the animals are placed under a heat lamp to maintain body temperature until they have recovered from the anesthesia.

Solid Tissue Transplants into Ventricles

The method of transplanting solid pieces of tissue into the ventricles has several advantages. First, CSF bathes the tissue and thus enhances its survival. Second, there is only modest damage to the host brain. Third, solid tissue grafts maintain their cellular organization and are easily distin-

[24] B. B. Mishell and S. M. Shiigi, "Selected Methods in Cellular Immunology," p. 21. Freeman, San Francisco, California, 1980.

[25] L. J. Pellegrino, A. S. Pellegrino, and A. J. Cushman, "A Stereotaxic Atlas of the Rat Brain." Plenum, New York, 1979.

guished from the host brain. A disadvantage, however, is that, unless placed in the fourth ventricle, grafts are not accessible for electrical stimulation and recording.

The dissected blocks of fetal tissue are cut into five to eight smaller pieces in CEM 2000 medium. The pieces along with 5 μl of medium are drawn into a syringe with a 21-gauge needle. The syringe needle is inserted into a guide cannula which has been stereotaxically positioned at the desired ventricular location. Tissue and medium are injected into the ventricle in 1-μl increments. The needle and guide cannula are removed, and the contents remaining in the syringe are inspected to ensure that tissue has been deposited into the ventricles. Gelfoam is placed in the opening of the skull and the scalp sutured.

Solid Tissue Transplants into Aspirated Cavities

Another approach is the grafting of solid tissue pieces into aspirated cavities. With this approach, a cavity is made in the parenchyma superficial to the surface of the brain. This permits the grafted tissue to be easily visualized for studies utilizing tracing, stimulation, recording, or lesions. This approach requires a vascular bed or choroidal plexus upon which to place the transplanted tissue. In sites which lack a vascular bed or choroid plexus, a delayed cavitation method is employed. By delaying the time between cavity formation and tissue grafting, neovascularization takes place within the aspirated cavity. This process typically occurs within 2 weeks. In addition, injury-induced tropic factors are thought to be released 4–7 days after cavitation and enhance the survival of the grafted tissue. The disadvantage of the cavitation method is the extent of tissue damage.

Aspirated cavities are made with a 20-gauge aspiration needle connected to a vacuum pump. Pressure is controlled with a ventilation hole made in the tubing. The aspiration needle should be gently placed on the tissue to be removed and not forced into the brain parenchyma. The sucking action created by aspiration collapses capillaries and minimizes bleeding. Some bleeding is inevitable, however, which can be controlled by placing Gelfoam into the cavity. It is essential that all bleeding into the cavity be stopped prior to placement of the transplant.

In nondelayed cavitation protocol, the dissection of the donor tissue and preparation of recipient cavity must be coordinated such that the donor tissue is available as soon as the cavity is formed. This minimizes the time between dissection of the donor tissue and its implantation and enhances its survival. In this case, it is best to remove individual fetuses from the uterine horn as needed rather than removing the entire string of fetuses at one time.

The dissected tissue is transferred within a bubble of medium poised between the blades of iridectomy scissors and placed within the aspirated cavity. In the nondelayed protocol, it is important to ensure that a vascular bed or choroid plexus is exposed at the base of the cavity. Excess fluid surrounding the tissue is removed, and the graft is immobilized with Gelfoam, filling the remainder of the cavity.

Conclusions

We have presented practical aspects of selected microsurgical techniques that have numerous applications for studying the dynamics of neuroendocrine interactions in the living animal. Although some of these techniques require specialized tools and fine surgical skills, they are not beyond the reach of many laboratories. Given that the demonstration of hormonal effects on *in vitro* systems always requires confirmation *in vivo*, it is likely that these microsurgical methods will continue to serve as indispensable tools for probing a variety of systems in the hypothalamopituitary complex.

Acknowledgments

This work was supported by National Institutes of Health Grants NS-13243 and HD-21135 to NBJ and American Heart Association Grant 87-1051 and NIH Grant NS-24464 to WCL.

[16] Superfusion and Static Culture Techniques for Measurement of Rapid Changes in Prolactin Secretion[1]

By GONZALO MARTÍNEZ DE LA ESCALERA, KAREN C. SWEARINGEN, and RICHARD I. WEINER

Multifactorial Regulation of Prolactin Release

The secretion of prolactin (PRL) is under the regulation of both inhibitory and stimulatory hypothalamic hormones. Dopamine (DA), a well-characterized inhibitory hormone,[2,3] tonically suppresses PRL secretion.[3]

[1] Supported by National Institutes of Health Grant HD09824 and by the Mellon Foundation.

[2] R. M. MacLeod, *Front. Neuroendocrinol.* **4,** 169 (1976).

[3] R. I. Weiner and W. F. Ganong, *Physiol. Rev.* **58,** 905 (1978).

Interruption of DA regulation increases the responsiveness of lactotropes to stimulation of PRL secretion by hypothalamic extracts and the putative PRL-releasing hormone thyrotropin-releasing hormone (TRH).[4] This sequential coding was originally proposed based on *in vivo* studies in which suckling was shown to potentiate the action of TRH,[4,5] and it was subsequently confirmed by DA infusion to rats depleted of endogenous DA,[6,7] pharmacological blockade of DA receptors,[8] and DA perifusion of pituitary fragments.[9] *In vivo* studies have serious experimental limitations associated with the complexity of regulation as well as the access to the analysis of intracellular events, whereas various *in vitro* methods generally lack the ability to monitor temporal aspects of secretory responses and interactions between multiple factors.[10] We were interested in investigating the cascade of biochemical and structural intracellular changes involved in the transduction of the actions of inhibitory and stimulatory hormones that regulate PRL release.[11,12] This chapter describes a superfusion method for cultured monolayers of anterior pituitary cells that permits the analysis of interactions between various regulatory factors, allows the monitoring of the temporal patterns of hormone release, and is compatible with the simultaneous determination of biochemical and structural changes correlated with secretion.

Animals

In the majority of experiments, we used Sprague–Dawley (Bantin & Kingman, Fremont, CA) cycling rats weighing 180–200 g. Rats were bilaterally ovariectomized and implanted with an estradiol (E_2)-filled silastic capsule (Dow Corning: 1 cm; i.d. 0.062 inch, o.d. 0.125 inch) for 14 days. The E_2 treatment increased both the yield of cells, from approximately 1.5 to 6–10 million cells/pituitary, and the relative proportion of lactotropes, from 30–50 to 70–80%. This enrichment of lactotropes allowed the determination of specific biochemical events to be correlated with PRL secre-

[4] C. E. Grosvenor and F. Mena, *Endocrinology* (*Baltimore*) **107,** 863 (1980).

[5] C. E. Grosvenor, G. T. Goodman, and F. Mena, *in* "Prolactin Secretion: A Multidisciplinary Approach" (F. Mena and C. M. Valverde-R, eds.), p. 275. Academic Press, Orlando, Florida, 1984.

[6] W. J. de Greef and T. J. Visser, *J. Endocrinol.* **91,** 213 (1981).

[7] P. M. Plotsky and J. D. Neill, *Endocrinology* (*Baltimore*) **111,** 168 (1982).

[8] D. J. Haisenleder, J. A. May, R. R. Gala, and D. M. Lawson, *Endocrinology* (*Baltimore*) **119,** 1980 (1986).

[9] K. D. Fagin and J. D. Neill, *Endocrinology* (*Baltimore*) **109,** 1835 (1981).

[10] W. S. Evans, M. J. Cronin, and M. O. Thorner, this series, Vol. 103, p. 294.

[11] G. Martínez de la Escalera, J. Guthrie, and R. I. Weiner, *Neuroendocrinology* **47,** 38 (1988).

[12] G. Martínez de la Escalera and R. I. Weiner, *Neuroendocrinology* **47,** 186 (1988).

tion. Cells from E_2-treated rats responded to DA, TRH, and the potentiating action of the transient removal of DA in a fashion very similar to that of cells from lactating rats.[11]

The ability of E_2 to increase the size of the anterior pituitary[13] varies among strains and, in fact, between the same strain obtained from different suppliers. Sprague–Dawley rats obtained from Bantin & Kingman showed a 3- to 4-fold increase in anterior pituitary weight following E_2 treatment. With the same treatment, however, the anterior pituitary of Sprague–Dawley Simonsen rats doubled in weight, and those of Fischer 344 increased 4- to 5-fold.

Dispersion and Plating of Cells

Two to four anterior pituitary glands were placed in Hanks' balanced salt solution, calcium- and magnesium-free (obtained, as well as the other media, sera, and antibiotics, through the UCSF Cell Culture Facility), containing penicillin–streptomycin (250 U/ml), ascorbic acid (AA) (100 μM), and DA (500 nM). Pituitaries were minced, washed twice, and incubated in 5 ml of the previous buffer containing collagenase (3 mg/ml, Type I; Worthington Biochemical Corporation, Freehold, NJ) and DNase (40 μg/ml, Type I; Sigma Chemical Co., St. Louis, MO) for 40 min in a metabolic shaker at 37°. The fragments were dissociated into individual cells by gentle trituration through siliconized Pasteur pipets. The cell suspension was harvested by centrifugation (200 g, 10 min) and washed 3 times with Dulbecco's modified Eagle's medium H-21 (DME H-21) containing calf serum (10%), gentamycin (100 μg/ml), fungizone (25 μg/ml), penicillin–streptomycin (250 U/ml), AA (100 μM), and DA (500 nM). One million cells were placed in 6-well plates containing 25-mm round coverslips (Thermanox, Miles Scientific, Naperville, IL) previously coated with extracellular matrix produced by bovine corneal endothelial cells.[14] Cells plated on extracellular matrix rapidly attach to the coverslips and remain firmly attached throughout the superfusion. The medium was replaced the following morning, and 4 hr later the coverslips were transferred to Sykes–Moore chambers (Bellco Glass, Inc., Vineland, NJ). Essentially, the chambers consist of two coverslips with the cells facing to the inside, pressed against either side of a rubber O ring. The coverslips are kept in place by two threaded metal plates which can be tightened to form a seal (Fig. 1, top). In some instances, we have used only one of the coverslips coated with cells. This procedure produces preparations in

[13] J. Wiklund, N. Wertz, and J. Gorski, *Endocrinology (Baltimore)* **109,** 1700 (1981).

[14] R. I. Weiner, C. L. Bethea, P. Jaquet, J. S. Ramsdell, and D. J. Gospodarowicz, this series, Vol. 103, p. 287.

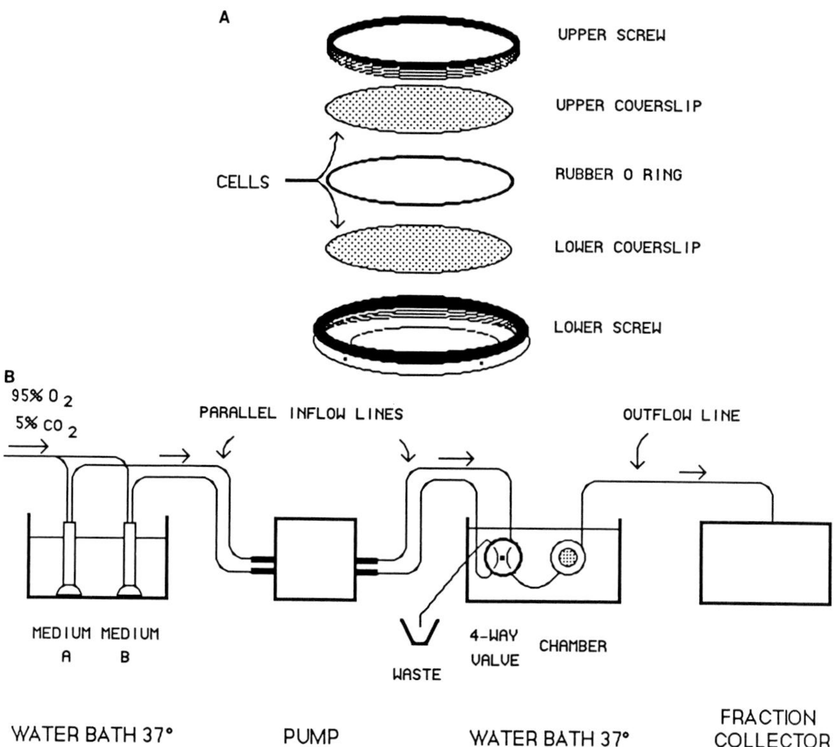

FIG. 1. Schematic representation of the Sykes–Moore chamber (A) and the superfusion system (B).

which the response to TRH or DA withdrawal was unchanged within 3–96 hr after plating the cells.

Superfusion

Four chambers were normally superfused in parallel. Figure 1 details the connections for one of these chambers. A double-channel four-port valve (Pharmacia Inc., Piscataway, NJ) is positioned just before each chamber.[15] In this way two inflow lines, supplying the chamber, converge at the valve, and a single line (PE 90 tubing, i.d. 0.034 inch, o.d. 0.050 inch) is attached to the chamber. This permits rapid changes of medium, with the medium in one inflow line supplying the chamber while medium from the other is discarded. Each inflow line carries medium from sepa-

[15] J. C. Kraicer and A. E. H. Chow, *Endocrinology (Baltimore)* **111,** 1173 (1982).

rate reservoirs placed in a 37° water bath and constantly gassed with a mixture of 95% O_2, 5% CO_2. Medium containing fresh DA is added to the reservoirs every 30 min. Medium is pumped through the system by a peristaltic pump (Cassette Multichannel Peristaltic Pump, Manostat, New York, NY). The last 25 cm of the inflow lines, the valve, the single connecting line, and the chamber are submerged in the 37° water bath. This arrangement allows switching of media with virtually no variation in pressure, flow rate, or temperature. The single inflow and the outflow (PE 90 tubing) lines are connected to the chamber by 25-gauge needles inserted through the rubber O-ring at opposite poles. The total volume of the chamber is 200 μl. At a flow rate of 0.3 ml/min, the transit time of medium from the valve to the chamber is only 6 sec, and the filling time of the chamber is approximately 40 sec. These times permit the study of signals separated by intervals of at least 1 minute. We found that the rate of PRL release was independent of flow rates ranging from 0.3 to 0.9 ml/min. We normally used the lowest rate, which allowed PRL to be measured by radioimmunoassay (RIA) in fractions collected over 2.5 min. Fractions from the four chambers were collected simultaneously with the aid of a fraction collector (Model 328, ISCO, Lincoln, NE). It took approximately 30 min after initiation of the superfusion for the secretion rate of PRL from cells to reach equilibrium (Fig. 2). PRL levels then remained steady for up to 5 hr. We routinely discarded the medium collected during the first 60 min.

PRL secretion was determined by radioimmunoassay using materials supplied by the National Pituitary Agency, NIAMDD. The intraassay coefficient of variation was 5%, and all fractions from each experiment were analyzed in the same assay. The results may be expressed as rate of

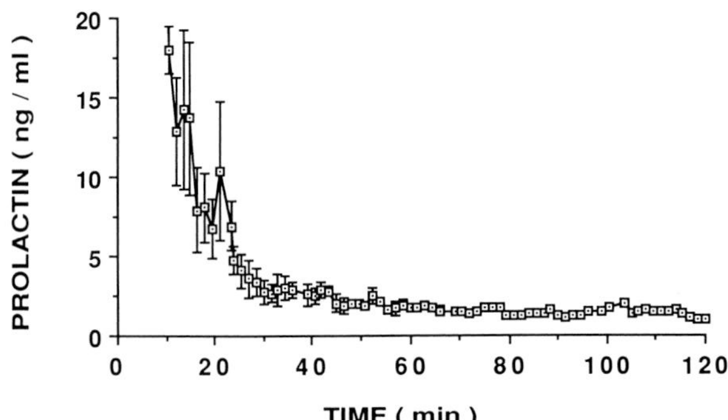

FIG. 2. Prolactin secretion measured by RIA in consecutive 2.5-min fractions depicting means ± SEM of three chambers containing 1.5 × 10^6 cells. Time 0 represents the start of the superfusion.

PRL release (ng/ml/2.5 min), as a percentage of the average release during an initial control period, or as the integrated area under the curve. The area under the curve can be simply measured by plotting the normalized release and cutting and weighing the area under the curve for a selected period of time. Statistical differences between groups were analyzed by ANOVA followed by the Fisher's multiple comparison test.

Figure 3 shows the release of PRL (both as rate and as percentage of the basal) from cells challenged with TRH (10 n*M*) for three consecutive 10-min periods after a transient suspension of DA (500 n*M*) infusion. The rate of PRL release was rapidly increased after either DA removal or TRH administration. The addition of DA back to the superfusion medium resulted in a rapid decrease in the rate of PRL release. The responses to three separate challenges with TRH did not differ.

Pulse–Chase Experiments

In order to monitor release of hormone of various ages we developed a technique for pulse labeling PRL for 1 hr with [^{35}S]methionine. Data presented in Fig. 4 are for 4-hr-old PRL. Cells were incubated with methionine-free DME H-21 (plus 500 n*M* DA and 100 μ*M* AA) containing 100 μCi/ml [^{35}S]methionine (specific activity 1103 Ci/mol, New England Nuclear Research Products, Boston, MA). After 1 hr this medium was replaced with complete DME H-21 (DA + AA). Coverslips were then placed in the superfusion chamber and samples collected every 2.5 min. The cells were exposed to DA inhibition for the first 30 min, following which DA was removed for 1 hr. DA was then added back to the medium for the final 30 min. The amount of total PRL released (ng/ml) was deter-

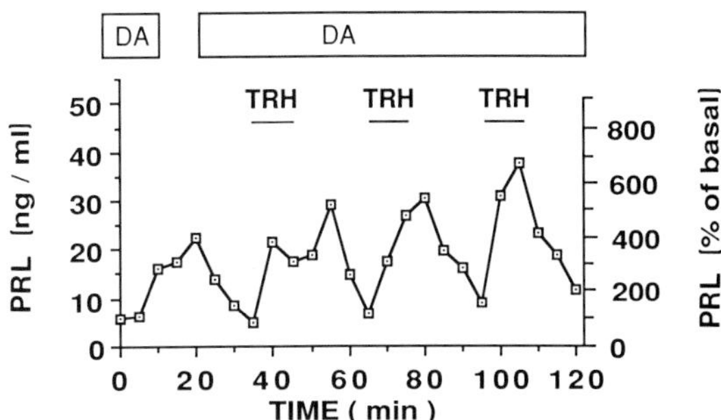

FIG. 3. Prolactin secretion measured by RIA in alternate 2.5-min fractions. Cells were challenged by either removing DA (500 n*M*) from or adding TRH (10 n*M*) to the medium. The rate of PRL release is expressed as ng/ml and as percentage of the basal release. Cells were tonically incubated with DA for 24 hr. Medium from the first 60 min in superfusion was discarded.

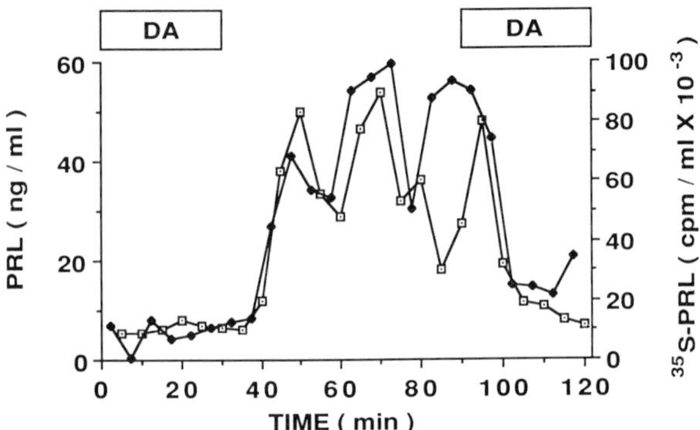

FIG. 4. Effect of removing DA (500 nM) from the superfusion medium for 60 min on radioimmunoassayable PRL (□) and [^{35}S]PRL release (◆). DA infusion was suspended at 30 min and reestablished at 90 min. PRL was measured by RIA on alternate fractions. [^{35}S]PRL was labeled 4 hr beforehand by a 1-hr pulse with [^{35}S]methionine. Cells were tonically incubated with DA for 24 hr.

mined in alternate fractions by RIA, whereas the amount of 4-hr-old PRL released was estimated by trichloroacetic acid (TCA) precipitation of an aliquot from alternate fractions.

Measurement of ^{35}S-Labeled Prolactin Released

The release of [^{35}S]PRL was estimated by the measurement of radiolabeled TCA-precipitable material. For this purpose, 0.5 ml of each fraction (with 0.1 mg bovine serum albumin added as a carrier protein) was mixed with 0.5 ml ice-cold 20% TCA. The precipitate was pelleted by centrifugation, redissolved in 0.1 ml 1 N NaOH, and counted in 12 ml Scinti Verse II in a Beckman LS7500 spectrometer. We confirmed that greater than 90% of labeled released protein was PRL by simultaneous estimation of labeled PRL by polyacrylamide gel electrophoresis and TCA precipitation.

Figure 4 shows the release of 4-hr-old [^{35}S]PRL (cpm/ml) and of total PRL (ng/ml). We have observed that, following the removal of DA inhibition, PRL release *in vitro* was episodic.[16] The release of 4-hr-old PRL was also episodic, showing the same general pattern as radioimmunoassayable PRL. When DA was returned to the superfusion medium, both total and [^{35}S]PRL release were rapidly inhibited.

[16] K. C. Swearingen, G. Martínez de la Escalera, and R. I. Weiner, *Reprod. Biol.* (submitted for publication).

Semidynamic Studies

The use of superfusion techniques, although providing a method for studying dynamic processes, produces a large number of samples and allows few replicates for biochemical studies. On the other hand, rapid manipulations of cells maintained in static culture give variable and unreliable results when medium is aspirated from the dishes. We observed that, when cells were cultured on coverslips, the coverslips could be moved from one well to another without causing artifactual changes in PRL secretion. Cells were plated on plastic coverslips (12 or 25 mm in diameter) previously coated with either ECM[14] or Matrigel (Collaborative Research, Inc., Lexington, MA). To coat the coverslips with Matrigel, the coverslips were dipped in a solution of Matrigel in culture medium DME H-21 (1 : 4, at 4°). The coverslips were then incubated for 60 min at 37° to allow gelation to take place before the cells were plated. The use of ECM and Matrigel gave identical results in experiments in which PRL release, cAMP accumulation, and phosphoinositide hydrolysis were assessed.

The coverslips with the attached cells are progressively transferred from well to well in seconds. The medium containing the specific agents to be tested, radioactive precursors to be incorporated in specific metabolic pathways, etc. had been previously added to the wells. This technique allowed reliable measurement of PRL release during periods of 1–10 min. A sequence of at least three preincubation changes of medium decreased the variability between replicates. Using this method, we found that the stimulation of PRL secretion by a 10-min challenge of TRH after a 10-min suspension of DA inhibition was essentially identical to the integrated response seen with superfused cells (Fig. 5). The cells of the experiment illustrated in Fig. 5A were plated at a density of 10^6 cells on 25-mm ECM-coated coverslips contained in 6-well plates. Cells treated with constant DA were transferred for three consecutive 10-min periods between wells containing DME H-21 with 500 n*M* DA; the cells were then challenged for 10 min with 100 n*M* TRH. In the presence of constant DA, TRH induced a 4-fold increase in PRL release (38.2 ± 2.17 versus 10.05 ± 1.28 ng/ml, $n = 3$). On the other hand, the cells incubated in medium without DA for 10 min released more PRL (18.9 ± 0.821 ng/ml PRL, $n = 3$) and hyperresponded to TRH (227 ± 41 ng/ml PRL, $n = 3$). This increased responsiveness to TRH was very similar in magnitude to that observed in superfused cells (Fig. 5B) when the release over 10 min of TRH administration was integrated as the area under the peak. Using this procedure we have performed experiments involving up to 48 wells. We correlated changes in PRL secretion with changes in second messengers (cAMP concentration),

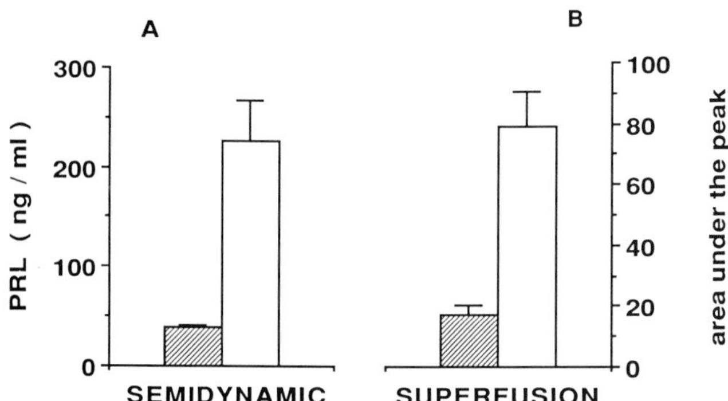

FIG. 5. Comparison of the effectiveness of TRH (100 nM for 10 min) to induce PRL release from cells maintained under constant DA administration (hatched bars) or after a transient 10-min removal of DA (open bars). (A) Cell-coated coverslips were transferred from well to well every 10 min. The open bars represent TRH-induced PRL released by cells incubated in medium without DA for 10 min prior to the incubation for 10 min in medium with DA plus TRH. (B) Cells were continuously superfused with DA except for 10 min (open bar), 20 min before adding TRH. The release of PRL was measured as the integrated area under the peak response.

enzymatic activity (phospholipase C), and cellular compartmentalization (translocation of protein kinase C). Attempts to perform these experiments by aspirating media from wells containing cultured cells were unsuccessful.

Conclusion

The multifactorial regulation of PRL secretion represents a specific but not unique process, in which the sequence and duration of exposure to multiple hypothalamic hormones determine the secretory response. Growth hormone,[17,18] thyroid-stimulating hormone (TSH),[19,20] luteinizing hormone (LH),[21,22] and probably all the pituitary hormones as well as secretory products from other glands are regulated in an analogous fashion.[5-9,11,12] The culture/superfusion technique we have described for dis-

[17] W. Vale, J. Vaughan, G. Yamamoto, J. Spiess, and J. Rivier, *Endocrinology (Baltimore)* **112,** 1553 (1983).

[18] J. Kraicer, J. S. Cowan, M. S. Sheppard, B. Lussier, and B. C. Moore, *Endocrinology (Baltimore)* **119,** 2047 (1986).

[19] C. Dieguez, S. M. Foord, J. R. Peeters, R. Hall, and M. F. Scanlon, *Endocrinology (Baltimore)* **114,** 957 (1984).

[20] M. Michalkiewicz, M. Suzuki, and M. Kato, *Endocrinology (Baltimore)* **121,** 371 (1987).

[21] A. Pickering and G. Fink, *J. Endocrinol.* **81,** 223 (1979).

[22] J. C. Hwan and M. E. Freeman, *Endocrinology (Baltimore)* **120,** 483 (1987).

sociated cells permits the unraveling of the hierarchy of hypothalamic regulation of the secretion of anterior pituitary hormones. Both rapid changes in hormone secretion and cellular events, i.e., events occurring within minutes, and slower modulatory effects, i.e., effects taking days, can be correlated by the coupling of perifusion techniques with those of cell culture.

[17] Intracellular Calcium Levels in Rat Anterior Pituitary Cells: Single-Cell Techniques

By DENIS A. LEONG

Introduction

A pivotal signaling role for intracellular calcium in a wide variety of tissues is frequently postulated without any direct measurement of the activity. Researchers have been limited to obtaining indirect evidence: typically, correlations of cell function with tracer fluxes, transmembrane currents, calcium content, alterations in extracellular calcium, or treatment with ionophores, inhibitors of calmodulin, or putative blockers of calcium movements. This large body of evidence suggesting the universal importance of intracellular calcium has provided the incentive to measure calcium directly in normal endocrine tissues.

Single-Cell Studies

The best opportunity to measure calcium accurately in specific cell types of the pituitary is offered by single-cell approaches. Single-cell studies provide an increasingly important strategy to unravel the mechanism of hormone secretion. Although measurements in bulk populations give an overview of the secretory response, single-cell studies give additional information normally obscured by taking the population average of the response. A good example has come from our recent studies of isolated rat anterior pituitary cells. Some of these cells generate repetitive calcium oscillations.[1] A striking pulsatile mode of calcium mobilization is observed asynchronously from cell to cell. Trains of calcium are distributed predominantly in a subpopulation of constitutively active growth hormone (GH) cells. Calcium responses measured using bulk populations

[1] R. W. Holl, M. O. Thorner, and D. A. Leong, *J. Biol. Chem.* **263,** 9682 (1988).

would spuriously average out the spontaneous rhythm of changing intracellular calcium.

On the other hand, there is a common disadvantage associated with this undertaking. Interpretation of single-cell studies is invariably more intricate owing to the response heterogeneity among apparently similar cells. Heterogeneity among the secretory responses of individual cells is a universal phenomenon existing even among established pituitary cell lines.[1a,2] There is particular difficulty in dealing with the case where some cells respond to a stimulus, but other cells, apparently of the same cell type, do not. Does this have anything to do with recent evidence suggesting that not all pituitary cells which store hormone can secrete it?[3]

To begin to deal with these issues we have devoted a considerable effort to measure concomitantly calcium and the secretory response of each individual cell. The reverse hemolytic plaque assay, a technique to measure cumulative hormone release in single cells, has provided a promising solution.[4–6] Thus coapplication of two single-cell techniques raises the possibility that intracellular calcium and hormone release may be measured essentially at once.

Intracellular Signaling and the Pituitary Gland

Intracellular signaling information is conveyed by a diverse group of primary messengers including free calcium, cAMP, cGMP, diacylglycerol, inositol 1,4,5-trisphosphate (IP_3), and arachidonate. The classic question of *which* messenger pathway is activated in response to a specific stimulus is almost certainly too narrowly phrased. Multiple messenger responses to a single stimulus are recognized and rapidly becoming commonplace. Mammalian cells are replete with feed-forward and feedback mechanisms that provide the cell with a complex network of interrelated messenger pathways. The mechanism(s) that links individual messenger pathways together are poorly understood. No unifying mechanism is likely since individual cell types clearly link messenger pathways together

[1a] M. C. Gershengorn, M. Cohen, and S. T. Hoffstein, *Endocrinology (Baltimore)* **103,** 648 (1978).

[2] F. R. Boockfor, J. P. Hoeffler, and L. S. Frawley, *Endocrinology (Baltimore)* **117,** 418 (1985).

[3] P. F. Smith, L. S. Frawley, and J. D. Neill, *Endocrinology (Baltimore)* **115,** 2484 (1984).

[4] P. F. Smith, E. H. Luque, and J. D. Neill, this series, Vol. 124, p. 443.

[5] J. D. Neill, P. F. Smith, E. H. Luque, M. Munoz de Toro, G. Nagy, and J. J. Mulchahey, *Recent Prog. Horm. Res.* **43,** 175 (1987).

[6] D. A. Leong, S. K. Lau, Y. N. Sinha, D. L. Kaiser, and M. O. Thorner, *Endocrinology (Baltimore)* **116,** 1371 (1985).

in different ways although a limited set of general interrelationships seems reasonable.[7] The focus in our laboratory has been to apply techniques that will directly measure secretagogue-induced intracellular calcium responses in normal pituitary corticotropin (ACTH), luteinizing hormone (LH), and GH cells.

The ever enduring problem is that the anterior pituitary gland comprises a mixed-cell population of at least six distinct cell types. Calcium-sensitive fluorescent probes like quin-2 have been available since 1982, yet few reports using normal pituitary cells have emerged.[8,9] The lack of success seems surprising since the apparent specificity of some secretagogues can be exploited to provide a measure of cell type specificity. For instance, relatively robust cAMP responses to corticotropin-releasing hormone (CRH) or growth hormone-releasing hormone (GHRH) can be measured in mixed-cell pituitary preparations without intervention of cell-enrichment procedures. Interpretation of these studies is more straightforward since CRH or GHRH seems to act on one pituitary cell type. In retrospect, a major reason calcium experiments in bulk populations have been problematic arises from the very high "background" generated from high-amplitude calcium oscillations in about 35% of the pituitary cell population.[1] This might generate particular difficulty for studies of the less abundant cell types [ACTH, LH, follicule-stimulating hormone (FSH), and thyroid-stimulating hormone (TSH)] since the calcium response will be relatively modest. A variety of cell-sorting techniques to enrich fractions of pituitary cell types have been tested; however, no single cell-sorting technique can solve the problem of all pituitary cell types. Some isolated successes, particularly with gonadotropes, have been reported.[9]

It is important not to overlook information derived from calcium measurements on relatively homogeneous pituitary tumor cell lines.[10–12] To place these findings in complete perspective, comparison with normal cells will be required. Interesting new evidence of major disturbances in the function of intracellular signaling pathways triggered by the products of particular oncogenes suggests a further cautionary note.[13] Thus, even though free calcium is recognized as a pivotal intracellular messenger,

[7] H. Rassmussen and P. Q. Barrett, *Physiol. Rev.* **64,** 938 (1984).

[8] D. L. Clapper and P. M. Conn, *Biol. Reprod.* **32,** 269 (1985).

[9] R. Limor, D. Ayalon, A. M. Capponi, G. V. Childs, and Z. Naor, *Endocrinology (Baltimore)* **120,** 497 (1987).

[10] J. Axelrod and T. D. Reisine, *Science* **224,** 452 (1984).

[11] M. Gershengorn, *Recent Prog. Horm. Res.* **41,** 607 (1985).

[12] P. R. Albert and A. H. Tashjian, *J. Biol. Chem.* **260,** 8746 (1985).

[13] M. R. Hanley and T. Jackson, *Nature (London)* **328,** 668 (1987).

accurate measurements of any pituitary cell type originating from normal tissue have been elusive.

These technical obstacles have been overcome by the coapplication of two existing single-cell techniques. This strategy eliminates the requirement for pituitary cell enrichment. Calcium measurements were performed using the calcium-sensitive fluorescent probe fura-2/AM and a digital imaging microscope. Identification of the pituitary cells in a mixed-cell population was accomplished using a reverse hemolytic plaque assay (RHPA). New versions of the RHPA permit quantitative measurements of net hormone secreted by single cells in culture. Thus the RHPA provided the means not only to identify individual pituitary cells but also to quantitate the amount of hormone released, allowing direct comparison with calcium measurements (Fig. 1).

Quantitative Measurement of Hormone Release from Single Cells

The ability to measure the secretory response of individual cells has been available for antibody-secreting B cells since development of the original plaque assays of Jerne *et al.*[14] This venerable assay evolved gradually into a form suitable to measure hormone release from single cells. Pituitary applications of RHPA were first developed in J. D. Neill's laboratory[15] for prolactin cells. Similar reports on GH, LH, ACTH, and TSH soon followed. The essential quantitative nature of the RHPA has been established.[3,15a] This is not to say that hormone release from single cells can be routinely expressed in absolute units as provided by radioimmunoassay. Routinely, it is the relative amount of hormone release measured for each cell that can be quantitated and expressed in arbitrary units. The technique has been well reviewed[5] and is considered here only briefly.

The principle of the assay is schematically represented in Fig. 2. Dissociated pituitary cells are plated together with an excess of specialized indicator cells. The indicator cells (sheep erythrocytes) are manipulated so as to lyse when exposed to a specific pituitary hormone. The inset in Fig. 2, representing a section of the erythrocyte membrane, explains the basis for the lytic reaction. Protein A has been conjugated to the erythrocyte membrane in order to attach the hormone antiserum that will be provided in the chamber. If ACTH antiserum is introduced, immune complexes will form only on the membrane of erythrocytes surrounding a

[14] N. K. Jerne, C. Henry, A. A. Noedin, H. Fuji, A. M. C. Koros, and I. Lefkovits, *Transplant. Rev.* **18,** 130 (1974).

[15] J. D. Neill and L. S. Frawley, *Endocrinology (Baltimore)* **112,** 1135 (1983).

[15a] D. A. Leong and M. O. Thorner, submitted for publication.

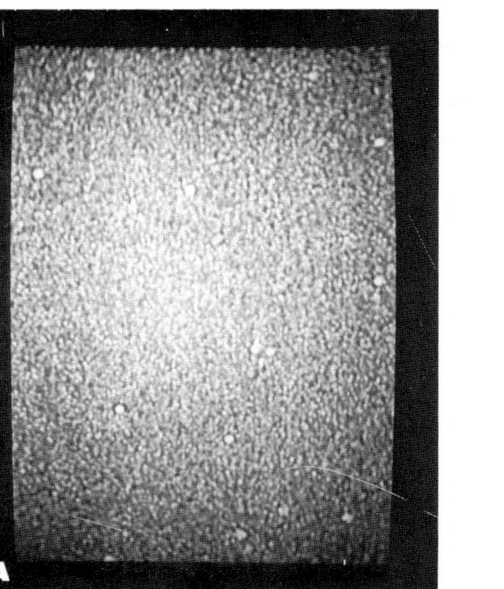

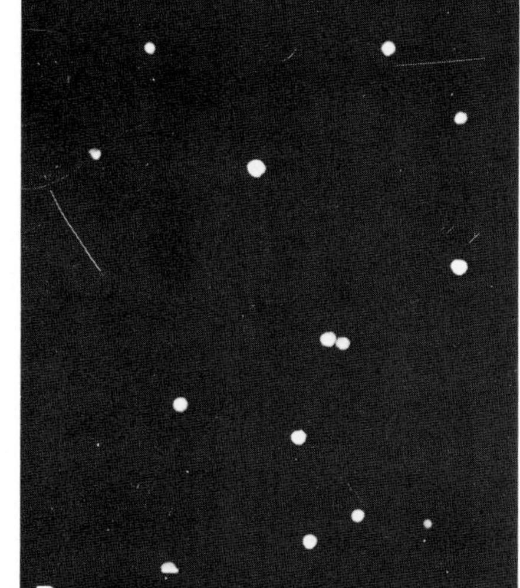

FIG. 1. Appearance of the same field of cells at different stages of the protocol. (A) Bright-field image of 14 pituitary cells (large cells) surrounded by protein A-coated erythrocytes (small cells). (B) Baseline fluorescence 340/380 ratio image. (C) Bright-field image after complement addition to visualize hemolytic plaques (note the large, medium, and small ACTH plaques formed in the field).

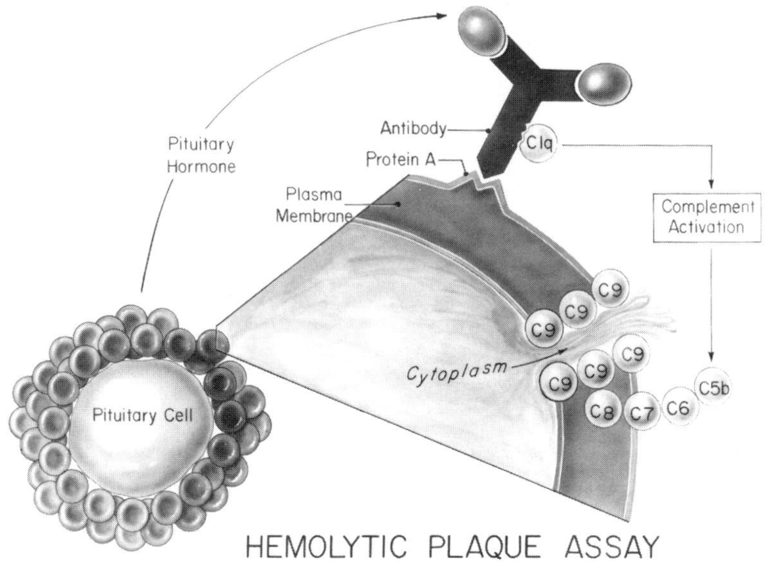

FIG. 2. The reverse hemolytic plaque assay. Reproduced with permission from D. A. Leong *et al.*, *Endocrinology* (*Baltimore*) **116,** 1371 (1985).

secretory corticotrope. Since hormone diffuses radially from the cell, a circular pattern of immune complexes is formed around each secretory corticotrope. This pattern can be visualized by exploiting the ability of complement to recognize immune complexes and initiate erythrocyte lysis via the classic pathway. A zone of hemolysis (a hemolytic plaque) forms around the secretory pituitary cell (Fig. 3). The central pituitary cell is spared from the hemolytic reaction as assessed by the ability to exclude trypan blue.

Anterior pituitary glands from female Sprague–Dawley rats (125–250 g) are enzymatically dispersed with 0.1% trypsin (Sigma)/0.1% bovine serum albumin (Sigma) for no more than 1.5 hr in a temperature-controlled (37°) spinner suspension flask. Dissociated cells are washed, suspended in culture medium, and mixed with protein A-coated ovine erythrocytes. Ovine erythrocytes (Colorado Serum Company) are coated with *Staphylococcus* protein A (Sigma) using chromium chloride as a coupling agent. A mixed suspension of pituitary and coated erythrocyte cells is introduced into a Cunningham chamber.[16] The cells are allowed 1 hr to plate on the poly(L-lysine)-treated surface of the Cunningham chamber (Fig. 4). Cunningham chambers provide optical qualities suitable for

[16] A. J. Cunningham and A. Szenberg, *Immunology* **14,** 559 (1968).

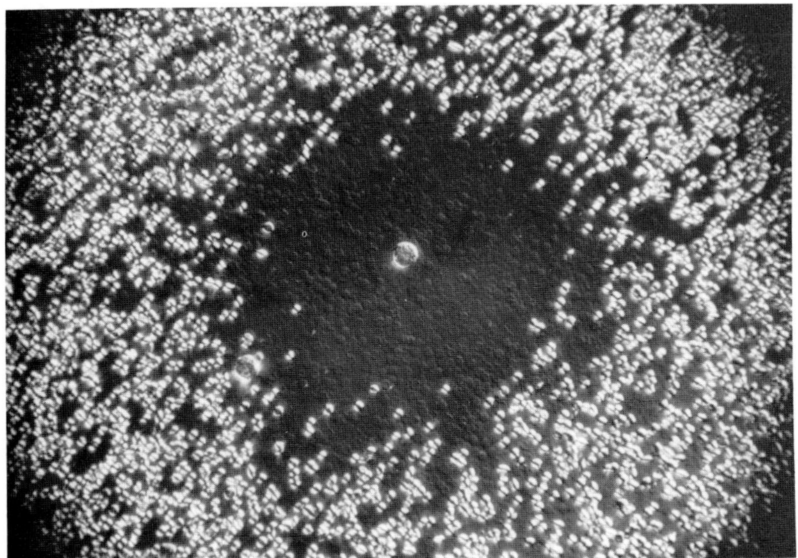

FIG. 3. A pituitary ACTH cell visualized by the reverse hemolytic plaque assay. Note the central pituitary cell surrounded by erythrocyte cell ghosts (a hemolytic plaque) bordered radially by the intact lawn of erythrocytes. One other large pituitary cell included in the field has not formed a plaque.

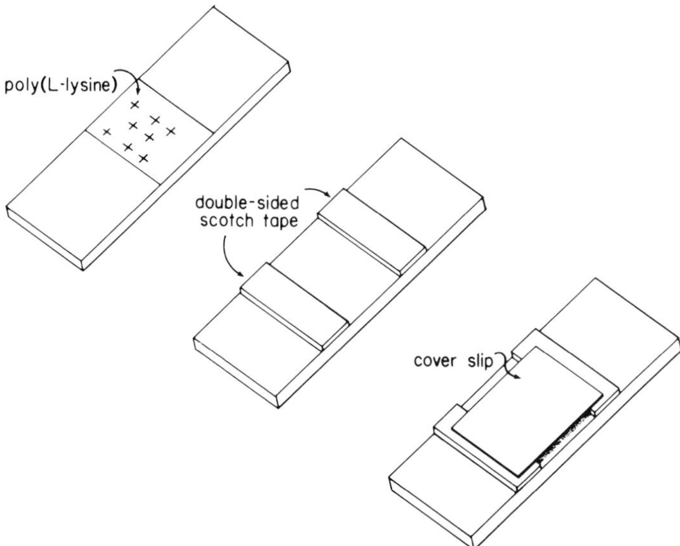

FIG. 4. The Cunningham chamber. See A. J. Cunningham, *Immunology* **14,** 539 (1968).

viewing pituitary cells with an upright microscope. The cells are covered with a thin film of culture medium confined by a thin glass coverslip. All drugs are delivered to the cells in a remote fashion by attaching polyethylene tubing to the Cunningham chamber and drawing the liquid through the chamber with the aid of a paper towel. This enables rapid and complete medium exchange in the chamber without interruption of calcium measurements in progress.

An important consideration is the order in which the measurements (calcium and hormone release) are combined. One approach is to first perform the RHPA and identify the cell of interest in the mixed population. The advantage here is that a high-power objective can subsequently be used to image the identified cell, permitting high-resolution spatial mapping of calcium in the cytosol. The disadvantage is the requirement to stimulate secretion in order to form a hemolytic plaque. In our RHPA for ACTH cells, few spontaneous plaques are formed. Thus RHPA preidentified cells are not in an appropriately naive form in terms of their prior exposure to secretagogue. Possibly another concern is that it has never been unequivocally demonstrated that each plaque-forming pituitary cell completely escapes the local effects of complement-induced lysis. While plaque-forming cells are plainly robust and viable, the possibility of subtle forms of cellular damage has not been eliminated. Indeed, the experience has been that "preplaquing" for ACTH cells prior to calcium measurements produces highly variable results. These and other considerations have resulted in the routine addition of complement to develop ACTH plaques after the calcium experiments are complete. It is not necessary to include complement during *de novo* formation of immune complexes. Indeed, the labile properties of complement make it optimal to delay complement addition until the incubation period desired is complete.

After the calcium measurements are recorded on video tape ready for processing, the final stages of the reverse hemolytic plaque assay are completed to identify and quantitate the relative amount of ACTH release from each pituitary cell. Guinea pig serum as a source of complement (Gibco, diluted 1 : 5 to 1 : 20) is added to promote plaque formation around secretory ACTH cells. The relative amount of hormone secreted from individual pituitary cells is quantitated by measuring the dimensions of the hemolytic area formed around each secretory cell with computer assistance (Zeiss Videoplan System). The specificity and quantitative nature of the RHPA have been reported previously.[7] Finally, it is important to note that RHPA methods measure cumulative hormone release in single cells over periods of hours. The RHPA has been critical first in the identification of secretory cells. Information about the cumulative amount of hormone released from individual cells has also been provided. However, new methods will ultimately be required that can measure hormone release on a second-to-second basis, to match the resolution presently achieved for calcium measurements.

Calcium-Sensitive Probes

The elegant probes designed by R. Y. Tsien have permitted direct observation of intracellular calcium in single cultured cells *in situ*. Repeated calcium measurements can be made in serial fashion essentially in real time. The superior probe fura-2 was introduced recently, and this chapter exclusively describes our experience. Fura-2 comprises a tetracarboxylic acid to form a chelation "cage" for one calcium ion much like the prototype quin-2. In fura-2, the calcium chelation component is combined with a chromophore based on a stilbene structure.[17] An important characteristic is the rapid on–off constants for calcium binding that provides exquisite time resolution in the 100-msec range.[18,18a] Our initial attempts to apply quin-2 for individual pituitary cells were frustrated by the very large amount of dye that needed to be loaded into cells to acquire an adequate fluorescent signal. Furthermore, quin-2 is highly susceptible to photobleaching, and calcium dynamics could be measured only for several minutes before the fluorescent signal completely decayed. Fura-2 has a very high extinction coefficient quantum yield, resulting in a probe about 30 times more fluorescent than quin-2. Thus, much less fura-2 needed to be loaded into cells to obtain a suitable fluorescent signal. This reduced the potential of fura-2 to buffer endogenous calcium transients and interfere with the physiologic response. Another improved quality of fura-2 is the much lower affinity for magnesium ions and heavy metals than quin-2.[17] The greatest operational advantage of fura-2 is its remarkable resistance to photobleaching.

The calcium-sensitive fluorescent indicators do not passively cross cell membranes. Based on principles first developed to enhance intestinal absorption of antibiotics, Tsien[19] masked the troublesome carboxylates with special esterifying groups which then hydrolyze inside the cells, regenerating and trapping the original indicators (Fig. 5).

The following loading procedure has been developed for pituitary cells. The acetoxymethyl ester derivative of the dye fura-2 (1.0 μM fura-2/AM, Calbiochem-Behring) is introduced to a Cunningham chamber

[17] G. Grynkiewicz, M. Poenie, and R. Y. Tsien, *J. Biol. Chem.* **260,** 3440 (1985).

[18] R. Y. Tsien, *Biochemistry* **19,** 2396 (1980).

[18a] J. P. Y. Kao and R. Y. Tsien, *Biophys. J.* **53,** 635 (1988).

[19] R. Y. Tsien, *Nature (London)* **290,** 527 (1981).

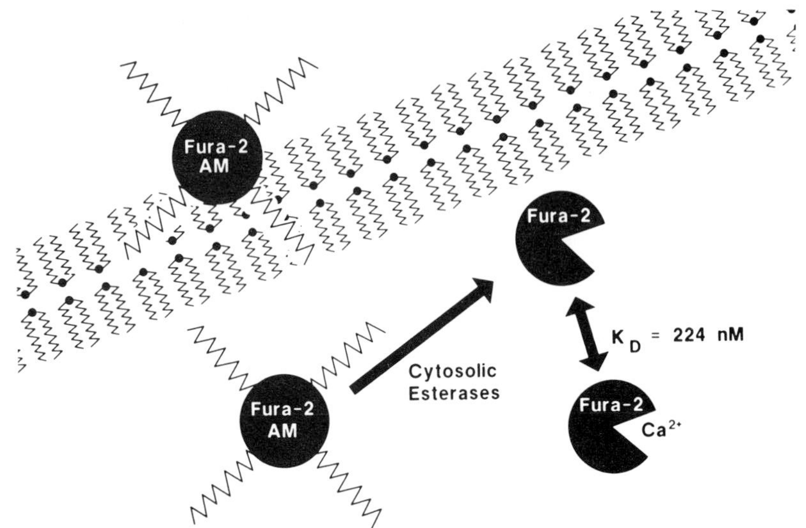

FIG. 5. Diagram of fura-2/AM entry and cleavage inside the cell to liberate the calcium-sensitive trapped form of the dye.

containing plated pituitary cells for 20 min and then washed with medium containing antiserum directed against ACTH. A period of 20–40 min at 37° is allowed to ensure intracellular cleavage of the dye's acetoxymethyl ester side groups to generate the calcium-sensitive form of the dye. This dye-loading procedure is followed in exactly the same way for each experiment. The fluorescence of the indicator is evenly distributed throughout the cell, and the dye does not appear to sequester in compartments other than cytoplasm. It has been reported that the magnitude of calcium increases is sometimes underestimated following presumed generation of a calcium-insensitive dye species resulting from incomplete ester cleavage.[20–24] Pituitary cells were discarded and a new set reloaded with 1.0 μM fura-2/AM on the rare occasions that intracellular compartmentalization of the dye was visualized. This potential problem has been assessed under the conditions of our loading procedure. Treatment with potent calcium-mobilizing drugs always reduce the fluorescence in the 380-nm image to near background levels. This argues that the primary species of regenerated fura-2 in pituitary cells is calcium sensitive.

[20] M. Scanlon, D. A. Williams, and F. S. Fay, *J. Biol. Chem.* **262,** 6308 (1987).

[21] W. Almers and E. Neher, *FEBS Lett.* **192,** 13 (1985).

[22] Q. Li, R. A. Altschuld, and B. T. Stokes, *Biochem. Biophys. Res. Commun.* **147,** 120 (1987).

[23] M. Poenie, R. Y. Tsien, and A. M. Schmitt-Verhulst, *EMBO J.* **6,** 2223 (1987).

[24] J. A. Connor, *Proc. Natl. Acad. Sci. U.S.A.* **83,** 6179 (1986).

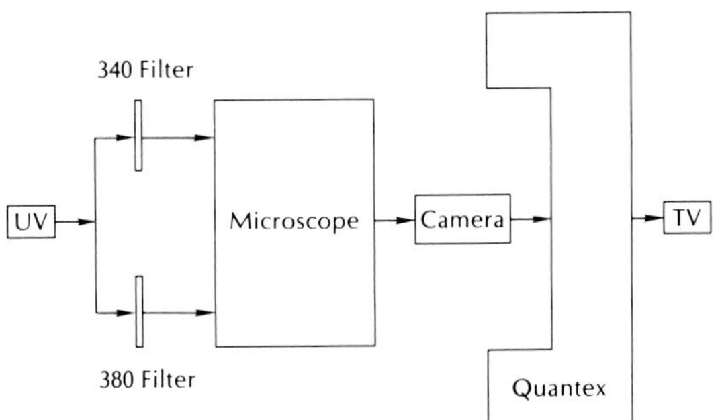

FIG. 6. Instrumentation used for digital imaging microscopy.

Pituitary cells are imaged using a Leitz Orthoplan microscope, equipped with epifluorescence and quartz optics in the epifluorescence light path, and a Nikon objective (Fig. 6). Images of fluorescent pituitary cells are obtained using a long-pass barrier filter that transmits emission wavelengths of 410–580 nm. Images are collected with a silicon-intensified target (DAGE Series 65) and stored for later processing on video tape (Ampex) using a professional quality video recorder (Sony VO-5600). More recently, there have been substantial improvements in the design of charge-coupled devices (CCD) that may offer particular advantages in terms of the quality of the images.[25] On the other hand, one limitation of CCD cameras concerns the relatively sluggish temporal response. SIT cameras offer good temporal resolution, an attribute we wish to retain to register the highly dynamic nature of calcium transients in pituitary cells. It is also important to demonstrate that the camera/video recorder unit measures the fluorescent response in a linear fashion over the entire range of interest (Fig. 7).

Methods to measure calcium, using indicators like quin-2 and fura-2 in single cells, have evolved rapidly. Measuring calcium transients using digital imaging microscopy is a new technique that continues to be refined as experience has consolidated. The calcium-bound form of fura-2 maximally emits fluorescence measured at 510 nm after excitation at 340 nm. Calcium changes are proportional to fluorescence intensity at 340 nm excitation. The calcium-free form of fura-2 maximally emits at 518 nm after excitation at 380 nm. Calcium changes measured with 380 nm excitation are inversely related to the fluorescence response, that is, an increase in calcium is mirrored by a decrease in the emission intensity. By forming the ratio image ($I_{340/380}$), a measure of calcium with a high dynamic range is provided that will change in the same direction as intracellular calcium.

[25] R. Y. Tsien and M. Poenie, *Trends Biochem. Sci.* **11,** 450 (1986).

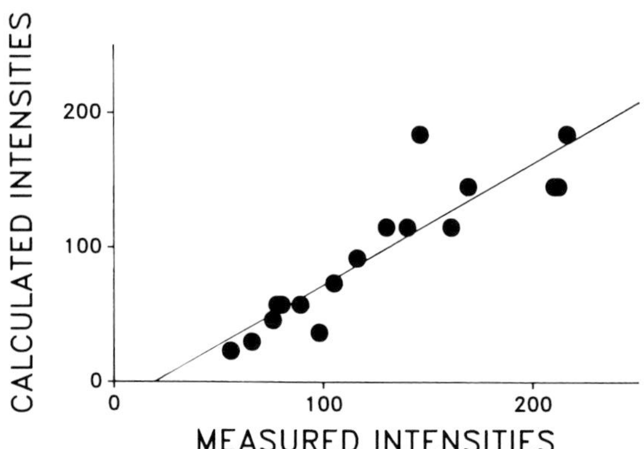

FIG. 7. Linearity of the instrument response. A bright-field image was recorded on video tape using the silicon-intensified target camera and video recorder unit. The image was digitized using a Quantex QX-7210 image analyzer. The intensity of the image was decreased in a stepwise fashion by placing calibrated neutral density filters in the light path. This procedure is also a test of the image analysis algorithm used to analyze the fluorescent images: the linear response obtained between the calculated and measured values suggests that an appropriate algorithm has been selected.

An important corollary of ratio imaging is that measurements are made essentially independent of uneven dye loading, cell height, and variations in instrumental efficiency.[25–27] Ratio imaging principles were originally applied to measure calcium with quin-2 using 340/360 as the ratio.[28,29]

The excitation spectrum of fura-2 as a function of calcium concentration is such that the exact choice of wavelengths is to some extent arbitrary. Other workers have monitored the bound form of the dye at 350 nm and the unbound form at 385 nm. The advantage of using 350 nm excitation is that glass optics can be used. The effect of glass is to prevent transmission of wavelengths below 350 nm. Expensive quartz optics are required to provide 340 nm excitation. On the other hand, quartz optics provide a lower level of "background" fluorescence generated within the optical system. Nikon objectives have become the "standard" objective of choice because of their unusual property of transmitting wavelengths as low as 340 nm. This is fortunate because the expense of quartz objectives is prohibitive.

The requirement to change excitation wavelengths rapidly is not presently available on commercial fluorescent microscopes. Monochromators

[26] R. Y. Tsien, T. J. Rink, and M. Poenie, *Cell Calcium* **6,** 145 (1985).

[27] G. R. Bright, J. Rogowska, G. W. Fisher, and D. L. Taylor, *BioTechniques* **5,** 556 (1987).

[28] B. A. Kruskal, C. H. Keith, and F. R. Maxfield, *J. Cell Biol.* **99,** 1167 (1984).

[29] D. W. Sawyer, J. A. Sullivan, and G. L. Mandell, *Science* **230,** 663 (1985).

can be used, but they are expensive and best suited to a dual-purpose instrument designed for both cuvette and microscope applications such as one marketed by Spex Industries (3880 Parke Avenue, Edison, NJ 08820). Pairs of excitation wavelengths can be more cheaply obtained with interference filters (Corion, Holliston, MA) mounted on a rotating wheel to provide specific excitation at 340 and 380 nm. A computer-driven custom-built filter wheel/shutter system is available from Kramer Scientific Corporation (1 Odell Plaza, Yonkers, NY 10701).

Cells are excited sequentially first at 340 nm (12-nm band half-width) and subsequently at 380 nm (12-nm band half-width). The 380 nm excitation filter is usually paired with a transmission neutral density filter to provide an approximately equal fluorescence image intensity for the two excitation wavelengths. The fluorescent images comprising the ratio pair are separated routinely by less than 0.5 sec.

For most pituitary cell applications about 20–30 pituitary cells are imaged together in one field using a 20× Nikon objective and the fluorescent responses recorded on video tape. For 40× and 100× resolution, glycerin-immersion objectives are required, and this sometimes causes problems. The occasional experiment can be ruined if glycerin smears and penetrates the Cunningham channel. The air interface using a 20× objective avoids this problem and provides a mathematical chance of there being at least one corticotrope (relative abundance about 6%) or one gonadotrope (relative abundance about 4%) in the field studied. It is important to select a field at random to avoid bias in the sampling of individual cells.

Our first approach used a high-powered objective (100×) to image a single pituitary cell. Cell identification was made possible by the relatively large size of gonadotropes. By studying cells selected on the basis of size (the very largest pituitary cell was selected) we obtained a secretory gonadotrope (confirmed by RHPA) at least 80% of the time. While these early studies produced great excitement about the potential of calcium imaging, the nagging possibility remained that calcium responses from gonadotropes selected in this way represented only a discrete subpopulation of pituitary gonadotropes. Indeed, our recent studies have demonstrated that the very largest pituitary gonadotropes appear to represent a subpopulation that secretes LH in a distinctive manner.[15a]

Quantitation of Fluorescence

There are two basic approaches to quantitation of fluorescence available. The average fluorescence of an individual cell can be measured directly using a photomultiplier tube equipped with pin holes to mask out all regions other than the cell of interest. Only one cell can be studied at a time. This approach is limiting since the RHPA would have to be applied first to identify the cell of interest prior to calcium measurements. Measuring calcium in a single cell and subsequently trying to identify it by RHPA would be like looking for a needle in a haystack. Furthermore, since photometric measurements give the average response of the cell, this method will not provide definitive answers because calcium rises may occur in concentration gradients without a net increase in cell calcium.

The second method to quantitate fluorescence makes use of digital imaging microscopy. This method enables calcium measurements on multiple numbers (15–25) of pituitary cells all imaged together in a single field using a 20× objective. This permits the RHPA to be performed either during or after calcium measurements, since the measurement of hormone release begins when hormone antiserum is introduced to the chamber. Some spatial information concerning the distribution of calcium inside the cell is retained.

Using an image processor (Quantex 9210), the raw fluorescence image is first converted to a pixel array (640 by 480 by 8 bits). A sequence of 340 nm and then 380 nm images is recorded directly on video tape. A box is carefully placed over each plaque-forming cell so that it approximately confines the cell. An area of 50–150 pixels per cell is formed, and the fluorescence intensity is calculated for the area defined by the box. The "box" method provides the average calcium value for each cell. Thus digital imaging microscopy is used much like a photometer, the difference being that images of multiple cells are acquired and noncellular regions are blocked out during image processing. It is important to note that some spatial information concerning the distribution of calcium gradients is still retained in the recorded images for later use. Thus, there is a mathematical probability that a corticotrope or gonadotrope is included in the randomly selected field which will subsequently be identified by RHPA. More recently, special software has been written that instructs the image analyzer to automatically identify each fluorescent pituitary cell in a field, outline the cell with a circle rather than a box, calculate the average fluorescence intensity of each cell, and store these values in a suitable file for later processing, all within a few seconds.

Average background values are subtracted for each cell. The background fluorescence was obtained by averaging multiple values for an intact area of erythrocytes flanking the cell of interest. In effect, the "background" is the RHPA red cell lawn. Even though erythrocytes are known to be able to load fura-2/AM[19] the fluorescent signal emitted by the red cells is only slightly greater than the background obtained in a Cunningham chamber filled with cell-free medium. We have determined that when erythrocytes loaded with fura-2/AM are plated together with nonfluorescent pituitary cells (not loaded with fura-2) the signal obtained from each pituitary cell is equal to the signal measured from the surrounding erythrocyte lawn. Parenthetically, our dual procedure (calcium and hormone measurements) would probably be impossible were it not for the very strong effect of hemoglobin absorbance. This quenches the otherwise high background obtained from intracellular calcium in erythrocytes.

The data are then loaded onto a floppy disk and transferred for further processing to an IBM-compatible personal computer (PC's Limited Model 286). The PC software includes a Lotus Symphony spread sheet and Sigma Plot (Jandel Scientific, 65 Koch Rd., Corte Madera, CA 94925) for graphics. High-quality plots of the data may be obtained using Sigma Plot graphics and a plotter. Each 340/380 ratio pair is calculated using the spread sheet, and an estimate of absolute calcium concentration is calculated according to the equation formulated by Tsien.[17] Separate computers for postimage processing have been important to protect the image analyzer from becoming the bottleneck in the processing system. The rate-limiting factor is the overwhelming volume of data. Using the existing system, intracellular calcium may be serially sampled at about 5-sec intervals (the time to record a ratio pair) for periods up to 45–60 min.

The estimation of absolute $[Ca^{2+}]_i$ from the 340/380 ratio in single cells is performed using the equation below. The absolute values of calcium estimated using this equation should presently be regarded as approximations, for reasons that are discussed below. More rigorous approaches to establishing the precise values in pituitary cell systems for some of the constants are presently being evaluated. K_D is the dissociation constant of

$$[Ca^{2+}]_i = K_D \left(\frac{R - R_{min}}{R_{max} - R}\right)\left(\frac{I_{380} \text{ in } Ca^{2+}\text{-depleted cells}}{I_{380} \text{ in } Ca^{2+}\text{-saturated cells}}\right)$$

fura-2 for Ca^{2+}. R is the ratio of the fluorescence intensity at 340 nm to that at 380 nm, R_{min} and R_{max} are the R values when probe is saturated with Ca^{2+} and depleted of Ca^{2+}, respectively. I_{380} is the fluorescence intensity at 380 nm. R_{max} and R_{min} in single pituitary cells have been determined with the calcium ionophore 8-bromo-A23187 (10 μM, Calbiochem-Behring), an ionophore derivative without significant autofluorescence, in medium containing high or low (5.0 mM EGTA added) amounts of calcium.

The K_D for fura-2 will need to be determined according to a calibration curve obtained for each cell system of interest. Experiments in cuvette spectrofluorometers suggest a K_D for fura-2 of about 200–225 nM.[17,30] The formula above is explicitly derived from ratios calibrated in a cuvette and cannot be directly applied to measurements on single cells. The equation also assumes that the dye behaves the same in cells as it does in calibration medium. Thus, it is prudent to confirm that R_{min} and R_{max} compare favorably to those of solutions of fura-2 under the same conditions with each experiment. Full calibration curves in both cells and liquid medium need to be constructed to confirm these assumptions and derive an accurate K_D. In our hands, R_{min} and R_{max} values are comparable whether derived from pituitary cells or calibration medium.

The original image analyzer used for pituitary studies had just two buffer memories. This limited the scope of the experiments and the precision with which background corrections could be made. A new model Quantex 710 with eight memory buffers permits more precise background subtraction. Multiple video frames (8–32) are routinely averaged sequentially to increase the signal-to-noise ratio in a given image. One approach to correct for spatial variations in dark current of the camera is to average multiple images, obtained without input to the camera (sample chamber filled with media but without cells), and subtract these values pixel by pixel from the image. To correct for spatial variations in the intensity of the excitation source, and similar variations in camera gain, a uniform fluorescent field (a well-mixed solution of fura-2-free acid) is imaged. It is assumed that the mean gray level is the correct value for this calibration image. For each pixel, the ratio for use in correction is simply the pixel's value divided by the mean. The correction can be performed by dividing each pixel in the measurement image by the corresponding one in the calibration image and multiplying by the constant mean calibration image value. In practice, Tsien uses a thin film of solution trapped between two coverslips containing 120 m*M* KCl, 10 m*M* EDTA (dipotassium dihydrate), 20 m*M* HEPES, pH 7.5, and 10 m*M* fura-2 pentopotassium salt (without acetoxymethyl ester side chains). This approach has the advantage of essentially normalizing the ratios against a relatively reproducible standard (''low'' calcium) so that experiments performed on different days can be compared.

Spatial Maps of Intracellular Calcium

Perhaps the most exciting application of digital imaging microscopy stems from the ability to visualize local gradients of calcium. Most recent reports have focused on the distribution of calcium gradients strikingly represented in color-coded maps.[24,25,29,30] For example, a frequent finding

[30] D. A. Williams, K. E. Fogarty, R. Y. Tsien, and F. S. Fay, *Nature (London)* **318,** 558 (1985).

has been the apparent difference in $[Ca^{2+}]_i$ for the nucleus versus the cytoplasm, which was first reported by Williams *et al.*[30] It is presently not clear whether this finding is applicable to pituitary cells since acutely plated cells are highly rounded. Thus, it is unusual to find a cell in which the cytoplasmic compartment is relatively distinct from the nuclear region. Nevertheless, preliminary findings in the gonadotrope suggest that the nuclear region is maintained at a lower ambient calcium level than the cytoplasm, in agreement with the findings of Williams *et al.*[30] This area is presently controversial since it is unclear whether fluorescence differences reflect calcium differences within cellular compartments rather than differences in fura-2 uptake into the compartment. For example, the different ionic and viscous environment of the nucleus may affect the fluorescence of fura-2 differently than in cytoplasm, making the calculations of $[Ca^{2+}]_i$ in each compartment difficult to compare directly.

Successful measurement of calcium using digital imaging microscopy in single pituitary cells involves several compromises. The method described here requires about 15–25 cells/field to be measured to ensure at least one ACTH or LH cell is studied. Thus much lower power objectives need to be used, providing limited spatial resolution. On the other hand, about 50–150 pixels of spatial information is obtained with pituitary cells depending on cell size. Thus, the method described here for pituitary cells measures the average calcium response from the cell. In addition, some spatial information is obtained that should be sufficient to assess whether dynamic regional changes in calcium are also occurring.

Disadvantages

One major problem revolves around possible incomplete deesterification of the dye within cells. This problem leads to gross underestimations of $[Ca^{2+}]_i$. A related problem is the tendency of the dye with time to sequester inside subcellular organelles and no longer report on calcium changes originating solely in the cytoplasm. Both of these problems have been shown to vary with different cell systems. The approach we have taken in pituitary cells is to determine empirically the best loading conditions using time and temperature as variables. There is still some variability, however, and the occasional cells with apparent dye compartmentation are discarded and a new set prepared. An important test of compartmentation is to permeabilize the cells' plasma membrane using a judicious amount of digitonin that does not compromise subcellular compartments. If all of the dye is in the cytoplasm, the intensity reading should be reduced to background. We are fortunate that we can adequately load pituitary cells with fura-2/AM with just a 20-min incubation. Care is taken to stagger the loading of the cells so that the sequence of fura-2 loading and the experimental calcium measurements are always comparable. In pituitary cells, calcium-mobilizing agents almost always reduce the intensity in the 380-nm wavelength image to near background levels. The dynamic range of our ratio values is very high, suggesting that the overwhelmingly predominant species of fura-2 is sensitive to calcium. Thus, it seems unlikely that we have dye compartmentation or a problem with deesterification. However, these are important issues that must be thoroughly evaluated in all cell systems. Some researchers have chosen to emphasize fluorescence ratios and acknowledge the possibility that the absolute values of $[Ca^{2+}]_i$ calculated from these ratios may have to be revised when the effect of intracellular environment is more completely evaluated and an accurate K_D for the indicator has been determined. This active area of investigation continues to provide more refined ways of accurately measuring intracellular calcium.

The greatest drawback is that the instrumentation for quantitative fluorescence microscopy is expensive. Image analyzers that operate in close to real time and a high-quality microscope are essential for the enterprise. An optical disk recorder is also useful for archival purposes, as the volume of data generated on video tape is overwhelming.

Advantages

Measurements of calcium at the single-cell level complement fluorometer studies on bulk populations. Calcium values derived from bulk cell populations are particularly appropriate for comparison with other biochemical responses also measured in the bulk population. In standard cuvette fluorometers, cell suspensions are used, and calcium measurements reflect the averaged response from millions of cells. However, there are a number of advantages that single-cell microscopy brings to the measurement of calcium. The single-cell procedure inherently requires a small number of animals. The single-cell approach emphasizes economy since an overwhelming number of cells can be provided by one animal per experimental day.

The microscopy procedure uses plated cells, and we believe this practice is fundamental since most cells are of the anchorage-dependent variety. It is well established that anchorage-dependent pituitary cells are poorly responsive to hypothalamic hormones when challenged in suspension. Some modified fluorometers are capable of determinations on plated cells, but these instruments are not common. The modified Cunningham chamber we use permits the medium bathing the cells to be rapidly replaced. Such flexibility is not always provided in the cuvette of a fluorometer. Thus, we can withdraw or change our stimulus at will without interrupting calcium measurements, whereas the lid to a cuvette must be removed to apply the stimulus, which interrupts calcium measurements because light is admitted. Consequently, in a cuvette the precise kinetics of the initial response are difficult to determine. Using microscopy the initial response can be monitored continuously at a single wavelength. It is noteworthy that we have made accurate measurements, with a resolution of hundreds of milliseconds, and demonstrated calcium responses, within 1–2 sec, to the classic releasing hormones in LH and ACTH cells. Furthermore, Sykes–Moore chambers can be used that permit the cells to be perfused with medium while being visualized in the microscope. In this way, pulsatile or sequential administration of drugs is feasible.

The single-cell technique allows selection of a healthy cell for study. Dying cells can be excluded, together with cells that have formed precipitates of dye and cells in which fura-2 has entered intracellular calcium stores. Thus there is less potential for artifacts originating from dye leakage or cell necrosis.

As discussed earlier, the single-cell procedure gives detailed information that is unique. Calcium can be accurately measured and shown to be sequestered in regional domains within the cytoplasm of the cell. For example, the ACTH studies described earlier suggest that CRH-induced calcium rises are episodic. The effect of averaging would be to conceal the true pattern of the calcium response occurring in each individual cell. Thus standard cuvette fluorometers, by virtue of measuring calcium simultaneously from millions of cells, cannot provide the level of precision provided by microscopy. Finally, calcium studies in many tissues are confounded by mixed-cell populations. The microscopy approach overcomes the need for cell purification, provided a method is available to identify the cells before or after microscopy. Techniques other than RHPA, such as immunocytochemistry, can be used to identify the pituitary cell type.

Preliminary Studies

Corticotropin-releasing hormone (CRH) is the prime hypothalamic regulator of pituitary ACTH secretion. The CRH receptor is clearly coupled to adenylate cyclase to generate cAMP as a second messenger. However, the signaling steps beyond this simple scheme are unknown. Recent studies[31] have demonstrated that CRH mobilizes calcium in secretory corticotropes in a highly specific manner (Fig. 8).

Three general mechanisms of CRH action to increase $[Ca^{2+}]_i$ are possible: (1) promotion of influx of extracellular calcium via membrane ion channels, (2) prevention of calcium efflux from the cytoplasm, and (3)

[31] D. A. Leong *et al.*, in preparation.

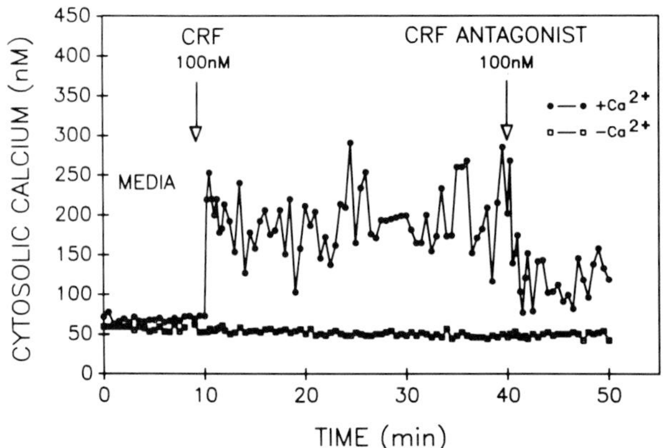

FIG. 8. Effect of 100 nM CRH-induced Ca^{2+} mobilization, illustrated as a representative response from a single ACTH cell. CRH-induced Ca^{2+} changes typical of plaque-forming corticotropes (n = 28) are shown, identified from about 500 pituitary cells analyzed. CRH induced a sustained rise in $[Ca^{2+}]_i$ initiated 5–40 sec after delivery of the stimulus. Treatment with a potent CRH antagonist rapidly lowered $[Ca^{2+}]_i$ to basal levels followed by a characteristic rebound to establish a moderately elevated baseline. Withdrawal of extracellular Ca^{2+} (at 10 min) completely abolished CRH-induced Ca^{2+} rises (n = 5).

promotion of calcium mobilization from intracellular stores. The most likely mechanism is that CRH opens calcium channels to promote influx from an extracellular source. Thus, CRH-induced rises in $[Ca^{2+}]_i$ are completely abolished when a source of extracellular calcium is depleted with 5.0 mM EGTA. This evidence cannot fully exclude mechanism 2 above since it is not known if the normal function of calcium efflux mechanisms requires extracellular calcium.

Thus, CRH promotes cAMP accumulation and calcium mobilization. We next asked whether forskolin, a pharmacologic agent that stimulates cAMP rises (bypassing the CRH receptor), might induce calcium mobilization in secretory ACTH cells. Forskolin induced a sustained rise of calcium that was abolished when the source of extracellular calcium was removed (Fig. 9). Thus CRH and forskolin each induce replica rises in $[Ca^{2+}]_i$ that are abolished when extracellular calcium is withdrawn. Based on these interesting findings, the following working hypothesis is proposed (Fig. 10). The scheme proposes a sequential and causal link between the cAMP and calcium signaling systems.

It is generally accepted that the metabolites of inositol lipid metabolism (particularly IP_3) regulate calcium mobilization from intracellular stores. More recently, the first examples of second messenger-regulated calcium ion channels have been established using direct patch clamp studies. In all three known cases the messengers (IP_3, IP_4, Ca^{2+}) that regulate calcium ion channels (in neutrophils, mast cells, and oocytes) are also

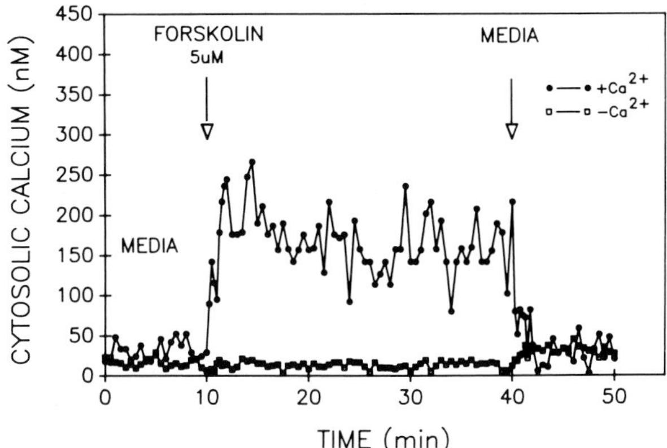

FIG. 9. Effect of 5 μM forskolin-induced Ca^{2+} mobilization shown as a representative response from a single ACTH cell. Forskolin induced this typical Ca^{2+} response in all plaque-forming corticotropes ($n = 20$). Medium replacement, to wash out forskolin, rapidly lowered Ca^{2+} to basal levels followed by a characteristic rebound to establish a new elevated baseline. Withdrawal of extracellular Ca^{2+} (at 5 min as shown, or 10 min) completely abolished forskolin-induced Ca^{2+} rises.

generated from activation of the inositol messenger pathway. Thus, to date, the inositol lipid signaling system (and not cAMP) is the only known pathway linked to *de novo* calcium influx. The mechanism for pituitary ACTH cells above proposes a novel sequential link between cAMP and calcium mobilization via a cAMP-regulated membrane ion channel.

Final Comments

Precise calcium measurements in single cells will tell us how the population response is built from discrete unitary responses. Novel findings of oscillatory calcium transients in pituitary cells raise the possibility that the "language" of intracellular signaling involves information coded in

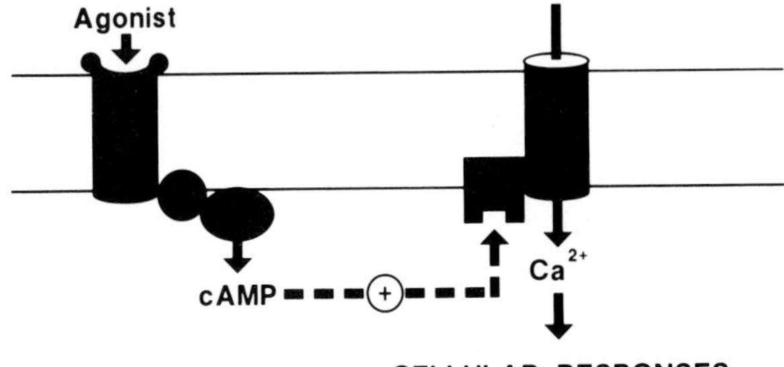

FIG. 10. Schematic model of the second messenger-operated Ca^{2+} channel.

both analog and digital modes. The temporal and spatial resolution obtained by digital imaging microscopy is unprecedented for any intracellular messenger. Just as summing the population response can obscure critical information resolved only by single-cell measurements, averaging the calcium response of single cells (Figs. 8 and 9) might also mask calcium gradients within responding cells. The digital imaging microscopy approach described here essentially solves these problems. The combination of RHPA methods, calcium-sensitive indicators, and digital imaging microscopy represents a powerful new tool to investigate intracellular calcium in pituitary cells.

An implicit strength of the technique is that questions can be asked about the fundamental interrelationships of individual messenger pathways. It is clear that each pituitary cell type is organized in a particular manner with distinct networks of feed-forward and feedback pathways. Yet very little information is known about the "hard wiring" of the multiple messenger systems. The messenger equivalents of integrated cellular responses must be built from these relationships. To characterize the possible feed-forward interdependence of these messenger pathways, one strategy will be to increase pharmacologically the concentration of a particular messenger (e.g., cAMP, cGMP, IP_3, diacylglycerol, arachidonate) and look for a calcium response (see Fig. 9). Similarly, feedback relationships can be revealed by performing the same experiments in combination with a secretagogue known to mobilize calcium (e.g., CRF in corticotropes). Again, the single-cell approach is emphasized since uninterpretable data would be obtained from mixed-cell populations. These advances promise new insights into the complex mechanisms of intracellular signaling.

Acknowledgments

I would like to thank Walter May for substantial contributions to the ACTH single cell studies. Studies with pituitary GH cells were performed in collaboration with Reinhard Holl and Michael Thorner. These studies were made possible with instrumentation provided by James Sullivan and Gerald Mandell (University of Virginia). David Orth (Vanderbilt University) supplied ACTH antiserum and the CRH antagonist was provided by Wylie Vale and Jean Rivier (Salk Institute). Thanks also to Kay Hancock for manuscript preparation. These studies were supported by grants from the NIH DK-35937, BRSG 5-SO7-RR 431-29, and the Thomas F. Jeffress and Kate Miller Jeffress Memorial Trust.

NOTE ADDED IN PROOF

Since this chapter was originally submitted, several examples of cAMP-regulated *de novo* calcium influx have been reported in primary sensory neurons.[32,33]

[32] P. Avenet, F. Hoffman, and B. Lindeman, *Nature* (*London*) **331,** 351 (1988).

[33] T. Nakamura and G. H. Gold, *Nature* (*London*) **325,** 442 (1987).

Section III

Preparation and Maintenance of Biological Materials

[18] Use of Protein Kinase C-Depleted Cells for Investigation of the Role of Protein Kinase C in Stimulus–Response Coupling in the Pituitary

By CRAIG A. MCARDLE and P. MICHAEL CONN

Introduction

Protein kinase C is a Ca^{2+}- and phospholipid-dependent enzyme, which is activated at a low Ca^{2+} concentration (<1 μM) in the presence of diacylglycerols of the sn-1,2 conformation.[1–6] Research into the role of this enzyme in stimulus–response coupling has been spurred greatly by the suggestion that diacylglycerols, produced as a consequence of phosphoinositide hydrolysis, influence cellular activity by activation of protein kinase C.[2,3] In pituitary gonadotropes, for example, the binding of gonadotropin-releasing hormone (GnRH) to cell surface receptors causes both an increase in the mass of cellular diacylglycerol[7] and a redistribution of protein kinase C from cytosolic to particulate fractions prepared from homogenates of pituitary tissue or cells.[8–10] The production of diacylglycerol and the association of this phospholipid-dependent enzyme with the phospholipid-rich particulate fraction provide strong evidence that the enzyme is activated by GnRH. Moreover, the demonstration that direct activation of protein kinase C (with exogenous phorbol esters or diacylglycerols) causes the release of luteinizing hormone (LH) from pituitary

[1] Y. Takai, A. Kishimoto, Y. Iwasa, Y. Kawahara, T. Mori, and Y. Nishizuka, *J. Biol. Chem.* **254,** 3692 (1979).

[2] Y. Takai, A. Kishimoto, U. Kikkawa, T. Mori, and Y. Nishizuka, *Biochem. Biophys. Res. Commun.* **91,** 1218 (1979).

[3] A. Kishimoto, Y. Takai, T. Mori, U. Kikkawa, and Y. Nishizuka, *J. Biol. Chem.* **255,** 2273 (1980).

[4] J. F. Kuo, R. G. G. Andersson, B. C. Wise, L. Mackerlova, I. Salomonsson, N. L. Brackett, N. Katoh, M. Shoji, and R. W. Wrenn, *Proc. Natl. Acad. Sci. U.S.A.* **77,** 7039 (1980).

[5] Y. Nishizuka, *Science* **233,** 305 (1986).

[6] T. Kitano, M. Go, U. Kikkawa, and Y. Nishizuka, *in* "Neuroendocrine Peptide Methodology" (P. M. Conn, ed.), pp. 371–374. Academic Press, San Diego, California, 1989.

[7] W. V. Andrews and P. M. Conn, *Endocrinology (Baltimore)* **118,** 1148 (1986).

[8] K. Hirota, T. Hirota, G. Aguilera, and K. J. Catt, *J. Biol. Chem.* **260,** 3243 (1985).

[9] Z. Naor, J. Zer, H. Zakut, and J. Herman, *Proc. Natl. Acad. Sci. U.S.A.* **82,** 8203 (1985).

[10] C. A. McArdle and P. M. Conn, *Mol. Pharmacol.* **29,** 570 (1986).

cell cultures[11-13] is compatible with a role for protein kinase C as a mediator of GnRH-stimulated LH release.

Recognizing the possibility that the effects produced by pharmacological activation of protein kinase C with phorbol esters might not adequately reflect the effects produced by physiological activation of the enzyme, we have sought means for testing the *requirement* of hormone-stimulated protein kinase C activation for hormone action in pituitary cells. To this end we have taken advantage of the observation that an apparent loss of cellular protein kinase C occurs in pituitary-derived clonal cell lines,[14-16] and in several other cell types,[17-22] after pretreatment with protein kinase C-activating phorbol esters. Here we describe methods for obtaining primary cultures of pituitary cells which are depleted of protein kinase C as indicated by a loss of responsiveness to exogenous protein kinase C activators,[23] a loss of extractable protein kinase C activity,[23] and a loss of phorbol ester binding sites. We also give an example of the manner in which such cells can be used for assessing whether hormonal activation of this enzyme is required for the mediation of other effects of the hormone.

Materials and Methods

Preparation and Phorbol Ester Pretreatment of Pituitary Cells

Primary cultures of dispersed pituitary cells from female weanling rats (60–75 g, Sprague–Dawley; Sasco, Omaha, NE) are prepared by enzymatic dispersion.[24] The cells are suspended in Medium 199 (M199; Gibco,

[11] M. A. Smith and W. W. Vale, *Endocrinology (Baltimore)* **107,** 1425 (1980).

[12] W. A. Smith and P. M. Conn, *Endocrinology (Baltimore)* **114,** 553 (1984).

[13] P. M. Conn, B. R. Ganong, J. Ebeling, D. Staley, J. E. Neidel, and R. M. Bell, *Biochem. Biophys. Res. Commun.* **126,** 532 (1985).

[14] S. Jaken, A. H. Tashjian, and P. M. Blumberg, *Cancer Res.* **41,** 2175 (1981).

[15] M. A. Phillips and S. Jaken, *J. Biol. Chem.* **258,** 2875 (1983).

[16] R. Ballester and O. M. Rosen, *J. Biol. Chem.* **260,** 15194 (1985).

[17] P. M. Tapley and A. W. Murray, *Eur. J. Biochem.* **151,** 419 (1985).

[18] P. J. Blackshear, L. A. Witters, P. R. Girard, J. F. Kuo, and S. N. Quamo, *J. Biol. Chem.* **260,** 13304 (1985).

[19] E. Melloni, S. Pontremoli, M. Michetti, O. Sacco, B. Sparatore, and B. L. Horecker, *J. Biol. Chem.* **261,** 4101 (1986).

[20] K. Chida, N. Kato, and T. Kuroki, *J. Biol. Chem.* **261,** 13013 (1986).

[21] H. J. G. Matthies, H. C. Palfrey, L. D. Hirning, and R. J. Miller, *J. Neurosci.* **7,** 1198 (1987).

[22] P. J. Blackshear, R. A. Nemenoff, J. G. Hovis, D. L. Halsey, D. J. Stumpo, and J.-K. Huang, *Mol. Endocrinol.* **1,** 44 (1987).

[23] C. A. McArdle, W. R. Huckle, and P. M. Conn, *J. Biol. Chem.* **262,** 5028 (1987).

[24] J. Marian and P. M. Conn, *Mol. Pharmacol.* **16,** 196 (1979).

Grand Island, NY) containing 0.3% bovine serum albumin (BSA, Fraction V; Sigma, St. Louis, MO), 10% horse serum, 2.5% fetal calf serum (Irvine Scientific, Santa Ana, CA), 20 μg/ml gentamicin sulfate (Sigma), and 10 m*M* HEPES, pH 7.4 (plating medium). Aliquots of this cell suspension are then maintained in culture in a water-saturated atmosphere at 37°. For LH-release studies, cells prepared from one pituitary are suspended in 4–5 ml of plating medium, and 1-ml aliquots of this suspension are incubated in 16-mm wells of 24-well culture plates (Costar, Cambridge, MA). For phorbol ester-binding experiments, the cells are suspended at a density of approximately one pituitary per 1.5 ml, and 2-ml aliquots of this suspension are placed in 22-mm wells of 12-well culture plates. For assays of extractable protein kinase C activity, cells are suspended at density of one pituitary per milliliter, and 10- to 12.5-ml aliquots of this suspension are placed in sterile 50-ml centrifuge tubes.

After 2 days in culture, the cells are washed with M199 containing 0.3% BSA and 10 m*M* HEPES, pH 7.4 (BSA/M199) and then pretreated by incubation for 6 hr in BSA/M199 containing phorbol ester or vehicle alone. The pretreatment is terminated by removal of the challenge medium and washing the cells with BSA/M199. After pretreatment the cells are incubated for 12 hr in fresh plating medium, then washed twice with BSA/M199, and are either used for quantification of extractable protein kinase C or phorbol ester binding, as described below, or incubated (3 hr) with 1 ml of BSA/M199 containing secretagogue or appropriate vehicle (treatment). For these studies media are collected at the end of both pretreatment and treatment periods for determination of LH release by radioimmunoassay (RIA).[25] In addition, cells which have received pretreatment and incubation in plating medium alone are solubilized by freezing and thawing in 1 ml of BSA/M199 containing 0.1% Triton X-100 (Sigma) for determination of cellular LH.

Extraction and Assay of Protein Kinase C

Pituitary cells are dispersed, maintained in culture, pretreated, and returned to plating medium for 12 hr as described above. The cells are then washed and collected by scraping with a plastic pipet followed by centrifugation at 120 *g* for 15 min. Detergent-extractable protein kinase C activity is then determined.[23] The cell pellets are homogenized in 10 ml of 25 m*M* Tris–HCl, pH 7.5, containing 0.25 *M* sucrose, 2.5 m*M* $MgCl_2$, 2.5 m*M* EGTA, 50 m*M* 2-mercaptoethanol, and 0.1 m*M* phenylmethylsulfonyl fluoride (homogenization buffer) with 0.3% Triton X-100. These suspensions are shaken for 30 min and applied to DEAE–cellulose

[25] W. A. Smith, R. L. Cooper, and P. M. Conn, *Endocrinology* (*Baltimore*) **111**, 1843 (1982).

columns which have been equilibrated with homogenization buffer. The columns are then washed with 10 ml of homogenization buffer followed by 2 ml of this buffer containing 20 mM NaCl. Samples are eluted from the column with 5 ml of homogenization buffer containing 100 mM NaCl, followed by 5 ml of this buffer containing 400 mM NaCl. The Trixon X-100 extraction and DEAE–cellulose purification procedures are performed at 4°.

Protein kinase activity is assayed by determination of the rate of transfer of ^{32}P from [γ-^{32}P]ATP to histone as previously described.[26] The standard assay is performed using 250 μl of 20 mM Tris–HCl, pH 7.5, containing 5 mM magnesium nitrate, 0.2 mg/ml histone (Sigma Type III-S), 10 μM [γ-^{32}P]ATP, and 50 μl of DEAE–cellulose column eluate. This solution is supplemented with 1 mM EGTA, with 1 mM $CaCl_2$, or with 1 mM $CaCl_2$ and lipids (40 μg/ml phosphatidylserine and 4 μg/ml 1,2-diolein). Alternatively, the solution is supplemented as indicated with 0 or 0.05–15 nM phorbol 12-myristate 13-acetate (PMA), 10 μM $CaCl_2$, and 4 μg/ml phosphatidylserine. The assay is started by addition of the [γ-^{32}P]ATP and is continued for 5–6 min at 30° before stopping by precipitation of protein in 10% trichloroacetic acid.[10] After extensive washing, ^{32}P incorporation into precipitated protein is determined by liquid scintillation spectroscopy. Kinase activity is expressed as picomoles of inorganic phosphate incorporated per minute per milligram protein applied to the column. Protein is assayed according to the method of Bradford,[27] using BSA as the protein standard. We routinely perform the kinase assays immediately after the DEAE–cellulose purification step. All buffers can be prepared in advance (store at 4°) except that the histone, lipids, phorbol esters, and phenylmethylsulfonyl fluoride are added on the day of use.

[^{3}H]Phorbol 12,13-Dibutyrate (PDBu) Binding

Dispersed pituitary cells are prepared, maintained in culture, pretreated for 6 hr and returned to plating medium for 12 hr as described above. After washing and equilibration (20 min in BSA/M199 at 23°) the medium is decanted and replaced with 0.5 ml of BSA/M199 at 23° containing 1.5–50 nM [^{3}H]PDBu and 0 or 1 μM PMA. Nonspecific binding is defined as the amount of tracer binding measured in the presence of 1 μM PMA. After 20 min the binding incubation is terminated by removal of the radioligand-containing medium and washing the cells rapidly in 1 ml of BSA/M199 at 4°. The cells are then collected by scraping from the wells in 2 ml of BSA/M199 containing 2.5 mM EGTA at 4°. The cell suspension is

[26] M. Castagna, Y. Takai, K. Kaibuchi, K. Sano, U. Kikkawa, and Y. Nishizuka, *J. Biol. Chem.* **257,** 7847 (1982).

[27] M. M. Bradford, *Anal. Biochem.* **72,** 248 (1976).

then layered over 1 ml of M199 containing 0.3 *M* sucrose, and the cells are collected by centrifugation (10 min, 3000 *g*, 4°). After resuspension in 0.4 ml of distilled water, the radioactivity in the cell pellet is determined by liquid scintillation spectroscopy.

Materials

Natural sequence GnRH is provided by the National Pituitary Agency. Phorbol esters (Sigma or LC Services, Woburn, MA), *sn*-1,2-dioctanoylglycerol (prepared as described in Ref. 12), and the Ca^{2+}-selective ionophore A23187 (Calbiochem-Behring, San Diego, CA) are prepared as concentrated stock solutions in dimethyl sulfoxide (DMSO) or ethanol.[28] The vehicle concentration should not exceed 1% (v/v) in the bioassay medium and has no measurable effect on LH release, [^{3}H]PDBu binding, or kinase activity at this concentration. 1,2-Diolein (1,2-dioleoyl-*rac*-glycerol) and phosphatidylserine (both from Sigma) are stored as stock solutions at 1 mg/ml in chloroform. Aliquots of these lipids are mixed, evaporated to dryness under N_2, and resuspended by sonication in the required buffer (typically 20 m*M* Tris–HCl, pH 7.5, containing 5 m*M* $CaCl_2$) on the day of use. [γ-^{32}P]ATP can be prepared according to Glynn and Chappel[29] or purchased from Amersham (Arlington Heights, IL) or NEN Research Products (Boston, MA). We now routinely store [γ-^{32}P]ATP at a specific activity of >5000 Ci/mmol and dilute this stock on the day of use in 20 m*M* Tris–HCl (pH 7.5) containing 50 μM unlabeled ATP (Sigma). Kinase assays are initiated by adding 50 μl of this diluted stock to the other constituents of the assay (final ATP concentration 10 μM, final specific activity 0.05–0.75 Ci/mmol). The [^{3}H]phorbol 12,13-dibutyrate (approximately 20 Ci/mmol) is purchased from Amersham.

Characterization of Protein Kinase C-Depleted Cells

As shown in Fig. 1, treatment of pituitary cells for 3 hr with PMA causes a concentration-dependent release of LH into the culture medium. This effect appears to be a consequence of activation of protein kinase C *in situ* because a wide range of phorbol esters and diacylglycerols provoke LH release from gonadotropes with a potency order similar to that with which they activate extracted protein kinase C.[11–13] The LH-releasing effect of PMA is greatly reduced in cells which have been pretreated for 6 hr with 150 n*M* PMA and then returned to plating medium for 12 hr prior to treatment with PMA. This effect occurs without any measurable change in the EC_{50} for PMA-stimulated LH release and provides a simple

[28] P. M. Conn, D. C. Rogers, and F. S. Sandhu, *Endocrinology (Baltimore)* **105,** 1122 (1979).
[29] I. M. Glynn and J. B. Chappel, *Biochem. J.* **90,** 147 (1964).

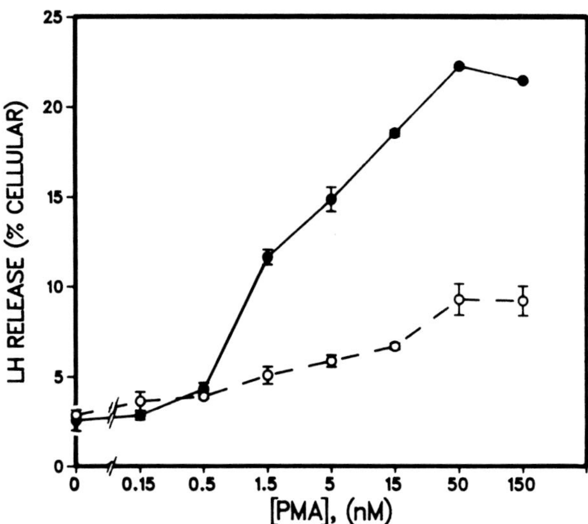

FIG. 1. Effect of PMA pretreatment on PMA-stimulated LH release. Pituitary cells were prepared and cultured for 2 days. The cells were then washed and pretreated for 6 hr with PMA (○, 150 n*M* in BSA/M199 with 0.5% DMSO) or with BSA/M199 containing vehicle alone (●). The cells were then washed and incubated for a further 12 hr in plating medium, prior to being washed and incubated for 3 hr with the indicated concentrations of PMA in BSA/M199. The response of PMA-pretreated and control cells differed significantly ($p < 0.05$) at 1.5–150 n*M* PMA. The values shown are the means ± SE of three duplicate determinations ($n = 3$) and are representative of those obtained in six similar experiments. Reprinted from Ref. 23 with permission from The American Society of Biological Chemists, Inc.

and functional demonstration of the loss of ability of these cells to respond to direct activation of protein kinase C. A further demonstration of the loss of responsiveness to PMA is provided by the observation that the inhibition of GnRH-stimulated inositol phosphate accumulation by PMA is also greatly reduced in PMA-pretreated gonadotropes.[23]

A more direct assessment of the effect of PMA pretreatment on protein kinase C activity can be obtained using *in vitro* kinase assays to quantify PMA-stimulable or Ca^{2+}- and phospholipid-stimulable kinase activity. For these studies the protein kinase C from cells pretreated with BSA/M199 containing PMA or vehicle alone is extracted with Triton X-100 and partially purified by DEAE–cellulose chromatography before assay. As shown in Fig. 2, the kinase activity in a DEAE–cellulose-purified detergent extract of pituitary cells is stimulated in a concentration-dependent manner by PMA. PMA-stimulated kinase activity, like PMA-stimulated LH release, is markedly reduced after pretreatment for 6 hr with 150

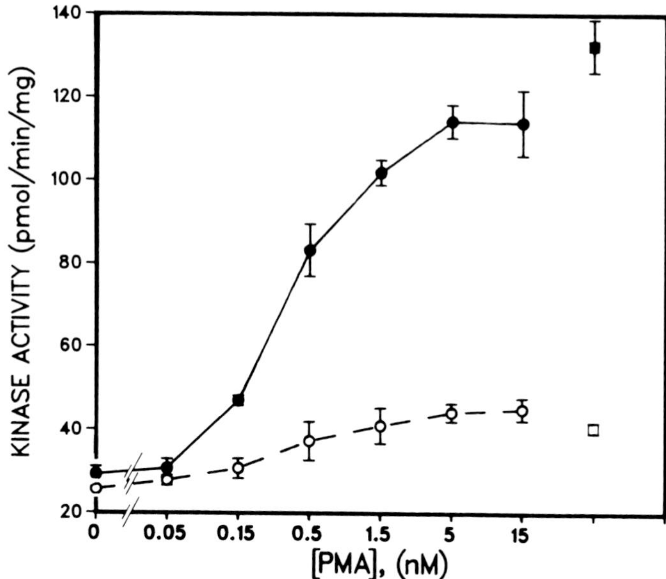

FIG. 2. PMA-stimulable kinase activity is reduced in PMA-pretreated pituitary cells. Pituitary cells were dispersed and challenged for 6 hr with PMA (○, 150 nM in BSA/M199 with 0.5% DMSO) or with BSA/M199 containing vehicle alone (●). The cells were then returned to plating medium for 12 hr prior to collection of cells, extraction, and purification on DEAE–cellulose. Samples eluted with homogenization buffer containing 20–100 mM NaCl were used as the enzyme source. Kinase activity was determined in the presence of 10 μM $CaCl_2$, 4 μg/ml phosphatidylserine, and the indicated concentrations of PMA. The kinase activity measured in the presence of 1 mM $CaCl_2$, 40 μg/ml phosphatidylserine, and 4 μg/ml 1,2-diolein is also shown for control (■) and PMA-pretreated cells (□). PMA-stimulable kinase activity of PMA-pretreated and control cells differed significantly ($p < 0.05$) at 0.15–15 nM PMA. The values shown are the means ± SE of triplicate determinations and are representative of those obtained in three similar experiments. Reprinted from Ref. 23 with permission from The American Society of Biological Chemists, Inc.

nM PMA. This loss of extractable protein kinase C occurs without any measurable change in the EC_{50} for PMA-stimulated kinase activity. Kinase activity in the presence of 1 mM $CaCl_2$, 40 μg/ml phosphatidylserine, and 4 μg/ml 1,2-diolein is also shown in Fig. 2. Under these conditions (which we routinely use for maximal activation of protein kinase C) the kinase activity is of similar magnitude to maximal PMA-stimulated kinase activity and, again, is greatly reduced after PMA pretreatment.

Pretreatment of pituitary cells for 6 hr with 1.5–1500 nM PMA (followed by 12 hr in plating medium) causes a concentration-dependent reduction in the proportion of cellular LH released in a subsequent 3-hr treatment with PMA (Fig. 3A). The specificity of this effect is indicated by the fact that no such reduction is seen in cells pretreated with phorbol esters which do not activate protein kinase C: 150 n*M* 4α-phorbol 12,13-didecanoate (4αPDD, Fig. 3) or 1 μ*M* 4αPMA (not shown). PMA pretreatment also causes a loss of extractable Ca^{2+}- and lipid-stimulated kinase activity (protein kinase C activity) over the same concentration range (Fig. 3B). Again, this effect appears to be specific since 4αPDD (Fig. 3) and 4αPMA (not shown) do not reduce extractable protein kinase C activ-

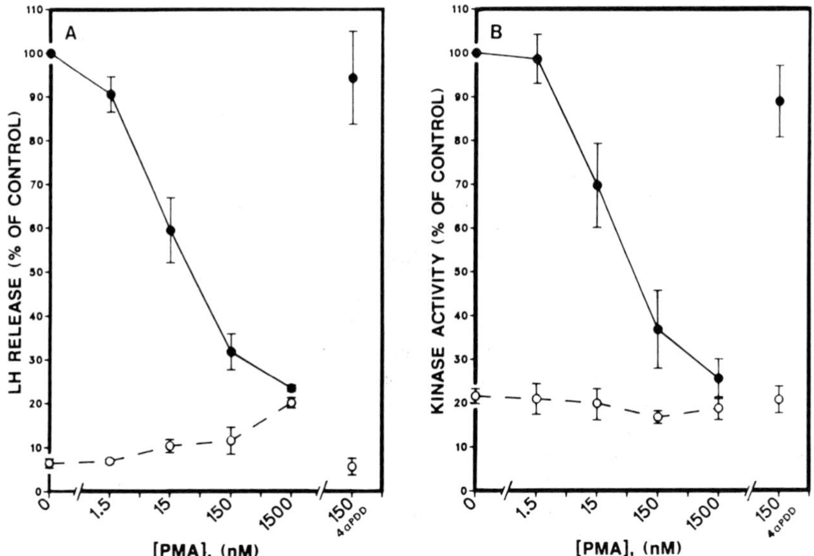

FIG. 3. Concentration dependence of the effect of PMA pretreatment on PMA-stimulated LH release and pituitary kinase activity. Pituitary cells were prepared, cultured for 2 days, and pretreated for 6 hr with the indicated concentrations of PMA or with 150 n*M* 4αPDD in BSA/M199 containing 0.5% DMSO. The cells were then washed and incubated for a further 12 hr in plating medium. (A) Cells were washed and incubated for 3 hr with 15 n*M* PMA (●, in BSA/M199 containing 0.5% DMSO) or with vehicle alone (○). LH release has been expressed as a percentage of that obtained in control (vehicle-pretreated) cells. The data shown are pooled from four separate experiments (means ± SE, *n* = 3–4), each experimental observation having been determined in triplicate. (B) After pretreatment as described above, cells were washed and collected for extraction, DEAE–cellulose purification, and assay of kinase activity. Kinase activity observed in the presence of 1 m*M* $CaCl_2$ (○) and 1 m*M* $CaCl_2$ plus lipid (●, 10 μg phosphatidylserine, 1 μg 1,2-diolein) is expressed as a percentage of that observed with $CaCl_2$ and lipid in control (vehicle-pretreated) cells. The values shown are pooled from six separate experiments (means ± SE, *n* = 3–6), each experimental observation having been determined in triplicate. Reprinted from Ref. 23 with permission from The American Society of Biological Chemists, Inc.

ity, and because PMA pretreatment does not measurably alter lipid-independent kinase activity.

Studies of the time course of the effect of PMA pretreatment reveal that Triton X-100-extractable protein kinase C activity is reduced to 50, 25, 35, and 75% of control 0, 12, 24, and 48 hr, respectively, after pretreatment for 6 hr with 150 n*M* PMA. For this reason we routinely use these protein kinase C-depleted cells 12-hr after the termination of the pretreatment.

Binding studies with phorbol esters provide a further means by which the effect of PMA pretreatment on cellular protein kinase C can be assessed. As shown in Fig. 4, [^{3}H]PDBu binds with high affinity to a specific and saturable binding site in pituitary cell cultures. This binding is reduced in a concentration-dependent manner by PMA pretreatment (see

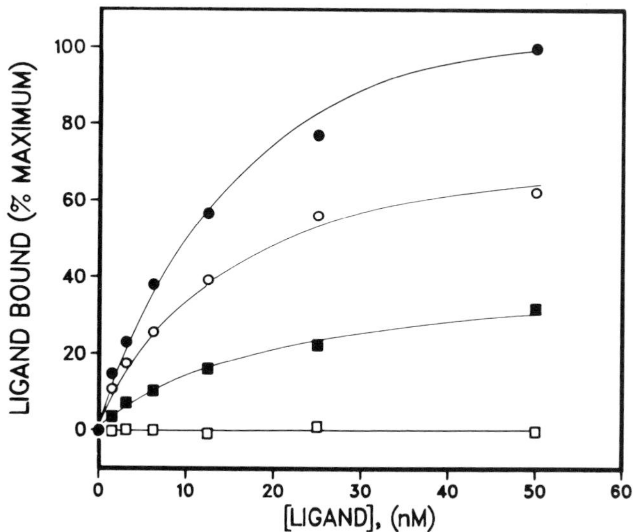

FIG. 4. Effect of PMA pretreatment on [^{3}H]PDBu binding. Pituitary cells were prepared, cultured for 2 days, and pretreated for 6 hr with BSA/M199 containing 0.5% DMSO and 0 (●), 10 (○), 100 (■), or 1000 n*M* PMA (□). The cells were then washed and returned to plating medium for 12 hr prior to assessment of [^{3}H]PDBu binding. Specific binding is shown expressed as a percentage of maximum binding. Maximum binding, defined as that observed in control cells in the presence of 50 n*M* [^{3}H]PDBu, was approximately 60 fmol/well or about 50,000–100,000 sites/cell. Nonspecific binding (in the presence of 1 μM PMA) was directly proportional to radioligand concentration and was 25% of total binding in the presence of 50 n*M* [^{3}H]PDBu. The values shown are means of those obtained in three to four separate experiments; standard errors, which have been omitted for clarity, are less than 6%.

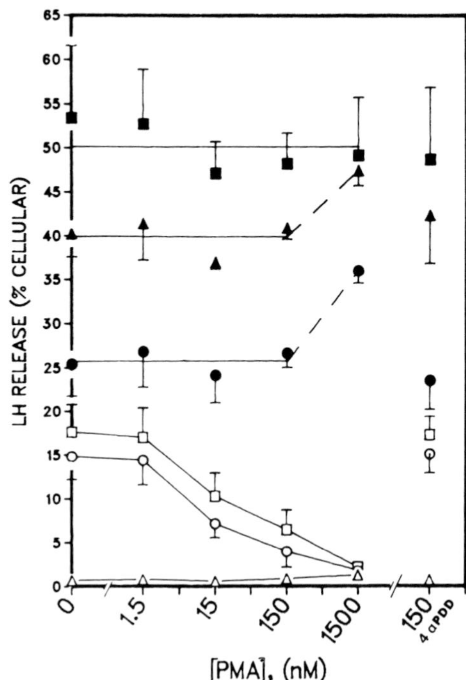

FIG. 5. Concentration dependence of the effect of PMA pretreatment on LH release in response to GnRH, A23187, and PMA. Pituitary cells were prepared, cultured for 2 days, and pretreated for 6 hr with the indicated concentrations of PMA or with 150 n*M* 4αPDD in BSA/M199 containing 0.5% DMSO. The cells were then washed and incubated for a further 12 hr in plating medium before being washed again and incubated for 3 hr with 15 n*M* PMA (○), 150 n*M* PMA (□), 10^{-9} *M* GnRH (●), 10^{-7} *M* GnRH (■), 40 μ*M* A23187 (▲), or with vehicle alone (△, BSA/M199 containing 0.5% DMSO). The data shown are pooled from five separate experiments (means ± SE, $n = 3$–5), each experimental observation having been determined in triplicate. Reprinted from Ref. 23 with permission from The American Society of Biological Chemists, Inc.

also Refs. 14 and 15 for studies in which clonal pituitary cell lines were used). Since protein kinase C is the major high-affinity cellular receptor for phorbol esters,[30–32] this loss of [^{3}H]PDBu binding is assumed to reflect the loss of cellular protein kinase C.

[30] J. E. Niedel, L. J. Kuhn, and G. R. Vanderbank, *Proc. Natl. Acad. Sci. U.S.A.* **80,** 36 (1983).

[31] C. L. Ashendel, J. M. Staller, and R. K. Boutwell, *Cancer Res.* **43,** 4333 (1983).

[32] U. Kikkawa, Y. Takai, Y. Tanaka, R. Miyake, and Y. Nishizuka, *J. Biol. Chem.* **258,** 11442 (1983).

Figure 5 shows the results of a series of experiments in which the effect of PMA pretreatment on the responsiveness of gonadotropes to protein kinase C-activating and Ca^{2+}-mobilizing secretagogues was compared. As shown (see also Fig. 3), pretreatment for 6 hr with 1.5–1500 n*M* PMA causes a concentration-dependent reduction in the proportion of cellular LH released in a subsequent 3-hr challenge with PMA. In contrast, no such reduction is seen in the proportion of cellular LH released on incubation with GnRH (a Ca^{2+}-mobilizing secretagogue) or A23187 (a Ca^{2+}-selective ionophore). The demonstration that the proportion of cellular LH released in response to GnRH is not reduced in cells which do not release measurable LH in response to PMA, do not possess a specific high-affinity phorbol ester binding site, and contain no measurable Triton X-100-extractable protein kinase C activity suggests that activation of this enzyme by GnRH is not required for sustained GnRH-stimulated LH release.

Comments

Methodology

For interpretation of studies in which LH release is quantified, it is important to note that release of LH also occurs during the period of pretreatment with PMA. This release of LH is associated with a reduction of as much as 35% in cellular LH content. Therefore, in order to normalize the data we routinely express LH release as a percentage of cellular LH (at the start of the 3-hr treatment with secretagogue) and consider the LH-releasing effect of secretagogues to be reduced only when the PMA pretreatment reduces the proportion of cellular LH released in a subsequent challenge.

In our preliminary studies pituitary cells were treated with secretagogue immediately after pretreatment with PMA. Under these conditions the basal rate of LH release (that seen in medium containing vehicle alone) was increased in PMA-pretreated cells. This effect, which is assumed to reflect incomplete washout of the lipophilic phorbol ester, prevents accurate assessment of maximum PMA-stimulable LH release at early time points after pretreatment. For this reason, and because the depleting effect of PMA pretreatment on Triton X-100-extractable protein kinase C activity is most pronounced approximately 12 hr after pretreatment, we typically return PMA-pretreated cells to plating medium for 12 hr before use. Pretreatment with a less lipophilic phorbol ester (such as PDBu) would presumably be more appropriate for experiments in which the time course for development of, or recovery from, effects of phorbol ester pretreatment is of central importance.

In order to estimate cellular protein kinase C activity we routinely purify the Triton X-100-extracted enzyme by DEAE–cellulose chromatography prior to assay. This procedure is necessary because of the presence of endogenous substances in crude homogenates of pituitary cells which alter protein kinase C activity (lipids, ions, etc.). We typically use Econocolumns from Bio-Rad (Richmond, CA; bed volume ~1 ml) and find it more convenient to pour fresh columns than to regenerate used columns. DEAE–BioGel A can be substituted for DEAE–cellulose. With the method described at least 90% of the elutable protein kinase C activity is found in the 20–100 m*M* NaCl fraction, and the activity in this fraction alone can be taken as a measure of cellular protein kinase C. For the experiments shown here, we also determined the kinase activity in a 100–400 m*M* NaCl fraction because it has been demonstrated that PMA-stimulated proteolysis of protein kinase C in platelets and neutrophils produces a Ca^{2+}- and phospholipid-independent kinase which elutes from DEAE–cellulose at a higher salt concentration than protein kinase C.[17,19] Our own experiments revealed no evidence for the PMA-stimulated production of a Ca^{2+}- and phospholipid-independent kinase and therefore suggested that any such kinase formed may itself have been metabolized prior to extraction.

For the purpose of the experiments shown here we define protein kinase C activity of DEAE–cellulose eluates as that seen in the presence of a maximally activating combination of Ca^{2+} and lipids (1 m*M* $CaCl_2$, 40 μg/ml phosphatidylserine, and 4 μg/ml 1,2-diolein) minus that in the presence of $CaCl_2$ alone. Alternatively, the enzyme can be activated by PMA in the presence of low concentrations of $CaCl_2$ (10 μM) and phosphatidylserine (4 μg/ml) which are sufficient to facilitate the action of PMA, but do not alone appreciably activate the enzyme.

Caution should be used when performing the procedures described herein because the investigator is exposed both to tumor-promoting phorbol esters and to a relatively high-energy β particle-emitting radionuclide (^{32}P). Precautions appropriate for using phorbol esters include handling the compounds in solution only and inactivation before disposal (e.g., by treatment with NaOH or household bleach). Further details on appropriate handling of these compounds are provided by the suppliers. Although the amount of ^{32}P used in the experiments described is typically less than 0.25 mCi, plexiglass shielding of approximately 1 cm thickness should be used, and personal exposure should be monitored with wrist and finger dose meters.

An alternative approach to probing the role of protein kinase C in hormone action is the use of protein kinase C inhibitors, several of which are commercially available. Although such compounds have been found to be of great utility by some groups,[33–38] other systems appear to be less sensitive to the effects of these inhibitors. We have been unable, for example, to inhibit phorbol ester-stimulated LH release from pituitary gonadotropes with compounds such as 1-(5-isoquinolinesulfonyl)-2-methylpiperazine dihydrochloride (H7; Seikagaku America, Inc., St. Petersburg, FL), sphingosine (Sigma), and phorbol diacetate (LC Services). Moreover, although several lipid-interacting compounds, including Ca^{2+} antagonists and calmodulin blockers, are known to inhibit protein kinase C at high concentrations,[39] the inhibitory effect of such compounds on GnRH-stimulated LH release appears independent of protein kinase C inhibition because they do not reduce phorbol ester- or diacylglycerol-stimulated LH release at the concentrations used in culture.[40]

Interpretation of Results

Although the mechanism underlying protein kinase C depletion is not well understood, phorbol ester-stimulated association of protein kinase C with cell membranes and consequent proteolysis of the membrane-associated enzyme provide a likely explanation for this phenomenon. Evidence in support of this possibility includes the demonstration that (1) PMA causes the association of protein kinase C with cell membranes,[41,42] (2) membrane-associated protein kinase C is susceptible to proteolysis,[16,43,44]

[33] R. J. Sturm, B. M. Smith, R. W. Lane, D. L. Laskin, L. S. Harris, and R. A. Carchman, *Cancer Res.* **43,** 4552 (1983).

[34] S. Kawamoto and H. Hidaka, *Biochem. Biophys. Res. Commun.* **125,** 258 (1984).

[35] M. Inagaki, S. Kawamoto, and H. Hidaka, *J. Biol. Chem.* **259,** 14321 (1984).

[36] C. Gerard, L. McPhail, A. Marfat, N. P. Stimler-Gerard, D. A. Bass, and C. E. McCall, *J. Clin. Invest.* **77,** 61 (1986).

[37] Y. A. Hannun, C. R. Loomis, A. H. Merrill, and R. M. Bell, *J. Biol. Chem.* **261,** 12604 (1986).

[38] Y. A. Hannun and R. M. Bell, *Science* **235,** 670 (1987).

[39] T. Mori, Y. Takai, R. Minakuchi, B. Yu, and Y. Nishizuka, *J. Biol. Chem.* **255,** 8378 (1980).

[40] C. E. Harris, D. Staley, and P. M. Conn, *Mol. Pharmacol.* **27,** 532 (1985).

[41] A. S. Kraft and W. B. Anderson, *Nature (London)* **301,** 621 (1983).

[42] M. W. Wooten and R. W. Wrenn, *FEBS Lett.* **171,** 183 (1984).

[43] A. Kishimoto, N. Kajikawa, M. Shiota, and Y. Nishizuka, *J. Biol. Chem.* **258,** 1156 (1983).

[44] A. Rodriquez-Pena and E. Rozengurt, *Biochem. Biophys. Res. Commun.* **120,** 1053 (1984).

(3) the PMA-stimulated loss of cellular protein kinase C is inhibited by selective inhibitors of Ca^{2+}-dependent protease activity,[17,19,20] and (4) in several systems the loss of protein kinase C (~80 kDa) is associated with the production of one fragment (~45–60 kDa) which has Ca^{2+}- and phospholipid-independent kinase activity[16,18,22] and another (~35 kDa) which specifically binds [^{3}H]PDBu.[45] In light of such observations we consider degradation of cellular protein kinase C to be the most likely explanation for the PMA-stimulated loss of protein kinase C activity in DEAE–cellulose-purified Triton X-100 extracts of pituitary cells. Accordingly we use the term "protein kinase C depletion" to describe this phenomenon.

An alternative explanation for the effect of PMA pretreatment is that it causes the conversion of protein kinase C to a "cryptic form" which is not measurable in DEAE–cellulose-purified Triton X-100 extracts, is not stimulable in intact cells, does not bind [^{3}H]PDBu, and is not recognized by specific antibodies to protein kinase C.[16,18,22] It is important to note that this caveat does not undermine the use of PMA-pretreated cells for studies of the requirement of protein kinase C for hormone action. Indeed, the availability of cells that have been made selectively refractory to maximally effective concentrations of exogenous phorbol esters provides a useful model for probing the requirement of activation of protein kinase C-mediated signal transduction pathways even in the absence of an identified mechanism for the production of such refractory cells.

When using cells depleted of protein kinase C by phorbol ester pretreatment to test whether activation of this enzyme is required for hormone action, it is important to recognize the possibility that this requirement could be met by sustained responses to protein kinase C activation initiated during the pretreatment. This possibility does not present a problem when testing the specific hypothesis that acute hormonal stimulation of protein kinase C is required to mediate acute hormone-stimulated responses. Furthermore, it is unlikely that such sustained responses to protein kinase C activation are pertinent for mediation of the effects of stimuli which are intermittent or pulsatile under physiological conditions (such as GnRH-stimulated LH release). However, in cells in which protein kinase C plays a physiological role in the production of sustained hormonal effects (e.g., by regulation of gene expression or by "priming" cellular responsiveness to stimulation), the *required* activation of protein kinase C could conceivably be achieved during the pretreatment with phorbol ester.

[45] M. Hoshijima, A. Kikuchi, T. Tamimoto, K. Kaibuchi, and Y. Takai, *Cancer Res.* **46,** 3000 (1986).

Concluding Remarks

In this chapter we describe a method for preparation of pituitary cell cultures which do not contain measurable protein kinase C activity, or specific phorbol ester binding sites, and do not respond to protein kinase C-activating phorbol esters. We describe these cells as "protein kinase C depleted" because an actual loss of this enzyme provides the most likely explanation for these effects of PMA pretreatment and for similar effects produced in several other cell types. Although our observations with this model do not exclude a role for protein kinase C in GnRH action, they do suggest that activation of protein kinase C by GnRH is not required for GnRH-stimulated LH release (Fig. 5 and Refs. 23 and 46) or for GnRH-induced homologous desensitization or receptor down-regulation.[47]

The demonstration that LH release from pituitary gonadotropes is provoked by exogenous protein kinase C activators, but that GnRH-stimulated LH release is not blocked in cells which are apparently protein kinase C-depleted, underlines the fact that the effects produced by pharmacological activation of protein kinase C need not necessarily reflect the effects produced by physiological activation of the enzyme. Studies in the presence of protein kinase C inhibitors, or in protein kinase C-depleted cells, can therefore constitute an important extension of studies with exogenous protein kinase C activators. In the absence of protein kinase C inhibitors, which produce a specific blockade of this enzyme in intact gonadotropes, protein kinase C-depleted pituitary cells provide a useful model for probing whether hormonal activation of this enzyme is required for GnRH action in these cells. Similar methods have been used in several other cell types[14–22] and would presumably be equally useful for studying stimulus–response coupling in the nongonadotrope cells (e.g., somatotropes, lactotropes, and corticotropes) of the pituitary cell cultures used here.

Acknowledgments

This work was supported by National Institutes of Health Grant HD 19899 and the Mellon Foundation.

[46] J. R. Hansen, C. A. McArdle, and P. M. Conn, *Mol. Endocrinol.* **1,** 808 (1987).

[47] C. A. McArdle, W. C. Gorospe, W. R. Huckle, and P. M. Conn, *Mol. Endocrinol.* **1,** 421 (1987).

[19] Purification of Neuropeptides: Cholecystokinin and Vasoactive Intestinal Peptide

By JOHN ENG and ROSALYN S. YALOW

Introduction

Cholecystokinin (CCK) and vasoactive intestinal peptide (VIP) are peptides common to the brain and gut. VIP is a neuropeptide in both regions and has an identical molecular form in brain and gut. In contrast, CCK is a neuropeptide in brain but is an endocrine peptide in gut. Unlike VIP, CCK has a distribution of molecular forms that differs between the two regions. In brain, CCK octapeptide (CCK8) predominates, while its intact precursor, CCK58, is only 10% as abundant as CCK8 on a molar basis. In gut, CCK39 and CCK33 are major CCK components along with CCK8 and CCK58. Brain concentrations of CCK8 average about 250 pmol/g in most species that have been studied. Concentrations of gut VIP are comparable to those of brain CCK8, but brain VIP concentrations are 5- to 10-fold lower. Because of these differences in concentration, the tissue of choice for purification of CCK8 is brain, and for VIP it is gut.

With the introduction of newer HPLC separation technologies and automated amino acid sequencing instrumentation, it is now possible to obtain complete sequence analyses of peptides from nanomole or even subnanomole quantities. This new capability for microsequencing permits the purification of bioactive peptides to be scaled down to levels that might be termed micropurification. An important objective in micropurification is to minimize the quantity of starting tissue. With reduced starting material, there is a nearly direct introduction of tissue extract into high-resolution analytical HPLC columns. The high-resolution separations provided by these columns permit the isolation of pure peptide in a minimum number of steps. In the purification protocols to be described, CCK8 and VIP can be purified and sequenced in a period of 1 to 2 weeks.

The purification process for CCK8 and VIP follows a series of well-defined steps. The target peptide is extracted from an appropriate tissue source, and the extract is subjected to a purification cycle which consists of a sequence of concentration, separation, and assay steps. One or more purification cycles are required to isolate pure peptide. The complexity of the purification strategy is dictated by a number of factors: the amount of tissue available, the tissue concentration of peptide, and the chemical characteristics of the peptide.

Brain CCK8

As little as 5 g of brain tissue is sufficient for purification of CCK8. Brain tissue is extracted with 5 volumes of methanol in a glass tissue grinder with Teflon pestle. CCK8 is quantitatively extracted by methanol. The suspension is filtered through Whatman No. 1 filter paper in a Büchner funnel under reduced pressure. The methanol extract filtrate is placed at −70° for 60 min. The white precipitate that forms is removed by centrifugation for 15 minutes at −20° and 6000 *g*. CCK8 in the methanol extract is concentrated by passing the extract through a strong anion-exchange column (QMA Sep-Pak, Waters Associates, Milford, MA) at 1 ml/min. Prior to use, the QMA Sep-Pak is prepared by sequential washes with 5 ml 50 m*M* Tris–HCl, pH 7 (Tris), containing 1 *M* NaCl followed by 10 ml Tris. The capacity of the cartridge is approximately 50 column volumes, i.e., as much as 50 ml of methanol extract can be passed through a QMA cartridge with quantitative adsorption of CCK8. The cartridge is washed with 10 ml Tris and eluted with 3 ml Tris–1 *M* NaCl. An aliquot is removed from the eluate for quantitation by radioimmunoassay (RIA), and the remainder is injected through a 10-ml injection loop in the HPLC system onto a Nova C_{18} radial-pak column (5 mm i.d., Waters Associates) which is preequilibrated in 0.1% trifluoroacetic acid (TFA). The column is eluted with a 60-min linear gradient from 0.1% TFA to 0.1% TFA–40% acetonitrile (ACN) at a flow rate of 1 ml/min with UV detector monitoring at 214 nm. One-minute fractions are collected. Purified CCK8 elutes as a prominent UV peak in the latter part of the elution (see Fig. 1). A portion of the peak is applied to a gas-phase sequencer (Applied Biosystems, Foster City, CA) for determination of amino acid sequence.

The purification of CCK8 from brain is accomplished in 1 day. The rapidity of this procedure is made possible by the use of methanol as the extraction solvent. Methanol yields a "clean" extract, i.e., CCK8 is extracted selectively relative to potential contaminants present in brain tissue. It is important to use freshly obtained and frozen brain tissue in order to minimize potential interference from protein degradation products in the HPLC separation. When tissue quality is not known, CCK8 purity in the QMA eluate can be assessed by injecting a portion of the eluate in a trial HPLC run. If a pure CCK UV peak is not present, then a multicycle purification procedure is employed.[1]

All reagents and solvents are HPLC or sequencer grade, including the methanol used for extraction. Tris solutions are purified by passage through a C_{18} Sep-Pak cartridge before use. A blank HPLC injection and

[1] Z.-W. Fan, J. Eng, M. Miedel, J. D. Hulmes, Y.-C. E. Pan, and R. S. Yalow, *Brain Res. Bull.* **18,** 757 (1987).

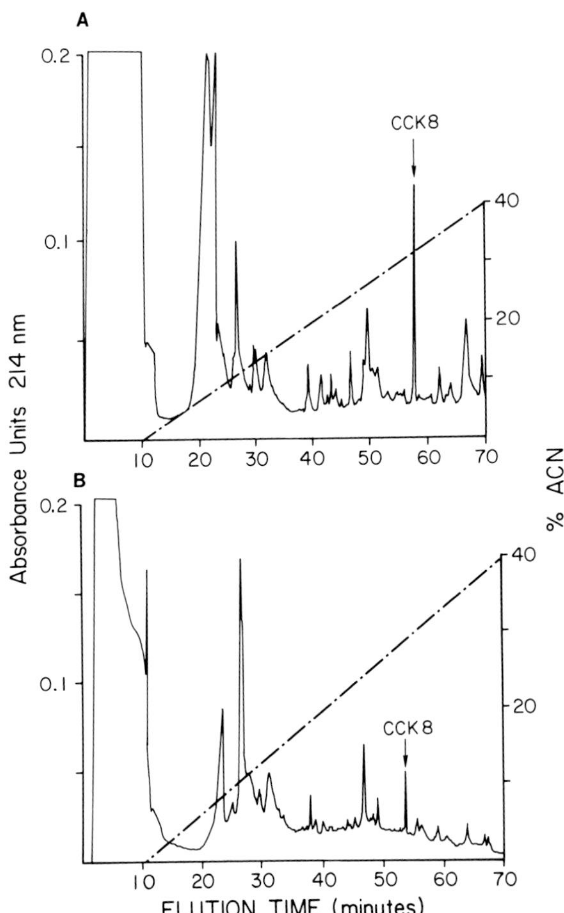

FIG. 1. Isolation of CCK8. A fresh, frozen brain with cerebellum and brain stem removed was obtained from a rabbit (brain weight, 7.4 g) or a guinea pig (3.1 g) and extracted with 5 volumes of methanol. The methanol extract was concentrated on a QMA Sep-Pak and the eluate injected onto a Nova C_{18} column. The column was eluted with a linear gradient from 0.1% TFA to 0.1% TFA–40% ACN between 10 and 70 min. (A) Rabbit CCK8: the UV peak eluting at 57 min has the sequence DYMGWMDF. From a total of 750 pmol CCK8 in the peak fraction, 150 pmol was used for sequence analysis. (B) Guinea pig CCK8: the UV peak eluting at 55 min has the sequence DYVGWMDF. Half of the 200 pmol of CCK8 in the peak fraction was used for sequence analysis.

run is routinely performed before sample runs to check for the presence of artifact peaks arising from HPLC solvents or columns.

Gut VIP

VIP is more difficult to purify than CCK8 because a comparably selective extraction solvent is not yet available for this peptide. Furthermore, VIP is a basic peptide, and basic peptides in general are more difficult to purify than acidic ones because they give broader peaks with lower resolution on most separations. Thus, multiple purification cycles are required to isolate pure VIP, and monitoring of its purification with RIA is necessary. Since VIP is present in higher concentrations in gut than in brain, intestinal tissue is chosen as starting material for VIP purification.

VIP concentrations in gut tissue are typically 200–500 pmol/g. Since the purification protocol gives a final yield of pure peptide of approximately 20% and since 1–2 nmol pure VIP is needed for full sequence analysis, at least 50 g of starting tissue is required. One of two extraction solvents may be used, acid–alcohol (ethanol containing 1% TFA and 0.1% 2-mercaptoethanol)[2] or aqueous acid (1% TFA containing 0.1% 2-mercaptoethanol).[3] A methanol cake prepared from gut tissue as previously described is extracted in a Waring blendor with either 5 volumes of acid–alcohol or 10 volumes of aqueous acid based on original tissue weight. The extract suspension is centrifuged at 6000 *g* for 30 min at 4°. The supernatant is removed, stored at −20° overnight, and centrifuged the next morning to remove additional precipitate. The aqueous extract is thawed before centrifugation. Next, SE53 cellulose (Whatman) is added at a concentration of 5 g/100 ml extract to acid–alcohol extract or 1 g/100 ml extract to aqueous acid extract. The cellulose suspension is stirred overnight at 4°. VIP is concentrated by adsorption onto this cation-exchange cellulose.

The SE53 cellulose is collected by centrifugation the next day and the supernatant discarded. The cellulose is washed once with distilled H_2O (5 ml/g cellulose) and twice with 50 m*M* sodium acetate, pH 5 (also 5 ml/g). VIP is eluted from the SE53 cellulose with a solution of 50 m*M* sodium acetate containing 1 *M* NaCl (5 ml/g cellulose). The elution is performed twice and the eluates pooled. VIP in the eluate is purified with a series of three or four HPLC steps.

[2] S.-C. Wang, B.-H. Du, J. Eng, M. Chang, J. D. Hulmes, Y.-C. E. Pan, and R. S. Yalow, *Life Sci.* **37,** 979 (1985).

[3] J. Eng, B.-H. Du, J.-P. Raufman, and R. S. Yalow, *Peptides* **7** (Suppl. 1), 17 (1986).

The first HPLC separation is performed on a MB C_{18} radial-pak column (8 mm i.d., Waters Associates) equilibrated in 0.13% heptafluorobutyric acid (HFBA) with UV monitoring at 280 nm. VIP in the pooled eluate is concentrated by pumping the pool directly through the HPLC column at 1 ml/min. The column is washed with 20 ml 0.13% HFBA and then eluted with a linear gradient from 0 to 60% ACN in 0.13% HFBA at a flow rate of 2 ml/min. One-minute fractions are collected and assayed for VIP by RIA.

The peak fractions of VIP as determined by RIA are pooled and further purified by separation on a Mono S HR5/5 strong cation-exchange column (Pharmacia, Piscataway, NJ). The pooled VIP fractions are injected onto a Mono S column equilibrated in 0.1% TFA–20% ACN, and the column is eluted with a gradient from 0 to 0.5 *M* NaCl in 0.1% TFA–20% ACN at a flow rate of 1 ml/min. The column effluent is monitored with UV absorbance at either 280 or 214 nm depending on the level of impurities still present. One-minute fractions are collected, and the fractions are assayed for VIP by RIA.

The peak fractions of VIP may be pooled or individually purified by a third step using a Nova C_{18} column equilibrated in 0.1% TFA. The column is eluted with a linear gradient from 25 to 35% ACN in 0.1% TFA at a flow rate of 1 ml/min. One-minute fractions are collected. With UV monitoring at 214 nm, a pure VIP peak is usually observed[2,3] from which a portion is taken for amino acid sequencing. If the VIP is not entirely pure at this point (see Fig. 2) then individual fractions are diluted with 5 volumes of 0.1% TFA, reinjected onto the column, and purified with an identical elution gradient.

It is apparent that the purification protocol for VIP is more complex than that for CCK8. The protocol requires approximately 1 week to complete: 1 day to prepare the tissue extract, 1 day to concentrate the extract on cation-exchange cellulose, and 1 day for each of the three to four purification cycles.

Several different modes of separation are illustrated in the purification of VIP. A separation based on charge difference is made on the Mono S column whereas separations based on differential hydrophobicity occur on the C_{18} columns. It has been reported by others[4] and it is our own experience that in the presence of heptafluorobutyric acid peptide mixtures exhibit a significantly different HPLC elution profile from that produced in the presence of trifluoroacetic acid. Thus, it is useful to perform differential separations on a C_{18} column using these two counterions,

[4] H. P. J. Bennett, C. A. Browne, and S. Solomon, *Biochemistry* **20**, 4530 (1981).

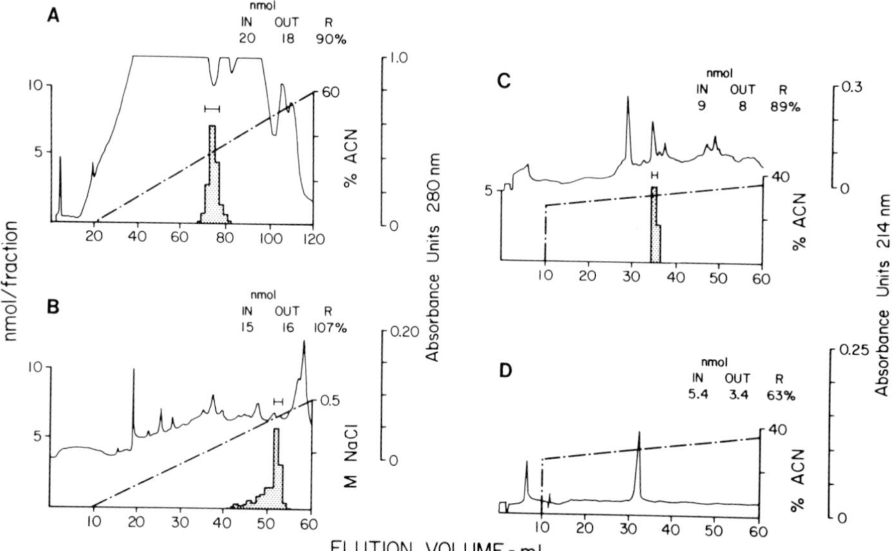

FIG. 2. Isolation of VIP. The entire colon was removed from a freshly sacrificed rabbit, opened lengthwise, washed free of fecal content under a stream of cold tap water, and stored frozen at $-70°$ until use. The tissue (100 g) was extracted with methanol and the methanol cake extracted with 1% TFA–0.1% mercaptoethanol to yield 27 nmol of VIP in 700 ml of extract. VIP in the acid extract was concentrated by adsorption onto 10 g SE53 cellulose with overnight stirring. The cation-exchange cellulose was washed and eluted with sodium acetate–1 *M* NaCl to yield 20 nmol VIP which was purified by a series of four HPLC separations with the following columns and elution conditions. Bars indicate the fraction or pool of fractions that was applied to the next separation step. (A) MB C_{18} column; 0.13% HFBA/0–60% ACN. (B) Mono S HR 5/5 column; 0.1% TFA/20% ACN/0–0.5 *M* NaCl. (C) Nova C_{18} column; 0.1% TFA/25–35% ACN. (D) Nova C_{18} column; 0.1% TFA/25–35% ACN. Five nanomoles pure VIP was obtained. The final purification step for the other VIP fraction in C is not shown. From the pure peak of VIP in D, 10% was used for amino acid sequencing. Rabbit VIP is identical to pig VIP in amino acid sequence.

respectively. We find that the presence of 20% acetonitrile throughout the Mono S separation produces higher recoveries of basic peptides and helps to increase peak sharpness and resolution.

We use dedicated HPLC columns, keeping columns used in the initial purification cycle separate from those used in the final purification cycle. We also avoid drying samples between runs or after the peptide is purified. Losses occur when nanomole amounts of peptide are dried and redissolved. Elution fractions are stored at 4° without problem since all of the HPLC runs contain acetonitrile, which is a good preservative.

Discussion

A key element in micropurification is the assay system used to monitor purification. Although it is only one of many types of assay systems that can be used, RIA has a number of ideal characteristics. First, it is sensitive. This minimizes cumulative losses from assay sampling, which, if large enough, could lead to insufficient yield of peptide. RIA typically requires only femtomole quantities of peptide for detection, which accounts for less than 1% of sample. Second, RIA is rapid. Multiple samples can be assayed overnight or in hours if necessary. This facilitates experimentation with strategies for extraction and separation prior to formulation of a final purification protocol.

Another important factor in micropurification is selectivity. Selectivity is the ability to separate desired from undesired material by physicochemical means. Successful micropurification depends on maximizing selectivity throughout the purification process. Beginning with the starting material, tissue is selected to contain high concentrations of target peptide. The peptide concentration is increased by dissecting free and eliminating those anatomical parts with low peptide concentrations. Next, a selective extraction solvent is used. The ideal extractant is one which extracts the target peptide with few other contaminants. A "clean" extract increases the ease of purification. This is illustrated by a comparison of the purification protocols for brain CCK8 and gut VIP. Maximum selectivity in the purification cycles is achieved by using different modes of separation when needed and by maintaining a high degree of resolution in each separation. High-resolution separations make it possible to have fewer purification cycles, shorter purification times, and higher yields of pure peptide. As improvements are made in these areas and incorporated into the purification scheme, the purification process will become simpler and the characterization of biologically significant peptides made quicker and easier.

Acknowledgment

This work was supported by the Medical Research Program of the Veterans Administration.

[20] Purification and Characterization of Insulin-Like Growth Factor-II Receptors

By CAROLYN D. SCOTT and ROBERT C. BAXTER

Characteristics of the IGF-II Receptor

Insulin-like growth factor-II (IGF-II), a small peptide (MW 7500) with structural similarity and biological activity in common with insulin, has so far been shown to exert most of its known mitogenic and anabolic effects through cross-reactivity at the insulin and IGF-I receptors.[1–3] However, a high affinity receptor with high specificity for IGF-II has been described in a number of cell types and tissues.[4,5] Called the type II IGF receptor, this receptor can be distinguished on a structural basis from the type I IGF receptor (IGF-I receptor) and the insulin receptor. Type I IGF and insulin receptors have a native tetrameric structure of MW 450,000 which breaks down to subunits of 90,000 and 135,000 on reduction, while the native type II receptor of MW 220,000–240,000 remains as a single species following reduction. IGF-II receptors show low cross-reactivity for IGF-I and no cross-reactivity for insulin. This contrasts with the type I or IGF-I receptor which has about 10% cross-reactivity for IGF-II and less than 1% cross-reactivity for insulin. The role of the IGF-II receptor is currently unknown although it is predicted to have a role in fetal growth and development as IGF-II has been found in high levels in the fetal circulation.[6] It is probably also important in brain and neuron function as high levels of both IGF-II[7] and its mRNA[8] have also been reported to be present in certain areas of the brain where the peptide may have an autocrine or paracrine action. IGF-II has been found to be more potent

[1] C. Mottola and M. P. Czech, *J. Biol. Chem.* **259,** 12705 (1984).

[2] R. W. Furlanetto, J. N. DiCarlo, and C. Wisehart, *J. Clin. Endocrinol. Metab.* **64,** 1142 (1987).

[3] N. L. Krett, J. H. Heaton, and T. D. Gelehrter, *Endocrinology (Baltimore)* **120,** 401 (1987).

[4] J. Massague and M. P. Czech, *J. Biol. Chem.* **257,** 5038 (1982).

[5] M. M. Rechler and S. P. Nissley, *Annu. Rev. Phsiol.* **47,** 425 (1985).

[6] A. C. Moses, S. P. Nissley, P. A. Short, M. M. Rechler, R. W. White, A. B. Knight, and O. Z. Higa, *Proc. Natl. Acad. Sci. U.S.A.* **77,** 3649 (1980).

[7] G. K. Haselbacher, M. E. Schwab, A. Pasi, and R. E. Humbel, *Proc. Natl. Acad. Sci. U.S.A.* **82,** 2153 (1985).

[8] A. L. Brown, D. E. Graham, S. P. Nissley, D. J. Hill, A. J. Strain, and M. M. Rechler, *J. Biol. Chem.* **261,** 13144 (1986).

than insulin in stimulating the growth of astroblasts[9] and in enhancing neurite outgrowth from cultured sensory neurons.[10] Examination of the regulation and distribution of IGF-II receptors should provide further information on the action of IGF-II, and several publications have recently appeared describing the purification of IGF-II receptors from a variety of sources[11–14] and the raising of antisera to the receptor.[13–15]

Purification Strategy

Successful purifications of IGF-II receptor to date have all employed IGF-II–agarose affinity chromatography. This technique has the advantage of being highly specific for the IGF-II receptor, resulting in high yields of essentially pure receptor after a single step. Most investigators have used preparations of rat IGF-II (otherwise known as multiplication stimulating activity or MSA) coupled to agarose to isolate small amounts of receptor (less than 50 μg) from small batches of membrane isolated from rat placenta,[11] chondrosarcoma cells,[12] or an IGF-II producing rat cell line.[13] However, as described in this chapter, the procedure has also been performed successfully using a column of immobilized human IGF-II, and it can be scaled up by using a column of higher capacity, resulting in the purification of milligram amounts of receptor from several grams of membrane.[14] While it should be possible to purify IGF-II receptor from any tissue or cells containing reasonable quantities, we have found rat liver to be an excellent starting material, being very rich in IGF-II receptors with little or no IGF-I receptors, thus minimizing the risk of copurification. This could be a serious problem in some tissues such as human placenta, where there is a large quantity of type I IGF receptor with significant IGF-II cross-reactivity.[16] However, antibodies raised to rat liver IGF-II receptor cross-react fully with receptors in all

[9] R. Lim, J. F. Miller, D. J. Hicklin, A. C. Holm, and B. H. Ginsberg, *Exp. Cell. Res.* **159,** 335 (1985).

[10] E. Recio-Pinto, M. M. Rechler, and D. N. Ishii, *J. Neurosci.* **6,** 1211 (1985).

[11] C. L. Oppenheimer and M. P. Czech, *J. Biol. Chem.* **258,** 8539 (1983).

[12] G. P. August, S. P. Nissley, M. Kasuga, L. Lee, L. Greenstein, and M. M. Rechler, *J. Biol. Chem.* **258,** 9033 (1983).

[13] R. G. Rosenfeld, D. Hodges, H. Pham, P. D. K. Lee, and D. R. Powell, *Biochem. Biophys. Res. Commun.* **138,** 304 (1986).

[14] C. D. Scott and R. C. Baxter, *Endocrinology* (*Baltimore*) **120,** 1 (1987).

[15] Y. Oka, C. Mottola, C. L. Oppenheimer, and M. P. Czech, *Proc. Natl. Acad. Sci. U.S.A.* **81,** 4028 (1984).

[16] S. J. Casella, V. K. Han, A. J. D'Ercole, M. E. Svoboda, and J. J. Van Wyk, *J. Biol. Chem.* **261,** 9268 (1986).

other rat tissues examined and therefore could be used in studies on IGF-II receptors in other rat tissues or cells.

Membrane Preparation

In order to prepare large quantities of purified receptor several batches of microsomal membranes can be prepared and pooled prior to purification. Rat liver membranes are prepared as described previously[17] from 50 g tissue (freshly excised or frozen in liquid nitrogen and stored at −80°) and three to four batches pooled for receptor purification, resulting in 3–4 g protein being used as starting material. Protease inhibitors are included in all buffers to prevent proteolytic degradation of the IGF-II receptor during membrane preparation.

Reagents

0.25 *M* Sucrose
25 m*M* Sodium 4-(2-hydroxyethyl)-1-piperazineethane sulfonate (Na-HEPES), pH 7.4

Both reagents are supplemented with protease inhibitors:

0.5 m*M* Bacitracin
50 mg/liter Soybean trypsin inhibitor
0.1 m*M* L-1-Tosyl-2-phenylethyl chloromethyl ketone (TPCK)

Procedure

1. Mince 50 g fresh or thawed tissue in a beaker containing a small quantity of ice-cold 0.25 *M* sucrose.
2. Homogenize with 4 volumes (200 ml) of 0.25 *M* sucrose using a motor-driven Teflon–glass homogenizer, followed by 30 sec with the large probe of an Ultra-Turrax (Junke and Kunkel, Staufen, FRG) or Polytron (Kinematica, Lucerne, Switzerland) tissue grinder (to disperse any large sheets of plasma membranes).
3. Centrifuge at 12,000 *g* for 20 min at 4°.
4. Pour off supernatant and recentrifuge at 12,000 *g* for 20 min at 4° to ensure complete removal of pellet.
5. Ultracentrifuge supernatant at 100,000 *g* for 60 min at 4°.
6. Aspirate supernatant and clean any residual fat from the sides of the tubes with a cotton swab.
7. Resuspend pellet by homogenization in same volume of 0.25 *M* sucrose (200 ml) and ultracentrifuge at 100,000 *g* for 45 min at 4°.

[17] R. C. Baxter and J. R. Turtle, *Biochem. Biophys. Res. Commun.* **84,** 350 (1978).

8. If supernatant is not yet clear, repeat washing and ultracentrifugation as in Step 7.
9. Aspirate supernatant and resuspend pellet by homogenization in 25 m*M* NaHEPES, pH 7.4.
10. Freeze in liquid nitrogen and store at −80°.
11. When required, membranes should be thawed rapidly and solubilized by adding Triton X-100 to a final concentration of 1% and mixing end over end for 30 min at 22°. Insoluble material is removed by centrifugation at 12,000 *g* for 20 min.

Assay of IGF-II Receptor Activity

IGF-II receptor activity is determined by radioligand binding.[14] The assay conditions described below are optimal for estimating rat liver IGF-II receptor, but as different tissues have different optima for pH, calcium, protein concentration, and time of incubation[18] these should be determined for the tissue under study. IGF-II tracer is incubated with solubilized membrane or purified receptor for 2 hr at 22° and bound receptor precipitated with polyethylene glycol. Nonspecific binding is calculated by the addition of sufficient excess IGF-II to cause total displacement of specifically bound tracer. In our studies 10 ng unlabeled human IGF-II has been sufficient to fully displace tracer from the rat liver receptor. This amount needs to be determined for every new tissue examined, however, and may be much higher (up to 1 μg/tube) in some systems. Binding of IGF-II tracer is proportional to membrane or protein concentration only over a limited range, and little additional binding is achieved beyond this. The protein concentration range over which proportionality is obtained should be determined for each membrane or receptor preparation studied. Using the purification protocol described here, starting material will require up to 10-fold dilution and purified receptor up to 100-fold dilution for the assay to be quantitative.

Reagents

25 m*M* NaHEPES, pH 7.4, 10 m*M* $CaCl_2$, 0.1% Triton X-100, 0.2% bovine serum albumin (BSA)

^{125}I-Labeled IGF-II (10,000 cpm/assay tube; specific activity 100–200 Ci/g)

IGF-II (unlabeled)

Bovine γ-globulin, 40 mg/ml in 0.15 *M* NaCl

Polyethylene glycol (PEG), 180 g/liter in 25 m*M* NaHEPES, pH 7.4, 0.15 *M* NaCl

[18] J. E. Taylor, C. D. Scott, and R. C. Baxter, *J. Endocrinol.* **115,** 35 (1987).

Procedure

1. Add receptor and buffer in a total volume of 390 μl to four replicate tubes per sample.
2. To two of the tubes add unlabeled IGF-II in 10 μl buffer (these are used to determine nonspecific binding). To the remaining two tubes add 10 μl buffer alone (these are used to determine total binding).
3. Add ^{125}I-labeled IGF-II in 100 μl buffer. Mix tubes well and incubate for 2 hr at 22°.
4. Add 20 μl 40 mg/ml bovine γ-globulin as carrier protein to each tube. Mix well. Add 1 ml cold 180 g/liter PEG and centrifuge immediately at 4000 g for 20 min. Drain or apsirate supernatant and count receptor-bound ^{125}I-labeled IGF-II in the pellet in a gamma counter.
5. Specific binding is calculated by subtracting the nonspecific binding from the total binding.

Preparation of IGF-II Affinity Column

General Considerations

Affinity chromatography relies on the immobilization of a particular ligand onto a cross-linked agarose or polyacrylamide gel which can then specifically bind the protein of interest and remove it from a mixture of other proteins. Various commercial gels are available which have been treated to produce charged groups which will couple rapidly to ligands by following the manufacturer's instructions. IGF-II has been successfully coupled to cyanogen bromide-activated agarose (CNBr-activated Sepharose 4B, Pharmacia, Uppsala, Sweden)[11–13] and to agarose *N*-hydroxysuccinimide ester (Affi-Gel 15, Bio-Rad, Richmond, CA),[14] both of which couple spontaneously with primary amino groups. Although both of these gels are capable of coupling 20–40 mg protein/ml gel it is advantageous to prepare a column with a lower concentration of IGF-II to increase the size of the column and thus the exposure time of the receptor to its ligand. Routinely, 1–5 mg IGF-II coupled to 5–10 g gel produces a high capacity affinity column. Excess active groups on the gel are blocked (after ligand coupling) by incubation with a primary amine such as ethanolamine or Tris buffer. As IGF-II is not yet widely available, it can either be isolated from human plasma[19] or from medium conditioned by IGF-II-producing cells (e.g., MSA from BRL cells[20]). Care should be taken to

[19] R. C. Baxter and J. S. M. De Mellow, *Clin. Endocrinol.* **24,** 267 (1986).

[20] A. C. Moses, S. P. Nissley, P. A. Short, M. M. Rechler, and J. M. Podskalny, *Eur. J. Biochem.* **103,** 387 (1980).

exclude primary amines such as Tris from the ligand mixture and coupling buffer.

Procedure

1. Wash gel of choice according to manufacturer's instructions.
2. Couple ligand to gel for 1–2 hr at 22° by mixing end over end in a small volume of buffer (0.1 *M* NaHEPES, pH 7.5, or 0.1 *M* $NaHCO_3$, pH 8.0, are suitable).
3. Block excess active sites with 100 μl of 1 *M* Tris–HCl, pH 8.0, or 1 *M* ethanolamine–HCl, pH 8.0, or other primary amine.
4. Pour the gel into a 1-cm-diameter column and wash with 1 liter 0.1 *M* NaHEPES, pH 7.5. Coupling efficiency can be monitored by assaying for IGF-II in the column wash or by including a small amount of ^{125}I-labeled IGF-II in the ligand mixture before coupling and monitoring the column wash for radioactivity.
5. To remove any remaining uncoupled ligand the column should be further washed in 100 ml 0.5 *M* NaCl, 100 ml 0.5 *M* sodium acetate, pH 3, 50 ml 2 *M* sodium acetate, pH 7, and 50 ml 0.1 *M* NaHEPES, pH 7.5. If the column is to be stored before use add 2 g/liter sodium azide to the last wash and store at 4° to prevent growth of microorganisms.

Purification of IGF-II Receptor

Successful purification of IGF-II receptor by affinity chromatography depends on sufficient exposure of the receptor to the immobilized ligand on the agarose followed by rapid elution and neutralization. Membranes prepared from a variety of rat tissues are all readily solubilized in Triton X-100[14]; however, we have found other detergents such as 3-[(3-cholamidopropyl)dimethylammonio]-1-propane sulfonate (CHAPS) (Calbiochem-Behring, La Jolla, CA) to solubilize equally well, and these may be more suitable for other tissues. At 22° rat IGF-II receptors reach equilibrium binding rapidly (90–120 min),[21] and 15–30 min of exposure to the column is sufficient to bind receptor, as determined by the absence of receptor activity in the column flow through. During elution of the receptor, fractions are neutralized immediately with 2 *M* Tris base. Several purifications of IGF-II receptor can be performed by a single affinity column before any deterioration of binding capacity can be detected. Solubilized membranes are passed through a guard column of Sephadex G-10 before

[21] J. M. Bryson and R. C. Baxter, *J. Endocrinol.* **113,** 27 (1987).

passing onto the affinity column to prevent particulate matter accumulating. Addition of protease inhibitors to column buffers and storage of the column in 2 g/liter sodium azide should also prolong the life of the gel.

Reagents

Triton X-100 (Boehringer, Mannheim, FRG)
Buffer A: 25 m*M* NaHEPES, pH 7.4, 10 m*M* $CaCl_2$
Buffer B: buffer A with 0.1% Triton X-100
Buffer C: buffer B with 0.15 *M* NaCl
Buffer D: buffer C adjusted to pH 4 with glacial acetic acid ($CaCl_2$ may be omitted)
2 *M* Tris base (unadjusted pH = 10.8)
Sephadex G-10

All buffers are supplemented with protease inhibitors:
0.5 m*M* Bacitracin
50 μg/ml Soybean trypsin inhibitor
0.1 m*M* TPCK

Procedure

1. Thaw microsomal membranes (3–4 g protein in 100 ml) rapidly and solubilize in Triton X-100 as described in *Membrane Preparation*.
2. Dilute solubilized membranes 10-fold with buffer A containing protease inhibitors to give a final Triton X-100 concentration of 0.1%.
3. Centrifuge membranes at 12,000 *g* for 20 min at 22° to remove any residual insoluble material. A minor white precipitate formed during dilution of the membranes (with no detectable IGF-II binding capacity) is also removed by this procedure.
4. Prepare a guard column of Sephadex G-10 (5 cm long × 2.5 cm diameter is a suitable size) and place in line before the IGF-II affinity column.
5. Prewash the IGF-II–agarose affinity column and guard column in 100 ml buffer B at 22°.
6. Apply solubilized membranes to column by pumping at the rate of 60 ml/hr at 22° and collect 10-min fractions.
7. Wash column with 200 ml buffer C at 120 ml/hr and collect 5-min fractions.
8. Elute IGF-II receptor with 50 ml buffer D at 60 ml/hr. Collect 1-ml fractions, measure their pH, and neutralize immediately with 2 *M* Tris base. IGF-II receptor activity is found in the first few fractions following the rapid drop in pH after the column void volume.

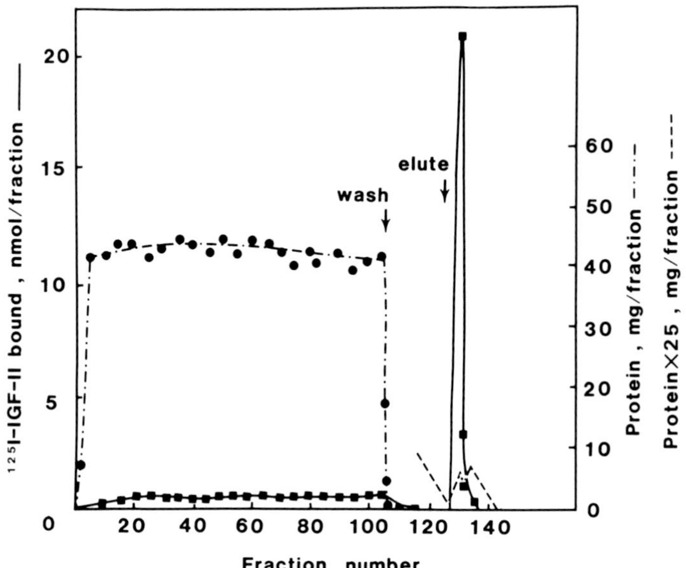

FIG. 1. Purification of rat liver IGF-II receptor. Microsomal membranes (7 g protein) solubilized in Triton X-100 were diluted to 1 liter and applied to an agarose–IGF-II affinity column (1 × 13 cm) as described in the text. Unbound material (fractions 1–105) and wash buffer (fractions 106–125) were collected in 10-ml fractions. Material eluted at pH 4 (fractions 126–140) was collected in 2-ml fractions. Fractions were then assayed for ^{125}I-labeled IGF-II binding activity (■) and protein content (●). [Reproduced with permission from C. D. Scott and R. C. Baxter, "Purification and immunological characterization of rat liver IGF-II receptor," *Endocrinology (Baltimore)*, **120,** 1–9, © by The Endocrine Society, 1987.]

9. Assay all fractions for IGF-II receptor activity as described previously and estimate protein by the Bradford method[22] using dye reagent from Bio-Rad.

Purification Profile

A typical purification profile of IGF-II receptor from solubilized rat liver microsomes is shown in Fig. 1. In this preparation 7 g membrane protein was applied to the IGF-II–agarose column. Less than 2% of receptor activity was found in the unbound fraction, and 27% was recovered in the fractions eluted at pH 4. Recoveries ranged from 20 to 45% in 10 similar preparations. Repeated use of the column with large amounts of protein later resulted in lower binding to the IGF-II–agarose, suggesting degradation or removal of IGF-II from the column. An initial purification of membrane glycoproteins on columns of immobilized wheat germ lec-

[22] M. Bradford, *Anal. Biochem.* **126,** 144 (1976).

tin[13] or concanavalin A[12] would possibly remove protease activity from tissues such as liver and result in a longer life for the IGF-II affinity column. The remainder of the binding activity not eluted is probably destroyed by the acid conditions of the elution and may be present in the eluate as denatured protein. However, denatured receptor is probably at least as antigenic as native receptor and should not hinder the raising of antibodies.

Characteristics of Purified Receptor

Purified IGF-II receptor showed similar properties to the microsomal membrane receptor. As shown in Fig. 2, specificity of the receptor is unaltered by purification, showing high affinity for IGF-II, 1% cross-reactivity with IGF-I, and no cross-reactivity with insulin. Sodium dodecyl sulfate–polyacrylamide gel electrophoresis (SDS–PAGE) of rat liver membranes and purified receptor cross-linked with ^{125}I-labeled IGF-II shows a single band of radioactivity at MW 250,000 under reducing conditions (Fig. 3). This band is absent in samples incubated with unlabeled IGF-II but present in samples incubated with insulin. A major band of Coomassie blue-staining material is present at MW 250,000 with a minor band occasionally seen at 70,000. Silver-stained gels also revealed the same two major and minor bands (results not shown). Radioiodinated

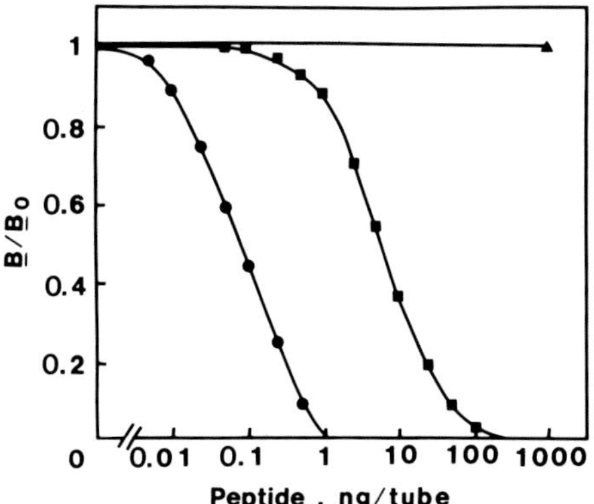

FIG. 2. Peptide specificity of purified IGF-II receptor. Purified receptor (3.4 ng) was assayed as described in the presence of varying concentrations of IGF-II (●), IGF-I (■), or insulin (▲). Displacement of ^{125}I-labeled IGF-II from the receptor is expressed as the amount bound in the presence of the peptide (B) divided by the amount bound in the absence of unlabeled peptide (B_0). [Reproduced with permission from C. D. Scott and R. C. Baxter, "Purification and immunological characterization of rat liver IGF-II receptor," *Endocrinology (Baltimore)*, **120,** 1–9, © by The Endocrine Society, 1987.]

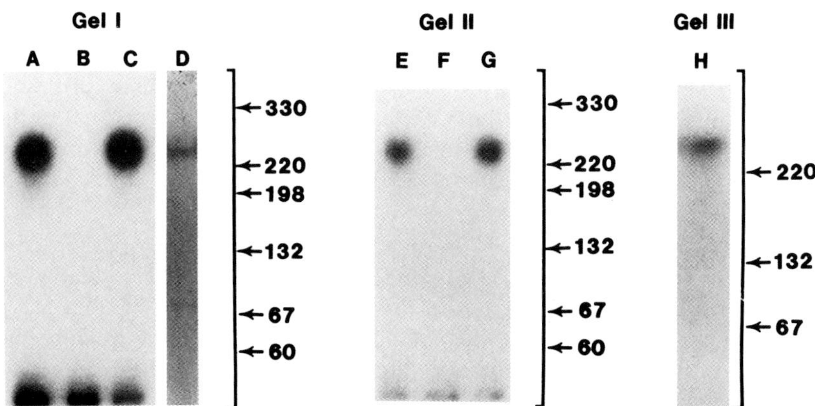

FIG. 3. SDS–PAGE of IGF-II receptor. Samples were prepared and reduced as previously described[14] and subjected to SDS–PAGE on 6–12% polyacrylamide gels. Radioactivity was detected by autoradiography. Purified receptor, 100 ng (Gel I, lanes A–C), and Triton X-100-extracted microsomal membranes, 5 μg (Gel II, lanes E–G), were affinity labeled with ^{125}I-labeled IGF-II in the absence of unlabeled peptide (lanes A and E), in the presence of 1 μg/ml unlabeled IGF-II (lanes B and F), or in the presence of 10 μg/ml unlabeled insulin (lanes C and G). Gel I, lane D, shows purified receptor (30 μg) stained with Coomassie blue, and Gel III, lane H, shows radioiodinated purified receptor (1 ng; 5 Ci/g). Arrows indicate molecular weight standards located by Coomassie blue staining and expressed as MW $\times$ 10^{-3}. [Reproduced with permission from C. D. Scott and R. C. Baxter, "Purification and immunological characterization of rat liver IGF-II receptor," *Endocrinology (Baltimore)*, **120,** 1–9, © by The Endocrine Society, 1987.]

purified receptor also exhibits a single band of radioactivity at MW 250,000 after electrophoresis.

Preparation and Assay of IGF-II Receptor Antibodies

Polyclonal antisera to IGF-II receptors appear to be easily raised by standard immunization procedures.[13–15] Sera can be assayed for antibodies capable of blocking IGF-II binding to its receptor (blocking antibodies) and for antibodies capable of precipitating IGF-II receptor already labeled with IGF-II (precipitating antibodies).

Immunization Procedure

1. Concentrate neutralized IGF-II receptor to approximately 200 μg/ml by ultrafiltration (YM100 membranes from Amicon, Danvers, MA, are suitable). Emulsify receptor with an equal volume of Freund's complete adjuvant and immunize two or three rabbits each with 1 ml (100 μg receptor) by injecting 100 μl into 10 subscapular sites.

2. Boost animals 2 weeks later with 100 μg purified receptor as described above, but using Freund's incomplete adjuvant.
3. After a further 2 weeks boost the rabbits by injection in the hind leg muscles with 100 μg purified receptor without adjuvant (in neutral isotonic buffer).
4. Test bleed rabbits 7 days later by collecting 1–2 ml blood from an ear vein and test for antibody activity.
5. If no antibody activity is detected, repeat boosts and test bleeds at 2-week intervals 2–3 more times or until antibodies are detected.

Assay for Blocking Antibodies

1. Using reagents as for *Assay of IGF-II Receptor Activity*, add receptor and buffer in a total volume of 390 μl to duplicate tubes for each serum concentration to be tested and also for total and nonspecific binding (with excess unlabeled IGF-II) in the absence of serum.
2. Make serial dilutions of sera from immunized and nonimmunized rabbits and add to assay tubes in 10 μl assay buffer. An initial range of 1 : 25 to 1 : 2,500 is suggested (to give a final concentration of 1 : 1,000 to 1 : 100,000).
3. Preincubate tubes for 2–6 hr at 22° or 16 hr at 4°.
4. Add 10,000 cpm ^{125}I-labeled IGF-II and incubate for 2 hr at 22° and then precipitate with bovine γ-globulin and PEG as described previously under *Assay of IGF-II Receptor Activity*.

In this assay the presence of blocking antibodies will be detected by a decrease in the specific binding of ^{125}I-labeled IGF-II to membrane incubated with serum from immunized rabbits compared to membrane incubated with serum from nonimmunized rabbits.

Assay for Precipitating Antibodies

1. Using reagents as described for *Assay of IGF-II Receptor Activity*, add receptor, buffer, and 10,000 cpm ^{125}I-labeled IGF-II in a total volume of 490 μl to duplicate tubes for each serum concentration to be tested and also for measuring total and nonspecific binding (with excess unlabeled IGF-II) in the absence of serum.
2. Incubate 2 hr at 22°.
3. Make serial dilutions of serum from immunized and nonimmunized rabbits in the range 1 : 20 to 1 : 2,000 and add to assay tubes in 10 μl assay buffer to give a final concentration of 1 : 1,000 to 1 : 100,000.
4. Incubate for 3–6 hr at 22° or 16 hr at 4°.
5. Receptor in tubes incubated without serum (for determining total

and nonspecific binding) is precipitated with bovine γ-globulin and 18 g/liter PEG as described in *Assay of IGF-II Receptor Activity*. Receptor–antibody complexes in tubes incubated with serum are precipitated by incubating with anti-rabbit γ-globulin for 30 min [we found 50 μl goat anti-rabbit γ-globulin (Anti-RGG serum, Bio-Mega, Montreal, Quebec) with 1 μl normal rabbit serum as carrier to give good precipitation; however, a number of commercial preparations should give equally good results]. Add 1 ml 60 g/liter PEG.

6. Centrifuge all tubes at 4500 *g* for 20 min, decant supernatants, drain tubes, and count bound radioactivity in pellets.

In this assay the presence of antibodies capable of precipitating IGF-II receptors occupied at their binding sites with ^{125}I-labeled IGF-II is detected by an increase in the amount of radioactivity precipitated by serum from immunized rabbits compared with serum from nonimmune rabbits. The proportion of total available receptor precipitated at each serum dilution is determined by comparison with receptor incubated in the absence of serum and precipitated by 180 g/liter PEG.

Antibody Characteristics

In three studies undertaken to date, antisera raised against rat IGF-II receptors have been found both to block binding of IGF-II to its receptor and to precipitate receptor already occupied with IGF-II.[13–15] In three rabbits immunized with purified rat liver IGF-II receptor as described, we found high-titer antibodies with both activities in all three rabbits at the first test bleed. The results for one of these antibodies, C-1, is shown in Fig. 4. This antiserum blocked ^{125}I-labeled IGF-II binding completely at a dilution of 1 : 10,000 and fully precipitated receptor at 1 : 10,000 in 4 hr (Fig. 4) or at 1 : 100,000 in 16 hr.[14] Antibody titers did not improve greatly with subsequent boosts. All three of the antibodies raised by us against rat IGF-II receptor cross-reacted with liver receptor from mouse but not from rabbit or human. Full cross-reactivity was seen to IGF-II receptors in all other rat tissues examined, as shown in Table I.

Applications of Receptor Antibodies

Studies of IGF-II Actions

IGF-II receptor antisera can be used to block binding of IGF-II to cells. As shown in Table II, ^{125}I-labeled IGF-II binding to rat hepatocytes and rat hepatoma cells (HTC, H-35, 5123) in culture can be totally blocked by preincubation of the cells with antisera. Thus, studies on

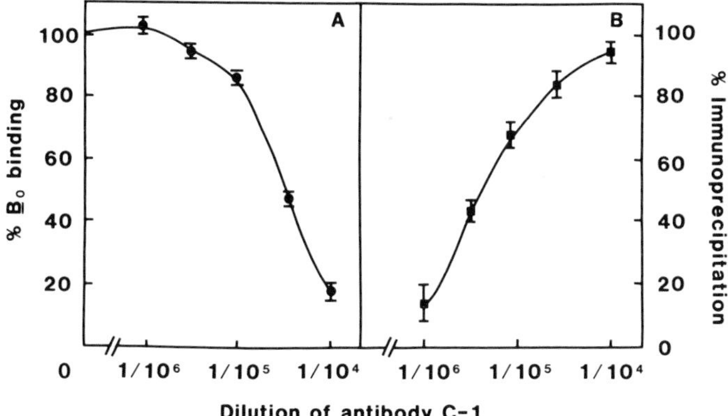

FIG. 4. Effect of antiserum raised to purified IGF-II receptor on rat liver microsomal IGF-II receptor. (A) Blocking activity of antiserum was assayed as described in the text and results expressed as a percentage of the binding measured in the absence of serum (B_0). (B) Precipitating activity of antiserum was assayed as described in the text and expressed as the percentage of total receptor-bound ^{125}I-labeled IGF-II precipitated at each antiserum dilution.

IGF-II actions on the cells can be performed to determine whether IGF-II is acting through the type II IGF receptor to effect its response or by cross-reactivity at the type I IGF or insulin receptor. Mottola and Czech[1] have demonstrated with IGF-II receptor antibodies that the

TABLE I
CROSS-REACTIVITY OF IGF-II RECEPTORS IN VARIOUS RAT TISSUES WITH RAT LIVER IGF-II RECEPTOR ANTISERUM

	Antiserum-blocking activity[a]		Antiserum-precipitating activity[b]	
Tissue[d]	B_0 binding[c] (%)	Inhibition by C-1 (% B_0)	B_0 binding (%)	Precipitation by C-1 (% B_0)
Liver	44.9 ± 2.9	96.0 ± 6.3	46.4 ± 4.2	72.3 ± 0.8
Brain	10.5 ± 1.0	88.7 ± 8.7	16.9 ± 1.7	49.0 ± 1.1
Kidney	44.4 ± 4.1	95.1 ± 8.8	51.3 ± 4.5	61.7 ± 0.9
Heart	20.2 ± 1.2	99.5 ± 6.0	24.6 ± 1.4	59.2 ± 4.9
Ovary	58.6 ± 1.5	91.2 ± 2.3	55.3 ± 7.1	64.1 ± 0.5
Muscle	10.8 ± 0.9	96.8 ± 7.9	15.7 ± 1.3	62.0 ± 1.4
Testes	33.2 ± 6.2	93.2 ± 7.4	49.1 ± 2.7	62.6 ± 1.4
Adrenal	43.9 ± 2.0	94.7 ± 4.2	44.0 ± 3.9	76.1 ± 3.2

[a] Assayed as described in the text with antiserum C-1 at 1 : 10,000.
[b] Assayed as described in the text with antiserum C-1 at 1 : 100,000.
[c] All results expressed as means ± SEM for three membrane preparations.
[d] Ten micrograms solubilized microsomal membranes prepared as described in the text for liver.

TABLE II
EFFECT OF IGF-II RECEPTOR ANTISERUM ON ^{125}I-LABELED IGF-II BINDING TO CULTURED CELLS

Serum	Dilution	Inhibition of ^{125}I-labeled IGF-II binding[a] (%) for cell type			
		Hepatocyte	HTC	H-35	5123
C-1	1:5000	47 ± 15	36 ± 11	56 ± 3	38 ± 18
	1:500	91 ± 10	84 ± 11	95 ± 8	85 ± 13
Nonimmune	1:5000	0	0	0	0
	1:500	5 ± 1	19 ± 1	6 ± 3	13 ± 2

[a] ^{125}I-labeled IGF-II binding was carried out in confluent cultures of 3–5 × 10^5 cells/4.5 cm^2 well in the presence and absence of 0.5 μg unlabeled IGF-II in serum-free medium (Eagle's minimum essential medium, Flow Laboratories, Irvine, Scotland) as described by C. D. Scott and R. C. Baxter [*J. Cell. Physiol.* **133,** 532 (1987)]. Immune (C-1) or nonimmune serum was preincubated with the cells for 1 hr before ^{125}I-labeled IGF-II binding. Results are expressed as means ± SD for three experiments.

IGF-II-stimulated increase in DNA synthesis in rat hepatoma cells is not mediated via the type II IGF receptor.

Regulation and Localization of IGF-II Receptors

Regulation of IGF-II receptors can be studied by a variety of techniques using receptor antibodies. The rapid translocation of IGF-II receptors from intracellular pools to the plasma membrane, as seen in adipocytes[23] and in H-35 hepatoma cells,[24] appears to be an important control of receptor activity at the cell surface. In investigations of the role of various factors on IGF-II receptor levels, antibodies can be used to quantitate receptors in various membrane fractions by radioimmunoassay[14] or by Western blotting as shown in Fig. 5. Figure 5 shows the higher proportion of IGF-II receptor present in Golgi membranes compared to the plasma membranes isolated from normal adult rat liver. Localization and regulation of the receptor can also be studied by immunohistochemistry at the light or electron microscopic level by using a variety of labeled second antibodies.

[23] L. J. Wardzala, I. A. Simpson, M. M. Rechler, and S. W. Cushman, *J. Biol. Chem.* **259,** 8378 (1984).

[24] J. Massagué, L. A. Blinderman, and M. P. Czech, *J. Biol. Chem.* **257,** 13948 (1982).

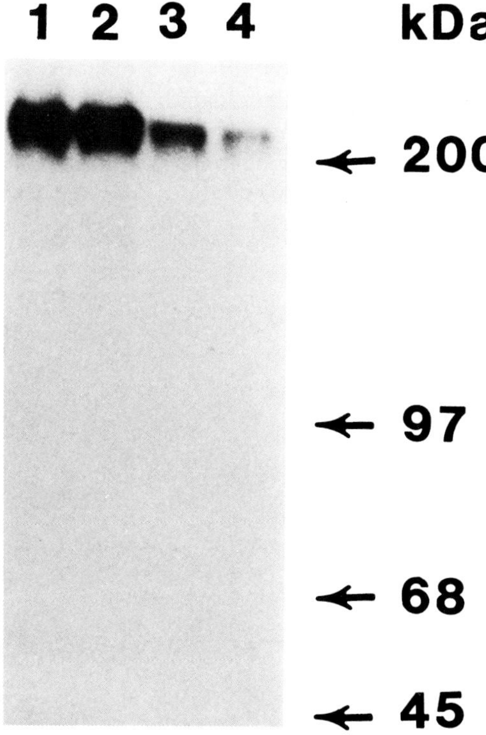

FIG. 5. Western blotting of rat liver membranes. Rat liver plasma membranes [A. L. Hubbard, D. A. Wall, and M. A. Anne, *J. Cell Biol.* **96,** 217 (1983)] and Golgi membranes [J. J. M. Bergeron, J. H. Ehrenreich, P. Siekeveitz, and G. E. Palade, *J. Cell Biol.* **59,** 73 (1973)] were prepared and 50 μg membrane protein subjected to SDS–PAGE without reduction on 6–12% gradient gels. Separated proteins were then electrotransferred onto nitrocellulose at 40 V for 6 hr. The nitrocellulose was blocked for 16 hr at 37° with 30 g/liter BSA, incubated with antiserum C-1 at 1 : 5000 for 4 hr at 22° and then with ^{125}I-labeled protein A (10^6 dpm) for 2 hr at 22°. Blots were washed extensively, dried, and autoradiographed. Golgi membranes (lanes 1 and 2) bound more ^{125}I-labeled protein A than plasma membranes (lanes 3 and 4), suggesting that a higher concentration of IGF-II receptors exists intracellularly in the Golgi than at the cell surface.

Relationship to the Mannose-6-phosphate Receptor

The IGF-II receptor gene has been cloned recently, and the amino acid sequence of the protein determined.[25] This sequence shows a strong homology with the cation-independent mannose-6-phosphate receptor,

[25] D. O. Morgan, J. C. Edman, D. N. Standring, V. A. Fried, M. C. Smith, R. A. Roth, and W. J. Rutter, *Nature* (*London*) **329,** 301 (1987).

suggesting a single protein with two independent ligand binding sites. This raises the possibility of purifying the IGF-II–mannose-6-phosphate receptor by affinity chromatography employing a mannose-6-phosphate-containing carbohydrate or glycoprotein ligand. A published purification using β-galactosidase affinity chromatography found very low recoveries of the receptor,[26] but methods using agarose–phosphomannan columns appear to give very favorable yields,[27,28] indicating that such ligands could be used as an alternative to IGF-II.

[26] G. G. Sahaglan, J. J. Distler, and G. W. Jourdian, *Proc. Natl. Acad. Sci. U.S.A.* **78,** 4289 (1981).

[27] G. G. Sahagian, J. J. Distler, and G. W. Jourdian, *in* "Methods in Enzymology" (V. Ginsburg, ed.), Vol. 83, p. 392. Academic Press, New York, 1982.

[28] A. W. Steiner and L. H. Rome, *Arch. Biochem. Biophys.* **214,** 681 (1982).

Section IV

Quantitation of Neuroendocrine Substances

[21] Measurement of Hormone Secretion from Individual Cells by Cell Blot Assay

By MARCIA E. KENDALL and W. C. HYMER

Introduction

The concept of functional heterogeneity in anterior pituitary cells was first inferred from cell separation approaches.[1] This heterogeneity has now been directly established by the reverse hemolytic plaque assay.[2,3] In this chapter a new method to evaluate heterogeneity at the cellular level is described. With this new method, the cell blot assay, visualization and quantification of hormone release from individual rat anterior pituitary cells are possible.[4]

In the cell blot assay, cells are incubated on, and bound to, a transfer membrane. Hormone released from the cells is captured on the membrane. The membrane is then processed according to Western blot methodology.[5] This includes incubation in (1) primary antiserum, (2) secondary antiserum to which an enzyme is conjugated, and (3) enzyme substrate. The released hormone product surrounding the hormone-secreting cells (e.g., prolactin, growth hormone) becomes visible as a zone of secretion immediately encompassing the cell. This secretion zone is then analyzed and quantified. Heterogeneity of hormone release per cell becomes apparent.

Cell Blot Methodology

Membrane

The membrane used in the cell blot assay is the hydrophobic transfer membrane Immobilon (polyvinyldiene difluoride, PVDF; Millipore, Bedford, MA). Immobilon has high mechanical strength and high protein binding capacity. Most importantly, the cells and their secretion products remain visible when the membrane is viewed with a light microscope using transmitted light. At low power, the membrane has a periodic struc-

[1] W. C. Hymer and J. M. Hatfield, this series, Vol. 103, p. 257.

[2] J. D. Neill and L. S. Frawley, *Endocrinology (Baltimore)* **112,** 1135 (1983).

[3] F. R. Boockfor and L. S. Frawley, *Endocrinology (Baltimore)* **120,** 874 (1987).

[4] M. E. Kendall and W. C. Hymer, *Endocrinology (Baltimore)* **121,** 2260 (1987).

[5] H. Towbin, T. Staehelin, and J. Gordon, *Proc. Natl. Acad. Sci. U.S.A.* **76,** 4350 (1979).

ture, which can best be described as intermittent white spots. The other side appears fibrous and does not have the periodic structure. Cells must be applied to the structured side.

Cell Application and Incubation

Single-cell suspensions are prepared from *individual* anterior pituitary glands. Cells (1×10^2 to 7.5×10^3) in 100 μl modified Eagle's medium (αMEM; Flow Laboratories, McLean, VA) containing 0.2% $NaHCO_3$, 0.025% bovine serum albumin (BSA, Fraction V, fatty acid free; Sigma), and 25 m*M* HEPES (pH 7.4) are applied to a 2×2 cm piece of Immobilon. Owing to the hydrophobic nature of the membrane, the cell-containing medium remains in a bead on the membrane during incubation on a microscope slide in a humidified petri dish. Incubation is usually at 37° in a 95% air/5% CO_2 incubator for 1–6 hr. After incubation, the medium is drawn off carefully without contacting the membrane. At this stage only a thin film of medium remains on the membrane. The membranes are then placed in individual wells of a 6-well tissue culture plate (Falcon; Becton Dickinson, Oxnard, CA). Reagents are added carefully to the membrane around (not directly on) the area of cell application. There are two reasons for care at this step: (1) to maintain cellular attachment and (2) to keep the membrane submerged during subsequent steps. The membrane is treated sequentially in various solutions with 2–3 intervening rinses of phosphate-buffered saline (PBS, 10 m*M*, pH 7.4) between each step. To accomplish rapid rinses, it is convenient to aspirate using a house vacuum and a Pasteur pipet.

Western Blotting

The procedure described below (see Fig. 1) is that used for measuring prolactin (PRL) secretion from single cells. After incubation as described above, the membranes are processed through the following reagents: (1) phenylhydrazine hydrochloride (0.15%, Fisher Scientific Company, Fair Lawn, NJ) in PBS, 30 min (to block endogenous peroxidase activity); (2) PBS/1.0% BSA, 2 hr (to block nonspecific binding sites); (3) primary antiserum (rabbit antiserum to rat RP-1 PRL),[6] at a final dilution of 1:80,000, overnight; (4) secondary antiserum (goat anti-rabbit immunoglobulin conjugated to horseradish peroxidase; Cappel Laboratories, Cochranville, PA), at a final dilution of 1 : 1000, 2 hr; (5) peroxidase–antiperoxidase serum (Cappel), final dilution 1 : 2000, 1 hr; and, finally, (6) substrate (3,3′-diaminobenzidine tetrahydrochloride, DAB, 0.05%, in 10

[6] J. M. Hatfield and W. C. Hymer, *Endocrinology (Baltimore)* **119,** 2670 (1986).

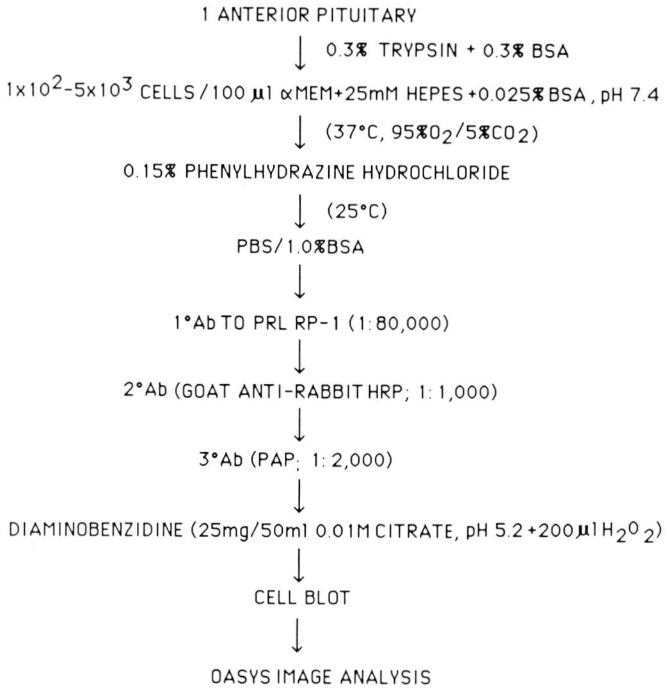

FIG. 1. Procedure for cell blot assay.

m*M* citrate buffer, pH 5.2; Sigma), initiated with 3.0% hydrogen peroxide (4 μl/ml), 30 min. In each of these steps, the membranes are gently agitated in a mechanical shaker at room temperature. After step 6, the membranes are rinsed with PBS and air dried. Antisera are diluted in PBS/1.0% BSA.

Standard Curves

Standard curves are generated by applying 0–1250 pg NIADDK RP-1 PRL in 100 nl of 10 m*M* $NaHCO_3$ to the Immobilon membrane with a sequencing pipet (Drummond Scientific Co., Broomall, PA). Incubation and Western blot processing is exactly as described above. Usually three replicates of each dose of hormone standard are used to generate a standard curve.

Quantification

Hormone release from individual cells is quantified using either the Oasys Image Analysis System (LeMont Scientific, State College, PA) or the Apple Macintosh ThunderScan system. Most of our data have been collected using the Oasys System. This digital color video image enhancement system, equipped with a microcomputer, processes gray levels (256) in the course of measuring both area and intensity of released hormone. Cell blots are analyzed directly with a light microscope and a video camera interfaced to the Oasys system. All images are digitized, stored, and analyzed as an array (512 × 480 pixels) of gray levels. Calculations of hormone released from standards and cells are made on the basis of area times average gray level. When quantifying cell secretion, the digitized cell body is subtracted from the zone of secretion.

An alternate method of analysis involves photography of cell blot images at 200× followed by insertion of the photograph into an Apple Imagewriter II printer and scanning with a laser digitizer (ThunderScan; ThunderWare, Inc., Orinda, CA). Quantification of the scanned blot can be achieved using Scan Analysis software (T. Burcham, Stanford Univ.) and an Apple Macintosh Plus computer. With either method of analysis, random fields are selected, and background is subtracted from all images.

Results

Examples of PRL secretion from single rat anterior pituitary gland cells are presented in this section. Cells were obtained from adult Fischer-344 rats (Charles River, Wilmington, MA) that had been ovariectomized (OVX) between 38 and 42 days of age. In some cases, they were injected (sc) with estradiol benzoate (EB; 0.5 μg day 1, 50.0 μg day 2)[7] between days 10 and 17 after surgery; others were implanted with a 5.0 mg diethylstilbesterol (DES)-containing Silastic capsule[8] for 50–70 days. All rats were sacrificed either 10–20 days after ovariectomy, the day after estradiol injection, or after 50–70 days of DES treatment.

Cells remain intact after 6 hr of incubation (Fig. 2A). An example of PRL secretion from individual cells which had been obtained from long-term estrogen-treated rats is shown in Fig. 2B. Note the heterogeneity in the size of the secretion zone. Secretion zones are larger and more intense at 37° as opposed to 25°. Secretion zones at 4° are diminished even further.

Hormone release from cells could not be detected after any of the following treatments: (1) replacement of primary antiserum with normal rabbit serum; (2) deletion of either primary or secondary antiserum; (3) incubation of the Immobilon membrane without cells; and, finally, (4)

[7] J. D. Neill, *Endocrinology (Baltimore)* **90,** 1154 (1972).

[8] J. Wiklund, N. Wertz, and J. Gorski, *Endocrinology (Baltimore)* **109,** 1700 (1981).

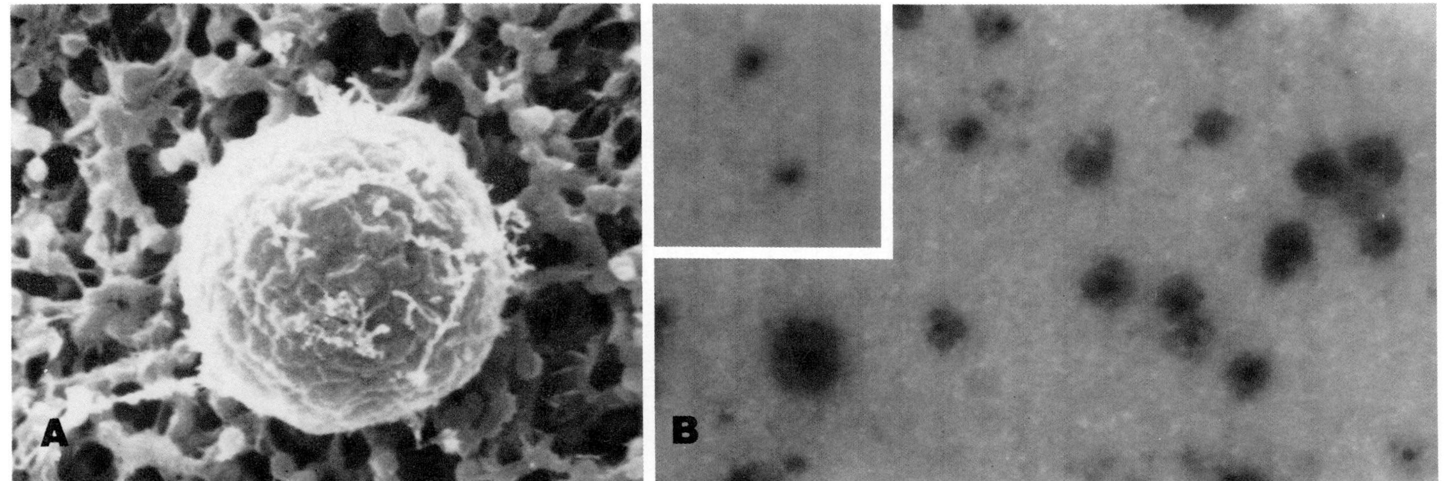

FIG. 2. (A) Scanning electron micrograph of an anterior pituitary cell incubated for 6 hr on Immobilon. ×3825. (B) PRL cell blot with cells (dark objects in center of images) and zones of secretion. ×170. These cells were prepared from DES-treated rats and were incubated for 2 hr prior to Western blotting. (B, insert) PRL cell blot from OVX rat incubated for 2 hr. Note that the size and intensity of secretion are less than those of cells from DES-treated rats.

preabsorption of PRL antiserum with PRL (0.3 μg/ml, NIADDK I-5) for 24 hr prior to use.

Examples of the digitized images of pituitary cells and their zones of secretion from OVX and OVX plus DES-treated rats are shown in Fig. 3. Note the tendency for the zone of secretion to be larger after 6 hr of incubation (compare Fig. 3B with A). Sometimes asymmetry of secretion is apparent, but the reason for this is unknown. Large zones of secretion from cells obtained from DES-treated rats (Fig. 3C and D) reflect intensive secretion activity.

A linear relationship exists between absorbance and dose of PRL standard applied to the Immobilon membrane (Fig. 4A). This standard curve is used to quantify hormone released from individual pituitary cells. Average release from 50–75 cells per point in each of two experiments is linear in the estrogen-treated group (Fig. 4C) but not in the other groups (Fig. 4B and D). There is variation in the amount of secretion during the 6-hr incubation period in the three treatment groups. In cells from OVX rats, the amount of hormone released ranges from 0.11 to 1.14 pg/cell; in cells from estrogen-treated rats, the range is 0.23–1.32 pg/cell; and in OVX plus DES-treated rats, 0.18–1.55 pg/cell. When dopamine is included in the incubation medium, suppression of PRL released per cell is seen (Fig. 4E). As dopamine concentration increases, PRL release decreases.

An example of the application of the cell blot assay to human tissue is shown in Fig. 5. In this case PRL released from cells prepared from a 31-year-old male with a prolactinoma shows easily detectable secretion signals after 2 hr of incubation. The antiserum to human PRL is hPRL-3 (AFP-C11580, NIADDK; final dilution 1 : 20,000).

Optimization of Assay

Membrane

In addition to the Immobilon PVDF membrane, several types of commercially available nitrocellulose membranes have been evaluated. These include Millipore, Bio-Rad, Gelman, and Schleicher and Schuell. In our experience the cells do not maintain integrity as well on nitrocellulose membranes. Furthermore, it is difficult for transmitted light to pass through these membranes. Finally, it is our impression that cells may not attach as well to nitrocellulose. The enhanced sensitivity, compatibility with immunostaining, and low background make Immobilon PVDF the membrane of choice for the cell blot assay.

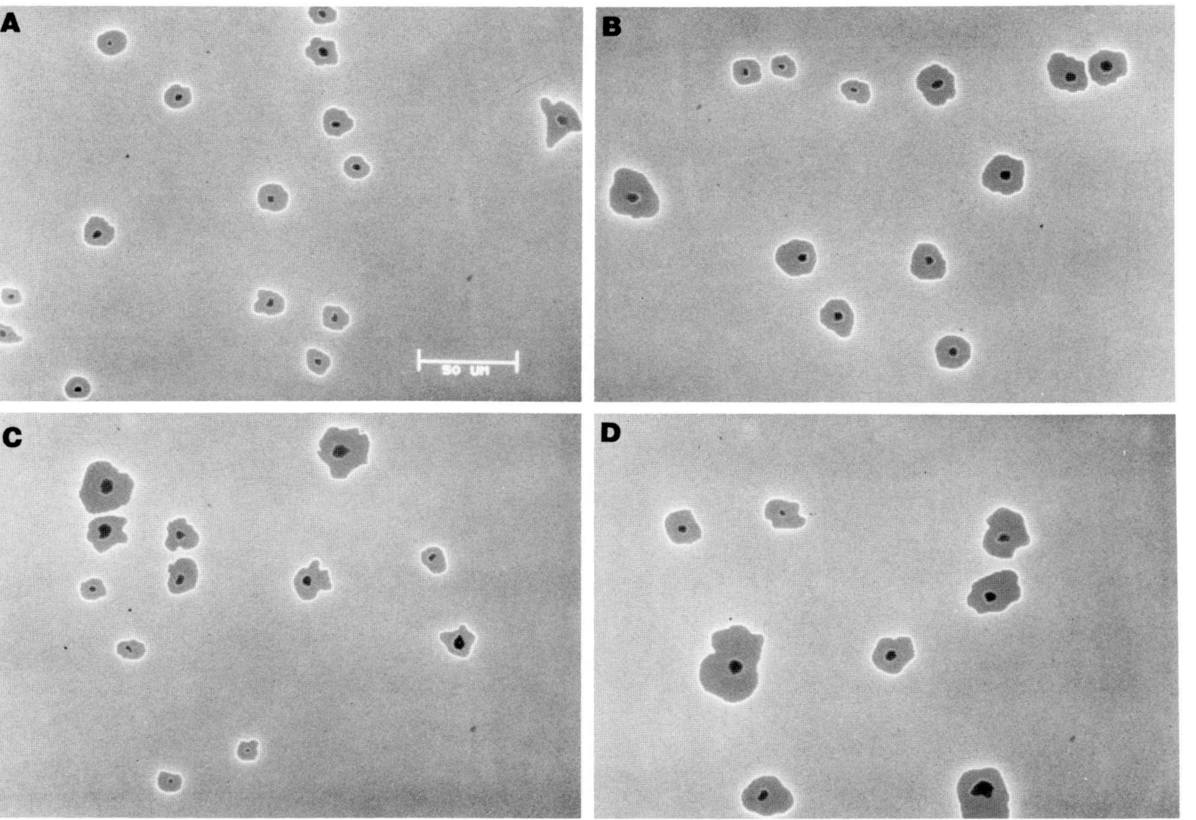

FIG. 3. (A) Digitized image (Oasys) of cells (center) and zones of secretion (surrounding the cell). The cells were from an OVX rat and were incubated 1 hr. (B) OVX rat, cells incubated 6 hr. (C) DES-treated rat, cells incubated 1 hr. (D) DES-treated rat, cells incubated 6 hr. ×204.

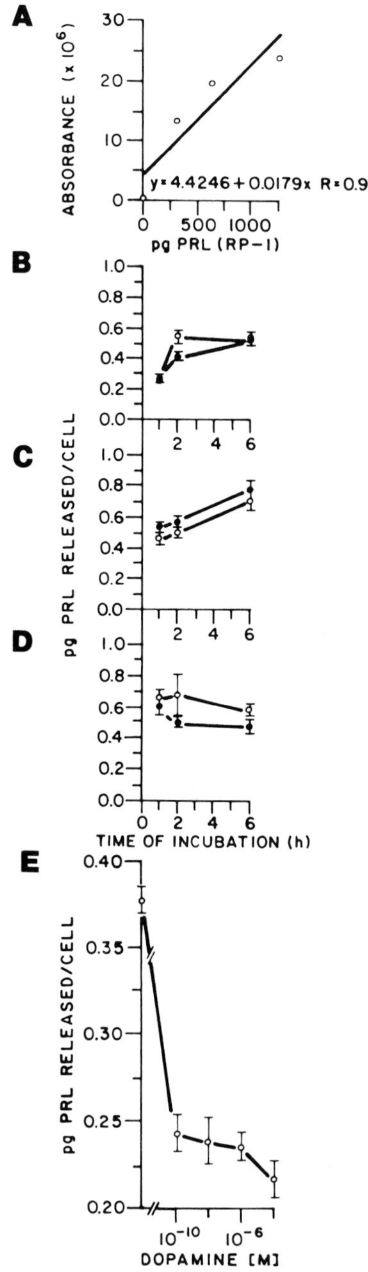

FIG. 4. (A) PRL (RP-1) standard curve developed on Immobilon. Each point represents the mean of three separate standard curves. The standard errors of the mean are small and are contained within each datum point. (B–D) Mean PRL release per cell after 1–6 hr of incubation of cells from OVX rats (B), OVX plus estrogen-treated rats (C), and OVX plus

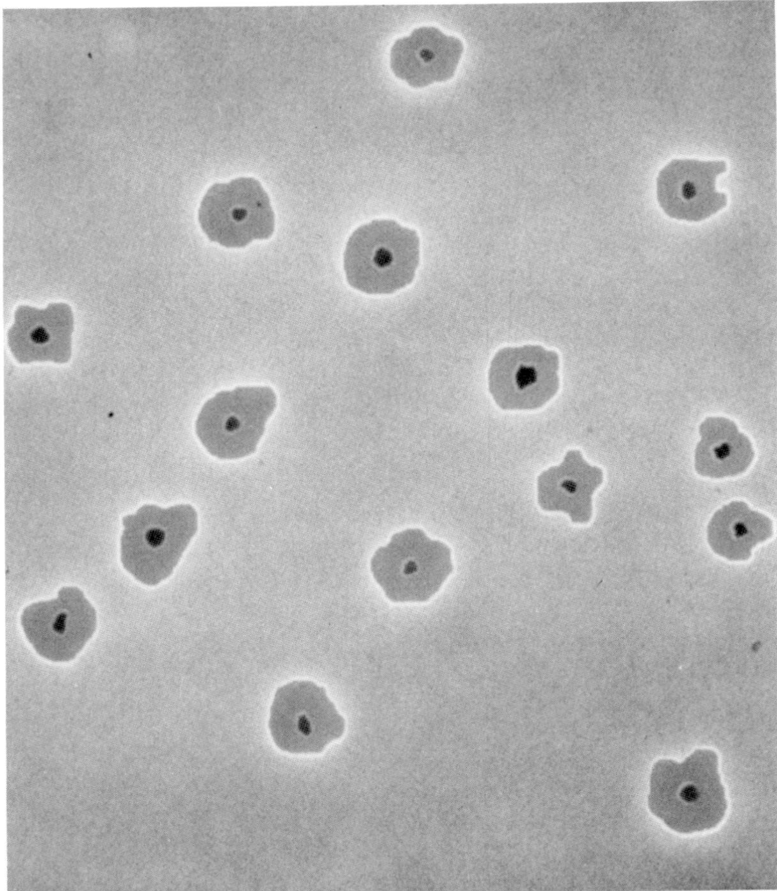

FIG. 5. Digitized image of cell blot obtained from human male prolactinoma. ×292. Cells were obtained by trypsinization[6] of biopsy tissue.

DES-treated rats (D). In all cases, the mean release ± SEM was obtained from at least 50 cells per point. Each curve represents one experiment. (E) Effect of dopamine on PRL release during a 2-hr incubation of cells from an OVX rat. Each datum point represents the mean ± SEM of release from 25–30 cells. Statistical analyses: B, (○) 1^{a}, 2^{b}, 6 hr^{b}, (●) 1^{a}, 2^{a}, 6 hr^{b}; C, (●) 1^{a}, 2^{a}, 6 hr^{b}, (○) 1^{a}, 2^{a}, 6 hr^{b}; D, (○) 1^{a}, 2^{a}, 6 hr^{a}, (●) 1^{a}, 2^{a}, 6 hr^{a}; E, 0^{a}, $10^{-10\,b}$, $10^{-8\,b}$, $10^{-6\,b}$, $10^{-4\,b}$ *M* dopamine. Means with different superscripts are significantly different at $p < 0.05$ on the basis of one-way analysis of variance.

Cell Attachment

Pretreatment of the membranes with (1) poly(L-lysine) (0.1%), (2) fibronectin (0.01%), (3) αMEM plus 0.2% $NaHCO_3$ plus 0.025% BSA plus 25 m*M* HEPES, or (4) Tween 20 (5%, Fisher) does not noticeably increase efficiency of pituitary cell attachment. Furthermore, pretreatment of the Immobilon with 100% methanol, as recommended by the manufacturer for opening membrane pores for protein attachment in the Western blot procedure, tends to lead to cell disruption. Time and temperature of incubation affect cell attachment. Although a small number of cells attach after 5 min, incubation times of at least 1 hr are required for significant attachment. Attachment is diminished when the temperature is reduced to 4°.

Cell Number

The number of cells incubated on the membrane has been varied from 2.5×10^2 to 1.0×10^4/100 μl. Our impression is that hormone release is independent of cell number. For image analysis and photography it is advantageous to have 5×10^3 cells in 100 μl of medium. Reduction of incubation medium volume to 25 μl has no qualitative effect on hormone secretion.

Incubation Medium

Several types of media have been tried in the development of this method. These include (1) αMEM plus 0.2% $NaHCO_3$ plus 0.025% BSA plus 25 m*M* HEPES, (2) αMEM plus 0.2% $NaHCO_3$ plus 0.1% BSA plus 25 m*M* HEPES, (3) Dulbecco's MEM plus 5% horse serum, (4) αMEM plus 0.1% horse serum, and (5) modified serum-free medium as described by Denef *et al.*[9] The amount of cellular secretion in the presence of serum tends to be decreased, but this has not been fully investigated. The medium that we use most frequently is αMEM plus 0.2% $NaHCO_3$ plus 0.025% BSA plus 25 m*M* HEPES.[10]

Western Blotting Reagents

Effects of reagent concentrations and incubation times on cell secretion have been examined but not quantitated.

[9] C. Denef, M. Baes, and C. Schramme, *Endocrinology (Baltimore)* **114,** 1371 (1984).
[10] W. W. Wilfinger, J. A. Davis, E. C. Augustine, and W. C. Hymer, *Endocrinology (Baltimore)* **105,** 530 (1979).

Primary Antiserum. A final dilution of 1 : 80,000 appears optimal. At this dilution cellular secretion is minimal after 1 hr, modest after 6 hr, and maximal after 12 hr of incubation.

Secondary Antiserum. A final dilution of 1 : 1000 appears optimal (range tested 1 : 1000 to 1 : 10,000).

Peroxidase–Antiperoxidase. A final dilution of 1 : 2000 appears optimal (range tested 1 : 2000 to 1 : 20,000). In some cases this step is omitted (e.g., rat growth hormone secretion).

3,3′-Diaminobenzidine Tetrahydrochloride. The secretion image appears sufficiently intense when the DAB concentration is 0.05% (range tested 0.02 to 0.1%) and the buffer is 10 m*M* citrate (pH 5.2) instead of 50 m*M* Tris (pH 7.6).

Limitations

One of the limitations of the cell blot assay is the inability to visualize the nonsecreting cells on the Immobilon membrane. Attempts to fix and stain nonprolactin cells with hematoxylin, Leukostat (Fisher), or toluidine blue have been unsuccessful. Fixatives tried to date include (1) methanol (100%, 30 min), (2) glutaraldehyde (3.0%, 30 min), (3) formalin vapors (30 min), (4) heat (60°, 15 min), (5) heat (60°, 15 min) plus formalin vapors (30 min), (6) graded methanol series (30, 50, 70, 80, 95, 100%, 2 min each step), (7) Bodian's (30 min), (8) methanol vapors (2 hr), and (9) methanol (70%, 30 min). In all cases cell integrity was poor. Single cells are essential in this assay. Trypsin techniques yield suspensions with the most single cells.[6]

3,3′-Diaminobenzidine tetrahydrochloride is a known carcinogen; consequently, precautions are required in its use. Preliminary experiments suggest that the avidin–biotin complex (ABC)[11] can successfully replace DAB (J. M. Connors, personal communication). The substrate 4-chloro-1-napthol is not satisfactory. Other potential limitations, e.g., incomplete capture of secreted product or secretagogue binding to Immobilon, have not been addressed to date.

Advantages

Advantages of the cell blot assay include (1) its simplicity and rapidity, (2) use of dilute preparations of antisera, (3) objectivity in measurement of hormone secretion, and (4) capability of quantification of hormone release

[11] G. V. Childs, *Stain Technol.* **58,** 281 (1983).

from single cells. The cell blot assay should prove useful for the quantitative evaluation of subpopulation heterogeneity of cells in secretory tissue for which an antibody to a secreted product is available. Ongoing studies include application of the cell blot assay to secretion of interleukin 1 and interleukin 2 from mammalian lymphocytes, growth hormone and thyroid-stimulating hormone from rat anterior pituitary gland cells, and alkaline phosphatase from bone cells.

Acknowledgments

This work was supported by grants PHS CA-23248, NASA NCC2-370, NASA 9-17416, and the NASA Center for Cell Research NAGW 1196. We thank LeMont Scientific, State College, PA, for use of their Oasys Image Analysis System. Prolactinoma tissue was kindly supplied by Dr. Robert Page, Hershey Medical Center, Hershey, PA.

[22] Measurement of Phosphoinositide Hydrolysis in Isolated Cell Membrane Preparations

By MILLIE M. CHIEN and JOHN C. CAMBIER

Introduction

The hydrolysis of inositol lipids by phospholipase C (PLC) is a integral element in signal transduction by cell surface receptors for a variety of hormones, neurotransmitters, and growth factors.[1-3] The primary substrate for this receptor-activated PLC is phosphatidylinositol 4,5-bisphosphate ($PtdInsP_2$) which is replenished by phosphorylation of phosphatidylinositol (PtdIns).[1] The breakdown products of $PtdInsP_2$, diaclglycerol (DAG) and inositol 1,4,5-trisphosphate ($InsP_3$) have been shown to have intracellular second-messenger function. DAG activates a Ca^{2+} and phospholipid-dependent protein kinase C (PKC), and $InsP_3$ stimulates Ca^{2+} release from intracellular stores into the cytosol. The effects of activation of PKC and Ca^{2+} calmodulin-dependent processes include a vast array of biological responses ranging from gene expression to secretion, depending on the tissue under study.

The biochemistry of ligand-induced perturbations in inositol lipid metabolism appears to have enormous complexity. Although PLC-mediated hydrolysis of phosphoinositides and phosphorylation and dephosphoryla-

[1] M. J. Berridge, *Biochem. J.* **220,** 345 (1984).

[2] M. J. Berridge and R. F. Irvine, *Nature (London)* **312,** 315 (1984).

[3] M. C. Sekar and L. E. Hokin, *J. Membr. Biol.* **89,** 193 (1986).

tion of inositol phosphates has been demonstrated in numerous cell systems, there has been only limited characterization of the operative molecular regulatory mechanisms. Of particular interest here is the precise mechanism whereby membrane immunoglobulin binding ligands initiate hydrolysis of inositol-containing lipids by PLC. In an effort to reduce the complexity inherent in studying the questions in whole cells, we have developed a cell-free system in which perturbations in phosphoinositide metabolism can be demonstrated. Here we describe this experimental system.

Materials

Inositol 1,4,5-trisphosphate, D-[2-^{3}H(N)]inositol ($InsP_3$) (specific activity, 4.0 Ci/mmol); phosphatidylinositol, L-α-[2-^{3}H(N)]*myo*-inositol (PtdIns) (specific activity, 10.0 Ci/mmol); and phosphatidylinositol 4,5,bisphosphate, [2-^{3}H(N)]inositol ($PtdInsP_2$) (specific activity, 3.5 Ci/mmol) were purchased from New England Nuclear (Boston, MA). Anion-exchange resin AG1-X8 (formate form, 100–200 mesh) and Bio-Rad protein assay reagent were obtained from Bio-Rad (Richmond, CA). Guanosine 5′-triphosphate (disodium salt) (GTP) was purchased from Boehringer Mannheim Biochemicals (Indianapolis, IN). D-Glucose, potassium chloride, and calcium chloride were purchased from J. T. Baker Chemical Co. (Phillipsburg, NJ). Phenylmethylsulfonyl fluoride (PMSF) was obtained from Calbiochem-Behring Corp. (La Jolla, CA). Sodium borate was from Fisher Scientific Co. (Fair Lawn, NJ). Adenosine 5′-triphosphate (disodium salt) (ATP), ammonium formate, ethylenediaminetetraacetic acid (EDTA), ethyleneglycol (bis(β-aminoethyl ether) N,N,N,′N′-tetraacetic acid (EGTA), lithium chloride, magnesium chloride, magnesium sulfate, sodium chloride, and sodium formate were obtained from Sigma (St. Louis, MO). Fetal calf serum (FCS) was purchased from M. A. Bioproducts (Walkersville, MD). *N*-2-Hydroxyethylpiperazine-*N*′-2-ethanesulfonic acid (HEPES) (1 *M* buffer) and rabbit complement were obtained from Gibco Lab (Grand Island, NY). Lymphocyte separation medium (LSM) was from Organon Teknika Corp. (Durham, NC). Rabbit anti-mouse Ig (RAMIG) and sheep anti-mouse Ig (SAMIG) were purified from the hyperimmune serum using mouse IgG–Sepharose affinity chromatography with elution using 3.5 *M* $MgCl_2$.

Formulations for solutions used were as follows: balanced salt solution (BSS) (5.6 m*M* glucose, 0.3 m*M* KH_2PO_4, 1.8 m*M* Na_2HPO_4, 1.3 m*M* $CaCl_2$, 5.3 m*M* KCl, 137 m*M* NaCl, 1.1 m*M* $MgCl_2$, and 1.4 m*M* $MgSO_4$); sonication buffer (0.25 *M* sucrose, 1 m*M* EDTA, and 10 m*M* HEPES, pH 7.0); incubation buffer (96 m*M* NaCl, 86 m*M* glucose,

0.1 mM EGTA, 5 mM EDTA, and 10 mM HEPES, pH 7.4); lysis buffer (25 mM sucrose, 0.1 mM EGTA, 0.1 mM PMSF, and 10 mM HEPES, pH 7.0).

Membrane Preparation

B lymphocytes were prepared from spleens of 6- to 15-week-old BDF1 mice as follows. Spleen cells are isolated and depleted of erythrocytes using Gey's solution and depleted of T lymphocytes by treatment with monoclonal anti-Thy 1.2 (HO13.4.9)[4] and anti-Thy 1 (T24/40)[5] antibodies and rabbit complement. Dead cells are removed by centrifugation over LSM. The isolated B cells consisted of more than 90% Ig-positive cells by immunofluorescence staining with fluorescein isothiocyanate-coupled rabbit anti-mouse Ig (FITC–RAMIG). A balanced salt solution (BSS) containing 5% heat-inactivated FCS was used to wash cells during their preparation.

For lysis membrane preparation, the isolated B cells (25×10^8/50 mice) were washed one time with incubation buffer, and the cells were resuspended in 12 ml incubation buffer plus 3.8 ml of 60% glycerol in incubation buffer.[6] Cells were allowed to swell at room temperature for 7 min before being centrifuged at 800 g for 10 min. The pellet was suspended in a small volume (10^{10}/ml) in incubation buffer. Cell disruption was affected by dropwise addition of this suspension to a 100× volume of the cold lysis buffer with constant mixing. Cells were further disrupted using a Dounce homogenizer (10 strokes) before the addition of DNase (10 μg/ml) and $MgCl_2$ (5 mM) and centrifugation at 800 g for 10 min to remove nuclei and any unbroken cells. The supernatant was centrifuged at 100,000 g for 30 min (Ti50 rotor) in a Beckman Model L5-50 ultracentrifuge to obtain a pellet enriched in plasma membranes. In some cases this fraction was resuspended in 10% sucrose, and this suspension layered on top of 50% sucrose cushion. Centrifugation at 100,000 g for 60 min (SW50.1 rotor) yielded an enriched plasma membrane fraction (based on ^{125}I-RAMIG binding activity) at the interface of 10 and 50% of sucrose. This interface was washed with a buffer consisting of HEPES (50 mM, pH 7.0) and KCl (100 mM). Electron microscopy revealed that these membranes were greater than 80% sheets with very few closed vesicles.

For membrane preparation by cell sonication, B lymphocytes were suspended at 10^8/ml in ice-cold sonication buffer. Cell disruption was

[4] A. Marshak-Rothstein, P. Fink, T. Gridley, D. H. Raulet, M. J. Bevan, and M. L. Gefter, *J. Immunol.* **122,** 2491 (1979).

[5] G. Dennert, R. Hyman, J. Lesley, and I. S. Trowbridge, *Cell. Immunol.* **53,** 350 (1980).

[6] A. J. Barber and G. A. Jamieson, *J. Biol. Chem.* **245,** 6357 (1970).

accomplished by sonication at 0° for a total of 20 sec (a 20-sec rest in the ice bath between 2 10-sec sonications) with a 1/8-inch diameter probe set at an output control of 4 in a Sonifier Cell Disruptor (Model W185D, Branson Sonic Power Co., Plainview, NY). Efficiency of disruption was monitored by microscopy. The sonicate was then centrifuged at 500 *g* for 10 min to remove unbroken cells and large particulate material. The supernatant was then centrifuged at 100,000 *g* for 30 min (Ti50 rotor) to isolate membrane (pellet) and cytosolic (supernatant) fractions.

Membranes isolated using osmotic lysis or sonication were suspended in HEPES (50 m*M*, pH 7.0) and KCl (100 m*M*) at a protein concentration of approximately 2 mg/ml and stored at −70° until use. Protein was determined by the Bio-Rad protein microassay.[7]

Incorporation of Labeled Precursors into the Isolated Membrane

The [^{3}H]PtdIns or [^{3}H]PtdInsP_2 in organic solvents were evaporated to dryness under a stream of nitrogen, and the residue was solubilized in 10% sodium cholate by overnight incubation at room temperature. Solutions were then brought to 5% sodium cholate with equal volume of water.[8] The [^{3}H]InsP_3 was diluted to 10 times its original volume with 50 m*M* HEPES (pH 7.0) before the experiment. The reaction mixtures consisted of B cell membrane (100–200 μg of protein), KCl (100 m*M*), EGTA (2.5 m*M*), LiCl (10 m*M*), with or without dialyzed cytosol (150 μg of protein), [^{3}H]PtdIns or [^{3}H]PtdInsP_2 (100,000 cpm, in sodium cholate 10 m*M* final concentration), or [^{3}H]InsP_3 (50,000 cpm, 60 μl of 10-fold diluted [^{3}H]InsP_3), and the appropriate ligand of anti-Ig (1 μM) and the volume brought to 500 μl with HEPES (50 m*M*, pH 7.0). Lithium chloride (10 m*M*) was added to the reaction mixture to block dephosphorylation of InsP.[9]

The reaction was initiated by adding ligand and labeled precursors at 37° and stopped by addition of 120 μl of 0.22 *N* HCl. The acidic conditions facilitated polyphosphoinositide extraction by preventing loss of lipids to the aqueous phase. Water-soluble inositol phosphates were extracted by the addition of 2.7 ml chloroform–methanol (1 : 2, v/v) to a reaction mixture in 13 × 100 mm screw-cap test tubes. After mixing, chloroform (0.9 ml) and 1 *M* KCl (0.9 ml) were added to separate phases. The mixtures were vortexed, and tubes were centrifuged at 800 *g* for 10 min.[10] The

[7] M. Bradford, *Anal. Biochem.* **72,** 248 (1976).

[8] S. Jackowski, C. W. Rettenmier, C. J. Sherr, and C. O. Rock, *J. Biol. Chem.* **261,** 4978 (1986).

[9] M. J. Berridge, C. P. Downes, and M. R. Hanley, *Biochem. J.* **206,** 587 (1982).

[10] M. K. Bijsterbosch and G. G. B. Klaus, *J. Exp. Med.* **162,** 1825 (1985).

lower phase was transferred to clean tubes and washed with 2 ml of CH_3OH–1 *N* HCl (1 : 1, v/v). The lipid extract was dried under a stream of nitrogen and kept at −20° for TLC analysis. The upper acidic aqueous phase was washed once with 1 ml of chloroform and immediately fractionated by anion-exchange chromatography.

Assay of Inositol Phosphates

The anion-exchange resin AG1-X8 (formate form, 100–200 mesh) was suspended and washed in deionized water to remove fine particles. The washed resin (1 ml packed volume) was put into small plastic columns (Bio-Rad Econo columns) and used for fractionation of inositol phosphates which occurred in the aqueous phase of reaction mixture. The acidic aqueous phase was diluted with 25 ml of water and then neutralized with 0.1 *N* NaOH. The diluted extracts were applied to columns and the

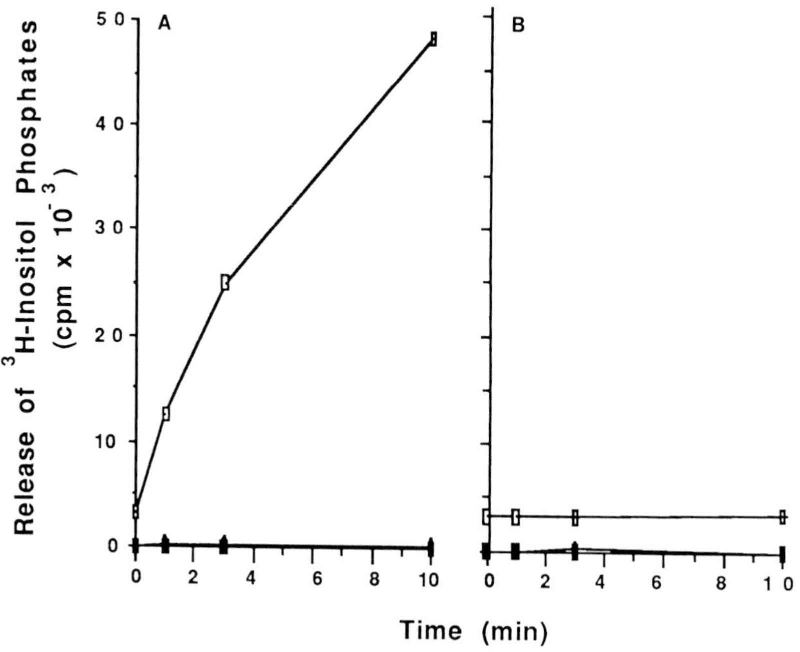

Fig. 1. Release of [^{3}H]inositol from [^{3}H]PtdIns-labeled membranes. The membranes (200 μg protein), obtained from sonication of B cells, were incubated with the dialyzed cytosol (150 μg protein) and [^{3}H]PtdIns (100,000 cpm) for varying times at 37° with (A) or without (B) RAMIG (1 μM). Samples were then extracted and analyzed for inositol phosphates. Values represent the amounts of the respective [^{3}H]inositol phosphates (cpm) released from the [^{3}H]PtdIns-labeled membranes. (□) InsP, (◆) $InsP_2$, (■) $InsP_3$.

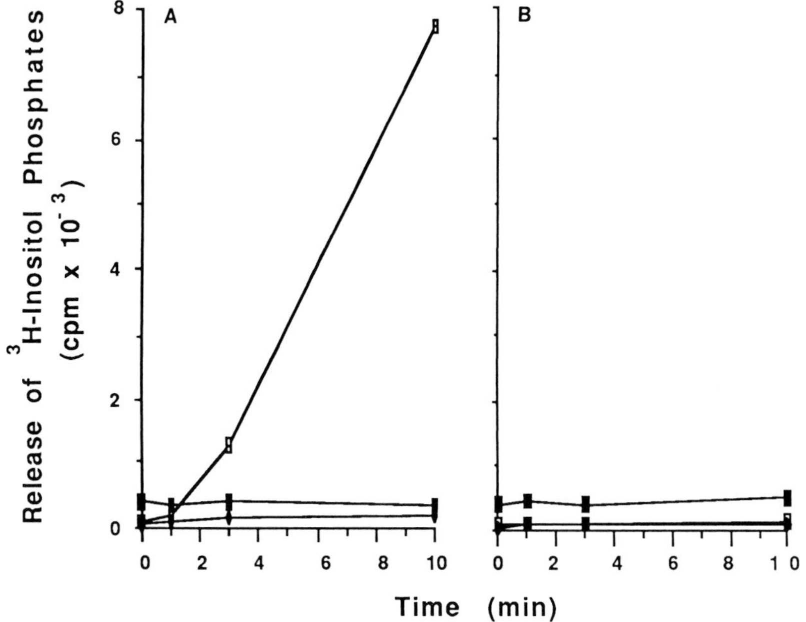

FIG. 2. Release of [^{3}H] inositol phosphates from [^{3}H]PtdInsP$_2$-labeled membranes. The sonicated B cell membranes (200 μg protein) were incubated with dialyzed cytosol and [^{3}H]PtdInsP$_2$ (100,000 cpm) for varying times at 37° with (A) or without (B) RAMIG (1 μM) before release of [^{3}H]InsP, [^{3}H]InsP$_2$, and [^{3}H]InsP$_3$ was assessed. (□) InsP, (◆) InsP$_2$, (■) InsP$_3$.

effluent (which contained free inositol) collected. Columns were then washed using 10 ml of water before applying 15 ml of sodium formate (30 mM)–sodium tetraborate (5 mM) for elution of glycerophosphoinositol. The inositol phosphates InsP, InsP$_2$, and InsP$_3$ were eluted in sequence using 15 ml of each of the following buffers: (1) ammonium formate (0.2 M)–formic acid (0.1 M); (2) ammonium formate (0.5 M)–formic acid (0.1 M); (3) ammonium formate (1.0 M)–formic acid (0.1 M).[10] The effluent and eluent were collected in 1-ml fractions and counted using a Beckman LS 3801 scintillation counter.

Inositol Lipid Hydrolysis and Inositol Phosphate Dephosphorylation

The effects of receptor Ig ligation on PLC and inositol polyphosphate phosphatase activity are illustrated in Figs. 1–5. As shown in Fig. 1, anti-Ig induced a rapid increase in the accumulation of [^{3}H]InsP when

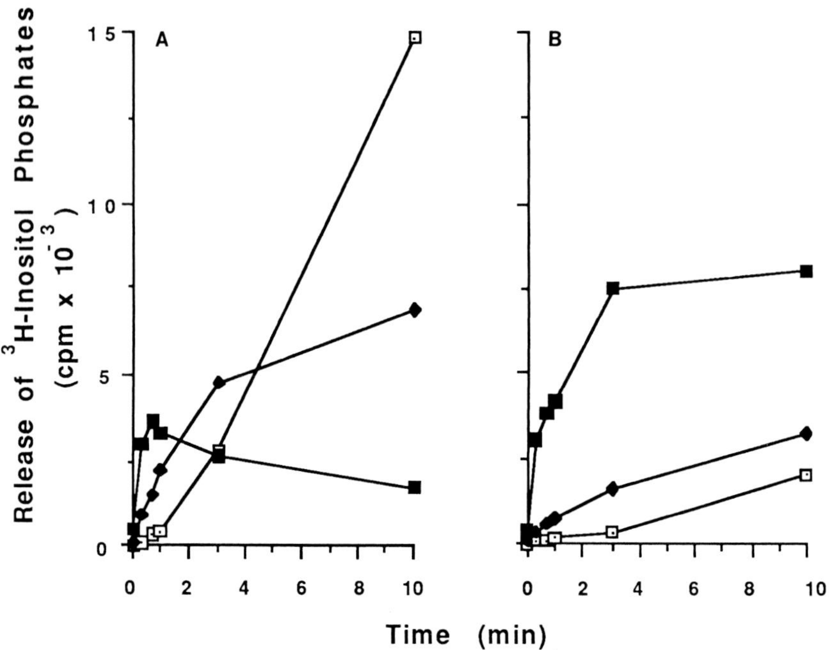

FIG. 3. (A) Release of [^{3}H]inositol phosphates from [^{3}H]PtdInsP$_2$-labeled membranes. The experiment was conducted as in Fig. 2 with addition of Ca^{2+} (1.0 μM) and GTP (10 μM). (B) Control. (□) InsP, (◆) InsP$_2$, (■) InsP$_3$.

[^{3}H]PtdIns was used as precursor. There were no significant changes in levels of [^{3}H]InsP, [^{3}H]InsP$_2$, and [^{3}H]InsP$_3$ during 10 min of incubation without anti-Ig. These findings indicate that anti-Ig stimulates the activation of a PLC which hydrolyzes PtdIns. As shown in Fig. 2, anti-Ig also induced release of [^{3}H]InsP when [^{3}H]PtdInsP$_2$ was used as precursor. This result is consistent with two distinct possibilities. First, anti-Ig may induce formation of [^{3}H]InsP$_3$ by PLC, but this [^{3}H]InsP$_3$ may be immediately dephosphorylated to InsP. Second, anti-Ig may induce dephosphorylation of PtdInsP$_2$, forming PtdIns which is then hydrolyzed by activated PLC to yield InsP.

We next examined the effect of Ca^{2+} and GTP on this response to determine if the lack of detectable release [^{3}H]InsP$_3$ from [^{3}H]PtdInsP$_2$ might reflect a requirement for Ca^{2+} and GTP in PtdInsP$_2$ hydrolysis as demonstrated in other systems.[11] As shown in Fig. 3, some hydrolysis of

[11] S. Cockcroft, *Biochem. J.* **240,** 503 (1986).

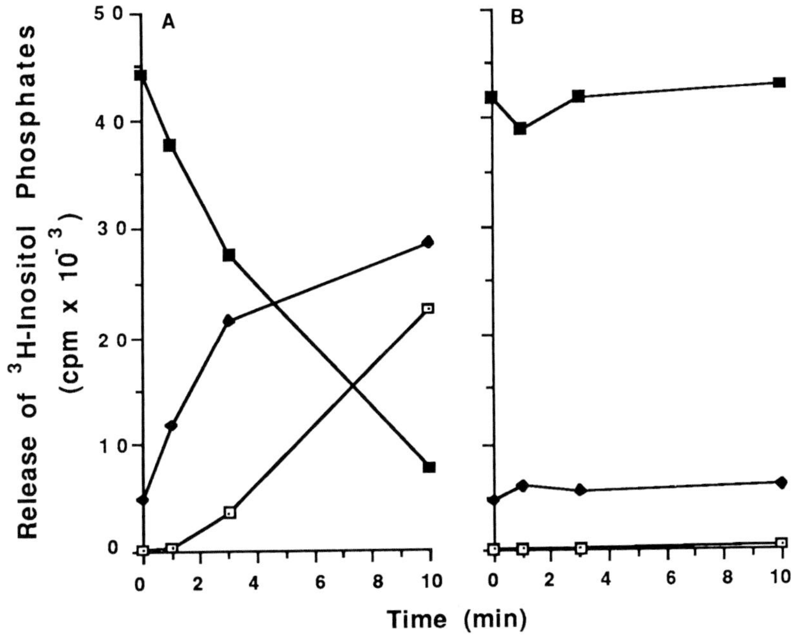

FIG. 4. [^{3}H]$InsP_3$ dephosphorylation. The sonicated B cell membranes (200 μg protein) were incubated with (A) and without (B) anti-Ig, with the dialyzed cytosol and [^{3}H]$InsP_3$ (50,000 cpm) at 37°. The reaction was stopped at varying times, and the amounts of [^{3}H]InsP, [^{3}H]$InsP_2$, and [^{3}H]$InsP_3$ were determined. (⊡) InsP, (◆) $InsP_2$, (■) $InsP_3$.

[^{3}H]$PtdInsP_2$ to yield [^{3}H]$InsP_3$ was seen in the presence of Ca^{2+} and GTP alone, suggesting the existence of a PLC activatable by Ca^{2+}/GTP. As also shown in Fig. 3, anti-Ig did not further stimulate formation of recoverable [^{3}H]$InsP_3$ but stimulated [^{3}H]InsP formation partially at the expense of background [^{3}H]$InsP_3$. These results were inconclusive in terms of the hypotheses stated earlier, but they provided suggestive evidence that anti-Ig stimulates activation of an $InsP_3$/$InsP_2$ phosphatase. To further address this possibility, we stimulated membranes with anti-Ig in the presence of [^{3}H]$InsP_3$ and assessed conversion of [^{3}H]$InsP_3$ to [^{3}H]$InsP_2$ and [^{3}H]InsP. As shown in Fig. 4, anti-Ig induced this conversion, confirming the activation of a phosphatase.

Conclusive proof that anti-Ig induces the hydrolysis of $PtdInsP_2$ was dependent on demonstration of enhanced accumulation of [^{3}H]$InsP_3$ following anti-Ig stimulation of membranes. We reasoned documentation of such an effect would require that we block activity of the phosphatase activated following stimulation with ligand. As shown in Fig. 5, ATP (or

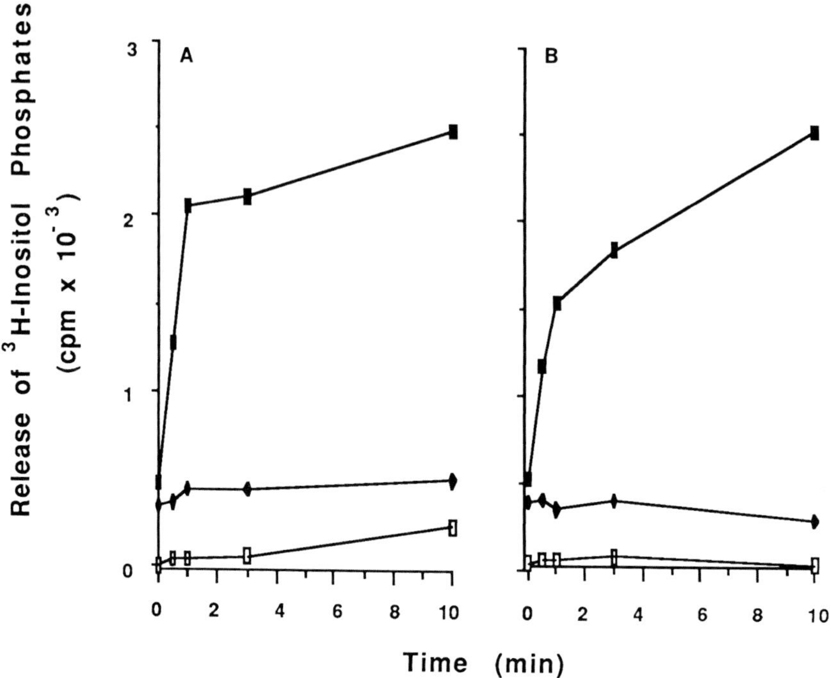

FIG. 5. (A) Release of [^{3}H]inositol phosphates from [^{3}H]PtdInsP$_2$-labeled membranes. The experiment was carried out as in Fig. 2, except SAMIG (1 μM) was used as stimulant and Ca^{2+} (0.1 μM), GTP (10 μM), and ATP (3 mM) were added to the reaction mixtures. (B) Unstimulated control. (□) InsP, (◆) InsP$_2$, (■) InsP$_3$.

ATPγS, data not shown) present in reaction mixtures at a concentration 3 mM blocked anti-Ig-induced InsP$_3$ dephosphorylation and allowed the increased accumulation of detectable [^{3}H]InsP$_3$ following anti-Ig stimulation of [^{3}H]PtdInsP$_2$-labeled membranes. Thus, the available evidence indicates that anti-Ig stimulated PLC-mediated hydrolyis of PtdIns (Fig. 1) and PtdInsP$_2$ (Fig. 5) in these isolated membranes and also stimulates dephosphorylation of InsP$_3$ and InsP$_2$ (Fig. 4).

Conclusion

Phosphoinositides, which constitute a small proportion of membrane phospholipids, are exquisitely sensitive to a phospholipase C activity stimulated by a number of external signals. Here, we have described techniques of membrane preparation, labeling, and analysis which allow documentation of hydrolysis of PtdIns and PtdInsP$_2$ and dephosphorylation of $InsP_3$ and $InsP_2$. These methods should be broadly applicable to other cell–receptor systems.

Acknowledgments

We thank Jan Schmid for assistance in constructing Ca^{2+} buffers, Holly Pickles for the preparation of the reagents, and Julia Gunnerson for secretarial assistance. Research supported by U.S. Public Health Service Grants AI 20519 and AI 21768.

Note Added in Proof

Further analysis has revealed that the anti-Ig preparations used in these studies were contaminated with ~1 μM $MgCl_2$ and that this $MgCl_2$ was responsible for the induction of PtdIns PLC and $InsP_3$ phosphatase activities reported here.

[23] A Rapid Method for the Resolution of Protein Kinase C Subspecies from Rat Brain Tissue

By Mark S. Shearman, Kouji Ogita, Ushio Kikkawa, and Yasutomi Nishizuka

Introduction

The importance of protein kinase C (PKC) in determining the response of cells to various stimuli, such as hormones, neurotransmitters, and growth factors, is well recognized.[1,2] The desire to study in more detail the biochemical and biophysical properties of this enzyme has led to the development of numerous purification procedures in our laboratory[3–6] and in many others.[7–10] Recently, however, molecular cloning analysis has revealed that the enzyme exists not as a single entity, but as a family of closely related subspecies. Chromatography on a hydroxyapatite column

[1] Y. Nishizuka, *Nature (London)* **308,** 693 (1984).

[2] Y. Nishizuka, *Science* **233,** 305 (1986).

[3] U. Kikkawa, Y. Takai, R. Minakuchi, S. Inohara, and Y. Nishizuka, *J. Biol. Chem.* **257,** 13341 (1982).

[4] U. Kikkawa, M. Go, J. Koumoto, and Y. Nishizuka, *Biochem. Biophys. Res. Commun.* **135,** 636 (1986).

[5] T. Kitano, M. Go, U. Kikkawa, and Y. Nishizuka, *in* "Neuroendocrine Peptide Methodology" (P. M. Conn, ed.), pp. 371–374. Academic Press, San Diego, California, 1989.

[6] M. Go, J. Koumoto, U. Kikkawa, and Y. Nishizuka, this series, Vol. 141, p. 424. See also references cited therein.

[7] S. E. Salama, *Thromb. Res.* **44,** 649 (1987).

[8] J. R. Woodgett and T. Hunter, *J. Biol. Chem.* **262,** 4836 (1987).

[9] G. W. Walton, P. J. Bertics, L. G. Hudson, T. S. Verdick, and G. N. Gill, *Anal. Biochem.* **161,** 425 (1987).

[10] M. W. Wooten, M. Vandenplas, and A. E. Nel, *Eur. J. Biochem.* **164,** 461 (1987).

has allowed the purified enzyme from whole rat brain to be resolved into three distinct fractions.[11,12] The method described herein is an adaptation of the aforementioned procedures that allows the rapid resolution of three distinct fractions enriched in PKC subspecies from a small amount (~1 g) of starting material.

Materials

DEAE–cellulose (DE-52) was obtained from Whatman. Hydroxyapatite columns (0.78 × 10 cm, type S) were purchased from Koken Co. Ltd. (Tokyo). 1,2-Diolein and phosphatidylserine (bovine brain) were obtained from Serdary Research Laboratories (London, ON, Canada) and employed directly without further purification. H1 histone was prepared from calf thymus as described previously.[13] [γ-^{32}P]ATP was purchased from Amersham.

Assay of Protein Kinase C

PKC subspecies are assayed by measuring the incorporation of ^{32}Pi into H1 histone from [γ-^{32}P]ATP, essentially as described previously.[4] The standard reaction mixture (0.25 ml) contains 20 m*M* Tris–HCl at pH 7.5, 200 μg/ml H1 histone, 10 μM ATP (containing ~10^5 cpm/assay tube), 5 m*M* magnesium acetate, 8 μg/ml phosphatidylserine, 0.8 μg/ml diolein, 0.3 m*M* calcium chloride, and the enzyme fraction from the hydroxyapatite column. For assays of nonspecific kinase activity, phospholipid and calcium are replaced by 0.5 m*M* ethyleneglycolbis(β-aminoethyl ether)-*N,N,N′,N′*-tetraacetic acid (EGTA). Assay tubes are incubated at 30° for 3 min. This incubation period is extended for samples containing low amounts of enzyme. The reaction is terminated by the addition of 2 ml of 25% trichloroacetic acid (TCA), and the acid-precipitable material is collected by filtration over nitrocellulose membrane disks (pore size 0.45 μm). Filters are rinsed 3 times with 25% TCA, and the radioactivity is quantitated by Cerenkov counting.

Experimental Procedures

Step 1: Preparation of the Crude Tissue Extract. Male Sprague–Dawley rats weighing between 150 and 200 g are used. All steps are carried out at 4°. Brain tissue (1–3 g wet weight) is homogenized in a Potter–Elveh-

[11] K.-P. Huang, H. Nakabayashi, and F. L. Huang, *Proc. Natl. Acad. Sci. U.S.A.* **83,** 8535 (1987).

[12] Y. Ono, U. Kikkawa, K. Ogita, T. Fujii, T. Kurokawa, Y. Asaoka, K. Sekiguchi, K. Ase, K. Igarashi, and Y. Nishizuka, *Science* **236,** 1116 (1987).

[13] E. Hashimoto, M. Takeda, Y. Nishizuka, K. Hamana, and K. Iwai, *J. Biol. Chem.* **251,** 6287 (1976).

jem Teflon–glass homogenizer with 10 volumes of 20 mM Tris–HCl at pH 7.5, containing 0.25 M sucrose, 10 mM EGTA, 2 mM ethylenediaminetetraacetic acid (EDTA), 1 mM phenylmethylsulfonyl fluoride, and 20 μg/ml leupeptin. The homogenate is centrifuged at 1500 g for 10 min to remove the nuclear fraction and cell debris, and the supernatant decanted off. The pellet is rehomogenized in 5 volumes of the above buffer, centrifuged at 1500 g for 10 min, and again the supernatant is retained. The combined supernatants are then centrifuged at 100,000 g for 60 min. The supernatant from this spin is employed as the crude tissue extract.

Step 2: DEAE–Cellulose Chromatography. The crude extract is applied to a DEAE–cellulose column (4 ml gel/g original wet weight of tissue), which has been equilibrated beforehand with 20 mM Tris–HCl, pH 7.5, containing 0.5 mM EGTA, 0.5 mM EDTA, and 10 mM 2-mercaptoethanol (buffer A). The column is then washed with 2 column volumes of buffer A, followed by 15 column volumes of buffer A containing 20 mM NaCl. Then, PKC activity is eluted from the column by the application of 3 column volumes of buffer A containing 120 mM NaCl, the first half-column volume of which is discarded.

Step 3: Hydroxyapatite Chromatography Coupled to a FPLC System. The DEAE–cellulose eluate is applied to a hydroxyapatite column (0.78 × 10 cm), coupled to a FPLC system (Pharmacia), which has been equilibrated beforehand with 20 mM potassium phosphate at pH 7.5, containing 0.5 mM EGTA, 0.5 mM EDTA, 10% glycerol, and 10 mM 2-mercaptoethanol (buffer B). PKC subspecies are resolved by the initiation of a preprogrammed phosphate buffer gradient elution, which, after washing the column with 48 ml of buffer B, raises the concentration of potassium phosphate linearly from 20 to 215 mM in a volume of 84 ml. The flow rate is 0.4 ml/min. Fractions of 1 ml each are collected, and the PKC activity present is determined as described above.

Figure 1 shows a representative elution profile of PKC activity from a hydroxyapatite column using rat hippocampus as the source of tissue. PKC is resolved into three distinct fractions, designated type I (γ), type II (β), and type III (α),[14] the peak activities eluting at approximately 70, 90,

[14] In the rat whole brain, type I enzyme consists of 697 amino acids and is encoded by γ sequence; type II enzyme is an unequal mixture of two enzymes determined by βI and βII cDNA sequences (encoding 671 and 673 amino acids, respectively), which differ from each other only in the carboxy-terminal region of about 50 amino acid residues. Partial genomic analysis indicated that βI and βII cDNAs result from alternative splicing of a common primary mRNA transcript[12]; type III enzyme consists of 672 amino acids and is encoded by α sequence. The nomenclature of α, βI, βII, and γ cDNAs is as described in U. Kikkawa, Y. Ono, K. Ogita, T. Fujii, Y. Asaoka, K. Sekiguchi, Y. Kosaka, K. Igarashi, and Y. Nishizuka, *FEBS Lett.* **217,** 227 (1987). It should be noted here, however, that the above designation may, in fact, be an oversimplification. A strong possibility exists that further heterogeneity will be found among the PKC subspecies so far detected.

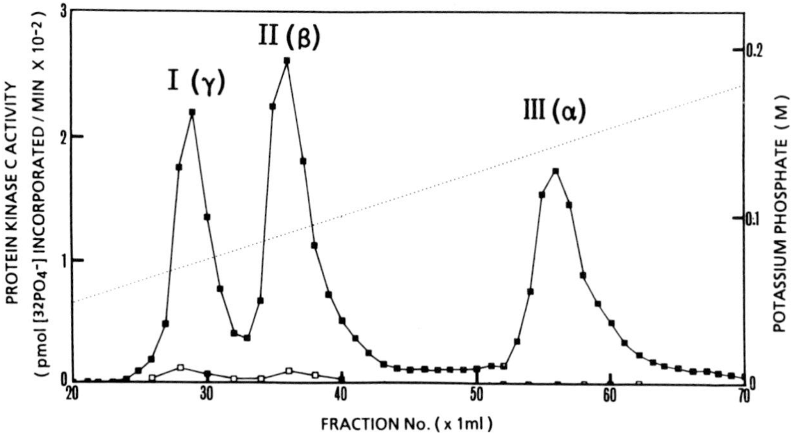

FIG. 1. Resolution of protein kinase C subspecies by hydroxyapatite column chromatography. Rat brain tissue crude extract (in this case hippocampus) was subjected to DEAE–cellulose followed by hydroxyapatite column chromatography, and the protein kinase activity in the column eluate fractions assayed as described in the text. Solid symbols, protein kinase activity in the presence of Ca^{2+}, phosphatidylserine, and 1,2-diolein; open symbols, protein kinase activity in the presence of EGTA. The nomenclature of the three enzyme fractions is as described in footnote 14.

and 140 m*M* potassium phosphate, respectively. (*N.B.* In practice, the absolute potassium phosphate concentration at which the three peaks elute shows some variability, which may in part be dependent on the condition of the hydroxyapatite column. The relative position of the three peaks, however, remains constant.) Similar elution profiles have been obtained when using tissue from other brain areas, although, as recently reported,[15] the relative amount of each PKC subspecies is dependent on the tissue area studied.

Comments

The method described above for the resolution of PKC subspecies has a number of advantageous features. First, the method is extremely rapid, the period of time from the initial processing of the tissue to obtaining separated fractions of PKC subspecies being less than 12 hr. Second, the method is sensitive enough to allow the use of small amounts of brain tissue (1 g or less), which facilitates the study of discrete brain areas.

[15] M. S. Shearman, Z. Naor, U. Kikkawa, and Y. Nishizuka, *Biochem. Biophys. Res. Commun.* **147,** 911 (1987).

Third, the simplicity of the method affords a high degree of reproducibility.

The PKC enzyme fractions eluted from the hydroxyapatite column, using the method described above, correspond closely to those reported by Huang *et al.*[11] and to those from a previous report from our laboratory[12] in which whole rat brain enzyme purified to apparent homogeneity was used. In addition, the enzyme subspecies show staining patterns very similar to the purified whole brain enzyme fractions when subjected to Western blotting analysis with type-specific monoclonal antibodies and polyclonal antisera. The latter evidence confirms the identity of the PKC subspecies obtained using the present method and attests to its suitability as a primary investigative approach to the study of PKC subspecies in brain, as well as in peripheral tissues.

[24] Assay of Peptidylglycine Monooxygenase: Glycine-Directed Amidating Enzyme

By THOMAS BROCK, JANE HUMM, and J. S. KIZER

Introduction

Over one-half of all biologically active secretory peptides and hormones are carboxamides. Analyses of the structure of the precursors of several polypeptide hormones led to the postulate that during the posttranslational hydrolysis of prohormones in the secretory vesicle, exposure of a carboxy-terminal glycine–COOH would result in the amidation of the preceding residue. This hypothesis was quickly proven correct by the discovery in pituitary secretory vesicles of an enzyme that catalyzed this reaction.[1] Subsequently, the enzyme, now designated as peptidylglycine monooxygenase, has been found in many different tissues and species, including frog skin.[2–7]

[1] A. F. Bradbury, M. D. Finnie, and D. G. Smyth, *Nature* (*London*) **278,** 686 (1982).

[2] B. A. Eipper, R. E. Mains, and C. C. Glembotksi, *Proc. Natl. Acad. Sci. U.S.A.* **80,** 5144 (1983).

[3] I. Husain and S. S. Tate, *FEBS Lett.* **152,** 277 (1983).

[4] J. S. Kizer, W. H. Busby, Jr., C. Cottle, and W. W. Youngblood, *Proc. Natl. Acad. Sci. U.S.A.* **81,** 3228 (1983).

[5] S. Gomez, C. di Bello, L. T. Hung, R. Genet, J.-L. Morgat, P. Fromageot, and P. Cohen, *FEBS Lett.* **167,** 160 (1984).

The molecular weight of the enzyme differs depending on the tissue source from which it is extracted. An M_r of 33,000 has been reported for the enzyme from *Xenopus laevis* skin,[6] 64,000 for the porcine pituitary enzyme, and 38,000, 54,000, and 60,000 for the proteins found in bovine pituitary.[1,8] The enzyme is a glycoprotein,[10] is maximally active at a pH of approximately 7.0,[1-10] and is a metalloenzyme, probably with a planar transition metal such as nickel or copper at the catalytic site.[2,6,9,10] Metal chelators, such as *o*-phenanthroline, imidazole, and EDTA, and thiol reducers, such as dithiothreitol and mercaptoethanol, readily inhibit the enzyme.[2,6,10]

The affinity of the catalytic pocket appears highly specific for substrates possessing a carboxy-terminal glycine, although artificial substrates ending in D-alanine appear to be competitive inhibitors.[11] Model peptides with charged amino acids preceding the glycine are poor substrates, whereas peptides with neutral or hydrophobic amino acids in this position are considerably better.[9,11] In addition, substrates with proline preceding the glycine also have a poorer affinity for the enzyme.[12]

By kinetic analysis, catalysis involves a two-step transfer or Ping-Pong mechanism, and the two substrates, peptidylglycine–COOH and ascorbate, are competitive inhibitors of one another.[10] This latter property of the enzyme complicates the interpretation of results of assays where changes in reaction velocity may be due to changes in affinity for either of the two substrates. (For a further discussion of problems relating to assay of enzymes demonstrating double-substrate inhibition, see below).

Enzymatic catalysis is thought to involve the generation of an hydroxyl radical by the reaction of ascorbate and copper at the substrate binding site.[1,2,10,13] Next, the hydroxyl radical could extract the α hydrogen from the glycine, generating a peptide free radical[10,13] or perhaps a dehydropeptide,[1] a postulate consistent with the observation that abstraction of an α hydrogen from glycine is the rate-limiting step.[10] A second

[6] K. Mizuno, J. Sakata, M. Kojima, K. Kangawa, and H. Matsuo, *Biochem. Biophys. Res. Commun.* **137,** 984 (1986).

[7] J. Sakata, K. Mizuno, and H. Matsuo, *Biochem. Biophys. Res. Commun.* **140,** 230 (1986).

[8] A. S. N. Murthy, R. E. Mains, and B. A. Eipper, *J. Biol. Chem.* **261,** 1815 (1986).

[9] A. F. Bradbury and D. G. Smyth, *Biochem. Biophys. Res. Commun.* **112,** 372 (1983).

[10] J. S. Kizer, R. C. Bateman, Jr., C. R. Miller, J. Humm, W. H. Busby, Jr., and W. W. Youngblood, *Endocrinology (Baltimore)* **118,** 2262 (1986).

[11] A. E. N. Landymore-Lim, A. F. Bradbury, and D. G. Smyth, *Biochem. Biophys. Res. Commun.* **117,** 289 (1983).

[12] L. J. Moray, C. R. Miller, W. H. Busby, Jr., J. Humm, R. C. Bateman, Jr. and J. S. Kizer, *J. Neurosci. Methods* **14,** 293 (1985).

[13] R. C. Bateman, Jr., W. W. Youngblood, W. H. Busby, Jr., and J. S. Kizer, *J. Biol. Chem.* **260,** 9088 (1985).

hydroxyl radical could react with the peptide radical generating a "hemiamidal" which would spontaneously degrade to the peptide amide and glyoxylic acid.[10,13] Alternatively, a dehydropeptide could undergo hydrolysis to the same two products.[1] Either postulate is compatible with the observation of Bradbury *et al.* that the glycine nitrogen is the ultimate source of the amide nitrogen.[1]

Principles of Assay

At present there are three useful assays for the measurement of peptidylglycine monooxygenase. All are based on the enzymatic conversion of a model peptide ending in glycine to the corresponding peptide amide. All also utilize short peptides containing the unnatural amino acid, D-tyrosine, to minimize proteolysis of substrate and product by other enzymes.

The first of the assays is based on the conversion of ^{125}I-D-Tyr-Val-Gly-OH to ^{125}I-D-Tyr-Val-NH_2 and the separation of product and substrate by either HPLC[1] or ion-exchange chromatography.[2] The second assay is based on the conversion of ^{125}I-*N*-acetyl-Tyr-Phe-Gly-OH to ^{125}I-*N*-acetyl-Tyr-Phe-NH_2 and the extraction of product into aqueous ethyl acetate at pH 7.0.[6] Product in both of these two assays is measured by direct counting.

The third assay, reported in this chapter, is based on the conversion of unlabeled D-Tyr-Val-Gly-OH to D-Tyr-Val-NH_2 and measurement of product in a highly sensitive radioimmunoassay which readily distinguishes substrate and product.[12] We believe the merits of this approach to be several: (1) A very large number of samples can easily be assayed at the same time since time-consuming steps to separate product and substrate are unnecessary; (2) the radioimmunoassay for product is highly sensitive, readily discriminates between substrate and product, and can reproducibly measure as little as 30 fmol of D-Tyr-Val-NH_2 (Fig. 1); (3) iodination of model peptides such as D-Tyr-Val-Gly-OH or ⟨Glu-His-Pro-Gly-OH considerably reduces their efficiency as substrates.[12] In principle, any substrate ending in glycine which does not contain a charged amino acid in the penultimate position may be used for a substrate provided that one has a reproducible means for distinguishing substrate and product, and can account for any degradative activity directed toward sample, product, or substrate.

Preparation of Tissues

Fresh or frozen tissues or subcellular fractions are homogenized in a hypotonic buffer (usually 50 m*M* sodium phosphate, pH 6.8–7.0) at a

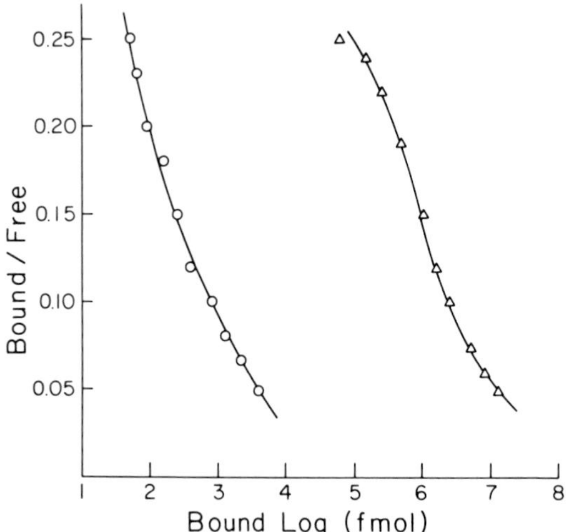

FIG. 1. Scatchard analysis comparing affinity of antiserum raised against BSA–D-Tyr-Val-NH_2 for D-Tyr-Val-NH_2 (○) and D-Tyr-Val-Gly-OH (△). (Reproduced with permission from Moray *et al.*[12])

concentration of 20%, w/v, or 2% protein/volume. Freshly prepared *N*-ethylmaleimide is quickly added to a final concentration of 5.0 m*M*. The homogenate is frozen and thawed twice, sonicated for 10–15 sec at 20,000 Hz at 150–200 W, and centrifuged at high speed for 20 min (average relative centrifugal force greater than 25,000 *g* is usually sufficient). The resulting high-speed supernatant may then be assayed directly or subjected to further purification as needed.

The addition of the *N*-ethylmaleimide is essential to stabilize the enzyme and insure maximal enzyme activity. In early work, crude enzyme preparations were observed to rapidly lose activity which could be restored on addition of transition metals such as copper or nickel.[2] Furthermore, maximal activity of the enzyme in crude preparations was obtained only after the addition of relatively high concentrations of these metals. These observations led to the postulate that a transition metal binding protein present in tissues inhibited the enzyme and that assay of the enzyme in crude samples required the addition of large quantities of exogenous metal to overcome this inhibition.[2,9] It was also believed that accurate measurement of enzyme activity required the determination of an optimal metal concentration for each tissue because of the observation that excessive metal ion itself inhibits the enzyme.[2,6,10]

Recent evidence argues that addition of *N*-ethylmaleimide throughout tissue preparation and assay removes the need for optimizing the metal concentration for each tissue, presumably by inactivating sulfhydryls on the interfering proteins or enzymes.[7,12] These observations are buttressed by the absence of a requirement for *N*-ethylmaleimide or of high concentrations of transition metals for assay of purified enzyme.[6,8,10]

Choice of Buffers

Any of several buffers may be chosen for tissue preparation and assay, including Tris, HEPES, MOPS, and phosphate. Certain buffers such as borate may inhibit the enzyme due to chelation of transition metals.

Enzyme Incubations

Twenty-five microliters of enzyme are added to 10 × 75 mm glass tubes followed by 25 μl of a cocktail containing nickel or copper chloride, *N*-ethylmaleimide, ascorbic acid, potassium iodide, catalase, and the substrate, D-Tyr-Val-Gly-OH (Bachem, Torrance, CA). The final concentration of reactants is as follows: copper, 5 μM (nickel, 10 μM); *N*-ethylmaleimide, 0.5 m*M*; ascorbic acid, 5.0–10 m*M*; potassium iodide, 25 m*M*; catalase, 100 μg/ml; and D-Tyr-Val-Gly-OH, 0.2 m*M*. The sample tubes are covered and incubated at 37°. Blank assays consisting of boiled tissue and assay cocktail must also be included. After 1–3 hr, 2 ml of 0.1 *M* HCl is added to stop the reaction. The samples are vortexed, and 15 to 40-μl aliquots are transferred in duplicate to a second set of 10 × 75 mm glass tubes. The samples are lyophilized and the dried product assayed by radioimmunoassay as described below.

Radioimmunoassay for Product (D-Tyr-Val-NH_2)[12]

Preparation of Immunogen and Raising of Antiserum. D-Tyr-Val-NH_2 (Bachem) is mixed with bovine serum albumin or other suitable carrier at a molar ratio of 20 : 1 in 0.4 ml of 0.1 *M* MES buffer, pH 4.9. Solid 1-ethyl-3-(3-dimethylaminopropyl)carbodiimide, a water-soluble carbodiimide, is added to a 5 molar excess with respect to the D-Tyr-Val-NH_2. The reaction is allowed to continue overnight at 4°. The conjugate is subsequently dialyzed exhaustively against 0.15 *M* NaCl. Approximately 200–300 μg of the conjugate in 1 ml of 0.15 NaCl are mixed with 1 ml of Freund's complete adjuvant (200–300 μg per animal) and injected into a New Zealand White rabbit by either the standard intramuscular route or by the multiple intradermal method of Vaitukaitis *et al.*[14] Booster injections using the same antigen preparation, but substituting incomplete for Freund's

[14] J. Vaitukaitis, J. B. Robbins, E. Nieschlag, and G. T. Ross, *J. Clin. Endocrinol.* **33,** 988 (1971).

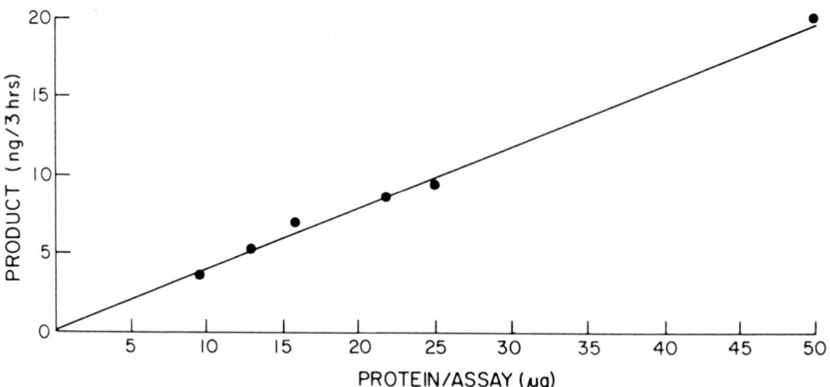

FIG. 2. Reaction product formed by crude rat brain supernatant as a function of protein content. (Reproduced with permission from Moray *et al.*[12])

complete adjuvant, are made every 2–3 weeks after the primary immunization. ^{125}I-Labeled-D-Tyr-Val-NH_2 is used both to screen the rabbits for a reactive antiserum and as tracer for the development of the radioimmunoassay.

Radioimmunoassay for D-*Tyr-Val-NH_2*. One hundred microliters of RIA buffer (0.154 *M* NaCl, 10 m*M* sodium phosphate, 0.01% sodium azide, 10 m*M* EDTA, and 0.4% bovine serum albumin, pH 7.5) is added to each of the lyophilized samples of product. Next, 100 μl of primary anti-D-Tyr-Val-NH_2 serum appropriately diluted with NRS buffer (RIA buffer containing 1.5% nonimmune rabbit serum) is added. Next, approximately 20,000 dpm of ^{125}I-D-Tyr-Val-NH_2 is added in 100 μl of RIA buffer. Sufficient antiserum against rabbit IgG is then added to ensure maximal precipitation of the antibody–ligand complex. The tubes are vortexed, covered, and incubated for at least 12 hr at 4°. Finally, the tubes are centrifuged at 900 *g* for 15 min, and the supernatants are removed by vacuum aspiration. The pelleted immunoglobulin–ligand complexes are then counted in a gamma counter. Standards of unlabeled D-Tyr-Val-NH_2 carried through the same procedure are used to quantitate the product.

Troubleshooting and Interpretation of Results

The addition of catalase and (potassium) iodide, a nucleophilic halogen, is essential and serves to inhibit the generation of excess hydrogen peroxide, hydroxyl radical, and superoxide ion by the interaction of ascorbic acid and copper in solution.[12,13,15] Omission of these two chemicals

[15] B. Halliwell and J. M. C. Gutteridge, *Biochem. J.* **219,** 1 (1984).

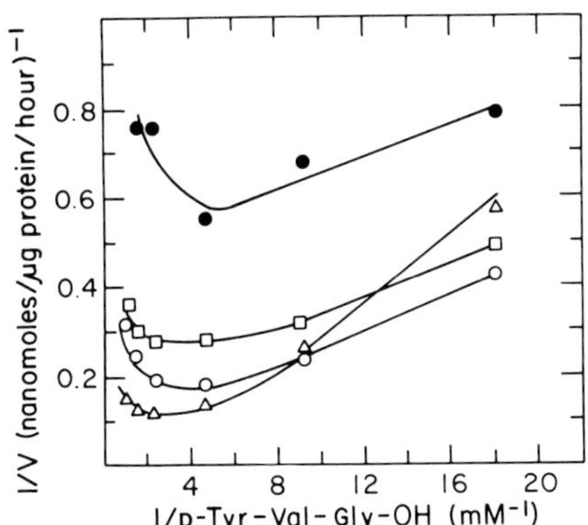

FIG. 3. Double-reciprocal plot of velocity (V) versus substrate concentration (D-Tyr-Val-Gly-OH) at varying concentrations of ascorbate [(●) 0.38 mM, (△) 4.0 mM, (□) 1.0 mM, (○) 2.0 mM]. (Reproduced with permission from J. S. Kizer, R. C. Bateman, Jr., C. R. Miller, J. Humm, W. H. Busby, Jr., and W. W. Youngblood, "Purification and characterization of a peptidyl glycine monooxygenase from porcine pituitary," *Endocrinology* (*Baltimore*), **118,** 2262–2267, 1986, © by The Endocrine Society.)

from the incubation results in the measurement of significantly less enzymatic activity, probably because of destruction of the enzyme by activated oxygen species.[12] It is also worth emphasizing that the N-ethylmaleimide should be freshly prepared since this active species is unstable in solution. Ascorbate is also unstable at neutral pH and should also be freshly prepared.

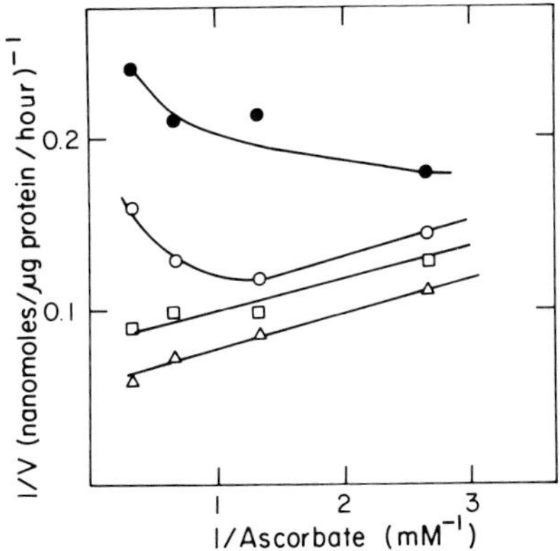

FIG. 4. Double-reciprocal plot of voleocity (V) versus substrate concentration (ascorbate) at varying concentrations of D-Tyr-Val-Gly-OH [(●) 0.05 mM, (○) 0.1 mM, (□) 0.2 mM, (△) 0.4 mM]. (Reproduced with permission from J. S. Kizer, R. C. Bateman, Jr., C. R. Miller, J. Humm, W. H. Busby, Jr., and W. W. Youngblood, "Purification and characterization of a peptidyl glycine monooxygenase from porcine pituitary," *Endocrinology* (*Baltimore*), **118,** 2262–2267, 1986, © by The Endocrine Society.)

The enzyme assay is linear for 2–3 hr, depending on the amount of enzyme in the samples, and can be used on relatively crude supernatants (Fig. 2). Optimal protein concentrations must be determined empirically for each tissue. In addition, it is necessary to determine the relative stability of product and substrate in homogenates from each tissue source under conditions in which the peptidylglycine monooxygenase is inactive, i.e., no added ascorbate.

The concentrations of ascorbate and substrate are critical to the sensitivity of the assay. The catalytic mechanism of the enzyme involves a two-step transfer, and, as is common for many enzymes of this type, each substrate is a competitive inhibitor of the other (Figs. 3 and 4).[10] This property of the enzyme makes it somewhat more difficult to choose appropriate concentrations of ascorbate and D-Tyr-Val-Gly-OH to obtain maximal activity. In practice, concentrations of ascorbate (5.0–10.0 mM) above its K_m (3.0 mM) are used, and the concentration of D-Tyr-Val-Gly-OH is adjusted to give maximal activity. In a compulsory ordered system in which ternary complexes are not formed, this technique will allow measurement of near maximum velocity.

Interpretation of the results of assays wherein changes in enzyme activity are being investigated are also complicated by the enzyme mechanism. Because the enzyme assay cannot be run at saturation, increases in measured activity may result from either increases in enzyme protein or *decreases* or *increases* in the affinity of the enzyme for either of its two substrates. Only detailed kinetic analysis will distinguish among these several possibilities.

[25] Assay of Glutaminylpeptide Cyclase

By JEANNE B. KOGER, JANE HUMM, and JOHN S. KIZER

Introduction

The amino-terminal amino acid of many hormones and neurotransmitter peptides is pyroglutamic acid. Recent evidence indicates that a pyroglutamyl peptide results from the enzymatic cyclization of an amino-terminal glutamine during the posttranslational processing of a prohormone.[1,2] Enzymes catalyzing this reaction have been found in secretory vesicles from pituitary, brain, adrenal medulla, and B lymphocytes.[1,2]

[1] W. H. Fischer and J. Spiess, *Proc. Natl. Acad. Sci. U.S.A.* **84,** 3628 (1987).

[2] W. H. Busby, Jr., G. E. Quackenbush, J. Humm, W. W. Youngblood, and J. S. Kizer, *J. Biol. Chem.* **262,** 8532 (1987).

The enzyme(s) found in porcine pituitary is perhaps the best characterized.[2] It is a glycoprotein exhibiting charge and size heterogeneity and is predominantly found in secretory vesicles. It also has a neutral pH optimum, is stimulated by high-salt concentrations, probably has catalytically important sulfhydryls, and requires no added cofactors. Because the cyclization of glutamine is thermodynamically favorable, wherein the nucleophilic α amine attacks the amide carbon with the release of ammonia, it is likely that the enzyme merely ensures a favorable conformation on the substrate, thereby enhancing the rate of reaction. The substrate specificity of the enzyme is currently uncertain, but substrates with an amino-terminal glutamine followed by two or more amino acids may have the highest affinity for the catalytic site. The influence of the chemical characteristics of the amino acids following glutamine on affinity or reaction velocity is unknown. In this chapter, we describe a simple assay for glutaminylpeptide cyclase applicable to a number of different tissues.

Principles of Assay

The assay is based on the conversion of the substrate, Gln-His-Pro-NH_2, to the product, ⟨Glu-His-Pro-NH_2 (thyrotropin-releasing hormone, TRH). The product is quantitated by radioimmunoassay using a unique TRH antiserum which has a higher affinity for product than substrate, or by differentially extracting substrate from product on an anion-exchange resin and measuring the product in a standard TRH radioimmunoassay (RIA).[2] Other means of assay based on separation of substrate and product by HPLC are also useful.[1]

Interpretation of the results of assay of crude tissue samples is complicated by the presence of pyroglutamyl peptide hydrolases (both soluble and particulate) and prolyl endopeptidase (EC 3.4.21.26), which degrade the product, in the case of the former, and both substrate and product, in the case of the latter.[3–10] Approaches to this difficulty are discussed in the following sections. In principle, any Gln–peptide substrate can be substituted for Gln-His-Pro-NH_2 provided that one has a means for distinguishing substrate and product and can account for any degradative activity in the samples.

[3] T. H. Friedman and S. Wilk, *J. Neurochem.* **46,** 1231 (1986).

[4] W. H. Busby, W. W. Youngblood, and J. S. Kizer, *Brain Res.* **242,** 261 (1987).

[5] B. O'Conner and G. O'Cuinn, *J. Neurochem.* **48,** 676 (1987).

[6] M. Orlowski, E. Wilk, S. Pearce, and S. Wilk, *J. Neurochem.* **33,** 461 (1987).

[7] E. C. Griffiths and J. A. Kelly, *Mol. Cell. Endocrinol.* **14,** 3 (1979).

[8] C. Prasad and A. Peterkofsky, *J. Biol. Chem.* **251,** 3229 (1976).

[9] K. Bauer and H. Kleinkauf, *Eur. J. Biochem.* **106,** 107 (1980).

[10] M. S. Kreider, A. Winokur, and N. R. Krieger, *Neuropeptides* **1,** 455 (1981).

Preparation of Tissues

Fresh or frozen tissues or subcellular fractions are homogenized in a hypotonic buffer (usually 10–50 m*M* MOPS, pH 7.0–7.5) at concentration of 10%, w/v (or 1% protein/volume). The homogenate is frozen and thawed twice, sonicated for 1–15 sec at 20,000 Hz at 150–200 W, and centrifuged at high speed (average relative centrifugal force greater than 50,000 *g* is usually sufficient) for 20–30 min. The resulting high-speed supernatant may then be assayed directly or subjected to further purification as needed.

Enzyme Incubations

Fifty microliters of enzyme is added to glass tubes followed by the addition of 30 μl of a mixture containing 10 nmol of Gln-His-Pro-NH_2 (final concentration, 0.125 m*M*), 4 μmol of EDTA (final concentration, 50 m*M*), and 32 μmol of NaCl (final concentration, 400 m*M*). (In practice, it is often easier to add solid NaCl to the enzyme to a final concentration of 400 m*M* before assay. This results in maximum activation and stability of the enzyme and avoids the use of high salt concentrations in the substrate cocktail.) The sample tubes are covered and incubated at 37°. Blank assays consisting of boiled tissue samples must also be included. After 60 min, the reaction is stopped by the addition of 2–4 ml of either ice-cold water or 10 m*M* MOPS, pH 7.2, depending on the method chosen for assay of product (see below). Under these conditions the assay is linear for up to 90 min (Fig. 1), provided that the amount of enzyme present converts less than 10% of the added substrate to product.

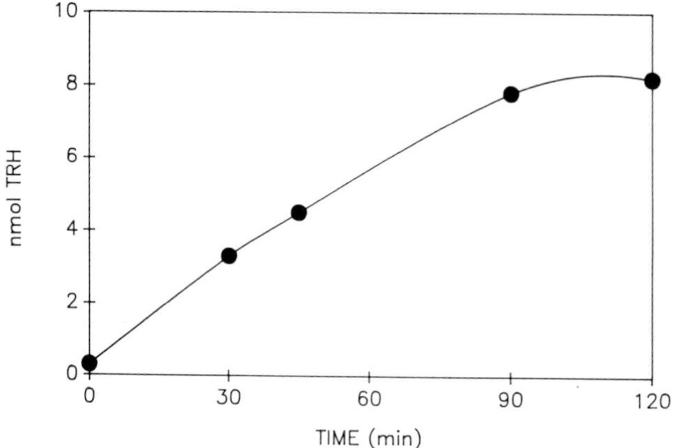

FIG. 1. Formation of product (nmol TRH formed/50 μg of protein) as a function of time by fractions obtained from gel filtration of a porcine pituitary homogenate as shown in Fig. 2.

Choice of Buffers

A number of different buffers can be used for the preparation of tissues and enzyme incubations, including MOPS, HEPES, and Tris. Phosphate buffers are to be avoided because they catalyze the spontaneous cyclization of amino-terminal glutamine, especially at higher temperatures.

Assay of Product

RIA of Product Using Amino-Terminal Directed Anti-TRH Serum. ⟨Glu-His-Gly-COOH (commercially available through Peninsula Labs, Belmont, CA) is mixed with a carrier protein such as bovine serum albumin at a molar ratio of 20 : 1 in 400 μl of 0.1 MES buffer, pH 5.0. Solid 1-ethyl-3-(3-dimethylaminopropyl)carbodiimide, a water-soluble carbodiimide, is added to a 5 molar excess with respect to ⟨Glu-His-Gly-COOH.[2] The reaction is incubated overnight at 4° and then dialyzed for 6–8 hr against 0.15 *M* saline. Approximately 200–300 μg of the conjugate in 1 ml of 0.15 *M* saline are mixed with 1 ml of Freund's complete adjuvant (200–300 μg per animal) and injected into New Zealand White rabbits by either the standard intramuscular route or by the multiple intradermal method of Vaitukaitis *et al.*[11] Booster injections using the same antigen preparation, but substituting incomplete for Freund's complete adjuvant, are made every 2–3 weeks after the primary immunization. ^{125}I-Labeled TRH is used both to screen the rabbits for a reactive antiserum and as tracer for the development of a TRH RIA (see below).

This amino-terminally directed antibody may be expected to have an affinity for the product, ⟨Glu-His-Pro-NH_2, approximately 10^3 liter/mol higher than for the substrate, Gln-His-Pro-NH_2, allowing measurement of product in the presence of substrate.[2] Antiserum obtained from early phlebotomies are usually of most use, since we have observed the frequent development of tolerance to this particular antigen after several immunizations. After enzyme incubation, the reaction is stopped by the addition of 2–4 ml ice-cold water, and 5 to 25-microliter aliquots removed for radioimmunoassay.

Assay of Product Based on Differential Anion-Exchange Chromatography. The theoretical pK_a of the product, ⟨Glu-His-Pro-NH_2, is approximately 6.3 (range 5.6–7.0). For the substrate, the pK_{a1} (imidazole) is approximately 6.3 (range 5.6–7.0), and the pK_{a2} is approximately 8.0 (range 7.6–8.4). Therefore, at a pH of 7.2, the greatest difference in charge between substrate and product will obtain, and the majority of the substrate will be retained by an anion-exchange resin while the product

[11] J. Vaitukaitis, J. B. Robbins, E. Nieschlag, and G. T. Ross, *J. Clin. Endocrinol.* **33**, 988 (1971).

will not. This permits the rapid separation of substrate from product by batch extraction, permitting the subsequent assay of the product, TRH, using the readily available, standard RIA for TRH.

Following the enzyme incubation, 2–4 ml of ice-cold 10 m*M* MOPS, pH 7.2, is added to each sample. Next is added 400 μl of a 1 : 1 suspension (constantly stirred) of Dowex 50 in 10 m*M* MOPS, pH 7.2 (prepared as outlined below). The sample is allowed to sit at 4° with occasional vortexing for 60 min after which it is centrifuged at low speed to sediment the resin. Samples, 5–20 μl, may then be removed for TRH RIA. It is not necessary to remove the supernatant from the resin as we have found the resin–substrate complex to be stable for at least 8 hr (i.e., ⟨Glu-His-Pro-NH_2 forms only very slowly under these conditions).

Preparation of Anion Exchanger

Dowex 50 (4% cross-linked, 200–400 mesh) is rinsed with several volumes of distilled water over a Büchner funnel. The resin cake is then heated to 100° for 10 min in 4 volumes of 4 *M* NaOH. The resin is allowed to settle, and the supernatant decanted. The resin cake is then rinsed repeatedly with distilled water until the pH is neutral. Next, the boiling and rinse steps are repeated with 4 *M* HCl in place of NaOH. Finally, the resin is washed exhaustively with distilled water until the wash gives a negative silver nitrate test for chloride ion.

The pH of the resin is adjusted to 7.2 by washing with 1.0 *M* MOPS, pH 7.2, followed by 5 washes with 4 volumes of 10 m*M* MOPS, pH 7.2, draining well between washes. The Dowex is now conditioned and is suspended in an equal volume of 10 m*M* MOPS, pH 7.2. In this form, it is stable for several months at 4°.

Radioimmunoassay of Product (TRH)[12]

Aliquots (5–25 μl) from the enzyme assay are added to 10 × 75 millimeter glass tubes, and a sufficient amount of RIA buffer (0.154 *M* NaCl, 10 m*M* sodium phosphate, 0.01% sodium azide, 0.4% bovine serum albumin, pH 7.5) is added to bring the sample volume to 100 μl. One hundred microliters of primary anti-TRH serum appropriately diluted with NRS buffer (RIA buffer containing 1.5% normal rabbit serum) is added. Next, approximately 25,000 dpm of ^{125}I-TRH is added in 100 μl of RIA buffer. Sufficient antiserum against rabbit IgG is then added to en-

[12] J. S. Kizer, M. Palkovits, M. Tappaz, J. Kebabian, and M. J. Brownstein, *Endocrinology (Baltimore)* **98,** 685 (1976).

sure maximal precipitation of the antibody–ligand complex. The tubes are vortexed, covered, and incubated for at least 12 hr at 4°. Last, the tubes are centrifuged at 900 *g* for 15 min, and the supernatants are removed by vacuum aspiration. The pelleted immunoglobulin–ligand complexes are then counted in a gamma counter. Standards carried through the same procedure are used to quantitate the product.

Preparation of Substrate

Because the substrate Gln-His-Pro-NH_2 (and any other Gln–peptide) is somewhat unstable (the amino-terminal Gln undergoes spontaneous cyclization to ⟨Glu) proper handling of the Gln–peptide is essential to ensure the lowest blanks and highest enzyme activity. Conditions favorable to spontaneous cyclization include heat, excess aqueous acid or base, and phosphate buffers.[13] Milder conditions of heat, acid, base or phosphate drive the reaction more efficiently in the presence of a surface such a chromatography resin. Although Gln–peptides may be synthesized or purchased with an unblocked α amine, we believe it is best to synthesize or purchase an N-blocked substrate and deblock it prior to use. The following is a standard protocol for use of the substrate N^{α}-tBoc-Gln-His-Pro-NH_2.

N^{α}-tBoc-Gln-His-Pro-NH_2 may be purchased as a custom synthesized peptide or synthesized by standard techniques. Gln-His-Pro-NH_2 is then freshly prepared as needed by reaction of the N^{α}-tBoc-Gln-His-Pro-NH_2 with 3.5 *M* HCl in *anhydrous* dioxane, followed by drying under a steam of nitrogen and dissolution of the product in 50 m*M* MOPS, pH 7.2. The Gln-His-Pro-NH_2 stored in this manner is stable for 2–3 weeks at −20°.[2]

Problems with Pyroglutamyl Peptide Hydrolases and Prolyl Endopeptidase

As mentioned previously, enzymes which degrade the product, TRH, are present in many tissues. Prolyl endopeptidase, an enzyme which hydrolyzes the carboxy-terminal prolylamide, is found in pituitary, brain, and other tissues, has a molecular weight of 50,000–70,000, and appears to be predominantly cytosolic.[3,4,6–10] It is probably a serine protease whose catalytic activity also depends on reactive sulfhydryls. Inhibitors of this enzyme include sulfhydryl inactivators such as *N*-ethylmaleimide or *p*-mercuribenzoate, heavy metals such as Zn or Hg, and possibly, the

[13] K. Folkers, J.-K. Chang, B. L. Currie, C. Y. Bowers, A. Weil, and A. V. Schally, *Biochem. Biophys. Res. Commun.* **39,** 110 (1970).

cystatins. The enzyme is activated by thiol-reducing agents such as dithiothreitol and metal chelators such as EDTA.[3,4,6–10]

A second enzyme, pyroglutamyl peptide hydrolase (broad specificity), is also found in many different tissues. It is both particulate and soluble, has a molecular weight of about 30,000, and depends on active sulfhydryls for catalytic activity. It is inhibited by the same sulfhydryl inactivators as prolyl endopeptidase and activated by thiol-reducing agents as well.[3,4,7–10]

A third enzyme, pyroglutamyl peptide hydrolase (narrow specificity), is abundant in brain. This enzyme is of rather narrow substrate specificity (appearing to have affinity only for ⟨Glu-His-Pro-NH_2 or closely related substrates). It has a high molecular weight, is a metalloenzyme, and is particulate, being released from membrane preparations only after treatment with papain. This enzyme is inhibited by metal chelators such as EDTA and thiol-reducing agents.[3,5]

There are several approaches to ridding assay samples of these interfering proteolytic enzymes.

1. Samples may be chromatographed to separate the degradative activity from the peptide glutaminyl cyclase. This is a reasonable approach if there are few samples. In gel filtration, for example, the molecular weight of the cyclase permits it to elute either leading the pyroglytamyl peptide hydrolase or following the prolyl endopeptidase (Fig. 2). Other chromatographic methods to separate the cyclase from the degradative enzymes may also be useful.

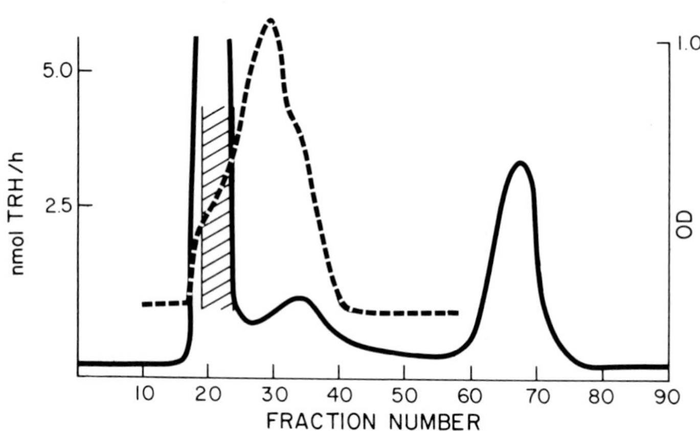

FIG. 2. Formation of ⟨Glu-His-Pro-NH_2 (TRH) from Gln-His-Pro-NH_2 and degradation of exogenous ⟨Glu-His-Pro-NH_2 by fractions of a gel permeation chromatography of porcine pituitary. (——) Absorbance (254 m*M*), (---) glutaminyl cyclase activity, (▨) region of TRH degradation. (Reproduced with permission from Busby *et al.*[2])

2. High-speed centrifugation of homogenates is useful for removing pyroglutamyl peptide hydrolase (narrow specificity) from assay samples. Usually, following centrifugation, there is little of this enzyme remaining to interfere with the assay of glutaminylpeptide cyclase. Also the routine use of EDTA in the assay system will completely inhibit this enzyme.

3. Because glutaminyl cyclase and the degradative enzymes, prolyl endopeptidase and pyroglutamyl peptide hydrolase (broad specificity), are all dependent on reactive sulfhydryl groups, thiol-reactive agents, such as Zn, Hg, Fe, *N*-acetylimidazole, and iodoacetamide, cannot be used in the assay cocktail to prevent degradation of product. We have observed, however, that the sensitivity of the glutaminyl cyclase to inhibition by *N*-ethylmaleimide (2–5 m*M*) appears less than that for either prolyl endopeptidase and pyroglutamyl peptide hydrolase. Consequently, in some tissues, glutaminyl cylase can be measured in the presence of the two degradative enzymes by adding appropriate amounts of *N*-ethylmaleimide to the incubation (1–2 m*M*). Alternatively, substrate analogs of the degradative enzymes such as *N*-benzyloxycarbonylprolylprolinal or *N*-acetylglycylprolinamide (inhibitors of prolyl endopeptidase) or pyroglutamylamide or pyroglutamyl diazomethyl ketone (inhibitors of pyroglutamyl peptide hydrolase) may be added to the incubation.[3,4]

[26] Primary Thyrotropin-Releasing Hormone-Degrading Enzymes

By CHARLES H. EMERSON

Introduction

Thyrotropin-releasing hormone (TRH) is unstable in serum and tissues due to the presence of both specific and nonspecific peptidases.[1] As noted elsewhere in this volume ([40]), two modes of primary enzymatic degradation of TRH are currently recognized. The first of these is due to TRH pyroglutamyl aminopeptidase (EC 3.4.19.3, 5-oxoprolyl-peptidase) activity, generating His-Pro-NH_2 from TRH. Brain contains at least two enzymes that exhibit this activity.[2]

[1] C. H. Emerson, *in* "The Thyroid" (S. H. Ingbar and L. E. Braverman, eds.), p. 1472. Lippincott, Philadelphia, Pennsylvania, 1986.

[2] C. H. Emerson and C. F. Wu, *Endocrinology (Baltimore)* **120,** 1215 (1987).

TRH Pyroglutamyl Aminopeptidases I and II

Liver, kidney, and other visceral organs contain TRH pyroglutamyl aminopeptidase (PAPase) activity that exhibits many of the properties of a thiol-activated TRH PAPase found in brain cytosol. The brain enzyme has been purified and its molecular weight is estimated to be 24,000.[3] It is referred to here and elsewhere[2] as TRH PAPase I. However, TRH PAPase I is probably similar to pyrrolidone-carboxylate peptidase, found in *Pseudomonas fluorescens* and purified from animal tissues.[4] TRH PAPase I is active against several other bioactive peptides including luteinizing hormone-releasing hormone (LHRH), neurotensin, and bombesin.[3] The product of TRH PAPase activity is His-Pro-NH_2. His-Pro-NH_2 itself is degraded by post-proline dipeptidyl aminopeptidase enzymes.[5] Post-proline dipeptidyl aminopeptidase activity is technically difficult to study using His-Pro-NH_2 as a substrate because this compound is very unstable and spontaneously converts to histidylproline diketopiperazine (HPD).

As noted below, brain homogenates also contain a second TRH PAPase referred to here as TRH PAPase II. In contrast to TRH PAPase I, TRH PAPase II exhibits marked substrate specificity for TRH and its estimated molecular weight, 230,000, is approximately 10 times greater than that of TRH PAPase I.[6] Whereas TRH PAPase I activity can be measured using the fluorogenic substrate pyroglutamate-7-amido-4-methylcoumarin,[3] TRH PAPase is currently best detected with authentic TRH as the substrate.

It is apparent that, when TRH is incubated in brain homogenates, measured His-Pro-NH_2 generation could be influenced by three enzymes. These are TRH PAPase I and TRH PAPase II, responsible for generating His-Pro-NH_2 from TRH, and the His-Pro-NH_2-degrading enzyme(s), post-proline dipeptidyl aminopeptidase. Fortunately, by employing saturating concentrations of substrate and incorporating inhibitors or activators into the incubation media, it is possible to obtain conditions in which the activity of either TRH PAPase I or II is the major determinant of His-Pro-NH_2 formation. Therefore, with respect to the measurement of TRH PAPase I and TRH PAPase II activities in brain, the following considerations are important:

1. TRH PAPase I is localized primarily in cytosol and, under conditions of Method 1 described below, requires sulfhydryl-reducing agents to

[3] P. Browne and G. O'Cuinn, *Eur. J. Biochem.* **137,** 75 (1983).
[4] T. C. Friedman, T. B. Kline, and S. Wilk, *Biochemistry* **24,** 3907 (1985).
[5] B. O'Connor and G. O'Cuinn, *Eur. J. Biochem.* **154,** 329 (1986).
[6] B. O'Connor and G. O'Cuinn, *Eur. J. Biochem.* **144,** 271 (1984).

demonstrate activity.[2] TRH PAPase I is not inhibited by 5 m*M* EDTA, 0.5 m*M* bacitracin, or 0.5 m*M* puromycin.[3]

2. TRH PAPase II is localized primarily, but not exclusively, in synaptosomal membranes. It is inhibited by 2 m*M* EDTA and does not require sulfhydryl-reducing agents to demonstrate activity in broken cell preparations. TRH PAPase II is not inhibited by 1.0 m*M* bacitracin or 1.0 m*M* puromycin.[6]

3. Post-proline dipeptidyl aminopeptidase activities in brain are variably inhibited by puromycin and bacitracin. Whereas synaptosomal membrane-associated post-proline dipeptidyl aminopeptidase is weakly inhibited by 0.5 m*M* puromycin, it is strongly inhibited by 0.5 m*M* bacitracin.[5] In contrast, post-proline dipeptidyl aminopeptidase associated with brain cytosol is strongly inhibited by 0.5 m*M* puromycin not by 0.5 m*M* bacitracin.[3]

Our method[2] utilizes these different properties of the enzymes in order to simultaneously measure brain TRH PAPase I and TRH PAPase II activities in rat brain.

Method: Simultaneous Assay of TRH PAPase I and TRH PAPase II

Basis. The basis of this method is the measurement of HPD generation, in the presence and absence of enzyme modifiers, when TRH is used as the substrate.

Procedure. Tissue samples (see below) are incubated at 37° for 30 min with synthetic TRH in a final concentration of 35 m*M*. The incubation buffer is 0.32 *M* sucrose, 100 m*M* imidazole, pH 7.4, and the final incubation volume is 0.4 ml. Bacitracin and puromycin should be added to the incubation buffer to obtain final concentrations of 0.5 m*M* for each compound. The protein content of samples for determination of enzyme activity is 0.1–1.0 mg per incubation tube. Parallel incubations are performed in tubes containing, or not containing, EDTA and dithiothreitol (DTT), each in a final concentration of 2 m*M*. Reactions are started by the addition of substrate, and terminated by the addition of 2 ml methanol. Blanks are generated by adding substrate at the end of the incubation. After termination of the incubation the tubes are vortexed and centrifuged at 3000 rpm. The supernatants are decanted and dried. After reconstitution in water they are covered and heated at 80° for 1 hr to ensure conversion of His-Pro-NH_2 to HPD. HPD content is then determined by radioimmunoassay (RIA) (see this volume [40]).

The amount of HPD generated in tubes containing EDTA and DTT is used to calculate TRH PAPase I activity. Conversely, the data from tubes lacking EDTA and DTT is used to calculate TRH PAPase II activity.

Results are expressed as micromoles HPD generated per minute per milligram protein. This transforms to units of enzyme activity per milligram protein. It must be emphasized that the concentration of enzyme units in a given tissue will change if the assay conditions are altered, as, for example, if the DTT concentration is changed (see below).

Comments. Several points should be emphasized concerning this method. First, 2 m*M* EDTA inhibits TRH PAPase II but does not enhance or inhibit TRH PAPase I activity. When adult rat brain cytosol is incubated in the absence of sulfhydryl-reducing agents (activators of TRH PAPase I), and in the presence of TRH PAPase II (an inhibitor of TRH PAPase II), no generation of HPD from TRH is observed. Rat brain cytosol can be shown to contain both TRH PAPase I and II activities by this method, but they can be separated on Sepharose CL-6B chromatography. These observations suggest that the use of DTT and EDTA completely distinguishes between the activities of TRH PAPase I and TRH PAPase II.

Second, the DTT concentration utilized in the method is not optimal for TRH PAPase activity in rat brain cytosol. Higher DTT concentrations can be employed, but the incubation time should be shortened to prevent substrate depletion which would result in a nonlinear relationship between incubation time and product formation. Third, puromycin and bacitracin should be added to inhibit, respectively, cytosol and membrane-associated post-proline dipeptidyl aminopeptidase activity. Finally, it should be noted that both cytosol and synaptosomal membranes contain TRH PAPase II-like activity by the methods described here. It is not clear if they result from the same enzymes. Therefore values for total TRH PAPase II will be obtained if whole homogenates, rather than subcellular fractions, are used as the tissue source.

Serum TRH Pyroglutamyl Aminopeptidase

TRH is rapidly degraded in the serum of many species. In rat, porcine, and human serum most, if not all, primary TRH degradation is due to TRH PAPase activity. Serum TRH PAPase is very similar to brain TRH PAPase II in that both enzymes have a narrow substrate specificity that does not include many other neuropeptides attacked by TRH PAPase I. In addition, both enzymes are inhibited by 2 m*M* EDTA and have similar molecular weights. We have found, however, that rat brain TRH PAPase II activity is not altered by thyroid status. In contrast, rat serum TRH PAPase activity is increased in hyperthyroidism and decreased in hypothyroidism.[2]

Method: Assay of Rat Serum TRH PAPase

Basis. The basis of the assay of rat serum TRH PAPase is the measurement of HPD generation, in the absence of enzyme modifiers, when TRH is used as the substrate.

Procedure. Identical incubation and analysis procedures are performed as those described for TRH PAPase II. That is, the incubation should be performed in the absence of DTT and EDTA. When rat serum is assayed the volume added to the incubation tubes should not exceed 12.5 μl.

Comments. No HPD generation is noted when incubations are performed in the presence of 2 m*M* EDTA, or 2 m*M* EDTA and 2 mM DTT. Volumes of serum greater than 12.5 μl give results in which HPD generation is not proportional to the amount of serum added to the incubation tubes. Physiological studies with the assay are consistent with previous studies concerning the effects of thyroid status on rat serum TRH-degrading activity.[2] In performing the assay, consideration should be given to including 1 m*M* diisopropyl fluorophosphate, to inhibit His-Pro-NH_2 degradation, as described by Bauer and Nowak.[7] The relatively large amounts of HPD that are generated in the method described above are due to the fact that the substrate concentrations employed are much greater than those used in previous studies in which HPD formation from TRH was difficult to detect in whole serum.[7,8]

TRH Deamidase

The second mode of TRH degradation is due to post-proline cleaving enzyme activity (prolyl endopeptidase activity). This catalyzes the conversion of TRH to deamido-TRH (acid TRH, TRH-OH). Post-proline cleaving enzyme(s) hydrolyze peptidylprolyl–peptide and peptidylprolyl–amino acid bonds of biologically active peptides including LHRH, angiotensin II, vasopressin, bradykinin, substance P, neurotensin, and TRH.[9,10] Post-proline cleaving enzymes differ from post-proline dipeptidyl aminopeptidases. The latter enzymes deamidate His-Pro-NH_2 but not pGlu-His-Pro-NH_2.[3] Post-proline cleaving activity with catalytic activity toward TRH (TRH deamidase activity) is found in the cytosol of many tissues, but not in serum. Post-proline cleaving enzyme and/or TRH

[7] K. Bauer and P. Nowak, *Eur. J. Biochem.* **99,** 239 (1979).

[8] W. L. Taylor and J. E. Dixon, *J. Biol. Chem.* **253,** 6934 (1978).

[9] H. Knisatschek and K. Bauer, *J. Biol. Chem.* **254,** 10936 (1979).

[10] S. S. Tate, *Eur. J. Biochem.* **118,** 17 (1981).

deamidase has been purified by a number of workers. Brain TRH deamidase exhibits immunological similarities with kidney post-proline cleaving enzyme.[11] Tate,[10] however, has purified TRH deamidase from brain that does not cleave oxytocin or vasopressin and exhibits substrate preference for TRH.

A variety of synthetic substrates including Gbz-Gly-Pro-MCA,[3] *N*-benzyloxycarbonylglycl-L-prolylsulfomethoxyazole,[12] pGlu-(*N*-benzyl-L-His)-Pro-β-naphthylamide,[11,13] and pyroglutamylhistidylprolyl-β-naphthylamide[13] have been employed to measure prolyl endopeptidase activity by fluorimetric methods. The latter two fluorogenic analogs are better substrates, when tested against purified TRH deamidase in the presence of 1 m*M* 2-mercaptoethanol, than TRH itself. They also exhibit similar pH–rate profiles as TRH.[13] These methods have been used to characterize the kinetic properties of the enzyme and its behavior in the presence of enzyme modifiers.

Method 1[13]*: Assay of TRH Deamidase*

Basis. The basis of the assay for TRH deamidase is the measurement of β-naphthylamine formation using the TRH analog, L-pyroglutamyl-*N*-benzyl-L-histidyl-L-prolyl-β-naphthylamide (TRH-A) as the substrate.

Procedure. TRH-A is prepared in a stock solution of 194 mg/dl in 1 *N* acetic acid. The pH of the stock solution is adjusted to 4.5 with sodium hydroxide, and an equal volume of 0.25 *M* sodium phosphate, 1 m*M* EDTA, 1 m*M* 2-mercaptoethanol, pH 7.5 (EDTA–PO_4 buffer), is added. Stock solution, 0.1 ml, is incubated at 37° with sample in the presence of EDTA–PO_4 buffer with the final incubation volume being 2 ml. β-Naphthylamide formation is monitored, and a β-naphthylamide standard curve is generated, using excitation and emission wavelengths of 335 and 410 nm, respectively.

Comments. Preliminary incubations must be performed to establish conditions under which substrate depletion is less than 10% and product formation is linear with time and directly proportional to the protein content of the test samples.

Method 2[14]*: Assay of TRH Deamidase*

Basis. Measurement of deamido-TRH by RIA using TRH as the substrate provides the basis for assaying TRH deamidase by Method 2.

[11] L. B. Hersh, *J. Neurochem.* **37,** 172 (1981).

[12] M. Orlowski, E. Wilk, S. Pearce, and S. Wilk, *J. Neurochem.* **33,** 461 (1979).

[13] P. C. Andrews, C. M. Hines, and J. E. Dixon, *Biochemistry* **19,** 5494 (1980).

[14] C. H. Emerson, W. Vogel, B. Currie, and T. Okal, *Endocrinology* (*Baltimore*) **107,** 443 (1980).

Procedure. Tissue samples are incubated at 37° with synthetic TRH at a final concentration of 0.46 m*M*. The incubation buffer is 0.15 *M* sodium maleate, pH 7, and the incubation volume is 0.4 ml. Reactions are started by the addition of substrate and terminated by the addition of 2 ml methanol. Blanks are generated by adding substrate at the end of the incubation. After termination of the incubation the incubation tubes are vortexed and centrifuged at 3000 rpm. The supernatants are decanted and dried, then reconstituted in water and assayed for deamido-TRH by RIA as described elsewhere in this volume ([40]). Results are expressed as micromoles HPD generated per minute per milligram protein. This transforms to units of enzyme activity per milligram protein.

Comments. This method has been used for physiological studies of TRH deamidase activity in tissue homogenates.[14] Under the assay conditions, deamido-TRH is stable, and, when cytosol from up to 10% of hamster cortex is assayed, deamido-TRH production is proportional to the protein content of the incubation tubes. The activity of TRH deamidase is enhanced by adding sulfhydryl-reducing agents. However, sulfhydryl-reducing agents also activate TRH PAPase I. Therefore, incubations are performed in the absence of DTT.

Acknowledgments

Supported by Research Grant AM27850 from the NIDDK, National Institutes of Health, Bethesda, Maryland.

[27] Quantification of the Mass of Tyrosine Monooxygenase in the Median Eminence and Superior Cervical Ganglion

By John C. Porter, Paulus S. Wang, Wojciech Kedzierski, and Hector A. Gonzalez

Introduction

The rate of biosynthesis of dopamine and norepinephrine is generally believed to be limited by the reaction that results in the conversion of L-tyrosine to L-3,4-dihydroxyphenylalanine (DOPA). This conversion in neurons is dependent on the catalytic activity of tyrosine hydroxylase (TH) [EC 1.14.16.2 tyrosine 3-monooxygenase, L-tyrosine, tetrahydropteridine : oxygen oxidoreductase (3-hydroxylating)].[1] However, an

[1] L. Levitt, S. Spector, A. Sjoerdsma, and S. Udenfriend, *J. Pharmacol. Exp. Ther.* **148,** 1 (1965).

analysis of this reaction is supportive of the conclusion that, in the formation of DOPA from L-tyrosine in catecholaminergic cells, other variables have significant roles. Of these, the following five appear to be especially important: (1) availability of L-tyrosine, (2) availability of tetrahydrobiopterin,[2] (3) mass of TH, (4) phosphorylation of TH,[3] and (5) end-product inhibition.[4]

The relative importance of each of these variables within neurons that synthesize dopamine or dopamine and norepinephrine, however, is not established since most studies on TH have been performed utilizing purified, partially purified, or broken cell preparations of the enzyme. The catalytic activity of TH under cell-free conditions is probably a poor prognosticator of this enzyme's activity within intact cells. Although it is difficult to critically study the relationship of any of these five variables to the *in situ* activity of TH, recent advances have made it possible to measure the quantity of TH in certain tissues, and thereby relate various indices of intracellular activity of TH to its mass.

In this chapter, we describe a specific and sensitive method for quantifying in rats the mass of TH in the median eminence (ME) of the hypothalamus as well as in one superior cervical ganglion (SCG). In doing so, an attempt is made to identify those steps of the procedure that seem to require caution as well as those that do not. When a particular manipulation is known to lead to an unsatisfactory result, the consequences will be identified and discussed. The procedure used in the determination of the mass of TH is called an immunoblot assay. Although this assay has been used earlier,[5] the procedure described here includes modifications and adaptations of the assay that are presently in use in our laboratory.

Purification of TH

TH is purified from tumor tissue harvested from rats of the New England Deaconess Hospital strain (Boston, MA). Tumors in these animals are grown from a subcutaneous inoculum of 10 to 20 million PC-12 cells.[6] The progenitor of these cells was isolated from a transplantable rat medullary pheochromocytoma.[7] PC-12 cells were a gift from Dr. Lloyd A. Greene (Department of Pharmacology, New York University). Tumor

[2] T. Nagatsu, M. Levitt, and S. Udenfriend, *J. Biol. Chem.* **239,** 2910 (1964).

[3] A. L. Cahill and R. L. Perlman, *Biochim. Biophys. Acta* **805,** 217 (1984).

[4] S. Udenfriend, P. Zaltman-Nirenberg, and T. Nagatsu, *Biochem. Pharmacol.* **14,** 837 (1965).

[5] J. C. Porter, *Endocrinology (Baltimore)* **118,** 1426 (1985).

[6] L. A. Greene and A. S. Tischler, *Proc. Natl. Acad. Sci. U.S.A.* **73,** 2424 (1976).

[7] S. Warren and R. N. Chute, *Cancer* **29,** 327 (1972).

TABLE I
PURIFICATION OF TH FROM TUMOR TISSUE HARVESTED FROM RATS INJECTED WITH PC-12 CELLS

Purification step	Total TH activity (units)[a]	Total protein (mg)	Specific activity of TH (units/mg protein)
1. 78,000 *g* Supernatant	15,196	1,100	13.8
2. 40% $(NH_4)_2SO_4$ precipitate	54,964	90.5	607
3. Ultragel AcA 22 chromatography	28,400	39.5	719
4. Precipitate formed between pH 5.2 and 5.8	13,335	28.5	468
5. Heparin–Sepharose chromatography	28,056	6.38	4,300
6. Phosphocellulose chromatography	6,920	1.33	5,200
7. Ultragel AcA 22 chromatography	9,875	0.99	9,970
8. Phosphocellulose chromatography	10,467	0.50	20,900

[a] A unit of TH activity is defined as 1 nmol of L-tyrosine oxidized per hour.

tissue is harvested when the mass of the tumor is 2–10 g, usually 5–6 weeks after the inoculation. Large spheroidal tumors should be avoided since such tumors frequently contain necrotized regions that are devoid of TH. The tumor tissue is stored at −80° until used. Under these storage conditions, TH in the tissue seems to be stable for an indefinite time.

In the purification of TH, we have found it convenient to process about 20 g of tissue at a time. Throughout the purification, the activity of TH is assayed according to the procedure of Nagatsu[8] as adapted by Foreman and Porter.[9] TH is purified according to the procedure described by Okuno and Fujisawa[10] for the isolation of this enzyme from rat adrenal tissue.

A representative result is given to Table I. Typically, a 1500-fold purification is achieved. The apparent increase in total TH activity after purification steps 2 and 5 may be the result of activation of TH by ammonium sulfate[11] and possibly heparin.[11,12] Alternatively, the assay conditions for TH activity at steps 1 and 4 may have been less than optimal, leading to low values for total TH activity.

After polyacrylamide gel electrophoresis under denaturing conditions using 4 μg of the final product (step 8), only one protein band is seen after

[8] T. Nagatsu, "Biochemistry of Catecholamines: The Biochemical Method," p. 162. University Park Press, Baltimore, Maryland, 1973.

[9] M. M. Foreman and J. C. Porter, *J. Neurochem.* **34,** 1175 (1980).

[10] S. Okuno and O. Fujisawa, *Eur. J. Biochem.* **122,** 49 (1982).

[11] R. T. Kuczenski and A. J. Mandrell, *J. Neurochem.* **19,** 131 (1972).

[12] I. R. Katz, T. Yamauchi, and S. Kaufman, *Biochim. Biophys. Acta* **429,** 84 (1976).

staining with Coomassie blue (Fig. 1). When calculated by the method of Weber and Osborn,[13] the molecular weight of the denatured product was found to be approximately 60,000, a value that is similar to that reported by others.[14,15] In eight purifications of TH using 22.5 ± 3.8 g (mean ± SD) of tumor tissue per purification, the average yield of TH was 496 μg. The specific activity of TH purified in these eight purifications was 19,250 ± 1,830 units/mg protein. (A unit of activity is defined as 1 nmol of L-tyrosine oxidized per hour.) The enzymatic activity of purified TH from tumor tissue is similar to that reported by Okuno and Fujisawa[10] for TH isolated from rat adrenal tissue.

Although the scheme followed in the purification of TH appeared to yield a homogeneous product as judged by polyacrylamide gel electrophoresis, the question persisted regarding the extent of contamination of purified TH with aromatic-L-amino-acid decarboxylase (AAAD) (aromatic-L-amino-acid carboxyl-lyase, EC 4.1.1.28) and dopamine β-hydroxylase (DBH) [dopamine β-monooxygenase, 3,4-dihydroxyphenylethylamine, ascorbate : oxygen oxidoreductase (β-hydroxylating), EC 1.14.17.1], two enzymes involved in the biosynthesis of norepinephrine by cells of tumors grown from PC-12 cells. To address this issue, freshly purified TH was assayed for AAAD and DBH activities. AAAD activity was assayed according to the method of Nagatsu,[16] and DBH activity was assayed using the procedure of Kato *et al.*[17] At the end of the incubation period, the incubation medium was adjusted to 300 m*M* with respect to perchloric acid (PCA) and centrifuged to sediment acid-insoluble proteins.

The supernatant fluid or the standard solution containing dopamine and norepinephrine in 300 m*M* PCA was neutralized with 5 volumes of 0.5 *M* Tris buffer (pH 8.6) containing 15 m*M* EDTA and extracted with 50 mg of acid-washed aluminum oxide. The catecholamines were eluted from the aluminum oxide with 200 μl of 300 m*M* PCA. Dopamine and norepinephrine in the eluate were quantified by means of HPLC with electrochemical detection as described by Felice *et al.*[18] with minor modification.[19] The assays for AAAD and DBH activities were sufficiently

[13] K. Weber and M. Osborn, *J. Biol. Chem.* **244,** 4406 (1969).

[14] T. H. Joh, D. H. Park, and D. J. Reis, *Proc. Natl. Acad. Sci. U.S.A.* **75,** 4744 (1978).

[15] A. M. Edelman, J. D. Raese, M. A. Lazar, and J. D. Barchas, *J. Pharmacol. Exp. Ther.* **216,** 647 (1981).

[16] T. Nagatsu, "Biochemistry of Catecholamines: The Biochemical Method," p. 166. University Park Press, Baltimore, Maryland, 1973.

[17] T. Kato, Y. Wakui, T. Nagatsu, and T. Ohnishi, *Biochem. Pharmacol.* **27,** 829 (1977).

[18] L. J. Felice, J. D. Felice, and P. T. Kissinger, *J. Neurochem.* **31,** 1461 (1978).

[19] J. C. Porter, *Mol. Cell. Endocrinol.* **46,** 21 (1986).

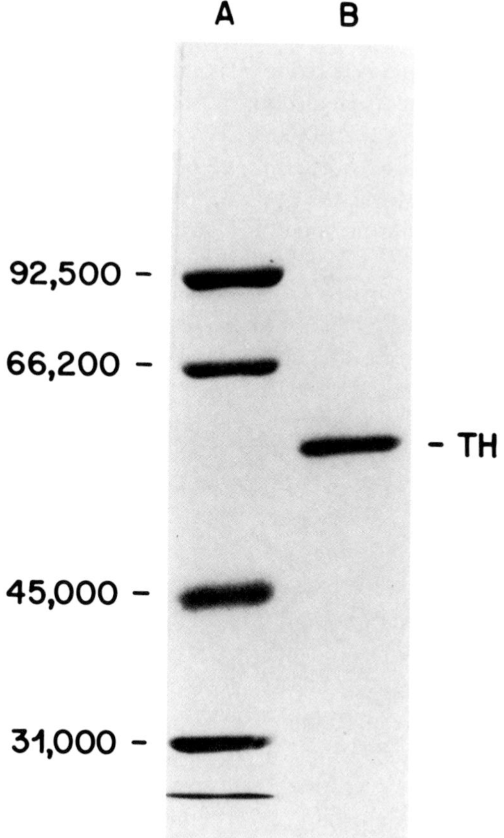

FIG. 1. Polyacrylamide gel (7.5% gel) slab electrophoresis in the presence of sodium dodecyl sulfate (0.1%) of TH purified from tumor tissue grown from PC-12 cells. The gels are calibrated with the following molecular weight standards: phosphorylase *b*, 92,500; bovine serum albumin, 66,200; ovalbumin, 45,000; carbonate dehydratase, 31,000; soybean trypsin inhibitor, 21,500; and lysozyme, 14,400 (Bio-Rad Laboratories). Molecular weight standards (lane A) and 4 μg purified TH (lane B) are stained with Coomassie blue. The molecular weight of denatured TH, calculated by the procedure of Weber and Osborn,[13] is approximately 60,000. [Reproduced with permission from J. C. Porter, "Relationship of age, sex, and reproductive status to the quantity of tyrosine hydroxylase in the median eminence and superior cervical ganglion of the rat," *Endocrinology* (*Baltimore*), **118,** 1426–1432, © by The Endocrine Society (1986).]

sensitive that the activity of each enzyme was readily measurable in a homogenate, equivalent to 4 μg protein, of rat adrenal tissue.

AAAD activity was not detected in 5.7 μg of freshly purified TH. On the basis of this finding, it is calculated that the specific activity of AAAD in highly purified TH is less than 15 nmol of dopamine formed per hour per milligram protein. Assuming the activity per unit mass of highly purified AAAD prepared from hog kidney[20] is the same as the activity per unit mass of AAAD in tumor tissue grown from PC-12 cells, we calculate that purified TH contains less than 0.003% AAAD as a contaminant.

Freshly purified TH (5.7 μg) contained a small amount of DBH activity, and, on the basis of the assay data, it is calculated that the specific activity of DBH in the purified TH preparation is 7 nmol of norepinephrine formed per hour per milligram protein. Assuming the activity per unit mass of highly purified DBH prepared from bovine brain[21] is the same as the activity per unit mass of DBH in tumors grown from PC-12 cells, we calculate that purified TH contains 0.08% DBH as a contaminant. These data support the view that the procedure described by Okuno and Fujisawa[10] for the isolation of TH from rat adrenal tissue is also suitable for isolating TH from tumor tissue grown from PC-12 cells.

Antiserum against TH

To prepare antiserum against rat TH, a suspension of highly purified TH, emulsified in Freund's complete adjuvant, was injected subcutaneously into female rabbits as described by Vaitukaitis *et al.*,[22] using 100–200 μg of enzyme per animal. The injections of TH emulsified in Freund's incomplete adjuvant were repeated at intervals of 3 weeks for 12–21 weeks. Successful immunization was achieved in four rabbits; after which each animal was bled every 2 weeks by way of an artery of an ear. The serum of this blood was stored at $-80°$ until used.

The antibody titer of serum from each of the many bleedings from these animals was estimated by means of an enzyme-linked immunosorbent assay.[23] When the antibody titer of one rabbit (S828) was maximal, the capacity of its serum to bind TH was quantified in this manner. Antiserum from rabbit S828 and preimmune rabbit serum were mixed in the following ratios: 0 : 5, 1 : 4, 2 : 3, 3 : 2, 4 : 1, and 5 : 0. Then, 5 μl of each of these serum mixtures was added to 55 μl of 5 m*M* sodium phosphate buffer, pH 7.5, containing 25% glycerol, 0.5% Tween 80, 0.1 m*M* EDTA,

[20] J. G. Christenson, W. Dairman, and S. Udenfriend, *Arch. Biochem. Biophys.* **141,** 356 (1970).

[21] H. Masui, C. Yamamoto, and T. Nagatsu, *J. Neurochem.* **39,** 1066 (1982).

[22] J. L. Vaitukaitis, G. D. Braunstein, and G. T. Ross, *Am. J. Obstet. Gynecol.* **113,** 751 (1972).

[23] A. Voller, D. E. Bidwell, and A. Bartlett, *Bull. W.H.O.* **53,** 55 (1976).

0.1% bovine serum albumin (BSA), and 200 ng of highly purified TH. After an incubation at room temperature for 20 min, 100 μl of a 10% suspension of *Staphylococcus aureus* (Staph A) was added to the solution containing TH and rabbit serum, and the incubation was continued for 1 hr with shaking. At the end of the incubation period, the mixture was centrifuged at 10,000 *g* for 2.5 min, and the supernatant fluid was assayed for TH activity.

The capacity of antibodies in serum from rabbit S828 to bind TH is illustrated in Fig. 2. On the basis of these binding studies, it was found that there is sufficient antibody in 1 μl of this antiserum to bind 2.4 units of enzyme activity. The catalytic activity of 200 ng of TH after incubation with 5 μl of preimmune serum and extraction of the incubation medium with Staph A was 3.5 units. On the basis of these results, we calculate that 1 ml of this antiserum has the capacity to bind 135 μg or 0.56 nmol of TH (assuming the molecular weight of native tetrameric TH to be 240,000[10]).

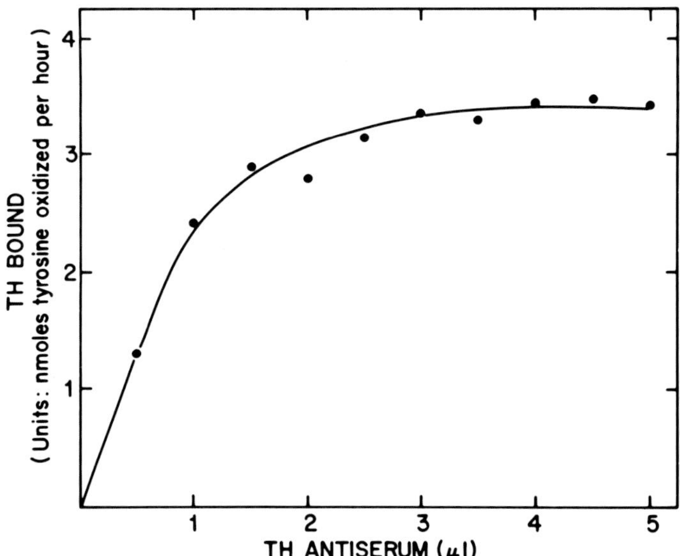

FIG. 2. Binding of TH by antibodies in antiserum from rabbit S828. TH bound by antibodies in a given volume of antiserum is calculated utilizing the difference between the activity of 200 ng of TH after immunoprecipitation with preimmune serum and the activity of TH after immunoprecipitation with antiserum against TH. [Reproduced with permission from J. C. Porter, "Relationship of age, sex, and reproductive status to the quantity of tyrosine hydroxylase in the median eminence and superior cervical ganglion of the rat," *Endocrinology (Baltimore)*, **118**, 1426–1432, © by The Endocrine Society (1986).]

Immunoblot Analysis

Preparation of Tissue

The ME of rats is excised according to the procedure of Arita and Kimura.[24] This dissection yields tissue that comprises the floor of the third ventricle and that has a protein content of approximately 25 μg, varying with the size of the donor. The SCG is excised with the aid of a dissection microscope. Each tissue sample is homogenized in ice-cold water (50 μl for the ME and 100 μl for an SCG). Glass microhomogenizers (Micro-Metric Instruments, Tampa, FL) are used for this purpose. The homogenate is centrifuged at 10,000 *g* for 1 min. A 20-μl aliquot of ME extract or a 40-μl aliquot of SCG extract is mixed with an equal volume of 2× Laemmli sample buffer.[25] The mixture is placed in a sealed tube, heated in a boiling water bath for 3 min, centrifuged at 10,000 *g* for 1 min, and stored at −80° to await analysis for TH.

Denatured TH appears to be quite stable. However, if tissue containing TH is allowed to stand for an extended period at room temperature or even cool conditions, fragments of the enzyme can be produced. For optimal results, such treatment of the tissue should be avoided.

Laboratory Standard

For the primary reference standard, freshly purified TH in Laemmli sample buffer,[25] pH 6.8, is placed in a sealed tube, heated in a boiling water bath for 3 min, and stored at −80°. Standards of 7.5–80 ng of TH constitute a useful range.

In the day-to-day operation, we have found it convenient to use a laboratory standard that is calibrated against the primary standard of purified TH. The laboratory standard is prepared as follows: PC-12 cells, grown in monolayer culture, are suspended in water, disrupted by sonication, and centrifuged to sediment cellular debris. The supernatant is mixed with an equal volume of 2× Laemmli sample buffer, pH 6.8 and heated for 3 min in a boiling water bath. After centrifugation, the concentration of TH in the supernatant fluid is standardized against purified TH by means of immunoblot analysis (described below). When stored at −80°, the laboratory standard preparation appears to be stable indefinitely.

[24] J. Arita and F. Kimura, *Neuroendocrinology* **39,** 524 (1984).

[25] U. K. Laemmli, *Nature* (*London*) **227,** 680 (1970).

Electrophoresis

Electrophoresis is performed on polyacrylamide gel slabs (7.5% gel) in the presence of 0.1% sodium dodecyl sulfate (SDS). The electrophoretic chamber is cooled by means of tap water (22–24°) flowing through the unit by way of a coiled conduit. The dimensions of the separating gel slab are as follows: height, 16.5 cm; width, 16 cm; and thickness, 1.5 mm. A stacking gel is formed using a 25-well comb where each well is 3.5 mm wide. A sample volume of 10–35 μl is used. The current conditions for electrophoresis are 30 mA per gel for 30 min for the stacking gel and 40 mA per gel for 2.5 hr for the separating gel. Throughout the electrophoresis, the buffer is stirred to maintain a uniform temperature in the electrophoretic chamber. Stirring is achieved by means of a magnetic bar.

Transfer to Membrane Matrix

After electrophoresis on SDS–polyacrylamide gel slabs, the gel below the dye front is removed to facilitate handling. The protein on the gel slab is transferred to nitrocellulose paper (Bio-Rad Laboratories, Richmond, CA), as described by Burnette,[26] or to a Biotrans nylon membrane (ICN, Irvine, CA) having a nominal pore size of 1.2 μm. Although nitrocellulose paper was used originally,[5] we now mostly use Biotrans nylon membrane because of ease of handling, and we describe its use in this chapter. The transfer is conducted in a Trans Blot cell (Bio-Rad Laboratories). The electrode buffer consists of 20 volumes of methanol and 80 volumes of an aqueous solution containing 25 m*M* Tris base and 192 m*M* glycine. The electrophoretic transfer of TH from the gel slab to the membrane is achieved at 4–6° using 300 mA for 2 hr. A temperature of 4–6° is maintained in the Trans Blot cell by means of a refrigerated recirculator (Bio-Rad Laboratories) that pumps coolant at 0° through a coil immersed in the cell. The solution in the Trans Blot cell is stirred continuously during the transfer using a magnetic stirrer.

TH–Antibody Reactions

At the end of the transfer of protein from the gel slab to the nylon membrane, the membrane is placed in 100 ml of prewarmed Tris-buffered saline (TBS: 20 m*M* Tris buffer, pH 7.5, 500 m*M* NaCl, and 0.01% Merthiolate) containing protein to saturate nonspecific binding sites, and incubated with shaking for 40 min at 50°. [All incubations and reactions in-

[26] W. N. Burnette, *Anal. Biochem.* **112,** 195 (1981).

volving the nylon membrane are carried out in rectangular staining dishes (20 × 10 × 7 cm).]

For saturation purposes, we have used a variety of protein solutions, including 3% gelatin, 3% BSA, and 8% nonfat dry milk. Among this assortment, nonfat dry milk is the most efficacious, as first noted by Johnson *et al.*[27] Routinely, we use Carnation nonfat dry milk purchased at a local supermarket. In addition to being an effective agent for saturation of nonspecific binding sites, nonfat dry milk is also inexpensive.

The membrane, saturated with protein, is washed for 15 min (three 5-min washes) with TBS buffer (100 ml per wash) containing 0.05% Tween 20, and then it is incubated overnight (~15 hr) with shaking at room temperature in 40 ml rabbit antiserum against TH (first antibody) diluted 1 : 500 with TBS containing 8% nonfat dry milk. The next morning, the membrane is washed for 30 min (three 10-min washes) in 100 ml TBS containing 0.05% Tween 20 and then incubated for 60 min in 100 ml of second antibody solution. The second antibody solution consists of horseradish peroxidase conjugated to affinity-purified goat antibodies against rabbit immunoglobulin (Bio-Rad Laboratories) diluted 1 : 2000 with TBS containing 8% nonfat dry milk. The membrane is washed for 30 min (three 10-min washes) with TBS containing 0.05% Tween 20 (100 ml per wash) and rinsed twice in 100 ml TBS. All manipulations and reactions are conducted at room temperature with continuous shaking using a platform shaker.

To localize the TH–antibody complex, the membrane is immersed in 480 ml of color development solution. This solution consists of 240 mg of 4-chloro-1-naphthol (Bio-Rad Laboratories) in 80 ml of cold methanol and 400 ml TBS containing 0.018% hydrogen peroxide. Color development is allowed to occur in a dimly lighted room. When the least TH standard becomes visible, the color development reaction is interrupted by immersing the membrane in distilled water for 5 min. The water wash is changed twice. (Note: The color continues to intensify somewhat in the water wash. With a little practice, one develops the ability to anticipate the intensification and to stop the development at a suitable level. Generally, overdevelopment is a greater obstacle to quantification than is underdevelopment.) The washed membrane is compressed between several sheets of Whatman No. 3 filter paper (Whatman, Clifton, NJ) and allowed to dry.

An immunoblot of purified TH using nitrocellulose paper saturated with 3% gelatin is shown in Fig. 3. It can be seen that the intensity of the

[27] D. A. Johnson, J. E. W. Gautsch, J. R. Sportman, and J. H. Elder, *Gene Anal. Technol.* **1,** 3 (1984).

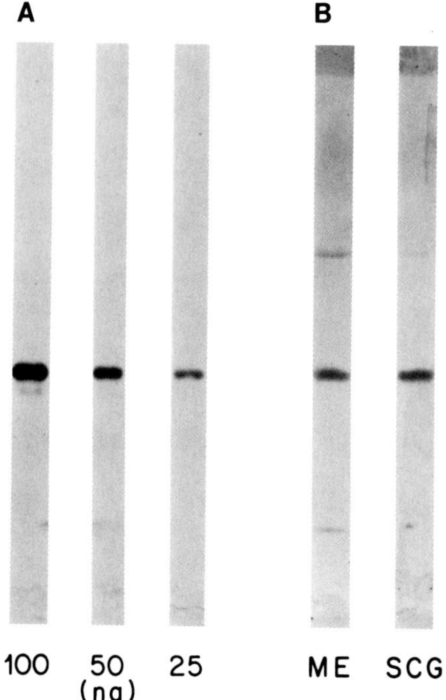

FIG. 3. Immunoblot of purified TH (A) and of TH extracted from ME tissue and SCG tissue (B), using nitrocellulose paper. The antiserum is from rabbit S828. The tissue is homogenized in 1× Laemmli sample buffer[25] and electrophoresed on an SDS–polyacrylamide gel slab. Nonspecific binding sites are saturated with 3% gelatin. [Reproduced with permission from J. C. Porter, "Relationship of age, sex, and reproductive status to the quantity of tyrosine hydroxylase in the median eminence and superior cervical ganglion of the rat," *Endocrinology (Baltimore)*, **118,** 1426–1432, © by The Endocrine Society (1986).]

colored band corresponding to TH increases progressively as the quantity of TH increases. Immunoblots of TH in an aqueous extract of rat ME and SCG are also shown in Fig. 3. The backgrounds on these immunoblots prepared in the presence of 3% gelatin are higher than desirable, and the presence of proteins other than TH are evident. These features of the immunoblot could be due to (1) inadequate saturation of binding sites on the nitrocellulose paper and/or (2) lack of specificity of the antiserum to TH.

Of these two possibilities, we believe the first is the most significant. This view is supported by results obtained using nylon membranes saturated with 8% nonfat dry milk. A representative immunoblot of TH in aqueous extracts of ME tissues from 12 rats is shown in Fig. 4. A compar-

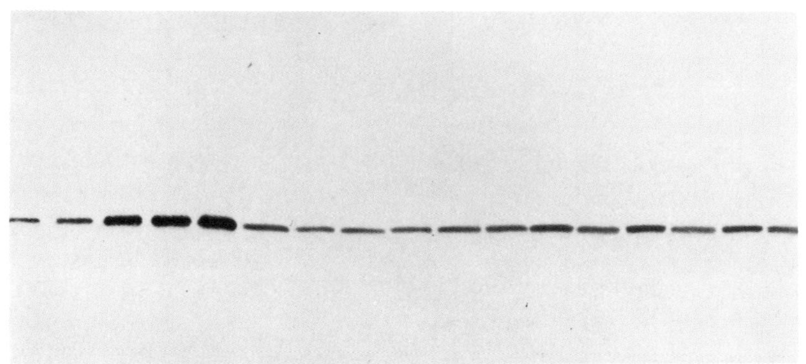

FIG. 4. Immunoblot of purified TH standard (lanes a–e) and of TH extracted from ME tissue from 12 rats (lanes 1–12), using a nylon membrane. Lane a corresponds to 12.5 ng of purified TH; lane b, 25 ng TH; lane c, 50 ng TH; lane d, 75 ng TH; and lane e, 100 ng TH. Nonspecific binding sites are saturated with 8% nonfat dry milk.

ison of the immunoblots in Figs. 3 and 4 shows a marked difference in background and nonspecific proteins.

Reflectance Analysis

The dry immunoblot is analyzed by means of a reflectometer/densitometer, using the reflectance mode. [The instrument supplied by Helena Laboratories (Beaumont, TX) has proved satisfactory in our hands.] The immunoblot should be analyzed within a few hours of development. With time the membrane becomes discolored, and discoloration impairs reflectance measurements.

An analysis of a representative immunoblot for TH is shown in Fig. 5. For optimal results, it is necessary that the colored band corresponding to TH be symmetrical and regular. Overloading the polyacrylamide gel, failure to keep the gel at a uniform temperature during electrophoresis, failure to keep the Trans Blot cell uniformly cold during the transfer, and overdevelopment will adversely affect the symmetry of each sample and the uniformity of the immunoblot.

Since each immunoblot is slightly different from another, it is necessary to include standards on each polyacrylamide gel slab. This necessity reduces the efficiency of the procedure, but we believe the cost in terms of chromatographic space is justified. We routinely include four or five standards with the unknowns on each gel slab, and the unknowns are quantified by comparison with the standards included on the same slab.

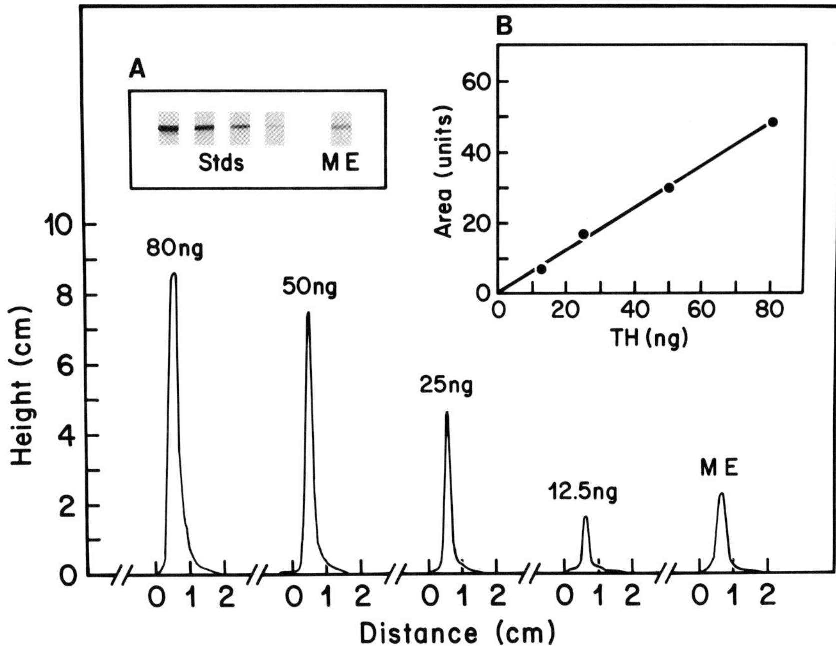

FIG. 5. Reflectance scans and analysis of an immunoblot of purified TH and TH in an aqueous extract of the ME of an immature male rat (5 weeks of age). The immunoblot of each TH standard and of TH from the ME is shown in inset A. A plot is shown in inset B of the area enclosed by each reflectance scan as a function of the quantity of TH. The area is given in arbitrary units. Slope in inset B equals 0.60; correlation coefficient in inset B equals 0.998. [Reproduced with permission from J. C. Porter, "Relationship of age, sex, and reproductive status to the quantity of tyrosine hydroxylase in the median eminence and superior cervical ganglion of the rat," *Endocrinology* (*Baltimore*), **118,** 1426–1432, © by The Endocrine Society (1986).]

Reflectance scans of four TH standards and of TH in an aqueous extract of the ME of a young male rat (5 weeks of age) are shown in Fig. 5. The area, in arbitrary units, defined by the envelope of the reflectance scans when plotted as a function of quantity of TH (right inset) reveals excellent proportionality. In this example, the correlation coefficient for the regression of Y on X is 0.998. The average correlation coefficient of 110 assays that we have performed to date is 0.996 ± 0.003 (mean $\pm$ SD). The intraassay coefficient of variation is approximately 8.5%.

When assaying TH in aqueous extracts of tissue, the concentration of TH relative to other proteins in the extract must be taken into consideration as protein load can limit the usefulness of the assay. In the analysis of ME or SCG tissue for TH, we estimate that the protein load per sample

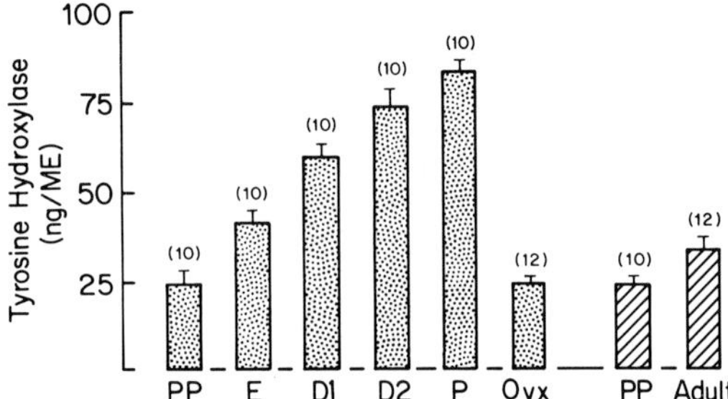

FIG. 6. TH in the ME of female rats (dotted bars) and male rats (hatched bars). Prepubertal rats (5–5.5 weeks of age) are identified as PP. Female animals on various days of the estrous cycle as identified as follows: E, estrus; D1, diestrus 1; D2, diestrus 2; P, proestrus. Ovx denotes mature ovariectomized animals that had been castrated at least 3 weeks. The adult male rats were mature animals. The height of the bar corresponds to the mean quantity of TH per ME. The vertical line denotes the magnitude of the SEM. The number of animals in each group is shown in parentheses. [Redrawn with permission from J. C. Porter, "Relationship of age, sex, and reproductive status to the quantity of tyrosine hydroxylase in the median eminence and superior cervical ganglion of the rat," *Endocrinology (Baltimore)*, **118,** 1426–1432, © by The Endocrine Society (1986).]

is between 10 and 20 μg. If larger amounts of protein are used, the efficacy of the immunoblot procedure should be reassessed.

Utility of Immunoblot Assay for TH

The quantity of TH, determined by immunoblot analysis, in the ME of immature and mature rats of both sexes is shown in Fig. 6. The amounts of TH in the ME of prepubertal male and female rats (5–5.5 weeks of age) are similar. In addition, the quantity of TH in the ME of mature ovariectomized female rats castrated at least 3 weeks before experiment is similar to that of prepubertal animals. The amount of TH in the ME is appreciably greater in intact, mature females than in mature males. In intact female rats, the amount of TH in the ME varies throughout the ovulatory cycle, being greatest on the day of proestrus and least on estrus. It seems probable that the marked reduction in the amount of TH in the ME between proestrus and estrus is a consequence of progesterone secretion at this time of the ovulatory cycle.[28] However, the basis for the large amount of TH in the ME of proestrous rats relative to that of diestrous and estrous animals is not known. It does not appear to be due to estrogen.[28]

[28] P. S. Wang and J. C. Porter, *Proc. Natl. Acad. Sci. U.S.A.* **83,** 9804 (1986).

Summary

A procedure is described that enables one to quantify the mass of TH in a fraction of the ME and SCG of rats. This procedure is specific and sensitive. It should be possible to study the biosynthetic activity of catecholaminergic neurons as a function of the mass of TH in such cells.

Acknowledgments

Supported by Research Grants AM01237, AG04344, and AG00306 from the National Institutes of Health, Bethesda, Maryland. P.S.W. was a Fogarty International postdoctoral fellow. H.A.G. is a Chilton Foundation postdoctoral fellow.

[28] Measurement of Vasopressin-Converting Aminopeptidase Activity and Vasopressin Metabolites

By J. P. H. BURBACH and BIN LIU

Introduction

Vasopressin (VP) is a nonapeptide with an internal disulfide bond (Fig. 1). It is well known as the hormone in the peripheral blood circulation regulating diuresis and blood pressure. VP is also present in the central nervous system, where it regulates brain functions.[1] Unlike the peripheral peptide, VP in the brain is not only biologically active like the intact nonapeptide entity, but it is also a precursor to smaller peptides with central activity.[2,3] These VP fragments are more selective and potent in their central effects than VP itself,[4] and they are endogenous in the brain.[5] In the brain VP is predominantly processed in a stepwise manner according to an aminopeptidase-like mechanism.[6] The enzyme or enzymes involved may have an essential role in regulating formation of active VP metabolites, thereby controlling the biological activities associated with

[1] D. De Wied, *Prog. Brain Res.* **60,** 217 (1983).

[2] J. P. H. Burbach, G. L. Kovács, D. De Wied, J. W. Van Nispen, H. M. Van Greven, *Science* **221,** 1310 (1983).

[3] J. P. H. Burbach, *in* "Vasopressin: Principles and Properties" (D. M. Gash and G. J. Boer, eds.), 497. Plenum, New York, 1987.

[4] D. De Wied, O. Gaffori, J. P. H. Burbach, G. L. Kovács, and J. M. Van Ree, *J. Pharmacol. Exp. Ther.* **241,** 268 (1987).

[5] J. P. H. Burbach, X.-C. Wang, J. A. Ten Haaf, and D. De Wied, *Brain Res.* **306,** 384 (1984).

[6] J. P. H. Burbach and J. L. M. Lebouille, *J. Biol. Chem.* **258,** 1487 (1983).

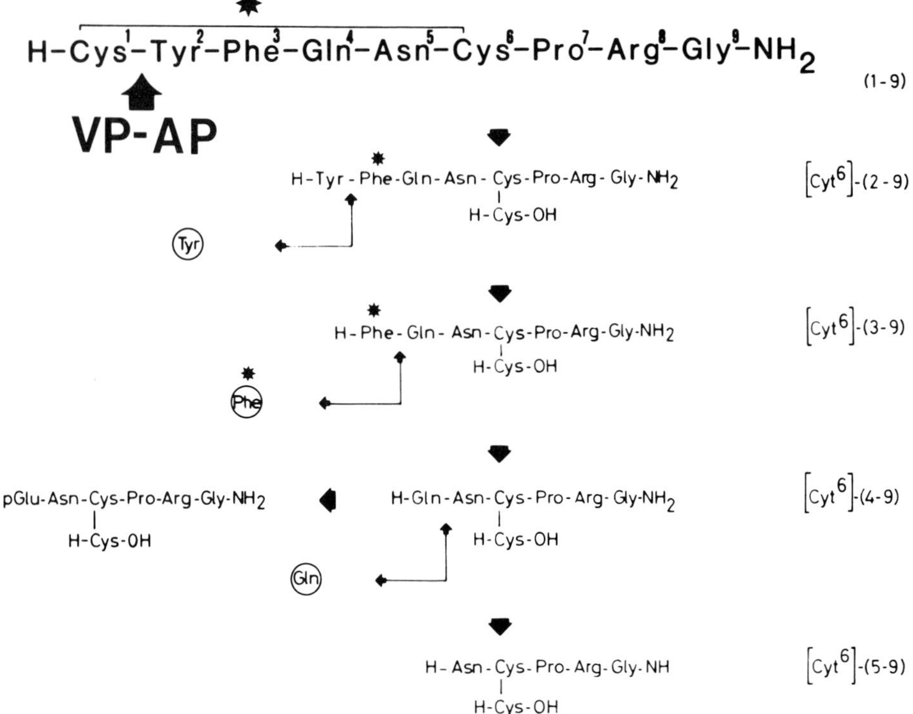

FIG. 1. Structure of [Arg^8]vasopressin [VP-(1–9)] and carboxy-terminal VP metabolites, which are formed by stepwise aminopeptidase cleavage. The enzyme activity responsible for initial cleavage is referred to as VP-converting aminopeptidase activity (VP-AP). The position of the labeled amino acid (Phe) is indicated by an asterisk.

them. In order to study biochemically the significance of VP metabolites we have devised methods for characterization and quantitation of VP metabolites as well as the aminopeptidase activity that converts vasopressin to active metabolites. This type of enzyme activity is briefly referred to as VP-converting aminopeptidase activity (VP-AP).

Materials and Methods

Measurement of VP-Converting Aminopeptidase Activity (VP-AP)

The principle of measurement of VP-AP is the radiometric quantitation of the ^{3}H-labeled amino acid which is released from [^{3}H]VP during VP-AP action (Fig. 1). Hydrophobic interaction chromatography on minicolumns is used to rapidly separate the ^{3}H-labeled amino acid from intact [^{3}H]VP. VP-AP can be determined in homogenates, membranes, and soluble fractions of various tissues. The assay method is described below for a typical brain membrane preparation used for current studies on VP-AP.[7]

Preparation of Brain Membranes. Rat brain tissue is homogenized in 20 m*M* Tris–HCl, pH 7.0 (10 ml/g tissue) by 10 strokes in a tightly fitting Teflon–glass homogenizer at 0°. The homogenate is centrifuged at 10,000 g_{av} for 60 min. The pellet is resuspended in the same volume of 150 m*M* NaCl, 5 m*M* Tris–HCl, pH 7.0, and spun down again under the same conditions. For the assay, membranes are resuspended in 20 m*M* Tris–HCl, pH 7.0, or in 40 m*M* Tris–HCl, 62 m*M* NaCl, pH 7.0.

A micropreparation for tissue samples as small as 1 mg (e.g., rat pineal glands[8]) is available. Tissue, approximately 1 mg, is homogenized in 50 μl of 150 m*M* NaCl in LP 2 tubes (Luckham, Ltd., Burgess Hill, UK) with a Teflon pastle. Samples are centrifuged at 1000 g_{av} for 10 min and the supernatant at 15,000 g_{av} for 30 min. The pellet is resuspended in 20 m*M* Tris–HCl, pH 7.0.

Membrane protein content is determined according to Bradford[9] by adding to 50 μl of membrane suspension 50 μl of 2 *M* NaOH and 900 μl of the Coomassie Brilliant Blue reagent.[9] Bovine serum albumin (BSA) in 1 *M* NaOH serves as standard. UV absorbance readings are obtained within 1 hr.

Incubation. [^{3}H-Phe^3]VP (New England Nuclear, Boston, MA; specific activity ~50 Ci/mmol) is used as substrate. Incubations are performed in 20 m*M* Tris–HCl, pH 7.0, or 40 m*M* Tris–HCl, 62 m*M* NaCl, pH 7.0. The VP-AP preparation, e.g., brain membranes at a protein concentration of 0.1–2 mg/ml, is first preincubated in 90 μl of this buffer at 37° for 10 min, and then 10 μl of [^{3}H]VP (0.01 μCi) is added to give a final concentration of 2.5 n*M*. Incubation is carried out for a further 30 min.

The enzyme reaction is instantaneously stopped by addition of 100 μl of 2 *M* acetic acid, and protein is denatured by boiling for 5 min. Membranes are removed by centrifugation at 10,000 g_{av} for 10 min. Samples can be stored for prolonged time at −20°.

Chromatography. A series of minicolumns are made in 1-ml pipet tips (plastic Eppendorf or Gilson pipet tips) and fitted in a device holding 10 rows of 10 columns above a box of 100 scintillation vials. Tips are packed with 500 mg polystyrene beads (Amberlite XAD-2; BDH, Poole, UK) in 96 or 100% (v/v) ethanol and equilibrated stepwise with, consecutively, 3

[7] J. P. H. Burbach, D. Terwel, and J. L. M. Lebouille, *Biochem. Biophys. Res. Commun.* **144,** 726 (1987).

[8] B. Liu and J. P. H. Burbach, *Endocrinology (Baltimore)* (in press) (1987).

[9] M. Bradford, *Anal. Biochem.* **72,** 248 (1976).

ml 100% ethanol, 3 ml 50% ethanol, and 3 ml 5% (v/v) ethanol in 0.1 *M* Tris base.[10] Aliquots of incubated samples (50–200 μl) are applied to the columns and left for 30 min and then eluted in two steps. The first step is 3 ml of 5% ethanol in 0.1 *M* Tris base, eluting [^{3}H]Phe. The second step is 3 ml of 100% ethanol, eluting [^{3}H]VP. The two fractions are collected separately in scintillation vials, and radioactivity is determined by scintillation counting. In each series of samples, chromatography is controlled by estimating the yields of known amounts of [^{3}H]Phe and [^{3}H]VP on separate columns. Typical yields are 75% for [^{3}H]Phe in the first step and 20% in the second, and less than 2% of [^{3}H]VP in the first step and 95% in the second step. Enzyme activity is calculated on the basis of the radioactivity in both fractions, taking into account the recovery of [^{3}H]Phe and [^{3}H]VP. Often scintillation counting of only the first (5% ethanol) fraction and correction for the yield of [^{3}H]Phe provide equally accurate data.

Measurement of VP Metabolites

The measurement of carboxy-terminal VP metabolites is achieved by virtue of an antiserum which recognizes the carboxy terminus of VP. It cross-reacts with the full series of centrally active VP metabolites, which are aminopeptidase products. An initial HPLC step is used to fractionate tissue extracts into fractions with separated VP metabolites, which are then quantitated by radioimmunoassay. Thus, all carboxyl-terminal VP metabolites are measured individually.

Tissue Extraction and Sample Preparation. Samples of rat brain tissue are collected and stored frozen at −80°. Frozen tissue samples are dropped in 1 *M* acetic acid at 100° in tubes standing in a boiling water bath and boiled for 10 min. Preferably, greater than 1 ml of acetic acid per 100 mg of tissue is used to achieve maximal recovery of peptides. After boiling the tubes are rapidly cooled to 0° in ice water. Small (<100 mg) tissues are disrupted by sonication at 0° (Branson sonifier, 20 KHz for 30 sec). Larger tissues are homogenized in a tightly fitting Teflon–glass homogenizer. The homogenates are centrifuged at 15,000 g_{av} for 30 min, and the clear supernatant is dried under reduced pressure at 50° in a Speed-Vac concentrator (Savant Instruments, Farmingdale, NY) or lyophilized. Dried samples are dissolved in 10 m*M* ammonium acetate, pH 4.15, and insoluble material is removed by centrifugation.

High-Performance Liquid Chromatography (HPLC). HPLC is performed on a two-pump gradient system (Millipore–Waters, Milford, MA) and μBondapak C_{18} columns. Two solvent systems with different selec-

[10] Minicolumns can be reused many times by washing with 5 ml 100% ethanol and reequilibration.

tivities have been employed. The reversed-phase system elutes the μBondapak C_{18} column with a concave gradient running from 0 to 40% (v/v) methanol [containing 0.15% (v/v) acetic acid] in 10 mM ammonium acetate, pH 4.15, in 30 min at a flow rate of 2 ml/min. The ion-pair reversed-phase system elutes the column with a concave gradient from 10 to 100% (v/v) solvent B in A over 30 min, at a flow rate of 2 ml/min [solvent A is 0.2% (v/v) heptanesulfonic acid (PIC-B7 reagent, Millipore–Waters) in water; solvent B is 0.1% heptanesulfonic acid in 50% methanol]. Fractions of 1 ml are collected in polystyrene LP3 tubes (Luckham, Ltd.) containing 100 μl of 0.1% (w/v) bovine serum albumin. Methanol is removed *in vacuo* at 50°, and the residue is lyophilized. Dried samples are dissolved in Veronal–HSA buffer (see below) for radioimmunoassay.

Radioimmunoassay (RIA). The RIA employs the polyclonal, carboxy-terminal VP-specific antiserum W1 (bleedings A or E) and ^{125}I-iodinated VP as tracer. Antiserum W1 was raised by Dr. W. B. J. Mens against a carbodiimide conjugate of synthetic [Arg^8]VP (Organon, Oss, The Netherlands) to bovine thyroglobulin injected intramuscularly in rabbits using standard procedures.[11]

^{125}I-Labeled VP was prepared by the iodogen method of Salacinski *et al.*[12] and purified or Sep-pak C_{18} cartridges (Millipore–Waters) eluted stepwise with methanol in 0.1% (v/v) trifluoroacetic acid (TFA). The effluents at around 40% methanol were tested for antiserum binding. Briefly, 10-μl aliquots of effluent were incubated with 100 μl of antiserum W1A or W1E (final dilution 1 : 2000), overnight. Binding was assessed after separation with charcoal (see below). Fractions with highest binding were used as tracer.

In the RIA the following Veronal–HSA working buffer was used: Veronal stock solution: 3.68 g diethylbarbituric acid (Veronal, Merck, Darmstadt, FRG), 3.72 g EDTA, 16 mg L-cystine in 1 liter water; Veronal–HSA working buffer: 8.12 g NaCl, 5 g human serum albumin (HSA, Sigma Chemical Co., St. Louis, MO) in 1 liter Veronal stock solution, adjusted to pH 8.0 with 4.5 ml of 5 M NaOH at room temperature. Synthetic [Arg^8]VP is used to construct standard curves.

In the standard protocol 50 μl of sample, 10 μl of ^{125}I-VP (4000 cpm), and 50 μl of antiserum W1E (final dilution 1 : 240,000) or W1A (final dilution 1 : 160,000) are subsequently pipetted to polystyrene LP_2 tubes (Luckham, Ltd.) on ice. After being vortexed, the samples are incubated at 4° for 48 hr. Then, 100 μl of a charcoal–dextran solution [4 g charcoal,

[11] H. Van Vunakis and J. J. Langone (eds.), this series, Vol. 70.

[12] P. R. P. Salacinski, C. McLean, J. E. C. Sykes, V. V. Clement-Jones, and P. J. Lowry, *Anal. Biochem.* **117,** 136 (1981).

750 mg dextran T70 (Pharmacia, Uppsala, Sweden) in 100 ml of 50 m*M* sodium phosphate buffer, pH 7.4, stirred for at least 30 min and kept at 4°] is added, and the mixture is vortexed and centrifuged at 7000 *g* for 10 min at 4°. Supernatants are siphoned off at room temperature, and the radioactivity in the pellet (free fraction) is determined in a gamma counter.

Evaluation and Application

VP-Converting Aminopeptidase Activity

VP-AP is a novel enzyme activity which is defined by the initial cleavage of the Cys^1—Tyr^2 bond in the amino terminus of VP (Fig. 1) and oxytocin (OT).[3,6,13,14] This cleavage is the limiting step in VP processing and converts the cyclic portion of VP to the linear peptide [Cyt^6]VP-(2–9). Subsequently, amino-terminal amino acids are removed, and smaller carboxyl-terminal metabolites are formed. The conversion of [Cyt^6]VP-(3–9) to -(4–9) involves the release of Phe^3 (Fig. 1). By using [^{3}H-Phe^3]VP as substrate, this step is detected by the VP-AP assay. [^{3}H-Tyr^2]VP, which is also commercially available (Amersham Radiochemical Centre, Amersham, UK), will also serve as a useful substrate for this VP-AP assay. It is not known whether VP-AP is also responsible for conversion of linear intermediate metabolites, or whether another aminopeptidase is required for these secondary steps. This may have consequences when purified VP-AP is assayed (see below).

Chromatography. The assay of VP-AP is based on the measurement of the amount of [^{3}H]Phe, which is released during VP-AP action on [^{3}H-Phe^3]VP (Fig. 1). A simple chromatographic procedure on minicolumns allows rapid separation of labeled product and substrate in large series of samples. Due to a low substrate concentration (2.5 n*M*), ^{3}H-labeled intermediates of VA-AP action, such as [Cyt^6]VP-(2–9) and -(3–9),[6] are absent in digests, and therefore [^{3}H]Phe and [^{3}H]VP are the only radioactive substances in the samples (Fig. 2). However, in samples containing enzymes that attack the carboxyl terminus of VP, like brain cytosol, the metabolites ^{3}H-labeled VP-(1–8) and -(1–7) can also be formed (Fig. 2). These do not interfere with the quantitation of [^{3}H]Phe, since they elute together with [^{3}H]VP in the 100% elution step. It has been shown by HPLC analysis that [^{3}H]Phe is also accurately determined in

[13] J. P. H. Burbach, *in* "Current Topics in Neuroendocrinology" (D. Ganten and D. Pfaff, eds.), p. 55. Springer-Verlag, Berlin and New York, 1986.

[14] W. H. Simmons and A. T. Arawski, *Proc. Int. Congr. Biochem., 13th,* Abstr. TU-276 (1985).

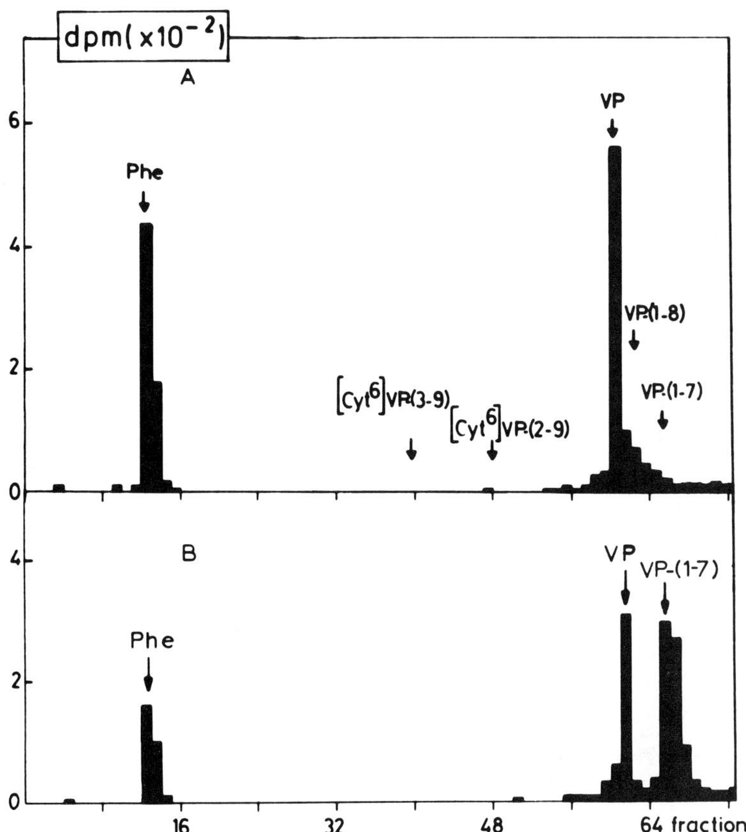

FIG. 2. ^{3}H-Labeled products in digests of [^{3}H-Phe^3]VP by brain membranes (A) and brain cytosol (B) as analyzed by HPLC using an ammonium acetate–methanol solvent system (see the text). Arrows indicate the elution position of synthetic peptides and Phe. Incubations were performed at a protein concentration of 0.5 mg/ml (A) or 0.25 mg/ml (B) for 40 min.

situations with high carboxyl-terminal cleaving enzymes.[7] It must be realized that VP-(1–8) and -(1–7) are also substrates for VP-AP and that the kinetics of their cleavage can be different from that of [^{3}H]VP.

Enzyme Reaction. VP-AP has a linear relationship with protein concentration and incubation time up to the point of approximately 50% conversion of the substrate [^{3}H]VP (Fig. 3). Since enzyme activity is assessed at a substrate concentration far below its K_m (12 μM^3), the concentration of intact [^{3}H]VP is limiting for the enzyme velocity. Depending on the specific activity of VP-AP, the experimental samples

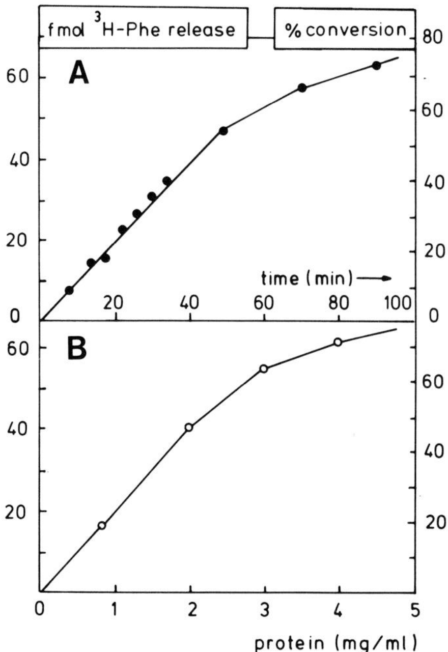

FIG. 3. Measurement of VP-AP as a function of incubation time (A) and brain membrane protein concentration (B). In A, a membrane protein concentration of 1 mg/ml was used. In B, the incubation time was 20 min. Conversion is expressed as femtomoles [^{3}H]Phe and is shown as the actual percentage of the initial amount of [^{3}H]VP.

should be titered in such a way that less than 50% conversion occurs. Preferably, conversion should be between 15 and 45% of [^{3}H]VP. Alternatively, the incubation time can be changed to obtain proper conversion of [^{3}H]VP (Fig. 3A).

A variety of buffer substances in the assay can be used without influencing the activity of VP-AP or the chromatographic separation of [^{3}H]Phe and [^{3}H]VP. For instance, Tris–HCl up to 100 m*M*, sodium phosphate up to 80 m*M*, and NaCl up to 250 m*M* are without effect on VP-AP as compared to the activity measured in 20 m*M* Tris–HCl, pH 7.0. However, since the VP-AP is a metallopeptidase,[3] citrate and phosphate buffers have been avoided in our studies for their metal-chelating properties. The pH of the incubation medium is a critical parameter, since VP-AP has a narrow optimum between pH 6.5 and 7.5. Triton X-100 concentrations up to 1% (v/v) can be used in the enzyme reaction without effect on the separation of [^{3}H]Phe and [^{3}H]VP on the Amberlite XAD-2

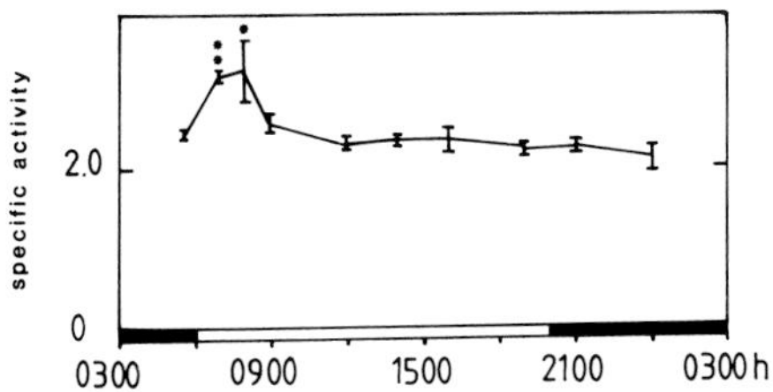

FIG. 4. Measurement of VP-AP in the rat pineal gland during the light–dark cycle. Enzyme activity is expressed as specific activity (pmol [^{3}H]Phe released/mg/hr). The clock time and light–dark regimen are shown on the *x* axis.

minicolumns. It has been noted that 0.2% Triton X-100 enhances the activity of VP-AP in brain membranes 3- to 4-fold. This phenomenon is probably caused by solubilization of the enzyme and can be used to increase the sensitivity of the assay on membranous samples. The assay of VP-AP is currently being used to study its distribution and endocrine regulation in the brain, to characterize its enzymatic properties, and to follow its purification. An example of the usefulness of the assay for application to small brain samples is the measurement of VP-AP in the rat pineal gland during the light–dark cycle.[8] The assay allows duplicate determination of VP-AP in membranes of single rat pineal glands (protein concentration in the assay is 0.1–0.6 mg/ml). Thus, a small, but highly significant cyclic variation in activity of VP-AP has been found, occurring at the onset of light[8] (Fig. 4).

The VP-AP assay is the basis for purification of VP–AP from rat brain. VP-AP is readily solubilized in 0.2% Triton X-100, and its activity can be followed after gel filtration as illustrated in Fig. 5. Two peaks of activity are observed, one eluting in the void volume, which may represent a protein aggregate, the other in the molecular mass range of 100–140 kDa. A potential hazard during purification work is the possibility that purified VP-AP, which is specialized in the initial cleavage of the cyclic amino terminus of VP (Fig. 1) is less or not active on the first linear metabolite to occur ([Cyt^6]VP-(2–9)^6), and thus does not release [^{3}H]Phe^3. In crude tissue fractions this problem does not arise owing to the abundance of a variety of peptidases, which attack the linear metabolites of VP. Whether VP-AP itself can also convert the linear metabolites and release [^{3}H]Phe is being investigated. Alternatively, a second aminopeptidase which is inactive on intact VP but acts on the linear metabolites could be used in a second incubation step to release all [^{3}H]Phe from the initial metabolite of VP-AP.

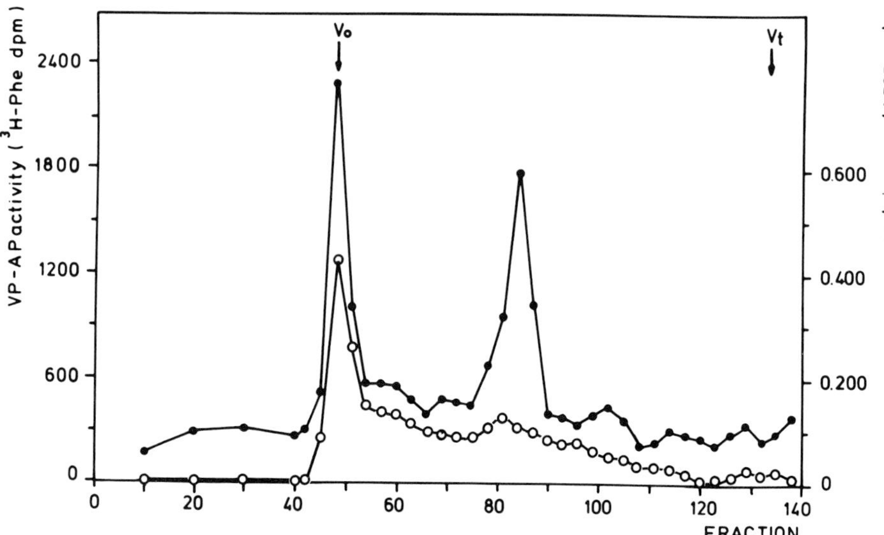

FIG. 5. Gel filtration profile of VP-AP (●) solubilized from brain membranes on Sephadex G-200. The column (1.6 × 94 cm) was run with 0.02% (v/v) Triton X-100 in 20 m*M* Tris–HCl, pH 7.0. Fractions of 1.3–1.6 ml were collected, and VP-AP was determined by incubation of 90 μl of fractions with 14,000 dpm of [^{3}H]VP for 60 min. The protein content of fractions was determined according to Bradford,[9] with 100-μl aliquots of the fractions, and expressed as $A_{595\text{ nm}}$ (○).

Carboxyl-Terminal VP Metabolites

Aminopeptidase products of VP like [pGlu^4,Cyt^6]VP-(4–9), [Cyt^6]VP-(5–9), and their desglycinamide^9 derivatives are the most potent VP fragments.[4] They are also the VP fragments which are the most selective in their effects.[2,4] Endogenous forms of these VP metabolites are measured by an antiserum specific for the carboxyl terminus of VP. Individual measurement of the different metabolites is achieved by HPLC fractionation prior to RIA.

Properties of the RIA. Antiserum W1 (bleedings A and E) recognizes VP and carboxyl-terminal metabolites as small as [Cyt^6]VP-(5–9). The affinity of the antiserum decreases with decreasing length of the peptide (Table I). This requires correction of data for cross-reaction and causes a decrease in sensitivity for the shorter peptides. The sensitivity is approximately 0.25–0.5 pg VP per tube (90% tracer displacement). Tracer displacement of 50% is achieved at 8 pg VP per tube (Fig. 6).

Tissue Extraction and HPLC. The recovery of [^{3}H]VP added to tissues extracts is overall approximately 40–50% as measured after the

TABLE I
IMMUNOREACTIVITY OF VP-RELATED PEPTIDES AND FRAGMENTS TO VP ANTISERUM W1[a]

Peptide	Cross-reactivity (%)	
	Bleeding A	Bleeding E
[Arg^8]Vasopressin [VP-(1–9)]	100	100
[Arg^8]Vasotocin	100	100
[Cyt^6]VP-(2–9)	50	Not tested
[Cyt^6]VP-(3–9)	31	50
[pGlu^4,Cyt^6]VP-(4–9)	23	25
[Cyt^6]VP-(5–9)	12.5	13
Oxytocin	<0.01	<0.01
Deamino[D-Arg^8]VP	Not tested	<0.01
VP-(1–8)	<0.01	<0.01
VP-(1–7)	<0.01	<0.01

[a] Antiserum W1A was used in a final dilution of 1 : 160,000; W1E in 1 : 240,000. ^{125}I-Labeled VP served as tracer. Cross-reactivities of synthetic peptides (courtesy of Dr. J. W. Van Nispen, Organon International, Oss, The Netherlands) were determined by comparison of tracer displacement curves with that of synthetic VP.

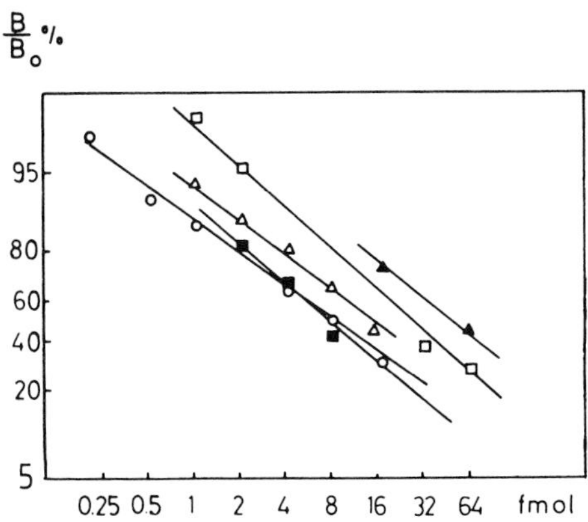

FIG. 6. Displacement of ^{125}I-labeled VP to antiserum W1E by [Arg^8]vasopressin (VP) (●), vasotocin (VT) (■), and [Cyt^6]VP fragments [VP-(3–9) (△), VP-(4–9) (□), VP-(5–9) (▲)] as indicated by the sequence numbers of VP (see Fig. 1). The relationship between percentage specific binding of ^{125}I-labeled VP ($B/B_0 \times 100$) and quantity of peptide is shown in a logit–log plot.

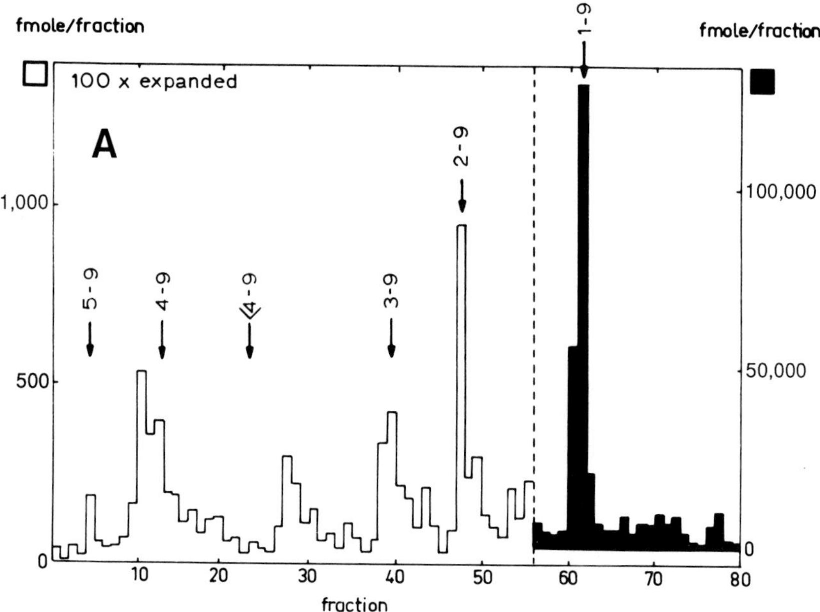

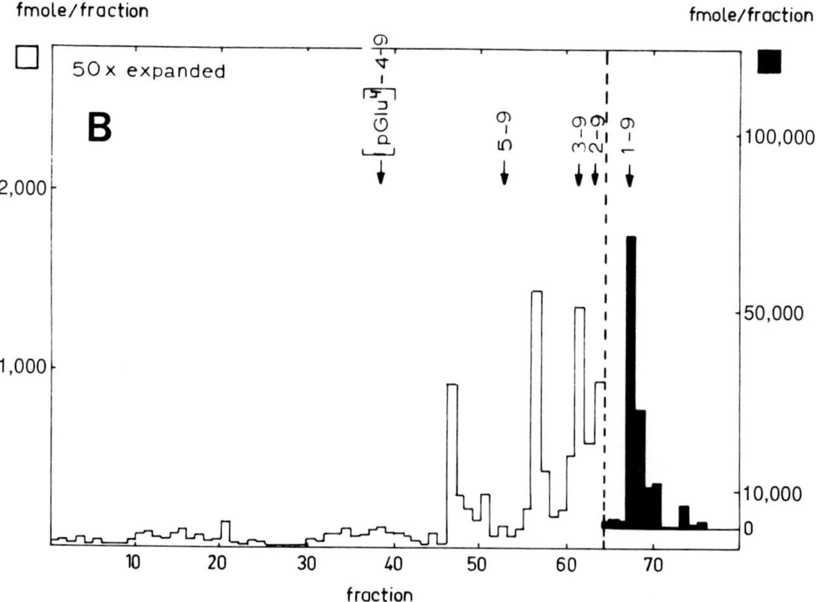

FIG. 7. HPLC analyses of VP-immunoreactive peptides in a rat hypothalamic extract. Ten hypothalami were pooled. Antiserum W1A was used in the RIA. (A) The HPLC solvent system was ammonium acetate–methanol; (B) heptanesulfonic acid and methanol were used. For chromatography conditions, see the text. The arrows indicate the elution positions of synthetic peptides. The bars represent the total immunoreactivity (in fmol) per fraction.

HPLC step.[15] The recovery is predominantly dependent on the ratio of tissue weight to volume of homogenization, with higher recoveries being achieved at higher ratios.

Two different HPLC solvent systems have been developed for separation of VP metabolites. The ammonium acetate reversed-phase system separates all RIA-measurable VP metabolites. The disadvantage, however, is the elution of [Cyt^6]VP-(4–9) and -(5–9) very near the void volume (Fig. 7A). The other heptanesulfonic acid ion-pair system[16] shifts all VP-related peptides to longer retention times and displays a different selectivity, as seen from the elution order of VP metabolites (Fig. 7B). In particular, [pGlu^4,Cyt^6]VP-(4–9) displays a characteristic shift in retention.

Measurements of VP metabolites have been performed in rat brain tissue samples.[3,5,15] RIA screening of HPLC fractions showed that a number of immunoreactive substances are eluting in the region of VP metabolites. Precise coelution with synthetic peptides has been observed only for a few metabolites (e.g., [Cyt^6]VP-(2–9) and -(3–9), Fig. 7), whereas others have the retention times of reduced forms of VP metabolites.[3,5]

The sensitivity of RIA together with the necessity of HPLC separation and the low level of VP metabolites in brain tissue require relatively large amounts of tissue. The levels of the various VP fragments in the hypothalamus are only 1–3% of the VP content. In extrahypothalamic areas, such as hippocampus, amygdala, septum, and brainstem, the levels of fragments are between approximately 5 and 30% of the VP content.[3,5,15] These levels imply that pools of tissue from several animals should be used in this application to obtain measurable amounts of peptides in RIA. Therefore, the present methodology cannot be applied as yet to measurement of the various VP metabolites in small tissue samples of individual animals. Improvements based on antisera with higher selectivity and without cross-reaction with intact VP are being developed in order to circumvent the HPLC step.

Acknowledgments

We are grateful to our colleagues Dr. Xin-Chang Wang (presently at Nanking University, People's Republic of China), Dr. Jos L. M. Lebouille, Mr. Dick Terwel, Ms. Agnes Tan, and Mr. Arno van der Kleij for the contribution of data to this chapter. The support of the Stichting Farmacologisch Studiefonds (to B.L.) and Medigon (to J.L.M.L.) is acknowledged.

[15] X.-C. Wang, J. P. H. Burbach, J. A. Ten Haaf, and D. De Wied, *Chin. Sci.* **3,** 257 (1986).

[16] X.-C. Wang and J. P. H. Burbach, *FEBS Lett.* **197,** 164 (1986).

[29] Quantitation of Vasopressin and Oxytocin mRNA Levels in the Brain

By HUBERT H. M. VAN TOL and J. PETER H. BURBACH

Introduction

Recent developments in molecular biology and the isolation of the two different but closely related genes for vasopressin (VP) and oxytocin (OT) made it possible to develop methods for quantitation of the mRNA levels of these two neuropeptides. These mRNA levels are a reflection of the gene transcription rates, and they also provide a useful index for the biosynthetic capacity of neurons to synthesize a given peptide. The use of the vasopressin and oxytocin systems for studies on neuropeptide gene regulation is very advantageous since the vasopressinergic and oxytocinergic neurons are clearly grouped into distinct nuclei of the brain which are linked to peripheral or central functions of these neuropeptides.[1] The separation of these neurons into distinct nuclei enables us to quantify VP and OT mRNA levels by rapid and simple techniques such as solution hybridization assay, dot-blot analysis, and Northern blot analysis combined with microdissection techniques like the punch technique of Palkovits.[2] Furthermore, changes in mRNA levels can be determined by quantitative *in situ* hybridization.

General Considerations

The methods for detection and measurement of VP and OT mRNAs described here were developed with the cloned exon C fragments from the VP gene (314 bp) and the OT gene (223 bp)[2] derived from the subclones pVλ*Pst*I and pOλBalI which were a kind gift from Drs. H. Schmale, R. Ivell, and D. Richter.[3,4] Exons C of both genes were chosen as hybridization probes in the different assays since they have the lowest sequence similarity.[3,4] As was apparent from Northern blot analysis (see below) no cross-hybridization could be detected between the exon C-derived probes and both mRNAs (Fig. 1). It should be noted that the

[1] R. M. Buijs, G. J. de Vries, F. W. van Leeuwen, and D. F. Swaab, *in* "Progress in Brain Research" (B. A. Gross and G. Leng, eds.), p. 115. Elsevier, Amsterdam, 1983.

[2] J. P. H. Burbach, H. H. M. Van Tol, M. H. C. Bakkus, H. Schmale, and R. Ivell, *J. Neurochem.* **47,** 1814 (1986).

[3] H. Schmale, S. Heinsohn, and D. Richter, *EMBO J.* **2,** 763 (1983).

[4] R. Ivell and D. Richter, *Proc. Natl. Acad. Sci. U.S.A.* **81,** 2006 (1984).

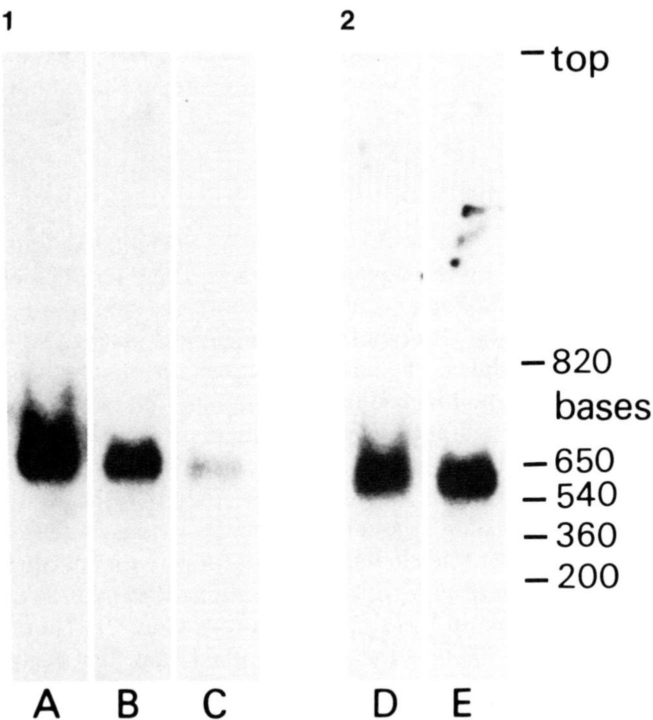

FIG. 1. Northern blot analysis for (1) VP mRNA in the nucleus supraopticus (SON), nucleus paraventricularis (PVN), and nucleus suprachiasmaticus (SCN) using the clone MPB-1 which contains exon C from the rat VP gene (lanes A, B, and C). For the detection of (2) OT mRNA in the SON and PVN (lanes D and E) the clone MPB-5, containing the exon C fragment derived from the rat OT gene, was used. RNA isolated from the SON, PVN, and SCN was pooled from two, two, and three rats, respectively. The length of the RNA was determined by using M13 digested with *Hpa*II as size markers.

conditions of these assays may be modified when probes with different characteristics, for example, in sequence and/or length, are used.

The following precautions should be taken to prevent RNA degradation. Surgical gloves should be worn during the experimental procedures. Glassware is incubated overnight at 280°, and plastic disposable materials are autoclaved for 20 min at 120°. Equipment that comes into contact with RNA preparations, such as parts of the apparatus for dot-blot hybridization or for gel electrophoresis in Northern blot analysis, is kept overnight in 0.1% diethyl pyrocarbonate (DEPC). Double-distilled water and solutions which do not themselves inhibit RNase are treated overnight with 0.1% (DEPC) and autoclaved thereafter. Since DEPC is not stable in Tris-containing solutions, Tris buffers are made with DEPC-treated water and then autoclaved. The chemicals used are kept apart from the general stocks and are preferably poured from their containers or handled with an autoclaved spatula.

Microdissection of Brain Tissue

For microdissection of brain tissue rats are decapitated, and the brain is removed and immediately frozen on dry ice. The microdissection technique developed by Palkovits is used.[5] In short, the brains are cut in 300-μm slices in a cryostat. By quickly thawing and freezing, the slices are mounted onto glass slides. The nuclei of interest are punched out from the frozen slices using a hollow needle of 1 mm inner diameter. The nucleus supraopticus (SON) and nucleus paraventricularis (PVN) are isolated by 10 punches over 5 slices and the nucleus suprachiasmaticus (SCN) by 3 punches over 3 slices. This way it is ascertained that the whole nucleus was isolated in a constant amount of tissue. A surplus of tissue which is not part of the nucleus itself does not interfere with the quantitation, since, as can be seen from *in situ* hybridization, VP and OT mRNAs are predominantly located in the soma of neurons (Fig. 2). Therefore, it is more appropriate to express the data obtained from the assays per nucleus than per milligram of protein or RNA. As negative controls to determine background hybridization, equal amounts of brain cortex tissue, which is devoid of VP and OT neurons, can be used.

RNA Preparation

The isolated nuclei are homogenized at room temperature in 100 μl of 4 *M* guanidine thiocyanate, 0.1 *M* 2-mercaptoethanol, 25 m*M* sodium citrate, pH 7.0 (GTC) by suction and ejection through a 21-gauge needle from a syringe (about 10 times). The GTC extraction solution is prepared by dissolving 50 g guanidinium thiocyanate (Fluka AG, Buchs, Switzerland) in 80 ml water, adding 2.5 ml 1 *M* sodium citrate (pH 7.0), and adjusting the pH to 7.0 with 1 *M* sodium hydroxide. The volume is made up to 99.3 ml with water, and the solution is filtered through a microfilter (Millipore, Bedford, MA; Type HAWP, pore size 0.45 μm). After filtration, 700 μl of 14 *M* 2-mercaptoethanol is added. The homogenate is extracted with an equal volume of phenol–chloroform–isoamyl alcohol (25 : 24 : 1, v/v/v), pH 8.0 (phenol–CIAA). Following centrifugation for 5 min at 10,000 *g*, 80 μl of the aqueous phase is removed and saved. Keep-

[5] M. Palkovits, *Brain Res.* **59,** 449 (1973).

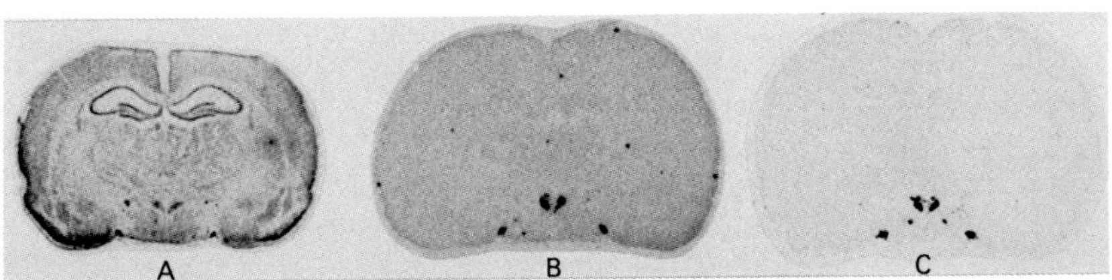

FIG. 2. *In situ* hybridization of VP mRNA (A and B) and OT mRNA (C) in the brain using a ^{35}S-labeled RNA probe with an exon C-containing fragment of the VP gene (A and B) and a single-stranded ^{35}S-labeled DNA probe containing the exon C sequence of the OT gene. In A the *in situ* hybridization technique using a RNA probe was performed without RNase treatment to reduce aspecific hybridization, while in B RNase treatment was included. Transverse sections of the rat brain are shown.

ing GTC and phenol–CIAA at 4° improves the formation of a clear interphase, particularly when larger amounts of tissue are used. The phenol–CIAA phase is reextracted with 100 μl GTC, and then the isolated GTC supernatant is combined with the previous one.

RNA is precipitated by addition of 0.6 volumes of 45 m*M* acetic acid in ethanol (0.28%, v/v) at −20°, overnight incubation at −20°, and centrifugation for 20 min at 10,000 *g*. The phenol–CIAA extraction step is included for removal of proteins since thereafter the pellet obtained following ethanol precipitation of the GTC extract can be more easily dissolved. The pellet is washed once by addition of 150 μl 70% ethanol (−20°) and centrifugation for 1 min at 10,000 *g*. The supernatant is discarded, and the pellet is dried under reduced pressure. Dissolving of the pellet in water or a solution of 10 m*M* Tris–HCl, 1 m*M* EDTA, pH 7.5 (TE), is enhanced by incubation at 60°. RNA isolated by this extraction procedure has a highly reproducible recovery (SEM ±1%) as determined with ^{35}S-labeled RNA.[2] The ratio of the UV absorbance profile of this isolated RNA for 240 : 260 : 280 nm is 0.50 : 1 : 0.52. This RNA is used in solution hybridization assays and Northern blot analysis.

Further purification, which is necessary for dot-blot analysis, is accomplished by a subsequent incubation of the RNA samples in 50 μl of 50 μg/ml proteinase K, 10 m*M* Tris–HCl, 5 m*M* EDTA, 0.5% sodium dodecyl sulfate (SDS), pH 7.8, at 37° for 1 hr. The proteinase K solution is preincubated for 15 min at 37° to degrade possible contaminating RNase activity. After the incubation, 50 μl of water is added to the sample, to obtain a volume which is easier to handle. This sample is extracted with 100 μl phenol–CIAA, and RNA from the isolated aqueous supernatant is precipitated by addition of 0.1 volumes of 3 *M* sodium acetate, pH 5.2, and 3 volumes of ethanol (−20°). After an incubation period of 30 min at

−80°, RNA is precipitated by centrifugation at 10,000 *g* for 20 min and washed once with 70% ethanol.

Hybridization Probes

Single-Stranded ^{35}S-Labeled DNA Probes for Solution and in Situ Hybridization

Isolation of single-stranded M13 DNA and plasmids is performed according to standard procedures.[6,7] The single-stranded M13 clones MPB-2 and MPB-4 containing the exon C sense sequence of the VP and OT genes, respectively, are used for the preparation of single-stranded ^{35}S-labeled DNA probes.[1] The 17-mer universal M13 sequencing primer (60 ng) (New England Biolabs, Beverly, MA) is hybridized to 1.5 μg MPB-2 or MPB-4 in a total of 100 μl solution containing 15 μl of 0.1 *M* Tris–HCl, pH 7.9, 0.6 *M* sodium chloride, 66 m*M* magnesium chloride (hybridization buffer) by heating in a water bath at 85° for 10 min and cooling slowly to room temperature overnight. To 10 μl of this solution are added 1 μl of 50 m*M* dithiothreitol (DTT); 1 μl 1 m*M* dTTP, dATP, and dGTP; 50 μCi [α-^{35}S]dCTP (1,000 Ci/mmol) (Amersham Radiochemical Centre, Amersham, UK); and 2.5 units DNA polymerase I Klenow fragment (Boehringer Mannheim, Mannheim, FRG). The final volume should be 15 μl; however, in case this is exceeded appropriate amounts of DTT and hybridization buffer should be added. After 90 min at room temperature an extra aliquot of 2.5 units DNA polymerase I Klenow fragment is added, and incubation is continued for another 90 min at room temperature. The ^{35}S-labeled inserts are cut from MPB-2 by addition of 12 units *Hin*dIII, and from MPB-4 by addition of 5 units *Eco*RI and 1 μl 1 *M* sodium chloride, and subsequently incubated at 37° for 3 hr.

DNA is precipitated by the addition of 0.1 volumes of sodium acetate, pH 5.2, 3 volumes of ethanol (−20°), followed by a 30-min incubation at −80° and centrifugation (0°) at 10,000 *g* for 20 min. The precipitate is washed once with ethanol, then quickly dried *in vacuo*. This pellet is dissolved in formamide (deionized) containing 0.3% bromphenol blue, 0.3% xylene cyanol, 10% 0.89 *M* Tris–borate, 0.89 *M* boric acid, 20 m*M* EDTA, pH 8.0, and heated to 100° for 10 min. The denatured samples are quickly cooled to 0° and then loaded and run on a 5% polyacrylamide–7 *M* urea gel. After electrophoresis the gel is exposed to X-ray film to locate

[6] J. Messing, this series, Vol. 104, p. 20.

[7] T. Maniatis, E. F. Fritsch, and J. Sambrook, "Molecular Cloning: A Laboratory Manual." Cold Spring Harbor Laboratory, Cold Spring Harbor, New York, 1982.

the single-stranded ^{35}S-labeled probe, which is then excised from the gel. The gel is minced by a glass rod, and the DNA is eluted overnight in 700 μl 0.5 M ammonium acetate, 10 mM magnesium chloride, 0.1% SDS, 0.1 mM EDTA, pH 8.0, and 1 μg tRNA. The gel is precipitated by centrifugation, and the probe in the supernatant is precipitated using sodium acetate and ethanol as described before. The DNA is dissolved in an appropriate volume of water and extracted with an equal volume of phenol–CIAA as described before, and remnants of phenol–CIAA are removed by three subsequent extractions with ether. In general a yield of 2–5 × 10^6 dpm of single-stranded ^{35}S-labeled probe can be achieved. These probes are used in the solution hybridization assays and for *in situ* hybridization.

Single-Stranded ^{32}P-Labeled DNA Probes for Northern and Dot-Blot Analysis

For hybridization of dot-blots and Northern blots we used the single-stranded M13 clones MPB-1 and MPB-5, containing the exon C antisense sequences of the VP and OT genes, respectively.[1] These probes are labeled by primer extension in the M13 vector arm. For this purpose the 17-mer hybridization probe primer (New England Biolabs) is hybridized to MPB-1 and MPB-5 in a similar manner as described before. To 10 μl of this solution are added 1 μl 50 mM DTT; 1 μl 1 mM dTTP, dATP, and dGTP; 15 μCi [α-^{32}P]dCTP (3000 Ci/mmol) (Amersham Radiochemical Centre); and 2.5 units of DNA polymerase I Klenow fragment. The volume is adjusted with water to 15 μl, and then the mixture is incubated at room temperature for approximately 15 min. Next, 60 μl 5 M ammonium acetate, 80 μl water, 1 μl 1 mg/ml salmon sperm DNA, and 90 μl 2-propanol are added and kept for 20 min at room temperature. DNA is precipitated by centrifugation at 10,000 g for 10 min. Only DNA which is larger than approximately 150 bases is precipitated by this procedure. The pellet, containing the probe, is washed once with 70% ethanol (−20°) and quickly dried *in vacuo*.

Single-Stranded ^{35}S-Labeled RNA Probes for in Situ Hybridization

In addition to single-stranded DNA probes, RNA probes are also used for the detection of VP mRNA by *in situ* hybridization. For this purpose the 211-bp *Eco*RI–*Taq*I exon C-containing fragment of MPB-2 was cloned into the *Eco*RI–*Acc*I sites of the vector pGEM-2 (Promega Biotec, Madison, WI). Transcription from the T7 or SP6 promotor, located on opposite sites from the cloned exon C insert on the pGEM-2 vector, can give RNA with the sense or antisense sequence, respectively. The transcript from the SP6 promotor containing the antisense sequence of exon C from the

VP gene is used for *in situ* hybridization. RNA synthesis is done according to the instructions of the manufacturer of the pGEM kit (Promega Biotec), using [^{35}S]UTP (1000 Ci/mmol, SP6 grade; Amersham Radiochemical Centre).

Solution Hybridization Assay

Basically the solution hybridization assay for VP mRNA and OT mRNA is performed according to principles for a solution hybridization assay for specific mRNAs as described by Durnam and Palmiter.[8]

Total RNA isolated as described before is dissolved by vortexing in 15 μl of 40% formamide, 0.6 *M* sodium chloride, 4 m*M* EDTA, 10 m*M* Tris–HCl, pH 7.4, and 3,000 dpm single-stranded [^{35}S]DNA probe. After a short centrifugation step to spin down the sample, it is covered with 20 μl paraffin oil and incubated at 50° for 30 hr. After 30 hr nonhybridized probe is digested by adding 150 μl 0.3 *M* sodium chloride, 30 m*M* sodium acetate, 3 m*M* zinc chloride, pH 4.2, 5 μg denatured salmon sperm DNA, and 400 units S_1 nuclease (Boehringer Mannheim), vortexing, and subsequently incubating at 50° for 1 hr. S_1 nuclease-resistant material is precipitated by addition of 1 ml of 11.5% (w/v) trichloroacetic acid, 1.66% (w/v) sodium pyrophosphate, and 10 μg salmon sperm DNA. After 20 min at 0° the precipitate is collected by filtration over a Whatman GF/B filter (Whatman, Maidstone, UK) which had been soaked in an aqueous solution of 2.5% trichloroacetic acid–1.66% sodium pyrophosphate at 0°. The Eppendorf tube, which contained the sample, is washed 3 times with 1 ml of 2.5% trichloroacetic acid–1.66% pyrophosphate at 0° and poured over the filter. Next the filter is washed with 5 ml of 2.5% trichloroacetic acid–1.66% sodium pyrophosphate (0°) followed by 3 ml 75% ethanol at 0°. The ^{35}S radioactivity of the precipitate on the filters is determined by scintillation counting.

Standard curves are made with known quantities (5–250 pg) of MPB-2 or MPB-4. To these standards is added 5 μg yeast RNA to simulate the composition of the RNA samples. The samples of the standard curve are processed similarly as the RNA samples. Furthermore, the conditions of the hybridization solution are chosen to minimize differences between DNA–DNA and DNA–RNA hybridization.[9]

The assay using MPB-2 and MPB-4 has been optimized as the procedure described above for the amount of S_1 nuclease as well as for the precipitation procedure to give an optimal signal-to-noise ratio. The assay

[8] D. M. Durnam and R. D. Palmiter, *Anal. Biochem.* **131,** 385 (1983).

[9] J. Casey and N. Davidson, *Nucleic Acids Res.* **4,** 1539 (1977).

can also be performed with single-stranded ^{32}P-labeled probes.[10] In the assay described here we used single-stranded ^{35}S-labeled DNA probes since they are more reliable for optimal S_1 nuclease degradation than single-stranded ^{32}P-labeled DNA. For degradation of single-stranded ^{32}P-labeled probes, S_1 nuclease can be used only in a small range of concentrations. This is probably due to strand breaks which accompany ^{32}P disintegration (Fig. 3). Improvement in the preparation of probe and standards, and possibly the sensitivity of this assay, might be achieved by the use of RNA probes and standards derived from an *in vitro* RNA transcription system. In such a solution hybridization assay using the principle of RNA–RNA hybridization instead of DNA–RNA hybridization, the nonhybridized RNA probe can be digested by RNase A. Successful application of such an assay has been described.[11]

Northern Blot Analysis

Denaturation of RNA and subsequent electrophoresis is done according to the methods described by McMaster and Carmichael.[12] The RNA pellet is dissolved in 5.5 μl water and 2.4 μl 100 m*M* sodium hydrogen phosphate buffer, pH 6.5. After dissolving the RNA 4 μl of 6 *M* glyoxal (deionized) and 12 μl dimethyl sulfoxide are added. This RNA solution is incubated at 50° for 60 min in a closed Eppendorf tube. Next, the sample is cooled on ice, and 6 μl of loading buffer, containing 50% glycerol, 10 m*M* sodium hydrogen phosphate buffer, pH 6.5, 0.4% bromphenol blue, and 0.4% xylene cyanol, is added. The denatured RNA is loaded and run at 3–4 V/cm through a 1.4% agarose gel containing 10 m*M* sodium hydrogen phosphate (pH 6.5) submerged in 10 m*M* sodium hydrogen phosphate buffer, pH 6.5. To avoid changes in the pH of the electrophoresis buffer, the buffer should be recirculated constantly or renewed every hour.

RNA in the gel is transferred by capillary blotting, using 25 m*M* sodium hydrogen phosphate buffer, pH 6.5, as transfer fluid, to a nylon membrane (Gene-Screen, NEN–Du Pont, Boston, MA, or Hybond-N, Amersham). After transfer the membrane is briefly rinsed in 25 m*M* sodium hydrogen phosphate buffer, pH 6.5. The membrane is dried at room temperature, and RNA is baked onto the membrane at 85° for 2–3 hr. The hybridization procedure and subsequent autoradiography is described below (see *Hybridization and Autoradiography of the Northern and Dot Blots*). Easily detectable bands for VP mRNA and/or OT mRNA can be

[10] J. P. H. Burbach, M. J. De Hoop, H. Schmale, D. Richter, E. R. De Kloet, J. A. Ten Haaf, and D. De Wied, *Neuroendocrinology* **39,** 582 (1984).

[11] L. Devi, J. Douglass, and E. Herbert, *Mol. Brain Res.* **388,** 173 (1987).

[12] G. K. McMaster and G. G. Carmichael, *Proc. Natl. Acad. Sci. U.S.A.* **47,** 4835 (1977).

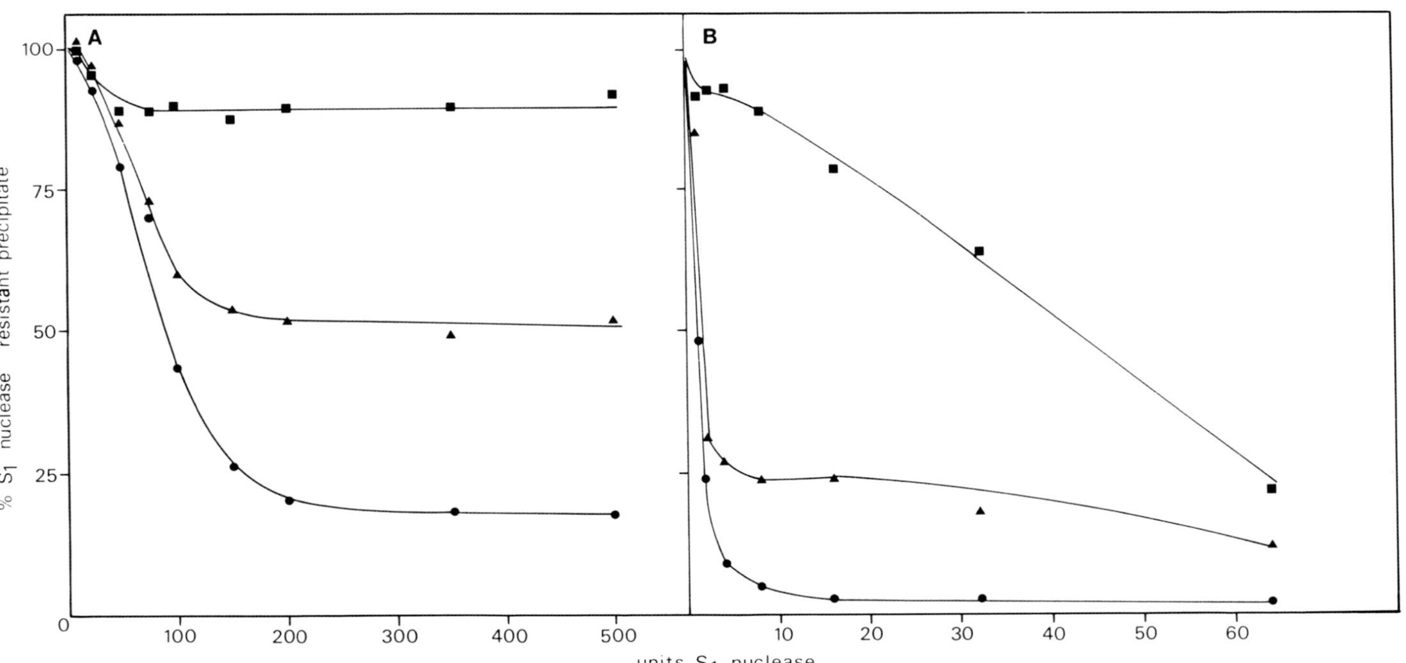

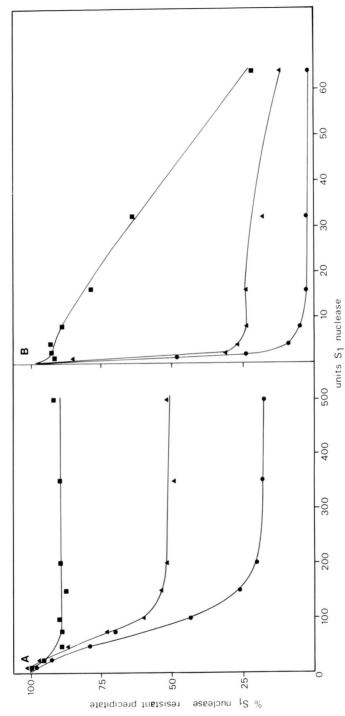

FIG. 3. Solution hybridization assay, using different concentrations of S_1 nuclease, with single-stranded ^{35}S-labeled (A) [reproduced with permission from Burbach *et al.*[2] (Fig. 2)] or ^{32}P-labeled (B) VP-specific DNA probes. In A the ^{35}S-labeled probe is hybridized to 0 pg (●), 200 pg (▲), and 1000 pg (■) MPB-2, respectively. In B the ^{32}P-labeled probe is hybridized to 0 pg (●), 200 pg (▲), and 2000 pg (■), respectively.

obtained for RNA isolated from the SON or PVN of one rat and for the SCN pooled from three rats (Fig. 1).

Dot-Blot Analysis

The dot-blot procedure is in essence the same as that which has been described by White *et al.*[13] RNA samples are dissolved in 50 μl 6.7× SSC (1× SSC: 150 m*M* sodium chloride–15 m*M* sodium citrate, pH 7.0) with 7.4% formaldehyde. The samples are incubated for 15 min at 65° and then cooled on ice. The denatured RNA is spotted onto a nylon membrane (Hybond-N, Amersham) (prewetted with 20× SSC) using a dot-blot apparatus (Bio-Dot microfiltration apparatus, Bio-Rad, Richmond, CA). The tubes which contained the sample are washed with 100 μl of 20× SSC and spotted on the membrane. Finally the slot of the dot-blot apparatus is washed with 100 μl 20× SSC. The membrane is removed from the dot-blot apparatus, briefly rinsed in 2× SSC, dried at room temperature, and then baked at 85° for 2–3 hr.

As standards for quantitation of the hybridization signal, known amounts (0.5–40 pg) of MPB-2 and MPB-4 are used and processed similarly as the RNA samples (Fig. 4). For adequate quantification 1/6, 1/4, and 1 part of RNA isolated from the SON, PVN, and SCN, respectively, of one rat are applied to the blots.

Hybridization and Autoradiography of the Northern and Dot Blots

The blots are prehybridized in plastic bags containing 10 ml 50% formamide (deionized), 0.2% polyvinylpyrrolidone, 0.2% bovine serum albumin, 0.2% ficoll, 0.1% sodium dodecyl sulfate (SDS), 0.1% sodium pyrophosphate, 50 m*M* Tris–HCl (pH 7.5), 1 *M* sodium chloride, 10% dextran sulfate, and 100 μg/ml denatured salmon sperm DNA at 55° for approximately 6 hr. After prehybridization, 2.5 ml of the prehybridization solution without dextran sulfate and containing 5–10 ng/ml ^{32}P-labeled MPB-1 or MPB-5 (see *Hybridization Probes*) is added to the prehybridization mixture. The blots are further incubated with the hybridization mixture at 55° for approximately 20 hr. To remove aspecific hybridization, blots are washed twice in 2× SSC, 0.1% SDS at room temperature for 5 min; once in 0.5× SSC, 0.1% SDS at 50–60° for 10–20 min; and once in 0.1× SSC, 0.1% SDS at 60–70° for 5–20 min. The durations of washing are variable and are modified according to the amount of radioactivity left on the blots checked by a hand-held ^{32}P counter. After washing, the blots

[13] B. A. White, T. Lufkin, G. M. Preston, and C. Bancroft, this series, Vol. 124, p. 269.

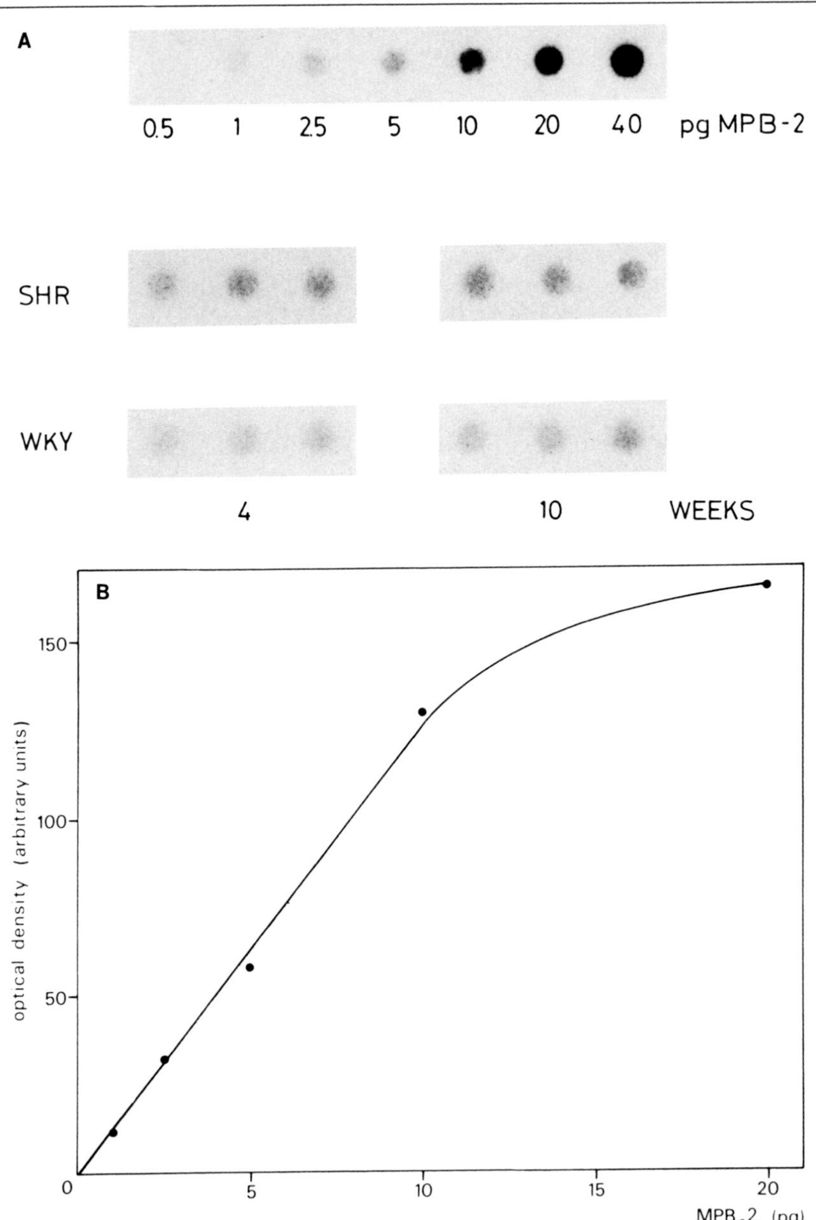

FIG. 4. Dot-blot analysis for VP mRNA isolated from the SON of 4- and 10-week-old spontaneously hypertensive rats (SHR) and Wistar–Kyoto controls (WKY). (A) Autoradiographic results of the standard curve using MPB-2 and the RNA samples of the SON (diluted 6-fold) of the different rat groups. (B) Standard curve obtained by image analysis.

are briefly air dried and then exposed to X-ray film (Kodak XAR-5, Eastman Kodak, Rochester, NY) with an intensifying screen at −70° for various lengths of time. For the given amount of RNA on the blots (see *Northern Blot Analysis* and *Dot-Blot Analysis*) the exposure time is usually between 12 hr and 3 days. If after autoradiogaphy it appears that there is still too much aspecific hybridization one can wash these blots once more under more stringent conditions.

For rehybridization we obtained good results with removal of the probe from the blots by washing them in 50% formamide (deionized), 10 m*M* Tris–HCl, 10 m*M* EDTA, pH 8.0, at 60° for 1 hr. However, other procedures described by the manufacturers of the nylon membranes can be used as well.

In Situ Hybridization

For *in situ* hybridization, rats are anesthetized with nembutal (0.1 ml/100 g body weight intraperitoneally), and brain tissue is fixed. In short, fixation is performed via intracardial perfusion during 3 min with 0.9% sodium chloride buffered with 10 m*M* sodium phosphate, pH 7.2 (PBS), followed by a perfusion with 4% paraformaldehyde in PBS for 10 min. The brains are removed and soaked in 4% paraformaldehyde in PBS for 15 min and left overnight at 4° in 15% sucrose in PBS. Next, the brains are frozen on dry ice and stored at −80°. The brains are cut into 16-μm slices in a cryostat (−12°). The sections are thaw-mounted onto slides which were precoated with 0.1% poly(L-lysine), stored overnight at −20°, and subsequently stored at −80°.

The *in situ* hybridization technique used is based on the procedure of Lawrence and Singer[14] and described previously.[15] The sections on the slides are preincubated at room temperature in 10 m*M* PBS, pH 7.4, 5 m*M* magnesium chloride for 10 min, followed by 0.2 *M* Tris–HCl, pH 7.4, 0.1 *M* glycine for 10 min, and finally for 10 min in 2× SSC, 50% formamide (deionized). The slides are dried with a paper towel, but drying out of the sections should be avoided. Each brain section is covered by 25 μl of 2× SSC, 50% formamide (deionized), 10% dextran sulfate, 0.5 mg/ml salmon sperm DNA, 2 mg/ml yeast tRNA, and 20,000 dpm of ^{35}S-labeled probe (see *Hybridization Probes*). Then the sections are protected with coverslips and incubated at 37° in a sealed humidified box for approximately 18 hr. When single-stranded DNA is used as a probe, aspecific

[14] J. B. Lawrence and R. H. Singer, *Nucleic Acids Res.* **5,** 1777 (1985).

[15] J. P. H. Burbach, Th. A. M. Voorhuis, H. H. M. Van Tol, and R. Ivell, *Biochem. Biophys. Res. Commun.* **145,** 10 (1987).

hybridization on the sections is removed by rinsing the slides in 50% formamide (deionized), 2× SSC at 37° for 30 min; 50% formamide (deionized), 1× SSC at 37° for 30 min, and 1× SSC at room temperature for 30 min (Fig. 2C). In case a RNA probe is used, the washing procedure has to be extended with an RNase treatment to remove aspecific hybridization. Without RNase treatment heavy nonspecific hybridization is evident (Figs. 2A and 2B). For this purpose, after washing in 50% formamide–1× SSC the sections are incubated in 2 μg/ml RNase A (Boehringer Mannheim) at 37° for 15 min. Next, the sections are rinsed in 1× SSC for 15 min, dehydrated in ethanol, and dried. The slides are exposed to X-ray film (Fuji RX, Fuji photo film, Tokyo, Japan) for 1–2 weeks.

Quantitative Measurement of Autoradiographic Signals by Image Analysis

Apparatus and Strategy

The determination of specific radioactivity on Northern blots, dot blots, and *in situ* hybridization is done by measurement of the total optical density (OD) of the whole autoradiographic signals. This is performed with a VIPER image analysis system (Gesotec, Darmstadt, FRG) equipped with a Hitachi-CCTV videocamera (Hitachi, Tokyo, Japan). Objects for analysis are maximally enlarged using a 50 mm lens and a bellows coupled to the videocamera. Illumination is constant, and the variation in the intensity of the field is adjusted by an automatic shading correction function. Illumination is chosen in such way that the camera is not overexposed by light. The area with a specific hybridization signal is automatically contoured using a preset OD limit which is kept constant throughout the measurements. Outputs are area size, minimum and maximum of the OD levels, and average OD of the contoured areas. Generally, the average OD of these contoured areas is determined. The OD of the background next to the measured areas is subtracted from the OD value obtained from these areas.

Measurement of Dot Blots

Quantitative measurement is done by dot-blot analysis in which the standards are processed like the RNA samples. The relationship between the amount of the standard MPB-2 and the average OD level obtained is shown in Fig. 3. Final results are unaffected when peak levels (maximum of OD) instead of average OD levels are used. For reliable measurements the RNA samples on the dot blots should be diluted such that their OD values are in the linear part of the standard curve (Fig. 3). Alternatively, when nonlinearity is inherent to the film response curve, the exposure time can be adapted. The OD of the RNA samples is converted by the standard curve to picograms of mRNA. The values obtained for cortex tissue are subtracted from the values of the different nuclei, resulting in values which represent the total amount mRNA per nucleus.

Measurement of Northern Blots

Northern blotting is primarily used for qualitative control of isolated RNA and provides a percent index of changes in mRNA levels. Percent changes in mRNA levels are based on average OD values of specific RNA bands on the autoradiograms. It is noted that generally the size of the autoradiographic signals vary in proportion to mRNA levels. Saturability of the film for the radioactive signal is deduced from autoradiograms of spots containing increasing amounts of ^{32}P radioactivity.

Measurement of in Situ Hybridization

For *in situ* hybridization the size of the area is dependent on the brain region. Therefore, peak levels of the OD values, which are not dependent on the area size, of these areas are used. Optical density values are interpolated from film response curves obtained from ^{35}S-labeled tissue standards, corrected for the background signal of the tissue section, and expressed as arbitrary units.

Discussion

The different assays have their own advantages with respect to quantitation, time, sensitivity, and control of mRNA quality. The solution hybridization assay does not require autoradiography nor advanced equipment for analysis. The assay is relatively fast and has a sensitivity of approximately 0.4 pg of VP mRNA/sample. The disadvantage of this assay in contrast to Northern blot analysis and dot-blot analysis is that samples can be measured only once with one probe. Therefore, the assay is most useful for mRNAs which are quite abundant as, for instance, proopiomelanocortin mRNA in the pituitary gland. Furthermore, there is no control for the nature of the mRNA, except when the precipitated material is analyzed on gels.[2]

Like the solution hybridization assay dot-blot analysis can provide accurate levels of VP mRNA and OT mRNA but not information on the nature and integrity of the RNA. Although the dot-blot assay requires a more extensive deproteination of RNA samples and takes more time to obtain results, it has the advantage that it has a higher sensitivity (0.1 pg/sample) due to the steep standard curve. The high sensitivity of this assay makes it possible to study, for example, the suprachiasmatic nucleus which has a low abundance of VP mRNA compared to the SON and PVN. Furthermore, the high precision provides the possibility to detect small differences, down to 20–25%, in VP mRNA and OT mRNA content. Another important advantage is that a dot blot can be hybridized several times with different probes, which opens the possibility for comparative studies for different mRNAs with the same RNA samples. The dot-blot analysis is illustrated by an experiment shown in Fig. 4. In this experiment VP mRNA levels of the SON of 4- and 10-week-old spontaneously hypertensive rats (SHR) were measured in comparison to age-matched Wistar–Kyoto controls (WKY). Using this method, it could be shown that in both age groups the SHR had higher VP mRNA levels.

Neither the solution hybridization assay nor the dot-blot assay gives information about the quality of the isolated RNA. Therefore, it is advisable to accompany the quantitative measurements using one of the two latter methods with a Northern blot analysis of the isolated RNA. This provides information about the length and integrity of the specific RNA and the specificity of the probe used. Using image analysis, differences in the signal strengths on the autoradiograms of the Northern blots can be quantified. Although not applied by us, the use of RNA standards run on the same agarose gel as the RNA samples might improve the accuracy of the quantitation of mRNAs by Northern blot analysis.

The previous three methods lack anatomical resolution and are dependent on the dissection of the tissue. Using the punch technique of Palkovits in combination with the mRNA detection techniques a relatively accurate dissection of brain nuclei can be established. However, more accurate anatomical information is obtained by the application of the *in situ* hybridization technique. Its greatest advantage over the other three techniques is its high sensitivity and the fact that it can be used to locate and quantitate mRNA at the cellular level (see chapters by Pfaff [59], Young [49], Uhl [53], and Lewis [58] in this volume). In this way one can get information about changes in mRNA levels in subpopulations of cells in a nucleus as, for example, can exist between the parvocellular and magnocellular divisions of the PVN. The method described in this chapter using X-ray film is useful for the VP and OT systems since neurons are grouped in compact clusters, which are visualized on the film (Fig. 2). Since cell groups are approximately equally dense, the OD of the hybridization signal can be used as an index for the average mRNA abundancy in a given nucleus. This approach has been successfully used to study the changes in OT mRNA in several distinct nuclei after salt loading.[16] However, application of liquid emulsions on such sections allows resolution at the cellular level. Standardization of the autoradiographic signals is still

[16] H. H. M. Van Tol, Th. A. M. Voorhuis, and J. P. H. Burbach, *Endocrinology (Baltimore)* **120,** 71 (1987).

not fully developed for this technique, but, in analogy to methods used in receptor autoradiography, the use of brain mash mixed with RNA standards could be promising. Future development in this field might be achieved by the introduction of nonradioactive probes[17,18] (see chapter [54] in this volume) which allows multiple stainings on one section for different mRNAs.

The techniques described have been applied in several experiments to study the regulation of the amount of VP mRNA and OT mRNA.[19] These experiments demonstrated that different stimuli which affect the release of the neurohypophyseal hormones also give changes in VP mRNA and OT mRNA levels in the hypothalamic nuclei.[2,16,20] Furthermore, alterations in the production of mRNAs in genetically different rat strains have been described.[21] Differences in regulatory control in the production of VP mRNA and OT mRNA between distinct hypothalamic nuclei have been studied.[10,19] These experiments indicated that the SON and PVN are responsive to several peripheral endocrine stimuli such as increased plasma osmolality induced by salt loading or lactation. This is in contrast to the SCN which projects only into the brain and seems to be nonresponsive to changes in plasma osmolality.[10,19,22,23] However, the SCN shows a clear circadian rhythm in VP mRNA levels with the highest levels at 5 o'clock in the afternoon. Such a circadian rhythm was not observed in the SON and PVN.[23,24] Further studies on the regulation of VP and OT mRNA production in distinct nuclei may help to unravel the basis for the differential regulation in mRNA production in these distinct nuclei.

Acknowledgments

The authors are grateful to Drs. H. Schmale, R. Ivell, and D. Richter (Hamburg, FRG) for the gift of the subclones pVλ*Pst*I and pOλ*Bal*I of the VP⁻ and OT genes. We thank Th. A. M. Voorhuis for her assistance in the development of the *in situ* hybridization technique. H. H. M. Van Tol is supported by the Netherlands Organization for Pure Research (Z.W.O., Medigon), Project 900-546-044.

[17] R. M. K. Dale and D. C. Ward, *Biochemistry* **14,** 2458 (1975).

[18] P. R. Langer, A. A. Waldrop, and D. C. Ward, *Proc. Natl. Acad. Sci. U.S.A.* **78,** 6633 (1981).

[19] H. H. M. Van Tol, Ph.D. thesis. Rudolf Magnus Institute for Pharmacology, State University of Utrecht, Utrecht, The Netherlands, 1987.

[20] H. H. M. Van Tol, E. L. M. Bolwerk, B. Liu, and J. P. H. Burbach, *Endocrinology (Baltimore)* **122,** 945 (1988).

[21] H. H. M. Van Tol, Th. A. M. Voorhuis, F. G. M. Snijdewint, G. J. Boer, and J. P. H. Burbach, *FEBS Lett.* **204,** 101 (1986).

[22] T. G. Sherman, O. Civelli, J. Douglass, E. Herbert, and S. J. Watson, *Neuroendocrinology* **44,** 222 (1986).

[23] J. P. H. Burbach, H. H. M. Van Tol, B. Liu, and M. A. Seger, *in* "Vasopressin; Cellular and Integrative Functions" (A. W. Cowley, J.-F. Liard, and D. A. Ausiello, eds.), p. 295. Raven, New York, 1988.

[24] G. R. Uhl and S. M. Reppert, *Science* **232,** 390 (1986).

[30] Granulosa Cell Aromatase Bioassay for Follicle-Stimulating Hormone

By KRISTINE D. DAHL, XIAO-CHI JIA, and AARON J. W. HSUEH

Introduction

Follicle-stimulating hormone (FSH)[1] is required for the maturation of ovarian follicles and testicular tubules during pubertal development.[2-5] In adult life, circulating FSH levels change in a cyclic manner during female reproductive cycles, and the gonadotropin regulates gonadal differentiation and steroidogenesis.[2,3] Serum levels of FSH have been measured by specific radioimmunoassay (RIA) to elucidate the role of FSH in various physiological and pathological conditions. However, increases in ovarian follicle growth are not always correlated with elevations in immunoreactive FSH levels in serum.[6] Similarly, changes in the biological activity of pituitary FSH are not always associated with variations in FSH immunoreactivity.[7,8] Furthermore, circulating FSH levels cannot be measured in many animal species owing to the lack of an established RIA. Although urinary FSH bioactivities have been measured, urinary immunoreactive FSH is difficult to quantitate by RIA because of the prior extraction required to minimize inhibitory factors.

Because of its low sensitivity, the classic Steelman–Pohley FSH bioassay[9] cannot be used to measure serum FSH levels. For urine samples, up to 1 liter of urine is extracted for the *in vivo* bioassay. We have recently used primary cultures of rat granulosa cells as the FSH target cell to

[1] Abbreviations: FSH, follicle-stimulating hormone; RIA, radioimmunoassay; GAB, granulosa cell aromatase bioassay; hCG, human chorionic gonadotropin; LH, luteinizing hormone; DES, diethylstilbesterol; MIX, 1-methyl-3-isobutylxanthine; PEG, polyethylene glycol; β-TGF, β-transforming growth factor; BSA, bovine serum albumin.

[2] J. H. Dorrington and D. T. Armstrong, *Recent Prog. Horm. Res.* **35,** 301 (1979).

[3] J. S. Richards, *Physiol. Rev.* **60,** 51 (1980).

[4] A. J. W. Hsueh, E. Y. Adashi, P. B. C. Jones, and T. H. Welsh, Jr., *Endocr. Rev.* **5,** 76 (1984).

[5] R. S. Swerdloff and W. D. Odell, *in* "The Testis in Normal and Infertile Men" (P. Troen and H. R. Nankin, eds.), p. 95. Raven, New York, 1977.

[6] V. D. Ramirez and C. H. Sawyer, *Endocrinology (Baltimore)* **94,** 475 (1974).

[7] W. D. Peckham, T. Yamaji, D. J. Dierschke, and E. Knobil, *Endocrinology (Baltimore)* **92,** 1660 (1973).

[8] E. M. Bogdanove, G. T. Campbell, and W. D. Peckham, *Endocr. Res. Commun.* **1,** 87 (1974).

[9] S. L. Steelman and F. M. Pohley, *Endocrinology (Baltimore)* **53,** 604 (1953).

design a sensitive and specific *in vitro* bioassay to quantitate FSH levels in body fluids.[10,11] In this chapter, we describe the granulosa cell aromatase (estrogen synthetase) bioassay (GAB) and its applications to the measurement of FSH bioactivities.

Materials

Human FSH [LER-907; FSH biopotency, 20 IU/mg; luteinizing hormone (LH) biopotency, 60 IU/mg] and human chorionic gonadotropin (hCG; CR121, 13,450 IU/mg) were the generous gifts of the National Hormone and Pituitary Distribution Program, NIDDKD. Pergonal (menotropins U.S.P.; FSH biopotency, 7.5 IU/mg; LH biopotency, 7.5 IU/mg; Second International Reference Preparation of human menopausal gonadotropin) is obtained from Serono Laboratories, Inc. (Randolph, MA). Porcine insulin (26.8 U/mg) is provided by Lilly Research Laboratories (Indianapolis, IN). McCoy's 5a medium (modified, without serum), penicillin–streptomycin solution, L-glutamine, and trypan blue stain (0.4%) are obtained from Gibco (Santa Clara, CA). Diethylstilbesterol (DES), androstenedione, 1-methyl-3-isobutylxanthine (MIX), and polyethylene glycol (PEG) (MW 8000) are obtained from Sigma Chemical Co. (St. Louis, MO). Tritiated estradiol (102 Ci/mmol) is obtained from New England Nuclear (Boston, MA). β-Transforming growth factor (β-TGF) is obtained from R and D Systems, Inc. (Minneapolis, MN). Cell culture plates (16 mm, 24 well) are obtained from Corning Glass Works (Corning, NY).

Isolation of Granulosa Cells

Intact female Sprague–Dawley rats (21–22 days old) are implanted with silastic capsules (10 mm) containing approximately 10 mg DES to stimulate granulosa cell proliferation. Four days after implantation, animals are sacrificed by cervical dislocation, and ovaries are dissected for granulosa cell collection.[12] Ovaries are decapsulated, follicles are punctured with 27-gauge hypodermic needles, and granulosa cells are carefully expressed into McCoy's 5a medium. The cells are centrifuged on low speed for 5 min in a tabletop Whisperfuge (Damon/IEC Co., Needham Heights, MA). The supernatant is discarded, and the cells are washed

[10] X.-C. Jia and A. J. W. Hsueh, *Endocrinology (Baltimore)* **119,** 1570 (1986).

[11] K. D. Dahl, N. M. Czekala, P. Lim, and A. J. W. Hsueh, *J. Clin. Endocrinol. Metab.* **64,** 486 (1987).

[12] X.-C. Jia and A. J. W. Hsueh, *Neuroendocrinology* **41,** 445 (1986).

with fresh medium and centrifuged. The cells are diluted in a final volume of medium which corresponded to 50,000–80,000 viable cells/60 μl. An aliquot is diluted with trypan blue stain, and viable cells are counted with a hemacytometer.

Assay Methods

Granulosa Cell Aromatase Bioassay (GAB)

Prior attempts at the development of a specific *in vitro* bioassay for FSH were hampered by a lack of sensitivity, or the assays could not be applied to serum samples owing to inhibitory factors.[13] Successful development of the GAB assay was, therefore, dependent on increasing granulosa cell sensitivity to FSH, as well as pretreating serum samples to remove interfering factors. This assay takes advantage of the fact that the combined actions of various hormones and factors enhance the responsiveness of granulosa cells to the aromatase-inducing action of FSH.

In view of the facts that granulosa cells are dependent on androgen substrates for aromatization and both estrogens and androgens enhance FSH-stimulated aromatase activity, androstenedione and DES are added to the granulosa cell cultures. In addition, a phosphodiesterase inhibitor, 1-methyl-3-isobutylxanthine (MIX), is added to minimize cAMP breakdown (Fig. 1, curve 1). When insulin and hCG are included in cultures containing FSH, androstenedione, DES, and MIX, a synergistic augmentation of estrogen production is observed, resulting in further increases in granulosa cell responsiveness to FSH (Fig. 1, curve 2). However, addition of serum to the cultures containing all these enhancing reagents (GAB assay medium) inhibits estrogen production (Fig. 1, curve 3). Therefore, the serum is pretreated with polyethylene glycol (PEG) to partially remove inhibitory substances. Pretreatment with the optimal concentration of 12% PEG partially removes serum inhibitory factors without removing FSH from the serum (Fig. 1, curve 4).

Reagents

DES capsules: diethylstilbesterol, 10 mm Silastic tubing containing approximately 10 mg DES

Penicillin: stock solution 10,000 U/ml; diluted to 100 U/ml in McCoy's 5a medium just prior to use

Streptomycin: stock solution 10 mg/ml; diluted to 100 μg/ml in McCoy's 5a medium just prior to use

[13] A. Baghdassarian, S. Fisher, H. Guyda, A. Johnson, and T. P. Foley, Jr., *Am. J. Obstet. Gynecol.* **108,** 1178 (1970).

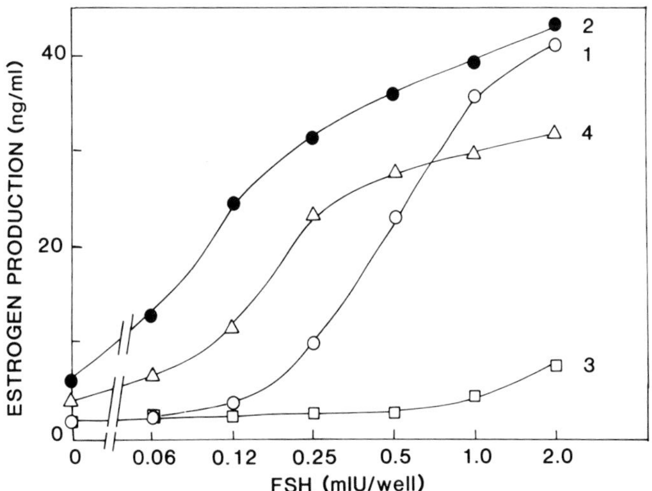

FIG. 1. Combined actions of various hormones and factors on FSH-stimulated estrogen production. Granulosa cell cultures were incubated with increasing concentrations of FSH and various hormones and factors. (1) Augmenting effect of MIX. (2) Synergistic effects of MIX, insulin, and hCG. (3) Serum (4% gonadotropin free) interference in GAB assay medium. (4) Effect of serum pretreatment with 12% PEG.

L-Glutamine: stock solution 200 m*M*; diluted to 2 m*M* in McCoy's 5a medium just prior to use

Androstenedione: stock solution 1 m*M*, 2.86 mg androstenedione dissolved in 10 ml 100% ethanol, tightly sealed to prevent evaporation, and stored at $-20°$; dilute 1 : 1000 to 1 μM in McCoy's 5a medium just prior to use

DES: diethylstilbesterol stock solution 1 m*M*, 2.68 mg DES dissolved in 10 ml 100% ethanol, diluted 1 : 10 to 10^{-4} *M* and tightly sealed to prevent evaporation, and stored at $-20°$; dilute 1 : 1000 to 10^{-7} *M* in McCoy's 5a medium just prior to use

MIX: isobutylmethylxanthine stock solution 1 mg/ml (4.5 m*M*), dissolved in McCoy's 5a medium by constant shaking at 37° for several hours and stored at $-20°$; dilute to 0.125 m*M* in McCoy's 5a medium just prior to use (because MIX acts optimally in a narrow range, titration of this agent may sometimes be necessary)

Human CG: stock solution 10 μg/100 μl, dissolved in phosphate-buffered saline (PBS) and stored at $-70°$; dilute to 30 ng/ml just prior to use

Insulin: stock solution 50 μg/50 μl, dissolved 1 mg initially in 50–100 μl of 0.01 *N* HCl and diluted to 50 μg/50 μl using McCoy's 5a medium for storage at −70°; dilute to 1 μg/ml in McCoy's 5a medium just prior to use

βTGF: β-transforming growth factor stock solution 100 ng/100 μl, dissolved in 4 m*M* HCl with 1 mg/ml BSA; dilute to 1 ng/ml in McCoy's 5a medium just prior to use

LER-907: stock solution 20 mIU/50 μl, dissolved in McCoy's 5a medium and stored at −70°; make 1 : 5 dilution to 4.0 mIU/50 μl in the GAB assay medium containing all the above-mentioned reagents and then make serial 1 : 2 dilutions to 2.0, 1.0, 0.5, 0.25, 0.125, and 0.06 mIU/50 μl in McCoy's 5a medium

Pergonal: stock solution 20 mIU/50 μl, dissolved in McCoy's 5a medium and stored at −70°; make 1 : 5 dilution to 4 mIU/50 μl and then make serial 1 : 2 dilutions to 2.0, 1.0, 0.5, 0.25, 0.125, and 0.06 mIU/50 μl in McCoy's 5a medium

PEG: polyethylene glycol (MW 8000); weigh 2.4 g and bring up to 10.0 ml (24%) using 8.4 ml McCoy's 5a medium by shaking in 37° water bath just prior to use

Preparation of the GAB Assay Medium. On the day that the bioassay is to be performed, an appropriate amount of GAB medium (McCoy's 5a medium containing various reagents) is mixed as described in Table I. Each serum sample requires 12 culture wells or 4 concentrations in triplicate. To calculate the amount of GAB medium required for a particular quantity of samples, divide the number of samples by 2. This would be the number of 16-mm, 24-well tissue culture plates necessary for the samples. Add 2 plates for the two FSH standard curves (8 concentrations in triplicate) and multiply this number by 10 (9.6 ml/plate or 400 μl/well) to estimate the amount of GAB assay medium necessary. Once the appropri-

TABLE I

DILUTION PROTOCOL FOR THE PREPARATION OF GAB MEDIUM

Total volume (ml)	Androstenedione, 1 m*M* (μl)	DES, 10^{-4} *M* (μl)	hCG, 10 μg/100 μl (μl)	MIX, 1 mg/ml (ml)	Insulin, 50 μg/50 μl (μl)
20	25	25	7.5	0.7	25
50	62.5	62.5	18.75	1.75	62.5
100	125	125	37.5	3.5	125
200	250	250	75	7	250
250	313	313	93.5	8.75	313
400	500	500	150	14	500

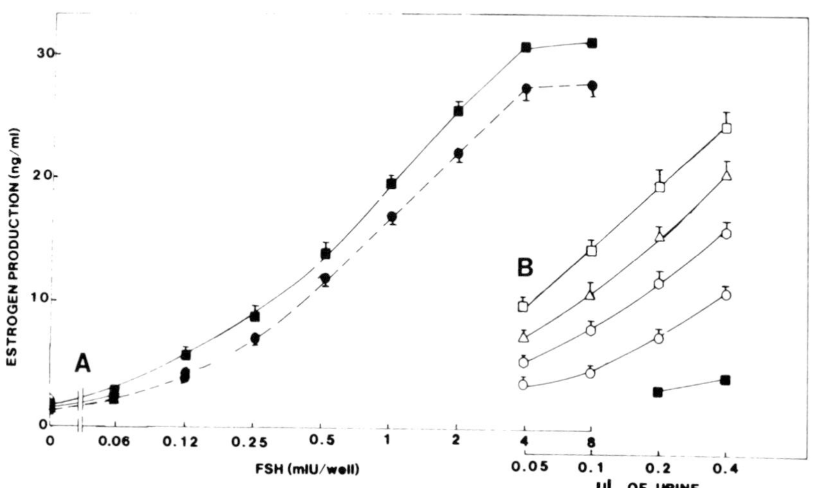

FIG. 2. Dose-dependent stimulation of estrogen production by a urinary FSH preparation and urine from women at different stages of the menstrual cycle. (A) Granulosa cell cultures were incubated with increasing concentrations of a human urinary FSH preparation (Pergonal) with or without (0.4 μl) gonadotropin-free urine. (●) Pergonal alone, (■) Pergonal plus 0.4 μl urine. (B) Increasing aliquots (0.05–0.4 μl) of urine obtained during the (△) early follicular phase (EFP), (⬡) late follicular phase (LFP), (□) surge, or (○) luteal phase (LP) were balanced (to a total of 0.4 μl) with urine from (■) oral contraceptive pill users (OCP) and added to the cultures.

ate amount of GAB assay medium is calculated, various reagents are added to McCoy's 5a medium to make the GAB medium. Based on the use of 400 μl of this medium for a final volume of 500 μl culture medium/well (by adding 40 μl of sample and 60 μl of cells), the final concentrations of the various reagents are the following: 100 U/ml penicillin, 100 μg/ml streptomycin, 2 m*M* L-glutamine, 1 μM andronstenedione, 10^{-7} *M* DES, 0.125 m*M* MIX, 30 ng/ml hCG, and 1 μg/ml insulin (Table I). To improve the sensitivity of the assay for samples containing low amounts of FSH, 1 ng/ml β-TGF can also be included.

Preparation of Standard. Depending on the type of samples that are assayed, an appropriate standard is necessary to calculate FSH concentrations. Two such standards, a human pituitary standard (LER-907) and a human urinary standard (Pergonal), are used. Contrary to the inhibitory effect serum has on the standard curve, addition of 0.4 μl of gonadotropin-free urine to the GAB assay medium results in a slight augmentation of FSH-stimulated estrogen production (Fig. 2). With respect to the use of standards from other sources, the GAB assay does not appear to be species specific, and, therefore, all other preparations that have been tested (e.g., rat, ovine, and monkey pituitary FSH preparations) display parallel dose–response curves. Although these preparations resulted in dose–response curves which are parallel to the above-mentioned preparations, new standard preparations should be initially tested for potency prior to use.

Pretreatment of Test Samples and Sample Balance. For serum samples, 200 μl of serum is pipetted into microfuge tubes with 200 μl of 24% PEG (final dilution 12%), vortexed, and incubated at 4° for 30 min. The tubes are then spun in an Eppendorf microcentrifuge (Model 5414; Fisher Scientific, Tustin, CA) for 2 min at room temperature, which results in a well-defined pellet and a clear supernatant. The supernatant is saved for measurement of FSH bioactivity. Urine samples do not require pretreatment with PEG but should be diluted 1 : 100 in McCoy's 5a medium just prior to use.

To ensure a constant volume of 20 μl of serum or 0.4 μl of urine in the total volume of 500 μl of culture medium, all samples and standards are balanced with the gonadotropin-free serum or urine from oral contraceptive pill users. Because some pools of serum or urine from the oral contraceptive pill users contain abnormally high levels of FSH bioactivity, individual pools should be prescreened and frozen in small aliquots to avoid repeated freezing and thawing. To balance the serum samples and standard curve, gonadotropin-free serum, as assessed by bioassay prior to use, from the oral contraceptive pill users is pretreated in the same fashion as above.

Granulosa Cell Cultures. Cells are cultured in 16-mm, 24-well culture plates. For the FSH standard curve, take 1 ml of gonadotropin-free, PEG-pretreated serum or 1 ml of prediluted urine balance (1 : 100) and add 9 ml of the GAB assay medium. Then add 400 μl of this mixture to each culture well (equivalent to 4% gonadotropin-free serum or 0.8% gonadotropin-free urine in final cell culture medium). Subsequently, add 50 μl of the appropriate standard in concentrations of 0, 0.06, 0.125, 0.25, 0.5, 1.0, 2.0, and 4.0 mIU/culture to triplicate culture wells.

Initially, 400 μl of the GAB assay medium is pipetted into all culture wells designated for samples. For each test sample, fixed amounts of serum or urine, balanced with gonadotropin-free serum or urine, are added in triplicate to culture wells according to Table II. Initially balanced serum is added to the respective wells, and then the corresponding test samples are added. For test samples with estimated high FSH levels, 2.5, 5, 10, and 15 μl of serum equivalents are used. For samples near sensitivity levels, 5, 10, 15, and 20 μl of serum equivalents are recommended. The resultant standard curves should display a dose-dependent increase

TABLE II
PROTOCOL FOR ADDITION OF TEST SAMPLES AND BALANCE

Sample equivalent (μl/culture)	Test sample (μl)	Gonadotropin-free balance (μl)
Serum	PEG-treated	PEG-treated
2.5	5	35
5	10	30
10	20	20
15	30	10
20	40	0
Total volume 40 μl		
Urine	Diluted (1 : 100)	Diluted (1 : 100)
0.05	5	35
0.1	10	30
0.2	20	20
0.4	40	0
Total volume 40 μl		

in estrogen production, while increasing concentrations of the test samples should result in parallel dose–response curves.

Granulosa cells (50,000–80,000/60 μl) are obtained after preparation of culture medium and added to all cultures. The cells are then cultured for 3 days at 37° in a humidified, 95% air–5% CO_2 incubator. At the end of the incubation, the culture plates are removed from the incubator and stored at −20° until the medium is assayed for estrogen content by RIA.

Estrogen RIA and Data Analysis

The estrogen RIA is performed using 10–20 μl of medium as previously described.[14] RIA data are analyzed with a program that utilizes a weighted logit–log regression analysis.[15] Calculation of FSH bioactivity in

[14] B. Kessel, Y. X. Liu, X.-C. Jia, and A. J. W. Hsueh, *Biol. Reprod.* **32,** 1038 (1983).

[15] S. E. Davis, M. L. Jaffe, P. J. Munson, and D. Rodbard, "Technical Report." National Institutes of Health, Bethseda, Maryland, 1979.

samples is performed using a standard curve fitted with a second-degree polynomial. The lowest dose of FSH capable of stimulating estrogen production greater than the mean plus 2 SE of the basal estrogen production by cultures without FSH is considered the minimal effective dose. The sensitivity of the assay for serum samples is about 0.12 mIU/culture, whereas the sensitivity for urine is approximately 0.06 mIU/culture.

In conclusion, the extreme sensitivity of the present *in vitro* bioassay allows for the measurement of circulating or urinary levels of bioactive FSH. Measurement of serum and urine levels of bioactive FSH should provide insight regarding the role of FSH in various physiological, pharmacological, and pathophysiological conditions. Since rat granulosa cells respond to FSH preparations from different species, this *in vitro* assay also provides valuable information on FSH levels in diverse animal species including those which lack a specific RIA.

Acknowledgments

This work was supported by National Institutes of Health Research Grant HD-23273. K.D.D. is the recipient of NIH Fellowship HD-06875.

[31] Neuroendocrine Regulation of Oocyte Tissue Plasminogen Activator

By THOMAS A. BICSAK, CARLA M. HEKMAN, and AARON J. W. HSUEH

Introduction

Plasminogen activators (PA)[1] have been implicated in a number of physiological events other than their classic role in fibrinolysis.[2] Two types of plasminogen activator are known to exist. Tissue-type plasminogen activator (tPA) differs from urokinase (uPA) in that the activity of tPA is greatly enhanced in the presence of fibrin[3] or substances which mimic

[1] Abbreviations: PA, plasminogen activator; tPA, tissue-type PA; uPA, urokinase-like PA; BSA, bovine serum albumin; DTNB, 5,5′-dithiobis(2-nitrobenzoic acid); Z-Lys-SBzl, thiobenzyl benzyloxycarbonyl-L-lysinate; PMSG, pregnant mare's serum gonadotropin; hCG, human chorionic gonadotropin.

[2] K. Danø, J. Andreasen, J. Grøndahl-Hansen, P. Kristensen, L. S. Nielsen, and L. Skriver, *Adv. Cancer Res.* **44,** 139 (1985).

[3] P. Wallen, *in* "Thrombosis and Urokinase" (R. Paoletti and S. Sherry, eds.), p. 91. Academic Press, London, 1977.

fibrin such as polylysine[4] or CNBr fragments of fibrinogen.[5] Numerous studies have demonstrated that ovarian somatic cells produce both uPA and tPA under different hormonal conditions both *in vivo* and *in vitro*.[6–11] In contrast, a recent report has shown that oocytes contain only tPA.[12] Oocytes synthesize tPA in response to gonadotropin treatment *in vivo*[13] or following spontaneous germinal vesicle breakdown *in vitro*.[14] These latter studies have used the qualitative method of fibrin autography[15] to detect tPA and have therefore provided only semiquantitative information. Our studies have been aimed at developing a sensitive, quantitative assay suitable for measuring tPA content of oocytes. An adaptation[16,16a] of the method of Coleman and Green[17,18] was found to be most suitable for this task, and, coupled with fibrin autographic methods, it provides a complete approach to the study of oocyte tPA regulation. In this chapter, we describe both of these methods in the context of their application to identification and measurement of oocyte tPA.

Materials

Human urokinase (Catalog No. 672123) and bovine fibrinogen (Catalog No. 341573) were obtained from Calbiochem-Behring (La Jolla, CA). Human single-chain tPA was a product of Biopool (Umea, Sweden). Thrombin (140 U/mg), hyaluronidase, bovine serum albumin (BSA), gelatin, 5,5′-dithiobis(2-nitrobenzoic acid) (DTNB), thiobenzyl benzyloxycarbonyl-L-lysinate (Z-Lys-SBzl), and poly(L-lysine) (average M_r 35,000) were from Sigma Chemical Co. (St. Louis, MO). Reagents for SDS–PAGE were from Bio-Rad (Richmond, CA). Sea-plaque agarose was from

[4] R. A. Allen, *Thromb. Haemostasis* **47,** 41 (1982).

[5] J. H. Verheijen, W. Nieuwenhuizen, and G. Wijngaards, *Thromb. Res.* **27,** 377 (1982).

[6] W. H. Beers, *Cell* **6,** 379 (1975).

[7] W. H. Beers, S. Strickland, and E. Reich, *Cell* **6,** 387 (1975).

[8] W. K. Liu, D. Burleigh, and D. N. Ward, *Mol. Cell. Endocrinol.* **21,** 63 (1981).

[9] C. Wang and A. Leung, *Endocrinology (Baltimore)* **112,** 1201 (1983).

[10] T. Ny, L. Bjersing, A. J. W. Hsueh, and D. J. Loskutoff, *Endocrinology (Baltimore)* **116,** 1666 (1985).

[11] R. Reich, R. Miskin, and A. Tsafriri, *Endocrinology (Baltimore)* **116,** 516 (1985).

[12] Y.-X. Liu, T. Ny, D. Sarkar, D. Loskutoff, and A. J. W. Hsueh, *Endocrinology (Baltimore)* **119,** 1578 (1986).

[13] Y.-X. Liu and A. J. W. Hsueh, *Biol. Reprod.* **36,** 1055 (1987).

[14] J. Huarte, D. Belin, and J.-D. Vassalli, *Cell* **43,** 551 (1985).

[15] A. Granelli-Piperno and E. Reich, *J. Exp. Med.* **147,** 223 (1978).

[16] C. M. Hekman and D. J. Loskutoff, *Biochemistry* **27,** 2911 (1988).

[16a] C. M. Hekman and D. J. Loskutoff, *Arch. Biochem. Biophys.* **262,** 199 (1988).

[17] P. L. Coleman and G. D. J. Green, *Ann. N.Y. Acad. Sci.* **370,** 617 (1981).

[18] P. L. Coleman and G. D. J. Green, this series, Vol. 80, p. 408.

FMC Corp. (Rockland, ME). Plasminogen was purified from human plasma by lysine–agarose affinity chromatography[19] followed by gel filtration. Ninety-six-well microtitration plates (Catalog No. 76-381-04) and sealing sheets were from Flow Laboratories (McLean, VA). All other chemicals used were of reagent grade.

Isolation of Oocytes

Oocytes were isolated from 26-day-old female rats which had been treated with 10 IU pregnant mare's serum gonadotropin (PMSG), or with PMSG followed by 10 IU (1 μg) human chorionic gonadotropin (hCG). Animals were sacrificed by cervical dislocation either 48 hr after PMSG treatment or 24 hr after the subsequent hCG treatment. Oocytes from PMSG-treated animals were collected by needle puncture of large follicles into McCoy's 5a medium, followed by repeated pipeting through a Pipetman (P-1000) to remove adhering granulosa cells.[12] This was followed by several transfers of denuded oocytes into fresh medium to remove remaining granulosa cells. Oocytes from PMSG–hCG-treated rats, obtained from the oviduct, were denuded by a combination of hyaluronidase treatment (50 μg/ml for 10 min) and repeated pipeting.[13] Control experiments demonstrated that hyaluronidase treatment did not affect the tPA content of the oocytes. Granulosa cell-free oocytes were collected in medium (1–2 μl/oocyte), and one-tenth volume 0.1% Tween 80 was added prior to sonication (15 sec on ice). Sonicated oocytes were typically frozen (−20°) until assayed.

Assay Methods

Fibrin Autography

Fibrin autography, which takes advantage of the stability of PAs to treatment with sodium dodecyl sulfate (SDS), was originally described by Granelli-Piperno and Reich[15] and has since been used extensively to qualitatively identify the type of PAs present in a given tissue or cell culture system. The first step in this assay is to fractionate the putative PAs on an SDS–PAGE slab gel (usually 9% polyacrylamide according to Laemmli[20]). Since this method is so commonly used, it is only briefly described here.

[19] D. G. Deutsch and E. T. Mertz, *Science* **170,** 1095 (1970).

[20] U. K. Laemmli, *Nature* (*London*) **227,** 680 (1970).

Reagents

Phosphate-buffered saline (PBS): 8 g NaCl, 0.2 g KCl, 0.92 g Na_2HPO_4, 0.2 g KH_2PO_4, dissolved in 1 liter distilled water
Thrombin: 1000 U/ml dissolved in PBS and stored at −20°
Plasminogen: 2 mg/ml in PBS and stored at −70°
Fibrinogen: 12.5 mg/ml dissolved in PBS and stored at −20°
Agarose: 2% (w/v) solution in water and stored at 4°
Triton X-100: 2.5% (v/v) solution in water and stored at room temperature

Procedure. After electrophoresis, the slab gel is soaked on a rotating table in 250 ml 2.5% Triton X-100 with two changes of buffer every 45 min. During the latter part of the second Triton wash, the PBS, thrombin, plasminogen, and fibrinogen are all equilibrated to 37°, while the agarose is melted (conveniently in a microwave oven) and allowed to cool to 45°. The 10 ml of PBS, 5 μl thrombin, and 100 μl plasminogen are mixed. To this is added 10 ml of agarose and 4 ml of fibrinogen. After mixing rapidly the agarose mixture is poured onto a glass plate on a leveling table and allowed to set undisturbed for at least 10 min. During this time, the fibrinogen polymerizes and forms an opaque fibrin film. To slow down the polymerization step, the glass plate may be preheated in a 56° oven prior to pouring the fibrin–agarose mixture. The SDS–PAGE slab gel is then gently blotted between two paper towels and carefully layered onto the fibrin–agarose film, taking care to avoid bubbles between the two layers. The gel overlay is then placed into a humidified box (with wet paper toweling in the bottom) and allowed to incubate at 37° until clear lytic zones appear in the opaque fibrin film. Depending on the amount of PA present, lytic zones appear in as little as 4 hr or as long as 48 hr. Gel overlays can then be photographed using dark-field photography. This is accomplished by placing the gel on an elevated stand on top of a fluorescent light box which is covered by black paper in the area directly underneath the gel. Using Polaroid Type 665 (P/N) film, a 1-sec exposure at f/8 should give a suitable photograph. Figure 1 shows a typical fibrin autograph of oocytes obtained from rats treated with either PMSG alone (lane 1) or PMSG followed by hCG (lane 2).

Z-Lys-SBzl Chromogenic Substrate Assay

The basis of the two-step Z-Lys-Bzl chromogenic substrate assay is described in detail by Coleman and Green.[18] As originally described, this assay was performed in individual tubes with absorbances read manually

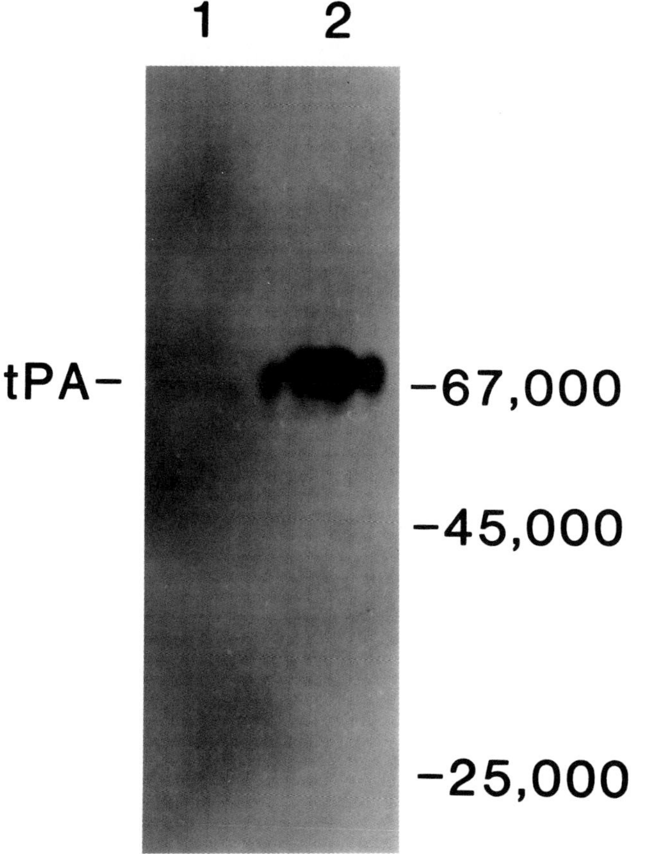

FIG. 1. Fibrin autograph of rat oocytes. Fifteen denuded oocytes collected from rats treated with PMSG alone (lane 1) or PMSG followed by hCG (lane 2) were electrophoresed on a 9% SDS–PAGE gel and subjected to fibrin autography as described in the text. The overlay was incubated at 37° for 20 hr, at which time the photograph was made. The migration positions of the molecular weight markers BSA (67,000), ovalbumin (45,000), and chymotrypsinogen (25,000) are indicated on the right.

on a spectrophotometer. We have adapted the basic method so as to perform the entire assay in a 96-well microtitration plate,[16,16a] thus simplifying the measurement and allowing for the convenient analysis of a large number of samples.

Reagents

PBS–BSA: 3 g BSA, 10 ml 10× PBS, 50 μl Tween 80, 19.8 mg sodium azide, all dissolved to 100 ml final volume with water

PBS–Tween: 0.01% (v/v) Tween 80 dissolved in PBS

PBS–Tween–gelatin: 1.25 g gelatin dissolved in 500 ml PBS–Tween and autoclaved for 90 min

tPA: 100 μg/ml dissolved in PBS–Tween–gelatin and stored in 20-μl aliquots at −70°; serially diluted from 1 : 5,000 to 1 : 40,000 in PBS–Tween–gelatin just prior to use

uPA: 1,000 U/ml dissolved in PBS and stored at −20°; serially diluted from 1 : 20,000 to 1 : 160,000 in PBS–Tween–gelatin just prior to use

Plasminogen: 2 mg/ml dissolved in PBS and stored in 50-μl aliquots at −70°; diluted 1 : 8 with PBS–Tween–gelatin immediately prior to use (this dilution may need to be adjusted depending on the particular plasminogen preparation used)

Polylysine: 1 mg/ml dissolved in PBS and stored in 50-μl aliquots at −20°; diluted 1 : 100 with PBS–Tween–gelatin immediately prior to use

Phosphate–NaCl: 0.4 M Na_2HPO_4, 0.4 M NaCl, 0.02% (v/v) Triton X-100, pH adjusted to 7.5 with HCl

DTNB: 220 μM dissolved in 50 mM Na_2HPO_4 and stored in 300-μl aliquots at −20°

Z-Lys-SBzl: 200 μM dissolved in water and stored in 300-μl aliquots at −20°

Plasmin substrate: 250 μl DTNB, 250 μl Z-Lys-SBzl, 12.5 ml water, and 12 ml phosphate–NaCl, mixed just prior to use

Procedure. Nonspecific binding sites on the microtitration plate are blocked by adding 300 μl PBS–BSA to each well and incubating at 37° for 1 hr. The plate is washed 3 times with PBS–Tween, shaken dry, and allowed to dry fully at 37°. The initial step of this two-step assay involves incubation of the PA together with plasminogen in a total volume of 30 μl, resulting in production of plasmin. This step is initiated by the addition of 10 μl of plasminogen to 10 μl of PA (either standard or unknown) premixed with 10 μl of polylysine (for tPA standard) or PBS–Tween–gelatin (for uPA standard and unknowns). Wells with PBS–Tween–gelatin substituted for PA (zero-dose well) are always included to assess the spontaneous hydrolysis of the plasmin substrate. For each unknown sample, control wells without plasminogen added are included to determine plasminogen-independent hydrolysis. After the reaction is initiated by addition of 10 μl of plasminogen to each well, the plate is covered with a sealing sheet and incubated in a humidified box at 37° for 90 min.

At the end of the first incubation, 200 μl of the plasmin substrate solution is added to each well and the plate incubated further at 37°. This addition effectively stops the action of PA. During the second incubation, the plasmin generated in the first part of the assay cleaves the Z-Lys-SBzl, which in turn undergoes disulfide exchange with DTNB, resulting in the formation of a yellow chromophore. The absorbance of the samples (at 405 nm) is measured at 20-min intervals after addition of plasmin substrate. Such readings are taken for at least 60 min but may proceed for up to 3 hr if absorbances are very low. Absorbance values greater than 1.0 are not linear with time.

The absorbances for each sample time point are corrected by subtracting the absorbance of the corresponding well which did not contain PA. These corrected absorbances are plotted as a function of the time of the second incubation. The slopes (ΔA_{405}/min) of these lines are directly proportional to the amount of plasmin which was generated during the first incubation. Standard curves for either tPA or uPA standards are constructed by plotting the ΔA_{405}/min from the primary plots as a function of the amount of PA added to the wells. Typical standard curves for a 90-min first incubation are shown in Fig. 2. Experiments with shorter initial incu-

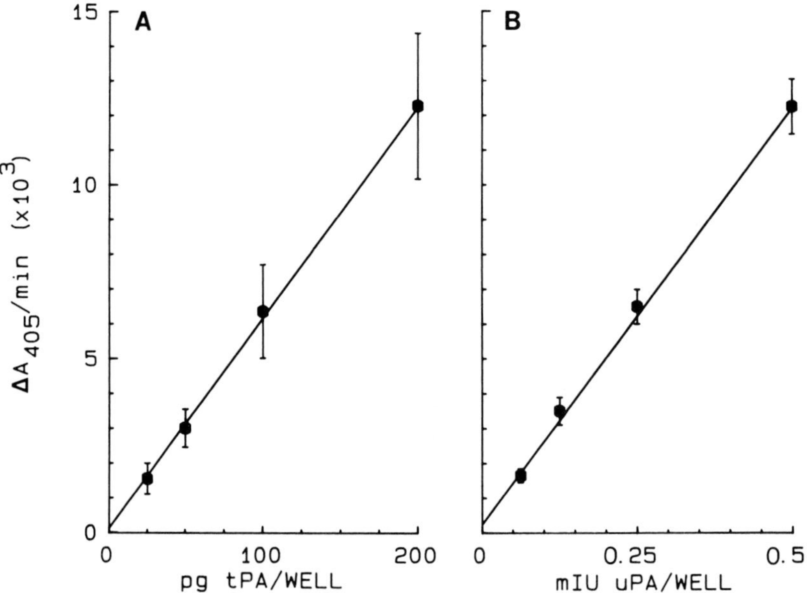

FIG. 2. Standard curves for tPA- and uPA-catalyzed plasminogen activation as measured by the Z-Lys-SBzl assay. Standard curves were generated using tPA (A) or uPA (B), using the assay conditions described in the text. Each point represents the average of five separate determinations, each assayed in duplicate, with the bars indicating the standard deviation for each point. The lines were fit by linear regression analysis and yielded the following lines: tPA, slope = 6.12×10^{-5} ΔA_{405}/pg min, intercept = 6.33×10^{-5} ΔA_{405}/min; uPA, slope = 0.024 ΔA_{405}/mIU min, intercept = 2.89×10^{-4} ΔA_{405}/min.

bations showed that the first reaction (conversion of plasminogen to plasmin) is a linear function of time for up to at least 90 min. The primary plot of ΔA_{405}/min for unknown samples can then be used to determine the amount of PA present by comparison with the appropriate standard curve.

Remarks

The assay as described here has a detection limit of approximately 10 pg (0.14 fmol) of purified tPA, if measured in the presence of the fibrin substitute poly(L-lysine). In the presence of 3.3 μg/ml polylysine, the activity of the purified tPA standard is stimulated approximately 5- to 10-fold. This was the maximal stimulation we observed with doses of polylysine ranging from 0.1 to 33 μg/ml (data not shown). In contrast, none of the concentrations of polylysine tested significantly increased the tPA activity of oocyte homogenates. In fact, the higher concentrations of polylysine (3.3–33 μg/ml) were somewhat inhibitory. This suggests that the oocytes contain a physiological stimulator of tPA activity, and therefore it is justifiable to compare an unstimulated oocyte tPA sample with a maximally stimulated tPA standard.

The applicability of the assay to measurement of oocyte tPA is demonstrated in Fig. 3. The assay detects tPA in less than one oocyte equivalent. Homogenates of oocytes obtained from oviducts of PMSG–hCG-treated rats show a dose-dependent increase in tPA activity as measured by a larger ΔA_{405}. Comparison of the data in Fig. 3 with the tPA standard curve in Fig. 2 reveals that ovulated oocytes contain approximately 30 pg tPA (0.095 mIU) per oocyte. This is in contrast with oocytes from rats treated only with PMSG, which contain the equivalent of roughly 1 pg tPA (0.0025 mIU) per oocyte. This result is consistent with the fibrin autograph shown in Fig. 1.

When using the Z-Lys-SBzl assay or the fibrin autograph with previously uncharacterized samples, it is imperative that controls containing all components except plasminogen be included in the assay. This control is necessary to rule out the presence of nonspecific proteases or peptidases which may be present and be capable of directly cleaving the Z-Lys-SBzl. In our hands, such controls for all oocyte samples assayed have revealed no such component (data not shown).

In conclusion, the Z-Lys-SBzl assay described here should prove to be of considerable use in the characterization of oocyte tPA activity under a number of physiological and possibly pathological conditions. In combination with the fibrin autograph, it provides a complete analysis of the PA activity in oocytes, or in any other system which contains a PA. Even samples which contain both uPA and tPA can be quantitated using this

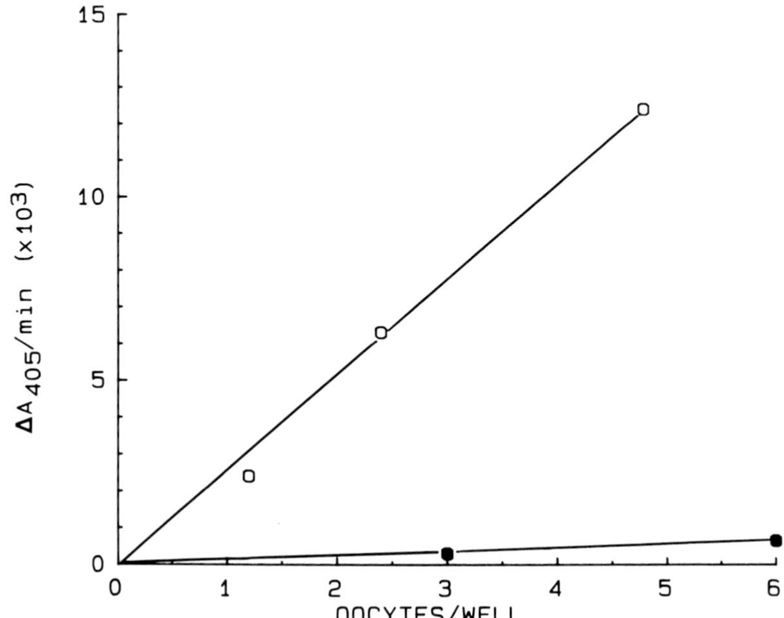

FIG. 3. Plasminogen activator content of oocytes collected from rats treated with gonadotropins. Oocytes were collected from rats treated with PMSG alone or from rats treated with PMSG followed by hCG and were assayed using the Z-Lys-SBzl assay, as described in the text. Oocytes from PMSG–hCG-treated rats showed a large increase in plasminogen activator (○) (slope = 0.0027 ΔA_{405}/oocyte min) when compared with those obtained from PMSG-treated rats (●) (slope = 0.00011 ΔA_{405}/oocyte min).

method, although the total PA activity should be expressed in units of uPA activity. Because the extreme sensitivity of the two-step enzymatic cascade allows the present chromogenic substrate assay to determine tPA activity in a small number of oocytes, this assay should provide the unique opportunity to study the neuroendocrine regulation of oocyte functions. Future quantitative analysis of secreted tPA from fertilized oocytes or early embryos may also serve as the basis for noninvasive monitoring of the well-being of these cells.

Acknowledgments

This work was supported by National Institutes of Health Grant HD-14084 and the Andrew W. Mellon Foundation. T.A.B. is the recipient of NIH Postdoctoral Fellowship HD-06939.

[32] Assessment of Peptide Regulation of the Autonomic Nervous System

By MARVIN R. BROWN, ROBERTA ALLEN, and LAUREL A. FISHER

Introduction

The autonomic nervous system (ANS) controls and coordinates virtually all visceral, vascular, and glandular activities, both tonically and in response to acute environmental demands. Accordingly, ANS outflow to various tissue sites is highly differentiated even under basal conditions and may reset rapidly to meet the changing needs of the organism. This nonuniformity of efferent nerve activity is present within both the sympathetic and parasympathetic divisions of the ANS. Furthermore, within a given tissue, reciprocal and nonreciprocal patterns of sympathetic and parasympathetic activation or withdrawal may be evoked. Such an extreme degree of fine control implies the existence of multiple complex regulatory processes for generating the observed viscerotropic fractionation of efferent ANS activity. Indeed, it is well appreciated that supraspinal, spinal, and ganglionic mechanisms all contribute to the specificity of ANS discharge patterns. However, it remains a great challenge to understand fully the neuroanatomical and neurochemical substrates involved in central nervous system (CNS) autonomic regulation.

A potentially important role of neuropeptides in mediating CNS regulation of ANS activity is suggested by numerous observations concerning the following: (1) the distribution of peptides and peptide receptors throughout CNS pathways that process autonomic and visceral information; (2) the ability to modify autonomic and visceral activities by CNS administration of peptides and/or agents that alter their synthesis, release, or action at receptors; and (3) the changes in CNS peptide levels occurring in disease states that have autonomic and visceral manifestations. The physiological involvement of peptides in regulating pituitary hormone secretion further supports the notion of their participation in CNS integrative mechanisms that coordinate and synchronize endocrine and autonomic outputs. If the capacity to generate extremely specialized and diverse patterns of ANS outflow depends in part on neurochemical specificity, certainly peptides are tenable candidates for encoding such information owing to the potentially enormous number of unique sequence structures. Outlined below are current research strategies employed to evaluate both ANS activity and its modulation by neuropep-

TABLE I
PEPTIDES REPORTED TO AFFECT THE AUTONOMIC NERVOUS SYSTEM

Angiotensin II	Met- and Leu-Enkephalin
Atrial natriuretic peptide	Galanin
Bombesin	Neurotensin
Calcitonin gene-related peptide	Neuropeptide Y
Cholecystokinin	Somatostatin
Corticotropin-releasing factor	Substance P
Dynorphin	Thyrotropin-releasing factor
β-Endorphin	Vasoactive intestinal peptide
	Vasopressin

tides. Neuropeptides reported to act within the CNS to influence ANS activity are listed in Table I.

Methods of Assessment of Autonomic Nervous System Function

The most meaningful assessment of ANS function would include a complete temporal analysis of the release, turnover, and biological effectiveness of all chemical transmitters involved in efferent and afferent autonomic neurotransmission under a variety of conditions. Such a measurement is impossible for a number of reasons, including the following: (1) incomplete identification of transmitters released from autonomic nerve terminals other than norepinephrine, epinephrine, dopamine, acetylcholine, and neuropeptide Y; (2) lack of methods to identify and measure turnover of all ANS transmitters; and (3) lack of methods to quantify effector responses and correlate these with ANS activity. Because of these limitations, the methods used to evaluate the ANS are frequently crude, indirect, and incomplete.

Table II summarizes some of the methods used by various investigators to assess ANS activity. In conscious animals, measurements of plasma catecholamine levels and end organ physiologic functions (e.g., cardiovascular, gastrointestinal, and metabolic parameters) are the most practical and widely used methods to assess ANS function.

Methods Used to Study Central Nervous System Actions of Peptides on Autonomic Nervous System Function

Putative neurotransmitters involved in the CNS regulation of the ANS may act at one or more of the multiple neuroanatomic loci believed to impart effects on autonomic efferent or afferent pathways. Knowledge of

TABLE II
METHODS OF ASSESSMENT OF AUTONOMIC NERVOUS SYSTEM FUNCTION

Adrenal medulla
Measurement of adrenal nerve electrical activity
Measurement of adrenal tyrosine monooxygenase activity
Measurement of plasma dopamine β-monooxygenase activity
Measurement of adrenal phenylethanolamine *N*-methyltransferase (PNMT) activity
Measurement of epinephrine and norepinephrine concentrations in adrenal vein or systemic blood
Measurement of end organ responses following administration of PNMT inhibitors, adrenal nerve transection, or adrenodemedullation
Sympathetic nervous system
Measurement of individual sympathetic nerve electrical activity
Measurement of ganglionic tyrosine monooxygenase activity
Measurement of norepinephrine or its precursors in tissue extracts of animals treated with inhibitors of catecholamine synthesis
Measurement of the uptake or release of labeled norepinephrine into noradrenergic terminals
Measurement of the rate of incorporation of radioactive precursors (e.g., tyrosine) into catecholamines
Measurement of plasma concentrations of catecholamines
Measurement of end organ responses following selective denervation or administration of drugs that interfere with the synthesis, storage, metabolism, release, or receptor binding of norepinephrine
Parasympathetic nervous system
Measurement of vagal nerve electrical activity
Measurement of tissue acetylcholine content
Measurement of the incorporation of radioactive choline into acetylcholine
Measurement of tissue acetylcholine turnover
Measurement of high affinity choline uptake
Measurement of end organ responses following selective denervation or administration of drugs that interfere with the metabolism, release, or receptor binding of acetylcholine

the neuroanatomic distribution of a particular peptide or neurotransmitter and/or its respective cellular receptors may provide the basis for a systematic evaluation of the brain sites of participation of such a substance in the regulation of ANS activity. To date, however, there is little evidence to support the concept that recognition of the anatomic distribution of a peptide or its cellular receptors is useful in predicting a functional site of action or physiologic significance of a peptide contained within the CNS. For instance, there is no apparent correlation between the distribution of peptide-containing nerve terminals and peptide receptor density.[1]

[1] M. Herkenham and S. McLean, *in* "Quantitative Receptor Autoradiography" (C. A. Boast, E. Q. Snowhill, and C. A. Altar, eds.), p. 137. Liss, New York, 1986.

One approach to answering the question of whether a particular peptide acts within the CNS to regulate ANS function involves the following steps. (1) Administration of chemically characterized peptide into brain ventricles, cisternal spaces, or parenchyma and concurrent measurement of some parameter of ANS function. (2) Demonstration that the peptide of interest does not leak from the CNS to produce effects outside the brain. This may be accomplished by systemic administration of receptor antagonists or by passive immunization against the peptide that is administered into the brain. Also, one may compare the qualitative or quantitative responses to the peptide when given into the brain versus periphery. (3) Determination of a CNS site of action and correlation of this information with the neuroanatomy and receptor distribution of the peptide being studied. (4) Once a site of action is identified, determination of the afferent and efferent pathways of this particular neuroanatomic locus. (5) If receptor antagonists are available, these should be tested against the agonist and in any animal model system in which the putative peptide is suspected of being involved as an endogenous transmitter. (6) If an animal model exists or can be developed in which the peptide of interest is a putative regulator of ANS function, efforts can be made to modify endogenous concentrations of this peptide and determine the effect of this manipulation on ANS function. (7) If methods are available, attempts can be made to quantitate the release or turnover of the peptide within the area of its site of action during changes of autonomic function under physiologic circumstances.

Animal Model

Owing to the depressant or stimulant actions of most anesthetic and sedative agents on CNS and ANS function, efforts should be made to utilize unanesthetized animal preparations. This can be achieved in most animals by prior placement of arterial and venous catheters for cardiovascular monitoring and/or obtaining blood samples. Chronic indwelling cannulae are placed into brain cisternal, ventricular, or parenchymal sites through which peptides can be administered as single doses or chronic infusions. Volumes of administration of peptides into any of these three sites should be kept as small as possible: 1–5 μl i.c. and i.c.v., and less than 200 nl for parenchymal injections. Controversy exists over the diffusion of substances injected into brain parenchyma. At present no adequate way exists to evaluate the extent of diffusion of a peptide administered into brain tissue. Functional characterization of diffusion can be achieved following the identification of a putative site of action by placing multiple injections around the preferred site to determine what degree of specificity is present.

Evaluation of the adequacy of delivery of a peptide can be achieved by noting the movement of air bubbles placed in the delivery line or by the distribution of a fluorescent or colored marker administered simultaneously with the peptide. Identification of cannulae placements can be made by injection of dye through the i.c. or i.c.v. cannulae, or by histologic examination of tissue damage produced by parenchymal cannulae.

Injection cannulae can be constructed from segments of stainless steel tubing ranging from 22 to 32 gauge, or one can purchase prefabricated injection cannulae. Vascular cannulae can be constructed using polyethylene, Silastic, Tygon, or other commercially available tubing. The patency of these cannulae can be enhanced through attention to sterile technique and leaving the cannula filled with heparinized saline between experiments.

Measurement of Plasma Catecholamine Concentrations

Plasma concentrations of epinephrine, norepinephrine, and dopamine can readily be measured using either a radioenzymatic or an HPLC method. In animals from which small blood samples are obtained, the radioenzymatic assay is preferable due to the limited sensitivity of the HPLC method. In humans or in large animals, where sufficient volumes of plasma can be obtained, the HPLC method for catecholamine determination is more economical and, with the use of an autosampler, allows assays to be performed more expediently.

The principle of the radioenzymatic assay is based on the conversion of the catecholamines to their respective methyl derivatives using the enzyme catechol *O*-methyltransferase (COMT, EC 2.1.1.6) in the presence of *S*-adenosylmethionine serving as an ^{3}H-methyl donor. Following the enzymatic reaction, unreacted *S*-adenosyl-L-[*methyl*-^{3}H]methionine ([^{3}H]SAM) is removed by extraction, and the products are separated using thin-layer chromatography (TLC) on silica plates. The areas corresponding to ^{3}H-methylated derivatives are cut out and placed in scintillation vials, eluted with aqueous solvent, extracted into nonpolar scintillation cocktail, and counted by liquid scintillation spectometry.

Details of the Radioenzymatic Catecholamine Assay

The method described below is a modified version of the assay described by Peuler and Johnson.[2]

[2] J. D. Peuler and G. A. Johnson, *Life Sci.* **21,** 625 (1977).

COMT Preparation

1. Fresh livers (~50 g) from rats fasted overnight are removed, rinsed, and then Polytron-homogenized in 3 volumes of ice-cold isotonic KCl for 2 min.
2. Centrifuge the homogenate at 40,000 *g* for 30 min at 4°. (In all subsequent steps keep the extract as cold as possible.) Discard pellet.
3. Adjust the supernatant to pH 5.0 by adding ice-cold 1 *M* acetic acid. Let stand 10 min, then centrifuge at 14,000 *g* for 10 min. Discard pellet; measure supernatant.
4. Slowly add ammonium sulfate to 30% saturation (17.7 g/100 ml) while stirring in an ice bath. Let stand 15 min.
5. Centrifuge at 40,000 *g* for 30 min. Discard pellet; measure supernatant.
6. Add ammonium sulfate to 55% saturation (16.2 g/100 ml). Let stand 15 min.
7. Centrifuge at 40,000 *g* for 30 min. Save pellet; discard supernatant.
8. Resuspend pellet in 0.1 *M* sodium phosphate buffer, pH 7.0 (0.5 ml/g starting material).
9. Dialyze the resuspended pellet overnight against 4 liters 0.1 *M* sodium phosphate buffer, pH 7.0, containing 1 m*M* dithiothreitol. Repeat dialysis next day.
10. Centrifuge dialyzate at 14,000 *g* for 10 min. Discard pellet; measure supernatant.
11. Add *O*-benzylhydroxylamine–HCl to 1 μM.
12. Divide into aliquots (0.5–1.0 ml). Store below −20°.

Collection of Plasma Unknowns

1. Collect blood into iced tubes containing the following mixture (10 μl/ml of blood): EGTA/glutathione (150/100 mg/ml, pH 7.4). EGTA chelates Ca^{2+} which inhibits COMT; glutathione prevents oxidation of catecholamines.
2. Centrifuge blood at 2000 *g* for 15 min.
3. Decant plasma into fresh tubes.
4. Cap, freeze at −20° until assay, then thaw and keep on ice.

Preparation of Catecholamine-Free Plasma (CFP)

1. Pool blood from several rats into a tube containing 20 μl/ml of blood of EGTA (60 mg/ml, pH 7.4).
2. Invert tube to mix.
3. Centrifuge and decant as above.

4. Dialyze plasma overnight (4°) against 100 volumes of phosphate-buffered saline (50 m*M* sodium phosphate, 0.15 *M* NaCl, pH 7.4).
5. Aliquot and store at −20°.

Reagents, Buffers, and Other Components

[^{3}H]SAM: *S*-Adenosyl-L-[*methyl*-^{3}H]methionine (New England Nuclear: NET 155H) is stored at −20°. The specific activity is usually approximately 70 Ci/mmol, adjusted to 500 Ci/ml. (If label is obtained from another source, make appropriate adjustments as assay sensitivity and linearity depend on the concentration and specific activity of SAM.)

NE, E, DA: Norepinephrine, epinephrine, and dopamine (Sigma) are stored in aliquots at −20° as a stock mixture, 100/100/100 μg/ml in 0.001 *N* HCl (make corrections for weight of salts).

NM/M/MT: Normetanephrine, metanephrine, and 3-methoxytyramine (Sigma) are stored in aliquots at −20° as a stock mixture, 250/250/250 m*M* in 0.001 *N* HCl.

Glutathione (reduced; Sigma) is prepared freshly for each assay, 100 m*M* in distilled H_2O.

Pargyline (Sigma) is stored in aliquots at −20°, 22 m*M* in distilled H_2O.

Tris/EGTA/Mg^{2+} 1400/160/410 m*M*, pH 8.3.

Boric acid/EDTA 1000/100 m*M*, pH 11.0.

TLC plates can be obtained from a variety of vendors. Purchase EM 5735-7, precoated TLC sheets, silica gel 60, F-254, 20 × 20 cm × 0.2 mm with plastic support. We typically spot 12–14 samples/plate.

Assay Procedure (Plasma). This assay is typically sensitive to detect 1 pg/sample (i.e., 20 pg/ml plasma) for E, NE, and DA. "Sensitivity" is defined as the amount of material required to yield twice the blank value. Our standard curve typically ranges from 1 to 250 pg per tube (i.e., 20 pg–5 ng/ml plasma), which is usually sufficient for quantitating normal and elevated levels of plasma catecholamines.

Samples. A typical assay contains four types of samples, prepared in duplicate 13 × 100 mm glass culture tubes in an ice bath: (1) blanks, (2)

Sample type	50 μl	10 μl
Blanks	CFP	0.001 *N* HCl
Standards	CFP	Standards in 0.001 *N* HCl
Internal control	Control rat plasma	0.001 *N* HCl
Unknowns	Unknown plasma	0.001 *N* HCl

standards, (3) internal control (to monitor intra- and interassay variation), and (4) unknowns. Each tube is normalized to contain 50 μl plasma ["catecholamine-free" (CFP) or unknown] and 10 μl 0.001 *N* HCl (in which standard is prepared), as tabulated on p. 437.

Reaction. The reaction mixture is prepared immediately before use. Calculate the amount needed for assay, then prepare the appropriate volume, allowing for pipetting error. Add each in the order given, vortexing between additions. Keep [^{3}H]SAM and COMT frozen until needed.

Glutathione	1 μl
Pargyline	1 μl
Tris–Mg^{2+}	12 μl
H_2O	11 μl
[^{3}H]SAM	5 μl
COMT	10 μl
	40 μl/tube

Add as rapidly as possible, as the mixture will precipitate. Vortex each tube, then incubate 60 min at 37° in a shaking water bath. While tubes are incubating, prepare another complete set of 13 × 100 mm glass tubes containing 100 μl 0.1 *M* acetic acid. Remove tubes to an ice bath and terminate the reaction by adding 50 μl of stopping/visualizing mixture to each tube. Vortex. While tubes are incubating, prepare the following:

Boric acid/EDTA, pH 11.0	48 μl
NM/M/MT	2 μl
	50 μl/tube

Extractions. Prepare a dry ice/95% ethanol bath for freezing phases.

First extraction. Add 2 ml toluene/isopentyl alcohol (3 : 2, v/v) to extract the methylated catecholamines into the organic phase. Vortex 15 sec. Centrifuge 2 min at 1300 *g* to separate phases. Quick-freeze the lower phase and decant the organic phase into the appropriate acetic acid-containing tube. Discard lower phase.

Second extraction. Vortex 15 sec. Centrifuge as above. Freeze the lower (acetic acid) phase and aspirate off the organic phase, keeping in mind that the unreacted [^{3}H]SAM is in this phase. At this point it is possible to stop for the day, leaving the acetic acid phase frozen. Parafilm the tubes and store at −20°.

Third extraction. Add 1 ml toluene/isopentyl alcohol to each tube. Vortex 15 sec. Centrifuge as above. Freeze the lower phase and aspirate off the organic phase. Add 100 μl 100% EtOH to each tube. Vortex briefly. Centrifuge as above.

Chromatography. Spot each sample (200 μl), using a TLC multispotter or by hand, onto 20 × 20 cm silica gel places (wash syringes in 100%

ethanol). Prepare solvent for developing plates: 80 ml chloroform, 15 ml 100% ethanol, and 10 ml 70% ethylamine (Adrich). Place plates in the TLC chamber, making sure the solvent level is below the spot level and the plates are not touching each other. Leave, covered, to develop until the solvent front reaches halfway up the plate (10 cm), usually about 1 hr. Remove plates and allow to dry. Using a pencil and UV light, outline the spots in preparation for cutting them out. Place each spot for E and NE into a corresponding scintillation vial, and each DA spot into a 16 × 125 mm glass tube. Add 1 ml 50 m*M* NH_4OH to elute the silica gel. Let stand 20 min or longer, shaking occasionally.

Periodate Oxidation. Greater sensitivity in the assay of E and NE can be obtained by oxidizing their ^{3}H-methylated derivatives to [^{3}H]vanillin.

For E and NE. Add 50 μl 4% sodium metaperiodate (w/v); after 5 min add 50 μl 10% glycerol; and then add 1 ml 0.1 *M* acetic acid to neutralize. (Shake the box containing vials after each addition.) Prepare scintillation fluid: toluene/Liquifluor (1000 : 50, v/v); add 10 ml to each vial; and vortex 15 sec, wipe clean, wait 60 min, and count.

For dopamine. Add 10 ml toluene/isopentyl alcohol (3 : 2, v/v); cap tubes with rubber stoppers and vortex 15 sec. Centrifuge to separate phases, then freeze lower phase; decant upper phase into scintillation vials. Add 300 μl Liquifluor and count.

HPLC Determination of Catecholamines

The principle of this method is based on the liquid–solid extraction of catecholamines onto alumina followed by their elution with dilute acid. Following this, separation can be achieved either on cation-exchange or reversed-phase columns using a variety of solvent systems.[3] Numerous manufacturers offer HPLC systems with electrochemical detectors.

Assessment of Norepinephrine Turnover

The adrenal medulla is the principal source of epinephrine that is present in plasma. Measurement of plasma concentrations of epinephrine provides an index of the functional activity of the adrenal medulla. Evaluation of adrenal epinephrine secretion can be further refined through the collection of adrenal venous blood samples. Measurement of plasma norepinephrine levels is believed to reflect the spillover of norepinephrine from peripheral sympathetic nerve terminals. The measurement of plasma concentrations of norepinephrine in the absence of evaluation of the arteriovenous (AV) difference across a selected group of noradrenergic termi-

[3] P. Hjemdahl, *Am. J. Physiol.* **247** (*Endocrinol. Metab.* **10**), E13 (1984).

nals does not indicate the anatomic site of origin of the norepinephrine measured.

Evaluation of AV differences of norepinephrine across various organ systems in small animals is technically difficult, if not impossible. Another approach to determine the activity of noradrenergic terminals in specific sites is to evaluate the turnover of norepinephrine within those sites. This objective can be achieved by one of four methods: (1) evaluation of the disappearance of tissue norepinephrine following the blockade of tyrosine monooxygenase (EC 1.14.16.2) with α-methyl-*p*-tyrosine[4,5]; (2) evaluation of the incorporation of labeled tyrosine into norepinephrine using a pulse–chase method[6]; (3) measurement of the neuronal uptake or release of labeled norepinephrine[7]; and (4) measurement of dopamine accumulation in the presence of blockade of the enzyme dopamine β-monooxygenase (EC 1.14.17.1).[8,9]

The use of α-methyl-*p*-tyrosine and measurement of tissue norepinephrine concentrations is associated with two major problems. First, α-methyl-*p*-tyrosine is extremely toxic to animals, and, second, owing to the large norepinephrine pool size this method does not reflect acute changes of norepinephrine concentrations. Several hours are required to detect changes in norepinephrine levels. The use of tritiated norepinephrine uptake and release as a method to assess noradrenergic activity also suffers from the problem of the large norepinephrine pool size, and thus the lack of sensitivity in acute experiments. The use of dopamine accumulation in dopamine β-monooxygenase-blocked animals is extremely useful in short time duration experiments[9]; however, this method has not been widely used. The most useful, albeit laborious, method for determination of norepinephrine turnover is to perform pulse–chase experiments with tritiated tyrosine.[6] This method requires the measurement of both labeled and unlabeled tyrosine and norepinephrine for calculation of specific activities. It is not known whether the pulse–chase method is useful for acute experiments.

Radioimmunoassay of Neuropeptide Y

The rationale for the measurement for neuropeptide Y (NPY) concentration in plasma is based on the observation that this peptide is secreted

[4] S. Spector, A. Sjoerdsma, and S. Udenfriend, *J. Pharmacol. Exp. Ther.* **147,** 86 (1965).
[5] B. B. Brodie, E. Costa, A. Dlabae, N. H. Neff, and H. H. Smookler, *J. Pharmacol. Exp. Ther.* **154,** 493 (1966).
[6] R. J. Wurtman and C. J. Watkins, *Nature (London)* **265,** 79 (1977).
[7] N. Weiner, *in* "Neuropsychopharmacology of Monoamines and Regulatory Enzymes" (E. Usdin, ed.), p. 143. Raven, New York, 1974.
[8] B. A. Bennett and D. K. Sundberg, *Life Sci.* **28,** 2811 (1981).
[9] M. Brown, R. Allen, and L. Fisher, *Brain Res.* **400,** 35 (1987).

concomitantly with norepinephrine from sympathetic nerve terminals and with epinephrine from the adrenal medulla.[10] Similar to catecholamines, NPY is a potent vasoconstrictor; however, in contrast to catecholamines, NPY produces negative inotropic effects on the heart.[11]

Antibodies against NPY have been prepared by coupling the peptide to human α-globulin (Research Plus Laboratories, Inc., Denville, NJ) using glutaraldehyde. Peptide and a 4 times weight excess of carrier protein are dissolved separately in 0.1 *M* sodium phosphate buffer, pH 7.0 (4 and 9 mg/ml, respectively) then added together on a magnetic stirrer. Glutaraldehyde, 25% (w/v) was diluted to 0.375% in the same buffer and added slowly while stirring to achieve a final dilution of 0.14%. The reaction was allowed to proceed for 5 hr at room temperature, then stopped by dialysis at 4° against 4 changes of 1 liter of glass-distilled water and finally 1 liter of 0.9% saline over a 24-hr period. Spectrapor dialysis tubing with a molecular weight cutoff of 12,000 is used. After dialysis, aliquots containing about 4 mg of NPY–human α-globulin conjugate are prepared and stored at −20°.

For immunization, an emulsion is made by mixing Freund's complete adjuvant-modified *Mycobacterium butyricum* (Calbiochem, La Jolla, CA) with an equal volume of the prepared conjugate and saline. Rabbits are initially immunized with 2 mg of immunogen and boosted monthly with 1 mg in a total volume of 1 ml injected into about 20 intradermal sites. Animals are bled 10 days after injection. Antisera are characterized by RIA to determine titer and specificity.

Plasma samples are assayed for NPY by RIA either unextracted or after cartridge purification using octadecyl (C_{18}) Bondelut columns containing 500 mg sorbent (Analytichem Intl., Harbor City, CA). Blood is collected into tubes containing EDTA to achieve a final concentration of 3.5 m*M*. Plasmas are decanted and stored at −20° and assayed within 2 months.

Prior to RIA or extraction, plasmas are centrifuged at low speed for 10–15 min and decanted. To extract NPY, columns are prewashed with 1 volume (2.5 ml) methanol (HPLC grade), then 2 volumes of triethylammonium formate (TEAF: 11.5 ml 88% HCOOH per liter of glass-distilled water, pH adjusted to 3.0 with triethylamine). A maximum of 1 ml of plasma is applied by gravity, and the columns are washed again with 2 volumes of TEAF. NPY is eluted with 2 ml of 75% acetonitrile (HPLC

[10] T. L. O'Donohue, B. M. Chronwall, R. M. Pruss, E. Mezey, J. Z. Kiss, L. E. Eiden, V. J. Massari, R. E. Tessel, V. M. Pickel, D. A. DiMaggio, A. J. Hotchkiss, W. R. Crowley, and Z. Zukowskagrojec, *Peptides* **6,** 755 (1985).

[11] J. M. Allen, P. M. M. Bircham, A. V. Edwards, K. Tatemoto, and S. R. Bloom, *Regul. Pep.* **6,** 247 (1983).

grade), 25% TEAF and lyophilized in a Speed Vac. Samples are reconstituted in RIA buffer prior to assay.

To assay NPY in unextracted plasma samples, "NPY-free" plasma is prepared to add to the RIA standard curve. Fresh frozen human plasma (200 ml) is applied to Bondelut columns as described above, except the plasma is collected, pooled, and adjusted to pH 7.4. Aliquots are made and frozen at −20° for use in the RIA at appropriate sample volumes (up to 100 μl/tube). Separation of free tracer from tracer bound to antibody is achieved by the addition of 0.5 ml charcoal–dextran. Recovery of NPY from Bondelut columns is routinely 80–85%, calculated by adding a known amount of the peptide to normal human plasma, which is then extracted and assayed.

Critique of Other Methods for Assessment of Autonomic Nervous System Function

Adrenal Medulla. As noted in Table II, several methods exist for the assessment of adrenal medullary function. It should be recognized that the adrenal medulla secretes not only epinephrine but also norepinephrine and a variety of peptides, including NPY, somatostatin, vasoactive intestinal peptide, substance P, Met- and Leu-enkephalin, corticotropin-releasing factor, neurotensin, and others.[12–15] To date, little effort has been made to measure the secretion of peptides from the adrenal medulla to determine their physiology or as an index of ANS function.

Measurement of adrenal nerve activity is limited by the necessity to perform such experiments in anesthetized animals. Furthermore, this method does not indicate the quantity of catecholamines or other substances that are secreted by the adrenal medulla. Measurement of tyrosine monooxygenase, dopamine β-monooxygenase, and phenylethanolamine *N*-methyltransferase (PNMT, EC 2.1.1.28) represents methods to assess adrenal activity based on the activity of enzymes involved in the synthesis of epinephrine.[16–18] Each of these methods is limited by the

[12] J. M. Allen, T. E. Adrian, J. M. Polak, and S. R. Bloom, *J. Auto. Nerv. Syst.* **9,** 559 (1983).

[13] F. Leboulenger, P. Leroux, C. Delarue, M. C. Tonon, Y. Charnay, P. M. Dubois, D. H. Coy, and H. Vaudry, *Life Sci.* **32,** 375 (1983).

[14] A. Bucsics, A. Saria, and F. Lembeck, *Neuropeptides* **1,** 329 (1981).

[15] R. Corder, D. F. J. Mason, D. Perrett, P. J. Lowry, V. Clement-Jones, E. A. Linton, G. M. Besser, and L. H. Rees, *Neuropeptides* **3,** 9 (1982).

[16] S. J. Fluharty, G. L. Snyder, E. M. Strickler, and M. J. Zigmond, *Brain Res.* **267,** 384 (1983).

[17] P. B. Molinoff, R. Weinshilboum, and J. Axelrod, *J. Pharmacol. Exp. Ther.* **178,** 425 (1971).

[18] K. E. Moore and O. T. Phillipson, *J. Neurochem.* **25,** 289 (1975).

inability to follow an animal's adrenal activity over time and by the lack of ways to correlate the activities of these enzymes with the actual amount of epinephrine secreted.

Other methods of assessment of adrenal medullary activity include measurement of end organ responses following administration of peripherally acting PNMT inhibitors,[19] adrenal nerve transection, or adrenal demedullation. Use of adrenal demedullation may be complicated by incomplete removal of chromaffin tissue and/or permanent changes of adrenal cortical function. A specific method to assess adrenal medullary activity is to measure epinephrine and norepinephrine concentrations in adrenal venous blood.

Sympathetic Nervous System. In contrast to assessment of adrenal medullary activity, where the measurement of plasma concentrations of epinephrine adequately reflects adrenal medullary activity, measurement of plasma concentrations of norepinephrine does not specifically indicate its site of origin. As noted above, methods are available for assessment of tissue norepinephrine turnover that may provide a better index of the viscerotropic activity of sympathetic nerves. As shown in Table II, several methods exist for the assessment of sympathetic nerve activity. Similar to the measurement of adrenal nerve activity, measurement of sympathetic nerve activity does not distinguish afferent from efferent function, nor does it necessarily correlate with the secretory patterns of norepinephrine. The use and limitation of norepinephrine turnover methods have already been mentioned. Finally, measurements of end organ responses following selective denervation or administration of drugs that interfere with the synthesis, storage, metabolism, release, or receptor binding of norepinephrine can be carried out. However, these methods are indirect and may be hampered by the toxicity of drugs used to produce these effects.

Parasympathetic Nervous System. In general, the assessment of parasympathetic nervous system activity relies on end organ responses. Methods are available for the measurement of tissue acetylcholine content, incorporation of radioactive choline into acetylcholine, measurement of tissue acetylcholine turnover, and measurement of high affinity choline uptake.[20-22] In general, the most useful of these methods *in vivo* is the assessment of end organ responses following selective denervation or administration of drugs that interfere with the metabolism release or receptor binding of acetylcholine.

[19] R. G. Pendleton, J. P. McCafferty, and J. M. Roesler, *Eur. J. Pharmacol.* **66,** 1 (1980).

[20] F. P. Bymaster, K. W. Perry, and D. T. Wong, *Life Sci.* **37,** 1775 (1985).

[21] O. M. Brown and J. J. Salata, *Life Sci.* **33,** 213 (1983).

[22] J. K. Blusztajn and R. J. Wurtman, *Science* **221,** 614 (1983).

[33] Recognition, Purification, and Structural Elucidation of Mammalian Physalaemin-Related Molecules

By LAWRENCE H. LAZARUS and WILLIAM E. WILSON

Introduction[1]

This chapter describes the results of a strategy which was originally designed to determine whether a peptide like physalaemin (PHY), a pharmacologically potent amphibian tachykinin,[2–6] might occur in mammalian tissue. Such an effort was considered reasonable because of the large number of precedents which led to the formulation of the brain–gut–skin triangle,[7,8] a relationship which evolved as a consequence of observations to the effect that structural analogs to peptides initially recognized in amphibian skin are frequently later found in the central nervous system and intestinal tract of many mammals.[3,7]

Historically, a number of biological and chemical methods have been utilized to facilitate the recognition of bioactive peptides. The classic pharmacological (e.g., smooth muscle contractility) and physiological (e.g., *in vivo* alterations in blood pressure) assessments of biological activities[2,3] enabled Erspamer to isolate and characterize nearly 40 bioactive peptides from amphibian skin.[9] An interesting chemical approach involving recognition of peptide carboxy-terminal amino acid amides, devised by Tatemoto and Mutt,[10] has been applied to the isolation and characterization of several bioactive peptides from mammalian tissues.[11–15] More

[1] We greatly appreciate collaboration with, and encouraging support of, the following individuals: D. L. Carlton, G. de Caro, R. P. DiAugustine, M. D. Erisman, V. Erspamer, A. Guglietta, C. W. Hamm, D. J. Harvan, O. Hernandez, T. Hökfelt, B. J. Irons, G. D. Jahnke, D. G. Klapper, R. I. Linnoila, V. Mutt, C. M. Soldato, L. Stone, and H. Yajima. The literature searches by R. J. Hester and L. L. Wright have proved invaluable.

[2] M. Bertaccini, *Pharmacol. Rev.* **28,** 127 (1976).

[3] V. Erspamer and P. Melchiorri, *Pure Appl. Chem.* **35,** 463 (1979).

[4] V. Erspamer, *Trends Neurosci.* (*Pers. Ed.*) **4,** 267 (1981).

[5] T. Nakajima, *Trends Pharmacol. Sci.* **2,** 202 (1981).

[6] V. Erspamer, *Trends Neurosci.* (*Pers. Ed.*) **6,** 200 (1983).

[7] V. Erspamer, P. Melchiorri, M. Broccardo, G. Falconieri Erspamer, P. Falashi, G. Improta, L. Negri, and T. Renda, *Peptides* **2** (Suppl. 2), 7 (1981).

[8] T. Fujita, R. Yui, T. Iwanaga, J. Nishiitsutsuji-Uwo, Y. Endo, and N. Yanaihara, *Peptides* **2** (Suppl. 2), 123 (1981).

[9] V. Erspamer, *Comp. Biochem. Physiol. C* **79,** 1 (1984).

[10] K. Tatemoto and V. Mutt, *Proc. Natl. Acad. Sci. U.S.A.* **75,** 4115 (1978).

[11] K. Tatemoto and V. Mutt, *Nature* (*London*) **285,** 417 (1980).

[12] K. Tatemoto, *Proc. Natl. Acad. Sci. U.S.A.* **79,** 5485 (1982).

recently, applications of complementary DNA methodology have also resulted in the isolation and structure determination of new peptides that possess sequence similarity with other known peptides, e.g., substance K (neurokinin A)[16,17] and gastrin-releasing peptide.[18]

Once an appropriate body of background peptide sequence information is available, another approach to the recognition and isolation of new peptides is to take advantage of specific site-directed antibodies to detect predetermined amino acid sequences and epitope domains.[19,20] This use of unique immunoglobulins as molecular probes provides a powerful tool which permits a high degree of selectivity for recognizing a particular peptide sequence. Utilization of a radioimmunoassay (RIA) to guide the isolation procedure permits quantitation of minute quantities of cross-reacting unknown material for which an appropriate pharmacological or physiological testing paradigm or tissue might not be anticipated. The use of antibodies to amphibian peptides also permits testing of hypotheses regarding the extent to which those amino acid sequences are conserved in an evolutionary context.[21]

Our approach to the recognition of physalaemin-like immunoreactive material (PHLIM) in mammalian tissues required that antisera be raised to the amino-terminal portion of PHY because the carboxy-terminal region contains sequence similarity with all members of the tachykinin family of peptides, including substance P.[3,4,9] The elicited antisera permitted recognition of mammalian material from normal tissue of several species,[22–24] and from small-cell carcinoma of the lung,[25,26] which cross-

[13] K. Tatemoto, M. Carlquist, and V. Mutt, *Nature (London)* **296,** 659 (1982).

[14] K. Tatemoto, R. Rokaeus, H. Jornvall, T. J. McDonald, and V. Mutt, *FEBS Lett.* **164,** 124 (1983).

[15] K. Tatemoto, J. M. Lundberg, H. Jornvall, and V. Mutt, *Biochem. Biophys. Res. Commun.* **128,** 947 (1983).

[16] H. Nawa, T. Hirose, H. Takashima, S. Inayama, and S. Nakanishi, *Nature (London)* **306,** 32 (1983).

[17] H. Nawa, H. Kotani, and S. Nakanishi, *Nature (London)* **312,** 729 (1984).

[18] E. Spindel, *Trends Neurosci. (Pers. Ed.)* **9,** 130 (1986).

[19] H. M. Geysen, J. A. Tainer, S. J. Rodda, T. J. Mason, H. Alexander, E. D. Getzoff, and R. A. Lerner, *Science* **235,** 1184 (1986).

[20] E. D. Getzoff, H. M. Geysen, S. J. Rodda, H. Alexander, J. A. Tainer, and R. A. Lerner, *Science* **235,** 1191 (1986).

[21] L. H. Lazarus, W. E. Wilson, G. Gaudino, B. J. Irons, and A. Guglietta, *Peptides* **6** (Suppl. 3), 295 (1985).

[22] L. H. Lazarus and R. P. DiAugustine, *Anal. Biochem.* **107,** 350 (1980).

[23] L. H. Lazarus, R. I. Linnoila, O. Hernandez, and R. P. DiAugustine, *Nature (London)* **287,** 555 (1980).

[24] L. H. Lazarus, R. P. DiAugustine, and C. M. Soldato, *Exp. Lung Res.* **3,** 329 (1982).

reacted immunohistochemically[25] and by RIA.[25,26] Isolation and structure elucidation of the physalaemin-like immunoreactive peptides (PHLIPs) is influenced by the following considerations: (1) relatively low levels of PHLIM occur in most tissues where it is present; only 40–50 ng equivalents PHY/g dry weight occur in rabbit stomach, the tissue with highest concentrations estimated by RIA; (2) the use of high affinity, low capacity antiserum is suitable for RIA, however, it is inadequate for construction of an immunoaffinity column; and (3) because the chosen antiserum recognizes the amino-terminal region of PHY, the amino-terminal residue of which is <Glu, it was anticipated that attempted removal of <Glu by pyroglutamate aminopeptidase treatment might significantly alter antiserum cross-reactivity and might also introduce other peptide contaminants which could interfere with structure determination efforts. Thus, mass spectrometry is employed to deduce the amino acid sequences of the PHLIPs.

An unexpected but very interesting result of this strategy of using site-specific antibody probes is that it subsequently led to the recognition that high molecular weight range (M_r) rabbit stomach glycoproteins contain sequences which cross-react with the PHY antiserum. Even though more extensive structural information on the mucin glycoproteins[27] is not presently available, it is quite likely that PHLIP-8 represents an amino-terminal sequence of such a macromolecule. Thus, this type of experimental design has the potential to yield valuable information on macromolecular, as well as peptide, structural[19,20] and conformational properties.[28]

Materials and Methods

Materials. Synthetic physalaemin is obtained from Bachem, Inc. or Peninsula Labs. 1-Ethyl-3-(3-dimethylaminopropyl)carbodiimide–HCl (EDCI), keyhole limpet hemocyanin, and Freund's complete adjuvant are obtained from Calbiochem. Bovine serum albumin (BSA) is from either Miles (Pentex) or Sigma Chemical Co. (RIA grade). Goat anti-rabbit antiserum is a product of Cappel Labs. (Cooper Ind.). Chloramine-T, sodium metabisulfite, and ammonium sulfate, Grade 1, are from Sigma.

[25] M. D. Erisman, R. I. Linnoila, O. Hernandez, R. P. DiAugustine, and L. H. Lazarus, *Proc. Natl. Acad. Sci. U.S.A.* **79,** 2379 (1982).

[26] L. H. Lazarus, R. P. DiAugustine, G. D. Jahnke, and O. Hernandez, *Science* **219,** 79 (1983).

[27] M. R. Neutra and J. F. Forstner, *in* "Physiology of the Gastrointestinal Tract" (L. R. Johnson, ed.), 2nd Ed., p. 975. Raven, New York, 1987.

[28] J. Gariepy, T. A. Mietzner, and G. K. Schoolnik, *Proc. Natl. Acad. Sci. U.S.A.* **83,** 8888 (1986).

Dowex analytical grade ion-exchange resins (Bio-Rad Labs.) are acid- and alkali-washed, then washed with water before use. After swelling in methanol, Sephadex LH-20 (Pharmacia) is equilibrated prior to use with *n*-butanol–acetic acid–1 m*M* aqueous phenol (10:2:1). Sixty grams of octyl-bonded silica gel (J. T. Baker), packed in two Michel–Miller columns (Ace Glass), is washed with acetonitrile then with water before use. HPLC grade organic solvents, absolute ethanol, and other chemicals are of highest purity commercially available. Phyllomedusin and uperolein are gifts from V. Erspamer (University of Rome, Italy). J. Rivier (Salk Institute, La Jolla, CA) donated α-MSH (melanocyte-stimulating hormone, melanotropin) and β-hMSH (human).

[Lys^5,Thr^6]Physalaemin, [Ile^5,Asn^6]PHLIP-8, PHLIP-8, and PHLIP-7 were synthesized in the laboratory of H. Yajima (Kyoto University, Kyoto, Japan) using reagents and methods which have been described elsewhere.[29–31]

Equipment. Gradient HPLC systems are from Du Pont Instrument Co. and Waters Associates. Ancillary HPLC instrumentation includes a Waters Associates Model 420 fluorescent detector, Du Pont column heater, and V4 variable wavelength absorbance monitors from ISCO.

Corning Pyrex culture tubes, 6 × 50 mm (Cat. No. 9820), are used for peptide hydrolysis. RIA is performed in polystyrene tubes from Sarstedt. HPLC eluant solutions are collected in test tubes which are lyophilized using a Speed Vac (Savant Instruments). Tissue disruption is facilitated with the aid of a Polytron device (Brinkmann). Amino acid analysis is conducted using the ion-exchange technique of Klapper,[32] which will be briefly described, as well as the Pico.Tag (Waters Associates) technique.[33]

Reversed-phase HPLC is performed using the following columns: 4.6 × 250 mm Zorbax C_{18} (Du Pont) column, a 4.6 × 250 mm LC 5DP (Supelco) column, a 7.8 × 300 mm μBondapak C_{18} (Waters Associates), and a 4 × 250 mm Aminex A9 cation-exchange resin column from Bio-Rad.

The tandem mass spectrometer (MS–MS) with a fast atom bombardment (FAB) source, a ZAB 4F obtained from Vacuum Group Anal. Ltd. (Manchester, UK) is housed in the Laboratory of Molecular Biophysics

[29] H. Yajima, S. Funakoshi, and K. Akaji, *Int. J. Pept. Protein Res.* **26,** 337 (1986).

[30] W. E. Wilson, D. J. Harvan, C. Hamm, L. H. Lazarus, D. G. Klapper, H. Yajima, and Y. Hayashi, *Int. J. Pept. Protein Res.* **28,** 58 (1986).

[31] N. Fugii, Y. Hayashi, K. Akaji, S. Funakoshi, M. Shimamura, S. Yuguchi, L. H. Lazarus, and H. Yajima, *Chem. Pharm. Bull.* **35,** 1266 (1987).

[32] D. G. Klapper, *in* "Methods in Protein Sequence Analysis" (M. Elzinga, ed.), p. 509. Humana, Clifton, New Jersey, 1982.

[33] B. A. Bidlingmeyer, S. A. Cohen, and T. L. Tarvin, *J. Chromatogr.* **336,** 93 (1984).

(National Institute of Environmental Health Sciences). The magnetic (B) and electrostatic (E) sectors of this FAB–MS–MS instrument are arranged in B–E–E–B geometry; a He collision chamber is positioned between the two electrostatic sectors for these experiments. Mass measurements are achieved using either a mass marker driven by a Hall-effect probe[34] or a linked-scan, computer-based technique.[35]

Physalaemin Antibody Selection and RIA Development

Antigen Synthesis. One-half milliliter saline (0.15 *M* NaCl) solution containing 4 mg PHY is mixed with 1 ml saline containing 25 mg keyhole limpet hemocyanin; then 1 ml of an aqueous solution of 20 mg ECDI[36] is added dropwise while stirring at 22°. The final molar ratios of peptide–carrier–coupling reagent are 54 : 1 : 1800.[26] Only the ε-amino group of Lys, at position 6 of PHY, is expected to couple to the carrier under these conditions.[37] After stirring at room temperature overnight, dialysis is performed against 4 changes of 4 liters saline containing 30 m*M* 2-mercaptoethanol (2-ME). Subsequent dialysis against deionized water permits removal of salt and thiol. The antigen is stored frozen at −20°. Studies using radiolabeled PHY indicate that over 70% of the peptide is coupled in this procedure.

Antibody Production. An aliquot of antigen containing 100 μg of PHY is diluted to 2.5 ml with sterile saline which is emulsified with an equal volume of Freund's complete adjuvant. The emulsified antigen is injected over 10 sites on the backs of three New Zealand White rabbits. After four monthly inoculations with freshly emulsified antigen, 25–30 ml blood is collected from an ear vein into polypropylene centrifuge tubes and allowed to coagulate at room temperature for 1 hr; the blood samples are then placed in a cold box at 4° overnight. After rimming the clots gently with a sterile applicator stick, the sera are separated by centrifugation at 2000 *g* for 10 min at 4°. The resulting antisera are stored at −20° in 1.5- to 5-ml NUNC freezer vials fitted with silicone rubber seals.

Iodination of PHY. PHY is iodinated, using chloramine-T as oxidant, at a molar ratio of peptide–^{125}I (radioiodine)–oxidant of 1 : 1 : 10.[38] Chloramine-T (1.05 μg in 10 μl of 0.4 *M* phosphate, pH 7.4) is added to 0.59 μg

[34] R. K. Boyd, P. A. Bott, D. J. Harvan, and J. R. Hass, *Int. J. Mass Spect. Ion Proc.* **69,** 251 (1986).

[35] R. K. Boyd, P. A. Bott, B. R. Beer, D. J. Harvan, and J. R. Hass, *Anal. Chem.* **59,** 189 (1987).

[36] T. L. Goodfriend, L. Levine, and G. Fasman, *Science* **144,** 1344 (1964).

[37] J. P. Briand, S. Muller, and H. M. V. van Regenmortel, *J. Immunol. Methods* **78,** 59 (1985).

[38] L. H. Lazarus, M. H. Perrin, and M. R. Brown, *J. Biol. Chem.* **252,** 7174 (1977).

PHY and 0.5 mCi $Na^{125}I$ (carrier-free) in 10 μl 0.4 *M* sodium phosphate, pH 7.4, then mixed. After 30 sec at room temperature, 10 μl sodium metabisulfite (10 mg/ml water) is added to terminate the reaction. The reaction mixture is transferred to a 7 × 45 mm column containing Sephadex LH-20 and eluted with an acidified *n*-butanol–water solution (see above), and fractions are collected. Labeled PHY elutes in the first radioactive peak; the shelf life of labeled PHY is about 3 weeks at 4°.

Development of the RIA. The RIA is performed in 12 × 75 mm polystyrene tubes as follows: 1–40 μl of standard peptide, unknown sample, or water is placed in the tube; next, 10 μl of diluted (1 : 40,000) antiserum is added; finally, 50 μl of 100 m*M* sodium cacodylate, pH 5.8, containing 1 mg BSA and 4000–5000 cpm of radiolabeled peptide is placed in the tube to obtain a final total volume of 100 μl. Duplicate analyses are performed with tubes placed in 90-tube racks. Tube contents are mixed by shaking the rack of tubes vigorously by hand for 10 sec, and then the racks are placed in sealed plastic bags (to prevent evaporative losses) at 4° for 12–18 hr. Next, to each tube is added 10 μl of a 1 : 200 dilution of nonimmune rabbit serum and 10 μl of a 1 : 4 dilution of goat anti-rabbit antiserum; the racks of tubes are shaken and allowed to stand at room temperature for 1–3 hr. Finally, 1 ml of deionized water is added to each tube; after mixing on a vortex mixer, the tubes are centrifuged at 2500 *g* for 30 min at 4°. This simultaneously washes the precipitate and tubes, thus eliminating a separate wash step. The supernatant is removed by aspiration and discarded (the vacuum pump is protected by in-line charcoal filters and Drierite in order to prevent contamination of the pump oil). The tubes with their residual radioactive films are placed in a gamma counter to estimate radioisotope content.

Selection of Antiserum Which Recognizes the Amino-Terminal Region of PHY. The antiserum which has the most potential for specific recognition of the amino-terminal region of PHY is selected as a result of a series of titrations of the three antisera with PHY and structurally related peptides over a concentration range of 10^{-3} to $>10^4$ pmol. Table I summarizes the cross-reactivities of each of the antisera to a number of tachykinins and to PHLIP-8. Although each antiserum recognizes the amino-terminal domain of PHY, PS-1 exhibits particular affinity for the amino-terminal hexapeptide sequence, <Glu-Ala-Asp-Pro-Asn-Lys. PS-1 antiserum appears to exhibit a strict requirement for the tripeptide sequence of Asp-Pro-Asn and, to a lesser degree, a requirement for <Glu^1 and Ala^2. The lack of recognition of Lys^6 by any of the antisera is anticipated since the ε-amino group of Lys was bound to the carrier in the antigen.

It is to be noted that none of the peptides in Table I cross-react with any of the antisera to give titration curves which exactly parallel that of

TABLE I
SEQUENCE SPECIFICITY OF PHYSALAEMIN POLYCLONAL ANTIBODIES[a]

Peptide	Sequence[d]	Percentage cross-reactivity[b] to antiserum[c]		
		PS-1	PS-2	PS-3
Physalaemin (PHY)	E-A-D-P-N-K-F-Y-G-L-M-NH_2	100	100	100
Uperolein	- -P- - - - - - -A- - - - - - - - - - - - - -	2.03	109	221
Phyllomedusin	E-N- - - - -R- - - - - - - - - - - - - -	0.36	0.026	0.071
PHLIP-8	- -V- - - - - - - -I -Q-A	0.74	—	—
[Ile^5,Asn^6]PHLIP-8	- -V- - - - - -I -N-Q-A	<0.0001	—	—
[Lys^5,Thr^6]PHY	- - - - - - - - -K-T- - - - - - - - - - - - - -	0.001	—	—
Substance P	R-P-K- - -Q-Q- - -F- - - - - - - - - -	<0.0001	0.02	0.013
Eledoisin	- -P-S-K-D-A- - -I- - - - - - - - - -	<0.0001	<0.001	0.005
Kassinin	D-V-P-K-S-D-E- - -V- - - - - - - - - -	0.013	0.037	0.22

[a] Modified and reproduced with permission of Academic Press.[22]

[b] Data based on 50% displacement values derived from log–logit plots. Other peptides tested that exhibited <0.0001% cross-reactivity are bombesin, litorin, neurotensin, VIP, bradykinin, oxytocin, α-MSH, β-hMSH, [Met^5]- and [Leu^5]enkephalin.

[c] Antisera dilutions for PS-1, -2, and -3 are 1 : 40,000, 1 : 30,000, and 1 : 20,000, respectively.

[d] Single letter amino acid codes: A, Ala; D, Asp; E, Glu or <Glu; F, Phe; G, Gly; I, Ile; K, Lys; L, Leu; M, Met; N, Asn; P, Pro; Q, Gln; S, Ser; T, Thr; V, Val; Y, Tyr.

PHY. The high affinities of the antisera for PHY (Table II) are as anticipated from the dosage of antigen and the inoculation schedule.[39,40] Moreover, the capacity (~100 ng PHY/ml antiserum) is inadequate to permit utilization of these antisera as ligands for an immunoaffinity column.

Purification of PHLIPs

Tissue Selection and Recovery. The selection of rabbit stomach as a tissue source for isolation of the PHLIPs is based on a consideration of the results of RIA studies which indicated that this tissue contained the greatest quantity of material which cross-reacted with PS-1 antiserum.[22–24] Rabbit stomachs are recovered within 5 min of sacrifice. PHLIM proteolysis is prevented by immediate removal of the stomach contents and washing of the tissue with cold water; subsequent immediate heating in a typical household-type microwave oven for 1.5–2 min inactivates tissue protein hydrolases. PHLIPs, on the other hand, are obtained from stomachs which are allowed to remain packed in ice for 20–60 min prior to cleaning and washing with cold tap water. Next, each stomach is frozen in

[39] H. N. Eisen and G. W. Siskind, *Biochemistry* **3,** 996 (1964).

[40] E. Ruoslahti, *Scand. J. Immunol. Suppl.* **3,** 3 (1976).

TABLE II
AFFINITIES OF PHYSALAEMIN POLYCLONAL ANTIBODIES

Antibody	IC_{50} (±SD)[a] pg	IC_{50} (±SD)[a] fmol	n
PS-1	6.79 ± 1.98	5.29 ± 1.54	34
PS-2	5.67 ± 1.82	4.42 ± 1.42	3
PS-3	10.21 ± 1.75	7.95 ± 1.36	3

[a] Titrations are carried out with ^{125}I-labeled physalaemin. Antisera dilutions are as in footnote *c* to Table I.

liquid nitrogen, lyophilized overnight or until dry, and then stored at room temperature in sealed jars prior to extraction.

Extraction of PHLIPs. Dry rabbit stomach is powdered by mixing 5- to 50-g batches in a Waring blendor at high speed for several minutes. A sample of 500 g dry tissue is extracted 3 times with 5 liters of a solution of 90% ethanol containing 1.2 *M* formic acid and 1 m*M* 2-ME. The volume of the extracts is reduced in a rotary evaporator with a water bath set at a temperature of 60°. After removal of 85–90% of the initial extraction liquid volume, 1 volume of water is added, and the resulting precipitate is removed by centrifugation at 15,000 *g* for 1 hr. The volume of the resulting supernatant solution is reduced to a convenient volume (e.g., 400 ml when starting with 500 g dry tissue) on the rotary evaporator. After adjusting the pH to 6.5 with ammonium hydroxide, this suspension is clarified by centrifugation at 35,000 *g* for 1 hr. The supernatant recovered from 500 g dry tissue contains the immunoreactive equivalent of 30–40 μg PHY.

Chromatographic Purification. Subsequent purification involves performance of several ion-exchange and preparative reversed-phase chromatography steps; then a number of analytical reversed-phase HPLC steps are performed to achieve final purification. Early efforts to adsorb the PHLIPs to reversed-phase chromatography supports reveal that aqueous solutions of these peptides have a finite, but extremely low, affinity for such phases; marked increases in peptide affinities for reversed-phase packings occur when 10–100 m*M* trifluoroacetic acid (TFA) or sodium sulfate is added to the mobile phase. (TFA is frequently presumed to enhance peptide affinity for reversed-phase surfaces as a consequence of ion pairing; however, this phenomenon is inappropriate for rationalization of chromatographic behavior of the PHLIPs as these peptides do not contain appropriate functionality, i.e., free amino groups.) The method described here utilizes sodium sulfate as the mobile phase modifier which facilitates peptide adsorption to the reversed-phase column packing material.

During the preparative and analytical reversed-phase HPLC steps, column effluent absorbance is monitored at 210–215 nm, and fractions of convenient size are collected. After each chromatography step, RIA is carried out to determine the location and recovery of the immunoreactive material.

Step 1. Initially, the stomach extract from 500 g dry rabbit stomach is passed through a column containing 450 g Dowex 50-X8 cation-exchange resin (H^+ form). An estimated >95% of the input PHLIM is recovered in the flow-through and wash solutions. After neutralization, this solution is reduced in volume to 300 ml with the aid of the rotary evaporator.

Step 2. The neutralized Dowex 50 flow-through plus wash solution is next passed through three tandemly arranged columns, each containing 450 g Dowex 1-X8 anion-exchange resin (OH^- form). The columns are washed with 1 liter water, followed by 1 *M* formic acid until a change occurs in the color of the resin at the bottom of the last column; an additional 2 liters of 1 *M* formic acid is passed through the columns to ensure desorption of PHLIPs. Any unadsorbed immunoreactive material which flows through the columns is combined with that which is recovered in the formic acid desorption step. After reduction in volume on a rotary evaporator, the resulting solution is adjusted to pH 6 by addition of ammonium hydroxide. The anion-exchange chromagraphy step results in recovery of only 5–6 μg equivalents out of about 36 μg equivalents of the input PHLIM.

Step 3. The neutralized flow-through plus wash solution from the previous step is pumped through two tandem Michel–Miller columns containing 60 g bonded-phase octyl stationary support equilibrated with water. The columns are washed with 200 ml 1 m*M* dithiothreitol (DTT) before initiating a step-gradient elution with ethanol solutions containing 1 m*M* DTT. Volumes and ethanol percentages of the elution solutions are as follows: 300 ml 10%; 300 ml 20%; 400 ml 40%; and 325 ml 95%. PHLIPs elute with 10% ethanol and are evaporated to near dryness.

Step 4. The PHLIP solution is adjusted to a 50 ml volume containing 1 m*M* sodium sulfate; then this solution is pumped, at 1 ml/min, onto a 7.8 × 300 mm μBondapak C_{18} column equilibrated with initial mobile phase (i-mp): 1 m*M* sodium sulfate containing 1 m*M* DDT. After washing the column with an additional 20 ml of i-mp, a step gradient is performed to 10% ethanol plus 1 m*M* DTT. PHLIP immunoreactivity and the major 215 nm absorbance band coelute over a 10-min (10 ml) interval immediately after this new mobile phase passes through the column void volume.

TABLE III
HPLC OPERATING CONDITIONS FOR PURIFICATION OF PHLIP-8[a]

HLPC run no.[d]	Mobile phase composition: Initial	Mobile phase composition: Final	Gradient interval[b] (min)	Retention times[c] (min)
1–3	1 mM Na_2SO_4 + 1 mM DTT	9.5% ethanol	15	12–19
4	5 mM Na_2SO_4 + 1 mM DTT	19% ethanol	30	19–22
5	5 mM Na_2SO_4 + 1 mM DTT	19% ethanol	30	23.0
6	5 mM Na_2SO_4 + 1 mM DTT	19% ethanol	30	25.4
7	5 mM Na_2SO_4	19% ethanol	30	22.5
8	5 mM Na_2SO_4	5 mM Na_2SO_4 + 15% ethanol	0	23.2
9	5 mM Na_2SO_4	5 mM Na_2SO_4 + 20% ethanol	30	32.8
10	5 mM Na_2SO_4	5 mM Na_2SO_4 + 10% CH_3CN	30	30.0
11	5 mM Na_2SO_4	5 mM Na_2SO_4 + 10% CH_3CN	30	33.0
12	5 mM Na_2SO_4	5 mM Na_2SO_4 + 20% 1-propanol	30	15.1
13	5 mM Na_2SO_4	5 mM Na_2SO_4 + 20% 1-propanol	30	15.0
14	5 mM Na_2SO_4	5 mM Na_2SO_4 + 20% 1-propanol	30	15.0

[a] From Wilson *et al.*,[30] © 1986 Munksgaard International Publishers Ltd., Copenhagen, Denmark.

[b] After sample injection, initial mobile phase is pumped through the column for 25 min prior to initiation of the linear gradients in runs 1–7, or for 10 min in runs 9–15; in run 8 a step gradient is initiated after 25 min of initial mobile phase. Mobile phase is pumped at 1 ml/min except in runs 6 and 7, in which the pumping rates are 0.4 and 0.5 ml/min, respectively.

[c] Retention times represent the interval after initiation of the gradient until elution of the peak concentration of PHLIP-8. In each run the immunoreactive material eluted in 2–3 ml mobile phase.

[d] HPLC runs 1–13 were performed with a Zorbax ODS (Du Pont) column, 4.6 × 250 mm. Run 14 used a Supelco LC 3DP column, 4.6 × 250 mm.

PHLIP-containing fractions are divided into three aliquots which are chromatographed as indicated in HPLC runs 1–3 of Table III. The PHLIP-containing fractions, eluted in runs 1–3, are combined and rechromatographed in run 4. This type of recovery procedure is repeated for the succeeding chromatographic steps (runs 5–14, Table III); PHLIP-containing fractions are combined and lyophilized to dryness in a Speed Vac, then each resulting residue is dissolved in 150–500 μl of the indicated i-mp prior to performance of the next HPLC run. Since DTT decomposition products contribute to interference with PHLIP absorbance at 215 nm, it is omitted from the mobile phases after run 6. Prior to amino acid analysis, the PHLIPs are desalted on a reversed-phase analytical column as indicated in Fig. 1.

Structure Elucidation of the PHLIPs

Amino Acid Analyses. The ultraviolet absorption spectra of the purified PHLIPs are devoid of an absorption band in the 275–280 nm region, indicating the absence of tryptophan. Five or ten percent of each desalted peptide is transferred to a 6 × 50 mm Corning Pyrex tube and dried. The

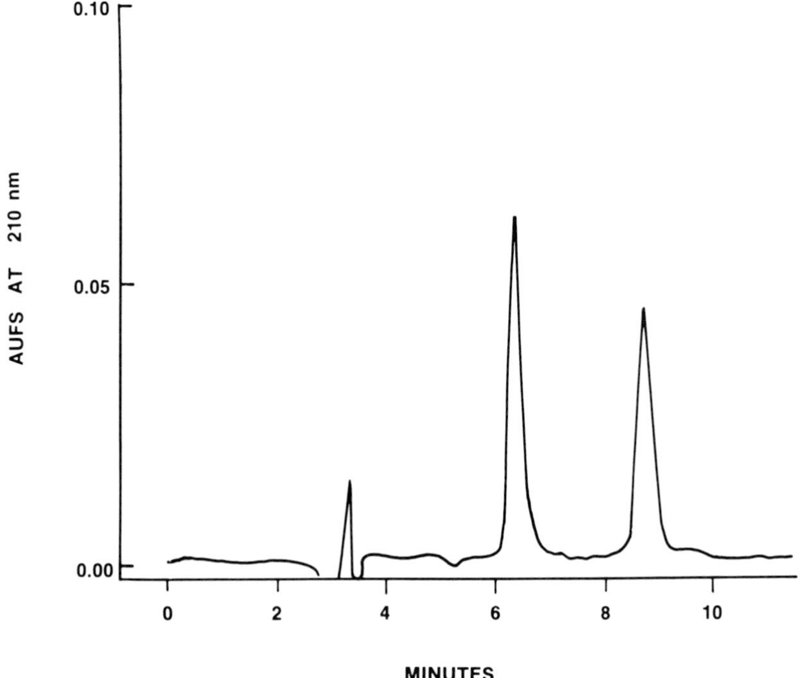

FIG. 1. HPLC of PHLIPs on a 4.6 × 250 mm Vydac 5 butyl column. The mobile phase is 8% 2-propanol in 9.2 m*M* TFA; flow rate, 1 ml/min; chart speed, 1 cm/min; 20 drops/fraction. Retention times were 6.20 min for PHLIP-7 (k' = 1.06) and 8.62 min for PHLIP-8 (k' = 1.87). Filters are in-line before and after the column. (From Wilson *et al.*,[30] © 1986 Munksgaard International Publishers Ltd., Copenhagen, Denmark.)

resulting residue is subjected to hydrolysis for 20 hr at 110° in the presence of 200 μl constant boiling HCl containing 1% phenol; hydrolysis is conducted in a nitrogen-purged, closed glass chamber at a reduced pressure of about 1 mtorr. Suitable amino acid analysis is anticipated from hydrolyzates of 50–100 pmol of peptide using either of two HPLC techniques: (1) ion-exchange chromatography followed by postcolumn derivatization with *o*-phthaldialdehyde (OPA) to yield the corresponding fluorescent isoindole derivatives[32] or (2) reversed-phase HPLC of the phenyl isothiocyanate derivatives of the amino acids using the Pico.Tag amino acid analysis protocol.[33]

The ion-exchange method of Klapper[32] involves injection of a 10% aliquot of the peptide hydrolyzate, in 20 μl of 0.2 *M* sodium citrate, onto a 4 × 250 mm Aminex A9 cation-exchange resin column equilibrated with 20 m*M* sodium citrate, pH 3.15. A 25-min linear gradient is initiated to 20 m*M* sodium borate buffer, pH 10, followed by a 20-min isochratic interval.

The column effluent is mixed with OPA reagent (at a flow rate of 0.1 ml/min) in a stainless steel tee, then this mixture is passed through a 10-foot coil of 1/16-inch stainless steel tubing to ensure completion of reaction before the solution enters the fluorescence detector. Column, tee, and coil are housed in a Du Pont heater compartment maintained at 62°. Fluorescence of amino acid isoindoles is determined using a 340-nm excitation filter and a 420-nm emission filter in the model 420E detector. Detector output for each amino acid derivative is integrated with the model 3390A reporting integrator. OPA reagent is prepared daily by mixing 6 ml of stock solution A (0.4% OPA in methanol) with 200 μl 2-ME and 100 μl of stock solution B (5% potassium borate, pH 10.5). The OPA reagent container is covered with Parafilm, degassed with He, and mixed before and during use.

The results of the amino acid analysis are that PHLIP-8 contains two Glx, two Asx, Ala, Ile, Pro, and Val; PHLIP-7 contains two Glx, two Asx, Ile, Pro, and Val. Results of the amino acid analyses permit estimates of recoveries of PHLIPs from the starting material (e.g., from two 500-g batches of dry stomach powder, ~20 μg of PHLIP-8 is recovered in one purification effort, while 5 μg of PHLIP-8 and 15 μg of PHLIP-7 are recovered in a second effort[30]).

Fast Atom Bombardment–Mass Spectrometry. The ZAB 4F tandem mass spectrometer with the "soft ionization" FAB source permits bombardment of the peptide, in an appropriate matrix, with primary fast neutral atoms generated with the aid of a saddle field ion source operated with xenon gas at 8 kV. A suitable matrix is obtained by mixing approximately 1 μl peptide solution in 50% isopropanol with either thioglycerol or a solution of 0.5% oxalic acid in glycerol on a stainless steel probe. In order to obtain satisfactory fragment ion spectra using a mass marker driven by a Hall-effect probe, 1–5 μg peptide is adequate; when the computer-based data system is used to collect the spectra, satisfactory B2–E2 fragment ion spectra may be obtained with 100–400 ng peptide.[41]

By peak matching at a mass resolution of 5000, the B1–E1 (FAB–MS) spectrum indicates that the protonated molecular ion (MH^+) of PHLIP-8 occurs at m/z (mass to charge ratio) 867.419. In view of the amino acid composition of PHLIP-8, the calculated exact mass of $C_{37}H_{59}N_{10}O_{14}$, equaling 867.421, is compatible with the possibilities that the octapeptide may be a cyclic peptide or a linear peptide in which the amino-terminal residue is <Glu. The B1–E1 spectrum of PHLIP-7 indicates a MH^+ at m/z 796.4. An elemental composition of $C_{34}H_{54}N_9O_{13}$, corresponding to an exact mass of 796.384, suggests that this heptapeptide could result from the loss of Ala from PHLIP-8.

While the MH^+ species is the most prominent ion in the B1–E1 spectrum of each peptide, the spectra of both peptides also indicate prominent MNa^+ (ions of m/z 889 for PHLIP-8 and m/z 818 for PHLIP-7). In addition, the most prominent fragment ion in the spectrum of each peptide occurs at m/z 650.

Analysis of B2–E2 Fragment Ion Spectra. Next, in order to obtain information about the sequence of amino acids in the two peptides, the MH^+ of each peptide is subjected to collision-induced decomposition (CID) in a chamber positioned between the electrostatic sectors of the FAB–tandem mass spectrometer. The resulting positively charged fragment ions that appear in the B2–E2 spectrum, and which are generally most useful in rationalization of the amino acid sequence of a new peptide, reflect cleavages between (1) the amide nitrogen and the acyl carbon (the peptide bond), (2) the acyl and α carbon in the amino-terminal fragment ion, and (3) an amide nitrogen and α carbon within a carboxy-terminal fragment of a peptide.[42,43] For example, the CID of PHLIP-8 results in cleavage of the peptide bond between Asn and Ile to give positively charged fragment ions at m/z 331 (a protonated tripeptide) and m/z 537 (a pentapeptide with a postulated carboxy-terminal acylium function which carries the charge); these two ions reflect the sums of the masses of the individual amino acid residues (total of values indicated in parentheses for each fragment) as shown in Scheme 1.

However, as can be seen in Figs. 2 and 3 and in Table IV,[41] the positively charged fragment ions obtained by CID of MH^+ of natural and

[41] D. J. Harvan, J. R. Hass, W. E. Wilson, C. Hamm, R. K. Boyd, H. Yajima, and D. G. Klapper, *Biomed. Environ. Mass Spectrom.* **14,** 281 (1987).

[42] H. R. Morris and M. Panico, *Biochem. Biophys. Res. Commun.* **101,** 623 (1981).

[43] I. Katakuse and D. M. Desiderio, *Int. J. Mass Spect. Ion Proc.* **54,** 1 (1983).

<Glu—Val—Asp—Pro—Asn$^+$ (m/z 537) Ile—Gln—Ala · H$^+$ (m/z 331)

(112) (99) (115) (97) (114) (113) (128) (89 + 1)

SCHEME 1

TABLE IV
FRAGMENT ION SPECTRA OF MH$^+$ IONS (m/z 867.4) FROM NATURAL AND SYNTHETIC PHLIP-8 PEPTIDES[a]

Natural peptide		Synthetic peptide	
m/z	R.I.[b]	*m/z*	R.I.[b]
852.5	12	852.6	3
850.6	24	850.5	22
849.5	24	849.6	19
848.5	7	848.5	7
778.5	71	778.4	66
777.4	12	777.4	2
657.5	27	657.4	10
650.5	100	650.5	100
649.4	29	649.3	10
622.3	11	622.4	6
		621.5	8
542.3	33	542.4	26
		541.4	15
537.4	48	537.3	11
509.2	14	509.1	4
492.2	7	492.3	6
453.2	10	453.2	4
423.3	5	423.1	5
		377.2	3
331.3	9	331.1	5
		327.1	3
326.2	8	326.2	2
325.3	4	325.1	3
297.2	5		
		299.2	2
218.0	7	218.0	4
212.0	16	212.0	11
211.0	2	211.0	3
199.9	5	200.0	4
166.6	5	166.7	2

[a] Obtained using the computer-based mass measurement technique. (From Harvan *et al.*[41] Copyright (1987). Reprinted by permission of John Wiley & Sons, Ltd.)

[b] Relative intensity of ion peaks (%).

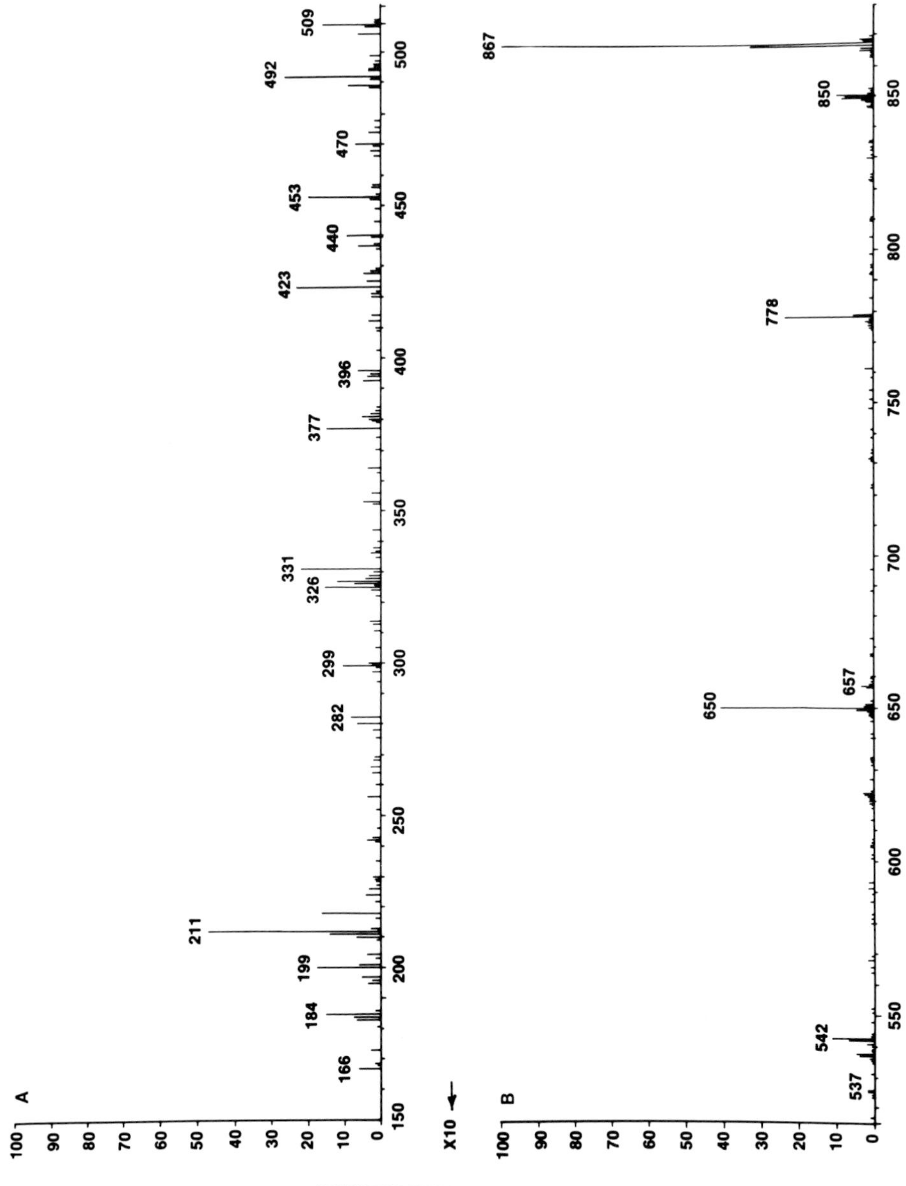
A
B
RELATIVE ABUNDANCE
X10
166
184
199
211
282
299
326
331
377
396
423
440
453
470
492
509
537
542
650
657
778
850
867

PHLIP-8

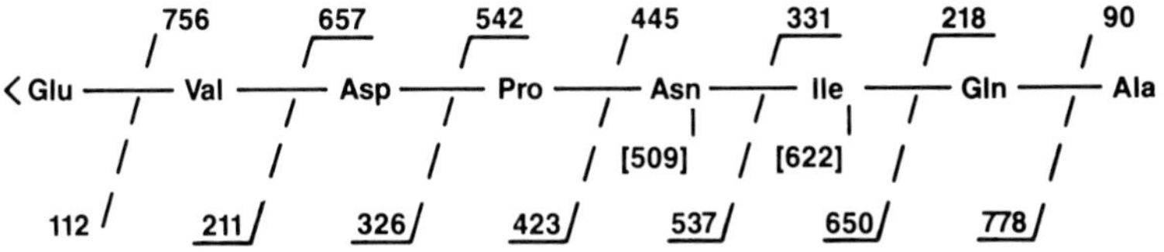

756 657 542 445 331 218 90
<Glu — Val — Asp — Pro — Asn — Ile — Gln — Ala
[509] [622]
112 211 326 423 537 650 778

PHLIP-7

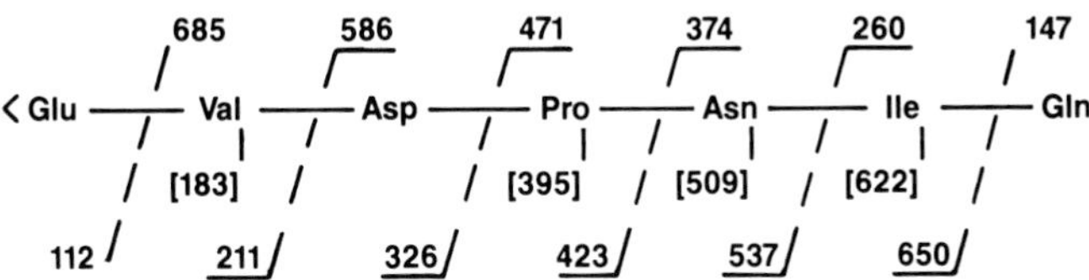

FIG. 3. Positive B2–E2 ions which result from cleavage of the amide bonds of MH^+ are underlined. The upper fragments, from the carboxy-terminal end, have a hydrogen added during the decomposition process and carry a positive charge on the nitrogen involved in the cleavage; the lower fragments, from the amino-terminal end, are postulated to carry the positive charge on the acylium carbon. The ions generated by cleavage between amino acid carbons 1 and 2 are indicated in square brackets; only amino-terminal fragments were detected. (From Wilson *et al.*,[30] © 1986 Munksgaard International Publishers Ltd., Copenhagen, Denmark.)

synthetic PHLIP-8 do not include all ions anticipated from the above-mentioned types of decomposition (e.g., there is no carboxy-terminal fragment ion resulting from cleavage of the peptide bond between <Glu-Val at m/z 756 for PHLIP-8 or at m/z 685 for PHLIP-7; nor is there a carboxy-terminal fragment ion resulting from cleavage of the peptide bond between Pro-Asn at m/z 445 in PHLIP-8). Other fragment ions resulting from multiple decomposition steps may be more prominent than those which are indicated in Fig. 3 [e.g., the relative intensity of the fragment ion Pro-Asn-H^+, at m/z 212 (Fig. 2), is severalfold greater than that of the ion at m/z 211 which arises from cleavage of the peptide bond between Val-Asp to yield <Glu-Val^+].

In order to accommodate the potential problems in attempting to interpret B2–E2 fragment ion spectra of unknown peptides, a computer pro-

FIG. 2. The FAB tandem MS (B2–E2) spectra of PHLIP-7 (A) and PHLIP-8 (B). The PHLIP-7 spectrum is amplified 5 times between 150 and 640 m/z; the spectrum of PHLIP-8 below 500 m/z is amplified 10 times. (From Wilson *et al.*,[30] © 1986 Munksgaard International Publishers Ltd., Copenhagen, Denmark.)

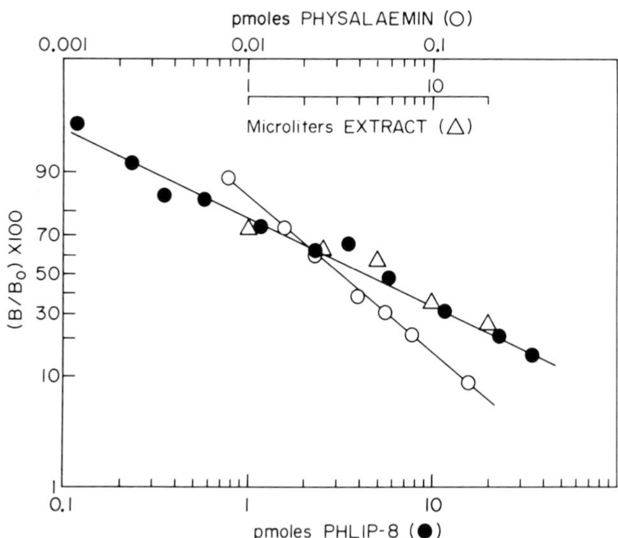

FIG. 4. Cross-reactivities of PHY, PHLIP-8, and unpurified extract components with PS-1 antiserum. The extract PHLIP(s) are in a 75% ethanol-soluble fraction of a neutralized formic acid extract of rabbit stomach powder; the powder is delipidated by prior extraction with chloroform–methanol (2 : 1, v/v). (From Wilson *et al.*,[30] © 1986 Munksgaard International Publishers Ltd., Copenhagen, Denmark.)

gram[44] is available to obtain the most probable fit(s) for those ions which are likely to be of greatest significance in determining the sequence of amino acid residues in a peptide. This program permits a high degree of flexibility in defining the potential forms in which nonpeptide-linked carboxyl (e.g., Asp versus Asn, Glu versus Gln or <Glu), amino (e.g., Lys versus *N*-acyl-Lys, etc.), or other groups may exist within the peptide; it also permits consideration of unusual or modified amino acids. When applied to analysis of the B2–E2 spectra of PHLIP-8 and PHLIP-7, this program permits prediction of unique, correct amino acid sequences.

Confirmation of Identities of Synthetic and Natural PHLIPs. Synthetic PHLIP-7, PHLIP-8, and [Ile^5,Asn^6]PHLIP-8[29–31] permit confirmation of the identity of natural and synthetic PHLIPs. Such evidence is indicated by (1) similarities in the fragment ion spectra for PHLIP-8 (Table IV) and PHLIP-7, (2) identical reversed-phase HPLC behavior under conditions indicated in Fig. 1, and (3) identical cross-reactivity with PS-1 antiserum as indicated in the titrations shown for synthetic and natural PHLIP-8 in Fig. 4.

[44] C. W. Hamm, W. E. Wilson, and D. J. Harvan, *Comput. Appl. Biosci.* **2,** 115 (1986).

Reexamination of the Nature of PHLIM

Although PHLIP-8 is found to be somewhat less than 1% as effective as physalaemin in stimulating contractility of guinea pig large intestine, the structures of the PHLIPs should not be expected to elicit characteristic tachykinin responsiveness. This leads to the question of whether the metabolic precursor of these peptides might be recoverable and whether such a compound might, in fact, possess a tachykinin amino acid sequence. In order to answer the first question, an effort is made to reexamine the technique for recovery of PHLIM. This recovery effort is performed using conditions which would ensure minimal proteolysis, i.e., heat denaturation of the tissue by microwave radiation for 1.5–2 min immediately following recovery as described above.

In this effort, 12.6 g dry, heat-denatured tissue is extracted with 10 ml of the following solutions per gram dry stomach: (1) chloroform–methanol (2 : 1, v/v), with the organic extract being back extracted with 22 m*M* ammonium acetate containing 5.7 m*M* 2-ME; (2) 22 m*M* ammonium acetate in 80% ethanol containing 5.7 m*M* 2-ME; (3) 1 *M* formic acid in 80% ethanol containing 5.7 m*M* 2-ME, with the resulting extract being neutralized prior to RIA; (4) 50% ethanol containing 5.7 m*M* 2-ME; (5) 20% ethanol containing 5.7 m*M* 2-ME; and (6) 5.7 m*M* 2-ME in water. In each extraction step, the tissue is initially homogenized for approximately 5 min with a Polytron apparatus at room temperature. Each tissue suspension is then heated at 60° for 15 min, sonicated for 10 min, and centrifuged at 7000 *g* for 30 min. RIA analysis of the aqueous supernatant solutions indicates that no PHLIM is recovered in any of these six extraction steps.

The sediment is next extracted with 1 *M* formic acid, using the above-mentioned mixing steps, and a large quantity of PHLIM is detected by RIA in the neutralized supernatant. The remaining sediment is exposed to 1 *M* formic acid at 95° for 1 hr; this results in extensive hydration of the sediment such that the aqueous phase is essentially filled with a gel. After cooling to 18°, the gel is centrifuged at 7000 *g* for 30 min; RIA of the neutralized opalescent supernatant indicates recovery of an even greater quantity of PHLIM than that recovered in the preceding step. Subsequent centrifugation of this supernatant material at 20,000 *g* for 3 hr at 18° removes some sediment; however, the supernatant retains essentially undiminished PHY antiserum cross-reactivity. Finally, centrifugation at 30,000 *g* for 20 hr at 18° yields a clear supernatant solution which does not cross-react with PHY antiserum, whereas the redispersed gelatinous sediment does cross-react with the antiserum. Thus, this PHLIM preparation exhibits properties which can only be attributed to a very large glycoprotein. With regard to the possibility that a tachykinin amino acid sequence might exist in the very high M_r PHLIM, an answer will require future efforts.

Summary

Although the PHLIPs are able to enhance gastrointestinal contractility to a very modest extent, the actual physiological significance of the PHLIPs is not completely understood at the present time. Should the high M_r PHLIM be the precursor molecule(s), the PHLIPs may represent partial enzymatic hydrolysis products of the amino-terminal region of the precursor(s). In this regard, it is interesting to note that immunohistochemical staining of rat duodenum indicates the presence of PHY antiserum cross-reacting material in Brunner's glands,[23] which appear to be involved in secretion of mucin.[45–47] The amphibian peptide, bombesin, has also been detected in amphibian skin mucous glands by immunochemistry[48]; other investigators have indicated that cerulein, an amphibian analog of the carboxy-terminal region of cholecystokinin, is discharged along with mucin from cutaneous glands in *Xenopus* after the injection of adrenalin.[3]

[45] W. Cochrane, D. V. Davies, A. J. Palfrey, and R. A. Stockwell, *J. Anat.* **98,** 1 (1964).
[46] C. R. Leeson and T. S. Leeson, *Anat. Rec.* **156,** 253 (1966).
[47] I. M. Lang and M. F. Tansy, *Life Sci.* **30,** 409 (1982).
[48] S. Yoshie, T. Iwanaga, and T. Fugiita, *Cell Tissue Res.* **239,** 25 (1985).

[34] Radioligand-Binding Assays for Study of Neurotensin Receptors

By MICHEL GOEDERT

Introduction

Neurotensin is a 13-amino-acid peptide first isolated and sequenced from bovine hypothalamus[1,2] and later from both bovine and human small intestine.[3,4] Xenopsin, a peptide isolated from the skin of *Xenopus laevis,*[5] and two peptides isolated from chicken intestine[6,7] share marked se-

[1] R. Carraway and S. E. Leeman, *J. Biol. Chem.* **248,** 6854 (1973).
[2] R. Carraway and S. E. Leeman, *J. Biol. Chem.* **250,** 1907 (1975).
[3] R. Carraway, P. Kitabgi, and S. E. Leeman, *J. Biol. Chem.* **253,** 7996 (1978).
[4] R. A. Hammer, S. E. Leeman, R. Carraway, and R. Williams, *J. Biol. Chem.* **255,** 2476 (1980).
[5] K. Araki, S. Tachibana, M. Uchiyama, T. Nakajima, and T. Yasuhara, *Chem. Pharm. Bull.* **21,** 2801 (1973).
[6] R. Carraway and Y. M. Bhatnagar, *Peptides* **1,** 167 (1980).
[7] R. Carraway and C. F. Ferris, *J. Biol. Chem.* **258,** 2475 (1983).

TABLE I
AMINO ACID SEQUENCES OF PEPTIDES OF THE NEUROTENSIN FAMILY

Peptide	Amino acid sequence
Mammalian neurotensin	pGlu-Leu-Tyr-Glu-Asn-Lys-Pro-Arg-Arg-Pro-Tyr-Ile-Leu-OH
Neuromedin N	H-Lys-Ile -Pro-Tyr-Ile-Leu-OH
Chicken neurotensin	pGlu-Leu-His-Val-Asn-Lys-Ala-Arg-Arg-Pro-Tyr-Ile-Leu-OH
[Lys^8,Asn^9]Neurotensin (8–13)	H-Lys-Asn-Pro-Tyr-Ile-Leu-OH
Xenopsin	pGlu-Gly-Lys-Arg-Pro-Trp-Ile-Leu-OH

quence similarities with the carboxy-terminal end of mammalian neurotensin (Table I). However, neither xenopsin nor [Lys^8,Asn^9]neurotensin$^{8-13}$ is present in central and peripheral tissues of various mammalian species.[8,9] A 6-amino-acid peptide called neuromedin N has been isolated and sequenced from porcine spinal cord[10]; it differs in only two amino acids from the carboxy-terminal end of mammalian neurotensin, and both peptides appear to be codistributed.[11,12] Neurotensin and neuromedin N are also part of the same precursor molecule.[13]

As is the case for most neuropeptides whose amino acid sequence is known and for which antibodies are available, far more is known about the distribution of neurotensin than about its physiological functions. The existing evidence for a neurotransmitter role of neurotensin in central and peripheral nervous systems and for an endocrine role in peripheral tissues has been recently reviewed.[14] Neurotensin receptors have been studied using biological assays, radioligand-binding studies, and, to a more limited extent, transmembrane events consequent to neurotensin receptor occupancy.[15] In general, there exists good agreement between the results obtained using these various assays. In this chapter, we summarize some of the results obtained using radioligand-binding assays for neurotensin receptors.

[8] M. Goedert, N. Sturmey, B. J. Williams, and P. C. Emson, *Brain Res.* **308,** 273 (1984).
[9] M. Goedert, W. N. Schwartz, and B. J. Williams, *Brain Res.* **342,** 259 (1985).
[10] N. Minamino, K. Kangawa, and H. Matsuo, *Biochem. Biophys. Res. Commun.* **122,** 542 (1984).
[11] R. Carraway and S. P. Mitra, *Endocrinology (Baltimore)* **120,** 2092 (1987).
[12] Y. C. Lee, J. A. Ball, D. Reece, and S. R. Bloom, *FEBS Lett.* **220,** 243 (1987).
[13] P. R. Dobner, D. L. Barber, L. Villa-Komaroff, and C. McKiernan, *Proc. Natl. Acad. Sci. U.S.A.* **84,** 3516 (1987).
[14] P. C. Emson, M. Goedert, and P. W. Mantyh, *Handb. Chem. Neuroanat.* **4,** 255 (1985).
[15] M. Goedert, R. D. Pinnock, C. P. Downes, P. W. Mantyh, and P. C. Emson, *Brain Res.* **323,** 193 (1984).

Tissue Membrane-Binding Assay

Tissues from central or peripheral tissues are dissected and homogenized in 10 volumes of cold 50 m*M* Tris–HCl, pH 7.4, with a Polytron at setting 7 for 15 sec. After a 20-min centrifugation at 50,000 *g*, the pellet is resuspended in 10 volumes of cold 50 m*M* Tris–HCl, pH 7.4, and incubated at 37° for 30 min. Following a 20-min centrifugation at 50,000 *g*, the pellet is resuspended in 50 m*M* Tris–HCl, pH 7.4, containing 0.1% bovine serum albumin (radioimmunoassay grade, Sigma Chemical Co.), 40 mg/liter bacitracin (Sigma), and 1 m*M* EDTA to yield a concentration equivalent to 10 mg original tissue/ml.

Freshly prepared tissue membranes must be used, as freezing and thawing the membranes results in a substantial loss of specific binding sites. One milliliter of membrane suspension, containing [^{3}H]neurotensin ([3,11-*tyrosyl*-3,5-^{3}H]neurotensin (1–13), New England Nuclear), is incubated at 25° for 10 min. Nonspecific binding is defined as that not displaceable by 1 μ*M* neurotensin (Cambridge Research Biochemicals), and specific binding is obtained by subtracting the nonspecific from the total binding. At the end of the incubation period the tubes are quickly transferred to ice and the mixtures filtered immediately under reduced pressure through GF/B glass fiber filters (Whatman) pretreated for more than 3 hr with 0.2% polyethyleneimine (Sigma) in water. Each tube is washed with 2 ml and each filter 4 times with 5 ml cold incubation buffer with a filtration time which does not exceed 20 sec per filter. Radioactivity is determined by liquid scintillation spectrometry, and protein concentrations are determined[16] using bovine serum albumin as the standard.

Tissue Section-Binding Assay

Tissues are dissected and 15–20 μm sections cut on a freezing cryostat. The sections are pressed onto gelatin-coated microscope slides, thawed in place, dried overnight at 4°, and stored at −20° for 24 hr. The thawed sections are then incubated for 10 min at 25° in 2 n*M* [^{3}H]neurotensin in 50 m*M* Tris–HCl, pH 7.4, containing 0.1% bovine serum albumin, 40 mg/liter bacitracin, and 1 m*M* EDTA. Nonspecific binding is estimated in parallel incubations in the presence of 1 μ*M* neurotensin (Cambridge Research Biochemicals). Following the incubation the sections are washed twice in cold buffer (2 min each time), dipped into cold water, and dried under a stream of cold air. They are then arranged in X-ray cassettes, covered in the darkroom with a tritium-sensitive film (^{3}H-

[16] O. H. Lowry, N. J. Rosebrough, A. L. Farr, and R. J. Randall, *J. Biol. Chem.* **193,** 265 (1951).

Ultrofilm, LKB Laboratories), and kept at −20° for 6 weeks. The film is developed under safe-light conditions using Kodak D-19 developer. Alternatively, the tissue sections can be apposed to emulsion-coated coverslips. They are apposed to the slides and attached at one side with cyanoacrylate glue. The assembly is held in place with a paper clip and stored at −20° for 6 weeks. The coverslips are then separated using a wooden spacer and the emulsion developed as above. The tissue sections are fixed in Carnoy's solution (60% absolute ethanol, 30% chloroform, and 10% glacial acetic acid) and stained with cresyl violet (0.05%, w/v, in sodium acetate buffer, pH 4.2) (Nissl stain). The sections are subsequently dehydrated through a series of graded alcohols and permanently mounted with Depex mounting medium in xylene.

Characterization of [^{3}H]Neurotensin, Neurotensin Fragments, and Analogs

The purity of [^{3}H]neurotensin is determined by subjecting an aliquot from each batch to reversed-phase high-performance liquid chromatography. The stability of [^{3}H]neurotensin is assessed by incubating 2 n*M* [^{3}H]neurotensin in the presence of 1 ml tissue membranes for 10 min at 25°. Following centrifugation (2000 *g* for 10 min) an aliquot of the supernatant is applied to a μBondapak C_{18} column (0.39 × 30 cm). A flow rate of 2 ml/min is used, and elution is achieved by using a 20 min linear gradient from 30 to 100% acetonitrile, with 10 m*M* ammonium acetate, pH 4.5, as the aqueous phase. The radioactivity present in 100 μl from each fraction is determined by liquid scintillation spectrometry. The purity and the concentration of neurotensin, neurotensin fragments, and analogs used are determined by amino acid analysis.

Characteristics of [^{3}H]Neurotensin-Binding Sites in Rat Brain

[^{3}H]Neurotensin adsorbs extensively to glass fiber filters from where it can be displaced by unlabeled peptide. Pretreatment of the filters with polyethyleneimine substantially reduces this nonspecific interaction, so that filter binding amounts to less than 3% of the total binding. When incubated with washed membranes for 10 min at 25° a substantial proportion of [^{3}H]neurotensin is degraded. Protease inhibitors, such as *p*-chloromercuribenzoylsulfonate and phenylmethylsulfonyl fluoride stabilize [^{3}H]neurotensin against degradation; however, they also markedly inhibit specific [^{3}H]neurotensin binding. The combination of bovine serum albumin, bacitracin, and EDTA was chosen,[17] because it protects

[17] M. Goedert, K. Pittaway, B. J. Williams, and P. C. Emson, *Brain Res.* **304,** 71 (1984).

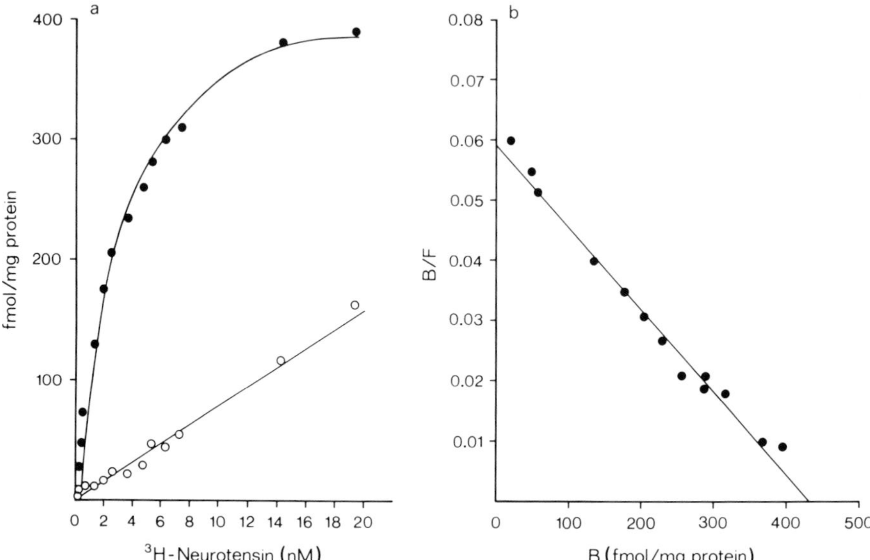

FIG. 1. Association time course of [^{3}H]neurotensin binding. Rat brain membranes were incubated for different time periods with 2 n*M* [^{3}H]neurotensin at 25°. (a) Specific (●) and nonspecific (○) binding as a function of time. (b) Specific binding data are represented as $\ln[B_{eq}/(B_{eq} - B)]$ versus time, where B_{eq} represents the amount of ligand bound at the time considered. (Reproduced with permission from Goedert *et al.*[17])

[^{3}H]neurotensin from degradation without interfering with specific binding. At a concentration of 2 n*M* the percentage of specific [^{3}H]neurotensin binding consistently amounts to more than 80% of the total binding. It shows a linear relationship to homogenate protein up to 1.4 mg protein/ml and a pH optimum between 7.0 and 8.0.[17] Specific [^{3}H]neurotensin binding is temperature dependent. At 4° specific binding increases slowly and does not fully plateau at 60 min. At 37°, following an initial increase, specific binding decreases with time, presumably due to degradation of the radioligand. Binding assays were therefore routinely performed at 25°.

The specific binding of 2 n*M* [^{3}H]neurotensin to rat brain membranes is time dependent and reversible, when performed at 25°. A plateau is reached after about 8 min (Fig. 1a), and a straight line with a slope of 6.38×10^{-3} sec^{-1} is obtained, when $\ln[B_{eq}/(B_{eq} - B)]$ is plotted as a function of time (Fig. 1b). The nonspecific binding remains constant throughout, and it does not exceed 20% of the specific binding at this radioligand concentration. The dissociation of [^{3}H]neurotensin is biphasic, with dissociation rate constants of 5.34×10^{-4} and 3.64×10^{-3} sec^{-1} (Fig. 2). The specific binding of [^{3}H]neurotensin in membranes from rat brain can be saturated with increasing concentrations of radioligand, whereas the nonspecific binding increases in a linear manner (Fig. 3a). Scatchard analysis of the specific binding data gives a straight line (Fig. 3b), indicating that under these conditions [^{3}H]neurotensin is bound to a single population of binding sites. The average maximum number of binding sites (B_{max}) amounted to 432 ± 26 fmol/mg protein, and the dissociation constant had a value of 2.85 ± 0.19 n*M*.

Neurotensin (1–13) competes with [^{3}H]neurotensin for its binding site with an IC_{50} of 3.09 n*M* (Table II). [^{3}H]Neurotensin appears to bind to a single population of noninteracting sites, since a Hill transformation of the specific binding data yielded a slope not significantly different from unity (Fig. 4). Of the neurotensin fragments and analogs tested, the carboxy-terminal fragment neurotensin (8–13) was equipotent with neurotensin

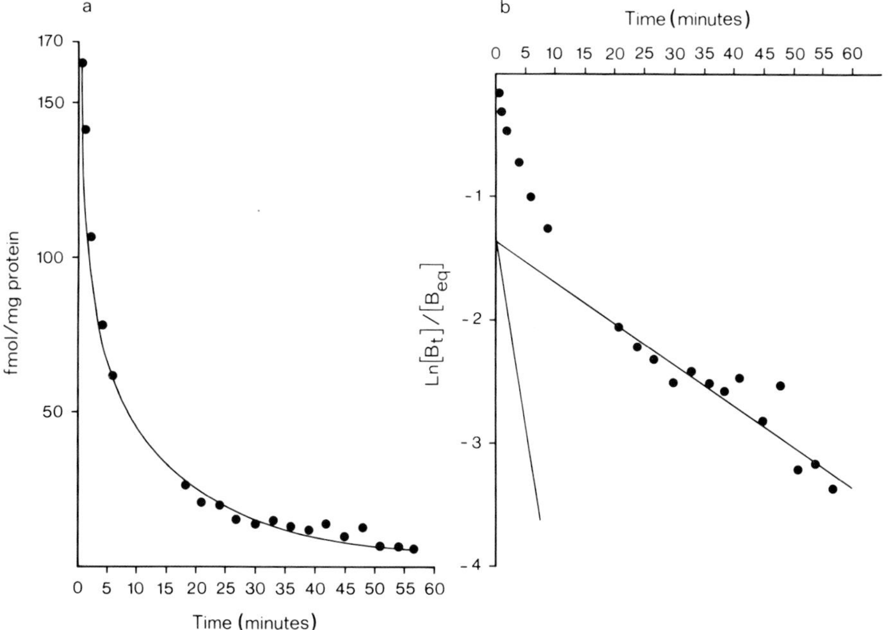

FIG. 2. Dissociation time course of specific [^{3}H]neurotensin binding. (a) [^{3}H]Neurotensin (2 n*M*) was incubated for 10 min at 25° with rat brain membranes. Neurotensin (1 μ*M*) was then added, and the concentration of [^{3}H]neurotensin specifically bound was determined at various times thereafter. Nonspecific binding (defined as binding in the presence of 1 μ*M* neurotensin) was determined separately for each time point. (b) Specific binding data are represented as $\ln(B_t/B_{eq})$ versus time, where B_{eq} represents the amount of ligand bound at equilibrium and B_t the amount of ligand bound at time t. The dissociation curve could be resolved into two different components with dissociation rate constants of 5.34×10^{-4} and 3.64×10^{-3} sec^{-1}. (Reproduced with permission from Goedert *et al.*[17])

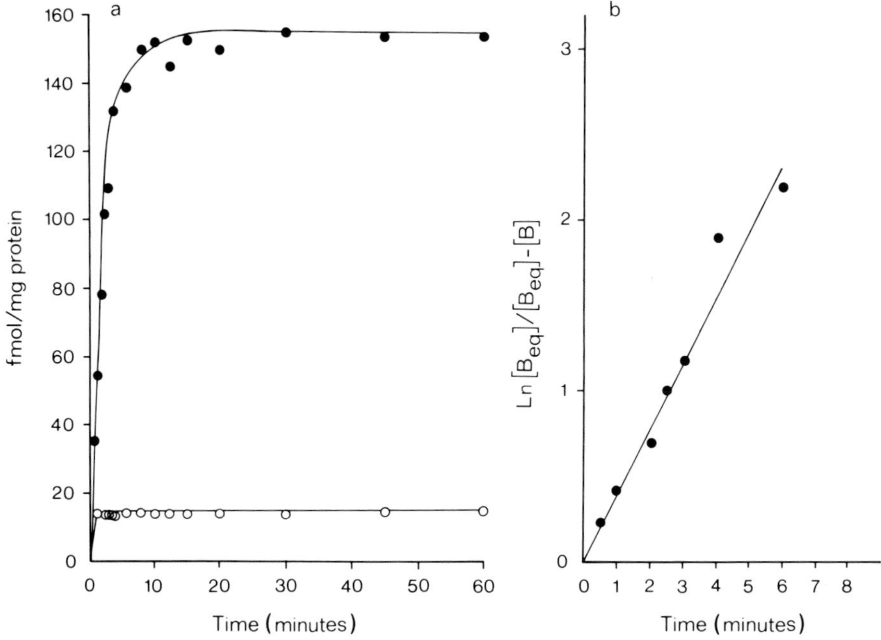

FIG. 3. [^{3}H]Neurotensin binding to rat brain membranes as a function of the radiolabeled peptide concentration. (a) Different concentrations of the radioligand were incubated for 10 min at 25° in the presence of brain membranes. Nonspecific binding (○) was defined as binding in the presence of 1 μM neurotensin, and specific binding (●) was obtained by subtracting nonspecific from total binding. (b) Scatchard transformation of specific [^{3}H]neurotensin binding data. (Reproduced with permission from Goedert *et al.*[17])

TABLE II
RELATIVE POTENCIES OF NEUROTENSIN FRAGMENTS AND RELATED PEPTIDES IN BINDING TO RAT BRAIN MEMBRANES

Peptide	Potency
Neurotensin (1–13)	1.00
Neurotensin (8–13)	1.16
Neurotensin (9–13)	0.51
Neurotensin (10–13)	<0.0004
Neurotensin (1–6)	<0.0004
Neurotensin (1–11)	<0.0004
Neuromedin N	0.14
[Lys^8,Asn^9]Neurotensin (8–13)	0.02
Chicken neurotensin (1–13)	0.96
Xenopsin	0.89

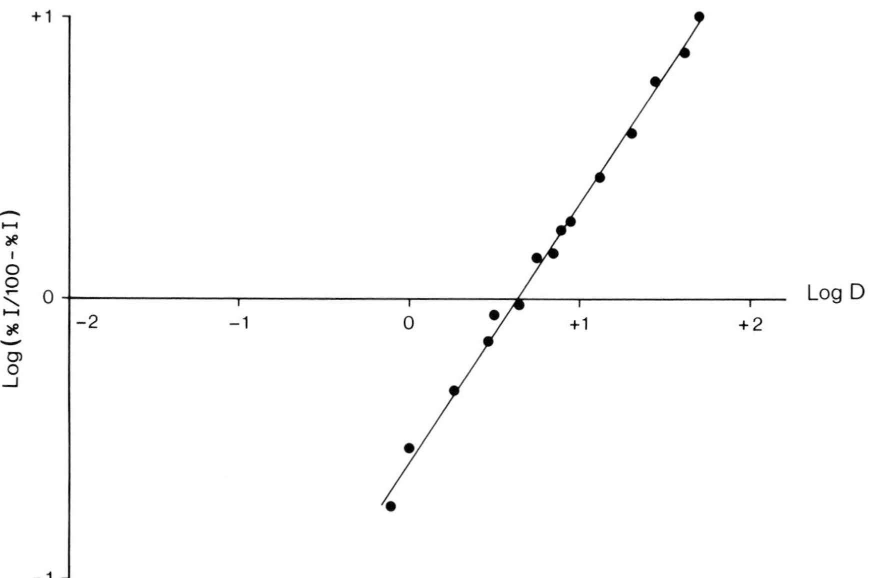

FIG. 4. Hill transformation of the competitive inhibition of [^{3}H]neurotensin binding to rat brain membranes by increasing concentrations of unlabeled neurotensin (1–13). (Reproduced with permission from Goedert *et al.*[17])

(1–13) (IC_{50} 2.66 n*M*), whereas neurotensin (9–13) showed little activity (IC_{50} 603 n*M*) and neurotensin (10–13) was inactive (IC_{50} > 10,000 n*M*). The amino-terminal fragments neurotensin (1–6), neurotensin (1–8), and neurotensin (1–11) were inactive (IC_{50} > 10,000 n*M*). The amphibian skin peptide xenopsin was equipotent with neurotensin (1–13) (IC_{50} 3.40 n*M*) and neuromedin N was roughly 5 times less potent than neurotensin (1–13) (IC_{50} 18.17 n*M*), whereas the chicken intestinal peptide [Lys^8,Asn^9]neurotensin (8–13) was only weakly active (IC_{50} 166 n*M*).

The amount of specifically bound [^{3}H]neurotensin and the dissociation constant were not significantly changed in the presence of $MgCl_2$, $MnCl_2$, $CaCl_2$ or KCl. However, physiological concentrations of NaCl resulted in a 35% reduction in the total number of binding sites and in a 2.6-fold increase in the dissociation constant. The addition of 100 μM of the guanyl nucleotides GTP, GDP, cGMP, or GMP–PNP did not result in significant changes in either the total number of specific [^{3}H]neurotensin-binding sites or in the dissociation constant. The latter results are in line with findings indicating that neurotensin stimulates inositol phospholipid hydrolysis in rat brain slices, while not affecting adenylate cyclase activity.[15]

The above results are consistent with [^{3}H]neurotensin binding to a physiological receptor. At present, it is unknown whether there exists a separate receptor for neuromedin N in mammalian tissues and for [Lys^8,Asn^9]neurotensin (8–13) in avian tissues. Binding studies using these peptides as radioligands have not yet been performed. However, in analogy with other neuropeptides, such as the tachykinins, it appears likely that there exist multiple receptors for the neurotensin family. It has been shown that it is possible to subdivide neurotensin receptors in rat, mouse, and hamster brains, but not in cat, dog, guinea pig, and human brains, on the basis of the inhibition of [^{3}H]neurotensin by levocabastine, a new histamine-1 antagonist that is structurally unrelated to neurotensin.[18] This has led to the classification into levocabastine-sensitive and levocabastine-insensitive sites. Future studies will reveal whether this subdivision reflects a physiological heterogeneity of neurotensin receptors. The structure–activity characteristics of neurotensin receptors observed in the rat brain are very similar to those seen in peripheral rat tissues and in central and peripheral tissues from other mammalian species. The only exception is xenopsin: it is equipotent with neurotensin$^{1-13}$ in the rat but has only 5–10% of its potency in cat and guinea pig brains.[17,19] This is probably of no physiological significance, as xenopsin-like immunoreactivity is not present in mammalian tissues.[8]

Distribution of Specific Neurotensin-Binding Sites in the Central Nervous System

The distribution of specific [^{3}H]neurotensin-binding sites in the rat brain has been investigated by both tissue homogenate binding and autoradiographic techniques. In tissue homogenates, the total number of binding sites varies among different brain regions, whereas the equilibrium dissociation constant does not differ significantly (Table III).[17] The density of binding sites is highest in hypothalamus, frontal cortex, and midbrain, followed by striatum, thalamus, hippocampus, and olfactory bulb, whereas only low levels are present in cerebellum and pons medulla (Table III). By autoradiography (Fig. 5)[20–22] the highest concentrations of binding sites are found in the external layer of the olfactory bulb. High

[18] A. Schotte, J. E. Leysen, and P. M. Laduron, *Naunyn-Schmiedeberg's Arch. Pharmacol.* **333,** 400 (1986).

[19] F. Checler, C. Labbé, G. Granier, J. van Rietschoten, P. Kitabgi, and J. P. Vincent, *Life Sci.* **31,** 1145 (1982).

[20] M. Goedert, *in* "Neuromethods" (A. Boulton, G. B. Baker, and P. D. Hrdina, eds.), Vol. 4, p. 251. Humana, Clifton, New Jersey, 1986.

[21] W. S. Young and M. J. Kuhar, *Brain Res.* **206,** 273 (1981).

[22] R. Quirion, P. Gaudreau, S. St. Pierre, F. Rioux, and C. B. Pert, *Peptides* **3,** 757 (1982).

TABLE III
DISTRIBUTION OF [^{3}H]NEUROTENSIN-BINDING SITES IN RAT BRAIN

Tissue	Binding sites (fmol/mg protein)	K_d (nM)
Hypothalamus	565 ± 62	2.43 ± 0.10
Frontal cortex	524 ± 56	3.09 ± 0.39
Midbrain	448 ± 25	3.72 ± 0.63
Striatum	418 ± 43	2.64 ± 0.40
Thalamus	392 ± 35	3.08 ± 0.63
Hippocampus	374 ± 22	2.37 ± 0.20
Olfactory bulb	366 ± 18	3.64 ± 0.21
Cerebellum	212 ± 18	3.71 ± 0.39
Pons medulla	202 ± 14	2.39 ± 0.25

levels are also present in the anterior cingulate cortex, the rhinal sulcus, the habenula, the zona incerta, the pars compacta of the substantia nigra, and the ventral tegmental area. Medium levels are present in the hypothalamus, the striatum, the amygdala, the superior colliculus, the periaqueductal gray, and the solitary tract nucleus. Low levels are observed in most cortical areas. A high concentration of neurotensin-binding sites is found in the substantia gelatinosa of the dorsal horn of the spinal cord.[23]

The injection of 6-hydroxydopamine into the rat substantia nigra results in a marked reduction in [^{3}H]neurotensin-binding sites, indicating that a large proportion of these binding sites is located on dopaminergic cell bodies.[24] Similarly, the number of [^{3}H]neurotensin-binding sites appears to be reduced in the substantia nigra of patients with Parkinson's disease.[25] Moreover, the chronic administration of neuroleptics has been shown to result in an apparent increase in [^{3}H]neurotensin-binding sites in the rat substantia nigra.[26]

[^{3}H]Neurotensin-binding sites are present in the internal plexiform layer of chicken and rat retinas,[27] in which neurotensin-like immunoreactivity is present in a population of amacrine cells.[28,29] A strikingly different

[23] M. Ninkovic, S. P. Hunt, and J. S. Kelly, *Brain Res.* **230,** 111 (1981).

[24] J. M. Palacios and M. J. Kuhar, *Nature (London)* **294,** 587 (1981).

[25] G. R. Uhl, P. J. Whitehouse, D. L. Price, W. W. Tourtelotte, and M. J. Kuhar, *Brain Res.* **308,** 186 (1984).

[26] G. R. Uhl and M. J. Kuhar, *Nature (London)* **309,** 350 (1984).

[27] P. W. Mantyh, M. Goedert, S. P. Hunt, and N. C. Brecha, *Invest. Ophthalmol. Vis. Sci. Suppl.* **26,** 277 (1985).

[28] N. Brecha, H. J. Karten, and C. Schenker, *Neuroscience* **6,** 1329 (1981).

[29] M. Fukuda, Y. Kuwayama, S. Shiosaka, S. Inagaki, I. Ishimoto, Y. Shimizu, H. Takagi, M. Sakanaka, K. Takatsuki, E. Senba, and M. Tokyama, *Curr. Eye Res.* **1,** 115 (1981).

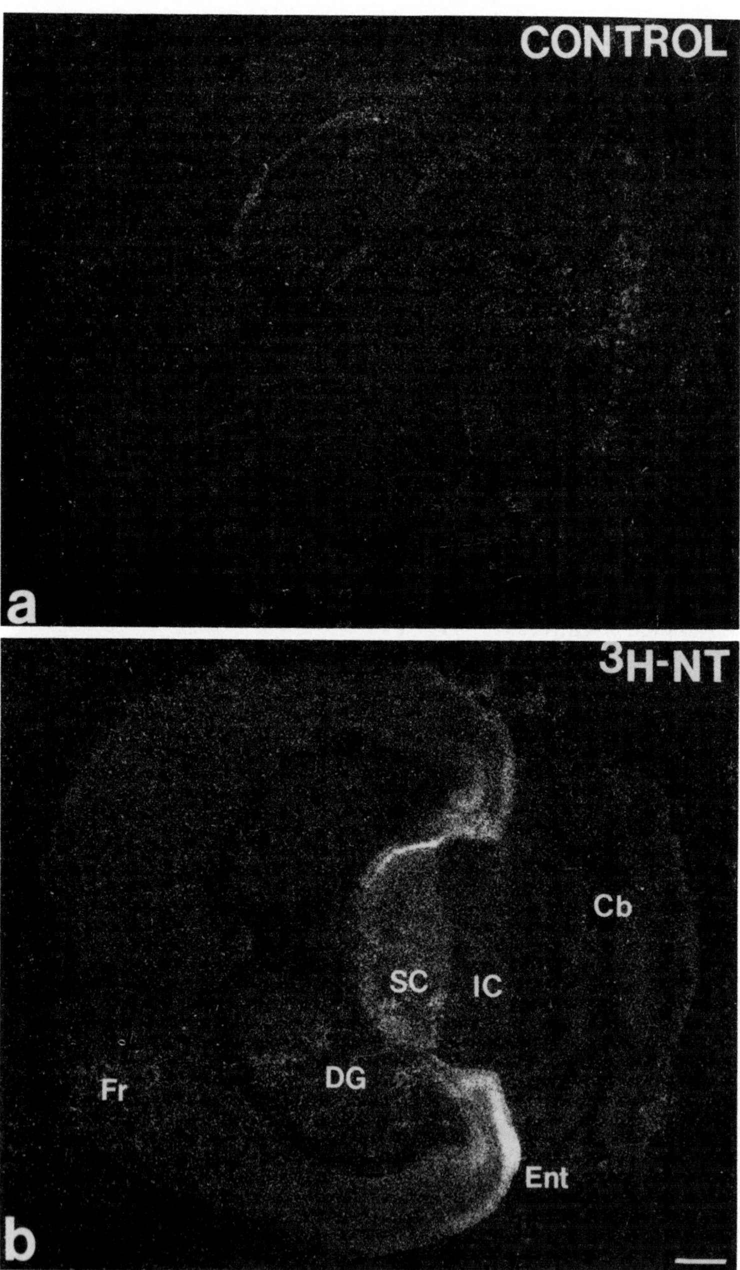

FIG. 5. Autoradiographic localization of [^{3}H]neurotensin-binding sites in rat brain. (a) Adjacent section to (b) incubated in the presence of 2 nM [^{3}H]neurotensin and 1 μM neurotensin (1–13). (b–f) Horizontal sections taken at different levels of the brain, with (b) representing the most dorsal and (f) the most ventral sections. All the sections were incubated in the presence of 2 nM [^{3}H]neurotensin. Scale bars, 2.5 mm. Fr, Frontal cortex; DG, dentate gyrus; SC, superior colliculus; IC, inferior colliculus; Ent, entorhinal cortex; Cb, cerebel-

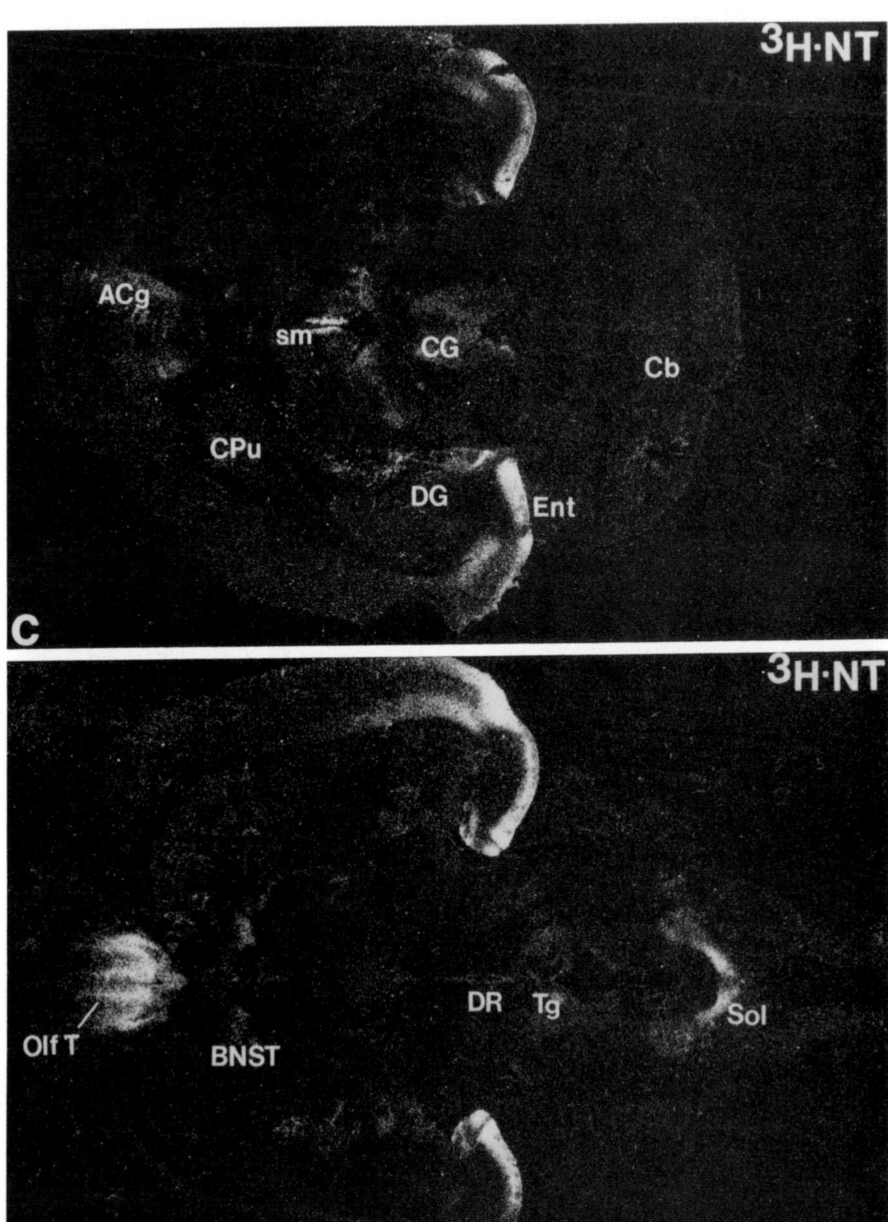

lum; ACg, anterior cingulate cortex; CPu, caudate putamen; sm, stria medullaris; CG, central gray; Olf T, olfactory tubercle; DR, dorsal raphe; Tg, tegmentum; Sol, solitary tract nucleus; RF, rhinal fissure; LS, lateral septum; ZI, zona incerta; PH, posterior hypothalamus; SNc, substantia nigra, pars compacta; Pa, paraventricular nucleus; VTA, ventral tegmental area; SNr, substantia nigra, pars reticulata. (Reproduced with permission from Goedert.[20])

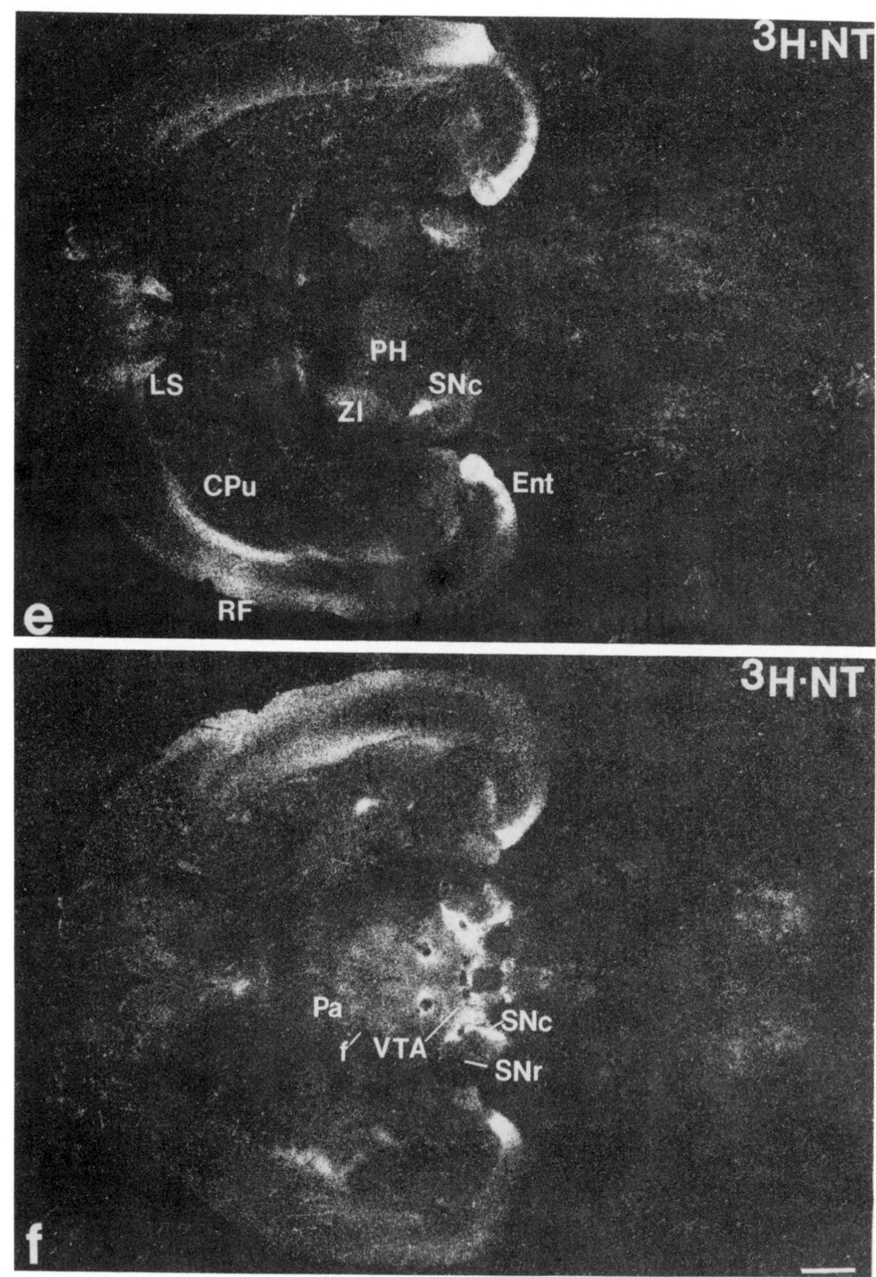

FIG. 5e and f. See legend on pp. 472–473.

topographical localization of neurotensin-like immunoreactivity[30] and [^{3}H]neurotensin-binding sites[31] within a structure characterized by distinct histochemical compartments is found in the case of the cat corpus striatum (Figs. 6 and 7). In the caudate nucleus, neurotensin-like immunoreactivity presents a mosaic distribution that coincides with the distribution of striosomes, defined by their high content of Met-enkephalin-like immunoreactivity and opiate receptor binding, as well as by their low acetylcholinesterase activity.[32] In contrast to Met-enkephalin-like immunoreactivity and opiate receptors, neurotensin-like immunoreactivity and [^{3}H]neurotensin-binding sites are inversely related (Figs. 6 and 7). The localization of striatal neurotensin receptors in the rat has been investigated by using kainic acid and 6-hydroxydopamine injections, as well as frontal cortex ablation.[33] More than 50% of the specific [^{3}H]neurotensin-binding sites are localized on intrinsic striatal neurons, approximately 35% on dopaminergic nerve terminals, and 20% on corticostriatal nerve fibers. This indicates that specific [^{3}H]neurotensin-binding sites in the rat striatum are localized on neurons and their processes, rather than on glial cells.

The reason for the inverse relationship between peptide-like immunoreactivity and peptide receptors is unknown. It could be caused by different turnover rates for a given peptide in different brain regions. An excess of peptide levels over binding density could also be caused by the presence of low-affinity binding sites not detected in binding studies operating at nanomolar range. An excess of binding sites over peptide levels could occur because the density of receptors is determined by the presence or absence of peptide-containing nerve fibers rather than by the number of these fibers. Alternatively, it could be because the endogenous ligands are able to bind to the binding sites but are not recognized by radioimmunoassay or immunohistochemistry. In the future it will be important to find an explanation for these discrepancies.

Distribution of Specific Neurotensin-Binding Sites in Peripheral Tissues

Comparatively little is known about neurotensin receptors in peripheral tissues. The binding sites on rat mast cells are clearly distinct from those in the central nervous system, with an equilibrium dissociation

[30] M. Goedert, P. W. Mantyh, S. P. Hunt, and P. C. Emson, *Brain Res.* **274,** 176 (1983).

[31] M. Goedert, P. W. Mantyh, P. C. Emson, and S. P. Hunt, *Nature (London)* **307,** 543 (1984).

[32] A. M. Graybiel and C. W. Ragsdale, *in* "Chemical Neuroanatomy" (P. C. Emson, ed.), p. 427. Raven, New York, 1983.

[33] M. Goedert, K. Pittaway, and P. C. Emson, *Brain Res.* **299,** 164 (1984).

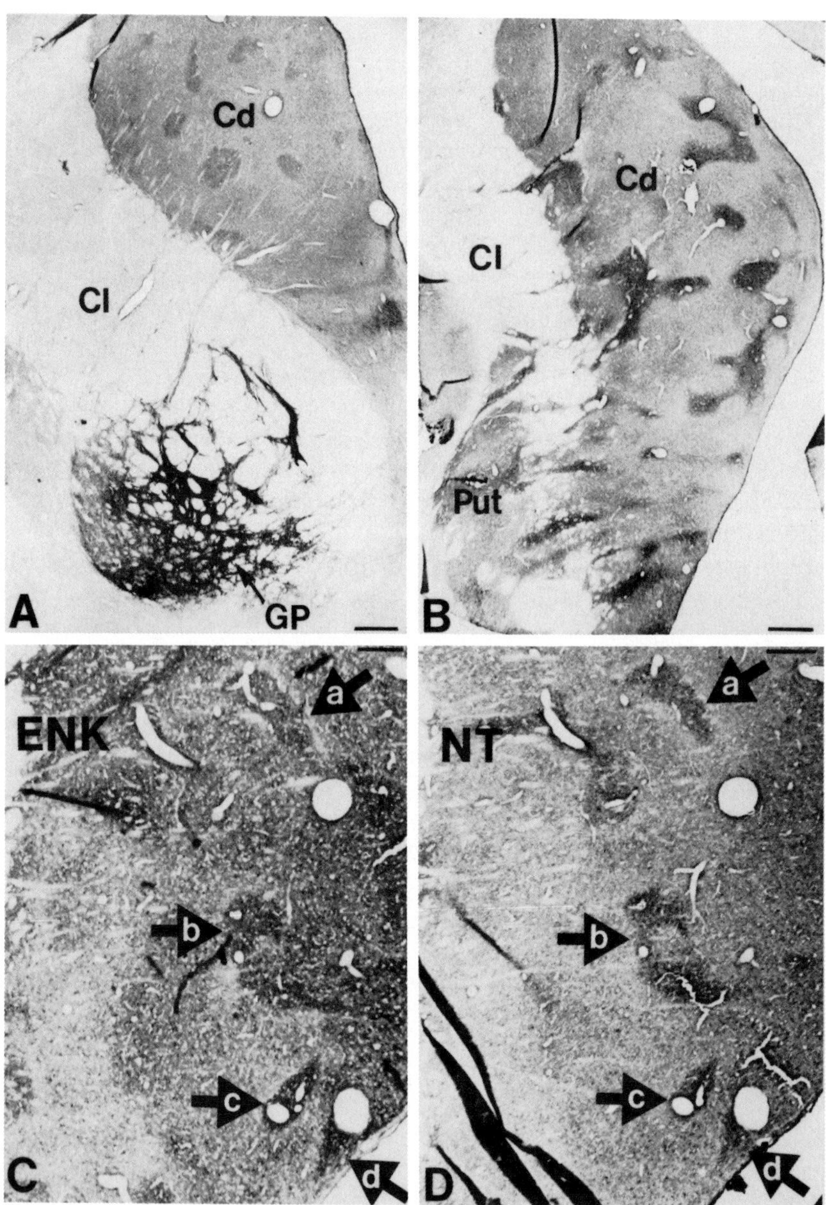

FIG. 6. Immunohistochemical localization of neurotensin-like immunoreactivity in the striatum of the cat (A, B). Note that in the caudate nucleus (Cd) there is a patchy distribution, whereas the globus pallidus (GP) has a homogeneous distribution broken up by myelinated fiber bundles. Scale bars, 450 μm. Photomicrographs C and D show Met-enkephalin-like immunoreactivity (ENK) and neurotensin-like immunoreactivity (NT) in adjacent sections of the caudate nucleus. The arrows indicate patches of immunoreactivity that appear to overlap in the two adjacent 30-μm brain sections. Scale bars, 250 μm. Cd, Caudate nucleus; CI, internal capsule; GP, globus pallidus; Put, putamen; ENK, enkephalin; NT, neurotensin. (Reproduced with permission from Goedert *et al.*[30])

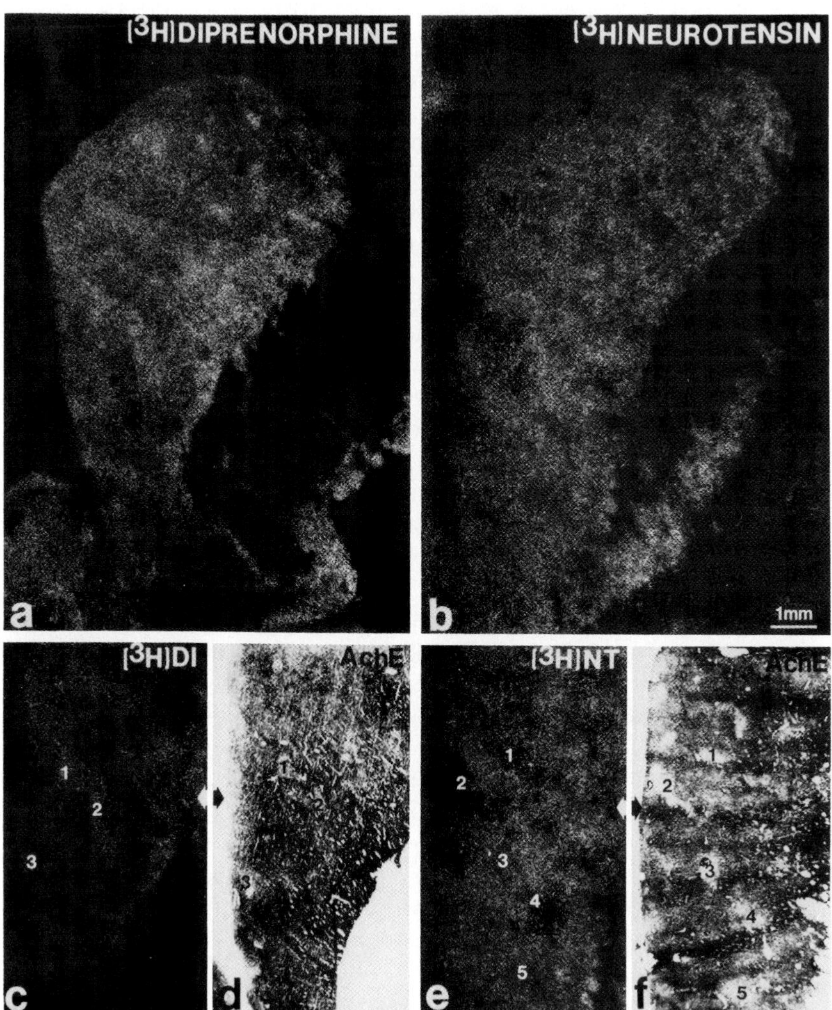

FIG. 7. Autoradiographic localization of [^{3}H]neurotensin- and [^{3}H]diprenorphine-binding sites in the cat caudate nucleus. The tissue sections were incubated in the presence of either 2 n*M* [^{3}H]neurotensin or 4 n*M* [^{3}H]diprenorphine. (a, b) The cat caudate nucleus labeled with [^{3}H]diprenorphine and [^{3}H]neurotensin, respectively. (c, d) Labeling of the sections with [^{3}H]diprenorphine ([^{3}H]DI) and visualization of acetylcholinesterase activity (AchE). (e, f) Labeling of the sections with [^{3}H]neurotensin ([^{3}H]NT) and visualization of acetylcholinesterase activity (AchE). The numbers in (c–f) refer to the acetylcholinesterase-poor striosomes. Note the correspondence between striosomes and dense [^{3}H]diprenorphine labeling and the inverse relationship between striosomes and [^{3}H]neurotensin-binding sites. Scale bar, 1 mm. [Reprinted by permission from *Nature* (*London*), Vol. 307, pp. 543–546. Copyright © Macmillan Magazines Limited.]

constant of 154 nM.[34] Autoradiographically, the majority of [^{3}H]neurotensin-binding sites in the guinea pig ileum is localized in the circular smooth muscle layer, with lower levels in the longitudinal muscle (Fig. 8).[35] The characteristics of the specific [^{3}H]neurotensin-binding sites in ileal homogenates are consistent with their binding to a physiological neurotensin receptor. The potencies of the various neurotensin fragments are in good agreement with their effects on the relaxation of the circular smooth muscle preparation and with their actions on the longitudinal smooth muscle layer.[34,35]

In endocrine tissues, [^{3}H]neurotensin-binding sites are found in the pituitary gland and the adrenal gland. [^{3}H]Neurotensin-binding sites are concentrated in the intermediate lobe of the rat (Fig. 9) and cat pituitary glands, with smaller amounts in both anterior and neural lobes. In the rat, the [^{3}H]neurotensin-binding sites are not affected by pituitary stalk transection, indicating that they are localized on cells of the intermediate lobe.[36] Using the test tube binding assay for [^{3}H]neurotensin, low levels of specific binding sites are found in whole rat pituitary glands. Considering that the intermediate lobe comprises less than 5% of the total pituitary gland, however, it appears that the local concentration of [^{3}H]neurotensin-binding sites is comparable with that found in the central nervous system. In the rat adrenal gland, high levels of [^{3}H]neurotensin-binding sites are present in the inner layer of the adrenal cortex and lower amounts in the adrenal medulla. The structure–activity profile of these specific binding sites in whole rat adrenal gland homogenates closely resembles the values obtained in peripheral bioassays and in binding assays in the central nervous system, therefore being consistent with binding to a physiological receptor.[37] The total number of specific [^{3}H]neurotensin binding sites is low; however, the fact that 75% of the binding is present in one layer of the adrenal cortex implies that the local concentration of neurotensin-binding sites is comparable with the central nervous system levels.

The physiological significance of the specific [^{3}H]neurotensin-binding sites in the zona reticularis of the adrenal cortex is unknown. It is possible that circulating neurotensin levels could influence steroid secretion from the adrenal cortex; alternatively, the neurotensin from the adrenal medulla could influence the adrenal cortex via retrograde blood flow, or the neurotensin-like immunoreactivity released from the splanchnic nerve

[34] L. H. Lazarus, M. H. Perrin, and M. R. Brown, *J. Biol. Chem.* **252,** 7174 (1977).

[35] M. Goedert, J. C. Hunter, and M. Ninkovic, *Nature (London)* **311,** 59 (1984).

[36] M. Goedert, S. L. Lightman, P. W. Mantyh, S. P. Hunt, and P. C. Emson, *Brain Res.* **358,** 59 (1985).

[37] M. Goedert, P. W. Mantyh, S. P. Hunt, and P. C. Emson, *Brain Res.* **299,** 389 (1984).

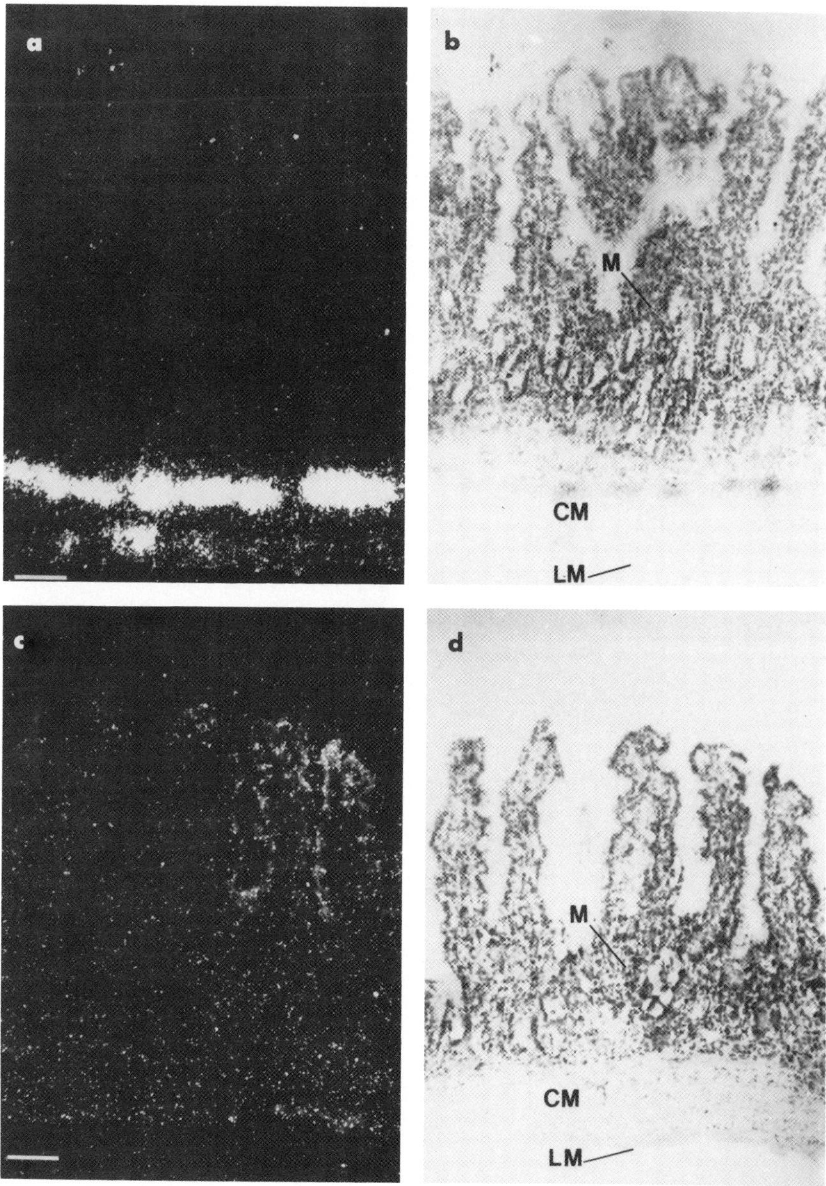

FIG. 8. Autoradiographic localization of [^{3}H]neurotensin-binding sites in guinea pig ileum. (a) Dark-field photomicrograph of the autoradiogram of a transverse section after incubation in the presence of 2 n*M* [^{3}H]neurotensin. (b) Light-field photomicrograph of the Nissl stain of the same section as in a. (c) Dark-field photomicrograph of the autoradiogram of a section adjacent to a after incubation in the presence of 2 n*M* [^{3}H]neurotensin and 1 μM neurotensin (1–13). (d) Light-field photomicrograph of the Nissl stain of the same section as in c. Scale bars, 100 μm. M, Mucosa; CM, circular muscle; LM, longitudinal muscle. [Reprinted by permission from *Nature* (*London*), Vol. 311, pp. 59–62. Copyright © Macmillan Magazines Limited.]

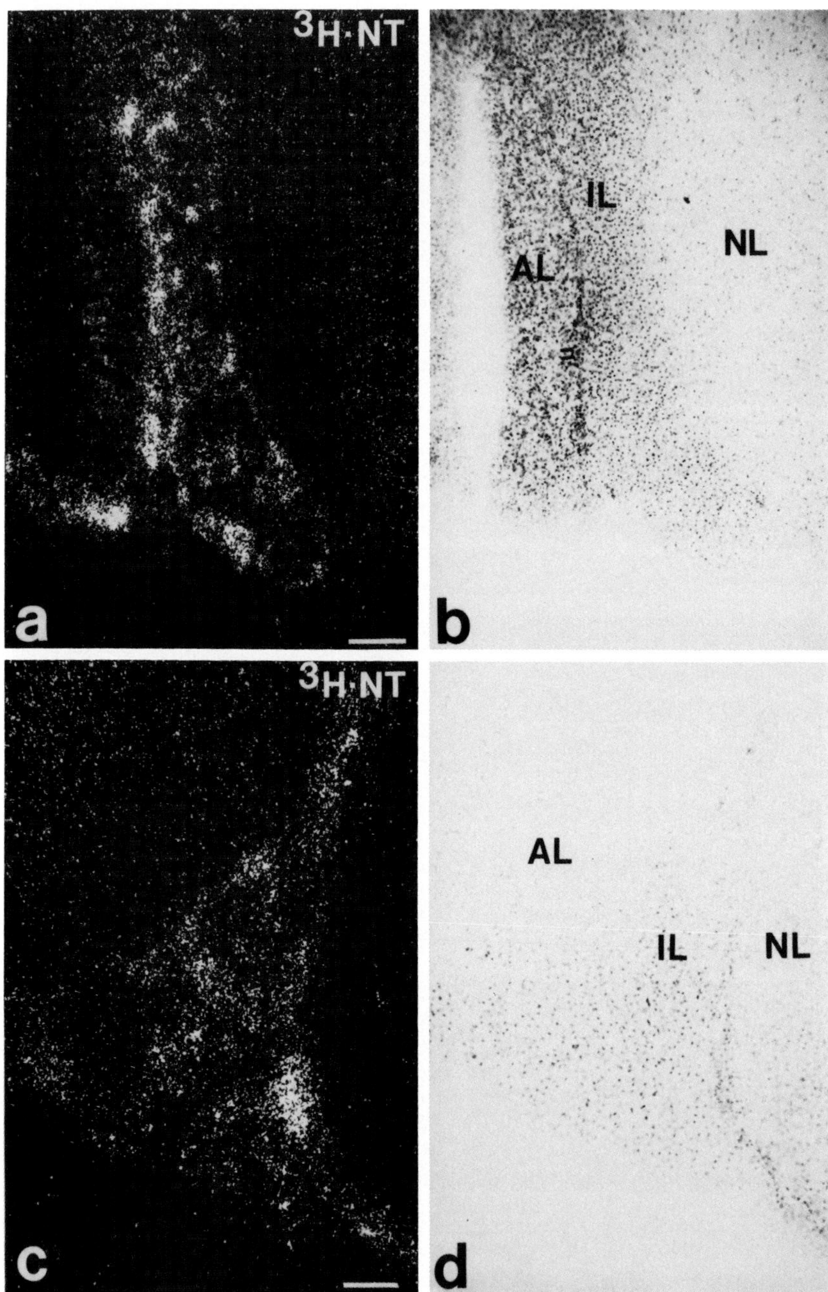

FIG. 9. Autoradiographic localization of [^{3}H]neurotensin-binding sites in rat pituitary. (a, c) Dark-field photomicrographs of the autoradiograms of transverse sections through the rat pituitary gland after incubation in the presence of 2 nM [^{3}H]neurotensin. (b, d) Light-field photomicrographs of the Nissl stains of the same sections as in a and c. Scale bars, 100 μm. AL, Anterior lobe; IL, intermediate lobe; NL, neural lobe. (Reproduced with permission from Goedert *et al.*[36])

could influence adrenal medullary function through the [^{3}H]neurotensin-binding sites.

Outlook

Radioligand-binding studies have provided valuable information concerning the characteristics, structure–activity relationships, and the cellular localization of neurotensin receptors. However, radioligand-binding studies are by their very nature indirect, deducing the existence of a receptor from the characteristics of membrane sites identified through radioligand binding. Therefore, the elucidation of the primary structure of neurotensin receptors constitutes the next important step. It has been reported that *X. laevis* oocytes injected with brain poly(A)$^+$ RNA exhibit an electrophysiological response to neurotensin.[38] This, in conjunction with a general technique for cloning neurotransmitter receptors without prior purification and partial amino acid sequence determination,[39] should permit the molecular cloning and sequencing of neurotensin receptors in the near future.

[38] I. Parker, K. Sumikawa, and R. Miledi, *Proc. R. Soc. London, Ser. B* **229,** 151 (1986).
[39] Y. Masu, K. Nakayama, H. Tamaki, Y. Harada, M. Kuno, and S. Nakanishi, *Nature (London)* **329,** 836 (1987).

[35] Characterization of Receptors for Bombesin/Gastrin-Releasing Peptide in Human and Murine Cells

By Terry W. Moody, Richard M. Kris, Gary Fiskum, Carol D. Linden, M. Berg, and Joseph Schlessinger

Introduction

Bombesin (BN), a 14-amino-acid peptide initially isolated from frog skin,[1] and gastrin-releasing peptide (GRP), a 27-amino-acid peptide isolated from porcine stomach tissue,[2] represent one class of peptides biologically active in the mammalian central nervous system (CNS), periphery, and in tumor cells. Each of these peptides has a common carboxy-termi-

[1] A. Anastasi, V. Erspamer, and M. Bucci, *Arch. Biochem. Biophys.* **148,** 443 (1973).
[2] T. J. McDonald, J. Jornvall, G. Nilsson, M. Vagne, M. Ghatei, S. R. Bloom, and V. Mutt, *Biochem. Biophys. Res. Commun.* **90,** 227 (1979).

TABLE I
STRUCTURE OF BN-LIKE PEPTIDES[a]

Peptide	Sequence
GRP	Ala-Pro-Val-Ser-Val-Gly-Gly-Gly-Thr-Val-Leu-Ala-Lys-Met-Tyr-Pro-Arg-Gly-Asn-His-Trp-Ala-Val-Gly-His-Leu-Met-NH_2
BN	pGln-Arg-Leu-Gly-Asn-Gln-Trp-Ala-Val-Gly-His-Leu-Met-NH_2
Neuromedin C	Gly-Asn-His-Trp-Ala-Val-Gly-His-Leu-Met-NH_2
Neuromedin B	Gly-Asn-Leu-Trp-Ala-Thr-Gly-His-Phe-Met-NH_2

[a] Sequence similarities are underlined.

nal heptapeptide essential for receptor binding and biological activity (Table I). Two other decapeptides, neuromedins B and C, isolated from porcine spinal cord by Minamino *et al.*,[3,4] have 7 and 10 of the same carboxy-terminal amino acids, respectively, as does BN or GRP. In the CNS these peptides may modulate neural activity and regulate behavior, food intake, and hormone (prolactin) secretion. In the periphery, these peptides may function in a paracrine manner and regulate hormone secretion (gastrin) and pancreatic secretion (amylase). In tumor cells BN/GRP may function as an autocrine growth factor.

Endogenous BN/GRP-like peptides have been detected in CNS neurons, peripheral neurons, ganglia, and endocrine as well as tumor cells.[5] GRP is derived from a 148-amino-acid precursor which has been cloned from a human lung carcinoid biopsy specimen.[6] This precursor contains one GRP sequence near the amino terminus and a carboxy-terminal GRP gene-associated peptide (GGAP). While the amino terminus of the prepro-GRP molecule is conserved in all clones thus far obtained, the carboxy terminus can vary due to alternative processing of the RNA, especially in small cell lung cancer (SCLC) tumors.[7] Thus, while each prepro-GRP yields a single GRP, the GGAP peptides may vary. GRP, but not prepro-GRP or the GGAP peptides, has demonstrable biological activity. While GRP is the predominant product in the porcine stomach, it can be further metabolized to GRP (18–27) (neuromedin C), which is found in the porcine spinal cord and is biologically active. The molecular processing of

[3] N. Minamino, K. Kangawa, and H. Matsuo, *Biochem. Biophys, Res. Commun.* **114,** 541 (1983).

[4] N. Minamino, K. Kangawa, and H. Matsuo, *Biochem. Biophys. Res. Commun.* **119,** 14 (1984).

[5] P. Panula, *Med. Biol.* **64,** 177 (1986).

[6] E. R. Spindel, W. W. Chin, J. Price, L. H. Rees, G. M. Besser, and J. F. Habener, *Proc. Natl. Acad. Sci. U.S.A.* **81,** 5699 (1984).

[7] E. A. Sausville, A. M. Lebacq-Verheyden, E. R. Spindel, F. Cuttitta, A. F. Gazdar, and J. F. Battey, *J. Biol. Chem.* **261,** 2451 (1986).

BN in the frog skin and of neuromedin B in mammals is not yet understood.

While much is known about the biosynthesis and biological activity of BN/GRP, we understand far less about the receptors. Initially, receptors for BN/GRP were discovered in rat brain.[8] [^{125}I-Tyr^4]BN bound with high affinity (K_d 3 nM) to a single class of sites in rat brain homogenate (B_{max} 80 fmol/mg protein). Subsequently, receptors which bound radiolabeled [Tyr^4]BN with high affinity (K_d 2 nM) to a single class of sites (B_{max} 5000/cell) were identified in dispersed guinea pig pancreatic acini.[9] [^{125}I-Tyr^4]BN also bound with high affinity (K_d 1 nM) to a single class of sites (B_{max} 3700/cell) on rat pituitary cells (GH_4C_1).[10] More recently, receptors for BN/GRP have been characterized in SCLC cells[11] and Swiss 3T3 cells.[12]

On SCLC cells the carboxy terminus of BN or GRP is essential for high-affinity binding to BN/GRP receptors (Table II). [Tyr^4]BN was the most potent peptide in that it inhibited 50% of the specific [^{125}I-Tyr^4]BN binding (IC_{50}) at a concentration of 0.7 nM. Ac-GRP (20–27), BN, GRP (14–27), and GRP were slightly less potent, with IC_{50} values of 1, 1.5, 2, and 5 nM, respectively. In contrast, GRP (1–16), [des-Leu^{13},Met^{14}]BN, and [D-Trp^8]BN had no measurable binding activity ($IC_{50} > 10{,}000$ nM). These data suggest that the carboxy terminus but not the amino terminus of BN or GRP may be essential for high-affinity binding activity. The putative BN/GRP receptor antagonist [D-Arg^1,D-Pro^2,D-Trp^{7,9},Leu^{11}]substance P has an IC_{50} value of 1 μM. Similar binding data have been obtained on Swiss 3T3 cells.[12]

In Swiss 3T3 cells, activation of receptors for BN/GRP stimulates phosphatidylinositol turnover,[13] elevates cytosolic Ca^{2+} levels,[14] and stimulates growth.[15] Because potent agonists such as BN or GRP stimulate each of these biological activities whereas inactive peptides such as

[8] T. W. Moody, C. B. Pert, J. Rivier, and M. R. Brown, *Proc. Natl. Acad. Sci. U.S.A.* **75,** 5372 (1978).

[9] R. T. Jensen, T. Moody, C. Pert, J. E. Rivier, and J. D. Gardner, *Proc. Natl. Acad. Sci. U.S.A.* **75,** 6139 (1978).

[10] J. Westendorf and A. Schonbrunn, *J. Biol. Chem.* **258,** 7527 (1983).

[11] T. W. Moody, D. N. Carney, F. Cuttitta, K. Quatrocchi, and J. D. Minna, *Life Sci.* **36,** 105 (1985).

[12] I. Zachary, J. W. Sinnett-Smith, and E. Rozengurt, *Proc. Natl. Acad. Sci. U.S.A.* **82,** 7616 (1985).

[13] K. D. Brown, J. Blay, R. F. Irvine, J. P. Heslop, and M. J. Berridge, *Biochem. Biophys. Res. Commun.* **123,** 377 (1984).

[14] S. A. Mendoza, J. A. Schneider, A. Lopez-Rivas, J. W. Sinnett-Smith, and E. Rozengurt, *J. Cell Biol.* **102,** 2223 (1986).

[15] E. Rozengurt and J. Sinnett-Smith, *Proc. Natl. Acad. Sci. U.S.A.* **80,** 2936 (1983).

TABLE II
SPECIFICITY OF BN-LIKE PEPTIDES USING SCLC CELLS[a]

Peptide	IC_{50} (nM)	Ca^{2+} response	Growth
[Tyr[4]]BN	0.7	++	n.d.
Ac-GRP (20–27)	1.0	++	n.d.
BN	1.0	++	++
GRP (14–27)	2.0	++	n.d.
GRP	5.0	++	++
SPant	1,000	−−	−−
[D-Trp[8]]BN	>10,000	−	n.d.
[des-Leu[13],Met[14]]BN	>10,000	n.d.	−
GRP (1–16)	>10,000	−	n.d.

[a] BN-like peptides were tested at varying doses for their ability to inhibit 50% of the specific binding using 1 nM [^{125}I-Tyr[4]]BN. Also, the ability of the peptides at a 1000 nM concentration to elevate cytosolic Ca^{2+} and the ability to alter growth at a 10 nM concentration were determined. The substance P antagonist (SPant) was [D-Arg[1],D-Pro[2],D-Trp[7,9],Leu[11]]substance P. Responses were measured as strong agonist (++), weak agonist (+), inactive (−), antagonist (−−), or not determined (n.d.).

GRP (1–16) do not, it is possible that phosphatidyl inositol turnover, alteration in Ca^{2+} levels, and growth are mediated by BN/GRP receptors. Here, the methods used in characterizing receptors for BN/GRP in human and murine cells are described.

Radiolabeling of Peptides

In order to conduct binding studies one must have an appropriate receptor probe. Routinely we iodinate [Tyr[4]]BN (Peninsula Laboratories Inc., San Carlos, CA) with Na^{125}I (Amersham Corp., Arlington Heights, IL). [Tyr[4]]BN (1.8 μg in 10 μl of 0.5 M $NaHPO_4$, pH 7.4) is incubated with ^{125}I (2 mCi of IMS 300), and 0.9 μg of chloramine-T (30 μl in H_2O) is added. After 60 sec the reaction is quenched with 2.1 μg of sodium metabisulfite (30 μl in H_2O). The radiolabeled peptide is separated from free ^{125}I by chromatography on a 0.7 × 15 cm Sephadex LH_{20} column using methanol–acetic acid–H_2O (10 : 2 : 1, v/v/v) as an eluant. When 1-min fractions are collected, the radiolabeled peptide elutes in fractions 9–11 and the free ^{125}I in fractions 19–21. Approximately 50% of the radioactivity is collected as [^{125}I-Tyr[4]]BN and 50% as free ^{125}I. The key to this radioiodination is the stoichiometry: 1.1 nmol each of [Tyr[4]]BN and ^{125}I, with a 3-fold excess of chloramine-T, and small reaction volume. This yields maximal incorporation of radioisotope with minimal oxidation of the peptide; oxidation of [Tyr^4]BN to [Tyr^4]BN sulfoxide reduces the biological activity by over two orders of magnitude.[16] $Na^{125}I$ from New England Nuclear has not been a suitable substitute in our laboratory as this isotope is kinetically slower, and, using the above procedure, it yields only approximately 10% of the radioactivity as [^{125}I-Tyr^4]BN.

Routinely, we pool the fractions containing [^{125}I-Tyr^4]BN (specific activity ~1100 Ci/mmol) and chemically reduce to convert any [^{125}I-Tyr^4]BN sulfoxide to [^{125}I-Tyr^4]BN. The sample is placed in a siliconized 12 × 75 mm borosilicate test tube and heated to 80° in a well-ventilated hood. After the methanol has evaporated (~45 min), 200 μl of 1 *M* dithiothreitol is added. After 2 hr the reduction is complete,[17] and the samples are stored in siliconized vials at −80° until use. Routinely, the frozen tracer is stable for 3 weeks in receptor binding assays, after which it promptly loses biological activity and the receptor binding activity is negligible. Repeated freezing and thawing of tracer is not recommended. The receptor binding activity of reduced tracer is 2- to 3-fold greater than that of unreduced tracer.

An alternative BN/GRP receptor probe is HPLC-purified ^{125}I-labeled GRP, available commercially from Amersham. In our hands this tracer yields almost identical binding data to that obtained using [^{125}I-Tyr^4]BN. One particular advantage of the ^{125}I-GRP is that it is relatively stable in lyophilized form. ^{125}I-GRP reconstituted in H_2O has a lifetime of 3 weeks when stored at −80°. While both [^{125}I-Tyr^4]BN and ^{125}I-GRP can be used for binding studies, ^{125}I-GRP also can be used for cross-linking studies because it contains a suitable free amino acid group, Lys^{13}.

Receptor Binding

SCLC receptor binding studies can be conducted using [^{125}I-Tyr^4]BN and the classic SCLC cell line NCI-H345. Two days after a change of the medium, cells are harvested by centrifugation and resuspended in serum-free HITES medium (RPMI-1640 containing 10^{-8} *M* hydrocortisone, 5 μl/ml bovine insulin, 10 μg/ml human transferrin, 10^{-8} *M* β-estradiol, and 3×10^{-8} *M* Na_2SeO_3). Routinely, 1×10^6 cells/assay are added to HITES medium, supplemented with 0.5% bovine serum albumin (BSA) and 100 μg/ml bacitracin, and 1 n*M* [^{125}I-Tyr^4]BN in the presence or absence of competitor (1 μ*M* BN) at 25° for 30 min. Bound peptide is separated from free by using centrifugation techniques. Unfortunately, it proved exceed-

[16] T. W. Moody, J. N. Crawley, and R. T. Jensen, *Peptides* **3,** 559 (1982).

[17] R. A. Houghten and C. H. Li, *Anal. Biochem.* **98,** 36 (1979).

ingly difficult to conduct binding studies with these cells because while [^{125}I-Tyr^4]BN bound with high affinity (K_d 0.5 nM) the density of receptors is low (1500/cell). SCLC cells routinely bound only 1000 cpm total, and the ratio of specific to nonspecific binding was 1 : 1.

It was reported that certain clones of Swiss 3T3 cells have 100,000 BN/GRP receptors/cell.[12] These cells, which are derived from mouse fibroblasts, are very different from the human SCLC cell lines. The 3T3 cells are adherent and are commonly propagated by 1 : 10 dilution every week, whereas the SCLC cells grow as floating aggregates and are divided every 2 weeks by 1 : 1 dilution. Thus, larger numbers of Swiss 3T3 cells can be generated readily for binding studies. Routinely, binding studies with 3T3 cells are conducted in 24-well tissue culture plates which are coated with fibronectin (5 μg/cm^2) to enhance cellular adherence. The wells are seeded with 10^5 cells/well and grown to confluence. Binding studies are conducted 3 days after seeding. The cells are washed 3 times with 500 μl of binding buffer [Dulbecco's modified Eagle's medium (DMEM) containing 20 mM HEPES–NaOH (pH 7.4), 0.2% BSA, and 100 μg/ml bacitracin]. Then the cells are incubated with 1 nM [^{125}I-Tyr^4]BN in the presence or absence of competitor (total volume 0.5 ml). After 30 min at 37°, the cells are washed twice with ice-cold binding buffer followed by 3 more washes with ice-cold phosphate-buffered saline (PBS). The cells are either solubilized with 0.5 N NaOH or scraped using cotton swabs and counted in a LKB gamma counter. Typically, total binding of 24,000 cpm/well is observed, and the ratio of specific to nonspecific binding is 12 : 1.[18]

On Swiss 3T3 cells ^{125}I-GRP binds with high affinity (K_d 1 nM) to a single class of sites (100,000/cell). High densities of BN receptors have been identified on a few additional cell lines. NIH 3T3-1 binds [^{125}I-Tyr^4]BN to a single class of sites (80,000/cell) with a K_d of 1.9 nM. Also, certain human glioma cells, such as U-138, bind radiolabeled [Tyr^4]BN with high affinity to a single class of sites (B_{max} 55,000/cell). The high density of BN receptors in these cell lines has facilitated further biochemical studies.

Cross-Linking

Cross-linking studies were performed to identify the cellular binding component for BN/GRP.[18] Because [Tyr^4]BN has no functional groups available for cross-linking, ^{125}I-GRP, with a free amino group at Lys^{13} and an unblocked amino terminus, was used as a ligand. The cross-linking studies were performed as follows. Swiss 3T3 cells, NIH 3T3 cells, or

[18] R. M. Kris, R. Hazan, J. Villines, T. W. Moody, and J. Schlessinger, *J. Biol. Chem.* **262,** 11215 (1987).

human glioma cells are seeded (10^5 cells/well) in six-well tissue culture dishes and grown to confluence. The cells are washed 3 times with DMEM medium containing 20 m*M* HEPES–NaOH (pH 7.4). Then ^{125}I-GRP (0.5 n*M*) is added in 1 ml of DMEM/HEPES. After 2 hr at 4°, the cells are washed 3 times with cold DMEM/HEPES to remove unbound radiolabeled GRP and rinsed once with cold PBS. The bound ^{125}I-GRP is cross-linked with the homobifunctional reagents (1 m*M*) disuccinimidyl suberate (DSS) or disuccinimidyl tartarate (DST) at 4° for 30 min. The cross-linking reaction is stopped by the addition of 10 μl of 2 *M* Tris–HCl (pH 8.0) and the cells rinsed once with cold PBS. Eighty microliters of Laemmli sample buffer[19] preheated to 95° is added to the cells and the cells scraped into an Eppendorf tube (1.5 ml). Samples are sonicated briefly, heated to boiling, and subjected to electrophoresis on an 8% sodium dodecyl sulfate (SDS) gel using the previously described buffer system.[19] Gels are dried and exposed on Kodak XAR-5 film for 3–4 days at −80° using image-intensifying screens.

^{125}I-GRP was cross-linked to a radioactive band of molecular weight 65,000, 75,000, and 75,000, using cell lines NIH 3T3-1, Swiss 3T3, and glioma line G-340, respectively.[18] No labeling was detected when 1 μM GRP or GRP (14–27) was present during the incubation (Fig. 1, lanes b and e). Also, Fig. 1 (lanes a and f) shows that the pharmacologically inactive peptide GRP (1–16) did not inhibit the cross-linking of ^{125}I-GRP to receptors on NIH 3T3-1 cells. These data indicated that the cross-linking specificity is consistent with the known pharmacology of BN/GRP receptors. Cross-linking to Swiss 3T3 cells with 6 m*M* ethylene glycol bis(succinimidyl succinate) yielded a radiolabeled protein with a molecular weight of 75,000–85,000 and an isoelectric point of 6.0–6.5.[20]

Glycoprotein Analysis

The BN/GRP receptor is a glycoprotein which contains amino-linked sugars. NIH 3T3-1 cells were cross-linked with ^{125}I-GRP and solubilized with 1% *N*-octylglucoside, 20 m*M* HEPES–NaOH (pH 7.4), 150 m*M* NaCl, 10% glycerol, and protease inhibitors, and the extract was applied to immobilized wheat germ agglutinin (WGA). In the absence of competing sugars, the MW 65,000 polypeptide bound to the WGA. If the solubilized cross-linked receptor solution was treated with *N*-acetylglucosamine, a sugar recognized by WGA, the MW 65,000 band was not retained. If the solubilized cross-linked receptor solution was treated with

[19] U. K. Laemmli, *Nature (London)* **277,** 680 (1970).

[20] I. Zachary and E. Rozengurt, *J. Biol. Chem.* **262,** 3947 (1987).

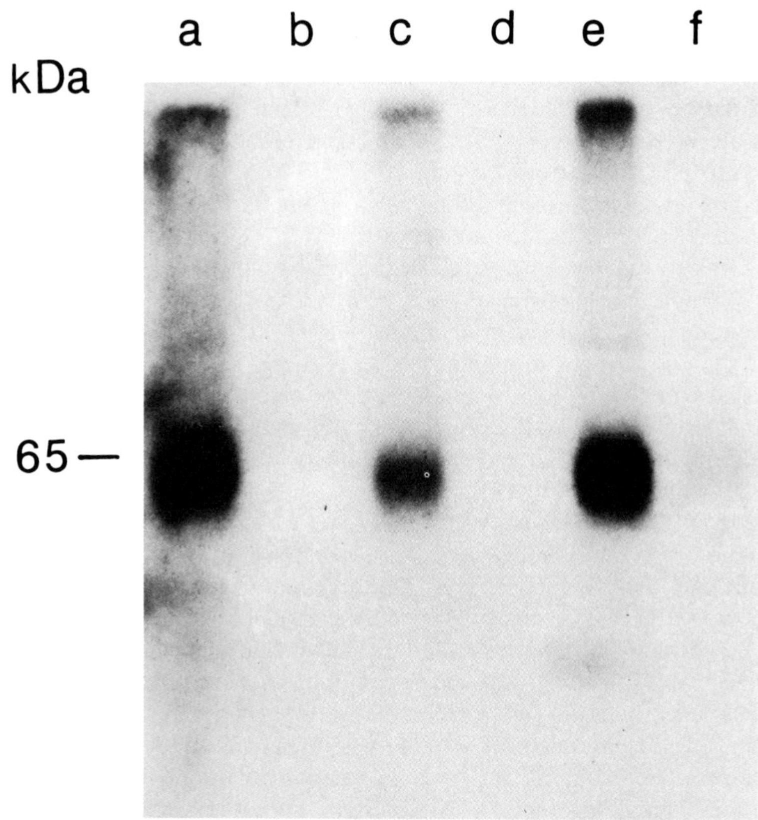

FIG. 1. Cross-linking reaction of ^{125}I-labeled GRP. ^{125}I-labeled GRP was cross-linked to NIH 3T3-1 cells using 1 mM DSS. ^{125}I-labeled GRP was incubated with the cells in the absence (a) and in the presence of 1 μM unlabeled GRP (b), 1 nM unlabeled GRP (c), 1 μM unlabeled GRP (1–16) (e), or 1 μM unlabeled GRP (14–27) (f) at 4°. Prior to the incubation of the cells with ^{125}I-labeled GRP, the cells were incubated with 1 nM unlabeled GRP for 30 min at 37° (d).

α-methyl mannoside, which is not recognized by WGA, the 65,000 band of radioactivity was retained by WGA.

The BN/GRP receptor appears to be heavily glycosylated. Swiss 3T3 cells were cross-linked with ^{125}I-GRP and then treated with the enzyme *N*-glycanase, which cleaves N-linked sugars. After the cross-linking, the cells were solubilized in 50 μl of 0.5% SDS and 0.1 M 2-mercaptoethanol, then heated at 95° for 3 min, and sonicated. A 40-μl aliquot of the solubilized extract was mixed with 43 μl of 0.5 M sodium phosphate (pH 8.6), 20 μl of 7.5% NP-40, and 2.4 μl of *N*-glycanase (250 units/ml). The mixture was incubated at 37° for 18 hr, and the reaction was terminated by the addition of Laemmli buffer and boiling. The proteins were then separated by SDS–PAGE under reducing conditions, after which the gels were dried and subjected to autoradiography for 3–4 days at −80°. The apparent molecular weight of the BN/GRP receptor decreased from 75,000 to 45,000. These data suggest that 40% of the apparent mass of the BN/GRP receptor is amino-linked oligosaccharide.

Internalization Studies

The intracellular fate of bound [^{125}I-Tyr^4]BN was determined using Swiss 3T3 cells. [^{125}I-Tyr^4]BN was incubated with Swiss 3T3 cells for 2 hr at 4° in the presence or absence of competitor. Free ^{125}I-GRP was removed by washing with buffer. The cells were then divided into two groups: one group was incubated for an additional 30 min at 4° and a second group incubated at 37° for 30 min. Internalized versus cell surface ^{125}I-GRP was assessed by dissociation of the peptide with acid. The cells containing bound ^{125}I-GRP were treated with 0.5 ml of 0.5 *M* acetic acid, 0.15 *M* NaCl (pH 2.5) at 4°. After 5 min the supernatant was removed and counted in a gamma counter. This material represented acid-dissociable [^{125}I-Tyr^4]BN which was presumably bound to the cell surface. The pellet was dissolved in NaOH and was also counted in a gamma counter. This fraction represented acid-resistant [^{125}I-Tyr^4]BN, which was internalized.

Figure 2 shows that, at 4°, approximately 99% of the specifically bound ^{125}I-GRP was acid dissociable and hence may represent binding to cell surface receptors. In contrast, after incubation at 37° for 30 min, approximately 41% of the specifically bound [^{125}I-Tyr^4]BN was acid resistant because it had been internalized. Similar experiments with rat pituitary GH_4C_1 cells showed approximately 90% of the specifically bound [^{125}I-Tyr^4]BN was on the cell surface at 4°, whereas approximately 55% was internalized at 37°. Internalized [^{125}I-Tyr^4]BN was degraded to yield free ^{125}I-labeled tyrosine at 37° but not at 4°. Because this degradation of ligand was inhibited by chloroquine, it was postulated that [^{125}I-Tyr^4]BN was internalized to lysosomes.[10] Additional experiments suggested that the receptor–peptide complex was internalized. If the cells were treated with 1 n*M* unlabeled GRP for 30 min at 37° but not 4°, then treated with ^{125}I-GRP at 4° for 2 hr and cross-linked, the MW 65,000 band of radioactivity was absent (Fig. 1, lanes c and d). Under these conditions the unlabeled GRP may trigger the internalization of the high-affinity BN/GRP receptors so that there are no longer unoccupied receptors left on the plasma membrane to bind and subsequently cross-link ^{125}I-GRP.

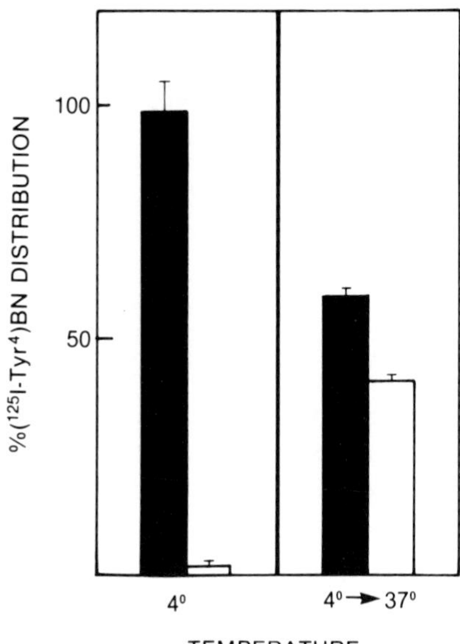

FIG. 2. Internalization of [^{125}I-Tyr^4]BN. [^{125}I-Tyr^4]BN was incubated with Swiss 3T3 cells for 2 hr at 4°, and the percentage of the specifically bound [^{125}I-Tyr^4]BN that was acid dissociable (solid bars) and not acid dissociable (open bars) was determined.

Phosphatidyl Inositol and Cytosolic Ca^{2+}

Ligand binding to BN/GRP receptors stimulates phosphatidylinositol turnover.[21] The inositol 1,4,5-trisphosphate produced may cause Ca^{2+} to be released from intracellular organelles such as the endoplasmic reticulum. SCLC cells were assayed for cytosolic Ca^{2+} as follows. One day after a medium change, SCLC cell line NCI-H345 is harvested and resuspended in HITES medium which contains 20 m*M* HEPES–NaOH (pH 7.4). Five milliliters of cells at a concentration of 2.5 × 10^6 cells/ml is incubated in flasks in the presence of 5 μM fura-2/AM.[22] After 30 min of shaking at 37° C, extracellular fura-2/AM is removed by centrifugation at 150 *g* for 10 min. The cells are resuspended at a concentration of 2.5 × 10^6 cells/ml and transferred to a spectrofluorimeter equipped with a magnetic stirring mechanism and 37° chamber. The excitation and emission wavelengths are 340 and 510 nm, respectively. The total volume is 2 ml, and

[21] J. P. Heslop, D. M. Blakeley, K. D. Brown, R. F. Irvine, and M. J. Berridge, *Cell* **47,** 703 (1986).

[22] G. Grynkiewicz, M. Poenie, and R. Y. Tsien, *J. Biol. Chem.* **260,** 3440 (1985).

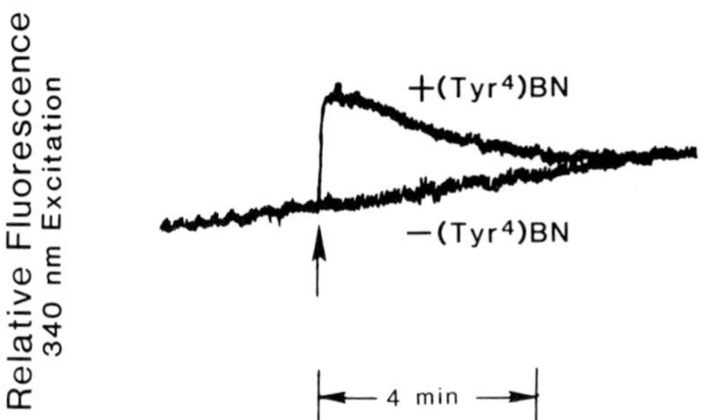

FIG. 3. Effect of [Tyr[4]]BN on cytosolic Ca^{2+} levels. Cell line NCI-H345 was loaded with fura-2, and the fluorescence intensity was determined in the absence and presence of 1 μM [Tyr[4]]BN.

concentrated peptide (10 μl in HITES medium) is rapidly added and the fluorescence intensity monitored for 4 min. All experiments are conducted within 1 hr after removal of the extracellular fura-2, as the sensitivity of the assay decreases after this time period.

The basal cytosolic free Ca^{2+} concentration of SCLC cell line NCI-H345 was 150 ± 40 nM.[23] After addition of a BN receptor agonist such as [Tyr[4]]BN, the fluorescence intensity rapidly increased within 15 sec to a steady-state value (Fig. 3), then slowly decreased. Similarly, those agonists which bound with high affinity such as BN, GRP, GRP (14–27), and Ac-GRP (20–27) caused elevation of the cytosolic Ca^{2+} (Table I), whereas [D-Trp[8]]BN and GRP (1–16) which have K_d values of greater than 10,000 nM did not alter the fluorescence intensity. Addition of the putative BN receptor antagonist [D-Arg[1],D-Pro[2],D-Trp[7,9],Leu[11]]substance P (100 μM) antagonized the ability to elevate the cytosolic Ca^{2+} levels. This assay can rapidly distinguish between BN receptor agonists, inactive compounds, and BN receptor antagonists.

In a related series of experiments, Swiss 3T3 cells were attached to Cytodex 2 beads, which float in aqueous solution, and the cytosolic Ca^{2+} levels were determined using the calcium-sensitive fluorescent dye quin2.[14] Thus in both Swiss 3T3 cells[14] and SCLC cells[23] BN/GRP receptors may regulate cytosolic Ca^{2+} levels. Because BN stimulates phosphatidyl inositol turnover, it also elevates diacylglycerol levels in Swiss 3T3 cells.[24]

[23] T. W. Moody, A. Murphy, S. Mahmoud, and G. Fiskum, *Biochem. Biophys. Res. Commun.* (in press).

[24] N. Tokuwa, Y. Tokuwa, W. E. Bollag, and H. Rassmussen, *J. Biol. Chem.* **262,** 182 (1987).

Growth Factors

BN functions as an autocrine growth factor in SCLC cells.[25] This was verified using a soft agarose clonal growth assay which is described below. SCLC cells (10^4) are washed 2 times in HITES medium to remove any residual serum and mixed with 0.3% agarose in culture medium. The cell viability is approximately 90% based on trypan blue exclusion. The cell suspension is then layered into 35-mm petri dishes containing a base layer of 0.5% agarose. Plates are examined microscopically to verify that only single cells were present initially. After 2 weeks, plates are scored for colony growth (cell aggregates of greater than 50 cells) using an inverted phase microscope.

Routinely, in HITES medium 30–80 colonies were present, whereas in serum-supplemented medium (RPMI-1640 which contained 10% heat-inactivated fetal bovine serum) 400–1300 colonies were observed, depending on the cell line used. Addition of exogenous BN (10 n*M*) increased the colony formation 8- to 12-fold for cell lines NCI-H345 and NCI-N592 using serum-free HITES medium, whereas BN had no effect on the number of colonies observed in the serum-supplemented medium.[26] It is possible that there are numerous other growth factors in the serum which stimulate the growth of the SCLC cells and mask the effect of BN. Because the inactive peptide [des-Leu^{13},Met^{14}]BN (10 n*M*) did not alter the colony formation rate for cell line NCI-H345 (Table II), the growth increase caused by BN is most likely mediated by BN receptors. The growth of SCLC cells was inhibited by an anti-BN monoclonal antibody which binds the carboxy terminus of BN or GRP such that the peptide was no longer able to bind to receptors.[25] Also, the BN receptor antagonist [D-Arg^1,D-Pro^2,D-Trp^{7,9},Leu^{11}]substance P (500 n*M*) reduced colony formation in the absence or presence of 10 n*M* BN using cell line NCI-N592.[27]

Stimulation of DNA synthesis by BN can be as assayed by measuring [^{3}H]thymidine uptake in Swiss 3T3 cells. Swiss 3T3 cells (1×10^5) are seeded in six-well plates containing 0.5 ml 10% fetal bovine serum in DMEM medium. After 7 days the cells are confluent, and each well contains approximately 6×10^5 cells. Each well is rinsed with 0.5 ml of DMEM medium before adding 0.5 ml of 0.5% fetal bovine serum in DMEM containing 2 μCi [^{3}H]thymidine with or without 10 μg/ml insulin.

[25] F. Cuttitta, D. N. Carney, J. Mulshine, T. W. Moody, J. Fedorko, A. Fischler, and J. D. Minna, *Nature (London)* **316,** 823 (1985).

[26] D. N. Carney, F. Cuttitta, T. W. Moody, and J. D. Minna, *Cancer Res.* **47,** 821 (1987).

[27] T. W. Moody, S. Mahmoud, A. Koros, F. Cuttitta, J. Willey, M. Rotsch, U. Zeymer, and G. Bepler, *Fed. Proc., Fed. Am. Soc. Exp. Biol.* **46,** 2201 (1987).

TABLE III
[³H]THYMIDINE UPTAKE IN SWISS 3T3 CELLS[a]

Compound	Uptake
Control	6,509 ± 261
Bombesin (1 n*M*)	52,509 ± 8,210
Bombesin (1 n*M*) + insulin (10 μg/ml)	683,480 ± 1,576
Fetal bovine serum (10%)	686,671 ± 3,934

[a] Mean value ± SE of two determinations is indicated.

Peptides (10 μl) are added in DMEM medium. After incubation at 37° for 28 hr, the trichloroacetic acid (TCA)-insoluble radioactivity is determined.[28] The cells are rinsed 3 times with PBS, and then 5% TCA is added. After 5 min at 4° the DNA is precipitated, and the TCA-insoluble material is washed with ethanol–diethyl ether (2 : 1, v/v). The precipitate is dried at 25°, and the cells are extracted with 0.5 *M* NaOH (0.5 ml). The extract is neutralized with 50 μl of 5 *M* HCl, scintillation fluid is added, and the samples are counted in a beta counter.

Table III shows that in the absence of mitogen, the baseline [^{3}H]thymidine incorporation was approximately 6500 cpm/well. Bombesin (1 n*M*) increased the radioactivity associated with the DNA approximately 8-fold, whereas insulin (10 μg/ml) slightly increased the [^{3}H]thymidine uptake relative to control values. Bombesin plus insulin, however, were synergistic, and thymidine uptake increased approximately 100-fold. Serum, which contains numerous growth factors, also increased the [^{3}H]thymidine uptake approximately 100-fold. Serum (10%) plus BN or serum plus BN and insulin did not increase the [^{3}H]thymidine uptake relative to serum alone. The data indicate that BN, especially in the presence of insulin, increased DNA synthesis in Swiss 3T3 cells to near maximal levels. This assay is particularly effective in Swiss 3T3 cells because they become arrested in the G_1/G_0 phase of the cell cycle when they deplete the nutrient medium of its growth-promoting activity. Addition of fresh serum or growth factors reinitiates DNA synthesis and cell division. The half-maximal effective dose for BN to stimulate radioactive thymidine uptake was 0.3 n*M*, and the uptake of ^{3}H-thymidine stimulated by BN was inhibited by [D-Arg^1,D-Pro^2,D-Trp^{7,9},Leu^{11}]substance P.[28,29] For the assay to work properly, however, the Swiss 3T3 cells must be quiescent prior to the addition of radioactive thymidine.

[28] A. N. Corps, L. H. Rees, and K. D. Brown, *Biochem. J.* **231,** 781 (1985).

[29] E. Rozengurt and J. Sinnett-Smith, *Proc. Natl. Acad. Sci. U.S.A.* **80,** 2936 (1983).

[36] Isolation and Identification of Neuroendocrine Peptides from Milk

By ELI HAZUM

Introduction

Milk, which is a mammal-specific biological fluid, is the first and sole food of the neonate for a certain period of time. Therefore, milk constituents are of great importance and have evolved differently, both in quality and quantity, to meet the specific requirements of neonatal development in various mammals. The composition of milk changes with age, stage of lactation (colostrum, transitional, and mature milk), environment, etc. Mature human milk contains 87.9% water, 0.9% protein (of which 40% is casein), 4.0% fat, 7% lactose, and 0.2% ash, whereas mature bovine milk contains more (3.3%) protein (of which 82% is casein) and less (4.7%) lactose.[1,2] The milk constituents can be synthesized in the mammalian glandular tissues from precursors in blood, they can be transferred directly from blood to milk through the cell membranes of the glandular tissue, or both.

Several neuropeptides, neurotransmitters, and hormones are present in milk from humans and other mammals.[3–7] Their concentrations in milk are significantly higher than that found in plasma, suggesting an active concentration mechanism in the mammary gland. In suckling mammals, the hormones and neuropeptides are absorbed through the gastrointestinal tract, and, when ingested by the neonate, they appear intact in the plasma. This absorption is age dependent and could have physiological and pharmacological significance in neonatal development.

The uniqueness of milk as a food for the neonate has resulted in the biochemical aspects of milk being extensively investigated since the early eighteenth century. The presence of hormones in milk was described 50 years ago, but only in the last decade has this subject been explored,

[1] B. Blanc, *World Rev. Nutr. Diet.* **36,** 1 (1981).

[2] J. T. Wilson, R. D. Brown, D. R. Cherek, J. W. Dailey, B. Hilman, P. C. Jobe, B. R. Manno, J. E. Manno, H. M. Redetzki, and J. J. Stewart, *Clin. Pharmacokin.* **5,** 1 (1980).

[3] T. Richardson and N. Mattarella, *J. Food Protect.* **40,** 57 (1977).

[4] A. T. Cowie and J. K. Swinburne, *Dairy Sci. Abstr.* **39,** 391 (1977).

[5] O. Koldovsky, *Life Sci.* **26,** 1833 (1980).

[6] E. Hazum, *Trends Pharmacol. Sci.* **4,** 454 (1983).

[7] V. Strbak, *in* "The Role of Maternal Milk in Endocrine Regulation of Sucklings." Vega, Bratislava, Czechoslovakia, 1985.

owing to the development of highly efficient procedures for the isolation of peptides and the methodological advances in hormone assays. The list of hormones and neuroendocrine peptides that have been found in milk is quite extensive.[3–7] Recently discovered peptides in milk include the following: gonadotropin-releasing hormone and thyrotropin-releasing hormone,[8–10] growth hormone-releasing hormone,[11] somatostatin,[12] vasoactive intestinal peptide,[13] bombesin,[14] prolactin,[15] opioid peptides and morphine,[16–18] neurotensin,[19] pro-γ-melanotropin,[19] δ-sleep-inducing peptide,[20] and epidermal growth factor.[21]

This chapter describes a general methodology to isolate neuroendocrine peptides from milk; the method of choice depends on the chemical and physical characteristics of the peptide of interest. This approach involves extraction of peptides from milk, a pre-HPLC concentration procedure, and HPLC purification of the neuroendocrine peptides. Purification steps are usually followed by a reliable and sensitive hormone assay such as biological activity, binding assay, and radioimmunoassay.

Extraction and Purification of Neuroendocrine Peptides

Extraction of Peptides from Milk

The efficient extraction of peptides from milk is conducted in acidic methanol. This extraction procedure has two major advantages: (1) precipitation of most proteins (especially casein), so that an enriched peptide extract is obtained, and (2) inactivation of endogenous proteases.

[8] T. Baram, Y. Koch, E. Hazum, and M. Fridkin, *Science* **198,** 300 (1977).

[9] A. K. Sarda and R. M. G. Nair, *J. Clin. Endocrinol. Metab.* **52,** 826 (1981).

[10] T. Amarant, M. Fridkin, and Y. Koch, *Eur. J. Biochem.* **127,** 647 (1982).

[11] H. Werner, T. Amarant, M. Fridkin, and Y. Koch, *Biochem. Biophys. Res. Commun.* **135,** 1084 (1986).

[12] H. Werner, T. Amarant, R. P. Millar, M. Fridkin, and Y. Koch, *Eur. J. Biochem.* **148,** 353 (1985).

[13] H. Werner, Y. Koch, M. Fridkin, J. Fahrenkrug, and I. Gozes, *Biochem. Biophys. Res. Commun.* **133,** 228 (1985).

[14] G. D. Jahnke and L. H. Lazarus, *Proc. Natl. Acad. Sci. U.S.A.* **81,** 578 (1984).

[15] P. V. Malven, *J. Anim. Sci.* **46,** 609 (1977).

[16] A. Henschen, F. Lottspeich, V. Brantl, and H. Teschemacher, *Hoppe-Seyler's Z. Physiol. Chem.* **360,** 1217 (1979).

[17] K. J. Chang, A. Killian, E. Hazum, P. Cuatrecasas, and J. K. Chang, *Science* **212,** 75 (1981).

[18] E. Hazum, J. J. Sabatka, K. J. Chang, D. A. Brent, J. W. A. Findlay, and P. Cuatrecasas, *Science* **213,** 1010 (1981).

[19] R. Ekman, S. Ivarsson, and L. Jansson, *Regul. Pept.* **10,** 99 (1985).

[20] M. V. Graf, C. A. Hunter, and A. J. Kastin, *J. Clin. Endocrinol. Metab.* **59,** 127 (1984).

[21] G. Carpenter, *Science* **210,** 198 (1980).

Milk Extracts. Powdered milk or lyophilized skim milk is stirred in one original volume of a mixture of acetic acid and methanol (1 : 10) for 24 hr at room temperature. Following filtration through Whatman folded filters (15.0 cm), the filtrate is evaporated under reduced pressure. The oily residue is centrifuged (10 min, 8000 *g*), and the crude milk extract is stored at −20° until a further purification step is performed. In case methionine or cysteine residues are expected to be present in the sequence, it is advised to add reducing agents, e.g., 2-mercaptoethanol or dithiothreitol, to avoid oxidation.

Pre-HPLC Purification Steps

Following extraction, initial purification steps of the milk extract can be accomplished using various methods. If a suitable antibody is available, affinity chromatography or immunoprecipitation of the crude extract may greatly enhance the purification of the peptide of interest. In our laboratory we have applied gel filtration chromatography or Sep-Pak cartridges as the first method of choice for purification.

Gel Filtration Chromatography. Gel filtration chromatography will result in an enriched peptide fraction with the appropriate molecular weight, thus removing salts and color pigments as well as other peptides and proteins. Typically, crude milk extract (equivalent to 500 ml milk) is loaded onto a previously equilibrated Sephadex G-25 or G-50 (Pharmacia) column (5 × 100 cm) and eluted with 1 *M* acetic acid. Fractions are assessed for optical density (at 210/280 nm) and activity. All active fractions are combined, lyophilized to dryness, and redissolved in the appropriate buffer for HPLC.

Sep-Pak Cartridges. The Sep-Pak method is used for small quantities of milk extracts. Aliquots of milk extracts are diluted 1 : 1 with 1% TFA (trifluoroacetic acid) and loaded onto a reversed-phase C_{18} cartridge (Sep-Pak, Waters Associates). Next, five successive fractions are collected using 1-ml rinses of this solvent each time. Finally, 2 ml of 80% acetonitrile in 0.1% TFA is applied. The fractions are then evaporated to dryness using a Speed Vac concentrator, redissolved in a small volume of buffer A of the next HPLC step (see below) and checked for activity. This procedure is rapid and can be repeated several times before application onto HPLC columns.

HPLC Purification

High-performance liquid chromatography (HPLC) is the most powerful method available for peptide purification and usually results in high recovery (>90%). In most studies with neuroendocrine peptides a reversed-phase column (μBondapak C_{18}, Waters Associates) has been used. The column is equilibrated with a solution of 0.1% TFA. The active fractions (after pre-HPLC purification steps) are applied and eluted using a linear gradient (from 0 to 80%) of acetonitrile in 0.1% TFA. Several other column types have been utilized in our laboratory, and elution was performed with gradients from 100% buffer A to 100% buffer B:

1. A nonpolar column (Lichrosorb RP-18, Merck), the elution buffers as follows: buffer A, 0.3% TFA in water; buffer B, 0.3% TFA in 20% 1-propanol.
2. A hydrophilic column (Lichrosorb Diol, Merck), the elution buffers being the following: buffer A, 1-propanol; buffer B, water.
3. An anion-exchange column (Mono Q, Pharmacia), with the elution buffers as follows: buffer A, 10 m*M* sodium phosphate, pH 7; buffer B, 0.3 *M* NaCl in buffer A.
4. A cation-exchange column (Mono S, Pharmacia), the elution buffers being the following: at pH 7, buffer A, 10 m*M* sodium phosphate and buffer B, 0.3 *M* NaCl in buffer A; at pH 5, buffer A, 20 m*M* acetate buffer and buffer B, 0.5 *M* NaCl in buffer A; at pH 2, buffer A, 0.38 *M* HCOOH and buffer B, 0.5 *M* NaCl in buffer A.

Summary

Hormone, hormone-like substances, and neuroendocrine peptides are natural constituents of milk, and they may have an important physiological and pharmacological role in neonate development. The concentration of these peptides in milk greatly exceeds those in serum and implies an active concentration mechanism in the mammary gland. The large quantities in which milk can be supplied and the development of highly efficient procedures for the purification of peptides suggest that milk may prove to be an excellent source for identifying as yet "unknown" hormones and neuroendocrine peptides.

Acknowledgments

I am grateful to Mrs. M. Kopelowitz for typing the manuscript and to Profs. C. Webb and M. Rubinstein for their useful suggestions. This work was supported by the U.S.–Israel Binational Science Foundation, the Fund for Basic Research administered by the Israel Academy of Sciences and Humanities, and the Minerva Foundation, Federal Republic of Germany.

[37] Passive Immunoneutralization: A Method for Studying the Regulation of Basal and Pulsatile Hormone Secretion

By MICHAEL D. CULLER and ANDRÉS NEGRO-VILAR

Introduction

During the last 20 years, it has become firmly established that all the anterior pituitary hormones, as well as many extrapituitary hormones, are secreted in a rhythmic, pulsatile manner. With this discovery has come the realization that the physiologic regulation of pituitary hormone secretion is far more complex than previously imagined. It is now believed that the basic pulses of a pituitary hormone's secretory pattern are the cumulative responses to stimulatory and/or inhibitory hypothalamic factors. Numerous secondary factors, both from the brain and from various target organs in the periphery, superimpose their influences on the basic pulse pattern to induce both subtle and dramatic alterations to the pattern in order to meet the needs of a given physiologic situation. One of the most effective approaches used to determine the contribution of a regulatory factor to the secretion of a pituitary hormone is to eliminate the factor, preferably without altering other regulatory elements, and determine either the effect on the normal secretory pattern or the manner in which the response to a given physiologic challenge is altered. Once the nature of an abolished factor's contribution is established, the next study, ideally, is to exogenously replace the missing factor to try to restore the normally observed secretory pattern or response. Various replacement regimens and patterns of administration may then also be tested to determine how the respective factor can be used to control and/or manipulate the particular aspect of secretion that it controls.

The first and surely one of the most elegant approaches to the study of pulsatile luteinizing hormone (LH) secretion was that used in the studies from the laboratory of Knobil (for a review, see Ref. 1). In these studies, hypothalamic luteinizing hormone-releasing hormone (LHRH) secretion was abolished by lesioning the hypothalamus and destroying the neurons which produce this peptide factor in order to determine its role in pulsatile gonadotropin secretion. Exogenous LHRH was then administered to study the effects of various pulsatile replacement regimens in stimulating and maintaining the pulsatile secretion of the gonadotropins. The information gained from these studies clearly illustrates the importance of

[1] E. Knobil, *Recent Prog. Horm. Res.* **36,** 53 (1980).

pulsatile hormone secretion in regulating the function of the target organ in both physiological and pathological situations. The only major disadvantage with this type of study is that by lesioning the brain, the secretion of not only the factor of choice but also of many other pituitary regulating factors, which may influence the physiological response under observation, is abolished or altered.

An alternative method of eliminating the influence of an endogenous factor that has been used to study the contribution of LHRH to pulsatile gonadotropin secretion[2] is the administration of a potent, long-lasting antagonist to block the biological receptors of the factor being studied. While this approach has the desired selectivity (assuming the antagonist is specific), it lacks the advantages of the Knobil model because replacement of the abolished biological activity is impossible while the biological receptors are occupied by the antagonist.

Another alternative is immunoneutralization, the use of antibodies or antisera to "eliminate" an endogenous factor within an organism. As is discussed in detail throughout the remainder of this chapter, immunoneutralization combines the advantages of both previously described methods by providing selectivity and specificity in eliminating an endogenous factor while still allowing the possibility of exogenously replacing the desired biological activity.

Passive versus Active Immunoneutralization

Immunoneutralization can either be active, in which an animal is sensitized to produce antibodies against an endogenous factor, or passive, in which an antiserum or antibody preparation generated in one animal is injected into another animal to neutralize a specific factor. While active immunoneutralization has been used to study the regulation of anterior pituitary secretion,[3,4] passive immunoneutralization is by far a more convenient procedure and offers a number of distinct advantages. Both methods are potentially highly specific and selective, removing only the desired factor without affecting other regulatory components. These requirements are much more easily assured, however, by passive immunoneutralization in which the antiserum to be utilized can be fully characterized and standardized. When active immunoneutralization is employed, the antiserum produced by each individual animal is different and must be individually characterized, thus greatly increasing the variation among

[2] L. V. DePaolo, *Endocrinology (Baltimore)* **117,** 1826 (1985).
[3] H. M. Fraser, A. Gunn, S. L. Jeffcoate, and D. T. Holland, *J. Endocrinol.* **63,** 399 (1974).
[4] I. J. Clarke, H. M. Fraser, and A. S. McNeilly, *J. Endocrinol.* **78,** 39 (1978).

the treated animals. Passive immunoneutralization also offers the advantages of causing minimal trauma to the animal, acting rapidly (thus allowing control of the onset of the neutralization effect), and of being cleared gradually from the circulation and thereby removing the neutralization effect. If, however, very long-term or permanent neutralization of an endogenous factor is required then active immunoneutralization is a more practical approach. One obvious limitation of passive immunoneutralization, as compared with active, is that in studies involving large animals, such as sheep and cattle, the quantity of exogenous antiserum necessary for passive immunoneutralization can be prohibitive. There have been studies, nevertheless, in which passive immunoneutralization has been successfully used to study the role of LHRH in LH and follicle-stimulating hormone (FSH) secretion in the sheep.[5,6]

Use of Passive Immunoneutralization for the Study of Pulsatile Hormone Secretion

The use of passive immunoneutralization to eliminate an endogenous factor is not new[7,8] and, in fact, predates the discovery that pituitary hormone secretion is pulsatile in nature. Soon after the elucidation of the structure of the first hypothalamic factors that regulate the pituitary, passive immunoneutralization began to be used as a means of determining their physiological function and significance.[9,10] Only recently, however, has immunoneutralization been used to determine the contribution of a given factor to pulsatile pituitary hormone secretion. Immunoneutralization was first combined with analysis of the secretory pattern by Ferland *et al.*[11] to study the role of somatostatin in the regulation of GH (growth hormone) secretion in the rat. Subsequent studies have also combined the techniques to study the role of LHRH in pulsatile LH and/or FSH secretion[5,6,12–14] and GHRH and/or somatostatin in pulsatile GH secretion.[15–17]

[5] G. A. Lincoln and H. M. Fraser, *Biol. Reprod.* **21,** 1239 (1979).

[6] H. M. Fraser and A. S. McNeilly, *J. Reprod. Fertil.* **69,** 569 (1983).

[7] W. A. Kelly, H. A. Robertson, and D. A. Stansfield, *J. Endocrinol.* **27,** 127 (1963).

[8] K. A. Laurence, O. Carpuk, and O. Wahby, *Excerpta Med. Int. Congr. Ser.* **99,** E145 (1965).

[9] Y. Koch, P. Chobsieng, U. Zor, M. Fridkin, and H. R. Lindner, *Biochem. Biophys. Res. Commun.* **55,** 623 (1973).

[10] A. Arimura, L. Debeljuk, and A. V. Schally, *Endocrinology (Baltimore)* **95,** 323 (1974).

[11] L. Ferland, F. Labrie, M. Jobin, A. Arimura, and A. V. Schally, *Biochem. Biophys. Res. Commun.* **68,** 149 (1976).

[12] G. B. Ellis, C. Desjardins, and H. M. Fraser, *Neuroendocrinology* **37,** 117 (1983).

[13] M. D. Culler and A. Negro-Vilar, *Endocrinology (Baltimore)* **118,** 609 (1986).

[14] M. D. Culler and A. Negro-Vilar, *Endocrinology (Baltimore)* **120,** 2011 (1987).

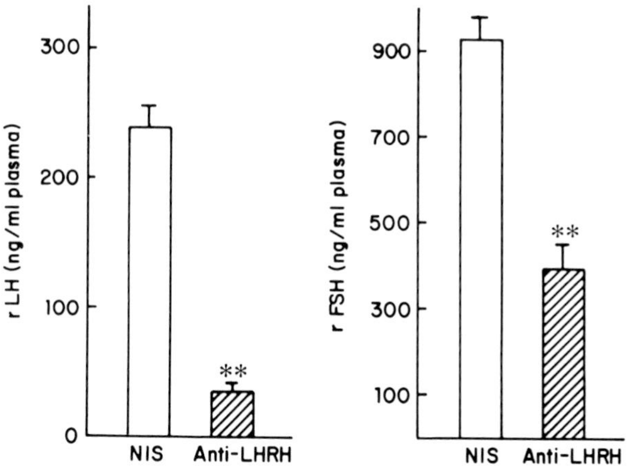

FIG. 1. Suppressive effect of ovine anti-LHRH antiserum (A-772; generous gift of Dr. A. Arimura) as compared with nonimmune ovine serum, on plasma LH and FSH in single blood samples taken 24 hr after treatment. **, $p < 0.01$. (M. D. Culler and A. Negro-Vilar, unpublished observations.)

While immunoneutralization provides a great deal of valuable information, the coupling of immunoneutralization with frequency sampling techniques that help to define the secretory pattern has revealed a new dimension of subtleties and interactions that are unobservable using either methodology alone. An example is illustrated in Fig. 1 that pertains to the role of LHRH regulating FSH secretion. The depicted results are from experiments that preceded pulsatility studies and clearly exemplify early immunoneutralization studies in which only single or infrequent samples were collected after injection of the antiserum. The results in Fig. 1 demonstrate that immunoneutralization of endogenous LHRH in the castrate rat causes a suppression of both LH and FSH secretion within 24 hr. The obvious conclusion is that LHRH is a (the?) physiological stimulator of both LH and FSH secretion. This conclusion, however, represents the extent of the information gained.

By taking frequent samples from chronically cannulated, unanesthetized, freely moving rats, in order to analyze the pattern of FSH secre-

[15] W. B. Wehrenberg, P. Brazeau, R. A. Luben, P. Böhlen, and R. Guillemin, *Endocrinology (Baltimore)* **111,** 2147 (1982).

[16] W. B. Wehrenberg, P. Brazeau, R. A. Luben, N. Ling, and R. Guillemin, *Neuroendocrinology* **36,** 489 (1983).

[17] P. M. Plotsky and W. Vale, *Science* **230,** 461 (1985).

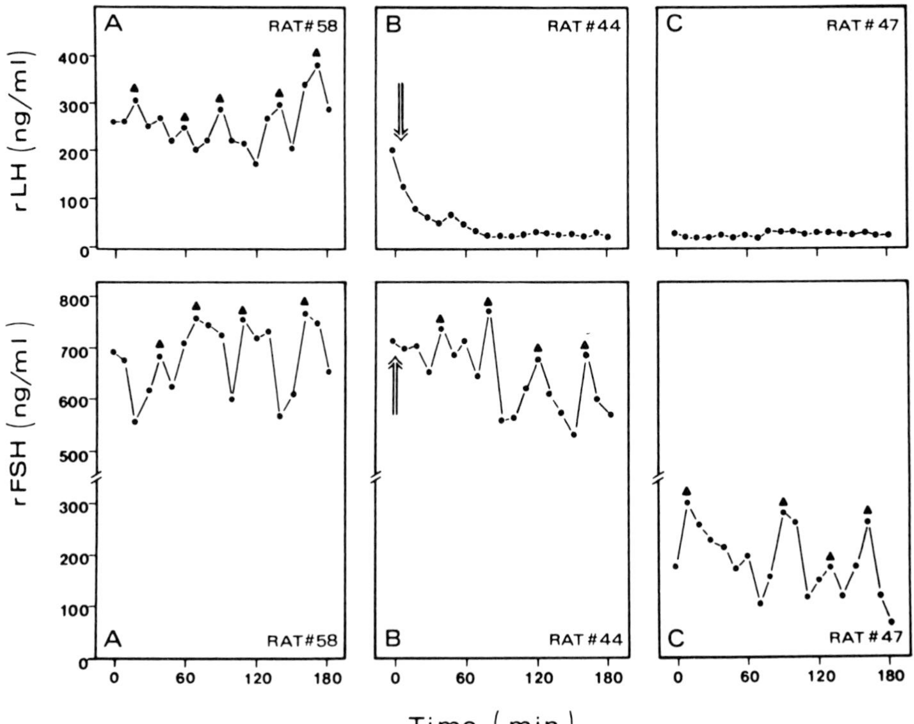

FIG. 2. Representative plasma LH and FSH patterns from chronically cannulated, unanesthetized, unrestrained castrate male rats after (A) injection of 500 μl nonimmune ovine serum, (B) 500 μl ovine anti-LHRH serum (A-772) immediately after withdrawing the first blood sample, and (C) 500 μl anti-LHRH serum 24 hr before initiation of sampling. Blood samples (200 μl) were withdrawn every 10 min for 4 hr. Arrows indicate times of injection. ▲, Pulse defined as a rise in plasma LH or FSH that is greater than twice the radioimmunoassay intraassay coefficient of variation at a comparable mean level of LH or FSH and that is followed by a trough. (From Culler and Negro-Vilar,[13] with permission.)

tion before and after immunoneutralization of endogenous LHRH (Fig. 2), the regulation of FSH secretion is found to be far more complex than previously suspected.[13,14] As illustrated in Fig. 2A, both LH and FSH are secreted in a pulsatile manner in either untreated or nonimmune serum-injected castrate male rats. Administration of anti-LHRH serum is immediately followed by an abolition of pulsatile LH secretion and a dramatic lowering of mean LH levels that results in a low, flat secretory pattern (Fig. 2B) which is still evident 24 hr after treatment (Fig. 2C). In contrast, the secretion of FSH is relatively unaffected within the first 3 hr after administration of the antiserum. Twenty-four hours after treatment with the antiserum, the mean, basal level of FSH has fallen significantly;

however, the pulsatile secretion of FSH, in terms of pulse frequency and amplitude, remains unaffected. Thus, not only has defining the secretory profile of FSH before and after immunoneutralization demonstrated a major difference between LH and FSH in the time course of the response to LHRH elimination, but strong evidence has been provided for an additional, LHRH-independent factor that regulates pulsatile FSH secretion.[13,14] This example clearly demonstrates that immunoneutralization, combined with frequent sampling techniques to analyze the pattern of secretion, represents an extremely powerful model for unraveling the mechanisms controlling pituitary hormone secretion.

Obstacles in Using Passive Immunoneutralization

While passive immunoneutralization is the most versatile and useful means of removing an endogenous factor, there are several drawbacks to the technique which should be considered. First, *in vivo* immunoneutralization of a factor requires a reasonably large quantity (up to several milliliters per animal in the case of the rat) of high-titer antiserum or purified antibodies; the exact amount depends on the titer and the ability of the antibodies to prevent the substance from interacting with its target receptors. Obtaining sufficient amounts of suitable antibodies or antisera for a given experiment can be extremely difficult or costly unless the researchers generate the antiserum themselves or have access to generous quantities of the material. The methods of generating antibodies, both polyclonal and monoclonal, are far too numerous and replete with individual modifications and idiosyncrasies to be considered here; however, reviews and descriptions of a number of basic antibody-generating techniques may be found in earlier volumes of this series.[18–20] Procuring sufficient quantities of certain factors (especially larger and noval peptides) in order to generate an antiserum can also be difficult and expensive.

A second major problem is that not all antisera are usable for immunoneutralization despite their having a significantly high antibody titer as demonstrated by radioimmunoassay or immunological methods. The reasons that certain antisera work poorly for immunoneutralization are unclear. Possible explanations are that the antibodies have lower affinities for the antigen than the biological receptors with which the antigen interacts or that the antibodies recognize and bind to the factor in such a way that the interaction of the biologically active portion of the factor with its receptor is not blocked.

[18] H. Van Vunakis and J. J. Langone (eds.), this series, Vol. 70.

[19] J. J. Langone and H. Van Vunakis (eds.), this series, Vol. 92.

[20] J. J. Langone and H. Van Vunakis (eds.), this series, Vol. 121.

Another possible reason why certain antisera are unusable for immunoneutralization is the presence of substances in the antiserum which counteract the immunoneutralizing properties of the antibodies either by interfering with the antibodies' ability to bind to the antigen or by introducing the very biological activity meant to be neutralized by the antiserum. The source of the interfering biological activity may include a large amount of the factor meant to be neutralized that is present in the antiserum. This may be quite common since the antiserum is produced by active immunization of the donor animal. If there is sufficient sequence similarity between the factor used as an antigen and the donor animal's own endogenous factor, then active immunoneutralization may occur. As the endogenous substance is neutralized, a common homeostatic response would be an increased secretion of either the factor itself or of other factors with similar biological activity in order to replace the deficiency. Thus, in addition to saturating many of the antibody binding sites with the endogenous factor, there may also be significant quantities of the free factor or of similarly acting factors present in the serum.

Another potential source of interfering biological activity is the production of anti-idiotypic antibodies in which a population of antibodies is generated against the binding site of the antibodies that recognize the antigen. Since the binding site of an antibody is believed to be composed of a peptide sequence that is complementary to the structure of the antigen, an antibody binding site produced against this site could conceivably contain a peptide sequence which would mimic the structure of the antigen.[21] If the similarity is sufficient, then theoretically the anti-idiotypic antibodies also could activate the biological receptors that normally interact with the antigen.

The presence of significant quantities of biologically active factors or of anti-idiotypic antibodies may explain, at least partially, why the researcher occasionally will encounter an antiserum that stimulates the biological activity meant to be neutralized or that partially immunoneutralizes at low concentrations but stimulates at higher concentrations. The need for initial screenings and preliminary trials of antiserum to be used for immunoneutralization cannot be overemphasized.

Validation of Immunoneutralization Experiments

Assuring Sufficient Antibody for Neutralization. While certain studies may not require complete blockade of an endogenous factor, most do in

[21] C. A. Bona and H. Kohler, *in* "Monoclonal and Anti-Idiotypic Antibodies: Probes for Receptor Structure and Function" (J. C. Venter, C. M. Fraser, and J. Lindstrom, eds.), p. 141. Liss, New York, 1984.

order to define the extent of its contribution to the physiologic response. One of the most important considerations for an immunoneutralization study must be the demonstration that the endogenous factor is completely neutralized for the required length of time. The approach to this problem depends on the needs of the particular study. If a study is designed to examine only the immediate effects of removing an endogenous factor, then a preliminary experiment testing various volumes of the antiserum (i.e., various amounts of antibody), to establish the minimal volume required to induce a maximal effect, is probably adequate to insure the presence of sufficient antibodies. This can be accomplished by intravenous (i.v.) or intraperitoneal (i.p.) injection of the antiserum and evaluation of the biological response under consideration at the appropriate time for the study. In studies in which a factor that regulates pituitary secretion is to be immunoneutralized, evaluation of the response usually consists of collecting blood at one or several appropriate time points and measuring the pituitary hormone under study by radioimmunoassay and/or bioassay.

If the immunoneutralization is to be more chronic in nature, then, in addition to establishing the minimal effective volume and a time course of the effects, an estimation of the rate of loss of available antibody should be made.

After injection, the antibodies will gradually lose their ability to neutralize. This is due to the combined effects of degradation, clearance from the circulation, and filling of available binding sites with the endogenous factor. Determining the rate at which the active, available antibodies are lost can be performed at the same time as determining the time course of effects. After injection of a known amount of antiserum, samples are taken at appropriate times to determine the antibody titer in the blood of the recipient animals. Ideally, any experimental procedures which may affect the availability of active antibodies should be performed as they will be executed during the actual study. This is particularly important in the assessment of the pattern of hormone secretion in which many blood samples are withdrawn over a relatively short period of time and a significant amount of the neutralizing antibodies may be removed.

An example of determining the rate at which antibody availability is lost is illustrated in Fig. 3. In this instance, antiserum against LHRH was injected either immediately or 24 hr before the initiation of the frequent bleeding period. After the bleeding period (4 hr of withdrawing 200-μl blood samples every 10 min) an additional sample of blood was withdrawn, and the titer of the LHRH antibody in the plasma was determined by the ability to bind ^{125}I-labeled LHRH. By comparing the antibody titer of the samples with the titer of the original antiserum used for immunoneutralization, an estimate of the rate of loss can be made as well as an estimate of the amount of antiserum that must be injected in order to

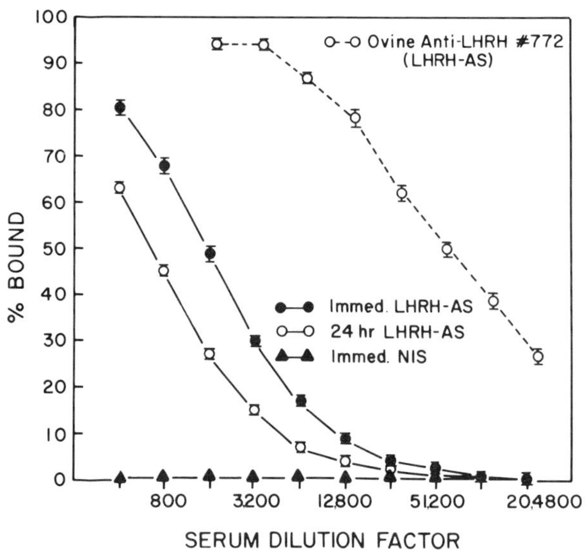

FIG. 3. ^{125}I-Labeled LHRH binding titers of rat plasma taken after a 4-hr sampling period from rats injected with nonimmune sheep serum or an ovine anti-LHRH serum (A-772) at the initiation of the sample period or with an ovine anti-LHRH serum 24 hr before the initiation of sampling as compared with the binding titer of the ovine anti-LHRH serum itself. Each point with brackets (except anti-LHRH serum alone, $n = 3$) represents the mean ± SEM of eight rats. (From Culler and Negro-Vilar,[14] with permission.)

ensure that more than the minimal effective concentration of antibodies will be present at the time of the experiment.

Completeness of Neutralization. Demonstrating the completeness of immunoneutralization, i.e., that the antiserum completely (or sufficiently) prevents the endogenous factor from interaction with its biological receptor and thereby stimulating some degree of response, can be a relatively easy or a very difficult task, depending on the particular factor under study. If synthetic antagonistic analogs of the factor are known and available, as is the case for several of the well-characterized hypothalamic peptides (LHRH, somatostatin, corticotropin-releasing factor, oxytocin, vasopressin), then demonstrating complete neutralization is relatively easy. Because the antagonists block the biological receptors, the injection of an antagonist when immunoneutralization is incomplete should produce a further reduction in the biological response. As illustrated in Fig. 4, to demonstrate the completeness of an anti-LHRH serum blockade, the LHRH antagonist, [D-pGlu^1,D-Phe^2,D-Trp^{3,6}]LHRH, was injected into previously (2 or 24 hr earlier) immunoneutralized rats at a concentration sufficient to completely block LH secretion. Because no

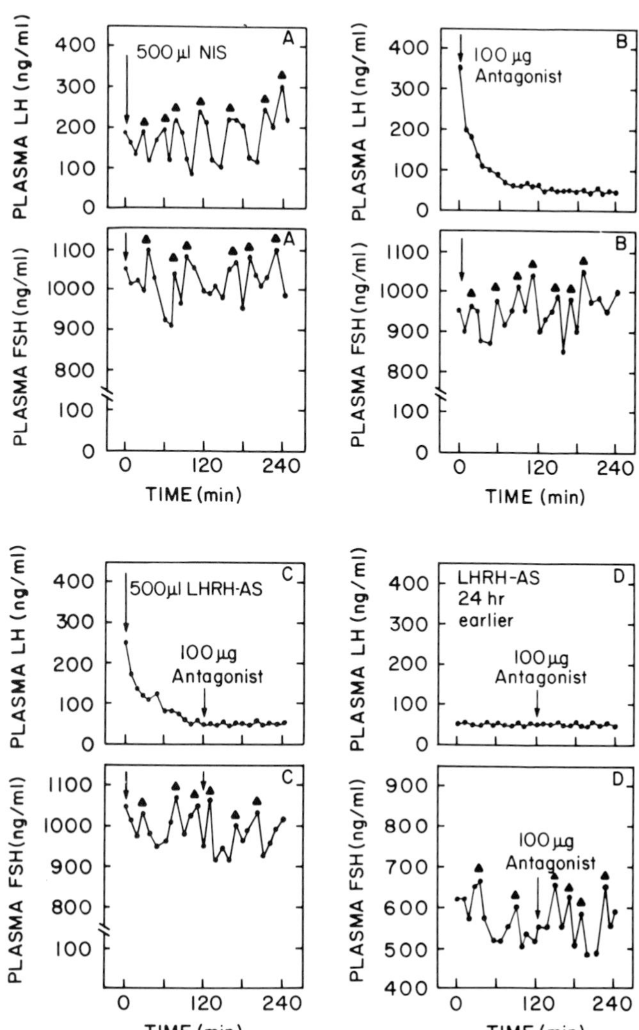

FIG. 4. Representative plasma LH and FSH patterns from unanesthetized, freely moving castrate male rats after (A) injection of nonimmune ovine serum (normal castrate gonadotropin secretion pattern), (B) injection of 100 μg LHRH antagonist [D-pGlu[1],D-Phe[2],D-Trp[3,6]]LHRH, (C) injection of 500 μl ovine anti-LHRH serum (A-772) at the initiation of sampling and 100 μg LHRH antagonist 2 hr later, and (D) injection of 500 μl ovine anti-LHRH serum 24 hr before and 100 μg LHRH antagonist 2 hr after the initiation of sampling. Arrows indicate times of injection. ▲, Pulse as defined in Fig. 2. (From Culler and Negro-Vilar,[14] with permission.)

further reduction in LH or FSH secretion occurred with the combination of antiserum and antagonist than had been observed with antiserum alone, the antiserum blockade was considered to be complete.

If antagonistic analogs are not available, then the problem of ascertaining the completeness of neutralization becomes more difficult. One alternative is to use a biological event other than the parameter under study to test the completeness of the antiserum blockade. The biological function used to test the blockade must be completely dependent on the factor that is to be abolished. If the event can be completely blocked by the amount of antiserum that is to be used in the actual study, then this is highly suggestive of a complete blockade. One example is the inhibition of suckling-induced milk ejection by a given concentration of oxytocin antibodies. The chief criticism of this method is that, unless demonstrated, there is no assurance that the biological receptors regulating the event to be studied have the same characteristics as those mediating the event used to verify neutralization. It may be argued that the threshold level of the endogenous factor under study, needed for a biological response, may be different for the two events. If data regarding the similarity of the receptors is unavailable and/or impractical to obtain, then it is best to help satisfy this legitimate concern by coupling this validation method with one of the other methods described in this chapter.

If the factor to be studied is available in a biologically active form that is recognized by the antiserum, then the completeness of neutralization can be tested by injecting a dose of the material normally considered sufficient to evoke a biological response. If the response is blocked, then this can be considered as reasonable evidence for a complete blockade. The problem with this test method is that not completely blocking the biological response to the "exogenous material" does not necessarily indicate an incomplete blockade of the endogenous factor. When exogenous factors are administered, especially in instances when hypothalamic–pituitary interactions are being studied, the amount of material that must be injected is relatively large compared with the amount of material normally secreted. The close proximity and relatively confined space of the hypophyseal portal system normally allows a relatively small amount of material (in terms of actual mass), secreted from the hypothalamus, to evoke a biological response. When administered from a distant site, such as the jugular vein where the material must travel through the heart before reaching the pituitary region, a comparatively large mass of material is required to assure that the factor will reach the pituitary in sufficient concentration to cause a response. When injected as a bolus, the exogenous material, at least initially, will travel through the circulation as a bolus. Thus, the concentration of antibodies within the segment of the bloodstream that contains this bolus of exogenous material may be inadequate to bind the mass injected, and sufficient material to elicit a biological response may still reach the pituitary.

If none of the above options are possible, or as an additional test for one of the less convincing methods, a second injection of antiserum can be administered at the end of the experimental period and monitoring continued to see if there is any further response. The problem with this approach is that high amounts of the antiserum may promote stimulating effects (as discussed earlier). A second drawback occurs when the effects of immunoneutralization do not immediately manifest themselves, such as with the immunoneutralization of endogenous LHRH on FSH secretion where several hours must pass before a significant effect is observed. In certain experiments it may not be feasible to continue monitoring the response for a sufficient period of time to test the completeness of the antiserum blockade owing to other experimental constraints (e.g., excessive bleeding may cause hypovolemia and, in turn, trigger the secretion of hormones such as ACTH and AVP).

Problems and Potential Solutions

Interference with Radioimmunoassays. Passive immunoneutralization of an endogenous factor can provide a wealth of information about the contribution of that factor to a physiologic response or function. It can also create several problems with data collection that must be recognized and dealt with from the onset of a study. The most serious problem when studying hormone secretion is the potential interference of the antiserum used for immunoneutralization with the radioimmunoassay used to monitor the biological response. When samples are collected from an immunoneutralized animal, it must always be remembered that they contain a relatively high concentration of the neutralizing antiserum.

One obvious problem occurs when the antiserum used for immunoneutralization recognizes (cross-reacts with) antigenic determinants on the radioimmunoassay reagents, particularly the tracer preparation. This possibility should be considered during the initial characterization of the antiserum before it is used for immunoneutralization. The most common problem arises when the RIA relies on a "second antibody reaction" to separate bound from free ligand. In this reaction, a "second" antiserum, generated against the γ-globulin of the species in which the primary antiserum was generated, is added to the RIA reaction mixture. The anti-γ-globulin antibodies bind not only to the specific antibodies of the primary antiserum that are against the antigen of interest but also to nonspecific (in terms of the RIA) γ-globulins of the primary antiserum and of normal or nonimmune serum from the same species as the primary antiserum, which is also added to the RIA reaction mixture. When conditions are optimal, the anti-γ-globulin antibodies bind to the γ-globulins from the primary antiserum and the normal serum and form a large lattice network which, because of its mass, precipitates out of solution and allows the primary antibodies with their bound ligand to be separated from the free ligand. If the ratio of anti-γ-globulin antibodies to total γ-globulins is not within a certain range (both high and low), the lattice network does not form or only partially forms. The concentration of normal serum in the assay in relation to the concentration of anti-γ-globulin serum (second antibody) is therefore critical.

If the antiserum used for immunoneutralization was generated in the same species as the primary antiserum of the RIA, then plasma samples from the immunoneutralized animal will greatly increase the total concentration of γ-globulins in the RIA reaction mixture and the precipitating second antibody reaction will be compromised or prevented. This problem can be circumvented in several ways. If possible, the antiserum used for immunoneutralization should be generated in a species other than that used to generate the primary antiserum of the RIA. Depending on the degree of sequence similarity between the structure of the γ-globulins of the two species and the antigenic determinants recognized by the anti-γ-globulin serum, this may reduce or completely eliminate the interference. If antiserum from the same species must be utilized or if the degree of cross-reaction between species is significant, then the extra γ-globulins carried in the sample must be compensated by adjusting the amount of normal serum and/or anti-γ-globulin serum added to the assay. In order to make this adjustment, it is necessary to know the critical concentration range of both anti-γ-globulin serum and total γ-globulin required for a complete precipitating reaction to occur. This can be accomplished by testing a range of normal serum concentrations with a range of second antibody concentrations.

Figure 5 illustrates the upper end of the critical range of normal sheep serum in relation to the concentration of rabbit anti-sheep γ-globulin necessary to fully precipitate the primary antiserum and bound ligand. In this example, the primary antiserum is sheep anti-human α-inhibin and the ligand is a ^{125}I-labeled fragment of the α chain of human inhibin. The amount by which the concentration of second antibody must be increased in order to accommodate increasing concentrations of normal sheep serum can easily be deduced. If, for example, a 350-g rat were injected with 1.0 ml of a sheep antiserum to immunoneutralize an endogenous factor, then, soon after injection, a 50-μl plasma sample from these animals would contain approximately 3.3 μl of the antiserum based on a 4.04%

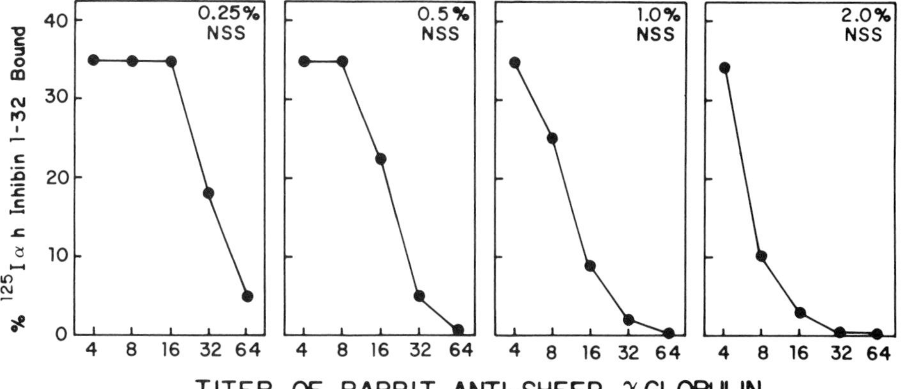

FIG. 5. Effect of increasing the concentration of normal sheep serum (NSS) on the ability of anti-sheep γ-globulin serum ("second antibody") to completely precipitate the antibodies (along with the bound ligand) of the primary antiserum. The concentration of both the anti-γ-globulin serum and the NSS must be within a critical range (both high and low) in order for a complete precipitation reaction to occur. In this example, illustrating the upper end of the critical range for NSS, increased concentrations of NSS can be compensated by increasing the concentration of anti-sheep γ-globulin. The primary RIA antiserum is ovine anti-inhibin, MC-4, and the tracer preparation used to measure binding is human ^{125}I-labeled α-inhibin (1–32). (M. D. Culler and A. Negro-Vilar, unpublished observations.)

blood volume per body weight.[22] When added to the inhibin RIA with a total reagent volume (before second antibody addition) of 400 μl, the 50-μl sample from these rats would increase the normal sheep serum content of the assay from 0 to 0.825%. As depicted in Fig. 5, omitting normal sheep serum from the assay and increasing the concentration of second antibody to 1 : 4 should allow the precipitation reaction to occur within the range of sheep serum carried by the samples. The conditions and concentrations will, of course, vary between different lots of second antibody and different assays and should be reevaluated as needed. Also, since the γ-globulins of several species may cross-react to some degree with the particular second antibody being used, this procedure should be performed even when the immunoneutralization antiserum and the RIA antiserum are from different species.

If the primary antiserum of the RIA was generated in the human, rabbit, or guinea pig, then an alternative method of separating bound from free ligand is the use of protein A. Protein A is a cell wall component of the bacterium *Staphylococcus aureus* that rapidly and specifically binds

[22] H. J. Baker, Lindsey, and S. H. Weisbroth (eds.), "The Laboratory Rat: Biology and Diseases," Vol. 1, Appendix 1, p. 411. Academic Press, New York, 1979.

γ-globulins of certain species.[23,24] By omitting normal serum from the assay (except that carried by the sample) the addition of a reasonably small amount of one of the commercially available protein A preparations (usually 50–100 μl) will precipitate all the γ-globulins present in the reaction mixture. Unfortunately, protein A does not bind all classes of γ-globulins in certain species, such as sheep, goat, mouse, and rat, and cannot be used as a precipitating agent when the primary RIA antiserum is from these species.[24]

If bound and free ligand in the RIA can be separated by one of the nonspecific methods, as in the use of ethanol or charcoal–dextran, then the only potential source of assay interference by the immunoneutralizing γ-globulins present in the sample, is cross-reaction with the tracer. Because these methods are not specific for precipitating γ-globulins, however, any substance in either the immunoneutralizing antiserum or even the primary antiserum of the RIA that can bind the tracer is a potential source of interference. This possibility should be thoroughly explored before relying on one of these methods of separation.

Interfering Substances in the Antiserum. Another potential source of problems is the possibility of substances, other than the γ-globulins, that are present in the antiserum or antibody preparation used for immunoneutralization that may either interfere with the RIA or alter some pertinent physiological factor when injected into the animal. Should this "nonspecific" interference be demonstrated or be perceived to occur at some phase of the study, it can often be eliminated by purifying the γ-globulins from the antiserum. Several methods have been used successfully for this procedure including ammonium sulfate precipitation, gel filtration, affinity chromatography, and DEAE ion-exchange chromatography. Reviews and descriptions of immunoglobulin purification methods may be found in Refs. 25 and 26 and in other volumes of this series.[27,28] The appropriate choice of procedures will depend, of course, on the requirements of the individual study. Regardless of the means used, the recovery of antibodies should be checked by comparing the antibody titer of the purified preparation with that of the original antiserum.

[23] S.-Y. Ying and R. Guillemin, *J. Clin. Endocrinol. Metab.* **48,** 360 (1979).

[24] S. W. Kessler, this series, Vol. 73, p. 442.

[25] J. S. Garvey, N. E. Cremer, and D. H. Sussdorf, "Methods in Immunology." Benjamin, Reading, Massachusetts, 1977.

[26] J. C. Jaton, D. C. Brandt, and P. Vassalli, *in* "Immunological Methods" (I. Kefkovits and P. Pernis, eds.), p. 43. Academic Press, New York, 1979.

[27] B. A. L. Hurn and S. M. Chantler, this series, Vol. 70, p. 104.

[28] P. D. Gorevic, F. C. Prelli, and B. Frangione, this series, Vol. 116, p. 3.

A convenient new development which may be considered for purifying γ-globulins is the Affi-Gel protein A MAPS II kit from Bio-Rad Laboratories (Richmond, CA) which provides protein A attached to an agarose matrix and a buffer system that can remove the γ-globulins from the protein A after they have been bound. While developed for monoclonal antibody purification, this kit can be used also to purify γ-globulins from rabbit serum with better than 90% recovery (M. D. Culler and A. Negro-Vilar, unpublished observations).

After purification by any of these techniques, the γ-globulin preparation must be desalted and concentrated before it can be used for immunoneutralization. The currently available methods include dialysis, gel filtration, and ultrafiltration for desalting and lyophilization, evaporation, and ultrafiltration for concentrating. Obviously, ultrafiltration offers the advantage of desalting and concentrating simultaneously. After concentration, the recovery of antibodies should be determined (as after purification) in order to establish the concentration available for immunoneutralization.

Use of Structurally Modified Agonists to Replace Immunoneutralized Biological Activity

The use of a specific antiserum to remove an endogenous factor has an added advantage in that structurally modified agonists that are poorly recognized by the antibodies can be used to study the exogenous replacement of the immunoneutralized, endogenous factor. An example illustrating pulsatile replacement of LHRH bioactivity in LHRH-immunoneutralized rats is shown in Fig. 6. In this study, the antigenic determinants recognized by the LHRH antiserum used were predominantly located at the amidated carboxy (C) terminus. By using the C-terminally modified agonist [des-Gly^{10}]LHRH ethylamide, the antiserum blockade was circumvented, and the effects of restoring LHRH biological activity on LH and FSH secretion were studied using various pulsatile replacement regimens.[14] Additionally, the replacement of pulsatile GHRH bioactivity in GHRH-immunoneutralized rats has been studied using GHRH analogs poorly recognized by the neutralizing monoclonal antibody preparation.[16] Replacement studies such as these are impossible with antagonist-blocked animals owing to the blockade of the biological receptors.

Two criteria must be met in order to successfully use a structurally modified analog to replace immunoneutralized biological activity: first, the analog must retain biological activity similar to that possessed by the parent compound and, second, the analog should be poorly recognized

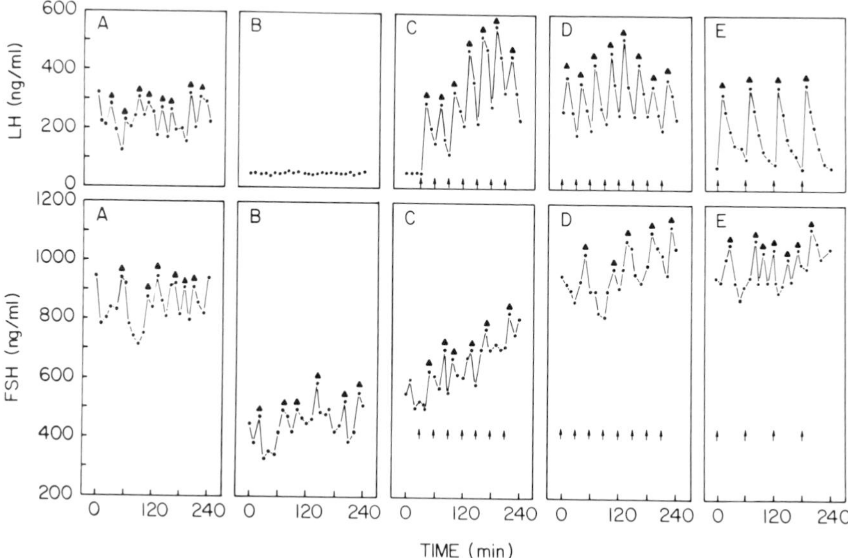

FIG. 6. Representative plasma LH and FSH patterns from unanesthetized, freely moving, castrate male rats after (A) injection of 500 μl nonimmune serum 24 hr earlier and giving bolus i.v. injections of saline every 30 min during the sample collection period or every 30 min for 10 hr before and continuing through the sampling period (vehicle control), (B) injection of 500 μl anti-LHRH serum A-772 (LHRH-AS) 24 hr earlier and giving saline as above (antiserum control), (C) injection of 500 μl LHRH-AS 24 hr earlier and giving 3 ng bolus i.v. injections of [des-Gly[10]]LHRH ethylamide (DG-LHRH) every 30 min during the collection period, (D) injection of 500 μl LHRH-AS 24 hr earlier and giving 3 ng bolus i.v. injections of DG-LHRH every 30 min for 10 hr before and continuing through the sampling period, and (E) injection of 500 μl LHRH-AS 24 hr earlier and giving 3 ng bolus i.v. injections of DG-LHRH every 60 min for 10 hr before and continuing through the sampling period. Arrows indicate times of DG-LHRH injections. ▲, Pulse as defined in Fig. 2. (From Culler and Negro-Vilar,[14] with permission.)

by the immunoneutralizing antiserum. Even if an agonist is highly characterized, such as the LHRH and somatostatin analogs, it should be tested for biological potency in the same model in which its endogenous parent compound is to be immunoneutralized. The biological potency of the [des-Gly[10]]LHRH ethylamide analog used in the previous replacement example was compared with LHRH for its ability to release LH in 4-week castrate male rates as illustrated in Fig. 7. In this model, it was found to be 2.6 times more potent than LHRH, and the concentration of agonist used to replace LHRH activity was appropriately adjusted. It should be kept in mind that altering the structure of a parent compound may not only affect potency but also may modify the biological activities and/or confer new

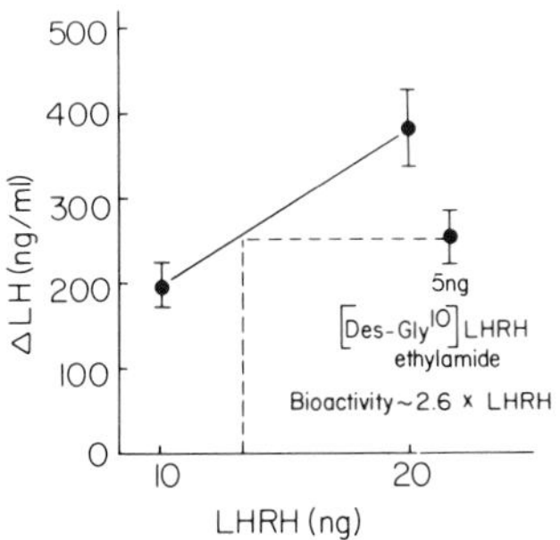

FIG. 7. LH-releasing activity of [des-Gly[10]]LHRH ethylamide compared with that of LHRH 15 min after i.v. injection into 4-week castrate male rats. Each point with brackets represents the mean ± SEM of six rats. (From Culler and Negro-Vilar,[14] with permission.)

activities. This could be an important consideration if the activity of a given analog has not been characterized previously.

The ability of an analog to circumvent an immunoneutralizing antiserum can be tested by determining its ability to compete with the parent compound in binding to the selected antiserum. As illustrated in Fig. 8, the [des-Gly[10]]LHRH ethylamide analog is able to completely inhibit the binding of the ^{125}I-labeled parent compound, LHRH, to the antiserum to be used for immunoneutralization. The mass of analog required to inhibit ^{125}I-labeled LHRH binding, however, is several orders of magnitude greater than the mass of LHRH required to induce the same degree of

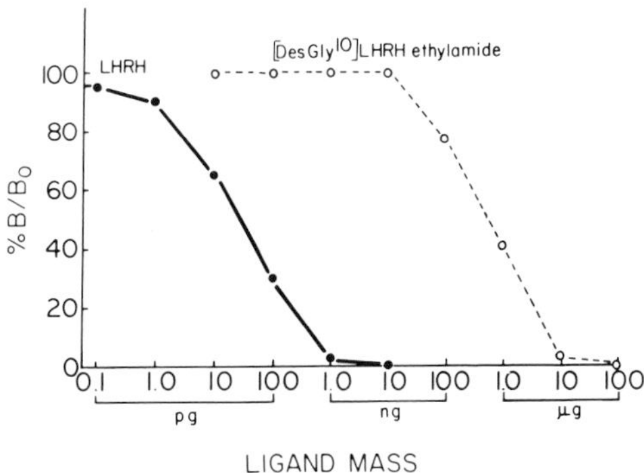

FIG. 8. Representative ^{125}I-labeled LHRH binding inhibition curves produced by serial dilutions of LHRH and [des-Gly[10]]LHRH ethylamide binding to ovine anti-LHRH serum A-772. The cross-reaction of the LHRH analog with this antiserum, directed against the carboxy-terminus of LHRH, is less than 0.0048% as compared with LHRH (100%). (From Culler and Negro-Vilar,[14] with permission.)

inhibition. The cross-reaction of the analog with the antiserum is less than 0.0048% as compared with LHRH (100%). This low degree of cross-reaction allows [des-Gly^{10}]LHRH ethylamide to circumvent the antiserum blockade and stimulate the pituitary LHRH receptors.

If the cross-reaction of an analog with the immunoneutralizing antiserum is too great to be used in a replacement study and if no other appropriate analogs are available, then it is usually easier to obtain or produce a different antiserum that will not recognize the analog than it is to create a new analog. This problem presents the main disadvantage in using structurally modified agonists for replacing biological activity in immunoneutralized animals. If there are no biologically active analogs available with sufficient structural modification so as to be poorly recognized by an antiserum, then replacement cannot be performed. In this instance the return of biological activity must be studied as the antiserum is gradually cleared from the blood or removed by transfusion. Fortunately, biologically active analogs have been discovered for many of the known pituitary-affecting factors, and the list of potential analogs increases markedly each year.

Summary

The combination of passive immunoneutralization of an endogenous factor with frequent sampling techniques to delineate the pattern of secretion represents a powerful tool for dissecting the contribution of that factor to normal and/or abnormal pituitary function. Passive immunoneutralization offers all of the advantages of other known means of abolishing endogenous factors. It has the specificity and the selectivity of antagonists while retaining the potential for replacement of biological activity that is possible with organ removal or selective brain cuts or lesions. The only major drawback to this approach is the potential difficulty in obtaining the appropriate reagents. The inherent flexibility of the model, however, increases the probability of obtaining useful data with those reagents that are available. At present, the number of studies using immunoneutralization is relatively small, but the knowledge gained from these studies has been substantial. As the number of immunoneutralization and replacement studies increases in the future, the usefulness of this powerful approach will be extended to its full potential.

[38] Combined Antibody–High-Performance Liquid Chromatography Approach to Assess Prohormone Processing

By W. WETSEL and A. NEGRO-VILAR

Introduction

In recent years there has been an ever-expanding number of peptides which have been identified within both the nervous system and the endocrine organs. Much of this progress rests on the application of molecular biology techniques to identify the underlying genes and transcripts which serve as templates for the biosynthesis of peptides. This approach has revealed an unexpected array of new proteins and peptides: those which have been deduced because they bear partial or complete nucleotide sequence homology to the gene or cDNA under study (e.g., the mineralocorticoid receptor and the glucocorticoid receptor[1]) and those which have been identified by their contiguous presence within the same precursor as the peptide of interest (e.g., neurotensin and neuromedin N[2]). In the past, peptides were identified by and purified according to their bioactivities or their ability to cross-react with a given antisera. Now, cDNA sequences are used to deduce the primary structure of precursor peptides and to predict the fragments which arise following processing of this precursor.

Despite the benefits of molecular biology, this approach creates special problems for the protein biochemist and physiologist because the amino acid sequences of the various peptides may not be known with certainty (the peptide has not been sequenced, only deduced from a cDNA), the regulatory steps involved in the biosynthesis, processing, and secretion of the peptides are unclear, and their functional roles are often unexplored. In addition, various tissues may metabolize the same precursor peptides by different routes, and these products may have different functional consequences depending on the locus of secretion and the physiological/pathological status of the organism. These circumstances emphasize the importance of adopting a coordinated and multifaceted approach in detailing a peptide's fate from biosynthesis to function.

[1] J. L. Arriza, C. Weinberger, G. Cerelli, T. M. Glaser, B. L. Handelin, D. E. Housman, and R. M. Evans, *Science* **237,** 268 (1987).

[2] P. R. Dobner, D. L. Barber, L. Villa-Komaroff, and C. McKierman, *Proc. Natl. Acad. Sci. U.S.A.* **84,** 8516 (1987).

Structure of the Precursor Peptide and Its Processing

An important consideration in studying peptide processing is the selection of tissues or cells. Various tissues may metabolize peptide precursors by different routes. This processing may be a function of the mRNA that is expressed, the molecular form(s) of the peptide precursor, and the nature of the metabolizing enzymes. Certain peptides such as somatostatin exist in multiple forms because their different mRNAs may be expressed from different genes.[3-5] Additional diversity is created by differential mRNA splicing as in the case of calcitonin and calcitonin gene-related peptide[6] or by transcription of different strands of DNA.[7] The mRNA may contain the coding sequence for a single-copy (e.g., insulin[8]), a multiple-copy (e.g., thyrotropin-releasing hormone[9]), or a multifunctional peptide (e.g., proopiomelanocortin[10]). Following translation, the precursor peptide may be subjected to proteolytic cleavage, glycosylation, methylation, acetylation, sulfation, amidation, phosphorylation, disulfide bonding, hydroxylation, and/or pyroglutamination.[11-14]

Precursor peptides are initially synthesized as prepropeptides. The prefix pre- refers to a secretory peptide or protein which contains a signal sequence at its amino terminus. This hydrophobic sequence is cleaved as the precursor or propeptide crosses the membrane of the endoplasmic reticulum and leaves the ribosome. It is this propeptide which undergoes a variety of posttranslational events leading to the final peptide product. Most frequently, the propeptide is cleaved at positions where two or more basic amino acid residues appear consecutively, but, as in the case of neurotensin or β-lipotropin (β-LPH), processing at such sites is not oblig-

[3] P. Hobart, R. Crawford, L. P. Shen, R. Pictet, and W. J. Rutter, *Nature (London)* **288,** 137 (1980).

[4] R. H. Goodman, J. W. Jacobs, W. W. Chin, P. K. Lund, P. C. Dee, and J. F. Habener, *Proc. Natl. Acad. Sci. U.S.A.* **77,** 5869 (1980).

[5] D. Shields, *J. Biol. Chem.* **255,** 11625 (1980).

[6] S. G. Amara, V. Jonas, M. G. Rosenfeld, E. S. Ong, and R. M. Evans, *Nature (London)* **298,** 240 (1982).

[7] J. P. Aldelman, C. T. Bond, J. Douglass, and E. Herbert, *Science* **235,** 1514 (1987).

[8] G. I. Bell, W. F. Swain, R. Pictet, B. Cordell, H. M. Goodman, and W. J. Rutter, *Nature (London)* **282,** 525 (1979).

[9] R. M. Lechan, P. Wu, I. M. D. Jackson, H. Wolf, S. Cooperman, G. Mandel, and R. M. Goodman, *Science* **231,** 159 (1986).

[10] S. Nakanishi, A. Inoue, T. Kita, M. Nakamura, A. C. Y. Chang, S. N. Cohen, and S. Numa, *Nature (London)* **278,** 423 (1979).

[11] B. A. Eipper and R. E. Mains, *Endocr. Rev.* **1,** 1 (1980).

[12] J. Spiess and B. D. Noe, *Proc. Natl. Acad. Sci. U.S.A.* **82,** 277 (1985).

[13] W. H. Fischer and J. Spiess, *Proc. Natl. Acad. Sci. U.S.A.* **84,** 3628 (1987).

[14] D. F. Steiner, *Diabetes* **26,** 322 (1977).

atory. Proteolytic cleavage can also occur at sites either of single basic amino acids or of other amino acids.

Propeptide processing events can be modified by hormones and neurotransmitters. For example, in the anterior pituitary, proteolytic cleavage of proopiomelanocortin (POMC) to adrenocorticotropin (ACTH) and β-LPH is regulated by corticotropin-releasing factor and glucocorticoids.[15] By contrast, in the neurointermediate lobe, dopamine stimulates the further cleavage both of ACTH to α-melanocyte-stimulating hormone and corticotropin-like intermediate lobe peptide (CLIP), and of β-LPH to γ-LPH and β-endorphin, respectively. Physiological status can also affect precursor processing. Intracellular degradation and secretion of parathyroid hormone (PTH) is influenced by plasma calcium concentrations.[16] During hypocalcemia, PTH is secreted primarily as an intact peptide which is bioactive. During hypercalcemia the bioactivity of PTH is reduced because most of it is cleaved into at least two peptide fragments which, by themselves, show little activity. Clearly, multiple strategies exist for peptide processing, and these strategies are a function of the tissue under study, the primary structure of the peptide, the presence of certain cleavage or modifying enzymes, and the physiological status of the animal.

Strategies Used to Evaluate Peptide Processing *in Vivo*

Preliminary Considerations

In order to study peptide processing, one can either examine the entire metabolic fate from propeptide to degradation products or focus on particular steps in the pathway. Both approaches require (1) a judicious selection of starting material, (2) the generation of multiple antisera that are specific and sensitive for the "precursor" and for the different fragments, and (3) the ability to chemically separate the different peptides from each other so that their immunological and chemical identities can be ensured and their concentrations quantitated. Each of these three aspects must be carefully evaluated before beginning a study.

In the choice of starting material, one must consider (1) whether the cells in the tissue actually biosynthesize the peptide or if the peptide is merely transported through or to the tissue, (2) whether the tissue con-

[15] E. Herbert, E. Oates, G. Martens, M. Comb, H. Rosen, and M. Uhler, *Cold Spring Harbor Symp. Quant. Biol.* **48,** 375 (1983).

[16] G. P. Mayer, J. A. Keaton, J. G. Hurst, and J. F. Habener, *Endocrinology* (*Baltimore*) **104,** 1778 (1979).

tains a homogeneous population of cells, and (3) whether there are sufficient quantities of cells that biosynthesize the peptide if radiolabeled experiments are to be conducted. Perhaps the most critical component of a processing study is the preparation of antisera which can specifically recognize the propeptide as well as different fragments of the precursor peptide. Because the propeptide is usually too long to chemically synthesize, it must be produced in bacteria, or one must chemically synthesize fragments of the propeptide and use them as antigens. The antisera that are produced may have low affinities for or fail to recognize the propeptide because they were generated against one of the terminal amino acids or they recognize a particular conformation of the peptide fragment which is different from that assumed by the propeptide. Finally, there should be some method that can separate the immunoactive peptides by chemical criteria. Chemical separation may reveal whether the antisera recognize more than one immunoreactive substance and whether the immunoreactive material has chemical properties that are similar to that of the synthetic radioimmunoassay (RIA) standard.

Immunological Approach

Three general approaches have been used to study processing of peptides *in vivo*. In the first, antisera that have been raised to several different domains of a "precursor" peptide are used to quantitate the respective immunoreactive peptides in biological samples. Typically, the antisera have been characterized to determine their specificity. Since the molecular weight (MW) of the chemically synthesized antigenic peptide is known and competitive inhibition curves for this peptide standard and tissue extracts may be parallel, molar ratios of immunoreactivities are often calculated. This procedure relies heavily on the immunological identity between the known standard and the antigen in the tissue sample. In order to calculate a molar ratio, the inference is made that the peptide standard and the unknown also share chemical identity. One may encounter several problems with this procedure, and they involve the extraction of peptides from tissues or fluids, determination of cross-reactivity of antisera, and assumptions involving the calculation of molar ratios. During the course of extraction of peptides from tissue, additional processing of the "precursor" and/or differential losses of the various processed products can occur.[17]

Another problem concerns the characterization of the antisera. Typically, competitive inhibition curves are generated using the synthetic pep-

[17] A. Miyata, K. Kangawa, T. Toshirmori, T. Hatoh, and H. Matsuo, *Biochem. Biophys. Res. Commun.* **129,** 248 (1985).

tide antigen as the standard, and its displacement is compared against serial dilutions of various different peptides and tissue extracts. The precise antigenic determinants can be analyzed by amino acid substitution and/or deletion of residues from the peptide to which the antisera were raised. This procedure has been applied to LHRH (luteinizing hormone-releasing hormone) antisera,[18,19] but it is very tedious and expensive when used to characterize larger peptides. Nonetheless, most investigators test selective peptides (same size, similar residues, similar conformation) to generate inhibition curves. If parallelism is seen between the synthetic peptide standard and tissue extracts, then both are considered immunologically identical. If any of the other synthetic peptides or tissue extracts show displacement of label but not parallelism, then cross-reactivity is inferred. These data must be interpreted with caution because investigators usually do not perform exhaustive tests on all synthetic peptides that may be cross-reactive and because many of the cross-reacting substances in tissue extracts remain to be purified and sequenced. This criticism can be partially overcome if several different antisera that recognize different antigenic determinants of the same peptide are used.

Expression of immunoreactivities as molar ratios can be helpful in proposing the pathway for processing and in speculating which products may have biological significance. However, these data can be misleading. For instance, the antisera may possess different affinities for the "precursor" peptide and the various fragments.[20] These separate cross-reactivities may be obscured in tests of parallelism of displacement curves from tissue extracts and peptide standards. Therefore, the pathway of metabolism may be incorrectly inferred, and the quantitation of final product may be under- or overestimated. As a result, the data expressed as molar ratios will not be completely accurate. In addition, if the cells that comprise the tissue are heterogeneous, there may be several different processing pathways for the "precursor."[11] Finally, care must be taken in inferring biological significance from the quantity of peptide in tissue. During the course of processing, different peptides are generated, and some of these may be active while others may be biologically inactive.

Chemical Characterization Approach

While the first approach to studying the processing of peptides *in vivo* relies on immunological criteria, the second expands on this model by

[18] K. C. Copeland, M. L. Aubert, J. Rivier, and P. C. Sizonenko, *Endocrinology (Baltimore)* **104,** 1504 (1979).

[19] A. Arimura, *Folia Endocrinol. Jpn.* **52,** 1159 (1976).

[20] R. Benoit, N. Ling, B. Alford, and R. Guillermin, *Biochem. Biophys. Res. Commun.* **107,** 944 (1982).

invoking chemical characterization. In this procedure, tissue extracts are submitted to chromatography or electrophoresis, the mixture is chemically separated, and the material in the fractions or that eluted from the gel is quantitated by RIA. Once the MW of the immunoreactive substance(s) is estimated, some determination of the size of the "precursor" and the processed products can be made. Molecular weight determination is particularly useful if the amino acid sequence of the precursor is known or has been deduced from the cDNA and if the antigenic determinants of the antisera have been described. With this information, the processing sites of the precursor can be inferred and the amino acid sequence of the immunoreactive peptides can be deduced.

One must exercise caution with this approach, however, because of certain potential problems. First, the recovery of IR peptides from the column or electrophoretic gel may be variable.[21] This problem is compounded if only one antibody is used to detect each immunoreactive peptide or if the antibody cross-reacts with other peptides or substances which contain some sequence similarity to the peptide under study. These circumstances could lead to incorrect assignment of one metabolic route over another, or some peptides may be missed in the pathway. Multiple antisera can help provide a more accurate estimation of recovery and help resolve different immunoreactive peptides, but if the antisera exhibit substantial cross-reactivity, then calculation of recovery can be difficult. Second, if the peptides are known to have sulfur-containing amino acids, then samples should be incubated and run under denaturing and dissociating conditions. Usually the sample must be desalted, whereby peptide loss may occur.

Finally, the separation system may not give reliable results. For instance, since many peptides are small and some may be glycosylated, estimation of MW by electrophoresis may be difficult or inaccurate.[11] If the peptide is not smaller than 1000 MW, some success can be achieved in separating low MW peptides by using a higher concentration of acrylamide in the gel. Similarly, if a peptide retains significant charge while in the mobile phase or if its conformation is globular, then MW determination by liquid chromatography may be incorrect. These difficulties can usually be corrected by incubating the peptide in dissociating and denaturing solutions and by altering the mobile phase. Finally, the estimation of MW by electrophoresis or gel-permeation chromatography can be imprecise because the resolution of immunoreactive species varies as a function of their size. This problem can be corrected through additional separation steps by isoelectric focusing in gels or reversed-phase chromatography.

[21] J. D. White, J. E. Krause, and J. F. McKelvy, *J. Neurosci.* **4,** 1262 (1984).

In reversed-phase separations in which the peptide contains a methionine residue, the sulfur group can have different oxidation states which will cause it to elute in positions different from that of the reduced peptide.[22] This difficulty can be overcome by oxidizing the sample prior to chromatography. It should be noted that the immunoreactivity of the oxidized peptide standard should be tested before implementing this procedure.

Biosynthetic Approach

A final approach to studying the processing of peptides *in vivo* combines the two previous paradigms and uses biosynthesis as a criterion. This approach is especially important because the fate of label incorporation into precursor and products can be followed and the kinetics of biosynthesis and processing can be studied.[23] In this paradigm, the precursor–product relationship can be clearly established, and the specificity of the antisera can be further verified. In addition, the radiolabeled peptide can be microsequenced, and the position of the labeled amino acid(s) within the peptide can be determined.[24]

While this procedure can provide invaluable information, care must be taken in implementing it. First, the labeled amino acid may be metabolized to another amino acid, it may be degraded, or it may be diluted in the amino acid pool.[25] One way to circumvent these difficulties is to administer several different labeled amino acids separately and compare their kinetics of incorporation. Second, few cells in the tissue may biosynthesize the peptide, the biosynthetic rate in the cells may be low, and/or the biosynthesizing cells may be distributed in diffuse clusters. These circumstances may preclude using the biosynthetic approach because the specific activity of the precursor and its products may be too low to quantitate (one often uses ^{3}H- or ^{14}C-labeled amino acids). This situation can be improved if sulfur-containing amino acids are present in both the precursor and the products so that higher energy isotopes such as [^{35}S] can be used. Finally, the labeled precursor and products should be purified to radioactive and immunological homogeneity such that the ratio of disintegrations/minute to immunoreactivity is constant for each peptide.[21] Since more than one separation is often required, losses of radioactive peptides can be substantial. In addition, since very low quantities of precursor may

[22] J. Rivier, J. Spiess, and W. Vale, *Proc. Natl. Acad. Sci. U.S.A.* **80,** 4851 (1983).

[23] B. D. Noe, D. J. Fletcher, and J. Spiess, *Diabetes* **28,** 724 (1979).

[24] L. Tan and P. Rousseau, *Biochem. Biophys. Res. Commun.* **109,** 1061 (1982).

[25] P. Lestage, M. Gonon, P. Lepetit, P. A. Vitte, G. Debilly, C. Rossatto, D. Lecestre, and P. Bobillier, *J. Neurochem.* **48,** 352 (1987).

exist in tissue, quantitation of label incorporation into the propeptide may be difficult.

Processing of LHRH Precursor

Introduction, General Protocol, and Tissue Dissection

Recently, the amino acid sequence for the luteinizing hormone-releasing hormone (LHRH) propeptide has been deduced from the cDNA from human placenta and human and rat hypothalamus (see Fig. 1A).[26,27] This 69-amino-acid peptide contains LHRH at its amino terminus, a combined dibasic processing and amidation site, and a 56-amino-acid peptide designated gonadotropin-releasing hormone-associated peptide (GAP). While the physiological effects of LHRH on anterior pituitary secretion of luteinizing hormone (LH) and follicle-stimulating hormone (FSH) are well known, the functional significance of GAP is currently being investigated. Different laboratories have indicated that an amino-terminal fragment of GAP[28] or GAP itself[29] may stimulate secretion of LH and FSH *in vitro*. In addition, GAP (1–56) has also been reported to inhibit prolactin secretion from the anterior pituitary.[29] This last report, however, has remained unconfirmed by other laboratories. While these data suggest that proLHRH is a polyfunctional peptide, the processing scheme for this precursor has not been elucidated. More importantly, neither proLHRH nor any of its fragments except LHRH have been purified and sequenced. In order to study processing of proLHRH, we have used high-performance gel-permeation chromatography and RIA to deduce the various cleavage and modification patterns of the propeptide.

The general scheme used in our study of proLHRH processing is shown in Fig. 2. Briefly, preoptic–hypothalamic tissue containing cells that biosynthesize, process, and/or secrete LHRH was dissected from rat brain. These tissues were defatted and homogenized, and the extracts were filtered prior to separation of peptides. Samples were separated according to MW by high-performance size-exclusion chromatography (HPSEC), and fractions were screened for LHRH- and GAP-like immunoreactivity. The fractions corresponding to GAP-like immunoreactivity that eluted at approximately 6500 MW were pooled separately and

[26] P. H. Seeburg and J. P. Adelman, *Nature (London)* **311,** 666 (1984).

[27] J. P. Adelman, A. J. Mason, J. S. Hayflick, and P. H. Seeburg, *Proc. Natl. Acad. Sci. U.S.A.* **83,** 179 (1986).

[28] R. P. Millar, P. J. Wormald, and R. C. D. Milton, *Science* **232,** 68 (1986).

[29] K. Nikolics, A. J. Mason, E. Szonyi, J. Ramachandran, and P. H. Seeburg, *Nature (London)* **316,** 511 (1985).

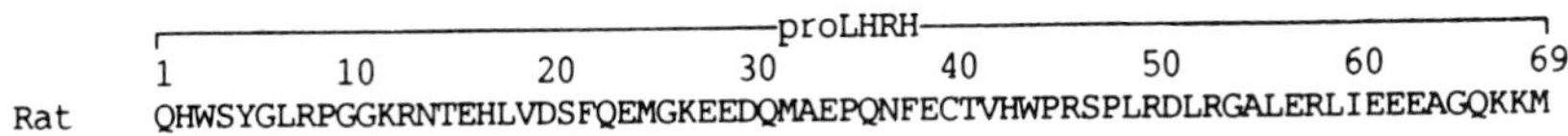

B

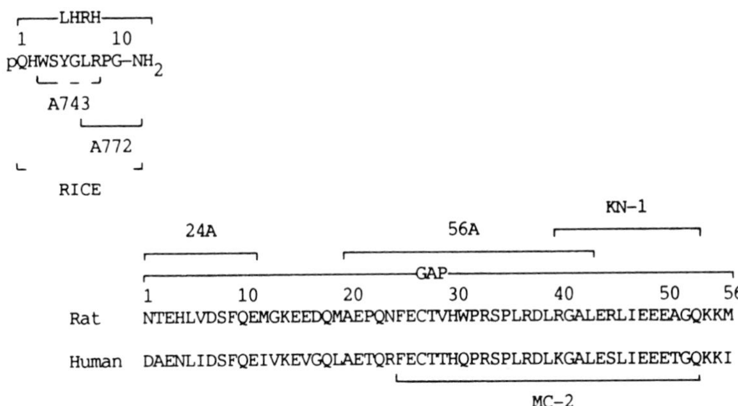

C

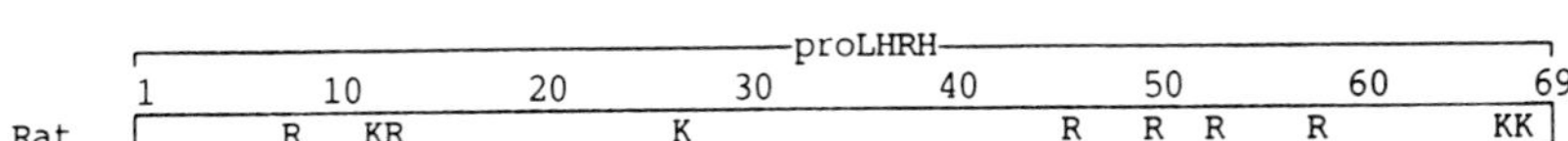

FIG. 1. The deduced amino acid sequence of proLHRH, the antigenic determinants of the LHRH and GAP antisera, and the location of basic amino acid residues in proLHRH are shown. (A) The deduced amino acid sequences of rat and human proLHRH contain 69 amino acids each. All amino acids are designated by a single-letter abbreviation and are numbered consecutively. (B) Three different LHRH and four different GAP antisera were used for the radioimmunoassays in the study. The antigenic amino acids of each of the LHRH antisera are underlined. The abbreviations pQ and G-NH_2 signify the pyroglutamate and the amidated glycine residues of the LHRH decapeptide. The antigenic amino acids for the GAP peptide have not been determined; however, the antisera have been characterized. The peptide fragments which were used to generate the four different GAP antisera are indicated. (C) Location of arginine and lysine residues within the proLHRH molecule is emphasized. The most likely sites for proteolytic cleavage in peptides are at dibasic residues or at single lysine (K) or arginine (R) residues.

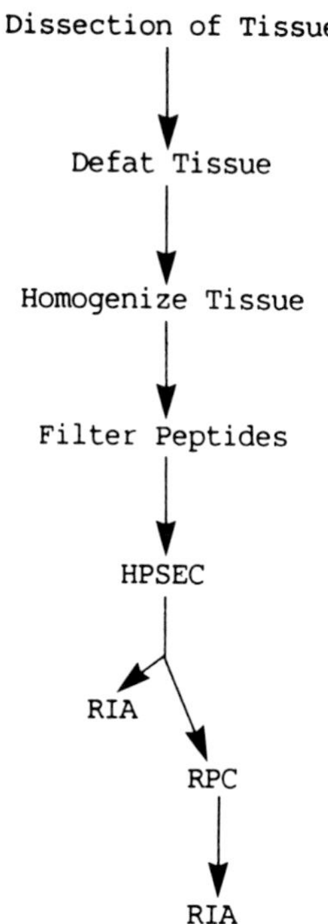

FIG. 2. Diagram outlining the approach used in characterization of the proLHRH metabolic pathway. Tissues from the preoptic area and hypothalamus of rat brain were dissected, defatted, and homogenized. The peptides were extracted and filtered prior to chromatography. Peptides were separated according to molecular weight (MW) using high-performance size-exclusion chromatography (HPSEC). Fractions were collected and screened for LHRH- and GAP-like immunoreactivity by RIA. The fractions containing GAP (1–56)-like immunoreactivity at approximately 6500 MW were combined and injected onto a high-performance reversed-phase column. Fractions were collected and examined by RIA to verify the homogeneity of the GAP-like peak seen on HPSEC.

injected onto a reversed-phase column, and the fractions were screened by RIA to determine the homogeneity of this HPSEC peak.

In our study, we incorporated two unique features by which to study processing of the LHRH precursor. In the first, we took advantage of the anatomical topography of the LHRH neuronal system. In the rat, the highest concentration of LHRH is found in a narrow band of cells that stretches from the preoptic area, through the hypothalamus, to nerve terminals in the median eminence (ME).[30] Since propeptides are biosynthesized in the neuronal cell body and the final processed products are

[30] I. Merchenthaler, G. Kovacs, G. Lovasz, and G. Setalo, *Brain Res.* **198,** 63 (1980).

found in the nerve terminals, we microdissected the ME[31] from the hypothalamus and then removed the remaining preoptic–hypothalamic area (POH) as a tissue block. The different molecular forms of LHRH and GAP were characterized separately from these two neuronal regions. Briefly, the brain was removed and placed ventral side up. The ME was dissected, and the POH was excised from the rest of the brain by making coronal cuts anterior to the optic chiasm and posterior to the mammallary bodies, vertical cuts along the hypothalamic sulcus on each side of the midline, and a horizontal cut at the anterior commissure. This approach is analogous to biosynthetic time course experiments. In biosynthetic experiments, label is given and its incorporation into precursor peptide and product are followed across time. In our study, the POH should contain the precursor peptide and partially processed products of the metabolic pathway (the "early" time point in biosynthetic experiments), while the ME should contain the final products (the "late" time point).

A second purpose of our study involved studying the physiological regulation of biosynthesis and processing of the LHRH precursor. Since orchidectomy (ORDX) has been reported to reduce the concentration of LHRH in hypothalamus, we compared the concentrations of different forms of LHRH- and GAP-like immunoreactivity in the POH and the ME from 20 adult male, intact and 20 ORDX rats. The ORDX animals were castrated 1 month prior to the experiment. At the time of decapitation, blood was collected for analysis of serum hormones. Serum FSH was 452 $\pm$ 28 and 2572 $\pm$ 103 ng/ml, serum LH was 7 $\pm$ 0.6 and 142 $\pm$ 9.5 ng/ml, and serum testosterone was 4 $\pm$ 0.7 and 0.013 $\pm$ 0.002 ng/ml for intact and ORDX animals, respectively. These serum concentrations were similar to other published values.[32] Inasmuch as serum FSH and LH levels are regulated by the LHRH that is released into the portal blood and this LHRH is a function of the amount of decapeptide that is biosynthesized and stored in LHRH neurons, then levels of LHRH in this neuronal system should be affected by this treatment. This approach should reveal whether biosynthesis or different processing steps in the proLHRH metabolic pathway can be regulated.

Sample Extraction

After the POH and ME were collected, they were pooled separately and frozen immediately. The frozen tissues were lyophilized and defatted with a solution of ether and petroleum ether (1 : 2, by volume). It should

[31] A. Negro-Vilar, S. R. Ojeda, and S. M. McCann, *Endocrinology (Baltimore)* **104,** 1749 (1979).

[32] S. P. Kalra, P. S. Kalra, and E. O. Mitchell, *Endocrinology (Baltimore)* **100,** 201 (1977).

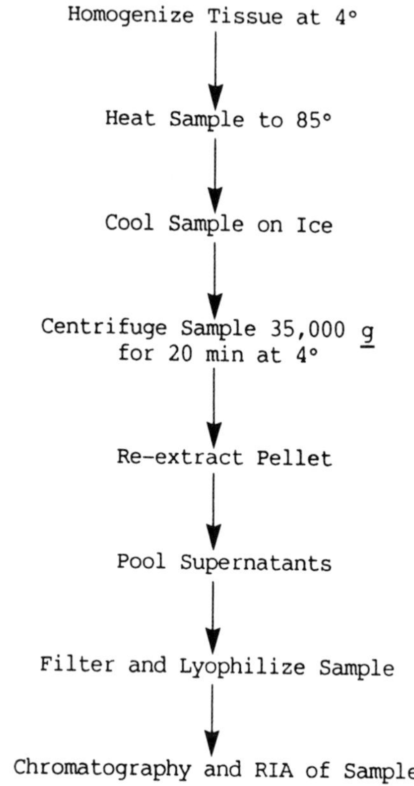
Homogenize Tissue at 4°
Heat Sample to 85°
Cool Sample on Ice
Centrifuge Sample 35,000 g for 20 min at 4°
Re-extract Pellet
Pool Supernatants
Filter and Lyophilize Sample
Chromatography and RIA of Samples

FIG. 3. Steps involved in extraction of peptides from tissue beginning with homogenization of the tissue sample through peptide separation by chromatography and quantification by RIA.

be noted that in preliminary experiments we found that recoveries of LHRH and GAP-like immunoreactivity were similar whether the tissue was defatted or not. However, we elected to defat the tissue because lipids can clog the frits on the HPSEC columns and cause loss of sample.

Following the defatting step, peptides were extracted from the POH and the ME. The general protocol for sample extraction is depicted in Fig. 3. After the reextraction step, pellets from the POH and the ME were saved for later protein analysis. The levels of protein in POH and ME from ORDX and intact rats were compared by the Lowry method[33] to discount any nonspecific effects of ORDX on protein biosynthesis and degradation. In the filtration step, supernatants from the POH and ME

[33] O. Lowry, M. Rosebrough, A. Farr, and R. Randall, *J. Biol. Chem.* **193,** 265 (1951).

TABLE I
COMPOSITION AND PREPARATION OF THE HOMOGENIZATION BUFFER[a]

I. Prepare 200 ml of buffer
  A. Prepare the 0.2 m*M* phenylmethylsulfonyl fluoride (PMSF)–10 μM pepstatin A solution in a 25-ml volumetric flask (buffer A)
    1. Weigh out 7.0 mg PMSF and 1.4 mg pepstatin A
    2. Bring to volume with methanol
  B. Prepare the 63 m*M* 2-mercaptoethanol (2-ME)–10 μM leupeptin–2 m*M* aprotinin–1 m*M* EDTA solution (buffer B)
    1. Add 125 ml deionized–distilled water to a beaker
    2. Weigh out 67.2 mg EDTA and 0.9 mg leupeptin
    3. Add 1 ml 2-ME and 2 ml aprotinin
  C. Addition of 1 *N* acetic acid and 0.1 *N* hydrochloric acid
    1. Add 11.49 ml acetic acid and 0.35 ml hydrochloric acid to buffer B
    2. Add buffer A to buffer B
    3. Bring to volume with deionized–distilled water
II. Cool buffer to 4° and use immediately

[a] The PMSF, EDTA, and pepstatin A were purchased from Sigma Chemical Co. (St. Louis, MO). The leupeptin was from Boehringer-Mannheim (Indianapolis, IN), the 2-ME from Fisher Scientific (Fair Lawn, NJ), and the aprotinin from Mobay Chemical Corporation (New York, NY).

pools were extracted separately over Waters Sep-Pak C_{18} cartridges using 50% 2-propanol, 40% acetonitrile, 0.1% trifluoroacetic acid as the eluting buffer.

In preliminary experiments we compared three different methods of peptide extraction and examined recovery of LHRH- and GAP-like immunoreactivity after homogenization, pooling the supernatants, and filtering–lyophilizing the sample. We compared extraction with 0.1 *N* acetic acid, with an acidic solution containing different classes of enzyme inhibitors as outlined by Barrett[34] (Table I), and with the buffer described by Bennett *et al.*[35] which contained 1 *M* hydrochloric acid, 5% formic acid (v/v), 1% sodium chloride (w/v), and 0.1% trifluoroacetic acid (v/v). The protocol for extraction with the last buffer was different from that of the other two, in that samples were not heated prior to centrifugation. Recoveries at the centrifugation and filtration–homogenization steps were similar under all three extraction methods. At the homogenization step, however, recovery of LHRH- and GAP-like immunoreactivity was highest with the acid–enzyme inhibitor buffer, and it was higher with the 0.1 *N* acetic acid buffer than with buffer of Bennett *et al.*[35] In addition, the levels

[34] A. J. Barrett, *Fed. Proc., Fed. Am. Soc. Exp. Biol.* **39,** 9 (1980).
[35] H. P. J. Bennett, C. A. Browne, and S. Solomon, *Biochemistry* **20,** 4530 (1981).

of LHRH-like immunoreactivity in the hypothalamus of the acid–enzyme inhibited extracts were similar to those reported in the literature.[32] For this reason, we elected to use the protocol in Fig. 3 and the buffer in Table I in all subsequent peptide extractions.

High-Performance Size-Exclusion Chromatography

Following peptide extraction and filtration, samples were submitted to HPSEC for separation of material by MW. HPSEC was selected over conventional Sephadex chromatography because HPSEC provides superior resolution, recoveries are usually higher, and the run time is much shorter. Since the deduced amino acid sequence for rat GAP contains one cysteine, dimers may form between other GAP peptides or with other sulfur-containing substances. Hence, samples were reconstituted in a 5 *M* guanidine hydrochloride, 50 m*M* sodium sulfate, 20 m*M* sodium phosphate (pH 6.8), and 6.0% 2-mercaptoethanol (2-ME) solution, heated to 85° for 10 min, cooled, and left at room temperature for at least 1 hr prior to chromatography to denature proteins and to reduce any disulfide-bridged substances. Samples were separated according to MW on a ToyoSoda TSK-G2000SW column (7.5 × 600 mm) using 5 *M* guanidine hydrochloride, 50 m*M* sodium sulfate, and 20 m*M* sodium phosphate (pH 6.8) as the eluting buffer. Absorbance was monitored at 280 nm, and the separation was run at 0.5 ml/min at room temperature while 0.5-min fractions were collected. Samples were collected in polypropylene tubes containing 10 μl 0.8% 2-ME and 100 μg bovine serum albumin. All fractions were desalted with Waters C_{18} Sep-Pak cartridges, as previously described, and lyophilized. Samples were reconstituted in RIA buffer[36] and analyzed for LHRH- and GAP-like immunoreactivity.

Radioimmunoassays

LHRH- and GAP-like immunoreactivities were quantitated with antisera which recognized different regions of the GAP and LHRH peptides. Figure 1B depicts the different antibodies and their antigenic determinants (LHRH) or the synthetic peptides to which they were generated (GAP). Three different antisera were used to quantitate LHRH-like immunoreactivity. The A772 and A743 antisera were generous gifts of Dr. A. Arimura, while the Rice #5 antisera was kindly provided by Dr. G. Rice and Dr. A. Barnea. The A743 antisera recognizes the midportion of LHRH,[19] while the A772 antisera requires the amidated carboxy terminus

[36] M. D. Culler and A. Negro-Vilar, *Brain Res. Bull.* **17,** 219 (1986).

of LHRH for recognition.[37] With the Rice #5 antisera, the amino terminus and the amidated carboxy terminus are required for detection.[38]

Four different antisera were employed to examine processing of GAP. The MC-2 antiserum was kindly provided by Dr. M. Culler, and the 24A, 56A, and KN-1 antisera were generously provided by Dr. K. Nikolics. The MC-2 antiserum was raised against the sequence contained in human (h) GAP (25–53), and it recognized mid-portion fragments of rat (r) GAP (20–43) and hGAP (25–53) and the entire rat and human GAP sequences.[36] The amino and carboxy termini of rGAP or hGAP are not detected by this antiserum. The 24A, 56A, and KN-1 antisera were raised against the amino terminus, mid-portion and carboxy terminus of rGAP, respectively. Each of these three antisera have a much higher affinity for the different peptide fragments to which they were generated than to either rGAP (1–56) or hGAP (1–56). Standard curves were run using MC-2 antiserum with ^{125}I-[Tyr^{0}]hGAP (1–56) and unlabeled hGAP (1–56). Additional GAP assays were conducted with 24A, 56A, and KN-1 antisera using rGAP (1–11), rGAP (20–43), and rGAP (39–53) as the respective standards against the corresponding radioiodinated *p*-hydroxyphenylmaleimide derivatives.[39] All iodinations were performed with the chloramine-T method.

Reversed-Phase High-Performance Liquid Chromatography

All fractions from the HPSEC separation were examined for LHRH-like immunoreactivity with the three antisera and for GAP-like immunoreactivity with the four antisera. When GAP-like immunoreactivity was detected by all four GAP antisera at the 6500 MW, ME fractions comprising and immediately surrounding this peak were pooled and submitted to a second separation by reversed-phase chromatography (RPC). This separation was performed to ensure that the immunoreactive peak from HPSEC was homogeneous. The RPC separation was performed on a 5-μm Vydac C_{18} column (4.5 × 250 mm) run at 1 ml/min at room temperature. Materials were separated using a linear gradient from 20 to 70% acetonitrile in 0.085% trifluoroacetic acid that was developed over a 60-min period. One-minute fractions were collected, lyophilized, reconstituted with RIA buffer and run in the GAP RIA using MC-2 antiserum.

[37] A. Arimura, N. Nishi, and A. V. Schally, *Proc. Soc. Exp. Biol. Med.* **152,** 71 (1976).

[38] M. P. Hedger, D. M. Robertson, C. A. Browne, and D. M. de Kretser, *Mol. Cell. Endocrinol.* **42,** 163 (1985).

[39] I. Clarke, J. T. Cummins, F. J. Karsch, P. H. Seeburg, and K. Nikolics, *Biochem. Biophys. Res. Commun.* **143,** 665 (1987).

Chromatographic and Radioimmunoassay Results

The protein concentrations were similar in the POH and in the ME from ORDX and intact rats, and, therefore, it is assumed that ORDX had no effect on the overall biosynthesis and degradation of proteins. In addition, recoveries for LHRH- and GAP-like immunoreactivity at each step from homogenization through chromatography were similar for ORDX and intact rats. Thus, any effect on the concentration of LHRH- and GAP-like immunoreactivity in ORDX or intact animals must be due to the treatment.

The chromatographic profiles for LHRH-like immunoreactivity in both groups are shown in Fig. 4 for the POH and in Fig. 5 for the ME. In Figs. 4 and 5, panels A, B, and C reflect the LHRH-like immunoreactivity that was measured with the respective antisera: A743, A772, and Rice. Despite these antisera having different antigenic determinants for the LHRH molecule (see Fig. 1B), only one LHRH-like immunoreactive form was detected in both the POH and the ME. This LHRH-like immunoreactive substance eluted at approximately 1200 MW, in the same position as synthetic LHRH.

Chromatograms depicting the hGAP (1–56)-, rGAP (1–11)-, rGAP (20–43)-, and rGAP (39–53)-like immunoreactivities are displayed in Figs. 6–9. In panel A of Figs. 6–9 can be seen the GAP-like immunoreactivity detected in the POH, while panel B shows the immunoreactivity in the ME. Some GAP-like immunoreactive forms were detected by all four antisera while other forms were recognized by some or only one of the antisera. All four antisera bound to GAP-like immunoreactive material at approximately 14,000–16,000 and at 6,500 MW. The three rGAP antisera also recognized immunoreactivity at approximately 8,200 MW. Although the MC-2 antiserum did not distinguish clearly this immunoreactive material, the 6,500 MW peak was broad and the higher MW substance could have been obscured. The amino terminal and mid-portion rGAP antisera detected immunoreactive material at approximately 3,500 MW, while the carboxy-terminal rGAP antisera recognized an immunoreactive substance at approximately 2,800 MW. Other, smaller MW material also bound to the amino- (Fig. 7) and carboxy-terminal (Fig. 9) antisera.

Proposed Processing Scheme for proLHRH

From our HPSEC and immunoreactivity results we have proposed a processing scheme for proLHRH. The processing model in Fig. 10 was devised (1) by estimating the MW of the immunoreactive substances rec-

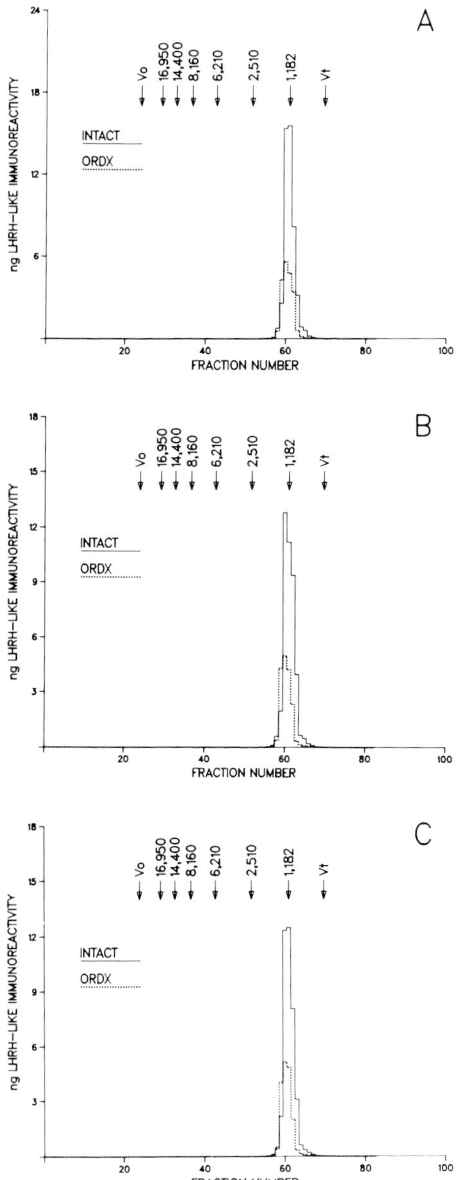

FIG. 4. High-performance, size-exclusion chromatograms of LHRH-like immunoreactivity in the preoptic hypothalamic area (POH) of intact and orchidectomized (ORDX) rats. Molecular weight markers which indicate the approximate MW of the immunoreactive material are shown at the top of each figure. V_0 is the void volume, and V_t is the total volume of the column. A, B, and C depict the chromatographic profiles of LHRH-like immunoreactivity detected with A743, A772, and Rice antisera, respectively.

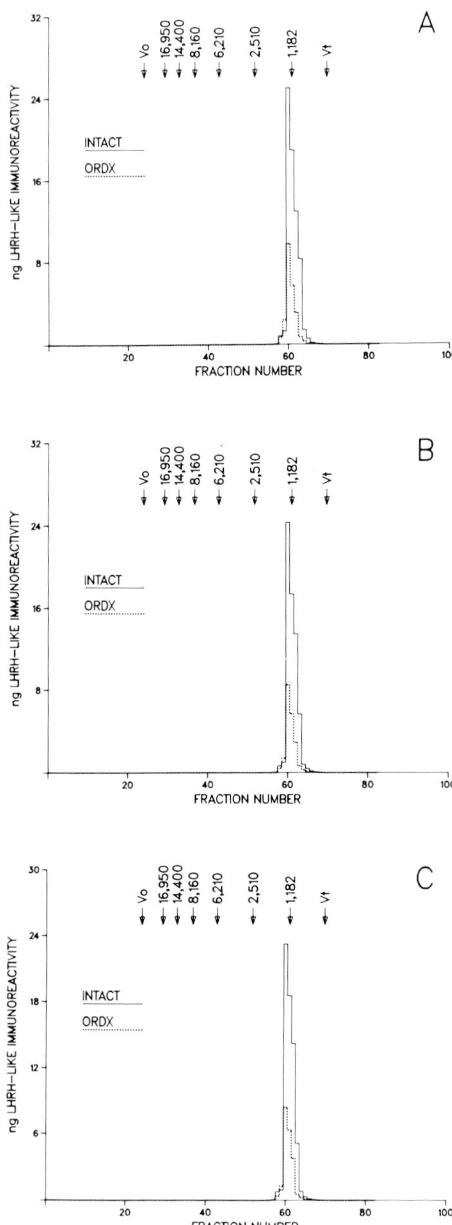

FIG. 5. Chromatograms of LHRH-like immunoreactivity in the median eminence (ME) of intact and ORDX rats. For details, see the legend to Fig. 4.

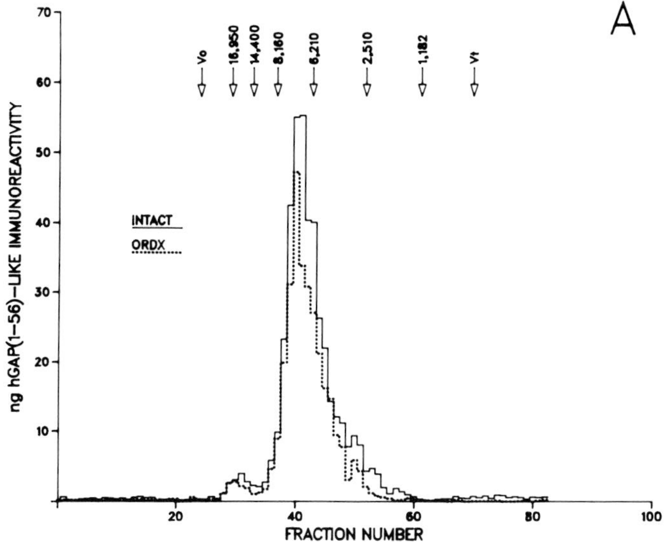

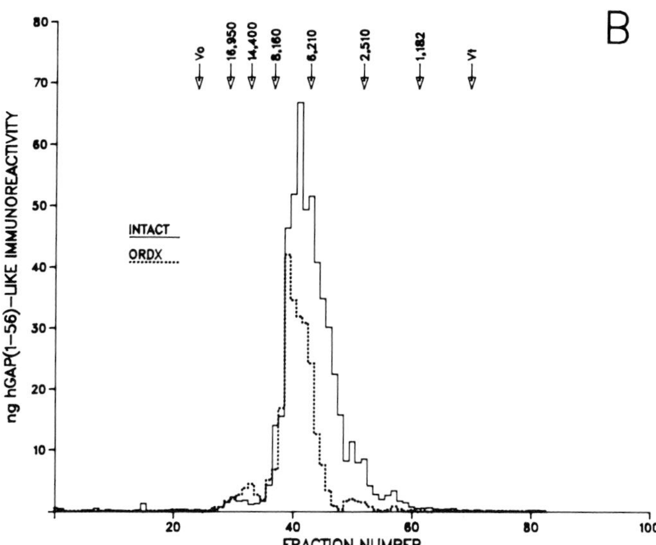

FIG. 6. High-performance, size-exclusion chromatograms of hGAP (1–56)-like immunoreactivity in the POH (A) and the ME (B) of intact and ORDX rats. MW markers are shown at the top of each figure.

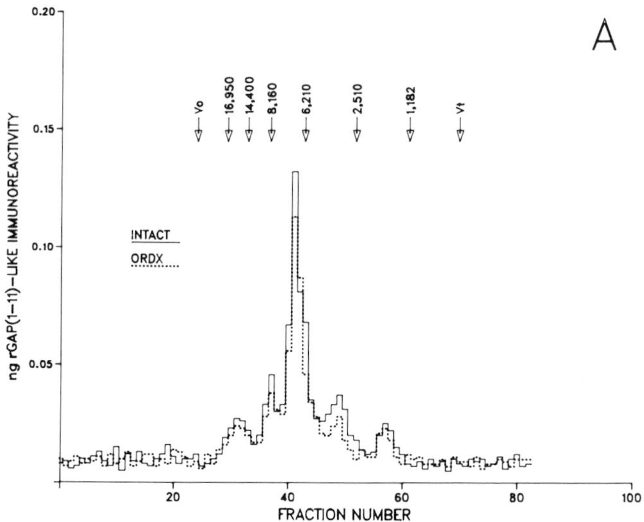

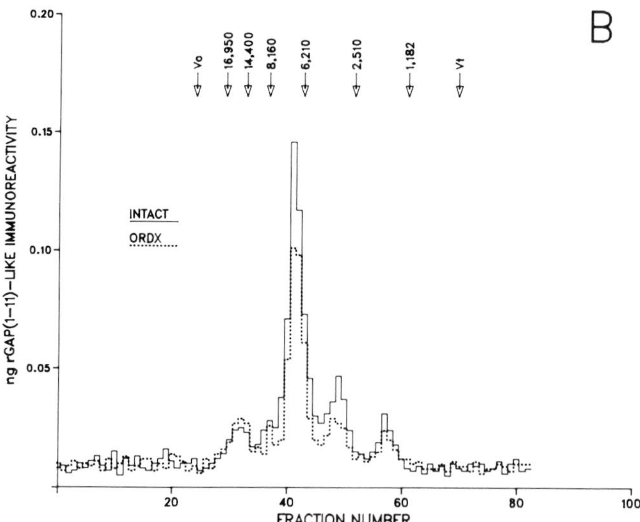

FIG. 7. Chromatograms of rGAP (1–11)-like immunoreactivity in the POH (A) and ME (B) from intact and ORDX rats. For details, see the legend to Fig. 6.

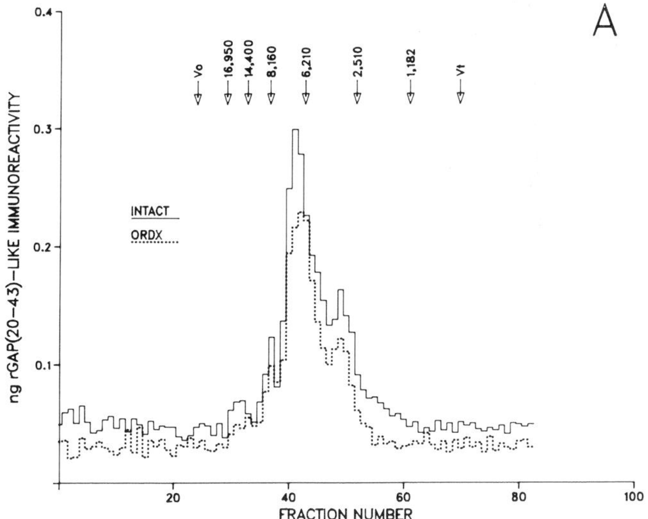

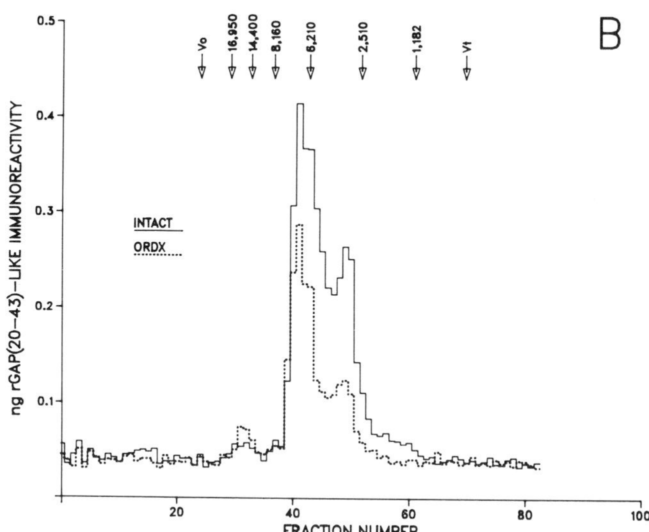

FIG. 8. Chromatograms depicting rGAP (20–43)-like immunoreactivity in the POH (A) and ME (B) from intact and ORDX rats. For details, see the legend to Fig. 6.

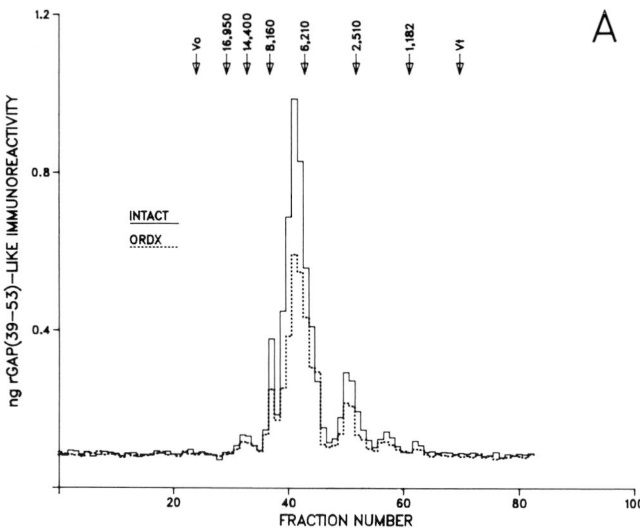

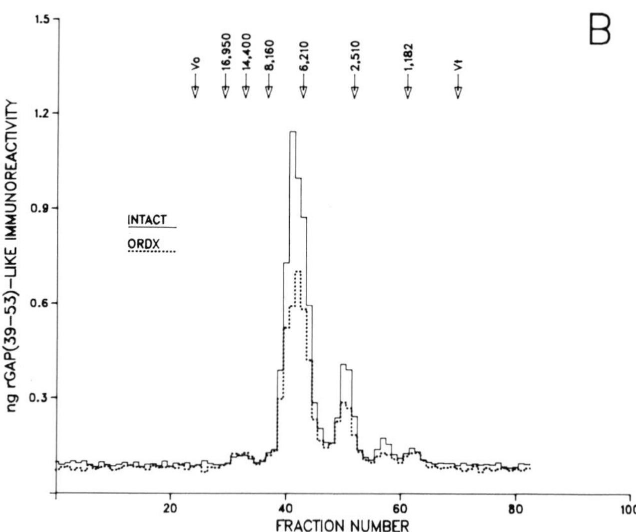

FIG. 9. Chromatograms of rGAP (39–53)-like immunoreactivity in the POH (A) and in the ME (B) from intact and ORDX rats. For details, see the legend to Fig. 6.

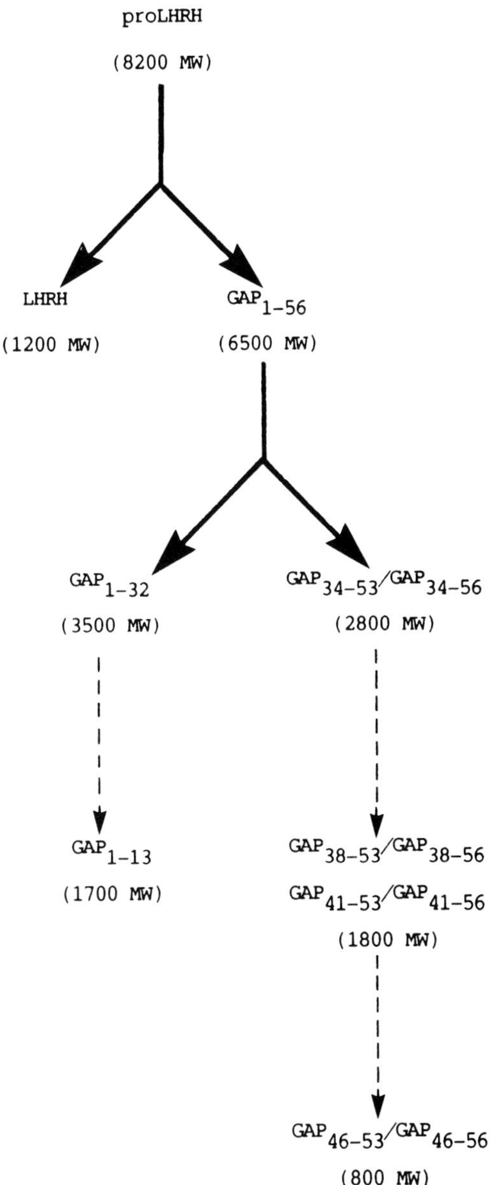

FIG. 10. Proposed processing scheme of the proLHRH peptide. The estimated MW from chromatography is located in parentheses under each immunoreactive form. The numbers assigned to each GAP represent the amino acids that comprise the peptide (see Fig. 1B). The solid lines connect immunoreactive species that comprise the major pathway. The dashed lines connect GAP species which are tentatively proposed as degradation products.

ognized by each GAP and LHRH antisera; (2) by deducing the "structure" of the immunoreactive material by noting which antisera (e.g., amino-, mid-portion, carboxy-terminal GAP antisera) bound to which immunoreactive substance; (3) by calculating the MW of the various possible peptides (e.g., those cleaved at basic residues) that would include and that might be derived from proLHRH; and, finally, (4) by comparing the estimated MW and proposed "structure" of the immunoreactive material with the calculated MW and deduced amino acid sequence of the peptide. This proposed processing scheme made the following assumptions: (1) that the deduced amino acid sequence and the size of the proLHRH peptide were correct; (2) that the HPSEC method separated material according to MW and that the estimation of the MW of the immunoreactive material was reliable; (3) that the different GAP antisera were specific, in that they detected only proLHRH and all of its metabolic products; (4) that proteolysis of the proLHRH peptide occurred only at basic amino acid residues by trypsin-like enzymes; (5) that this carboxy-terminal basic amino acid was removed by a carboxypeptidase B-like enzyme; and (6) that all LHRH neurons biosynthesized the same proLHRH and processed it by the same pathway. A diagram of the different basic residues found in the deduced sequence for proLHRH is displayed in Fig. 1C, which demonstrates that there are many potential cleavage sites (basic amino acids) in proLHRH.

All four GAP antisera recognized immunoreactive material at approximately 14,000 to 16,000 MW (see Figs. 6–9). The amount of this substance determined by the area under the curve declined from the POH to the ME. This was the highest MW material detected, and all four GAP antisera bound to it. Since the deduced amino acid sequence for rat proLHRH should have a MW of 8,200, the observed high MW material is too large to be proLHRH alone. It should be noted that proLHRH and rGAP (1–56) contain one cysteine residue which may have permitted disulfide bridge formation between the GAP-containing peptides themselves [proLHRH or GAP (1–56)] or other cysteine-containing material. However, we attempted to circumvent this problem by incubating and chromatographing our samples under dissociating and denaturing conditions. At this time, we can only speculate that the 14,000–16,000 MW material may be a dimer of proLHRH, a dimer of GAP (1–56), or a heterodimer of proLHRH and GAP (1–56), or proLHRH or GAP may be linked to some other cysteine-containing fragment(s).

After the immunoreactive material at 14,000–16,000 MW, the next smaller MW material was detected at approximately 8,200. More of this material was detected in the POH than in the ME of intact or ORDX rats. While all three of the rGAP fragment antisera recognized this substance (see Figs. 7–9), the only GAP-derived peptide in this MW range is proLHRH. However, none of the LHRH antisera bound to this material. The Rice and A772 antisera could not be expected to bind to the material because, together, they required LHRH to be cleaved from the proLHRH peptide and Gly[10] to be α-amidated.[37,38] By contrast, the A743 antiserum, which detected the mid-portion of LHRH, would be expected to bind to proLHRH,[19] but no peptide in this MW range was recognized by this antiserum. Perhaps the conformation of proLHRH precluded its detection by the A743 antiserum.

Following the 8200 MW material, a substance of approximately 6500 MW was recognized by all four GAP antisera (see Figs. 6–9). Hence, this immunoreactive material must contain the amino-terminal, mid-portion, and carboxy-terminal regions of rGAP. The 6500 MW substance eluted in the same position as synthetic rGAP (1–56) from the size-exclusion column. However, resolution on a gel-permeation column is not sharp, and rGAP (1–56) contains a dibasic residue at its carboxy terminus that could give rise to rGAP (1–53) which would also be detected by all four GAP antisera. In order to more accurately determine the nature of this 6500 MW material, fractions containing and immediately surrounding the peak were combined and rechromatographed on a C_{18} column. Since rGAP (1–56) contains two lysine and one methionine residue at its carboxy terminus, it should be more hydrophobic than rGAP (1–53) and, hence, should elute later in the chromatogram. For this reason, rGAP (1–53) should not elute in the same position as a synthetic rGAP (1–56) standard. When the 6500 MW material was rechromatographed, it coeluted with synthetic rGAP (1–56) (see Fig. 11). The immunological and chemical criteria applied to the 6500 MW substance strongly suggest that it is rGAP (1–56). In addition, since all four GAP antisera recognized the 6500 MW substance as the most abundant of all GAP-like immunoreactive material, and because the quantity of this material was greater in the ME than in the POH, this 6500 MW substance must be a major metabolic product in the proLHRH pathway.

If rGAP (1–56) was cleaved from proLHRH, then a LHRH-like peptide that eluted at 1100–1500 MW should be present (see Fig. 10). All three LHRH antisera recognized similar quantities of a substance that eluted at approximately 1200 MW in both the POH and ME. In addition, this substance coeluted with synthetic LHRH. It should be noted that some investigators,[40–42] using antisera different from our own, have re-

[40] R. P. Millar, C. Aehnelt, and G. Rossier, *Biochem. Biophys. Res. Commun.* **74,** 720 (1977).

[41] J. P. Gautron, E. Pattou, and C. Kordon, *Mol. Cell. Endocrinol.* **24,** 1 (1981).

[42] A. Curtis and G. Fink, *Endocrinology (Baltimore)* **112,** 390 (1983).

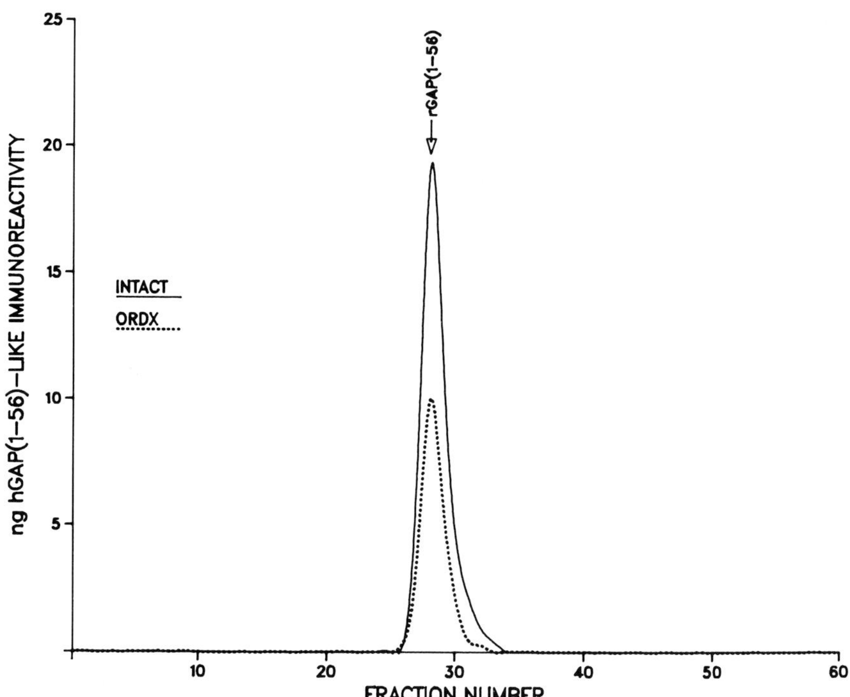

Fig. 11. High-performance reversed-phase chromatogram of the 6500 MW immunoreactive material from the ME of intact and ORDX rats that was obtained from a previous separation on a size-exclusion column (see Fig. 6). The material was separated on a C_{18} column with an acetonitrile gradient from 20 to 70% in 0.085% trifluoroacetic acid over a 60-min period. One-minute fractions were collected and assayed with MC-2 antiserum (see Fig. 1B) and expressed in terms of hGAP (1–56)-like immunoreactivity. The elution position of synthetic rGAP (1–56) is indicated.

ported high MW forms of LHRH. While our three antisera recognized only one form of LHRH, there may be others that were not detected.

In order for the three antisera to detect LHRH in the POH, LHRH must be cleaved from proLHRH, Gln^1 must be converted to pyroglutamyl-LHRH, and Gly^{10} must be α-amidated. While our data could not distinguish events which occurred in the cell body from those in the axon (because the POH contains both regions of the LHRH neuron), our results indicated that these three processing events must occur soon after proLHRH biosynthesis.

Since high quantities of LHRH-like immunoreactive material were detected by all three antisera and no other immunoreactive substances were recognized and because high levels of GAP-like immunoreactivity were detected by all four GAP antisera at approximately 6500 MW, LHRH and GAP (1–56) must be major metabolic products in the pro-LHRH pathway. Indeed, when the molar ratios of the area under the curves for GAP-like immunoreactivity eluting at 6500 MW (using MC-2 antiserum) and LHRH-like immunoreactivity eluting at 1200 MW (using A772 antiserum) were calculated from intact ME, the ratio was 1.0. This calculation indicates that LHRH and rGAP (1–56) are the major metabolites in the proLHRH pathway of the rat.

Although we had no evidence that LHRH was processed further, additional GAP-like immunoreactive forms were found. The amino-terminal and mid-portion antisera both detected GAP-like material at approximately 3500 MW (see Figs. 7 and 8), while the carboxy-terminal antiserum recognized material at 2800 MW. These immunoreactive substances would correspond to GAP (1–32) and GAP (34–53) or GAP (34–56) (see Fig. 10). Interestingly, a synthetic peptide containing part of the amino-terminal region of GAP has been reported to possess LH and FSH releasing activity.[28,29]

Several additional GAP-like substances were also detected with the amino- and carboxy-terminal antisera (see Figs. 7 and 9). However, because the binding of these materials was very low and they were not consistently detected in some of our other experiments, they probably represent degradation products of GAP. Some rGAP (1–11)-like immunoreactivity was seen at approximately 1700 MW. This material may be GAP (1–13) which has been reported to possess biological activity.[28] If GAP (1–13) is produced, then GAP (15–32) should also be present. Neither the MC-2 nor the 56A antiserum detected any material at this low MW (see Figs. 6 and 8).

Additional GAP-like material was also found with the carboxy-terminal antiserum (see Fig. 9). These substances were poorly recognized by this antiserum, and they probably represent degradation products of GAP. The material at approximately 1800 MW might be one of the carboxy-terminal fragments: GAP (38–53), GAP (38–56), GAP (41–53), or GAP (41–56). The substance at approximately 800 MW might be GAP (46–53) or GAP (46–56). However, the latter fragment would be specific to the rat since the hGAP (1–56) does not contain a basic residue at this position. The resolution of the HPSEC and the specificity of the KN-1 antiserum did not permit us to clearly distinguish among these fragments to determine which might actually be present.

The presence of additional GAP-like immunoreactive material beyond the 6500 MW species created special problems for the proposed pathway in Fig. 10. Since the deduced sequence of proLHRH contained 1 mol of LHRH and 1 mol of GAP (1–56) and a molar ratio of approximately 1 was observed, the appearance of additional processed forms of GAP should be accompanied by additional LHRH. We did not see this latter relationship. For this reason, there might be additional LHRH–GAP forms, or LHRH might be degraded[43] to forms that escaped detection by our antisera.

When the processing pathway for proLHRH was examined with respect to castration, the same processed forms were seen as in the intact POH and ME. The absolute quantities of LHRH- and GAP-like immunoreactivities were at least 2-fold lower in the POH and ME of ORDX rats than in the same regions of intact rats. In both groups, the ME contained more LHRH- and GAP-like material than the POH. When the proportion of high MW material (14,000–16,000 and 8,200 MW) to the total GAP-like immunoreactive material was calculated, a higher proportion was found in the ME of ORDX rats than in the same region from intact rats. This difference was reflected in the different molar ratios of GAP to LHRH in the ME of intacts (1.0) compared to that in ORDX (1.5) rats. Taken together, these results demonstrated that ORDX rats may be deficient in processing of the LHRH precursor in at least two respects. First, the lower levels of LHRH- and GAP-like immunoreactive forms in both the POH and ME of ORDX animals indicated that biosynthesis of proLHRH was decreased, or that degradation and/or secretion of its products was increased relative to intact controls. The depression in biosynthesis of proLHRH might be the more parsimonious explanation since the forebrains of intact rats contained more LHRH mRNA *in situ* than that found in castrated rats.[44] Second, the higher molar ratio of GAP to LHRH in the ME of ORDX rats indicated that processing of proLHRH was retarded or that axoplasmic transport of processed products was slowed relative to intacts. While some investigators[45,46] have reported that the *in vivo* release of LHRH from hypothalamus is increased by castration, we[47] have shown that secretion of LHRH *in vivo* into the hypophyseal portal circulation and *in vitro* using ME explants is reduced relative to intact controls. Other work seems to support this view.[32] Therefore, ORDX must exert its effects at biosynthesis, processing, and transport of LHRH.

The combined use of HPLC and antisera that are specific for different fragments of the proLHRH molecule has permitted us to identify multi-

[43] J. E. Krause, J. P. Advis, and J. F. McKelvy, *Biochem. Biophys. Res. Commun.* **108,** 1475 (1982).

[44] J. M. Rothfeld, J. F. Hejtmancik, P. M. Conn, and D. W. Pfaff, *Exp. Brain Res.* **67,** 113 (1987).

[45] R. L. Eskay, R. S. Mical, and J. C. Porter, *Endocrinology (Baltimore)* **100,** 263 (1977).

[46] N. Ben-Jonathan, R. S. Mical, and J. C. Porter, *Endocrinology (Baltimore)* **93,** 497 (1973).

[47] M. M. Valenca, M. Ching, C. Masotto, and A. Negro-Vilar, *in* "Regulation of Ovarian and Testicular Function" (V. B. Mahesh, E. Anderson, D. Bhindsa, and S. Kalra, eds.), pp. 617–621. Raven, New York, 1987.

ple processed products of this precursor. This approach became unique when we characterized the different processed products in terms of their location along the LHRH neuron and when we examined the regulation of this pathway by gonadal factors. Perhaps, through additional biochemical and physiological manipulations, we will be able to dissect more clearly the contribution of different gonadal factors to the biosynthesis, processing, transport, secretion, and/or degradation of these various processed products from LHRH neurons.

[39] Chromatographic Methods for Characterization of Angiotensin in Brain Tissue

By KLAUS HERMANN, MOHAN K. RAIZADA, and M. IAN PHILLIPS

Introduction

A variety of previously unknown peptides such as atrial natriuretic peptide[1] or head activating peptide[2] have been discovered in the last few years. Likewise, a great number of peptides first described in plasma or in the gastrointestinal tract has been identified in other tissues such as heart, kidney, adrenal gland, gonads, and the central nervous system (for a review, see Ref. 3). A representative of such a peptide is angiotensin, which was first isolated and purified from plasma[4,5] but which now has been identified in adrenal gland, testis, heart, and brain.[6-9] The production of antiangiotensin antibodies made it possible to localize the peptide in different tissues using immunocytochemistry[10-12] or to identify it in tissue

[1] H. Matsuo and K. Kangawa, *Clin. Exp. Ther. Pract.* **A6,** 1717 (1984).

[2] H. Bodenmüller and H. C. Schaller, *Nature (London)* **293,** 579 (1981).

[3] D. T. Krieger, M. J. Brownstein, and J. B. Martin (eds.), "Brain Peptides." Wiley, New York, 1983.

[4] W. S. Peart, *Biochem. J.* **62,** 520 (1956).

[5] L. T. Skeggs, J. R. Kahn, and N. E. Shumway, *J. Exp. Med.* **103,** 301 (1956).

[6] G. Aguilera, A. Schirar, A. Baukal, and K. J. Catt, *Nature (London)* **289,** 507 (1981).

[7] V. J. Dzau, *Am. J. Cardiol.* **59,** 59A (1987).

[8] R. E. Richard, *Am. J. Cardiol.* **59,** 56A (1987).

[9] K. Hermann, D. Ganten, C. Bayer, T. Unger, R. E. Lang, and W. Rascher, *in* "Experimental Brain Research: The Renin–Angiotensin System in the Brain" (D. Ganten, M. Printz, M. I. Phillips, and B. A. Schölkens, eds.), Suppl. 4, p. 192. Springer-Verlag, Berlin and New York, 1982.

[10] K. Fuxe, D. Ganten, T. Hökfelt, and P. Bolme, *Neurosci. Lett.* **2,** 229 (1976).

[11] M. I. Phillips, J. Weyhenmeyer, D. Felix, D. Ganten, and W. E. Hoffman, *Fed. Proc., Fed. Am. Soc. Exp. Biol.* **38,** 2260 (1979).

[12] R. W. Lind, L. W. Swanson, and G. Ganten, *Neuroendocrinology* **40,** 2 (1985).

extracts.[13,14] Both techniques, however, possess inherent problems such as specificity of the antibodies, nonspecific interference or interference with chemically related peptides, or cross-reactivity with degradation products. This had led to controversies which still need to be resolved. For a further and more conclusive characterization, methodologies were needed to overcome these difficulties.

In this chapter we describe the use of chromatographic methods in combination with immunological procedures for the characterization of angiotensin peptides in brain tissue. The production of angiotensin II antibody, its purification by affinity chromatography and characterization, as well as the isolation and purification of angiotensin II with affinity and high-performance liquid chromatography (HPLC) are presented. The methodology is also applied to brain cells in culture for the study of the pathways involved in the synthesis and degradation of angiotensin peptides in neuronal and glial cells. Taking angiotensin as a model peptide, this methodology offers a general approach for the characterization of other peptides generated and localized in brain and other tissues.

Angiotensin II Antibody Purification

For the measurement of angiotensin II (Ang II) in brain tissue a specific and sensitive antiserum is needed. Because of the necessity of this analytical tool we produced polyclonal Ang II antibodies.

Antibody Preparation

Antiserum against [Ile^5]Ang II is produced in New Zealand White rabbits by coupling the peptide to either bovine serum albumin or bovine thyroglobulin using carbodiimide or glutaraldehyde. After purification of the conjugates on a Sephadex G-50 column developed with 50 mM Tris buffer, pH 7.4, containing 10 mM NaCl, the conjugates are emulsified with Freund's complete adjuvant and administered intradermally. Several booster injections are given. The antibody titers are monitored periodically, and after they reached high plateau levels the carotid artery is catheterized and blood withdrawn. Sera are kept frozen in aliquots at −80°.[14,15]

[13] D. Ganten, K. Hermann, C. Bayer, T. Unger, and R. E. Lang, *Science* **221,** 869 (1983).

[14] M. I. Phillips and B. Stenstrom, *Circ. Res.* **56,** 212 (1985).

[15] K. Hermann, B. Kimura, and M. I. Phillips, *Biochem. Biophys. Res. Commun.* **136,** 685 (1986).

Coupling of Ang II to Affi-Gel 102

Two milliliters Affi-Gel 102 suspended in 2 ml distilled water is mixed with 20 mg [Ile^5]Ang II corresponding to 18.2 μmol dissolved in 2 ml distilled water in the presence of 20 mg 1-ethyl-3-(3-dimethylaminopropyl)carbodiimide. The pH is adjusted with 1 *N* HCl to between 4.7 and 5.0, and the reaction mixture is shaken overnight at room temperature. The gel suspension is transferred to a plastic column (6 $\times$ 0.8 cm), and the gel with covalently bound [Ile^5]Ang II (Affi-Gel 102–Ang II) is successively washed 5 times with 1 ml 0.1 *M* sodium acetate, pH 4.0, 5 times with 1 ml 0.1 *M* sodium borate, pH 8.0, and stored in 50 m*M* 3-(*N*-morpholino)propanesulfonic acid buffer (MOPS), pH 7.4, containing 0.02% sodium azide at 4°.

The coupling efficiency of [Ile^5]Ang II to Affi-Gel 102 is monitored radioimmunologically. Routinely, more than 95% of the [Ile^5]Ang II was covalently bound to the gel. The column did not show any significant loss in binding capacity after being used for several months when treated as indicated. From time to time it is advisable to remove nonspecifically bound protein with 5 bed volumes of 1 *M* urea containing 1 *M* NaCl.

Affinity Chromatography on Affi-Gel 102–Ang II

Ang II antiserum is purified on Affi-Gel 102–Ang II at 4°. The column is equilibrated with 10 ml 50 m*M* phosphate buffer, pH 7.4, at a flow rate of 0.8 ml/min, and 2 ml antiserum is applied. The column is washed 3 times with 5 ml 50 m*M* phosphate buffer, pH 7.4, and the elution is carried out with 0.1 *M* sodium citrate buffer containing 1 *M* NaCl, using a stepwise pH gradient from pH 5.0, 4.0, 3.0 to 2.0. Fractions of 5 ml are collected in plastic tubes, each containing 1 ml 0.5 *M* phosphate buffer, pH 7.4. Aliquots of each fraction are taken to measure protein according to the method of Lowry *et al.*[16] and to measure the binding of ^{125}I-Ang II. For this purpose, 0.05 ml of each fraction is withdrawn and mixed with 0.45 ml 50 m*M* Tris–HCl buffer, pH 7.4, containing 0.2% bovine serum albumin (BSA). This solution is further diluted in the same buffer to yield about 50% binding of added ^{125}I-Ang II. To determine the specificity of the eluting proteins for Ang II, 1.7–455 fmol synthetic [Ile^5]Ang II is added to displace bound ^{125}I-Ang II.

Native Ang II antiserum could be applied directly to the column without any preceding treatment (Fig. 1). The majority of the protein could be

[16] O. H. Lowry, N. J. Rosebrough, A. L. Farr, and R. J. Randall, *J. Biol. Chem.* **193,** 265 (1951).

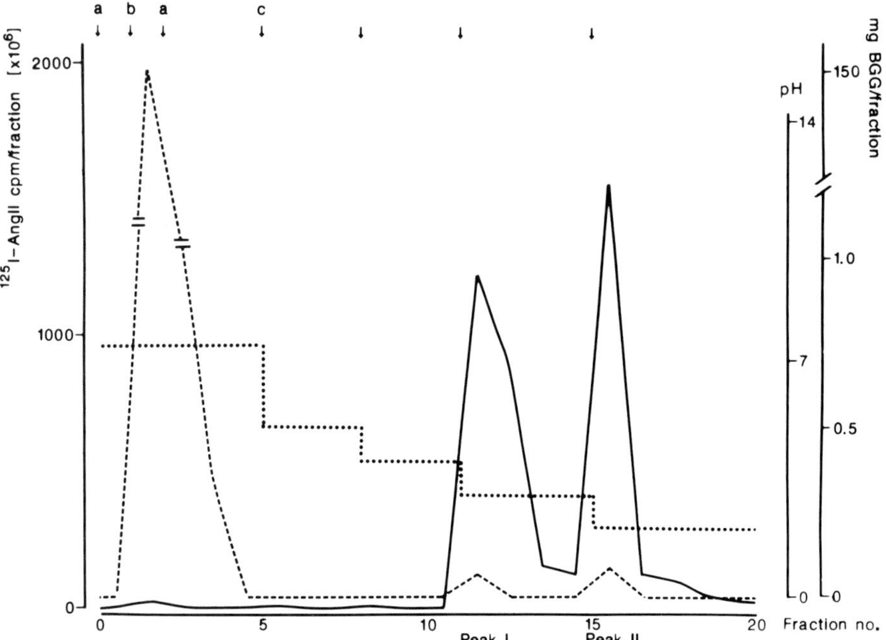

FIG. 1. Affinity purification of Ang II antiserum on Affi-Gel 102–Ang II. The column was equilibrated with 50 mM phosphate buffer, pH 7.4. Antiserum was applied, and the column was washed with 50 mM phosphate buffer. The elution was performed with 0.1 M sodium citrate buffer containing 1 M NaCl, using a stepwise pH gradient from pH 5.0 to 2.0. For neutralization the fractions contained 1 ml 0.5 M phosphate buffer, pH 7.4. Aliquots of each fraction were withdrawn to measure protein and binding of ^{125}I-Ang II. Most of the protein with very low binding of ^{125}I-Ang II eluted in fractions 2, 3, and 4, whereas two peaks with a very low protein concentration but very high binding capacity for ^{125}I-Ang II could be eluted at pH 3.0 and 2.0. —, Binding of ^{125}I-Ang II; ---, protein; ····, pH gradient profile; a, equilibration, respective wash; b, sample application; c, elution.

identified in fractions 2, 3, and 4, which showed negligible binding of ^{125}I-Ang II. At pH 3.0 (peak I) and 2.0 (peak II), two distinct protein species with high binding affinity to ^{125}I-Ang II could be eluted. In contrast to the unretained protein peak, these two peaks revealed a very low protein concentration. The low protein concentration and the high specific binding capacity speak in favor of a very pure preparation of both peaks. Calculations of the ratio between the specific binding capacity and the protein concentration showed values of 0.54 for the unretained protein peak, whereas for peaks I and II values of 10,400 and 10,500, respectively, were obtained, representing a 19,000-fold purification in a single step. That the binding was specific could be demonstrated by displacement of bound ^{125}I-Ang II with increasing concentrations of unlabeled synthetic [Ile^5]Ang II.

Figure 1 shows the purification of 2 ml antiserum, but the column could also be used for large-scale purification without any loss in resolution. Loading of the column with about 11 ml antiserum corresponding to about 800 mg protein in three steps with intermediate washings produced no significant differences in the elution profile and binding characteristics. From the total amount of 808.8 mg protein applied to the column, 806 mg was unretained and 2.8 mg (corresponding to 0.4%) could be eluted with the gradient. Referring to peak I (Fig. 1), eluting at pH 3.0, the protein concentration of this peak was 1.4 mg, which was 50% of the elutable and 0.2% of the total amount of protein added to the column.

Characterization of Purified Ang II Antiserum

Peaks I and II were further characterized to determine their molecular weights and their cross-reactivities. A Sephadex G-200 column (1.5 × 30 cm) was flushed with 10 m*M* ammonium acetate buffer, pH 5.4, containing 0.1 *M* NaCl at a flow rate of 0.140 ml/min. The column was calibrated with bovine thyroglobulin (MW 669,000), horse apoferritin (MW 443,000), yeast alcohol dehydrogenase (MW 150,000), bovine serum albumin (MW 69,000), and carbonate dehydratase (MW 29,000). The elution was monitored with a UV detector at a wavelength of 280 nm and a sensitivity of 0.1 absorbance units full scale. Under identical conditions, 0.5 ml of fraction 12 (peak I) and 16 (peak II) were separated. Fractions were collected at 10-min intervals, and 0.05-ml aliquots of each fraction were withdrawn to determine binding of ^{125}I-Ang II. Peak I eluted with two maxima, the first (I_1) corresponding to a molecular weight of 150,000 and the second (I_2) of 60,000. The elution of peak II was coincident with that of the 150,000 MW protein marker. No protein with a higher molecular weight such as immunoglobulin M (IgM) with a MW of 900,000 was present, suggesting that the two purified proteins were γ-globulin (immunoglobulin G, IgG) and probably an IgG degradation product.

The cross-reactivity of native antiserum and the purified Ang II-binding proteins (peaks I and II) was tested in a radioimmunoassay against a standard curve of [Ile^5]Ang II (1.7–455 fmol). Native and purified antisera were diluted to yield 50% binding of ^{125}I-Ang II, and different angiotensin peptides as well as oxytocin (OXT) and Arg-vasopressin (AVP), ranging from 1 to 5000 fmol, were screened for their ability to displace bound ^{125}I-Ang II. The results are summarized in Table I. For peptides chemically unrelated to angiotensin peptides, such as OXT and AVP, native antiserum as well as peaks I and II showed no cross-reaction. Native antiserum

TABLE I
CROSS-REACTIVITY OF NATIVE Ang II ANTISERUM AND PEAKS I AND II WITH VARIOUS PEPTIDES

Peptide	MW	Cross-reactivity (%)		
		Antiserum	Peak I	Peak II
[Ile^5]Ang II	1046	100	100	100
[Ile^5]Ang III	931	100	36	36
[Ile^5]Ang II (3–8) hexapeptide	757	100	0.06	0.14
[Ile^5]Ang II (5–8) tetrapeptide	513	0	0	0
[Ile^5]Ang II (6–8) tripeptide	381	0	0	0
[Val^5]Ang II	1032	26	41	33
[Ile^5]Ang I	1296	0.5	5.5	5.9
TDP	1756	1.5	1.89	1.49
AVP	1116	0	0	0
OXT	1007	0	0	0

cross-reacted only 0.5% with [Ile^5]Ang I, the direct biosynthetic precursor of [Ile^5]Ang II, whereas the cross-reactivity with peaks I and II was 5.5 and 5.9%, respectively. Peaks I, II, and untreated antiserum did not reveal any cross-reactivity with [Ile^5]Ang II tri- and tetrapeptide. Substitution of the amino acid Ile in position 5 with Val leads to [Val^5]Ang II, found in bovine. The antiserum directed against [Ile^5]Ang II crossreacted 26%, peak II, 33%, and peak I, 41% with this peptide. [Ile^5]Ang III showed the same affinity as [Ile^5]Ang II when tested with unpurified antiserum. Peaks I and II cross-reacted only 36% with [Ile^5]Ang III. The cross-reaction of peaks I and II with [Ile^5]Ang II hexapeptide was 0.06 and 0.14%, whereas for native antiserum 100% was found. Tetradecapeptide (TDP), a synthetic renin substrate, cross-reacted 1.89% with peak I, 1.49% with peak II, and 1.5% with unpurified antiserum.

Isolation and Purification of Angiotensin II Using Affinity and High-Performance Liquid Chromatography

To purify Ang II, a three-step procedure has been developed using extraction, affinity chromatography, and high-performance liquid chromatography (HPLC). The efficiency and usefulness of the methodology for the purification of Ang II from biological sources were tested with ^{125}I-labeled and ^{3}H-labeled [Ile^5]Ang II and synthetic [Ile^5]-Ang II added to rat brains prior to extraction. Likewise, this procedure was also used for the purification of endogenous Ang II from pig brain.

Extraction

Whole rat brains or pig brain hypothalami are boiled in 1 *M* acetic acid in a ratio of 1 : 10 tissue to extraction medium, followed by homogenization using a Polytron. The homogenates are centrifuged for 30 min at 30,000 *g* at 4°. The supernatant is collected, and the pellet is reextracted with half of the original volume of 1 *M* acetic acid and recentrifuged under the same conditions. Both supernatants are pooled and defatted with equal volumes of petroleum ether. The upper petroleum phase is taken off, and petroleum ether is removed on a hot plate at 40° with a stream of air in the hood and then lyophilized. The aqueous phase is frozen and lyophilized.

Typical recoveries of added [^{125}I]Ang II and [^{3}H]Ang II after extraction were 73.78 ± 6.20% (n = 3) and 74.84 ± 1.85% (n = 5), respectively. In the petroleum ether extract 10.58 ± 1.69% (n = 3) of ^{125}I-Ang II and 7.90 ± 0.64% (n = 5) of [^{3}H]Ang II could be recovered. The recovery of synthetic [Ile^{5}]Ang II after extraction was 74.13%.

Coupling of Purified Antiserum to Affi-Gel 10

Affi-Gel 10, 1.2 ml, is washed 10 times with 1 ml ice-cold 2-propanol followed by 10 washes with 1 ml ice-cold distilled water. The prepared gel is added to fraction 12 (peak I) from the purification step of Ang II antiserum after Affi-Gel 102–Ang II in a ratio of 1 ml gel to 0.5 mg protein. The pH is adjusted to 9.0 with the addition of NaOH solution. The mixture is shaken for 24 hr at 4°. After the coupling is completed the gel is allowed to settle, and the supernatant is taken off. The supernatant is checked for binding of ^{125}I-Ang II as described for the purification of Ang II antiserum on Affi-Gel 102–Ang II to determine the coupling efficiency. To the gel suspension 2 ml of 1 *M* ethanolamine solution, pH 8.5, is added, and the mixture is shaken again for 24 hr at 4°. The gel suspension is transferred to a plastic column (6 × 0.7 cm) precoated with 0.1% BSA solution. The gel with the covalently bound Ang II antibody (Affi-Gel 10–AB) is successively washed 5 times with 1 ml 50 m*M* phosphate buffer, pH 7.4, 5 times with 1 ml 0.5 *M* sodium borate buffer, pH 8.0, and 5 times with 1 ml 0.1 *M* sodium citrate buffer, pH 4.0, containing 1 *M* NaCl. Aliquots of the washing solutions as well as the 1 *M* ethanolamine treatment are taken for the measurement of ^{125}I-Ang II binding. The column is stored in 50 m*M* MOPS buffer, pH 7.4, at 4°.

The coupling efficiency of the purified antiserum fraction to Affi-Gel 10 was typically greater than 97%. After treatment of the Affi-Gel 10–AB column with 1 *M* ethanolamine and the subsequent washing procedures with 50 m*M* phosphate buffer, pH 7.4, 0.5 *M* sodium borate buffer, pH

8.0, and 0.1 *M* sodium citrate buffer, pH 4.0, no binding capacity for ^{125}I-Ang II could be found in these fractions. This indicates that the conjugation was complete. The column could be used for several months without any loss in the separation performance and reproducibility when used as reported.

Affinity Chromatography on Affi-Gel 10–AB

Prior to sample application the Affi-Gel 10–AB column is equilibrated with 15 ml 50 m*M* phosphate buffer, pH 7.4. Separations are performed at room temperature at a flow rate of 0.26 ml/min. Brain extracts are dissolved in 2 ml 50 m*M* phosphate buffer, pH 7.4, and centrifuged at room temperature for 2 min at 10,000 *g*. The samples are allowed to penetrate into the gel, then the outlet of the column is closed and the mixture incubated at room temperature for 20 hr. The outlet of the column is opened again, and the gel is washed 3 times with 1 ml 50 m*M* phosphate buffer, pH 7.4. The elution is performed with 0.1 *M* sodium citrate buffer containing 1 *M* NaCl in a stepwise pH gradient ranging from 5.0 to 2.0. Fractions of 1 ml are collected. The elution of radioactivity is monitored directly for ^{125}I-Ang II in a gamma counter (Beckman Gamma 5500). For ^{3}H-labeled Ang II, aliquots of 0.1 ml of each fraction are taken, mixed with 3 ml Liquiscint, and the radioactivity counted in a liquid scintillation (LS) counter (1215 Rackbeta II, LKB). The elution of endogenous Ang II from pig brain is measured radioimmunologically. Typical elution patterns of ^{125}I-Ang II and ^{3}H-Ang II are shown in Fig. 2.

In both cases the chromatogram revealed two peaks (Fig. 2), the first one detectable in fractions 2, 3, and 4, representing unbound and unretained peptide. At pH 5.0 small amounts of radioactivity could be monitored in fractions 6, 7, and 8, but most of the radioactivity was found in fractions 9, 10, 11, and 12. The total recovered radioactivity as well as the ratio between unretained and elutable radioactivity are summarized in Table II. In experiment 1 with the use of ^{125}I-Ang II, 20.10% was unretained. At pH 5.0, 16.59% and at pH 4.0, 46.76% of ^{125}I-Ang II could be eluted. For the other experiments [^{3}H]Ang II was used. The amount of unretained radioactivity was $31.81 \pm 2.31\%$ ($n = 5$). At pH 5.0, $10.55 \pm 3.94\%$ ($n = 5$) and, at pH 4.0, $39.44 \pm 2.70\%$ ($n = 5$) of radioactive material was elutable. With the use of synthetic [Ile^{5}]Ang II 35.38% was unretained; at pH 5.0, 8.97%, at pH 4.0, 38.81, and at pH 3.0, 5.18 could be eluted.

The specific enrichment of Ang II extracted from pig brain of Affi-Gel 10–AB is illustrated in Fig. 3. Ang II eluted in two peaks at pH 3.0 and 4.0, and only minor quantities of Ang II were unretained.

For the determination of the loading capacity of the Affi-Gel 10–AB

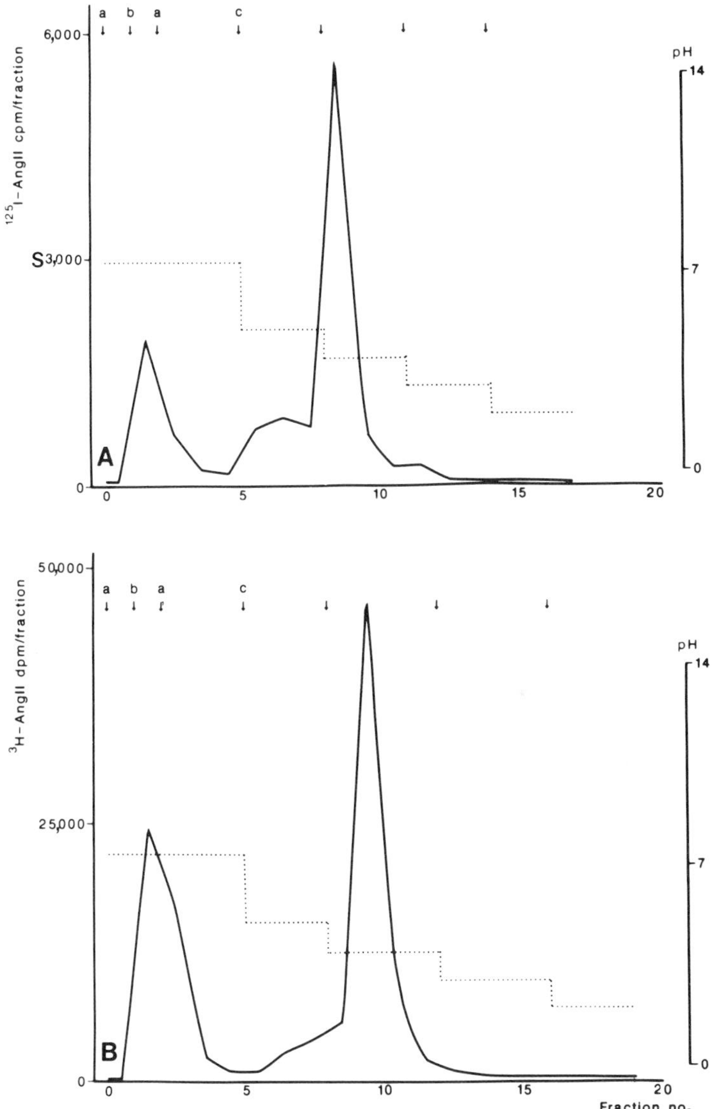

FIG. 2. Affinity purification of Ang II on Affi-Gel 10–AB. Rat brain extracts with the addition of ^{125}I-Ang II (A) and [^{3}H]Ang II (B) were applied to the Affi-Gel 10–AB column. Unretained radioactivity could be identified in fractions 2, 3, and 4. With the stepwise pH gradient small amounts of radioactivity could be monitored at pH 5.0 in fractions 6, 7, and 8. The majority of radioactivity could be eluted in fractions 9, 10, 11, and 12 at pH 4.0. For chromatographic conditions, see the text. ····, pH gradient profile; a, equilibration and wash; b, sample application; c, elution.

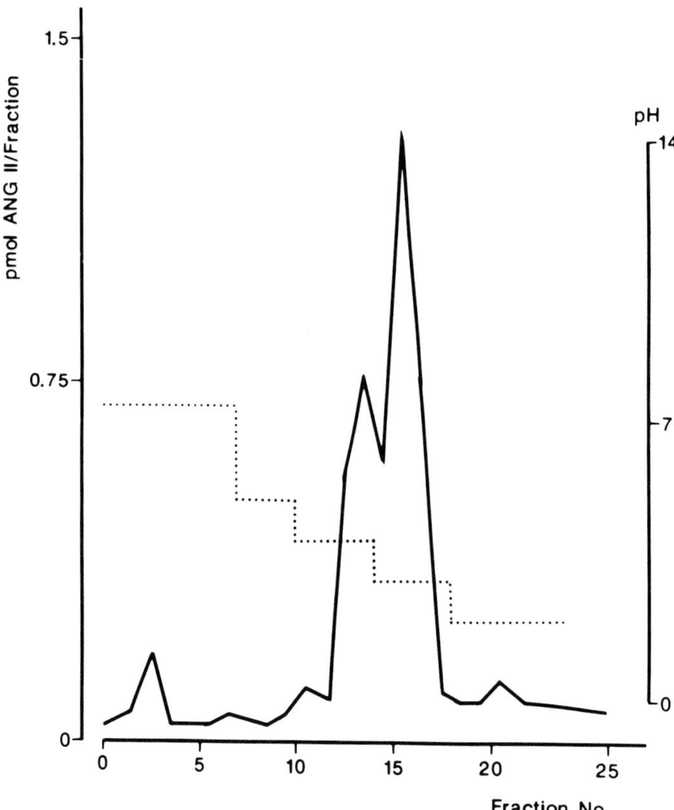

FIG. 3. Purification of pig brain extracts on an Affi-Gel 10–AB column. Extract was applied to the column and incubated for 20 hr. After washing the column with phosphate buffer, pH 7.4, Ang II was eluted with 0.1 *M* sodium citrate and 1 *M* NaCl using a pH gradient from pH 5.0, 4.0, 3.0 to 2.0. Ang II eluted in two peaks at pH 4.0 and 3.0. ····, pH gradient profile.

column, 2 μg, corresponding to 1.9 nmol synthetic [Ile^5]Ang II, was used and processed through the entire procedure, and Ang II was monitored radioimmunologically for each individual step. The amount of [Ile^5]Ang II retained and elutable from the Affi-Gel 10–AB column was 0.556 μg, corresponding to 0.54 nmol.

High-Performance Liquid Chromatography

The fractions after affinity purification on Affi-Gel 10–AB are pooled, lyophilized, and further characterized on a reversed-phase C_{18} column (Ultrasil ODS, 10 μm, 250 × 4.6 mm), using a methanol gradient system. The mobile phase is composed of methanol–water–1 *M* ammonium ace-

TABLE II
RELATION BETWEEN UNRETAINED AND ELUTABLE PEPTIDE AND TOTAL RECOVERY ON THE AFFI-GEL 10–AB COLUMN[a]

	Percentage ^{125}I-Ang II, [^{3}H]Ang II, or [Ile^5]Ang II				
		Retained and elutable			
Experiment	Unretained	pH 5.0	pH 4.0	pH 3.0	Total recovery
1	20.10	16.59	46.76	—	83.45
2	26.45	25.99	41.17	—	92.61
3	39.41	4.30	39.95	—	83.66
4	31.50	5.63	44.56	—	81.69
5	27.79	7.35	29.08	—	64.22
6	33.88	9.55	42.44	—	87.87
7	35.98	8.97	38.81	5.18	88.94

[a] In experiment 1, ^{125}I-Ang II was used. In experiments 2–6, [^{3}H]Ang II was used. In experiment 7, synthetic [Ile^5]Ang II was used.

tate, pH 5.4, 30 : 69 : 1 (v/v/v) [solution A (sol A)] and methanol–water–1 *M* ammonium acetate, pH 5.4, 80 : 19 : 1 (v/v/v) [solution B (sol B)]. A linear gradient of sol A to sol B within 35 min, starting 4 min after sample injection, is performed. Separations are carried out at room temperature at a flow rate of 1.0 ml/min, and fractions of 1.0 ml are collected.

The column is calibrated with synthetic angiotensin standard peptides. This HPLC system was used in the past for the baseline separation of angiotensin peptides and other peptides and is described in detail elsewhere.[17] In both cases, with the use of either ^{125}I-Ang II or [^{3}H]Ang II, the HPLC analysis demonstrated that the radioactive material eluting at pH 4.0 on Affi-Gel 10–AB was coincident with the elution of the synthetic ^{125}I-Ang II or [^{3}H]Ang II standards, yielding only one peak which underlines their high purity (Fig. 4). For the pig brain Ang II, following specific enrichment on Affi-Gel 10–AB, the HPLC chromatogram revealed a single peak with the same retention time as synthetic [Ile^5]Ang II standard peptide (Fig. 5).

Biosynthesis of Angiotensin I and II in Cultured Rat Brain Cells

Since Ang II is a circulating plasma peptide its presence in the brain and other tissue raises the question of whether it is autochthon or whether

[17] K. Hermann, R. E. Lang, T. Unger, C. Bayer, and D. Ganten, *J. Chromatatogr.* **312,** 273 (1984).

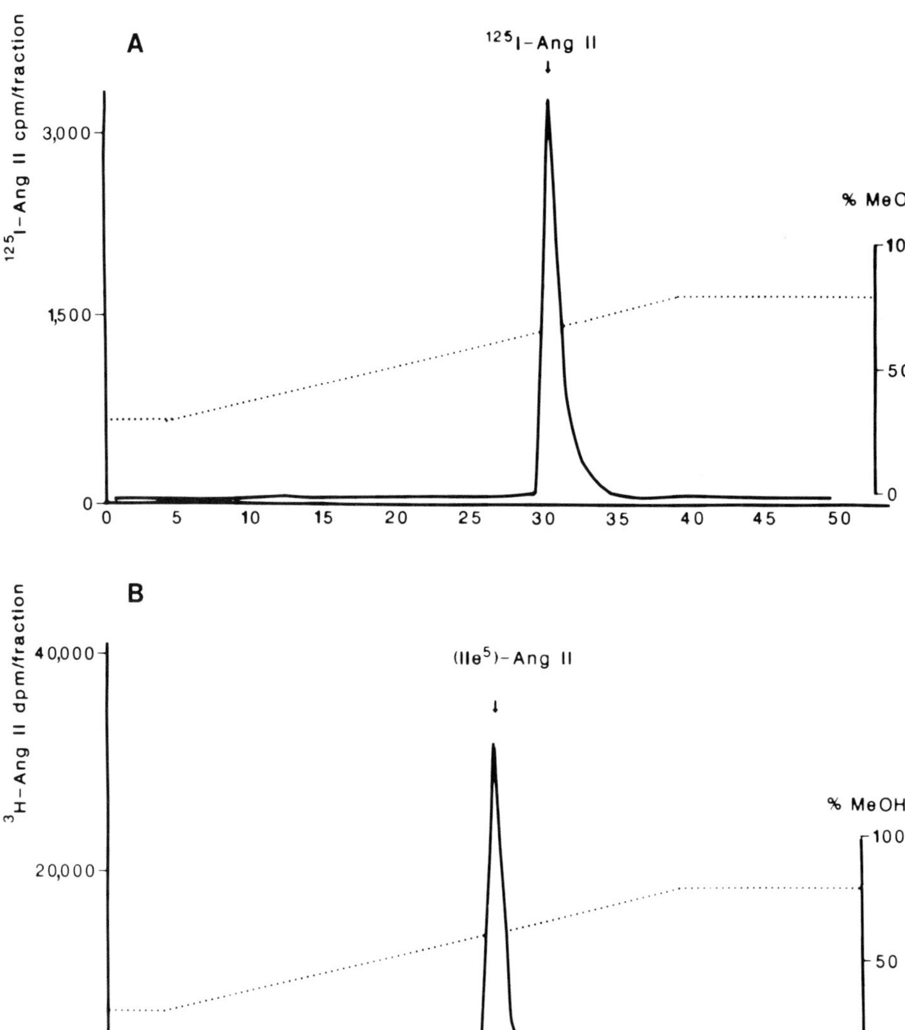

FIG. 4. HPLC purification of ^{125}I-Ang II (A) and [^{3}H]Ang II (B) on a reversed-phase C_{18} column in the methanol gradient system. Radioactivity obtained at pH 4.0 on Affi-Gel 10–AB could be identified as one peak with the same retention time as ^{125}I-Ang II standard or synthetic [Ile^{5}]Ang II standard peptide.

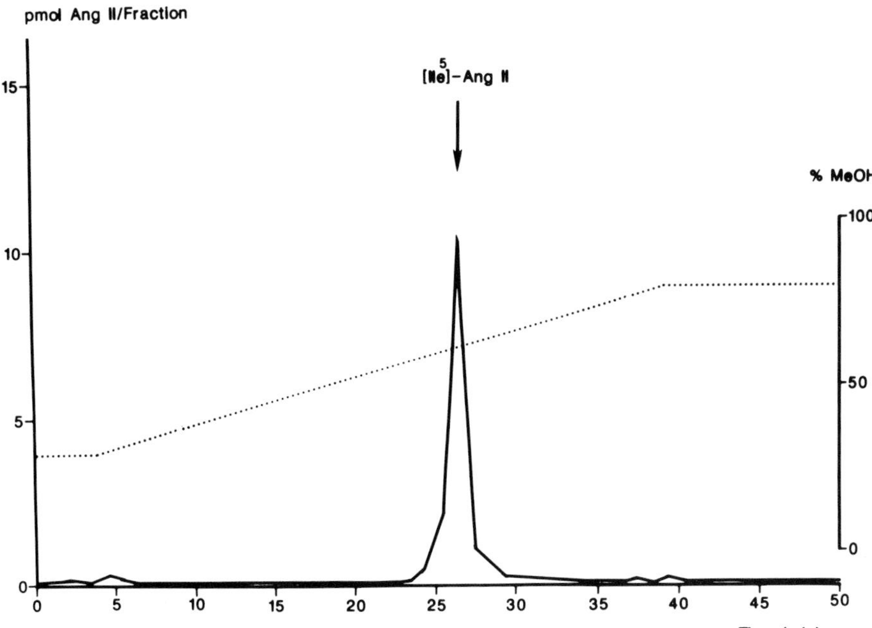

FIG. 5. HPLC purification of pig brain Ang II on a reversed-phase C_{18} column in the methanol gradient system. Ang II-containing fractions from the Affi-Gel 10–AB purification step were pooled, lyophilized, and further purified on HPLC. Ang II revealed the same retention time as synthetic [Ile^5]Ang II standard peptide and eluted in a single peak. The arrow indicates retention time of synthetic [Ile^5]Ang II.

it is taken up from the blood. To exclude the influence of the peripheral renin angiotensin system (RAS) neuronal and glial cells in primary culture have been used in our laboratory in the past for the localization of various neuropeptides and their receptors.[18–21]

The presence of renin, Ang I, and Ang II has been demonstrated in both glial and neuronal cells cultured from rat brain.[22,23] Despite the fact that primary brain cells in culture lack the influence of the peripheral RAS, the presence of these two peptides alone does not necessarily prove

[18] M. K. Raizada, M. I. Phillips, and J. S. Gerndt, *Neuroendocrinology*, **36,** 64 (1983).

[19] D. W. Clarke, F. T. Boyd, M. S. Kappy, and M. K. Raizada, *J. Biol. Chem.* **259,** 11672 (1984).

[20] J. Weyhenmeyer, M. K. Raizada, M. I. Phillips, and R. E. Fellows, *Neurosci. Lett.* **16,** 41 (1980).

[21] M. K. Raizada, J. W. Yang, M. I. Phillips, and R. E. Fellows, *Brain Res.* **207,** 343 (1981).

[22] K. Hermann, M. K. Raizada, C. Sumners, and M. I. Phillips, *Brain Res.* **437,** 205 (1987).

[23] K. Hermann, M. K. Raizada, C. Sumners, and M. I. Phillips, *Neuroendocrinology* **47,** 125 (1988).

that they can be synthesized by these cells. Therefore, we focused on the capability of primary rat brain cells in culture to synthesize Ang I and II after incubation with [^{3}H]isoleucine.

Cell Culture Preparation

Neuronal and glial cell cultures are prepared from either whole brains or cocultures of brain stem and hypothalamic areas of 1-day-old rats as described elsewhere.[18,19,24,25] Briefly, whole brains, brainstem (medulla oblongata and pons), and hypothalamic areas including supraoptic, paraventricular, anterior, lateral, dorsomedial, ventromedial, and posterior nuclei are chopped into 2-mm pieces after removal of blood vessels and pia mater, and cells are dissociated by trypsin and deoxyribonuclease I treatment. The dissociated cells are centrifuged, washed in Dulbecco's modified Eagle's medium (DMEM) containing 10% plasma-derived horse serum (PDHS), and plated in 100-mm Falcon tissue culture dishes precoated with poly(L-lysine). After 3 days of growth at 37° in a humidified incubator with 5% CO_2–95% air, plates are treated for 3 days with cytosine arabinoside to obtain neuron-enriched cells in culture. For the preparation of glial cells, the treatment with cytosine arabinsoide is omitted. Cultures are grown for 9–11 days in DMEM containing 10% PDHS before being used for experiments. Phase-contrast microscopy and immunocytochemical analysis with the use of antibodies to neuron-specific enolase or glial fibrillary acidic protein are used to show that neuronal cultures contain 75–85% neuronal cells and that glial cultures contain more than 95% astrocytic glial cells.[18,19,24,25]

Incubation of Brain Cell Cultures

Neuronal or glial cells from 10 tissue cultures dishes (100 mm in diameter) are pooled for one experiment. The growth medium is aspirated, and the cultures are washed 3 times with 10 ml phosphate-buffered saline (PBS), pH 7.4. Each culture dish is incubated in 5 ml $NaHCO_3$-buffered Krebs–Ringer solution in the presence of 0.02 mCi [^{3}H]isoleucine (Ile) for 6 hr at 37° in a humidified incubator with 5% CO_2–95% air.

Extraction of the Brain Cell Cultures

After 6 hr, the incubation medium is removed. The cells are washed 3 times with 10 ml PBS, pH 7.4, and scraped from the dishes with the help

[24] C. Sumners and M. K. Raizada, *Am. J. Physiol.* **246,** C502 (1984).

[25] M. K. Raizada, M. I. Phillips, F. T. Crews, and C. Sumners, *Proc. Natl. Acad. Sci. U.S.A.* **84,** 4655 (1987).

of a rubber policeman in 1 ml ice-cold 1 *M* acetic acid per dish. The acidic cell suspension is homogenized using ultrasonification for 10 sec at 4° followed by centrifugation at 2200 *g* for 15 min at 4°. The supernatant is collected, and the pellet is reextracted with 5 ml of ice-cold 1 *M* acetic acid. Both supernatants are pooled, frozen, and lyophilized.

HPLC Gel Filtration

Glial and neuronal extracts are subjected to HPLC gel filtration on Bio-Sil TSK 250 and Bio-Sil TSK 125 in tandem using 50 m*M* Na_2SO_4–20 m*M* NaH_2PO_4, pH 6.8, containing 50 ml HPLC-grade acetonitrile per 1000 ml as mobile phase. Separations are performed at room temperature at a flow rate of 0.7 ml/min, and fractions of 0.7 ml are collected. The columns are calibrated with markers of different molecular weights under identical chromatographic conditions. Antiserum against [Ile^5]Ang I and [Ile^5]Ang II prepared in our laboratory[23] is used to detect radioactive labeled Ang I and II synthesized by the cells. For both glial and neuronal cells the majority of radioactivity detectable with the Ang I and Ang II antibodies elutes in the low molecular weight range. Additionally, proteins cross-reacting with the Ang I and Ang II antibodies with high and intermediate molecular weights could be identified (Fig. 6).

Reversed-Phase HPLC

Fractions from the HPLC gel filtration containing low molecular weight substances identified by their cross-reaction with the Ang I and Ang II antibodies are pooled and lyophilized for glial and neuronal cells. Further characterization is achieved by reversed-phase C_{18} HPLC as described above. For glial as well as neuronal cells two peaks of radioactivity could be identified with the same retention times as synthetic [Ile^5]Ang I and [Ile^5]Ang II (Fig. 7). Measurement with the Ang I and Ang II antibodies revealed a peak for Ang I and Ang II. Both peaks were not only coincident with peaks of radioactivity but showed retention times identical to those of synthetic [Ile^5]Ang I and [Ile^5]Ang II standard peptides. The majority of radioactive material found in the void volume of the column could be identified as unincorporated [^{3}H]Ile. Compared to the total radioactivity present in the cell extracts, 0.04 and 0.05% were incorporated in Ang I and Ang II in neuronal cells. For glial cells, values of 0.07 and 0.08% were found for Ang I and Ang II. This clearly demonstrates the presence of [Ile^5]Ang I and [Ile^5]Ang II by *de novo* synthesis in glial and neuronal cells.

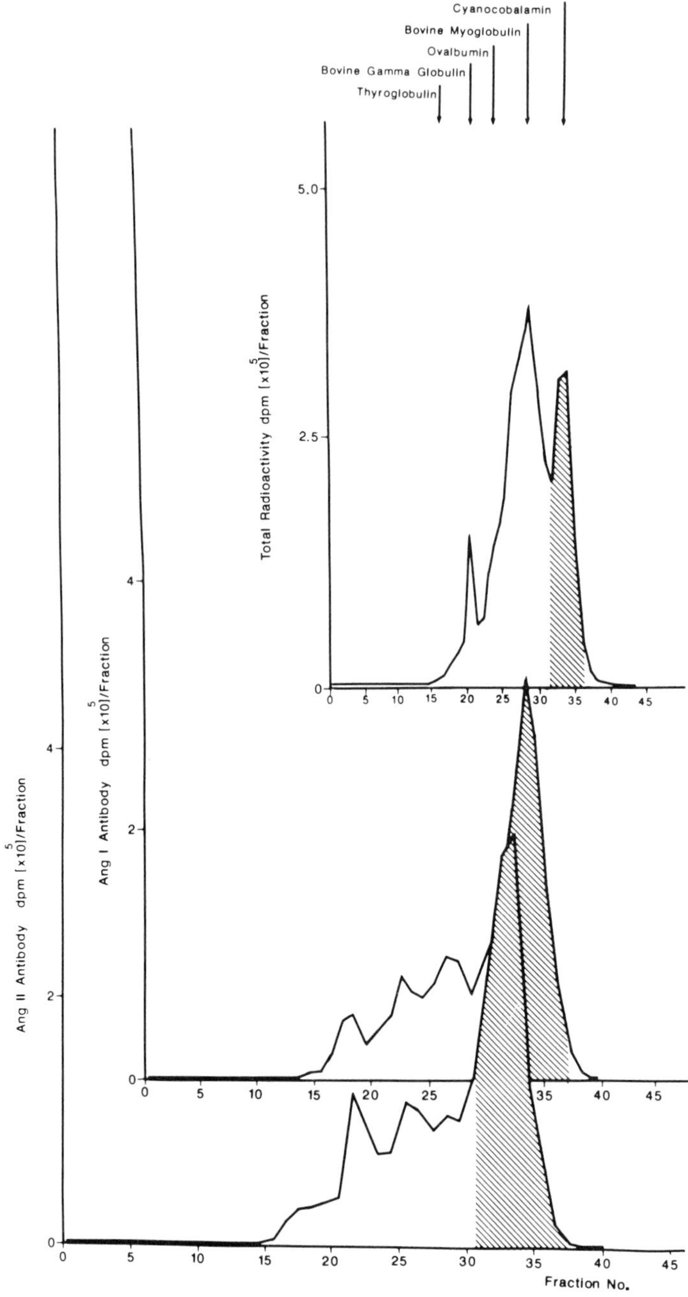
Cyanocobalamin
Bovine Myoglobulin
Ovalbumin
Bovine Gamma Globulin
Thyroglobulin
Total Radioactivity dpm [x10^5]/Fraction
Ang I Antibody dpm [x10^5]/Fraction
Ang II Antibody dpm [x10^5]/Fraction
Fraction No.

Degradation of Angiotensin Peptides in Cultured Brain Cells

Studies of the degradation of angiotensin peptides were in the past conducted in brain tissue homogenates.[26,27] In these preparations, composed of blood vessels, neuron-, and glia-derived tissue, the source of the enzymatic activity is difficult to define. In this regard it is advantageous to work with preparations lacking contaminations of plasma- and vascular tissue-derived enzymatic activity. An approach to overcome these problems is the use of primary cell cultures from rat brain. Based on the processing of these cell cultures, it is possible to differentiate between glia-enriched and neuron-enriched cells in culture.[18,19] The aim of this study was to investigate the degradation pattern of angiotensin peptides in neuron- and glia-enriched cells in culture using HPLC to separate and purify the metabolites prior to analysis of the amino acid composition.

Incubation of Brain Cell Cultures

Neuronal or glial cells from five tissue culture dishes (100 mm diameter) are pooled for one experiment. The growth medium is aspirated, and the cultures are washed 3 times with 10 ml PBS, pH 7.4. The cells are scraped from the dishes in 0.5 ml ice-cold 50 m*M* phosphate buffer, pH 7.4, containing 0.3 *M* NaCl and 0.1% Triton X-100 per dish, with the help of a rubber policeman. Cells are homogenized using ultrasonification for 10 sec at 4°. Aliquots of the cell homogenate are taken for protein measurement according to the method of Lowry *et al.*, using BSA as a standard.[16] Synthetic [Ile^5]Ang I, [Ile^5]Ang II, and [Ile^5]Ang III are incubated individually in 2.5 ml cell homogenate at 37° for 0, 1, 2, 5, 10, 15, 20, 30, and 60 min in a ratio of 100 nmol of each peptide per 2–3 mg protein. At the individual time periods, 0.2 ml of the incubate is withdrawn, mixed

[26] J. A. D. M. Tonnaer, G. M. H. Engels, V. M. Wiegant, J. P. H. Burbach, W. Dejong, and D. De Wied, *Eur. J. Biochem.* **131,** 415 (1983).

[27] G. Simmonet, A. Carayon, M. Alard, F. Cesselin, and A. Lagoguey, *Brain Res.* **304,** 93 (1984).

FIG. 6. HPLC gel filtration of an extract of neuronal cells after incubation with [^{3}H]isoleucine on Bio-Sil TSK 250 and Bio-Sil TSK 125 columns in tandem. The columns were developed in 50 m*M* Na_2SO_4–20 m*M* NaH_2PO_4, pH 7.4, containing 50 ml of HPLC-grade acetonitrile per 1000 ml at a flow rate of 0.7 ml/min. Fractions of 0.7 ml were collected, and total radioactivity, Ang I, and Ang II were measured in the HPLC fractions. The majority of the radioactive material detectable with the Ang I and Ang II antibodies eluted in the low MW range. The fractions containing radioactive material cross-reacting with the Ang I and Ang II antibodies (shaded) were pooled, lyophilized, and further characterized on a reversed-phase C_{18} column. The elution profiles of several MW markers are indicated by arrows.

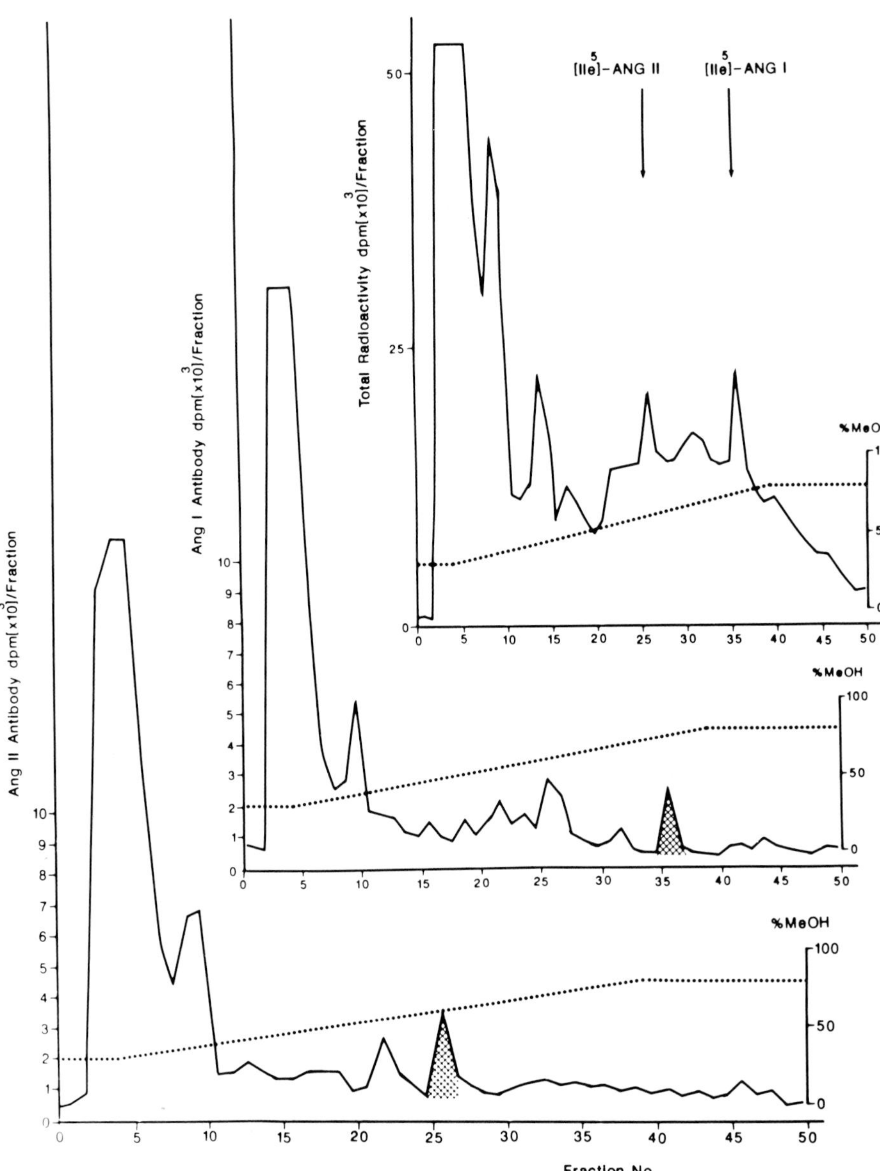

FIG. 7. HPLC separation of low MW radioactive material cross-reacting with the Ang I and Ang II antibodies from Fig. 6 on a reversed-phase C_{18} column in the methanol gradient system. For chromatographic conditions, see the text. Total radioactivity, Ang I, and Ang II were measured in the HPLC fractions. The majority of the radioactive material eluted in the void volume of the column and could be identified as unincorporated [^{3}H]Ile. The chromatogram shows two radioactive peaks with the same retention times as synthetic [Ile^5]Ang I and [Ile^5]Ang II standard peptides. Both peaks were also obtained when fractions were subjected to radioimmunological measurement with the Ang I and Ang II antibodies (shaded area). Arrows indicate the retention times of synthetic [Ile^5]Ang I and [Ile^5]Ang II.

with 0.2 ml 1 *N* HCl, and boiled for 5 min. After centrifugation at 10,000 *g* for 2 min at ambient temperature, aliquots of 0.2 ml are analyzed on HPLC.

HPLC Separation

Angiotensin metabolites are separated on a reversed-phase C_{18} column (Ultrasil ODS, 10 μm, 250 × 4.6 mm), using the methanol gradient system as described above. In addition, a 2-propanol gradient system is used. The mobile phase of this system is composed of 2-propanol–water–1 *M* triethylammonium phosphate, pH 3.0, 5 : 94 : 1 (v/v/v) [solution A (sol A)] and 2-propanol–water–1 *M* triethylammonium phosphate, pH 3.0, 40 : 59 : 1 (v/v/v) [solution B (sol B)]. A linear gradient of sol A to sol B within 45 min, starting 3 min after sample application, is performed. In both cases the flow rate is 1.0 ml/min, and fractions of 1.0 ml are collected. Metabolites are monitored with UV detection at 220 nm and a sensitivity of 0.1 absorbance units full scale. Fractions containing angiotensin metabolites are pooled, lyophilized, and rechromatographied first in the 2-propanol system. A second rechromatography in the methanol gradient system is performed prior to analysis of the amino acid composition of the different metabolites.

The degradation patterns of [Ile^5]Ang I, [Ile^5]Ang II, and [Ile^5]Ang III are the same in neuronal and in glial cell extracts. A representative example of the degradation of [Ile^5]Ang I in neuronal cells after different incubation times is shown in Fig. 8. After 0 min of incubation (Fig. 8A) only [Ile^5]Ang I is present in the chromatogram. After 10 min of incubation (Fig. 8B) the peak of [Ile^5]Ang I has declined, and, in addition to [Ile^5]Ang I, des-Asp^1-[Ile^5]Ang I, [Ile^5]Ang II, and [Ile^5]Ang III are visible. By monitoring and analyzing the metabolites of [Ile^5]Ang I, [Ile^5]Ang II, and [Ile^5]Ang III the degradation pattern of angiotensin peptides in cultured brain cells can be formulated as follows (Fig. 9). Converting enzyme (CE) cleaves the dipeptide His-Leu from the carboxy-terminal end of [Ile^5]Ang I to yield [Ile^5]Ang II. Ang II is metabolized by the action of aminopeptidase A to [Ile^5]Ang III. An alternative pathway is the cleavage of the amino-terminal amino acid Asp of [Ile^5]Ang I by aminopeptidase A. The generated peptide des-Asp^1-[Ile^5]Ang I is then hydrolyzed by CE to [Ile^5]Ang III, which seems to play a key role in the degradation scheme. Proceeding from Ang III, different aminopeptidases are involved in the subsequent steps of degradation, hydrolyzing the amino-terminal amino acids to successively generate [Ile^5]Ang II (3–8) hexapeptide, [Ile^5]Ang II (4–8) pentapeptide, and [Ile^5]Ang II (5–8) tetrapeptide. Except for the dipeptidyl carboxypeptidase CE, aminopeptidases seem to play a major role in the degradation of angiotensin peptides. No evidence for the in-

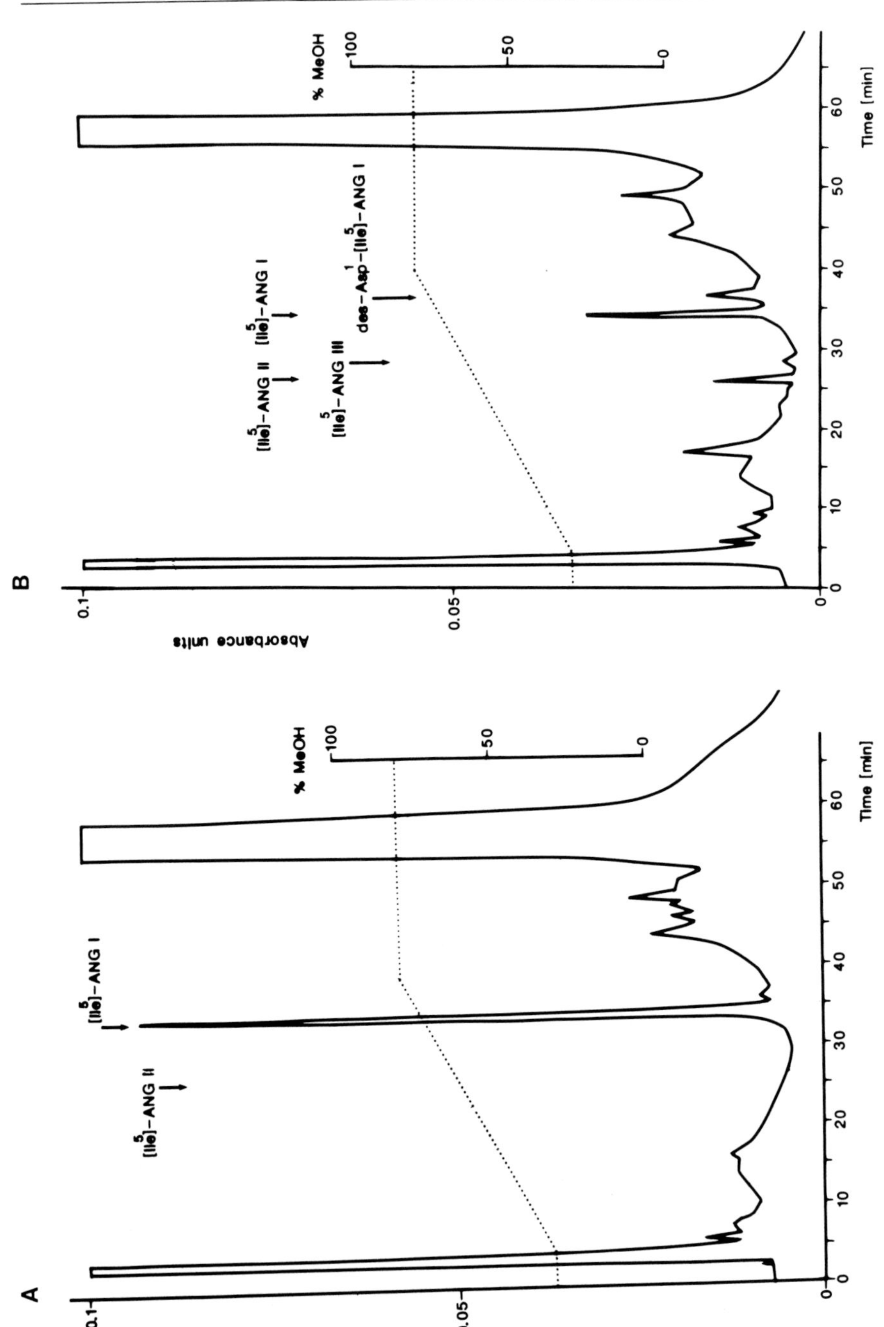

A
B
Absorbance units
0.1
0.05
0
% MeOH
100
50
0
Time [min]
[Ile5]-ANG II
[Ile5]-ANG I
[Ile5]-ANG III
des-Asp1-[Ile5]-ANG I

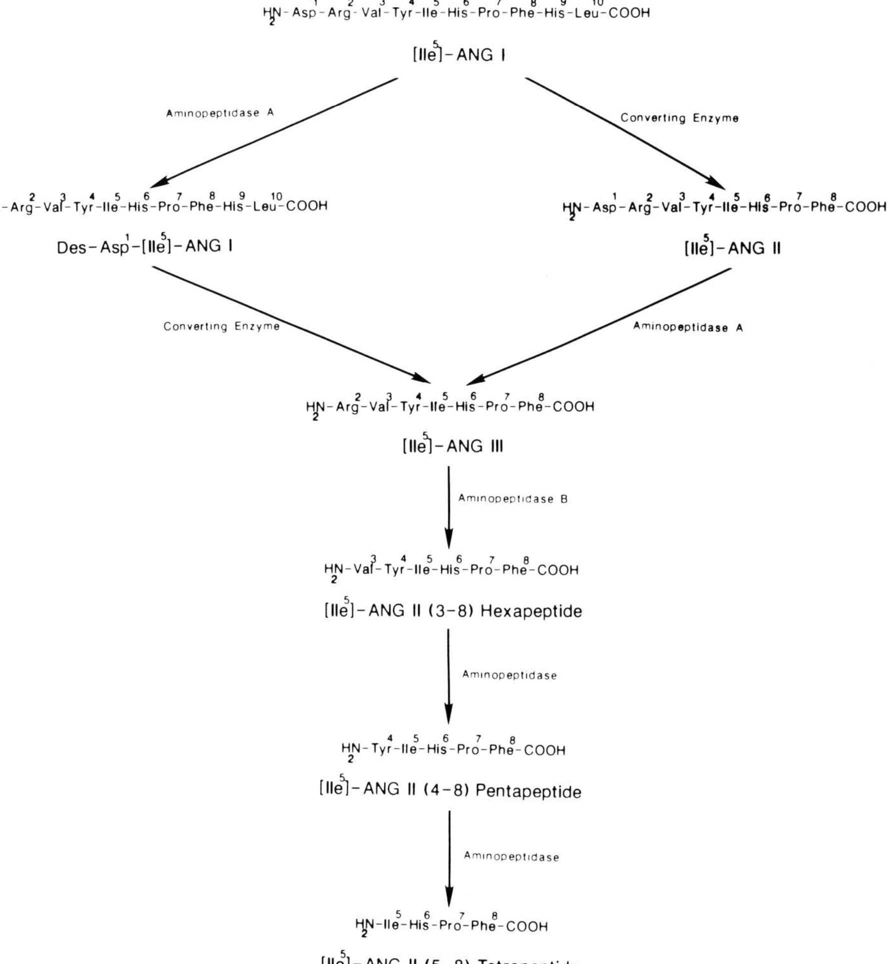

FIG. 9. Degradation pattern of angiotensin peptides in cultured rat brain cells. [Ile[5]]Ang I is hydrolyzed by converting enzyme to [Ile[5]]Ang II, which is metabolized on the action of aminopeptidase A to [Ile[5]]Ang III. An alternative pathway is the cleavage of the amino-terminal amino acid Asp of [Ile[5]]Ang I to generate des-Asp[1]-[Ile[5]]Ang I, which then is hydrolyzed by converting enzyme to [Ile[5]]Ang III. [Ile[5]]Ang III is metabolized in a step-by-step cleavage from the amino-terminal end to [Ile[5]]Ang II (3–8) hexapeptide, [Ile[5]]Ang II (4–8) pentapeptide, and [Ile[5]]Ang II (5–8) tetrapeptide.

FIG. 8. HPLC separation of a neuronal cell extract incubated with [Ile[5]]Ang I on a reversed-phase C_{18} column in the methanol gradient system with UV detection. After 0 min of incubation only [Ile[5]]Ang I is present (A). After 10 min of incubation, the concentration of Ang I is decreased, and des-Asp[1]-[Ile[5]]Ang I, [Ile[5]]Ang II, and [Ile[5]]Ang III have been generated (B).

volvement of endopeptidases in the metabolism of angiotensin peptides was obtained.

Summary and Conclusions

Ang II antiserum with high sensitivity and specificity was produced. The native Ang II antiserum was purified by affinity chromatography on Affi-Gel 102 with covalently coupled [Ile^5]Ang II, and purified Ang II antiserum was covalently coupled to Affi-Gel 10. The column with the covalently coupled Ang II antiserum was used for the specific enrichment of Ang II from brain extracts. The efficiency and usefulness of affinity chromatography for the purification of Ang II from biological sources were tested with ^{125}I-labeled, ^{3}H-labeled, and synthetic [Ile^5]Ang II added to rat brains prior to extraction. In addition, the methodology was used for the purification of endogenous Ang II from pig brain. The described three-step procedure for the isolation and purification of Ang II including extraction, affinity chromatography, and HPLC is rapid and highly specific with high loading capacity. We have applied the method to the peptide Ang II in brain, but the methodology may also be used in general for the rapid purification of other neuropeptides.

A combination of HPLC with specific radioimmunoassays for Ang I and Ang II was utilized to demonstrate that rat brain cells in culture devoid of the influence of the peripheral RAS were able to synthesize radioactively labeled Ang I and Ang II after incubation with [^{3}H]isoleucine. And, finally, an HPLC system capable of separating Ang I, Ang II, and its metabolites was used to obtain insight into the degradation pattern of angiotensin peptides in the brain. Aminopeptidases appear to be the major angiotensin-degrading enzymes, and endopeptidases do not appear to be involved.

Acknowledgment

These series of experiments were carried out in Dr. Phillips' laboratory while Klaus Hermann was a Feodor Lynen fellow of the Humboldt Foundation, Federal Republic of Germany. Support for this work came from NIH grants and the American Heart Association. We are grateful to Birgitta Kimura for technical assistance and to Elizabeth Albert for production of the cells.

[40] Measurement of Thyrotropin-Releasing Hormone and Its Metabolites

By CHARLES H. EMERSON

Introduction

Thyrotropin-releasing hormone (TRH) was the first pituitary releasing factor of hypothalamic origin to be characterized. This chapter describes methods for the measurement of TRH, and the putative TRH metabolites deamido-TRH and histidylproline diketopiperazine. A previous chapter (this volume [26]) reviews the enzymes that catalyze the primary degradation of TRH and describes some of the methods used to measure the activity of these enzymes.

Current methods for measuring TRH and TRH metabolites are based on radioimmunoassay (RIA). This technique provides the ability to process a large number of samples in a relatively rapid fashion. However, when working at the threshold sensitivity of the assay, "noise" and lack of precision are major problems. In addition, the presently available RIAs for TRH lack specificity with respect to analogs that contain substituted histidyl residues. RIAs for TRH and TRH metabolites have been used for two general applications. The first is to measure endogenous TRH or TRH metabolites in extracts of tissues, or after release from organ cultures. It is in this application that limitations in sensitivity and specificity pose major problems. The second application is to quantitate the loss of TRH, or the appearance of TRH metabolites, when synthetic TRH is infused *in vivo*, perfused into intact organs, or incubated with broken cell preparations. These are the applications that are more pertinent to the assessment of TRH-degrading enzymes, and in this setting assay sensitivity and specificity are less of a problem.

The structure of TRH has been established, in both ovine and porcine tissue, to be pyroglutamylhistidylprolineamide.[1,2] Two modes of primary enzymatic degradation of TRH have been described.[3]

[1] R. Burgus, T. Dunn, D. Desiderio, D. N. Ward, W. Vale, and R. Guillemin, *Nature (London)* **226,** 321 (1970).

[2] J. Boler, F. Enzmann, K. Folkers, C. Y. Bowers, and A. Schally, *Biochem. Biophys. Res. Commun.* **37,** 705 (1969).

[3] C. H. Emerson, in "The Thyroid" (S. H. Ingbar and L. E. Braverman, eds.), p. 1472. Lippincott, Philadelphia, Pennsylvania, 1986.

$$\underset{\text{(TRH)}}{\text{pGlu-His-Pro-NH}_2} \xrightarrow{\text{TRH deamidase}} \underset{\text{(deamido-TRH)}}{\text{pGlu-His-Pro}} \quad (1)$$

$$\underset{\text{(TRH)}}{\text{pGlu-His-Pro-NH}_2} \xrightarrow[\text{(5-oxoprolyl-peptidase, EC 3.4.19.3)}]{\text{pyroglutamyl aminopeptidase}} \text{His-Pro-NH}_2 \quad (2)$$

The compounds of interest that these enzyme activities generate are deamido-TRH (prolyl endopeptidase activity) and His-Pro-NH_2 (pyroglutamyl aminopeptidase activity). His-Pro-NH_2 is very unstable and spontaneously cyclizes to histidylproline diketopiperazine (HPD), otherwise known as cyclo[His-Pro].

$$\text{His-Pro-NH}_2 \xrightarrow{\text{nonenzymatic}} \underset{\text{(HPD)}}{\text{His-Pro-NH}}$$

HPD resists enzymatic degradation, and concentrations of HPD can, with certain reservations (see this volume [26]), be used as an index of His-Pro-NH_2 concentrations.

Materials

The following materials are required for most radioimmunoassays: (1) standard, i.e., a known quantity of the substance in pure form; (2) label, i.e., a labeled form of the substance; and (3) first antibody, namely, an antibody directed against the substance being measured. Pure synthetic standards for the TRH, deamido-TRH, and HPD RIAs are commercially available. In contrast, most investigators prepare their own label and first antibody for these RIAs. TRH, deamido-TRH, and HPD contain a histidyl residue. Because of this common feature the methods for developing antibodies against these compounds are similar in that the antigen consists of a protein conjugated to the peptide through the histidyl residue. In addition, the method of preparing label often consists of iodination of the histidyl group. The TRH RIA is the prototype for the assays discussed in this chapter and is described in greatest detail.

TRH Radioimmunoassay

Standard and Label

Highly purified synthetic TRH is available for use as the standard and to prepare labeled TRH. ^{125}I-TRH has generally been employed as the label for the TRH RIA[4,5]; however, some investigators have used [^{3}H]TRH.[6] The method for TRH iodination is adapted from Greenwood *et al.*[7] Synthetic TRH is dissolved in water in a concentration of 500 μg/ml, and 0.010 ml is added to 1 mCi carrier-free [^{125}I]sodium iodide having a specific activity above 15 mCi/μg. Next 0.05 ml of 0.5 *M* sodium phosphate, pH 7.5, is added. The iodination reaction is started by adding 0.025 ml of chloramine-T, prepared just prior to starting the iodination reaction by mixing 15 mg dry, light-shielded chloramine-T with 10 ml 10 m*M* sodium phosphate, 0.15 *M* NaCl, pH 7.4. The iodination reaction is allowed to proceed for 30 sec and is terminated by adding 0.1 ml of a 12.6 m*M* solution of $Na_2S_2O_5$. Following this, 0.05 ml of 2.5% bovine serum albumin (BSA) is added, and the iodination mixture is applied to a 1 × 15 cm Sephadex G-10 (Pharmacia, Inc., Piscataway, NJ) column equilibrated in 10 m*M* sodium phosphate, 0.15 *M* NaCl, pH 7.4. One-milliliter fractions are collected and counted.

Monoiodinated ^{125}I-TRH elutes in fraction 11 (peak B) and [^{125}I]iodide in fraction 17 (peak C). Iodinated material, presumably diiodinated ^{125}I-TRH, elutes in fraction 5 (peak A). Under the iodination conditions described, the amount of radioactivity in peak A is small. However, greater amounts of peak A material are generated if the iodination time, the chloramine-T concentration, or the molar ratio of iodide to TRH is increased[8] (unpublished observations). Whereas TRH–Ab binds monoiodinated peak B material we have observed no binding of peak A material by TRH–Ab. As reported by Martino *et al.*,[8] stable TRH elutes prior to the bulk of peak B and carrier-free ^{125}I-TRH should be enriched in the later fractions of peak B. Although it is theoretically desirable to use these later fractions for the TRH RIA, we have not found major variations in assay sensitivity when different portions of peak B material were used for the TRH RIA.

First Antibody

Because of its relatively low molecular weight, TRH must be coupled to a protein in order to produce anti-TRH antibodies (TRH–Ab) for use in the TRH RIA. Most TRH–Ab have been generated by immunizing rabbits

[4] R. M. Bassiri and R. Utiger, *Endocrinology (Baltimore)* **90,** 722 (1972).

[5] C. H. Emerson, L. A. Frohman, M. Szabo, and I. Thakkar, *J. Clin. Endocrinol. Metab.* **45,** 329 (1977).

[6] G. F. Bryce, *Immunochemistry* **11,** 507 (1974).

[7] F. C. Greenwood, W. H. Hunter, and J. S. Glover, *Biochem. J.* **89,** 114 (1963).

[8] E. Martino, H. Seo, and S. Refetoff, *Endocrinology (Baltimore)* **103,** 246 (1978).

with BSA–TRH conjugates, although some workers have used other proteins such as thyroglobulin.[9]

Bassiri and Utiger[4] described the first preparation of polyclonal TRH–Ab. Their method was very similar to previously described methods[10] for preparing protein erythrocyte conjugates with bisdiazotized benzidine (BDB). BDB is prepared, at 4°, by dissolving 0.23 g benzidine dihydrochloride in 45 ml 0.2 *N* HCl. $NaNO_2$, 0.175 g, previously dissolved in 5 ml H_2O, is then added to the benzidine dihydrochloride solution, and the sample is stirred for 1 hr at 4°. Aliquots of the BDB are then frozen in dry ice–acetone and stored at −60°. Since certain dry diazonium compounds may be explosive, BDB should not be dried or heated.[11] The coupling reaction is performed at 4°. Fifty milligrams of BSA is dissolved in 5 ml 0.16 *M* borate, 0.13 *M* NaCl, pH 9.0 (borate buffer), and stirred in a 50-ml beaker. To this is added 5 mg TRH, previously dissolved in 5 ml borate buffer. Finally 1.6 ml BDB is dissolved in 5 ml borate buffer, and this is added to the TRH–BSA mixture. The sample is stirred slowly at 5° and dialyzed overnight at 5° against water. The BSA–TRH conjugate is then stored at −20° until used for immunization. It has been estimated that of the 5 mg TRH committed to the reaction, 1.9 mg (38%) is incorporated into the BSA–TRH conjugate.[4]

In order to be reasonably certain that suitable TRH–Ab will be obtained, it is best to immunize a group of at least six rabbits. Initially, for each rabbit, 1 mg BSA–TRH is brought up to a volume of 1 ml with H_2O and emulsified with 0.8 ml Freund's complete adjuvant and 0.3 ml diphtheria–tetanus–pertussis (DPT) vaccine. The emulsion is injected subcutaneously in divided amounts in different regions. The goal of this type of maneuver[12] is to expose as many lymphatic drainage sites as possible to the BSA–TRH conjugate. Four weeks after primary immunization the rabbits are boosted with half the amount of material used for primary immunization. In this and subsequent immunizations Freund's incomplete adjuvant is substituted for Freund's complete adjuvant. Rabbits are bled 10–14 days after the second booster immunization.

To determine which rabbits are producing TRH–Ab, sera are tested by incubating individual samples with ^{125}I-TRH. Any of a number of standard methodologies can be employed to separate antibody-bound ^{125}I-TRH

[9] I. M. D. Jackson and S. Reichlin, *Endocrinology* (*Baltimore*) **95,** 854 (1974).

[10] V. Likhite and A. Sehon, *in* "Methods in Immunology and Immunochemistry" (C. A. Williams and M. W. Chase, eds.), Vol. 1, p. 165. Academic Press, New York, 1967.

[11] A. Nisonoff, *in* "Methods in Immunology and Immunochemistry" (C. A. Williams and M. W. Chase, eds.), Vol. 1, p. 121. Academic Press, New York, 1967.

[12] J. Viatukaitis, J. B. Robbins, E. Nieschlag, and G. T. Ross, *J. Clin. Endocrinol. Metab.* **33,** 988 (1971).

from the free label. We generally screen antisera by incubating them overnight at 4°, at a final dilution of 1 : 100, with 10,000 cpm ^{125}I-TRH. The bound and the free labeled TRH are separated by adding polyethylene glycol in at a final concentration of 12.5%.[13] Rabbits that do not produce TRH–Ab after three immunizations are removed from the study, and, if necessary, additional rabbits are immunized. Rabbits that are producing TRH–Ab are boosted at 1- to 2-month intervals until a satisfactory antibody titer is obtained.

Procedure and Comments

Standard RIA methodologies[14,15] are employed to set up the TRH assay. The sensitivity and specificity of the RIA will, of course, depend on the nature of the antibody generated. TRH RIAs generally utilize TRH–Ab in a final dilution of 1 : 3,000 to 1 : 12,000. The goal is to obtain 30–40% specific binding of the label. For many of the published assays the minimum detectable dose is 2–20 pg TRH per assay tube. The conditions for the assay, such as the antibody dilution, the range of the standard curve, and the method of separating bound from free label should be established first. Once this is accomplished data should be obtained regarding the within and between assay variability, the slope of the dose–response curve, the midrange of the assay, and the least detectable concentration of TRH.

Several aspects of RIA methodology are important to emphasize. First, tubes used to generate the standard curve should contain, as near as possible, the same material as found in the sample tubes. For example, if the samples being analyzed for TRH content are 0.1 ml volumes of extracts of tissue homogenates that have been incubated with synthetic TRH, then the standard curve tubes should contain 0.1 ml volumes of extracts of tissue homogenates that have not been incubated with synthetic TRH. This practice will minimize nonspecific effects that could otherwise result in either under- or overestimation of TRH content.

Second, peptides and TRH analogs should be tested to determine the degree to which they cross-react in the TRH RIA. The specificity of the RIA is dependent on the specificity of those antibodies in the antiserum employed that react with labeled TRH, and not the specificity of all of the antibodies in the antiserum employed. It is evident, for example, that TRH–Ab prepared as described above also contains IgG directed against BSA. However, this IgG does not bind ^{125}I-TRH, and, therefore, BSA

[13] B. Desbuquois and G. D. Aurbach, *J. Clin. Endocrinol. Metab.* **33,** 732 (1971).

[14] W. D. Odell and W. H. Daughaday (eds.), "Principles of Competitive Protein-Binding Assays." Lippincott, Philadelphia, Pennsylvania, 1971.

[15] H. V. Vunakis, this series, Vol. 70, p. 201.

does not cross-react in the TRH RIA. In this sense RIA is more specific than procedures that are influenced by all the antibodies present in a particular antiserum, such as immunocytochemistry. The results of cross-reaction studies with the TRH RIA used in our laboratory are typical of most assays using the methodology described here. Constituent amino acids and most neuropeptides have little or no cross-reaction in the assay. Carboxy-terminal substituted analogs such as pGlu-His-Pro and pGlu-His-Pro-Gly-NH_2 are from 0.2 to 2% as potent as TRH. In contrast, histidyl-substituted analogs may be highly reactive, having as much as half the potency (50% cross-reaction) of TRH itself. It is likely that analogs with histidyl-substituted amino acids have a high degree of cross-reaction in the TRH RIA because the method of conjugation for antigen production and the labeling procedure involve alteration of the histidyl residue.

Deamido-TRH Radioimmunoassay

The methodology for developing and performing the deamido-TRH RIA is identical to that for the TRH RIA except that equimolar doses of deamido-TRH are substituted for TRH in the standard curve and in the production of deamido-TRH–Ab and ^{125}I-deamido-TRH. Purification of iodinated deamido-TRH on Sephadex G-10 results in similar profiles as those obtained after iodination of TRH. In general, deamido-TRH–Ab is more difficult to obtain than TRH–Ab, and the titers and affinity of the antiserum generated are less than that obtained for TRH Ab. In contrast to the TRH RIA, the minimal detectable dose of deamido-TRH is generally 50–100 pg/tube. In most deamido-TRH RIAs the cross-reaction of TRH and TRH analogs containing an amide group is less than 0.5% that of TRH.[16]

Histidylproline Diketopiperazine (HPD) Radioimmunoassay

Standard and Label

Pure HPD for use in the HPD RIA is commercially available. Iodination of stable HPD is performed in a manner described for TRH except that equimolar concentrations of HPD are substituted for TRH.[17,18] Un-

[16] C. H. Emerson, A. Mishal, H. L. Mahabeer, and B. L. Currie, *J. Clin. Endocrinol. Metab.* **49,** 138 (1979).

[17] M. Safran, C. F. Wu, R. Matys, S. Alex, and C. H. Emerson, *Endocrinology* (*Baltimore*) **115,** 1031 (1984).

[18] T. Nogimori, S. Alex, S. Baker, and C. H. Emerson, *Endocrinology* (*Baltimore*) **117,** 565 (1985).

fortunately, iodination of HPD is far more complex than iodination of TRH. In our experience iodination of HPD frequently generates multiple labeled peaks when analyzed by Sephadex G-10 chromatography. Moreover, even if efforts are made to reproduce the iodination procedure in the same manner, the profile of iodinated products, as detected by Sephadex G-10 chromatography, frequently differs. Finally, variation of the iodination procedure in terms of altering the iodination time or the chloramine-T concentration has not solved this problem. Therefore the iodination peaks must be tested against HPD–Ab and then subjected to further purification on SP-Sephadex G-25 ion-exchange chromatography (Pharmacia). A 1.5 × 25 cm column is employed with a starting buffer of 10 m*M* ammonium acetate, pH 5.2. A 200-ml linear gradient is run to 0.2 *M* ammonium acetate, pH 7.0. When 1-ml fractions are collected, iodinated material that binds HPD–Ab elutes at fraction 88. In some cases we have observed more than one peak of binding when the iodination products are analyzed by ion-exchange chromatography.

Because of the difficulties obtaining consistent iodination, we currently use[18] custom synthesized prolyl-labeled [^{3}H]HPD (New England Nuclear, Boston, MA) for the HPD RIA. This is more convenient than preparing ^{125}I-HPD, and the sensitivity of the HPD RIA is improved. However, if [^{3}H]-HPD is difficult to obtain, ^{125}I-HPD can be used.

First Antibody

HPD–Ab is generated in an identical manner as TRH–Ab except that equimolar concentrations of HPD are substituted for TRH.

Procedure and Comments

The HPD RIA is performed in an identical manner as the TRH RIA except that appropriate label and first antibody is used. In general, the HPD RIA is highly specific for HPD. TRH, deamido-TRH, and HPD analogs usually have less than 0.3% cross-reaction as compared to HPD. Unfortunately, the titer and affinity of most HPD–Ab are less than that for the Ab used for the TRH and deamido-TRH RIAs. Therefore, the sensitivity of the RIA is usually not better than 50 pg/tube.

Acknowledgments

Supported by Research Grant AM27850 from the NIDDK, National Institutes of Health, Bethesda, Maryland.

[41] Monoclonal Antibodies: Uses in Studies on Vasopressin

By GUY VALIQUETTE and SIMON NEUBORT

Introduction

Since their development over a decade ago,[1] hybridomas have been used to produce monoclonal antibodies for the study of widely different substances. We first decided to develop monoclonal antibodies to neurohypophyseal hormones several years ago. In this project, we have experienced a number of unexpected problems, particularly in the use of radioimmunoassay (RIA) techniques with these antibodies. This chapter describes our experience, mostly with vasopressin antibodies.

The neurohypophyseal hormones vasopressin (VP) and oxytocin (OT) have been known since the very beginning of the twentieth century,[2-5] yet the first RIAs to these hormones were not developed until 10 years after the description of this technique by Berson and Yalow.[6-8] Even in the 1970s, VP RIAs were still considered difficult and few were available.[9] In retrospect, there are a number of reasons for this delay. Neurohypophyseal hormones are small peptides, 9 amino acids long, and are poorly antigenic. Many different peptides have been identified in various vertebrates, although any given species generally has only two, a pressor or VP-like peptide and a neutral or OT-like peptide (Table I). They also circulate at extremely low concentrations, typically in the 1–10 pM range, and have a short half-life in circulation, 5–10 min.

For these reasons, most researchers have conjugated the peptides to larger proteins to enhance their antigenicity.[10] Standard labeling techniques have also been modified, as have the related purification techniques, and RIAs have had to be highly optimized to detect these hormones in circulation. It is a testimony to the ingenuity of the researchers that highly reliable RIAs for these peptides are now available.

[1] G. Köhler and C. Milstein, *Nature* (*London*) **256,** 495 (1975).
[2] G. Oliver and E. A. Schäfer, *J. Physiol.* (*London*) **18,** 277 (1895).
[3] H. H. Dale, *J. Physiol.* (*London*) **34,** 163 (1906).
[4] I. Ott and J. C. Scott, *Proc. Soc. Exp. Biol. Med.* **8,** 48 (1910).
[5] V. du Vigneaud, *Harvey Lect.* **1** (1956).
[6] S. A. Berson and R. S. Yalow, *J. Clin. Invest.* **38,** 1996 (1959).
[7] J. Roth, S. M. Glick, L. A. Klein, and M. J. Petersen, *J. Clin. Endocrinol.* **26,** 671 (1966).
[8] M. A. Permutt, C. W. Parker, and R. D. Utiger, *Endocrinology* (*Baltimore*) **78,** 809 (1966).
[9] G. L. Robertson, *Recent Prog. Horm. Res.* **33,** 333 (1977).
[10] W. R. Skowsky and D. A. Fisher, *J. Lab. Clin. Med.* **80,** 134 (1972).

TABLE I

NATURALLY OCCURRING NEUROHYPOPHYSEAL PEPTIDES

Common structure:

1 2 3 4 5 6 7 8 9

Cys- - - - Asn - Cys - Pro - - Gly(NH_2)

(disulfide bridge between Cys 1 and Cys 6)

Peptide	Amino acid in position 2	3	4	8	Animals where present
Vasopressor (basic)					
Arginine-vasotocin (AVT)	Tyr	Ile	Gln	Arg	All vertebrates except some suina
Arginine-vasopressin (AVP)	Tyr	Phe	Gln	Arg	All mammals except some suina and Australian marsupials
Lysine-vasotocin (LVT)	Tyr	Ile	Gln	Lys	Suina (and some marsupials?)
Lysine-vasopressin (LVP)	Tyr	Phe	Gln	Lys	Suina and marsupials
Phenypressin (PVP)	Phe	Phe	Gln	Arg	Australian marsupials
Oxytocin-like (neutral)					
Oxytocin (OT)	Tyr	Ile	Gln	Leu	Mammals, American marsupials, holocephali (chimeras), [birds, reptiles, amphibians (disputed)]
Mesotocin (MT)	Tyr	Ile	Gln	Ile	Australian marsupials, birds, reptiles, amphibians, sarcopterygii (lungfishes)
Isotocin (IT)	Tyr	Ile	Ser	Ile	Actinopterygii (bony fishes)
Glumitocin (GT)	Tyr	Ile	Ser	Glu	Some elasmobranchii (skates)
Aspartocin (AT)	Tyr	Ile	Asn	Leu	Some elasmobranchii (sharks)
Valitocin (VT)	Tyr	Ile	Gln	Val	Some elasmobranchii (sharks)

Methods

Antigen Conjugation to Carrier Protein and Immunization

Although there have been rare exceptions, neurohypophyseal hormones must be conjugated to larger proteins to become antigenic. Unconjugated peptides generally behave as haptens. The following procedure was used to prepare the conjugate[10]: ten milligrams of the peptide and 50 mg of bovine thyroglobulin (Sigma) are dissolved in 1.5 ml of distilled water, pH 7.5; 2.3 mg of carbodiimide (Sigma) is dissolved in 0.5 ml of distilled water, pH 7.5. The two solutions are mixed and allowed to stand at room temperature overnight. Two milliliters of 1 *M* hydroxylamine in water is added and allowed to stand at room temperature for 5 hr. The resulting mixture is dialyzed against normal saline and kept frozen at −20° until use.

Immunization Procedure and Cell Fusion

These procedures were performed as described in the literature, and did not require any modification from the standard methods. They have been described elsewhere[11] and will not be repeated here. Any proven immunization schedule and cell fusion technique should work equally well.

In our first fusion experiments, cell lines were tested for antibody production by both liquid-phase and solid-phase radiobinding assays (see Ref. 11). With the experience we have accumulated, it is now clear that at least equally good results can be obtained with less labor by using liquid-phase radiobinding tests only (see *Recommendations for Monoclonal Antibody Screening Tests* below).

Radiolabeling and Tracer Purification

Neurohypophyseal hormones are cyclic nonapeptides with generally nonreactive or blocked side chains, limiting the number of radiolabeling methods available. Two widely known radioiodination techniques have been used, with some modification.

Chloramine-T Radioiodination. The chloramine-T method, initially described by Hunter and Greenwood,[12] substitutes an iodine atom for an hydrogen on the phenol ring of a tyrosine (position 2 in neurohypophyseal peptides) and can therefore be used with all neurohypophyseal hormones (except PVP). The only modification required from the prototype proce-

[11] A. Hou-Yu, P. H. Ehrlich, G. Valiquette, D. L. Engelhardt, W. H. Sawyer, G. Nilaver, and E. A. Zimmerman, *J. Histochem. Cytochem.* **30,** 1249 (1982).

[12] W. M. Hunter and F. C. Greenwood, *Nature* (*London*) **194,** 495 (1962).

dure[13] is the omission of the reduction step used to stop the iodinating reaction (it would open the disulfide bond between Cys^1 and Cys^6); this step is replaced by quenching with an excess of another protein [normally bovine serum albumin (BSA)]. It is routinely performed in our laboratory as follows. The following solutions are added, in order, to a 12×75 mm polystyrene test tube: 0.5 *M* phosphate buffer, pH 7.5, 50 μl; peptide, from a 1 mg/ml stock solution in 0.2 *M* acetic acid, 10 μl; ^{125}I, 1 mCi/10 μl, 10 μl; freshly prepared chloramine-T, 4 mg/ml in 50 m*M* phosphate buffer, 10 μl. The contents are mixed by a few gentle finger flicks. Sixty seconds after having added the chloramine-T solution, add the following: BSA, 10% solution in 50 m*M* phosphate buffer, 100 μl, and again gently agitate by flicking the tube. The tracer is then purified as described below.

"Peptide" can be essentially any neurohypophyseal peptide. In our laboratory, the method described above has been applied to AVP, LVP, OT, MT, and AVT. We have used both ^{125}I and ^{131}I from both Amersham and New England Nuclear with the same technique and results. We have also used "old" iodine, even more than 1 half-life old, with equally consistent results: we simply increase the volume of the radioiodine solution to provide 1 mCi. In fact, the method is so reliable that, when we have an iodination failure, the first question usually is, "Did we use OT instead of AVP?" if the gel elution pattern is as expected or, "Did we forget to add the peptide?" if it is not! We have had no reason to use alternate radioiodinating methods for labeling in the same positions, such as the peroxidase method or the iodogen method.

Bolton–Hunter Radioiodination. The Bolton–Hunter radioiodination method, as initially described,[14] was designed to provide a more gentle alternative to the chloramine-T method. We have used it to provide a chemically (as well as immunologically, see below) different tracer. The method is based on the conjugation of preradioiodinated *N*-succinimidyl-3-(4-hydroxyphenyl)propionate to free amino groups (amino terminal or lysine side chains) on the protein to be labeled. The ester hydrolyzes to release the 3-(4-hydroxyphenyl)propionic acid in aqueous solutions, which then reacts with the amino groups.

The Bolton–Hunter method, although reasonably reliable, is not quite as foolproof as the chloramine-T method. It is important to obtain, and use, the Bolton–Hunter reagent as fresh as possible. The preparation schedule of the supplier (Amersham or New England Nuclear) should be consulted before ordering, and the conjugation reaction conducted as

[13] A. E. Bolton, this series, Vol. 124, p. 18.

[14] A. E. Bolton and W. M. Hunter, *Biochem. J.* **133,** 529 (1973).

soon as possible after the reagent is received. The Bolton–Hunter reagent should be stored at 4° until used. Most of our failures have been associated with the use of less than fresh reagent (more than 1 week old). In our laboratory, the labeling reaction is performed as follows.

The reagent is commercially available as a solution in benzene (1 mCi in ~0.5 ml, Amersham) in a conical-tipped glass tube which is inserted in the neck of a glass vial. The benzene should be evaporated by a gentle stream of dry nitrogen. It is most convenient to evaporate the benzene and conduct the reaction in the shipping vial. Under a fume hood, two 18-gauge hypodermic needles are inserted through the Teflon-lined rubber septum of the vial and attached to intravenous fluid administration set tubings. The outlet side tubing is taped to the side of the fume hood to the upper part of the hood. The inlet side brings nitrogen, dried by a silica gel column, into the vial. The nitrogen stream should be very gentle, to avoid splashing. The radioactive reagent is not volatile but may be carried out into the tubing as droplets if the nitrogen stream is too strong. Several hours of drying, at room temperature, should be allowed so that no benzene remains.

In a separate tube, the following are mixed: peptide, from a 1 mg/ml stock solution in 0.2 *M* acetic acid, 10 μl; 0.1 *M* borate buffer, pH 9.0, 50 μl. The rubber septum of the shipment vial is removed, and the solution of peptide in borate buffer is added to the shipping vial. The vial is closed with the supplied screw cap. The vial is vigorously agitated for 15 min at 4°. In actual practice, it is agitated on a vortex mixer for 30 sec, returned to the ice bath for 30 sec, back to the vortex mixer, etc. After the initial 15 min of vigorous agitation, the vial is left in the ice bath for a further hour with occasional agitation (30 sec every 3–5 min). The reaction is terminated by adding 0.5 ml of 0.2 *M* glycine in 0.1 *M* borate buffer. The vial is agitated every few minutes for 15–30 min. The tracer is then purified as described below.

The vigorous agitation and long reaction times described above are essential for success in the iodination of neurohypophyseal peptides. When we first used this method for LVP, we allowed the conjugation reaction to proceed for 15 min, as described by Bolton and Hunter,[13,14] and yields were only 5–10%. With the above modifications, 50–80% of the radioactive reagent is conjugated to the peptide.

Since the Bolton–Hunter radiolabeling method depends on the availability of an amino group on the peptide to be labeled, it might be expected that all neurohypophyseal hormones can be labeled by this method. Strictly speaking, this is true, although neurohypophyseal peptides without lysine in position 8 are labeled at very low efficiency. Peptides with lysine in position 8, [deamino-Cys_1]LVP (dLVP), LVP, and LVT, have been labeled by this method and used as tracers. In the case of LVP and LVT, the vast majority of the Bolton–Hunter reagent appears to react with the ε-amino group of Lys^8.

Although amino-terminal Bolton–Hunter tracers can be prepared from any neurohypophyseal peptide, the efficiency of labeling in this position is very low, less than 10%, compared with the 50–80% obtained from the ε-amino group of lysine. Given the low conjugation efficiency, further purification of such a tracer by affinity chromatography with neurophysins (see below) is prudent, to remove unlabeled peptide. Nevertheless, we have found that neurohypophyseal peptides labeled at the amino-terminal cysteine by the Bolton–Hunter method lose essentially all their immunoreactivity (see below).

Tracer Purification

Owing to the abnormal behavior of neurohypophyseal peptides in gel filtration, simple Sephadex gel filtration is sufficient to obtain pure mono- or diiodinated tracer. Neurohypophyseal peptides interact hydrophobically with the dextrans of the Sephadex gel and elute beyond the position expected from their molecular weights. Furthermore, this hydrophobic interaction is amplified by the presence of iodine on Tyr^2 or the 3-(4-hydroxyphenyl)propionic acid moiety, and the tracers elute beyond the unlabeled peptide.

We routinely use a Sephadex G-25 Superfine (Pharmacia) column with a bed volume of approximately 80 ml. If newly prepared, the column should be "coated" by loading 2–3 ml of 10% BSA in 0.2 *M* acetic acid and eluting with at least one bed volume of 0.2 *M* acetic acid, before use. The column can be reused for years without repouring if it is rinsed with one or two bed volumes of 0.2 *M* acetic acid after each use. The column is left in 0.2 *M* acetic acid between uses, and no preservative is necessary.

The radiolabeling reaction mixture is brought to approximately 1 ml with 2 *M* acetic acid and applied to the Sephadex column. It is eluted overnight at room temperature with 0.2 *M* acetic acid at a flow rate of 7–10 ml/hr. A fraction collector (LKB) is used to collect 160 fractions. One milliliter/fraction is collected for a chloramine-T radioiodination, and 1.5 ml/fraction for a Bolton–Hunter radioiodination. In either case, 50 μl of a 10% BSA solution in 0.2 *M* acetic acid is added to each collection tube. A 10-μl aliquot of each fraction is counted to identify the peaks. Figure 1 shows typical elution profiles for both chloramine-T and Bolton–Hunter radioiodinations. The peak tubes of the tracer fractions are kept for use in

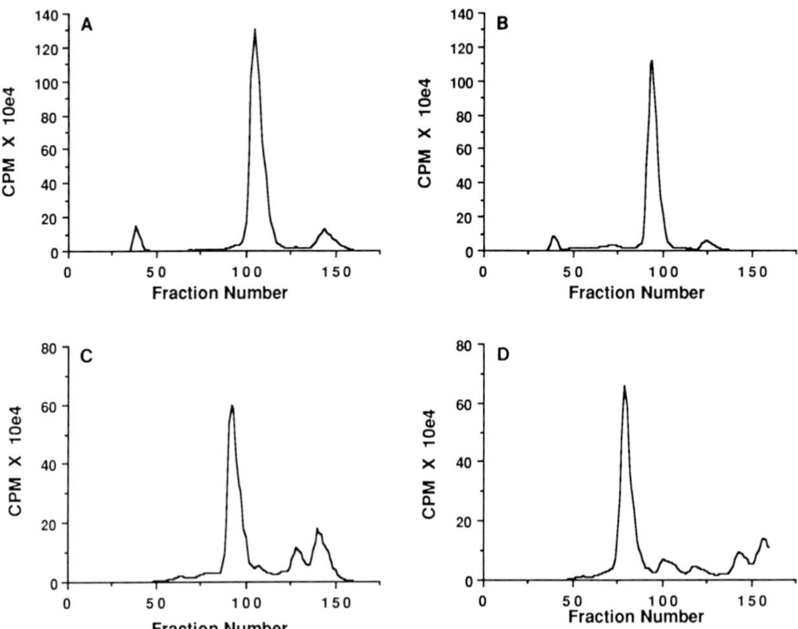

FIG. 1. Actual elution profiles of four neurohypophyseal hormone tracers prepared as described in the text. (A) Chloramine-T labeled AVP: the first peak (tube 40) represents ^{125}I-BSA, the second (tube 110) ^{125}I-AVP, and the third (tube 140) di-[^{125}I]iodo-AVP. (B) Chloramine-T labeled OT: in addition to the peaks identified in A, a very small peak is occasionally seen between ^{125}I-BSA and the tracer (tube 70); this represents free ^{125}I. (C) Bolton–Hunter labeled dLVP: the tracer constitutes the major peak. (D) Bolton–Hunter labeled LVT: note the similarity to the elution profile in C.

RIA. Contrary to other authors,[15–20] we have not found that tubes on the descending side of the peak bind better than the peak tube itself. This seems to be the result of the large bed volume that we use.[21] The tracer

[15] M. K. Husain, N. Fernando, M. Shapiro, A. Kagan, and S. M. Glick, *J. Endocrinol. Metab.* **37,** 616 (1973).

[16] F. Fyhrquist, M. Wallenius, and H. J. G. Hollemans, *Scand. J. Clin. Lab. Invest.* **36,** 841 (1976).

[17] P. H. Baylis and D. A. Heath, *Clin. Endocrinol.* **7,** 91 (1977).

[18] G. Moore, A. Lutterodt, G. Burford, and K. Lederis, *Endocrinology (Baltimore)* **101,** 1421 (1977).

[19] M. A. Moulin, R. Camsonne, M. C. Bigot, and D. A. Debruyne, *Clin. Chim. Acta* **88,** 363 (1978).

[20] W. G. North, F. T. LaRochelle, J. Haldar, W. H. Sawyer, and H. Valtin, *Endocrinology (Baltimore)* **103,** 1976 (1978).

[21] W. A. Sadler, C. P. Wright, and J. H. Livesey, *Clin. Chim. Acta* **155,** 61 (1986).

tubes are kept frozen at −20° and can be used for more than one half-life of the isotope, although the specific activity of the tracer, and the sensitivity of the RIA, decrease proportionately. The tracer does not suffer from freeze–thaw cycles, nor do stock solutions of the peptides.

The radioactive diiodo-CT-AVP tracer (or diiodo-CT-OT, etc.) is also immunoreactive with most antisera whose epitopes are directed away from Tyr^2 of the neurohypophyseal peptide molecule, and it can be used as a tracer for these antisera. Although this increases the sensitivity of the RIA, because of the doubled specific activity of the tracer, the diiodo tracer degrades rapidly and we have not found its use necessary. Diiodinated Bolton–Hunter reagent is also available commercially. We do not have any experience with this reagent and can only speculate that the tracer thus obtained should be comparable in stability to the diiodo chloramine-T tracer.

Tracers and Binding Affinities

It was mentioned in the introduction that unexpected problems were met in the development of monoclonal antibodies to AVP. These first surfaced when we noticed discrepancies between the results of a solid-phase radiobinding assay and those of liquid-phase radiobinding assays. To clarify this situation, we prepared a number of different tracers and tested all positive cell lines with these tracers.

Eleven of 177 hybridomas obtained from one fusion experiment were considered positive for anti-AVP secretion from the initial screening tests. Spent culture medium was used as the source of the antibody. Sufficient culture medium was pooled to conduct these experiments, and these pools were tested in liquid-phase radiobinding tests. Table II summarizes the results obtained. A conventional rabbit antiserum (AVP-5) was also included for comparison. All binding tests were conducted as follows. Incubation buffer contained the following: 0.1 *M* phosphate, pH 7.5; 10 m*M* EDTA; 0.1% NaN_3 (sodium azide); 0.1% BSA. All incubations were performed at 4° for 48 hr. The final incubation volume was 0.5 ml. Bound tracer was separated from free tracer by the polyethylene glycol (PEG) method.[22] Fifty microliters of a 2% solution of bovine γ-globulin in incubation buffer was added to the incubation mixture, followed by 1 ml of a solution of 250 g of PEG in 750 ml of distilled water. The assay tubes (12 × 75 mm, polystyrene) were agitated on a vortex mixer and immediately centrifuged at approximately 2000 *g* at 4° for 30 min. Only as many tubes as can be centrifuged in one batch should receive

[22] B. Desbuquois and G. D. Aurbach, *J. Clin. Endocrinol.* **33,** 732 (1971).

TABLE II
APPARENT TITERS OF MONOCLONAL ANTIBODIES WITH VARIOUS TRACERS[a]

Antibody	CT-AVP	BH-dLVP	BH-AVP	[^{14}C]AVP
I D-5	1:2,050	1:3,225	40%	
I G-2	0.0%	0.5%	0.4%	
II G-2	0.3%	0.5%	0.4%	
II G-5	27%	1:80	37%	
III D-7	10%	4.5%	13%	1:98
III D-9	1:113	1:425	41%	
III G-7	41%	0.6%	5.5%	
III G-11	1:1,560	33%	16%	
IV B-3	1:45	1:1,295	39%	18%
IV G-3	6.5%	17%	7.4%	
IV G-8	4.0%	1.0%	0.0%	
AVP-5[b]	1:253,950	1:21,870	43%	1:815

[a] CT-AVP, Chloramine-T labeled AVP; BH-dLVP, Bolton–Hunter labeled dLVP; BH-AVP, Bolton–Hunter labeled AVP; [^{14}C]AVP, intrinsically labeled AVP. Data are expressed as apparent titers for 50% binding of 1 pg of tracer (iodinated tracers only), if reached, or percentage binding at a 1 : 10 dilution of the culture medium.
[b] At a 1 : 100 dilution.

the PEG, as the bound tracer will slowly dissociate after the PEG is added. After centrifugation, the supernatant is aspirated and the precipitate is counted. The results are shown in Table II. Some cell lines, such as I G-2 and II G-2, seem to have been falsely positive in the initial screening test. The most remarkable feature, however, is how the apparent titer is dependent on the tracer used. This is most notable for cell lines II G-5, III G-11, and IV B-3.

Bolton–Hunter labeled AVP was found to be a universally poor tracer. This is most probably due to the fact that the introduction of a third aromatic ring (the iodophenylpropionic acid residue) in close proximity to the two aromatic side chains of AVP, Tyr^2 and Phe^3 (see Fig. 2), leads to such distortion of the cyclic part of AVP (residues 1–6) through π-electron interactions[23,24] that the resulting molecule is no longer recognized as AVP. The importance of tracer choice is not surprising when one considers the degree of molecular distortion that some labeling methods entail (Fig. 2). Probably more important, though, is whether the distortion

[23] D. W. Mason and A. F. Williams, *Biochem. J.* **187,** 1 (1980).
[24] M. Kobayashi, M. Nakanishi, and M. Tsuboi, *J. Biochem. (Tokyo)* **91,** 407 (1982).

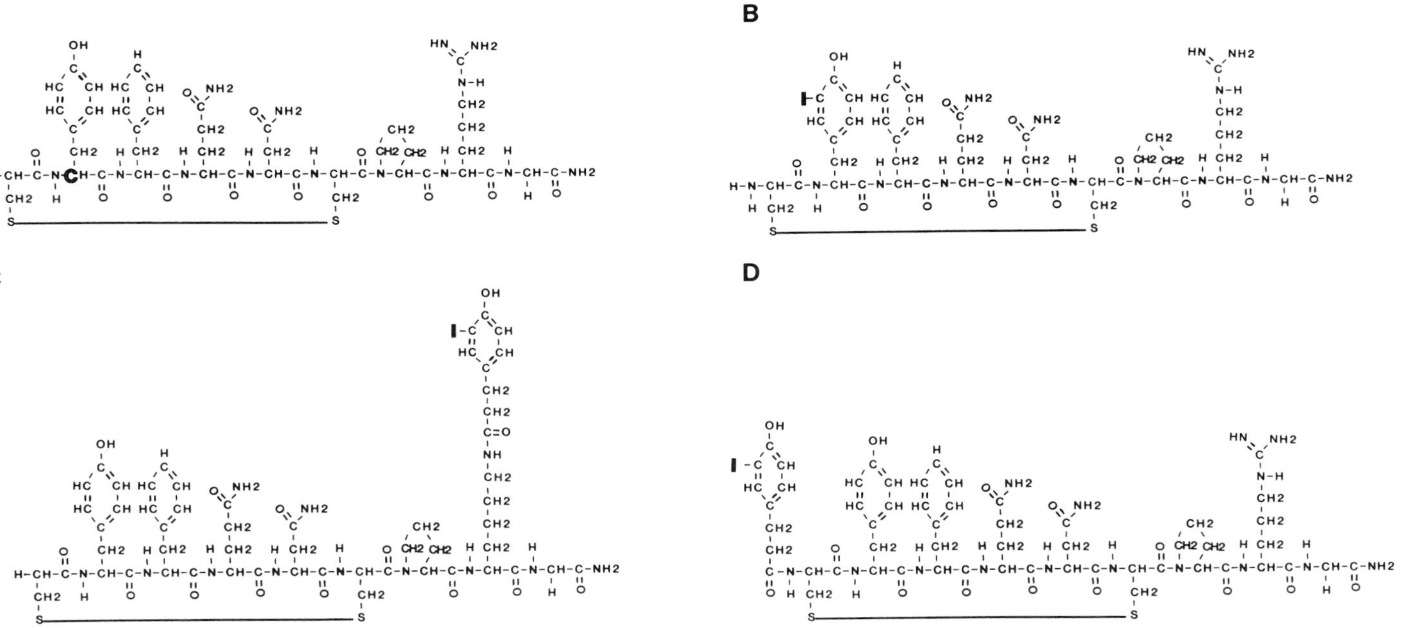

FIG. 2. Structural formulas of four tracers. (A) [^{14}C]AVP, intrinsically labeled. (B) Chloramine-T labeled AVP. (C) Bolton–Hunter labeled dLVP. (D) Bolton–Hunter labeled AVP. In all formulas, the radioactive atom is identified by bold type.

is in the part of the molecule encompassed by the epitope recognized by a particular monoclonal antibody.

An extreme example of how extrinsic radiolabeling can influence the immunoreactivity of a tracer can be seen in the behavior of the antibody produced by cell line III D-7 (Table II). Monoclonal antibody III D-7 does not bind any extrinsically labeled tracer to any significant extent, yet at that same time, other members of our group had used it with remarkable success in immunohistochemistry.[11] This suggested that the antibody recognizes an epitope covering most of the structure of the peptide and that extrinsic labeling destroys the immunoreactivity of the tracer. Intrinsically tritiated neurohypophyseal peptides were not available when this work was conducted, but we were fortunate to receive a gift of intrinsically labeled [^{14}C]AVP from Dr. J. McKelvey. This tracer allowed us to elucidate the atypical behavior of III D-7. The tracer was purified by neurophysin affinity chromatography (see below) before use. Given the low specific activity of the ^{14}C tracer, it became evident that III D-7 had a high apparent titer. By using the apparent titers of rabbit antiserum AVP-5 to extrapolate the expected titer of III D-7 with extrinsically labeled tracers, one obtains an apparent titer of approximately 1 : 25,000, while the observed titer with all iodinated tracers is less than 1 : 10!

Recommendations for Monoclonal Antibody Screening Tests

There are, unfortunately, no totally error-free screening methods. Screening methods should therefore strive to minimize false-negative results even if some false-positive results must be accepted: it is preferable to carry forward some nonproducing hybridomas and to discard them later in the project than to lose a potentially precious cell line.

Solid-phase radiobinding assays, as we initially performed them,[11] require large amounts of antigen, which may be expensive or scarce. They also require a labeled second antibody and are more labor intensive than liquid-phase radiobinding assays. Finally, the specificity information they provide is limited. However, they do provide a testing environment more akin to immunocytochemistry. Given our experience, we feel that carefully designed liquid-phase binding assays are easier and more cost effective for screening the few hundred cell lines resulting from a single fusion. Adaptability of an antibody for a particular application can be more easily assessed later on each of the dozen or so hybridomas producing the desired antibody.

Liquid-phase radiobinding tests with intrinsically labeled tracers can approximate the ideal screening test. However, the lower specific activity of tritium may limit sensitivity, and such tracers are not available for all antigens. A mixture of different extrinsically labeled tracers, possibly using different isotopes (e.g., ^{131}I-CT-AVP with ^{125}I-BH-LVP) provides more sensitivity, owing to the higher specific activity of the iodine isotopes. This provides some protection against false-negative results, but there is no absolute guarantee.

It should also be noted that an arbitrary choice must be made of the buffer used in the screening test. Although a middle-of-the-road approach, such as the buffer described above, is generally a reasonable choice, it must be kept in mind that maximum affinity may be obtained in conditions far removed from physiological.

Neurophysin Affinity Chromatography

Neurohypophyseal peptides have a naturally occurring ligand for affinity chromatography: neurophysins. These proteins, of approximately 10,000 molecular weight, are synthesized from the same mRNA molecule that encodes the neurohypophyseal peptide itself.[25] Although there are two distinct neurophysins, associated with the two neurohypophyseal peptides, either neurophysin will bind either peptide indiscriminately. The physiological role of these proteins is unclear. Nevertheless, the high degree of sequence conservation through evolution[26] and the fact that the mutation in the Brattleboro rat, which does not produce AVP, is a single base deletion (and subsequent frameshift) in the latter third of the invariant portion of the neurophysin sequence, while the AVP coding sequence, toward the 5′ end of the mRNA, is intact,[27] suggest that the neurophysins play an important role in the biosynthesis of neurohypophyseal hormones.

Although neurophysins will bind all naturally occurring neurohypophyseal hormones indiscriminately, there obviously are minimal structural requirements for binding. A free α-terminal amino group is essential; thus, dLVP and BH-labeled tracers with conjugation at the α-terminal amino group of Cys^1 will not bind. Iodinated peptides with iodine on Tyr^2 (chloramine-T method) bind very poorly and can be considered as nonbound in the batchlike techniques used in affinity chromatography. Synthetic ligands for neurophysins have been constructed, and these di- or tripeptides define the minimum requirements for binding.[28,29] Binding is reversible, with an optimum at pH 5.8 and essentially complete elution in

[25] H. Land, G. Schütz, H. Schmale, and D. Richter, *Nature (London)* **295,** 299 (1982).

[26] G. Valiquette, E. A. Zimmerman, and J. L. Roberts, *J. Theor. Biol.* **112,** 445 (1985).

[27] H. Schmale and D. Richter, *Nature (London)* **308,** 705 (1984).

[28] E. Breslow, *Annu. Rev. Biochem.* **48,** 251 (1979).

[29] J. Carlson and E. Breslow, *Biochem. Biophys. Res. Commun.* **100,** 455 (1981).

0.1 *M* formic acid. Neurophysin affinity columns can be cycled an indefinite number of times.

A neurophysin affinity column is easily prepared, and the neurophysin–gel conjugate is stable for years. Any laboratory working with neurohypophyseal peptides on a regular basis should consider using neurophysin affinity columns and not shy away from their preparation. The methods we used to purify the neurophysins from bovine pituitary glands and prepare the affinity gel are described below.

Neurophysin Extraction from Pituitaries and Crude Purification. A limited degree of purification is all that is required to prepare neurophysins for affinity chromatography. As long as most of the conjugated protein is neurophysin, a perfectly satisfactory immobilized ligand will be obtained, even if some of the protein conjugated is albumin. As neurophysins represent a large proportion of posterior pituitary proteins, a simple purification scheme is perfectly adequate. The following scheme (adapted from Refs. 30 and 31) was used in our laboratory.

Fresh-frozen bovine pituitary glands were obtained from a slaughterhouse. Thawed whole pituitary glands (900 g) were dried in several (4–5) changes (24 hr each) of acetone at 4°. The dried glands should feel hard and somewhat brittle. The posterior lobes were separated from the adenohypophysis. We found that it was easier to do this with acetone-dried glands than with freshly thawed glands. The acetone was evaporated under a nitrogen stream. Thirty-five grams of acetone-dried posterior pituitary tissue was obtained. All the following steps were conducted at 4°. The posterior pituitary tissue was homogenized in 400 ml of 0.1 *N* HCl, and the homogenate was centrifuged at 1000 *g* for 30 min. The pellet was reextracted with a further 400 ml of 0.1 *N* HCl, and this mixture was centrifuged again, the pellet discarded, and the supernatants combined. The combined supernatants were brought to pH 7.0 with NaOH. The resulting precipitate was centrifuged at 1000 *g* for 30 min and discarded. The supernatant was reacidified to pH 3.9 with HCl. NaCl was then added to 10 g/100 ml to "salt-out" the neurophysins. The mixture was agitated for 24 hr with a magnetic stirrer to allow complete precipitation of the neurophysins. The precipitate was collected by centrifugation at 2000 *g* for 30 min and then dialyzed in several changes of 0.1 *M* ammonium acetate, pH 7.0. This "crude neurophysins" preparation can be kept frozen or lyophilized. A yield of 900 mg of crude neurophysins was obtained.

Neurophysin Conjugation to Cyanogen Bromide-Activated Sepharose. The conjugation of neurophysins to cyanogen bromide-activated

[30] A. G. Robinson, *J. Clin. Invest.* **55,** 360 (1975).

[31] R. Acher, J. Chauvet, and G. Olivry, *Biochim. Biophys. Acta* **22,** 421 (1956).

Sepharose was performed essentially as suggested by the manufacturer of the gel[32] (Pharmacia Fine Chemicals). It is described here for completeness and to give an indication of expected yields. Many of the deviations from the method suggested by the manufacturer consisted of the use of the same buffers used for radiolabeling by the Bolton–Hunter method.

The processing of the gel was conducted in a batch-wise fashion by gently centrifugating the gel (500 g for 1 min) at the end of each step. Two grams of cyanogen bromide-activated Sepharose was allowed to swell in 25 ml of 0.2 M acetic acid for 1 hr at room temperature. The gel was rinsed in 2 further changes of 25 ml of 0.2 M acetic acid. *Working quickly*, the gel was washed with 2 changes of 25 ml of 0.1 M sodium borate with 0.5 M NaCl, pH 9.0. Fifty milligrams of "crude neurophysin" in 25 ml of 0.1 M sodium borate–0.5 M NaCl, pH 9.0, was added to the gel and agitated in a reciprocating mixer for 2 hr at room temperature. Residual conjugation sites on the gel were inactivated by adding 25 ml of 0.1 M glycine in 0.1 M sodium borate–0.5 M NaCl, pH 9.0, to the gel and agitating on a reciprocating mixer overnight at room temperature. Adsorbed proteins and reagents were eluted by 3 cycles of high/low pH rinses with 25 ml of 0.1 M sodium borate–1 M NaCl, pH 9.0, and 0.1 N acetic acid–1 M NaCl. Storage of the neurophysin affinity gel was best at 4° in 0.1 N formic acid. As determined by Lowry protein assays on the neurophysin solutions before and after conjugation to the cyanogen bromide-activated Sepharose, the efficiency of conjugation of the protein was 41.6%.

Conclusions

Monoclonal antibodies to AVP and OT have been used in immunocytochemistry, radioimmunoassay, and for passive immunization *in vivo*.[11,33,34] In all these applications, they have distinct advantages over conventional polyclonal antibodies. However, the small size of the neurohypophyseal hormones and the monospecificity of the antibodies require careful attention to details.

Acknowledgments

The authors thank Dr. J. McKelvey for the [^{14}C]AVP and Dr. W. H. Sawyer for many of the neurohypophyseal peptides used in this work as well as helpful advice.

[32] Pharmacia Fine Chemicals, "Affinity Chromatography: Principles and Methods," p. 8. Pharmacia Fine Chemicals, Uppsala, Sweden, 1974.

[33] V. M. Tennyson, G. Nilaver, A. Hou-Yu, G. Valiquette, and E. A. Zimmerman, *Cell. Tissue Res.* **243,** 15 (1986).

[34] G. Valiquette, A. Hou-Yu, and E. A. Zimmerman, *Proc. Serono Symp.* **30,** 65 (1986).

[42] Detection and Purification of Inhibin Using Antisera Generated against Synthetic Peptide Fragments

By JOAN M. VAUGHAN, JEAN RIVIER, ANNE Z. CORRIGAN, RICHARD MCCLINTOCK, CAROLYN A. CAMPEN, DIANE JOLLEY, JOSEF K. VOGLMAYR, C. WAYNE BARDIN, CATHERINE RIVIER, and WYLIE VALE

Introduction

Inhibin is a hormone whose best established physiological role is selective suppression of the release of follicle-stimulating hormone (FSH) from the pituitary. Inhibin has been isolated from the gonadal fluids of several species[1-5] and characterized as a heterodimer consisting of an α subunit and one of two distinct β subunits, β_A or β_B, linked by disulfide bridges. The complete amino acid sequences of the precursors of porcine,[6,7] human,[7,8] bovine,[9] and rat[10,11] inhibins have been predicted from complementary DNA sequences. The two inhibin β subunit precursor proteins predicted from the cDNA sequence show structural and sequence similarity to an emerging family of hormones possessing growth or

[1] D. M. Robertson, L. M. Foulds, L. Leversha, F. J. Morgan, M. T. W. Hearn, H. G. Burger, R. E. H. Wettenhall, and D. M. de Kretser, *Biochem. Biophys. Res. Commun.* **126,** 220 (1985).

[2] K. Miyamoto, Y. Hasegawa, M. Fukuda, M. Nomura, M. Igarashi, K. Kangawa, and H. Matsuo, *Biochem. Biophys. Res. Commun.* **129,** 396 (1985).

[3] J. Rivier, J. Spiess, R. McClintock, J. Vaughan, and W. Vale, *Biochem. Biophys. Res. Commun.* **133,** 120 (1985).

[4] N. Ling, S.-Y. Ying, N. Ueno, F. Esch, L. Denoroy, and R. Guillemin, *Proc. Natl. Acad. Sci. U.S.A.* **82,** 7217 (1985).

[5] D. M. Robertson, F. L. de Vos, L. M. Foulds, R. I. McLachlan, H. Burger, F. J. Morgan, M. T. W. Hearn, and D. M. de Kretser, *Mol. Cell. Endocrinol.* **44,** 271 (1986).

[6] A. J. Mason, J. S. Hayflick, N. Ling, F. Esch, N. Ueno, S.-Y. Ying, R. Guillemin, H. Niall, and P. H. Seeburg, *Nature (London)* **318,** 659 (1985).

[7] K. E. Mayo, G. M. Cerelli, J. Spiess, J. Rivier, M. G. Rosenfeld, R. M. Evans, and W. Vale, *Proc. Natl. Acad. Sci. U.S.A.* **83,** 5849 (1986).

[8] A. J. Mason, H. D. Niall, and P. H. Seeburg, *Biochem. Biophys. Res. Commun.* **135,** 957 (1986).

[9] R. G. Forage, J. M. Ring, R. W. Brown, B. V. McInerney, G. S. Cobon, R. P. Gregson, D. M. Robertson, F. J. Morgan, M. T. W. Hearn, J. K. Findlay, R. E. H. Wettenhall, H. G. Burger, and D. M. de Kretser, *Proc. Natl. Acad. Sci. U.S.A.* **83,** 3091 (1986).

[10] F. S. Esch, S. Shimasaki, K. Cooksey, M. Mercado, A. J. Mason, S.-Y. Ying, N. Ueno, and N. Ling, *Mol. Endocrinol.* **1,** 388 (1987).

[11] T. K. Woodruff, H. Meunier, A. J. W. Hsueh, and K. E. Mayo, *Mol. Endocrinol.* **1,** 561 (1987).

differentiation-regulating properties. Members of this family include transforming growth factor β (TGF-β),[12] known to exert a wide range of cellular responses in a variety of cell types, Mullerian duct inhibiting substance (MIS),[13] which causes regression of the Mullerian duct during development of the mammalian testis, and fly decapentaplegic gene complex (DPP-C),[14] responsible for important developmental changes in *Drosophila*.

Subsequent to the identification of inhibin, two laboratories isolated a protein from porcine follicular fluid which selectively released FSH from pituitary cell cultures.[15,16] Characterization of this protein, named FSH releasing protein (FRP)[15] or activin,[16] showed that it was a homo-[15] or heterodimer[16] of inhibin β subunits. Thus it appears that α and β subunits of inhibin associate to form dimers with very different biological functions.

S_1 nuclease analysis has been used to investigate the pattern of inhibin subunit expression in the rat.[17] Expression of the inhibin subunits (α, β_A, and β_B) is found in a variety of gonadal and extra gonadal tissues including the ovary, testis, placenta, pituitary, adrenal, bone marrow, kidney, spinal cord, and brain. In addition to the actions of inhibin and FRP on the pituitary, effects have been seen in the gonad, placenta, brain, and bone marrow (see Ref. 18 for a review).

Study of this complex regulatory system requires both adequate amounts of inhibin and FRP and antibodies specific to both proteins. We describe here the production of antibodies to the α, β_A, and β_B subunits of inhibin; development of a radioimmunoassay (RIA) specific for inhibin, using an antibody directed to the α subunit; use of these antibodies for Western blot analysis; and a method for concentrating inhibin and FRP from biological fluids. In addition we describe a method which we have developed for rapid isolation of inhibin from ram rete testis fluid (RTF),

[12] R. Derynck, J. A. Jarrett, E. Y. Chen, D. H. Eaton, J. R. Bell, R. K. Assoian, A. B. Roberts, M. P. Sporn, and D. V. Goeddel, *Nature (London)* **316,** 701 (1985).

[13] R. L. Cate, R. J. Mattaliano, C. Hession, R. Tizard, N. M. Farber, A. Cheung, E. G. Ninfa, A. Z. Frey, D. J. Gash, E. P. Chow, R. A. Fisher, J. M. Bertonis, G. Torres, B. P. Wallner, K. L. Ramachandran, R. C. Ragin, T. F. Manganaro, D. T. MacLaughlin, and P. K. Donahue, *Cell* **45,** 685 (1986).

[14] R. W. Padgett, R. D. St-Johnston, and W. M. Gelbart, *Nature (London)* **325,** 81 (1987).

[15] W. Vale, J. Rivier, J. Vaughan, R. McClintock, A. Corrigan, W. Woo, D. Karr, and J. Spiess, *Nature (London)* **321,** 776 (1986).

[16] N. Ling, S.-Y. Ying, N. Ueno, S. Shimasaki, F. Esch, M. Hotta, and R. Guillemin, *Nature (London)* **321,** 779 (1986).

[17] H. Meunier, C. Rivier, R. M. Evans, and W. Vale, *Proc. Natl. Acad. Sci. U.S.A.* **85,** 247 (1988).

[18] W. Vale, C. Rivier, J. Spiess, A. Hsueh, J. Vaughan, H. Meunier, T. Biczak, J. Yu, P. Sawchenko, A. Corrigan, R. McClintock, C. Campen, W. Bardin, and J. Rivier, *Recent Prog. Horm. Res.* **44,** in press (1988).

using immunoaffinity chromatography with an antibody directed against the α subunit.

Inhibin Radioimmunoassays

Development of Specific Antisera

Antibodies can be raised to inhibin and FRP using several different strategies. McLachlan *et al.*[19] developed a radioimmunoassay using polyclonal antisera raised in rabbits against purified 58-kDa bovine inhibin. In this assay system, human follicular fluid (FF) cross-reacted 30%, while ovine FF and rete testis fluid (RTF) and rat ovarian extracts had no significant cross-reactivity. Miyamoto *et al.*[20] used purified bovine 32-kDa inhibin to prepare monoclonal antibodies specific for the 20-kDa α and the 13-kDa β subunits and used these antibodies in immunoblotting procedures to elucidate the forms of inhibin present in bovine FF. Another strategy is to use synthetic peptides predicted from the DNA sequence to raise antibodies to the gene product. This technology, reviewed by Lerner,[21] has proven successful for producing antibodies to a variety of DNA and RNA tumor virus proteins. We chose to use this last approach to raise polyclonal antisera in rabbits against various synthetic fragments of inhibin α, β_A, and β_B subunits (see Table I).

The choice of synthetic fragments for use as immunogens is complicated by the lack of knowledge of the position of intra- and intermolecular disulfide bridges. In the native protein, antigenic determinants are often constructed from discontinuous regions of the protein chain brought into proximity by disulfide bridges. To avoid this complication, we chose mainly regions of the amino acid sequence lacking cysteine residues and which had a tyrosine at the amino or carboxy terminus. In cases where the peptide sequence of interest lacked a tyrosine at the amino or carboxy terminus, a tyrosine connected by a spacer arm was added. The tyrosine moiety can then be used as an iodination as well as a coupling site.

We were able to obtain useful antisera directed toward the cysteine-free amino terminus of the porcine inhibin α subunit which also recognizes bovine, human, ovine, rat, and rhesus monkey gonadal fluids. Using these antisera, we have developed a radioimmunoassay and have mea-

[19] R. I. McLachlan, D. M. Robertson, H. G. Burger, and D. M. de Kretser, *Mol. Cell. Endocrinol.* **46,** 175 (1986).

[20] K. Miyamoto, Y. Hasegawa, M. Fukuda, and M. Igarashi, *Biochem. Biophys. Res. Commun.* **136,** 1103 (1986).

[21] R. A. Lerner, *Nature (London)* **299,** 592 (1982).

TABLE I
SYNTHETIC FRAGMENTS OF INHIBIN CHAINS USED AS IMMUNOGENS[a]

Rabbit/antisera	Synthetic fragment conjugated to human α-globulins for use as immunogen
104, 105, 106	Porcine inhibin α (1–6)-Gly-Tyr
119, 120, 121	Porcine inhibin α (1–26)-Gly-Tyr
122, 123, 124	Porcine inhibin β_B (1–10)-Gly-Tyr
140, 141, 142	Porcine inhibin α (41–58)
143, 144, 145	Porcine inhibin β_B (47–65)
146, 147, 148	Porcine inhibin β_A (66–79)
153, 154, 155	Ac-porcine inhibin β_A (81–113)-NH_2 (cyclized)
179, 180, 181	Human inhibin α (1–25)-Gly-Tyr
185, 186, 187	Human inhibin β_A (85–102)
197, 198, 199	Ac-human inhibin β_B (80–112)-NH_2 (cyclized)

[a] All synthetic fragments were coupled using the bisdiazotized benzidine method, except for human inhibin β_A (85–102) which was coupled using the glutaraldehyde method.

sured (1) the production of inhibin by cultured rat granulosa cells[22]; (2) the production of inhibin by cultured Sertoli cells[23]; and (3) levels of inhibin in rat plasma.[24] We have also injected these antisera into rats and have shown a subsequent rise in plasma FSH levels, which can most easily be explained by neutralization of endogenous inhibin.[24] Ying *et al.*[25] produced a similar antisera against porcine inhibin α (1–29)-Tyr.

Procedure for Synthetic Fragment Coupling and Injection in Rabbits

Fragments are synthesized as described by Rivier *et al.*[26] and are conjugated to human α-globulins using either bisdiazotized benzidine (BDB), which couples through tyrosine and histidine residues, or glutaraldehyde, which will couple primary amines. BDB·2Cl is prepared by dissolving 0.23 g benzidine (Sigma) in 45 ml 0.2 *N* HCl. This solution and all solutions and reactions for BDB coupling are maintained in an ice bath

[22] T. A. Biczak, E. M. Tucker, S. Cappel, J. Vaughan, J. Rivier, W. Vale, and A. J. W. Hsueh, *Endocrinology (Baltimore)* **119,** 2711 (1986).

[23] T. A. Biczak, W. Vale, J. Vaughan, E. M. Tucker, S. Cappel, and A. J. W. Hsueh, *Mol. Cell. Endocrinol.* **49,** 211 (1987).

[24] C. Rivier, J. Rivier, and W. Vale, *Science* **234,** 205 (1986).

[25] S.-Y. Ying, J. Czvik, A. Becker, N. Ling, N. Ueno, and R. Guillemin, *Proc. Natl. Acad. Sci. U.S.A.* **84,** 4631 (1987).

[26] J. Rivier, C. Rivier, and W. Vale, *Science* **224,** 889 (1984).

at 0°. Sodium nitrite (0.175 g) (Sigma) is dissolved in 5 ml distilled H_2O and added dropwise to the benzidine solution with constant stirring. The reaction is continued for 30 min. Aliquots are snap frozen in dry ice–acetone and stored for up to 6 months at −20°. The final stock concentration of BDB·2Cl is 25 μmol/ml.

For coupling synthetic fragments to human α-globulins via BDB, the molar ratio of tyrosine plus histidines in the synthetic fragment to BDB·2Cl to tyrosines plus histidines in human α-globulins is 2 : 2 : 1. Human α-globulins contain approximately 100 μmol of tyrosine plus histidine/μmol protein. The coupling buffer is 0.13 *M* NaCl, 0.16 *M* boric acid, pH adjusted to 9.0 with NaOH. The human α-globulins (Fraction IV, Research Plus, Denville, NJ) are dissolved in coupling buffer in a glass vial, put in an ice bath, and stirred with a magnetic spin bar. The synthetic fragment is dissolved in 1.0 ml distilled H_2O and added to the reaction vial. BDB·2Cl, prepared as described above, is thawed and added dropwise to the reaction vial with rapid stirring, and the reaction is allowed to proceed for 3 hr in an ice bath. The total reaction volume is kept under 10 ml. The reaction is stopped by dialysis against 8 liters of 0.9% NaCl at 4° with 3 changes using Spectrapor dialysis tubing (Spectrum Medical) with a molecular weight cutoff of 1000.

For glutaraldehyde coupling, a 3 times weight excess of human α-globulins to synthetic fragment is used. Human α-globulins are dissolved in 4 ml coupling buffer (0.1 *M* sodium phosphate, pH 7.0) in a glass vial. The synthetic fragment is dissolved in 1.0 ml buffer and added to the reaction vial. Glutaraldehyde, 25% (w/v) (Sigma, Grade II), is diluted to 0.2% (w/v) in coupling buffer. Five milliliters of 0.2% glutaraldehyde is added to the reaction vial with constant stirring. The reaction is allowed to proceed for 4 hr at room temperature. The final concentration of glutaraldehyde in the reaction is 0.1% (w/v). The reaction is stopped by dialysis as described above for the BDB coupling.

After dialysis the synthetic fragment–human α-globulins conjugates are diluted with physiological saline (0.9%, w/v) to a final concentration of 1 mg of total protein per milliliter. The immunogens are prepared by emulsification with a Polytron. Freund's complete adjuvant-modified *Mycobacterium butyricum* (Calbiochem) is mixed with an equal volume of saline containing either 1.0 mg conjugate/ml (for initial injections) or 0.5 mg conjugate/ml (for boosters). For each immunization, a rabbit receives a total of 1 ml emulsion in 20–30 intradermal sites. Animals are injected every 2 weeks and bled through an ear vein 7 days after each booster. Blood is allowed to clot, and serum is separated from cells by centrifugation. The antiserum from each bleeding is characterized with respect to titer and affinity using several candidate tracers.

Development of Precipitating Antisera

Precipitating antisera (second antibody) were produced in sheep using rabbit γ-globulins (United States Biochemical Co.). Freund's complete adjuvant-modified *M. butyricum* (Calbiochem) with an equal volume of physiological saline (0.9% w/v) containing either 1.0 mg rabbit γ-globulins/ml (for initial injections) or 0.5 mg rabbit γ-globulins/ml (for boosters) is emulsified with a Polytron to prepare the immunogen. For each immunization, each sheep receives a total of 5 ml of emulsion in approximately 30 intradermal sites. Animals are injected every 2 weeks and bled through the jugular, using an evacuated bottle, 7 days after each booster. Blood is allowed to clot, and serum is separated by centrifugation. The antiserum from each bleeding is characterized with respect to titer using 0.1% normal rabbit serum in 0.5 ml volume per tube.

Preparation of Inhibin Tracers

Tracers are prepared using synthetic fragments or purified ovine inhibin using $Na^{125}I$ and the chloramine-T method. The molar ratio of synthetic peptide to $Na^{125}I$ to chloramine-T is 1 : 1 : 10. For natural products, we use 1 μg of inhibin or FRP and 0.4 nmol $Na^{125}I$ and 4 nmol chloramine-T for each iodination. The reaction is stopped with sodium metabisulfite for peptide fragments not containing cysteine residues or with excess tyrosine for synthetic peptide fragments containing cysteine residues or for natural products containing disulfide bridges. All tracers are purified using high-performance liquid chromatography (HPLC) with a 0.1% trifluoroacetic acid (TFA)–acetonitrile solvent system.

Reagents

- Phosphate (PO_4) buffer: 0.5 *M* Na_2HPO_4, pH adjusted to 7.4 with 0.5 *M* NaH_2PO_4
- Phosphate (PO_4) buffer: 50 m*M* pH 7.4; dilute above buffer 1 : 10 with distilled H_2O
- Chloramine-T: 100 μg/ml in 50 m*M* PO_4 buffer; diluted just prior to iodination
- Sodium metabisulfite: 250 μg/ml in 50 m*M* PO_4 buffer; diluted just prior to iodination (only for synthetic fragments not containing cysteine residues)
- Tyrosine: Dissolve first in 1 drop 10 *N* NaOH and then further dilute to 400 μg/ml with 50 m*M* phosphate buffer (only for synthetic fragments containing cysteine residues or for purified natural products)

Bovine serum albumin (BSA; Miles Labs., Pentex crystalline grade): 10% in 50 m*M* PO_4 buffer (for synthetic fragments only).

$Na^{125}I$: 1 mCi vial in aqueous solution, pH 8–10 (New England Nuclear, NEZ 033 L)

BondElut C_{18} cartridge: 200 mg sorbent in 3-cm^3 syringe (Analytichem International, Harbor City, CA) (for synthetic fragments only)

Triethylammonium formate (TEAF): 11.5 ml of formic acid (88%) per liter of distilled H_2O, pH adjusted to 3.0 with triethylamine (MCB reagents); filter through 0.45-μm Millipore type HA filter (for synthetic fragments only)

Trifluoroacetic acid (TFA) (Pierce Chemical)

Acetonitrile, HPLC grade (Burdick and Jackson)

Methanol, HPLC grade (Burdick and Jackson) (for synthetic fragments only).

2-Propanol, HPLC grade (Burdick and Jackson) (for synthetic fragments only).

Iodination and Purification of Synthetic Fragments

The synthetic fragment used for iodination is dissolved in 10 m*M* acetic acid to 1 mg/ml and further diluted to 150 μg/ml in 50 m*M* PO_4 buffer. To a 1 mCi vial of $Na^{125}I$ is added 20 μl of 0.5 *M* PO_4 buffer, pH 7.4, and the appropriate amount of peptide at 150 μg/ml to give 0.4 nmol. This mixture is vortexed, and 10 μl of chloramine-T (100 μg/ml) is added. The iodination is allowed to proceed for 30–45 sec, at which point it is immediately quenched with either 10 μl of sodium metabisulfite solution (250 μg/ml) for peptides not containing cysteine residues or with 20 μl of tyrosine (400 μg/ml) for peptides with cysteine residues. After vortexing for 30 sec, 100 μl of 10% BSA is added to adsorb any excess free iodine, and the reaction vial is again vortexed for 30 sec. The mixture is then diluted with TEAF to the top of the iodination vial (~1 ml).

A BondElut C_{18} column is used to separate the iodinated peptide from the free iodine and BSA. The BondElut column is prewetted with 3 ml methanol, then washed twice with 3 ml of TEAF, and subsequently the iodination mixture is applied. The iodination vial is washed twice with TEAF, and the wash is applied to the column. The column is washed again with 3 ml TEAF. The majority of the free iodine and BSA will not be retained by the column; iodinated and free peptide will bind to the C_{18} sorbent. The iodinated peptide is eluted from the column into a glass tube with the addition of 2 ml of 50% TEAF–50% 2-propanol. The spent column is then discarded. The volume of the peptide fraction is reduced to approximately 0.2–0.5 ml with a Speed Vac rotary evaporator (Savant). It is necessary to remove most of the 2-propanol for the peptide to adsorb to the HPLC column. TFA (0.1% in HPLC-grade H_2O) (1.5 ml) is then added to the tube containing the peptide, and the iodinated peptide is purified using an Altex Model 332 HPLC system with a Rheodyne injector equipped with a 2-ml sample loop. The column used is a Vydac C_{18}, 5 μm, 300 Å pore size. The buffer for pump A is 0.1% TFA; pump B contains 60% acetonitrile in 0.1% TFA. The HPLC system is equilibrated to the desired start concentration of buffer B. The iodination mixture is applied, and a gradient is run to the desired final concentration of buffer B in 30 min at a flow rate of 1.5 ml/min. Fractions are collected every 0.5 min, and 5-μl aliquots of these fractions are counted for radioactivity. The peak tubes of radioactivity are saved, and 10 μl of 10% BSA is added to each peak tube to stabilize the iodinated peptide. Iodinated peptides stored in TFA–acetonitrile with added BSA at 4° are generally stable for at least 2 months. In the absence of standard peptides synthesized with cold iodide, all major peaks are screened with antisera. By employing consistent iodination and purification conditions, similar peaks can be produced repeatedly and therefore be available for radioimmunoassays.

For our two most commonly used tracers, human inhibin α (1–25)-Gly-Tyr and porcine inhibin α (1–26)-Gly-Tyr, the gradient is 50–80% B in 30 min followed by a gradient to 95% B in 5 min to wash the column. The optimal conditions for purification of other iodinated synthetic fragments can be determined using the method described in the following paragraph.

The first time a synthetic peptide fragment is iodinated, the peptide-containing fraction from the BondElut C_{18} column is concentrated and then diluted with buffer A of the HPLC system as outlined above. Small amounts (1% of total cpm) are applied to the Vydac C_{18} column using various gradient conditions until a gradient is found which optimizes separation of peaks containing radioactivity. Generally, the first run for any synthetic peptide is 20–95% B in 30 min at a flow rate of 1.5 ml/min. Fractions are collected every 0.5 min, and 5-μl aliquots of these fractions are counted for radioactivity. The radioactivity profile is assessed for adsorption of iodinated material to the reversed-phase column, recovery of total counts applied, separation of peaks, and approximate concentration of solvent B required to elute the iodinated material. A second trial run is made using a shallower gradient bracketing the percentage B required to elute the iodinated material. Fractions are collected, and the radioactivity profile is again evaluated. Further trial runs using different gradient conditions are made, if necessary, to optimize separation of radioactive peaks. Finally, the remainder of the radioactive mixture is applied using the optimized gradient conditions. Any subsequent iodinations of the same synthetic peptide are purified using the determined conditions.

Iodination and Purification of Natural Products

Purified 32-kDa ovine inhibin or 28-kDa porcine FRP $\beta_A\beta_A$ is concentrated from HPLC buffer using a Speed Vac for approximately 10 min to remove acetonitrile but is not concentrated to dryness. Generally, 1 μg inhibin or FRP in about 35 μl TFA–acetonitrile buffer is reduced to around 10 μl and then diluted with 100 μl 0.5 *M* phosphate buffer, pH 7.5. The inhibin or FRP is added to a vial containing 1 mCi $Na^{125}I$. This mixture is vortexed, and 10 μl of chloramine-T (100 μg/ml) is added. The iodination is allowed to proceed for 90 sec, at which point it is immediately quenched with 20 μl of tyrosine (400 μg/ml) and vortexed for an additional 30 sec. The iodination should not be stopped with large amounts of BSA as the iodination mixture is applied directly to a reversed-phase column. The iodination mixture is diluted to a final volume of 1500 μl with starting buffer of the HPLC system and immediately purified using an Altex Model 332 HPLC system with a Rheodyne injector equipped with a 2-ml sample loop. The column used is a Vydac C_4, 5 μm, 300 Å pore size. The buffer for pump A is 0.1% TFA; pump B contains 80% acetonitrile in 0.1% TFA. The HPLC system is equilibrated to 20% B prior to application of the iodination mixture. The sample is applied and a gradient is run to 95% B in 60 min at a flow rate of 0.75 ml/min for inhibin. Iodinated FRP is run from 10 to 95% B in 60 min at a flow rate of 0.7 ml/min. Fractions are collected every 1 min, and 5-μl aliquots of these fractions are counted for radioactivity. There are several major peaks of radioactivity. The very large peak of radioactivity which elutes with the injection spike is ^{125}I-labeled tyrosine and is discarded. BSA is added to all other peak tubes to a final concentration of 0.1%. Small aliquots of each peak, with and without 5% 2-mercaptoethanol, are examined using polyacrylamide–sodium dodecyl sulfate (SDS) slab gel electrophoresis, according to the method of Laemmli,[27] followed by autoradiography. Peaks with correct molecular weight with and without reducing are screened with antisera. Iodinated inhibin or FRP stored in TFA–acetonitrile with added BSA is stable at least 3–4 weeks at 4°.

Procedure for Titration of Antisera and Radioimmunoassay

Reagents

SPEA buffer: 0.1 *M* NaCl, 50 m*M* Na_2HPO_4–NaH_2PO_4, pH 7.4, 25 m*M* EDTA, 0.1% (w/v) sodium azide. To prepare two liters, weigh out the following:

2.62 g $NaH_2PO_4 \cdot H_2O$ [monobasic sodium phosphate, monohydrate, formula weight (FW) = 138]

[27] U. K. Laemmli, *Nature (London)* **227,** 680 (1970).

11.5 g Na_2HPO_4 (dibasic sodium phosphate, anhydrous, MW = 142)
11.69 g NaCl (sodium chloride, MW = 58.44)
18.61 g EDTA·Na_2·$2H_2O$ (ethylenediaminetetraacetic acid, disodium salt, dihydrate, FW = 372.2)
2.0 g NaN_3 (sodium azide, FW = 65.02).

Dissolve in approximately 1900 ml distilled H_2O and adjust the pH to between 7.3 and 7.5 with HCl or NaOH. Bring volume to exactly 2000 ml with distilled H_2O.

RIA buffer: SPEA buffer with 0.1% (w/v) BSA and either 1 m*M* CHAPS (3-[(3-cholamidopropyl)dimethylammonio]-1-propane sulfonate) (Pierce Chemicals) for α chain RIAs or 0.1% (v/v) Triton X-100 for β chain RIAs to reduce nonspecific binding.

Standard: Either synthetic peptide fragment or purified natural product. Synthetic peptide fragments are stored in frozen aliquots of 10 μg/ml (peptides are first dissolved to 1 mg/ml with 10 m*M* acetic acid and then diluted 1 : 100 in RIA buffer). Fresh aliquots of synthetic peptide standards are made every 2 months. Aliquots of 1 μg purified natural product/ml are diluted directly from HPLC buffer of the final purification step in RIA assay buffer. Just prior to assay, the standard is serially diluted with RIA buffer.

Antibody: A list of our current high-affinity antisera is given in Table I. Each antibody is used at a final dilution that gives 35–50% maximal tracer binding. Antibody is diluted 1 : 100 in RIA buffer, aliquoted, and stored frozen. A fresh aliquot is used for each assay.

Tracer: ^{125}I-labeled synthetic peptide fragment of α, β_A, or β_B chain or ^{125}I-labeled purified natural product, prepared and stored as described above, is diluted in RIA buffer for titration of antiserum or in RIA buffer with 0.5% normal rabbit serum (NRS; Colorado Serum Co.) for radioimmunoassays.

Sheep anti-rabbit second antibody: Produced as described above, although we have also tried several commercially available antisera. We find that antisera produced by Vanguard has consistently high titer and gives excellent results. Dilute appropriately in RIA buffer (generally 1 : 40).

Polyethylene glycol (PEG): 10% (w/v) of 6000–8000 molecular weight (Sigma) in RIA buffer.

RIA wash buffer: SPEA buffer only.

Antisera Titration Protocol

Antisera titrations are carried out using chilled reagents and with tubes partially immersed in ice water. On day one, tracer for testing is diluted in

RIA buffer to 20,000 cpm/100 μl. Tracer (100 μl) is mixed with (1) 400 μl of RIA buffer and normal rabbit serum (Colorado Serum Co.) only or (2) 400 μl RIA buffer containing antiserum for testing or (3) 400 μl RIA buffer containing antiserum for testing and additional normal rabbit serum. The assay is done in duplicate. Specific antiserum is tested at final dilutions ranging from 1 : 1,000 to 1 : 30,000,000. Normal rabbit serum is added in appropriate amounts so that each tube contains a final 0.1% (v/v) concentration of serum in 500 μl. After these additions, tubes are shaken and placed in the cold at 4° for 2–3 days.

On day 3 or 4, free tracer is separated from tracer bound to antibody by precipitating the tracer–antibody complex with sheep anti-rabbit γ-globulins and 10% PEG. To all tubes, except "total counts" tubes, add 100 μl of sheep anti-rabbit γ-globulins at the proper dilution (generally 1 : 40) and 0.5 ml of 10% PEG. Vortex all tubes and incubate for 15–30 min at room temperature. To all tubes except "total counts" tubes add 1.0 ml wash buffer and centrifuge at 4° for 30–45 min at 2000 g. Decant supernatants and count the pellets in a gamma counter. Percentage specific binding is calculated using the following equation:

$$\% \text{ Binding} = \frac{\text{cpm(antibody dilution)} - \text{cpm(background)}}{\text{cpm(total counts)} - \text{cpm(background)}} \times 100$$

The percentage specific binding versus final antibody dilution is plotted, and the maximum binding and EC_{50} are noted.

We have been able to raise high-titer antisera to all synthetic fragments attempted (see Table I) as determined by antibody titration assay. When measured by RIA, using the relative tracer and cold synthetic fragment in a competitive assay, all our antisera are of high affinity. Unfortunately only antibodies raised against synthetic fragments corresponding to extreme amino-terminal or carboxy-terminal regions of the inhibin or FRP chains can sufficiently bind the natural product tracers. Therefore, only competitive assays against extreme amino- or carboxy-terminal fragments work with natural products. Other synthetic fragments do not seem to correspond to regions of inhibin or FRP that are available to antibody binding when the protein is in its native conformation since antiserum with high titer for synthetic fragment tracers can recognize the various chains in Western blots of reduced but not nonreduced natural products. Because an RIA for either β chain would not be specific for inhibin, we developed an radioimmunoassay for inhibin using antiserum made against a synthetic fragment to porcine inhibin α. And because only our terminally directed antibody will recognize natural inhibin, we developed an RIA using antibody to porcine inhibin α (1–26)-Gly-Tyr.

Procedure for Radioimmunoassay of Inhibin α Chain

The radioimmunoassay is carried out using chilled reagents and with tubes partially immersed in ice water. On day 1, 100 μl of buffer with antibody at the proper dilution is added to borosilicate glass tubes containing standard or test samples or buffer only in a volume of 300 μl. All treatments are tested in duplicate. Standards ranging from 1 to 5,000 pg of synthetic peptide fragment or from 10 pg to 100 ng of purified natural product are used. Samples are generally tested at two to five dose levels, and standard curves are included at the beginning and the end of the assays. After these additions, tubes are vortexed and placed in the cold at 4° for approximately 24 hr. On day 2, 20,000 cpm of ^{125}I-labeled tracer with 0.5% normal rabbit serum (as a carrier) is added in a volume of 100 μl to all tubes. The tubes are vortexed and returned to the cold for approximately 24 hr.

On day 3, tracer bound to antibody is precipitated with sheep anti-rabbit γ-globulins and 10% PEG. To all tubes, except "total counts" tubes, add 100 μl of sheep anti-rabbit γ-globulins at proper dilution (generally 1 : 40) and 0.5 ml of 10% PEG. Vortex all tubes and incubate for 15–30 min at room temperature. Wash down all tubes, except "total counts" tubes, with 1 ml wash buffer and centrifuge at 4° for 30–45 min at 2000 *g*. Decant supernatants which contain tracer not bound to antibody and count the pellets in a gamma counter. Results are calculated using the logit–log radioimmunoassay data processing program of Faden, Huston, Munson, and Rodbard (NICHD RRB, NIH).

Biological fluids from bovine, human, ovine, porcine, rat, and rhesus monkey have been tested for total displacement of tracer and parallelism with standards using various combinations of antibody, synthetic peptide fragments, or purified natural products as tracer, and synthetic peptide fragments or purified natural products as standard. Owing to the limited supply of purified natural products, we routinely use synthetic fragment tracers and appropriate synthetic fragments as standards for measurement of inhibin α chain.

For rat plasma samples, the radioimmunoassay uses antibody #119 (10/17/85 bleed) at 1 : 325,000 final dilution with porcine inhibin α (1–26)-Gly-^{125}I-Tyr as a tracer and porcine inhibin α (1–26)-Gly-Tyr as a standard. The EC_{50} and minimum detectable for the synthetic peptide standard are 27.1 ± 7.0 and 1.4 ± 0.6 pg per dose, respectively. As shown in Fig. 1, plasma samples from female rats treated with one intravenous injection of 20 international units (IU) pregnant mare serum gonadotropin (PMSG) are parallel to the synthetic peptide using this system. As expected, plasma from ovariectomized rats treated in the same manner have

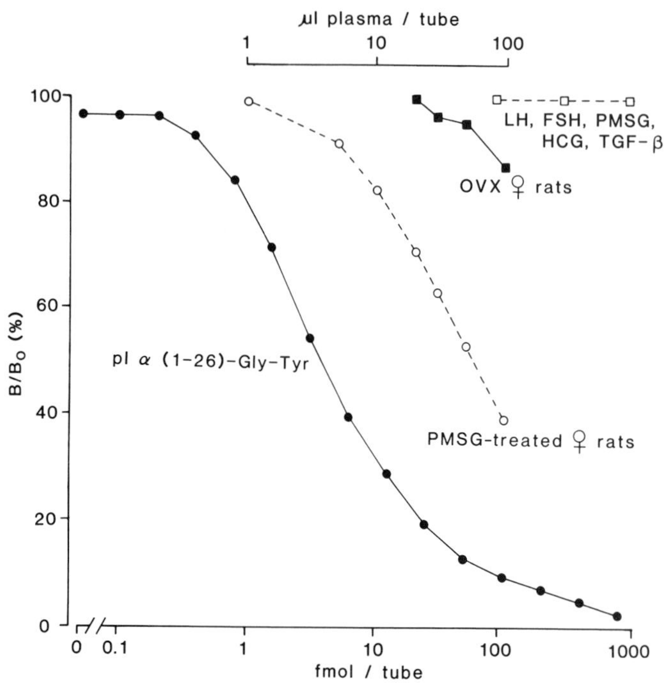

FIG. 1. Displacement of ^{125}I-labeled porcine inhibin α (1–26)-Gly-Tyr binding to antibody #119 (10/17/85 bleed) at a 1 : 275,000 final dilution by porcine inhibin (pI) α (1–26)-Gly-Tyr (●), by plasma from female rats treated with one intravenous injection of 20 IU PMSG 72 hr prior to bleeding (○), by plasma from ovariectomized (OVX) female rats treated with one intravenous injection of 20 IU PMSG 72 hr prior to bleeding (■), or by LH, FSH, PMSG, HCG, and TGF-β (□).

greatly reduced levels of inhibin α chain. Luteinizing hormone (LH), FSH, PMSG, human chorionic gonadotropin (HCG), and TGF-β will not displace the tracer, confirming the specificity of the antibody for inhibin. As shown in Fig. 2, purified 32-kDa ovine inhibin compared to porcine inhibin α (1–26)-Gly-Tyr is 10-fold less effective on a molar basis in displacing porcine inhibin α (1–26)-Gly-^{125}I-Tyr tracer. Therefore we multiply all molar amounts of inhibin by 10 when using the peptide standard to obtain relative total amounts of inhibin in rat plasma. But because we do not have purified rat inhibin available for comparison in this radioimmunoassay system, we can not be certain of absolute amounts of inhibin in the rat plasma samples measured. We are currently evaluating the use of cold and iodinated rat inhibin α (1–26)-Tyr for the assay of rat inhibin.

Samples generated during the fractionation of porcine follicular fluid are assayed by RIA employing antibody #120 (3/6/86 bleed) at

1 : 1,000,000 final dilution, porcine inhibin α (1–26)-Gly-^{125}I-Tyr as a tracer, and porcine inhibin α (1–26)-Gly-Tyr as a standard. The EC_{50} and minimum detectable dose for the synthetic peptide standard are 32.6 ± 4.5 and 3.3 ± 0.5 pg, respectively. For human, bovine, and rhesus monkey samples the radioimmunoassay uses antibody #120 (3/6/86 bleed) at a 1 : 275,000 final dilution, D-Ser^1-Nle^5 human inhibin α (1–25)-Gly-^{125}I-Tyr as a tracer, and human inhibin α (1–25)-Gly-Tyr as a standard. The EC_{50} and minimum detectable for the synthetic peptide standard are 43.8 ± 5.6 and 2.6 ± 1.8 pg, respectively.

Radioimmunoassay of ovine samples uses antibody #120 (3/6/86 bleed) at 1 : 275,000 final dilution, human inhibin α (1–25)-Gly-Tyr as a tracer, and ovine inhibin α (1–25)-Gly-Tyr as a standard. As shown in Fig. 2, the EC_{50} and minimum detectable for the synthetic peptide standard are 9.5 and 1.5 fmol per dose, respectively. Both unpurified ram rete testis fluid and purified 32-kDa ovine inhibin displacement curves are parallel to

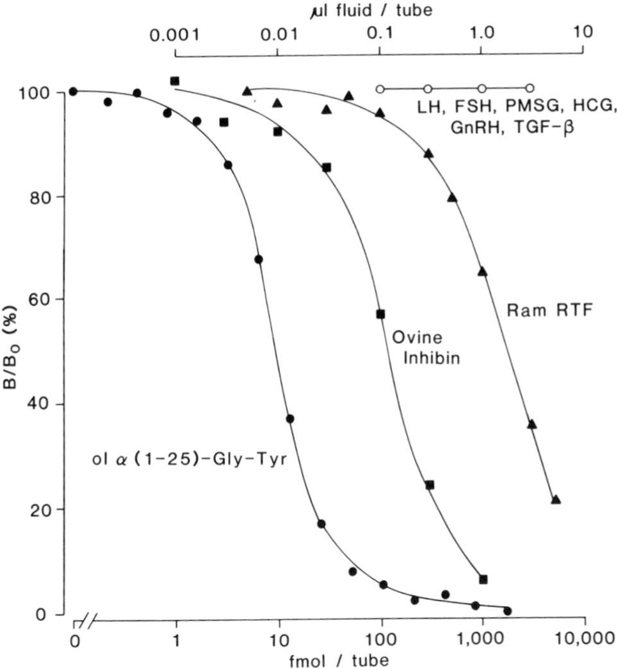

FIG. 2. Displacement of ^{125}I-labeled human inhibin α (1–25)-Gly-Tyr binding to antibody #120 (3/6/86 bleed) at a 1 : 275,000 final dilution by ovine inhibin (oI) α (1–25)-Gly-Tyr (●), by purified 32-kDa ovine inhibin (■), by ram rete testis fluid (RTF) (▲), or by LH, FSH, PMSG, HCG, GnRH, and TGF-β (○).

the synthetic peptide standard curve. Additionally, it can be seen that the purified natural product is less effective in displacing the tracer than the synthetic fragment. The EC_{50} and minimum detectable for the 32-kDa ovine inhibin standard are 122 and 16 fmol, respectively. We have repeated this many times with different combinations of antibodies and porcine, ovine, and human synthetic fragments and various batches of 32-kDa ovine inhibin, and the natural product is always 10-fold less effective than the synthetic fragment.

In Vitro Bioassay

Rat Anterior Pituitary Cell Culture

Rat pituitary cells are dissociated and established in culture essentially as described by Vale *et al.*[28] Reagents are as per Vale *et al.*[28] with the exception that cells are cultured in β-PJ culture medium containing 2% fetal bovine serum (FBS; Hyclone) or 2% FBS stripped of steroids by charcoal adsorption for studies involving interaction of inhibin and steroids.

Dissociation Procedure

Immediately following decapitation of adult (220–220 g) male Sprague–Dawley rats, anterior pituitary lobes are removed and collected in HEPES dissociation buffer (HDB).[28] The lobes are washed multiple times with HDB and then placed in a sterile, siliconized, water-jacketed double-sidearm Celstir (Wheaton) flask with a solution of 0.4% collagenase (Worthington, Type II), 80 μg/ml DNase II (Sigma), 0.4% BSA, and 0.2% glucose in HDB. FBS (5.0%) may be added to this enzymatic solution if the tryptic activity of the particular lot of collagenase is relatively high. Approximately 1 ml of this enzyme solution is used per pituitary lobe. The Celstir flask is attached by latex tubing to a Lauda MS (Brinkman) circulating water bath (37°) and placed on a Biostir (Wheaton) magnetic stir plate set to stir the lobes at approximately 200 rpm. Dissociation of the pituitary fragments is aided by gently drawing the fragments in and out of a siliconized Pasteur pipet every 20–40 min during the 1.5–2 hr procedure. Following the collagenase treatment, the cells are poured into a polypropylene tube and are pelleted in a centrifuge. The cell pellet is resuspended in approximately 0.5 ml of 0.25% (v/v) Viokase (Gibco) in HDB per lobe and returned to the Celstir flask and stirred for 5–10 min.

[28] W. Vale, J. Vaughan, G. Yamamoto, T. Bruhn, C. Douglas, D. Dalton, C. Rivier, and J. Rivier, this series, Vol. 103, p. 565.

The Viokase reaction is inactivated with 10% FBS, and once again the cells are pelleted. The cells are then resuspended and pelleted 3 times in culture medium β-PJ[28] containing 2% FBS. After the third wash, the cells are counted by hemacytometer and resuspended at a density of 3.3×10^5 cells/ml in β-PJ culture medium containing 2% FBS. The cell suspension is aliquoted (1 ml per well) into 24-well Linbro (Flow Laboratories, #76-033-05) tissue culture plates and incubated at 37° in 92.5% air–7.5% CO_2.

Experimental Incubation Procedure

After 3 days, the cultured cells are firmly anchored to the wells and can be washed by multiple medium change. The basis of the bioassay is the comparison of the abilities of crude porcine follicular fluid, crude ram rete testis fluid, or purified ovine inhibin standard and the test samples to affect LH or FSH secretion by the cultured pituitary cells. Multiple concentrations of the appropriate standard and usually two to five different concentrations of the unknown sample are tested. Cells are washed 2 times with 1.0 ml of the culture medium, and treatments are added in small volumes ($\leq$50 μl) to triplicate wells. After 48–72 hr of incubation, medium is removed from the wells with Pasteur pipets to 12×75 mm polystyrene tubes and stored for future radioimmunoassays. When intracellular contents must be obtained, cells are disrupted with 1 ml of 0.1% (v/v) Nonidet P-40 (NP-40; Calbiochem) in culture medium. After 1–2 min of incubation, samples are transferred to tubes. Then each well is washed with an additional 1 ml of medium, and the wash is transferred to the appropriate tube. FSH and/or LH are measured using an RIA kit provided by the National Hormone and Pituitary Program of NIDDK. Parallel dose–response curves (3–5 points) give relative potencies versus an appropriate internal standard.

Preparation of Purified Antisera and Immunoaffinity Columns

Preparation of Affinity-Purified Antisera

Antisera to inhibin α and β_A chains were affinity purified for preparation of immunoaffinity columns and for use in Western blot analysis and immunocytochemistry. Although unpurified antiserum can be used for these purposes, we find all methods benefit greatly from the use of affinity-purified antisera. Western blots and immunocytochemistry have much reduced background. Immunoaffinity columns prepared using purified antiserum give superior results to those made using the total IgG fraction of serum.

Antiserum #120 produced in rabbits (see Table I) is purified using an affinity column to which porcine inhibin α (1–26)-Gly-Tyr is covalently attached. Affi-Gel 10 (Bio-Rad), a cross-linked agarose gel with a 10-atom neutral spacer arm and active *N*-hydroxysuccinimide ester, is used for coupling with 100 m*M* sodium HEPES (Calbiochem), pH 7.5, as coupling buffer. Porcine inhibin α (1–26)-Gly-Tyr (10 mg, ~2.3 μmol) is dissolved in 1 ml distilled H_2O and further diluted to 20 ml in coupling buffer. Affi-Gel 10 (15 ml bed volume) is transferred to a scintered glass funnel and washed with ice-cold distilled H_2O. The Affi-Gel 10 is mixed with the synthetic peptide and allowed to rotate for 4 hr at 4° using a Roto-Torque rotator (Cole-Parmer). The gel is spun down, and supernatant containing uncoupled porcine inhibin α (1–26)-Gly-Tyr is removed. The remaining active ester is blocked by the addition of 15 ml 1 *M* ethanolamine–HCl, pH 8, followed by rotation at 4° for 1 hr. The gel is spun down and supernatant is removed. Noncovalently bound peptide is stripped from the gel by washing with 20 column volumes (300 ml) of 1 *N* acetic acid. Affi-Gel 10–porcine inhibin α (1–26)-Gly-Tyr is then equilibrated with 200 ml sodium HEPES, pH 7.5. Coupling efficiency of porcine inhibin α (1–26)-Gly-Tyr to Affi-Gel 10 is 90%.

Antiserum #120 (40 ml, 40 nmol binding sites/ml antiserum) is rotated overnight with Affi-Gel 10–porcine inhibin α (1–26)-Gly-Tyr at 4°. Antiserum and gel are then packed into a 1.5 × 10 cm column (Bio-Rad) at 30 ml/hr. The column is washed extensively with 50 m*M* sodium HEPES containing 1 m*M* CHAPS, pH 7.5, at 30 ml/hr to remove nonspecifically bound material. Specific immunoglobulins to porcine inhibin α (1–26)-Gly-Tyr are then eluted using 1 *N* acetic acid. The purified immunoglobulin fraction is neutralized and dialyzed against 150 m*M* NaCl, 50 m*M* sodium HEPES, 0.05% sodium azide, pH 7.5, at 4° with 3 buffer changes using Spectrapor (Spectrum Medical) dialysis tubing with molecular weight cutoff 12,000–14,000.

Antiserum #153 (see Table I) is purified with an affinity column to which 10 mg porcine inhibin β_A Ac (81–113)-NH_2 is covalently attached to 10 ml bed volume Affi-Gel 10, using the method outlined above for coupling porcine inhibin α (1–26)-Gly-Tyr to Affi-Gel 10. The coupling efficiency of porcine inhibin β_A Ac (81–113)-NH_2 is 80%. Antiserum #153 (65 ml, 10 nmol of binding sites/ml antiserum) is rotated overnight at 4° with Affi-Gel 10–porcine inhibin β_A Ac (81–113)-NH_2. The antiserum–gel mixture is then packed into a 1.5 × 10 cm column (Bio-Rad) at 30 ml/hr. The column is washed extensively with 50 m*M* sodium HEPES containing 1 m*M* CHAPS, pH 7.5, at 30 ml/hr to remove nonspecifically bound material. Specific immunoglobulins to porcine inhibin β_A Ac (81–113)-NH_2 are then eluted with 1 *N* acetic acid at 30 ml/hr. The purified immunoglobulin fraction is then concentrated 10-fold using Centriprep 30 concentrators

(Amicon) with 30,000 molecular weight cutoff filters. The concentrate is neutralized with NaOH and diluted to the original volume of antiserum (65 ml) with 50 m*M* sodium HEPES, 150 m*M* NaCl, 0.02% sodium azide, pH 7.5.

Coupling Purified Antibody to Agarose Gel

Affinity-purified antiserum #120 (45 ml equivalents, 73 mg total protein) is coupled to 25 ml bed volume of Affi-Gel 10. The Affi-Gel 10 is washed with distilled H_2O, and the antiserum in 150 m*M* NaCl, 50 m*M* sodium HEPES, 0.05% sodium azide, pH 7.5, is added. The gel is allowed to rotate with antiserum for 24 hr at 4° using a Roto-Torque (Cole-Parmer). The gel is spun down, and the supernatant containing uncoupled antiserum is removed. Remaining active ester is blocked by the addition of 25 ml of 1 *M* ethanolamine–HCl, pH 8.0, and the gel is again rotated for 1 hr at 4°. The gel is spun down and the supernatant is removed. The gel is stripped of noncovalently bound antibody by successive washes with (1) two bed volumes distilled H_2O, (2) two bed volumes 2 *M* guanidine thiocyanate, (3) two bed volumes distilled H_2O, (4) 2 times two bed volumes 1 *N* acetic acid. For each wash, the gel slurry is rotated at 4° for 30 min. The gel is then spun down and the supernatant removed. The gel containing covalently bound antiserum #120 is equilibrated with 100 m*M* sodium HEPES, pH 7.5. Approximately 70–80% of the affinity-purified antibody is covalently coupled to the Affi-Gel 10.

Preparation of Antibody–Protein A Matrix

An affinity column is made using unpurified antiserum #120 (Table I) coupled directly to protein A–Sepharose CL-4B (Pharmacia). Thirty milliliters of antiserum #120 is adsorbed with 60 mg of human α-globulins at 4° for 24 hr to remove any antibodies made against the carrier protein. The antiserum is spun down and the pellet discarded. The adsorbed antiserum is then rotated for 45 min at room temperature with 30 ml bed volume protein A–Sepharose CL-4B, previously swollen and washed with 250 ml 100 m*M* sodium phosphate, pH 7.4. The protein A–Sepharose CL-4B beads are spun down and the supernatant is removed. The beads are washed once with 100 m*M* sodium phosphate, pH 7.4, and once with 0.2 *M* triethanolamine–HCl, pH 8.2, to remove nonspecifically bound proteins. The immunoglobulins bound to protein A–Sepharose CL-4B are then covalently cross-linked using dimethylpimelimidate dihydrochloride (DMP; Pierce) essentially as described by Schneider *et al.*[29] The beads are

[29] C. Schneider, R. A. Newman, D. R. Sutherland, U. Asser, and M. F. Greaves, *J. Biol. Chem.* **257,** 10766 (1982).

resuspended in 10–20 volumes of 20 m*M* DMP freshly made in 0.2 *M* triethanolamine–HCl, pH 8.2, and rotated at room temperature for 60 min. The beads are centrifuged, the supernatant removed, and the reaction stopped by resuspending the beads in 10–20 volumes of 20 m*M* ethanolamine–HCl, pH 8.2. The antibody–protein A beads are then washed twice with 1 *N* acetic acid to remove any immunoglobulins which are not covalently linked and finally washed and equilibrated with 100 m*M* sodium phosphate, pH 7.4. The coupling efficiency of the immunoglobulin fraction to protein A–Sepharose CL-4B is 90%.

The antibody–protein A–Sepharose CL-4B method is useful because the immunoglobulin fraction of antiserum can be purified and coupled with high efficiency in one step. Furthermore, the yield of antigen binding sites is quite high; we find that 20–25% of the total number of sites in the unpurified antiserum are still usable when bound to the gel.

Purification of Inhibin from RTF Using Affinity Chromatography

Fluid is collected using cannulae inserted into the rete testes of adult rams as previously described.[30] Fluid from several animals is pooled and stored at −70°. The RTF averages 700 μg protein per milliliter and possesses inhibin-like bioactivity and immunoactivity.

Throughout the purification, fractions are monitored for inhibin activity using either or both the *in vitro* bioassay or the radioimmunoassay. In the bioassay, parallel dose–response curves (3–5 points) give relative potencies versus an internal standard of crude RTF. Approximate amounts of immunoreactive inhibin are measured by radioimmunoassay using the synthetic peptide as standard. The radioimmunoassay uses antibody #120 and human inhibin α (1–25)-Gly-Tyr as a tracer because crude RTF is able to completely displace the human tracer but not the porcine tracer from the antibody. The first immunoaffinity purification used human inhibin α (1–25)-Gly-Tyr as standard; we have since found that purified 32-kDa ovine inhibin is completely parallel to ovine inhibin α (1–25)-Gly-Tyr standard (Fig. 2) but not to human inhibin α (1–25)-Gly-Tyr standard. Subsequent batches of RTF purified by immunoaffinity chromatography were screened by RIA using ovine inhibin α (1–25)-Gly-Tyr as a standard.

For the immunoaffinity columns and reversed-phase HPLC columns, aliquots for assays (0.01–1.0% of the total fraction volume but never less than 5 μl) are transferred to polypropylene tubes containing bovine serum albumin (10 μl of 10 mg/ml) and dried in a Speed Vac (Savant). For FPLC

[30] J. K. Voglmayr, G. M. H. Waites, and B. P. Setchell, *Nature (London)* **210,** 861 (1966).

gel permeation colums where desalting is necessary before assay, small aliquots are removed and transferred to glass tubes containing 0.5 ml of 10 m*M* sodium HEPES plus 0.1% bovine serum albumin, pH 7.5. Squares of dialysis tubing with a molecular weight cutoff of approximately 1000 are secured over the tops of the glass tubes with rubber bands. These covered tubes are inverted and dialyzed against 10 m*M* sodium HEPES, pH 7.5. The retentates are then dried using a Savant Speed Vac. All fractions are resuspended in cell culture medium for the *in vitro* bioassay or in RIA buffer for the radioimmunoassay.

Use of Immunoaffinity Matrix

Ram RTF, 2000 ml per batch, is processed using either the affinity-purified antibody #120–Affi-Gel 10 or the antibody #120–protein A–Sepharose CL-4B immunoaffinity matrix. For each batch, the RTF is allowed to rotate with the immunoaffinity matrix for 48 hr at 4°. The mixture is then packed into a 1.5×10 cm column at 25 ml/hr. The column is then washed at 30 ml/hr with 1 m*M* CHAPS in 100 m*M* sodium phosphate buffer, pH 7.4. Proteins specifically bound to the immunoaffinity matrix are then eluted using 1 *N* acetic acid at a flow rate of 30 ml/hr. An example of immunoaffinity chromatography using the affinity-purified antibody #120–Affi-Gel 10 resin is show in Fig. 3. Either immunoaffinity matrix works well, but the purification factor is approximately 2.5-fold higher for the affinity-purified antibody #120–Affi-Gel 10 than for the antibody #120–protein A–Sepharose CL-4B. This is due to adsorption of RTF to the protein A–Sepharose CL-4B backbone of the antibody–protein A–Sepharose CL-4B resin and possibly adsorption of RTF to immunoglobulins not specific to the α chain. This can be greatly reduced, however, by preadsorbing RTF with protein A–Sepharose CL-4B. No bioactive or immunoreactive inhibin can be detected in the material retained on protein A–Sepharose CL-4B alone. For the immunoaffinity chromatography, recovery of immunoactivity is 85% and recovery of bioactivity is 40%. This difference is due to material not retained by the column which contains bioactivity but not immunoreactivity using this amino-terminal directed antibody when not reduced. Western blot analysis shows only material with α chains of 19,000–21,000 and 44,000 is retained by the immunoaffinity column. Unretained material has truncated α chains. Furthermore, an earlier fractionation of ram RTF solely by reversed-phase HPLC[31] resulted in isolation of biologically active ma-

[31] J. Rivier, R. McClintock, J. Vaughan, G. Yamamoto, H. Anderson, J. Spiess, W. Vale, J. Voglmayr, C. Y. Cheng, and C. W. Bardin, *in* "Male Contraception: Advances and Future Prospects" (G. I. Zatuchni, A. Goldsmith, J. M. Spieler, and J. J. Sciarra, eds.), pp. 401–407. Harper, New York, 1986.

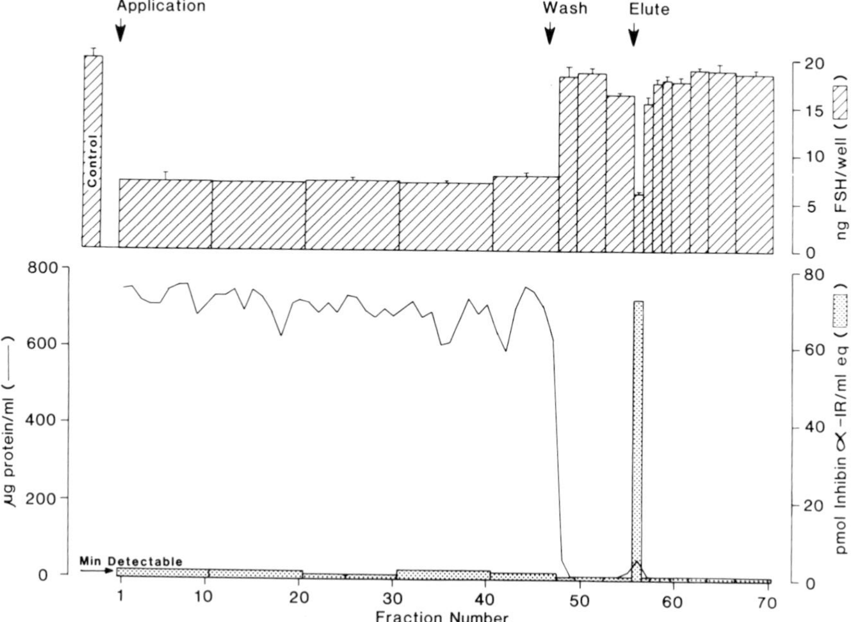

FIG. 3. Immunoaffinity chromatography of RTF using affinity-purified antibody #120 covalently coupled to Affi-Gel 10. Specifically bound material is eluted with 1 *N* acetic acid. Aliquots of fractions are measured for FSH inhibition using the *in vitro* bioassay (3-day incubation) and inhibin α chain immunoreactivity (IR) using the radioimmunoassay as described for ovine samples.

terial with an intact β_A chain and an α chain lacking the amino-terminal 15 residues.

FPLC Gel Filtration

The active fraction from the immunoaffinity column is pooled, lyophilized, resuspended in gel filtration column eluant, and further purified using an FPLC system (Pharmacia) equipped with tandem Superose 12B columns, 10 μm, 10 × 300 mm each. The column eluant is 6 *M* guanidine–HCl, 0.1 *M* ammonium acetate, pH 4.75; the flow rate is 0.4 ml/min. Fractions are collected every 1 min. The active fraction from the immunoaffinity column is generally processed in two batches of not more than 5 mg total protein per run to optimize resolution. Two zones of immunoactivity are eluted, but only the later eluting zone is biologically active (Fig. 4).

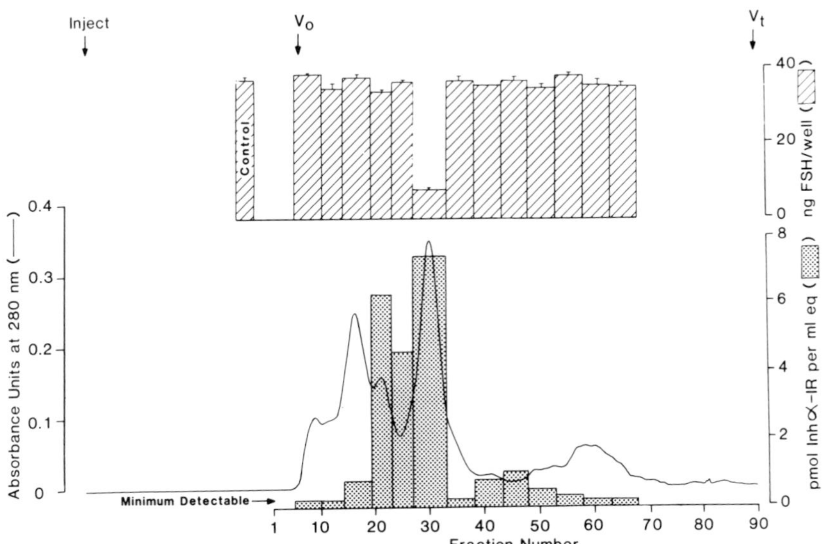

FIG. 4. FPLC gel permeation of material specifically adsorbed to the inhibin α chain immunoaffinity column. Tandem Superose 12B columns, 10 μm, 10 × 300 mm each, are used at a flow rate of 0.4 ml/min. Eluant is 6 *M* guanidine–HCl, 0.1 *M* ammonium acetate, pH 4.75. Aliquots of fractions are assayed for FSH inhibition using the *in vitro* bioassay (3-day incubation) and α chain immunoreactivity using the radioimmunoassay as described for ovine samples.

Analytical Reversed-Phase HPLC

The two active zones from the FPLC gel filtration are separately purified using a Beckman HPLC system with reversed-phase columns. The zone with both immunoactivity and bioactivity is purified in three batches using a Vydac C_8 column, 0.46 × 25 cm, 5 μm particle size, at 40°. Buffer A is 0.1% trifluoroacetic acid (TFA); buffer B is 80% acetonitrile in 0.1% TFA. Columns are loaded at 25% B at a flow rate of 1.2 ml/min, and then a gradient is run to 95% B in 45 min at a flow rate of 0.7 ml/min. The active zone from the three runs is pooled and then concentrated using a gradient of 25 to 95% B in 70 min at a flow rate of 0.7 ml/min. The active fraction from this column is analyzed for purity using SDS–polyacrylamide gel electrophoresis and for composition using Western blot analysis.

The zone with immunoactivity only is purified in two batches using a Vydac diphenyl column, 0.46 × 25 cm, 5 μm particle size, at room temperature. Columns are loaded at 10% B at a flow rate of 1.2 ml/min, and then a gradient is run to 75% B in 25 min. The active zone from the two runs is pooled and analyzed for purity using SDS–polyacrylamide gel electrophoresis and for composition using Western blot analysis.

Analysis of the highly purified fraction with both immunoactivity and bioactivity by SDS–PAGE using a gradient of 8–25% acrylamide shows a single band of 32 kDa unreduced and bands of 21 and 14 kDa on reduction with 5% 2-mercaptoethanol (Fig. 5). Depending on the batch of ram rete testis fluid used, the yield is 32–45 ng of 32-kDa inhibin/ml of crude RTF processed as determined by amino acid analysis. The purified 32-kDa ovine inhibin was 1000 times more potent on a weight basis than crude RTF in the *in vitro* bioassay (Fig. 6). The recovery of FSH inhibiting activity was 20% when compared to an internal standard of crude RTF in the *in vitro* bioassay. Because some of the bioactivity of crude ram RTF may be due to material with truncated α chain and steroids, it is difficult to assess the true recovery of the 32-kDa inhibin from the various chromatographic procedures used. From a sample of 32-kDa ovine inhibin purified earlier by conventional chromatography,[32] we determined the first 24 residues of the amino terminus of the α chain and the first 29 residues of the amino terminus of the β chain and found them to be closely related but not identical to the α chain and β_A of porcine inhibin. β_A subunits from all other species sequenced to date are identical to one another, whereas the β_A chain of ovine inhibin has a tyrosine for phenylalanine substitution at position 17.

Analysis of the highly purified fraction with immunoreactivity but not bioactivity by SDS–PAGE using a gradient of 8–25% acrylamide shows a band of 43 and a band of 44 kDa on reduction with 2-mercaptoethanol. The yield from 1 ml of crude RTF is 90 ng of this 44-kDa material as determined by amino acid analysis. Western blot analysis, using antibody #120 directed against the amino terminus of the α chain, shows that this material is 44-kDa α chain without either β chain.

Western Blot Analysis

Procedure for Western Blot Analysis

Samples to be analyzed are dissolved in sample buffer (0.1 *M* Tris, pH 6.8, 1.0% SDS, 20% glycerol, 0.01% bromphenol blue) either with or without 5% 2-mercaptoethanol and heated in a boiling water bath for 5 min. Samples are electrophoresed in a 12.5% polyacrylamide–SDS slab

[32] C. W. Bardin, P. L. Morris, C.-L. Chen, J. Shaha, J. Voglmayr, J. Rivier, J. Spiess, and W. W. Vale, *in* "Inhibin–Non-steroidal Regulation of Follicle Stimulating Hormone Secretion" (H. G. Burger *et al.*, eds.), Vol. 42, pp. 179–190. Raven, New York, 1987.

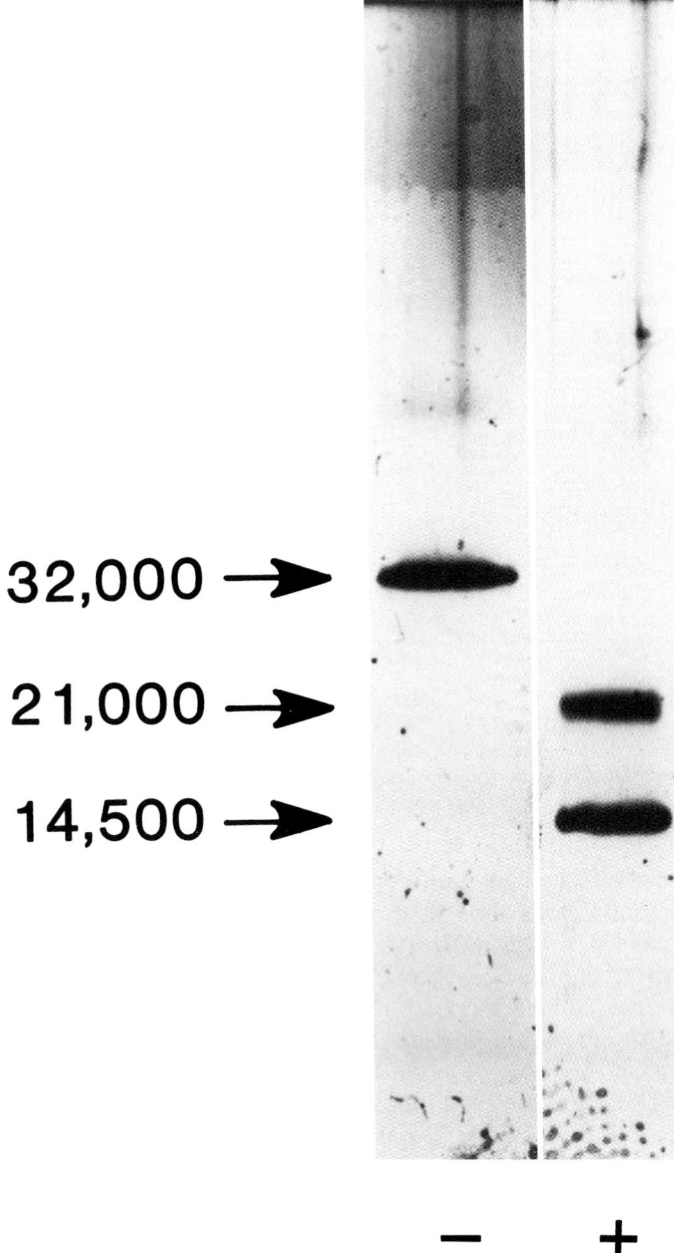

FIG. 5. SDS–polyacrylamide gel electrophoresis of ovine inhibin purified using immunoaffinity chromatography, gel permeation FPLC, and reversed-phase HPLC. A Pharmacia Phastgel system is used with an 8–25% gradient gel and without (−) or with (+) 5% 2-mercaptoethanol in the sample buffer. The gel is subsequently silver stained.

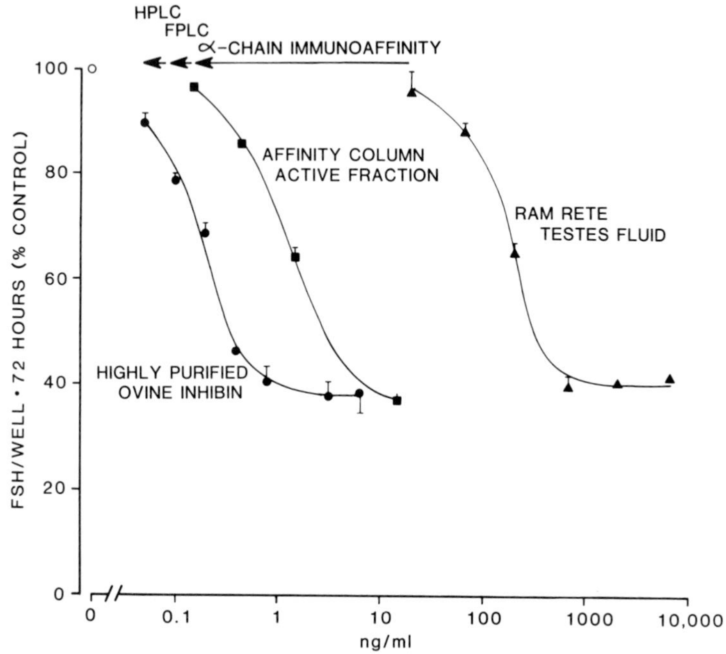

FIG. 6. *In vitro* bioassay of RTF (▲), specifically adsorbed material from inhibin α chain immunoaffinity chromatography of RTF (■), and 32-kDa ovine inhibin purified using immunoaffinity chromatography, gel permeation FPLC, and reversed-phase HPLC (●).

gel with a 4% stacking gel, 0.75 mm thickness, using a Bio-Rad Mini Protean II apparatus according to the method of Laemmli.[27] Proteins are transferred from SDS–polyacrylamide gels to a nitrocellulose membrane using a Bio-Rad Mini TransBlot Cell with transfer buffer (25 m*M* Tris, 192 m*M* glycine, 20% methanol) for 1 hr at 100 V constant voltage with cooling. All incubations and washes are done at room temperature using an American Variable Rotator (American Hospital Supply Company). Unbound sites on the membrane are blocked by incubation with 3% gelatin in Tris-buffered saline (50 m*M* Tris, 500 m*M* NaCl, pH 7.5) for 30 min. The membrane is washed for 5 min twice with 0.05% Tween 20 in Tris-buffered saline. The membrane is then incubated overnight with affinity-purified antiserum #120, affinity-purified antiserum #153 (to α or β chains of inhibin), or antiserum blocked by preincubation with the appropriate α or β chain peptide fragment as a control. Antiserum is diluted 1 : 1000 in 50 ml 0.05% Tween 20 Tris-buffered saline containing 1% gelatin. For blocked antiserum, 2 mg peptide is dissolved in 100 μl 10 mM acetic acid and further diluted to 0.5 ml with SPEA buffer (described in the RIA section). This peptide solution is mixed with 1 ml equivalent of affinity-purified antiserum and incubated 24–48 hr at 4°. The mixture is centrifuged and the supernatant removed. This supernatant (100 μl) is then diluted in 50 ml 0.05% Tween 20 Tris-buffered saline and is used as a antiserum control. After two 5-min washes with 0.05% Tween 20 in Tris-buffered saline, the nitrocellulose membrane is incubated with one of various second antibodies, washed twice with 0.05% Tween 20 Tris-buffered saline and detected appropriately.

Detection methods we use include the following: (1) goat anti-rabbit IgG–horseradish peroxidase conjugate (Bio-Rad, 1 : 3000 final dilution), 2-hr incubation followed by staining with 4-chloro-1-naphthol substrate; (2) goat anti-rabbit IgG–alkaline phosphatase conjugate (Bio-Rad, 1 : 3000 final dilution), 2-hr incubation followed by staining with BCIT (5-bromo-4-chloro-3-indolyl phosphate *p*-toluidine salt) and NPT (*p*-nitro blue tetrazolium chloride) substrates; (3) goat anti-rabbit IgG–colloidal gold conjugate (Bio-Rad, diluted according to directions), 4-hr incubation followed by silver enhancement (Bio-Rad kit); (4) goat anti-rabbit ^{125}I-labeled IgG (New England Nuclear, 50–100 μCi per gel), 1-hr incubation followed by autoradiography. We routinely use the goat anti-rabbit IgG–alkaline phosphatase system for samples containing low levels of inhibin α and/or β chains because it has high sensitivity, low background, and is easy to use. For samples where sensitivity is not a problem, we routinely use the horseradish peroxidase system because of the availability of biotinylated standards which can be detected using avidin–horseradish peroxidase and 4-chloro-1-naphthol substrate. Using the biotinylated standards, we can accurately determine the molecular weight of the various α and β inhibin subunits.

Results of Western Blot Analysis

Western blots are useful for analysis of fractions generated during the purification of RTF. Western blots of reduced RTF show up to six different forms of α chains in varying amounts depending on the batch of RTF. There are single bands of molecular weights 44,000 and 29,000 and a doublet of molecular weights 20,200 and 19,500, and a doublet of molecular weights of 15,000 and 14,700 (Fig. 7). Similar analysis reveals a β_A chain of 15,500 for reduced RTF (Fig. 7). Western blot analysis of 32,000 ovine inhibin, purified by immunoaffinity chromatography, shows a single band at 32,000 unreduced and a single band at 20,200 reduced, using

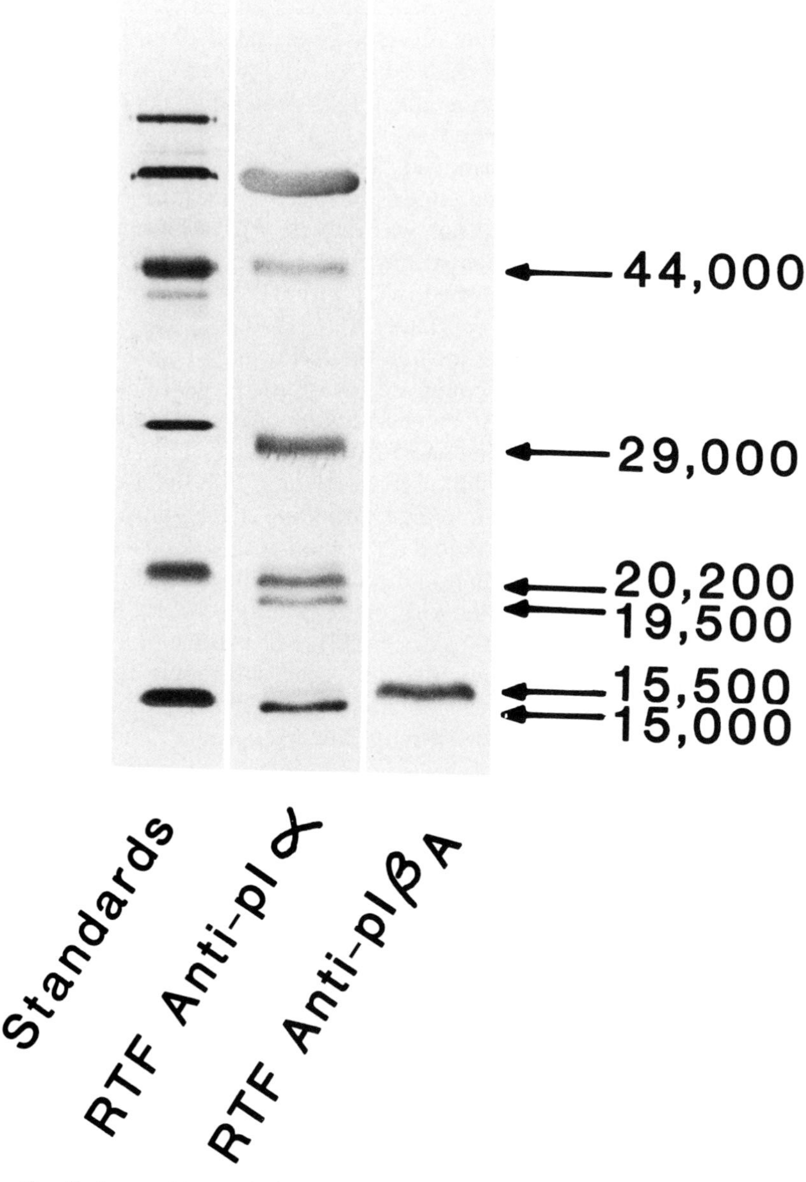

FIG. 7. Immunoblot analysis of ram rete testis fluid (RTF) electrophoresed with 5% 2-mercaptoethanol in the sample buffer. Proteins are transferred onto nitrocellulose membranes and incubated with either affinity-purified antiserum #120 to inhibin α chain (RTF anti-pI α) or affinity-purified antiserum #153 to β_A chain (RTF anti-pIβ_A). Biotinylated standards are included in the analysis. Nitrocellulose membranes are incubated with goat anti-rabbit IgG–horseradish peroxidase (1:3000) and avidin–horseradish peroxidase (1:3000). Bands are colormetrically developed using 4-chloro-1-naphthol substrate. [The highest molecular weight band in the middle lane (~60,000) is not specific as this band is still apparent when antiserum is blocked by preincubation with porcine inhibin α (1–26)-Gly-Tyr.]

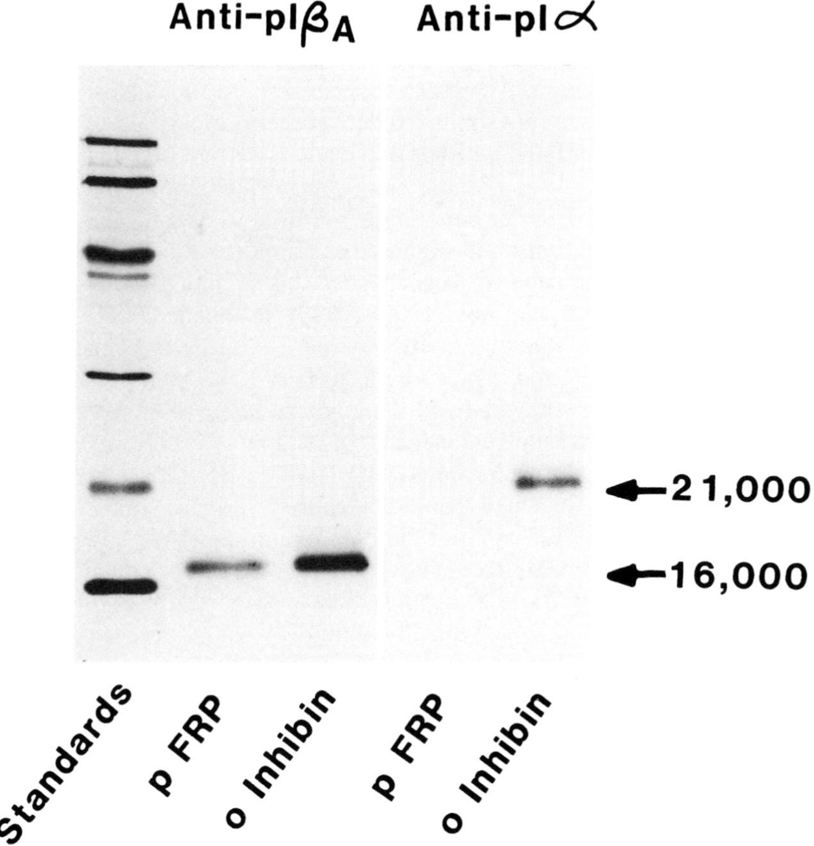

FIG. 8. Immunoblot analysis of 28-kDa $\beta_A\beta_A$ porcine FRP and 32-kDa ovine inhibin electrophoresed with 5% 2-mercaptoethanol in the sample buffer. Biotinylated standards of molecular weights 97,400, 66,200, 43,000, 31,000, 21,500, and 14,400 are also subjected to analysis. Proteins are transferred to nitrocellulose membranes which are then incubated with affinity-purified antiserum #153 to inhibin β_A chain or affinity-purified antiserum #120 to inhibin α chain. Membranes are incubated with goat anti-rabbit IgG–horseradish peroxidase (1 : 3000) and avidin–horseradish peroxidase (1 : 3000). Bands are colormetrically developed using 4-chloro-1-naphthol substrate.

rabbit anti-porcine inhibin α (1–26)-Gly-Tyr as primary antibody (Fig. 8). Using rabbit anti-porcine β_A Ac (81–113)-NH_2 as primary antibody, a single band at 32,000 was observed when ovine inhibin was unreduced and a single band of 14,700 when reduced (Fig. 8). No bands could be detected using our β_B chain antibodies.

Concentration of Inhibin from Biological Fluids

The routine concentration of small amounts of inhibin or FRP from biological fluid samples is complicated because both molecules possess physical properties (size, p*I*) similar to many other proteins, and there are many forms of both inhibin and FRP. Jansen *et al.*[33] used Matrex Gel Red A (Amicon) as a first step for the purification of inhibin from bovine follicular fluid. Using a buffer containing 1 *M* urea as eluant, they recovered 90% of the bioactivity and obtained a 20-fold purification with this one step. Later, Matsuo and co-workers used this method for the isolation of both porcine inhibin[2] and bovine inhibin[34] from follicular fluids. They recovered 81% of the bioactivity with porcine FF[2] and 91% of the bioactivity with bovine FF[34] and achieved a 5- to 6-fold purification factor for this one step. Further, they showed the material retained on Matrex Gel Red A had four distinct molecular forms of inhibin with porcine FF[2] and at least three distinct molecular forms with bovine FF.[34] McLachlan *et al.* adapted this method for small samples to purify 58-kDa and 31-kDa iodinated bovine inhibin for use in a radioimmunoassay.[19] We have developed an effective, although maybe not optimal, method for concentrating inhibin and FRP from biological fluids based on a modification of the original Jansen method. We typically obtain 80–100% recovery of all forms of inhibin and FRP from both RTF and porcine FF and a 20-fold purification factor using this one step. Recoveries are highest using 1 ml or more of resin per 30 mg total protein and by incubating the resin with the sample for sufficient periods of time (30 min to 1 hr).

Reagents

Matrex Gel Red A (Amicon): Washed extensively with alternating 8 *M* urea and distilled H_2O using a scintered glass funnel until uncoupled dye is no longer visible in the eluant

Column: Disposable polypropylene column with a 70-μm frit (Evergreen Scientific) packed with 2 ml washed Matrex Gel Red A

Buffer A: 50 m*M* sodium HEPES plus 1 m*M* CHAPS, pH 7.5

Buffer B: 50 m*M* sodium HEPES plus 1 m*M* CHAPS plus 0.15 *M* NaCl, pH 7.5

Buffer C: 50 m*M* sodium HEPES plus 1 m*M* CHAPS plus 0.5 *M* NaCl plus 0.5 *M* urea, pH 7.5

[33] E. H. J. M. Jansen, J. Steenbergen, F. H. de Jong, and H. J. van der Molen, *Mol. Cell. Endocrinol.* **21,** 109 (1981).

[34] M. Fukuda, K. Miyamoto, Y. Hasegawa, M. Nomura, M. Igarashi, K. Kangawa, and H. Matsuo, *Mol. Cell. Endocrinol.* **44,** 55 (1986).

Buffer D: 50 m*M* sodium HEPES plus 1 m*M* CHAPS plus 1.5 *M* NaCl plus 1.0 *M* urea, pH 7.5

Procedure for Matrex Red A Chromatography

Matrex Gel Red A chromatography is carried out at 4°. Columns are equilibrated with 10 ml buffer A. Samples are mixed with an equal volume of buffer A and not more than 65 mg total protein is applied to each 2-ml column. Plasma (1 ml), other biological fluids or tissue extracts can be used provided they are at neutral pH and are not cloudy. They may be centrifuged to remove any precipitate. For comparisons of samples in the same experiment, equal amounts of total protein are loaded per column for each sample. Samples are applied to the columns at a low flow rate (~1 drop in 15 sec). For optimum recovery, columns can be capped at both ends and rotated with the sample for 1 hr at 4°. The columns are washed once with 5 ml of buffer B followed by 5 ml of buffer C. The inhibin- and FRP-containing fraction is then eluted with 5 ml of buffer D. To decrease salt and urea concentrations, each eluant is brought up to 15 ml with distilled H_2O and then concentrated to 1 ml using Centriprep 10 concentrators (Amicon) at 4°. If further desalting is necessary, all eluants are again brought up to 15 ml and concentrated to 1 ml a second time. All concentrated eluants are measured for total protein by the Bradford assay (Bio-Rad kit) to assure the approximate total protein recovery is the same for all columns. Concentrated eluants can then be evaluated for inhibin and FRP using the radioimmunoassay, the *in vitro* bioassay, and/or Western blot analysis.

Acknowledgments

Research supported by National Institutes of Health Grants HD13527 and DK26741 and The Population Council. Research conducted in part by the Clayton Foundation for Research, California Division. C. Rivier and W. Vale are Clayton Foundation Investigators. We wish to thank Gayle Yamamoto for helpful comments.

[43] Regulation of Neuroendocrine Peptide Gene Expression

By M. Blum

Introduction

In eukaryotes, mRNA is transcribed from DNA by RNA polymerase II. The resulting primary transcript must go through posttranscriptional modifications before the mature mRNA is transported to the cytoplasm. In addition to the "capping" of the 5′ end of the mRNA with a 7-methylguanosine and the "tailing" of most mRNAs on the 3′ end with a polyadenosine tract, there are also internal regions within primary transcripts called intervening sequences or introns which must be removed. The remaining segments, which must be precisely religated together, constitute the mature mRNA and are called exons. It is generally believed that the half-life of a primary transcript is only about 10–20 min. It is either rapidly degraded or processed into a mature mRNA and transported to the cytoplasm. Each of these steps in the biosynthesis of mRNA (shown schematically in Fig. 1) is a potential site for regulation of gene expression. The protocols outlined in this chapter are designed to enable one to study possible hormonal regulation of neuroendocrine peptide genes at each of these steps of mRNA biosynthesis.

This chapter describes a "run-on" transcription assay, which measures the number of RNA polymerase complexes transcribing a specific gene, and an S_1 nuclease protection assay, which can be used to quantitate the absolute amount of either primary transcript or mature mRNA. Transcriptional analysis of neuroendocrine genes can be performed with fresh tissue or primary cell or clonal cell line cultures. The basis of the assay is to treat either the animal *in vivo* or the culture *in vitro* with the desired hormone for a specified time and then to isolate cell nuclei by homogenizing the tissue in a cold buffer containing Triton X-100, which solubilizes the cell membrane and allows for the nuclei to be pelleted at a low speed centrifugation. The resulting postnuclear supernatant contains the cytoplasmic RNA and also peptides or proteins of interest and therefore should be saved for further analysis. During the isolation of cell nuclei, the RNA polymerases remain bound to the genes being transcribed. When the nuclei are then incubated *in vitro* under the appropriate conditions along with radiolabeled nucleotide triphosphates, the polymerases will continue to transcribe the genes for a couple hundred nucleotides. In this way one may "end label" *in vitro* the newly synthesized

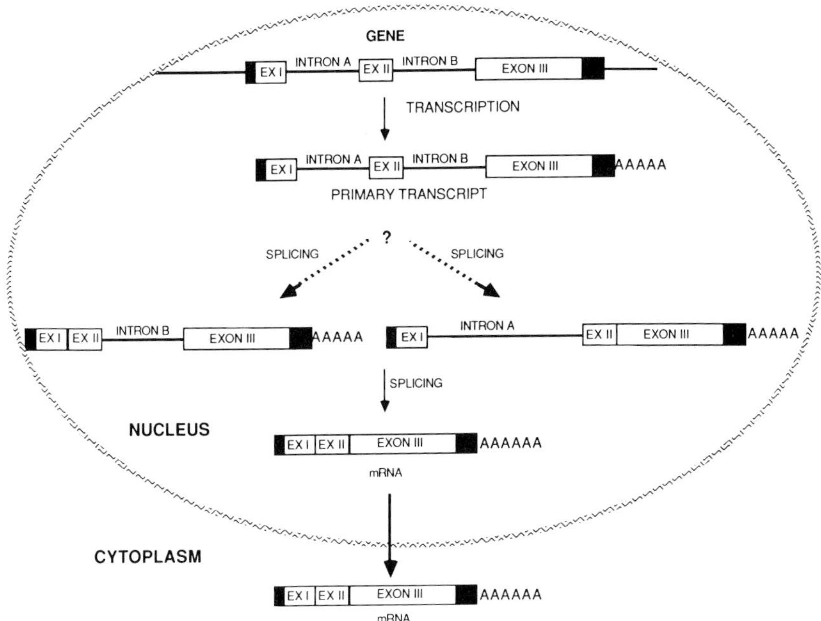

FIG. 1. Schematic representation of mRNA biosynthesis.

transcripts that were being transcribed *in vivo* under a particular hormonal or activity state. Then the newly synthesized RNA transcripts are purified and hybridized to specific cDNA probes bound to nitrocellulose filters. After hybridization the filters are washed and counted in a scintillation counter. (See Fig. 2 for a flow diagram of the assay.)

Alternatively, rather than performing transcriptional analysis, the primary transcripts, processing intermediates, mature mRNA from the pelleted nuclei, and mRNA from the supernatant (cytoplasm) can be purified so that one can determine if hormonal treatment affects RNA processing. To quantitate the amounts of primary transcript, processing intermediates, and mature mRNA, a high specific activity (10^9 cpm/μg) RNA probe complementary to an intron–exon junction (see Fig. 5) is synthesized for use in S_1 nuclease analysis. The RNA samples are then hybridized to the probe in solution. A standard curve is set up with a known amount of coding RNA so that the number of counts/minute protected in the samples can be converted to picograms of a specific RNA species per microgram of total RNA. After the hybridization reaction goes to completion, any nonhybridized RNA remaining is digested away with S_1 nuclease.

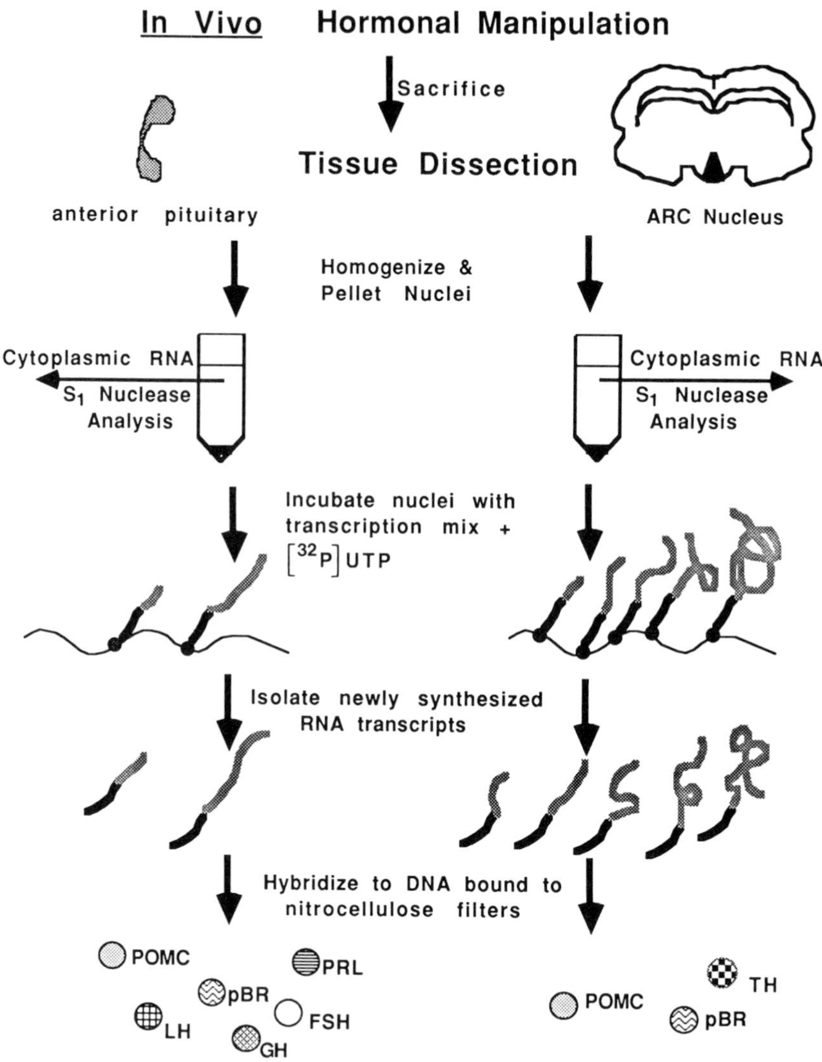

FIG. 2. Flow diagram of the nuclear transcription run-on assay.

The protected double-stranded RNA is then electrophoresed on an acrylamide gel so that each of the protected bands can be resolved. The bands are then cut out of the gel and counted in a scintillation counter. The counts/minute are converted to picograms from the standard curve.

Isolation of Cell Nuclei

When preparing nuclei for transcriptional analysis I have found that the less they are manipulated the better. This protocol was developed so that a large number of samples may be rapidly processed with little transfer of the nuclei.[1] This is in contrast to several of the published protocols which require an ultracentrifugation step and at least two transfers of the nuclei.

To isolate cell nuclei from fresh tissue, homogenize between 25 and 50 mg wet weight/ml of cold RNase-free AT buffer [10 m*M* Tris–HCl, pH 8, 3 m*M* $CaCl_2$, 2 m*M* $MgCl_2$, 0.5 m*M* dithiothreitol (DTT), 0.3 *M* sucrose, 0.15% Triton X-100] in a glass/Teflon homogenizer. Transcriptional analysis requires about 10^6 nuclei; therefore for genes expressed in the anterior pituitary around 1 to 3 pituitaries per point may be used. For the neuroendocrine peptide genes thus far analyzed from brain tissue, one usually needs to pool together microdissected tissue taken from three animals per point ($\sim 10^6$ nuclei). Complete homogenization of the tissue usually requires only 10 strokes by hand.

Carefully layer the homogenate over 400 μl of 0.4 *M* AT buffer (same as AT buffer except containing 0.4 *M* sucrose rather than 0.3 *M* sucrose) in an 1.5-ml microcentrifuge tube. Centrifuge at 2500 *g* for 10 min at 4°. Remove the supernatant, leaving behind the 0.4 *M* AT cushion, and transfer it to a 2-ml microcentrifuge tube. If one is interested in quantitating the peptide content in the tissue dissection, an aliquot may be taken for radioimmunoassay and for total protein determination. The rest of the supernatant may be frozen (−20° short term or −70° long term) until one is ready to purify the cytoplasmic RNA from this postnuclear supernatant as discussed below. Separate the 0.4 *M* AT cushion from the nuclei pellet and discard; the cushion serves to protect the nuclear fraction from contamination by any RNase present in the cytosol fraction. Add 1 ml of 0.3 *M* AT buffer to the nuclear pellet and very gently resuspend the nuclei by flicking the tube with a finger. Do not vortex the nuclei for this has been found to greatly reduce nucleotide incorporation. Centrifuge once again at 2500 *g* for 10 min at 4°. After the last centrifugation step remove the supernatant.

For most genes expressed in the pituitary, it has been found that transcriptional analysis must be performed on freshly prepared nuclei, in contrast to other tissues, such as liver, where nuclei can be stored in glycerol at −70° for later study. What has been observed is that nuclei which have been stored in glycerol will incorporate radiolabeled nucleo-

[1] M. Blum, B. S. McEwen, and J. L. Roberts, *J. Biol. Chem.* **262,** 817 (1987).

tides but hybridization above background cannot be detected for some specific genes.[2] For the analysis performed with brain tissue, one has to determine whether fresh tissue is required for the particular gene one is interested in analyzing. I have found that for some genes frozen nuclei work fine while for others they do not. To freeze the nuclei, add 60 μl (for 10^6–10^7 nuclei) of buffer B (50 mM Tris–HCl, pH 8.3, 5 mM $MgCl_2$, 0.1 mM EDTA, 40% glycerol) and store at $-70°$. If using fresh nuclei, proceed to the section describing the transcription assay. Do not add any buffer B to the nuclei when using them immediately in order to allow addition of the transcription mix without having to dry down the radioactive label.

To isolate nuclei from cells in culture merely remove the tissue culture medium, which you may want to save to quantitate hormone release, and add cold AT buffer. For a 100-mm confluent dish ($\sim 5 \times 10^6$ cells) add 2 ml of AT buffer, scrape the cells off with a rubber policeman, and divide them between two tubes each containing 400 μl of 0.4 M AT as described above; there are probably enough cells present to perform transcriptional analysis in duplicate if desired.

Transcription Assay

The transcription reaction mixture that is used to "run-on" the RNA polymerases was optimized for analyzing RNA polymerase II transcripts[1] based on the optimal metal ion, pH, and ionic strength described by Roeder and Rutter[3] for the purified enzyme. It was also determined which of the labeled nucleotide triphosphates gave the best results; use of radiolabeled UTP at a final concentration of 12 μM (400 Ci/mmol) worked the best. The transcription mix is made fresh just before use by diluting concentrated RNase-free stock solutions (see Table I).

To each nuclei sample add 100 μl of the transcription mix [40 mM Tris–HCl, pH 8.3, 5 mM $MgCl_2$, 2 mM $MnCl_2$, 2.5 mM DTT, 32% glycerol, 80 mM $(NH_4)_2SO_4$, 0.6 mM ATP, 0.25 mM CTP, 0.25 mM GTP, 500 μCi [α-^{32}P]UTP (400 Ci/mmol), 7 U/ml nucleoside-5′-diphosphate kinase, 2 mM creatine phosphate, 35 μg/ml creatine phosphokinase, 0.1 mM EDTA]. After adding the mixture, gently resuspend the nuclei as described above. Again, it is important not to vortex the nuclei. Incubate the reaction at 29° for 30 min. To stop the reaction add 1 μl of DNase I (20 U/μl, Worthington ultrapure) and incubate for 5 min at room temperature. (After the incubation, the nuclei may be frozen.)

[2] J. H. Eberwine and J. L. Roberts, unpublished observation (1984).

[3] R. G. Roeder and W. J. Rutter, *Nature (London)* **224,** 234 (1969).

TABLE I
TRANSCRIPTION MIX[a]

Stock	Volume/tube (μl)	Final concentration
1 *M* Tris–HCl, pH 8.3	4	40 m*M*
1 *M* $MgCl_2$	0.5	5 m*M*
1 *M* $MnCl_2$	0.2	2 m*M*
1 *M* Dithiothreitol	0.25	2.5 m*M*
Glycerol (ultrapure)	32	32%
1 *M* $(NH_4)_2SO_4$	8	80 m*M*
50 m*M* ATP	1.2	0.6 m*M*
50 m*M* CTP	0.5	0.25 m*M*
50 m*M* GTP	0.5	0.25 m*M*
24 μ*M* [α-^{32}P]UTP	49	12 μ*M* (0.5 mCi, 400 Ci/mmol)
1 U/μl Nucleoside-5′-diphosphate kinase	0.7	7 U/ml
200 m*M* Creatine phosphate	1	2 m*M*
3.5 mg/ml Creatine phosphokinase (made immediately before use)	1	35 μg/ml
24 U/μl RNasin	1	250 U/ml
66 m*M* EDTA	0.15	0.1 m*M*

[a] Usually make up extra transcription mix, excluding the label. Then add 51 μl of mix and 49 μl of label to each assay tube.

The newly synthesized RNA is then purified as described by Evans *et al.*[4] I have tried several different protocols[5,6] and have found this one to give the best results. In addition to being a reliable method for purification, it is also very rapid compared to other protocols. To purify the RNA as described by Evans *et al.*,[4] add 500 μl of 1× SET [1% sodium dodecyl sulfate (SDS), 5 m*M* EDTA, 10 m*M* Tris–HCl, pH 8] and 2.5 μl of proteinase K (10 mg/ml) and incubate for 30 min at 45°. Add 15 μl of 5 *M* NaCl, 10 μl of yeast RNA (10 mg/ml), and 200 μl of phenol (saturated with TE: 10 m*M* Tris–HCl, pH 8, 1 m*M* EDTA) and vortex the samples. Then add 200 μl of chloroform, vortex the samples again, and centrifuge for 10 min at 16,000 *g* at room temperature. Remove the aqueous phase and transfer to a 1.5-ml microcentrifuge tube. Add 60 μl each of 0.5 *M* sodium pyrophosphate and 50% trichloroacetic acid (TCA). Incubate on ice for 5 min.

Collect the precipitated nucleic acid with a gentle vacuum on nitrocellulose filters that have been previously boiled for 10 min in 1× SET. I use

[4] M. I. Evans, L. J. Hager, and G. S. McKnight, *Cell (Cambridge, Mass.)* **25,** 187 (1981).

[5] G. S. McKnight and R. D. Palmiter, *J. Biol. Chem.* **254,** 9050 (1979).

[6] J. H. Eberwine and J. L. Roberts, *J. Biol. Chem.* **259,** 2166 (1984).

a 12-place manifold (Millipore) to collect the TCA precipitates. Wash the filters with at least 20 ml of 3% TCA, 40 m*M* sodium pyrophosphate. Transfer the nitrocellulose filters to a 20-ml scintillation vial. Add 1 ml of DNase buffer (100 m*M* Tris–HCl, pH 8, 25 m*M* HEPES, pH 8, 5 m*M* $MgCl_2$, 1 m*M* $MnCl_2$, 1 m*M* $CaCl_2$) and 1 μl DNase (20 U/ul) and incubate at 37° for 30 min. To elute the RNA from the nitrocellulose filters, add 100 μl of 10× SET and incubate at 65° for 10 min. Transfer the buffer to a 5-ml polypropylene tube. Add 500 μl of 1× SET to the scintillation vial containing the filters, incubate at 65° for 10 min to extract any remaining RNA, and then combine with the other RNA in the 5-ml tube.

To the combined RNA sample add 36 μl of 5 *M* NaCl and 750 μl phenol (saturated with TE) and vortex. Then add 750 μl chloroform, vortex, and centrifuge for 10 min. Transfer the aqueous phase to another 5-ml tube, add 3 ml of 100% ethanol, and store at −20° overnight. Spin down the pellet (10,000 *g* for 20 min) and pour off the ethanol, being careful not to lose the pellet. Then add 2 ml of 70% ethanol and vortex the samples to dissolve any salt that may have come down with the pellet. Centrifuge (10,000 *g*) for 10 min and pour off the supernatant. After allowing the pellet to dry, resuspend the pellet in 23 μl of sterile H_2O. Take two 1-μl aliquots for determination of the radioactivity incorporated by spotting them on DE-81 ion-exchange paper and then washing 2 times for 10–15 min with 0.5 *M* $NaPO_4$, pH 6.5. After drying the filters, count them in a scintillation counter.

To set up the hybridization, add to the resuspended RNA pellet 1 μl of sense [^{3}H]RNA (~5,000–10,000 cpm, ~10^6 cpm/μg) synthesized with SP6 RNA polymerase (or one of the other RNA polymerases, depending on vector and orientation of your insert) for each gene being analyzed in order to determine the efficiency of hybridization. This is important because the efficiency of hybridization can vary considerably (>50%). It is possible to analyze more than six different genes per hybridization reaction, duplicate filters for the same gene may be used if desired. Then add 14 μl 50% dextran sulfate and about 35 μl of 2× hybridization buffer (1 *M* NaCl, 100 m*M* PIPES, pH 7, 66% formamide, 0.8% SDS, 4 m*M* EDTA), depending on the volume of [^{3}H]RNA that was added, so that the final concentration is 0.5 *M* NaCl, 50 m*M* PIPES, pH 7, 33% formamide, 0.4% SDS, and 2 m*M* EDTA in a volume between 75 and 80 μl.

Add 7-mm nitrocellulose filter circles that have 1 μg of DNA insert bound to them and, as a control, add a filter that has 1 μg of pBR322 (or any other nonspecific DNA). To make the nitrocellulose filters (modification of McKnight and Palmiter[5]) isolate 50 μg of DNA insert and resuspend it in 5000 μl of 2 *M* NaCl, 0.2 *M* NH_4OH. I have found that use of DNA insert versus the whole plasmid-containing insert can result in as much as 10 times higher efficiency of hybridization. Heat denature at 100° for 5 min and quickly cool on ice. Spot onto nitrocellulose which has been prewetted with 10× SSC (1× SSC: 150 m*M* NaCl, 6 m*M* sodium citrate, pH adjusted to 7.0 with HCl). Using a dot blot manifold, spot 100 μl (~1 μg) of the heat-denatured DNA insert into each well and allow the DNA to adsorb to the nitrocellulose by gravity or with a very gentle vacuum. The nitrocellulose is then taken out of the manifold and each of the "dots" is punched out with a hole punch (7 mm). The filters are baked for 2 hr under vacuum at 80° and then washed with 2× SSC for about 2 hr, with changing of the 2× SSC 2 or 3 times. The filters are then air dried and stored until ready for use. The pBR322 filters are made the same way, using 1 μg of linearized vector per filter. When preparing to hybridize, the filters should be prehybridized for at least 2 hr at 42° in 1× hybridization buffer before being added to the hybridization reaction. Once the filters have been added to the hybridization reaction, they are covered with mineral oil to prevent evaporation and incubated for 3.5 days at 42°.

After the hybridization the filters are washed with 2× SSC, 0.1% SDS twice for 30 min each at room temperature, 1× SSC, 0.1% SDS twice for 30 min each at room temperature, and with 0.1× SSC, 0.1% SDS twice for 30 min each at room temperature. The final wash is with 0.1× SSC, 0.1% SDS twice for 30 min at 55°. The filters are then counted in a scintillation counter. First the nonspecific counts/minute bound to the pBR322 filters are subtracted, then the counts/minute are corrected for the efficiency of hybridization by determining the percentage of the [^{3}H]RNA, included in the hybridization reaction, that hybridized to the filters. Then the counts/minute are divided by the total counts/minute incorporated. The background counts/minute on the pBR322 filters usually average between 30 and 60 parts per million (ppm). For example, the average background, corrected for efficiency of hybridization, for proopiomelanocortin gene transcription in the anterior lobe is 300 ppm, in the intermediate lobe 1000 ppm, and in the arcuate region is 200–400 ppm.

When the transcription rate of the gene being analyzed is low compared to background, I have used solution hybridization rather than filter hybridization because in solution one can drive the hybridization reaction to completion (100%) whereas with filter hybridization efficiency ranges from 5 to 50%. Therefore, by being able to drive the hybridization to 100% one can gain a severalfold increase in the signal. This is done by resuspending the newly synthesized RNA in 12 μl of H_2O, taking two 1-μl aliquots to determine total incorporation, and then adding 20 μl of the hybridization buffer described in the next section which includes 1 μg of unlabeled antisense RNA. If one is interested in analyzing more than one gene, this can be accomplished by using antisense RNA probes of different lengths so that one may distinguish between them on an acrylamide gel. The remainder of the protocol is as described in the next section, except that the counts/minute in the protected bands are expressed as counts/minute per total counts/minute incorporated.

Quantitation of hnRNA and mRNA

This section describes how to quantitate RNA using a very sensitive solution hybridization S_1 protection assay (Blum *et al.*,[1] modified from Durnam and Palmiter[7]). With the availability of a new generation of vectors which contain promoters for specific bacteriophage RNA polymerases, it is possible to synthesize antisense RNA probes[8] of high specific activity. In addition, large quantities of sense RNA can also be synthesized with these vectors. Having both of these tools, one may hybridize RNA samples in solution with a high specific activity antisense RNA probe and set up a standard curve with synthetic sense strand. A standard curve obtained with a known amount of RNA makes it possible to then express the number of hybridized counts/minute in the unknown samples as a mass amount rather than a relative density number or equivalents as done in the past when doing either Northern or dot blot analysis. This allows one to compare across experiments or with other laboratories. With the high specific activity antisense probe, it is possible to detect routinely 0.5 pg of a specific RNA.

Since this assay is so sensitive, one does not need so much total RNA as with a Northern or dot blot to measure a specific mRNA. This has several advantages. For one, it is possible to analyze the amount of mRNA from total RNA isolated from individual animals. Therefore, it is possible to analyze a large number of samples per treatment group, so one may perform statistical analysis. Each sample can also be analyzed multiple times to ensure an accurate determination. This allows one to detect small changes (~20%) in the levels of RNA. For example, using this assay we have been able to quantitate the amount of gonadotropin-releasing hormone (GnRH) mRNA isolated from a single rat preoptic area,[9] whereas using Northern analysis one would have to use 10 μg of poly(A)-selected RNA isolated from roughly 20 animals to detect a signal. Therefore, to measure a hormonal effect on RNA levels that is statistically

[7] D. M. Durnam and R. D. Palmiter, *Anal. Biochem.* **131,** 385 (1983).

[8] D. A. Melton, P. A. Krieg, M. R. Rebagliati, T. Maniatis, K. Zinnard, and M. R. Green, *Nucleic Acids Res.* **12,** 671 (1984).

[9] J. L. Roberts, M. Blum, R. T. Fremeau, Jr., C. M. Dutlow, R. P. Millar, and P. H. Seeburg, submitted for publication.

significant, Northern analysis might not be feasible owing to the prohibitively large number of animals required. In addition, by using the protocol developed for isolating cell nuclei and cytoplasmic RNA in combination with S_1 nuclease analysis, one may measure regulatory effects not only on the steady-state levels of mRNA in the cytoplasm but also on the levels of primary transcript, processing intermediates, and mature mRNA in the nucleus.

To purify the primary transcript, processing intermediates, and mature mRNA in the nucleus, first isolate the nuclear pellet as described above. Resuspend the pellet in 300 μl of DNase buffer (50 m*M* Tris–HCl, pH 8, 5 m*M* $MgCl_2$, 1 m*M* $MnCl_2$, 1 m*M* DTT), add 1 μl of RNasin (20 U/μl), 2 μl of DNase I (20 U/μl, Worthington ultrapure), and incubate for 5 min at 37°. Then add 30 μl of 10× SET (1× SET: 1% SDS, 5 m*M* EDTA, 10 m*M* Tris–HCl, pH 8) and 3 μl of proteinase K (10 mg/ml) and incubate for 1 hr at 45°. Add 175 μl of phenol (saturated with TE), vortex, and then add 175 μl of chloroform. Vortex again, centrifuge (16,000 *g*) for 5 min, remove the aqueous phase, and transfer it to a 1.5-ml microcentrifuge tube. Add 40 μl of 3 *M* ammonium acetate and 1 ml of 100% ethanol and store at −20° until ready to use.

To purify the mRNA from the postnuclear supernatant, if starting with approximately 30 mg wet weight of tissue that was homogenized in 1 ml of AT buffer as described above and removing a 200-μl aliquot for peptide content determination, thus leaving 800 μl, add 800 μl of 10× SET and 8 μl of proteinase K (10 mg/ml) and incubate for 1 hr at 45°. Add 450 μl of phenol (saturated with TE), vortex, and add 450 μl of chloroform. Vortex and centrifuge (16,000 *g*) for 5 min. Remove the aqueous layer and transfer to a 2-ml microcentrifuge tube. If the aqueous phase is cloudy, extract with phenol : chloroform once more. Then add 90 μl of 3 *M* ammonium acetate and 900 μl of 2-propanol. Store at −20° until ready to use.

This purification of RNA is simple, rapid, and results in a high yield of very clean RNA. From a single dissection of a rat preoptic area that was done by making a 2-mm-thick coronal section by cutting first anterior and then posterior to the optic chiasm and then cutting out a 2-mm square at midline using the anterior commissure as the dorsal landmark, we were able to isolate between 7 and 8.5 μg of total RNA.[9] Because this procedure is so simple and does result in a high yield, it is a very good method to use to isolate RNA when one wants to quantitate RNA from microdissected or punched brain regions taken from individual animals. This isolation protocol is especially good when one wishes to measure the effect of a hormone treatment on the levels of RNA and has 7 or 8 samples per treatment group. As mentioned above, by using this procedure for purify-

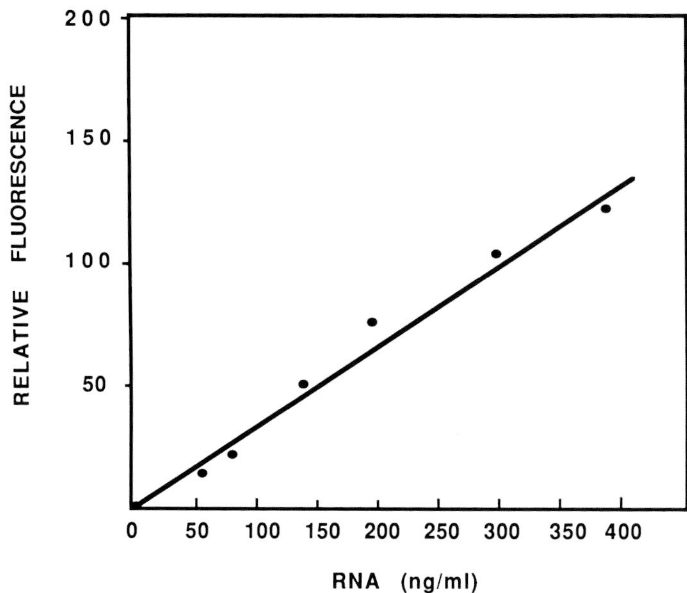

FIG. 3. Quantitation of total RNA by ethidium bromide (EtBr) fluorescence. A standard curve was generated by measuring the amount of fluorescence (λ_{ex} 286 nm, λ_{em} 588 nm) of 0–450 ng/ml of total brain RNA in 0.1 μg EtBr/ml TEA (see text) with a Perkin Elmer LS-5B spectrofluorometer.

ing RNA in combination with the solution hybridization protection assay, we have been able to successfully quantitate the amount of GnRH mRNA, which is a very rare mRNA (~0.5 pg/μg of total RNA or 0.00005% of total RNA), from one preoptic area dissection.[9]

In some experiments, to quantitate the amount of a specific mRNA from RNA isolated from a brain microdissection of a single animal, most of the RNA is needed to quantitate the specific mRNA, and therefore there is not much sample to spare in order to quantitate accurately the total amount of RNA in the sample. When there is not enough sample to read the optical density (OD) on a spectrophotometer (at least 500 ng/ml) one may determine the amount of fluorescence in the presence of ethidium bromide (EtBr).[10] Even though ethidium bromide, which works by intercalating with double-stranded nucleic acid,[11] is more sensitive in detecting DNA, we have been able to detect RNA concentrations down to 50 ng/ml. Figure 3 is a standard curve generated using a known amount total brain RNA diluted in 0.1 μg EtBr/ml TEA (40 m*M* Tris base, 20 m*M*

[10] J.-B. LePecq and C. Paoletti, *Anal. Biochem.* **17,** 100 (1966).

[11] J.-B. LePecq and C. Paoletti, *J. Mol. Biol.* **27,** 87 (1967).

sodium acetate, 20 m*M* NaCl, 2 m*M* EDTA, adjusted to pH 8.5 with acetic acid) and reading the amount of fluorescence with a spectrofluorometer. To quantitate the amount of RNA in our samples, we took 1 μl from each sample, diluted it in 250 μl of 0.1 μg/ml EtBr in TEA, and measured the fluorescence. The amount of total RNA in each sample was determined from the standard curve. In this way we could accurately measure as little as 12.5 ng/μl of sample. We found that at higher concentrations the results obtained with the spectrofluorometer were comparable to results obtained with the spectrophotometer.

To quantitate the amount of RNA isolated from nuclei preparations, one must ensure that the fluorescence detected in the sample is due to RNA and not to any DNA that may be remaining in the sample. This is especially important because DNA binds EtBr more efficiently, such that a small amount of DNA contamination in the sample can be a real problem. To determine if the samples contain DNA, one may check for fluorescence in the samples with DNA-specific dyes (4′,6′-diamidino-2-phenylindole or Hoescht 33258 dye). Therefore, if the samples are not fluorescent with the DNA-specific dyes, the fluorescence detected with EtBr must be due to RNA. Conversely, one can measure the fluorescence in the presence of EtBr before and after RNase digestion.

To quantitate the amount of a specific RNA by the protection assay, first synthesize the sense strand to be used for a standard, so it is possible to express the data in picograms of a specific RNA per amount of total RNA put into the assay. The synthesis is done as outlined by Promega. Briefly, linearize your probe 3′ to the insert. If using a vector which has a promoter on either side, linearize such that you can make the coding sequence. Set up the reaction with the following: 28 μl H_2O, 10 μl 5× transcription buffer (200 m*M* Tris–HCl, pH 8, 30 m*M* $MgCl_2$, 10 m*M* spermidine, 50 m*M* NaCl), 0.5 μl of 1 *M* DTT, 2 μl RNasin, 0.5 μl 50 m*M* ATP, 0.5 μl 50 m*M* CTP, 0.5 μl 50 m*M* GTP, 0.5 μl 50 m*M* UTP, 5 μl of linearized DNA template (1 μg/μl), and 2 μl (of the appropriate polymerase; SP6, T3, or T7) RNA polymerase (5 U/μl). Incubate for 1 hr at 37°. Then add 50 μl of H_2O, 5 μl of 1 *M* Tris–HCl, pH 8, 1 μl of 1 *M* $MgCl_2$, 1 μl RNasin, 1.5 μl of DNase I (10 ng/μl, Worthington ultrapure) and incubate at 37° for 30 min. To purify the probe from any unincorporated nucleotides run the sample on a Sephadex G-100 column. Usually around 10–20 μg are synthesized in this reaction. To quantitate the amount of RNA synthesized one can either determine the OD or include in the reaction 0.5 μl of [α-^{32}P]UTP (400 Ci/mmol), calculate the percentage incorporation, and thus determine the amount synthesized. The resulting specific activity is only about 1 cpm/pg so it will not interfere with the assay. Once you determine the concentration of the probe, make several 1 : 10 serial dilutions until you get to the desired working concentration. I have found that the RNA synthesized for standard is stable. Usually I make several aliquots of the lower dilutions and freeze them. For stability, the highest concentration may be stored precipitated under ethanol.

To quantitate the amount of RNA in the 0.250–5 pg range synthesize the antisense strand at a specific activity of ~2×10^9 cpm/μg. To do this, dry down 200 μCi of [α-^{32}P]UTP (~800 Ci/mmol), then add 3.5 μl of H_2O, 2 μl of 5× transcription buffer, 1 μl of 100 m*M* DTT, 0.5 μl of RNasin, 0.5 μl of 50 m*M* ATP, 0.5 μl of 50 m*M* CTP, 0.5 μl of 50 m*M* GTP, 1 μl of linearized DNA template (1 μg/μl), and 0.5 μl of the specific RNA polymerase (10 U/μl), and incubate at 37° for 1 hr. After the incubation add 92 μl of H_2O and take two 1-μl aliquots to determine the total counts/minute incorporated and the total radioactivity in the reaction. In this way determine the fraction of incorporation and calculate the amount synthesized. Add 5 μl of 1 *M* Tris–HCl, pH 8, 1 μl of 1 *M* $MgCl_2$, 1 μl of yeast RNA (10 μg/μl), 1 μl of RNasin (20 U/μl), and 1 μl of DNase I (10 ng/μl, Worthington ultrapure) and incubate at 37° for 30 min to remove the DNA template. Purify the probe by running on a Sephadex G-100 column. Probes synthesized at this specific activity are not very stable and can only be used for protection assays for 3 days, the reason being that wherever the ^{32}P decays, the probe breaks into smaller and smaller pieces until finally the probe gives only a smear instead of a protected band. One should check to see whether the probe synthesized is full length by running it on an acrylamide gel before use.

To set up a protection assay, first set up a series of tubes with a standard curve; for example, add no RNA to tube 1, 0.5 pg of GnRH RNA to tube 2, 1 pg to tube 3, 2 pg to tube 4, 3 pg to tube 5, and 4 pg to tube 6. Add TE to the standard tubes to bring the final volume to 5 μl. Add the unknown samples to the next set of tubes, also in a final volume of 5 μl. Then to each tube add ~100 pg in a volume of 5 μl of the newly synthesized antisense GnRH RNA probe (~10^9 cpm/μg). One should add at least 10-fold excess probe so that the hybridization will go to completion; therefore, when analyzing in the 0.5–5 pg range one should add between 50 and 100 pg of probe. Finally add 20 μl of the hybridization mix (60% formamide, 0.9 *M* NaCl, 6 m*M* EDTA, 60 m*M* Tris–HCl, pH 7.4, 2.5 mg/ml yeast RNA) and two drops of mineral oil over each sample to prevent evaporation. Heat denature at 85° for 5 min and incubate overnight (~16 hr) at 68°.

Transfer the samples to fresh microcentrifuge tubes, leaving behind the oil. Add 300 μl of 1× S_1 nuclease buffer (0.3 *M* NaCl, 30 m*M* sodium acetate, pH 4.8, 3 m*M* ZnCl) and add the S_1 nuclease, which you will have to titer. For S_1 nuclease purchased from Sigma usually about 1500–2500 U is required to digest completely the unprotected material. Some investigators like to use RNase A and RNase T1 rather than S_1 nuclease. RNase works well, is cheaper, and may even give a cleaner signal; however, I prefer not to use it because it is harder to control than S_1 nuclease, which requires an acidic pH and Zn^{2+}, whereas RNase can work under any condition, i.e., it is hard to get rid of. So, when working with precious RNA samples, I find it safer not to work with large quantities of RNase. However, if RNase is used, it should be titrated rather than just using the 40 μg/ml of RNase A and 2 μg/ml of RNase T1 as described in the Promega protocol, to avoid overdigestion of samples.

After adding the S_1 nuclease to the samples incubate for 1 hr at 56°. Add 40 μl of 5 *M* ammonium acetate, pH 4.8, 1 μl of 0.5 *M* EDTA, 1 μl of yeast RNA (10 μg/ml), 150 μl of phenol (saturated with TE), vortex the samples, and add 150 μl of chloroform. Vortex again and centrifuge (16,000 *g*) for 5 min. Extract the samples, transfer to 1.5-ml microcentrifuge tubes, and add 600 μl of 2-propanol and store at −20° for at least 1 hr. Then spin down the samples in a microfuge for 10 min, pour off the supernatant, and add 0.5 ml of 70% ethanol, vortex, and centrifuge for 5 min. Carefully remove the supernatant, for the pellets can be a bit slippery, and then dry the pellets. Resuspend them in 4 μl of 1× TE and 1 μl of 5× dye mix. Heat denature at 65° for 5 min and then load the samples on an acrylamide gel. After running the gel, dry it and expose it to X-ray film with an intensifying screen at −70°. Then, using the X-ray film as a guide, cut out the protected bands and determine the radioactivity. Some investigators[7] merely precipitate the protected material with TCA, collect the precipitate onto filters, and count the filters, rather than running the gel. I have found, however, that running the protected material on a gel gives much greater sensitivity and provides additional information which makes it easier to troubleshoot when something goes wrong with the assay.

In Fig. 4 is an autoradiograph (see insert) of a representative standard curve obtained with GnRH and the resulting plot obtained from the counts/minute measured in each of the bands. It is possible to quantitate 500 fg, and, if very fresh probe is used, it is possible to detect down to 200 fg. When analyzing samples one must always run a standard curve. The curve is very reproducible; however, the counts/minute protected for each point varies as the probe gets older. Usually we analyze each unknown sample 2 to 3 times and have found the assay to be very reproducible despite the numerous transfers.

To analyze the primary transcript, processing intermediates, and mature mRNA in the nucleus, the assay is done very similarly. As shown in

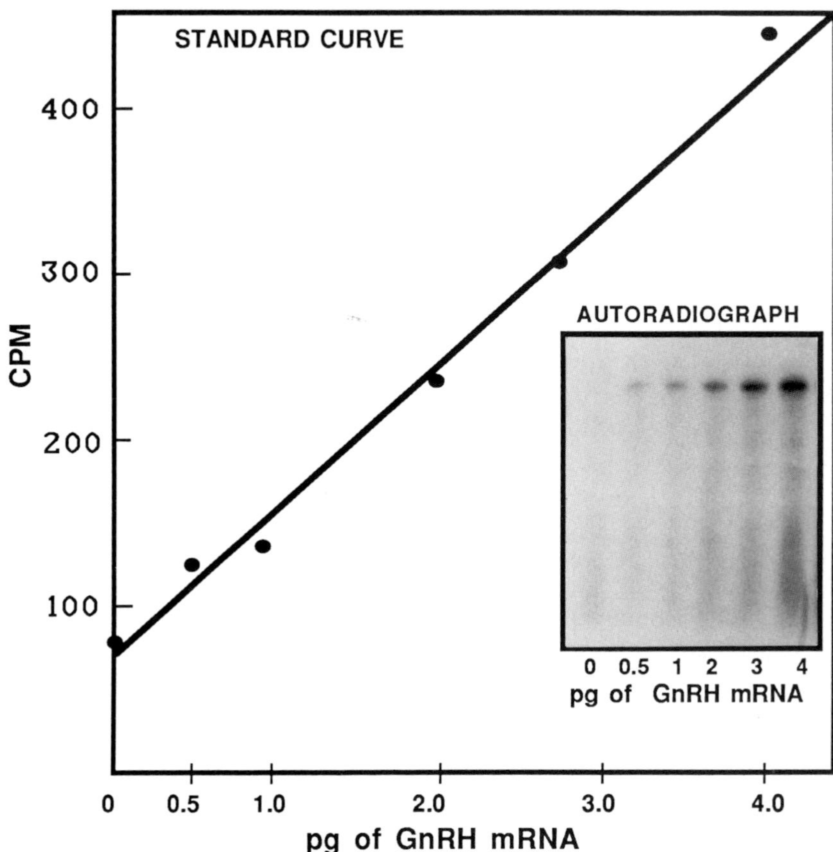

FIG. 4. (Insert) Representative autoradiograph of a standard curve set up with synthetic sense GnRH mRNA hybridized with antisense GnRH mRNA (~10^9 cpm/μg). The plot was generated by cutting out the protected bands from the acrylamide gel and determining the radioactivity.

Fig. 5, a probe may be synthesized that is complementary to an intron–exon–intron junction and used to detect the primary transcript which will protect the full-length probe. It is also possible to detect processing intermediates with this probe. By synthesizing the probe such that the regions which are complementary to the two different introns are of unequal lengths, it may be determined which of two possible pathways processing takes. And finally, with this same probe, the amount of mature mRNA in the nucleus can be measured. One thing to be considered when measuring nuclear RNA compared to cytoplasmic RNA is to make sure that the

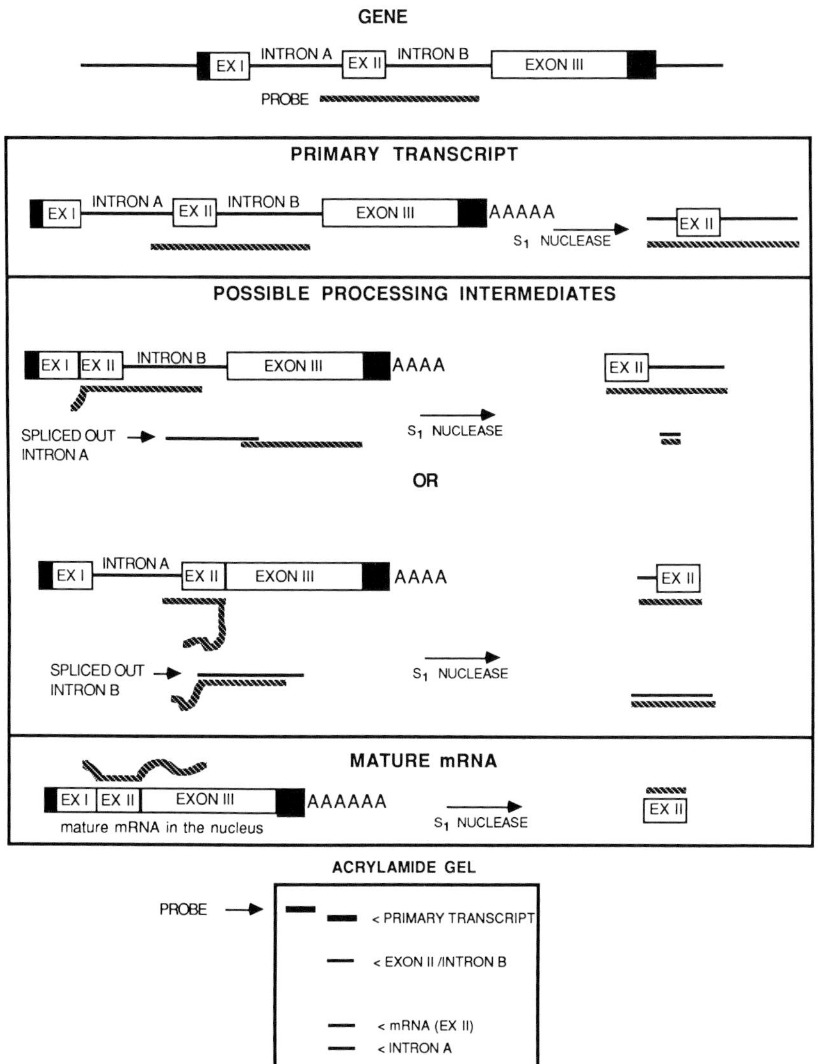

FIG. 5. Diagram demonstrating an intron–exon–intron antisense probe that may be used to measure, by S_1 nuclease analysis, the primary transcript, possible processing intermediates, and mature mRNA from RNA isolated from the cell nucleus.

probe is in 10-fold excess for all protected bands being measured. For example, the amount of primary transcript may be in 10-fold excess of the processing intermediates and the mature mRNA. In addition, the standard curve would be required to span a wider range.

[44] Solution Hybridization–Nuclease Protection Assays for Sensitive Detection of Differentially Spliced Substance P- and Neurokinin A-Encoding Messenger Ribonucleic Acids

By JAMES E. KRAUSE, JEAN D. CREMINS, MARK S. CARTER, ELAINE R. BROWN, and MARGARET R. MACDONALD

Introduction

In recent years it has been observed that many eukaryotic genes are transcribed in the nucleus of cells and that the precursor RNA is subsequently processed by intron removal which may include differential or alternative exon usage.[1,2] These mRNAs are then transported into the cytoplasm of cells where they may be translated into functional proteins. Since in many instances the differentially or alternatively spliced exons correspond to the protein-coding region of the mRNA species, it is important to differentiate between the various types of mRNAs that can potentially be expressed. Many methods are available to investigate the mRNAs expressed, including the following: (1) RNA blot (i.e., Northern) analysis[3] with exon-specific or spliced mRNA-specific complementary DNA (cDNA), genomic, or oligonucleotide probes; (2) end-labeled or uniformly labeled cDNAs used in solution hybridizations[4,5] followed by digestion with the single-strand-specific nuclease, S_1, derived from *Aspergillus oryzae*[6]; and (3) uniformly labeled complementary RNA (cRNA) probes[7] used in solution hybridizations followed by digestion with nuclease S_1 or with a combination of ribonucleases, A and T_1.[8] In this chapter, we describe experimental methods useful for both the documentation and routine quantitation of differentially spliced mRNA species of low abundance by using solution hybridization and subsequent nuclease protection (methods 2 and 3 above). These solution hybridization–nuclease protection methods are especially useful in that more information with regard to

[1] R. A. Padgett, P. J. Grabowski, M. M. Konarska, S. Seiler, and P. A. Sharp, *Annu. Rev. Biochem.* **55,** 1119 (1986).

[2] S. E. Leff, M. G. Rosenfeld, and R. M. Evans, *Annu. Rev. Biochem.* **55,** 1091 (1986).

[3] P. S. Thomas, *Proc. Natl. Acad. Sci. U.S.A.* **77,** 5201 (1980).

[4] A. J. Berk and P. A. Sharp, *Cell* **12,** 721 (1977).

[5] J. Favaloro, R. Treisman, and R. Kamen, this series, Vol. 65, p. 718.

[6] V. M. Vogt, this series, Vol. 65, p. 248.

[7] D. A. Melton, P. A. Krieg, M. R. Rebagliati, T. Maniatis, K. Zinn, and M. R. Green, *Nucleic Acids Res.* **12,** 7035 (1984).

[8] J. N. Davidson, *in* "The Biochemistry of Nucleic Acid," 7th Ed. Academic Press, New York, 1972.

METHODS IN ENZYMOLOGY, VOL. 168

both the qualitative and quantitative nature of the mRNAs can be attained. Moreover, the last method using uniformly labeled cRNA probes can be at least 2- to 10-fold more sensitive than Northern blots in detecting specific mRNAs.

In this chapter we use as an example the detection and quantitation of multiple mRNAs derived from the gene that encodes the tachykinin neuropeptides substance P (SP) and neurokinin A (NKA; also known as substance K). The structures of three rat preprotachykinin (PPT) mRNAs (α-, β-, and γ-PPT mRNA) have been determined from cDNA cloning and nuclease protection experiments.[9–13] α-PPT mRNA encodes a PPT of 112 amino acids which lacks the NKA coding sequence (exon 6). β-PPT mRNA encodes a PPT of 130 amino acids (exons 1–7), and γ-PPT mRNA encodes a PPT of 115 amino acids which lacks the pentadecapeptide sequence between SP and NKA (exon 4). Both tachykinins SP and NKA can be produced from β- and γ-PPT, as can amino-terminal extensions of NKA. Thus, different peptides may be produced by the expression of each type of PPT mRNA.

Nuclease Protection Assays in the Context of Gene and mRNA Identification and Analysis. Information on the intron/exon organization of the gene of interest will greatly facilitate an understanding of results obtained in nuclease protection experiments, although this information is not initially necessary. Prior to the use of nuclease protection assays for the analysis of gene expression, a cDNA clone (or in some cases, a genomic clone) must be available. The cDNA can be used for the isolation of the gene encoding the mRNA(s) from which the cDNA has been derived. From an analysis of the genomic clones isolated, an understanding of the number and size of nucleotide sequences corresponding to the exons can be achieved. It is clear that the ultimate characterization of any putative cellular RNA species necessitates the determination of the complete nucleotide sequence of individual RNA species by cDNA cloning and nucleotide sequence analysis.

General Strategy of Solution Hybridization–Nuclease Protection Assays. Berk and Sharp[4] originally introduced an S_1 nuclease protection assay for the analysis of adenovirus-derived mRNAs. The procedure consisted of hybridization of unlabeled RNA to ^{32}P-labeled DNA. The hybrid-

[9] H. Nawa, H. Kotani, and S. Nakanishi, *Nature (London)* **312,** 729 (1984).

[10] J. E. Krause and J. M. Chirgwin, *Soc. Neurosci. Abstr.* **11,** 1115 (1985).

[11] Y. Kawaguchi, M. Hoshimoru, H. Nawa, and S. Nakanishi, *Biochem. Biophys. Res. Commun.* **139,** 1040 (1986).

[12] J. E. Krause, J. M. Chirgwin, M. S. Carter, Z. S. Xu, and A. D. Hershey, *Proc. Natl. Acad. Sci. U.S.A.* **84,** 881 (1987).

[13] M. S. Carter and J. E. Krause, *J. Biol. Chem.*, submitted for publication.

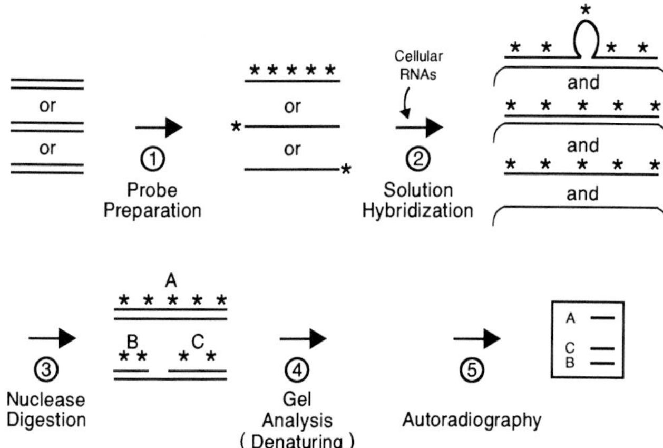

FIG. 1. Schematic illustration of the various experimental procedures used in solution hybridization–nuclease protection assays. As discussed in the text, some of the procedures differ somewhat depending on whether end-labeled cDNA or uniformly labeled cRNA probes are to be used. The scale used for the nucleic acid probes at the hybridization step differs from that used in the remainder of the figure.

ized molecules were subsequently treated with S_1 nuclease under conditions in which single-stranded nucleic acid is hydrolyzed without introduction of strand breaks into hybridized molecules.[6] The size of the single-stranded DNA protected fragment(s) (after strand separation) then was determined by alkaline agarose gel electrophoresis. Favaloro *et al.*[5] summarized methods for nuclease protection assays using DNA probes for the analysis of polyoma virus gene transcription. In particular, they used a two-dimensional (neutral and alkaline) agarose gel mapping of the S_1 digests.

Recently, efficient *in vitro* systems have been described for the synthesis of single-stranded cRNAs.[7,14–17] These methods utilize transcription vectors which contain phage promoter sequences upstream of multiple cloning sites. With these vectors and the cognate RNA polymerases, cRNA probes can be generated for use in nuclease protection assays. The overall nuclease protection assay procedure using either DNA or RNA probes is schematically illustrated in Fig. 1. The assay procedure can be divided into five steps, including (1) preparation and radiolabeling of the

[14] S. J. Stahl and K. Zinn, *J. Mol. Biol.* **148,** 481 (1981).

[15] P. Davenloo, A. H. Rosenberg, J. J. Dunn, and F. W. Studier, *Proc. Natl. Acad. Sci. U.S.A.* **81,** 2035 (1984).

[16] M. Golomb and M. J. Chamberlin, *J. Virol.* **21,** 743 (1977).

[17] C. E. Morris, J. F. Klement, and W. T. McAllister, *Gene* **41,** 193 (1986).

hybridization probe, (2) annealing of the radiolabeled probe with cellular RNAs, (3) digestion of the nonannealed nucleic acid with nuclease(s), (4) urea–sodium dodecyl sulfate (SDS) polyacrylamide or other denaturing gel analysis of the protected RNA–probe hybrid after strand separation, and (5) autoradiographic analysis. The significant differences between protection assays using end-labeled cDNA probes compared with cRNA probes lie in steps number 1, 2, and possibly 3 listed above. Detailed descriptions and protocols of the two assays, as well as their uses and limitations, are presented below.

Detailed Protocol of Solution Hybridization–Nuclease Protection Assays

Buffers and Reagents

T7 RNA polymerase (Promega Biotec)
SP6 RNA polymerase (Promega Biotec)
T3 RNA polymerase (Stratagene)
Klenow fragment of *Escherichia coli* DNA polymerase I (Boehringer Mannheim)
T4 DNA polymerase (Bethesda Research Labs.)
S_1 nuclease (Sigma)
Ribonuclease A (Sigma)
Ribonuclease T_1 (Sigma)
Proteinase K (Sigma)
RNasin (i.e., placental ribonuclease inhibitor, Sigma)
RQ_1 DNase (Promega Biotec)
T7 polymerase transcription buffer: 200 m*M* Tris–HCl, pH 7.5, 1 m*M* dithiothreitol, 30 m*M* $MgCl_2$, 10 μM spermidine, 50 m*M* NaCl; 500 μM ATP, 500 μM GTP, 500 μM UTP, 12 μM CTP, 50 μCi [α-^{32}P]CTP (3000 Ci/mmol), 1 U/μl RNasin, 200 ng *Hin*dIII linearized pG1β-PPT, 10 U T7 RNA polymerase
Formamide hybridization buffer: 40 m*M* PIPES, pH 6.4, 400 m*M* NaCl, 1 m*M* EDTA, 80% (v/v) formamide
S_1 nuclease digestion buffer: 30 m*M* NaOAc, pH 4.4, 280 m*M* NaCl, 4.5 m*M* $ZnSO_4$, 100–1000 U/ml S_1 nuclease
Ribonuclease A/T_1 digestion buffer: 10 m*M* Tris–HCl, pH 7.5, 5 m*M* EDTA, 300 m*M* NaCl; 40 μg/ml RNase A, 2 μg/ml RNase T1
Termination mix: 2.5 *M* NH_4OAc, 50 m*M* EDTA

Preparation of β-PPT Hybridization Probe

1. [^{32}P]β-PPT is prepared with linearized pG1β-PPT in a 30-min transcription reaction at 40° (incorporation of ^{32}P into product is usually 50–85% input ^{32}P).

2. After transcription, the DNA template is removed by digestion with RQ_1 DNase at a concentration of 1 U/μg DNA at 37° for 15 min.

3. The reaction mixture is extracted with a phenol–chloroform (1 : 1) mixture and is then extracted with chloroform.

4. The ^{32}P probe is purified by 2–3 ethanol precipitations from 2.5 *M* NH_4OAc. The extract is made 2.5 *M* in NH_4OAc (by adding an equal volume, 20 μl, 5 *M* NH_4OAc), and 2 volumes of 100% ethanol are added. The solution is chilled at −80° for at least 30 min.

5. The ethanol solution is brought to room temperature (~5 min) and the ^{32}P probe is recovered by centrifugation at 10,000 *g* for 15 min. This procedure, which separates unincorporated ^{32}P from ^{32}P-labeled probe, is repeated 1–2 more times.

6. The ^{32}P-labeled probe is resuspended in 100–200 μl diethyl pyrocarbonate-treated H_2O containing 10–20 U RNasin, and an aliquot is taken for ^{32}P determination by liquid scintillation spectrometry. A working solution of 2×10^5 cpm ^{32}P probe/μl is used.

α-, β-, and γ-Preprotachykinin mRNA Protection Assay

1. [^{32}P]β-PPT cRNA (2×10^5 cpm) or ^{32}P-end-labeled β-PPT cDNA (3 ng) is ethanol coprecipitated with 25 μg total RNA (if lower amounts of specific tissue RNA are to be used, then total liver RNA or *E. coli* ribosomal RNA is added as carrier to attain a total of 25 μg).

2. The ^{32}P probe–RNA precipitate is briefly dried at room temperature (~10 min) and resuspended in 10 μl hybridization buffer as follows. The hybridization buffer is placed on top of the ^{32}P probe–RNA pellet and mixed briefly by vortex. The solution is heated to 65° for 10 min in the 1.5-ml microcentrifuge tubes. Complete resuspension of the ^{32}P probe–RNA pellet is ascertained by repeated pipetting of the hybridization buffer and checking by Geiger counter. Finally, the samples are briefly spun to collect the hybridization buffer at the bottom of the tube, denatured in a boiling water bath for 5 min, and incubated at 45° for at least 6 hr, usually overnight. If cDNA hybridization probes are used, shorter hybridization times are needed.

3. The nonannealed ^{32}P probe is digested for 30 min at 37° in S_1 nuclease digestion buffer, containing from 100 to 1000 U S_1 nuclease/ml (300 μl buffer). Note that the appropriate S_1 nuclease concentration must be determined by titration for each lot of S_1 nuclease.[18]

[18] If ribonucleases A and T_1 are to be used for digestion (step 3), then step 4 is replaced with a proteinase K digestion (10 μl of a 10 μg/ml solution at 37° for 15 min) followed by a phenol–chloroform (1 : 1) extraction prior to the addition of carrier RNA.

4. Seventy-five microliters termination mix is added to each tube, followed by 5 μg carrier RNA [usually poly(A)$^-$ RNA or total eukaryotic RNA].

5. The nucleic acids are precipitated from an equal volume of 2-propanol after 30 min at $-80°$ and allowed to air dry.

6. The samples are resuspended in 2–5 μl 99% formamide sequencing dye,[17] boiled for 5 min, and quick chilled on ice prior to electrophoresis on a 6% polyacrylamide gel containing 7 *M* urea.

7. The gel is run at 500 V total (25 V/cm) until the xylene cyanol dye is near the bottom of the gel. This dye comigrates with oligonucleotides of approximately 106 bases, whereas the bromphenol blue dye comigrates with oligonucleotides of about 26 bases. ^{32}P-Labeled DNA markers ranging from 75 to 700 bases are used as size standards, either *Msp*I-digested pBR322 (labeled with [α-^{32}P]dCTP and Klenow fragment of DNA polymerase I) or *Dde*I-digested pUC19 (labeled with [γ-^{32}P]ATP and T4 polynucleotide kinase).

8. The gel is rinsed in 10% acetic acid for 10 min, dried onto Whatman paper on a gel dryer, and exposed to XAR-5 X-ray film (Kodak).

End-Labeled cDNAs in Nuclease Protection Assays

When cDNA(s) of interest are isolated from libraries of plasmid or phage recombinants, the excised cDNA after appropriate preparation and radiolabeling can be used in nuclease protection assays. The cDNA sequence to be used as the solution hybridization probe should possess the full complement of exons possible, and preferably should be full length at both the 5′ and 3′ corresponding end of the mRNA. If the cDNA is lacking one or more exons due to splicing, it will be difficult to use as a probe in protection assays since annealing of an mRNA which contains an additional exon(s) will produce a hybrid whereby the additional exon encoded in the mRNA will loop out from the cDNA probe. This situation will most likely result in the nucleolytic scission of the mRNA single-strand loop but only limited "cut through" of the cDNA owing to inaccessibility of the nuclease to this region, or to the lack of a significant probe–mRNA mismatch. A schematic illustration of the generation of misleading results arising from the use of an incomplete solution hybridization probe is displayed in Fig. 2. Thus, if this type of incorrect probe were used, the interpretation of the results regarding the type and abundance of the putative mRNA(s) would be wrong. For similar reasons, a uniformly labeled cRNA probe should also possess the full complement of exons.

The choice of restriction sites for excising the cDNA that will be radiolabeled is also crucial. For the detection of mRNA(s) that lacks an

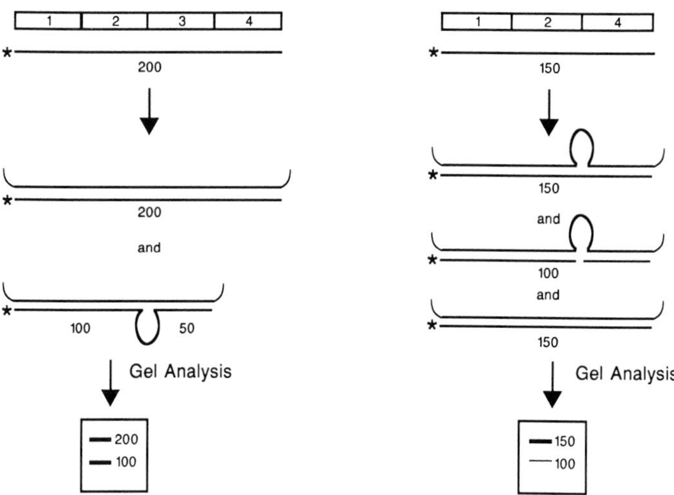

FIG. 2. Comparison of the use of a "correct" probe compared with an "incorrect" probe used in nuclease protection assays. The left-hand scheme illustrates the use of a "correct" end-labeled nuclease protection probe that is derived from a cDNA that contains a full complement of exon-corresponding sequence. In this hypothetical situation the probe is hybridized to an equimolar amount of mRNAs, one of which contains sequences corresponding to exons 1–4 while the other lacks the sequence corresponding to exon 3. After hybridization and nuclease digestion, two autoradiographic signals of equal intensity corresponding to the protected fragments are observed on denaturing gel electrophoresis and autoradiography. The interpretation of the results from this experiment, i.e., an equimolar amount of two mRNAs, one lacking the exon 3-corresponding sequence, would be correct. The right-hand scheme displays the use of an "incorrect" end-labeled nuclease protection probe that is derived from a cDNA lacking the exon 3-corresponding sequence. This probe is also hybridized to an equimolar amount of mRNAs as described above for the "correct" probe. After hybridization and nuclease digestion, two hybridization signals differing greatly in intensity corresponding to the protected fragments are observed on denaturing gel electrophoresis and autoradiography. The major signal at 150 bases corresponds both to the mRNA lacking the exon 3 sequence and in part to the mRNA possessing sequences corresponding to the four exons. The minor band observed at 100 bases corresponds to an mRNA species possessing the full complement of exons; however, incomplete "cut-through" of the probe has occurred, as described in the text. This situation would result in an incorrect interpretation of the relative amount and type of mRNAs expressed.

exon (or exons), both 5′ and 3′ end-labeled fragments are necessarily used since one or the other will allow for the identification of only one end of the mRNA and both ends must be identified. Restriction sites should be located near the 5′ end and 3′ end of the cDNA that are amenable to radiolabeling to a high specific radioactivity. For example, on the antisense-corresponding strand of the cDNA fragment to be radiolabeled, the 3′ end should be recessed so that a "fill in" reaction can be used with a nucleoside [α-^{32}P]triphosphate and either Klenow fragment of DNA polymerase[19] or T4 DNA polymerase I.[20] A different strategy is taken when the 5′ end of the cDNA is to be labeled. On the antisense-corresponding strand of the cDNA fragment to be radiolabeled, the 5′ end should overhang (or at least be blunt-ended) so that the 5′ phosphate can be enzymatically cleaved and the terminal nucleotide can be labeled with [γ-^{32}P]ATP and T4 polynucleotide kinase.[21] Both recessed and blunt 5′ termini are less efficiently labeled than are 5′ overhangs.[22]

It is also important to include some vector sequence on one end of the radiolabeled DNA so that the mRNA with the full complement of exons that protects the cDNA from nucleotide scission will migrate differently (i.e., smaller) than the reannealed cDNA probe. Though theoretically it is possible to adjust the annealing temperature so that RNA/DNA hybrids are favored, in practice it is difficult to prevent DNA/DNA reannealing. This is important because the reannealed cDNA/cDNA strands without any vector sequence would on denaturing gel analysis comigrate with the protected species that corresponded to the mRNA with the full complement of exons present in the probe.

Preparation of End-Labeled β-Preprotachykinin cDNA Solution Hybridization Probes. The cloning of rat cDNAs for multiple SP- and NKA-encoding mRNAs[9,10] provided the appropriate tools for solution hybridization–nuclease protection assays. This has allowed for the additional documentation and quantitation of mRNAs derived from the preprotachykinin I gene. A nearly full-length cDNA corresponding to β-PPT mRNA, which contains the sequences of all seven exons present in the gene, was used as the hybridization probe in these experiments.

The strategy for restricting the β-PPT cDNA that was contained in the pUC19 plasmid vector, as well as an illustration of the radiolabeling of its termini, is shown in Fig. 3. The plasmid pSP27-4[12] has a cDNA insert size of approximately 1130 bases and contains the sequences of exons 1–7 as displayed. Restriction sites for *Rsa*I are found at the 3′ end of the cDNA as well as in the pUC19 vector at about 140 bases upstream from the 5′ end of the cDNA. The strategy for labeling the 5′ end of this cDNA fragment can be seen in Fig. 3C. After excision of the *Rsa*I fragment, the terminal phosphate is removed by digestion with calf intestinal alkaline phosphatase,[19] and the fragment is purified and extracted from low melting temperature agarose (Seaplaque, FMC Corp.). The isolated cDNA

[19] T. Maniatis, E. F. Fritsch, and J. Sambrook, "Molecular Cloning: A Laboratory Manual." Cold Spring Harbor Laboratory, Cold Spring Harbor, New York, 1982.

[20] P. O'Farrell, *Focus (Bethesda Res. Labs.)* **3,** 1 (1981).

[21] G. Chaconas and J. H. Van De Sande, this series, Vol. 65, p. 75.

[22] J. R. Lillehaug, R. K. Kleppe, and K. Kleppe, *Biochemistry* **15,** 1858 (1976).

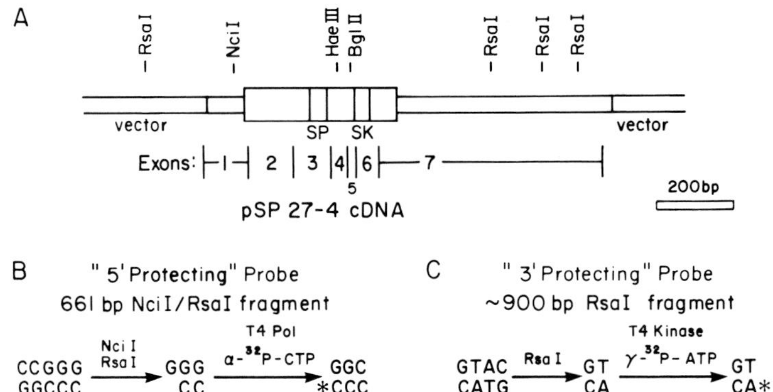

FIG. 3. Preparation of end-labeled β-preprotachykinin cDNA probes for the detection of differentially spliced mRNAs encoding substance P and neurokinin A. (A) cDNA insert derived from pSP27-4 along with some 5′ and 3′ vector sequence. Various restriction enzyme sites are displayed above the illustrated cDNA, and a representation of the cDNA that corresponds to the complement of exons (1–7) present in the PPT gene is shown below. (B) Strategy for radiolabeling the *Nci*I/*Rsa*I fragment with T4 DNA polymerase and [α-^{32}P]dCTP. This probe can be used for the detection of 5′ ends of PPT mRNAs that lack the full complement of exons. (C) Strategy for radiolabeling the *Rsa*I fragment with [γ-^{32}P]ATP and T4 kinase. This probe can be used for the detection of 3′ ends of PPT mRNAs that lack the full complement of exons.

fragment is quantitated by spectrophotometry at 260 nm and by agarose gel electrophoresis adjacent to standards of known mass. The 900-bp *Rsa*I fragment is 5′ end-labeled with [γ-^{32}P]ATP and T4 polynucleotide kinase. The ^{32}P-labeled *Rsa*I fragment is separated from unincorporated [γ-^{32}P]ATP by gel filtration and is used directly in the solution hybridization–nuclease protection assay as described below. The approach taken for labeling the 3′ end of this cDNA can be seen in Fig. 3B. The unlabeled *Rsa*I fragment derived from pSP27-4 is further digested with *Nci*I to produce a fragment of 560 bases that is similarly gel purified, extracted, and quantitated as described above for the *Rsa*I fragment. The *Nci*I/*Rsa*I fragment is amenable to "fill in" labeling with either T4 DNA polymerase[20] or with the Klenow fragment of *E. coli* DNA polymerase I[19] in the presence of [α-^{32}P]dCTP. This ^{32}P probe is separated from unincorporated [α-^{32}P]dCTP by gel filtration and is also used directly in the solution hybridization–nuclease protection assay as described below.

Solution Hybridization–Nuclease Protection Assays with End-Labeled β-Preprotachykinin Probes. Total RNA prepared by the guanidinium isothiocyanate–cesium chloride method of Chirgwin *et al.*[23] is co-

[23] J. M. Chirgwin, A. E. Przbyla, R. J. MacDonald, and W. J. Rutter, *Biochemistry* **18,** 5294 (1979).

precipitated with 1–5 ng ^{32}P probe in a 1.5-ml Eppendorf tube. The precipitate is rinsed with 70% ethanol, and the damp precipitate is resuspended in 10 μl 80% formamide hybridization buffer. If the precipitate is allowed to dry completely, trouble is sometimes encountered in its subsequent resuspension. The cDNA and RNA are denatured by placing in a boiling water bath for 10 min. Complete resuspension of the nucleic acid is ascertained by visual inspection as well as by obtaining a defined volume in a pipet tip and examining the amount of radioactivity with a Geiger counter. If the precipitate is completely resuspended in the 80% formamide hybridization buffer, the tube is removed to an open air incubator at the predetermined hybridization temperature. If the precipitate does not appear to be resuspended, the solution is repeatedly pipetted and is denatured a second time in a boiling water bath.

The temperature and time of hybridization must be empirically determined, and the following formula[24] is useful for estimating the T_m of a DNA/DNA hybrid of defined length:

$$T_m = 16.6 \log[Na^+] + 0.41(\%G + C) + 81.5$$

Note also that for every 1% of formamide present, the T_m is lowered by 0.65°. Since DNA/RNA hybrids are more stable than DNA/DNA hybrids, a hybridization temperature slightly above (2–4°) that calculated for a DNA/DNA hybrid has been used for the β-PPT probes. The hybridization temperatures may have to be determined empirically once a theoretical T_m is calculated in order to maximize RNA/DNA annealing and minimize probe reannealing. Temperatures of 45 and 46° are used for the *Rsa*I and *Nci*I/*Rsa*I β-PPT probes, respectively.

Hybridizations are carried out for 3 hr in order to minimize probe reannealing. After hybridization, the solutions are subjected to digestion with S_1 nuclease for 30 min at 37° after dilution with 300 μl S_1 nuclease digestion buffer as described above. The protected species are analyzed on a 6% polyacrylamide gel containing 7 *M* urea[25] adjacent to an aliquot of the undigested probe and appropriate size standards. Figures 4A and 4B display the results using the end-labeled β-PPT probes while Figs. 4C and 4D provide a schematic illustration of the annealing of the probes to the three PPT mRNAs.[12] With the 3′ end-labeled 900-base *Rsa*I probe, a fragment of 738 bases is protected, which corresponds to β-PPT mRNA. With the 5′ end-labeled *Nci*I probe, a species of 661 bases, corresponding to β-PPT mRNA, is protected. It should be pointed out that this is also the size of the reannealed *Nci*I probe. The RNA species corresponding to α-PPT mRNA results in the protection of fragments 296 bases and 311

[24] J. Marmus and P. Doty, *J. Mol. Biol.* **5,** 109 (1962).

[25] A. M. Maxam and W. Gilbert, this series, Vol. 6, p. 499.

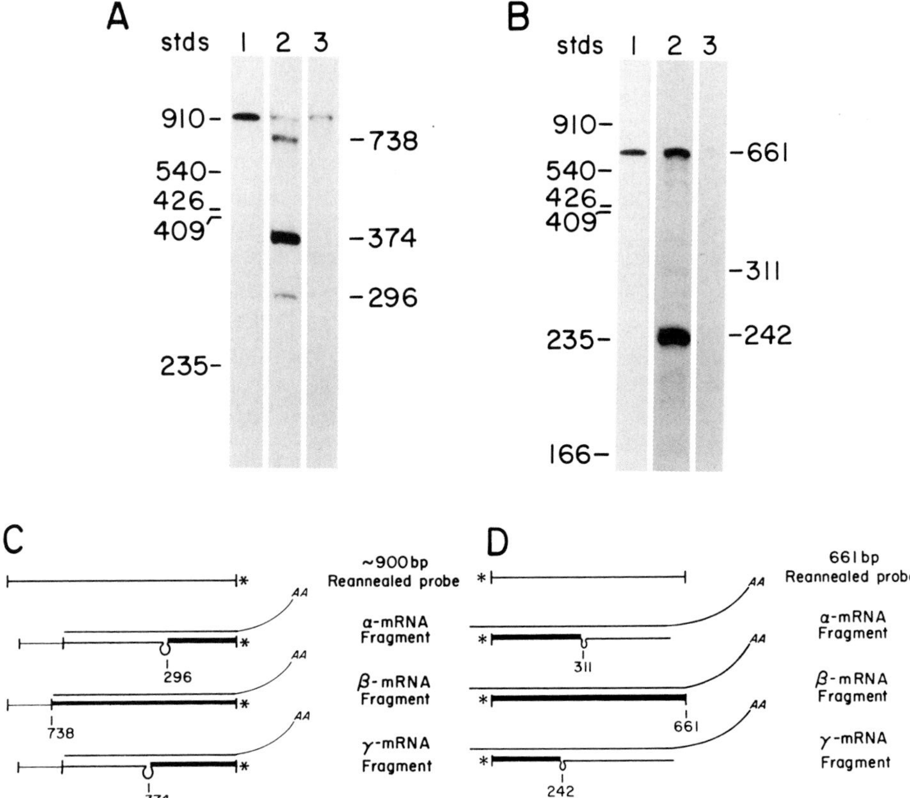

FIG. 4. Solution hybridization analysis of rat preprotachykinin mRNAs using end-labeled cDNA probes. (A) Autoradiogram of an S_1 nuclease protection experiment in which 50 μg rat striatal (lane A2) or liver (lane A3) total RNA was annealed to the ^{32}P-labeled *Rsa*I fragment isolated from pSP27-4, as shown in Fig. 3, and subsequently digested with S_1 nuclease as discussed in the text. Lane A1 shows the undigested probe at 2.5% of the amount used in the protection assay. The size standards are a ^{32}P-labeled *Dde*I digest of pUC19, and the sizes of the protected fragments in bases are displayed to the right of the autoradiogram. (B) Autoradiogram of an S_1 nuclease protection experiment in which 50 μg striatal (lane B2) or liver (lane B3) total RNA was annealed to the ^{32}P-labeled *Nci*I–*Rsa*I fragment isolated from pSP27-4, as shown in Fig. 3, and subsequently digested with S_1 nuclease as discussed in the text. Lane B1 shows the undigested probe at 1% of the amount used in the protection assay. The size standards are the same as in A above, and the sizes of the protected fragments in bases are displayed to the right of the autoradiogram. (C and D) Schematic depictions of the striatal PPT mRNAs protected in the S_1 nuclease experiments shown in A and B.

bases with the end-labeled *Rsa*I and *Nci*I probes, respectively. The γ-PPT mRNA species results in the protection of fragments of 374 and 242 bases with the end-labeled *Rsa*I and *Nci*I probes, respectively.

Uniformly Labeled Complementary RNAs in Solution Hybridization–Nuclease Protection Assays

Melton and co-workers[7] and others[14–17] have developed methods for the preparation of cRNAs by *in vitro* transcription. The DNA of interest is inserted downstream from bacteriophage (i.e., SP6, T7, or T3) promoters for the cognate RNA polymerases. The radiolabeled cRNAs generated can be used in RNA measurements. These prokaryotic promoters are strong and very specific, and the methods used for the transcription reactions are straightforward. Furthermore, the probes can be prepared of uniform length and desired specific radioactivity. These probes have seen widespread use for RNA blots (Northern) and, more recently, for solution hybridization–nuclease protection assays as well as for *in situ* hybridization analyses. Since the specific radioactivity of the probe can be adjusted by the input of radiolabeled and cold nucleoside triphosphates in the transcription reaction, these probes can be used for the detection and quantitation of low-abundance RNAs.[26]

In our hands this solution hybridization–nuclease protection assay is at least 2- to 10-fold more sensitive than Northern blot analyses due to (1) the fact that the hybridization can be driven to completion and (2) the absence of probe reannealing. Furthermore, as pointed out by Quarless and Heinrich,[26] total RNA concentrations used can be higher than that used in RNA blot analyses, and problems arising from the limited capacity of the hybridization membrane to bind nucleic acid are not encountered. Northern blot analysis of low-abundance RNAs may also be limited by the relatively low efficiency of RNA transfer from the gel to the membrane. Thus, these cRNA probes can be ideal for the detection and routine quantitation of differentially or alternatively spliced mRNAs in complex RNA mixtures. In this section, we describe methods for the preparation of a β-PPT cRNA solution hybridization probe and demonstrate its specificity and sensitivity for the detection of SP- and/or NKA-encoding mRNAs.

Preparation of a β-Preprotachykinin cRNA Solution Hybridization Probe. As part of our ongoing studies on the cell-free translation and posttranslational processing of α-, β-, and γ-PPTs,[27] a construction was

[26] S. A. Quarless and G. Heinrich, *Biotechniques* **4,** 434 (1986).

[27] M. R. MacDonald and J. E. Krause, *Soc. Neurosci. Abstr.* **12,** 1042 (1986).

made with pGEM1 that contained the coding region of β-PPT as well as some 5′ and 3′ untranslated sequence. The pGEM vector system (obtained from Promega Biotec) has T7 and SP6 promoter sites on opposite strands of the plasmid DNA adjacent to a multiple cloning site. The constructed plasmid, called pG1β-PPT, proved to be highly desirable for solution hybridization–nuclease protection experiments as it contains part of the exon 1 sequence, exons 2–6, and part of the exon 7 sequence. An illustration of pG1β-PPT is displayed in Fig. 5A. This plasmid was constructed by isolating the *Fnu*D2 fragment of pSP27-4[12] and blunt-end ligating it into *Bam*HI-digested pGEM after the *Bam*HI ends were "filled in" with the aid of the Klenow fragment of DNA polymerase I and deoxynucleoside triphosphates. Figure 5B presents the β-PPT cDNA along with some 5′ and 3′ vector sequence and the antisense β-PPT cRNA which can be transcribed. This transcript is derived from *Hin*dIII-linearized pG1β-PPT as a result of *in vitro* transcription from the T7 promoter. Also indicated is the exon number and sizes (in bases) of the appropriate vector and exon sequences. Figure 5B also presents schematically the annealing of α-, β-, and γ-PPT mRNAs to the antisense β-PPT cRNA transcript. The size in bases of the undigested probe and the potential protected fragments are indicated above the appropriate cRNA and mRNAs displayed.

Solution Hybridization–Nuclease Protection Assay with the β-Preprotachykinin cRNA Probe. As indicated above in the discussion of the general strategy of nuclease protection assays, steps number 1, 2, and potentially 3 of the assays using end-labeled cDNA probes compared with uniformly labeled cRNA probes may differ (see Fig. 1). These include (1) preparation and radiolabeling of the hybridization probe, (2) annealing of the radiolabeled probe with cellular RNAs, and (3) digestion of the nonannealed probe with nucleases. Below we discuss our procedures for nuclease protection assays using the β-PPT cRNA probe. We shall also discuss some results obtained with this assay with either S_1 nuclease or a combination of ribonucleases A and T_1 that demonstrate both its specificity and sensitivity.

Note that either S_1 nuclease or ribonucleases A and T_1 can be used (after optimization of each) with similar results. Nuclease S_1 nonspecifically degrades single-stranded RNA (as well as DNA) to yield 5′-phosphoryl mono- or oligonucleotides. Ribonucleases A and T_1, however, do exhibit some specificity. Ribonuclease A (bovine pancreas) attacks pyrimidine nucleotides at the 3′ phosphate group and leaves the phosphate linked to the adjacent nucleotide. Ribonuclease T_1 (from *Aspergillus oryzae*) specifically attacks the 3′-phosphate groups of guanosine nucleotides and leaves a 5′ hydroxyl group on the adjacent nucleotide.

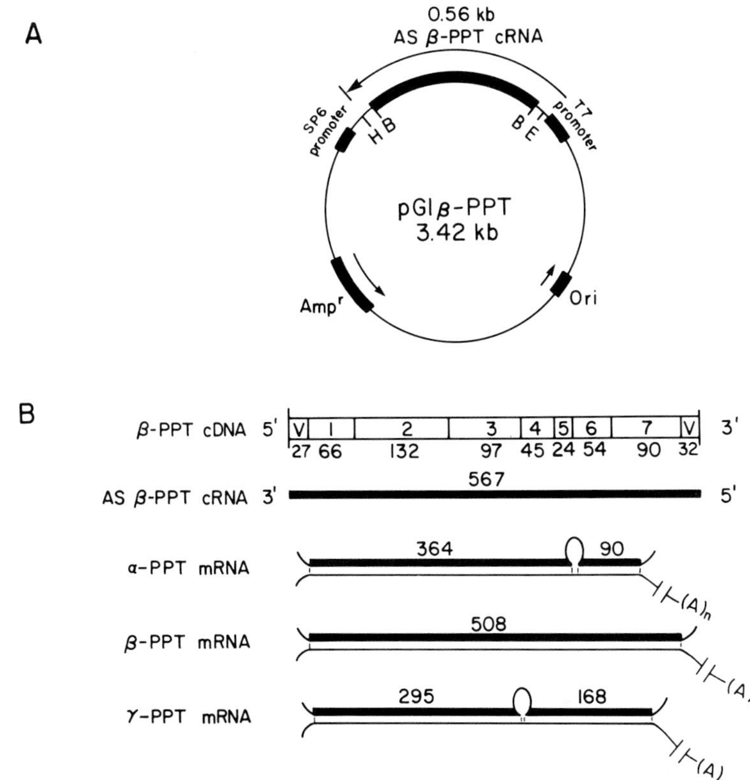

FIG. 5. Schematic illustration of the hybridization of a uniformly labeled β-preprotachykinin cRNA to α-, β-, and γ-preprotachykinin mRNAs encoding substance P and neurokinin A. (A) The plasmid pG1β-PPT used to produce the antisense β-PPT cRNA. The plasmid is linearized by digestion with *Hin*dIII, and transcription is carried out with T7 polymerase as described in the text. H, *Hin*dIII; B, *Bam*HI; E, *Eco*RI; Amp^r, ampicillin resistance; Ori, origin of replication. (B) The β-PPT cDNA insert present in pG1β-PPT with the 5′ and 3′ sequence from the vector as is present in the transcript obtained with T7 polymerase as described above. The exon number corresponding to the regions present in the cDNA is shown, and the size of these regions, in nucleotides, is displayed beneath the β-PPT cDNA. Also shown are the 567-base antisense β-PPT cRNA and a schematic illustration of the annealing of α-, β-, and γ-PPT mRNA to the antisense β-PPT cRNA probe. The size of the homologous segments that are protected after annealing and nuclease digestion is shown above the annealed structures. v, Vector.

Full-length α-, β-, and γ-PPT cRNA expressing vectors were used in the pGEM system so that the message-sense cRNAs produced could be used as size and quantitation standards for the assay. Figure 6A displays the plasmids constructed for the *in vitro* transcription of α-, β-, and γ-PPT

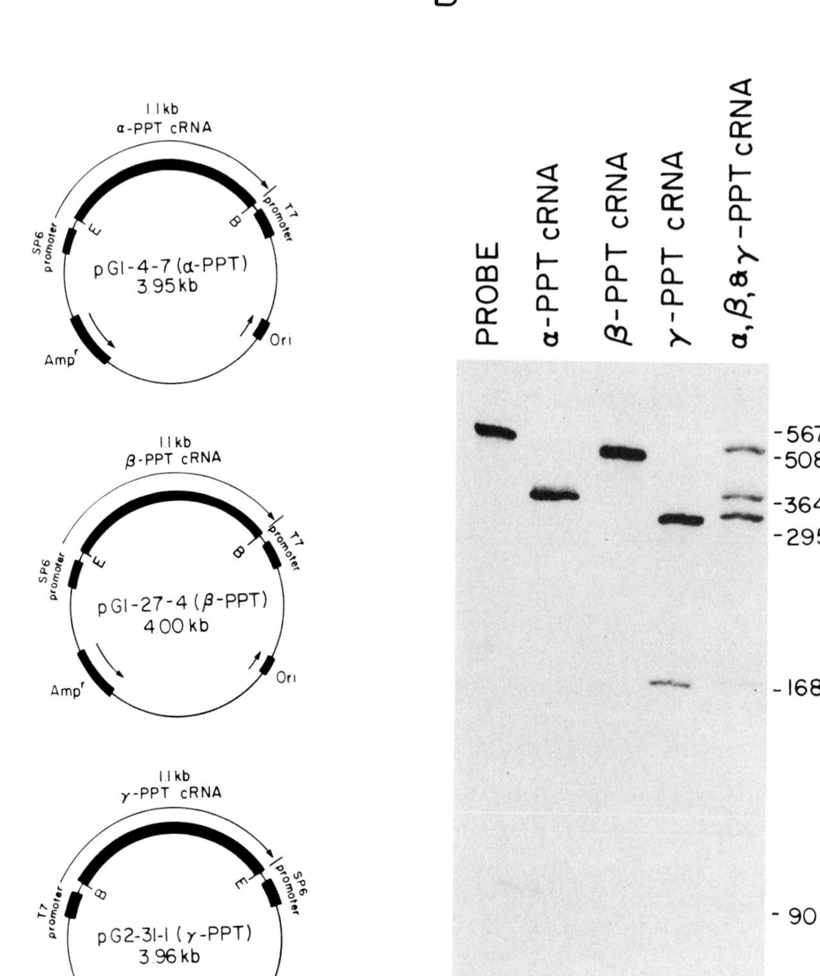

FIG. 6. Solution hybridization–nuclease protection analysis of α-, β-, and γ-preprotachykinin cRNAs with a uniformly labeled β-preprotachykinin cRNA probe. (A) Three plasmids that were constructed for the *in vitro* production of α-, β-, and γ-PPT cRNAs. The promoter site used for sense strand transcription is shown, as is the site used for plasmid linearization. The names and sizes in kilobases of the plasmids are displayed within the plasmid illustration. B, *Bam*HI; E, *Eco*RI. (B) Autoradiogram of the results of the solution hybridization analysis performed with 10 pg α-, β-, and γ-PPT cRNAs or a combination of all three at 5 pg adjacent to an aliquot of undigested probe. The sizes indicated are in bases, based on the nucleotide sequence of the cDNAs.

cRNAs from the promoters as indicated. Figure 6B shows the results from a typical experiment in which 10 pg synthetic α-, β-, or γ-PPT cRNA, or a combination of all three at 5 pg, was used in an S_1 nuclease protection assay. Note that a single band is observed on gel electrophoresis with the β-PPT cRNA, whereas two bands are observed as expected both with the α-PPT cRNA and with the γ-PPT cRNA. The assay is quantitative as indicated by the results shown in Fig. 7. From 1 to 1000 pg synthetic cRNA corresponding to α-, β-, and γ-PPT cRNA was examined in the assay, and the autoradiographic signals were scanned densitometrically to evaluate the linearity of signals observed at various doses. The log–log plots of Fig. 7 illustrate that all three cRNAs produce a linear response in the nuclease protection assay with a correlation coefficient (r) of at least 0.90. Furthermore, with the α-PPT cRNA and γ-PPT cRNA, the ratio of autoradiographic signals between the larger and smaller band is consistent with the guanosine content of the cRNAs (the probe is labeled with [^{32}P]CTP), demonstrating that both ends of the cRNAs can be quantitated accurately and either (or both) can be used as an estimate of the abundance of α-PPT mRNA or γ-PPT mRNA.

We have used either S_1 nuclease or a combination of ribonucleases A and T_1 with equally satisfactory results as evidenced by the data displayed in Fig. 8, where liver, striatal, and intestinal RNAs were used in the β-PPT cRNA nuclease protection assay. Some concern has been raised over the controllability and efficiency of digestion with the ribonuclease A and T_1 combination.[26] We have found that S_1 nuclease preparations from different suppliers as well as different lots from the same suppliers show substantial variability in activity and must be titrated for optimal results. In a similar vein, the temperature of RNase A and T1 digestion has to be empirically determined since a 37° digestion with some probes and hybrids yields decreased signals corresponding to protected species when compared with digestions at lower temperatures (i.e., 30 or 22°). However, optimal conditions can be found for both S_1 nuclease and a combination of ribonucleases A and T_1 with the antisense β-PPT cRNA probe as illustrated in Fig. 8 for striatal and intestinal total RNA. Note the presence of β- and γ-PPT mRNAs in these tissues and barely detectable levels of α-PPT mRNA. The abundance of the PPT mRNAs in striatum is some 50- to 100-fold greater than that in intestine. Thus, it is apparent that either digestion procedure can be used after optimization with comparable results.

Summary

In this chapter we discussed methods that can be used for the sensitive detection and quantitation of differentially or alternatively spliced

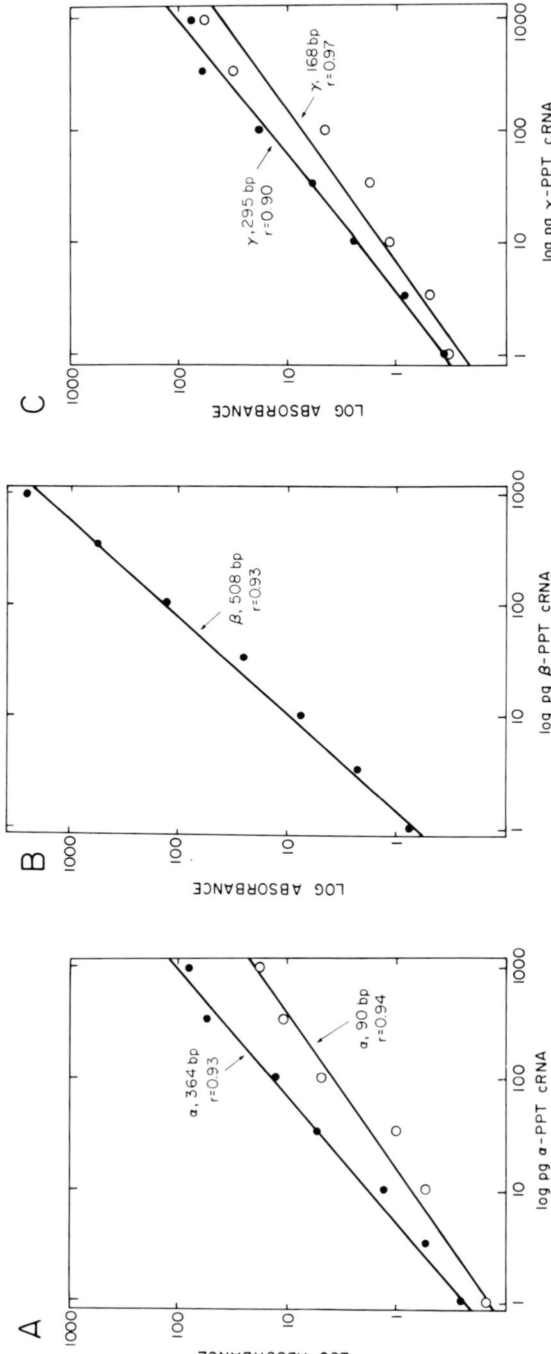

FIG. 7. Plots of log absorbance versus log picogram cRNA for α-, β-, and γ-preprotachykinin cRNAs in the solution hybridization assay with the ^{32}P-labeled pG1/β-PPT cRNA probe. *In vitro* transcribed cRNAs from 1 to 1000 pg were subjected to solution hybridization and S_1 nuclease digestion as described in the text. The observed correlation coefficient (*r*) of the data points and the size of the protected fragments are displayed adjacent to the plotted data. See Fig. 5 and the text for further details.

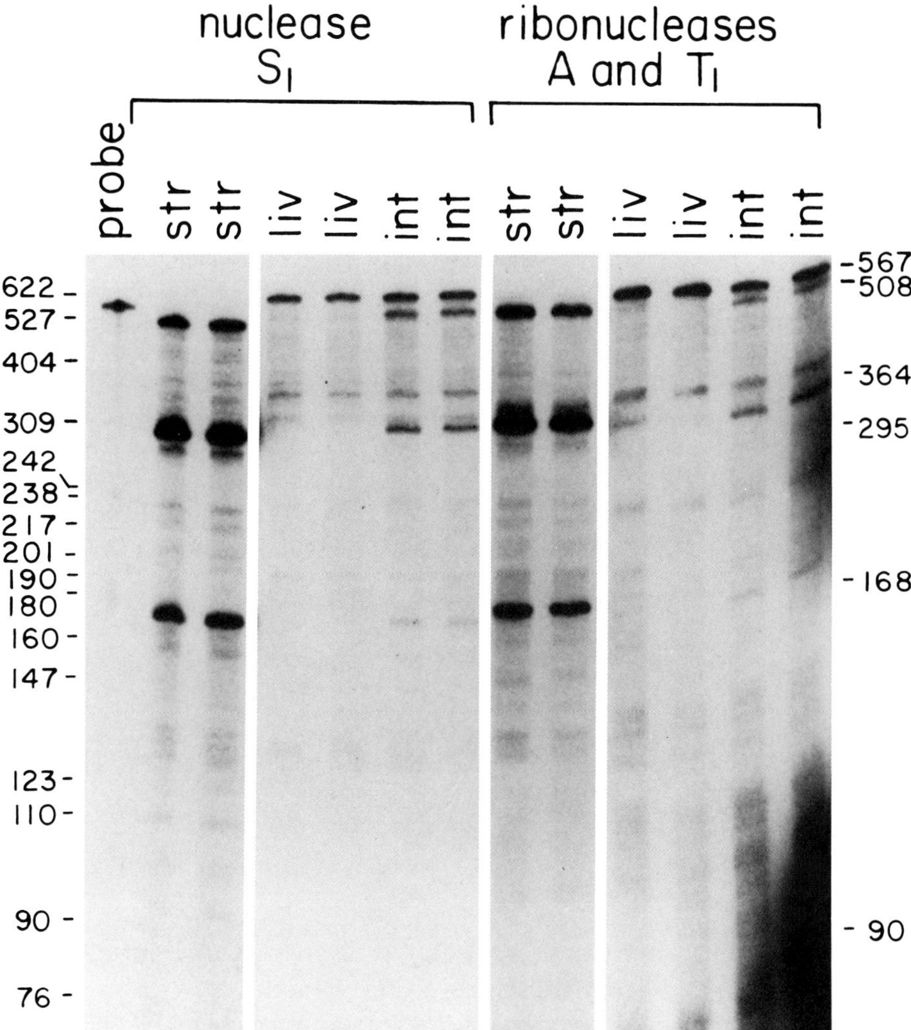

FIG. 8. Solution hybridization–nuclease protection analysis of preprotachykinin mRNAs present in total RNA isolated from rat striatum and intestine. Either S_1 nuclease or a combination of ribonucleases A and T_1 were used to hydrolyze nonannealed nucleic acids in hybridizations of rat striatal, intestinal, or liver total RNA (25 μg each) with 2×10^5 dpm [^{32}P]β-PPT antisense cRNA (prepared as described in Fig. 5 and in the text). The autoradiographic exposure time for intestine (int) and liver (liv) was 15 hr and was 7.5 times that for striatum (str). The signals barely visible at 364 and 90 bases correspond to α-PPT mRNA, whereas the signals observed at 508 bases and at 295 and 168 bases correspond to β-PPT mRNA and γ-PPT mRNA, respectively. None of the PPT mRNAs are present in the liver, which serves as a negative control. Note the similarity of results observed with replicate samples that were subjected to nuclease digestion either with S_1 nuclease or with a combination of ribonucleases A and T_1.

mRNAs as well as mRNAs of low abundance. Although mechanisms responsible for splicing (and differential splicing in particular) have not been fully determined, many RNAs derived from a variety of genes have been observed to undergo the process. The impact of splicing with regard to the expanded potential of gene expression emphasizes the usefulness of the solution hybridization–nuclease digestion technique described here, compared to Northern blot analysis. The use of radiolabeled cRNA(s) provides for an assay of both high specificity and high sensitivity. While end-labeled cDNA probes can be used, they do not have the sensitivity inherent in the assay performed with uniformly radiolabeled cRNAs. If multiple mRNAs are derived from a single gene as a result of differential or alternative precursor RNA splicing, however, the results with a cRNA probe may initially appear to be quite complicated, and end-labeled cDNAs may yield more easily interpretable results. Nonetheless, both types of probes are useful in the context of gene expression analysis, and it is clear that for routine purposes of quantitation cRNA probes in solution hybridization–nuclease protection assays are clearly more desirable than RNA blot analyses due to their truly quantitative nature as well as ease of assay.

Acknowledgments

This work was supported in part by National Institutes of Health Grant NS21937 and the Pew Memorial Trust. JEK is a Pew Scholar in the Biomedical Sciences.

[45] Quantifying Carrier-Mediated Transport of Peptides from the Brain to the Blood

By WILLIAM A. BANKS and ABBA J. KASTIN

Introduction

Peptides are found both in peripheral tissues and in the central nervous system (CNS).[1,2] They play important roles in the regulation of physiologic events and, when levels are altered in pathological states, can be responsible for symptoms of the disease. It is well established that

[1] D. T. Krieger and J. B. Martin, *N. Engl. J. Med.* **304,** 876 and 944 (1981).

[2] T. Hokfelt, O. Johansson, A. Ljungdahl, J. M. Lundberg, and M. Schultzberg, *Nature (London)* **284,** 515 (1980).

peptides administered peripherally can affect CNS function.[3,4] Peptides injected into the CNS can also alter the function of peripheral tissues.[5–8] It is clear from such findings that the administration of peptides results in the transfer of information across the blood–brain barrier (BBB). This transfer probably involves several mechanisms,[3] but the one in which we have been most interested is the most direct, that is, the movement of the peptide itself across the BBB.

The early studies evaluating movement of peptides across the BBB occurred in the 1970s. The conclusions were variable among laboratories, probably because of methodological shortcomings in those studies and unsupported preconceptions. Despite contradictory findings in the literature, a consensus arose that subscribed to the belief repeated by a publication of the National Institutes of Health, "It is unlikely that significant entry [of peptides] occurs into brain."[9] This belief currently persists, and perhaps represents the majority opinion, as evidenced by the numerous reviews and articles still making such pronouncements as: "Subsequent studies have confirmed that the BBB excludes peptides,"[10] "Circulating peptides cannot penetrate the BBB,"[11] "Studies show . . . a blood–CSF barrier, and, presumably, a blood–brain barrier (BBB), for peptides,"[12] "Thus it is unlikely that distribution of peptides in brain will occur after systemic administration,"[13] and, "The fact that circulating peptides do not effectively cross the BBB. . . ."[13] The premature conclusion that peptides do not cross the BBB was made despite the early finding that small amounts of peptide could be shown to cross the BBB in intact form.[14] The mid-1980s have seen a resurgence of interest in the movement of peptides across the BBB, with many laboratories now investigating this topic. From this body of work, it now appears that many peptides cross the BBB, many by diffusion across the cell membranes

[3] A. J. Kastin, R. D. Olson, A. V. Schally, and D. H. Coy, *Life Sci.* **25,** 401 (1979).

[4] A. J. Kastin, W. A. Banks, J. E. Zadina, and M. Graf, *Life Sci.* **32,** 295 (1983).

[5] P. J. Kulkosky, J. Gibbs, and G. O. Smith, *Physiol. Behav.* **28,** 505 (1982).

[6] Y. Tache, W. Vale, J. Rivier, and M. Brown, *Peptides* **2,** 51 (1981).

[7] M. A. Petty and W. de Jong, *Brain Res.* **260,** 322 (1983).

[8] G. Feurerstein, E. Powell, and A. I. Faden, *Peptides* **6,** 11 (1985).

[9] Task Force 11, "Neuroendocrinology," NIH Publ. 81-2394, p. 495. National Institutes of Health, Bethesda, Maryland, 1981.

[10] E. M. Cornford, *Mol. Physiol.* **7,** 242 (1985).

[11] W. M. Pardridge, *in* "The Neuronal Microenvironment" (H. F. Cserr, ed.), p. 231. NY Acad. Sci., New York, 1986.

[12] J. B. Martin and S. Reichlin, *in* "Clinical Neuroendocrinology," p. 626. Davis, Philadelphia, Pennsylvania, 1987.

[13] W. M. Pardridge, *Fed. Proc., Fed. Am. Soc. Exp. Biol.* **43,** 201 (1984).

[14] A. J. Kastin, C. Nissen, K. Nikolics, K. Medzihradszky, D. H. Coy, I. Teplan, and A. V. Schally, *Brain Res. Bull.* **1,** 10 (1976).

that comprise the BBB.[15–17] Some classes of peptides depart from this general principle and so suggest the existence of saturable transport systems. The best studied of these is the system that has been shown to transport Tyr-MIF-1 (Tyr-Pro-Leu-Gly-amide) and the enkephalins from the brain to the blood.[18–20] Different methodologies have lent support to the idea that members of this group of peptides may cross the BBB, including demonstration of saturable binding to the choroid plexus[21,22] and to capillaries of the brain.[23] This chapter describes a method extensively used by our group that allows the quantitative measurement of the rate of transport of peptides from the brain to the blood. We have used this method to study thyroid hormones, Tyr-MIF-1, methionine (Met)-enkephalin, arginine (Arg)-vasopressin (AVP), iodide, technetium pertechnetate, albumin, and peptide T.[19,20,24–27]

Procedure

The first part of the method that we use to study brain-to-blood transport rates in the mouse was modified from that developed by Noble *et al.*[28] to inject radioactive norepinephrine into the lateral ventricle of the brain. Both procedures take advantage of the relationship in rodents between the external morphology of the cranium and the position of the lateral ventricle. Noble *et al.* showed with their technique that injection of radioactive norepinephrine resulted in reliable delivery into the lateral ventricle, subsequent distribution throughout the rat brain within an hour, and results under various experimental regimens that were indistinguishable from experiments in which the material had been delivered with the use of stereotactic apparatus.

[15] W. A. Banks and A. J. Kastin, *Brain Res. Bull.* **15,** 287 (1985).

[16] W. A. Banks and A. J. Kastin, *Psychoneuroendocrinology* **10,** 385 (1985).

[17] W. A. Banks, A. J. Kastin, D. H. Coy, and E. Angulo, *Brain Res. Bull.* **17,** 155 (1986).

[18] W. A. Banks and A. J. Kastin, *Pharmacol. Biochem. Behav.* **21,** 943 (1984).

[19] W. A. Banks, A. J. Kastin, A. J. Fischman, D. H. Coy, and S. L. Strauss, *Am. J. Physiol.* **251,** E477 (1986).

[20] W. A. Banks, A. J. Kastin, and E. A. Michals, *Peptides* **8,** 899 (1987).

[21] J. T. Huang and A. Lajtha, *Neuropharmacology* **17,** 1075 (1978).

[22] H. M. Firemark, *in* "Neurobiology of Cerebrospinal Fluid" (J. H. Wood, ed.), p. 77. Plenum, New York, 1983.

[23] H. T. Pretorious, A. J. Kastin, and W. A. Banks, *Brain Res. Bull.* **17,** 829 (1986).

[24] W. A. Banks, A. J. Kastin, and E. A. Michals, *Life Sci.* **37,** 2407 (1985).

[25] W. A. Banks and A. J. Kastin, *J. Pharmacol. Exp. Ther.* **239,** 668 (1986).

[26] W. A. Banks, A. J. Kastin, A. Horvath, and E. A. Michals, *J. Neurosci. Res.* **18,** 326 (1987).

[27] C. M. Barrera, A. J. Kastin, and W. A. Banks, *Brain Res. Bull.* **19,** 629 (1987).

[28] E. P. Noble, A. J. Wurtman, and J. Axelrod, *Life Sci.* **6,** 281 (1967).

Mice are anesthetized with urethane (ethyl carbamate), 2 g/kg injected i.p. in a volume of 10 ml/kg. Sodium pentobarbital has also been used, but we have found that urethane seems to be a superior general anesthetic for the mouse. The scalp is removed to expose the skull. The bregma, the intersection of the coronal and sagittal sutures, is located and a hole made through the skull 1 mm lateral and one 1 mm posterior to it. This is accomplished with a 26-gauge needle that has a cuff of polyethylene tubing covering all but the terminal 3.5 mm of its length. The polyethylene tubing acts as a guard, so that when pressure is applied as the needle enters the skull to a depth of 3.5 mm, it is stopped by the tubing. At this depth, the needle pierces the roof but not the floor of the lateral ventricle.

A 1.0-μl Hamilton syringe (Hamilton Co., Reno, NV) is used to make the injection. Hamilton syringes have several types of needles that are probably suitable, but we use the cemented needle with an electrotaper and polished tip (Catalog No. 7101 N). The accuracy of the syringe or the operator's technique can be checked by delivery of the injectate into a test tube that is then counted in the gamma counter. Typically, an experienced technician will have an average delivery of 0.998 to 1.002 μl/injection with a coefficient of variation of 5–6%. It is useful to practice by injecting a dye that will stain the brain tissue so that the needle tract and, if the unguarded portion of the needle is too long, an indentation in the floor of the ventricle can be seen. The technique is so easy that few people miss the ventricle even with the first injections. It does take some practice, however, to attain the high degree of reproducibility that is possible with this technique.

Mice are then decapitated at the desired time, and the entire brain except for the pineal and pituitary, which lie outside the BBB, is removed, rinsed in saline, and counted for 3 min in a gamma counter. The transport rate (T) in mol/g of brain-min is determined by the equation:

$$T = (A - M)C/itw$$

where M is the number of counts remaining in the individual brain, C the amount of material injected expressed in moles, i the amount of injected material expressed in counts per minute (cpm), t is the time in minutes from injection to decapitation, and w the weight of the brain in grams. A is the amount of material expressed in cpm that is available for transport; its derivation is discussed below. To express A in terms of mol/g of brain (Cb) the following equation is used:

$$Cb = AC/iw$$

Kinetic parameters such as $V_{\max}$ or K_{m} can be obtained by determination of the value of T with varying values for C. The relationship between T

and *Cb* can then be determined and the kinetic parameters derived with the use of a computer program such as ALLFIT.[29]

Determination of *A*

It is critical that *A*, the amount of material expressed in cpm available in the brain for transport, be accurately determined. This value is a fraction of *i*, the amount of material expressed in cpm that was injected, because the material is distributed throughout the CNS and is not limited to the brain only. The ratio (*r*) of *A* divided by *i* is usually between 0.4 and 0.6 as measured for thyroxine, ^{125}I-Tyr-MIF-1, ^{131}I-labeled Met-enkephalin, ^{99m}Tc-labeled albumin, and ^{125}I-AVP. Although *A* can be determined by several methods, they all give very similar values.

One simple method for the determination of *A* is to kill mice with an overdose of anesthetic, inject the labeled test material into the lateral ventricle, and remove the brain for counting after the same amount of time has elapsed as used in the alive mice. Mice dead between 10 and 30 min are usually used in this technique. It is assumed that during this interval, transport systems and bulk flow of CSF have stopped but no meaningful change in CSF or other CNS fluid compartments has occurred, so that the overriding factor in the disappearance of injected counts from the brain is distribution by diffusion within the CNS.

Another technique to determine *A* is by removal of brains at varying times (usually ranging from 2 to 30 min) after ventricular injection and construction of a plot between the log cpm remaining in the brain and time elapsed between injection and decapitation. The antilog of the intersection of the line with the ordinate is taken as the value of *A*. Time points taken too early may not have allowed sufficient time for distribution, and time points taken later may be inaccurate due to departures of the curve from linearity, accumulation of degradation products, or blood-to-brain influx of material.

A third technique for determining *A* involves injection of a substance that selectively inhibits the saturable transport, usually an excess of the unlabeled peptide. The value for *A* may be taken as the cpm remaining in the brain at the end of the desired time of study, or a time curve may be constructed and *A* taken to be the intersection of the ordinate. These two variations on the method will disagree to the extent that the system has been incompletely inhibited and to the extent that bulk flow and nonsaturable transport is important. This method for the determination of *A* may not be useful if the substance used for inhibition alters nonsaturable parameters important in the distribution of peptide within the CNS (e.g.,

[29] A. De Lean, P. J. Munson, and D. Rodbard, *Am. J. Physiol.* **235,** E97 (1978).

distribution space) or exit from the brain (e.g., bulk flow). If the inhibitor does not affect nonsaturable parameters but does totally inhibit the saturable transport being studied, then this technique may yield some of the most accurate estimates of A.

A technique that does not seem to work well for the estimation of A involves the simultaneous injection of a marker, such as albumin, that is thought to exit the CNS primarily or exclusively by bulk flow. Albumin seems to have a smaller value for r, so that A for albumin does not equal A for the N-tyrosinated peptides. Such differences are possibly due to different rates of diffusion or a different space of distribution. This approach might be effective if a substance that had a diffusion profile similar to the peptide of interest were used. For example, simultaneous injection of ^{125}I-Tyr-MIF-1 and ^{131}I-D-Tyr-MIF-1 (Tyr-MIF-1 synthesized with a D- rather than an L-tyrosine results in a peptide that is not transported by the stereospecific transport system) might allow for the direct determination of the actual cpm transported by the saturable, stereospecific component in each animal.

Probably the most accurate of these techniques for determination of A is the third method that uses the variation of an excess of unlabeled inhibitor. This method is to be recommended when exceedingly precise estimates of A are needed, as in the determination of kinetic parameters. The easiest method for determining A, however, is the use of animals killed by an overdose of anesthetic. The values obtained by this method have often agreed very closely with those obtained by extrapolation or by use of excess inhibitor, especially for AVP, Tyr-MIF-1, thyroxine, and albumin.

Validation of the Procedure

Radioiodinated peptides are readily susceptible to degradation to amino acids and iodide. Some amino acids and iodide are transported out of the CNS by saturable transport systems. Validation of a candidate transport system for peptides must rule out the possibility that radiolabeled nonpeptide fragments are being transported. Two methods that we have used are competition with fragments of the peptide being tested and chromatography of the radioactivity in the brain or blood.

The use of fragments in the validation of a transport system for peptides is best illustrated for Tyr-MIF-1,[19] which has the sequence Tyr-Pro-Leu-Gly-amide. The radioactive label for this peptide is attached to the tyrosine, so only two radioactively labeled nonpeptides (iodide and iodotyrosine) and three radiolabled peptides (^{125}I-Tyr-Pro, ^{125}I-Tyr-Pro-Leu, ^{125}I-Tyr-MIF-1) can be produced. If a fragment were being transported, then the nonradioactive version of that fragment would be a good inhibitor. For example, if iodotyrosine were being generated and transported, then unlabeled tyrosine or iodotyrosine (synthesized with nonradioactive ^{127}I) should be able to inhibit the transport of the radioactive moiety. Tyr-MIF-1, but not tyrosine, iodotyrosine, Tyr-Pro, or Tyr-Pro-Leu, inhibits the disappearance of radioactivity from the brain after injection of ^{125}I-Tyr-MIF-1 into the lateral ventricle. Therefore, the entire peptide must be the moiety transported since the entire peptide is required to inhibit transport.

The use of chromatography to validate peptide transport is illustrated by work with AVP.[26] AVP consists of nine amino acids, six of which form a ring. The radioactive label attaches to the tyrosine, which is one of the amino acids that constitutes the ring; thus, while iodide and iodotyrosine are the only nonpeptide fragments that can occur, many different peptide fragments containing radioiodinated tyrosine could be generated. Rather than synthesizing and testing for competition all the possible tyrosine-containing peptide fragments that could be generated from AVP, we examined by high-performance liquid chromatography (HPLC) the radioactivity that appeared in the blood after the intraventricular injection of radioiodinated AVP. Enzymatic degradation in blood is often so robust that radioiodinated peptides added to blood are often almost entirely degraded by the time that processing of the sample is completed, even if the blood is cooled to 4° and contains enzyme inhibitors such as aprotinin. A method is needed, then, that immediately halts enzyme activity at the time of sample collection.

The method that we used for AVP was to collect the blood directly into a beaker containing 30% trifluoroacetic acid (TFA). This denatures proteins immediately, but does not alter the elution pattern of radioiodinated AVP on HPLC. To determine the degree of degradation of peptide that occurs in the blood after transport out of the CNS but before collection into TFA, other animals receive an infusion directly into the jugular vein over a period equal to that which elapsed from the time of central injection to decapitation. An infusion rather than a bolus is used because this more closely mimics the brain-to-blood transport which occurs over time. A bolus could overestimate the rate of peripheral degradation by exposing the entire injectate to blood enzymes for the full length of time from i.v. injection to decapitation. We found with this technique that 59.2% of the radioactivity appearing in the blood after central injection eluted at the same position as radioiodinated AVP, while 68.8% eluted at this position after peripheral infusion. This shows that not only does intact AVP enter the peripheral circulation from the CNS, but that almost all of the degradation of transported radioactivity that does occur takes place in the circulation and not in the CNS.

This work with centrally injected AVP shows that the radioactivity appearing in the peripheral circulation represents intact peptide. However, one would expect some radioactivity to appear eventually in the circulation even in the absence of a saturable transport system owing to nonsaturable membrane diffusion and bulk flow. The final step in validation, then, must be to determine whether the amount of radioactivity appearing in the circulation is similar to that predicted by the rate of disappearance from the brain. To do this, the half-time disappearance and volume of distribution of the compound being studied are determined. Then, the radioactive peptide is injected centrally as before, and, at the desired time, an arterial blood sample is obtained. The rate of brain-to-blood transport based on the appearance of radioactivity in the peripheral circulation can then be determined with the equation[30]:

$$T = P(K)(V_d)/(1 - e^{-KT})w$$

where P is mol/ml, K the inverse of the half-time disappearance from blood multiplied by 0.693, and V_d the volume of distribution after injection into the blood. The rate is also divided by w, brain weight, in order to express T in terms of mol/g-min. For AVP, the transport rate based on the disappearance of radioactivity from the brain showed that 42.8% of available material had been transported, while the transport rate based on the appearance of radioactivity in the blood showed that 56.2% of available material had been transported. Furthermore, both the disappearance rate from the brain and the appearance rate in the blood were inhibited by about 50% when unlabeled peptide acting as an inhibitor was added to the lateral ventricular injection. These findings show that the appearance of radioactivity in the blood is due to the saturable system transporting peptide out of the brain.

Interpretation of Results

Accurate interpretation of the results obtained by this method depends on an understanding of the underlying assumptions. One must, for example, be able to accurately and precisely deliver material into the lateral ventricle. Use of the Hamilton syringe allows accurate delivery of 1 μl of material, an amount unlikely to seriously influence CNS physiology.

The importance of the precise determination of the value for A, the amount (cpm) of material available in the brain for transport, has also been discussed in detail. The value of A is less than the amount of material injected (usually 40–60%) because only the brain is counted and not the

[30] A. Goldstein, L. Aronow, and S. M. Kalman, "Principles of Drug Action: The Basis of Pharmacology," p. 311. Wiley, New York, 1974.

cerebrospinal fluid or the spinal cord, which are also potential compartments for distribution. It does not have to be assumed that instantaneous distribution, homogeneous mixing, or complete equilibrium has been achieved any more than such assumptions are necessary in the determination of half-time disappearance or volume of distribution of substances in the periphery.

Transport kinetics are based on the inhibitable portion of the disappearance curve. As such, bulk flow, passive diffusion, and BBB leakage or disruption do not contribute directly to the transport rate. The method does not assume the location of the transport site(s) within the CNS. It also does not make assumptions about the various possible components of the transport rate (brain to blood, brain to CSF, blood to brain, blood to CSF, CSF to brain, or CSF to blood) but rather measures net CNS-to-blood transport. For this reason, the results are expressed in terms of grams per whole brain. The values for transport rate and the kinetic parameters approximate the actual values depending on the degree of diffusion and ubiquity of location of transport sites within the CNS. The amount of material within the CNS decreases logarithmically with time, and so the transport rate is an average over the time studied. Keeping these limitations in mind will help to prevent misinterpretation of the results and will enhance the use of the technique in the quantitative determination of the brain-to-blood transport of peptides or other materials.

Acknowledgments

Supported by the Veterans Administration and the Office of Naval Research.

[46] Characterizing Molecular Heterogeneity of Gastrin-Releasing Peptide and Related Peptides

By JOSEPH R. REEVE, JR. and JOHN H. WALSH

Introduction

Structures of Gastrin-Releasing Peptides

Gastrin-releasing peptide (GRP) is a member of the mammalian bombesin family.[1] Another chemically characterized member of this mamma-

[1] T. J. McDonald, H. Jornvall, K. Tatemoto, and V. Mutt, *FEBS Lett.* **156,** 349 (1983).

lian neuropeptide family is neuromedin B.[2] Other members of the family may be characterized in the future. Erspamer and colleagues have identified numerous members of this family in amphibians.[3] Substance P and other tachykinins also have a structural resemblance to the bombesin family in that they share a biologically active carboxyl-terminal dipeptide amide sequence.

Approaches to Identification of Molecular Variants

Two approaches that have been combined for identification and quantification of various molecular forms of mammalian bombesin peptides are radioimmunoassay detection and high-resolution methods of separation. Region-specific antibodies provide additional information that assists in prediction of structure. Several methods of peptide separation are available. This chapter discusses the merits and difficulties encountered with gel permeation chromatography, ion-exchange chromatography, and reversed-phase high performance liquid chromatography (HPLC) for the purification of gastrin-releasing peptide.

Naturally Occurring Gastrin-Releasing Peptides

Gastrin-releasing peptide was first characterized from porcine stomach as a heptacosapeptide.[4] The same sized gastrin-releasing peptide has also been chemically characterized from chicken,[5] dog,[6] and man[7] (Table I and Refs. 4–9 cited therein). Two other carboxyl-terminal fragments of the 27 amino acid peptide have been isolated and characterized from dog intestine.[6] The second largest form, a tricosapeptide, probably arises from post-proline dipeptidyl aminopeptidase action on the larger peptide. This form has only been characterized in dog. The smallest form characterized is the carboxyl-terminal decapeptide. It could arise from the larger form

[2] N. Minamino, K. Kangawa, and H. Matasuo, *Biochem. Biophys. Res. Commun.* **114,** 541 (1983).

[3] P. Melchiorri, *in* "Gut Hormones" (S. R. Bloom, ed.), p. 534. Livingstone, Edinburgh, 1978.

[4] T. J. McDonald, H. Jornvall, G. Nilsson, M. Vagne, M. Ghatei, S. R. Bloom, and V. Mutt, *Biochem. Biophys. Res. Commun.* **90,** 227 (1979).

[5] T. J. McDonald, H. Jornvall, M. Ghatei, S. R. Bloom, and V. Mutt, *FEBS Lett.* **122,** 45 (1980).

[6] J. R. Reeve, Jr., J. H. Walsh, P. Chew, B. Clark, D. Hawke, and J. E. Shively, *J. Biol. Chem.* **258,** 5582 (1983).

[7] M. S. Orloff, J. R. Reeve, Jr., C. M. Ben-Avram, J. E. Shively, and J. H. Walsh, *Peptides* **5,** 865 (1984).

[8] A. Anastasi, V. Erspamer, and M. Bucci, *Arch. Biochem. Biophys.* **148,** 443 (1972).

[9] N. Minamino, K. Kangawa, and H. Matsuo, *Biochem. Biophys. Res. Commun.* **119,** 14 (1984).

TABLE I
CHEMICALLY CHARACTERIZED BOMBESIN AND GASTRIN-RELEASING PEPTIDES[a]

Source	Sequence
Frog	p Q Q R L G N Q W A V G H L M*
Pig	A P V S V G G G T V L A K M Y P R G N H W A V G H L M*
	G N H W A V G H L M*
Chicken	A P L Q P G G S P A L T K I Y P R G S H W A V G H L M*
Dog	A P V P G G Q G T V L D K M Y P R G N H W A V G H L M*
	G G Q G T V L D K M Y P R G N H W A V G H L M*
	G N H W A V G H L M*
Human	V P L P A G G G T V L T K M Y P R G N H W A V G H L M*
	G N H W A V G H L M*

[a] References for sequences are as follows: frog bombesin,[8] pig GRP-27,[4] pig GRP-10,[9] chicken GRP-27,[5] dog GRP-27,[6] dog GRP-23,[6] dog GRP-10,[6] human GRP-27,[7] and human GRP-10. The asterisk at the end of each sequence indicates that the carboxyl-terminal methionine is amidated.

by cleavage at a single basic residue. It has also been isolated from a human endocrine tumor[7] and from porcine spinal cord.[9]

Other Gastrin-Releasing Peptide Gene Products

Gastrin-releasing peptide represents only 27 of 147 amino acids in preprogastrin-releasing peptide.[10] In man, the gastrin-releasing peptide sequence immediately follows the 23-amino-acid signal peptide. The carboxyl-terminal extension peptide of progastrin-releasing peptide is coded by three different types of mRNA produced by alternate mRNA splicing.[11] It has recently been shown that there are three different forms of preprogastrin-releasing peptide messenger RNA, referred to as forms I, II, and III.[12] The translated peptides also have the same designation, i.e., gastrin-releasing peptide gene-related peptides forms I, II, and III. The

[10] E. R. Spindel, W. W. Chin, J. Price, L. H. Rees, G. M. Besser, and J. F. Habener, *Proc. Natl. Acad. Sci. U.S.A.* **81,** 5699 (1984).

[11] E. R. Spindel, M. D. Zilberberg, and W. W. Chin, *Mol. Endocrinol.* **1,** 224 (1987).

[12] E. A. Sausville, A. M. Lebacq-Verheyden, E. R. Spindel, F. Cuttitta, A. F. Gazdar, and J. F. Battey, *J. Biol. Chem.* **261,** 2451 (1986).

mRNA structures indicate that all three forms have the same signal peptide, followed by the gastrin-releasing peptide sequence and amidation site and a 68-amino-acid common extension peptide sequence that follows immediately after the amidation site. The three forms vary near the carboxyl terminus of preprogastrin-releasing peptide (Table II). Form I contains an additional 27-amino-acid carboxyl-terminal extension. Form II contains the same extension, but it has a 7-amino-acid deletion caused by a 21-base deletion in the mRNA. Form III is a unique heptadecapeptide sequence at the carboxyl terminus resulting from a frameshift in translation caused by a 19-base deletion in the mRNA after the bases that code for amino acid 98 in the common mRNA region. Characterization of the unique regions of these three peptide forms is discussed.

Methods of Detecting Gastrin-Releasing Peptides

Biological Assay

Gastrin-releasing peptide initially was isolated from porcine stomach and intestine based on bioassay of gastrin release in conscious dogs.[13] Similar gastrin-releasing activity can be monitored in anesthetized rats.[14] A group of bioassays including hormone release and contraction of various smooth muscle preparations was used by Erspamer and co-workers to characterize bombesin-like activity in mammalian tissues.[15] Full potency of the decapeptide compared with the heptacosapeptide was shown in strips of canine gastric smooth muscle.[16] A wide spectrum of biological actions also has been found when bombesin-like peptides were injected into the brain.[17,18] Despite these potent biological actions, measurement of bombesin-like activity now is done more conveniently and accurately by radioimmunoassay using an antibody specific for the biologically active portion of the molecule. However, biological activity remains a useful property for monitoring purification of other peptides with bombesin-like biological activity, such as neuromedin B.[2]

[13] T. J. McDonald, G. Nilsson, M. Vagne, M. Ghatei, S. R. Bloom, and V. Mutt, *Gut* **19,** 767 (1978).

[14] J. H. Walsh, H. C. Wong, and G. J. Dockray, *Fed. Proc., Fed. Am. Soc. Exp. Biol.* **38,** 2315 (1979).

[15] V. Erspamer, G. F. Erspamer, P. Melchiorri, and L. Negri, *Gut* **20,** 1047 (1979).

[16] E. A. Mayer, J. R. Reeve, Jr., S. Khawaja, P. Chew, J. Elashoff, B. Clark, and J. H. Walsh, *Am. J. Physiol.* **250,** G581 (1986).

[17] Y. Tache, W. Marki, J. Rivier, W. Vale, and M. Brown, *Gastroenterology* **81,** 298 (1981).

[18] M. Brown, W. Marki, and J. Rivier, *Life Sci.* **27,** 125 (1980).

TABLE II

PREPROGASTRIN-RELEASING PEPTIDE[a]

Common region forms I, II, and III

−23 −20 −15 −10
Met - Arg - Gly - Ser - Glu - Leu - Pro - Leu - Val - Leu - Leu - Ala - Leu - Val - Leu -

−5 −1 +1 5
Cys - Leu - Ala - Pro - Arg - Gly - Arg - Ala - Val - Pro - Leu - Pro - Ala - Gly - Gly -

10 15 20
Gly - Thr - Val - Leu - Thr - Lys - Met - Tyr - Pro - Arg - Gly - Asn - His - Trp - Ala -

25 30 35
Val - Gly - His - Leu - Met - Gly - Lys - Lys - Ser - Thr - Gly - Glu - Ser - Ser - Ser -

40 45 50
Val - Ser - Glu - Arg - Gly - Ser - Leu - Lys - Gln - Gln - Leu - Arg - Glu - Tyr - Ile -

55 60 65
Arg - Trp - Glu - Glu - Ala - Ala - Arg - Asn - Leu - Leu - Gly - Leu - Ile - Glu - Ala -

70 75 80
Lys - Glu - Asn - Arg - Asn - His - Gln - Pro - Pro - Gln - Pro - Lys - Ala - Leu - Gly -

85 90 95
Asn - Gln - Gln - Pro - Ser - Trp - Asp - Ser - Glu - Asp - Ser - Ser - Asn - Phe - Lys - Asp -

Variable regions forms I, II, and III

	100 105 110 115 120 125
I	Val - Gly - Ser - Lys - Gly - Lys - Val - Gly - Arg - Leu - Ser - Ala - Pro - Gly - Ser - Gln - Arg - Glu - Gly - Arg - Asn - Pro - Gln - Leu - Asn - Gln - Gln
	100 105 110 115
II	Val - Gly - Ser - Lys - Gly - Lys - * * * * * * * - Gly - Ser - Gln - Arg - Glu - Gly - Arg - Asn - Pro - Gln - Leu - Asn - Gln - Gln
	100 105 110
III	Leu - Val - Asp - Ser - Leu - Leu - Gln - Val - Leu - Asn - Val - Lys - Glu - Gly - Thr - Pro - Ser

[a] Human GRP forms I, II, and III are identical for 98 residues. The amino acids corresponding to the 21-nucleotide deletion in form II are indicated with asterisks. The unique sequence in form III is underlined.

Radioimmunoassay

Specific and sensitive radioimmunoassays have been developed for measurement of bombesin-like and gastrin-releasing peptide immunoreactivity.[15,19,20] A monoclonal antibody produced by immunizing a mouse with conjugated bombesin has been used to produce a solid-phase radioimmunoassay suitable for measurement of carboxyl-terminal fragments of gastrin-releasing peptide.[21] Antibodies that are specific for the carboxyl-terminal region of the GRP or bombesin molecules usually produce results that agree well with chemical or bioassay measurement of peptide content. However, antibodies have been described that discriminate between bombesin and gastrin releasing peptide.[22] Region-specific antisera also have been raised against predicted carboxyl-terminal gene products of the alternately processed mRNAs and used to detect production of these peptides in mammalian tissues (F. Cuttitta, personal communication).

Methods of Separating Gastrin-Releasing Peptide Molecular Forms

Tissue Source

The first step in separating various molecular forms of gastrin-releasing peptide was determination of tissue distribution. Figure 1 shows the concentrations of gastrin-releasing peptide per unit weight of tissue, determined as bombesin immunoreactive equivalents in 2% trifluoroacetic acid extracts[6] of canine gastrointestinal and brain tissues. The regions with highest concentrations were gastric fundic mucosa and small intestinal smooth muscle layer. These tissues each contained approximately 40 pmol bombesin-like immunoreactivity per gram.

Molecular Heterogeneity in Tissues

Acetic acid extracts of gastrointestinal tissues and brain contain different proportions of small and large forms of bombesin-like peptides as revealed by gel filtration elution profiles. Proportions of small and large

[19] J. H. Walsh and J. H. Wong, "Methods of Hormone Radioimmunoassay," p. 581. Academic Press, New York, 1979.

[20] S. Knuhtsen, J. J. Holst, U. Knigge, M. Olesen, and O. V. Nielsen, *Gastroenterology* **87,** 372 (1984).

[21] F. Cuttitta, S. Rosen, A. F. Gazdar, and J. D. Minna, *Proc. Natl. Acad. Sci. U.S.A.* **78,** 4591 (1981).

[22] C. Yanaihara, A. Inoue, T. Mochizuki, N. Sakura, H. Sato, and N. Yanaihara, *in* "Peptide Chemistry 1978" (N. Izumiya, ed.), p. 183. Protein Research Foundation, Osaka, Japan, 1979.

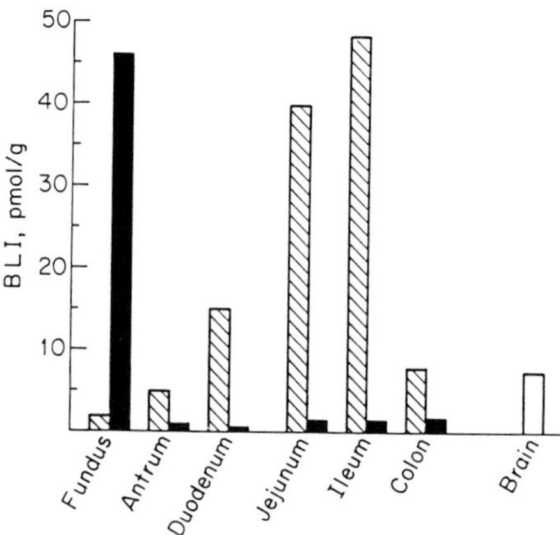

FIG. 1. Tissue levels of bombesin-like immunoreactivity (BLI) in canine gastrointestinal tract and brain. Tissues were boiled in water, and the muscle and mucosa layers were separated and then extracted with 2% trifluoroacetate (10 ml/g). The extracts were centrifuged and the supernatants measured for bombesin-like immunoreactivity using antibody 1078.[19] The muscle levels are designated by hatched bars, the mucosa levels by solid bars, and the brain levels by open bars.

forms also may vary among mammalian species. The rat acid-secreting gastric mucosa contains approximately equal portions of small and large forms, while rat brain[23] contains mainly small forms (Fig. 2). Acetic acid extracts of human acid-secreting gastric mucosa contain nearly 80% large forms (Fig. 2). Canine brain contains approximately equal portions of large and small forms, but predominantly small forms are extractable from rat brain and porcine spinal cord. Thus the relative proportions of large and small forms depend on the tissue extracted and the species used for the studies. Some of this variation is caused by species differences, but some must be caused by differences in extraction methods, separation methods, or detection methods.

Method of Extraction

The recovery of bombesin-like peptides from tissue is greatly influenced by the method of extraction. Boiling the tissue to denature proteolytic enzymes and then extracting the peptides in weak acids (2% trifluoroacetic acid or 3% acetic acid) has yielded high recovery. Neutral or basic extraction conditions yield much lower recoveries. This preferen-

[23] The structure of rat prepro-GRP gene has been recently published [A.-M. Lebacq-Verheyden, G. Krystal, O. Sator, J. Way, and J. F. Battey, *Mol. Endocrinol.* **6,** 556 (1988).]

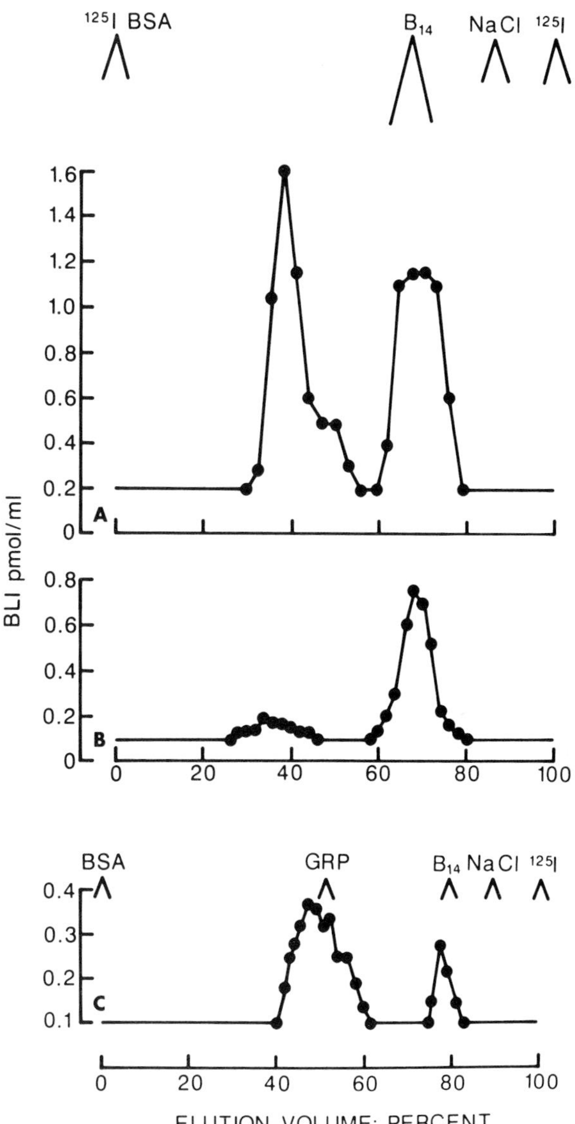

FIG. 2. Molecular forms of GRP or bombesin-like immunoreactivity (BLI) in various mammalian tissues. The molecular forms (A) in rat fundus extracts, (B) in rat brain extracts, and (C) in human fundus extracts. Separations were on Sephadex G-50 columns (1 × 90 cm) eluted with 1% acetic acid. Radiolabeled bovine serum albumin (BSA) was used to mark the void volume (0% elution) and radioiodide the total volume (100% elution). Synthetic porcine GRP-27, bombesin-14, and NaCl were run separately to allow comparisons with the immunoreactive elution profile of the rat and human extracts. Tissues were extracted with 3% boiling acetic acid (10 ml/g).

tial extraction may be explained by the fact that gastrin-releasing peptides are basic and thus more soluble in acidic buffers. The method of extraction will also greatly influence the amount of artifactual postmortem processing of gastrin-releasing peptide. Tissue processed within a matter of minutes should provide a more representative picture of naturally stored forms than tissue that is collected over a long period of time and frozen for future extraction and characterization.

Concentration of Gastrin-Releasing Peptides

Similar to some other basic peptides, it is very difficult to concentrate nanomole and subnanomole amounts of gastrin-releasing peptides from large volumes of initial extracts. Over 80% of the total bombesin-like immunoreactivity was lost from canine intestinal extracts if they were dried by rotary evaporation, lyophilization, or by blowing dry nitrogen over the extract. Freezing Sephadex G-50 fractions of concentrated intestinal extracts at −20° for 2 weeks caused more than 50% loss of immunoreactivity. Solutions containing picomolar concentrations of gastrin-releasing peptide should contain a carrier protein such as albumin to prevent adsorption of the peptide onto glass. During the purification of canine gastrin-releasing peptides, good recovery was obtained with concentration by ion-exchange chromatography at pH 5 or by HPLC using buffers containing 0.1% trifluoroacetate.

Gel Permeation Chromatography of Gastrin-Releasing Peptides

Crude extracts of canine small intestine were concentrated by batch absorption onto CM-Sephadex equilibrated in ammonium acetate, pH 5, and elution with 1.5 *M* ammonium acetate, pH 5. Up to 100 ml of the concentrated solution was loaded onto a Sephadex G-50 SF column (5 × 90 cm) equilibrated in 3% acetic acid. The column was eluted with the same buffer, and two major immunoreactive peaks were detected (Fig. 3). The first eluted before bombesin-14 standards and was subsequently shown to contain GRP-23 and GRP-27.[6] The second immunoreactive peak eluted between bombesin-14 and iodinated tyrosine bombesin-10 standards. This second peak has been shown to contain GRP-10. Natural GRP-10 and synthetic iodinated bombesin-10 contain the same number of amino acids. The labeled peptide is larger due to the presence of an iodide residue. However, the labeled peptide consistently eluted later than the smaller, natural peptide, indicating that modes of separation other than size exclusion were operative. It is likely that hydrophobic interaction between iodide and the Sephadex gel dextran impedes elution of the labeled peptide. The 27- and 23-amino-acid forms of canine gastrin-releas-

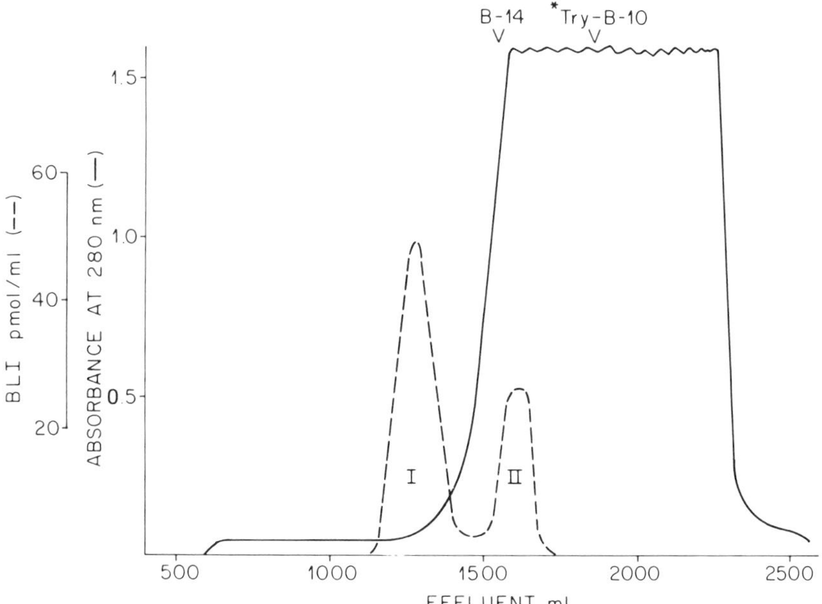

FIG. 3. Gel permeation chromatography of canine intestinal bombesin-like immunoreactivity (BLI). Extracts of intestinal muscle were concentrated as described in the text and loaded onto a Sephadex G-50 column (5 × 90 cm) equilibrated in 3% acetic acid. After loading the column was eluted with the same buffer. The effluent was measured for bombesin-like immunoreactivity and absorbance at 280 nm. Synthetic bombesin tetradecapeptide (B-14) and radioiodinated tyrosine bombesin-10 (*Tyr-B-10) were chromatographed on separate runs for column calibration.

ing peptide cannot be resolved by gel permeation chromatography under these conditions.

Good recoveries (70–90%) were obtained during gel permeation chromatography of intestinal muscle extracts. It was found that storage of gel permeation column fractions was best at 2–4°. Immunoreactive fractions stored at this temperature, in acidic buffers, for 0.5–6 months contained over 90% of their original immunoreactivity.

Gel permeation chromatography is an excellent step to use after initial concentration of the crude extract. It has excellent resolution properties for large amounts of protein and produces quantitative yields. Relatively small sample volumes must be applied, and samples are diluted 2- to 5-fold during the chromatography. Dilution during chromatography is a disadvantage during the final stages of purification when concentrated samples are needed for chemical characterization.

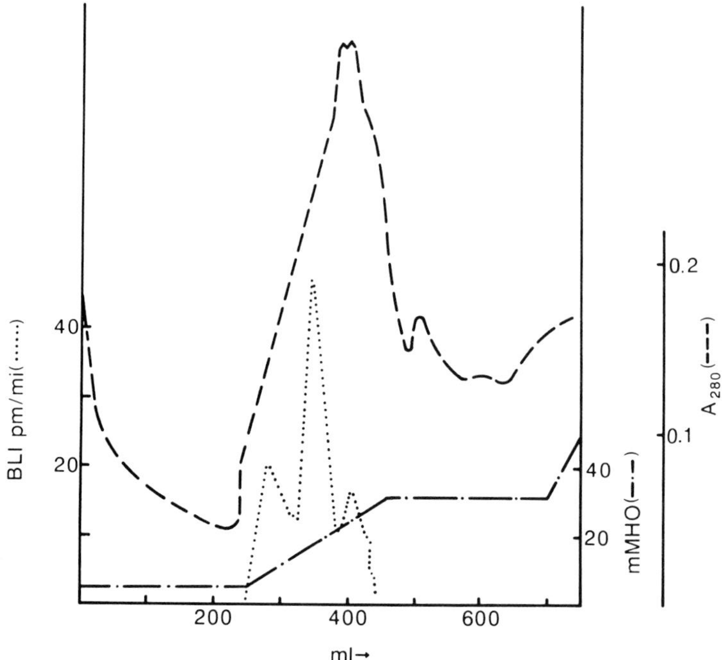

FIG. 4. Ion-exchange chromatography of canine intestinal muscle bombesin-like immunoreactivity. Extraction and chromatographic details are given in the text. After loading and rinsing, the BLI was eluted with a gradient from 0.1 to 1 *M* ammonium acetate, pH 5, and the absorbance at 280 nm was measured.

Low-Pressure Ion-Exchange Chromatography

Canine intestinal muscle was extracted in trifluoroacetic acid (2%). The pH was adjusted to 5 with concentrated ammonium hydroxide, the conductivity adjusted with water to equal that of 0.1 *M* ammonium acetate, pH 5, and the extract loaded onto a CM-Sephadex column (1.5 × 30 cm) equilibrated in 0.1 *M* ammonium acetate. The column was rinsed with 0.1 *M* ammonium acetate, pH 5, until the absorbance at 280 nm stabilized, and then the bombesin-like immunoreactivity was eluted with a linear gradient to 1 *M* ammonium acetate over 200 ml. Three immunoreactive peaks were eluted (Fig. 4), the earliest eluting just after the bombesin-14 standard, the other two eluting in the same region as procine GRP-27 (results not shown). Over 75% of the loaded immunoreactivity was detected in the column eluates.

Low-pressure ion-exchange chromatography resolves various gastrin-releasing peptides. It also separates the various forms from other major absorbance peaks, but the products of separation are usually in relatively large volumes containing high concentrations of salt.

High-Performance Ion-Exchange Chromatography

Separation of synthetic peptides on a high-performance ion-exchange column has been studied. Synthetic peptides were dissolved in buffer A (0.1% trifluoroacetate brought to pH 3 with ammonium hydroxide, containing 20% acetonitrile). A sulfoethylaspartamide SCX column (5 m, 4.6 × 200 mm, The Nest Group, Southboro, MA) was equilibrated in buffer A. As the sample was injected, a gradient was started to 50% buffer B (5% ammonium hydroxide brought to pH 3 with trifluoroacetic acid, containing 20% acetonitrile) over 50 min. Individual components of mixtures containing approximately 3 nmol each of GRP-10, −23, and −27 were well separated (Fig. 5A). When the same mixture was chromatographed under identical conditions using the same column, buffers, and detectors, but with the flow rate changed from 1 to 2 ml/min, the peaks appeared earlier owing to more rapid gradient development (Fig. 5B). The peak heights were smaller, but the total peak areas (flow rate × width at half-height × height) were approximately the same.

The advantage of high-performance over low-pressure ion-exchange chromatography is that better separation of peptides can be achieved in less time. Elution volumes are smaller than with low-pressure systems, but high salt concentrations in the separated products cannot be avoided.

Reversed-Phase High-Performance Liquid Chromatography

Reversed-phase high-performance liquid chromatography (HPLC) is extremely useful for purification of small peptides. The ability to separate very similar peptides in a very short period of time has made HPLC the method of choice for several analytical studies on the molecular forms of gastrin-releasing peptide. However, caution must be applied to conclusions based entirely upon HPLC results. In the following section the strengths of HPLC are described along with examples of deviations from expected profiles obtained in purification of gastrin-releasing peptides.

The chromatography of human tumor gastrin-releasing peptide illustrates unexpected HPLC results. The crude extract (40 ml 4% trifluoroacetic acid) from 8 g of disseminated metastases in the liver of a primary bronchial carcinoid was loaded onto an HPLC column after centrifugation.[7] The column (Waters C_{18} Z module, 10 μM, 8 × 100 mm) was equilibrated in buffer A (0.1% trifluoroacetate). After loading, the column was eluted with a gradient to 30% buffer B (50% acetonitrile containing 0.1% trifluoroacetate) over 5 min, then a gradient from 30 to 75% buffer B

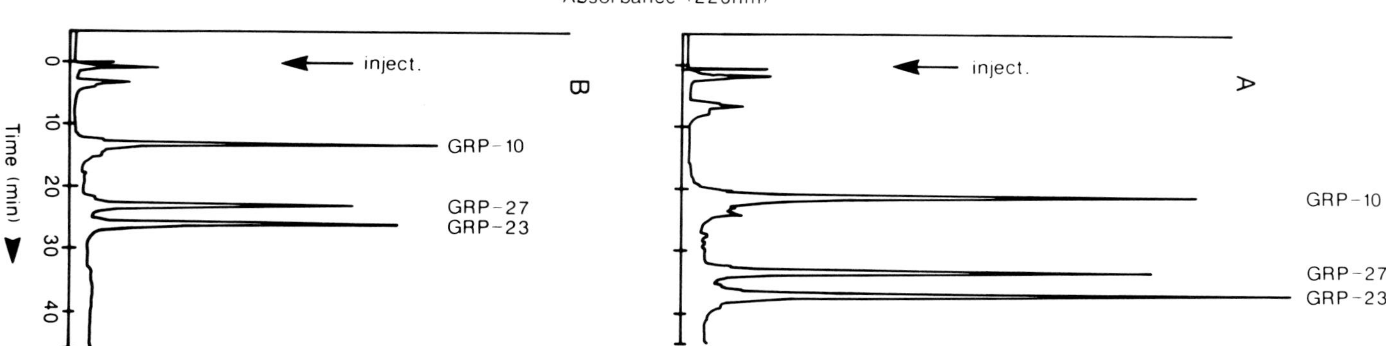
Absorbance (220nm)
A
inject.
GRP-10
GRP-27
GRP-23
B
inject.
GRP-10
GRP-27
GRP-23
0
10
20
30
40
Time (min)

over 90 min. Three peaks of immunoreactivity eluted from this HPLC column. A diffuse, high level of immunoreactivity eluted between peaks 2 and 3 (Fig. 6A). The first two peaks were pooled (pool a), and in subsequent chromatography of this pool only one peak was detected. The diffuse area of immunoreactivity was pooled (pool b), and chromatography on a Waters alkyphenyl reversed-phase column (10 μM, 3.9 × 300 mm) yielded two peaks (Fig. 6B). The earlier eluting peak appeared to coelute with the immunoreactivity from pool a and with synthetic GRP-10. The pooled third peak from the C_{18} column (pool c) produced a similar chromatographic elution pattern as shown for pool b. The earliest eluting material from pools a, b, and c were all purified to apparent homogeneity. They all had the same absorbance to immunoreactivity ratios, the same sequence by microsequence analysis, and they all coeluted with synthetic GRP-10 during isocratic elutions (Fig. 6C shows the coelution of GRP-10 with the early eluting material from pool b). Therefore, the same peptide initially had eluted in three different peaks and diffusely during chromatography of the crude extract. This variable elution probably resulted from association of gastrin-releasing decapeptide with other peptides or proteins during the initial chromatography. The multiple elution positions of GRP-10 from this crude extract illustrate that control experiments with coinjections of natural peptides with standards are necessary to validate conclusions about molecular forms based on HPLC elution profiles.

Several different column supports are available from many manufacturers. Experience during the isolation of gastrin-releasing peptides and other gastrointestinal peptides has shown that columns can be overloaded readily to a point that very little separation is achieved. Rat gastrin-releasing peptide was extracted from 250 g of intestine with 4% trifluoroacetate, and the entire 2.5 liters of extract was centrifuged and loaded onto a single HPLC column (Dynamax C_{18}, 4.2 × 10 cm). The gastrin-releasing peptide immunoreactivity was eluted in approximately 120 ml, but no separation of the various molecular forms was obtained. The fractions containing the immunoreactivity were diluted 3-fold and loaded

FIG. 5. High-performance ion-exchange chromatography of synthetic canine GRP-10, GRP-23, and GRP-27. The three peptides were loaded on a sulfoethylaspartamide SCX column equilibrated in 0.1% trifluoroacetate brought to pH 3 with ammonium hydroxide and containing 20% acetonitrile. The column was eluted with a gradient between this buffer and 50% buffer B (5% ammonium hydroxide brought to pH 3 with trifluoroacetic acid, containing 20% acetonitrile) over 50 min. This gradient was started at the time of injection. (A) Absorbance profile when the column was eluted at 1 ml/min. (B) Absorbance profile of an identical sample when the column was eluted at 2 ml/min. The position of each peptide was identical to the position when the peptides were chromatographed separately.

through the pump onto another HPLC column (Dynamax C_8, 2.1 × 25 cm). Two immunoreactive peaks were well separated during the elution of this column with a gradient similar to that used for the column just before this step where no resolution was observed.

Another feature of reversed-phase chromatography is that changes in the column matrix or the manufacturing process can alter the relative elution position of various peptides. Use of different columns allows purifications that are not possible by use of a single column. With gastrin-releasing peptide it has been observed that Vydac C_{18} and C_4 columns give slightly different relative elution positions, but Waters alkylphenyl columns give very different relative elution positions.

It has also been observed that there is a limit to the degree of purification that can be obtained through reversed-phase HPLC. Rat GRP carboxyl-terminal decapeptide did not separate from another unidentified

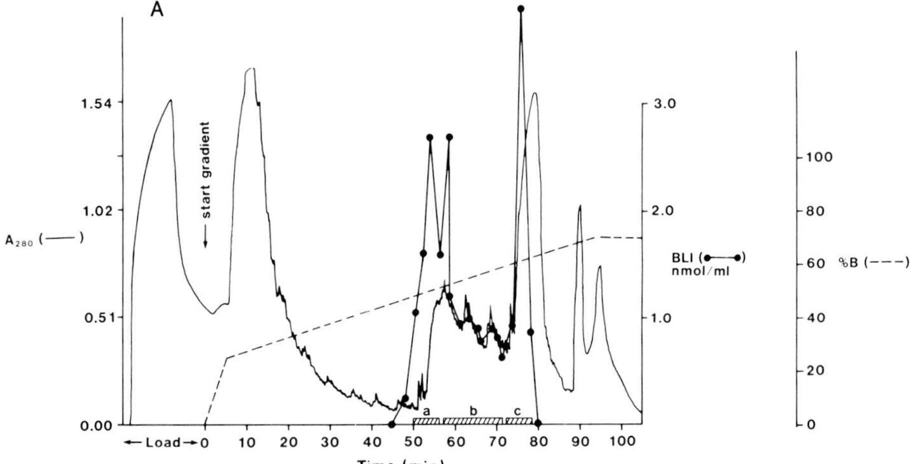

FIG. 6. Reversed-phase high-performance liquid chromatography of human tumor bombesin-like immunoreactivity (BLI). (A) The centrifuged extract of the tumor was loaded directly on the C_{18} column equilibrated in buffer A (0.1% trifluoroacetic acid). The bombesin-like immunoreactivity was eluted with a gradient of buffer B (50% acetonitrile containing 0.1% trifluoroacetic acid). The absorbance profile at 280 nm was also recorded. (B) The immunoreactivity from pool b in A was diluted 3-fold with buffer A and loaded onto an alkylphenyl column equilibrated in buffer A. The bombesin-like immunoreactivity was eluted with a gradient of buffer B. The absorbance profiles at 280 and 220 nm are also shown. (C) The earlier eluting immunoreactivity from B was further purified by reversed-phase chromatography steps[7] until a single absorbance peak was associated with the bombesin-like immunoreactivity. This purified tumor peptide appeared to elute in the same position as synthetic GRP-10 when the peptides were chromatographed separately. This coelution was confirmed when the two peptides were cochromatographed.

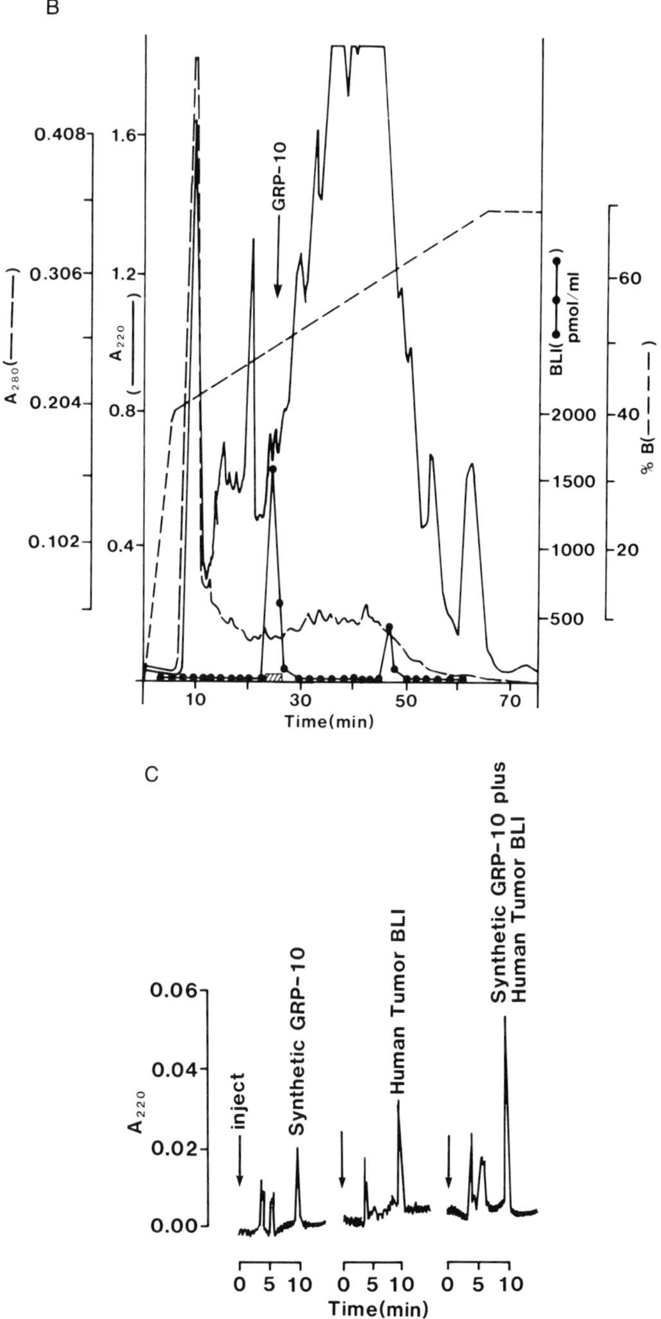
B
0.408
0.306
0.204
0.102
A_{280}(– – –)
1.6
1.2
0.8
0.4
A_{220}(——)
GRP-10
BLI(●—●—●) pmol/ml
2000
1500
1000
500
60
40
20
% B(– – – –)
10
30
50
70
Time(min)
C
0.06
0.04
0.02
0.00
A_{220}
inject
Synthetic GRP-10
Human Tumor BLI
Synthetic GRP-10 plus Human Tumor BLI
0 5 10
0 5 10
0 5 10
Time(min)

FIG. 6B and C.

component during its isolation. On Vydac C_4, Vydac C_{18}, Dynamax C_8, and Waters alkylphenyl columns one absorbance peak was associated with the GRP immunoreactivity, but three peptides were identified by microsequence and mass spectral analysis.

As already shown for human tumor GRP a single peptide can elute over a very broad range of acetonitrile concentrations. Chromatographic characterization of gastrin-releasing peptides is done best by cochromatography of highly purified natural peptide with synthetic standards. In summary, reversed-phase HPLC is an ideal final step in the purification of gastrin-releasing peptides. Recovery in this step usually exceeds 80%, the final volume is small compared to other methods, and many HPLC buffers are volatile. Peptides are obtained under conditions suitable for microsequence analysis, mass spectral analysis, or bioassay.

Isolation Strategy for Gastrin-Releasing Peptides

In our isolation strategies for various gastrin-releasing peptides two considerations were employed. The first was to always keep the peptide from being adsorbed to surfaces (plastic or glass), omitting freezing steps and always working at a pH below 5. The second was to use as many different modes of separation as possible.

The initial extracts were concentrated rapidly and the products rapidly purified in order to minimize contact with proteolytic enzymes also present in the extracts. Crude, batch-wise ion-exchange chromatography or reversed-phase chromatography can concentrate several liters of extract quickly, but little resolution is obtained. Both of these methods are suitable for concentrating 1–5 liters of extract in preparation for gel permeation chromatography. Gel permeation chromatography was a good early step that separated gastrin-releasing peptides on the basis of size and quickly separated enzymes from gastrin-releasing peptide. It gave good yields, good resolution, and was suitable for large amounts of protein.

After separation of the various forms by size, high-resolution ion-exchange or reversed-phase HPLC steps proved to be ideal purification modes. High-performance ion-exchange chromatography separates peptides on the basis of charge, and it gives better resolution of gastrin-releasing peptides than low-pressure ion-exchange chromatography and is also much faster. Preparative columns are able to handle peptides from a few kilograms of tissue extracts that had been purified by gel permeation chromatography. Reversed-phase HPLC separates peptides based on their hydrophobicity. Two HPLC steps on different columns lead to complete purification. It is an ideal last step for GRP purification since it is rapid, gives good yields, and produces small volumes.

Translation of Multiple Preprogastrin-Releasing Peptide Forms

As discussed in the detection section, once the cDNA was determined for preprogastrin-releasing peptide, synthetic analogs could be made for nongastrin-releasing peptide regions of the deduced propeptide. This was done for form I GRP, and antibodies corresponding to the carboxyl terminus of this form were raised by Dr. Frank Cuttitta. This material was isolated from culture media from a small cell lung cancer line. A peptide corresponding the carboxyl terminus of form I was isolated based on its immunoreactivity, and it was partially sequenced (F. Cuttitta, unpublished observations). Another peptide was isolated from the culture media on the basis of a major absorbance peak eluting from the HPLC columns. Its structural characterization showed that it was the carboxyl terminus of from III. Thus, multiple forms of preprogastrin-releasing peptide are translated in this small cell tumor line.

Conclusions

The characterization of multiple gastrin-releasing peptide molecular variants, and gene-related peptides, has been achieved utilizing region-specific radioimmunoassays and high-resolution purification techniques. The posttranslational processing sites that yield these various forms are double basic residues and several other types of bonds. For example, processing occurs after second position prolines at the amino terminus, at single basic residues, and between aspartic and serine residues.

Coinjections of similar amounts of synthetic and natural peptide should be done to avoid false conclusions about elution positions caused by variability in pump speed, chart recorder speed, or gradient formation. The various forms of gastrin-releasing peptide are well separated in several different chromatography systems. A combination of more than one separation technique is suggested to validate any conclusions about molecular forms based on immunoreactivity coeluting with standard peptides.

Section V

Use of Chemical Probes

[47] Neuropeptide Gene Transcription in Central and Peripheral Nervous System Tissue by Nuclear Run-On Assay

By JEFFREY D. WHITE and EDMUND F. LAGAMMA

Introduction

Rapid progress in molecular biological techniques has enabled the study of gene expression in a wide variety of neuroendocrine tissues. Through the use of dot blot,[1] Northern gel,[2] and *in situ*[3] hybridization techniques, the mechanisms of accumulation or depletion of mRNA coding for specific neuroendocrine precursor proteins can now be elucidated. Similarly, insights into the synthesis, processing, and secretion of neuroendocrine hormones has been advanced through the use of high-performance liquid chromatography.[4] Presently, it has become increasingly evident that a complete molecular investigation of neuroendocrine regulation must also involve an examination of neuroendocrine mRNAs at the level of gene transcription. For example, with respect to the production of the mRNA coding for the cellular protooncogene c-*myc*, several laboratories have demonstrated that some cells maintain a constant level of c-*myc* gene transcription but regulate the level of c-*myc* mRNA by controlling c-*myc* mRNA degradation.[5] Similarly, the level of c-*fos* mRNA has been shown to be regulated at the stages of both transcription and mRNA degradation.[6] Thus, it can no longer be assumed in neuroendocrine systems that increased mRNA content is the direct consequence of increased gene transcription.

There are two general methods for determining transcription of a specific mRNA. In both methods, transcriptionally active nuclei are isolated from tissue, or cultured cells, incubated in the presence of labeled nucleo-

[1] B. A. White, T. Lufkin, G. M. Preston, and C. Bancroft, this series, Vol. 124, p. 269.

[2] J. D. White, K. D. Stewart, and J. F. McKelvy, this series, Vol. 124, p. 548.

[3] J. N. Wilcox, C. E. Gee, and J. L. Roberts, this series, Vol. 124, p. 510.

[4] J. F. McKelvy, J. E. Krause, and J. D. White, this series, Vol. 103, p. 511.

[5] M. Dean, R. A. Levine, W. Ran, M. S. Kindy, G. E. Sonenshein, and J. Campisi, *J. Biol. Chem.* **261,** 9161 (1986).

[6] R. Bravo, M. Neuberg, J. Burckhardt, J. Almendral, R. Wallich, and R. Muller, *Cell (Cambridge, Mass.)* **48,** 251 (1987).

side triphosphate, and newly synthesized RNA is then isolated. Both methods involve the use of a complementary nucleotide sequence to remove selectively the specific labeled mRNA of interest from the pool of transcripts for quantitation. In the first method, hybridization is performed in solution, and the hybrids are isolated by filter binding[7]; in the second method filter hybridization is used. The advantages of the first method (solution hybridization) are that the hybridization can be driven to completion and the data can be expressed in terms of counts/minute bound versus counts/minute added (parts per million) to arrive at an absolute number for quantitation. The primary disadvantages to this method are that the gene under study must be expressed at a level sufficient to permit accurate liquid scintillation spectrometry and appropriate stringency controls (see discussion of controls below) are difficult to ascertain. The methods for this type of analysis have been extensively described[7] and will not be discussed further here. In the second method (filter hybridization), the unlabeled nucleotide strand is bound to a filter, and hybridization is conducted by incubating the solution of labeled RNA with the filter. Quantitation is then performed by film autoradiography and densitometric analysis of the autoradiographic band. This method has the advantages of permitting the investigation of transcription of several genes simultaneously and the increased sensitivity possible with prolonged exposure of the X-ray film. Although bound counts can be cut from the filter and counted, the chief disadvantage is that quantitation must be expressed in relative densitometric terms. In addition, if hybridization does not proceed to completion absolute counts are not accurate; thus, inclusion of an internal standard, i.e., a gene whose transcription does not change in response to the stimulus, is necessary to permit comparison between stimuli. The filter hybridization method is described below with examples of its use in investigating neuropeptide gene transcription.

Methods

Reagents

Sucrose I: 0.32 *M* sucrose–0.1% Triton X-100 in 5 m*M* HEPES, 1 m*M* $MgCl_2$, 1 m*M* dithiothreitol (DTT), pH 7.6 (HMD buffer) (DTT should be added to the buffer immediately prior to use)

Sucrose II: 2.1 *M* sucrose in HMD

Sucrose III: 0.25 *M* sucrose in HMD

[7] M. E. Greenberg and E. Ziff, *Nature (London)* **311,** 433 (1984).

Cold wash buffer: 20 mM Tris, 20% glycerol, 140 mM KCl, 10 mM $MgCl_2$, 1 mM DTT, pH 7.9

Complete reaction buffer: 1× cold wash buffer containing 0.1 mg/ml creatine phosphokinase, 8.5 mM phosphocreatine, 2 mM each ATP, CTP, and GTP, 1 U/μl RNasin, and 250 μCi [^{32}P]UTP per reaction

High-salt buffer: 500 mM NaCl, 50 mM $MgCl_2$, 2 mM $CaCl_2$, 10 mM Tris, pH 7.4

Proteinase K buffer: 20 mM EDTA, 0.5% sodium dodecyl sulfate (SDS), 10 mM Tris, 50 μg/ml proteinase K, pH 8

Sephadex G-50 column buffer: 0.1% SDS, 50 mM NaCl, 1 mM EDTA, 10 mM Tris, pH 8

Sephadex G-50 slurry: 1 g/15 ml column buffer, autoclaved together

5 M NaCl

1 M NaOH

2 M HEPES

Hybridization buffer: 0.2% SDS, 50 mM TES (or MOPS), pH 7.4, 0.3 M NaCl, 10 mM EDTA, 0.02% polyvinylpyrrolidone, 0.02% Ficoll, 0.1% sodium pyrophosphate, 0.1 mg/ml yeast tRNA

Prehybridization buffer: as above with 0.2% polyvinylpyrrolidone and 0.2% Ficoll

Isolation of Nuclei

It is important that nuclei be prepared with minimal damage to maintain them in the transcriptionally active state. Procedures will vary slightly from tissue to tissue and must be derived empirically.[8] Moreover, care must be taken to avoid disruption of lysosomes and the release of ribonucleases. In these and all subsequent steps it should be remembered that the procedure involves the isolation of RNA and therefore all appropriate precautions must be taken to avoid RNase contamination of the sample and of the buffers used, e.g., use of sterile plasticware or baked glassware, continual wearing of gloves, preparation of buffers in RNase-free containers using RNase-free salts and RNase-free water. If buffer solutions are treated with diethyl pyrocarbonate (DEPC) to inactivate RNase, *all* traces of the DEPC *must* be removed from the solutions to prevent inactivation of RNA polymerase, e.g., autoclave buffers for 2 cycles.

For this assay we have successfully used between 10^6 and 10^8 cells or about 1–10 mg of tissue. The tissue is homogenized in 1 ml sucrose I

[8] B. S. McEwen and R. E. Zigmond, *Res. Methods Neurochem.* **1,** 139 (1978).

buffer in a dounce homogenizer (type B pestle) with 10 strokes by hand. If necessary, the tissue can be allowed to swell in the buffer for up to 30 min prior to homogenization. Two milliliters of sucrose II is added, and the mixture is layered over 2 ml sucrose II in a 5-ml ultracentrifuge tube. The homogenate is centrifuged at 31,000 *g* for 45 min at 4° in an SW 50.1 Beckman ultracentrifuge rotor. The nuclei are recovered as the pellet, resuspended in 0.5 ml sucrose III, and recentrifuged at 2500 *g* for 5 min at 4°. The nuclei are resuspended in 1.0 ml cold wash buffer, and the yield is estimated either using a hemocytometer or by phase-contrast microscopy with counterstaining with 0.1% cresyl violet. The nuclei are again washed by centrifuging at 2500 *g*. The function of the washes is to remove residual Triton X-100, which was added to aid cell lysis and strip away outer nuclear membrane.[8] Attempts to homogenize tissue without Triton X-100 should be made as these may yield higher incorporated counts. As an alternative, "crude" nuclei can be isolated by homogenization in sucrose I, washing twice in cold wash buffer, and pelleting at 1000 *g* in a variable speed microcentrifuge. Following the final wash the nuclei are resuspended in 100 μl of complete reaction buffer. In our hands, concentrations of nuclei of about 10^7/100 μl reaction volume have worked well. If necessary, it is possible to freeze the nuclei in 2× cold wash buffer following the pelleting in sucrose III; however, the resulting yield of transcribed RNA will decline slightly compared to freshly prepared nuclei.

Transcription and RNA Isolation

When the nuclei have been resuspended in complete reaction buffer, they are incubated at 30° for 30 min with gentle swirling every 10 min. The salt conditions of the assay mix enhance RNA polymerase II transcripts while suppressing polymerase I and polymerase III activity.[9] The reaction is terminated by centrifuging the mixture at 2500 *g* for 5 min and resuspending the pellet in 250 μl high-salt buffer containing 55 U DNase I. The sample is briefly vortexed and incubated at 37° for 30 min. Thirty-six microliters of proteinase K buffer is added, to digest protein and inactivate the DNase, with further incubation at 37° for 30 min. For highest purity, the sample can then be transferred to a 1.5-ml microcentrifuge tube and extracted with an equal volume of phenol/chloroform[10] (optional). The aqueous and organic phases are separated with a 5–10 min centrifugation step in a microcentrifuge. The aqueous (upper) phase is recovered and loaded onto the Sephadex G-50 column.

[9] R. G. Roeder and W. J. Rutter, *Proc. Natl. Acad. Sci. U.S.A.* **65,** 675 (1970).

[10] T. Maniatis, E. F. Fritsch, and J. Sambrook, "Molecular Cloning: A Laboratory Manual." Cold Spring Harbor Laboratory, Cold Spring Harbor, New York, 1982.

For the column, we have found it convenient to use a 10-ml plastic pipet cut at the 4-ml mark and filled with Sephadex G-50 slurry. The aqueous phase from the phenol/chloroform extraction is loaded onto the column and eluted with column buffer, collecting 200-μl fractions. The early eluting peak (at the void volume) corresponds to the labeled RNA and the "hotter," late eluting peak represents unincorporated tracer. The labeled RNA is precipitated by adding sufficient 5 *M* NaCl to bring the final NaCl concentration to 0.2 *M* and adding carrier tRNA and 2-propanol. For example, to 600 μl of pooled fractions, add 20 μl 5 *M* NaCl, 10 μg carrier tRNA, and 622 μl 2-propanol. The samples are then stored at $-20°$ for 1 hr and pelleted in a microcentrifuge at 4°, after which the supernatant is removed and the pellet gently washed with 70% ethanol to remove salt.

Preparation of Membrane and Hybridization

Following the ethanol wash, the sample is vacuum dried and resuspended in a small volume of RNase-free water, e.g., 50 μl. To facilitate the formation of stable hybrids, the sample is fragmented with 0.2 *M* NaOH, i.e., add 12.5 μl of 1 *M* NaOH to the 50 μl of water. The sample is mixed and incubated on ice for 15 min. HEPES (2 *M*) is added to a final concentration of 0.33 *M* (12.5 μl added to the mix) to neutralize the NaOH, and then the sample can be added to the hybridization mixture.

In parallel with the preparation of the RNA transcripts, the membrane for hybridization can be prepared. For each probe used it is advisable first to establish a dose response for the plasmid used (see discussion of controls below and Fig. 1). We have been successful in using from 1 to 5 μg plasmid DNA/sample using a slot blot apparatus. Larger amounts of DNA may be required when using a dot blot apparatus. The DNA samples are made 0.1 *N* in NaOH, boiled for 5 min, then rapidly chilled on ice, to denature the strands, and finally made 2 *M* in NaCl.

The nitrocellulose, or nylon, membrane is loaded into the manifold and washed once with RNase-free water followed by two washes with 6× SSC.[10] The plasmid DNA samples are loaded onto the membrane then washed twice with 6× SSC. The membrane is blotted dry then vacuum baked at 75–80° for 2 hr to bind the DNA to the membrane. Prior to hybridization, the membrane is prehybridized by first wetting in 10× SSC then in prehybridization buffer in a Seal-a-Meal bag with 0.5–1 ml buffer/slot. Prehybridization is for 4–24 hr at the hybridization temperature.

Following the prehybridization, the membrane is carefully removed from the bag and placed in a fresh bag with an equal volume of hybridization buffer containing 1–2 × 10^6 cpm of transcribed RNA/ml of hybridiza-

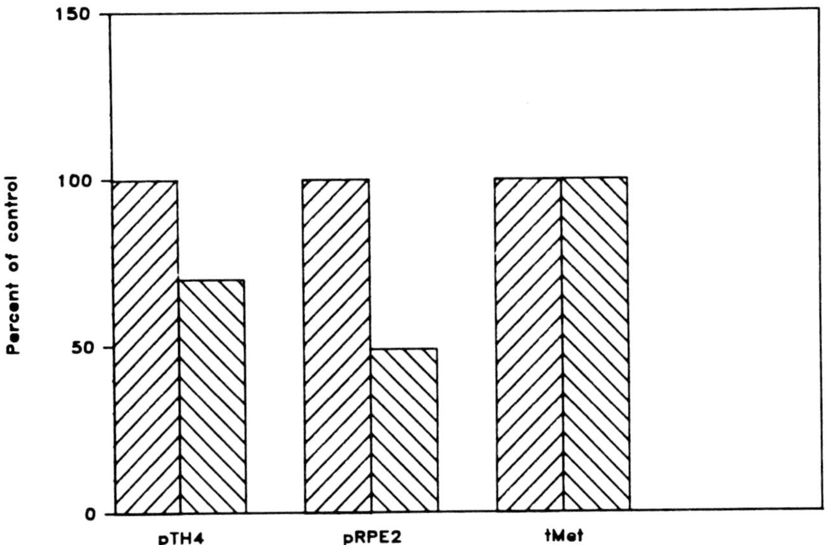

FIG. 1. Histogram plot of densitometric scan data from "run-on" results of 10 control (▨) adrenal medullas or 10 medullas assayed in the presence of the RNA polymerase II inhibitor, α-amanitin (▧) (1 μg/ml). Data are normalized to the polymerase III transcript tMet.

tion solution. We generally allow membranes to hybridize for 40 hr with shaking. The optimal temperature for hybridization and washing must be determined for each probe. We have successfully used temperatures between 42 and 65°. Following hybridization, the membrane is washed in 3 changes of 2× SSC for 30 min each at the hybridization temperature. If nonspecific hybridization or high overall background is a problem, the membranes can be treated with pancreatic RNase A (0.45 U/ml) in 2× SSC for 1 hr at 37° and washed with 2× SSC twice at room temperature. The membrane is then apposed to X-ray film with an intensifying screen and the film exposed. Generally, for rare neuroendocrine mRNA transcripts, exposures of 2–10 days will be required. Providing the exposure is in the linear range of both the film and the densitometer, the autoradiographic bands can then be quantified by densitometry.

To date, we have been successful in detecting neuropeptide mRNA transcripts using the entire insert-containing plasmid on a slot blot apparatus, although purified insert may be used as well. To reduce possible background, however, it may be advisable to produce antisense RNA using a vector containing the SP6 or T7 bacteriophage promoter. In this case the resulting RNA–RNA hybrids would be more stable than the

DNA–RNA hybrids, allowing more stringent hybridization and washing conditions to be used. Naturally, in this case the plasmid denaturation steps prior to blotting would be unnecessary.

Requisite Control Experiments

Following successful hybridization, several control experiments are recommended to demonstrate the specificity of the transcription reaction and hybridization product. Although assay conditions favor polymerase II transcription, selective inhibition of this activity can be demonstrated using a polymerase II inhibitor such as α-amanitin (0.1–10 μg/ml) in the incubation mix. In addition, data can then be normalized to genes transcribed by polymerase I or polymerase III, using appropriate clones (e.g., the gene for methionine tRNA) to verify preferential inhibition (see example in Fig. 1).

Linearity of incorporated radioactivity can be ascertained by extending the incubation time of a large homogenate from 15 to 30, 45, then 60 min. Aliquots of labeled RNA can then be hybridized to equal amounts of immobilized cloned gene fragments for comparison. Our experience indicates incubation for up to 30 min yields maximal incorporation, failing to increase thereafter. However, since individual tissue variations exist, optimal incubation times must be determined for each new system. Nevertheless, long incubations are not recommended since it is not known whether reinitiation of transcription can occur, leading to false estimates of transcriptional activity. Including heparin in the reaction mixture may limit this effect.

In addition to linearity of incorporation, increasing amounts of plasmid (or insert) must be bound to filters to determine optimal hybridization conditions and maximal sensitivity. Thus, it must be determined that the hybridization conditions used provide sufficient excess plasmid DNA to permit maximal binding of transcribed RNA such that comparisons between samples will be valid.

Consideration must also be given to the issue of binding specificity. Similar to any hybridization protocol, nonspecific binding is limited by hybridization conditions and washing procedures. Routinely, nonspecific binding to the plasmid and insert is determined by incubating the vector plasmid without an insert in an additional slot. As an additional control, it may be advisable to test hybridization from a tissue sample that is not known to transcribe the gene of interest; for a neuroendocrine gene product such a control sample might be a fibroblast cell line. If high backgrounds are a problem, the first step is to increase stringency for washing membranes (e.g., 1% SDS, higher temperatures, lower salt). If labeled

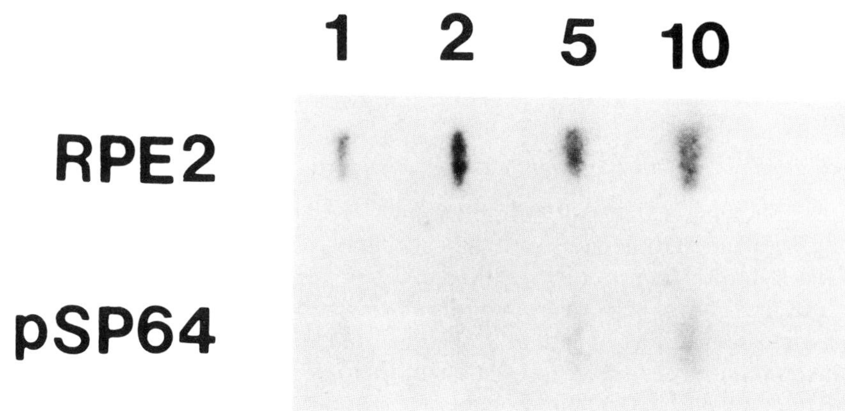

FIG. 2. Autoradiogram of preproenkephalin mRNA transcription from rat corpus striatum hybridized with increasing amounts of plasmid DNA. The numbers above the lanes represent the amounts (μg) of plasmid blotted. RPE2, cDNA clone for preproenkephalin; pSP64, plasmid vector.

RNA is still detected, background signal can be further limited by including increased amounts of tRNA in the prehybridization solution or adding poly(C) or poly(dC) (100–200 μg/ml).

Detection of Preproenkephalin mRNA Transcripts in Rat Brain

As an example of detecting neuropeptide precursor protein mRNA transcripts in the central nervous system, preproenkephalin mRNA transcription was assayed in the corpus striatum, the richest source of enkephalin in the rat central nervous system. For this study to test the linearity of signal versus plasmid concentration, the paired striata from a single rat brain were used in the reaction mixture, and 1.5×10^7 cpm was used for hybridization. One, two, five, and ten micrograms of plasmid insert DNA were blotted onto the nitrocellulose membrane. As can be seen in Fig. 2, the hybridization signal reached saturation by 5 μg.

Detection of Preproenkephalin and Tyrosine Monooxygenase Transcripts in Cultured Adrenal Medullas

In an earlier study,[11] we demonstrated that explanted rat adrenal medullas dramatically increase their content of preproenkephalin mRNA with time in culture and that this increase can be modulated by depolarization. Moreover, the activity of the catecholamine biosynthetic enzyme

[11] E. F. LaGamma, J. D. White, J. E. Adler, J. E. Krause, J. F. McKelvy, and I. B. Black, *Proc. Natl. Acad. Sci. U.S.A.* **82,** 8252 (1985).

tyrosine monooxygenase did not change under these same culture conditions. Therefore it became of interest to determine if the transcription of these two genes was similarly regulated by these cells. Explanation of medullas to culture resulted in a 2-fold increase in preproenkephalin mRNA transcription, paralleling the previously described rise in cellular

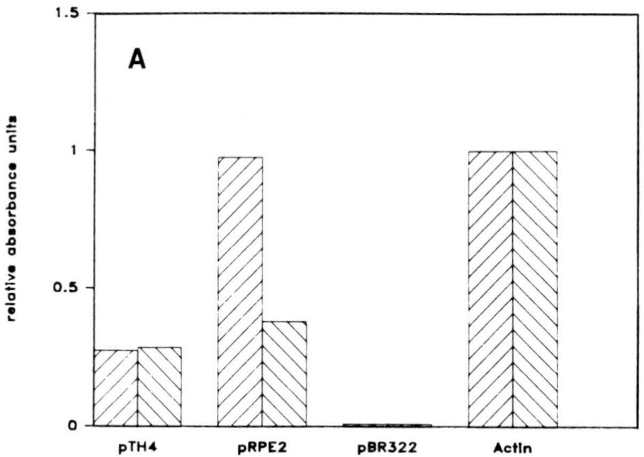

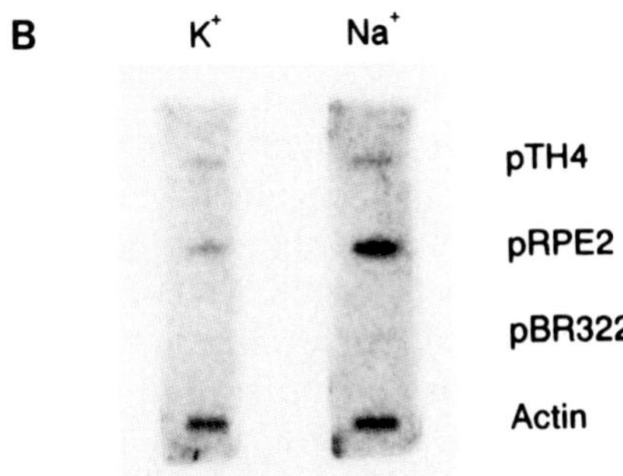

FIG. 3. (A) Densitometric scan data of an autoradiogram of a run-on transcriptional analysis of adrenal medullas in the absence or presence of depolarizing concentrations of potassium chloride. (▨) Results from sodium-treated explants. (▧) Results from potassium-treated 2-day explant cultures. Results were obtained by adding equal amounts of incorporated radioactivity to each blot then normalized to data from analysis of β-actin transcription. (B) Autoradiogram of the hybridizing material from potassium- and sodium-treated cultures. pTH4, cDNA clone for tyrosine monooxygenase mRNA; pRPE2, cDNA clone for preproenkephalin mRNA; pBR322, plasmid vector; Actin, cDNA clone for β-actin mRNA.

preproenkephalin mRNA content and in enkephalin peptide; tyrosine monooxygenase transcription was unchanged (Fig. 3). Additionally, depolarizing stimuli, known to decrease enkephalin peptide levels in rat medullas, elicited a 2.5-fold decrease in enkephalin gene transcription. These data support the conclusion that depolarizing stimuli alter neuroendocrine gene activity.[12]

Acknowledgments

This work was supported by National Institutes of Health Grant MH 42074 and a fellowship from the Aaron Diamond Foundation to J. D. W. and by a Grant-in-Aid from the American Heart Association and a Basil O'Conner Award from the March of Dimes Birth Defects Foundation to E. F. L.

[12] E. F. LaGamma and I. B. Black, submitted for publication (1988).

[48] Assaying the Reporter Gene Chloramphenicol Acetyltransferase

By David W. Crabb, Carolyn D. Minth, and Jack E. Dixon

Introduction

Chloramphenicol acetyltransferase (CAT, acetyl-CoA:chloramphenicol O^3-acetyltransferase, EC 2.3.1.28) is perhaps the most widely used marker gene for studies on the regulation of gene expression. Because this enzyme is absent from all eukaryotic cell lines, there is no background activity in transfected cells.[1] The enzyme activity can be readily measured by a sensitive radiochemical method. We have found, however, that extracts of certain cells contain several enzymatic activities which may interfere with the CAT assay.[2] If this interference is not recognized, the results of the assay may suggest that the cells have not been efficiently transfected. Moreover, the degree of interference with the assay varies between extracts of different cell lines. Differences in CAT activity observed after transfecting various cells with the same promoter–CAT gene constructs may suggest that the promoter under study exhibits tissue-

[1] C. M. Gorman, G. T. Merlino, M. C. Willingham, I. Pastan, and B. H. Howard, *Proc. Natl. Acad. Sci. U.S.A.* **79,** 6777 (1982).

[2] D. W. Crabb and J. E. Dixon, *Anal. Biochem.* **163,** 88 (1987).

specific expression. We have examined, therefore, the causes of this interference and surveyed a variety of cell types for the presence of interfering activities. We also describe a simple method which reliably eliminates this problem.

Methods and Materials

Cell Lines and Culture Conditions

H4IIE, H4IIE C3 (rat hepatomas), HepG2 (human hepatoblastoma), HeLa, COS, CHO-K1, and BSC-40 cell lines were obtained from the American Type Culture Collection (Rockville, MD). CA-77 cells (rat medullary thyroid carcinoma) were a gift from Dr. Bernard Roos (American Lake VA Medical Center, Tacoma, WA). PC-12 cells were obtained from Dr. G. E. Landreth (Department of Neurology, Medical University of South Carolina, Charleston, SC). LAN-5 cells were from Dr. Robert C. Seeger (Department of Pediatrics, UCLA School of Medicine, Los Angeles, CA). PLC/PRF/5, HuH6, HA22T, and Hep3B cells (all derived from human hepatomas) were kindly provided by Dr. Chao-Hung Lee (Department of Pathology, Indiana University School of Medicine, Indianapolis, IN). The MCF-7 cells were provided by Dr. Sam Brooks of the Michigan Cancer Foundation. These cells were generally grown in Dulbecco's modified Eagle's medium supplemented with 10% heat-inactivated fetal bovine serum with 10 μg/ml gentamicin. The hepatoma cells were cultured in Waymouth's MB 752/1 medium supplemented with 10% fetal bovine serum plus ascorbate, linoleic acid, tocopherol, trace metals, streptomycin, and penicillin.[3] Cell extracts were prepared from confluent monolayers from 100-mm culture dishes.

The following were provided by members of the faculty of Purdue University (West Lafayette, IN): *Xenopus laevis* oocytes, from Dr. D. Smith, Department of Biology; WE-4 cells (rat medullary thyroid carcinoma), from Randy Haun, Department of Biochemistry; 10T1/2 cells, from Dr. E. J. Taparowsky, Department of Biology; yeast (*Saccharomyces cerevisiae*), from Dr. G. Kohlhaw, Department of Biochemistry; and corn (Black Mexican sweet corn B73 × A188) and rice (IR 54) cells, from Dr. Tom Hodges, Department of Botany and Plant Pathology. Two insect cell lines, *Trichoplusia ni* (TN-368) and *Spodoptera frugiperda* (IPLB-SF-21AE), were the gift of Dr. James Maruniak (University of Florida, Gainesville, FL).

[3] D. W. Crabb and J. E. Roepke, *In Vitro Cell. Dev. Biol.* **23,** 303 (1987).

Preparation of Cell Extracts

The cell monolayer is scraped with a rubber policeman from the plates into medium and centrifuged at 2000 *g* for 4 min at room temperature. The cell pellet is resuspended in 0.25 *M* Tris, pH 8, and recentrifuged. The supernatant is decanted, and the cell pellet is resuspended in 100 μl of the Tris buffer. The suspension is then sonicated using a Branson Sonifier set at 50% maximum power for three 5-sec bursts. In order to break the corn and rice cells, the suspension is homogenized with 10–15 passes of a glass Dounce homogenizer. The mixture is then sonicated for 20 sec. Yeast cells are broken by 6 cycles of vortexing with glass beads for 15 sec followed by cooling on ice. To heat-treat the extracts, 1 μl of 0.5 *M* EDTA is added, and the suspension is then incubated in a water bath at 60° (unless otherwise noted) for 10 min. The suspension is then centrifuged at full speed in a tabletop Eppendorf microfuge (14,000 *g*) for 15 min at 4°. One hundred microliters of the supernatants is tested for ability to interfere with the CAT assay.

CAT Assay

The CAT assay is modified from that originally described by Gorman *et al.*[4] The assay mixture contains 100 μl of cell extract, 1 m*M* acetyl-CoA (Sigma Chemical Co., St. Louis, MO), and 0.2 μCi of [^{14}C]chloramphenicol (Du Pont–New England Nuclear, Boston, MA). The assay is carried out in a 1.5-ml Eppendorf tube. One unit of pure CAT (PL-Pharmacia, Piscataway, NJ) is mixed with the extract, and the assay mixture is incubated at 37° for 2 hr. The tubes are then transferred to ice, and the reaction products are extracted with 1 ml of cold ethyl acetate. The solution is dried under a stream of air, resuspended in 50 μl of ethyl acetate, and spotted on a silica thin-layer plate (Eastman Kodak Co., Rochester, NY). The chromatogram is developed in 95% chloroform–5% methanol and autoradiographed. The conversion of chloramphenicol to acetylated derivatives is quantified by cutting out the radioactive spots and counting them in a liquid scintillation counter. CAT activity is expressed as the percentage of the parent compound converted to mono- and diacetylated derivatives. The degree of interference with the reaction by untreated extracts was calculated as the percent reduction of CAT activity in the untreated extract compared to the heat-treated extract.

Assay of Acetyl-CoA Hydrolysis by Cell Extracts

The cell extracts are incubated with 1 m*M* acetyl-CoA at 37°. Aliquots are removed at intervals and added to a solution of 5,5′-dithiobis(2-nitro-

[4] C. Gorman, L. Moffat, and B. Howard, *Mol. Cell. Biol.* **2,** 1044 (1982).

benzoic acid) (1 mg/ml) in 50 mM sodium phosphate (pH 7.5)–methanol (1 : 1, v/v). The rate of appearance of free CoA-SH was calculated from the increase in absorbance at 412 nm,[5] using a molar extinction coefficient for the disulfide adduct of 13.6 M^{-1} cm^{-1}.

Hepatoma Cell Extracts Interfering with the CAT Assay

We initially found that H4IIE C3 hepatoma cells appeared to be difficult to transfect with pRSV-CAT. We subsequently found that we were unable to assay pure CAT added directly to the cell extract (Fig. 1). This interference may be due to at least three factors, described in greater detail below: deacetylation of the newly synthesized acetylchloramphenicol, proteolytic degradation of the CAT enzyme, and hydrolysis of acetyl-CoA in the assay mixture.

The deacetylation reaction was detected when we allowed pure CAT to acetylate [^{14}C]chloramphenicol in buffer, then added hepatoma cell extract and continued the incubation for an additional 2 hr. Nearly all of the acetylated chloramphenicol was converted to the parent compound (Fig. 1). This strongly suggests that the cell extract contains a carboxylesterase activity capable of removing the 1- and 3-acetoxy groups from acetylchloramphenicol. Thus, in the hepatoma extracts, a futile cycle is established during the assay, with the labeled chloramphenicol repeatedly acetylated and deacetylated. This may also deplete the acetyl-CoA in the mixture.

Proteases present in crude tissue or cell extracts are a commonly encountered problem in enzyme purification schemes. We therefore examined whether the addition of protease inhibitors to the cell extract improved the CAT assay. None of the compounds tested inhibited pure CAT assayed in buffer alone (Fig. 2). The addition of EDTA, a mixture of serine protease inhibitors, or bacitracin each increased the apparent activity of added CAT in hepatoma extracts. The addition of pepstatin had no effect on the interference (Fig. 2). These data suggest that serine proteases or metalloproteases may degrade CAT during the 2-hr assay. However, it is also possible that EDTA, serine protease inhibitors, or bacitracin inhibit the deacetylase activity or the thioesterase (see below). In particular, it would not be surprising if the phenylmethylsulfonyl fluoride included in the mixture of protease inhibitors also inhibited a serine esterase.

The third enzymatic activity which could interfere with the CAT assay is a thioesterase which hydrolyzes acetyl-CoA. This was identified by

[5] W. V. Shaw, this series, Vol. XLIII, p. 737.

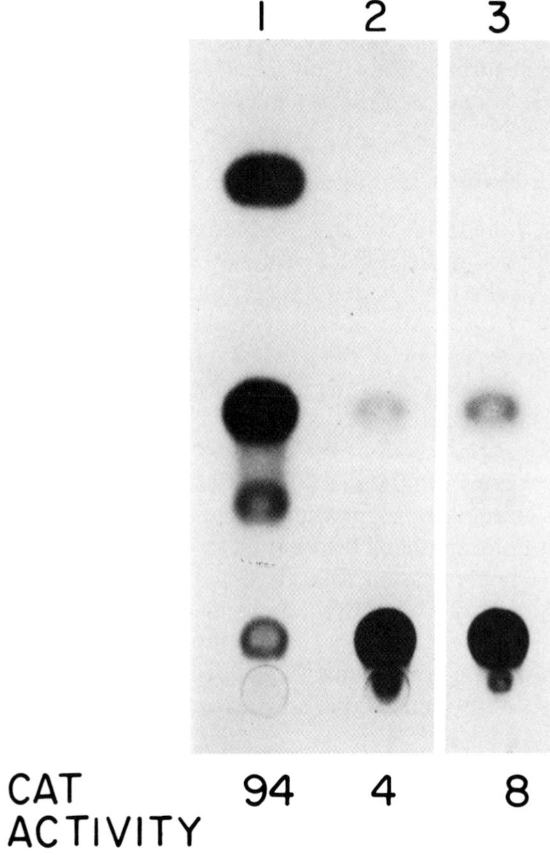

FIG. 1. Interference with the CAT assay by hepatoma cell extracts. One unit of pure CAT was assayed for 2 hr in buffer (lane 1) or in extract prepared from H4IIE C3 hepatoma cells (lane 2). In lane 3, the CAT was allowed to acetylate chloramphenicol for 2 hr, then hepatoma extract was added and the incubation continued for an additional 2 hr. CAT activity is expressed as the percentage of total radioactivity recovered from the thin-layer chromatogram as mono- and diacetylated chloramphenicol.

following the formation of sulfhydryl groups (presumably CoA-SH) which react with 5,5′-dithiobis(2-nitrobenzoic acid) (Fig. 3). A much slower rate of hydrolysis is observed in heat-treated or boiled extracts (Fig. 3A), suggesting that the hydrolysis of acetyl-CoA is enzymatically catalyzed. This thioesterase is present in extracts of several other cell lines (Fig. 3B). The Michaelis constant of CAT for acetyl-CoA is about 60 μM.[5] Therefore, even at the end of the 2-hr assay, the acetyl-CoA concentration was

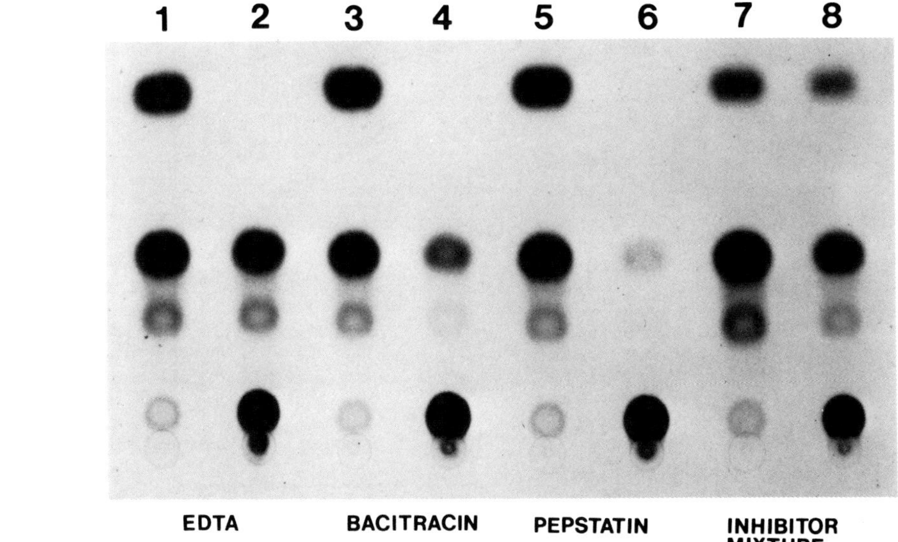

FIG. 2. Effects of protease inhibitors on CAT activity in hepatoma extracts. One unit of CAT was assayed in buffer (lanes 1, 3, 5, and 7) or in hepatoma extract (lanes 2, 4, 6, and 8) with the noted compounds. The concentrations used were as follows: EDTA, 5 mM; bacitracin, 100 μM; and pepstatin, 10 μg/ml. The mixture of protease inhibitors included (final concentrations) the following: EDTA, 1 mM; phenylmethylsulfonyl fluoride, 1 mM; tosyl-L-lysine chloromethyl ketone, 10 μg/ml; leupeptin, 10 μg/ml; and soybean trypsin inhibitor, 50 μg/ml. CAT activity is expressed as described in Fig. 1. (From Crabb and Dixon,[2] with permission.)

5- to 10-fold above the K_m for this substrate in our assay. Addition of acetyl-CoA to the assay at 30-min intervals did not prevent the interference (not shown). We therefore believe that the thioesterase activity is not a major contributor to the interference. However, other CAT assays utilize considerably lower concentrations of acetyl-CoA,[6] and under these conditions the thioesterase activity may be more important.

EDTA plus Heat Treatment to Abolish Interference with the CAT Assay

The initial experiments suggested several mechanisms by which the hepatoma cell extracts might interfere with the CAT assay. We then

[6] M. J. Sleigh, *Anal. Biochem.* **156,** 251 (1986).

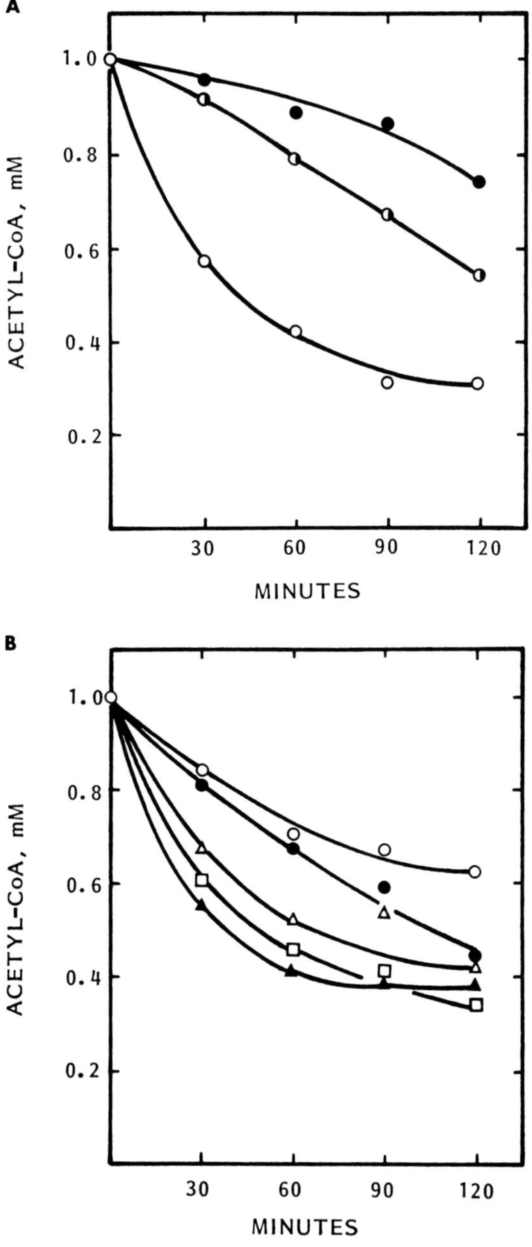

FIG. 3. Hydrolysis of acetyl-CoA by cell extracts. (A) Hepatoma extracts were incubated with 1 mM acetyl-CoA at 37°. The hydrolysis of acetyl-CoA was determined by measuring the appearance of sulfhydryl groups which reacted with 5,5′-dithiobis(2-nitrobenzoic acid). ○, Hepatoma extract; ◑, heat-treated extract (60° for 10 min); ●, boiled extract. (B) The rate of acetyl-CoA hydrolysis was determined in extracts of the following: ○, HeLa; ●, BSC-40; □, COS; △, PC-12; and ▲, CA-77 cells. (From Crabb and Dixon,[2] with permission.)

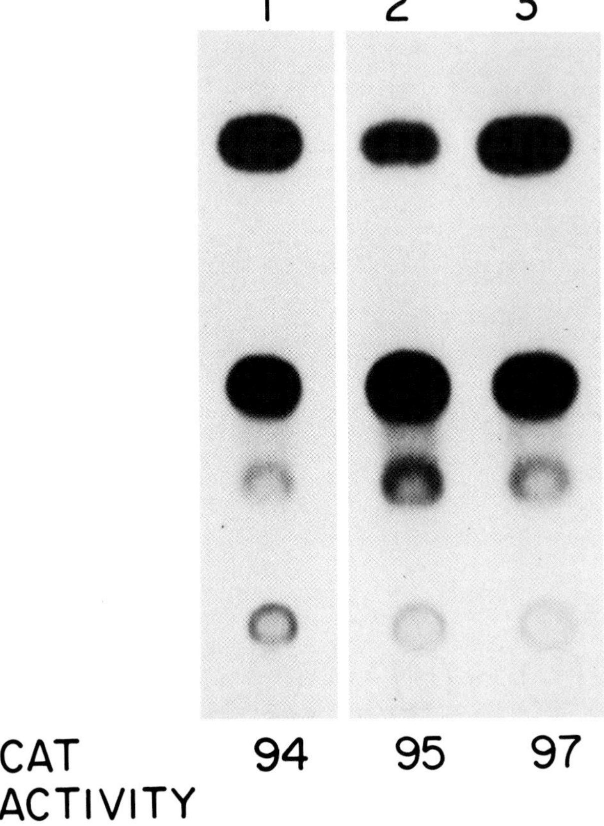

FIG. 4. CAT assays in heat-treated hepatoma extracts. One unit of CAT was assayed in boiled hepatoma extract (lane 1) or heat-treated extract (lane 2). CAT activity was stable to heating in buffer at 60° for 10 min (lane 3). CAT activity is expressed as in Fig. 1.

sought a simple way to prevent this interference. CAT is known to be rather heat-stable. Earlier purification schemes for CAT utilized a 60° heat-inactivation step to precipitate other proteins, leaving CAT in the supernatant.[5] We therefore tried to inactivate the proteases, deacetylase, and thioesterase by heat treatment (Fig. 4). Boiling the extract or addition of EDTA to 5 m*M* final concentration followed by a 10-min incubation of the extract at 60° permitted virtually quantitative recovery of CAT activity.

The effect of incubating the hepatoma extract, to which CAT had been added, at different temperatures is shown in Fig. 5. Heating the extract at 50–75° allowed us to assay a higher activity of added CAT than was possible in untreated extracts. The highest CAT activity was observed in

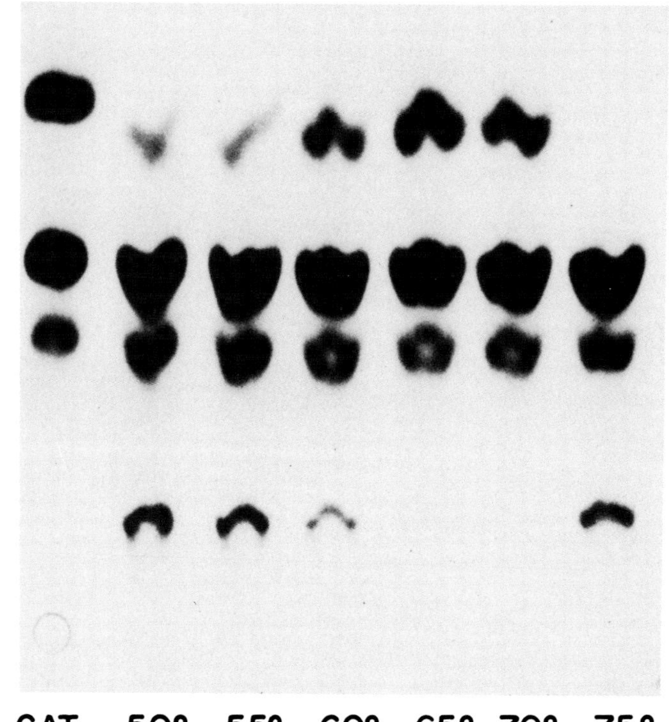

FIG. 5. Heat stability of CAT in hepatoma extracts. One unit of CAT was added to 100 μl of hepatoma extract containing 5 m*M* EDTA. The extract was then heated for 10 min at the noted temperatures, and CAT activity was assayed at 37° for 2 hr. CAT indicates enzyme activity assayed in buffer alone without heat treatment.

extracts which had been treated at 65–70°. We surmise that at lower temperatures some of the interfering activities are not inactivated, and that at temperatures over 70° CAT itself is inactivated. It is obvious that the temperature for the heat treatment is not critical, but it should be carefully controlled to optimize reproducibility of the assay.

Interfering Activity Observed with Many Different Cell Lines

We then undertook a survey of different cell lines to see how commonly the interference was encountered (Fig. 6 and Table I). Both rat hepatomas (H4IIE and H4IIE C3) and three of five human hepatoma cell lines exhibited substantial interference with the assay. In addition, there was interference by extracts of several neuroendocrine cell lines (PC-12, LAN-5, WE-4, and CA-77 cells), several other mammalian cell lines

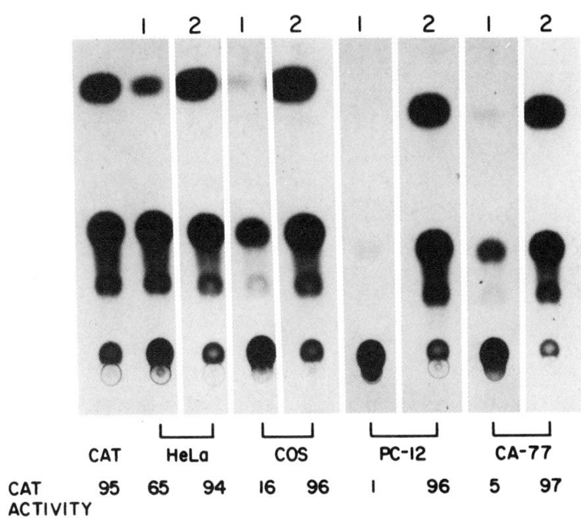

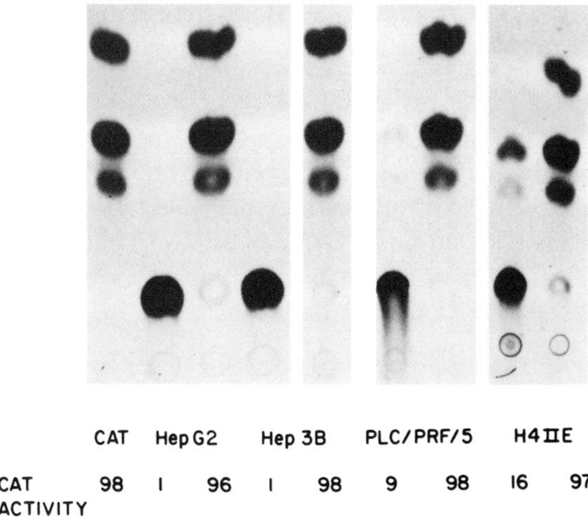

FIG. 6. Interference with the CAT assay in various cell lines. One unit of CAT was added to extracts of the noted cell lines and assayed for CAT activity (lanes 1), or extracts were EDTA- and heat-treated before the assay (lanes 2). CAT activity is expressed as in Fig. 1, and the lanes marked CAT indicate enzyme activity assayed in buffer alone without heat treatment.

TABLE I
INTERFERENCE WITH THE CAT ASSAY BY EXTRACTS OF DIFFERENT CELLS[a]

Cell type	Interference	Cell type	Interference
Neuroendocrine		Miscellaneous	
LAN-5	++	mammalian cells	
WE-4	+	HeLa	+
CA-77	+++	MCF-7	++
PC-12	+++	CHO-K1	++
		BSC-40	−
Hepatoma		COS	+++
H4IIE	+++	10T1/2	−
H4IIE C3	+++		
HepG2	+++	Amphibian cells	
Hep3B	+++	*Xenopus*	−
HuH6	−	oocytes	
HA22T	−		
PLC/PRF/5	+++	Insect cells	
		IPLB-SF-21AE	−
Plant cells		TN-368	−
Rice	+		
Corn	+	Yeast	−

[a] Extracts of the cells were prepared as described in the *Methods and Materials* section. The activity of 1 unit of pure CAT (determined as in Fig. 1) was measured in untreated or in EDTA- and heat-treated extracts. The degree of interference, calculated as described in *Methods and Materials*, is indicated as follows: −, less than 10%; +, 10–40%; ++, 40–70%; and +++, greater than 70%.

(HeLa, COS, MCF-7, and CHO-K1), and two plant cell extracts. In each case, the interfering activity could be abolished by EDTA plus heat treatment of the extract as described above. There was little interference in extracts of the BSC-40 or 10T1/2 cells, two of the human hepatoma cells (HA22T or HuH6), *Xenopus* oocytes, yeast, or two insect cell lines (Table I).

Summary

These experiments document the presence of enzymatic activities in extracts of commonly used cell lines which interfere with the determination of CAT activity. We suspect that the deacetylase activity is the most important, as the extract of the H4IIE C3 cells was capable of completely deacetylating the mono- and diacetylchloramphenicol formed during a 2-hr incubation of CAT with chloramphenicol and acetyl-CoA. The results of the inhibitor experiments are consistent with the presence of proteases which degrade CAT, or a serine carboxylesterase. The interference was also reduced by about half by EDTA; a metalloenzyme (either a protease or esterase) may therefore be involved.

This interference appears to be a common phenomenon. We have surveyed 23 different cell types for the presence of the interfering activity and found it in 15. The interference was particularly prominent in several neuroendocrine and hepatoma cells. We took advantage of the effect of EDTA and the heat stability of CAT to eliminate the interference. Addition of 5 m*M* EDTA and a 10-min incubation of the sonicated cell suspension at 60° prior to centrifugation abolished the interference in all cell lines tested. It is important to note that in order to reveal any CAT activity in some of the extracts (e.g., PC-12 or Hep3B, Fig. 6), it was necessary to run the CAT assay for 2 hr. The control assays were therefore run almost to completion, and were well beyond the linear range of the assay. Therefore, the small differences which we observed between the heat-treated and control samples in some instances (e.g., rice, corn, or HeLa cells) will be dramatically amplified when the CAT assay is performed under conditions in which only a small percentage of the substrate is converted to product.

After these studies had been performed, we found that others have also recommended heat treatment of the cell extract prior to CAT assay.[6–8] We concur with this recommendation. We suggest that EDTA plus heat treatment of the cell extract should be incorporated into all CAT assay protocols, unless it has been previously determined that extracts of the cells used do not interfere. Furthermore, the heat treatment step should be used whenever the activity of promoter–CAT constructs is compared among different cell lines, as is often done to define tissue-specific expression.

Acknowledgments

The authors wish to thank Ms. Ruth Ann Ross for her help with the cell cultures and Ms. Maggie Cox for her expert assistance in the preparation of the manuscript. Drs. O. Andrisani and R. Haun provided helpful suggestions. We are indebted to the many investigators who provided us with the different cells to test.

[7] M. Fromm, L. P. Taylor, and V. Walbot, *Proc. Natl. Acad. Sci. U.S.A.* **82,** 5824 (1985).

[8] L. G. Davis, M. D. Dibner, and J. F. Battey, "Basic Methods in Molecular Biology," p. 298. Elsevier, Amsterdam, 1986.

[49] *In Situ* Hybridization Histochemical Detection of Neuropeptide mRNA Using DNA and RNA Probes

By W. SCOTT YOUNG III

Introduction

In situ hybridization histochemistry (ISHH) can be used to detect the transcribed mRNAs coding for neuropeptides (or for enzymes or other proteins) in cells that express the corresponding genes. Tissue sections are incubated with labeled complementary DNA (cDNA) or cRNA probes which form hybrids with the cellular mRNA, and the location of labeled cells is then determined—usually by autoradiography for radioactively tagged probes. At present, there is no generally recognized best procedure as there are nearly as many approaches as laboratories that use ISHH. We have used two separate approaches to ISHH[1,2] that offer their own particular advantages and are relatively straightforward.

The first method employs cRNA probes that are synthesized using vectors that contain bacteriophage RNA polymerase promotors, as described by Melton and co-workers.[3] The lengths of the cRNA probes can be varied using the sites present in the cDNA insert that are recognized by restriction endonucleases. The RNA polymerase then produces single-stranded copies of cRNA of defined length and specific activity. The use of cRNA probes for ISHH was pioneered by Cox *et al.*[4] and modified by Harper *et al.*[5] This method allows one to immediately examine tissues for expression of a gene once its cDNA is isolated and placed into the "riboprobe" vector. Because full-length probes can be made, various splicing alternatives may be detected, and the gene's expression in other species may be studied by adjusting the conditions of the hybridization (i.e., stringency[4]). Length may be a drawback, as well, however, because the probe may also detect other genes' transcripts if the cDNA insert contains regions of sequence similarity to them.

The second approach employs oligodeoxyribonucleotide probes. Our probes are usually 48 bases long and offer several advantages over the

[1] R. E. Siegel and W. S. Young III, *Neuropeptides* **6,** 573 (1985).

[2] W. S. Young III, T. I. Bonner, and M. R. Brann, *Proc. Natl. Acad. Sci. U.S.A.* **83,** 9827 (1986).

[3] D. A. Melton, P. A. Krieg, M. R. Rebagliati, T. Maniatis, K. Zinn, and M. R. Green, *Nucleic Acids Res.* **12,** 7035 (1984).

[4] K. H. Cox, D. V. DeLeon, L. M. Angerer, and R. C. Angerer, *Dev. Biol.* **101,** 485 (1984).

[5] M. E. Harper, L. M. Marselle, R. C. Gallo, and F. Wong-Staal, *Proc. Natl. Acad. Sci. U.S.A.* **83,** 772 (1986).

cRNA probes. They can be synthesized and labeled to a defined specific activity once a sequence is published. These probes can be designed to distinguish between a gene's splicing alternatives or between related genes' mRNAs. They also readily penetrate processed tissue sections (see below) and yield good signal-to-noise ratios without resorting to nucleases (e.g., RNase for the cRNA probes) to reduce background. Currently, this approach with cDNA probes is our preferred method.

Experimental Procedures

Tissue Preparation

We find that freshly frozen tissues that are subsequently fixed display the greatest signal and lowest background. We freeze the tissues in powdered dry ice and store them at −70° (in tubes with water to prevent desiccation) until used. The tissues are then cut in a cryostat–microtome, and the sections (usually 12 μm) are thaw-mounted onto twice gelatin-coated slides. The slides are then warmed at 37° for 1–2 min to dry the sections, after which they are placed back in the freezer until hybridized.

Prior to hybridization the slides are warmed for 10 min at room temperature by placing them on aluminum foil. They are then placed in polyethylene Coplin jars (previously sterilized by soaking in 0.1% diethyl pyrocarbonate overnight and then autoclaved; Thomas Scientific) containing 4% formaldehyde (10% formalin) in phosphate-buffered saline (PBS) (9 g NaCl, 0.122 g KH_2PO_4, 0.815 g Na_2PO_4 per liter) for 5 min at room temperature. The sections are then rinsed twice with PBS and placed into 0.25% acetic anhydride[6] in 0.1 *M* triethanolamine hydrochloride–0.9% NaCl, pH 8, for 10 min at room temperature. Next, the sections are transferred through 70 (1 min), 80 (1 min), 95 (2 min), and 100% ethanol (1 min); 100% chloroform (5 min); and 100 (1 min) and 95% ethanol (1 min) before standing upright to air-dry. Tissues fixed by perfusion with 4% paraformaldehyde are processed similarly except the sections go directly into the acetic anhydride from the warming step. These tissues may be used simultaneously for immunohistochemistry[7–9] and tract tracing,[9,10] but they display decreased signal and increased background.

[6] S. Hayashi, I. C. Gillam, A. D. Delaney, and G. M. Tener, *J. Histochem. Cytochem.* **26,** 677 (1978).

[7] C. E. Gee, C.-L. C. Chen, J. L. Roberts, R. Thompson, and S. J. Watson, *Nature (London)* **306,** 374 (1983).

[8] W. S. Young III, É. Mezey, and R. E. Siegel, *Mol. Brain Res.* **1,** 231 (1986).

[9] M. Schalling, T. Hökfelt, B. Wallace, M. Goldstein, D. Filer, C. Yamin, and D. H. Schlesinger, *Proc. Natl. Acad. Sci. U.S.A.* **83,** 6208 (1986).

[10] J. N. Wilcox, J. L. Roberts, B. M. Chronwall, J. F. Bishop, and T. L. O'Donohue, *J. Neurosci.* **16,** 89 (1986).

Our tissue processing leads to a low background which we find very satisfactory for localizing most neuropeptide mRNAs. Occasionally, however, a probe seems to have greater background or the mRNA is less abundant, necessitating longer exposures (more than a few months). In these cases, a prehybridization for 2 hr at 37° in the hybridization buffer (see below) containing 50 μM α-thiodeoxyadenosine triphosphate (New England Nuclear), adapted from a similar approach for cRNA probes,[11] further reduces the background.

Hybridization with Oligodeoxyribonucleotide Probes

A typical rat brain coronal section is next hybridized with about 25–50 μl of hybridization buffer containing 1–2 $\times$ 10^6 dpm of labeled probe under a Parafilm coverslip. If there was a preliminary prehybridization, then the coverslip is lifted, the old buffer removed by aspiration, and new buffer with the labeled probe is applied. The amount of buffer and probe is increased for larger sections. The hybridization buffer contains 4$\times$ SSC (1$\times$ SSC is 0.15 *M* NaCl, 15 m*M* sodium citrate, pH 7.2), 50% formamide (BDH), 10% dextran sulfate 500 (Sigma), 100 m*M* dithiothreitol, 500 μg/ml herring testes DNA (sheared with a 20-gauge needle and boiled; Sigma), 250 μg/ml yeast tRNA (BRL), and 0.02% each of polyvinylpyrrolidone 360 (Sigma), bovine serum albumin, and Ficoll 400 (Pharmacia). The sections are incubated in a humid chamber for 20–24 hr at 37°. Dr. G. Mengod (Sandoz, Basel, Switzerland) has suggested another hybridization buffer containing 600 m*M* NaCl, 80 m*M* Tris–HCl (pH 7.5), 4 m*M* EDTA, 0.1% sodium pyrophosphate, 0.2% sodium dodecyl sulfate, 0.2 mg/ml heparin, and 50% formamide. We have used this buffer plus 100 m*M* dithiothreitol with good preliminary results.

The coverslips are removed by floating them off in 2$\times$ SSC. Then the sections are washed 4 times for 15 min in 2$\times$ SSC–50% formamide at 37° (or 1$\times$ SSC at 55°) followed by two more washes for 30 min in 1$\times$ SSC at room temperature. The sections are dipped quickly in water and then 70% ethanol and dried standing up.

The oligonucleotide probes are made on an Applied Biosystems DNA synthesizer according to the manufacturer's instructions. They can be purified prior to labeling on 8 *M* urea–8% polyacrylamide preparative sequencing gels. The oligonucleotides are labeled using terminal deoxynucleotidyltransferase (Boehringer-Mannheim)[12] and [α-^{35}S]thiodeoxyadenosine triphosphate (>1000 Ci/mmol, New England Nuclear) to spe-

[11] C. E. Bandtlow, R. Heumann, M. E. Schwab, and H. Thoenen, *EMBO J.* **6,** 891 (1987).

[12] T. Maniatis, E. F. Fritsch, and J. Sambrook, "Molecular Cloning: A Laboratory Manual." Cold Spring Harbor Laboratory, Cold Spring Harbor, New York, 1982.

cific activities of greater than 5000 Ci/mol. Each lot of enzyme is checked by running the probes on a sequencing gel to determine the incubation time necessary to produce an average tail length of 10–12 bases. Longer tail lengths produce higher backgrounds as do labelings with other nucleoside triphosphates.

Hybridization with Ribonucleotide Probes

The slides are placed on a slide warmer at 50° and 25–50 μl of hybridization buffer (as above) containing $1-2 \times 10^6$ dpm of probe is placed on each section under a Parafilm coverslip. The hybridizations proceed for 3 hr in a humid chamber, and then the coverslips are floated off in 2× SSC. The sections are rinsed in 2× SSC–50% formamide at 50° for 5 min, and again for 20 min. The slides are then placed on a 37° slide warmer and incubated under coverslips for 30 min with 40 μl of an ribonuclease solution (100 μg/ml RNase A and 1 μg/ml RNase T_1 in 2× SSC). The coverslips are removed, and the sections are rinsed twice for 5 min in 2× SSC–50% formamide at 50° and twice for 5 min in 2× SSC at room temperature. The sections are dipped briefly in water and then 70% ethanol and air-dried.

Our cRNA probes are generated using the SP6 plasmids (Promega Biotech) according to the manufacturer's instructions. Concentrations of the ribonucleotide triphosphates during the synthesis are 500 μM except for 25 μM UTP and 1–3 μM[α-^{35}S]UTP (>1000 Ci/mmol, New England Nuclear). The resultant probes have specific activities greater than 10^4 Ci/mmol (>10^8 dpm/μg) and are checked for size and purity on denaturing acrylamide or agarose gels.

Autoradiography

The sections are first examined by apposing them to X-ray film (Kodak X-Omat). This preliminary screening provides information on the strength of the hybridization and the amount of background. Based on the strength of the signal on the film, estimates are made for the subsequent exposure with nuclear emulsion (NTB, 1 : 1 with water, Kodak). The sections can be dipped directly into the emulsion at 42° and hung for 3 hr to dry, or they can be apposed to emulsion-coated coverslips.[13] The emulsions are developed in Dektol (half-strength, Kodak) for 2 min at 18° and the tissues stained with 0.4% toluidine blue or desired stain. The film and coated coverslip approaches offer freedom from two potential artifacts. One is edge artifact commonly seen with dipping. The other, which is more

[13] W. S. Young III and M. J. Kuhar, *Brain Res.* **179,** 255 (1979).

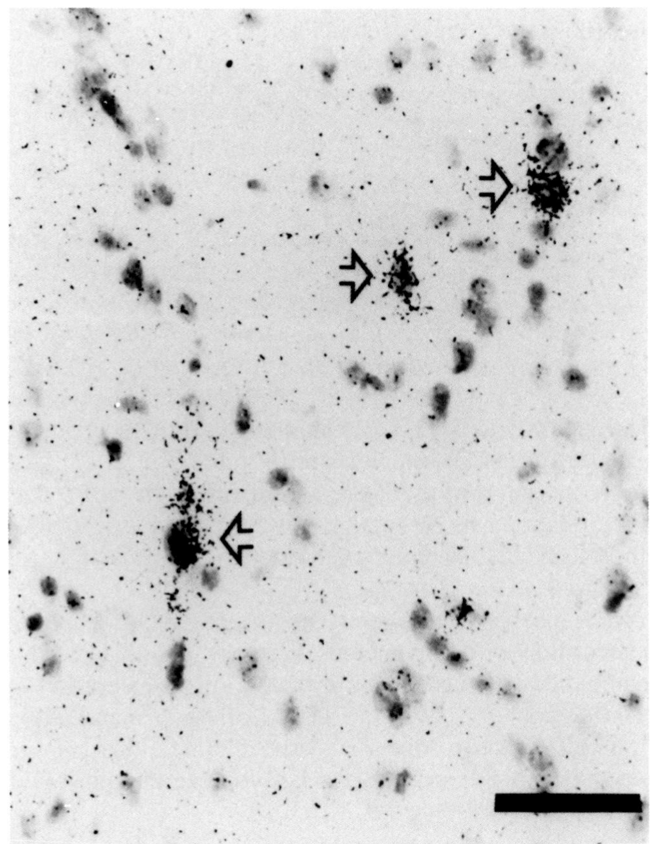

FIG. 1. Vasopressin mRNA-containing cells in the rat bed nucleus of the stria terminalis, medial subdivision. Labeled cells (arrows) appear with clusters of dark grains over their soma. Exposure of dipped slide was for 3 weeks. Bar, 50 μm. (Produced in collaboration with R. T. Zoeller.)

theoretical than apparent in our studies, is the production of grain density differences due to variation in emulsion thickness over similar levels of radioactivity. Since the β emissions of ^{35}S are an order of magnitude greater in energy than those of ^{3}H, the density of grains is proportional to the emulsion thickness with the former isotope. Figure 1 shows cells in the rat bed nucleus of the stria terminalis identified with an ^{35}S-labeled complementary oligodeoxyribonucleotide probe to the mRNA coding for the last 16 amino acids of the glycopeptide region of vasopressin.[14]

[14] R. Ivell and D. Richter, *Proc. Natl. Acad. Sci. U.S.A.* **81,** 2006 (1984).

Discussion

We discuss a few aspects of the oliogodeoxyribonucleotide method primarily, including tissue preparation, probe labeling, controls, and quantitation. Our approach to tissue preparation initially was based on the assumption that perfusion fixation would reduce potential ribonuclease activity within cells. However, we were obliged to use freshly frozen tissues from primates and other large mammals and noted good signal and low background with postfixation of those sections in formaldehyde. The same situation held in direct comparisons in rats, and, furthermore, the light microscopic tissue morphology was as good as we had obtained with perfusion and sucrose cryoprotection. Presumably, the ribonucleases remain apart from the mRNA in the thaw-mounted tissue sections that rapidly dry on the warm surface.

We have not found it necessary to treat our sections with proteases or HCl, which worsen tissue morphology, to increase penetration, perhaps because the oligodeoxynucleotides are generally smaller than the probes derived from cloned inserts and because we delipidate our tissue sections. The probes are able to penetrate at least as far as 28 μm through tissue (Fig. 2A), and the signal is proportional to section thickness (Fig. 2B). The delipidation, along with the dextran sulfate in the hybridization buffer, greatly reduces background while increasing signal. Without these two facets, there is significant background over white matter and cell-dense areas.

Labeling the 5′ ends of oligodeoxyribonucleotide probes was initially performed using T_4 polynucleotide kinase and [γ-^{32}P]ATP. Lewis *et al.*[15] demonstrated that the oligonucleotides labeled at the 3′ end using terminal deoxynucleotidyltransferase and [α-^{32}P]dCTP (or ^{125}I- or [^{3}H]dCTP) to higher specific activities were quite useful for ISHH. Although ^{32}P has a greater initial specific activity than ^{35}S (3000 versus 1000 Ci/mmol), its 6-fold shorter half-life makes ^{35}S our choice for label, especially as many exposures are for greater than 1 month. Also, the resolution of the hybridization signal on X-ray film is considerably better with ^{35}S- than with ^{32}P-labeled probes.

The choice of controls is not straightforward. The best approach, of course, is to choose as many as possible. Comparisons with the hybridizations using sense probes and probes for other transcripts help evaluate nonspecific binding as a complicating factor. Probes for different portions of the same mRNA should produce identical patterns of hybridization. If a well-characterized immunocytochemical distribution of the translation product is available, comparison with the ISHH distribution of the encod-

[15] M. E. Lewis, T. G. Sherman, and S. J. Watson, *Peptides* **6,** 75 (1985).

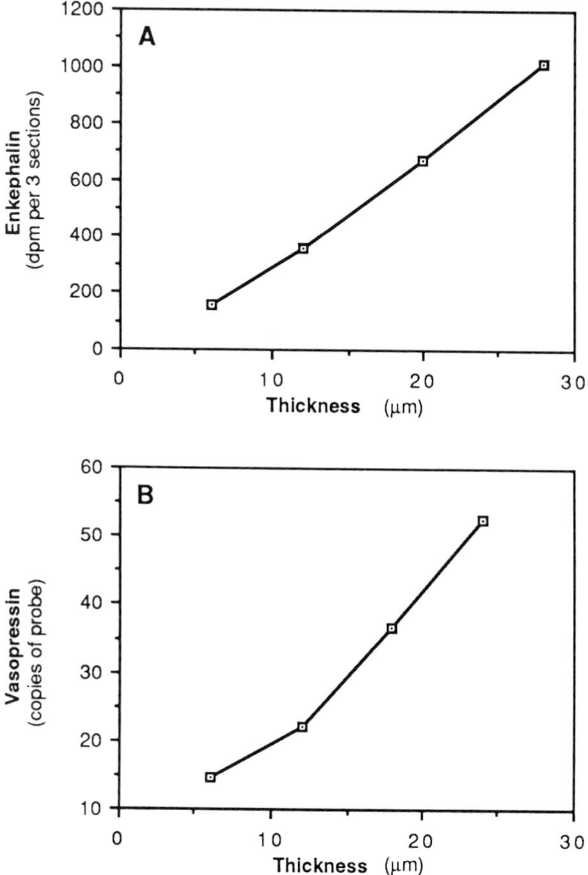

FIG. 2. Oligodeoxyribonucleotides, about 60 bases long after labeling, readily penetrate sections of various thicknesses (A), and their ^{35}S β emissions are also detected through the sections (B). The curves intersect the abscissa a few micrometers from the origin, suggesting that the initial few micrometers of tissue against the glass slide are rendered inaccessible to the probe. (A) Rat coronal sections of various thicknesses containing the caudate putamen were hybridized with a probe for preproenkephalin A,[2] and, after washing, the sections were scraped off and the radioactivity measured in a scintillation counter. (B) A vasopressin probe[8] was hybridized to rat coronal sections of various thicknesses containing the hypothalamic paraventricular nuclei. After exposure of these sections and accompanying ^{35}S-impregnated brain paste standards to X-ray film, the optical densities over the paraventricular nuclei were converted to copies of probe present in the tissue directly below a 1 μm^2 area.

ing mRNA is appropriate. It is useful to compare the sizes of RNA detected upon Northern analysis of total RNA using the probe under identical degrees of stringency as used for ISHH. This analysis can provide further evidence that the probe detects only the appropriate sized transcript. The use of ribonucleases to treat sections prior to hybridization is similar to treating sections with proteases prior to immunohistochemistry and is of questionable value. Similarly, adding excess unlabeled probe to the hybridization simply reduces the specific activity of the probe. A more valid competition could use two other, unlabeled probes (of similar size to the labeled probe) which together span the region targeted by the labeled probe.

The greatest area of controversy is quantitation of ISHH. At least three chapters in this volume,[16–18] as well as a recent book,[19] address this topic. The general approach[7] is analogous to quantitation of receptor autoradiography through the use of calibrated radioisotope-impregnated brain paste standards.[20] Knowing the specific activity of the probe enables one to determine the number of copies of the probe that are hybridized to a measured portion of the tissue section. It is important to stress that values determined in this fashion do *not* reflect total mRNA copies—only the amount of probe hybridized. No assumption is necessarily made about the efficiency of hybridization. Quantitation is most useful when comparing different physiological or developmental states, for example.

Quantitation of ISHH compares favorably with other approaches to measuring mRNA levels in tissues, such as solution hybridization and Northern analyses. Indeed, because ISHH does not entail extraction and purification (and transfer to a filter for Northerns) of mRNA prior to hybridization, it would not necessarily be unexpected for ISHH to be a quite accurate quantitative method. An example of the concordance between ISHH and assays that use extracted mRNA is provided by the effect of drinking saline on vasopressin mRNA levels in the paraventricular and supraoptic nuclei of the rat hypothalamus. McCabe *et al.*[21] and our laboratory[22] found 2- to 4-fold increases by ISHH compared with 1.6- to 3-fold increases in recent studies using paraventricular punches[23] or dissected hypothalami.[24] A study by Sherman *et al.*,[25] although finding larger increases, was internally consistent between ISHH and Northern analy-

[16] J. T. McCabe, R. A. Desharnais, and D. W. Pfaff, this volume [59].

[17] G. R. Uhl, this volume [53].

[18] M. E. Lewis, W. T. Rogers, R. G. Krause II, and J. S. Schwaber, this volume [58].

[19] G. R. Uhl (ed.), "*In Situ* Hybridization in Brain." Plenum, New York, 1986.

[20] J. R. Unnerstall, M. J. Kuhar, D. L. Niehoff, and J. M. Palacios, *J. Pharmacol. Exp. Ther.* **218,** 797 (1981).

[21] J. T. McCabe, J. I. Morrell, and D. W. Pfaff, *in* "Neuroendocrine Molecular Biology" (G. Fink, A. J. Harmar, and K. W. McKerns, eds.), p. 219. Plenum, New York, 1986.

[22] S. L. Lightman and W. S. Young III, *J. Physiol.* (*London*) **394,** 23 (1987).

[23] J. P. H. Burbach, H. H. M. Van Tol, M. H. C. Bakkus, H. Schmale, and R. Ivell, *J. Neurochem.* **47,** 1814 (1986).

[24] H. H. Zingg, D. Lefebvre, and G. Almazan, *J. Biol. Chem.* **261,** 12956 (1986).

[25] T. G. Sherman, J. F. McKelvy, and S. J. Watson, *J. Neurosci.* **6,** 1685 (1986).

ses. Nevertheless, more experience is necessary before there can be general confidence in quantitative ISHH.

Acknowledgments

The author greatly appreciates the many helpful discussions with Ruth Siegel, Michael Brownstein, Thomas Zoeller, and other researchers in the field and the technical assistance of Emily Shepard and Marjorie Warden.

[50] Use of Metal Complexes in Neuroendocrine Studies

By AYALLA BARNEA

Why Study Metal Complexes in Neuroendocrinology?

The following lines of evidence are strongly supportive of the view that metal complexes, particularly those of copper (Cu) and zinc (Zn) with some amines, peptides, or polypeptides, are present in the extracellular space of neuroendocrine cells in sufficient amounts to modulate the function of these cells. Cu and Zn are essential for numerous enzymatic reactions crucial for the survival and metabolic functions of all cells. In secretory, neuroendocrine, and some exocrine cells, Cu or Zn is essential for the execution of enzymatic reactions that are intimately involved in specific cell function, i.e., secretion. These "secretion-related" enzymatic reactions are integral to the formation or posttranslational modification of the substances destined to be secreted. For example, Cu is essential for dopamine β-monooxygenase (EC 1.14.17.1), which executes the conversion of dopamine to norepinephrine,[1,2] and for α-amidating enzymes,[3–5] which execute the posttranslational amidation of peptides (this modification is very common for peptides/polypeptides secreted from neuroendocrine as well as exocrine cells); Zn is important for the processing of proinsulin to insulin.[6,7] The common feature of all these enzymes is that

[1] J. H. Phillips and Y. P. Allison, *Neuroscience* **2,** 147 (1977).

[2] D. F. Kirksey, R. L. Klein, J. M. Baggett, and M. S. Gasparis, *Neuroscience* **3,** 71 (1978).

[3] B. A. Eipper, R. E. Mains, and C. C. Glembotski, *Proc. Natl. Acad. Sci. U.S.A.* **80,** 5144 (1983).

[4] R. E. Mains, A. C. Myers, and B. A. Eipper, *Endocrinology* (*Baltimore*) **116,** 2505 (1985).

[5] J. Sakata, K. Mizuno, and H. Matsuo, *Biochem. Biophys. Res. Commun.* **140,** 230 (1986).

[6] D. F. Steiner, W. Kemmler, H. S. Tager, and A. H. Rubenstein, *Adv. Cytopharmacol.* **2,** 195 (1974).

[7] S. O. Emdin, G. G. Dodson, J. M. Cutfield, and S. M. Cutfield, *Diabetologia* **19,** 174 (1980).

they are intragranular, which implies a similar localization of the metal. In addition, some secretory granules, e.g., those of mast cells, are enriched with Zn.[8]

It is known that the soluble components of secretory granules,[9–11] including dopamine β-monooxygenase and α-amidating enzymes, are coreleased during exocytosis. Thus, it is conceivable that the metals are released into the extracellular space during exocytosis. In support of this proposition are the findings that Zn is coreleased with insulin[12] and histamine and that depolarizing concentrations of K^+ release newly taken-up ^{65}Zn or ^{67}Cu from neuronal tissues incubated *in vitro*.[13,14] Although the chemical form of the intragranule-stored metal is not known, one could be certain that it is a complex and not an ion, since there are numerous substances that could complex the metal. It is usually thought that the secretory substances, e.g., norepinephrine (NE), histamine, or peptides, are released as such into the extracellular space. However, one should consider the possibility that some of the amines/peptides, particularly those peptides containing histidine (His) and/or cysteine (Cys), form complexes with Cu or Zn. Such complexes are known to have extremely high stability constants[15]; e.g., the stability constant of NE for Cu complexation is 10^{16} liter/mol, that of His is 10^{18}, and that of Cys is 10^{19}. Histamine could complex Zn; the stability constant for such a complex is $10^{8.7}$. Thus, it is conceivable that some of the secretory amines/peptides exist as metal complexes within and are secreted as such from the secretory granules. Moreover, one should consider the possibility that other complexes of Cu or Zn are secreted. Thus, the question arises: What are the actions of such metal complexes?

Preparation of Metal Complexes: Cu–His Complex as an Example

The most desirable approach would be to evaluate the action of the metal complex under *in vivo* conditions. However, owing to the presence in the circulation of numerous substances that can complex Cu with very

[8] B. Uvnas, C.-H. Aborg, and U. Bergqvist, *Acta Physiol. Scand.* **93,** 401 (1975).

[9] F. Konings and W. De Potter, *Biochem. Biophys. Res. Commun.* **104,** 254 (1982).

[10] M. Levine, A. Asher, H. Pollard, and O. Zinder, *J. Biol. Chem.* **258,** 13111 (1983).

[11] R. E. Mains, C. C. Glembotski, and B. A. Eipper, *Endocrinology* (*Baltimore*) **114,** 1522 (1984).

[12] G. M. Grodsky and F. Schmid-Formby, *Endocrinology* (*Baltimore*) **117,** 704 (1985).

[13] G. A. Howell, M. G. Welch, and C. J. Frederickson, *Nature* (*London*) **308,** 736 (1984).

[14] D. E. Hartter and A. Barnea, *Fed. Proc., Fed. Am. Soc. Exp. Biol.* **46,** 2107 (1987) (Abstr.).

[15] A. E. Martell, "Organic Ligands, in Stability Constants of Metal-Ions." The Chemical Society, London, 1964.

high avidity, there is a problem in interpretation of the results. The most problematic question is: Which complex exerts the biological effect? For this reason, an *in vitro* approach and the use of balanced salts solutions that do not contain metal chelators such as amino acids or proteins (e.g., Krebs–Ringer–phosphate buffer, KRP) are recommended. If inclusion of a peptidase inhibitor in the incubation buffer is needed, choose one that does not complex the metal with high affinity, for example, leupeptin.

Solutions

KRP buffer:
- 127 m*M* NaCl
- 5 m*M* KCl
- 1.8 m*M* $CaCl_2$
- 10 m*M* glucose
- 1.27 m*M* $MgSO_4 \cdot 7H_2O$
- 1.28 m*M* KH_2PO_4

Dissolve in 10 m*M* phosphate buffer, pH 7.4. This solution is kept refrigerated and used within 10 days.

Cu 20× stock solution: 20 m*M* $CuCl_2$ in deionized–distilled water.

Histidine 20× stock solution: 20 m*M* L-histidine–HCl in 0.1 *N* HCl.

The stock solutions are kept refrigerated for several months.

Dilute 20× His solution in KRP buffer to the desired final concentration of His, add 20× stock of Cu, and adjust pH to 7.4 with NaOH. The choice of the molar ratio of Cu to His is critical, and some considerations and examples are outlined below. Unless otherwise desired, the molar ratio optimal for the formation of the complex is used: in the case of His, a ratio of 1 : 2 is optimal[16]; for Cys, glutathione, some peptides such as Gly-His-Lys, pGlu-His-Pro-NH_2 (thyrotropin-releasing hormone), and albumin, a ratio of 1 : 1 is recommended. When radiolabeled Cu (^{67}Cu) is used for the formation of the complex, a 20× stock solution of ^{67}Cu–His complex is prepared by mixing His in KRP and the commercially available $^{67}CuCl_2$ (~1×10^6 Ci/mol; Oak Ridge National Laboratories, Oak Ridge, TN) at a ratio of 1 : 2. This is then diluted with the pH-adjusted nonradiolabeled Cu–His solution, prepared as described, to the desired final concentration of the Cu–His complex. Some Cu complexes can be used for several days, e.g., Cu–His, Cu–albumin; those that are comprised of unstable ligands such as cysteine or glutathione are freshly made on the day of the experiment.

[16] L. Casella and M. Gullotti, *J. Inorg. Biochem.* **18,** 19 (1983).

Some Considerations for the Choice of Ligand and the Molar Ratio of Cu to Ligand

It is generally believed that Cu, particularly in high concentrations, is deleterious to membranes. While this is correct when cells or membranes are exposed for prolonged periods of time to Cu, one should consider the fact that high concentrations of complexed Cu exist within cells in specific subcellular compartments, and that it may exist in high concentrations extracellularly, albeit for short periods of time. We have conducted a series of studies on the effects of Cu on the secretory function of a peptidergic neuron, using explants of brain tissues (the median eminence area, MEA) incubated under *in vitro* conditions, and found that concentrations as high as 100–200 μM Cu modulate the release of luteinizing hormone-releasing hormone (LHRH) in a fashion consistent with a physiological process exhibiting a limited number of Cu interactive sites and a requirement for Na^+/Cl^- transport.[17–20] In these studies, the longest exposure period to Cu was 15 min. In another series of studies using this range of [Cu] and a 5-min exposure period, we demonstrated that Cu amplifies prostaglandin E_2 (PGE_2) stimulation of LHRH release and that amplification is a post-PGE_2–receptor event.[21–24] Importantly, these effects of Cu are specific for this metal.[18,25] Thus, the proper use of the Cu complex can unravel important aspects of metal–tissue interactions.

Example 1. Importance of Complexation of Cu. The most important consideration is that ionic and complexed Cu may have different effects in some biological systems. As shown in Table I, complexation of Cu is essential for both Cu uptake by MEA explants and Cu action (amplification of PGE_2 stimulation of LHRH release) on MEA explants.

Example 2. Choice of Ligand to Determine Cu Uptake/Action. As shown in Table II, Cu uptake by MEA explants and Cu action (amplification of PGE_2 stimulation of LHRH release) are dependent on the ligand species: Cu complexed to histamine is the most effective; Cu complexed to Glu-His-Lys is less effective, and Cu complexed to albumin is ineffective.

[17] A. Barnea and M. Colombani-Vidal, *Proc. Natl. Acad. Sci. U.S.A.* **81,** 7656 (1984).
[18] M. Colombani-Vidal and A. Barnea, *Neuroendocrinology* **43,** 664 (1986).
[19] M. Colombani-Vidal and A. Barnea, *Neuroendocrinology* **44,** 276 (1986).
[20] M. Colombani-Vidal and A. Barnea, *Neuroendocrinology* **44,** 283 (1986).
[21] A. Barnea, G. Cho, and M. Colombani-Vidal, *Endocrinology (Baltimore)* **117,** 415 (1985).
[22] A. Barnea, G. Cho, and M. Colombani-Vidal, *Endocrinology (Baltimore)* **119,** 1254 (1986).
[23] A. Barnea, G. Cho, and M. Colombani-Vidal, *Endocrinology (Baltimore)* **119,** 1262 (1986).
[24] A. Barnea and G. Cho, *Proc. Natl. Acad. Sci. U.S.A.* **84,** 580 (1987).
[25] A. Barnea, G. Cho, and M. Colombani-Vidal, *Brain Res.* **384,** 101 (1986).

TABLE I
DEPENDENCE OF ^{67}Cu UPTAKE AND Cu ACTION ON THE MOLAR RATIO OF Cu TO His[a]

Complex	Cu-to-ligand molar ratio	^{67}Cu uptake (pmol/15 min/mg protein)	LHRH release (pg/15 min/MEA)
Cu–His	1:0	3.1 ± 0.50 (9)	1.7 ± 0.4 (8)
Cu–His	1:2	7.3 ± 0.30 (9)	8.9 ± 1.2 (17)
Cu–His	1:200	2.0 ± 0.10 (8)	1.2 ± 0.4 (8)

[a] For ^{67}Cu uptake, MEA explants, obtained from immature female rats, were incubated for 15 min with ^{67}Cu complex (100 μM Cu, 5 μCi/ml); for Cu action, MEA explants were incubated for 5 min with 150 μM Cu–His and then for 15 min with 10 μM PGE_2. LHRH release into the medium was evaluated by radioimmunoassay.

TABLE II
DEPENDENCE OF ^{67}Cu UPTAKE AND Cu ACTION ON THE LIGAND STRUCTURE[a]

Complex	^{67}Cu uptake (pmol/min/mg protein)	LHRH release (pg/15 min/MEA)
Cu–histamine	8.2 ± 0.5 (6)	10.3 ± 1.0 (11)
Cu–Gly-His-Lys	2.5 ± 0.1 (9)	2.6 ± 0.7 (6)
Cu–albumin	0.7 ± 0.02 (6)	Undetectable (7)

[a] For ^{67}Cu uptake, MEA explants were incubated for 15 min with 150 μM complexed ^{67}Cu (Cu-to-ligand ratio was 1:2 for Cu–histamine and 1:1 for the other complexes); for Cu action, MEA explants were incubated for 5 min with the Cu complex and then for 15 min with 10 μM PGE_2.

TABLE III
DIFFERENTIAL EFFECTS OF EXCESS His ON ^{67}Cu UPTAKE BY HYPOTHALAMIC TISSUE: DEPENDENCE ON [Cu][a]

Complex	[Cu] (μM)	Molar ratio	^{67}Cu uptake (pmol/min/mg protein)	Molar ratio	^{67}Cu uptake (pmol/min/mg protein)
Cu–His	0.2	1:2	0.16 ± 0.01 (22)	1:2000	0.63 ± 0.03 (23)[b]
Cu–His	0.6	1:2	0.90 ± 0.10 (12)	1:2000	1.80 ± 0.20 (7)[b]
Cu–His	5.0	1:2	18.50 ± 2.10 (16)	1:2000	9.30 ± 0.30 (17)[b]
Cu–His	10.0	1:2	53.60 ± 5.00 (14)	1:2000	16.60 ± 0.40 (9)[b]

[a] Hypothalamic tissue slices were incubated with ^{67}Cu–His (5 μCi/ml) for 30 min.
[b] t test of molar ratio 1:2 versus 1:2000, $p < 0.001$.

Example 3. Choice of the Molar Ratio of Cu to Ligand. Complexation of Cu is a reversible reaction.[26] Thus, one could assume that increasing the molar ratio of Cu to ligand would shift the reaction toward the formation of the complex and, hence, ensure complex formation. While within a limited range of ratios of Cu to ligand this consideration is correct, one should be aware of the possibility of an interaction between the ligand itself and the Cu complex interactive sites, i.e., competition of excess ligand with the Cu–ligand complex. As shown in Table I, complexation of Cu with His at a molar ratio of 1 : 2 facilitated both uptake and action of Cu, whereas complexation at a molar ratio of 1 : 200 (excess His over the Cu–His complex) inhibited Cu uptake/action. This result provides two important pieces of information: one, it gives us the range of molar ratios of the active metal complex, and, two, it tells us that the Cu–ligand complex and the ligand have common recognition sites.

Example 4. Benefits of Excess Ligand. From Example 3 one might get the impression that excess ligand always inhibits the Cu–ligand complex. However, one cannot presume such an inhibition in every aspect of metal complex action. We have found that ^{67}Cu is taken up by hypothalamic tissue by two saturable, ligand-dependent processes: a high and a low affinity process. When the effect of excess ligand on ^{67}Cu uptake was evaluated, it was noted that excess His facilitated ^{67}Cu uptake by the high affinity process, but it inhibited ^{67}Cu uptake by the low affinity process (Table III). Such a finding is suggestive that the interactive sites operative in the high affinity process have a higher affinity for the Cu–ligand complex than for the ligand.

Acknowledgments

The technical assistance of Gloria Cho and editorial assistance of Sandy Nance are highly appreciated. These studies were supported by National Institutes of Health Research Grants DK25692 and HD09988.

[26] F. A. Cotton and G. Wilkinson, "Advanced Inorganic Chemistry: A Comprehensive Text." Wiley, New York, 1980.

[51] Methods for the Study of Somatostatin

By M. C. AGUILA and S. M. MCCANN

Introduction

The methodology used for the study of the control, regulation, and mechanism of release of somatostatin from the hypothalamus is similar to that used for other hypothalamic peptides. In this chapter we mention all of the various ways that have been used for the study of the release of somatostatin; however, we place most of the attention on the method of static incubation of stalk median eminence fragments since this is the method which we have employed most extensively and which has yielded very clear, interpretable results.

Somatotropin release-inhibiting factor (SRIF) is the hypothalamic peptide identified originally as the growth hormone (GH) release-inhibiting factor by Krulich *et al.*, on the basis of the inhibition of GH release by the anterior pituitary incubated *in vitro* in the presence of partially purified hypothalamic extracts.[1] This neuropeptide was isolated and characterized by Brazeau *et al.*[2] and renamed somatostatin. It inhibits GH release both *in vivo* and *in vitro*.[2,3] Somatostatin is a tetradecapeptide which is widely distributed in the mammalian central nervous system and peripheral tissues. SRIF has been shown to be present in cell bodies and nerve fibers in hypothalamic and extrahypothalamic regions, with the highest concentration occurring in axons and nerve terminals of the median eminence (ME).[4,5]

In Vivo Methods of Study

The hypothalamic control of GH release can be studied indirectly by microinjecting putative transmitters or neuromodulators into the third

[1] L. Krulich, A. P. S. Dhariwal, and S. M. McCann, *Endocrinology (Baltimore)* **83,** 783 (1968).

[2] P. Brazeau, W. Vale, R. Burgus, N. Ling, M. Butcher, J. Rivier, and R. Guillemin, *Science* **179,** 77 (1973).

[3] R. Hall, M. Snow, M. Scanlon, B. Mora, and A. Gomez-Pan, *Metabolism* **27,** 1257 (1978).

[4] T. Hokfelt, R. Elde, K. Fuxe, O. Johannson, A. Ljungdahl, M. Goldstein, R. Luft, S. Efendic, G. Nilsson, L. Terenius, D. Ganten, S. L. Jeffcoate, J. Rehfeld, S. Said, M. Perez de la Mora, L. Possani, R. Tapia, L. Teran, and R. Palacios, "The Hypothalamus." Raven, New York, 1978.

[5] M. Brownstein, A. Arimura, H. Sato, A. V. Schally, and J. S. Kizer, *Endocrinology (Baltimore)* **96,** 1456 (1975).

brain ventricle of conscious rats.[6] If a peptide or other neurotransmitter alters growth hormone release following third ventricular injection into conscious, freely moving rats, it can be presumed to have an action within the brain or pituitary to alter GH. A pituitary site of action can be ruled in or out by *in vitro* experiments incubating hemipituitaries and/or dispersed pituitary cells *in vitro*. If there is no action of the putative transmitter to modify GH release from pituitary cells *in vitro* then the action obtained following intraventricular injection can be assumed to be within the brain. If, indeed, the action *in vitro* is to produce the same alteration in release of GH as occurs following intraventricular injection, then it is not possible to assign a site of action in the brain. If the action is of opposite sign obviously there is a brain action. Even if the action can be assumed to be in the brain as a result of these experiments, still one does not know the precise site of action. Presumably it would be on hypothalamic structures in the vicinity of the ventricle. To determine more precisely the site, it would be necessary to microinject the agent into various hypothalamic areas known to be concerned with the control of GH release. Even when this has been done, one does not know whether the action is via an alteration in the release of GH-releasing factor (GRF) or, alternatively, via an alteration of SRIF release or by both actions.

Most brain peptides have been found to have an action either to elevate or to lower GH following their intraventricular injection. Only in a few cases has it been possible to determine the mechanism of these effects. For example, in the case of gastrin-releasing peptide, it is now clear on the basis of immunoneutralization studies with intraventricular injection of SRIF antiserum that at least part of the action which is to inhibit GH release is mediated via release of SRIF within the hypothalamus which passes down the portal vessels and suppresses GH release by the somatotropes.[7]

So far, it has not been possible to assign any of the hypothalamic actions of these various peptides to an alteration in GRF discharge. The reason for this was the lack of available radioimmunoassays for GRF and also the failure so far to employ antibodies or antagonists against the peptide in order to identify the role of GRF. Presumably, such studies will be forthcoming shortly.

The most direct method for studying SRIF release *in vivo* is the collection of hypophyseal portal blood and measurement of the SRIF content by radioimmunoassay.[8–10] Most studies of SRIF in rat hypophyseal portal

[6] M. D. Lumpkin, A. Negro-Vilar, and S. M. McCann, *Science* **211,** 1072 (1981).

[7] S. Kentroti, M. C. Aguila, and S. M. McCann, *Endocrinology (Baltimore)* **122,** 1 (1988).

[8] K. Chihara, A. Arimura, and A. V. Schally, *Endocrinology (Baltimore)* **104,** 1434 (1979).

blood have involved anesthetized animals. Experiments with anesthetized animals are flawed by the use of anesthesia, which can alter the response of the neurons to putative transmitters. Furthermore, this acute procedure involves extensive surgical trauma so that the results are also confused by the effects of stress. One paper has reported the response of GH and TSH (thyroid-stimulating hormone) to TRH (thyrotropin-releasing hormone) in freely behaving as well as urethane-anesthesized rats.[9]

Several transmitters have been shown to influence the release of somatostatin into portal blood. Dopamine,[11] norepinephrine,[11] acetylcholine,[11] and γ-aminobutyric acid[12] have been reported to increase SRIF-like immunoreactivity in the portal blood, whereas a decreased release[11] has been reported for serotonin. Neuropeptides have also been examined, and both neurotensin[13] and bombesin[14] increase SRIF release into the portal blood. However, opioid peptides[13] and substance P did not alter the content of portal blood SRIF.[13]

Another *in vivo* method which gives great promise in the study of SRIF release is the intrahypothalamic push–pull cannula technique to measure the release of SRIF into the push–pull cannulae of freely moving rats.[15] This technique should allow the measurement of the release of SRIF into various regions of the hypothalamus in response to the microinjection of various transmitters or neuromodulators into the push–pull cannula. Work in this area will certainly soon be forthcoming.

In a recent additional modification, Ramirez has introduced the intrapituitary push–pull cannula technique.[16] In this technique the push–pull cannula is inserted directly into the anterior pituitary, and, after a recovery period, it can be used to perfuse the pituitary in the freely moving unanesthesized rat. This promises to give an index of portal blood concentrations of putative transmitters reaching the pituitary gland in the unanesthetized animal. Consequently, it should be superior to the portal

[9] I. Wakabayashi, M. Kanda, N. Miki, P. Demura, and K. Shizume, *Life Sci.* **24,** 2119 (1979).

[10] K. Chihara, A. Arimura, and A. V. Schally, *Fed. Proc., Fed. Am. Soc. Exp. Biol.* **37,** 638 (1978).

[11] K. Chihara, A. Arimura, and A. V. Schally, *Endocrinology (Baltimore)* **104,** 1656 (1979).

[12] G. M. Turkelson, K. Chihara, C. Kubli-Garfias, and A. Arimura, *Prog. Annu. Meet. Endocrine Soc.,* 144 (1979).

[13] H. Abe, K. Chihara, T. Chiba, S. Matsukura, and T. Fujita, *Endocrinology (Baltimore)* **108,** 1939 (1981).

[14] H. Abe, K. Chihara, N. Minamitani, J. Iwasaki, T. Chiba, S. Matsukura, and T. Fujita, *Endocrinology (Baltimore)* **109,** 229 (1981).

[15] V. D. Ramirez, J. C. Chan, E. Nudka, W. Lynn, and A. D. Ramirez, *Ann. N.Y. Acad Sci.* **473,** 434 (1986).

[16] D. Dluzen and V. D. Ramirez, *Neuroendocrinology* **45,** 328 (1987).

blood collection technique which involves anesthesia, surgical trauma, and usually the removal of the pituitary gland itself which could alter short-loop feedback relationships. We have carried out preliminary studies using the pituitary push–pull cannula and have observed that the release of SRIF into the cannula is increased following the initiation of ether anesthesia. Stress is known to suppress GH release in the rat.[17] Therefore, one would expect an enhanced release of SRIF following the initiation of ether anesthesia, but it had not yet been documented previously.

The disadvantages of the pituitary push–pull cannula technique are that it may not actually measure the intrapituitary concentration accurately because of time required for diffusion of the transmitter or peptide into the push–pull cannula. Furthermore, the results may well vary considerably depending on the site of the cannula in the gland. This can be checked, of course, by histological control following the completion of the experiment. It would seem likely that this will be a big advance in the *in vivo* study of SRIF release. The method is also equally applicable to other transmitters and neuropeptides.

In Vitro Methods for the Study of Somatostatin Release

Several *in vitro* techniques have been used to study SRIF release as well as the release of other transmitters and peptides. Initial experiments along these lines involved the coincubation of stalk–median eminence tissue with anterior pituitaries incubated *in vitro*. Following the addition of putative transmitters to the incubation medium, the output of pituitary hormones measured by bioassay was altered.[18] With the development of radioimmunoassays for the measurement of the peptides, the conditions could be altered such that only ventral hypothalamic fragments were incubated and the effect on the output of peptide or various transmitters and neuromodulators assessed.

In early studies we found that the use of a large fragment of tissue including the preoptic and anterior hypothalamic regions led to a poor response in terms of LHRH release to putative transmitters and prostaglandins. On the other hand, responsiveness of a median eminence–stalk (ME) fragment was excellent.[19,20] At this point in time, the dissection was also modified so as to incubate almost exclusively median eminence

[17] L. Krulich, E. Hefco, P. Illner, and C. B. Read, *Neuroendocrinology* **16,** 293 (1974).

[18] H. P. G. Schneider and S. M. McCann, *Endocrinology (Baltimore)* **85,** 121 (1969).

[19] A. Negro-Vilar, S. R. Ojeda, A. Arimura, and S. M. McCann, *Life Sci.* **23,** 1493 (1978).

[20] S. R. Ojeda, A. Negro-Vilar, and S. M. McCann, *Endocrinology (Baltimore)* **104,** 617 (1979).

and the proximal stump of the pituitary stalk. The earlier incubations had included almost certainly a considerable amount of the arcuate nucleus and possibly some ventromedial nucleus as well. By carrying out the dissection under a dissecting microscope one could obtain almost completely median eminence tissue. This structure contains mainly axonal fibers and nerve terminals, and therefore it was thought that actions of transmitters on this tissue *in vitro* were almost certainly interactions among the nerve terminals. The median eminence itself only has a few cell bodies. More careful evaluation of the remaining tissue indicates that probably even with this dissection there may be variable amounts of arcuate nucleus included. This would contain cell bodies of tuberoinfundibular dopaminergic neurons and of other peptidergic neurons which are now known to have cell bodies in the arcuate nucleus and axons which project to the external layer of the median eminence, presumably to release their product into the hypophyseal portal vessels.[21] Consequently, it is not possible to interpret these results simply in terms of interactions among nerve terminals; however, these are certainly the predominant structures present.

Once the median eminence fragments have been prepared they can be utilized in either a static or perifusion system to evaluate the effects of various secretagogues. We have used the perifusion system and in cases where we have used it we have found similar results to those obtained with the static incubation. The perifusion system has the advantage of allowing a determination of the latency and the complete time course of the effect of a transmitter; however, it is more cumbersome to use because of the very large number of samples generated. Consequently, for screening purposes and for initial studies we have used the static system exclusively. Obviously, more studies should be carried out with the perifusion system in the future.

Other *in vitro* methods have involved the culture of fetal rat hypothalamic cells.[22,23] This technique is very laborious and time consuming. It uses newborn or embryonic tissue and has the disadvantage that the neurons are now no longer in their normal interrelationships with the rest of the central nervous system. Having been derived from immature animals they may not have the same responsiveness as tissue from adult animals, and consequently the results may have limited significance *in*

[21] T. Hökfelt, B. Meister, B. Everin, W. Staines, T. Melander, M. Schalling, V. Mutt, A.-L. Hulting, S. Werner, T. Bartfai, O. Nordstrom, J. Fahrenkrug, and M. Goldstein, *in* "Integrative Neuroendocrinology: Molecular, Cellular and Clinical Aspects" (S. M. McCann and R. I. Weiner, eds.), pp. 1–34. Karger, Basel, 1987.

[22] R. A. Peterfreund and W. W. Vale, *Brain Res.* **239,** 463 (1982).

[23] R. A. Peterfreund and W. W. Vale, *Endocrinology (Baltimore)* **112,** 526 (1983).

vivo. However, in cases where the cultured neurons[24,25] have been used, results have been similar to those which have been obtained in the short-term *in vitro* incubation system.[19,26]

In vitro procedures have also included the use of synaptosome preparations,[27] dispersed hypothalamic cells,[23,28] and whole or partial hypothalamic fragments *in vitro*.[29] These types of preparations have been used to examine the effect of several neurotransmitters on SRIF release. Dopamine has been reported to increase SRIF release from hypothalamic fragments[30] and synaptosomal preparations.[27] Norepinephrine (NE) stimulated SRIF release from hypothalamic fragments[30,31] or slices[32]; however, other studies found NE to have no effect on SRIF release.[33,34] Acetylcholine (Ach) has been reported to inhibit[25] or have no effect[31] on SRIF release from hypothalamic fragments and to enhance SRIF release from dispersed cell cultures.[23] γ-Aminobutyric acid has been shown to inhibit SRIF release[24] from hypothalamic cells in culture, but it had no effect on release of SRIF from hypothalamic fragments.[31]

Serotonin also has been reported to inhibit[35] or to have no effect on SRIF release from hypothalamic fragments.[30] Several neuropeptides have been examined. Vasoactive intestinal polypeptide (VIP) has been reported to inhibit SRIF release from mediobasal hypothalamus.[36] Substance P[29,37] and neurotensin[29,37] were reported to increase SRIF release from hypothalamic fragments, whereas either an inhibitory[28,38] or lack of

[24] R. Gamse, D. E. Vaccaro, G. Gamse, M. Dipace, T. O. Fox, and S. E. Leeman, *Proc. Natl. Acad. Sci. U.S.A.* **77,** 5552 (1980).

[25] S. B. Richardson, C. S. Hollander, R. D'Eletto, P. W. Greenleaf, and C. Thaw, *Endocrinology (Baltimore)* **107,** 122 (1980).

[26] M. C. Aguila and S. M. McCann, *Endocrinology (Baltimore)* **117,** 762 (1985).

[27] I. Wakabayashi, Y. Myazawa, M. Kordon, N. Miki, R. Demura, H. Demura, and K. Shizume, *Endocrinol. Jpn.* **24,** 601 (1977).

[28] R. Moldow and C. H. Hollander, *Peptides* **2,** 489 (1981).

[29] M. C. Sheppard, S. Kronheim, and B. L. Pimstone, *J. Neurochem.* **32,** 647 (1979).

[30] K. Maede and L. A. Frohman, *Endocrinology (Baltimore)* **106,** 1837 (1980).

[31] L. C. Terry, O. P. Rorstad, and J. B. Martin, *Endocrinology (Baltimore)* **107,** 794 (1980).

[32] J. Epelbaum, L. Tapia-Arancibia, and C. Kordon, *Brain Res.* **215,** 393 (1981).

[33] R. M. MacLeod, A. M. Judd, W. D. Jarvis, and I. S. Login, *in* "Neuroendocrine Perspectives" (E. E. Muller and R. M. MacLeod, eds.), p. 45. Elsevier, Amsterdam, 1986.

[34] L. C. Terry and J. B. Martin, *Endocrinology (Baltimore)* **107,** 794 (1980).

[35] S. B. Richardson, C. S. Hollander, J. A. Prasad, and Y. Hirooka, *Endocrinology (Baltimore)* **109,** 602 (1981).

[36] J. Epelbaum, L. Arancibia, J. Besson, W. H. Rotsztejn, and C. Kordon, *Eur. J. Pharmacol.* **58,** 493 (1979).

[37] A. Shimatsu, Y. Kato, N. Matsushita, H. Katakami, and N. Yamaihara, *63rd Annu. Meet. Endocrine Soc., 63rd,* #198 (1981).

[38] S. V. Drouva, J. Epelbaum, Tapia-Arancibia, E. Laplante, and C. Kordon, *Neuroendocrinology* **32,** 163 (1981).

effect[29] has been reported for endogenous opioid peptides. However, the inhibition could only be demonstrated if SRIF release had been artificially elevated by depolarizing concentrations of potassium.

On the other hand, agonist SRIF analogs which did not interfere with the radioimmunoassay of SRIF inhibited SRIF release from cultured hypothalamic cells.[39] This result is consistent with the earlier *in vivo* studies in which it was shown that intraventricular injection of SRIF results in an elevation of GH presumably mediated, at least in part, by a reduction in SRIF release.[6]

In Vitro Studies with Stalk–Median Eminence (ME) Fragments

Because of the simplicity of the preparation and its good responsiveness to a variety of secretatogues, we have been extensively examining the release of SRIF from ME fragments. As indicated earlier the highest concentration of SRIF is present in the ME, which also contains all other known neurotransmitters or neuromodulators, be they peptides, monoamines, histamine, or acetylcholine.[40] We have used the preparation to study the effects of agonists, antagonists, receptor blockers, and even specific antibodies against peptides to determine their effect on the release of SRIF. Finally we have gone on to examine the mechanisms underlying the release of SRIF, exploring the involvement of cyclic nucleotides, calcium, and calmodulin in the process.

Methods

Hypothalamic Tissue Dissection. Adult male rats are used as tissue donors. Following decapitation ME fragments are dissected out and incubated in 0.4 ml of Krebs Ringer bicarbonate glucose buffer (pH 7.4).

The brains are removed and the ME is dissected from the brain under a stereomicroscope as described previously,[41] with modifications introduced by Negro-Vilar *et al.*,[42] to obtain a sample suitable for incubation purposes (see Fig. 1). Holding the proximal stump of the stalk with a pair of fine forceps, the tip of delicate iris scissors is introduced into the third ventricle, and, using the lateral limits of the infundibular recess as reference, two longitudinal cuts are performed in a forward direction. Last, a

[39] R. A. Peterfreund and W. W. Vale, *Neuroendocrinology* **39,** 397 (1984).

[40] M. Palkovits, *in* "Integrative Neuroendocrinology: Molecular, Cellular and Clinical Aspects" (S. M. McCann and R. I. Weiner, eds.), pp. 35–45. Karger, Basel, 1987.

[41] H. P. G. Schneider, D. B. Crighton, and S. M. McCann, *Neuroendocrinology* **5,** 271 (1969).

[42] A. Negro-Vilar, S. R. Ojeda, and S. M. McCann, *Fed. Proc., Fed. Am. Soc. Exp. Biol.* **37,** 296 (1978).

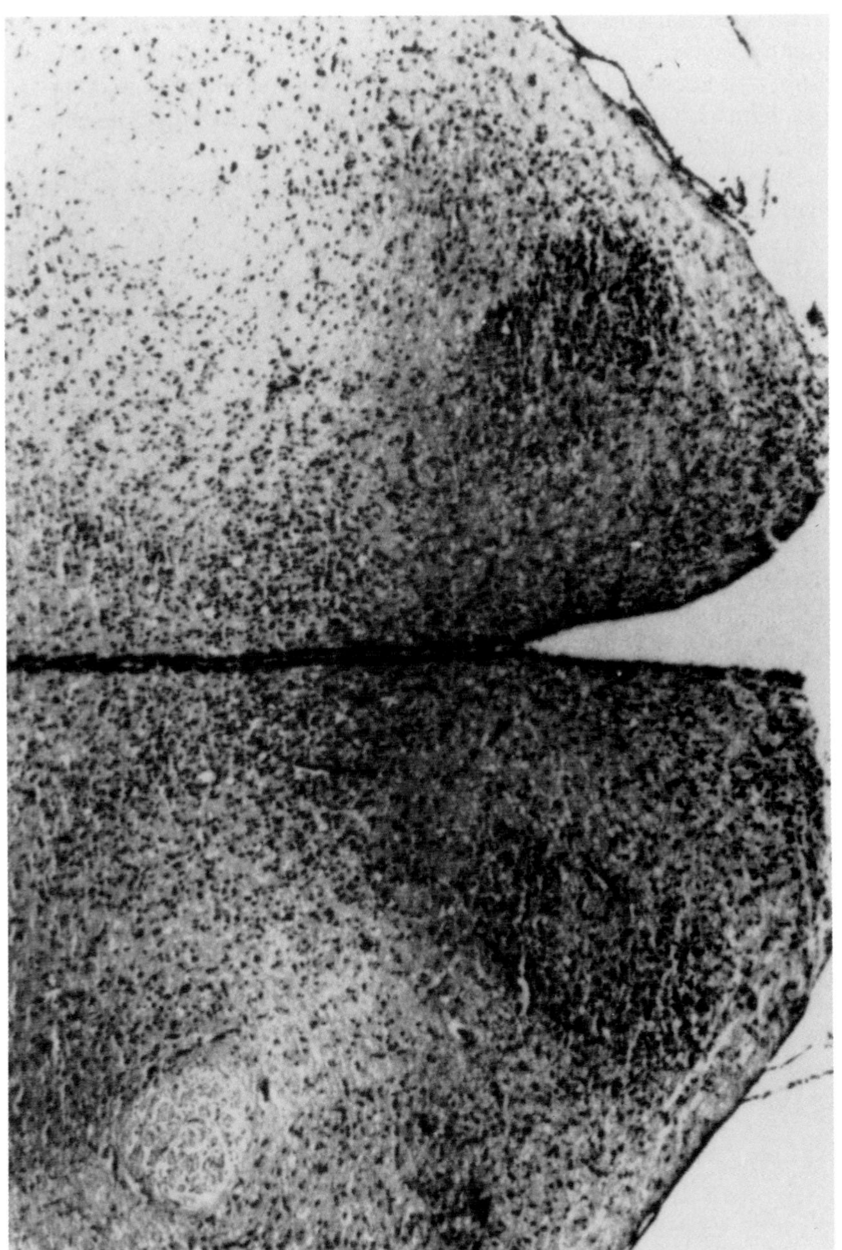

FIG. 1. Microphotograph from an animal in which the ME had been removed. Tissue was stained with Luxol fast blue–cresyl violet. ×20. (Reproduced from Ojeda *et al.*,[20] with permission of the publisher.)

frontal cut is made at a point behind the optic chiasm. The samples thus obtained contain the ME and a small part of the stalk.

In Vitro Incubation System. ME fragments are incubated in plastic vials with constant shaking (55–60 cycles/min) in an atmosphere of 95% O_2–5% CO_2. Each vial contains one ME in 0.4 ml incubation medium. The medium is Krebs–Ringer bicarbonate glucose buffer, pH 7.4, containing 2×10^{-5} *M* bacitracin.

Each experiment has a control group and several concentrations of the substance to be tested, with at least six flasks per group. In all the experiments, the ME fragments are preincubated for 30 min. At the end of that time, the medium is replaced by fresh medium containing the substances to be tested. Incubation of the tissue is then carried out for another 30 min. Immediately after incubation the samples are kept on ice at 4° until the medium of the last group is collected. Then the tubes are centrifuged at low speed for 10 min at 4°. After centrifugation, aliquots of the medium are assayed for SRIF. If the samples are not assayed immediately, they can be stored in 0.1 *M* acetic acid (100 μl of sample with 25 μl of acetic acid).

Results

Low doses of dopamine released SRIF from the ME, and the effects were blocked by the dopamine receptor blocker, pimozide. On the other hand, higher doses of norepinephrine were stimulatory, and these effects were blocked by the α receptor blocker, phentolamine.[43]

Several neuropeptides have been reported to influence SRIF secretion *in vitro*. Among the peptides found initially in the gastrointestinal tract and subsequently found to be localized in the brain, gastrin-releasing factor has been found to stimulate SRIF release from ME terminals *in vitro*.[7] Likewise, vasoactive intestinal polypeptide inhibited the release[29] (unpublished observations). Delta sleep-inducing peptide has also been demonstrated to inhibit SRIF release by a dopaminergic action.[44]

The effect of the various hypothalamic releasing and inhibiting hormones on the release of SRIF has been studied. TRH and LHRH had no effect on the release of SRIF.[34] Corticotropin-releasing factor stimulated the release of SRIF from ME terminals *in vitro*.[45,46] This result agrees well with *in vivo* studies in which the intraventricular injection of CRF was

[43] A. Negro-Vilar, S. R. Ojeda, A. Arimura, and S. M. McCann, *Life Sci.* **23,** 1493 (1978).

[44] K. S. Iyer and S. M. McCann, *Neuroendocrinology* **46,** 93 (1987).

[45] M. C. Aguila and S. M. McCann, *Brain Res.* **348,** 180 (1985).

[46] R. A. Peterfreund and W. W. Vale, *Endocrinology (Baltimore)* **112,** 1274 (1983).

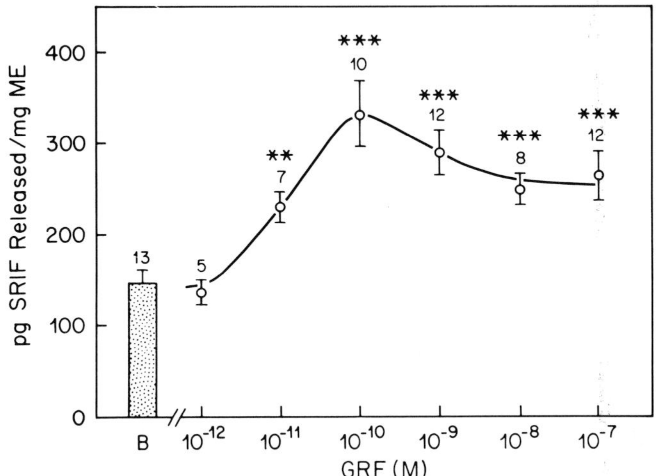

FIG. 2. Somatostatin (SRIF) release from rat median eminences incubated *in vitro* in the presence of synthetic human GH-releasing factor. $^{**}p < 0.01$, $^{***}p < 0.001$ versus basal (B), control release. (Reproduced from Aguila and McCann,[26] with permission of the publisher.)

found to lower plasma growth hormone.[47] That this action may have physiologic significance is suggested by the fact that stress suppresses GH release in the rat as already indicated. Furthermore, intraventricular injection of antibodies directed against CRF blocked stress-induced suppression of GH release.[48]

Growth hormone-releasing factor (GRF) was also capable of stimulating SRIF release from ME terminals *in vitro* in a dose-related fashion[26,45] (Fig. 2). The minimal effective dose of GRF to stimulate was an order of magnitude less than that for CRF. These *in vitro* results agree well with previous *in vivo* findings in which intraventricular injection of GRF was capable of suppressing GH release, presumably by stimulating SRIF release and/or also inhibiting GRF release.[49]

Since the stimulatory effect of GRF mimicked that of DA,[43] we then examined the possibility that endogenous dopamine mediated the effect *in vitro*. However, pimozide, a dopaminergic blocker, did not alter the stimulatory effect of GRF. Furthermore, submaximal concentrations of both

[47] N. Ono, M. D. Lumpkin, W. K. Samson, J. K. McDonald, and S. M. McCann, *Life Sci.* **35,** 1117 (1984).

[48] N. Ono, W. K. Samson, J. K. McDonald, M. D. Lumpkin, J. C. Bedran de Castro, and S. M. McCann, *Proc. Natl. Acad. Sci. U.S.A.* **82,** 7787 (1985).

[49] M. D. Lumpkin, W. K. Samson, and S. M. McCann, *Endocrinology* (*Baltimore*) **116,** 2070 (1985).

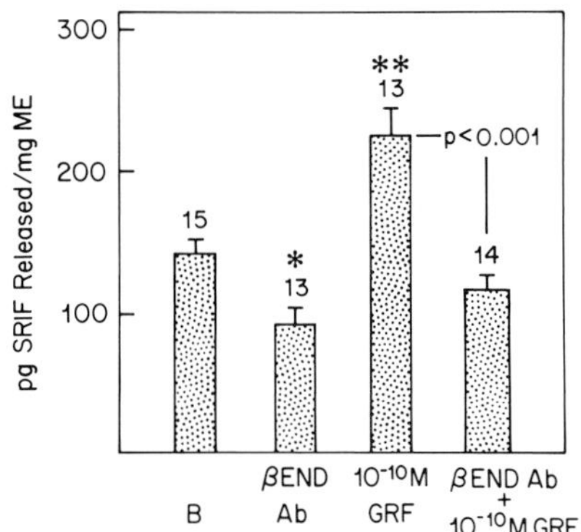

FIG. 3. Effects of β-endorphin antiserum (βEND Ab) on hGRF-induced SRIF release. β-Endorphin antiserum was added *in vitro* in a volume of 5 μl in 400 μl of medium; hGRF was used at a dose of 10^{-10} *M*. $^{**}p < 0.001$ versus basal (B). (Reproduced from Aguila and McCann,[50] with permission of the publisher.)

dopamine and GRF added together to the medium produced additive and not synergistic release, although by themselves each of these concentrations had little effect on SRIF release.[26] Thus, it is likely that DA and GRF activate separate receptors on the somatostatinergic terminals in the ME. Cholinergic and adrenergic mediations of the GRF action were ruled out in similar experiments in which we were unable to block the stimulated release of SRIF with atropine or phentolamine. However, in contrast with these negative results, the stimulatory effect of GRF on SRIF release was completely annulled by the opiate receptor blocker, naloxone.[50] The stimulatory effect of GRF was also completely blocked by anti-β-endorphin serum but was unaltered by anti-α-melanocyte stimulating hormone (αMSH) serum[50] (Fig. 3). These results indicate that endogenous β-endorphin acting through specific opiate receptors, located on SRIF nerve endings, mediates the release of SRIF induced by GRF.

In addition, we have examined the possible role of calcium and calmodulin in GRF-induced SRIF release from the ME. In these studies the stimulation of SRIF release induced by GRF was not inhibited by omission of extracellular calcium (Ca^{2+}) or when the remaining Ca^{2+} was chelated with EGTA. The calcium channel blockers nifedipine and verapamil failed to alter the increase of SRIF release induced by GRF. These results suggest that the effect of GRF is absolutely independent from Ca^{2+} influx. Calmodulin inhibitors (trifluoperazine, triflupromazine, and penfluridol) had an inhibitory effect on the stimulation of SRIF release

[50] M. C. Aguila and S. M. McCann, *Endocrinology (Baltimore)* **120,** 341 (1986).

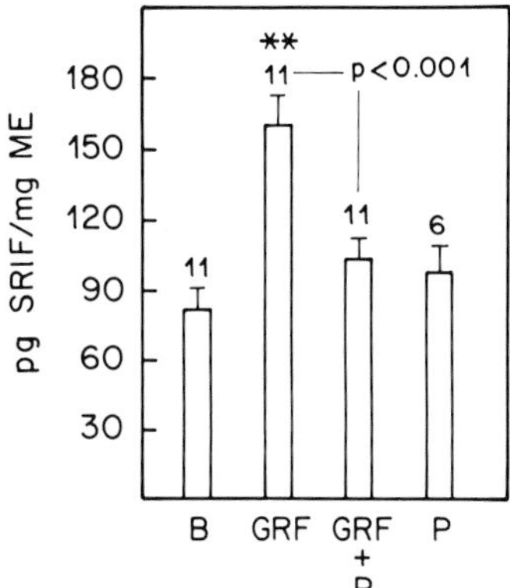

FIG. 4. Effect of penfluridol (P), a calmodulin inhibitor, on SRIF release evoked by rGRF (10^{-7} *M*). Penfluridol was employed during preincubation and incubation. Rat GRF (10^{-10} *M*) was present during the incubation period. $^{**}p < 0.005$ versus basal (B). (Reproduced from Aguila and McCann,[51] with permission of the publisher.)

induced by GRF and failed to alter resting release. Thus, the inhibitory effect of calmodulin inhibitors on GRF-induced SRIF release suggests that calmodulin may be involved in this process[51] (Fig. 4).

In conclusion we propose that GRF releases β-endorphin from the terminals of β-endorphinergic neurons in the median eminence. β-Endorphin combines with its receptors on the somatostatinergic terminals to stimulate release of SRIF. The mechanism for release appears to require mobilization of intracellular calcium stores. The rise in Ca^{2+} leads to the activation of calmodulin. The binding of Ca^{2+} by calmodulin results in increased SRIF release from the ME.

Summary

It is clear from the above that there are a number of methods for study of SRIF release. From the standpoint of convenience, the *in vitro* static incubation of ME is the most practical technique at the present time. Using this preparation SRIF release has been found to be modified by a number of neurotransmitters and peptides, and studies on the mechanism of release of the peptide have been initiated. There is no doubt that such studies should be complemented by perifusion studies, by studies involving larger pieces of the hypothalamus which encompass the entire somatostatinergic neuron, and by *in vivo* studies to determine the correlation of

[51] M. C. Aguila and S. M. McCann, *Endocrinology (Baltimore)* **123,** 305 (1988).

TABLE I

EFFECT OF NEUROTRANSMITTERS AND NEUROPEPTIDES ON SOMATOSTATIN SECRETION FROM HYPOTHALAMIC PREPARATIONS[a]

Substance	Effect	References
Neurotransmitters		
Dopamine	↑ ME terminals	43
	↑ Hypothalamic fragment	30
	↑ Portal blood	11
	↑ Synaptosome release	27
Norepinephrine	↑ ME terminal	43
	↑ Hypothalamic fragment	30, 31
	↑ Hypothalamic slices	32
	↑ Portal blood	11
Acetylcholine	→ Hypothalamic fragment	33, 34
	↑ Hypothalamic cells	23
	↓ Hypothalamic fragment	25
	→ Hypothalamic fragment	31
	↑ Portal blood	11
γ-Aminobutyric acid	↓ Hypothalamic cells	24
	↑ Portal blood	12
	→ Hypothalamic fragment	31
Serotonin	↓ Hypothalamic fragment	35
	→ Hypothalamic fragment	30
	→ Portal blood	11
Neuropeptides		
Growth hormone-releasing factor	↑ ME fragment	26
Corticotropin-releasing factor	↑ ME fragment	45
	↑ Hypothalamic cells	46
Luteinizing hormone-releasing hormone	→ ME fragments	45
Vasoactive intestinal polypeptide	↓ Mediobasal hypothalamic slices	36
	↓ ME fragments	Unpublished observations
Gastrin-releasing factor	↑ ME fragments	7
Delta sleep-inducing peptide	↓ ME fragments	44
Substance P	↑ Hypothalamic fragment	29, 37
	→ Hypothalamic fragment	36
	→ Portal blood	13
Neurotensin	↑ Hypothalamic fragment	29, 37
	↑ Portal blood	13
Endogenous opioid peptide	→ Hypothalamic fragment	29
	→ ME fragment	52
	↓ Hypothalamic fragment	28, 38
	→ Portal blood	13
Bombesin	↑ Portal blood	14

[a] Key to symbols: ↑, increased release; ↓, decreased release; →, unaltered release. (Adapted from Reichlin.[53])

in vivo and *in vitro* release. Among the *in vivo* techniques which have been utilized, the push–pull cannula technique employing cannulae implanted in hypothalamus or anterior pituitary gland offers the most promise. A summary of the effects of some neurotransmitters and neuropeptides on hypothalamic SRIF secretion is reported in Table. I.[52,53]

Acknowledgments

Supported by National Institutes of Health Grant AM10076.

[52] M. C. Aguila, O. Khorram, and S. M. McCann, *Brain Res.* **417,** 127 (1987).

[53] S. Reichlin, *in* "Brain Peptides" (D. T. Krieger, M. O. Brownstein, and J. B. Martin, eds.), p. 711. Wiley, New York, 1983.

[52] Mapping of Gonadotropin-Releasing Hormone–Receptor Binding Site Using Selective Chemical Modifications

By ELI HAZUM

Introduction

Chemical modification of proteins provides an important experimental tool for studying the relationships between their biological properties and specific amino acid residues. This approach has been widely applied to identify essential groups in the catalytic sites of enzymes. Similarly, chemical modification can be used to investigate biological interactions that involve the specific binding between a receptor protein and a hormone. The many applications of chemical modification in hormone–receptor studies include the development of affinity reagents, the introduction of reporter groups for receptor localization (e.g., fluorescent, electron-dense, and enzyme-conjugated groups), the use of cross-linking reagents, and the determination of the side-chain groups in the receptor essential for binding. The use of selective chemical reagents to probe and modify specific residues in the binding site of a given receptor molecule can serve as a supplementary procedure to the technique of *in vitro* mutagenesis. Chemical modification of receptors for structure–function analysis can be performed with membrane-associated receptors as well as with solubilized and purified receptors. This chapter summarizes the properties of a few selective chemical reagents and the applications of these reagents to identify the amino acid residues participating in the gonadotropin-releasing hormone (GnRH)–receptor binding site.

Selective Chemical Reagents for Protein Modification

The different chemical properties of the various amino acid side chains provide a basis for their selective modification. Most protein reagents react with more than one side chain, and the lack of specificity limits the usefulness of many reagents. However, by taking advantage of the differential properties of the available groups under various conditions, it is often possible to bring about selective modification. Thus, variation in reactivity as a function of pH provides a convenient way of controlling the course of many modification reactions. However, the relative positions of various groups, whether buried or exposed, obviously affect the chemical and biological properties of proteins. A selective list of chemical reagents useful for modifying different functional groups in proteins is given in Table I. Protein modification procedures, information on specificity, and applications are discussed in much greater detail in a recent review,[1] in books,[2,3] and in this series.[4]

Chemical Modification of Receptors

The principal approach to the study of structure–function relationships in receptors involves an irreversible and specific modification of amino acid residues located at or near the binding site. Such modifications can alter the characteristics of binding (B_{max} and K_d) and biological activity as well as other phenomena such as desensitization and internalization of receptors and the existence of low- and high-affinity binding sites. Typical of this approach are protection studies, i.e., prevention of the modification effect in the presence of the hormone or hormone analogs. Additional evidence that changes in properties of a receptor are primary effects of chemical modification can be obtained by removing the modifying groups and demonstrating restoration of original properties. For such studies, it is desirable to use reversible chemical reagents.[2-4] Using these combined techniques, it is then possible to identify the functional groups participating in the binding site. This information is of great importance

[1] R. E. Feeney, *Int. J. Pept. Prot. Res.* **29,** 145 (1987).

[2] G. E. Means and R. E. Feeney, "Chemical Modification of Proteins." Holden-Day, San Francisco, California, 1971.

[3] R. L. Lundblad and C. M. Noyes, "Chemical Reagents for Protein Modification," Vols. 1 and 2. CRC Press, Boca Raton, Florida, 1984.

[4] W. B. Jakoby and M. Wilchek (eds.), this series, Vol. 46; C. W. H. Hirs and S. N. Timasheff (eds.), this series, Vol. 91.

TABLE I
CHEMICAL REAGENTS USEFUL FOR MODIFYING SPECIFIC FUNCTIONAL GROUPS

Functional group	Residue	pK_a	Reagent
Carboxyl	Aspartate, glutamate, and carboxy terminus	4.6	Water-soluble carbodiimide plus nucleophiles and diazoacetyl compounds
α-Amino and ε-amino	Amino terminus and lysine	7.8 10.2	Acetic anhydride, active esters of carboxylic groups; ethyl acetimidate and trinitrobenzenesulfonic acid
Sulfhydryl[a]	Cysteine	8.5	Iodoacetate or iodoacetamide; *N*-ethylmaleimide and 5,5′-dithiobis(2-nitrobenzoic acid)
Imidazole	Histidine	7.0	Diazonium salts; *p*-bromophenacyl bromide and diethyl pyrocarbonate
Phenol	Tyrosine	10.5	Iodine[b]; aromatic diazonium compounds[b]; tetranitromethane and *N*-acetylimidazole
Indole	Tryptophan	—	2-Hydroxy-5-nitrobenzyl bromide; sulfenyl halides[c] and *N*-bromosuccinimide
Guanidino	Arginine	12–13	Phenylglyoxal; 2,3-butanedione and 1,2-cyclohexanedione

[a] Sulfhydryl groups can also be derived from the reduction of disulfide bonds with mild reducing agents such as 2-mercaptoethanol and dithiothreitol.
[b] React also with histidine.
[c] React also with sulfhydryl groups.

for the efficient use of cross-linking reagents as well as in the design of appropriate affinity-labeled hormones. The main advantage of affinity labeling is that large local concentration of the reagent enhances the specificity for the binding site of a receptor.

The following describes the methodology we have used to analyze the components comprising the binding site of the membrane-associated GnRH receptor.[5] This technique of chemical modification can also be applied to purified GnRH receptor.[6]

[5] D. Keinan and E. Hazum, *Biochemistry* **24,** 7728 (1985).
[6] E. Hazum, I. Schvartz, Y. Waksman, and D. Keinan, *J. Biol. Chem.* **261,** 13043 (1986).

Mapping of GnRH–Receptor Binding Site[5]

Treatment with Different Reagents. Methods for obtaining pituitary membrane preparations, iodination of buserelin [D-Ser (tBu)^6-des-Gly^{10}-GnRH-ethylamide], and the binding assay were previously described.[5,6] The reagents are of the highest quality available from commercial sources and are dissolved immediately before use.

Pituitary membrane preparations are incubated with increasing concentrations of reagent for 60 min at room temperature in 1 ml of 10 mM phosphate buffer, pH 7.4. At the end of the incubation period, the membranes are precipitated and washed twice. Each wash consists of a 15 min incubation with 20 ml of 10 mM Tris–HCl, pH 7.4, containing 0.1% bovine serum albumin (BSA) (assay buffer) followed by centrifugation for 20 min at 20,000 g at 4°. The membranes are resuspended in assay buffer and the binding measured.[5]

p-Diazobenzenesulfonic acid is prepared immediately before use as follows: *p*-aminosulfanilic acid (10 μmol) in 25 μl of cold 2 N HCl is diazotized by the addition of sodium nitrite (10 μmol) in 25 μl of cold water (4°). After standing for 8 min at 4°, 75 μmol of $NaHCO_3$ dissolved in 150 μl of ice-cold water (pH 8.5) is added. This reaction mixture at the appropriate concentration is immediately reacted with pituitary membranes. The colorless solution turns orange–brown within a few minutes; the reaction is allowed to proceed for an additional 1 hr.

The effects of various concentrations of 2-methoxy-5-nitrobenzyl bromide, *p*-diazobenzenesulfonic acid, and 1-ethyl-3-(3-dimethylaminopropyl)carbodiimide plus or minus glycine ethyl ester on ^{125}I-labeled buserelin binding to pituitary membranes are shown in Figs. 1, 2, and 3, respectively. Control membranes are treated under the same conditions without the reagent. The IC_{50} values for inhibition by various reagents and the residues modified are summarized in Table II.

Protection Experiment. To confirm that the inhibition of binding is a result of a direct modification of the binding site, the protective effect of GnRH against inactivation by one of the reagents was tested. The inactivation by 2-methoxy-5-nitrobenzyl bromide as a function of time (Fig. 4) indicates that the hormone-bound receptors are less accessible to the reagent than unoccupied receptors. For this experiment, pituitary membranes are suspended in 2 ml of 10 mM phosphate buffer, pH 7.4, with or without 10^{-6} M GnRH and incubated for 90 min at 4°. At the end of the incubation (time 0) 200 μl of each sample is centrifuged (3 min, 20,000 g). The membrane pellets are then incubated with 0.5 mM 2-methoxy-5-nitrobenzyl bromide; at 5, 10, 20, and 30 min, samples of 200 μl are centrifuged. The membrane pellets are extensively washed with assay buffer and the binding measured.

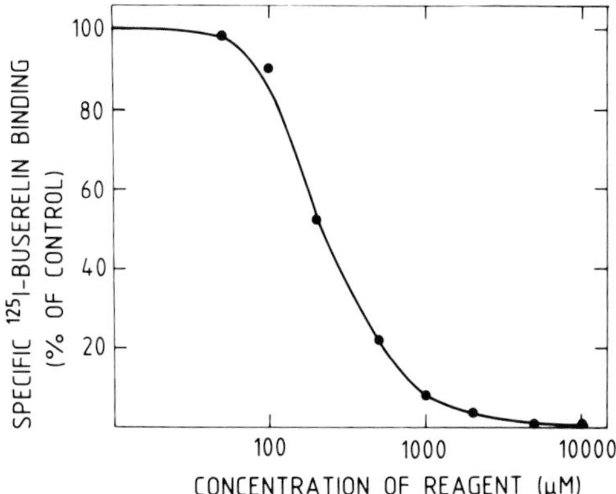

FIG. 1. Dose–response curve for the inhibition of ^{125}I-labeled buserelin binding to pituitary membrane preparations by 2-methoxy-5-nitrobenzyl bromide. The membranes are washed and incubated in 10 m*M* phosphate buffer with increasing concentrations of 2-methoxy-5-nitrobenzyl bromide (0.1–10 m*M*) dissolved in dimethylformamide. The membranes are washed twice with 10 m*M* Tris–0.1% BSA, and binding is assayed. [Reprinted with permission from D. Keinan and E. Hazum,[5] *Biochemistry* **24,** 7728–7732. Copyright (1985) American Chemical Society.]

TABLE II

IC_{50}[a] VALUES FOR THE INHIBITORY EFFECT OF VARIOUS REAGENTS ON ^{125}I-LABELED BUSERELIN BINDING

Reagent	IC_{50} (m*M*)	Site specificity
2-Methoxy-5-nitrobenzyl bromide	0.22	Indole
p-Diazobenzenesulfonic acid	0.1	Phenol and imidazole
1-Ethyl-3-(3-dimethylaminopropyl)carbodiimide plus glycine ethyl ester	25	Carboxyl
Sodium periodate	0.5	Sugar moiety
Dithiothreitol	14	Disulfide
Iodoacetamide	NE[b]	Sulfhydryl and imidazole (slightly)
N-Ethylmaleimide	NE[b]	Sulfhydryl
N-Hydroxysuccinimide ester of acetic acid	NE[b]	Amino

[a] IC_{50} is the concentration of reagent that inhibits the specific binding of ^{125}I-labeled buserelin to pituitary membrane preparations by 50%.

[b] NE, No effect at a concentration range of 0.1–100 m*M*. [Modified with permission from D. Keinan and E. Hazum,[5] *Biochemistry* **24,** 7728–7732. Copyright (1985) American Chemical Society.]

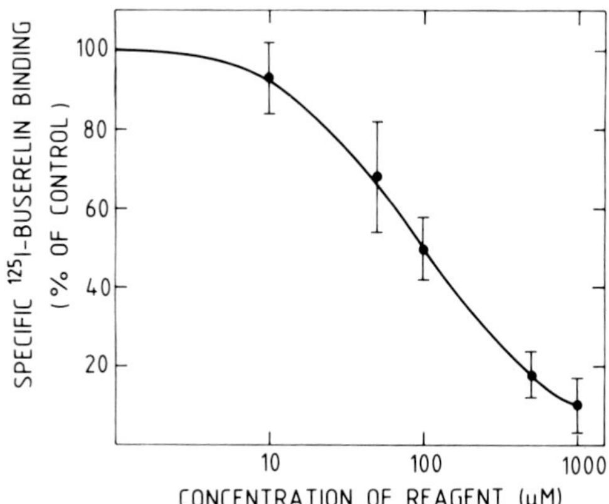

FIG. 2. Dose–response curve for the inhibition of ^{125}I-labeled buserelin binding to pituitary membrane preparations by *p*-diazobenzenesulfonic acid. *p*-Diazobenzenesulfonic acid is prepared before use as described in the text. The reaction mixture at the appropriate concentration is immediately reacted with pituitary membranes and incubated for 1 hr at room temperature. Following extensive washing, binding is assayed. [Reprinted with permission from D. Keinan and E. Hazum,[5] *Biochemistry* **24,** 7728–7732. Copyright (1985) American Chemical Society.]

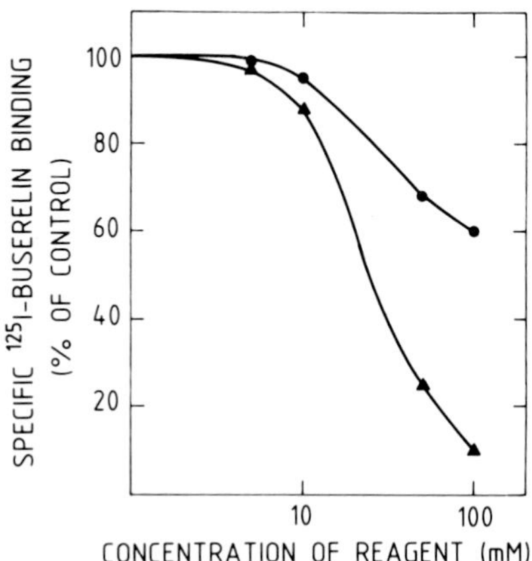

FIG. 3. Dose–response curve for the inhibition of ^{125}I-labeled buserelin binding to pituitary membranes by 1-ethyl-3-(3-dimethylaminopropyl)carbodiimide (DCI, 5–100 m*M*) (●) and DCI plus glycine ethyl ester (5–100 m*M*) (▲). The membranes, suspended in 10 m*M* phosphate buffer, are allowed to react with increasing concentrations of the reagents for 1 hr at room temperature. Following extensive washing with 10 m*M* Tris–0.1% BSA, binding is assayed. [Reprinted with permission from D. Keinan and E. Hazum,[5] *Biochemistry* **24,** 7728–7732. Copyright (1985) American Chemical Society.]

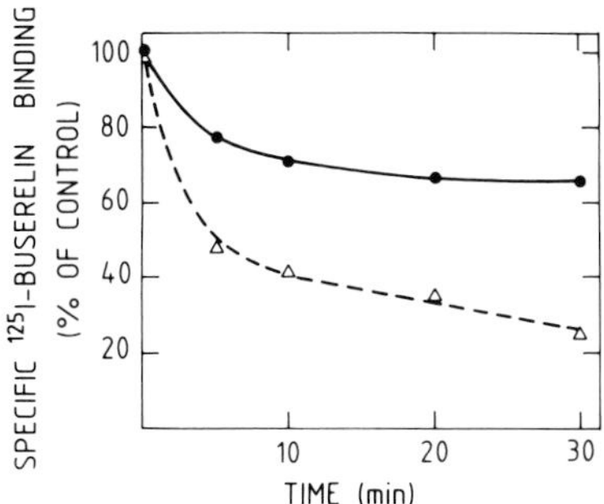

FIG. 4. Time course of inactivation by 2-methoxy-5-nitrobenzyl bromide of GnRH receptors in control membranes (△) and GnRH-protected membranes (●). Inactivation is performed as described in the text. [Reprinted with permission from D. Keinan and E. Hazum,[5] *Biochemistry* **24,** 7728–7732. Copyright (1985) American Chemical Society.]

Competition Binding Experiments and Scatchard Analysis. Pituitary membrane preparations are treated with 0.5 m*M* 2-methoxy-5-nitrobenzyl bromide (60 min, 24°), 10 m*M* dithiothreitol (30 min, 24°), or 0.5 m*M* sodium periodate (60 min, 24°) in 1 ml of 10 m*M* phosphate buffer, pH 7.4. Control (untreated) membranes are incubated with 1 ml of 10 m*M* phosphate buffer, pH 7.4, for 30 or 60 min at 24°. Following extensive washing (as described above), competition binding experiments with treated and untreated membranes are carried out by incubating radioactive buserelin (40,000 cpm) with increasing concentrations of unlabeled buserelin (10^{-10} to 10^{-7} *M*) in 0.5 ml assay buffer. After 90 min at 4°, binding is assayed, and the results are plotted according to Scatchard[7] and analyzed statistically by the method of Wilkinson.[8] The effects of reagent treatment on receptor binding parameters (B_{max} and K_d) are summarized in Table III.

Model for GnRH–Receptor Complex Formation[5]

Gonadotropin-releasing hormone is a decapeptide with the following amino acid sequence: pGlu-His-Trp-Ser-Tyr-Gly-Leu-Arg-Pro-Gly-NH_2. Three-dimensional analysis of the hormone in solution has indicated that there is a β turn in position 6 which brings the carboxy and the amino termini of the hormone into close proximity.[9,10] It has been suggested that the side chains of histidine, tyrosine, and arginine form a packed unit

[7] G. Scatchard, *Ann. N.Y. Acad. Sci.* **51,** 660 (1949).
[8] G. N. Wilkinson, *Biochem. J.* **80,** 324 (1961).
[9] M. Shinitzky and M. Fridkin, *Biochim. Biophys. Acta* **434,** 137 (1976).
[10] M. Shinitzky, E. Hazum, and M. Fridkin, *Biochim. Biophys. Acta* **453,** 533 (1976).

TABLE III
EFFECTS OF REAGENT–TREATMENT ON BINDING AFFINITY AND NUMBER OF GnRH RECEPTORS[a]

Reagent	K_d (nM)	B_{max} (fmol/pituitary)
Control	0.12 ± 0.01	165 ± 4
2-Methoxy-5-nitrobenzyl bromide (0.5 mM)	0.29 ± 0.04	135 ± 6
Sodium periodate (0.5 mM)	0.18 ± 0.01	180 ± 3
Dithiothreitol (10 mM)	0.22 ± 0.02	210 ± 7

[a] B_{max} and K_d values (means ± SEM) are calculated according to Wilkinson.[8] [Reprinted with permission from D. Keinan and E. Hazum,[5] *Biochemistry* **24,** 7728–7732. Copyright (1985) American Chemical Society.]

which may play an active role in the action of the hormone. Tryptophan, however, is at a maximal distance from this unit and thus may act as an independent active entity.[9,10] According to the spatial conformation of the native hormone and the participation of carboxyl, phenol, and indole groups in the binding site (Table II), we can postulate a model for the formation of the hormone–receptor complex. It is well known that arginyl residues on proteins may serve as positively charged loci for recognition of negatively charged anions. Thus, the driving force for the formation of the hormone–receptor complex is probably an ionic interaction between the amino acid arginine in position 8, which is positively charged, and the carboxyl groups in the binding site.[5,11] In addition to the ionic interaction, the hormone–receptor complex is stabilized by aromatic π–π interactions between the histidine, tryptophan, and tyrosine residues of the hormone and tyrosine and tryptophan in the receptor-binding site.

Conclusions

Chemical modification of proteins should continue to serve as an important method for studying biological interactions that involve the specific binding between a hormone and a receptor. This technique represents a complementary approach in such studies and is sometimes a potential replacement for the two powerful tools of *in vitro* mutagenesis and direct chemical synthesis.

[11] P. M. Conn, D. C. Rogers, S. G. Seay, and D. Staley, *Endocrinology* (*Baltimore*) **115,** 1913 (1984).

Acknowledgments

I am grateful to Mrs. M. Kopelowitz for preparing the manuscript and to Profs. A. M. Kaye and M. Fridkin for their useful suggestions. This work was supported by the U.S.–Israel Binational Science Foundation, the Fund for Basic Research administered by the Israel Academy of Sciences and Humanities, and the Minerva Foundation, Federal Republic of Germany.

Section VI

Localization of Neuroendocrine Substances

[53] *In Situ* Hybridization: Quantitation Using Radiolabeled Hybridization Probes

By GEORGE R. UHL

Introduction

In situ hybridization techniques are finding increasing use when localization of gene expression is of importance, and especially when regulated gene expression is studied in heterogeneous cell populations. Recent reviews document approaches to localizing specific nucleic acid sequences to particular cell populations by *in situ* hybridization.[1-5] Here, we explore some of the issues and caveats involved when the techniques are used in a quantitated fashion. How can *in situ* hybridization yield data that can be interpreted quantitatively? What are the limitations of interpretation of these data?

Studies of the behavior of individual cells in regulating gene expression provide impetus for these quantitative concerns. In the brain, for example, large numbers of genes are expressed, but the expression of many of these important genes is highly localized to discrete cellular subpopulations.[1,6] Thus, studies of nucleic acids extracted from whole tissue can fail to elucidate features of gene regulation that may have great functional significance. If the differential distribution of gene expression from one cell type to another is important, or if regulated changes in levels of cellular expression are of interest, quantitated *in situ* hybridization techniques are the approaches of choice.

Two sorts of information can be sought in these studies. First, comparisons of the cellular expression of specific nucleic acid sequences in two particular states often yield biologically important information. These questions most frequently concern regulation of the expression of specific messenger RNAs; we focus on these mRNAs as targets of hybridization. In these settings, *relative* determinations of the cellular levels of hybridiz-

[1] G. R. Uhl, ed., "*In Situ* Hybridization in Brain." Plenum, New York, 1986.

[2] B. D. Shivers, B. S. Schachter, and D. W. Pfaff, this series, Vol. 124, p. 497.

[3] J. N. Wilcox, C. E. Gee, and J. L. Roberts, this series, Vol. 124, p. 510.

[4] J. D. Penschow, J. Haralambidis, P. Aldred, G. W. Tregear, and J. P. Coghlan, this series, Vol. 124, p. 534.

[5] K. Valentino, J. Eberwine, and J. Barchas, "In-Situ Hybridization: Neurobiological Applications." Raven, New York, 1987.

[6] D. M. Chikaraishi, S. S. Deeb, and N. Sueko, *Cell* (*Cambridge, Mass.*) **13,** 111 (1978).

able mRNAs are adequate to address many biological questions. In this chapter, we describe several approaches to determination of relative cellular hybridization densities using *in situ* hybridization. In other circumstances we might wish to determine the *absolute* cellular content of mRNA. As we shall see below, however, accurate determination of this second quantity may be more difficult without the use of additional adjunctive studies of extracted RNAs.

Issues in Quantitation of *in Situ* Hybridization: Classic Kinetics and *in Situ* Hybridization

Classic descriptions of nucleic acid association kinetics emphasize hybridization in solution.[7,8] Hybridization occurs in two phases. In an initial nucleation event, a relatively short nucleic acid sequence recognizes a complementary sequence. This initial recognition aligns stretches of similar sequence and allows subsequent rapid "zippering" of longer complementary sequences. The affinity of hybridization of one nucleic acid to another is classically described in terms of concentration- and time-dependent association parameters (e.g., C_0t). Hybridization affinity can also be determined by study of hybridization stringency: the temperature, salt, and formamide conditions that can prevent hybridization or "melt" already formed hybrids.

Three characteristics of the target of hybridization in *in situ* studies, however, combine to produce cautions about using this C_0t modeling to exactly describe the results of these studies. First, the target of hybridization is immobilized. This feature invalidates one of the assumptions of classic nucleic acid association kinetics, that each species is free to diffuse toward the other. Second, the hybridization target lies behind a diffusion barrier. The cellular constituents located between the target and the externally applied labeled hybridization probe can serve to "sieve" heterogeneously sized probe molecules, to delay their access, and to provide a substrate for degradation of the probe. These elements can thus induce both qualitative and quantitative changes in hybridization, in comparison to solution hybridization. Third, messenger RNA *in vivo* is not necessarily accessible for hybridization. Polysomes, nascent polypeptide

[7] R. J. Britten and E. H. Davidson, *in* "Nucleic Acid Hybridization. A Practical Approach" (B. D. Hames and S. J. Higgins, eds.), p. 3. IRL Press, Washington, D.C., 1985.

[8] D. C. Campbell, *in* "*In Situ* Hybridization in Brain" (G. R. Uhl, ed.), p. 239. Plenum, New York, 1986.

chains, and other cellular constituents fixed in apposition to the RNA could provide substantial steric hindrance for hybridization. These factors could result in qualitative changes in the stringency of hybridization (based on interrupting part of the hybridizing sequence) and could also yield quantitative changes in the total amount of hybridization.

Each of these features raises concerns about direct application of simple kinetic analyses to *in situ* hybridization. Fortunately, several empiric approaches can validate *in situ* quantitation. First, studies in a growing number of biological systems reveal that *in situ* hybridization signals do vary in a fashion that makes biological sense and that correlates with studies of mRNAs extracted from the same tissues (e.g., Refs. 9–19). When such hybridization differences are found, they can be further validated through attention to features of the hybridization that can influence quantitative results, care with autoradiographic quantitation, and use of either of two tools to approach "absolute" quantitation. In saturation analyses, hybridization responses to increasing probe concentrations are assessed.[20,21] This can yield interpretable results in relatively uncomplicated test situations. Alternatively, use of mRNA-sense standards can define the relationship between hybridization densities and the concentration of hybridizable mRNA target molecules (J. Palacios, personal communication).[22] We discuss methodologic considerations related to quantitation in the order that they occur in practice: those related to probe, hybridization and standardization, autoradiography, and analysis.

[9] G. R. Uhl, H. H. Zingg, and J. F. Habener, *Proc. Natl. Acad. Sci. U.S.A.* **82,** 5555 (1985).

[10] J. M. Rothfield, J. F. Hejtmancik, P. M. Conn, and D. W. Pfaff, *Exp. Brain Res.* **67,** 113 (1987).

[11] B. Wolfson, R. W. Manning, L. G. Davis, R. Arentzen, and F. Baldino, Jr., *Nature (London)* **315,** 59 (1985).

[12] R. T. Fremeau, Jr., J. R. Lundblad, D. B. Pritchett, J. N. Wilcox, and J. L. Roberts, *Science* **234,** 1265 (1986).

[13] L. G. Davis, R. Arentzen, J. M. Reid, R. W. Manning, B. Wolfson, K. L. Lawrence, and F. Baldino, Jr., *Proc. Natl. Acad. Sci. U.S.A.* **83,** 1145 (1986).

[14] W. S. Young III, E. Mezey, and R. E. Siegel, *Neurosci. Lett.* **70,** 198 (1986).

[15] S. A. Lewis and N. J. Cowan, *J. Neurochem.* **45,** 913 (1985).

[16] S. M. Reppert and G. R. Uhl, *Endocrinology* **120,** 2483 (1987).

[17] G. R. Uhl and S. M. Reppert, *Science* **232,** 390 (1986).

[18] B. D. Shivers, R. E. Harlan, G. J. Romano, R. D. Howells, and D. W. Pfaff, *in* "*In Situ* Hybridization in Brain" (G. R. Uhl, ed.), p. 3. Plenum, New York, 1986.

[19] S. L. Lightman and W. S. Young III, *Nature (London)* **328,** 643 (1987).

[20] J. E. Kelsey, S. J. Watson, S. Burke, H. Akil, and J. L. Roberts, *J. Neurosci.* **6,** 38 (1986).

[21] J. B. Lawrence and R. H. Singer, *Nucleic Acids Res.* **13,** 1777 (1985).

[22] G. R. Uhl, B. Navia, and J. Douglas, *J. Neurosci.*, in press (1988).

Approaches to Quantitation: Practical Considerations When Hybridizing

Probes

An ideal probe for quantitated *in situ* hybridization studies should have several characteristics that can facilitate interpretation of results.[23] If the probe has a known specific activity and if it is stable under conditions of hybridization, then the amount of radioactivity detected in the tissue can be simply related back to the mass of hybridized probe. Substantial degradation of probe molecules during the *in situ* hybridization procedures increases uncertainty about the specific activity of the hybridized molecules. In addition, the probe should have relatively ready access to the RNA species targeted. If tissue constituents exert substantial "sieving," hybridization of smaller probe fragments may be favored. Even if these fragments represent a small fraction of the total probe applied to tissue, they might in fact be selected during hybridization. Specific activity of the material in the tissue, then, would be substantially different from that of the starting materials. These considerations are of special concern for nick-translated probes, frequently used as a mixture of different species, and for cRNA probes used after basic hydrolyses or under hybridization conditions that allow breakdown of these easily degraded molecules.

Single Probe Concentration

Rationale. In instances where the biological question to be addressed concerns the relative amount of hybridizable mRNA in one state compared to another, a single concentration of radiolabeled probe is frequently utilized. These studies have successfully documented changes in the cellular and regional expression of specific mRNAs induced by a large number of stimuli.[9–19] The approach has empiric validation: Under the conditions employed, variation in hybridization densities do correlate with variation in the amount of mRNA that can be extracted from tissue.

Further means to buttress the validity of this approach involve use of mRNA-sense RNA or DNA standards, procedures that we and others have recently developed (J. Palacios, personal communication).[22] Using SP6/T7 vector systems, mRNA-sense RNA corresponding to the gene of interest can be synthesized, quantitated spectrophotometrically, diluted appropriately, and applied in known amounts to specific regions of nylon filters or other suitable supports. Alternatively, known amounts of this

[23] G. R. Uhl, *in* "*In Situ* Hybridization in Brain" (G. R. Uhl, ed.), p. 227. Plenum, New York, 1986.

RNA or mRNA-sense DNA oligonucleotides can be mixed with brain paste or embedding medium, frozen, and sectioned at the same thickness as the experimental material. These standards can be hybridized in parallel with tissue sections, under the same conditions. The ability of increasing amounts of RNA to provide an increasing hybridization signal and the relationship between the increasing amounts of RNA and increasing hybridization signal can both be assessed. If these interactions are not linear, they can provide estimates that, for example, 2-fold changes in hybridization density may reflect 3-fold changes in mRNA content.

mRNA-Sense Standards. We construct mRNA-sense standards using cDNAs cloned into plasmids such as pGEM (Promega) that also contain promoters for the active RNA polymerases SP6 and T7. When these plasmids are cut at the end of the cDNA, transcription with the appropriate RNA polymerase yields full-length RNA, as assessed by Northern analyses. RNA concentrations can be determined by absorbance at 260 nm,[24] and a range of dilutions set up. One-microliter aliquots of these RNA solutions can be heated to 65° and spotted onto 1 × 1 cm nylon filter squares (Nytran, Schleicher and Schuell). Application of a stream of warm air during spotting results in more uniform distribution of the RNA. Under these circumstances, 8–12 standard spots can easily be applied to each filter square. The filters are baked for 2 hr at 80° to fix the RNA to the filter, after which they can be stored at −70°. Before use, the filter squares can be glued to slides with super-glue. They are then hybridized, washed, and subjected to film autoradiography under the same condition used for experimental tissue sections.

mRNA-sense oligonucleotide DNA standards can also be synthesized chemically, quantitated spectrophotometrically, and used for the same purposes.[25] These oligonucleotides may be too short to adhere effectively to nylon or nitrocellulose filters, but they can be embedded in mounting medium or carefully prepared brain paste at known concentrations. Frozen sections of these standards can be thaw-mounted onto slides and hybridized in parallel with the experimental unknown sections. Loss of the standardizing mRNA sense oligonucleotide from the matrix by elution during hybridization and washing can be reduced by fixation of the standards prior to hybridization (J. Palacios, personal communication).

These strategies can define the relationship between the density of applied mRNA-sense standardizing material and the accessible mRNA in tissue. They cannot account as easily for variable mRNA retention by standards or variable access to the mRNA in unknown tissue standards.

[24] T. Maniatis, E. F. Fritsch, and J. Sambrook, "Molecular Cloning." Cold Spring Harbor Laboratory, Cold Spring Harbor, New York, 1982.

[25] M. J. Gait, "Oligonucleotide Synthesis." IRL Press, Oxford, 1984.

Multiple Probe Concentrations

Limitations of a Single Probe Concentration. Use of a single concentration of hybridizing probe could have several limitations. If, under the conditions utilized, virtually all of the probe molecules were hybridized to target, then no change in hybridization signal would be seen with increasing concentrations of target mRNA. The use of mRNA-sense standards can provide one control for this feature, if the hybridization densities for the standards encompass the range of hybridization values found in the unknown samples. In addition, it is conceivable that different physiological states examined could be accompanied by differences in accessibility of the target RNA to probe. Use of several different probe concentrations could help to elucidate such differences. Differences both in the barrier to diffusion from hybridization solution to the tissue and in the extent of mRNA obstruction by riboproteins could be overcome in part by increasing probe concentration. Conversely, if a hybridization difference between experimental and control tissues is noted at several different probe concentrations, there is greater assurance that it relates to bona fide differences in the number of molecules of hybridizable target mRNA species.

Saturation Analyses. Quantitation of the number of hybridizable mRNA molecules present may be approached using a single probe concentration in conjunction with mRNA-sense standards as noted above. Another approach to determine the maximal number of hybridizable molecules involves saturation analyses.[20,21] With addition of increasing amounts of radiolabeled probe, hybridization values typically increase up to a point, and then plateau. At this plateau, each accessible molecule of target mRNA is presumed to be hybridized; determination of this plateau hybridization level can then lead to an estimate of the density of the target mRNA.

In practice, these studies require relatively large amounts of hybridization probe. They are most readily performed in cultured cells or homogeneous cell populations, where radioactivity can be effectively detected by liquid scintillation counting over a broad range of values.[20,21]

Autoradiographic quantitation of these results requires multiple exposures, in conjunction with careful autoradiographic standardization (Uhl and Parta, unpublished observations, 1985). Generation of these sorts of data at the level of individual scattered cells requires a substantial grain-counting effort. Use of the high probe concentrations necessary to obtain saturation can frequently raise the fraction of probe that is nonspecifically associated with tissue, decrease signal-to-noise ratios, and increase the variability of estimates of the true saturation value.

Approaches to Quantitation: Quantitation of Radioactivity

In simple systems, it may be possible to model features of *in situ* hybridization reactions by detecting radioactivity through scintillation counting as noted above.[26] In most circumstances, however, the requirement for high anatomic resolution demands utilization of autoradiography. Quantitation of the results of these techniques in turn requires careful attention to specific technical concerns. Autoradiographic methods can yield film or emulsion autoradiograms, but virtually all of the same basic issues pertain to each approach. Each method determines the density of radioactivity over a specific area, which can be related back to the density of hybridized probe and even to the effective concentration of hybridizable mRNA, as noted above. Because of difficulties with anatomic scattering of the emissions from high-energy isotopes and with quenching of low-energy particles, absolute quantitation of the number of molecules in a certain area may be difficult, as noted below.

Saturation

Each autoradiographic system has only a finite capacity to display the consequences of radioactive decay. Saturation occurs when most of the silver grains available are converted to a latent image; at this point the image can get no blacker.[26–29] If two regions in the same experiment both show emulsion saturation, it is still possible that the amounts of radioactivity in each may be different. Thus, it is important to demonstrate that the radioactivity from each of two unknown samples falls within but not at the top of a range of possible values using autoradiographic standards before concluding that the radioactivity emitted by each is reliably sampled autoradiographically. Otherwise, the same values can be attributed to samples that are not in fact equally radioactive.

Determining the Operating Characteristic

The operating characteristic of an autoradiographic system determines the relationship between increasing amounts of radioactivity and grain

[26] A. W. Rogers, "Techniques of Autoradiography." Elsevier, Amsterdam, 1973.

[27] P. Dormer, "Molecular Biology, Biochemistry and Biophysics," p. 347. Springer-Verlag, Berlin and New York, 1973.

[28] R. P. Perry, *in* "Methods in Cell Physiology" (D. M. Prescott, ed.), p. 305. Academic Press, New York, 1969.

[29] R. J. Przybylski, *in* "Introduction to Quantitative Cytochemistry" (G. L. Wied and G. F. Bahr, eds.), p. 477. Academic Press, New York, 1970.

density (for emulsion autoradiography) or film optical density.[26] This need not be linear, but is typically a sigmoid curve whose midsection may approximate a linear function. This operating characteristic provides the most convenient means for approaching data analysis.

After film or emulsion hybridization autoradiograms are generated, the amount of film optical density, or grain density, must be related back to the amount of radioactivity. These studies can reveal the nature of the relationship between radioactivity and grain density, on the one hand, and also allow for quantitation of the amount of probe hybridized. This can be performed by coexposing autoradiographic standards of known activity with the unknown samples. Commercially available standards now incorporate tritium and iodine-125 in a medium whose quenching/absorbance parameters resemble those of brain tissue. These standards are available in blocks that can be sectioned at the same thickness as experimental materials and exposed to autoradiographic films or emulsions along with the experimental sections. Alternatively, standards can be constructed by incorporating known amounts of isotope in aliquots of tissue homogenate that can be frozen, cut at the same thickness as experimental samples, and coexposed with unknown samples.[30] In our hands, relatively good ranges of extended nearly linear film and emulsion response to ^{3}H, ^{125}I, or ^{35}S radioisotopes can be obtained. Quantitation corresponding to ^{32}P can yield more strikingly nonlinear responses.

Artifacts: Section and Emulsion Thickness Variation

Several factors can produce artifactual variation in autoradiographic signals. When using relatively higher energy radioisotopes, such as ^{35}S and ^{32}P, the intensity of the autoradiographic signal will depend not only on the relative cellular and regional concentration of target RNA or DNA, but also on the thickness of the sections and of the emulsion.[26] Thus, tissue areas to which more emulsion adheres can display increased grain numbers based solely on the increased emulsion thickness. For these high-energy isotopes, the amount of radioactivity over a particular area will also be proportional to the section thickness. Thus, attention to section evenness is of substantial importance. One means for minimizing tissue-induced variability in emulsion thickness employs emulsion-coated coverslips.[31]

[30] G. Uhl and K. Hill, *in* "*In Situ* Hybridization in Brain" (G. R. Uhl, ed.), p. 287. Plenum, New York, 1986.

[31] M. J. Kuhar, *in* "Neurotransmitter Receptor Binding" (H. I. Yamamura, S. J. Enna, and M. J. Kuhar, eds.), p. 153. Raven, New York, 1985.

Artifact: Quenching

When using tritium, however, a different set of considerations prevails. Different tissue constituents can exert differential "quenching."[26,32] Differing degrees of tissue self-absorbance of weaker tritium β emissions can yield artifactual differences in apparent hybridization densities. Conversely, the relatively low mean free path length of these tritium β emissions makes small variations in emulsion and section thickness less crucial, because only the top few micrometers of section thickness and the adjacent few micrometers of emulsion thickness are involved in exposures using this isotope.

Artifacts: Chemography

Emulsions can react with nonradioactive chemical tissue constituents. These interactions can produce either artificially high or artificially low grain densities (positive and negative chemography).[26] Tissue sections free from radioactivity or those mounted over a uniform slide coating of isotope can be utilized to check for positive and negative chemography, respectively.

Approaches to Quantitation: Analyses

Film

Film autoradiograms are best analyzed by using an image analysis system that allows calibrated determination of optical densities over geometrically irregular regions of interest. The borders of these regions can be determined by careful comparison of the film autoradiographic images to cell-stained representations of the tissue section used to generate the autoradiograms. The film optical density will fall off more or less gradually at the edge of an area rich in radioactivity. Thus, standardized conventions should be employed to define the edge of a region, so that the same sampling strategy can be reproduced on each tissue section analyzed. This is of greater concern if the biological system under study can change its cellular characteristics or size.

When positively hybridizing cells are scattered over a region, care must be taken to avoid saturation of the small film areas that overlie the cells themselves. Regional film optical densities reflect both the high densities over hybridizing cells and the low densities between the cells. Thus, local film saturation can take place when the optical density values of the

[32] W. A. Geary III, A. W. Toga, and G. F. Wooten, *Brain Res.*, in press (1988).

entire region are substantially below saturation. Determination of hybridization densities over individual cells in emulsion autoradiograms can obviate this potential difficulty.

Background optical density values are typically subtracted from each observation. Careful attention to selection of tissue areas known to be free of the mRNA of interest can provide a background determination that takes into account radioactivity that is nonspecifically absorbed or trapped by the tissue. Alternatively, an area of film away from the tissue can provide an estimate of the autoradiographic background that is attributable to the density of the film alone. Optical densities over mRNA-sense and autoradiographic standards coexposed to the same film can also be determined. With these values in hand, the optical density values corresponding to regions of the unknown sample can be related more accurately to the density of radioactivity, the density of hybridized probe, and the density of accessible mRNA.

Emulsion

Signal. In studies where the relative cellular content of hybridizable mRNA is of interest, determining the density of autoradiographic grains over an area of a cell can provide a good measure of hybridization density. However, if absolute measurements of total cellular mRNA content are undertaken, then all of the grains associated with the cell must be counted.[26] The distribution of grains from a point source falls off with distance from the source. Thus, the distance from a cell of a grain attributable to a radioactive decay event within the cell will vary with the isotope used. Furthermore, if cells are close together, some of the grains lying over the second cell may be attributable to radioactive events in the first cell, and vice versa. These considerations make absolute determinations of total cellular grains difficult unless (1) the cells are widely separated and (2) the autoradiographic background is so low that virtually all of the grains in the vicinity of the cells can be confidently attributable to specific hybridization.

Background. Autoradiographic background values can again be determined over tissue zones free of the mRNA of interest, or from emulsion that overlies no tissue. These grain densities are typically subtracted from the signal observed over positively hybridizing zones.

Partial Volume Effects. In studies comparing hybridization to different cell types, it is important that the cells be represented to the same extent in the tissue section.[26] One approach to this problem involves quantitating autoradiographic signals only over cells whose nucleus is present in this section.[33] This technique avoids "partial voluming," the inclusion of only the edge of a cell within a small fraction of total section thickness.

Data Analysis

Regional Values: Film Optical Densities

When the cells of interest lie close together, and when they behave in a homogeneous fashion, the average optical density over the region is the parameter of interest. This value is corrected for background, typically by subtracting the optical density of an adjacent film area that did not lie over a tissue section. Finally, the position of the value on two standard curves is noted. In comparison with the radioactivity standards constructed with the same isotope used to label the hybridization probe, the relation of the value to film saturation and to other radioactivity levels can be assessed. In comparison with cohybridized mRNA-sense standards, the relation between the unknown value and the amount of added mRNA can be determined. If both of these comparisons yield values within a nearly linear range, then the unknown data can be validly compared directly to other background-corrected unknown samples without fear of improper interpretation due to saturation of the film or hybridization of all available probe molecules. Statements about the absolute density or copy number of mRNA present in the tissue, however, must still be tempered by the realization that tissue factors hindering full hybridization may not be present in the mRNA-sense standards.

Statistical treatment of such data is straightforward. Although there is no reason to believe *a priori* that the data need be parametric, these tests are often used in practice.

Cellular Values: Grain Densities from Emulsion Autoradiograms

If quantitation of hybridization to individual neurons is required, the density of autoradiographic grains above the cells of interest is easily determined. This can be done usually using a calibrated eyepiece grid or can be approached using an automated image analysis system. We routinely assess grain counts in a 10×10 μm grid box positioned over neuronal cytoplasm at 100× objective magnification.[33] Alternatively, several image analysis systems allow assessment of grain densities overlying cells of certain size classes. When these figures are corrected for autora-

[33] G. R. Uhl and C. A. Sasek, *J. Neurosci.* **6,** 3258 (1986).

diographic background, they can be displayed as frequency distribution histograms. The fraction of all cells displaying certain grain densities can be plotted against ascending grain density.

The numerical data obtained in this way can be analyzed based on such questions as (1) What is the fraction of total cells that express the mRNA of interest, and how does it change with a biological stimulus? and (2) What is the intensity (and range) of expression of the mRNA of interest, and how does this change with various stimuli? Determination of the threshold for considering a cell to be positively expressing can be based on its displaying several times (e.g., 3–5) more grains than background values or on an assessment of the shape of the cell labeling density–frequency distribution histograms. These numerical values can again be compared directly by simple statistical tests. If one wishes to compare statistically two population hybridization density distributions, more complex statistics such as the Kolmogorov–Smirnov test must be used. These methods are also discussed by McCabe *et al.* (this volume [59]).

Conclusions

As approaches to *in situ* hybridization mature, and as more questions arise about gene regulation in individual cells, quantitative *in situ* hybridization analyses will assume increasing importance. With careful attention to specific technical details, this approach should yield data that can help to address many of these biological questions.

Acknowledgments

I thank Drs. Robert Singer and Gerhard Heinrich for valuable comments, and Ms. Janice Caniff for assistance with the manuscript. This work was supported by the Howard Hughes Medical Institute, McKnight Foundation, Sloan Foundation, American Parkinson's Disease Association, and the National Institutes of Health.

[54] Nonradioactive Methods of *in Situ* Hybridization: Visualization of Neuroendocrine mRNA

By P. C. Emson, H. Arai, S. Agrawal, C. Christodoulou, and M. J. Gait

Introduction

Current methods for detecting and localizing neuropeptide mRNAs in the pituitary or brain have commonly used radioactive complementary DNA or RNA probes (either single- or double-stranded). Hitherto, nonradioactive methods have been generally unsuccessful in detecting peptide mRNAs, especially when used for *in situ* hybridization.[1] As part of a series of experiments designed to develop improved nucleic acid probes for nonradioactive hybridization procedures, we have chemically synthesized 5′-biotin-labeled oligodeoxyribonucleotides.[2] In this chapter we describe a procedure for the synthesis of 5′-biotin-labeled oligodeoxyribonucleotide probes and show that these can be used to detect the presence of vasopressin or oxytocin mRNAs in the hypothalamus and to follow their physiological regulation.

The principal advantage of using chemically synthesized oligonucleotides that are labeled with biotin at the 5′ position is that the labeling does not significantly alter the binding characteristics of the oligonucleotide. In contrast, incorporation of a biotinylated heterocyclic base into the oligonucleotide sequence may lower the melting temperature (T_m) of the probe–target duplex.[3-5] Similarly, biotin labeling with a photoactivatable analog of biotin (photobiotin) produces modifications in the heterocyclic bases that also result in a substantial reduction of T_m.[6] Further, chemical labeling of the 5′ position allows a spacer arm to be introduced to separate the oligonucleotide sequence from the biotin group, allowing easier access of avidin–biotin complexes to the biotin on the oligonucleotide.

[1] G. Uhl, ed., "*In Situ* Hybridization in Brain." Plenum, New York, 1986.
[2] S. Agrawal, C. Christodoulou, and M. J. Gait, *Nucleic Acids Res.* **15,** 6227 (1986).
[3] P. R. Langer, A. A. Waldrop, and D. C. Ward, *Proc. Natl. Acad. Sci. U.S.A.* **78,** 6633 (1981).
[4] D. J. Brigati, D. Myerson, J. J. Leary, B. Spalhoz, S. Travis, C. K. Y. Fang, G. D. Hsiung, and D. C. Ward, *Virology* **126,** 32 (1982).
[5] J. L. Leary, D. J. Brigati, and D. C. Ward, *Proc. Natl. Acad. Sci. U.S.A.* **80,** 4045 (1983).
[6] A. C. Forster, J. L. McInnes, D. C. Skingle, and R. H. Symons, *Nucleic Acids Res.* **13,** 745 (1985).

Methods

Oligonucleotide Synthesis

Oligonucleotide synthesis is carried out by the phosphoramidite method using an Applied Biosystems 380B DNA synthesizer and β-cyanoethyl phosphoramidites (BDH, Poole, UK). A preproduction sample of linker phosphoramite (see below) is also obtained from BDH.

High-Performance Liquid Chromatography

HPLC is carried out using a Waters gradient system consisting of two Model 510 pumps, a 680 gradient programmer, 481 variable wavelength UV detector, 730 data module, and a Rheodyne 7125 injector. The columns are either the radial PAK (used with a Z module system) or the stainless-steel type of μBondapak C_{18} reversed phase. Chromatography is carried out at a flow rate of 1.5 ml/min using a buffer of 0.1 *M* ammonium acetate (pH not adjusted) and gradients of acetonitrile.

Synthesis of Linker Molecule

A linker suitable for attachment of amino-specific labeling agents is prepared by reaction of 6-amino-1-hexanol with 9-fluorenylmethyl chloroformate in aqueous acetone containing sodium carbonate as previously described[2] to produce 2-(9-fluorenylmethoxycarbonyl)aminohexanol, which is then reacted with 2-cyanoethyl-*N*,*N*-diisopropylamino-chlorophosphite to give the linker phosphoramidite, 2-(9-fluorenylmethoxycarbonyl)aminohexyl-2-cyanoethyl-*N*,*N*-diisopropylaminophosphite. This linker phosphoramidite is used in the final stage of the assembly of the relevant oligonucleotide (in this case oligonucleotides complementary to the mRNA coding for part of the vasopressin glycopeptide or oxytocin sequence[7,8]), to give the relevant amino-functionalized oligonucleotide.[2] The amino oligonucleotide is purified by reversed-phase HPLC (Fig. 1a) and reacted overnight with *N*-hydroxysuccinimidobiotin (15 m*M*) in 125 m*M* Tris-HCl buffer (pH 7.4) in dimethylformamide/water (1 : 1) at room temperature. Alternatively, the crude amino oligonucleotide can be similarly biotinylated. The biotinylated product elutes slightly later from the reversed-phase column (Fig. 1b) and is collected and dialyzed against water. The 5′-aminohexylbiotinylated oligonucleotide can now be used directly in hybridization experiments. Biotinylated oligonucleotides are

[7] H. Land, M. Grez, S. Ruppert, H. Schmale, M. Rehbein, D. Richter, and G. Shultz, *Nature (London)* **302,** 342 (1983).

[8] S. Ruppert, G. Scherer, and G. Shultz, *Nature (London)* **308,** 554 (1984).

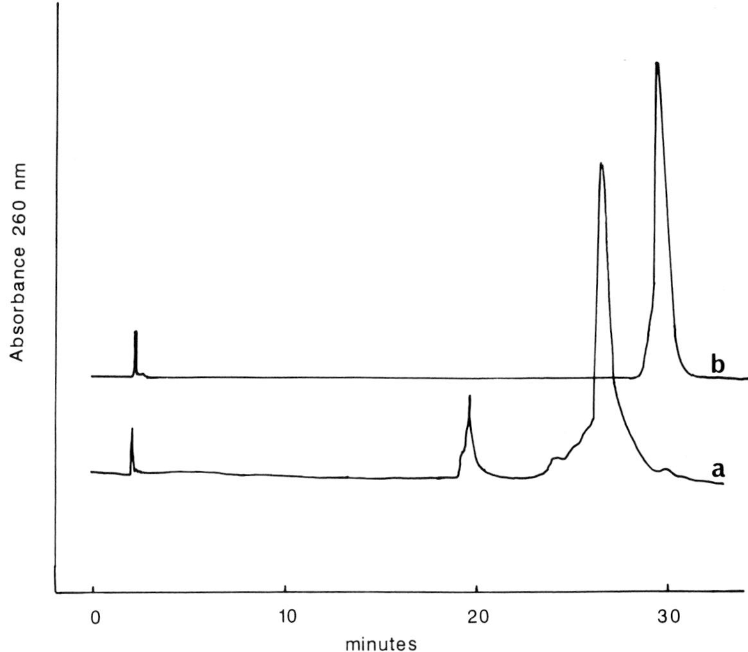

FIG. 1. Reversed-phase HPLC traces of (a) the aminohexyl oligonucleotide complementary to the vasopressin glycopeptide sequence and (b) the same oligonucleotide after biotinylation.

stable for at least 3 months when stored in aqueous solution at −20°. We have observed sensitivities of between 0.1 and 1.0 fmol for detection of sense vasopressin mRNA sequences spotted onto nitrocellulose (H. Arai, unpublished observation) using streptavidin and an alkaline phosphatase detection system, such as that marketed by BRL (Blue-Gene).

In Situ Hybridization

There are a number of variations on recipes and fixation conditions for *in situ* hybridization. The method described here works well with biotinylated probes and uses 20-μm free-floating sections which can be readily processed further for electron microscopy.

Fixation. Anesthetized rats are perfused via the ascending aorta with 300 ml of ice-cold 4% paraformaldehyde (buffered to pH 7.4 with 100 m*M* sodium phosphate) dissolved in autoclaved distilled water pretreated with the RNase inhibitor diethyl pyrocarbonate (DEP) (0.1%). DEP-treated, autoclaved water is used routinely for all solutions. After perfusion the brain is removed and left for at least 24 hr in 30% sucrose (the brain is completely impregnated with sucrose when it sinks to the bottom of the storage jar). Once impregnated, sections of 20 μm are cut on a freezing microtome or cryostat and collected into sterile PBS (phosphate-buffered saline, 100 m*M* $NaPO_4$, pH 7.4, 0.9% saline) made up in DEP-treated water.

Sections containing the hypothalamic nuclei of interest (supraoptic or paraventricular) are treated for 30 min with proteinase K (1 μg/ml in 100 m*M* Tris–HCl, pH 7.6, containing 50 m*M* EDTA) to enhance penetration of the biotinylated oligonucleotide. After proteinase K digestion the sections are briefly fixed again (5 min in 4% buffered paraformaldehyde) and rinsed 3 times in PBS/DEP, after which they are ready for prehybridization.

Hybridization and Prehybridization. Hybridization and prehybridization are both carried out at room temperature. The prehybridization solution consists of 50% deionized formamide, 5× SSC (1× SSC: 0.15 *M* sodium chloride, 15 m*M* sodium citrate, pH 7.4), 20 m*M* sodium phosphate, pH 6.5, 0.2% bovine serum albumin (BSA, molecular biology grade), 0.2% polyvinylpyrrolidone, 0.2% Ficoll 400, 5% dextran sulfate, 100 μg/ml herring sperm DNA, and 40 μg/ml yeast tRNA. Sections in sterile plastic centrifuge tubes are prehybridized in this solution for 3 hr at room temperature. The aim of prehybridization is to allow any material (DNA, RNA, proteins) that might nonselectively bind biotinylated probe to react with excess herring sperm DNA, yeast tRNA, or BSA. After prehybridization, the hybridization solution, which is identical to the prehybridization solution except that it contains 1–2 μg/ml biotinylated oligonucleotide, is added to the tubes containing the sections. Hybridization is carried out overnight at room temperature, and after this sections are rinsed twice in 4× SSC, for 30 min at 37°, and then in sterile PBS for 5 min at room temperature before development of the hybridization signal.

Development of Hybridization Signal. The biotinylated oligonucleotide can potentially be detected by antibiotin antibodies and/or by using conjugated avidin or streptavidin.[3–5] We have found that a considerable amount of "signal" amplification is required, and for this we have used two avidin–biotin complexes (ABC) linked by a biotinylated antiavidin antibody (Fig. 2). This method allows the use of a number of biotinylated enzymes as part of the ABC complex, and we have successfully used both alkaline phosphatase (Fig. 3A) and horseradish peroxidase (Fig. 3D). Sections are incubated in freshly prepared ABC complex (Vector Laboratories, Burlingame, California) for 15 min, rinsed in PBS/DEP for 15 min, and then incubated in PBS/DEP containing 1% BSA and biotinylated

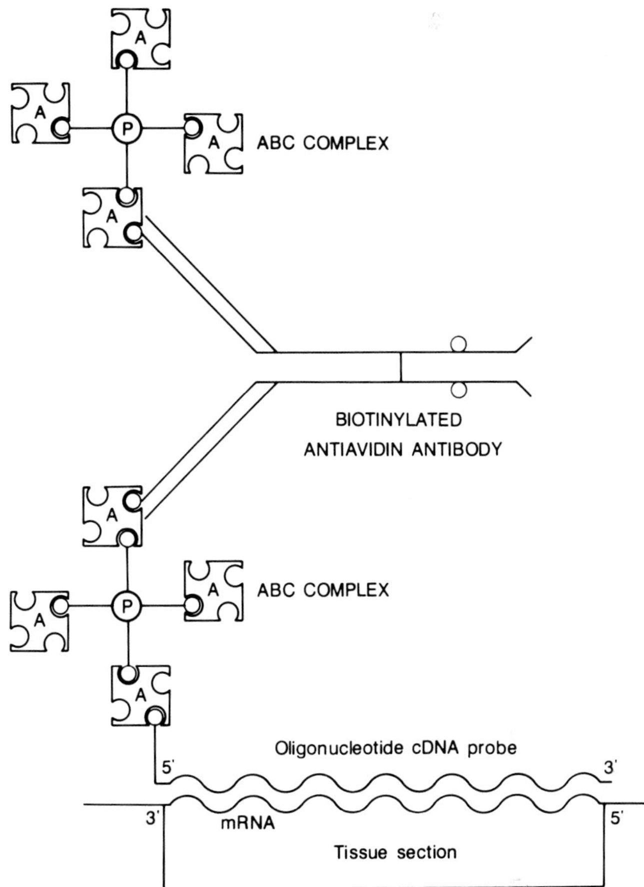

FIG. 2. Schematic representation of the ABC–antiavidin–ABC complex used to detect the biotinylated oligonucleotide on the tissue section. The ABC complex represented is between biotinylated horseradish peroxidase (P) and avidin (A). The linker antibody is biotinylated antiavidin. Note that the ABC complex will also detect biotin sites on the antiavidin antibody, but to avoid over complicating the figure this has been omitted.

antiavidin antiserum (1 : 20 dilution, Vector) for 30 min at 37°. After further washing in PBS/DEP at room temperature, sections are reincubated with ABC complex, and after a final rinse the ABC complex is detected using either diaminobenzidine (1 mg/ml) in PBS containing 0.03% hydrogen peroxide or an alkaline phosphatase substrate (Vector). For comparison of signals between brains, sections are developed for the same length of time so that differences in expression can be at least partially related to

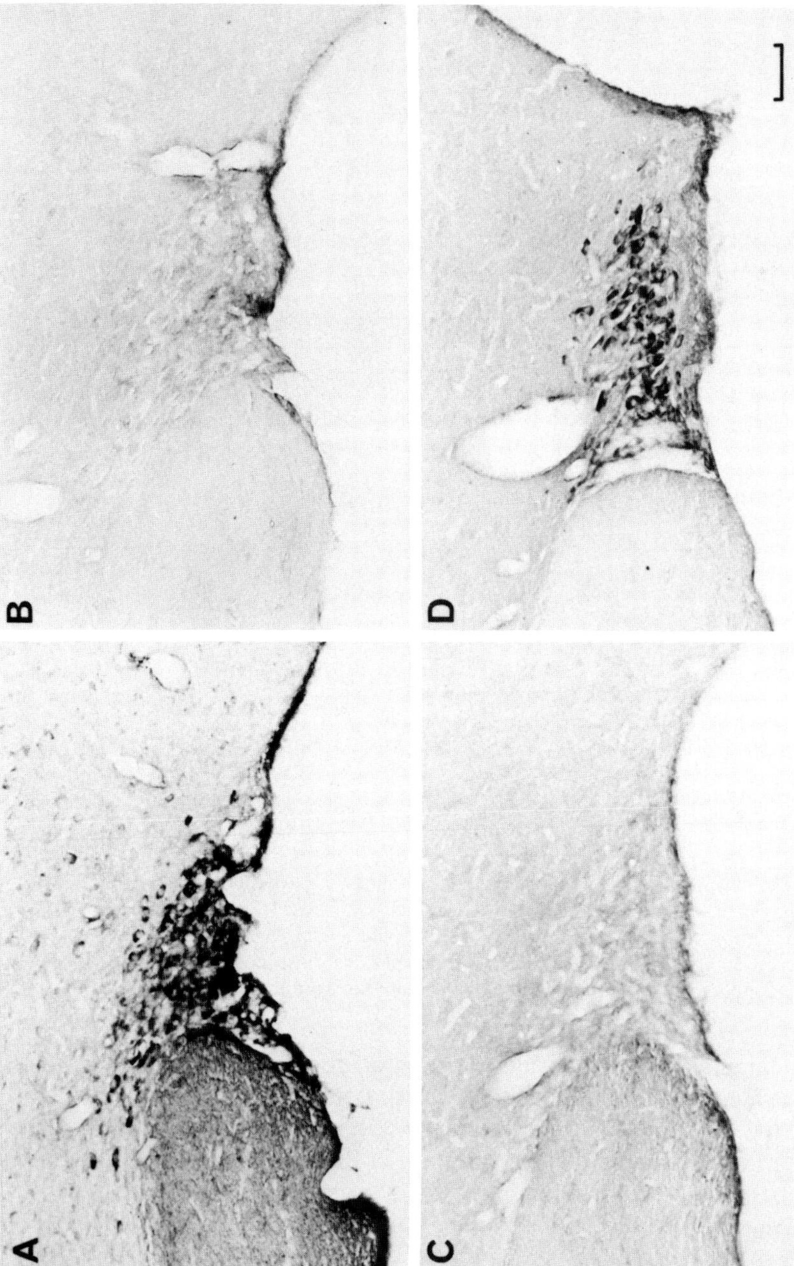

FIG. 3. Localization of vasopressin mRNA in supraoptic neurons using an alkaline phosphatase–ABC complex (A) or horseradish peroxidase (D). RNase pretreatment of the sections (B) or use of an excess (10 μg/ml) of complementary sense oligonucleotide probe (C) prevented specific staining. Bar, 50 μm.

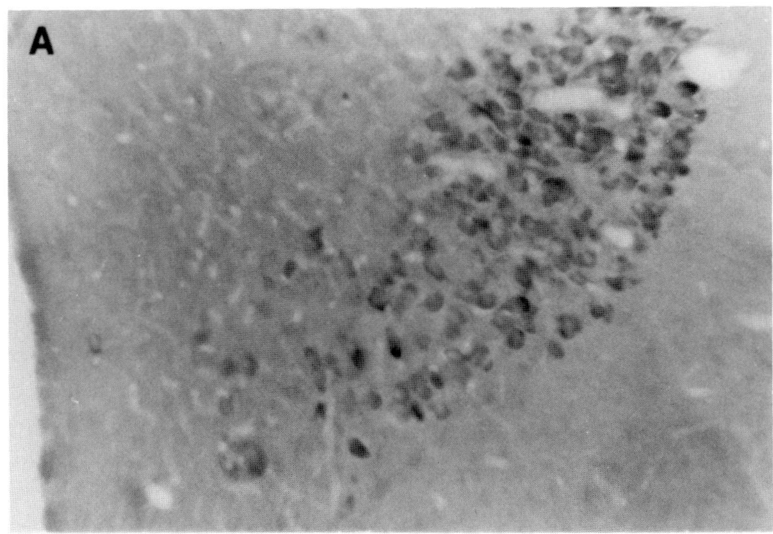

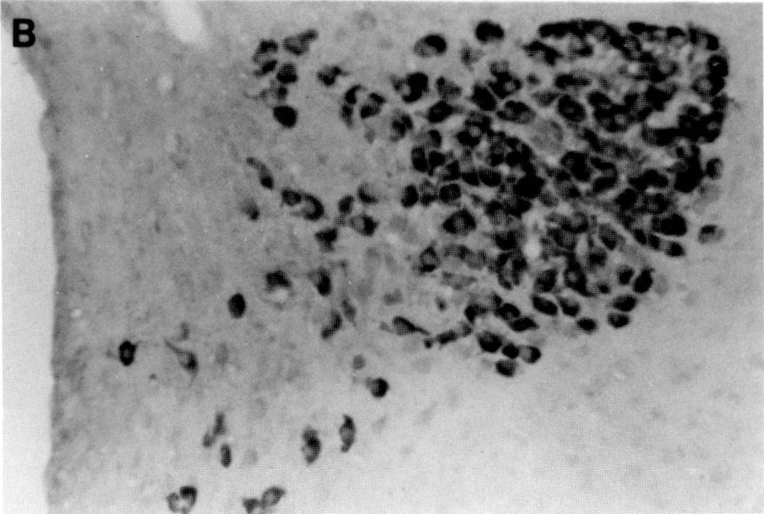

FIG. 4. Localization of vasopressin mRNA in the paraventricular nucleus of a control rat (A) and a rat given saline for 2 days before processing (B). Sections were incubated for the same amount of time. Note the increase in specific signal in the paraventricular nucleus of the salt-loaded rat. ×450.

intensity of color developed. In our case the increase in vasopressin expression produced by saline treatment of rats[9-11] can be readily visualized (Fig. 4).

Specificity. In any hybridization reaction or color reaction carried out on tissue sections, the possibility of artifacts or nonspecific binding of the biotinylated probe needs to be considered. In this case the signal detected is clearly localized over cells known to express vasopressin peptide and neurophysin I, and *competition* with excess (10 μg/ml) sense oligonucleotide probe complementary to the biotinylated oligonucleotide (Fig. 3C) removes the signal. This competition control is the best test of specificity. An additional useful control is the use of RNase A (20 μg/ml) to digest tissue mRNA (Fig. 3B). This RNase control, however, must be conducted carefully since use of excess RNase can also degrade ribosomal RNA and result in a reduction in background signal. It is also appropriate to note that, although the final wash of hybridized sections is in 4× SSC at 37° (a relatively low degree of stringency), the extensive washes in low-salt PBS and ABC solutions and in the color development solution introduce a very high degree of stringency to the procedure.

Conclusions

The method illustrated allows the detection of relatively abundant neuroendocrine mRNAs in the hypothalamus. The method should be particularly useful for studies of coexpression (detection of two mRNAs in one tissue section) and the localization of mRNA intracellularly. Indeed our initial results show that peroxidase product is localized to the rough endoplasmic reticulum of oxytocin-containing neurons (H. Arai, unpublished). The reason for the method's success probably lies in the use of a biotin group removed by a linker arm from the oligonucleotide and the considerable amplification of the original signal provided by two ABC complexes joined by an antiavidin antibody.

For the future we are currently exploring chemical labeling of cRNA probes with biotin and improving the chemical synthesis of biotinylated (and multibiotinylated) RNA and DNA probes. Success in the use of multibiotinylated probes will probably depend on ensuring that the biotins are sufficiently sterically separated so that access of conjugated avidin,

[9] J. T. McCabe, J. I. Morrell, R. Ivell, H. Schmale, D. Richter, and D. W. Pfaff, *J. Histochem. Cytochem.* **34,** 45 (1986).

[10] B. Wolfson, R. W. Manning, L. G. Davis, R. Arentzen, and F. Baldino, *Nature (London)* **315,** 59 (1985).

[11] W. S. Young III, E. Mezey, and R. E. Siegel, *Mol. Brain Res.* **1,** 231 (1986).

streptavidin, or antibiotin antibodies is not impeded while preserving the specificity of the DNA or RNA probe.

Acknowledgments

H. A. was a visiting fellow from the Institute of Psychiatry, Tokyo, Japan, supported by the Mental Health Foundation, UK.

[55] High-Resolution *in Situ* Hybridization Histochemistry

By FRANK BALDINO, JR., MARIE-FRANCOISE CHESSELET, and MICHAEL E. LEWIS

Introduction

The application of recombinant DNA technology to the study of the nervous system has permitted neuroscientists to accurately address questions relating to the transcription of specific genes within the central nervous system (CNS). The cloning and identification of a variety of genes and gene families which are expressed in nervous tissue have fostered the development of an impressive array of cDNAs to probe the nervous system for specific sequences of nucleic acids. These technologies take advantage of the electronic complementarity and thermodynamic stability of Watson–Crick base pairs. Under the appropriate conditions, single strands of complementary pairs of nucleotides interact by hydrogen bonding and base-pair stacking, resulting in the formation of stable hybrids. Hybridization can thus be achieved with virtually any complementary DNA or RNA sequence of sufficient length. These probes can be radiolabeled to a degree of specific activity that permits the autoradiographic localization of specific genomic or RNA species. Although hybridization techniques have been routinely employed by molecular biologists for a number of years, the application of these techniques to the CNS requires special considerations.

The CNS consists of a heterogeneous population of biochemically distinct neuronal and glial elements, each possessing different morphological, chemical, and functional attributes. Thus, the use of Northern blot hybridizations, solution hybridizations, or dot/slot blot hybridizations to assay regions of the nervous system comprised of multiple diverse cellular populations does not permit the consideration of singular neuronal elements within that population. Although these hybridization assays have provided valuable insights into the mechanisms responsible for the regulation of genes and primary transcripts in the CNS, they do not permit a determination of the factors mediating the regulation of these events in individual neurons or glial cells. Because of this limitation, methods have been developed to hybridize specific species of mRNA *in situ*.

In situ hybridization histochemistry (ISHH) has become a powerful tool to study the regulation of selected mRNA species in single neurons. Several laboratories have successfully used *in situ* hybridization to localize relatively rare mRNA species within individual neurons throughout many regions of the CNS.[1-11] The localization of these transcripts to individual neurons has allowed, for the first time, the identification of which neurons actually synthesize a given protein or neuropeptide. In addition, refinement of this hybridization technology has yielded insights into the regulation of neuropeptides, enzymes, and other CNS-related proteins at the level of single neurons. The purpose of this chapter is to provide a detailed guide for the reader to perform *in situ* hybridization histochemistry in nervous tissue with the high degree of resolution required to assay these hybrids within the cytoplasm of individual neurons. Oligonucleotide and RNA probes are most commonly used for ISHH. Each type of probe has specific advantages, but in most cases either probe can be used with similar success. Several aspects of the ISHH procedure are common to both types of probes and are described in more detail in the section on oligonucleotides.

[1] C. E. Gee, C. Chen, J. E. Roberts, R. Thompson, and S. J. Watson, *Nature* (*London*) **306,** 374 (1983).

[2] W. S. T. Griffin, M. Alejos, G. Nilaver, and M. R. Morrison, *Br. Res. Bull.* **10,** 597 (1983).

[3] B. Wolfson, R. W. Manning, L. G. Davis, R. Arentzen, and F. Baldino, Jr., *Nature* (*London*) **315,** 59 (1985).

[4] G. R. Uhl, H. H. Zingg, and J. F. Habener, *Proc. Natl. Acad. Sci. U.S.A.* **82,** 555 (1985).

[5] L. G. Davis, R. Arentzen, J. M. Reid, R. W. Manning, B. Wolfson, K. L. Lawrence, and F. Baldino, Jr., *Proc. Natl. Acad. Sci. U.S.A.* **83,** 1145 (1986).

[6] R. E. Siegel and W. S. Young III, *Neuropeptides* **6,** 574 (1985).

[7] B. D. Shivers, R. E. Harlan, G. J. Romano, R. D. Howells, and D. W. Pfaff, *Proc. Natl. Acad. Sci. U.S.A.* **83,** 6221 (1986).

[8] M. Schalling, T. Hökfelt, B. Wallace, M. Goldstein, D. Filer, C. Yamin, and D. H. Schlesinger, *Proc. Natl. Acad. Sci. U.S.A.* **83,** 6208 (1986).

[9] M. E. Lewis, R. Arentzen, and F. Baldino, Jr., *J. Neurosci. Res.* **16,** 117 (1986).

[10] B. Bloch, T. Popovici, D. Le Guetlec, E. Normond, S. Chouham, A. F. Guitteny, and P. Bohlen, *J. Neurosci. Res.* **16,** 183 (1986).

[11] D. R. Gehlert, B. M. Chronwall, M. P. Schafer, and T. L. O'Donohue, *Synapse* **1,** 25 (1987).

Synthetic Oligonucleotide Probes

General Considerations

The use of synthetic oligodeoxyribonucleotide (oligonucleotide) probes to detect known mRNA sequences provides significant advantages over the use of cloned cDNAs for *in situ* hybridization studies.[12] Unlike cDNAs, oligonucleotides are single-stranded and are easily labeled on either the 3′ or 5′ end with a variety of radioactive and nonradioactive tags (see below). Furthermore, the synthesis and labeling of these probes is readily accomplished *in vitro* by an enzymatic reaction without the need for sophisticated cloning procedures. The relatively short size of these probes (20–50 bases) makes them less susceptible to many of the methodological limitations often associated with long DNA sequences. They readily penetrate the cell membrane and diffuse into the cytoplasm. DNA is extremely stable within the cytoplasm and therefore has a high probability of forming a stable DNA–RNA hybrid.

One of the most important advantages of short oligonucleotide probes is that they permit a wide range of control experiments to be performed (see below). In this respect, the most significant feature of these probes is that their short length permits the predicted thermal denaturation of the DNA–RNA hybrid to be used as a sensitive index of hybridization specificity. Another unique feature of synthetic DNA is that complementary sequences can be synthesized to highly restricted regions of the RNA, thus permitting the cellular resolution of individual intron and exon regions of the RNA. Therefore, despite the limitation of their use to detection of sequenced mRNA species, the use of oligonucleotide probes to detect RNA within individual cells provides a powerful tool to study factors which regulate gene expression in the CNS.

Oligonucleotide Synthesis and Labeling

Recent years have seen the emergence of rapid and relatively simple methods for the solid-phase synthesis of oligonucleotides. Syntheses that were once the province of experienced organic chemists can now be carried out by chemically naive biologists using automated DNA synthesizers. Solid-phase synthetic methods have been discussed elsewhere[13] and will not be described here. For investigators who cannot or prefer not

[12] J. N. Wilcox, C. E. Gee, and J. L. Roberts, this series, Vol. 124, p. 510.

[13] M. J. Gait, ed., "Oligonucleotide Synthesis, A Practical Approach." IRL Press, New York, 1986.

to carry out their own syntheses, many universities and research centers now have central DNA synthesis facilities, and a number of companies (e.g., OCS Laboratories, Denton, TX; Molecular Biosystems, Inc., La Jolla, CA) will also undertake custom syntheses.

After the oligonucleotide has been synthesized and purified, it is then labeled to permit its detection following hybridization to the target mRNA. Four radioisotopes have been used for DNA probe labeling: ^{3}H, ^{35}S, ^{32}P, and ^{125}I. Each of these radioisotopes has particular advantages and disadvantages based on the energy of their emitted particles. Tritium, which emits low-energy β particles, provides excellent resolution, but requires long exposures, and is compromised for quantitative work by a high degree of tissue quenching. At the other extreme, ^{32}P emits very energetic β particles, yielding short exposure times, but generally at the cost of poor resolution. The "intermediate" solution is to use ^{35}S or ^{125}I, which both offer relatively good resolution with moderate exposure times.

There are essentially three methods for radiolabeling synthetic oligonucleotides: (1) 5′ end labeling,[14,15] in which T4 polynucleotide kinase is used to transfer the terminal phosphate from [γ-^{32}P]ATP to the free 5′ OH group of the oligonucleotide; (2) primer extension,[4,9,16] using the Klenow fragment of DNA polymerase I to catalyze the extension of the primer (with radioactive nucleotides) across a message sense oligonucleotide template; and (3) 3′ end labeling,[17] in which terminal deoxynucleotidyltransferase is used to catalyze the sequential addition of radioactive nucleotides to the 3′ terminus of the oligonucleotide. The primer extension and 3′ end labeling methods share the advantage of incorporating multiple radioactive nucleotides into each molecule of probe, but the latter procedure is technically much easier to perform and has been used to label DNA probes for both *in situ* hybridization[9,11,17–21]

[14] L. G. Davis, M. D. Dibner, and J. F. Battey, "Basic Methods in Molecular Biology." Elsevier, New York, 1986.

[15] T. Maniatis, E. F. Fritsch, and J. Sambrook, "Molecular Cloning: A Laboratory Manual." Cold Spring Harbor Laboratory, Cold Spring Harbor, New York, 1982.

[16] A. B. Studencki and R. B. Wallace, *DNA* **32,** 7 (1984).

[17] M. E. Lewis, T. G. Sherman, and S. J. Watson, *Peptides (N.Y.)* **6,** 75 (1985).

[18] M. E. Lewis, T. G. Sherman, S. Burke, H. Akil, L. G. Davis, R. Arentzen, and S. J. Watson, *Proc. Natl. Acad. Sci. U.S.A.* **83,** 5419 (1986).

[19] J. P. Card, S. Fitzpatrick-McElligott, I. Gozes, and F. Baldino, Jr., *Cell Tissue Res.*, in press (1988).

[20] B. D. Shivers, R. E. Harlan, J. F. Hejtmancik, P. M. Conn, and D. W. Pfaff, *Endocrinology (Baltimore)* **118,** 883 (1986).

[21] W. S. Young III, T. I. Bonner, and M. R. Brann, *Proc. Natl. Acad. Sci. U.S.A.* **83,** 9827 (1986).

and Southern blotting.[22] The 3′-tailing reaction is carried out by incubating 15 pmol of oligonucleotide in a 25-μl volume containing 60–100 pmol of radiolabeled deoxyribonucleotide triphosphate (e.g., ^{35}S-labeled dATP or ^{125}I-labeled dCTP, which must first be dried), and 20 U of terminal deoxynucleotidyltransferase in 100 m*M* potassium cacodylate (pH 7.2), 0.2 m*M* dithiothreitol (DTT), and 2 m*M* $CoCl_2$. The reaction is carried out for 1.5 hr and is terminated by purifying the labeled probe from the other reaction components by Nensorb-20 (Du Pont NEN) column chromatography according to the manufacturer's directions. The probe is eluted in 20% ethanol and can be used for hybridization without further steps except that ^{35}S-labeled probes should be stored at 4° in the presence of 10 m*M* DTT.

Determination of Oligonucleotide Probe Specificity

The usefulness of *in situ* hybridization experiments depends critically on probe specificity, and several criteria for specificity have emerged.[17,18] (1) *In situ* hybridization and immunocytochemistry can be carried out on adjacent sections[1,23] or sequentially on the same section[2,3,5,8,24] to provide evidence that the same cells contain both the mRNA and the peptide or protein it encodes. (2) To discriminate saturable hybridization from nonspecific probe binding, the tissue sections can be hybridized with unlabeled probe prior to the addition of labeled probe, resulting in elimination of the specific hybridization signal.[4,7,18] (3) Probes complementary to different regions of the same mRNA should label the same cells in adjacent sections or yield similar results in the detection of treatment effects.[18] (4) Treatment of tissue sections with RNase A prior to hybridization should eliminate specific hybridization signal,[4,7,25,26] although it should be noted that spurious hybridization, e.g., to ribosomal RNA, would also be eliminated. (5) Message sense (identical to cellular RNA) probes should not hybridize to tissue containing the target mRNA. (6) The complementary probe should detect RNA of the correct size following the extraction, electrophoresis, and transfer to nitrocellulose of the appropriate tissue RNA for probe hybridization.[5,17–19] (7) Finally, the probe–mRNA hybrid

[22] M. L. Collins and W. R. Hunsaker, *Anal. Biochem.* **151,** 211 (1985).

[23] D. Henken, A. Tessler, M.-F. Chesselet, A. Hudson, F. Baldino, Jr., and M. Murray, *Anat. Rec.* **218,** 60A (1987).

[24] M. Brahic, A. T. Haase, and E. Case, *Proc. Natl. Acad. Sci. U.S.A.* **81,** 5445 (1985).

[25] R. Arentzen, F. Baldino, Jr., L. G. Davis, G. A. Higgins, Y. Lin, R. W. Manning, and B. Wolfson, *J. Cell. Biochem.* **27,** 415 (1985).

[26] F. Baldino, Jr. and L. G. Davis, *in* "*In Situ* Hybridization in Brain" (G. Uhl, ed.), p. 97. Plenum, New York, 1986.

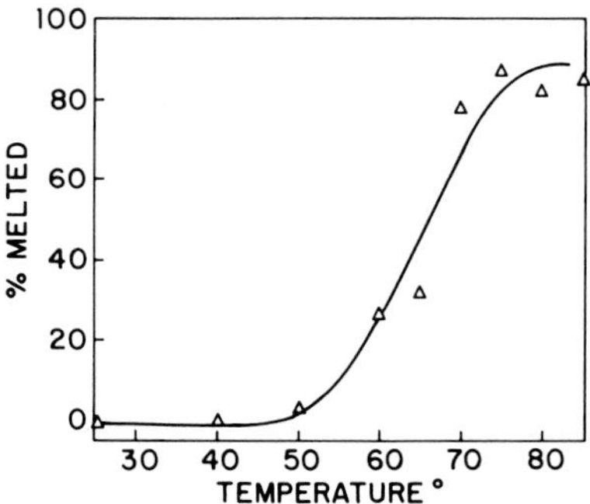

FIG. 1. Thermal denaturation curve for a 39-base somatostatin probe determined *in situ*. The data are reported as mean optical densities (OD) plotted as a function of temperature. The results were obtained from washing hybridized tissue sections in 0.5× SSC at different temperatures between 20 and 85°. The sections were exposed to X-ray film for 3 days, and the suprachiasmatic nucleus ODs were determined with a DUMAS image analysis system. Each datum point is a mean of four sections. (Redrawn from Card *et al.*[19])

has a predicted thermal stability (T_m value) which would be reduced by base-pair mismatching, i.e., inappropriate hybridization.

The T_m value is determined principally by G–C (guanine–cytosine base) content, length, number of base-pair mismatches, and sodium concentration.[27–29] For oligonucleotides T_m is given by Eq. (1), where the molar concentration of Na^+ is a maximum of 0.5 (1× SSC contains 0.165 *M* Na^+), the percentage of G + C bases is between 30 and 70, and the probe length does not exceed 100 bases.[14,26] The T_m study is carried out by exposing a similar series of hybridized sections to X-ray film, incubating the sections at different temperatures (e.g., 30–80° in 5° increments) in 0.5× SSC (1× SSC: 0.15 *M* NaCl, 15 m*M* sodium citrate, pH 7.2) for 1 hr, reexposing the sections to film for the same length of time, and then calculating and plotting percentage melted as a function of temperature. The interpolated "50% melted" point should be close to the calculated T_m [Eq. (1)], which has been found to be the case for oligonucleotides used in

$$\begin{aligned} T_m = 81.5 &+ 16.6(\log[\mathrm{Na}^+]) + 0.41[\%(\mathrm{G} + \mathrm{C})] \\ &- 675/(\text{probe length in bases}) - 1.0(\%\ \text{mismatch}) \\ &- 0.65(\%\ \text{formamide}) \end{aligned} \quad (1)$$

ISHH[17,18,26] (see Fig. 1). A T_m study will also provide the optimal hybridization temperature, which is often considered to be about 20° lower than

[27] T. I. Bonner, D. J. Brenner, B. R. Neufeld, and R. J. Britten, *J. Mol. Biol.* **81,** 123 (1973).

[28] R. J. Britten, D. E. Graham, and B. R. Neufeld, this series, Vol. 29, p. 363.

[29] C. R. Cantor and P. R. Schimmel, "Biophysical Chemistry," Part III. Freeman, San Francisco, California, 1980.

the T_m value. This information is especially critical if the investigator is using a probe based on the sequence from one species to carry out studies on the homologous mRNA from another species: base-pair mismatches will result in a lower actual T_m value in comparison to the value calculated without taking any mismatches into account.

Tissue Preparation

Hybridization can be performed on tissue sections prepared in a variety of ways. For perfusion fixation, an initial transcardiac flush with 25 ml 0.1 *M* phosphate-buffered saline (PBS) is usually followed by 200–300 ml of either sterile 4% paraformaldehyde in 0.1 *M* phosphate buffer (pH 7.4) containing 2% sodium periodate and 1.4% lysine–HCl, or 4% paraformaldehyde in 0.1 *M* PBS (pH 7.3). All perfusion solutions are sterile and may contain 0.02% of the RNase inhibitor diethyl pyrocarbonate (DEP). The entire brain is then removed, postfixed in sterile fixative for 1 hr at 4°, and transferred to sterile 10% 0.1 *M* phosphate-buffered sucrose for 1 hr followed by an additional hour in 20% phosphate-buffered sucrose and overnight at 4° in 30% phosphate-buffered sucrose. The tissue can then be sectioned on a cryostat (usually between 10 and 20 μm), collected on gelatin-coated slides, and stored at −20 or −80° until processed for ISHH. Because of the frequent use of stringent temperatures and proteases, slides should not be subbed with protein preparations [e.g., bovine serum albumin (BSA)]. For subbing, slides should be washed in hot soapy water and rinsed well in distilled water, soaked in 8% nitric acid for 20 min, rinsed well in distilled water, and dried. Dissolve 1.25 g gelatin (Fisher 0.275 bloom) in 500 ml preheated (55°) water; cool to room temperature and add 0.125 g chromium potassium sulfate, mixing well by stirring. The slides are subbed in this solution by dipping (in racks) several times slowly and allowing them to dry vertically in a dust-free area. Fresh frozen brains can be sectioned and postfixed for 5 min in freshly prepared 3% paraformaldehyde in 0.1 *M* phosphate-buffered saline containing 0.02% DEP.

In Situ Hybridization

Tissue sections can be delipidated by transferring them through a graded series of alcohols (50, 70, 100%), rehydrated, and then rinsed in DEP-treated sterile water. The value of this procedure depends on the hydrophobicity of the probe and the lipid content of the tissue. Partial digestion of the tissue sections with protease can be of some value in improving probe penetration across the cell membrane and for removing proteins and ribosomes from the mRNA. Incubation with proteinase K (1 μg/ml in 20 m*M* Tris–HCl, 2 m*M* $CaCl_2$, pH 7.4) for 15 min at 37°, followed by a 30-min rinse in 2× SSC (0.3 *M* NaCl, 30 m*M* sodium citrate, pH 7.2) is adequate for this purpose. However, improved penetration with proteinase K must be balanced against the risk of losing cellular mRNA by this procedure.[30]

The sections are then incubated in hybridization buffer for 1 hr, at the same temperature that hybridization is to be performed. The hybridization buffer contains 0.6 *M* sodium chloride, 10 m*M* Tris–HCl (pH 7.5), 0.02% Ficoll, 0.02% polyvinylpyrrolidone, 0.02% BSA, 1 m*M* EDTA, 0.05% yeast tRNA, 0.05% salmon sperm DNA, 10% (w/v) dextran sulfate, and 50% formamide. For ^{35}S-labeled probes, 10 m*M* DTT is also included in the hybridization buffer. The sections are then hybridized overnight with 1–5 ng of labeled oligonucleotide probe diluted in 50 μl of hybridization buffer per section. The hybridization temperature depends on the T_m of the oligonucleotide probe, and hybridization is usually performed at 20° below the T_m. Washing stringency may have to be varied for each probe. The salt concentration varies as function of probe length, G–C content, and T_m. Although one can theoretically predict the appropriate salt concentration for each hybridization condition, in general the sections are rinsed in decreasing concentrations of SSC to a final rinse of 0.5× SSC at the hybridization temperature. For ^{35}S-labeled probes, 14 m*M* 2-mercaptoethanol and 1% sodium thiosulfate are included in the wash solutions.

Autoradiography

For preliminary observations, the slides are apposed directly to X-ray film (Kodak XAR-5) in standard cassettes and exposed for 3–5 days (see Fig. 2A). X-Ray films can be developed manually or in a Kodak X-Omat processor. For single-cell resolution, the slides are subsequently dipped in photographic emulsion (Kodak NTB2, NTB3, or the Ilford series of emulsions) and kept in the dark at 4° for 7–60 days depending on the probe label and message abundance. The grain size of the emulsion selected depends on the degree of resolution required and the particular radiolabel. The emulsion-coated slides are then developed according to the manufacturer's protocol (e.g., for NTB2, with Kodak D-19 for 2 min at 16° and fixed for 5 min, dehydrated, and cleared) and examined by both dark-field and bright-field light microscopy.

After the autoradiographic procedure is completed, the data can be analyzed by several methods (see Lewis *et al.*[31]). Image analysis can be performed on either X-ray film or emulsion-coated slides (see Fig. 3).

[30] J. B. Lawrence and R. J. Singer, *Nucleic Acids Res.* **13,** 1777 (1985).

[31] M. E. Lewis, W. T. Rogers, R. G. Krause II, and J. S. Schwaber, this volume [58].

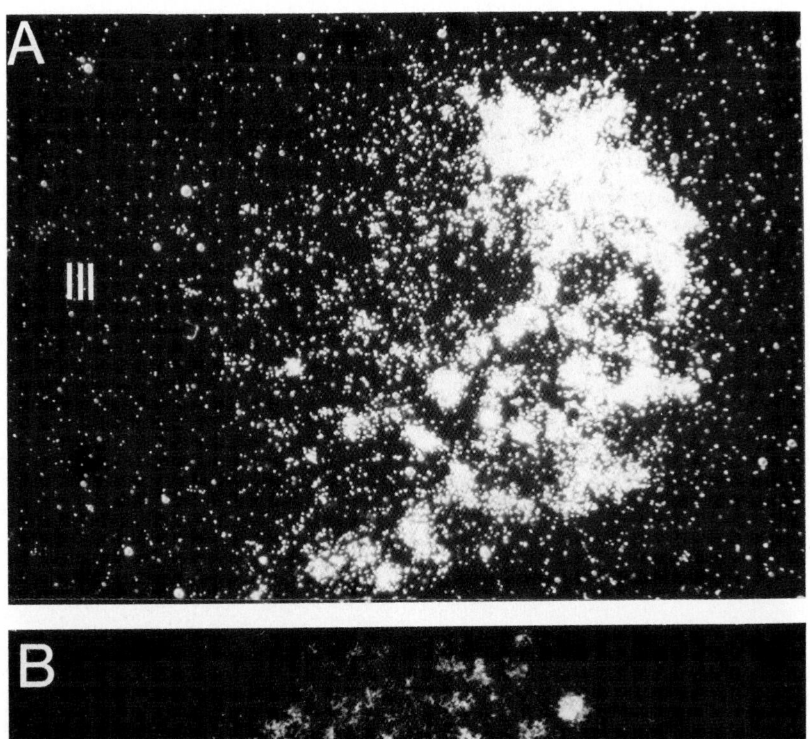

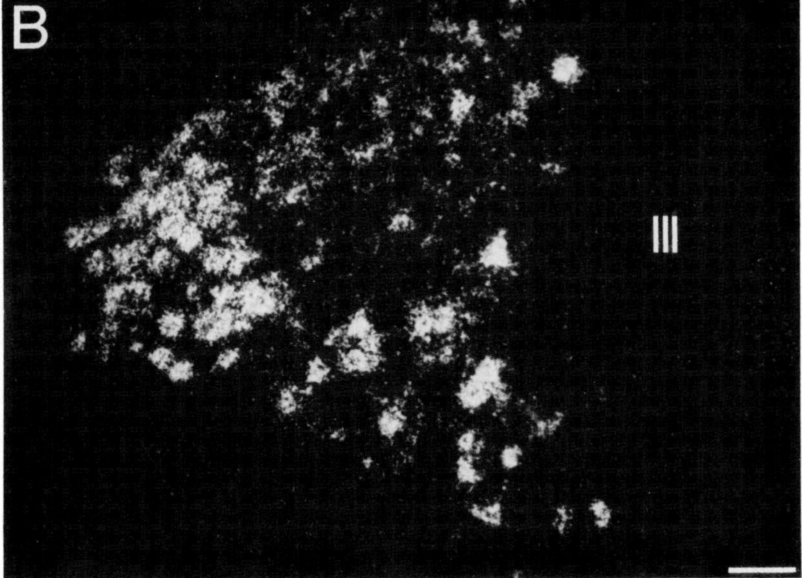

FIG. 2. Dark-field photomicrographs demonstrating the distribution of vasopressin mRNA within the paraventricular nucleus of the rat. (A) Hybridization with a 24-base oligonucleotide probe 3′ end labeled with ^{125}I-labeled dCTP. This probe was complementary to the glycoprotein-coding region of prepropressophysin. Sections (third ventricle, III) were dipped in Kodak NTB2 and exposed for 3 days. (B) Hybridization with the same 24-base probe 3′ labeled with ^{35}S-labeled dATP. Sections were dipped in Kodak NTB2 and exposed for 6 weeks. Note the differences in the resolution and time required for development with each procedure. Bar, 100 μm.

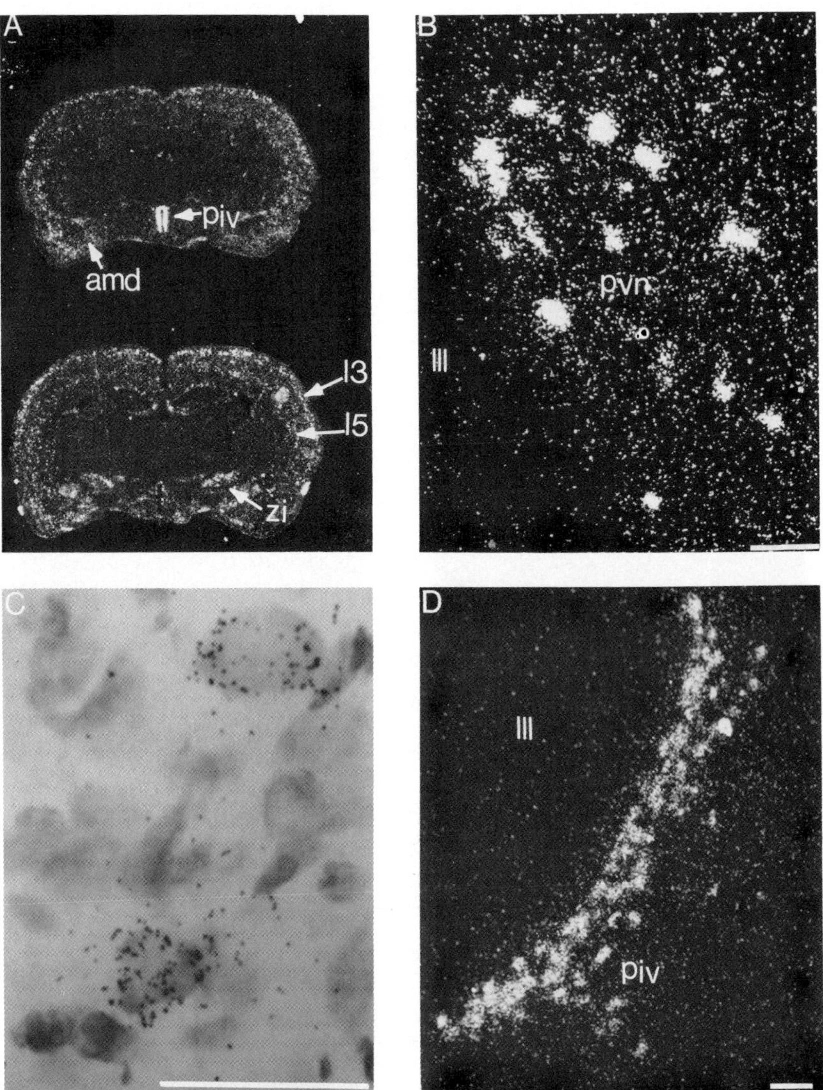

FIG. 3. Photomicrographs demonstrating hybridization with a 39-base oligonucleotide probe for somatostatin labeled on its 3′ end with ^{35}S. (A) Photograph of a X-ray autoradiograph showing the distribution of somatostatin mRNA throughout many regions within the same plane of a section: periventricular hypothalamus (piv), amygdala (amd), zona incerta (zi), cortical layer III (l3), and cortical layer V (l5). (B) Dark-field photomicrograph of hybridized neurons in the paraventricular nucleus (pvn) and third ventricle (III). Bar, 100 μm. (C) Bright-field photomicrograph showing large Nissl-stained neurons in the piv containing silver grains (hybridization signal). Bar, 50 μm. (D) Dark-field photomicrograph demonstrating the distribution of hybridized cells in the piv. Bar, 100 μm. These photographs are of material prepared in our laboratory in collaboration with Dr. Sandra Fitzpatrick-McElligott.

Referencing appropriate internal standards, the data can then be expressed as optical density or quantity of probe hybridized. Individual neurons can also be visually counted as can the grains within each defined cell profile. These values can then be compared against appropriate control data.

After fixation and washing (45 min), the sections can be counterstained with a variety of agents including cresyl violet, hematoxylin–eosin, and toluidine blue. The sections are then dehydrated through a graded series of alcohols (50, 70, 95, 100%), cleared in xylene, and coverslipped.

Combining Immunohistochemistry with Hybridization

In many experimental situations, it is desirable to localize the hybridization signal and an immunohistochemical label within the same cell. For this purpose, we have found it preferable to perform the immunohistochemical procedure prior to *in situ* hybridization since many of the elements in the latter procedure preclude the development of a high-quality immunohistochemical label. The peroxidase–antiperoxidase method of Sternberger[32] and the biotin–avidin method of Hsu *et al.*[33] have been successfully applied with oligonucleotide, cDNA, and RNA probes.[2,3,8,24,26] For best results, it is advantageous to terminate the immunochemical procedure in 2× SSC just prior to peroxidation and addition of the chromogen. At this critical point in the procedure, ISHH is performed to completion. After the final stringent wash (usually 0.5× SSC), the H_2O_2 and chromogen are added, and color is developed to completion. In this way the oxidation of the chromogen does not interfere with the hybridization procedure.

RNA Probes

General Considerations

Owing to the development of a simple and efficient system for the synthesis of labeled single-stranded RNA,[34] the use of RNA probes has recently become a useful alternative to the use of DNA probes for ISHH. RNA probes allow for a much higher sensitivity than nick-translated DNA.[34,35] RNA probes, therefore, are the method of choice to detect

[32] L. A. Sternberger and S. A. Joseph, *J. Histochem. Cytochem.* **27,** 1424 (1979).

[33] S. M. Hsu, L. Raine, and H. Fanger, *J. Histochem. Cytochem.* **29,** 577 (1981).

[34] D. A. Melton, M. R. Krieg, T. Rebogliat, K. Maniatis, K. Ziom, and M. R. Green, *Nucleic Acids Res.* **12,** 7035 (1984).

[35] K. H. Cox, D. V. DeLeon, L. M. Angerer, and R. C. Angerer, *Dev. Biol.* **101,** 485 (1984).

mRNAs encoded by genes for which no nucleic acid sequence data are available, thus precluding the use of synthetic oligonucleotides. RNA probes can be labeled to very high specific activity,[36] and several factors contribute to the excellent signal over background ratio usually achieved with RNA probes (Fig. 4). The higher stability of RNA–RNA hybrids compared to DNA–RNA hybrids, as well as the longer length of RNA probes relative to oligonucleotide probes, permit the use of more stringent hybridization conditions. In addition, unhybridized labeled RNA molecules can be destroyed by posthybridization treatment with RNase, which spares double-stranded RNA.

Labeled RNA probes are usually transcribed from the cDNA inserted into vectors containing an RNA polymerase initiation site upstream to the cloned DNA. Specific promoter–RNA polymerase combinations from phage T7 and from the *Salmonella typhimurium* phage SP6 are currently available in these vectors. cDNAs can be inserted in either orientation with regard to the SP6 or T7 promoter, allowing for the transcription of either antisense (complementary to the cellular mRNA) or sense (identical to the cellular mRNA) RNA probes.[34] Antisense RNA is used for detection of the cellular mRNA, and the sense probe is used as a control for the specificity of the hybridization (see below).

Protection against contamination with RNase, a concern common to all types of ISHH experiments, is even more crucial when the probe itself is a RNA. As a general rule, taking excessive precautions is probably the safest attitude. Gloves are worn at all times, including during thaw-mounting the cryostat-cut sections on microscope slides, and all solutions are treated with DEP.[15] All glassware and reusable plasticware are similarly treated. Disposable plasticware is autoclaved. Solutions containing Tris buffer cannot be treated with DEP but are made with DEP-treated water and autoclaved. Particular care must be taken in experiments using RNA probes to keep containers for pre- and posthybridization steps separated, as the posthybridization treatments include an incubation in RNase-containing solution.

In Vitro Transcription of RNA Probes

Prior to transcription, plasmids containing either the SP6 or the T7 promoter and the inserted cDNA must be linearized with a restriction endonuclease.[15] An enzyme which cuts not within the cDNA but between the insert and the plasmid must be chosen to allow for the transcription of full-length probes including only a minimal amount of plasmid sequence.

[36] M. F. Chesselet, L. Weiss, C. Wuenschell, A. J. Tobin, and H. U. Affolter, *J. Comp. Neurol.* **242,** 125 (1987).

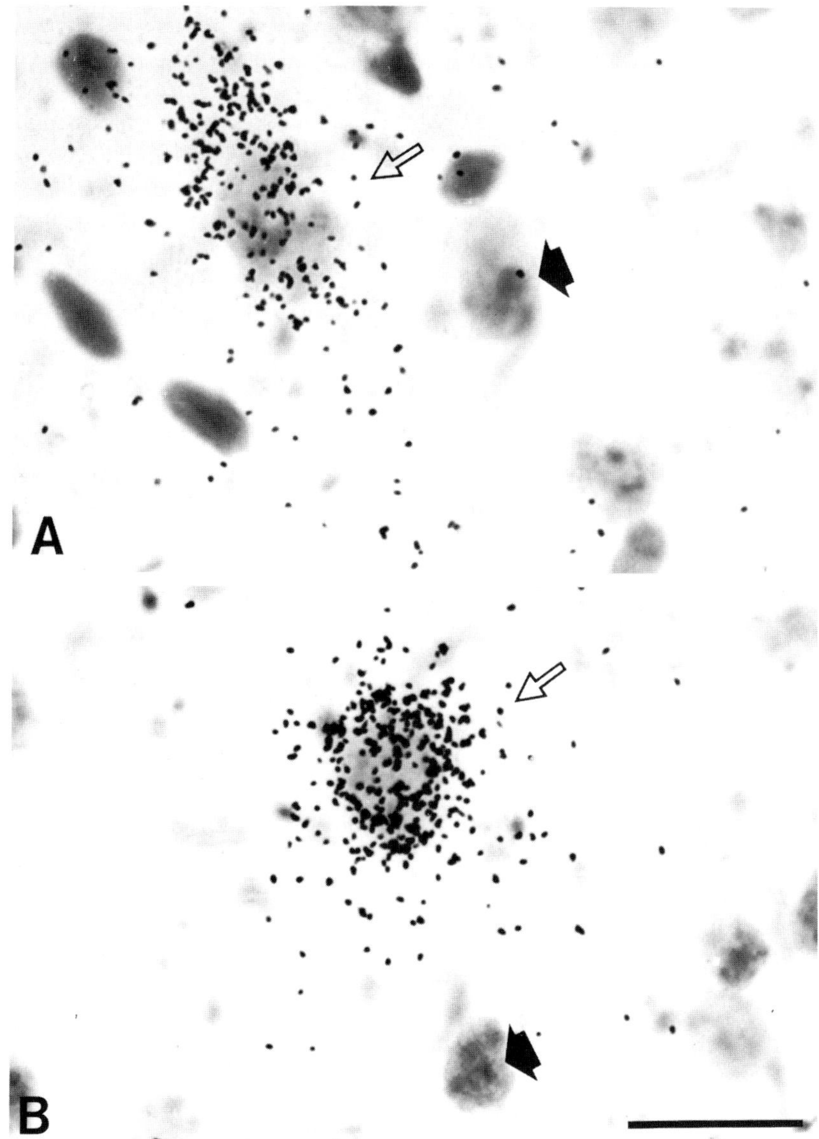

FIG. 4. *In situ* hybridization histochemistry with RNA probes. Sections of mouse brain were incubated with ^{35}S-labeled RNA probe complementary to the mRNA coding for glutamic acid decarboxylase (cDNA from A. Tobin, UCLA). The open arrows point to labeled cells in the substantia nigra pars reticulata (A) and the cerebral cortex (B). Note the counterstained nuclei and the dense accumulation of silver grains over the unstained cytoplasm of the labeled cells, contrasting with the low background and the presence of unlabeled cells (solid arrows). Bar, 20 μm.

Linearized plasmids are transcribed with the appropriate polymerase in the presence of one or more radiolabeled nucleotides. The use of only one labeled nucleotide generates enough specific activity for most applications and is therefore preferred. Both tritium- and ^{35}S-labeled nucleotides can be used for the synthesis of RNA probes. The choice of radioisotope is based on the same criteria as described for oligonucleotides (see above). The amount of labeled and unlabeled UTP must be chosen to provide the enzyme with 12 μM of substrate, a requirement for the production of full-length probes.[15] Specific activities achieved with the following protocol approximate 2×10^8 dpm/μg, depending on the batch of ^{35}S-labeled UTP.

Linearized templates (2 μg) are incubated in 40 mM Tris–HCl (pH 7.5), 6 mM $MgCl_2$, and 2 mM spermidine ("transcription buffer," usually provided in a concentrated form with the enzyme), containing 10 mM DTT, 25 U RNasin (an RNase inhibitor, Promega Biotech), 500 μM of ATP, GTP, and CTP, 10 μM of UTP, 2.5 μM of ^{35}S-labeled UTP (specific activity 1000 Ci/mmol or higher), and approximately 15 U polymerase. After a 1-hr incubation at 37°, 15 U more of polymerase is added, and the incubation is carried out for an additional hour. At this point, approximately 2 U of RNase-free RQ-1 DNase (Promega Biotech) and 25 U of RNasin are added to the reaction in order to remove the DNA template. The reaction is carried out for 10 min at 37° and stopped by addition of 10 mM EDTA and sodium/Tris/EDTA buffer (10 mM Tris–HCl, 100 mM NaCl, 1 mM EDTA, pH 8.0); this is done by adding aliquots of 10× concentrated solution and bringing the total volume of the reaction to 100 μl with DEP-treated water containing 20 mM DTT. After phenol/chloroform extractions and ethanol precipitation,[35] the ^{35}S-labeled RNA probes are usually resuspended in DEP-treated distilled water containing 20 mM DTT, 2% sodium dodecyl sulfate (SDS), and 2% ethanol and stored at 4°. Probes kept this way proved to be more stable than when stored as an ethanol precipitate at −70°. Prior to the experiment, the RNA is concentrated by ethanol precipitation and diluted appropriately in the hybridization mixture (see below).

Some investigators have suggested that partial alkaline hydrolysis of the probe into fragments of 100–150 bases greatly improves the signal, probably by improving probe penetration.[35] Using our conditions, such modification leads to only a weak (5–10%) increase of the hybridization signal. This may vary, however, with the conditions of the experiment, in particular the type of tissue and fixation. Partial alkaline hydrolysis can be performed in 40 mM $NaHCO_2$/60 mM Na_2CO_3, pH 10.2, at 60° for a time (in minutes) equal to $L_0 - L_f/KL_0L_f$, where L_0 and L_f are the initial and final fragment lengths in kilobases (kb) and K is the rate constant for hydrolysis (0.11 kb/min). The reaction is terminated by addition of sodium acetate (pH 6.0) to 0.1 *M* and of glacial acetic acid to 0.5%. The samples are then ethanol precipitated and stored as described above.

In Situ Hybridization

Tissue preparation for ISHH with RNA probes is identical to that described for oligonucleotides (see above). The following protocol has been used successfully with RNA probes.[36] Sections are rinsed for 2 min in 2× SSC; acetylated (10 min) in 0.1 *M* triethanolamine (pH 8.0) to which 0.25% acetic anhydride is added just prior to the addition of tissue sections; washed twice (1 min each) in 2× SSC followed by a rinse in PBS (1 min); rinsed for 30 min in 0.1 *M* Tris–glycine, pH 7.0; rinsed twice in 2× SSC (1 min each); and dehydrated in graded ethanol (70, 80, 95%, 1 min each).

After dehydration, the sections are air dried and covered with a drop of hybridization buffer containing the ^{35}S-labeled probe. The composition of the hybridization buffer is as follows: 40% deionized formamide, 10% dextran sulfate, 1× Denhardt's solution (0.02% Ficoll, 0.02% polyvinylpyrrolidone, 10 mg/ml RNase-free BSA), 1 mg/ml *Escherichia coli* tRNA, 1 mg/ml denatured salmon sperm DNA, and 4× SSC containing 10 m*M* DTT. The hybridization buffer can be prepared ahead of time and stored frozen at −70° in aliquots. The deionized formamide, tRNA, and DNA must be prepared and stored according to standard protocols.[13] Dextran sulfate must be dissolved in the formamide by ultrasound. Stock solutions of DTT are kept frozen and thawed just before use. The radiolabeled probe, precipitated by ethanol just prior to the experiment, is dissolved in DEP-treated water containing 20 m*M* DTT. The dilution is determined for each experiment such that 10× dilution in the hybridization buffer leads to an adequate probe concentration. The amount of probe per section (usually in a 20-μl drop) must be determined empirically for each probe and tissue. In our experience, amounts of probe of 1–5 ng/section usually result in a good signal with very little background. Much higher concentrations appear to be needed to reach saturation of the signal.

Hybridization is performed at 50° in humid boxes for 3.5 hr. The sections are subsequently washed in 50% deionized formamide/2× SCC at 52° for 5, 2, and 5 min. Between the second and third washes, the sections are usually treated with RNase A (100 μg/ml) for 30 min at 37°. The sections are finally rinsed in 2× SSC at room temperature and left overnight under mild agitation in 2× SSC containing 0.05% Triton X-100. The next day, sections are rinsed in 2× SSC, dipped in ammonium acetate (300 m*M*), dehydrated in graded ethanol solutions prepared with 300 m*M* ammonium acetate, delipidated in xylene (5 and 30 min), rinsed in 100% ethanol, and air dried. Several variations of the protocol for ISHH with

RNA probes have been successfully used by various authors. In particular, posthybridization washes can be performed as follows[37]: 2× SSC for 30–60 min at room temperature, 0.1× SSC, 60 min, 45°. Optimal stringency and duration of the hybridization and posthybridization washes should be ultimately determined for each probe.

Autoradiography and Counterstaining

Detection of the autoradiographic signal generated by the radiolabeled RNA–RNA hybrid (see Fig. 4) is done by film and emulsion autoradiography as described for oligonucleotide probes. However, one problem, specific to the use of RNA probes, is the loss of optimal conditions for Nissl stain due to the treatment with RNase. Counterstaining with toluidine blue, hematoxylin–eosin, or ethidium bromide is thus preferable for RNase-treated sections.

Determination of RNA Probe Specificity

As discussed earlier, the length of the probes and the stability of the RNA–RNA hybrid allows for the use of very stringent conditions for ISHH with RNA probes. This usually leads to specific hybridization, provided that a posthybridization treatment with RNase is performed. To ensure the specificity of the autoradiographic signal, controls described for the oligonucleotide probes also apply to the RNA probes (see controls 1, 2, 5, and 6 described above on p. 765). In particular, sections must be incubated under the same conditions as the experimental sections with radiolabeled RNA probes which do not hybridize to cellular mRNA. RNA transcribed from the plasmid DNA or from the cDNA strand coding for the cellular mRNA (sense RNA) can be used for this purpose. Some controls commonly used with oligonucleotide probes, however, are not appropriate for use with RNA probes. Pretreatment with RNase should abolish the signal, but this could be misleading as residual enzyme could hydrolyze the probe itself. Melting curves cannot be determined with full-length RNA probes because the length of the probes would require melting temperatures too high to be compatible with tissue integrity.

Concluding Remarks

In situ hybridization with synthetic oligonucleotide and RNA probes provides direct histochemical methods to assay the expression of a vari-

[37] C. Wuenschell, R. S. Fisher, D. L. Kaufman, and A. J. Tobin, *Proc. Natl. Acad. Sci. U.S.A.* **83,** 6193 (1986).

ety of mRNAs within individual cells. These methods are of sufficient sensitivity and resolution to detect relatively rare transcripts within individual neurons in the CNS, a tissue that requires this degree of resolution in order to study the regulation of neuropeptides, enzymes, and other brain-related proteins. RNA probes are particularly advantageous for detecting rare transcripts since these probes can be labeled to a very high specific activity.[36,38] Application of hybridization technology to neurobiology has already yielded insights into the cellular mechanisms mediating plasticity of mRNA expression induced by humoral and synaptic manipulation in the CNS and neuroendocrine tissues.[3,5,18,21,39–42]

Future studies using oligonucleotide probes to detect restricted regions of mRNA will be an especially valuable tool to decipher alterations in the processing of mRNA *in situ*. For example, the ability to evaluate specific exon-coding segments of any mRNA species *in situ* will provide a convenient histochemical method for distinguishing those mRNAs which are alternatively processed to translate proteins in a tissue-specific manner.[43] Moreover, the detection of intron sequences within heteronuclear mRNA species[44] permits, for the first time, the evaluation of the rate of transcription at the level of the single cell. Thus, high-resolution *in situ* hybridization provides a histochemical technique to study the molecular and cellular basis of genomic plasticity.

Acknowledgments

We gratefully acknowledge the technical expertise of Rudolph G. Krause II, Betty Wolfson, Terry O'Kane, and Elaine Robbins. We also wish to acknowledge Dr. Sandra Fitzpatrick-McElligott and Linda T. Weiss for their collaboration in the ISHH for somatostatin and glutamic acid decarboxylase, respectively.

[38] M. T. Johnson and S. A. Johnson, *BioTechniques* **2,** 156 (1984).

[39] T. G. Sherman, J. F. McKelvy, and J. J. Watson, *J. Neurosci.* **6,** 1685 (1986).

[40] J. Angulo, L. G. Davis, B. Burkhart, and G. Christoph, *Eur. J. Pharmacol.* **130,** 341 (1986).

[41] B. M. Chronwall, W. R. Millington, S. W. T. Griffin, J. R. Unnerstall, and T. L. O'Donohue, *Endocrinology (Baltimore)* **120,** 1201 (1987).

[42] F. Baldino, Jr., T. M. O'Kane, S. Fitzpatrick-McElligott, B. Wolfson, W. T. Rogers, and J. S. Schwaber, *Anat. Rec.* **218,** 13A (1987).

[43] M. G. Rosenfeld, J. J. Mermod, S. G. Amara, L. W. Swanson, P. E. Sawchenko, J. Rivier, W. W. Vale, and R. M. Evans, *Nature (London)* **304,** 129 (1983).

[44] R. T. Fremeau, Jr., J. R. Lundblad, D. B. Pritchett, J. N. Wilcox, and J. L. Roberts, *Science* **235,** 1265 (1986).

[56] *In Situ* Hybridization Histochemistry Combined with Markers of Neuronal Connectivity

By JAMES S. SCHWABER, BIBIE M. CHRONWALL, and MICHAEL E. LEWIS

Introduction

Analysis of the structure and organization of the central nervous system depends on differentiating specific neuronal types. Until recently, neural typology has been based (1) on cytoarchitecture (cell shape, size, orientation, and staining properties) and (2) on a neuron's axonal projections to other brain regions, which define its participation in specific functional circuitry. In the past decade, interest has increasingly been directed toward identifying neurons on the basis of chemical phenotype, such as transmitter or receptor type. Recently, the application of *in situ* hybridization histochemistry to the detection of brain mRNAs has extended the ability to detect and type neurons on the basis of their chemistry. The combination of these various techniques for differentiating neurons, for example, by combining immunohistochemistry with markers of connectivity (reviewed in Ref. 1), has greatly increased the analytical power of anatomical studies. This chapter describes new methods by which *in situ* hybridization histochemistry can be combined with contemporary retrograde fluorescent tracer techniques.

Retrograde tracing of neuronal connections was revolutionized in the early 1970s by the introduction of methods based on active axonal transport. It was initially shown that injection of the fluorescent dye Evans blue or the enzyme horseradish peroxidase into the terminal field of neurons resulted in active axonal uptake and transport of these markers to the neuronal cell bodies.[2,3] Subsequently, there has been rapid development of new tract-tracing methods as several generations of markers have rapidly superseded one another, such that the ease of use, resolving power, and sensitivity of these methods are now quite good. These markers of connectivity have often been combined with other methods, such as immunohistochemistry,[1] and several offer the possibility of being used for combination with *in situ* hybridization histochemistry.

[1] T. Hökfelt, G. Skagerberg, L. Skirboll, and A. Björklund, *in* "Methods in Chemical Neuroanatomy" (A. Björklund and T. Hökfelt, eds.), p. 228. Elsevier, Amsterdam, 1983.
[2] K. Kristensson, *Acta Neuropathol.* **16,** 293 (1970).
[3] J. H. LaVail and M. M. LaVail, *Science* **176,** 1416 (1972).

Although it is only in the last few years that the technique of *in situ* hybridization histochemistry has been extended to the mammalian brain and pituitary,[4,5] it has now become greatly simplified and can be used as a standard histological technique much as immunohistochemistry is used. The method was originally used for localization of specific DNAs in metaphase chromosomes.[6,7] Later, globin mRNA was detected in dispersed mammalian cells.[8] However, the greatest advantages and opportunities for the technique appear to be in the brain, since the complexity and heterogeneity of brain organization and function render identification of the specific cells of interest of great importance: (1) compared to Northern or dot-blot mRNA analyses the method offers the advantage of cellular specificity in histological context within the brain; (2) compared to immunohistochemistry, *in situ* hybridization offers the advantage of locating cells in which protein synthesis is taking place, not merely detection of the presence of the protein; (3) *in situ* hybridization appears to offer higher specificity in comparison to immunohistochemistry, where unexpected antigens regularly cross-react with antibodies; (4) quantitation of mRNA levels in specific cells is possible when radioactively labeled probes are used[9–11]; (5) in experiments where gene expression has been pharmacologically manipulated, *in situ* hybridization has become an invaluable analytical tool[12]; (6) changes in mRNA level during brain development[13] as well as changes in mRNA levels in neurological disease states[14] have also been elucidated with *in situ* hybridization. One of the most exciting concepts to come from this work is that a neuropeptide's mRNA level reflects the level of functional activity for that neuropeptide.

The combination of retrograde tracing and *in situ* hybridization methods can relate information on gene expression and gene regulation to

[4] C. E. Gee, C.-L. C. Chen, J. L. Roberts, R. Thompson, and S. J. Watson, *Nature (London)* **306,** 374 (1983).

[5] R. Pochet, H. Brocas, G. Vassart, G. Toubeau, H. Seo, S. Refetoff, J. E. Dumont, and J. L. Pasteels, *Brain Res.* **211,** 433 (1981).

[6] J. G. Gall and M. L. Pardue, *Proc. Natl. Acad. Sci. U.S.A.* **63,** 378 (1969).

[7] J. G. Gall and M. L. Pardue, this series, Vol. 38, p. 470.

[8] P. R. Harrison, D. Conkie, J. Paul, and K. Jones, *FEBS Lett.* **32,** 109 (1973).

[9] P. Szabo, R. Elde, D. M. Steffensen, and O. C. Ohlenbeck, *J. Mol. Biol.* **115,** 539 (1977).

[10] W. S. Young, E. Mezey, and R. E. Siegel, *Neurosci. Lett.* **70,** 198 (1985).

[11] W. T. Rogers, J. S. Schwaber, and M. E. Lewis, *Neurosci. Lett.* **82,** 315 (1987).

[12] L. G. Davis, R. Arentzen, J. M. Reid, R. W. Manning, B. Wolfson, K. L. Lawrence, and F. Baldino, Jr., *Proc. Natl. Acad. Sci. U.S.A.* **83,** 1145 (1986).

[13] W. S. T. Griffin, M. A. Alejos, E. J. Cox, and M. R. Morrison, *J. Cell. Biochem.* **27,** 205 (1985).

[14] W. S. T. Griffin, M. A. Alejos, E. J. Cox, and M. R. Morrison, *J. Cell. Biochem.* **79,** 4783 (1982).

specific neural circuitry.[15–18] This chapter describes new ways in which these methods may be combined. These combined methods are two-step procedures in which the axonal tracing is performed first and after some survival time the tissue is prepared and hybridized with a probe. Thus, the chapter is organized in terms of this sequence, first presenting tract-tracing methods, then *in situ* hybridization techniques and considerations necessary to combining the approaches. The result of this combination should be the simultaneous identification in tissue sections of neurons projecting to the injection location and neurons expressing a specific mRNA, some of which may be the same neurons, double-labeled. Thus, this technique permits the study of mRNA regulation in anatomically defined subsets of neurons within complexly organized brain nuclei. A section on how to examine and document these results concludes the chapter.

Retrograde Tracing Methods

General Properties of Retrograde Tracers

Tracer Uptake, Uptake by Fibers of Passage, and Effective Injection Site. Numerous axonal tracing methods based on the active axonal transport of enzymes, viruses, lectins, fluorescent dyes, and latex microspheres have been described. The uptake of these tracers is mediated by fluid-phase endocytosis, by which extraneuronal fluid is continually sampled by invagination of the cell membrane, particularly at terminals, creating internalized vesicles.[19] In some cases, where the tracer binds to the cell surface, adsorptive endocytosis also plays a role in the incorporation of tracer. It is important that both the endocytosis and the binding of the tracer be nonspecific to ensure the generality of tracer incorporation in all terminals independent of their function or chemistry. Although endocytosis occurs throughout the neuronal membrane, including along axons, it is rare along axons and most active at terminals.[20] Therefore, for most tracers, significant uptake by intact fibers passing through the injection site is

[15] B. M. Chronwall, *Peptides* **6** (Suppl. 2), 1 (1985).

[16] M. Schalling, T. Hökfelt, B. Wallace, M. Goldstein, D. Filer, C. Yamin, D. H. Schlesinger, and J. Mallet, *Proc. Natl. Acad. Sci. U.S.A.* **83,** 6208 (1986).

[17] B. M. Chronwall, M. E. Lewis, J. S. Schwaber, and T. L. O'Donohue, *in* "Neuroanatomical Tract-Tracing Methods II" (L. Heimer and L. Zaborsky, eds.). Plenum, New York, 1988.

[18] J. N. Wilcox, J. L. Robers, B. M. Chronwall, J. F. Bishop, and T. L. O'Donohue, *J. Neurosci. Res.* **16,** 89 (1986).

[19] S. Teichberg, E. Holtzman, S. M. Crain, and E. R. Peterson, *J. Cell Biol.* **67,** 215 (1975).

[20] P. T. Turner and A. B. Harris, *Brain Res.* **74,** 305 (1974).

not a problem. On the other hand, at sites of axonal injury passive diffusion of tracer might lead to internalization into vesicles. Thus, the effective injection site for retrograde tracers depends on several factors, including the necessary extracellular concentration of tracer, the affinity of the tracer for neuronal membranes, tracer diffusion and the possibility of uptake by damaged fibers of passage.

Active Axonal Transport. The endocytotic vesicles are subsequently taken up into retrograde vesicular transport[21] and transported toward the perikaryon at rates of 50–400 mm per day.[22,23] Consequently, the minimum survival time following injection of the retrograde tracer depends on the length of the pathway of interest and the rate of transport in the pathway. However, in the case of the contemporary fluorescent tracers recommended in this chapter, the degradation of the tracer in the cell body is very slow, and the tracers do not diffuse out of the cell to label other neurons. Thus, since tracer continues to arrive in the cell body for some period of time, it is generally best to allow some time past the minimum survival period for accumulation of tracer to occur.

Choice of Retrograde Tracer. The use of fluorescent tracers in combination with *in situ* hybridization offers the advantage of requiring no separate histochemical processing for visualization of the tracer. This eliminates the possibility of such processing affecting the hybridization. Further, fluorescent tracers are easily differentiable from markers of *in situ* hybridization, such as silver grains. It is equally important, however, to choose tracers not harmed by the requirements of the hybridization histochemistry. Several fluorescent dyes have been of particular interest, many of them originally introduced by Kuypers and associates (reviewed in Ref. 1). The dyes initially available suffered disadvantages, however, including rapid fading, uptake by fibers of passage, diffuse, undefined injection locations, and leakage from labeled neurons. The search for improved dyes has led to the continual, rapid introduction of new fluorescent markers. The most recent in this series are fluorescent latex microspheres (rhodamine beads[24]) and Fluoro-gold.[25] Undoubtedly these will not be the last word in tracer development, but they do appear to overcome most of the earlier problems and are presently the tracers of choice for retrograde transport studies. As tracers, they are nonselective, bright, chemically stable and resistant to fading, do not diffuse from the cell *in*

[21] B. Grafstein and D. S. Foreman, *Physiol. Rev.* **60,** 168 (1980).

[22] M. W. Brightman, *J. Cell Biol.* **26,** 99 (1965).

[23] J. H. LaVail, *in* "The Use of Axonal Transport for Studies of Neuronal Connectivity" (W. M. Cowan and M. Cuenod, eds.), p. 217. Elsevier, Amsterdam, 1975.

[24] L. C. Katz, A. Burkhalter, and W. J. Dreyer, *Nature* (*London*) **310,** 498 (1984).

[25] L. C. Schmued and J. H. Fallon, *Brain Res.* **377,** 147 (1986).

vivo or in tissue sections, and do not label intact fibers of passage. For combination with *in situ* hybridization, they appear to be nondisruptive to normal cell functions, can be stored without significant loss of signal (e.g., during exposure of emulsion-dipped material), and withstand the processing for *in situ* hybridization.

The rhodamine-labeled acrylic microspheres (beads; Luma Sleilor Inc., 50 New Valley Rd., New City, NY 10956) are 0.02–0.2 μm in diameter. The hydrophobic surfaces of the beads, i.e., their tendency to stick to cell membranes, and small size probably contribute to their uptake and transport. The hydrophobic polymers in the beads may also restrict access of oxygen to rhodamine dye trapped within the beads, attenuating fading. Bead-filled neurons show no degeneration and have normal action potentials as shown by intracellular recording.[24] Fluoro-gold (Fluorochrome, Inc., P.O. Box 4983, Englewood, CO 80155) is reported to be chemically a stilbene that offers the advantage of being easily separated to near purity, offering consistency for experimental use.[25] Fluoro-gold, as it back-fills a neuron, will completely fill the soma and proximal dendrites. Our experiments have shown no apparent experimental effect on hybridization pattern in cell groups as a consequence of labeling with either rhodamine beads or fluoro-gold. However, each tracer does have pros and cons: fluoro-gold gives consistent success in injections and fills cells and proximal dendrites completely but produces a relatively diffuse injection site; the beads sometimes bind together and clog the pipet tip, causing failure to inject, but the injection site is confined, well defined, and can be seen when sectioning. The beads are soluble in alcohol and clearing agents and thus are restricted to use in protocols that minimize exposure to these reagents.

Procedure for Retrograde Tracers

Tracer Preparation. Fluoro-gold is supplied lyophilized and can be stored dry at 4° for at least several months. The dye can be dissolved in water, physiological saline, or 0.2 *M* neutral phosphate buffer. Concentrations of 4 and 2% have been successful for use with *in situ* hybridization processing; lower concentrations produce somewhat smaller injection sites and somewhat reduced labeling. We have used dye dissolved in buffer and stored in the refrigerator for several months without loss of fluorescent intensity or amount of labeling. Rhodamine beads are supplied in an aqueous solution. This solution may be injected, or diluted up to 1 : 4 in distilled H_2O, NaCl, or KCl. We have used 1 : 4 dilutions in combination with *in situ* hybridization processing. The beads can be stored for at least 1 year in the refrigerator in a humidified container without loss of fluorescent intensity.

Tracer Injection. Pressure injections are performed under anesthesia, and are made into the projection field or the neural structure of interest. Volumes of 50–75 nl delivered via glass pipets have seemed optimum for both tracers. Pipets are pulled on a pipet puller (Kopf) from 1.0 mm i.d. glass capillary tubes (World Precision Instruments, 1b100F-4, with filament) and the tip broken back to the desired size while viewed under the microscope. Since the beads have a tendency to clump and block the tip of the glass pipet, tip sizes of 35–50 μm should be used, while 10–20 μm works well for fluoro-gold pressure injections. Sonicating the bead solution briefly prior to filling the pipet helps reduce bead clumping. There are numerous imaginative devices for performing these injections, ranging from simply gluing the pipet to the end of a Hamilton syringe to the use of a Nano Pump (World Precision Instruments). The pipet and injection instrument are first filled with light paraffin oil, and the tracer is then drawn into the pipet tip. Injections are made as continuous as possible over a 10-min period, and the pipet is then left in place for 10 min to reduce the spread of tracer up the pipet track. Since bead injection sites remain highly confined, several injections may be required to fill a desired projection field. Fluoro-gold can also be iontophoresed by a positive current in the range of 5 μA.

Survival Period. Survival times of 48 hr appear adequate for most pathways for both dyes in the rat, but longer times up to 7 days may lead to increased labeling. Optimal survival time should be uniquely determined for each system under study. Survival time, tracer dilution, and injection size all interact to produce an apparent injection site and amount of retrograde filling. Survival times up to 3 weeks have produced no diminution of label or leakage from filled cells, but such extended periods do appear to gradually decrease the apparent size of the injection site. Longer survival times may be desirable if the possible effect of surgical stress on mRNA level is an important consideration.

Fixation. Fixation is required both to cross-link the mRNA and thus retain it in its appropriate cellular compartment and to preserve tissue structure. The results in this chapter are from both tissue fixed post-sectioning and tissue fixed by vascular perfusion. The perfusion procedure is preferable for better tissue preservation. Transcardial perfusion is performed under deep anesthesia at the rate of 10–15 ml/min delivered by a peristaltic pump (Masterflex pump; Cole-Parmer) via a cannula placed into the aorta. In a standard procedure, perfusion is begun with phosphate-buffered saline (PBS) until the perfusate clears of erythrocytes, and this is followed by 350–400 ml of cold (4°) paraformaldehyde–lysine–periodate (PLP) fixative or 4% paraformaldehyde in PBS prepared according to Pease[26] over about 20 min. To make PLP fixative,[27] warm 1 liter 0.1 *M* PBS to 70°, stir 40 g of paraformaldehyde into the solution, filter, and allow to cool below 30°, then add 2.14 g sodium periodate and 13.7 g lysine. Cool to 4°. A T junction is useful to switch solutions without losing pressure on the vasculature. For some hybridization probes a lower concentration (1%) of formaldehyde might be tried.[28] Bouin's solution has been successfully used,[29] and 0.05–0.1% glutaraldehyde can be added, if resectioning for electron microscopy is desired. The brain is then removed from the skull, and immersed in the same fixative for an additional 90 min, and incubated 1 hr each in 10% and 20% sucrose and overnight (or until the brain sinks) in 30% sucrose in ribonuclease-free PBS prior to freezing for cryostat sectioning.

Postsectioning fixation gives greater control over the fixation, and appears to yield somewhat lower background for *in situ* hybridization. For this method, animals are sacrificed by decapitation, the brains are removed and frozen in crushed dry ice. The brains are then cryostat-sectioned, sections are picked up onto subbed slides and air dried as discussed below, and fixation is achieved by immersing slides into 4% paraformaldehyde in phosphate-buffered saline (PBS) at 4° for 5 min.

Preparation of Microscope Slides. In situ hybridization involves somewhat harsh tissue treatments and lengthy incubations and washes. Therefore, it is imperative to carefully clean and sub the microscope slides to ensure that the tissue does not float off the slides during the procedure. Slides are placed in slide racks, washed in hot, soapy water, rinsed, soaked in 8% nitric acid for 20 min, rinsed, dipped several times in a subbing solution, dried in a dust-free location, and stored in dust-free slide boxes. Subbing solution here has been made by dissolving 1.87 g gelatin (Fisher G-8) in 750 ml sterile water at 55°, cooling the solution to room temperature, and adding 0.187 g chromium potassium sulfate. This solution can be stored for up to a week at 4°.

Sectioning and Storage. The brains are cut on a cryostat (Hacker, Fairfield, NJ) at 5 to 30-μm section thickness and thaw-mounted onto chrome alum–gelatin subbed slides at room temperature. Wrinkles are smoothed with an autoclaved camel's hair brush wetted with sterile PBS. Gloves are worn, and the sections are touched only with heat-sterilized brushes to keep RNases from contaminating the sections. The slides are

[26] D. C. Pease, *Anat. Rec.* **142,** 342 (1962).

[27] I. W. McClean and P. K. Nakane, *J. Histochem. Cytochem.* **22,** 1077 (1974).

[28] K. H. Cox, D. V. DeLeon, L. M. Angerer, and R. C. Angerer, *Dev. Biol.* **101,** 485 (1984).

[29] M. E. Lewis, T. G. Sherman, and S. J. Watson, *Peptides* **6** (Suppl. 2), 75 (1985).

placed into slide boxes, and the sections are permitted to thoroughly dry onto the slides in order to prevent their floating off in later processing. Sections mounted on slides (or unsectioned brains) can be stored at −20 or −70°. The slides must be stored dry (with desiccant) to prevent RNase activity, and in the dark to prevent fading of the fluorescent compounds. Brains are frozen by immersion in powdered dry ice. Brains and sections will retain their tract tracing fluorescence for long periods, at least up to a year, and have been successfully hybridized after several months.

In Situ Hybridization Histochemistry

In situ hybridization histochemistry subsumes a wide diversity of procedures for localizing specific nucleic acid sequences within cells. All of these procedures require (1) a probe, which is a nucleic acid sequence partially or fully complementary to the sequence to be detected; (2) an incubation step in which the probe is hybridized to the endogenous nucleic acid; and (3) a method for detecting the location of the probe in a tissue section. Each of these procedures is briefly outlined below; more detailed discussion of our approach is available in other chapters in this volume ([55], [58]).

Choice of Probe and Probe Label

Several types of probes are currently in use for *in situ* hybridization: double-stranded cDNA probes, RNA probes, and synthetic oligonucleotides. It presently appears that the latter two offer clear advantages for *in situ* hybridization. RNA probes are an increasingly popular method for obtaining hybridization probes by the use of plasmids containing promoter sequences (e.g., SP6 and T_7) for transcription *in vitro* by DNA-dependent RNA polymerases. The RNA polymerase will repeatedly transcribe the cloned sequence (in the presence of radiolabeled precursor), resulting in the generation of large quantities of single-stranded RNA probes of very high specific activity. These cRNA probes are up to 8 times more sensitive than cDNA probes,[28] and hybridization background can be markedly reduced by the use of (1) posthybridization RNase treatment to digest unhybridized cRNA and mRNA and (2) elevated washing temperatures which are permitted by the greater thermal stability of RNA–RNA duplexes compared to DNA–RNA hybrids.

Despite the above benefits of RNA probes, this chapter focuses on the use of synthetic oligonucleotide probes because of their simplicity of acquisition and use. Synthetic oligonucleotides are prepared in a sequence complementary to the mRNA of interest (see Ref. 29 for a review). These probes are short (e.g., 20–50 nucleotides), single-stranded segments of DNA which are readily synthesized by automated apparatus (e.g., the Applied Biosystems Model 380A or the Vega Coder 300 DNA synthesizer) and then purified by HPLC for subsequent radiolabeling. Nucleotide or amino acid sequence information is obviously required to design the probe, and different design strategies are required, depending on which type of information is available.[29] Fortunately, many research institutions and universities now have oligonucleotide synthesis facilities, and a number of companies will synthesize and perform purification of a specified DNA sequence. Furthermore, synthetic oligonucleotide probes for several neuronal and nonneuronal mRNAs are now commercially available, and more should become available as increasing numbers of histochemists add *in situ* hybridization to their technical repertoire.

Probe Labeling. Among radioisotopic methods of labeling in general use, 3′ end-labeling offers the advantages for *in situ* hybridization of simplicity and of permitting multiple labeled nucleotides to be incorporated into the probe. The method is based on the ability of terminal deoxynucleotidyltransferase to catalyze the sequential addition of radioactive bases to the 3′ end of oligonucleotides,[30] resulting in the formation of a radioactive "tail" on the probe. This method was used[29,31] to detect proopiomelanocortin and vasopressin mRNA by *in situ* hybridization. It has the advantage of being able to label probes with multiple, lower energy radioisotopes to enhance anatomical resolution. While both ^{3}H- and ^{125}I-labeled probes have been useful, the former requires long exposure times, while the latter can lead to high background labeling with some probes. Labeling of oligonucleotides by 3′-tailing, using [^{35}S]dATP as substrate in the reaction, followed by a chromatographic purification step, has yielded probes which give excellent anatomical (i.e., cellular) resolution after relatively short exposure times.

Hybridization

The hybridization conditions have been derived from published methods[4,29] and aim to facilitate the interaction of the labeled probe with the appropriate mRNA in the tissue while suppressing nonspecific interactions of the probe with other molecules. The method has been further evolved empirically to simplify the procedure, and several conventional steps have been eliminated or collapsed where they have been found to not affect the results.

In brief, before the probe is added, the tissue is warmed to room

[30] F. J. Bollum, *Enzymes* **10,** 145 (1974).

[31] M. E. Lewis, S. Burke, T. G. Sherman, R. Arentzen, and S. J. Watson, *Soc. Neurosci. Abstr.* **10,** 358 (1984).

temperature (about 30 min) and incubated in 2× SSC (diluted from Sigma S-6639) for 10 min. The slides are then blotted dry with filter paper and placed in a Nunc culture dish lined with wetted filter paper backing. Radioactive probe in 500 μl of hybridization buffer is then pipetted onto each section. The hybridization buffer contains (1) formamide, which facilitates hybrid formation at lower temperatures; (2) dextran sulfate, an inert polymer which appears to accelerate the rate of hybridization by exclusion of DNA from the volume taken up by the polymer; (3) 3× SSC (450 m*M* NaCl, 45 m*M* trisodium citrate, pH 7.0), which maintains an ionic strength consistent with a rapid rate of hybrid formation; (4) Denhardt's solution, consisting of Ficoll, polyvinylpyrrolidone, and bovine serum albumin, originally designed to saturate nonspecific binding sites on nitrocellulose; and (5) yeast tRNA and denatured salmon sperm DNA, which may occupy nonspecific binding sites and serve as "decoy" substrates for endogenous or contaminating nucleases. The probe is then added to the tissue in this buffer at a temperature about 20–25° below the T_m,[29] between 37 and 45°, for a period of about 16 hr. The slides are then washed at decreasing concentrations of SSC and at high temperature (e.g., hybridization temperature) to destabilize nonspecific hybrids, thus reducing the background.

Autoradiography

Following hybridization the sections are evaluated for effective labeling. They are air dried, placed into an X-ray cassette with Kodak XAR-5 film, exposed for 3 days at room temperature, developed, and examined. Successful cases are dipped in liquid emulsion (Kodak NTB2, diluted 1 : 1) according to standard techniques.[32,33] The appropriate exposure time depends on the effectiveness of probe labeling, the efficiency of hybridization, and the amount of mRNA contained in the cells of interest. In our work 5 weeks has proved to be a good starting point.

Following exposure and development of the emulsion, the tissue must be cleared and coverslipped in order to detect the silver grains indicating hybridization. If the rhodamine beads have been used as the retrograde tracer, the exposure to alcohols and xylenes must be minimized to avoid solubilizing, and thus losing, the beads. A total of 2 min of exposure appears tolerable, but an alternative is to simply air dry the tissue, directly apply mounting medium, and then coverslip the tissue. There are several mounting media that are nonfluorescent, retain fluorescence in the

[32] A. W. Rogers, "Techniques of Autoradiography," 2nd Ed. Elsevier, Amsterdam, 1973.

[33] S. T. Edwards and A. Hendrickson, *in* "Neuroanatomical Tract-Tracing Methods" (L. Heimer and M. J. Robards, eds.), p. 171. Plenum, New York, 1981.

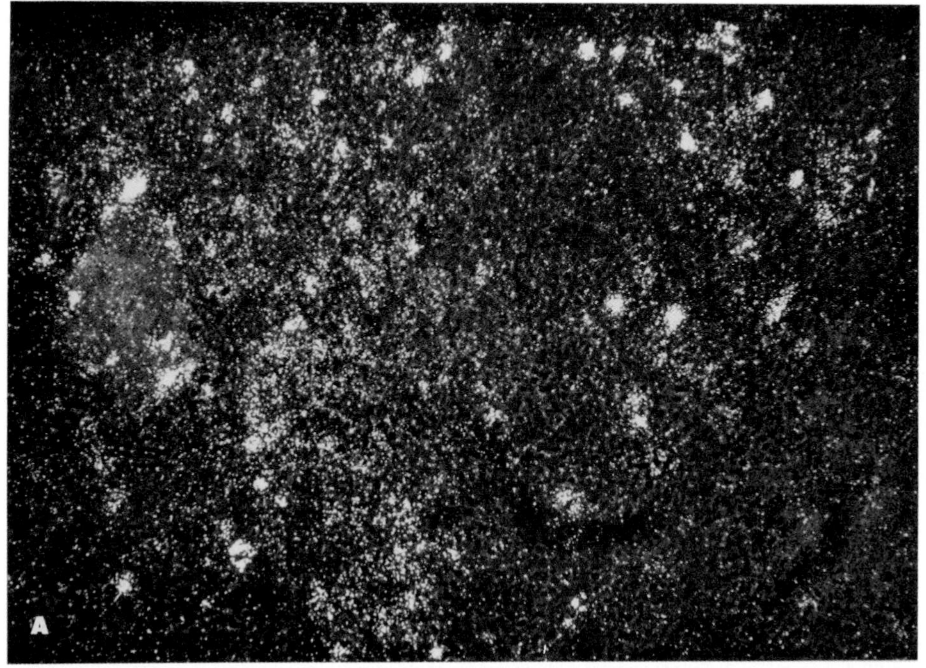
A

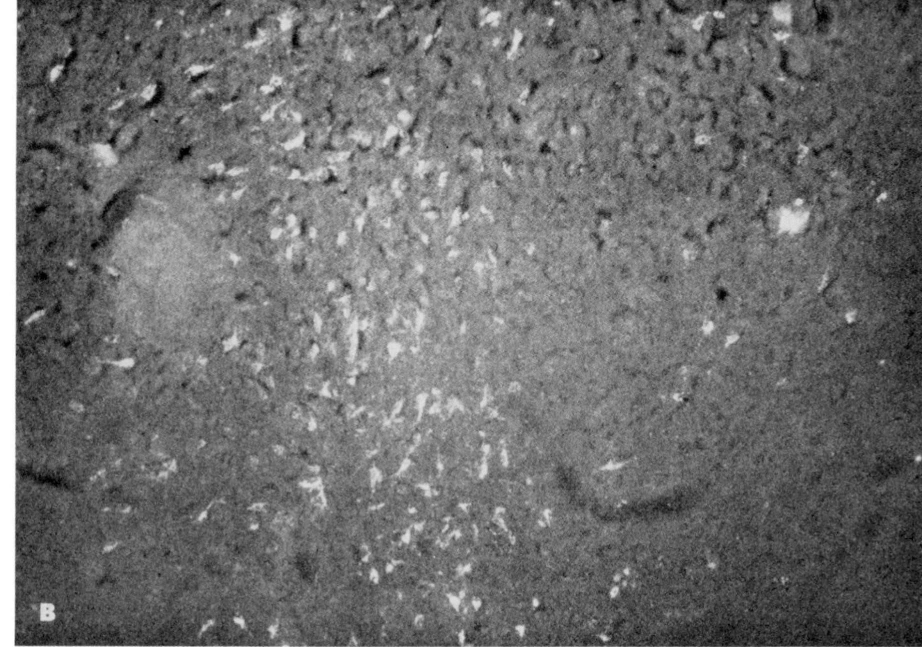
B

tissue, and in addition give a semipermanent preparation: Entellan (Merck, Darmstadt, FRG), Fluoromount (Southern Biotechnology Associates, Inc., Birmingham, AL 35226), Immumount (Accurate Chemicals), and Permafluor (Immunon, Troy, MI 48087). Polyvinyl alcohols in glycerol has proven to be very good mounting medium that retains fluorescence well and hardens, which prevents the coverslip from sliding around. A polyvinyl alcohol–glycerol medium can also be obtained commercially (Aqua-Mount, Lerner Lab., New Haven, CT 06513).

Microscopy and Photography

Cellular resolution, and thus microscopic examination, is a requirement of combined *in situ* hybridization and retrograde fluorescent connection tracing. The silver grains produced by the *in situ* hybridization label are most easily seen in dark-field microscopy. Clusters of silver grains about the size of neurons and located over neurons suggest specific labeling. Clustering of silver grains of this kind are most easily detected and discriminated from background grains at relatively low magnification, using a 4× or 10× objective (final magnification 40–100×), for example. The silver grains are, of course, visible at higher magnification for photography or simultaneous viewing with other neuronal labels. Standard fluorescence microscopy is used for the dyes. The beads require any rhodamine filter set, such as 510–560 nm excitation, 580 nm dichroic mirror, and 590 nm barrier filter. For fluoro-gold 323 nm is the excitation maximum (UV) and 408 nm the emission maximum at neutral pH, which is compatible with a standard UV filter set. The beads are visible using a 10× objective, but lightly labeled cells may only be seen with 20× or even 40× objectives. The fluorescent tracers and *in situ* hybridization label can be viewed simultaneously using epifluorescence for the dyes and dark-field transmission microscopy for the autoradiographic grains. By dodging your hand through the subcondenser light path, the field can be rapidly scanned for double-labeled cells. Alternating between the 10× and 20× objectives optimizes the viewing conditions for each label.

For photography, color documentation of the fluorescent tracers is most informative, although results are adequate with black and white film (Fig. 1). Kodak 200 Ektachrome slide film has proven good, as has the 800

FIG. 1. (A) A dark-field photomicrograph (10× objective) showing clusters of silver grains reflecting somatostatin mRNA *in situ* hybridization in the right central amygdaloid nucleus of the rat. (B) The identical microscopic field taken for rhodamine fluorescence. In this rat 50 nl of rhodamine beads had been injected into the ipsilateral nucleus tractus solitarii 10 days prior to sacrifice. Bead-labeled cells can be seen concentrated in the medial aspect of the nucleus but also extending laterally into the region with numerous neurons heavily labeled for somatostatin mRNA.

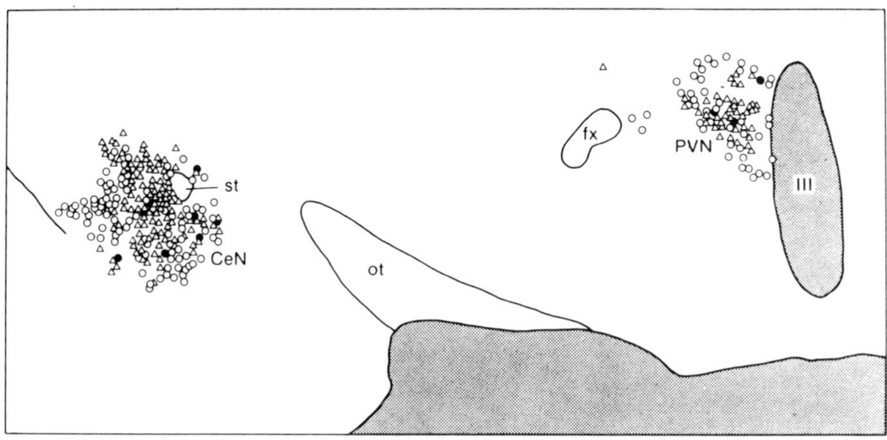

FIG. 2. Computer-generated line drawing of the left basal forebrain in a rat treated as in Fig. 1. The relative distribution of neurons labeled for somatostatin mRNA, for rhodamine beads, and for both is represented by circles, triangles, and solid circles, respectively. CeN, Central nucleus of the amygdala; fx, fornix; ot, optic tract; PVN, paraventricular nucleus; st, stria terminalis; III, third ventricle.

ASA Ektachrome professional film. In either case, it is necessary to increase the ASA to maintain good signal-to-noise ratios for the fluorescent tracers, by two f stops for the Fluoro-gold, and by two to three f stops for the rhodamine beads. To photograph double-labeled cells, the same field can be photographed under each condition separately without moving the stage. At least 20× and usually 40× objectives (200–400× final magnification) are typically required, and an attempt should be made to include "landmarks" when possible to confirm the relative positions of the cells in photomicrographs (Fig. 1). When photographing for dark-field conditions, the fluorescent filters should be removed from the light path to the camera. To convincingly show double labeling, double exposures are valuable, as it is possible for adjacent cells to be singly labeled, but color photomicrographs are necessary to separate the markers. For these photomicrographs the dark-field exposure has to be reduced by approximately two to three f stops to keep the brightness of each marker proportional. Depending on background, an increase of up to four f stops may be necessary for rhodamine fluorescence in order to keep background brightness low. At higher magnifications it is necessary to refocus between exposures, since the emulsion and the cells are in different focal planes. In order to compare photomicrographs it is important that slide film be mounted carefully to ensure registration of the frames using glass slides or other mounts that use pinhole registration.

Expected Benefits

The results will simultaneously show, in the same tissue section, the distribution of neurons both participating in a functional circuit and expressing a certain mRNA. By the use of appropriate computer-based data acquisition (e.g., Ref. 34) it is possible to obtain quantitative information on the spatial distribution of each label and the location and extent of double labeling, as in Fig. 2, thus examining gene expression in subpopulations of neurons defined by their axonal projections. By the addition of quantitation of *in situ* hybridization labeling,[11] the relationship between gene expression, regulation of gene expression, and function can be approached.

[34] W. T. Rogers and J. S. Schwaber, *Appl. Optics* **26,** 3384 (1987).

[57] Cytochemical Techniques for Studying the Diffuse Neuroendocrine System

By A. E. Bishop and J. M. Polak

Introduction

In recent years, it has become apparent that the endocrine and nervous systems, far from being two entirely separate units, are actually unified in the diffuse neuroendocrine system.[1] It is now recognized that the diffuse neuroendocrine system is a massive complex of endocrine cells, nerves, and intermediate cellular forms, such as neuroendocrine[2] and paracrine cells,[3] which permeates the entire body. The various components are capable of producing a vast range of biologically active peptides and amines which have potent effects on bodily functions.

The morphological study of the diffuse neuroendocrine system has centered in the past on the application of specialized dyes, e.g., toluidine blue, histochemical reactions, e.g., argyrophilia and argentaffinity,[4,5]

[1] A. G. E. Pearse, *in* "Interdisciplinary Neuroendocrinology" (M. Ratzenhofer, H. Höfler, and G. F. Walter, eds.), p. 1. Karger, Basel, 1984.

[2] A. G. E. Pearse, *Med. Biol.* **55,** 115 (1977).

[3] F. Feyrter, "Uber Diffuse Endokrine Epithelial Organe." Le Barth, Leipzig, 1937.

[4] A. G. E. Pearse, "Histochemistry: Theoretical and Applied." Livingstone, Edinburgh, 1985.

[5] L. Grimelius and E. Wilander, *in* "Endocrine Tumours" (J. M. Polak and S. R. Bloom, eds.), p. 95. Livingstone, Edinburgh, 1985.

ultrastructural analysis,[6] or, within the past 15 years, immunocytochemistry.[7] The aim of this chapter is to report on the latest techniques to be used in this field, including electron immunocytochemistry, *in situ* hybridization, and binding site analysis by *in vitro* autoradiography. Current application of these methods, in conjunction with the more established techniques, is allowing thorough investigation of the functional morphology of the diffuse neuroendocrine system.

Localization of Components of the Diffuse Neuroendocrine System

Immunocytochemistry remains the method of choice for localizing the various structures within the diffuse neuroendocrine system (DNES). At present, a variety of sophisticated immunocytochemical methods is available, and one should select the technique most suited to the particular area of interest.[7] However, there are some general advances which may be of use to those involved in various aspects of the study of the diffuse neuroendocrine system.

Specific Neuroendocrine Cell Products

For many years, immunostains of neural or endocrine cells have been made using antisera raised against the main characterized form of the cellular product. More recently, the structures of precursors of many peptide hormones and neurotransmitters have been disclosed by biochemists and molecular biologists. Thus, antisera are now produced to various portions of peptide precursor molecules. These are particularly useful in the investigation of neuroendocrine tumors as they can increase the efficiency of diagnosis of cases where little bioactive peptide may be stored or abnormal molecular forms of a hormone may be produced. For example, it is known that pancreatic glucagon is derived from a molecule, the so-called preproglucagon, which also gives rise to glicentin (or gut glucagon), glucagon-related peptides, and, at the carboxy terminus, glucagon-like peptides 1 and 2[8] (Fig. 1A). An immunocytochemical study of 10 pancreatic A cell tumors associated with the glucagonoma syndrome failed to detect pancreatic glucagon in 3 cases. However, all tumors were positive for one or another of the derivatives of preproglucagon[9] (Figs. 1B and 1C).

[6] E. Solcia, C. Capella, R. Buffa, L. Usellini, R. Fiocca, and F. Sessa, *in* "Physiology of the Gastrointestinal Tract" (L. R. Johnson, ed.), p. 39. Raven, New York, 1981.

[7] J. M. Polak and S. Van Noorden (eds.), "Immunocytochemistry: Modern Methods and Applications." Wright, Bristol, England, 1986.

[8] G. I. Bell, R. Sanchez-Pescador, P. J. Laybourn, and R. C. Najarian, *Nature (London)* **304,** 368 (1983).

[9] Q. A. Hamid, A. E. Bishop, K. L. Sikri, I. M. Varndell, S. R. Bloom, and J. M. Polak, *Histopathology* **10,** 119 (1986).

Immunostains of different precursor derivatives in neuroendocrine tumors may also aid analysis of tumor biology. Pulmonary and extrapulmonary small cell carcinomas and lung carcinoids have been associated for some time with the production of mammalian bombesin (or gastrin-releasing peptide), which has been described as an autocrine stim-

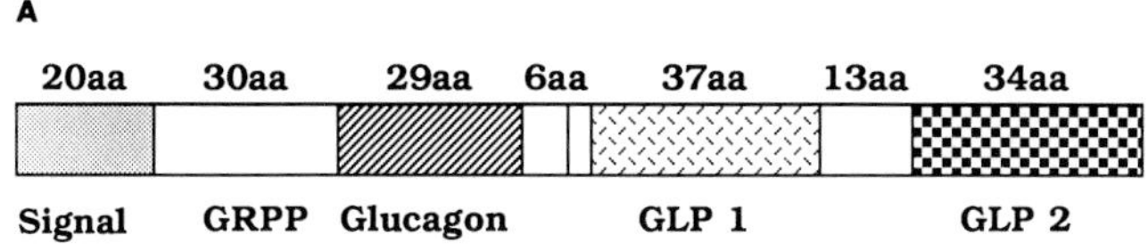

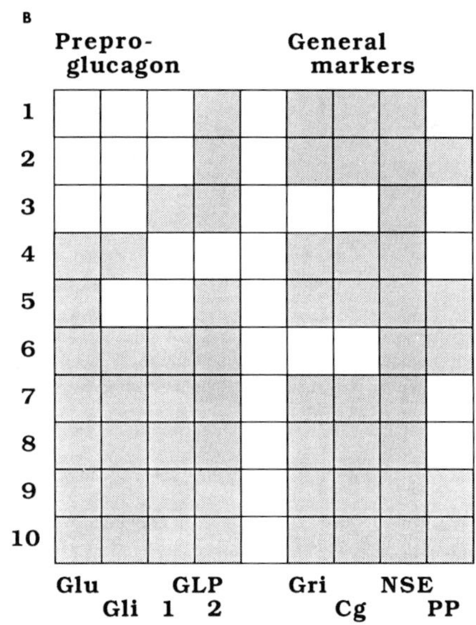

FIG. 1. (A) Diagram showing the structure of preproglucagon and the positions of its derivative peptides. (B) Tabulated results of an immunocytochemical study of 10 cases of glucagonoma. At left are shown the results for immunostains of preproglucagon derivatives. At right the results for the general markers Grimelius' argyrophilia (Gri), chromogranin (Cg), neuron-specific enolase (NSE), and pancreatic polypeptide (PP) are seen. A shaded square denotes a positive result. (C) Pancreatic glucagon immunostained in an A cell tumor (glucagonoma) of the human pancreas using the peroxidase–antiperoxidase technique. ×120.

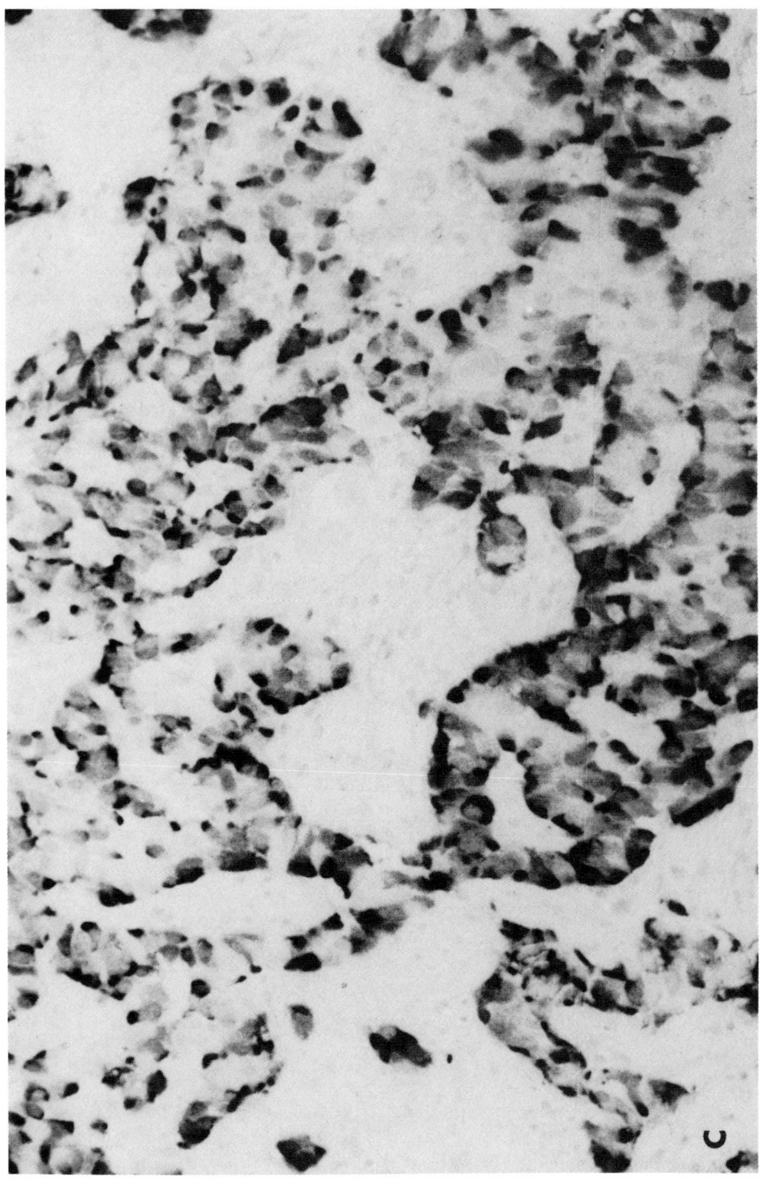

FIG. 1C. See legend on p. 793.

ulator of tumor growth.[10] Human preprobombesin has been shown to consist of a signal peptide, bombesin itself, and a carboxy-terminal extension.[11] Immunostaining of bombesin and its carboxy-terminal flanking peptide in small cell carcinomas and lung carcinoids has shown stronger immunostaining for bombesin in the more slow growing, well-differentiated carcinoids than in the carcinomas. Conversely, the rapidly growing, malignant small cell cancers contain much more immunoreactivity for the carboxy-flanking peptide.[12,13]

General Markers for Cells of the Diffuse Neuroendocrine System

Several proteins have been cited as components exclusively of neurons and/or endocrine cells. At present, a number of antisera are available to these so-called general neuroendocrine markers which allow immunostaining of all endocrine cells, all nerves, or both components of the diffuse neuroendocrine system at the same time.

Neuron-Specific Enolase (NSE). One of the first general neuroendocrine markers to be described was NSE, an isoenzyme of enolase first extracted from bovine brain[14] and later shown to occur in all types of cells of the diffuse neuroendocrine system[15,16] (Fig. 2A). As well as revealing the diffuse neuroendocrine system in its entirety, immunostaining for NSE allows visualization of poorly granulated cells as the enzyme has a cytoplasmic location. Also, its close chemical relationship with an enzyme involved in glycolysis means that the density of NSE within a cell may reflect the cellular metabolic state.

Chromogranins. A family of structurally related, large, acidic proteins was discovered in the catecholamine-storing secretory granules of the adrenal medulla.[17–19] This family consists mainly of the three chromo-

[10] F. Cuttitta, D. N. Carney, J. Mulshine, T. W. Moody, J. Fedorko, A. Fischler, and J. D. Minna, *Nature (London)* **316,** 823 (1985).

[11] E. R. Spindel, W. W. Chin, J. Price, L. H. Rees, G. M. Besser, and J. F. Habener, *Proc. Natl. Acad. Sci. U.S.A.* **81,** 5699 (1984).

[12] D. R. Springall, N. Ibrahim, J. Rode, M. Sharp, S. R. Bloom, and J. M. Polak, *J. Pathol.* **150,** 151 (1986).

[13] Q. Hamid, B. J. Addis, D. R. Springall, N. B. N. Ibrahim, M. A. Ghatei, S. R. Bloom, and J. M. Polak, *Virchows Arch. A* **411,** 185 (1987).

[14] E. Bock and J. Dissing, *Scand. J. Immunol.* **4,** 31 (1975).

[15] D. Schmechel, P. J. Marangos, and M. Brightman, *Nature (London)* **276,** 834 (1978).

[16] A. E. Bishop, J. M. Polak, P. Facer, G. L. Ferri, P. J. Marangos, and A. G. E. Pearse, *Gastroenterology* **83,** 902 (1982).

[17] H. Blaschko, R. S. Comline, F. H. Schneider, M. Silver, and A. D. Smith, *Nature (London)* **215,** 58 (1967).

[18] F. H. Schneider, A. D. Smith, and H. Winkler, *Br. J. Pharmacol. Chemother.* **31,** 94 (1967).

[19] D. T. O'Connor and R. P. Frigon, *J. Biol. Chem.* **259,** 237 (1984).

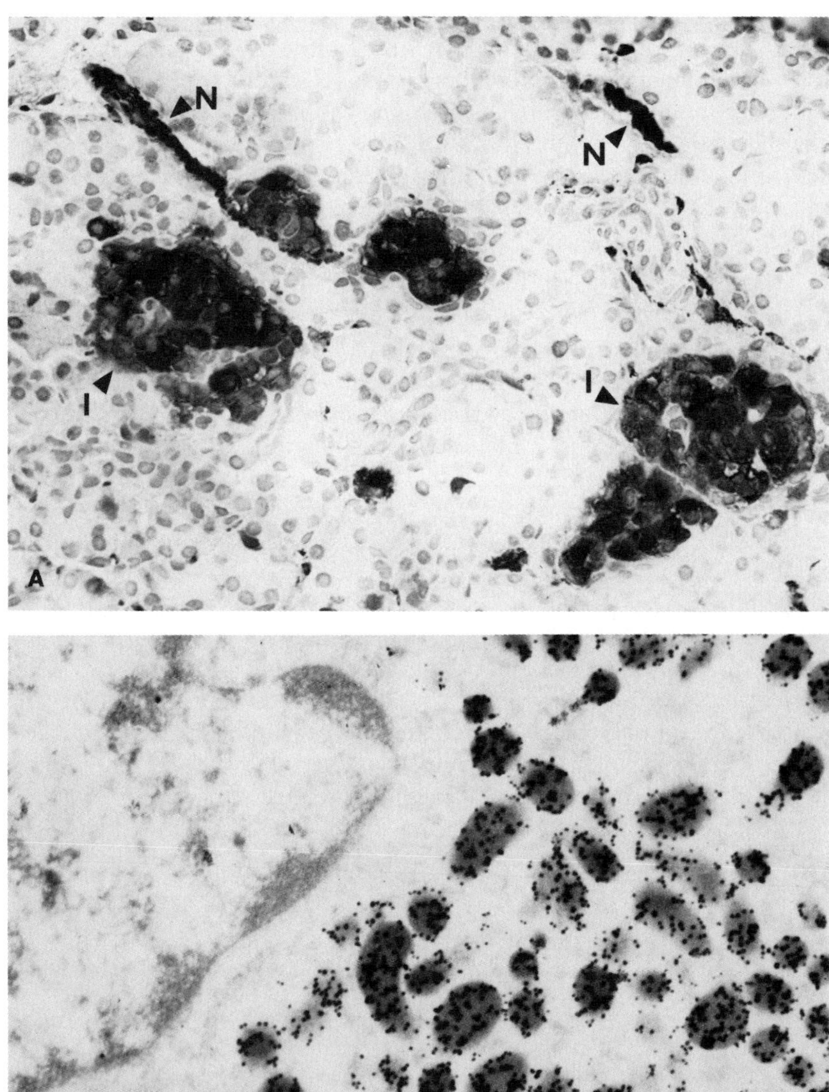

FIG. 2. (A) Section of human pancreas immunostained for NSE. The enzyme is present in both nerves (N) and the endocrine cells of the pancreatic islets (I). ×183. (B) Electron micrograph showing chromogranin immunostained in secretory granules of an endocrine cell using the immunogold staining method. ×10,950. (C) High-power micrograph of rat lung showing PGP 9.5 immunostained in a ganglion next to a bronchial blood vessel. ×329. (D) 7B2 immunostained in an islet of Langerhans of human pancreas. ×234.

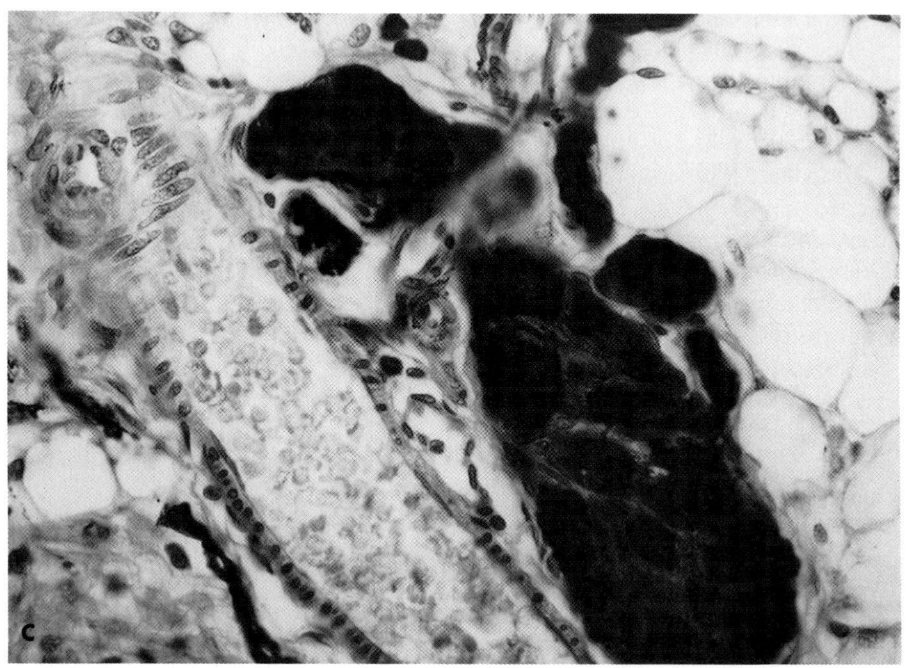

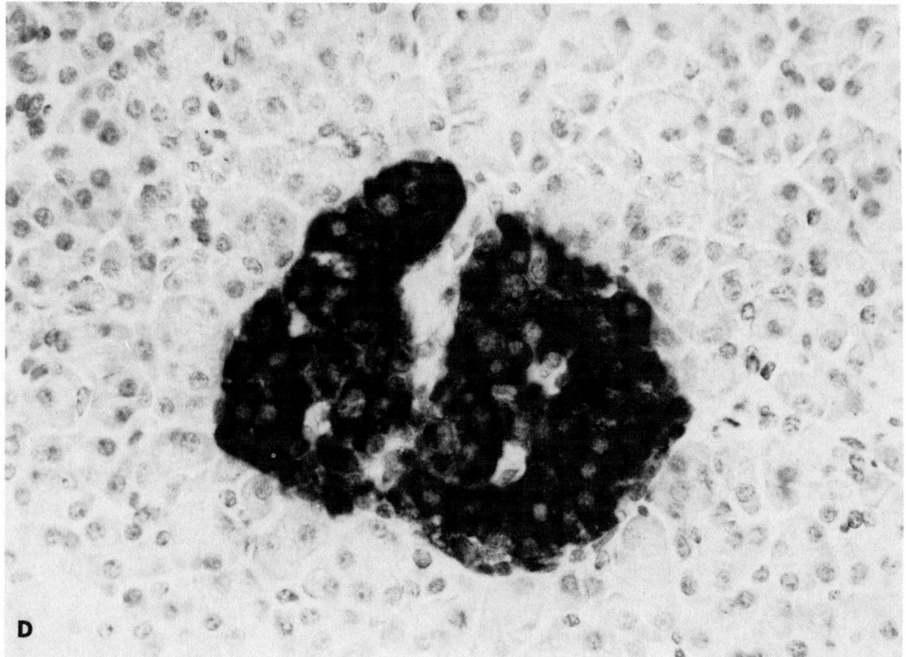

FIG. 2C and D.

granin proteins: chromogranin A, 70,000–75,000 MW and originally isolated from chromaffin granules; B, 100,000 MW also first found in chromaffin granules; C, 86,000 MW discovered in the anterior pituitary. Chromogranins have since been shown to be present in most types of endocrine cells,[20–23] and, although they are known to be released from sympathetic nerves,[19] consistent immunostaining of neural tissue has only been reported with certain polyclonal antisera. Unlike NSE, chromogranins are contained in secretory granules[24] (Fig. 2B), and, thus, the density of chromogranin immunostaining reflects granule storage, making application of NSE and chromogranin antisera in combination a useful means to ascertain the metabolic status of neuroendocrine cells. A recent development in the study of chromogranins has been the production of a cDNA probe for chromogranin A using recombinant DNA technology.[25] Using the clone, the primary structure of chromogranin A has been elucidated. Also, it has provided a probe for use in Northern blots, and this has given evidence that chromogranins may arise from more than one precursor protein.

Protein Gene Product 9.5. Another soluble protein, like NSE first isolated from the brain, is protein gene product 9.5 (PGP 9.5).[26] Antibodies to PGP 9.5 stain all types of nerves and appear to be particularly good markers for the peripheral nervous system[27] (Fig. 2C). The usefulness of PGP 9.5 antibodies as markers for all components of the diffuse neuroendocrine system is, however, debatable. The protein has been reported to occur in neuroendocrine tumors,[28] but its presence in normal endocrine cells has not been widely demonstrated.

Synaptophysin (or P38 Protein). Synaptophysin, or P38 protein, is a calcium-binding glycosylated peptide (MW 38,000) which has been detected in certain neuroendocrine cells.[29] Synaptophysin is a component of secretory vesicles but appears to be specific for the small, transparent

[20] D. T. O'Connor, D. Burton, and L. J. Deftos, *J. Clin. Endocrinol. Metab.* **57,** 1084 (1983).

[21] B. S. Wilson and R. V. Lloyd, *Am. J. Pathol.* **115,** 458 (1984).

[22] P. Facer, A. E. Bishop, R. V. Lloyd, B. S. Wilson, R. J. Hennessey, and J. M. Polak, *Gastroenterology* **89,** 1366 (1985).

[23] G. Rindi, R. Buffa, F. Sessa, O. Tortora, and E. Solcia, *Histochemistry* **85,** 19 (1986).

[24] I. M. Varndell, R. V. Lloyd, B. S. Wilson, and J. M. Polak, *Histochem. J.* **17,** 981 (1985).

[25] M. Grimes, A. Iacangelo, L. E. Eiden, B. Godfrey, and E. Herbert, *Ann. N. Y. Acad. Sci.* **493,** 351 (1987).

[26] R. J. Thompson, J. F. Doran, P. Jackson, A. P. Dhillon, and J. Rode, *Brain Res.* **278,** 224 (1983).

[27] S. Gulbenkian, J. Wharton, and J. M. Polak, *J. Auton. Nerv. Syst.* **18,** 235 (1987).

[28] J. Rode, A. P. Dhillon, J. F. Doran, P. Jackson, and R. J. Thompson, *Histopathology* **9,** 147 (1985).

[29] E. Navone, R. Jahn, G. Di Dioia, H. Stukenbrok, P. Greengard, and P. De Camilla, *J. Cell Biol.* **193,** 2511 (1986).

vesicles that resemble presynaptic vesicles[30] and are found only in certain types of neuroendocrine cells.[31,32] Thus, the protein can be localized to cells of the adrenal medulla, anterior pituitary, and endocrine pancreas, for example, but cannot be found in intestinal or parathyroid endocrine cells.[29,33]

7B2 (or APPG). Porcine and human pituitary glands were the original source of this large protein, which was first called APPG (anterior pituitary pig) but more recently termed 7B2 because of its chromatographic profile.[34] The protein can be found in several types of endocrine cells and in central and peripheral nerves and neuroendocrine tumors.[35] However, the largest concentrations of 7B2 can be found in pancreatic β-cells (Fig. 2D) and their neoplasms,[35] an observation which is interesting in view of the reported structural similarities between 7B2 and insulin-like growth factor.[34]

Neurofilaments. Neurofilaments are cytoskeletal elements composed of a triplet of proteins of different molecular weights (68,000, 150,000, and 200,000).[36] Polyclonal antisera recognizing various epitopes of the triplet proteins can be used to immunostain all classes of innervation.[37,38] In addition, these antisera can be useful in the investigation of neural tumors.[39]

Electron Immunocytochemistry

In the past, electron microscopy of endocrine cells has made a major contribution to their identification and classification.[6] Ultrastructural analysis demonstrates the typical secretory granules of each cell type, and the features of these granules, e.g., size, electron density, presence of limiting membrane, allow some prediction of the substance produced by the cell. The application of immunocytochemistry at the electron micro-

[30] B. Wiedenmann and W. W. Franke, *Cell* **45,** 1017 (1985).

[31] C. Capella, E. Hage, E. Solcia, and L. Usellini, *Cell Tissue Res.* **186,** 25 (1978).

[32] S. Kobayashi and R. E. Coupland, *Arch. Histol. Jpn.* **40,** 251 (1977).

[33] B. Wiedenmann, W. W. Franke, C. Kuhn, R. Moll, and V. E. Gould, *Proc. Natl. Acad. Sci. U.S.A.* **83,** 3500 (1986).

[34] K. L. Hsi, N. G. Seidah, G. De Serves, and M. Chretien, *FEBS Lett.* **147,** 261 (1982).

[35] H. Suzuki, M. A. Ghatei, S. J. Williams, L. O. Uttenthal, P. Facer, A. E. Bishop, J. M. Polak, and S. R. Bloom, *J. Clin. Endocrinol. Metab.* **63,** 147 (1985).

[36] P. N. Hoffman and R. J. Lasek, *J. Cell Biol.* **66,** 351 (1975).

[37] A. E. Bishop, F. Carlei, V. Lee, J. Q. Trojanowski, P. J. Marangos, D. Dahl, and J. M. Polak, *Histochemistry* **82,** 93 (1985).

[38] J. Q. Trojanowski, *in* "Immunocytochemistry: Modern Methods and Applications" (J. M. Polak and S. Van Noorden, eds.), p. 413. Wright, Bristol, England, 1986.

[39] F. Carlei, J. M. Polak, and A. Ceccamea, *Virchows Arch. A: Pathol. Anat.* **404,** 313 (1985).

scopical level, however, provides direct evidence of the product of individual granules. The immunocytochemical methods used in electron microscopy are preembedding or postembedding techniques. Preembedding methods using peroxidase–antiperoxidase complexes provided the first immunostains which could be visualized under the electron microscope. This was possible as the diaminobenzidine coupled to peroxidase could be made electron dense by osmication. There are certain disadvantages to this approach, one of the most important being spreading of the final reaction product which can obscure the structure of the granules.

In postembedding techniques, a variety of gold-labeling methods are now available which provide much better resolution of immunostaining.[40] Gold particles are seen as very electron-dense spheres which attach to secretory granules but do not mask their morphology. There are a number of procedures which use gold labeling, but one of the most widely applied is probably the immunogold staining method[41,42] which is an indirect immunostain, with the second layer antibody being labeled with colloidal gold, which absorbs onto proteins (Fig. 2B). As gold particles can be made in different sizes, the basic postembedding techniques which first utilized colloidal gold have been refined so that multiple immunostains can be made on single tissue sections, to show costorage of different substances or different derivatives of a single hormone precursor in one secretory granule[40,43–45] (Fig. 3).

Characterization of Peptide-Containing Nerves

A number of experimental procedures are now being used to establish the origins and nature of peptide-containing nerves.[46] For example, establishing whether a particular type of peptide-containing nerve has an intrinsic or extrinsic origin within a certain organ can be achieved first by administering chemicals like colchicine or vinblastine.[47] These agents block axonal transport of neurotransmitters and thus cause a buildup of

[40] J. M. Polak and I. M. Varndell, "Immunolabelling for Electron Microscopy." Elsevier, Amsterdam, 1984.

[41] J. De Mey, M. Moeremans, G. Geuens, R. Nuydens, and M. De Brabander, *Cell Biol. Int. Rep.* **5,** 889 (1981).

[42] I. M. Varndell, F. J. Tapia, L. Probert, A. M. J. Buchan, J. Gu, J. De Mey, S. R. Bloom, and J. M. Polak, *Peptides* **3,** 259 (1982).

[43] L. I. Larsson, *Nature (London)* **282,** 743 (1979).

[44] M. Bendayan, *J. Histochem. Cytochem.* **30,** 81 (1982).

[45] F. J. Tapia, I. M. Varndell, L. Probert, J. De Mey, and J. M. Polak, *J. Histochem. Cytochem.* **31,** 977 (1983).

[46] P. C. Emson, "Chemical Neuroanatomy." Raven, New York, 1983.

[47] A. Björklund and T. Hökfelt, "Methods in Chemical Neuroanatomy," Vol. 1. Elsevier, Amsterdam, 1983.

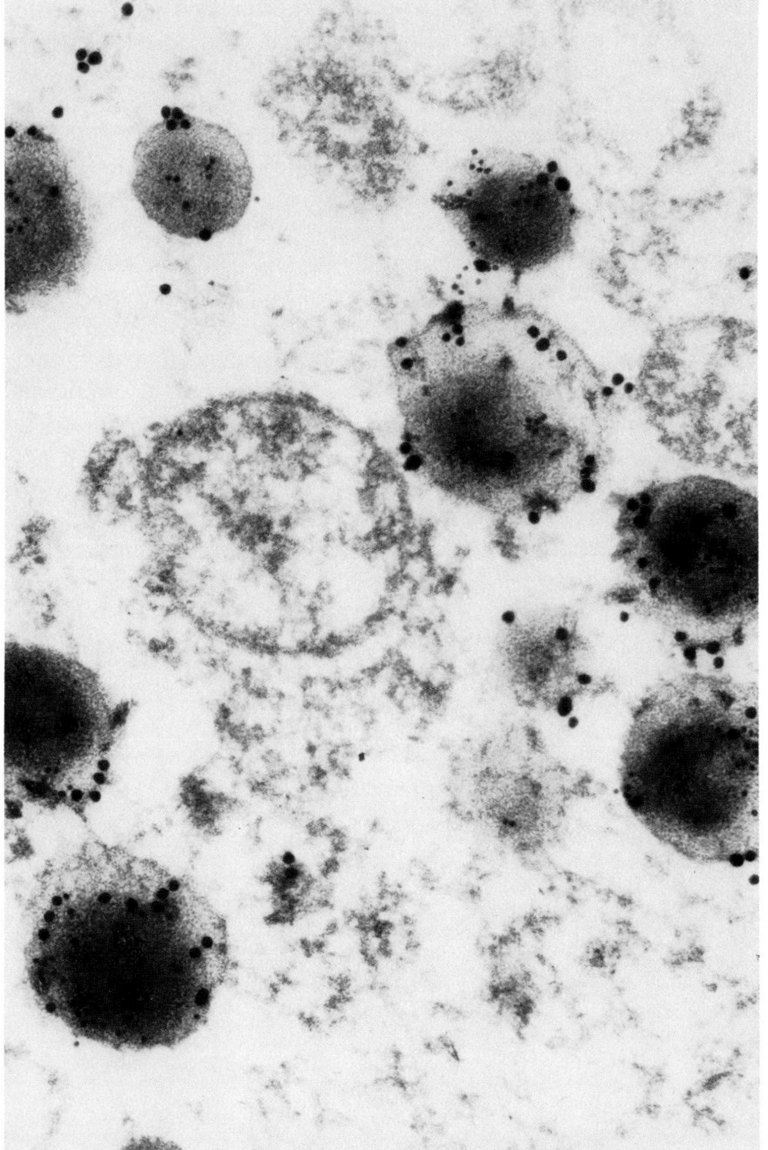

FIG. 3. Electron micrograph of a pancreatic A cell immunostained for derivatives of preproglucagon using the immunogold staining method. The core of the secretory granules contains pancreatic glucagon (small gold particles) whereas glicentin is immunostained in the halo (large gold particles). ×20,000.

peptide content within perikarya, allowing a more accurate analysis of the numbers of intrinsic peptide-containing ganglion cells than would be obtained in untreated tissue. Extrinsic denervation of the organ under study, followed by its maintenance *in vivo* for a number of days, or extirpation and culture of intrinsic ganglion cells give information on the origins of peptide-containing nerves.[48] In addition, the chemical identity of nerves can be studied by administration of, for example, 6-hydroxydopamine, which damages noradrenergic nerves,[49] or capsaicin, a chemical first extracted from red peppers and now known to cause release of neurotransmitters from primary afferent fibers.[50]

For those peptide-containing nerves where cell bodies lie outside the organ they innervate, there are retrograde tracing methods which allow extrinsic neural origins to be mapped.[51] In these methods, an appropriate substance (such as True blue dye, horesradish peroxidase, wheat germ agglutinin) is applied to the terminals of the nerve being studied by injection into or application to the surface of the innervated organ. Uptake and retrograde transport of the substance *in vivo* means that the cell bodies can be visualized. These can be identified as producing a particular peptide by serial sectioning or double-immunostaining techniques (Figs. 4A and 4B). A problem which has been encountered in the application of retrograde tracing techniques is the leakage of dye at the point of tracer application.[52,53] This can lead to inaccuracies but has been largely overcome recently by the use of barriers of dye diffusion, such as plastic skin[54] or gel foam.[55] Combined application of these surgical and chemical manipulations is now providing the means for establishing the neurochemical anatomy of the peptide-containing nervous system (cf. Ref. 56).

Demonstration of Peptide Gene Expression

Methods for the hybridization of complimentary nucleotide sequences have been applied for some time in the field of molecular biology. These

[48] K. R. Jessen, M. J. Saffrey, S. Van Noorden, S. R. Bloom, J. M. Polak, and G. Burnstock, *Neuroscience* **5,** 1717 (1980).

[49] H. Thoenen and J. P. Tranzer, *Arch. Pharmacol.* **261,** 271 (1968).

[50] M. Fitzgerald, *Pain* **15,** 109 (1983).

[51] L. Heimer and K. J. Robards, "Neuroanatomical Tract-Tracing Methods." Plenum, New York, 1981.

[52] Y. Amy, *J. Neurosci. Methods* **2,** 95 (1980).

[53] A. Van der Krans and P. V. Hoogland, *J. Neurosci. Methods* **9,** 95 (1982).

[54] E. A. Fox and T. L. Powley, *J. Auton. Nerv. Syst.* **15,** 55 (1986).

[55] B. Lindh, T. Hökfelt, L. G. Elfvin, L. Terenius, J. Fahrenkrug, R. Elde, and M. Goldstein, *J. Neurosci.* **6,** 2371 (1986).

[56] H. C. Su, A. E. Bishop, R. F. Power, Y. Hamada, and J. M. Polak, *J. Neurosci.* **7,** 2674 (1987).

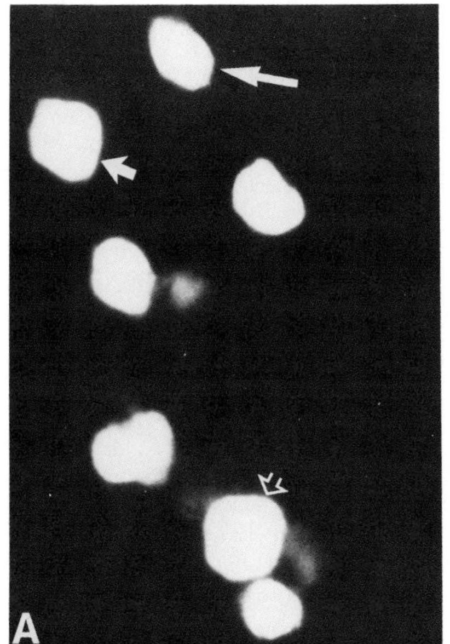

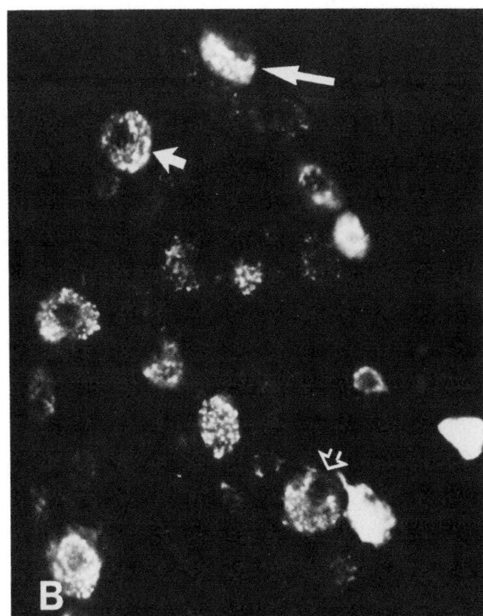

FIG. 4. Serial sections of rat dorsal root ganglia. (A) Injected True blue is present in some of the ganglion cells. (B) In the serial section, an immunostain for calcitonin gene-related peptide shows that it is present in some True blue-labeled cells (open, small solid, and long solid arrows show the pairs of serially sectioned cells). ×114.

methods reveal specific mRNAs directing the synthesis of a particular substance. Recently, these techniques have been adapted for use on tissue sections, and the method of *in situ* hybridization[57] is now adding a new dimension to the cytochemical investigation of the diffuse neuroendocrine system by allowing analysis of peptide gene expression.

A number of techniques of *in situ* hybridization have been proposed and are based on the use of single- or double-stranded DNA or RNA probes complementary to the intracellular mRNA (cf. Ref. 58). Different labels can be attached to the probes in order to visualize their distribution in tissue sections, and these can be radioactive or nonradioactive. Radioactive labels include ^{3}H, ^{35}S, and ^{32}P whereas most nonradioactive methods make use of the avidin–biotin system (Fig. 5A). Current opinion holds that the use of cRNA probes provides greater specificity and sensitivity of hybridization than other types of probe[59] and that radioactive

[57] P. Hudson, J. Penschow, J. Shine, G. Ryan, H. Niall, and J. Coghlan, *Endocrinology (Baltimore)* **108,** 353 (1980).

[58] J. D. Penschow, J. Haralambidis, P. E. Darling, I. A. Darby, E. M. Wintour, G. W. Tregear, and J. P. Coghlan, *Experientia* **43,** 741 (1987).

[59] R. E. Siegel and W. S. Young, *Neuropeptides* **6,** 573 (1986).

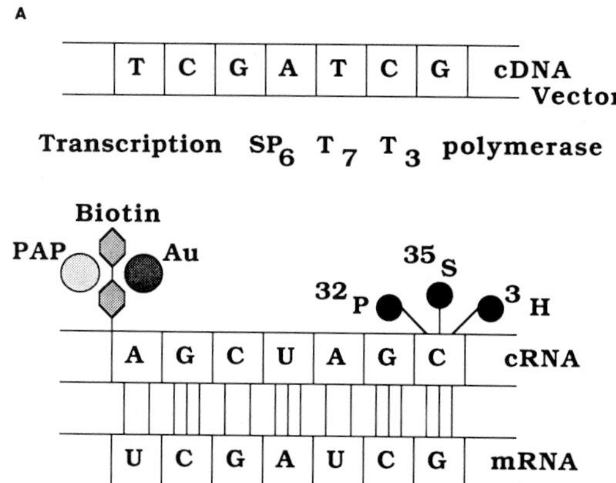

FIG. 5. (A) Diagrammatic representation of the basis of the *in situ* hybridization technique showing the various probe labels that can be used. (B) Localization of insulin mRNA in rat pancreas using a ^{32}P-labeled cRNA probe. The strong signal is seen as black grains overlying an islet. ×350.

labels are preferable to the avidin–biotin system as the sensitivity of the latter has been questioned.

Combined with immunocytochemistry and RNA blot analysis, *in situ* hybridization provides a means for studying peptide gene expression at the cellular level. Unlike immunocytochemistry, *in situ* hybridization can identify components of the diffuse neuroendocrine system irrespective of whether the cell contains stored product (Fig. 5B). It thus allows greater insight into the functional morphology of the diffuse neuroendocrine system.

Binding Site Analysis

Many studies have been aimed at identifying and characterizing peptide binding sites in preparations of isolated cells (cf. Ref. 60). However, although such preparations allow investigation of binding kinetics and receptor specificity, they cannot provide information on the exact tissue distribution of binding sites and cells or nerves producing peptides. Morphological techniques are now being used in this field.

Methods Using Antisera

Although the immunocytochemical demonstration of specific binding sites has been achieved using antiidiotypic antibodies,[61] this method has

[60] A. Undén, L. L. Peterson, and T. Bartfai, *Int. Rev. Neurobiol.* **27,** 141 (1985).
[61] G. N. Gaulton and M. I. Greene, *Annu. Rev. Immunol.* **4,** 253 (1986).

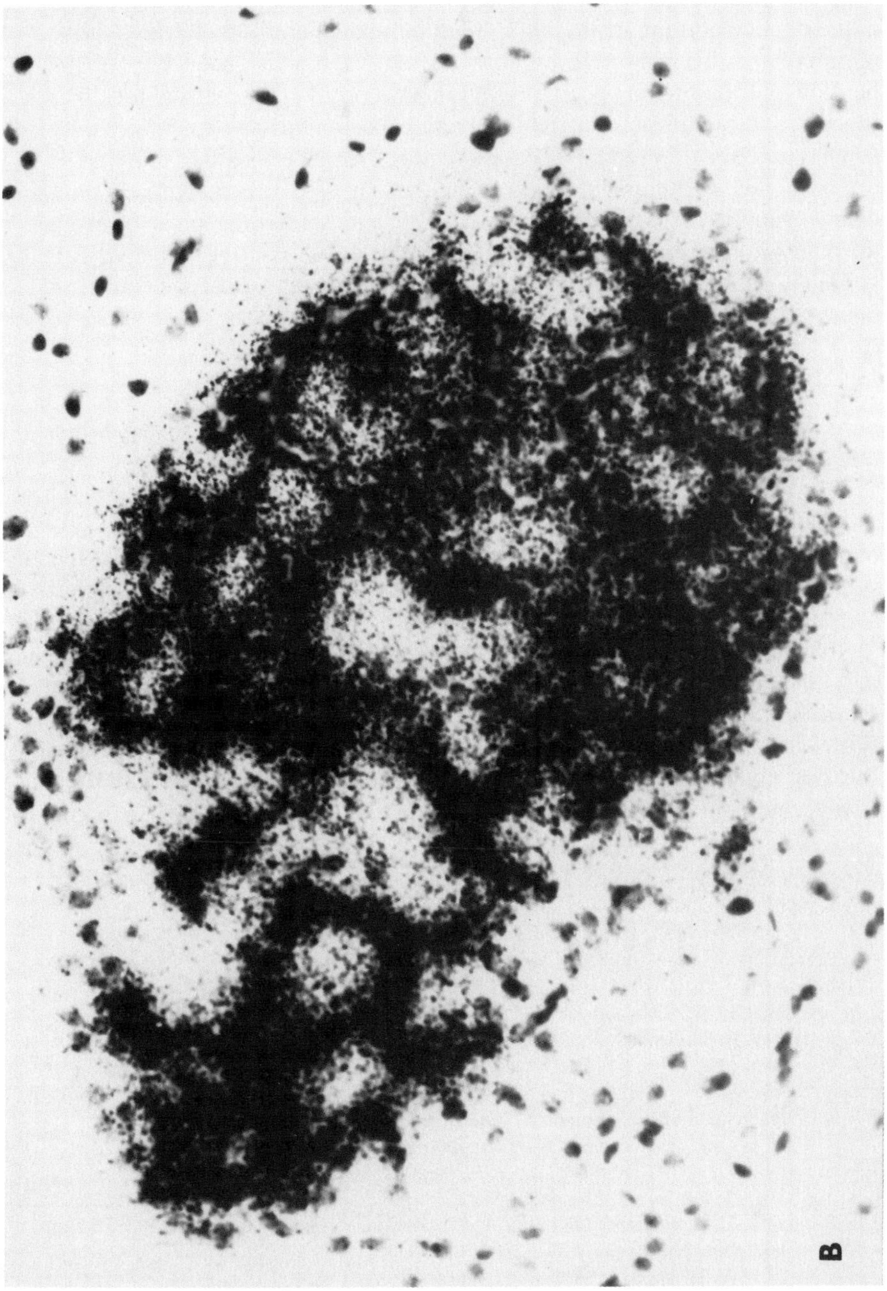

FIG. 5B.

not been widely applied to the study of peptide receptors. Another technique depends on the availability of a monoclonal antibody to the peptide of interest and knowledge of both the active binding sequence of the peptide and the region specificity of the antibody.[62] This method localizes receptors at the ultrastructural level and is particularly useful where only small numbers of binding sites exist and a highly sensitive method is required. The technique consists of incubation of fresh cells with divalent ligand prior to fixation in glutaraldehyde and sectioning on a freezing ultramicrotome. Application of an antibody to the peptide, labeled with colloidal gold, allows visualization of binding sites under the electron microscope.

In Vitro Autoradiography

A technical approach that combines binding site localization with histological analysis is *in vitro* autoradiography.[63] This method has been used successfully to study the distribution of peptide binding sites in various tissues including the central nervous system,[64,65] cardiovascular system,[66–68] respiratory tract,[69–71] and gastrointestinal tract.[72,73] In basis, the technique consists of incubation of tissue sections with a radiolabeled peptide. Sites of attachment of ligand can be established by apposing sections to film or emulsion-coated coverslips. Using film, it is possible to obtain a sharp image of binding site attachment (Fig. 6A), but the use of coverslips has the added advantage that the sections can be stained to show tissue histology or even immunostained to show the spatial relationship between components of the diffuse neuroendocrine system and their displacement of radiolabeled ligand by the addition of either the same

[62] P. M. Lackie, F. Cuttitta, J. D. Minna, S. R. Bloom, and J. M. Polak, *Histochemistry* **83,** 57 (1984).

[63] W. S. Young and M. J. Kuhar, *Brain Res.* **179,** 255 (1979).

[64] J. Besson, M. Dussaillant, J. C. Marie, W. Rostene, and G. Rosselin, *Peptides* **5,** 339 (1984).

[65] M. M. Schaffer and T. W. Moody, *Peptides* **7,** 283 (1986).

[66] P. J. Barnes, A. Cadieux, J. R. Carstairs, B. Greenberg, J. M. Polak, and K. Rhoden, *Br. J. Pharmacol.* **89,** 157 (1986).

[67] J. A. Stephenson, E. Burcher, and R. J. Summers, *Eur. J. Pharmacol.* **124,** 377 (1986).

[68] J. A. Stephenson and R. J. Summers, *Eur. J. Pharmacol.* **134,** 35 (1986).

[69] J. R. Carstairs and P. J. Barnes, *J. Pharmacol. Exp. Ther.* **239,** 249 (1986).

[70] J. R. Carstairs and P. J. Barnes, *Eur. J. Pharmacol.* **127,** 295 (1986).

[71] K. Leys, A. H. Morice, O. Madonna, and P. S. Sever, *FEBS Lett.* **199,** 198 (1986).

[72] R. F. Power, A. E. Bishop, J. Wharton, C. O. Inyama, R. H. Jackson, S. R. Bloom, and J. M. Polak, *Ann. N.Y. Acad. Sci.* **527,** 314 (1987).

[73] H. Sayadi, J. W. Harmon, T. W. Moody, and L. Y. Koorman, *Gastroenterology* **92,** 1617 (1987).

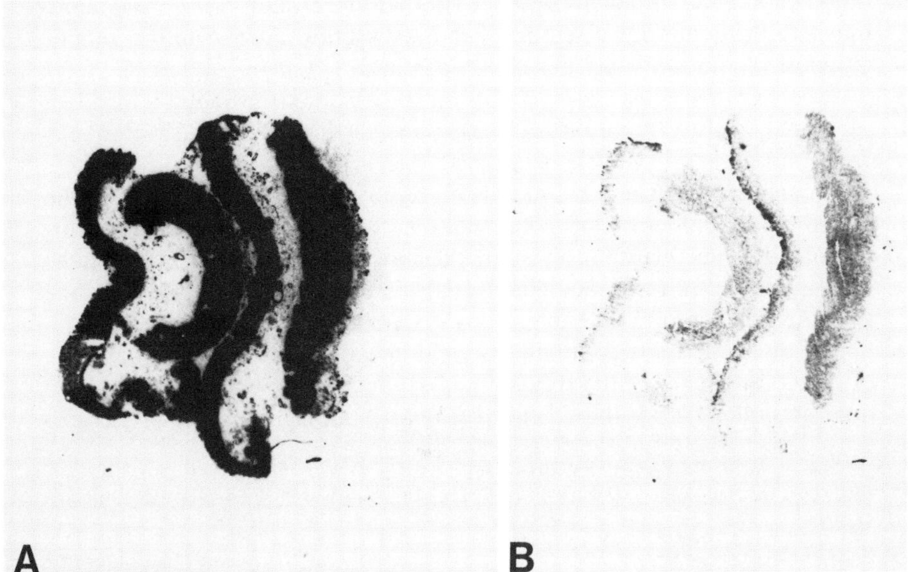

FIG. 6. Serial cryostat sections of human duodenum. (A) Autoradiograph of two sections which had been incubated with ^{125}I-labeled peptide histidine methionine (PHM). Binding is seen in the mucosa and muscle. (B) Control section, serial to that in A, which had been incubated with ^{125}I-labeled PHM in the presence of excess (1 μM) unlabeled PHM. ×31.

target cells. Specificity of binding can be determined by studying the peptide without radiolabel (Fig. 6B) or another peptide possibly active for the same binding sites.

It is possible to carry out quantitative biochemical analysis of the binding sites in tissue sections and to perform standard assays of their kinetics, e.g., association, dissociation, and saturation curves, with values for the dissociation constant (K_D) and maximal binding capacity (B_{max}) for each receptor type being determined from Scatchard plots. Such quantitative analysis can be achieved by two main methods. Labeled sections can be scraped off, counted in a gamma emission counter, and the concentration of bound ligand derived.[69,70] Alternatively, image analysis can be applied using a computerized system. An image of a section incubated with radiolabeled ligand in the presence of excess unlabeled peptide shows only nonspecific binding. Subtracting the image of this section from that of a serial radiolabeled section gives an automatically corrected image of specific binding. Quantification of the amount of binding can then be achieved by comparison of test autoradiographs with those generated by standards.[74,75] This latter method is not only more sensitive than emission counting of scraped sections, it also allows quantitation of binding to small tissue components.

Conclusions

Numerous cytochemical methods are being used for the study of the diffuse neuroendocrine system. Unlike early methods which gave information only on the static system, the newer morphological techniques described here allow concomitant analysis of the anatomy and dynamic status of this complex and ubiquitous system of cells and nerves. Application of these methods continues to yield new data on the functional roles of the components of the diffuse neuroendocrine system, their distributions and interrelationships, and their contributions to disease processes.

[74] P. Slater, *Br. J. Pharmacol.* **85,** 391 (1985).

[75] K. Zilles, A. Schleicher, M. Rath, T. Glaser, and J. Traber, *J. Neurosci. Methods* **18,** 207 (1986).

[58] Quantitation and Digital Representation of *in Situ* Hybridization Histochemistry

By Michael E. Lewis, Wade T. Rogers, Rudolph G. Krause II, and James S. Schwaber

Introduction

In situ hybridization histochemistry has frequently been used as a qualitative technique to detect and localize specific nucleic acid sequences in an anatomical context, without particular regard to the abundance of the detected sequence. For example, in studies of gene mapping in chromosomes[1] or the identification of a peptide-containing cell as a site of biosynthesis via mRNA colocalization,[2–6] quantitation of the relevant

[1] J. G. Gall and M. L. Pardue, *Proc. Natl. Acad. Sci. U.S.A.* **63,** 378 (1969).

[2] C. E. Gee, C. L. C. Chen, J. L. Roberts, R. Thompson, and S. J. Watson, *Nature (London)* **306,** 374 (1983).

[3] W. S. T. Griffin, M. Alejos, G. Nilaver, and M. Morrison, *Brain Res. Bull.* **10,** 597 (1983).

[4] M. Brahic, A. T. Haase, and E. Cash, *Proc. Natl. Acad. Sci. U.S.A.* **81,** 5445 (1984).

[5] B. Wolfson, R. W. Manning, L. G. Davis, R. Arentzen, and F. Baldino, Jr., *Nature (London)* **315,** 59 (1985).

[6] B. D. Shivers, R. E. Harlan, D. W. Pfaff, and B. S. Schachter, *J. Histochem. Cytochem.* **34,** 39 (1986).

nucleic acid sequence is unimportant for the purpose of the experiment. However, many investigators are now beginning to use *in situ* hybridization histochemistry as a tool to study the regulation of gene expression, e.g., during development or after experimental manipulations. Since such studies require comparisons among tissues from organisms of differing ages or treatment groups, some procedure to quantify the changes must be employed.

Quantitation of *in situ* hybridization histochemical material depends on the type of signal detection system employed. For radioactively labeled probes, hybrids will be detected autoradiographically using either liquid emulsions or emulsion-coated film. For hybrid detection using nonradioactively labeled probes, the signal can be generated by an enzyme (e.g., alkaline phosphatase or horseradish peroxidase) which is coupled to the probe and allowed to form some visible reaction product. However, since relatively few investigators use nonradioactive probes, and since their use has generally been for qualitative rather than quantitative purposes, the quantitation of nonradioactive probe hybridization is not discussed here. It should be noted, however, that the potential for quantitation exists, based on previous studies that measured histochemically detected enzyme activity in tissue sections.[7,8] The purpose of this chapter is to identify some of the issues relating to autoradiographic quantitation of *in situ* hybridization histochemistry, as well as to describe our approach to this problem.

Quantitative Film Autoradiography

The principles of quantitative film autoradiography are well established and have been reviewed elsewhere.[9,10] Since autoradiographic film images of heterogeneous tissues, such as brain, are highly complex, effective quantitative studies usually require the application of computerized image analysis techniques. After *in situ* hybridization is carried out with a radiolabeled probe, the slide-mounted sections can be exposed to film (Kodak X-Omat AR for ^{125}I, ^{35}S, or ^{32}P; LKB Ultrofilm or Amersham Hyperfilm for ^{3}H) prior to emulsion dipping. The distribution of radioactivity in the tissue is thus converted to a pattern of optical densities on the film. This image is then digitized and translated into gray values by the passage of light through the film into a video camera which is appropriately interfaced to an image analysis system. We have found the DUMAS

[7] R. H. Benno, L. W. Tucker, T. H. Joh, and D. J. Reis, *Brain Res.* **246,** 225 (1982).

[8] A. Bigeon and M. Wolff, *J. Neurosci. Methods* **16,** 39 (1986).

[9] M. J. Kuhar and J. R. Unnerstall, *Trends Neurosci.* **8,** 49 (1985).

[10] M. Herkenham, *NIDA Res. Monogr.* **62,** 13 (1985).

system[11] developed at Drexel University to be efficient and economical; however, other low-cost systems can be constructed.[12] Gray values are converted to optical density values with the aid of a photographic step tablet, and optical density values can also be converted to units of radioactivity by reference to radioactive standards[13] which are exposed to the same film as the tissue. The distribution of radioactivity in the tissue can then be represented digitally by pseudocolor via color lookup tables in the image memory.

The ease of carrying out film autoradiographic analysis of *in situ* hybridization histochemistry is offset by the lack of cellular resolution of this technique. Given the complex internal organization of brain nuclei, it is not reasonable to expect all the cells containing a particular mRNA to respond uniformly to a stimulus. For example, cells may respond differentially in relation to their connections with other neuronal systems, a situation which can be resolved by combining *in situ* hybridization with tract-tracing techniques.[14–17] Thus, it is important to be able to quantitate mRNA at the cellular level and then to have some means to represent these data in an anatomical context. The following sections describe our approach to this problem.

Cellular Grain-Counting Method

Our approach to the quantitation of *in situ* hybridization histochemistry is based on a novel, computerized procedure for counting the number of silver grains in the autoradiographic emulsion over an individual cell, rather than on a measurement of integrated optical density. With this computerized system, which is described in the following sections, a number proportional to the count of silver grains is stored along with the x,y coordinates for each cell. This allows quantitative data to be printed as a list or displayed as a map which shows marks that are gray-scale or color-scale intensity coded to display graphically the relative

[11] C. L. Chu, S. B. Seshadri, E. Feingold, and O. J. Tretiak, *IEEE Microprocessor Forum,* 93 (1985).

[12] A. Isseroff and D. Lancet, *J. Neurosci. Methods* **12,** 265 (1985).

[13] J. R. Unnerstall, M. J. Kuhar, D. L. Niehoff, and J. M. Palacios, *J. Pharmacol. Exp. Ther.* **218,** 797 (1981).

[14] B. M. Chronwall, *Peptides (N.Y.)* **6,** 1 (1985).

[15] J. N. Wilcox, J. L. Roberts, B. M. Chronwall, J. F. Bishop, and T. O'Donohue, *J. Neurosci. Res.* **16,** 89 (1986).

[16] B. M. Chronwall, M. E. Lewis, J. S. Schwaber, and T. L. O'Donohue, *in* "Neuroanatomical Tract-Tracing Methods" (L. Heimer and M. J. Robards, eds.), Vol. 2. Plenum, New York, 1988.

[17] J. S. Schwaber, B. M. Chronwall, and M. E. Lewis, this volume [56].

expression of mRNA for each cell in its true anatomical location. Since the absolute coordinates of each cell are stored, if the same analysis is repeated with a new probe in the same tissue, this procedure allows for the analysis of multiple species of mRNA or generation of hybridization melting point data for individual cells. Finally, this method also lends itself readily to computer graphic methods of three-dimensional reconstruction to show spatial distributions of cellular mRNA levels.

Digital Microscopy System

Figure 1 shows a diagram of the digital microscopy system used for the cellular grain-counting method. A microscope, two high-resolution color video monitors, monochrome monitor, modified drawing tube, graphics tablet, high-performance computer-controlled microscope stage, video camera, video digitizer, 4-megabyte digital image memory with high-speed data-transfer electronics, and computer comprise the system. Modifications to the camera lucida were made by replacing the 45° mirror with a dove prism in order to project an erect image from the mapping display monitor no. 1 to the microscope. Also, provision was made to easily

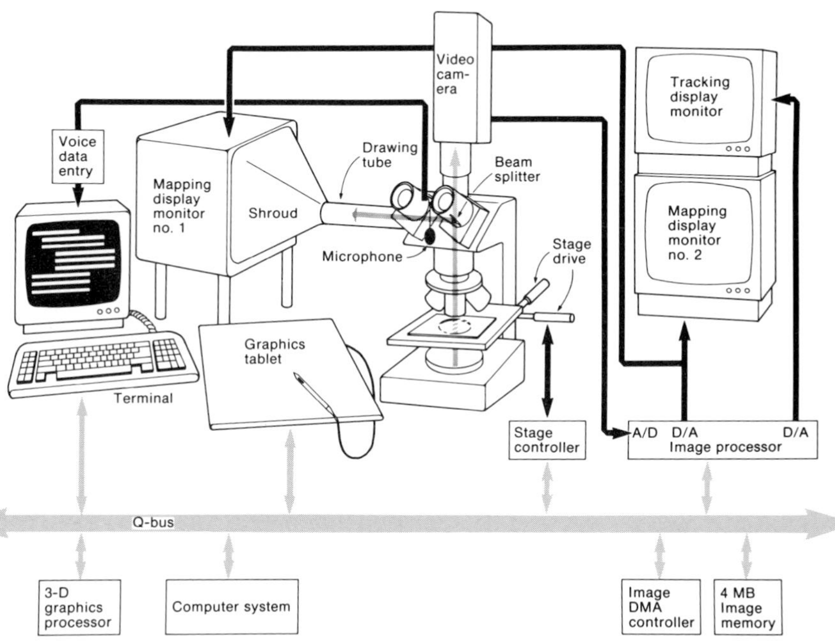

FIG. 1. Block diagram of the digital microscopy system. Reproduced from Ref. 18 with permission of the Optical Society of America.

change the beam splitter cube in the camera lucida to accommodate a wide range of objective illumination conditions, ranging from bright-field to very low light fluorescence. With the exception of the stage controller, the electronics and computer systems were assembled from commercially available components. The operational software, described below, was entirely custom written for specific applications.

Mapping

A record of the location of labeled cells is created using a method called flying field mapping,[18,19] in which the paper in a camera lucida drawing is replaced by a high-resolution CRT screen and the pen by a stylus and graphics tablet. Playing the role of ink marks on paper, luminous marks on the CRT screen appear to the microscopist to be superimposed on the section image in the microscope oculars. An important improvement over conventional camera lucida drawings or other computer-aided methods[20–22] is the fact that the stylus is used directly to control the microscope stage position via computer, rather than to control the position of a cursor on the video screen. Thus, rather than mapping a stationary tissue image, the system allows the microscopist essentially to "fly" continuously over the tissue, directing the system to mark features viewed in the microscope as they go by. This approach is much like that used by Capowski,[23] except that provisions are made to project back into the microscope a representation of the developing map superimposed on the tissue, so that the microscopist may take advantage of visual feedback to determine the status of the map.

In the operation of the main data-acquisition program (MAP), the camera lucida projects the image from the high-resolution color monitor to the microscope so that its image appears in the microscope oculars superimposed on that of the magnified anatomical section. A video cross-hair cursor is displayed on the monitor at the center of the field of view. A 512 × 480 pixel window is also displayed as a small subset of the 2048 × 2048 pixel digital image memory. This window serves as a "microscopic" graphic viewport, providing a movable high-resolution view into the digital image memory. This image memory may be conceptualized as a large

[18] W. T. Rogers and J. S. Schwaber, *Appl. Opt.* **26,** 3384 (1987).

[19] P. Jansson, W. T. Rogers, and J. S. Schwaber, U.S. Patent 4,673,988 (1987).

[20] D. E. Hillman, *Comput. Med.* **5,** 1 (1977).

[21] E. M. Glaser and H. Van Der Loos, *IEEE Trans. Biomed. Eng.* **12,** 22 (1965).

[22] R. D. Lindsay, *in* "Computer Analysis of Neuronal Structures" (R. D. Lindsay, ed.), p. 1. Plenum, New York, 1977.

[23] J. J. Capowski, *Comput. Biomed. Res.* **10,** 617 (1977).

two-dimensional array of locations, each of which maps to a unique x,y coordinate on the tissue section.

One of the difficulties microscopists face is that of "getting lost" in the tissue while working at high magnification. This problem was circumvented by creation of a "navigational" aid, a static, low-magnification image of the tissue section which is displayed on the monochrome monitor (tracking display). A video marker can then be superimposed on this tracking display to indicate at a glance the current location of the microscopic field of view. Thus, the first step in establishing a map is to digitize a video image of the entire tissue section using a low-power objective.

To create a map the operator switches to a higher power objective. While looking into the microscope oculars, with one hand holding the stylus of the graphics tablet and the other hand on the fine-focus control of the microscope, the operator moves the stylus, transmitting to the computer its x,y coordinates. For each new coordinate pair, the computer does four things: (1) causes the microscope stage to move under the objective in a corresponding way; (2) drives the display window through the large image memory so that the display appears to move in perfect registry with the optical image of the anatomical section, (3) moves the video marker on the tracking display, and (4) determines if a button on the stylus is depressed by the operator. If so, the computer stores a suitable mapping mark in the 4-megabyte image memory as well as in a vector list in the main memory and updates the current contents of the mapping display. The process is repeated rapidly enough to make the motion appear continuous. The correspondence between stylus and stage motions is carefully chosen so that the microscopist's perception is one of flying over the tissue; thus, the term "flying field mapping" is applied to this procedure.

Several benefits accrue from this method. First, because all coordinates are digitized from the optical axis of the microscope, the data are not corrupted by geometric distortion in the microscope optics or CRT graphical display. Second, the apparent magnification of the graphical display can readily be altered to accommodate a wide range of objective magnifications. The microscopist can thus change magnification at will during acquisition of a map, choosing conditions for optimum viewing of the tissue section. Third, the fact that objectives are not parcentric is easily corrected by storing (in stage coordinates) a table of x,y offsets for each objective. Last, microscope stage motion is inherent in the creation of the map. The microscopist need not periodically divert attention away from the analysis to choose a new fixed field of view, thus improving speed and accuracy of map acquisition. Another aid in avoiding interruption of the continuity of the microscopist's observation of the tissue during mapping, a voice data entry system is provided to allow verbal commands to be issued, obviating the need to look away from the microscope to type commands on a keyboard.

The mapping method allows data to be entered randomly. For example, the microscopist can map clearly defined anatomical landmarks for later reference, then at higher magnification enter distributions of cells and fibers. Other local landmarks can be mapped as they are encountered during the process of flying over the tissue. The operator may indicate different cell classes with different colors, which in turn are stored in the digital data base with distinguishable codes. By identifying a particular code for locations containing no cells, the observer may select small regions distributed widely over the tissue to be used to compute the autoradiographic nonspecific background, so that cell grain counts may be corrected. We note that, depending on the aim of the experiment, it is possible to map the actual appearance of the autoradiographic emulsion, thus locating only labeled cells, or to map counterstained (Nissl) cells, or even to map cells labeled by injection or immunohistochemical preparation. Creating a digital data base by this mapping process facilitates further analysis, including flexible two-dimensional display of map data, three-dimensional reconstruction, and a wide variety of quantitative analyses in both two and three dimensions, including the *in situ* hybridization grain-counting analysis described below.

Grain Counting by Digital Microscopy

Once the coordinate data base for a tissue section has been created, the data are passed to another program which performs the autoradiographic quantitation. At this point the microscope is set up for oil-immersion dark-field illumination at rather high (40–60×) objective magnification. This method forms the highest contrast image of silver grains in the emulsion.

The system automatically recalls the sequence of x,y coordinates for each cell and then brings the cells to the optical axis of the microscope one at a time. The live video image is then focused, and, for each cell, the image is digitized using a flash analog-to-digital converter and stored as a 512 × 480 pixel digital image in a frame buffer memory. Sixteen successive video frames are digitized in real time and summed into a frame buffer using a high-speed arithmetic-logic unit (ALU) to improve the signal-to-noise ratio in the video signal (frame averaging). A video threshold value is found by "flipping" between a gray-level display (Fig. 2A) and a binary black/white display (Fig. 2B), where the boundary between black

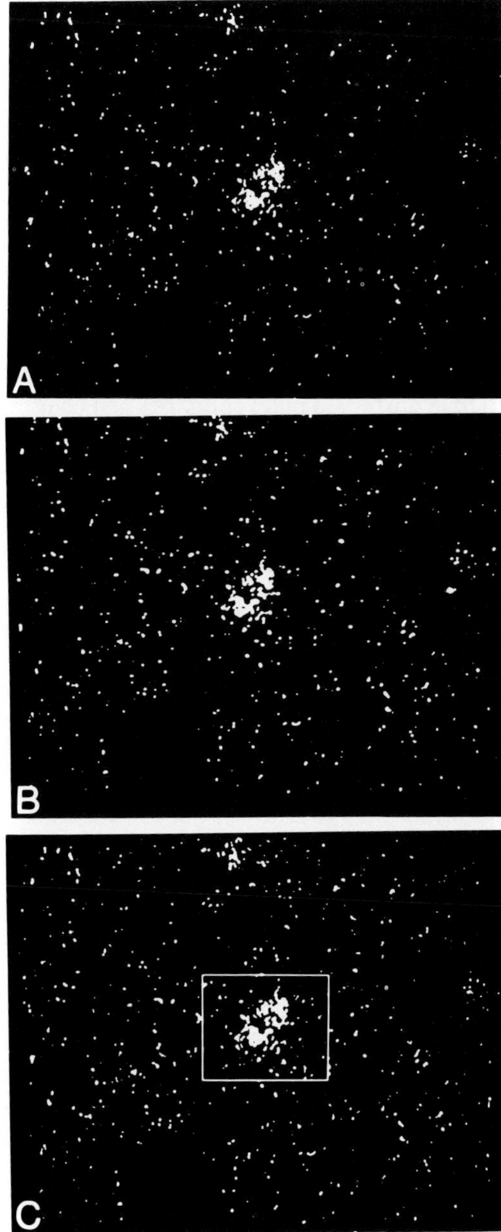

FIG. 2. Video images of a typical hybridized cell analyzed by the method described in the text. (A) Gray-level video image and (B) binary image after thresholding. The system allows the operator to rapidly "blink" back and forth between A and B while adjusting the threshold value via a graphics tablet. (C) A box indicates the region in which the image analysis system will look for grains after thresholding.

and white is the threshold value. The threshold value is manually adjusted until the threshold image accurately represents the gray-level image in terms of the area subtended by grains. The central portion of the field fully containing the cell (Fig. 2C) is then sent to a routine which counts the number of pixels above threshold. This number is stored in a file along with the x,y coordinates from the data base for the given cell.

The process is repeated sequentially for every cell stored in the data base, requiring approximately 5–10 sec/cell. The system then visits all locations chosen by the operator to represent background fields. Once all background fields have been analyzed, the average number of background pixels is computed and then subtracted from the stored numbers for each cell to arrive at a final corrected pixel count per cell. This method assumes uniform background grain density over the section, which has been confirmed in a number of cases.

The resulting data file is then printed out in tabular form for further analysis and displayed in graphical form on a CRT. Figure 3A shows the distribution of cells labeled for somatostatin mRNA in the right hemisphere of the forebrain of a rat. The intensity scale, shown at the left of Fig. 3A, going from black at the bottom to white at the top, indicates the relative level of hybridization in units of pixels above threshold. The range of the scale may be interactively adjusted such that the top of the scale (white) corresponds to lower levels of expression, thus expanding the dynamic range of the display for more lightly labeled cells (Fig. 3B). Any portion of the display may be spatially zoomed (Fig. 3C) to examine a particular region in more detail.

Calibration of Hybridization Signal

Calibration of the hybridization signal in the present procedure involves relating the number of pixels above threshold to the number of silver grains per cell. It is possible to obtain accurate relative values presuming that hybridization, dipping, exposure, and developing procedures are carefully controlled and carried out concurrently for all sections to be compared. Calibration within each experiment is achieved by photographing cells representing the range of labeling and manually counting the images of grains on the photomicrographs within an area corresponding to the area analyzed by the computer. A scattergram (Fig. 4) is produced, with computer-counted pixels plotted against manually counted grains. Linear regression analysis yields the slope of the corresponding best-fit line, which corresponds to the average number of pixels per grain. This conversion factor can then be used for all cells within each experi-

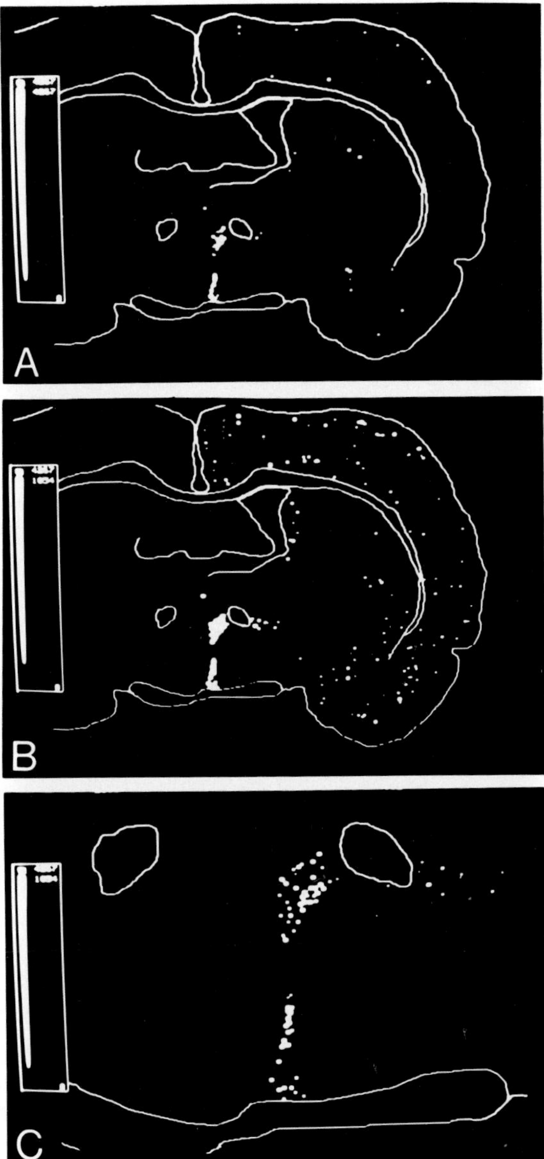

FIG. 3. Examples of data and maps resulting from a typical *in situ* hybridization experiment (^{125}I-labeled somatostatin oligonucleotide). (A) Low-magnification map, with lines indicating major anatomical boundaries and intensity-coded marks indicating relative hybridization per cell. The gray-scale wedge at the left of each panel indicates the intensity code in units of pixels above threshold. Color code is readily substituted for the gray scale.[24] (B) Same as A but with the range of the intensity code expanded to enhance the more lightly labeled cells, for example, in cortex. (C) Zoomed view showing a small region containing the paraventricular nucleus of the hypothalamus.

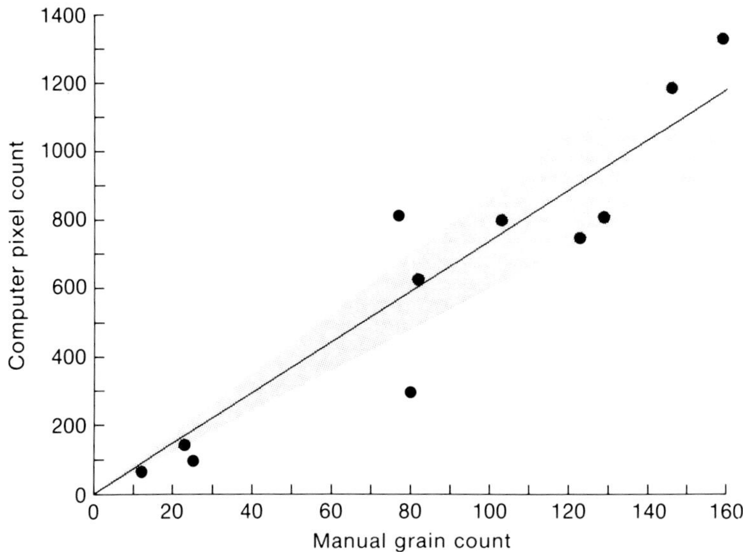

FIG. 4. Calibration curve comparing the number of pixels per cell counted by the computer (ordinate) against the number of grains counted manually (abscissa) from high-power photomicrographs of the same cells. The line represents the least-squares best fit (r = 0.98), with a slope of 7.38 pixels/grain, and the shaded area encompasses three standard errors on the slope of the best-fit line. Reproduced from Ref. 24 with permission of Elsevier, Amsterdam.

ment. We have found that a range of at least 10 to 150 grains/cell can be analyzed by the method without significant departures from linearity.[24]

Various experimental conditions were compared in order to optimize the accuracy and robustness of the measurements. For example, bright-field versus dark-field illumination was compared, showing that, while the resolution of bright-field images was slightly better than that of the corresponding dark-field images using the same objective magnification, the dark-field illumination produced more reliable results since the contrast of punctate opaque objects in bright-field microscopy is severely degraded by flare in the optics. Another comparison was made between fixed versus variable thresholds. In the first case, a threshold was selected on the basis of the first cell encountered in a tissue section and subsequently used for all cells in the section. In the second case, the operator was given

[24] W. T. Rogers, J. S. Schwaber, and M. E. Lewis, *Neurosci. Lett.* **82,** 315 (1987).

the ability to interactively choose a threshold value for each cell. Statistical comparison showed the latter to produce a lower variance of the residuals in the calibration least-squares fit due to the fact that background absorbance in the tissue under the emulsion causes the fixed threshold to be appropriate only in an average sense, while the operator is able to compensate for this variability in the variable threshold method. This ability to reliably compensate for background absorbance also allows the analysis of counterstained tissue with little loss of quantitation accuracy.

Data Analysis

Following mapping and quantitation as described here, several procedures can be used to analyze the data obtained in a hybridization study, e.g., the comparison of tissue sections from animals receiving different treatments. The most direct procedure is by visual observation and descriptive comparison of the graphical displays created by the present methods (Fig. 3). The relative distribution of labeled cells and of the amount of label in the cells is immediately obvious in these graphs and can be observed at low resolution across the tissue or, by zooming, observed regionally or within a specific nucleus. By selecting comparable sections across animals, the maps may be overlaid and aligned for blink comparison of distribution and amount of probe labeling. A second analytic method uses the figures generated for x, y, z location and grain number to perform statistics of average grain count, variability, distribution, and proportion of label between regions. These results can be displayed in histogram form, relating amount of label to anatomical location. Tests of statistical significance of differences between treatment groups can then be performed.

Issues in Quantitative *in Situ* Hybridization Histochemistry

The methods described here provide quantitative information about grain density per cell, permitting the measurement of relative mRNA levels by comparison of different brain areas, developmental stages, or the effects of pharmacological or physiological treatments. Although grain counts provide only a relative measure of mRNA level, it is important to note the conditions under which this measure is on a linear scale. As Rogers[25] has noted, the relationship between radiation dose and num-

[25] A. W. Rogers, "Techniques in Autoradiography," 2nd Ed. Elsevier, Amsterdam, 1973.

ber of grains per unit of emulsion is not linear, but logarithmic, due to (1) coincidence effects, i.e., as the number of β particles increases, the probability of a crystal being hit more than once increases although these additional hits are not detected, and (2) the difficulty in accurate grain counting when many grains overlap or even fuse during development. However, if grain density is plotted against the amount of radioactive disintegrations in the source, the first part of the logarithmic curve is essentially linear.[25] Thus, grain density can be considered proportional to radioactivity up to a limiting grain density, i.e., when coincidence effects or grain coalescence become significant (generally, more than 10% of the total grains). These factors are related to the size of the undeveloped silver halide crystals in the emulsion, such that smaller crystal diameters permit higher grain densities to be proportional to radioactivity.[25] Accordingly, we use a fine grain emulsion (Kodak NTB2, 0.26 μm) and carefully regulated exposure times to help ensure that differences in cellular grain densities are proportional to differences in hybrid formation.[26]

In order to begin extending this method to the absolute quantitation of cellular mRNAs, it is necessary to convert the measure of grains/cell to hybrids/cell, and this value may be calculated[27] using the specific activity of the probe (dpm/fmol) and the efficiency of autoradiography (e.g., 0.44 = mean grain yield/γ disintegration for Kodak NTB2 emulsion[28]). Standard curves, prepared by exposing radioactive tissue standards to emulsion, have been used for quantitating *in situ* hybridization,[29] and are especially important if prolonged exposures are employed to enhance the photogenic qualities of the hybridized cells.

Although quantitation of hybrids/cell is therefore possible, relating this value to the absolute amount of mRNA present is difficult. The problem is analogous, in receptor autoradiography, to comparing femtomoles of radioligand bound at one concentration to the true B_{max} value which can only be obtained by proper analysis of saturation experiments. Unfortunately, saturation experiments have often proved difficult in hybridization studies, and their quantitative interpretation depends further on the assumptions that no mRNA is lost during the numerous tissue processing steps and that all the mRNA present is available for hybridization. There is evidence that both assumptions may be untenable.[26,30–32] In addition,

[26] P. Szabo, R. Elder, D. M. Steffensen, and O. C. Uhlenbeck, *J. Mol. Biol.* **115,** 539 (1977).

[27] E. Cash and M. Brahic, *Anal. Biochem.* **157,** 236 (1986).

[28] G. L. Ada, J. H. Humphrey, B. A. Askonas, H. O. McDevitt, and G. J. V. Nossal, *Exp. Cell Res.* **41,** 557 (1966).

[29] W. S. Young III, E. Mezey, and R. E. Siegel, *Mol. Brain Res.* **1,** 231 (1986).

[30] P. R. Harrison, D. Conkie, N. Affara, and J. Paul, *J. Cell Biol.* **63,** 402 (1974).

[31] J. B. Lawrence and R. H. Singer, *Nucleic Acids Res.* **13,** 1777 (1985).

[32] J. E. Kelsey, S. J. Watson, S. Burke, H. Akil, and J. L. Roberts, *J. Neurosci.* **6,** 38 (1986).

the autoradiographic method itself imposes limitations on quantitative accuracy. Tissue absorption of β particles by cells, a severe problem with tritiated probes, will affect grain density in a complex manner depending on cell shape and the proximity of the cell to the emulsion. Although this problem can be avoided by the use of higher energy radioisotopes (e.g., ^{125}I or ^{35}S), the measurement of grains from transected cells will still give an underestimation of mRNA levels, which can be circumvented to some extent by counting grains only in cells which have a visible nucleus.

Given these barriers to the absolute quantitation of *in situ* hybridization histochemistry, it is worth pointing out that for most experimental needs, e.g., comparing the effects of different treatments on mRNA levels, relative quantitation of the material, if done accurately, should generally suffice to answer the question under investigation. This conclusion is reinforced by previous findings that local relative concentrations of target sequences in a tissue can be accurately quantitated even with subsaturating concentrations of probe.[33]

In summary, we believe that quantitative analysis is necessary for evaluating changes in gene expression with *in situ* hybridization histochemistry. The structural and functional complexity of the brain requires that this analysis be represented in an anatomical context in order to be meaningful. With autoradiographic film analysis, this process is efficiently carried out by computerized densitometry and pseudocolor representation of the film images of the hybridization signal. In contrast, a quantitative analysis of autoradiographic grain density at the cellular level is more difficult. However, the high degree of anatomical specificity and resolution obtained with emulsion autoradiography should facilitate any study of the distribution and regulation of mRNA levels in heterogeneous tissues such as the brain.

Acknowledgments

We wish to thank Ms. Alicia Walsh for assisting in the writing of the quantitation code, Dr. J. Patrick Card for assistance with photography, and Dr. Jill M. Roberts-Lewis for helpful comments on the manuscript.

[33] L. M. Angerer and R. C. Angerer, *Nucleic Acids Res.* **9,** 2819 (1981).

[59] Graphical and Statistical Approaches to Data Analysis for *in Situ* Hybridization

By JOSEPH T. MCCABE, ROBERT A. DESHARNAIS, and DONALD W. PFAFF

Introduction

Quantitative cytochemistry permits measurement of biological activity at the level of the single cell. In tissue as complex and heterogeneous as the nervous system, this level of analysis is imperative for understanding the pattern of gene expression in neurons following developmental and physiological challenges. We discuss methods of data analysis to explore changes in mRNA levels in single neurons. These analyses go beyond the usual, single description of experimental data, such as mean or median cell response, to explore what morphologically derived data can tell us about the pattern of individual cell response in a defined neuron population. By drawing on concepts and methods from quantitative autoradiography, *in situ* hybridization, and graphical and statistical analysis, this approach may allow a further understanding of the role of the single neuron within constellations of cells considered to mediate specific functions.

The tenets of this approach should be recognized at the outset: these assumptions outline part of the rationale for this view and keep us cognizant of interpretive limitations (see below). (1) It is assumed that neurons can be viewed as the elementary units of brain function, where each cell is an individual participant of the class of cells which we define functionally, or identify by means of phenotype or other morphological characteristics. (2) Each cell may exhibit measurable differences in its magnitude of response (here, number of copies/cell of a particular mRNA). (3) The antecedent conditions for this variation in cell response are determinable. It may be that a single cell's response exhibits differences in magnitude due to connectivity, size, position, or other controlling factors. These other factors include cell synaptology, receptor type and densities, as well as internal modulatory functions such as the nature of second messengers and sensitivity of genomic mechanisms. (4) It is assumed that rigorous application of experimental technique permits valid quantitative data. Graphical and statistical analysis, in the context of other work, can then provide a guide to further investigations. For quantitative application, a fixed, controlled relationship must exist between the number of mRNA molecules of interest in a single cell, the amount of nucleotide probe that hybridizes in that cell, and the number of autoradiographic grains that are generated in the photographic emulsion by the labeled probe. Variation in grain density (number of grains detected above a cell divided by cell profile area) over a single cell is then a meaningful quantity. It reflects not just variation resulting from grain production by "background" grain formation,[1] nonspecific binding of probe, variation due to the underlying randomness of β-particle decay, but also meaningful perturbations in mRNA copy number per cell.

The Neurohypophyseal System

The examples are drawn from work with the magnocellular neurosecretory or hypothalamic–neurohypophyseal system.[2,3] This system is best known for its synthesis of the hormones vasopressin (VP) (antidiuretic hormone) and oxytocin (OT). These hormones, as well as many other recently discovered neuropeptide substances, are transported down the axons of these neurons to the neurohypophysis. The substances are then released from the pituitary in a controlled manner. From an historical perspective, this cell system has played an important role in experimental and conceptual developments in neuroendocrinology as well as in neurobiology in general.[4,5] For example, the hypothalamic–neurohypophyseal system provided the first spectacular example of neurosecretion,[6] was an amenable subject for later investigations of how hormones are synthesized, processed, packaged, transported, and released from the pituitary,[7,8] and has been the subject of intense study of how bodily state, hormone condition, and electrophysiological and chemical stimulation affects neuron morphology[9,10] and hormone synthesis and release.[11–13]

[1] A. W. Rogers, "Techniques of Autoradiography," 3rd Ed. Elsevier, Amsterdam, 1979.

[2] B. Scharrer, *Annu. Rev. Neurosci.* **10,** 1 (1987).

[3] A.-J. Silverman and E. A. Zimmerman, *Annu. Rev. Neurosci.* **6,** 357 (1983).

[4] B. A. Cross and G. Leng, eds., "The Neurohypophysis: Structure, Function and Control." Elsevier, Amsterdam, 1983.

[5] S. Reichlin, ed., "The Neurohypophysis: Physiological and Clinical Aspects," p. 1. Plenum, New York, 1984.

[6] E. Scharrer and B. Scharrer, *Recent Prog. Horm. Res.* **10,** 183 (1954).

[7] M. J. Brownstein, J. T. Russell, and H. Gainer, *Science* **207,** 373 (1980).

[8] B. T. Pickering, *Essays Biochem.* **14,** 45 (1978).

[9] M. Castel, H. Gainer, and H. D. Dellman, *Int. Rev. Cytol.* **88,** 304 (1984).

[10] J. F. Morris, J. J. Nordmann, and R. E. J. Dyball, *Int. Rev. Exp. Pathol.* **18,** 1 (1978).

[11] J. A. Amico and A. G. Robinson, eds., "Oxytocin." Excerpta Medica, Amsterdam, 1985.

[12] L. P. Renaud, C. W. Bourque, T. A. Day, A. V. Ferguson, and J. C. R. Randle, *in* "The Secretory Process" (A. M. Poisner and J. M. Trifaro, eds.), Vol. 2, p. 165. Elsevier, Amsterdam, 1985.

[13] R. W. Schrier, ed., "Vasopressin." Raven Press, New York, 1985.

The cells of the neurosecretory system are located in the hypothalamus, predominantly in the supraoptic and hypothalamic paraventricular nuclei, although almost half of the neurons comprising this system are not localized in these brain nuclei, but are distributed in the anterior and midlateral hypothalamus.[14] These cells provide an excellent model for understanding neuronal function, and the examples discussed draw on this system's relative simplicity. These neurons produce large quantities of neurosecretory material for transport and release from a single source (the neurohypophysis). The cells contain relatively large amounts of VP or OT mRNA (in terms of copies per cell), and the cells are large, allowing easy identification and quantification.[15,16] In addition, a great deal of information already exists regarding the neuronal connectivity of these cells, the neuroendocrine functions mediated by these hormones, and what factors stimulate hormone synthesis and release.

Alterations in vasopressin and oxytocin mRNA levels have been observed following physiologically meaningful stimulation. Previous investigations have shown maintenance on salt water as drinking solution effectively increases plasma osmotic pressure. Fluid imbalance requires increased vasopressin secretion for kidney antidiuretic function. This stimulus, which can almost entirely deplete pituitary stores of vasopressin and oxytocin, induces increased production of hormone in magnocellular perikarya.[7,17] One component of the response to osmotic stimulation is increased levels of vasopressin and oxytocin mRNA[18,19] and cellular content of hybridizable ribosomal RNA.[20] These alterations are detectable with *in situ* hybridization methods.[20–24]

[14] C. H. Rhodes, J. I. Morrell, and D. W. Pfaff, *J. Comp. Neurol.* **198,** 45 (1981).

[15] J. T. McCabe, J. I. Morrell, and D. W. Pfaff, *in* "Neuroendocrine Molecular Biology" (G. Fink, A. J. Harmer, and K. W. McKerns, eds.), p. 219. Plenum, New York, 1986.

[16] W. S. Young III, É. Mezey, and R. E. Siegel, *Mol. Brain Res.* **1,** 231 (1986).

[17] C. W. Jones and B. T. Pickering, *J. Physiol.* (*London*) **203,** 449 (1969).

[18] J. P. H. Burbach, H. H. M. Van Tol, M. H. C. Bakkus, H. Schmale, and R. Ivell, *J. Neurochem.* **47,** 1814 (1986).

[19] J. A. Majzoub, A. Rich, J. van Boom, and J. F. Habener, *J. Biol. Chem.* **258,** 14061 (1983).

[20] M. Kawata, J. T. McCabe, C. Harrington, D. Chikaraishi, and D. W. Pfaff, *J. Comp. Neurol.* **270,** 528 (1988).

[21] J. T. McCabe, J. I. Morrell, D. Richter, and D. W. Pfaff, *Front. Neuroendocrinol.* **9,** 145 (1986).

[22] T. Sherman, J. F. McKelvy, and S. J. Watson, *J. Neurosci.* **6,** 1685 (1986).

[23] G. R. Uhl, H. H. Zingg, and J. F. Habener, *Proc. Natl. Acad. Sci. U.S.A.* **82,** 5555 (1985).

[24] H. H. M. Van Tol, D. T. A. M. Voorhuis, and J. P. H. Burbach, *Endocrinology* (*Baltimore*) **120,** 71 (1987).

In Situ Hybridization and Quantitative Autoradiography

In situ hybridization (or hybridization histochemistry) provides a method for visualizing the location of specific mRNA molecules in cells from sectioned tissue, as well as in cells cultured on glass coverslips. Two recent publications provide extensive background concerning the application of this technique to different tissue types and outline methodological issues.[25,26] Several others discuss strategies for application of this methodology to estimating the amount of hybridized probe and mRNA.[27–30]

Technically, the most demanding requirement of this form of hybridization is that reaction conditions be so well controlled and constant among experimental groups that quantitative analyses can be applied meaningfully. Four of our experiments show that such histochemical reliability can be achieved. (1) Quantitatively, induction of the preproenkephalin gene by haloperidol, measured with *in situ* hybridization, agrees with measurement by slot blot.[31] (2) The 3.1-fold induction of enkephalin mRNA by estradiol, determined by slot blots, closely matches the 3.1-fold estimate based on *in situ* hybridization.[32] (3) The increase in luteinizing hormone-releasing hormone (LHRH) message measured by *in situ* hybridization following long-term gonadal steroid treatment,[33,34] agrees with results from an S_1 nuclease protection assay.[35] (4) The measurement of hypothalamic vasopressin mRNA content in five strains of rats (Long–Evans, Wistar–Kyoto, Brattleboro, spontaneously hypertensive stroke-prone, and crossbred Brattleboro × hypertensive strains) by *in situ* hybridization agreed with results obtained by solution hybridiza-

[25] G. R. Uhl, ed., "*In Situ* Hybridization in Brain." Plenum, New York, 1986.

[26] K. Valentino, J. Eberwine, and J. Barchas, eds., "*In Situ* Hybridization in Neurobiology." Oxford Univ. Press, London and New York, 1987.

[27] E. Cash and M. Brahic, *Anal. Biochem.* **157,** 236 (1986).

[28] W. S. T. Griffin, *in* "*In Situ* Hybridization in Neurobiology" (K. Valentino, J. Eberwine, and J. Barchas, eds.), p. 97. Oxford Univ. Press, London and New York, 1987.

[29] J. T. McCabe, J. I. Morrell, and D. W. Pfaff, *in* "*In Situ* Hybridization in Brain" (G. R. Uhl, ed.), p. 73. Plenum, New York, 1986.

[30] W. S. Young III, *in* "*In Situ* Hybridization in Brain" (G. R. Uhl, ed.), p. 243. Plenum, New York, 1986.

[31] G. J. Romano, B. D. Shivers, R. E. Harlan, R. D. Howells, and D. W. Pfaff, *Mol. Brain Res.* **2,** 33 (1987).

[32] G. Romano, R. Howells, and D. W. Pfaff, *Mol. Endocrinol.*, in press (1988).

[33] J. M. Rothfield, J. F. Hejtmancik, P. M. Conn, and D. W. Pfaff, *Exp. Brain Res.* **67,** 113 (1987).

[34] J. M. Rothfeld, J. F. Hejtmancik, P. M. Conn, and D. W. Pfaff, *Mol. Brain Res.*, in press (1988).

[35] J. L. Roberts, personal communication.

tion assay.[36] In summary, the data show that quantitative *in situ* results can justify detailed mathematical interpretation.

The study of alterations in neuron mRNA content by *in situ* hybridization provides a way of measuring a single cell's response. This in turn may permit one to comprehend how a neuronal system (for example, the neurosecretory system) responds *in toto*. A serious limitation of *in situ* hybridization is that, while the method is quantifiable and can potentially provide a wealth of information concerning the specific cells which exhibit alterations in mRNA content, it does not give one a picture of response dynamics. Akin to Northern and dot blot techniques, hybridization with nucleotide probes on tissue sections estimates momentary cell mRNA content and disregards important parameters of mRNA synthesis such as the rate of mRNA synthesis, transport, and turnover.[37–39] The degree of correspondence between results derived from *in situ* hybridization and those from other hybridization methods, however, demonstrates the reliability of measures with *in situ* hybridization. Measurement of steady-state mRNA levels, though products of a series of complex proximal and distal "processing" steps,[38,39] provides an overall index of neuropeptide synthetic activity.[40,41]

Fremeau and colleagues[42] have recently applied a unique approach for detecting alterations in transcription by means of *in situ* hybridization. Utilizing an intervening (intron) sequence to the proopiomelanocortin (POMC) gene as a probe, these investigators labeled cell nuclei that contained POMC gene transcription product. This work demonstrated that within 1 hr of adrenalectomy the number of silver grains over corticotrope cell nuclei increased dramatically. By using a probe complementary to an intron sequence which is rapidly spliced from heterogeneous nuclear RNA transcripts, the latter observation suggests adrenalectomy increased POMC gene transcription rate.

The application of tissue autoradiography in neurobiology has a long tradition,[43] and early studies outline how its application to quantitative

[36] J. T. McCabe, K. Almasan, E. Lehmann, J. Hänze, R. E. Lang, D. W. Pfaff, and D. Ganten, *Neuroscience*, in press (1988).

[37] M. Blum, B. S. McEwen, and J. L. Roberts, *J. Biol. Chem.* **262**, 8817 (1987).

[38] J. E. Darnell, Jr., *Nature (London)* **297**, 365 (1982).

[39] B. Lewin, "Gene Expression," 2nd Ed., Vol. 2, p. 728. Wiley (Interscience), New York, 1980.

[40] B. S. Schachter, L. K. Johnson, J. D. Baxter, and J. L. Roberts, *Endocrinology (Baltimore)* **110**, 1442 (1982).

[41] J. P. Schwartz and E. Costa, *Annu. Rev. Neurosci.* **9**, 277 (1986).

[42] R. T. Fremeau, Jr., J. R. Lundblad, D. B. Pritchett, J. N. Wilcox, and J. L. Roberts, *Science* **234**, 1265 (1986).

[43] C. Pilgrim and W. E. Stumpf, *J. Histochem. Cytochem.* **35**, 917 (1987).

investigations is fraught with methodological difficulties.[1,44,45] The work described here applies tritium-labeled probes for determination of cellular content of vasopressin mRNA, using this isotope's excellent cellular resolution. This is a consequence of the low maximum energy of tritium that results in an extremely short scatter of grains around labeled cells observed with light microscopy (~0.3-μm resolution). The low energy of ^{3}H β particles (E_{max} 18.5 keV) also means that a significant number of particles which are emitted in labeled tissue are "self-absorbed" by cellular constituents.[46] For all intents and purposes, tritium-labeled probes must be within approximately 1000 Å of the tissue surface to generate autoradiographic grains effectively.[1] Consequently, the collision of β particles with nuclear track emulsion occurs essentially in the monolayer of halide crystals adjacent to the section.[1] In applications where relatively low amounts of radioactivity are employed for short exposure periods (to avoid the problem of "double hits"[1,44,47]), emulsion-dipped autoradiograms produce a linear relationship between the magnitude of cellular radioactivity and grain production in spite of perturbations in section and emulsion thickness across the tissue section.[48,49] For investigations where isotopes other than tritium are used, standards are critical for absolute quantification.[50] There are other problems when directly applying standards as an accurate measure of absolute radioactivity. Are brain paste and plastic standards, which may have different degrees of quench, equivalent to the tissue sample employed? Does a proportion of the isotope elute from the standards when slides are processed? Nevertheless, standards provide a means for detection of grain saturation of photographic emulsion and the related problem of double hits.[1,29,47]

Approaches to Data Analysis

The aforementioned experimental manipulation of chronic salt drinking was used to alter vasopressin mRNA production in magnocellular neurosecretory neurons. For this experiment, brain tissue was obtained from a rat maintained on water with 2% NaCl for 2 weeks and a control rat maintained on tap water. The tissue was processed for *in situ* hybridiza-

[44] P. Dörmer, *in* "Micromethods in Molecular Biology" (V. Neuhoff, ed.), p. 347. Springer-Verlag, Berlin and New York, 1973.

[45] H. Levi, *Scand. J. Haematol.* **1,** 138 (1964).

[46] H. Korr, *Histochemistry* **83,** 65 (1985).

[47] R. P. Perry, *Methods Cell Physiol.* **1,** 305 (1964).

[48] B. M. Kopriwa and C. P. Leblond, *J. Histochem. Cytochem.* **10,** 269 (1962).

[49] C. P. Leblond, B. Kopriwa, and B. Messier, *in* "Histochemistry and Cytochemistry" (R. Wegmann, ed.), p. 1. Macmillan, New York, 1963.

[50] G. R. Uhl, this volume [53].

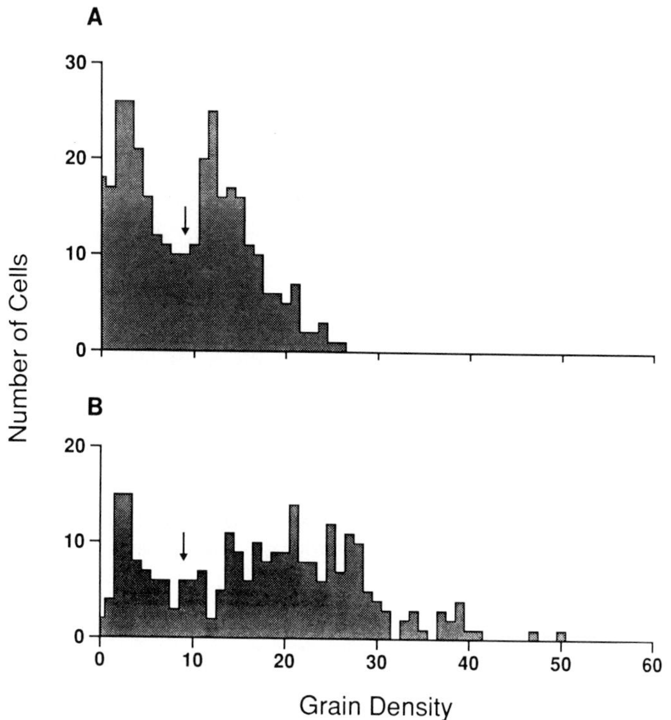

FIG. 1. Frequency histograms summarize cell grain counts over supraoptic nucleus neurons following *in situ* hybridization to label vasopressin mRNA. A brain tissue section was sampled from an ovariectomized female, Long–Evans strain rat maintained on tap water (control) (A) and from a rat maintained on 2% NaCl water for 2 weeks (B). The grain density of each cell profile was determined by visual counting of overlying silver grains and dividing this value by cell area (×1000). Results were taken from McCabe *et al.*[15]

tion[51] with a double-stranded nick-translated probe encoding the vasopressin gene.[52] Subsequent to standard autoradiographic procedures,[51] sections were stained with cresyl violet and analyzed by preparing camera lucida drawings of a tissue section containing a sample of the supraoptic nucleus. Grains over each cell were then counted by eye, and cell area profiles were quantified utilizing the Bio-Quant morphometric system (R & M Biometrics, Nashville, TN). Cell grain density for each neuron is then calculated (grains/cell divided by cell unit area, multiplied by 1000, where 1 cell unit area = 0.87 μm^2).

Figure 1 summarizes data obtained from a salt-loading experiment, where a vasopressin probe was used to detect changes in vasopressin

[51] J. T. McCabe, J. I. Morrell, R. Ivell, H. Schmale, D. Richter, and D. W. Pfaff, *J. Histochem. Cytochem.* **34,** 45 (1986).

[52] H. Schmale, S. Heinshohn, and D. Richter, *EMBO J.* **2,** 763 (1983).

mRNA.[15] Data are collected by grain counting over individual supraoptic nucleus magnocellular neurons in a single tissue section from an animal maintained on tap water and one maintained on salt water for 2 weeks. A total of 326 cells were counted in a single supraoptic nucleus of the control; 260 in the salt-loaded rat. With respect to mean grain density, the tissue from the control rat had an average of 9.17 ± 0.4 [± standard error of the mean (SEM)] grains/unit area in the overlying nuclear track emulsion, and the tissue from the experimental rat had a mean of 17.4 ± 0.7. By conventional statistical tests, such as the Student t-test, these samples are significantly different ($t_{584} = 11.74$, $p < 0.001$).

This analysis considers all supraoptic nucleus cells as suitable for grain counting. The traditional approach in quantitative autoradiography is to utilize a criterion, set prior to grain counting, to alleviate problems concerning what cells to analyze, and what data to include for numerical analysis and summary statistics. This method was not used here. Rather, we deliberately counted and included all observed cells in this analysis to illustrate the applicability of our methods of data analysis to cases where one does not have an independent criterion for deciding whether a cell should or should not be counted.

Several immunocytochemical papers have described the anatomy of this system in detail,[14,53] and we assume magnocellular neurons synthesize either VP or OT hormone exclusively. Preliminary visual study of the shapes of the histograms in Fig. 1 strongly suggests each data set is comprised of two subpopulations of cells. Although the subpopulations overlap, two modes in each histogram are easily recognized. The first mode presumably describes average grain density for the population of unlabeled (oxytocinergic) cells, while the second mode could be used to describe average grain density of labeled (vasopressinergic) cells. As noted earlier, several publications suggest a criterion for elimination of unlabeled cells from further consideration. For example, the mean grain density observed over neuropil can be designated nonspecific "background." Assuming a Poisson function describes the background, 3 times mean background[54,55] excludes most unlabeled cells from the data sets illustrated in the histograms. The arrows in Fig. 1 refer to this criterion, and they indeed lie very near the trough between the two modes in each histogram.

In this chapter we discuss two methods for analysis of autoradiographic data. Graphical methods provide a simple, fast, paper-and-pencil

[53] A. Hou-Yu, A. T. Lamme, E. A. Zimmerman, and A.-J. Silverman, *Neuroendocrinology* **44,** 235 (1986).

[54] A. P. Arnold, *J. Histochem. Cytochem.* **29,** 207 (1981).

[55] J. I. Morrell, M. S. Krieger, and D. W. Pfaff, *Exp. Brain Res.* **62,** 343 (1986).

(or personal computer graphics-generated) exploratory approach for data summarization and interpretation. Visual pictures of large data sets can be easily comprehended and allow the investigator flexibility in posing questions about alterations in mRNA content across a cell population. Several recent publications offer examples of exploratory techniques for data analysis.[56–59] The second method utilizes a parametric statistical approach. This approach applies maximum likelihood estimation to determine cell population parameters and allows one to "fit" specific probability distributions that describe quantitatively vasopressin gene expression (measured by mRNA content) of hypothalamic neurosecretory cells.

Quantile Plots

Quantile (Q) plots are classic graphical tools that continue to prove their utility.[56] Numerous modifications of this plotting technique have been described,[58,60–62] but all of these methods emphasize how Q plots allow rapid and versatile data exploration.

Figure 2 summarizes the same data shown in the histograms of Fig. 1 in terms of empirical quantile plots. To generate the quantile plot, the raw data (grain density/cell) were listed in rank order from smallest value to largest value. The sorted data sets now describe their respective quantile distributions. In Fig. 2, the ordered data points were plotted against what we shall call their corresponding *fraction value*. This value is derived from the data point's rank. The fraction value is computed using $(i - 0.5)/n$, where i is the datum's rank and n equals the sample size. A quantile point's fraction value then corresponds to the fraction of data that are less than or equal to the data point, and it is an empirical analog of the cumulative frequency used in probability distributions. For example, for the quantile data set of grain counts from the tissue section from the control rat (Fig. 1), the lowest grain density rank gives $i = 1$ and the highest gives $i = 326$. The corresponding fraction values are 0.0015337 and 0.99847, respectively ($n = 326$).

[56] J. M. Chambers, W. S. Cleveland, B. Kleiner, and P. A. Tukey, "Graphical Methods for Data Analysis." Wadsworth, Belmont, California, 1983.

[57] W. S. Cleveland, "The Elements of Graphing Data." Wadsworth Advanced Book and Software, Monterey, California, 1985.

[58] B. S. Everitt and D. J. Hand, "Finite Mixture Distributions." Chapman & Hall, London, 1981.

[59] J. W. Tukey, "Exploratory Data Analysis." Addison-Wesley, Reading, Massachusetts, 1977.

[60] K. Kafadar and C. H. Spiegelman, *Comput. Stat. Data Anal.* **4,**167 (1986).

[61] J. C. Lindsey, A. M. Herzberg, and D. G. Watts, *Biometrics* **43,** 327 (1987).

[62] M. B. Wilk and R. Gnanadesikan, *Biometrika* **55,** 1 (1968).

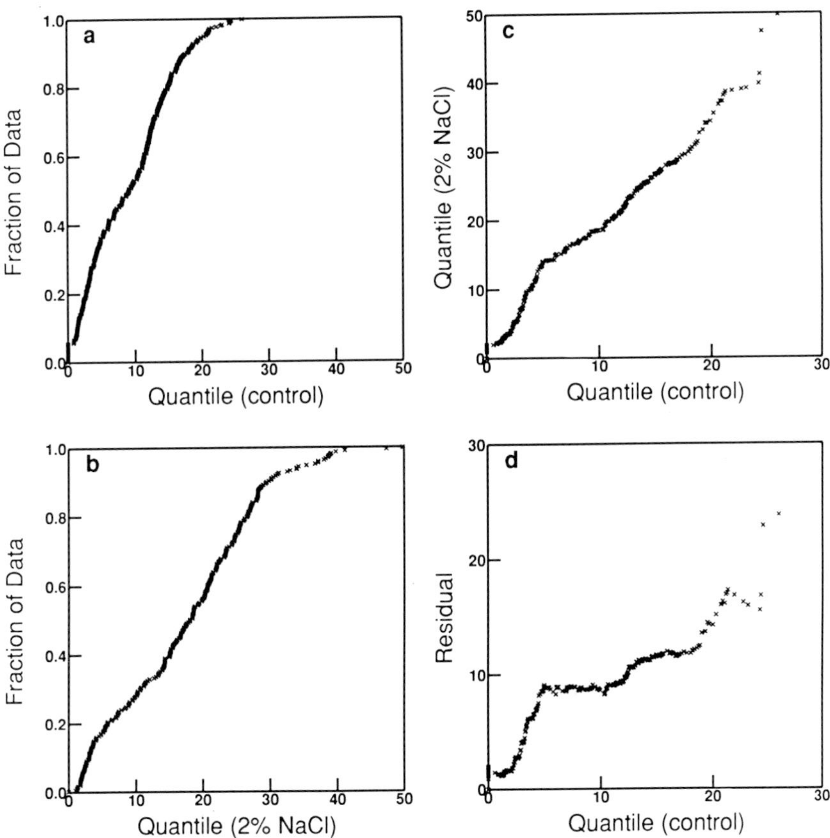

FIG. 2. Quantile (Q) plots of data summarized in histogram form in Fig. 1. (a) Quantile plot of the set of data obtained from the control rat tissue. (b) Quantile plot of the data set obtained from grain counts of a tissue section from the salt-loaded rat. (c) Empirical quantile–quantile (Q–Q) plot of these same data sets. (d) Residual Q plot. Values on the ordinate of the residual Q plot were calculated by subtracting correspondingly ranked values of the quantile data sets. The residual quantiles were then plotted against the quantile data set of the control rat. When quantile data sets are unequal in size, as in c and d, an interpolation step is required (see the text).

By plotting the quantile grain density values against their fraction value, one then has a direct, detailed visual summary of large data sets. Figure 2a and b are Q plots of the grain counts obtained for the control and salt-loading treatment conditions, respectively. Visual comparison of each quantile distribution now enables one to quickly determine if the quantile distributions possess any similarities in overall shape or location. The first salient feature of these plots is that the data sets do not traverse the same general location within the plot. (Note that the same scales for abscissa and ordinate are used in Fig. 2a and b to permit visual comparison.) Had the Q plots differed only in terms of location and not shape, this would suggest that these data sets differ by a constant and that cellular response function is "additive." The second easily detected feature of the Q plots is that neither falls on a straight line. Linear plots would suggest each data set is composed of equally spaced data points where each consecutive point increases by a constant (a rectangular or uniform distribution). Closer visual inspection of Fig. 2a and b suggests each plot may consist of four components. In Fig. 2a, the lowest segment of the plot (fraction value from 0 to about 0.4) is a straight line. A second component, with a different slope, describes fraction values from 0.4 to 0.6, a third component exists for fraction values from 0.6 to 0.8, and finally a curved line describes the highest points. In Fig. 2b, a similar situation appears to exist, except the "first component" of Fig. 2b extends only from fraction values 0 to perhaps 0.15. With respect to the relationship between the quantile plots and the histograms that summarized the data sets, it appears the "trough" regions of the histograms in Fig. 1, which describe the region of overlap between labeled and unlabeled cells, correspond to the middle region of the "second component" segments of the quantile plots. The most obvious difference in the shapes of these quantile plots, however, is the large shift in the general location of the upper portions of the quantile plot of the salt-loaded data set (Fig. 2b) compared with the control data set (Fig. 2a). The large difference in slope indicates that the data sets do not differ by a simple constant but possess a more complex relationship which may involve a multiplicative factor (see below).

Several alternative plots of these data are also possible. For example, instead of plotting the data sets against their fraction values, one could plot the quantiles against the log probability for normal (Gaussian) distributions. One would expect a linear relationship between the two axes if the data set fit a Gaussian function (see below). In addition, the data sets could be transformed (logarithmic, trigonometric, square root, or other power transformations) to determine the simplest linear relationship among quantile values or theoretical functions.

Two plots for making further detailed comparisons of two quantile distributions are the empirical quantile–quantile plot (Q–Q)[62] and the residual quantile plot. The Q–Q plot combines the information displayed in Fig. 2a and b directly. The simplest case for generating Q–Q plots is where the data sets to be compared are equal in size; equivalently ranked quantile points are then plotted as paired data. For cases where the data sets are unequal in size, a more complicated methodology is needed.[56] (Generating Q–Q plots underlies the rationale for calculating the fraction values of each data point of quantile distributions.) Let n_1 equal the number of elements of the larger data set and n_2 equal the number of elements of the smaller data set. Let $j = 1, 2, 3, \ldots, n_1$ for the larger data set plotted on the x axis, and $i = 1, 2, 3, \ldots,$ to n_2 for the smaller data set plotted on the y axis. The size of the larger data set must be interpolated so that a set of quantile data points are formed from the original data set which contains the same number of elements as the smaller data set plotted on the y axis. Let $v = (n_1/n_2)(i - 0.5) + 0.5$, for the values $i = 1, 2, 3, \ldots, n_1$. The value v designates which jth element of the x-axis quantile data set is plotted against the corresponding ith quantile of the y-axis data set. If v is an integer, then v designates the jth element of the x-axis data set to be plotted against the ith element of the y-axis data set. If v is not an integer, let j equal the integer portion of v, and let θ equal the decimal portion (e.g., for $v = 6.25, j = 6, \theta = 0.25$). An interpolated vector of quantile data points for the quantile data set originally plotted along the x axis is then calculated by determining $(1 - \theta)x_j + \theta x_{j+1}$. One now can plot the original y-axis set of quantiles against the corresponding set of interpolated quantile points determined for the quantile distribution plotted on the x axis.

The empirical Q–Q plot has been constructed for direct comparison of data sets in Fig. 2c. If the quantile distributions for the salt-loaded and control tissue were equal, Fig. 2c would describe a straight line with slope equal to one. The divergence in this Q–Q plot, however, is complex. First, there is wavering in the line, indicating no smooth, equal relative progression of changes in mRNA cell content across the data sets. Second, the plot is not merely parallel to the line with a slope of 1 that intersects the origin. This would have suggested a constant difference between the quantile distributions (altering just the y intercept). The slope in fact exceeds 1—approximately 1.25 as determined by a simple least square regression fit—and again implies a multiplicative difference between these quantile distributions. (Note that the axes in Fig. 2c are not equal.)

A further useful variant of the quantile plot compares the data of the control group, plotted along the x axis, to the *residuals* of the quantiles of the experimental group minus the control group (interpolated) data set. A plot of residuals is important since it greatly amplifies differences that are less detectable to the eye in the Q and Q–Q plots. In residual Q plots, a line equal to 0 is expected for instances where there are no differences between the experimental and control quantile distributions. Figure 2d, however, exhibits several complex departures from a line equal to 0. The lower data points exhibit a slow rise, then an initial plateau, a second rise, a second plateau, and then a further rise, with the final portion of the plot being composed of a scatter of points. One interpretation of the "plateau" segments of this plot would be that there are regions of each data set where the cell groups apparently differ by a constant. In this case one observes a slope equal to 0, but the y intercept of the plot is greater than 0. The most salient aspects of the residual plot in Fig. 2d are the second and third "plateau" regions. They imply the highest cell grain densities in the experimental group differ to a great extent from the control group such that they comprise a subgroup of cells with a very robust change in vasopressin mRNA content.

One difficulty with interpretation of both Q plots and Q–Q plots is that the shapes can be inordinately influenced by the proportion of unlabeled cells in the two samples. In such a case, there is a serious hindrance to interpretation. To explore how this factor may have altered graphical presentation of the data, and to simply examine data sets consisting primarily of labeled cells, quantile distribution plots were generated that discard the unlabeled cells. To further explore the relationships between the two data sets, an arbitrary decision to plot only data points where grain density is greater or equal to 9 (to the right of the arrows in Fig. 1) was used since this is the general location where the "trough" between the two modes is evident in the histograms.

The modified quantile plots, derived from truncation of the original data sets, are summarized in Fig. 3. Analogous to the quantile plots illustrated in Fig. 2a and b, the shape of the Q plots in Fig. 3 clearly depict the gross differences in grain densities of the control and experimental quantile distributions. The quantile distributions now appear to have just two components: a large portion of each quantile distribution approximates a straight line, and each data set contains a proportion of heavily labeled cells that depart from the linear trend. It should be noted that a straight line segment within a Q plot indicates an interval of the data which has a relatively uniform distribution.

Further data exploration by graphical technique with the truncated distribution now permits us to "see" significant relationships that may have gone undetected. The Q–Q plot (Fig. 3c) appears to suggest that there is a linear relationship between the quantile distributions of the data sets. The slope in this plot again is greater than 1, indicating that the distributions are not equal, and the slope suggests there may exist a multiplicative function that underlies cellular response in terms of mRNA content. The residual quantile plot (Fig. 3d) most clearly illustrates the relationship between the two data sets, and this plot describes how cellular content of vasopressin mRNA is altered by osmotic challenges. It was expected that the slope of the residual plot in Fig. 3d would approximate 0. Rather surprisingly, a slope of approximately 1 is observed. This suggests that the distribution of vasopressin mRNA content in individual

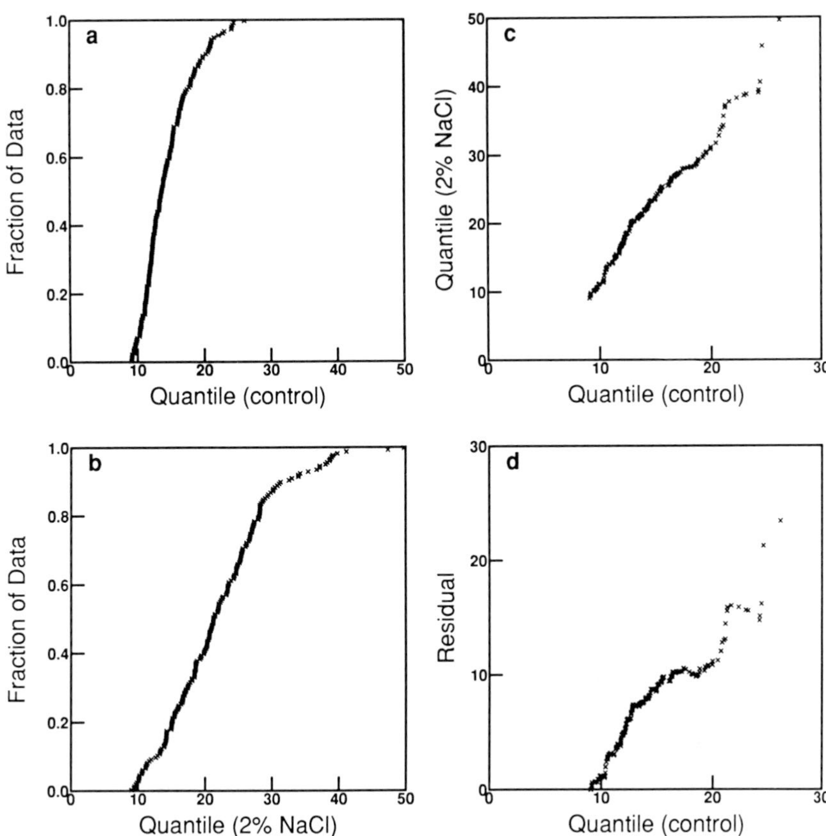

FIG. 3. Quantile plots were generated by recalculating the quantile distributions where cells with grain density less than 9.0 were first eliminated. (a and b) Q plots for the truncated data sets obtained from control and salt-treated rats, respectively. (c) The Q–Q plot indicates that the quantile distributions from the control and experimental tissues are different, since the slope of the plot is greater than 1.0. (d) The residual plot strongly suggests that mRNA content increases as a linear function across the cell population consequent to physiological stimulation of hormone production.

supraoptic nucleus neurons, under stimulatory conditions, changes as a function of their "initial state." That is, instead of all vasopressin cells exhibiting an additive, approximately constant, increase in message level following osmotic stimulation, the cells exhibited proportional or linear increments in message content. This linear relationship, therefore, suggests how neuron population response (increase in cellular mRNA content) is a function of individual cell phenotype. Under basal conditions, the population of cells contains an inherent steady-state level of message that depends on the integrated total of message synthetic and degradative mechanisms. (This steady-state level, of course, is constantly altered by physiologic demands.) With stimulation, however, message level across the cell population increases linearly: cells containing, for example, 1000 copies/cell increase their vasopressin mRNA content by 1000 copies to meet the demands for more peptide hormone. A cell containing 4000 copies, however, can increase its message level by 4000!

Statistical Model

Although graphical techniques are extremely useful for initial data exploration, estimation and hypothesis testing must be based on a statistical model. For autoradiographic data, the Poisson probability distribution is the model most frequently used.[54,63] This is because statistical variation in the disintegration rate of identical quantities of a radioisotope involves rare events within a large population of atoms, the basic circumstances which give rise to the Poisson.[64] In the context of quantitative autoradiography, however, one must assume a relatively uniform distribution of labeled sources among cells to use the Poisson, and in many cases this restriction may not apply. In particular, if the cells themselves show intrinsic variation in the amount of substance which binds the radioactive probe, or if the tissue consists of a heterogeneous mixture of cell types with different response levels, then the Poisson model may be inadequate. We now describe an alternative statistical model which takes these possibilities into account.

Our statistical model builds on the Poisson distribution. Let λ represent the expected number of autoradiographic grains for a uniformly labeled population of cells. The probability of observing exactly x grains per unit cell area is given by

$$p(x) = \frac{e^{-\lambda}\lambda^x}{x!}, \qquad x = 0, 1, 2, \ldots \tag{1}$$

Here, $p(x)$ is the Poisson probability distribution. If the amount of labeled probe varies among cells, then the parameter λ itself will be a random variable and have a probability distribution associated with it. A fairly flexible distribution for describing the variation in λ is the gamma distribution, which is defined for nonnegative real values and subsumes the expo-

[63] J. M. England and A. W. Rogers, *J. Microsc.* (*Oxford*) **92,** 159 (1970).

[64] N. L. Johnson and S. Kotz, "Distributions in Statistics: Discrete Distributions." Houghton, Boston, Massachusetts, 1969.

nential, χ^2, and normal distributions as special cases.[65] The density function for the gamma distribution is

$$g(\lambda) = \frac{\lambda^{\alpha-1}e^{-\lambda/\beta}}{\beta^{\alpha}\Gamma(\alpha)}, \qquad 0 \leq \lambda < \infty \tag{2}$$

where $\Gamma(\cdot)$ is the gamma function. The parameter α determines the shape of the gamma distribution (skewness decreases with increases in α), and β is a scaling parameter. Compounding the distribution (2) for λ with the Poisson Eq. (1) leads to the following expression for the probability of observing x grains per unit cell area[64]:

$$b(x) = \binom{\alpha + x - 1}{\alpha - 1} \beta^{x}(1 - \beta)^{-\alpha-x}, \qquad x = 0, 1, 2, \ldots \tag{3}$$

Equation (3) is a negative binomial probability distribution. It has a mean of $\alpha\beta$ and a variance of $\alpha\beta(\beta + 1)$.

In order to make contact with the experimental data of Fig. 1, we assume that the tissue is composed of two populations of cells: those exhibiting background labeling only and cells which produce meaningful quantities of mRNA. We also assume that background labeling is a relatively uniform process which can be described by a Poisson distribution, while the response levels of mRNA-producing cells is variable and requires a negative binomial distribution. If we let ϕ denote the proportion of background labeled cells and $(1 - \phi)$ the proportion of mRNA-producing cells, then the probability distribution for the number of grains per cell area for the combined cell populations is

$$f(x) = \phi p(x) + (1 - \phi)b(x), \qquad x = 0, 1, 2, \ldots \tag{4}$$

The functions $p(x)$ and $b(x)$ are given by Eqs. (1) and (3), respectively. We refer to Eq. (4) as a *mixture distribution*.

Maximum Likelihood Estimation

The utility of describing the data with a specific statistical model is that its parameters can be estimated and compared. A powerful technique for parameter estimation is the method of maximum likelihood.[66]

Let the integer $y(x)$ represent the observed number of occurrences of exactly x grains per unit cell area. The probability of observing exactly x grains is given by $f(x)$. If the observations are independent, then the

[65] N. L. Johnson and S. Kotz, "Distributions in Statistics: Continuous Univariate Distributions," Vol. 1. Houghton, Boston, Massachusetts, 1970.

[66] M. G. Kendall and A. Stuart, "The Advanced Theory of Statistics," Vol. 2, p. 35. Hafner, New York, 1961.

probability of $y(x)$ occurrences of x grains is equal to $f(x)$ raised to the power $y(x)$. The probability of obtaining the entire set of observations $y(0), y(1), y(2), \ldots, y(m)$ is the product

$$L(\phi, \lambda, \alpha, \beta) = \prod_{x=0}^{m} f(x)^{y(x)} \tag{5}$$

where m is the largest value of x for which there exists an observation. This probability is called the *likelihood function* because it measures the likelihood of obtaining the observed data from the statistical model. The method of maximum likelihood simply involves choosing the parameters of the model so as to maximize this probability. An equivalent but simpler problem is to maximize the logarithm of this probability:

$$\log L = \sum_{x=0}^{m} y(x) \log f(x) \tag{6}$$

Equation (6) is usually referred to as the *log-likelihood function*.

The maximization of a nonlinear function of several variables is a classic problem in numerical analysis.[67] Many powerful algorithms involve the use of first and second derivatives and have been implemented as commercially available software. In practice, one usually tries to solve the system of equations obtained by setting the first partial derivatives equal to 0. In the case of the log-likelihood function [Eq. (6)], the first partial derivatives are given by

$$\begin{aligned}
\frac{\partial \log L}{\partial \phi} &= \sum_{x=0}^{m} \left(\frac{y(x)}{f(x)}\right)\left(\frac{\partial f}{\partial \phi}\right) = 0 \\
\frac{\partial \log L}{\partial \lambda} &= \sum_{x=0}^{m} \left(\frac{y(x)}{f(x)}\right)\left(\frac{\partial f}{\partial \lambda}\right) = 0 \\
\frac{\partial \log L}{\partial \alpha} &= \sum_{x=0}^{m} \left(\frac{y(x)}{f(x)}\right)\left(\frac{\partial f}{\partial \alpha}\right) = 0 \\
\frac{\partial \log L}{\partial \beta} &= \sum_{x=0}^{m} \left(\frac{y(x)}{f(x)}\right)\left(\frac{\partial f}{\partial \beta}\right) = 0
\end{aligned} \tag{7}$$

Explicit expressions for the derivatives of $f(x)$ were derived and are available in the Appendix. (Many professionally written computer programs

[67] J. Stoer and R. Bulirsch, "Introduction to Numerical Analysis." Springer-Verlag, Berlin and New York, 1980.

can use numerical derivatives in lieu of analytical expressions.) At the maximum likelihood solution, the estimates of ϕ, λ, α, and β satisfy Eq. (7).

An important statistical property of the maximum likelihood procedure is that, for large samples, the parameter estimates will have a multivariate normal (MVN) distribution.[66] Furthermore, the *variance–covariance matrix,* which is the MVN analog of the variance, can be obtained from the second partial derivatives of the log-likelihood function evaluated at the maximum likelihood solution. Specifically, the parameter estimates of the mixture distribution will have a variance–covariance matrix given by

$$\Sigma = -\begin{bmatrix} \dfrac{\partial^2 \log L}{\partial \phi^2} & \dfrac{\partial^2 \log L}{\partial \phi\, \partial \lambda} & \dfrac{\partial^2 \log L}{\partial \phi\, \partial \alpha} & \dfrac{\partial^2 \log L}{\partial \phi\, \partial \beta} \\ \dfrac{\partial^2 \log L}{\partial \phi\, \partial \lambda} & \dfrac{\partial^2 \log L}{\partial \lambda^2} & \dfrac{\partial^2 \log L}{\partial \lambda\, \partial \alpha} & \dfrac{\partial^2 \log L}{\partial \lambda\, \partial \beta} \\ \dfrac{\partial^2 \log L}{\partial \phi\, \partial \alpha} & \dfrac{\partial^2 \log L}{\partial \lambda\, \partial \alpha} & \dfrac{\partial^2 \log L}{\partial \alpha^2} & \dfrac{\partial^2 \log L}{\partial \alpha\, \partial \beta} \\ \dfrac{\partial^2 \log L}{\partial \phi\, \partial \beta} & \dfrac{\partial^2 \log L}{\partial \lambda\, \partial \beta} & \dfrac{\partial^2 \log L}{\partial \alpha\, \partial \beta} & \dfrac{\partial^2 \log L}{\partial \beta^2} \end{bmatrix}^{-1} \tag{8}$$

Explicit expressions for these second partial derivatives were also derived and are available in the Appendix. (These second derivatives can also be estimated numerically, and the variance–covariance matrices themselves are often included in the output of commercial software for maximizing nonlinear functions.) The square roots of the diagonal elements of the matrix Σ can be used as large sample standard errors. Multiplying the standard error by ± 1.96 yields an approximate 95% confidence interval. If two or more data sets are to be compared, the standard errors can be combined to test for statistically significant differences. To illustrate, let λ_1 and λ_2 be two parameter estimates, and let σ_1 and σ_2 be their estimated standard errors. If n_1 and n_2 are the (large) sample sizes for two data sets, then an approximate 95% confidence on the difference $\lambda_1 - \lambda_2$ is given by

$$95\%\ \text{Confidence interval} = (\lambda_1 - \lambda_2) \pm 1.96 \left(\frac{n_1}{n_2} \sigma_1^2 + \frac{n_2}{n_1} \sigma_2^2 \right)^{1/2} \tag{9}$$

If this confidence interval contained 0, then one would conclude that there is no evidence for a statistically significant difference at the 5% probability level.

In many cases, the parameters of the mixture distribution are of less interest than some function which combines these parameters. If $z(\phi, \lambda, \alpha, \beta)$ is any differentiable function of the four parameters, then a large sample estimate is obtained by evaluating z using the maximum likelihood parameter values. To compute the standard error of z, let ω_z be the vector of first partial derivatives of z, evaluated at the maximum likelihood solution, i.e.,

$$\omega_z = \begin{bmatrix} \partial z/\partial\phi \\ \partial z/\partial\lambda \\ \partial z/\partial\alpha \\ \partial z/\partial\beta \end{bmatrix} \tag{10}$$

Denoting the standard error as σ_z, one uses

$$\sigma_z = (\omega_z' \textstyle\sum \omega_z)^{1/2} \tag{11}$$

where the prime indicates matrix transposition. This standard error can also be used to derive confidence intervals and to test for differences between data sets as described above.

Application to Autoradiograms

We now illustrate the parametric procedures described above by applying them to the data presented in Fig. 1. In what follows, the subscript 1 refers to the control, and the subscript 2 refers to the salt-loading treatment.

In the context of the preceding notation, the x variable represents the abscissa values of the histograms (grains per area), and $y(x)$ represents the height of the histogram (number of cells) at location x. The log-likelihood function [Eq. (6)] was maximized with respect to the four parameters ϕ, λ, α, and β of the mixture distribution [Eq. (4)]. A simple "grid search" was conducted to find an initial point near the maximum, and then a modification of Newton's method[67] was used to solve the nonlinear equations [Eqs. (7)]. The maximum likelihood estimates appear in Table I.

The parameter estimates can be used to construct fitted distributions. The estimated values were substituted in Eqs. (1), (3), and (4) to obtain expected mixture distributions for the control and salt-loaded treatments. Multiplying these distributions by the sample sizes gives an expected number of observations. These observed and expected histograms are plotted together in Fig. 4. The fitted functions do a reasonable job of reproducing the general shape of the observed histograms. A χ^2 test for

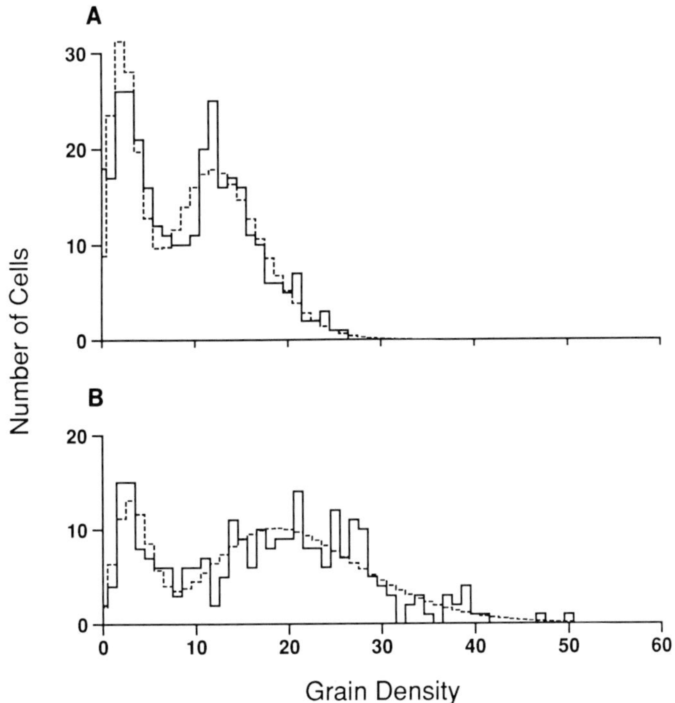

FIG. 4. The observed histograms from Fig. 1 (solid lines) were compared to the expected number of grain counts (dashed lines) obtained through multiplying the fitted theoretical probability distributions by the sample sizes. For both the control (A) and salt-loaded (2% NaCl) (B) treatments, the theoretical distributions reproduce the general shape of the observed histograms and do not differ significantly at the 5% probability level (see the text).

goodness of fit uses the test statistic

$$\chi^2 = \sum_{x=0}^{m} \frac{[y(x) - nf(x)]^2}{[nf(x)]^2}$$

where m is the largest value of x for which there are observations. The degrees of freedom (df) for this test statistic is 1 less than the number of terms in the χ^2 summation minus the number of estimated parameters. In many cases it is necessary to combine the results of two or more adjacent values of x if the expected numbers are less than 1.[68] For the data in Fig. 4, we pooled the right tails of the distributions for $x \geq 25$ in the control and

[68] G. W. Snedecor and W. G. Cochran, "Statistical Methods," 6th Ed., p. 235. Iowa State Univ. Press, Ames, 1967.

TABLE I
MAXIMUM LIKELIHOOD PARAMETER ESTIMATES

Parameter	Description	Control (± SE)	2% NaCl (± SE)	95% Confidence interval for difference[a]
ϕ	Proportion of background-labeled cells	0.3828 ± 0.0342	0.2290 ± 0.0296	(−0.2451, −0.0626)
λ	Mean grain density of background-labeled cells	2.642 ± 0.212	3.489 ± 0.324	(0.112, 1.581)
α	Shape parameter for variation in amount of labeled probe	21.783 ± 8.712	9.626 ± 1.819	(−31.507, 7.192)
β	Scale parameter for variation in amount of labeled probe	0.6033 ± 0.2330	2.224 ± 0.408	(0.742, 2.499)

[a] A 95% confidence interval on the difference between the parameter estimates for the salt-loaded treatment and the control was computed using Eq. (9). If this confidence interval contains 0, there is no statistically significant difference at the 5% probability level.

$x \geq 41$ in the salt-loaded treatment. We computed a χ^2 of 26.078 for the control ($df = 21$, $p > 0.2$) and 40.758 for the salt-loaded treatment ($df = 37$, $p > 0.3$). In neither case do we find sufficient evidence to reject the mixture distribution.

The variance–covariance matrices were computed using Eq. (8) and the analytical expressions in the Appendix. For the control and salt-loaded treatments, respectively, we obtained

$$\Sigma_1 = \begin{bmatrix} 0.00117 & 0.00316 & 0.13176 & -0.00339 \\ 0.00316 & 0.04507 & 0.87097 & -0.02234 \\ 0.13176 & 0.87097 & 75.90818 & -2.02437 \\ -0.00339 & -0.02234 & -2.02437 & 0.05428 \end{bmatrix}$$

$$\Sigma_2 = \begin{bmatrix} 0.00088 & 0.00292 & 0.01598 & -0.00329 \\ 0.00292 & 0.10522 & 0.20914 & -0.04271 \\ 0.01598 & 0.20914 & 3.30789 & -0.73222 \\ -0.00329 & -0.04271 & -7.73222 & 0.16627 \end{bmatrix}$$

The standard errors of the estimated parameters were computed from the square roots of the diagonal elements of these matrices. Equation (9) was used to construct a 95% confidence interval on the difference between the control and salt-loaded treatments. These results also appear in Table I.

It is interesting to note that there are small but statistically significant differences in both the estimated proportion of background labeled cells (ϕ) and the mean level of background labeling (λ). One of the strengths of this parametric approach is that these quantities can be estimated and compared. Experimental procedures which ignore potential differences in the level and frequency of background labeling could yield questionable conclusions.

The parameters α and β can be used directly to make inferences about the variability of labeled probe in the cells producing vasopressin mRNA. Recall that in the gamma distribution of Eq. (2) a large value of α implies a small skewness; this suggests distributions which look much like the normal. Both estimated values of α are relatively large (Table I). The confidence intervals for the parameter differences (Table I) suggest that the changes in amount of labeled probe were changes of scale (β) but not of shape (α). In other words, increasing the amount of labeled probe in the control cells by a constant (multiplicative) factor could yield data that are similar to those obtained for the salt-loaded treatment. This result is consistent with our observation of a linear relationship in the residual quantile plot of Fig. 3.

There are several functions of the original parameters which are of value in the interpretation of the experimental results. These functions are listed in Table II. Their estimates and standard errors were obtained using Eqs. (10) and (11). The first two functions, $\alpha\beta$ and $\alpha\beta(1 + \beta)$, represent the mean and variance of the negative binomial component [Eq. (3)] of the mixture distribution [Eq. (4)]. Since the negative binomial represents the distribution of grain density for vasopressin mRNA-producing cells, these functions can be used to characterize the data with the effects of background labeling removed. The increase in response above background level can be defined as the difference between the expected values of the negative binomial [Eq. (3)] and the Poisson [Eq. (1)], i.e., $\alpha\beta - \lambda$. This quantity was also estimated and is presented in Table II. The confidence interval for the difference between the salt-loaded and control treatments can be used as a statistical test for treatment effects with the complications of background labeling removed. For these data, this interval is located well above 0, indicating a statistically significant difference.

Finally, in light of the possible "scaling" hypothesis mentioned above, it is worthwhile to examine the coefficients of variation for the estimated negative binomial distributions. For any distribution, the coefficient of variation is defined as the ratio of the standard error to the mean, and it is a measure of relative variation. If the difference between two distributions is simply a matter of scale, then they will have the same coefficients of variation. In the case of the negative binomial, the coeffi-

TABLE II
ESTIMATES FOR SELECTED FUNCTIONS OF THE PARAMETERS

Function	Description	Control (± SE)	2% NaCl (± SE)	95% Confidence interval for difference[a]
$\alpha\beta$	Mean grain density for mRNA-labeled cells	13.143 ± 0.420	21.409 ± 0.645	(6.808, 9.726)
$\alpha\beta(1 + \beta)$	Variance in grain density for mRNA-labeled cells	21.072 ± 2.861	69.028 ± 8.756	(31.371, 64.642)
$\alpha\beta - \lambda$	Mean increase in grain density above background	10.500 ± 0.379	17.920 ± 0.644	(6.019, 8.822)
$[(1 + \beta)/\alpha\beta]^{1/2}$	Coefficient of variation for mRNA-labeled cells	0.3493 ± 0.0281	0.3881 ± 0.0258	(−0.0376, 0.1152)

[a] A 95% confidence interval on the difference between the function estimates for the salt-loaded treatment and the control was computed using Eq. (9). If this confidence interval contains 0, there is no statistically significant difference at the 5% probability level.

cient of variation is a function of α and β, and this function is given in Table II. The estimated values for the control and salt-loaded treatments are very similar, and the confidence interval for their difference contains 0. This observation lends further support to the hypothesis of a proportional, "scaling," increase in vasopressin mRNA levels due to salt loading.

Summary

Quantification of gene expression in a morphological context is an invaluable tool for neurobiological investigation. The ability to measure the quantity of specific mRNA molecules at the level of the single neuron permits one to monitor the modulation of complex cell synthetic activity of intact neuron populations. The cells of interest can be contiguous or dispersed in functionally significant patterns throughout a broad anatomical region of the brain.

The application of *quantitative in situ* hybridization is technically difficult and labor intensive. Nevertheless, it has great utility for investigating gene expression from a structural perspective. (1) *In situ* hybridization permits one to ask questions concerning the anatomical pattern of neuronal gene expression. (2) It permits analyses concerning the initiation of expression, cell location, cell type, and alterations of level of expression within a spatial and temporal context. (3) In cases where blotting methods suggest a message exists at low copy, *in situ* hybridization permits queries at the single-cell level. For example, *in situ* hybridization can determine if very few cells are expressing the gene product or if many neurons dispersed throughout a brain region exhibit low mRNA copy number/cell.

Quantitative analyses also allow detailed investigation of cell response to physiologically meaningful stimulation. Our application of statistical and numerical methods is a demonstration of the utility of probabilistic models; the mixture distribution accounted for data from both labeled and unlabeled sources. In agreement with many previous investigations, grain density over an unlabeled uniform source (oxytocinergic cells) was suitably described by the Poisson distribution.[63] The population of labeled vasopressinergic cells, however, was best described by the negative binomial distribution. Previous investigations from different fields of biology show that the negative binomial can be used to describe many biological phenomena,[69–72] and this distribution was considered in at least two previous investigations to evaluate autoradiographic data which did not fit the Poisson function.[73,74] From a theoretical perspective, the probabilistic relationship between β-particle decay (a Poisson function) and the distribution of message levels among individual neurons in a cell group (gamma distribution) prompts consideration of the negative binomial. For both data sets the observed variances were larger than the mean, and the labeled portion of the data sets exhibited positive skewness. On a practical level, the negative binomial seems appropriate for describing our empirical observations.

The analysis also demonstrated the importance of graphical inspection of data sets. Visual summaries with histograms and quantile plots can be useful for initial decisions about what statistical calculations are most suitable. Graphical exploration may also lead to further revelations concerning cell response characteristics. In the present case, the residual quantile plot of the truncated data set (Fig. 3d), provided an example of how these cells react to stimulation: the alteration in message levels was a linear function of "resting state" message levels. This inference prompted us to reexamine the parametric estimates. As would be ex-

[69] R. I. Baxter and C. D. Blake, *Nature* (*London*) **215,** 1168 (1967).

[70] C. I. Bliss, *Biometrics* **9,** 176 (1953).

[71] Student, *Biometrika* **5,** 351 (1907).

[72] C. B. Williams, *J. Ecol.* **32,** 1 (1944).

[73] B. Chernick and A. Evans, *Exp. Cell Res.* **53,** 94 (1968).

[74] J. Reddingius and O. M. H. DeVries, *J. Microsc.* (*Oxford*) **143,** 299 (1986).

pected based on the results in the residual quantile plot, the coefficients of variation for the control and salt-loaded data sets are equal.

The graphical and statistical analyses demonstrated here suggest further experiments on the mechanisms by which neuronal gene expression responds to physiological challenge. We hope future tests of the applicability of this approach, as well as examination of its suitability with other probes, will result in further methodological and theoretical developments.

Appendix

Here we present explicit expressions for the first and second partial derivatives used in Eqs. (7) and (8) (results derived by RAD).

Since the mixture distribution $f(x)$ is a composite of the Poisson $p(x)$ and the negative binomial $b(x)$, we begin by listing the first and second derivatives of these two latter distributions. In the case of the negative binomial, we have made frequent use of the following definition:

$$\binom{\alpha + x - 1}{\alpha - 1} = \frac{\alpha(\alpha + 1)(\alpha + 2) \ldots (\alpha + x - 1)}{x!}$$

The derivatives of $p(x)$ and $b(x)$ are given by

$$\frac{\partial p}{\partial \lambda} = p(x)\,\frac{x - \lambda}{\lambda}$$

$$\frac{\partial b}{\partial \alpha} = b(x)\left[\sum_{j=0}^{x-1} (\alpha + j)^{-1} - \log(1 + \beta)\right]$$

$$\frac{\partial b}{\partial \beta} = b(x)\left[\frac{x}{\beta} - \frac{\alpha + x}{1 + \beta}\right]$$

$$\frac{\partial^2 p}{\partial \lambda^2} = \left(\frac{\partial p}{\partial \lambda}\right)^2\left(\frac{1}{p(x)}\right) - \left(\frac{xp(x)}{\lambda^2}\right)$$

$$\frac{\partial^2 b}{\partial \alpha^2} = \left(\frac{\partial b}{\partial \alpha}\right)^2\left(\frac{1}{b(x)}\right) - b(x)\sum_{j=0}^{x-1} (\alpha + j)^{-2}$$

$$\frac{\partial^2 b}{\partial \alpha\, \partial \beta} = \left(\frac{\partial b}{\partial \alpha}\right)\left(\frac{\partial b}{\partial \beta}\right)\left(\frac{1}{b(x)}\right) - \left(\frac{b(x)}{1 + \beta}\right)$$

$$\frac{\partial^2 b}{\partial \beta^2} = \left(\frac{\partial b}{\partial \beta}\right)^2\left(\frac{1}{b(x)}\right) - b(x)\left[\frac{x}{\beta^2} - \frac{\alpha + x}{(1 + \beta)^2}\right]$$

These expressions can be substituted into the following equations to obtain the derivatives of the mixture distribution:

$$\frac{\partial f}{\partial \phi} = p(x) - b(x)$$

$$\frac{\partial f}{\partial \lambda} = \phi \left(\frac{\partial p}{\partial \lambda}\right)$$

$$\frac{\partial f}{\partial \alpha} = (1 - \phi) \left(\frac{\partial b}{\partial \alpha}\right)$$

$$\frac{\partial f}{\partial \beta} = (1 - \phi) \left(\frac{\partial b}{\partial \beta}\right)$$

Likewise, the first and second derivatives of $p(x)$ and $b(x)$ can be used in the following equations to compute the elements of the variance–covariance matrix:

$$\frac{\partial^2 \log L}{\partial \phi^2} = -\sum_{x=0}^{m} \left(\frac{y(x)}{f(x)}\right)\left(\frac{1}{f(x)}\right)\left(\frac{\partial f}{\partial \phi}\right)$$

$$\frac{\partial^2 \log L}{\partial \phi\, \partial \lambda} = -\sum_{x=0}^{m} \left(\frac{y(x)}{f(x)}\right)\left(\frac{\partial p}{\partial \lambda}\right)\left[\left(\frac{\phi}{f(x)}\right)\left(\frac{\partial f}{\partial \phi}\right) - 1\right]$$

$$\frac{\partial^2 \log L}{\partial \phi\, \partial \alpha} = -\sum_{x=0}^{m} \left(\frac{y(x)}{f(x)}\right)\left(\frac{\partial b}{\partial \alpha}\right)\left[\left(\frac{1 - \phi}{f(x)}\right)\left(\frac{\partial f}{\partial \phi}\right) + 1\right]$$

$$\frac{\partial^2 \log L}{\partial \phi\, \partial \beta} = -\sum_{x=0}^{m} \left(\frac{y(x)}{f(x)}\right)\left(\frac{\partial b}{\partial \beta}\right)\left[\left(\frac{1 - \phi}{f(x)}\right)\left(\frac{\partial f}{\partial \phi}\right) + 1\right]$$

$$\frac{\partial^2 \log L}{\partial \lambda^2} = -\sum_{x=0}^{m} \left(\frac{y(x)}{f(x)}\right) \phi \left[\left(\frac{\phi}{f(x)}\right)\left(\frac{\partial p}{\partial \lambda}\right)^2 - \left(\frac{\partial^2 p}{\partial \lambda^2}\right)\right]$$

$$\frac{\partial^2 \log L}{\partial \lambda\, \partial \alpha} = -\sum_{x=0}^{m} \left(\frac{y(x)}{f(x)}\right)\left(\frac{\phi(1 - \phi)}{f(x)}\right)\left(\frac{\partial p}{\partial \lambda}\right)\left(\frac{\partial b}{\partial \alpha}\right)$$

$$\frac{\partial^2 \log L}{\partial \lambda\, \partial \beta} = -\sum_{x=0}^{m} \left(\frac{y(x)}{f(x)}\right)\left(\frac{\phi(1 - \phi)}{f(x)}\right)\left(\frac{\partial p}{\partial \lambda}\right)\left(\frac{\partial b}{\partial \beta}\right)$$

$$\frac{\partial^2 \log L}{\partial \alpha^2} = -\sum_{x=0}^{m} \left(\frac{y(x)}{f(x)}\right) (1 - \phi) \left[\left(\frac{1 - \phi}{f(x)}\right)\left(\frac{\partial b}{\partial \alpha}\right)^2 - \left(\frac{\partial^2 b}{\partial \alpha^2}\right)\right]$$

$$\frac{\partial^2 \log L}{\partial \alpha\, \partial \beta} = -\sum_{x=0}^{m} \left(\frac{y(x)}{f(x)}\right)(1-\phi)\left[\left(\frac{1-\phi}{f(x)}\right)\left(\frac{\partial b}{\partial \alpha}\right)\left(\frac{\partial b}{\partial \beta}\right) - \left(\frac{\partial^2 b}{\partial \alpha\, \partial \beta}\right)\right]$$

$$\frac{\partial^2 \log L}{\partial \beta^2} = -\sum_{x=0}^{m} \left(\frac{y(x)}{f(x)}\right)(1-\phi)\left[\left(\frac{1-\phi}{f(x)}\right)\left(\frac{\partial b}{\partial \beta}\right)^2 - \left(\frac{\partial^2 b}{\partial \beta^2}\right)\right]$$

We checked the accuracy of these expressions by substituting the maximum likelihood estimates into the equations and comparing the results to approximations obtained from numerical derivatives. In every case, there is agreement within the error limits of the numerical derivatives.

Acknowledgments

The authors gratefully acknowledge the excellent technical assistance of Florence Lowe. We thank Drs. Dietmar Richter, Hartwig Schmale, and Richard Ivell (University of Hamburg) for the supply of probe, and Dr. Marlene Schwanzel-Fukuda (Rockefeller University) for the use of her image analysis system. R. A. D. was supported by a grant from the U.S. Environmental Protection Agency (R-812239-01-1).

Author Index

Numbers in parentheses are footnote reference numbers and indicate that an author's work is referred to although the name is not cited in the text.

A

C

D

E

F

H

I

J

L

M

O

P

Q

R

S

T

X

Y

Subject Index

A

D

E

F

G

K

L

M

O

P

Q

R

S

T

V

W

Z